Lippincott®
Illustrated
Reviews

Biochemistry

Ninth Edition

Lippincott® Illustrated Reviews

Biochemistry

Ninth Edition

Emine Ercikan Abali, PhD

Adjunct Professor of Biochemistry and Molecular Biology

Rutgers Robert Wood Johnson Medical School

Piscataway, New Jersey

Susan D. Cline, PhD

Professor of Biochemistry

Department of Biomedical Sciences

Mercer University School of Medicine

Macon, Georgia

David S. Franklin, PhD

Professor of Biochemistry & Molecular Biology

Tulane University School of Medicine

New Orleans, Louisiana

Susan M. Viselli, PhD

Professor of Biochemistry & Molecular Genetics

College of Graduate Studies

Midwestern University

Downers Grove, Illinois

Wolters Kluwer

Philadelphia • Baltimore • New York • London
Buenos Aires • Hong Kong • Sydney • Tokyo

Acquisitions Editor: Crystal Taylor
Development Editor: Deborah Bordeaux
Editorial Coordinator: Janet Jayne
Editorial Assistant: Parisa Saranj
Marketing Manager: Kirsten Watrud
Senior Production Project Manager: Alicia Jackson
Manager, Graphic Arts & Design: Stephen Druding
Art Director: Jennifer Clements
Senior Manufacturing Coordinator: Margie Orzech-Zeranko
Prepress Vendor: Aptara, Inc.

9th edition

9 8 7 6 5 4 3 2 1

Printed in Mexico

Library of Congress Cataloging-in-Publication Data

Names: Abali, Emine Ercikan, author. | Cline, Susan D., author. | Franklin, David S., author. | Viselli, Susan, author.
Title: Biochemistry / Emine Ercikan Abali, Susan D. Cline, David S. Franklin, Susan M. Viselli.
Other titles: Lippincott's illustrated reviews
Description: Ninth edition. | Philadelphia, PA : Wolters Kluwer, [2025] | Series: Lippincott illustrated reviews | Includes bibliographical references and index. | Summary: "Praised by faculty and students for more than two decades, Lippincott® Illustrated Reviews: Biochemistry is the long-established go-to resource for mastering the essentials of biochemistry. This best-selling text helps students quickly review, assimilate, and integrate large amounts of critical and complex information, with unparalleled illustrations that bring concepts to life"–Provided by publisher.
Identifiers: LCCN 2024039840 (print) | LCCN 2024039841 (ebook) | ISBN 9781975220495 (paperback) | ISBN 9781975220518 (epub)
Subjects: MESH: Biochemistry | Examination Questions
Classification: LCC QP514.2 (print) | LCC QP514.2 (ebook) | NLM QU 18.2 | DDC 572–dc23/eng/20240917
LC record available at https://lccn.loc.gov/2024039840
LC ebook record available at https://lccn.loc.gov/2024039841

shop.lww.com

QUADM0125

Dedication

This edition is dedicated to those we teach and to those who taught us.

Emine Ercikan Abali, PhD
Susan D. Cline, PhD
David S. Franklin, PhD
Susan M. Viselli, PhD

Acknowledgments

We extend gratitude to our families and loved ones, without their unwavering support this book would not have been possible. We also honor the legacy of the founding authors of this title, the late Dr. Pamela Champe and the late Dr. Richard Harvey, who created the first four editions, and to Dr. Denise Ferrier, who coauthored or authored the subsequent three editions. We have strived to carry on their tradition of excellence with the 8th and now the current 9th editions.

We value the many members of the Association of Biochemistry Educators who provided critical peer review of the new materials produced for this edition.

We are grateful to the team at Wolters Kluwer. We thank Crystal Taylor for her support throughout this project, Debbie Bordeaux for her editorial skills and guidance, Janet Jayne for her expert editorial coordination, and Kelly Horvath for her expert developmental editing.

Cover image: A depiction of the structure of a nucleosome.

CONTRIBUTING EDITOR, ONLINE UNIT REVIEW QUESTIONS

Jana M. Simmons, PhD
Associate Professor
Department of Biochemistry and Molecular Biology
Michigan State University, College of Human Medicine
Grand Rapids, Michigan

REVIEWERS

Ashley Bradley, DO Candidate
A.T. Still University Osteopathic Medical
    School
Mesa, Arizona

Tameka A. Clemons, PhD
Vanderbilt University Medical Center
Nashville, Tennessee

Vanessa De La Rosa, PhD
Burrell College of Osteopathic Medicine
Las Cruces, New Mexico

Sarah A. Evans, PhD
Kansas College of Osteopathic Medicine
Wichita, Kansas

Sheri F. T. Fong, MD, PhD
John A. Burns School of Medicine
Honolulu, Hawaii

Jack G. Goldsmith, PhD
University of South Carolina School
    of Medicine
Columbia, South Carolina

Zeynep Gromley, PhD
Lincoln Memorial University — DeBusk
    College of Osteopathic Medicine
Harrogate, Tennessee

Mark E. Hemric, PhD
Liberty University College of Osteopathic
    Medicine
Lynchburg, Virginia

Zachary I. Merhavy, MD Candidate
Ross University School of Medicine
Saint Michael, Barbados

Cristine Smoczer, MD, PhD
Detroit Mercy School of Dentistry
Detroit, Michigan

Luigi Strizzi, MD, PhD
Midwestern University College of Graduate
    Studies
Downers Grove, Illinois

Preface

Biochemistry is the study of chemical processes in living organisms, at the cellular and molecular levels. It includes how our bodies utilize nutritional substances in the diet to harness energy and to produce proteins to carry out biologic functions. This book provides a succinct and illustrative review of the complex mechanisms in carbohydrate, protein, and lipid biochemistry. In doing so, the book also offers examples of a useful organizational tool called a concept map. Here is an explanation of concept maps so that you may use them as you study biochemistry, and perhaps create your own concept maps in your studies.

CONCEPT MAPS

Biochemistry is a body of concepts to be understood in context of the whole person, and the use of concept maps illustrates the interconnectedness of biologic processes. Concept maps serve as guides, or road maps—to provide the student with an understanding of the context of how various topics fit together to tell a story. In this text, a series of biochemical concept maps have been created to graphically illustrate relationships between ideas and connections between concepts. These are presented near the end of each chapter to show how the information can be grouped or organized. A concept map is, thus, a tool for visualizing the connections between concepts. Material is represented in a hierarchical fashion, with the most inclusive, most general concepts at the top of the map, and the more specific, less general concepts arranged beneath. The concept maps ideally function as templates or guides for organizing information, so the student can readily find the best ways to help with the integration of new information with knowledge they already possessed. Concept map construction is described below.

A: Concept boxes and links

In the biochemical maps, concepts include abstractions (e.g., free energy), processes (e.g., oxidative phosphorylation), and compounds (e.g., glucose 6-phosphate). These broadly defined concepts are prioritized with the central idea positioned at the top of the page. The concepts that follow from this central idea are then drawn in boxes (see figure, part A). The size of the type indicates the relative importance of each idea. Lines are drawn between concept boxes to show which are related. The label on the line defines the relationship between two concepts, so that it reads as a valid statement (i.e., the connection creates meaning). The lines with arrowheads indicate in which direction the connection should be read.

B: Links to other parts of a map

Concept maps may contain cross-links that allow the reader to visualize complex relationships between ideas represented in different parts of the map (see figure, part B) or between the map and other chapters in this book (see figure, part C) or to other books in the Lippincott® Illustrated Reviews series (e.g., *Lippincott® Illustrated Reviews: Cell and Molecular Biology*). These links can help identify concepts that are central to more than one topic in biochemistry, empowering students to be effective in clinical situations and on professional licensure examinations that require integration of material. These maps with links provide a visual aid to represent nonlinear relationships between facts, in contrast to cross-referencing within linear text and concepts. The first example of a complete concept map can be found at the end of Chapter 2 (Fig. 2.15).

A **Linked concept boxes**

Amino acids
(fully protonated)

can

Release protons (H$^+$)

B **Concepts cross-linked within a map**

Degradation of body protein ← *is produced by* — **Amino acid pool**

Simultaneous synthesis and degradation — *leads to* → **Protein turnover**

Synthesis of body protein ← *is consumed by* — **Amino acid pool**

C **Concepts cross-linked to other chapters in the book**

. . . how the protein folds into its native conformation

Protein Structure **2**

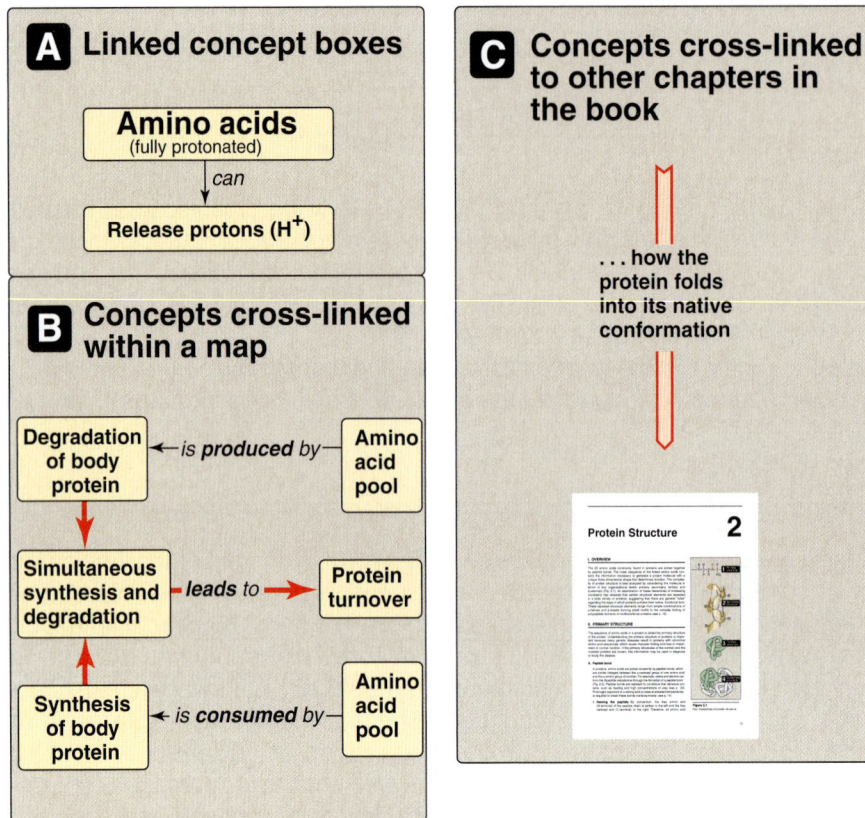

RECOMMENDED USE OF THIS TEXTBOOK AND OTHER RESOURCES

This book provides a comprehensive review of medical biochemistry. In addition to concept maps and illustrative figures, clinical boxes are included to offer applications of concepts. Students are also encouraged to challenge their understanding of the information that they have read through the completion of study questions at the end of each chapter and in the larger question bank available online.

Contents

UNIT VI Medical Nutrition

UNIT VII Storage and Expression of Genetic Information

Amino Acids and the Role of pH

1

I. OVERVIEW

Amino acids are the building blocks or subunits of proteins, an abundant and functionally diverse category of macromolecules in living systems. Proteins exist as structural supports, as enzymes, hormones, and as carriers. For example, enzymes and polypeptide hormones direct and regulate metabolism in the body. In the digestive system, enzymes break down food so the nutrients can be absorbed. Contractile proteins in muscle permit movement. In bone, collagen is a protein that forms a framework for the deposition of calcium phosphate crystals, strengthening the skeleton. In the bloodstream, hemoglobin transports oxygen, and albumin carries hormones and vitamins, while immunoglobulins fight infectious bacteria and viruses. Proteins associated with membranes facilitate movement of molecules into and out of cells. In short, proteins display an incredible diversity of functions, yet all share the common structural feature of being linear polymers of amino acids. This chapter describes the properties of amino acids that are linked together to form proteins. It also describes the importance of pH to normal protein and body function. Chapter 2 explores how these simple building blocks are joined to form proteins that have unique three-dimensional structures, giving them the capacity to perform specific biologic functions.

II. STRUCTURE

More than 300 different amino acids have been described in nature but only 20 are commonly found in mammalian proteins. These 20 standard amino acids are the only amino acids encoded by DNA, the genetic material in the cell. Nonstandard amino acids are produced by chemical modification of standard amino acids. Each amino acid has a carboxyl group, a primary amino group (except for proline, which has a secondary amino group), and a distinctive side chain or R group bonded to the α-carbon atom.

At physiologic pH ($\sim$7.4), the carboxyl group of an amino acid is dissociated, forming the negatively charged carboxylate ion ($-COO^-$), and the amino group is protonated ($-NH_3^+$) (Fig. 1.1A). This form of amino acid

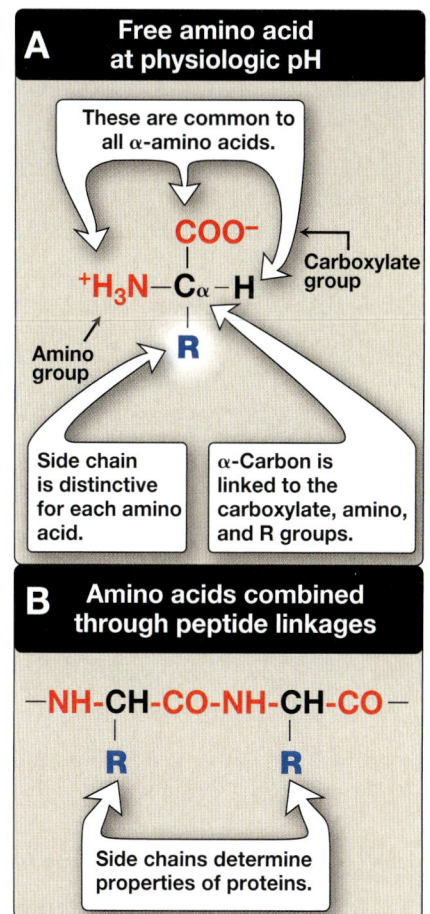

A Free amino acid at physiologic pH

These are common to all α-amino acids.

COO^-

Carboxylate group

$^+H_3N-C_\alpha-H$

R

Amino group

Side chain is distinctive for each amino acid.

α-Carbon is linked to the carboxylate, amino, and R groups.

B Amino acids combined through peptide linkages

$-NH-CH-CO-NH-CH-CO-$

R R

Side chains determine properties of proteins.

Figure 1.1
A, B. Structural features of amino acids.

1

is called a zwitterion. In proteins, almost all of these carboxyl and amino groups are combined through peptide bonds and, in general, are not available for chemical reactions except for hydrogen bond or ionic bond formation (Fig. 1.1B). Amino acids within proteins are referred to as **residues** in reference to the residual structure remaining after peptide bond formation between consecutive amino acids within a peptide chain. The nature of the side chains ultimately dictates the role an amino acid plays within a protein. Therefore, it is useful to classify the amino acids according to the properties of their side chains, that is, whether they are nonpolar, with an even distribution of electrons, or polar with an uneven distribution of electrons, such as acids and bases (Figs. 1.2 and 1.3).

A. Amino acids with nonpolar side chains

Each of the amino acids in this category has a side chain that does not gain or lose protons or participate in hydrogen or ionic bonds (see Fig. 1.2). The side chains of these amino acids can be thought of as "oily" or lipid-like, a property that promotes hydrophobic interactions (see Fig. 2.10).

NONPOLAR SIDE CHAINS

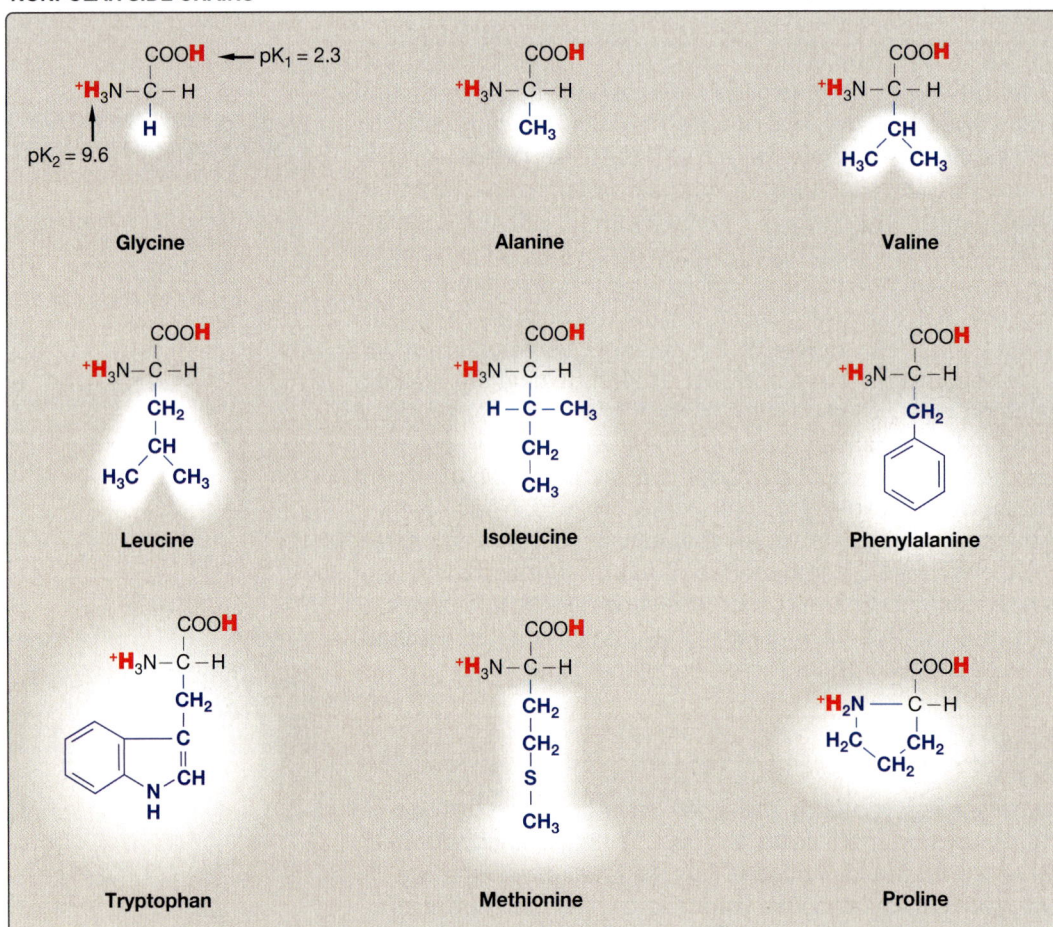

Figure 1.2
Classification of the 20 standard amino acids, according to the charge and polarity of their side chains at acidic pH, is shown here and continues in Figure 1.3. Each amino acid is shown in its fully protonated form, with dissociable hydrogen ions represented in *red*. The pK values for the α-carboxyl and α-amino groups of the nonpolar amino acids are similar to those shown for glycine.

UNCHARGED POLAR SIDE CHAINS

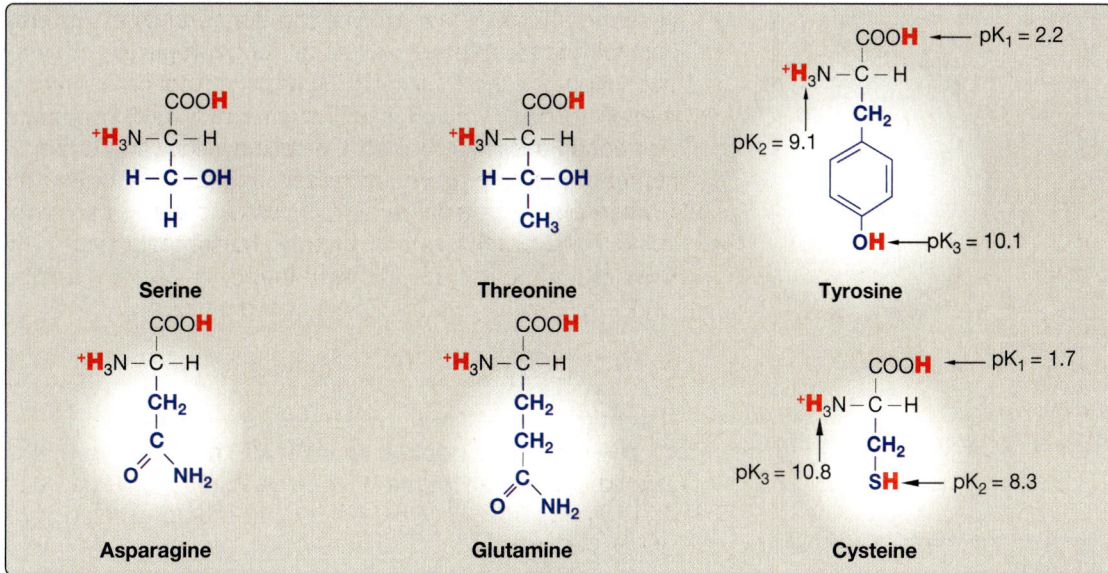

Serine

Threonine

Tyrosine — $pK_1 = 2.2$, $pK_2 = 9.1$, $pK_3 = 10.1$

Asparagine

Glutamine

Cysteine — $pK_1 = 1.7$, $pK_2 = 8.3$, $pK_3 = 10.8$

ACIDIC SIDE CHAINS

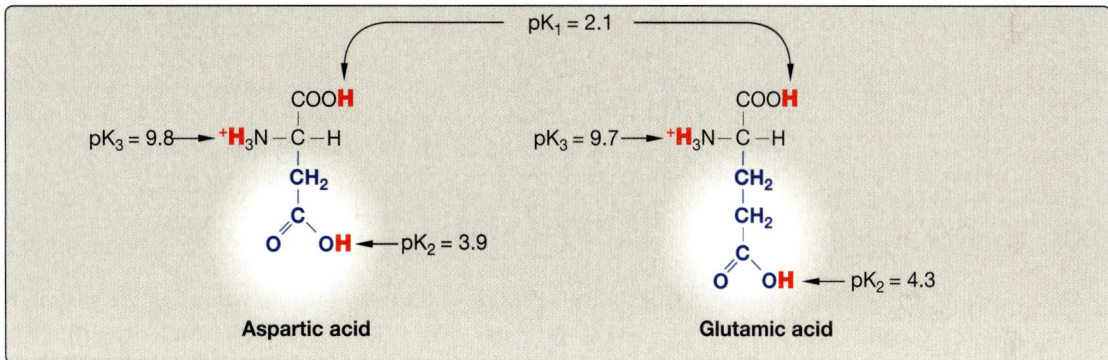

Aspartic acid — $pK_1 = 2.1$, $pK_2 = 3.9$, $pK_3 = 9.8$

Glutamic acid — $pK_1 = 2.1$, $pK_2 = 4.3$, $pK_3 = 9.7$

BASIC SIDE CHAINS

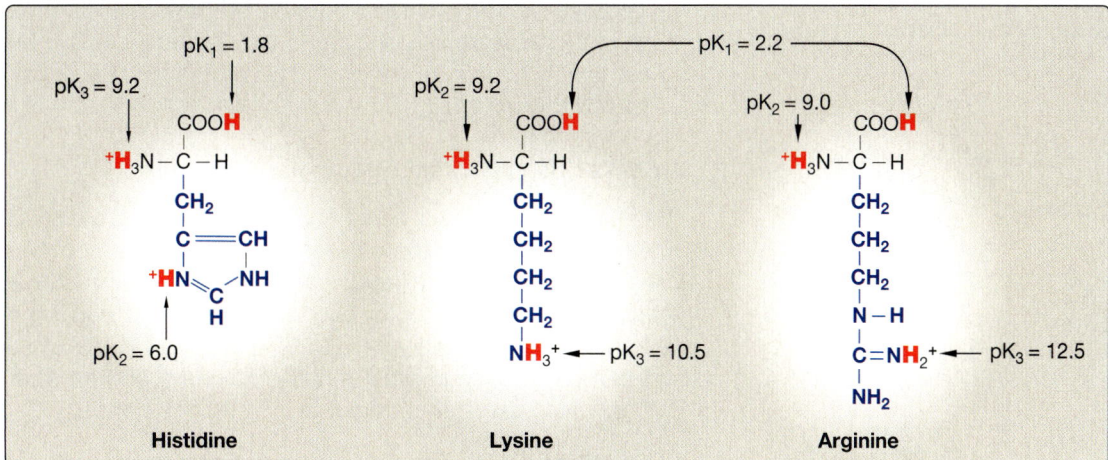

Histidine — $pK_1 = 1.8$, $pK_2 = 6.0$, $pK_3 = 9.2$

Lysine — $pK_2 = 9.2$, $pK_3 = 10.5$

Arginine — $pK_1 = 2.2$, $pK_2 = 9.0$, $pK_3 = 12.5$

Figure 1.3
Classification of the 20 standard amino acids, according to the charge and polarity of their side chains at acidic pH (continued from Fig. 1.2). [Note: At physiologic pH (7.35 to 7.45), the α-carboxyl groups, the acidic side chains, and the side chain of free histidine are deprotonated.]

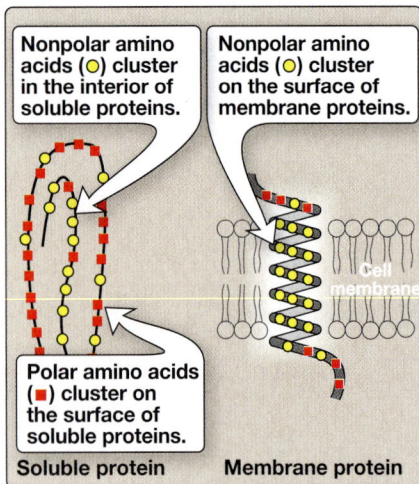

Figure 1.4
Location of nonpolar amino acids in soluble and membrane proteins.

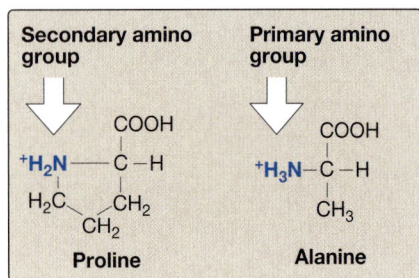

Figure 1.5
Comparison of the secondary amino group found in proline with the primary amino group found in other amino acids such as alanine.

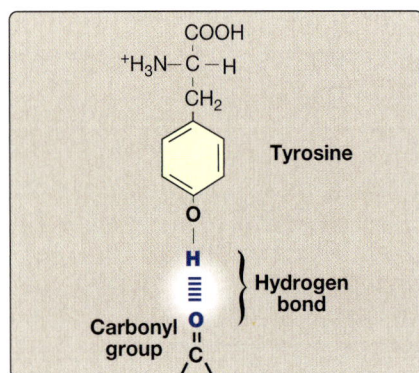

Figure 1.6
Hydrogen bond between the phenolic hydroxyl group of tyrosine and another molecule containing a carbonyl group.

1. **Location in proteins:** In proteins found in polar environments such as aqueous solutions, the side chains of nonpolar amino acids tend to cluster together in the interior of the protein (Fig. 1.4). This phenomenon is known as the hydrophobic effect and is the result of the hydrophobicity of the nonpolar R groups, which act much like droplets of oil that coalesce in an aqueous environment. By filling up the interior of the folded protein, these nonpolar R groups help give proteins their three-dimensional shape. The exclusion of water from the hydrophobic interior of a protein results in the entropy increase that drives protein folding.

   For proteins located in a hydrophobic environment, such as within the hydrophobic core of a phospholipid membrane, nonpolar R groups are found on the outside surface of the protein, interacting with the lipid environment (see Fig. 1.4). The importance of these hydrophobic interactions in stabilizing protein structure is discussed in Chapter 2.

   Sickle cell anemia, a disease of red blood cells (RBCs) that results from replacement of polar glutamate with nonpolar valine at the sixth position in the β subunit of hemoglobin A (see Chapter 3), causes the hemoglobin to precipitate under low oxygen conditions. A consequence is a change in shape from the expected round disc of an RBC to a C-shape or sickled appearance. Sickled erythrocytes are fragile and lyse open, causing hemolytic anemia.

2. **Proline:** Proline differs from other amino acids because its side chain and α-amino nitrogen form a rigid, five-membered ring (Fig. 1.5). Proline, then, has a secondary (rather than a primary) amino group and is referred to as an "imino acid." The unique geometry of proline contributes to the formation of the extended fibrous structure of collagen (see Chapter 4, II. B.), but it interrupts the α-helices found in more compact globular proteins (see Chapter 3, III.).

B. Amino acids with uncharged polar side chains

These amino acids have zero net charge at physiologic pH of approximately 7.4, although the side chains of cysteine and tyrosine can lose a proton at an alkaline pH (see Fig. 1.3). Serine, threonine, and tyrosine each have a polar hydroxyl group that can participate in hydrogen bond formation (Fig. 1.6). The side chains of asparagine and glutamine each contain a carbonyl group and an amide group, both of which can also participate in hydrogen bonds.

1. **Disulfide bond:** The side chain of cysteine contains a sulfhydryl (thiol) group (–SH), which is an important component within the active site of many enzymes. In proteins, the –SH groups of two cysteines can be oxidized to form a covalent cross-link called a disulfide bond (–S–S–). Two disulfide-linked cysteines are referred to as cystine. (See Chapter 2, IV. B. for a further discussion of disulfide bond formation.)

> ## Disulfide Bonds Stabilize Proteins
> Many extracellular proteins are stabilized by disulfide bonds. Albumin, a protein that functions in the transport of a variety of molecules in the blood, is one example. Fibrinogen, a blood protein converted to fibrin to stabilize blood clots, is another. Also, chains of the important hormone insulin are held together by disulfide bonds.

2. **Side chains as attachment sites for other compounds:** The polar hydroxyl group of serine, threonine, and tyrosine can serve as a site of attachment for phosphate groups. Kinases are enzymes that catalyze phosphorylation reactions with the result that serine, threonine, or tyrosine residues within their substrates are phosphorylated. Phosphatases are enzymes that remove the phosphate group. The changes in phosphorylation status of proteins (whether phosphorylated or not), especially of enzymes, alter their activity; some enzymes are more active when phosphorylated, while others are less active. In addition, the amide group of asparagine, as well as the hydroxyl group of serine and threonine, can serve as a site of attachment for oligosaccharide chains in glycoproteins (see also Chapter 14, VII.).

C. Amino acids with acidic side chains

The amino acids aspartic acid and glutamic acid are proton donors. At physiologic pH, the side chains of these amino acids are fully ionized, containing a negatively charged carboxylate group ($-COO^-$). The fully ionized forms are called aspartate and glutamate.

D. Amino acids with basic side chains

The side chains of the basic amino acids accept protons (see Fig. 1.3). At physiologic pH, the R groups of lysine and arginine are fully ionized and positively charged. In contrast, the free amino acid histidine

---

Clinical Application 1.1: Slower, Longer Acting Insulin Created by Substituting Amino Acids

Insulin glargine was first approved for use in the United States in 2000. It is a slower acting form of insulin created in the laboratory by replacing the asparagine normally at position 21 on the A chain of insulin with glycine and by extending the carboxy terminus by 2 additional arginine residues. The result of these changes is a less water-soluble form of insulin with a net charge of +0.2, which is closer to 0. Insulin glargine is absorbed more slowly from the site of injection. The glycine substitution prevents deamidation of the asparagine at acidic pH in the neutral, subcutaneous space. The additional arginine residues shift the isoelectric point from pH 5.4 to pH 6.7, making the molecule more soluble at acidic pH and less soluble at neutral pH. Insulin glargine is therefore a form of insulin that dissolves slowly, has longer activity, and requires less frequent injection. This form of insulin can be useful in the treatment of diabetes mellitus and help patients achieve better glycemic control. (See Chapter 23 for the structure of insulin.)

1 Unique first letter:

Cysteine	=	Cys	=	**C**
Histidine	=	His	=	**H**
Isoleucine	=	Ile	=	**I**
Methionine	=	Met	=	**M**
Serine	=	Ser	=	**S**
Valine	=	Val	=	**V**

2 Most commonly occurring amino acids have priority:

Alanine	=	Ala	=	**A**
Glycine	=	Gly	=	**G**
Leucine	=	Leu	=	**L**
Proline	=	Pro	=	**P**
Threonine	=	Thr	=	**T**

3 Similar sounding names:

A**r**ginine	=	Arg	=	**R** ("a**R**ginine")
Asparagine	=	Asn	=	**N** (contains N)
Aspartate	=	Asp	=	**D** ("aspar**D**ic")
Glutamate	=	Glu	=	**E** ("glut**E**mate")
Glutamine	=	Gln	=	**Q** ("**Q**-tamine")
Phenylalanine	=	Phe	=	**F** ("**F**enylalanine")
T**y**rosine	=	Tyr	=	**Y** ("t**Y**rosine")
Tryptophan	=	Trp	=	**W** (double ring in the molecule)

4 Letter close to initial letter:

Aspartate or asparagine	=	Asx	=	**B** (near A)
Glutamate or glutamine	=	Glx	=	**Z**
Lys**i**ne	=	Lys	=	**K** (near L)
Undetermined amino acid	=			**X**

Figure 1.7
Abbreviations and symbols for the standard amino acids.

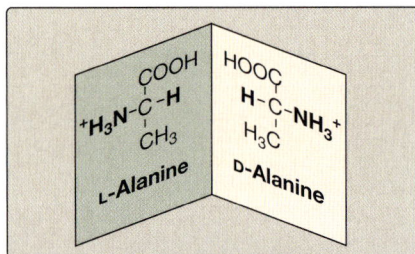

Figure 1.8
D and L forms of alanine are mirror images (enantiomers).

is weakly basic and largely uncharged at physiologic pH. However, when histidine is incorporated into a protein, its R group can be either positively charged (protonated) or neutral, depending on the ionic environment provided by the protein. This important property of histidine contributes to the buffering role it plays in the functioning of such proteins as hemoglobin (see Chapter 3). Histidine is the only amino acid with a side chain that can ionize within the physiologic pH range.

E. Abbreviations and symbols for commonly occurring amino acids

Each amino acid name has an associated three-letter abbreviation and a one-letter symbol (Fig. 1.7). The one-letter codes are determined by the following rules.

1. **Unique first letter:** If only one amino acid begins with a given letter, then that letter is used as its symbol. For example, V = valine.

2. **Most commonly occurring amino acids have priority:** If more than one amino acid begins with a particular letter, the most common of these amino acids receives this letter as its symbol. For example, glycine is more common than glutamate, so G = glycine.

3. **Similar-sounding names:** Some one-letter symbols sound like the amino acid they represent. For example, F = phenylalanine.

4. **Letter close to initial letter:** For the remaining amino acids, a one-letter symbol is assigned that is as close in the alphabet as possible to the initial letter of the amino acid, for example, K = lysine. Furthermore, B is assigned to Asx, signifying either aspartic acid or asparagine; Z is assigned to Glx, signifying either glutamic acid or glutamine; W is used for tryptophan and an X is used to represent an unidentified amino acid.

F. Amino acid isomers

Because the α-carbon of an amino acid is attached to four different chemical groups, it is an asymmetric or chiral atom. Glycine is the exception because its α-carbon has two hydrogen substituents. Amino acids with a chiral α-carbon exist in two different isomeric forms, designated D and L, which are enantiomers, or mirror images (Fig. 1.8). [Note: Enantiomers are optically active. If an isomer, either D or L, causes the plane of polarized light to rotate clockwise, it is designated the (+) form.] All amino acids found in mammalian proteins are of the L configuration. However, D-amino acids are found in some antibiotics and in bacterial cell walls. [Note: *Racemases* enzymatically interconvert the D- and L-isomers of free amino acids.]

III. ACIDIC AND BASIC PROPERTIES

Amino acids in aqueous solution contain weakly acidic α-carboxyl groups and weakly basic α-amino groups. In addition, each of the acidic and basic amino acids contains an ionizable group in its side chain. Thus, both free amino acids and some amino acids combined in peptide linkages can act as buffers. Acids may be defined as proton donors and bases as proton acceptors. Acids (or bases) described as weak ionize to only a limited extent.

A. pH

The concentration of protons ($[H^+]$) in aqueous solution is expressed as pH.

$$pH = \log 1/[H^+] \text{ or } -\log [H^+]$$

1. **Dissociation constants:** The salt or conjugate base, A^-, is the ionized form of a weak acid. By definition, the dissociation constant of the acid, K_a, is:

$$K_a = \frac{[H^+][A^-]}{[HA]}$$

The larger the K_a, the stronger the acid, because most of the HA has dissociated into H^+ and A^-. Conversely, the smaller the K_a, the less acid has dissociated and, therefore, the weaker the acid.

2. **Henderson–Hasselbalch equation:** By solving for the $[H^+]$ in the above equation, taking the logarithm of both sides of the equation, multiplying both sides of the equation by -1, and substituting $pH = -\log [H^+]$ and $pK_a = -\log K_a$, we obtain the Henderson–Hasselbalch equation:

$$pH = pKa + \log [A^-]/[HA]$$

This equation demonstrates the quantitative relationship between the pH of the solution and concentration of a weak acid (HA) and its conjugate base (A^-).

B. Buffers

A buffer is a solution that resists change in pH following the addition of an acid or base and can be created by mixing a weak acid (HA) with its conjugate base (A^-). If an acid is added to a buffer, A^- can neutralize it, being converted to HA in the process. If a base is added, HA can likewise neutralize it, being converted to A^- in the process.

Maximum buffering capacity occurs at a pH equal to the pK_a, but a conjugate acid–base pair can still serve as an effective buffer when the pH of a solution is within approximately ± 1 pH unit of the pK_a. If the amounts of HA and A^- are equal, the pH is equal to the pK_a. As shown in Figure 1.9, a solution containing acetic acid (HA = $CH_3 - COOH$) and acetate ($A^- = CH_3 - COO^-$) with a pKa of 4.8 resists a change in pH from 3.8 to 5.8, with maximum buffering at pH 4.8. At pH values less than the pKa, the protonated acid form ($CH_3 - COOH$) is the predominant species in solution. At pH greater than the pK_a, the deprotonated base form ($CH_3 - COO^-$) is the predominant species.

1. **Dissociation of the carboxyl group:** The dissociation constant of the carboxyl group of an amino acid is called K_1, rather than K_a, because the molecule contains a second titratable group.

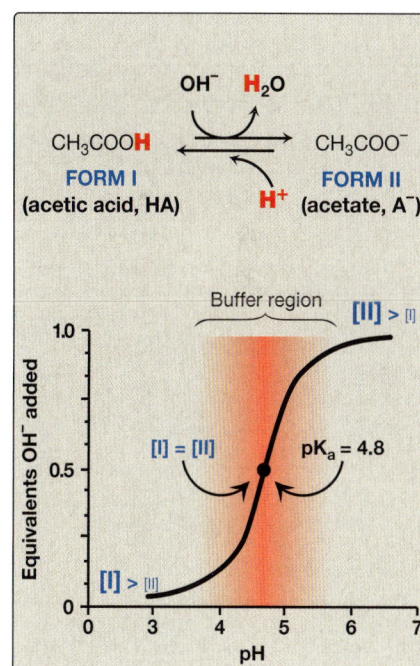

Figure 1.9
Titration curve of acetic acid.

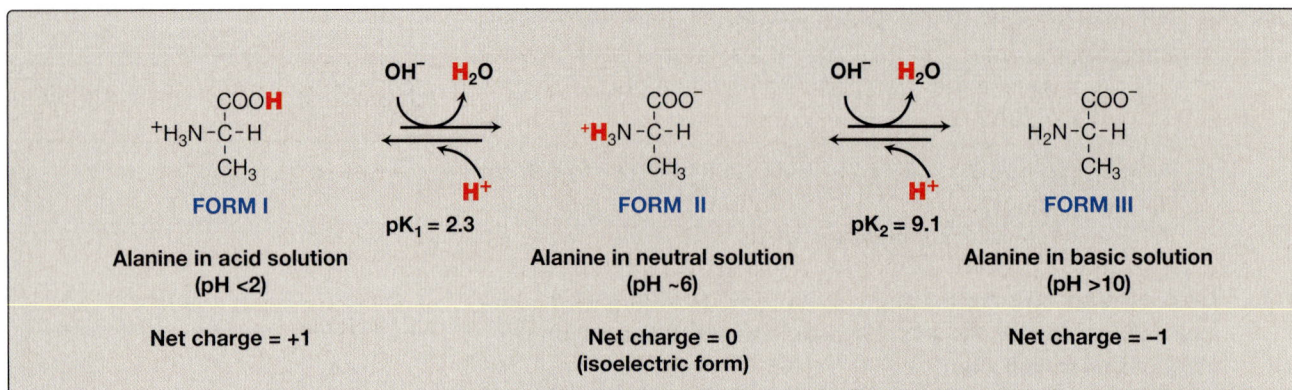

Figure 1.10
Ionic forms of alanine in acidic, neutral, and basic solutions.

The Henderson–Hasselbalch equation can be used to analyze the dissociation of the carboxyl group of alanine:

$$K_1 = [H^+][II]/[I]$$

where I is the fully protonated form of alanine and II is the isoelectric form of alanine (Fig. 1.10). This equation can be rearranged and converted to its logarithmic form to yield:

$$pH = pK_1 + \log [II]/[I]$$

2. **Amino group dissociation:** The second titratable group of alanine is the amino ($-NH_3^+$) group. Because this is a much weaker acid than the $-COOH$ group, it has a much smaller dissociation constant, K_2. [Note: Its pK_a is, therefore, larger.] Release of a H^+ from the protonated amino group of form II results in the fully deprotonated form of alanine, form III.

3. **pKs and sequential dissociation:** The sequential dissociation of H^+ from the carboxyl and amino groups is summarized in Figure 1.10 using alanine as an example. Each titratable group has a pK_a that is numerically equal to the pH at which exactly one-half of the H^+ have been removed from that group. The pK_a for the most acidic group ($-COOH$) is pK_1, whereas the pK_a for the next most acidic group ($-NH_3^+$) is pK_2. [Note: The pK_a of the α-carboxyl group of amino acids is ~2, whereas the pK_a of the α-amino group is ~9.]

By applying the Henderson–Hasselbalch equation to each dissociable acid group, it is possible to calculate the complete titration curve of a weak acid. Figure 1.11 shows the change in pH that occurs during the addition of base to the fully protonated form of alanine (I) to produce the complete deprotonated form (III).

a. **Buffer pairs:** The $-COOH/-COO^-$ pair can serve as a buffer in the pH region around pK_1, and the $-NH_3 \pm NH_2$ pair can buffer in the region around pK_2.

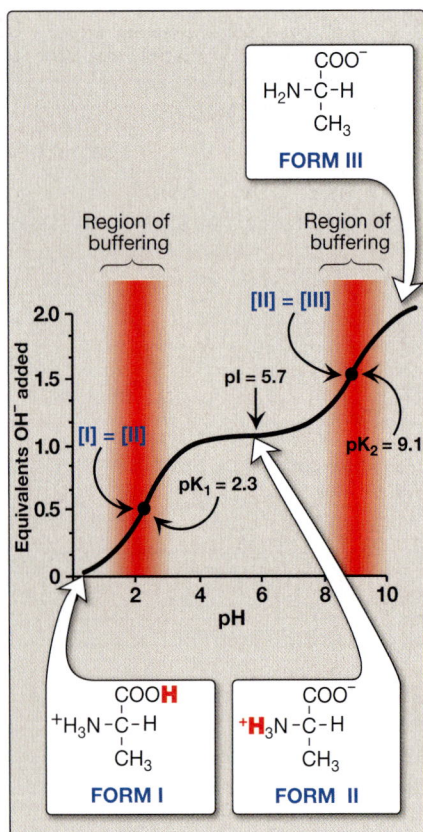

Figure 1.11
Titration curve of alanine.

b. When pH = pK: When the pH is equal to pK_1 (2.3), equal amounts of forms I and II of alanine exist in solution. When the pH is equal to pK_2 (9.1), equal amounts of forms II and III exist in solution.

c. Isoelectric point pI: At neutral pH, alanine exists predominantly as the dipolar form II in which the amino and carboxyl groups are ionized, but the net charge is zero. The isoelectric point (pI) is the pH at which an amino acid is electrically neutral, when the sum of the positive charges equals the sum of the negative charges. For alanine, with only two dissociable hydrogens (one from the α-carboxyl and one from the α-amino group), the pI is the average of pK_1 and pK_2 (pI = [2.3 + 9.1]/2 = 5.7) as shown in Figure 1.11. The pI is, thus, midway between pK_1 (2.3) and pK_2 (9.1). pI corresponds to the pH at which the form II (with a net charge of zero) predominates and at which there are also equal amounts of forms I (net charge of +1) and III (net charge of –1).

> In the laboratory, separation of plasma proteins by charge typically is done at a pH above the pI of the major proteins. Therefore, at a high alkaline pH, the charge on the proteins is negative. In an electric field, the proteins will move toward the positive electrode at a rate determined by their net negative charge. Variations in the mobility pattern are suggestive of certain diseases.

4. Net charge at neutral pH: At physiologic pH, amino acids have a negatively charged group ($-COO^-$) and a positively charged group ($-NH_3^+$), both attached to the α-carbon. Glutamate, aspartate, histidine, arginine, and lysine have additional potentially charged groups in their side chains. Substances such as amino acids that can act either as an acid or a base are described as amphoteric.

C. Buffering the blood, the bicarbonate buffer system

Blood pH is maintained in the slightly alkaline range of 7.35 to 7.45 by the bicarbonate buffer system. Most proteins function optimally at this physiologic pH, and their amino acid constituents exist in the chemical form described for their category; exceptions include some digestive enzymes that function at acidic pH of the stomach between pH 1.5 and 3.5. Lysosomal enzymes also function at an acidic pH range between pH 4.5 and pH 5.0. Maintaining arterial pH at 7.40 ± 0.5 is important for health; normally the bicarbonate buffer system is able to keep pH within the acceptable range.

The bicarbonate ion concentration [HCO_3^-] and the carbon dioxide concentration [CO_2] influence blood pH, as depicted in Figure 1.12A. The need for a buffering system can be appreciated by considering that organic acids (such as lactic acid) are generated during metabolism and glucose and fatty acid oxidation generate CO_2, the

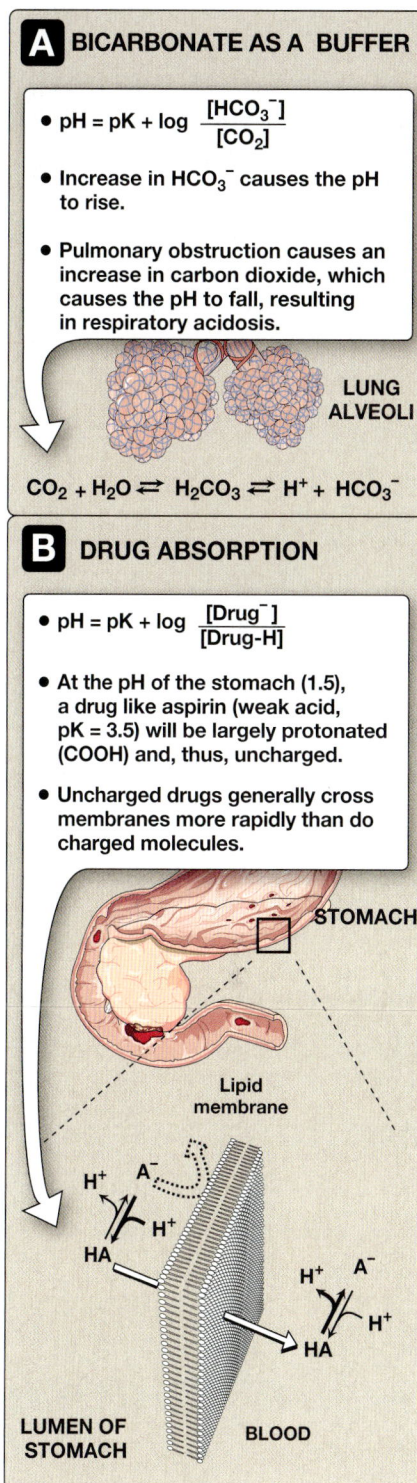

A BICARBONATE AS A BUFFER

- $pH = pK + \log \dfrac{[HCO_3^-]}{[CO_2]}$

- Increase in HCO_3^- causes the pH to rise.

- Pulmonary obstruction causes an increase in carbon dioxide, which causes the pH to fall, resulting in respiratory acidosis.

LUNG ALVEOLI

$CO_2 + H_2O \rightleftarrows H_2CO_3 \rightleftarrows H^+ + HCO_3^-$

B DRUG ABSORPTION

- $pH = pK + \log \dfrac{[Drug^-]}{[Drug\text{-}H]}$

- At the pH of the stomach (1.5), a drug like aspirin (weak acid, pK = 3.5) will be largely protonated (COOH) and, thus, uncharged.

- Uncharged drugs generally cross membranes more rapidly than do charged molecules.

STOMACH

Lipid membrane

H^+ A^-

H^+

HA

H^+ A^-

H^+

HA

LUMEN OF STOMACH   BLOOD

Figure 1.12
Henderson–Hasselbalch equation is used to predict: (**A**) changes in pH as the concentrations of bicarbonate (HCO_3^-) or carbon dioxide (CO_2) are altered and (**B**) the ionic forms of drugs.

anhydrous form of H_2CO_3 (carbonic acid), which is also an acid. The relatively water-insoluble CO_2 is converted by carbonic anhydrase to the water-soluble HCO_3^- (bicarbonate), which is carried through the blood to the lungs where dissolved CO_2 is exhaled. Therefore, lungs regulate loss and retention of CO_2 by altering the breathing rate. The kidneys are also important in regulating acid–base balance by retaining or excreting bicarbonate, H^+, ammonia, and other acids/bases that appear in the blood.

D. pH and drug absorption

Many drugs are administered orally and are transported across intestinal epithelial cells to be absorbed into blood. Most drugs are either weak acids or weak bases. Acid drugs (HA) release a H^+, causing a charged anion (A^-) to form. Weak bases (BH^+) can also release a H^+; however, the protonated form of basic drugs is usually charged, and the loss of a proton produces the uncharged base (B).

$$HA \leftarrow \rightarrow H^+ + A^-$$
$$BH^+ \leftarrow \rightarrow B + H^+$$

Drugs are best absorbed at the pH where dissociation of their side chains results in the most neutral molecule. The effective concentration of the permeable form of each drug at its absorption site is determined by the relative concentrations of the charged and uncharged forms (Fig. 1.12B). Transport of drugs into cells occurs via membrane transport proteins and often through active transport processes, although the systems are not well characterized. (See also *LIR Cell and Molecular Biology*, 3e Chapter 16.)

E. Blood gases and pH

As a consequence of certain disease processes or poisons, blood pH can become abnormal. Acidemia is arterial pH lower than 7.35, and alkalemia is an arterial pH higher than 7.45. In the bicarbonate buffer system, CO_2 is an acid, and bicarbonate a base. Because the bicarbonate buffer is an open system and CO_2 is released in the breath, changes in breathing impact the acid–base balance in the body. Hyperventilation can cause release of too much acid, causing alkalotic conditions; generation of excess metabolic acids (e.g., lactic acidosis or ketoacidosis that can accompany type 1 diabetes mellitus) can cause acidosis. Loss of excess acid through vomiting can also cause an acid–base disturbance. Neither renal compensation nor compensation by changing breathing rate (respiratory compensation) will bring pH back into physiologic range if excess metabolic acids have been generated. Neither the lungs nor the kidneys can fully compensate for and can never overcompensate for pH imbalances. Measuring CO_2 and bicarbonate along with pH can help to determine a patient's acid–base imbalance (Table 1.1).

Table 1.1: Disturbances in Acid–Base Balance

pH	[H$^+$]	Initial Issue	Response	Disorder
Decreased	Increased	Hypoventilation; increased retention of CO_2 (more acid)	Increased renal retention of HCO_3^- (more base)	**Respiratory acidosis**; lungs not excreting enough acid as CO_2 such as in Chronic obstructive pulmonary disease (COPD)
Increased	Decreased	Hyperventilation; increased release of CO_2 (less acid)	Decreased renal retention of HCO_3^- (less base)	**Respiratory alkalosis**; lungs excreting too much acid as CO_2 such as in hyperventilation and asthma
Decreased	Increased	More acid generated	More CO_2 released in breath (hyperventilation); renal excretion of HCO_3^-	**Metabolic acidosis**; body generates acid that cannot be excreted by lungs such as lactic acidosis, diabetic ketoacidosis, ingestion of acid
Increased	Decreased	HCO_3^- increases	Less CO_2 released in breath (hypoventilation); HCO_3^- will be low to attempt to buffer the acid	**Metabolic alkalosis**; blood is alkaline and not caused by respiratory imbalance such as excess loss of acid in vomiting or ingestion of a base

IV. CHAPTER SUMMARY

- Each amino acid has an α-**carboxyl group** and a primary α-**amino group** (except for **proline**, which has a **secondary amino group**).
- α-Carbons of each amino acid (except glycine) attach to four different chemical groups and are asymmetric (**chiral**), and amino acids exist in D- and L-isomeric forms that are optically active mirror images (**enantiomers**). The L-form of amino acids is found in proteins synthesized by humans. Amino acids within proteins are called **residues**.
- At physiologic pH, the α-carboxyl group is dissociated, forming the negatively charged carboxylate ion (–COO$^-$), and the α-amino group is protonated (–NH$_3^+$).
- Each amino acid also contains one of 20 distinctive **side chains** attached to the α-carbon atom.
- The chemical nature of this **R group** determines the function of an amino acid in a protein and provides the basis for classification of the amino acids as **nonpolar**, **uncharged polar**, **acidic (polar negative)**, or **basic (polar positive)**.
- All free amino acids, plus charged amino acids in peptide chains, can serve as **buffers**.
- The quantitative relationship between the pH of a solution and the concentration of a weak acid (HA) and its conjugate base (A$^-$) is described by the **Henderson–Hasselbalch equation**. Buffering occurs within ±1 pH unit of the pK_a and is maximal when pH = pK_a, at which [A$^-$] = [HA].
- The pH within blood is maintained in the slightly alkaline range of 7.4 ± 0.5 by the bicarbonate buffer system; the lungs regulate the acid CO_2 by altering breathing rate and the kidneys retain or release acids and bases.

Study Questions

Choose the ONE best answer.

1.1. The peptide Val-Cys-Glu-Ser-Asp-Arg-Cys:
- A. contains asparagine.
- B. contains a side chain with a secondary amino group.
- C. contains a side chain that can be phosphorylated.
- D. cannot form an internal disulfide bond.
- E. cannot move toward the cathode during electrophoresis at pH 5.

Correct answer = C. The hydroxyl group of serine can accept a phosphate group. Asp is aspartate, not asparagine. Proline contains a secondary amino group and is not within this peptide. The two cysteine residues can, under oxidizing conditions, form a disulfide bond. The net charge on the peptide at pH 5 is negative, and it would move to the anode.

1.2. An amino acid with a secondary amino group is geometrically incompatible with a right-handed spiral of an α helix; is observed to insert a kink in the amino acid chain and to interfere with the normally smooth, helical structure of the α helix; and is found in high concentration in collagen. The amino acid described is:

A. Ala
B. Cys
C. Gly
D. Pro
E. Ser

Correct answer = D. Proline differs from other amino acids in that its side chain and α-amino nitrogen form a rigid, 5-membered ring structure and, therefore, contains a secondary amino group. It interrupts α helices in globular proteins, contributes to the structure of collagen, and is found in high concentration in collagen. None of the other amino acids have these properties.

1.3. An amino acid that may have its side chain phosphorylated by the action of a kinase is:

A. Arg
B. Cys
C. Gly
D. Ser
E. Val

Correct answer = D. The polar hydroxyl group found within Ser, Thr, and Tyr can serve as a site of attachment for phosphate groups when acted on by kinases. None of the other amino acids contains a hydroxyl group susceptible to phosphorylation by a kinase.

1.4. The titration curve for a nonpolar amino acid is shown below. Which of the following descriptions about the letters A through D is correct?

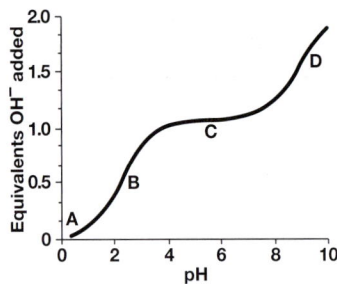

A. Point A represents the region where the amino acid is deprotonated.
B. Point B represents a region of minimal buffering.
C. Point C represents the region where the net charge on the amino acid is zero.
D. Point D represents the pK of the amino acid's carboxyl group.
E. The amino acid could be lysine.

Correct answer = C. Point C represents the isoelectric point, or pI, and as such is midway between pK_1 and pK_2 for a nonpolar amino acid. The amino acid is fully protonated at Point A. Point B represents a region of maximum buffering, as does Point D. Lysine is a basic amino acid, and free lysine has an ionizable side chain in addition to the ionizable α-amino and α-carboxyl groups.

1.5. An 18-year-old woman with a 15-year history of type 1 diabetes mellitus is brought to the emergency department for evaluation of nausea, vomiting, and altered consciousness. Her blood glucose is 560 mg/dL (reference range for random glucose, <200 mg/dL). Her arterial blood pH is 7.15 (reference range 7.35 to 7.45) and bicarbonate is 12 mEq/L (22 to 28 mEq/L). Which of the following would be the expected type of compensation in her body in response to this acid–base imbalance?

A. Increased respiration
B. Decreased respiration
C. Increased renal release of acid
D. Increased renal retention of base

Correct answer = A. In response to a metabolic acidosis, compensation is respiratory. Increased respiration removes acid in the form of CO_2 from the body. Since the acid is being generated metabolically (diabetic ketoacidosis suspected), renal release of acid or retention of base would not be compensatory.

1.6. A 5-year-old boy with an unremarkable past medical history was seen in the emergency department for evaluation of excessive vomiting and poor oral intake of 2 days' duration. He takes no medications. His abdomen appears distended. His blood glucose is 80 mg/dL (reference range 70 to 100 mg/dL). Arterial blood gases reveal a pH of 7.55 (reference range 7.35 to 7.45) and bicarbonate 35.5 mmol/L (reference range 22 to 26 mmol/L). pCO_2 results are not available. The most likely cause of this patient's acid–base imbalance is:

A. constipation
B. diabetes mellitus type 1
C. hypoventilation
D. renal failure
E. vomiting

Correct answer = E. In response to vomiting, acid is lost, and metabolic alkalosis can result as in this situation. Only the distended abdomen might support constipation, but his altered blood gases do not. His fasting blood glucose is normal, and, coupled with his unremarkable medical history and taking no medications, rules out diabetes mellitus type 1. Decreased respiration in hypoventilation would cause more retention of acid in the form of CO_2 in the body. Hypoventilation is likely occurring as compensation for the loss of acids in vomiting but is not a cause of the acid–base imbalance. There is no evidence that he has kidney failure. Renal compensation for this excess loss of acid is most likely occurring, but we cannot fully compensate for acid–base imbalances.

1.7. Replacement of which of the following would represent a potential modification to human insulin that would prolong its action?

A. Ala with Val
B. Asn with Gly
C. Asp with Glu
D. Cys with Ser
E. Met with Phe

Correct answer = B. Replacement of Asn with Gly represents a change from an uncharged polar side chain to a nonpolar side chain. This is the substitution made at position 21 on the A chain of insulin to produce insulin glargine, which has a smaller positive charge and slower absorption, prolonging its action. All the other substitutions represent replacement of an amino acid with another within the same category and would have little effect on the net charge and be unable to prolong the action of the insulin.

2 Protein Structure

I. OVERVIEW

Proteins are composed of amino acids that are joined together by peptide bonds in a linear sequence and then folded into a unique three-dimensional shape that determines their function. The complexity of protein structure is considered in terms of four organizational levels: primary, secondary, tertiary, and quaternary (Fig. 2.1). Examination of these levels of increasing structural complexity has revealed that certain structural elements are repeated in a wide variety of proteins and range from simple combinations of α-helices and β-sheets forming small motifs to the complex folding of polypeptide domains of multifunctional proteins (see Section IV.).

II. PRIMARY STRUCTURE

The primary structure of a protein is its linear sequence of amino acids. Many genetic diseases are caused by missense mutations that cause one amino acid to be replaced with another. The results are abnormal proteins that may be improperly folded and/or have lost or impaired function. If the primary structures of the normal and abnormal proteins are known, this information may be used to diagnose or study the disease.

A. Peptide bond

In proteins, adjacent amino acids are joined covalently by peptide bonds, which are amide linkages between the α-carboxyl group of one amino acid and the α-amino group of the next amino acid. For example, valine and alanine can form the dipeptide valylalanine through the formation of a peptide bond (Fig. 2.2). Peptide bonds are resistant to conditions that denature proteins, such as heat and high concentrations of urea. Prolonged exposure to a strong acid or base at elevated temperatures is required to nonenzymatically break these bonds.

1. **Naming the peptide:** By convention, the free amino end (N-terminal) of the peptide chain is written to the left and the free carboxyl end (C-terminal) to the right. Therefore, all amino acid sequences are read from the N- to the C-terminal end. For example, in Figure 2.2A, the order of the amino acids in the dipeptide is valine, alanine. Linkage of 50 or more amino acids through

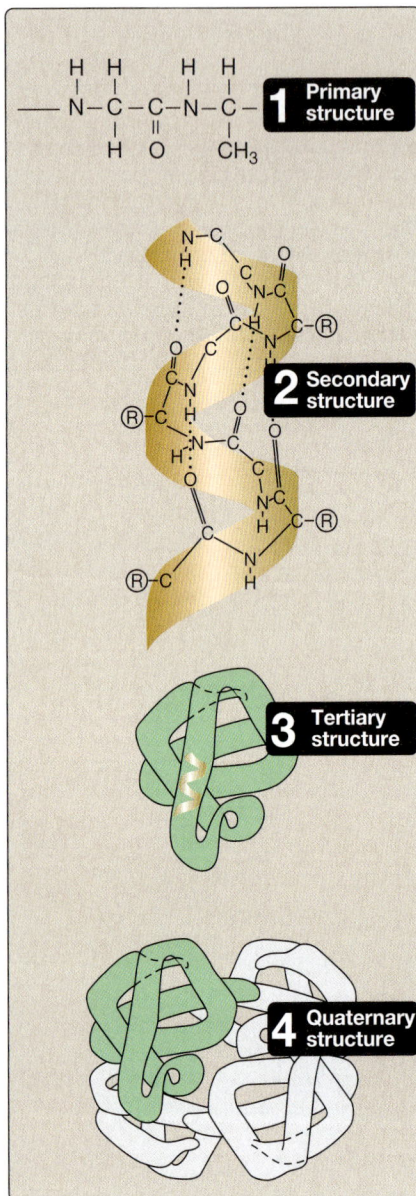

Figure 2.1
Four hierarchies of protein structure.

peptide bonds results in an unbranched chain called a polypeptide, or protein. Each component amino acid is called a residue because it is the portion of the amino acid remaining after the water atoms are lost during the formation of the peptide bond. When a peptide is named, all amino acid residues have their suffixes (-ine, -an, -ic, or -ate) changed to -yl, with the exception of the C-terminal amino acid. For example, a tripeptide composed of an N-terminal valine, a glycine, and a C-terminal leucine is called valylglycylleucine.

2. **Peptide bond characteristics:** The peptide bond has a partial double-bond character; it is shorter than a single bond and is rigid and planar (Fig. 2.2B). This prevents free rotation around the bond between the carbonyl carbon and the nitrogen of the peptide bond. However, the bonds between the α-carbons and the α-amino or α-carboxyl groups can rotate freely (although they are limited by the size and character of the R groups). This allows the polypeptide chain to assume a variety of possible conformations. The peptide bond is almost always in the *trans* configuration (instead of the *cis*; see Fig. 2.2B), largely because of steric interference of the R groups (side chains) when in the *cis* position.

3. **Peptide bond polarity:** Like all amide linkages, the $-C = O$ and $-NH$ groups of the peptide bond are uncharged, and neither accept nor release protons over the pH range of 2 to 12. The charged groups present in polypeptides consist solely of the N-terminal (α-amino) group, the C-terminal (α-carboxyl) group, and any ionized groups present in the side chains of the constituent amino acids. However, the $-C = O$ and $-NH$ groups of the peptide bond are polar, and are involved in hydrogen bonds (e.g., in α-helices and β-sheets), as described on page 17.

B. Determining the amino acid composition of a polypeptide

The first step in determining the primary structure of a polypeptide is to identify and measure its constituent amino acids. Many polypeptides have a unique amino acid profile. A purified sample of the polypeptide to be analyzed is first hydrolyzed by strong acid to cleave the peptide bonds and release the individual amino acids. These can then be separated by cation-exchange chromatography. Amino acids bind to the chromatography column with different affinities, depending on their charges, hydrophobicity, and other characteristics. Each amino acid is then sequentially released from the chromatography column by eluting with solutions of increasing ionic strength and pH (Fig. 2.3), and the separated amino acids are quantitated spectrophotometrically. The analysis described above is performed using an automated amino acid analyzer, whose components are depicted in Figure 2.3.

C. Sequencing the peptide from its N-terminal end

Sequencing is a stepwise process of identifying the specific amino acid at each position in the peptide chain, beginning at the N-terminal end. Automated sequencers are used now; the historical process to produce amino acid derivatives is shown in Figure 2.4.

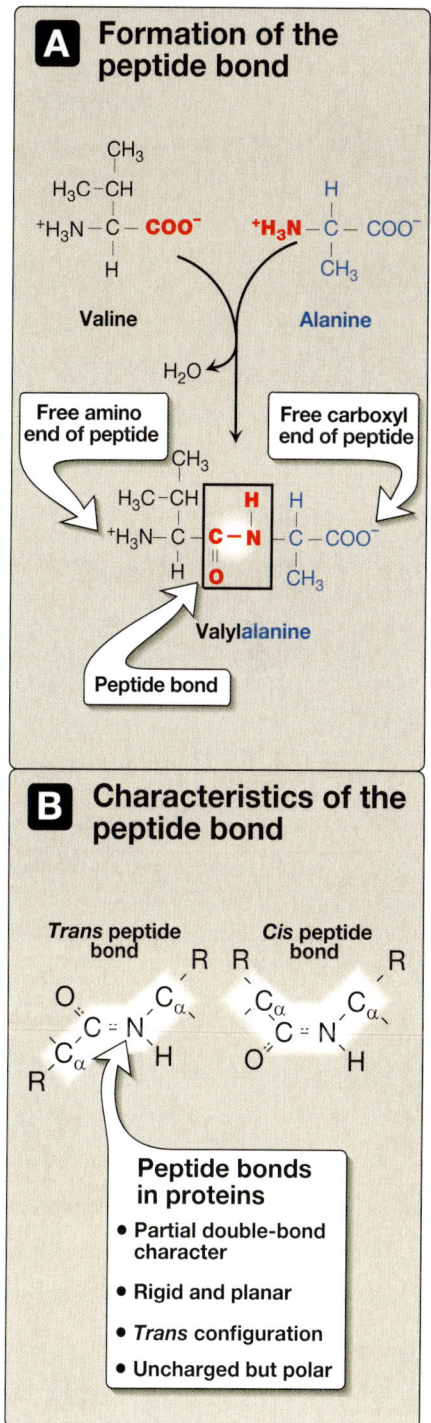

A Formation of the peptide bond

B Characteristics of the peptide bond

Peptide bonds in proteins
- Partial double-bond character
- Rigid and planar
- *Trans* configuration
- Uncharged but polar

Figure 2.2
A. Formation of a peptide bond, showing the structure of the dipeptide valylalanine. **B.** Characteristics of the peptide bond. [Note: Peptide bonds involving proline may have a *cis* configuration.]

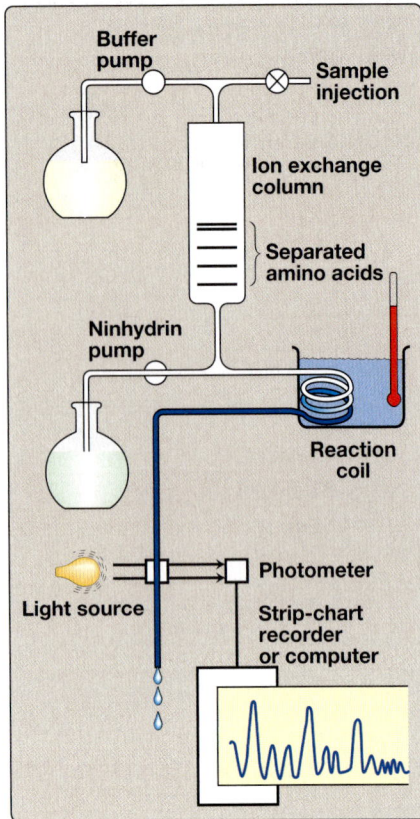

Figure 2.3
Determination of the amino acid composition of a polypeptide using an amino acid analyzer.

D. Cleaving the polypeptide into smaller fragments

Many polypeptides have a primary structure composed of more than 100 amino acids. Such molecules cannot be sequenced directly from end to end. However, these large molecules can be cleaved at specific sites and the resulting fragments sequenced (Fig. 2.5). Enzymes that hydrolyze peptide bonds are known as peptidases or proteases, which can be categorized by the location of the bond they cleave. Exopeptidases cut at the ends of proteins and are classified into aminopeptidases and carboxypeptidases. Carboxypeptidases are used in determining the C-terminal amino acid. Endopeptidases cleave within a protein.

E. Determining a protein's primary structure by DNA sequencing

The sequence of nucleotides in a protein-coding region of the DNA specifies the amino acid sequence of a polypeptide. If the nucleotide sequence can be determined, then the genetic code allows the sequence of nucleotides to be translated into the corresponding amino acid sequence of that polypeptide. This indirect process, used to obtain the amino acid sequences of proteins, has the limitations of not being able to predict positions of disulfide bonds in the folded chain and of not identifying amino acids that are modified after their incorporation into the polypeptide (posttranslational modification). Therefore, direct protein sequencing is an extremely important tool for determining the true character of the primary sequence of many polypeptides. (See also Chapter 34 for discussion of related techniques.)

III. SECONDARY STRUCTURE

The polypeptide backbone generally forms regular arrangements of amino acids that are located near each other in the linear sequence. These arrangements are known as the secondary structure of the polypeptide. α-Helices, β-sheets, and β-bends (or, β-turns) are examples of secondary structures commonly encountered in proteins. Each of these structures is stabilized by hydrogen bonds between atoms of the peptide backbone. [Note: The collagen α-chain triple helix, another example of secondary structure, is discussed in Chapter 4.]

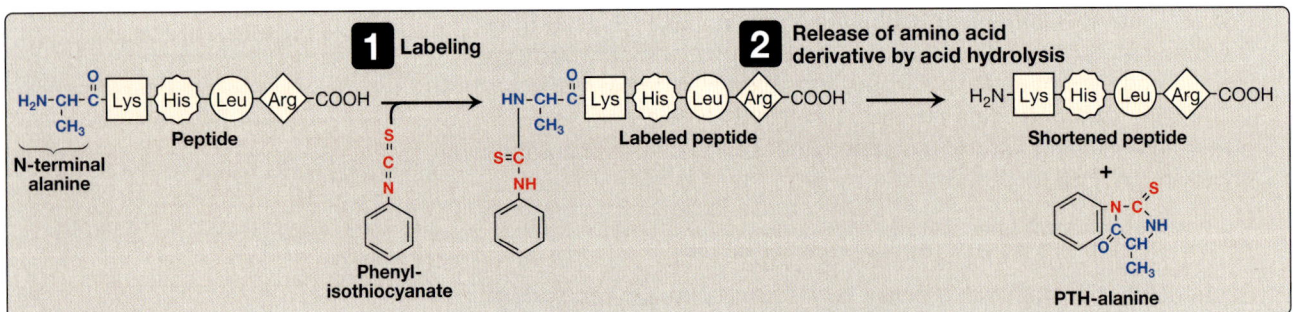

Figure 2.4
Determination of the amino (N)-terminal residue of a polypeptide by Edman degradation. PTH, phenylthiohydantoin.

A. α-Helix

Several different polypeptide helices are found in nature, but the α-helix is the most common. It is a rigid, right-handed spiral structure, consisting of a tightly packed, coiled polypeptide backbone core, with the side chains of the component L-amino acids extending outward from the central axis to avoid interfering sterically with each other (Fig. 2.6). A diverse group of proteins contain α-helices. For example, the keratins are a family of closely related, rigid, fibrous proteins whose structure is nearly entirely α-helical. They are a major component of tissues such as hair and skin. In contrast to keratin, myoglobin, whose structure is also highly α-helical, is a globular, flexible oxygen-binding molecule found in muscles (see Chapter 3, II. B.).

1. **Hydrogen bonds:** An α-helix is stabilized by extensive hydrogen bonding between the peptide bond carbonyl oxygens and amide hydrogens that are part of the polypeptide backbone (see Fig. 2.6). The hydrogen bonds extend up and are parallel to the spiral from the carbonyl oxygen of one peptide bond to the –NH group of a peptide linkage four residues ahead in the polypeptide. This ensures that all but the first and last peptide bond components are linked to each other through intrachain hydrogen bonds. Hydrogen bonds are individually weak but collectively serve to stabilize the helix.

2. **Amino acids per turn:** Each turn of an α-helix contains 3.6 amino acids. Thus, amino acids spaced three or four residues apart in the primary sequence are spatially close together when folded in the α-helix.

3. **Amino acids that disrupt an α-helix:** The R group of an amino acid determines its propensity to be in an α-helix. Proline disrupts α-helices because its rigid secondary amino group is not geometrically compatible with the right-handed spiral of the α-helix. It inserts a kink in the chain, interfering with the smooth, helical structure. Glycine, with hydrogen as its R group, confers high flexibility. Additionally, amino acids with charged or bulky R groups, such as glutamate and tryptophan, respectively, and those with a branch at the β-carbon, the first carbon in the R group (e.g., valine), are less likely to be found in an α-helix.

B. β-Sheet

β-sheets are another form of secondary structure in which all the peptide bond components are involved in hydrogen bonding (Fig. 2.7A). Because the surfaces of β-sheets appear to be folded or to form "pleats," they are often called β-pleated sheets. Pleating results from successive α-carbons being slightly above or below the plane of the sheet. Illustrations of protein structure often show β-strands as broad arrows (Fig. 2.7B).

1. **Formation:** A β-sheet is formed by two or more peptide chains (β-strands) aligned laterally and stabilized by hydrogen bonds between the carboxyl and amino groups of amino acids that either are far apart in a single polypeptide (intrachain bonds) or are in different polypeptide chains (interchain bonds). The adjacent β-strands are arranged either antiparallel to each other (with

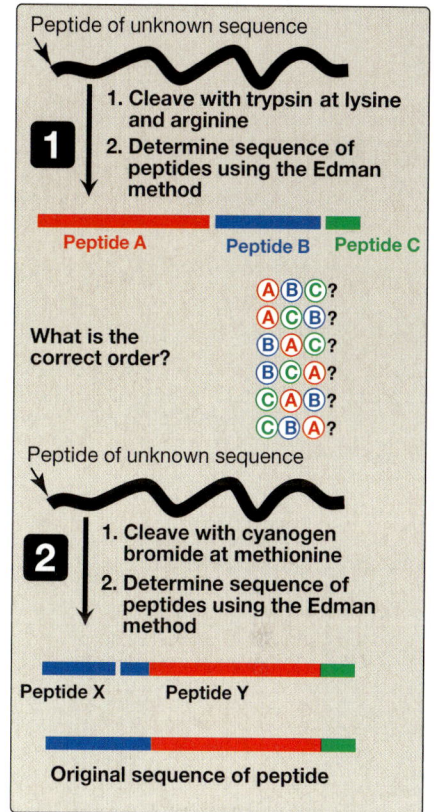

Figure 2.5
Overlapping of peptides produced by the cleavage action of trypsin and cyanogen bromide.

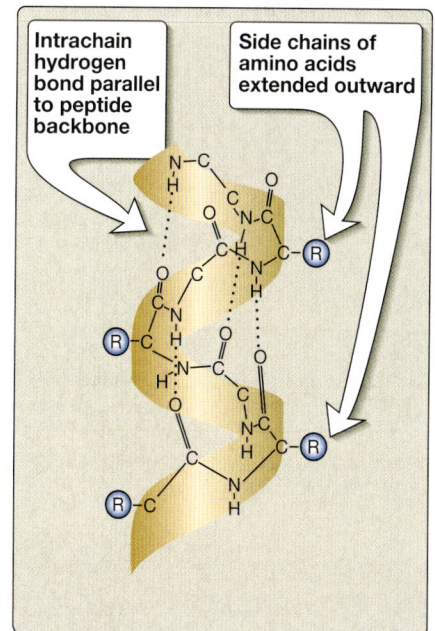

Figure 2.6
Structure of an α-helix.

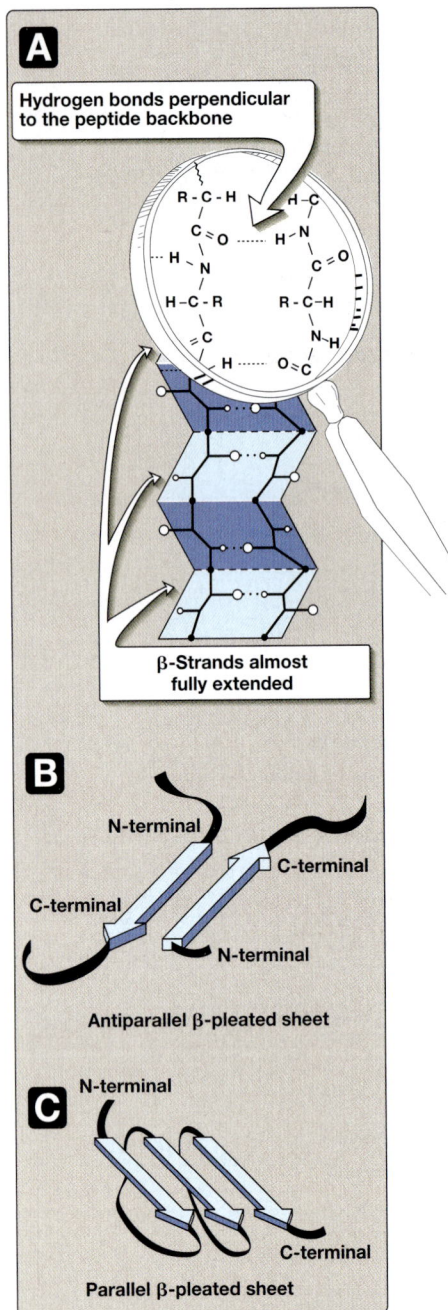

A. Hydrogen bonds perpendicular to the peptide backbone

β-Strands almost fully extended

B

N-terminal

C-terminal

C-terminal

N-terminal

Antiparallel β-pleated sheet

C N-terminal

C-terminal

Parallel β-pleated sheet

Figure 2.7
A. Structure of a β-sheet. **B.** Antiparallel β-sheet with the β-strands represented as *broad arrows*. **C.** Parallel β-sheet formed from a single polypeptide chain folding back on itself.

the N-termini alternating, as shown in Fig. 2.7B) or parallel to each other (with the N-termini together, as shown in Fig. 2.7C). On each β-strand, the R groups of adjacent amino acids extend in opposite directions, above and below the plane of the β-sheet. β-Sheets are not flat and have a right-handed curl (twist) when viewed along the polypeptide backbone.

2. **Comparing α-helices and β-sheets:** In β-sheets, the β-strands are almost fully extended, and the hydrogen bonds between the strands are perpendicular to the polypeptide backbone (see Fig. 2.7A). In contrast, in α-helices, the polypeptide is coiled, and the hydrogen bonds are parallel to the backbone (see Fig. 2.6).

> The orientation of R groups of amino acid residues in both α-helices and the β-sheets can result in formation of polar and nonpolar sides in these secondary structures, thereby making them amphipathic.

C. β-Bends

β-bends, also called reverse turns and β-turns, reverse the direction of a polypeptide chain, helping it form a compact, globular shape. They are usually found on the surface of protein molecules and often include charged residues. β-Bends were given this name because they often connect successive strands of antiparallel β-sheets. They are generally composed of four amino acids, one of which may be proline, the amino acid that causes a kink in the polypeptide chain. Glycine, the amino acid with the smallest R group, is also frequently found in β-bends. β-Bends are stabilized by the formation of hydrogen bonds between the first and last residues in the bend.

D. Nonrepetitive secondary structure

Approximately half of an average globular protein is organized into repetitive structures, such as the α-helix and β-sheet. The remainder of the polypeptide chain is described as having a loop or coil conformation. These nonrepetitive secondary structures are not random but rather simply have a less regular structure than those described above. "Random coil" refers to the disordered structure obtained when proteins are denatured (see Chapter 2, IV. D.).

E. Supersecondary structures (motifs)

Globular proteins are constructed by combining secondary structural elements including α-helices, β-sheets, and coils, producing specific geometric patterns, or motifs. These form primarily the core (interior) region of the molecule. They are connected by loop regions (e.g., β-bends) at the surface of the protein. Supersecondary structures are usually produced by the close packing of side chains from adjacent secondary structural elements. For example, α-helices and β-sheets that are adjacent in the amino acid sequence are also usually (but not always) adjacent in the final, folded protein. Some of the more common motifs are illustrated in Figure 2.8.

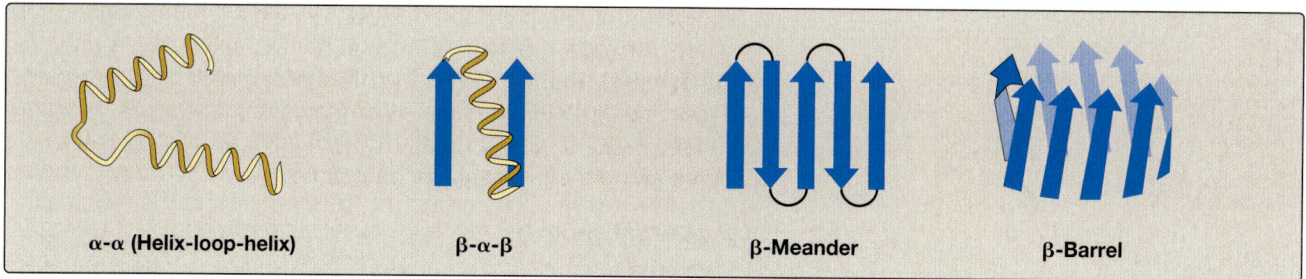

Figure 2.8
Common structural motifs involving α-helices and β-sheets. Their names describe their schematic appearance.

> Motifs are structural features that may be associated with particular functions. Proteins that bind to DNA contain a limited number of motifs. The helix–loop–helix motif is an example found in many proteins that function as transcription factors (see Chapter 31).

IV. TERTIARY STRUCTURE

The primary structure of a polypeptide chain determines its tertiary structure. Tertiary refers both to the folding of domains (the basic units of structure and function; see Section A) and to the final arrangement of domains in the polypeptide. The tertiary structure of globular proteins in aqueous solution is compact, with a high density (close packing) of the atoms in the core of the molecule. Hydrophobic side chains are buried in the interior, whereas hydrophilic groups are generally found on the surface of the molecule.

A. Domains

Domains are the fundamental functional and three-dimensional structural units of polypeptides. Polypeptide chains that are more than 200 amino acids in length generally consist of two or more domains. The core of a domain is built from combinations of supersecondary structural elements (motifs). Folding of the peptide chain within a domain usually occurs independently of folding in other domains. Therefore, each domain has the characteristics of a small, compact globular protein that is structurally independent of the other domains in the polypeptide chain.

B. Stabilizing interactions

The unique three-dimensional structure of each polypeptide is determined by its amino acid sequence. Interactions between the amino acid side chains guide the folding of the polypeptide to form a compact structure. The following four types of interactions cooperate in stabilizing the tertiary structures of globular proteins.

1. **Disulfide bonds:** A disulfide bond (–S–S–) is a covalent linkage formed from the sulfhydryl group (–SH) of each of two cysteine residues to produce a cystine residue (Fig. 2.9). The two cysteines

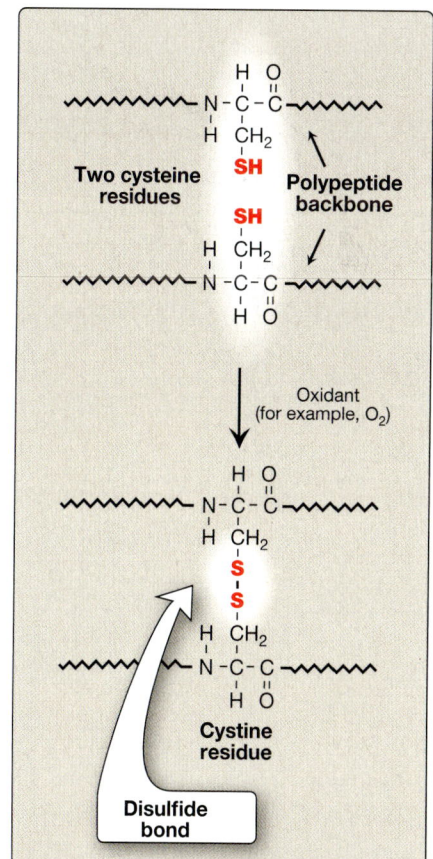

Figure 2.9
Formation of a disulfide bond by the oxidation of two cysteine residues, producing one cystine residue. O_2, oxygen.

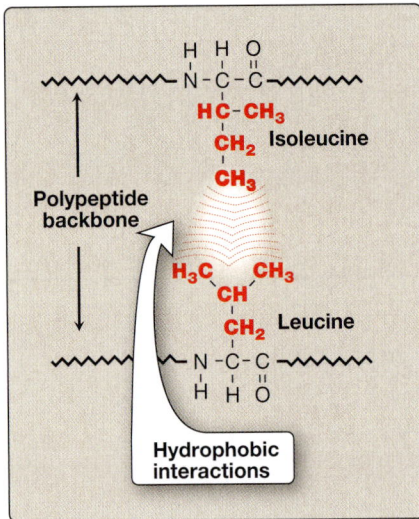

Figure 2.10
Hydrophobic interactions between amino acids with nonpolar side chains.

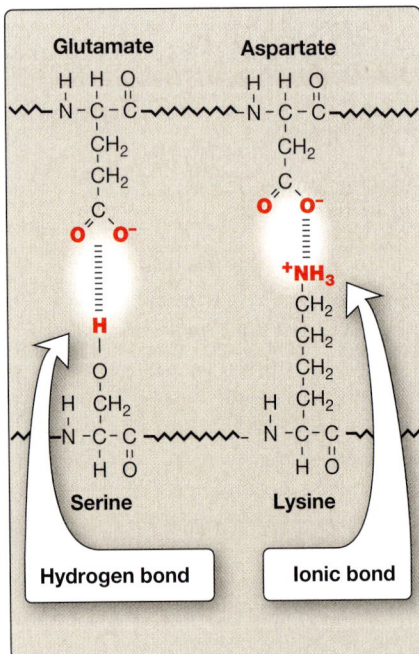

Figure 2.11
Interactions of side chains of amino acids through hydrogen bonds and ionic bonds (salt bridges).

may be separated from each other by many amino acids in the primary sequence of a polypeptide or may even be located on two different polypeptides. The folding of the polypeptide(s) brings the cysteine residues into proximity and permits covalent bonding of their side chains. A disulfide bond contributes to the stability of the three-dimensional shape of the protein molecule and prevents it from becoming denatured in the extracellular environment. For example, many disulfide bonds are found in proteins such as immunoglobulins that are secreted by cells. [Note: Protein disulfide isomerase breaks and reforms disulfide bonds during folding.]

2. **Hydrophobic interactions:** Amino acids with nonpolar side chains tend to be located in the interior of the polypeptide molecule, where they associate with other hydrophobic amino acids (Fig. 2.10). In contrast, amino acids with polar or charged side chains tend to be located on the surface of the molecule in contact with the polar solvent. In each case, a segregation of R groups occurs that is energetically most favorable.

3. **Hydrogen bonds:** Amino acid side chains containing oxygen- or nitrogen-bound hydrogen, such as in the alcohol groups of serine and threonine, can form hydrogen bonds with electron-rich atoms, such as the oxygen of a carboxyl group or carbonyl group of a peptide bond (Fig. 2.11; see also Fig. 1.6). Formation of hydrogen bonds between polar groups on the surface of proteins and the aqueous solvent enhances the solubility of the protein.

4. **Ionic interactions:** Negatively charged groups, such as the carboxylate group ($-COO^-$) in the side chain of aspartate or glutamate, can interact with positively charged groups such as the amino group ($-NH_3^+$) in the side chain of lysine (see Fig. 2.11).

C. Protein folding

Interactions between the side chains of amino acids determine how a linear polypeptide chain folds into the intricate three-dimensional shape of the functional protein. Protein folding, which occurs within the cell in seconds to minutes, involves nonrandom, ordered pathways. As a peptide folds, secondary structures form, driven by the hydrophobic effect; hydrophobic groups come together as water is released. These small structures combine to form larger structures. Additional events stabilize secondary structure and initiate formation of tertiary structure. In the last stage, the peptide achieves its fully folded, native (functional) form characterized by a low-energy state (Fig. 2.12). Some biologically active proteins or segments thereof lack a stable tertiary structure and are referred to as intrinsically disordered proteins.

D. Protein denaturation

Denaturation results in the unfolding and disorganization of a protein's secondary and tertiary structures without the hydrolysis of peptide bonds. When denatured proteins precipitate in a cell, this disrupts cell function. Denaturing agents include heat, urea, organic solvents, strong acids or bases, detergents, and ions of heavy metals such as lead. Denaturation may, under ideal conditions, be reversible, such that the protein refolds into its original native structure when the denaturing agent is removed. However, most proteins remain permanently disordered once denatured. Denatured proteins are often insoluble and precipitate from solution.

E. Chaperones in protein folding

The information needed for correct protein folding is contained in the primary structure of the polypeptide. However, most denatured proteins do not resume their native conformations even under favorable environmental conditions. This is because, for many proteins, folding is a facilitated process that requires ATP hydrolysis and a specialized group of proteins, referred to as molecular chaperones. The chaperones, also known as heat-shock proteins (HSPs), interact with a polypeptide at various stages during the folding process. Some chaperones bind hydrophobic regions of an extended polypeptide and are important in keeping the protein unfolded until its synthesis is completed (e.g., Hsp70). Others form cage-like macromolecular structures composed of two stacked rings. The partially folded protein enters the cage, binds the central cavity through hydrophobic interactions, folds, and is released (e.g., mitochondrial Hsp60).

Chaperones, then, facilitate correct protein folding by binding to and stabilizing exposed, aggregation-prone hydrophobic regions in nascent and denatured polypeptides, preventing premature folding.

V. QUATERNARY STRUCTURE

While many proteins consist of a single polypeptide chain and are defined as monomeric proteins, other proteins consist of two or more polypeptide chains that may be structurally identical or totally unrelated. The arrangement of these polypeptide subunits is the quaternary structure of the protein. Subunits are held together primarily by noncovalent interactions including hydrogen bonds, ionic bonds, and hydrophobic interactions. Subunits may function independently of each other or may work cooperatively, such as in hemoglobin, where the binding of oxygen to one subunit of the tetramer increases the oxygen affinity of the other subunits (see Chapter 3, II. E. 1.).

> Isoforms of particular proteins all perform the same function but have different primary structures. They can arise from different genes or from tissue-specific processing of the product of a single gene. If the proteins function as enzymes, they are referred to as isoenzymes.

VI. PROTEIN MISFOLDING

Protein folding is a complex process that can sometimes result in improperly folded molecules that are usually degraded within the cell (see Chapter 19, II). However, this quality control system is not perfect, and intracellular or extracellular aggregates of misfolded proteins can

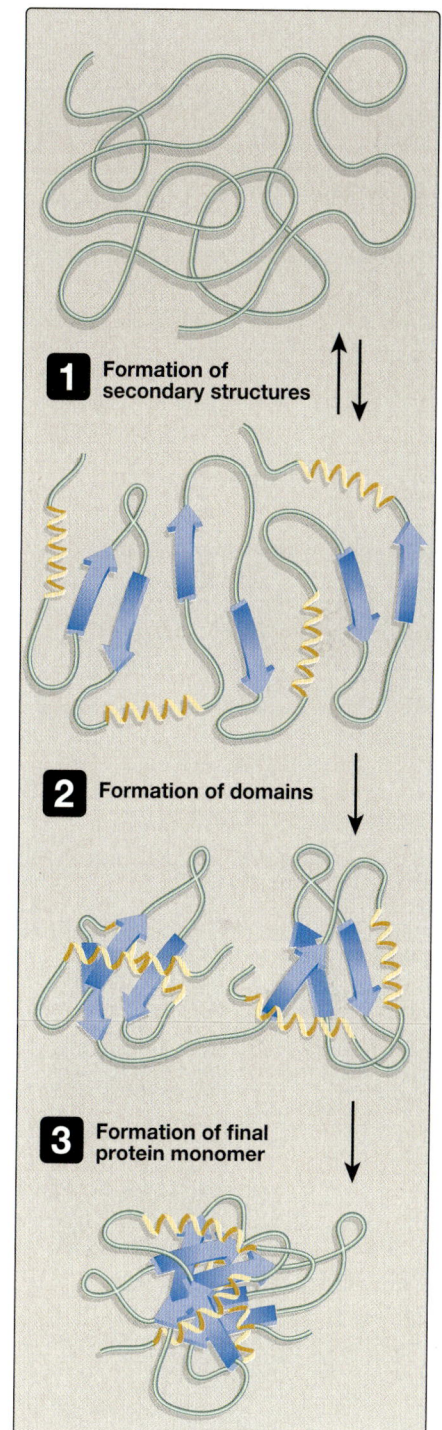

1 Formation of secondary structures

2 Formation of domains

3 Formation of final protein monomer

Figure 2.12
Steps in protein folding (simplified).

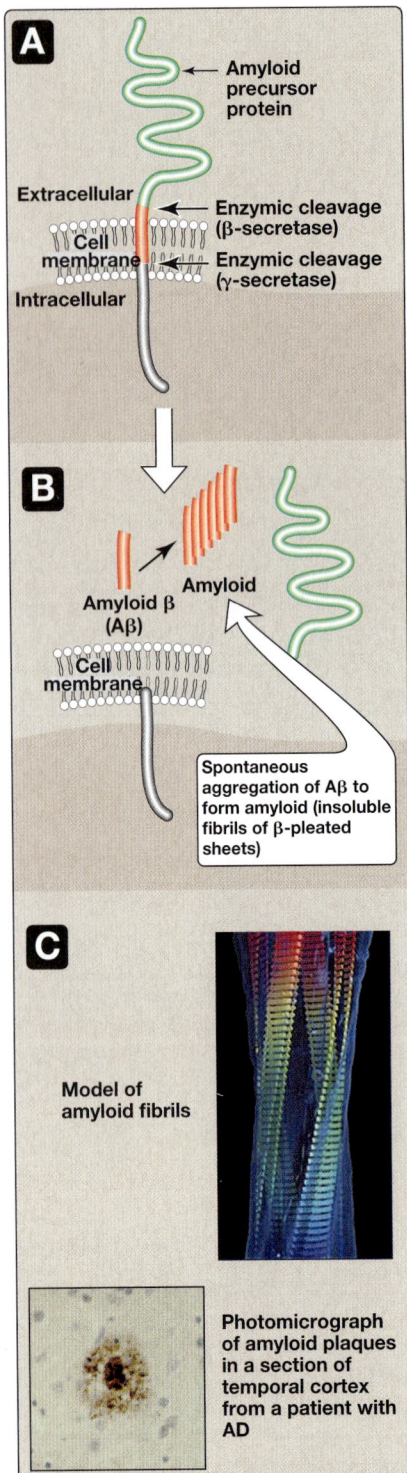

Figure 2.13
A–C. Formation of amyloid plaques found in Alzheimer disease (*AD*). [Note: Mutations to presenilin, the catalytic subunit of γ-secretase, are the most common cause of familial AD.]

accumulate, particularly as individuals age. Deposits of misfolded proteins are associated with a number of diseases.

A. Amyloid diseases

Protein misfolding may occur spontaneously or be caused by a mutation in a particular gene, which then produces an altered protein. In addition, some apparently normal proteins can, after abnormal proteolytic cleavage, take on a unique conformation that leads to the spontaneous formation of long, fibrillar protein assemblies consisting of β-pleated sheets. Accumulation of these insoluble fibrous protein aggregates, called amyloids, has been implicated in neurodegenerative disorders such as Alzheimer disease (AD) and Parkinson disease (PD), although some evidence suggests these insoluble plaques could be protective since a significant number of older adults have plaques but not dementia. The soluble oligomer may be involved with dementia in AD. The dominant component of the amyloid plaque that accumulates in AD is amyloid β (Aβ), an extracellular peptide containing 40 to 42 amino acid residues with a β-pleated sheet secondary structure in nonbranching fibrils. Aβ deposited in the brain in AD is derived by enzymatic cleavages by secretases from the amyloid precursor, a single transmembrane protein expressed on the cell surface in the brain and other tissues (Fig. 2.13).

The Aβ peptides aggregate, generating the amyloid found in the brain parenchyma and around blood vessels. Most cases of AD are not genetically based, although at least 5% of cases are familial. A second biologic factor involved in the development of AD is the accumulation of neurofibrillary tangles inside neurons. A key component of these tangled fibers is an abnormal tau (τ) protein that is hyperphosphorylated and insoluble. In its healthy version, τ helps assemble and stabilize microtubule structure. Defective τ blocks actions of its normal counterpart. In PD, amyloid is formed from α-synuclein protein.

B. Prion diseases

Prions or proteinaceous infectious particles are associated with certain diseases. The prion protein (PrP) is the causative agent of transmissible spongiform encephalopathies (TSEs), including Creutzfeldt–Jakob disease in humans; scrapie in sheep; and bovine spongiform encephalopathy, popularly called "mad cow disease," in cattle. The infectivity of the agent causing scrapie in sheep is associated with a single-protein species not complexed with detectable nucleic acid. This infectious protein is designated PrP^{Sc} (Sc = scrapie). It is highly resistant to proteolytic degradation and tends to form insoluble aggregates of fibrils, similar to the amyloid found in some other diseases of the brain. A noninfectious form of PrP^{C} (C = cellular), encoded by the same gene as the infectious agent, is present in normal mammalian brains on the surface of neurons and glial cells. Thus, PrP^{C} is a host protein. No primary structure differences or alternate post-translational modifications have been found between normal and infectious forms of the protein. The key to becoming infectious lies in changes in the three-dimensional conformation of PrP^{C}. Research has demonstrated that a number of α-helices present in noninfectious PrP^{C} are replaced by β-sheets in the infectious form (Fig. 2.14). This conformational difference is presumably what confers relative resistance to proteolytic degradation of infectious prions and permits them to be distinguished from the normal PrP^C in infected tissue. The infective agent is, thus, an altered version of a normal protein, which acts as a template for converting the normal protein to the pathogenic conformation. The TSEs are invariably fatal, and no treatment is currently available that can alter this outcome.

VII. CHAPTER SUMMARY

- A protein's unique three-dimensional structure is determined by its amino acid sequence or **primary structure**.
- A protein's native confirmation is its functional, fully folded structure (Fig. 2.15).
- Interactions between the amino acid side chains guide the folding of the polypeptide chain to form **secondary**, **tertiary**, and sometimes **quaternary** structures, which cooperate in stabilizing the native conformation of the protein.
- **Chaperones** are specialized proteins required for proper folding of many species of proteins.
- **Protein denaturation** results in unfolding and disorganization of the protein's structure, not accompanied by hydrolysis of peptide bonds.
- Disease can occur when an apparently normal protein assumes a conformation that is cytotoxic, as in the case of the **transmissible spongiform encephalopathies** (**TSEs**), including **Creutzfeldt–Jakob disease**.
- In Alzheimer disease, normal proteins undergo abnormal processing and take on a unique conformation that leads to formation of neurotoxic **amyloid β peptide** (**Aβ**) assemblies. In TSE, the infective agent is an altered version of a normal **prion protein (PrP)** that acts as a template for converting normal protein to the pathogenic conformation.

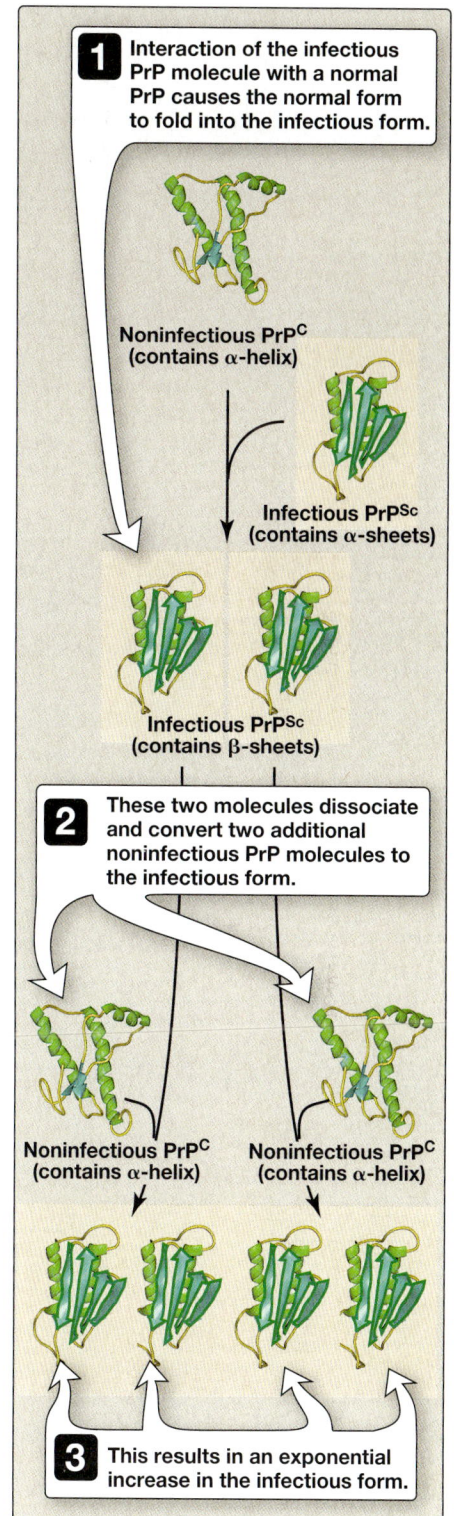

1. Interaction of the infectious PrP molecule with a normal PrP causes the normal form to fold into the infectious form.

Noninfectious PrP^C (contains α-helix)

Infectious PrP^{Sc} (contains α-sheets)

Infectious PrP^{Sc} (contains β-sheets)

2. These two molecules dissociate and convert two additional noninfectious PrP molecules to the infectious form.

Noninfectious PrP^C (contains α-helix)

Noninfectious PrP^C (contains α-helix)

3. This results in an exponential increase in the infectious form.

Figure 2.14
One proposed mechanism for multiplication of infectious prions. PrP, prion protein; PrP^c, prion protein cellular; PrP^{Sc}, prion protein scrapie.

Hierarchy of protein (polypeptide) structure

composed of

Primary
is
sequence of
amino acids

contributes to

leads to

α-Helix

β-Sheet

β-Bends (reverse turns)

Nonrepetitive structures

Supersecondary structures

consists of

Secondary
is
regular arrangements
of amino acids located
near each other
in primary structure

contributes to

can be

Fibrous
or
globular

Chaperones

*folding
assisted by*

leads to

Hydrophobic interactions

Hydrogen bonds

Electrostatic interactions

Disulfide bonds

stabilized by

Tertiary
is
the three-
dimensional
shape of
the folded chain

contributes to

**Native
conformation**

determines

**Biologic
function**

For example:
- Catalysis
- Defense
- Regulation
- Signal transduction
- Storage
- Structure
- Transport

may lead to

Hydrophobic interactions

Hydrogen bonds

Electrostatic interactions

stabilized by

Quaternary
is
the arrangement
of polypeptide
subunits in the
protein

may contribute to

*some
may
regain*

*unfolding
caused by*

Denaturants
For example:
- Urea
- Extremes of pH,
  temperature
- Organic solvents

lead to

**Loss of
secondary and
tertiary structure**

leads to

Loss of function

*can
undergo*

*most
proteins
cannot refold
upon removal
of denaturant*

**Creutzfeldt–Jakob
disease**

lead to

Infectious
prions

leads to

Altered folding
(conformational change)

Alzheimer disease
(most common cause of
dementia in older adults)

lead to

Amyloid proteins

leads to

**Irreversible
denaturation**

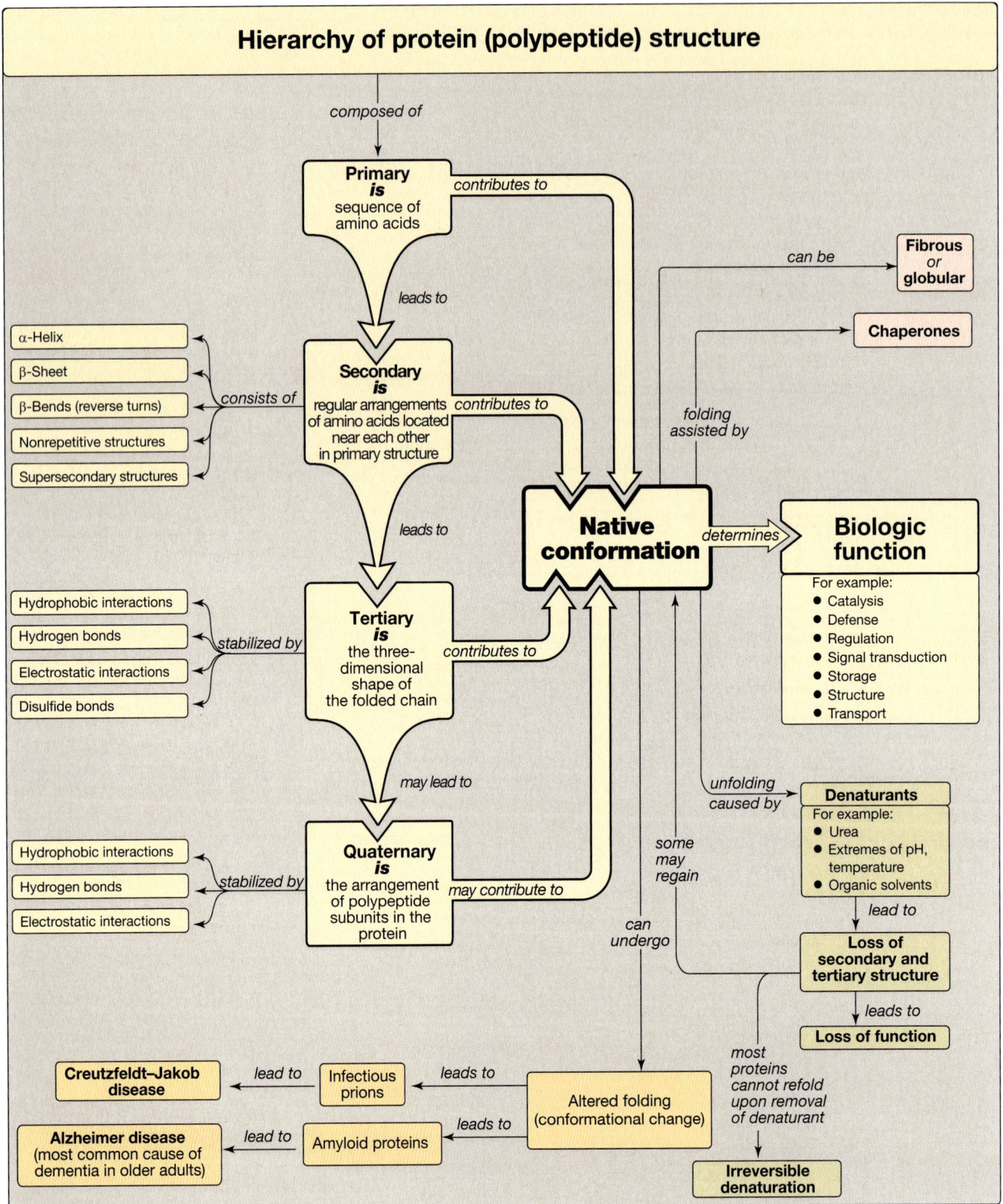

Figure 2.15
Key concept map for protein structure.

Study Questions

Choose the ONE best answer.

2.1. When considering protein structure, which of the following statements is true?

    A. Proteins with one polypeptide have quaternary structure stabilized by covalent bonds.

    B. Peptide bonds that link amino acids most commonly occur in the *cis* configuration.

    C. Disulfide bonds in proteins are between cysteine residues adjacent in the primary structure.

    D. Denaturation of proteins leads to irreversible loss of secondary structural elements.

    E. The primary driving force for protein folding is the hydrophobic effect.

Correct answer = E. The hydrophobic effect, or the tendency of nonpolar entities to associate in a polar environment, is the primary driving force of protein folding. Quaternary structure requires more than one polypeptide, and, when present, it is stabilized primarily by noncovalent bonds. The peptide bond is almost always *trans*. The two cysteine residues participating in disulfide bond formation may be a great distance apart in the amino acid sequence of a polypeptide (or on two separate polypeptides) but are brought into close proximity by the three-dimensional folding of the polypeptide. Denaturation may be reversible or irreversible.

2.2. A particular point mutation results in disruption of the α-helical structure in a segment of the resulting abnormal protein. The most likely change in the primary structure of the mutant protein is:

    A. glutamate to aspartate

    B. lysine to arginine

    C. methionine to proline

    D. valine to alanine

Correct answer = C. Proline, because of its secondary amino group, is incompatible with an α-helix. Glutamate, aspartate, lysine, and arginine are charged amino acids, and valine is a branched amino acid. Charged and branched (bulky) amino acids may disrupt an α-helix. The flexibility of glycine's R group (an H) can also disrupt an α-helix.

2.3. Adjacent amino acids in a polypeptide are joined together by peptide bonds that involve a/an:

    A. carboxyl group and amino group

    B. carboxyl group and hydrogen atom

    C. carboxyl group and R group

    D. R group and amino group

Correct answer = A. Peptide bonds are covalent bonds formed between consecutive amino acid residues between the carboxyl group of the first and the amino group of the next. A molecule of water is removed during the process. None of the other choices correctly describes peptide bond formation.

2.4. Quaternary structure of a protein refers to:

    A. folding into subunits driven by the way the R groups interact

    B. local arrangements of amino acids including α helices and β sheets

    C. multiple protein subunits, properly folded, assembled together

    D. the linear order of amino acids as encoded by DNA and translated from mRNA

Correct answer = C. Quaternary structure exists only in those proteins composed of multiple subunits and when they are assembled properly. Choices A, B, and C describe tertiary, secondary, and primary structure of a protein, respectively.

2.5. Which statement is true of β-sheets only, and not α-helices?

    A. They may be found in typical globular proteins.

    B. They are stabilized by interchain hydrogen bonds.

    C. They are examples of secondary structure.

    D. They may be found in supersecondary structures.

Correct answer = B. The β-sheet is stabilized by interchain hydrogen bonds formed between separate polypeptide chains and by intrachain hydrogen bonds formed between regions of a single polypeptide. The α-helix, however, is stabilized only by intrachain hydrogen bonds. Choices A, C, and D are true statements for both of these secondary structural elements.

2.6. Stability of tertiary protein structure is provided in part by:

    A. α helices.

    B. aminopeptidases.

    C. β meanders.

    D. disulfide bond formation.

> Correct answer = D. Disulfide bonds along with hydrophobic interactions, hydrogen bonds, and ionic interactions are used to stabilize tertiary structure of proteins. α Helices and β meanders are examples of secondary structures. Aminopeptidases are enzymes that cleave amino acids from the N-terminus of proteins and do not stabilize tertiary structure.

2.7. An 80-year-old man presents for evaluation of impairment of intellectual function, memory loss, and progressive disorientation over the past 6 months. He was tentatively diagnosed with Alzheimer disease (AD). If this diagnosis is correct, then his condition most likely:

    A. involves β-amyloid, an abnormal protein with an altered amino acid sequence.

    B. is associated with soluble oligomeric amyloid β peptide aggregates.

    C. occurred as the result of accumulation of amyloid precursor protein.

    D. resulted from accumulation of denatured proteins with random conformations.

    E. was acquired from environmental damage unrelated to his genetics.

> Correct answer = B. AD is associated with soluble assemblies consisting of β-pleated sheets found in the brain and elsewhere as well as abnormal processing of a normal protein. The accumulated altered protein occurs in a β-pleated sheet conformation that is neurotoxic. The amyloid β in the brain in patients with AD is derived by proteolytic cleavages from the larger amyloid precursor protein.

Globular Proteins

3

I. OVERVIEW

Chapter 2 described the types of secondary and tertiary structures that are the bricks and mortar of protein architecture. By arranging these fundamental structural elements in different combinations, widely diverse proteins capable of various specialized functions can be constructed. Two important protein structures are globular proteins and fibrous proteins (or scleroproteins). As the name implies, globular proteins are spherical (or "globe-like") in overall shape. They are usually somewhat water-soluble, possessing many hydrophilic amino acids on their outer surface, facing the aqueous environment. More nonpolar amino acids face the interior of the protein, providing hydrophobic interactions to further stabilize the globular structure. In contrast, fibrous proteins form long rod-like filamentous and more complex sheet-like structures, are relatively inert or water-insoluble, and often provide structural support in the extracellular environment. This chapter examines the relationship between structure and function of clinically important globular hemeproteins, such as hemoglobin and myoglobin. Fibrous structural proteins, such as collagen and elastin, are discussed in Chapter 4.

II. GLOBULAR HEMEPROTEINS

Hemeproteins are a group of specialized globular proteins that contain heme as a tightly bound prosthetic group. The function of the heme group is dictated by the three-dimensional structure of the protein. In the mitochondrial electron transport chain, the cytochrome protein structure allows for rapid and reversible oxidation–reduction electron transfer of the heme-coordinated iron, reversibly transitioning between its ferrous (Fe^{2+}) and ferric (Fe^{3+}) states. In the enzyme catalase, the heme group is structurally part of the enzyme's active site, which catalyzes the breakdown of hydrogen peroxide. In this chapter, we discuss the protein structure of hemoglobin and how it affects the alignment of Fe^{2+} iron with respect to the plane of the heme prosthetic group. Changes in this alignment affect the binding affinity and functional transport of oxygen by hemoglobin from the lungs, through the capillaries to our tissues.

A. Heme structure

Heme is a planar structure, comprised of a porphyrin ring with Fe^{2+} iron coordinated in the porphyrin ring center, as shown in Figure 3.1. The iron is held in the center of the heme molecule by bonds to four

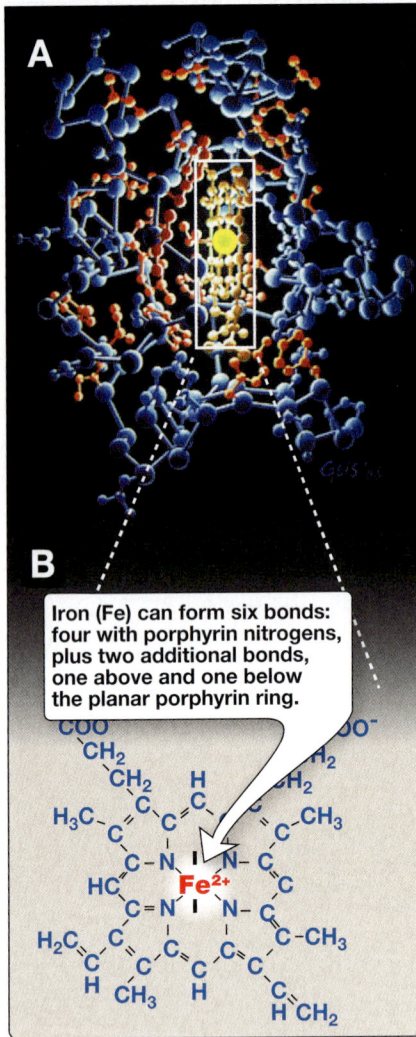

Figure 3.1
A. Hemeprotein (cytochrome c).
B. Structure of heme.

nitrogens of the porphyrin ring. The heme Fe^{2+} can form two additional bonds, one on each side of the planar porphyrin ring. In hemoglobin, one of these positions is coordinated to the side chain of a histidine residue of a globin molecule (α-globin or β-globin), whereas the other position is available to bind O_2 (Fig. 3.2).

B. Myoglobin structure and function

Myoglobin, a hemeprotein present in heart and skeletal muscle, functions both as an oxygen reservoir and as an oxygen carrier that increases the rate of oxygen transport within the muscle cell. Myoglobin consists of a single polypeptide chain that is structurally similar to the individual polypeptide chains of the tetrameric hemoglobin molecule. This homology makes myoglobin a useful model for interpreting some of the more complex structural and functional properties of hemoglobin.

1. **α-Helical content:** Myoglobin is a compact globular molecule, with ~80% of its polypeptide chain folded into eight stretches of α-helix. These α-helical regions, labeled A to H in Figure 3.2A, are terminated either by the presence of proline, whose five-membered ring cannot be accommodated in an α-helix or by β-bends and loops stabilized by hydrogen bonds and ionic bonds. [Note: Ionic bonds are also termed electrostatic interactions or salt bridges.]

2. **Location of polar and nonpolar amino acid residues:** The interior of the globular myoglobin molecule is composed almost entirely of nonpolar amino acids. Nonpolar amino acids are packed closely together, forming a structure stabilized by hydrophobic interactions between these clustered residues. In contrast, polar amino acids are located almost exclusively on the outer surface, where they can form hydrogen bonds, both with each other and with water in the aqueous environment.

3. **Binding of the heme group:** The heme prosthetic group of the myoglobin molecule sits in a crevice, which is lined with nonpolar amino acids. Notable exceptions are two histidine residues, which are basic amino acids (Fig. 3.2B). One of the two histidine residues, the proximal histidine (F8), binds directly to the Fe^{2+} of heme. The second, or distal histidine (E7), does not directly

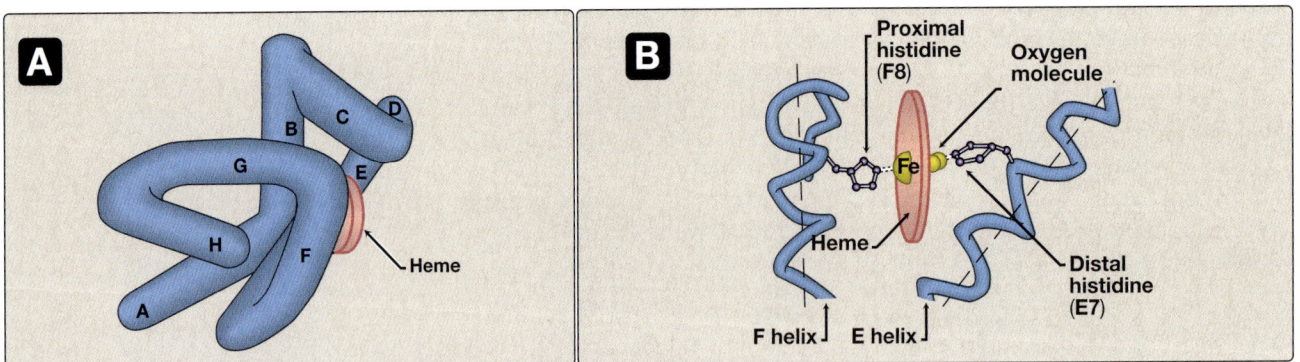

Figure 3.2
A. Model of myoglobin showing α-helices A to H. **B.** Oxygen-binding site of myoglobin.

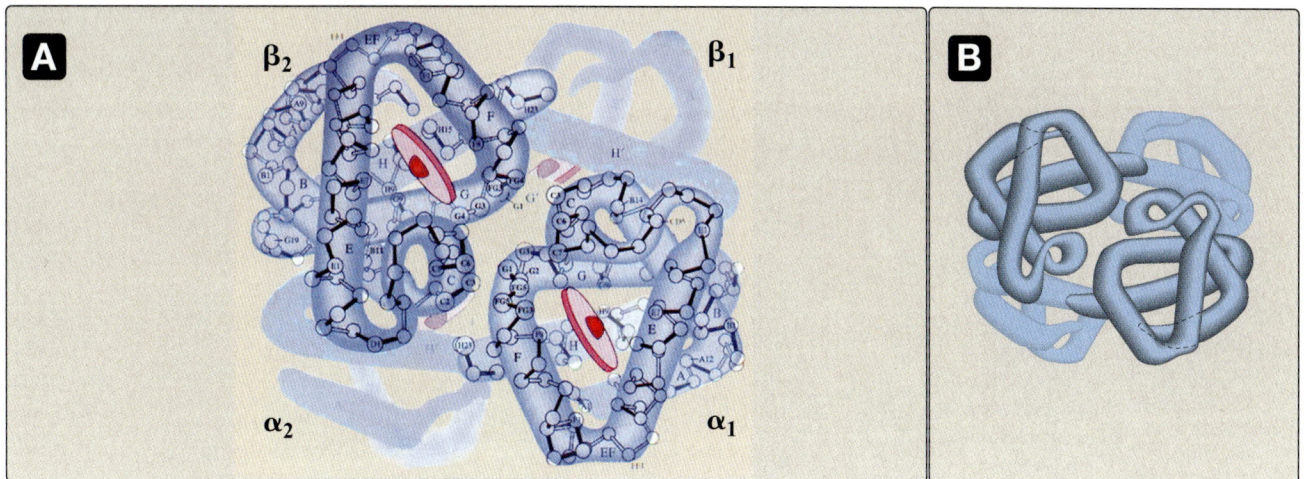

Figure 3.3
A. Structure of hemoglobin showing the polypeptide backbones. **B.** Simplified drawing showing the α-helices.

interact with the heme group but helps stabilize the binding of O_2 to Fe^{2+}. Thus, the protein, or globin, portion of myoglobin creates a special microenvironment for the heme that permits oxygenation, the reversible binding of one oxygen molecule. The simultaneous loss of electrons by Fe^{2+} (oxidation to the ferric Fe^{3+} form) occurs only rarely.

C. Hemoglobin structure and function

Hemoglobin is found exclusively in red blood cells (RBCs), where its main function is to transport O_2 from the lungs to the capillaries of the tissues. Hemoglobin A (HbA), the major hemoglobin in adults, is composed of four polypeptide chains (two α-globin chains and two β-globin chains) held together by noncovalent interactions (Fig. 3.3). Each chain (subunit) has stretches of α-helical structure and a hydrophobic heme-binding pocket similar to that described for myoglobin. However, the tetrameric hemoglobin molecule is structurally and functionally more complex than myoglobin. For example, hemoglobin can transport protons (H+) and carbon dioxide (CO_2) from the tissues to the lungs and can carry four molecules of O_2 from the lungs to the cells of the body. Furthermore, the oxygen-binding properties of hemoglobin are regulated by interaction with allosteric effectors.

1. **Quaternary structure:** The hemoglobin tetramer can be envisioned as composed of two identical dimers, $\alpha\beta_1$ and $\alpha\beta_2$. The two polypeptide chains within each dimer are held tightly together primarily by hydrophobic interactions (Fig. 3.4). [Note: In this instance, hydrophobic amino acid residues are localized not only in the interior of the molecule but also in a region on the surface of each subunit. Multiple interchain hydrophobic interactions form strong associations between the α-globin subunit and the β-globin subunit in each of the dimers.] In contrast, the two dimers are held together primarily by polar bonds (ionic and hydrogen bonds). The weaker interactions between dimers allow them to move with respect to one another. This movement results

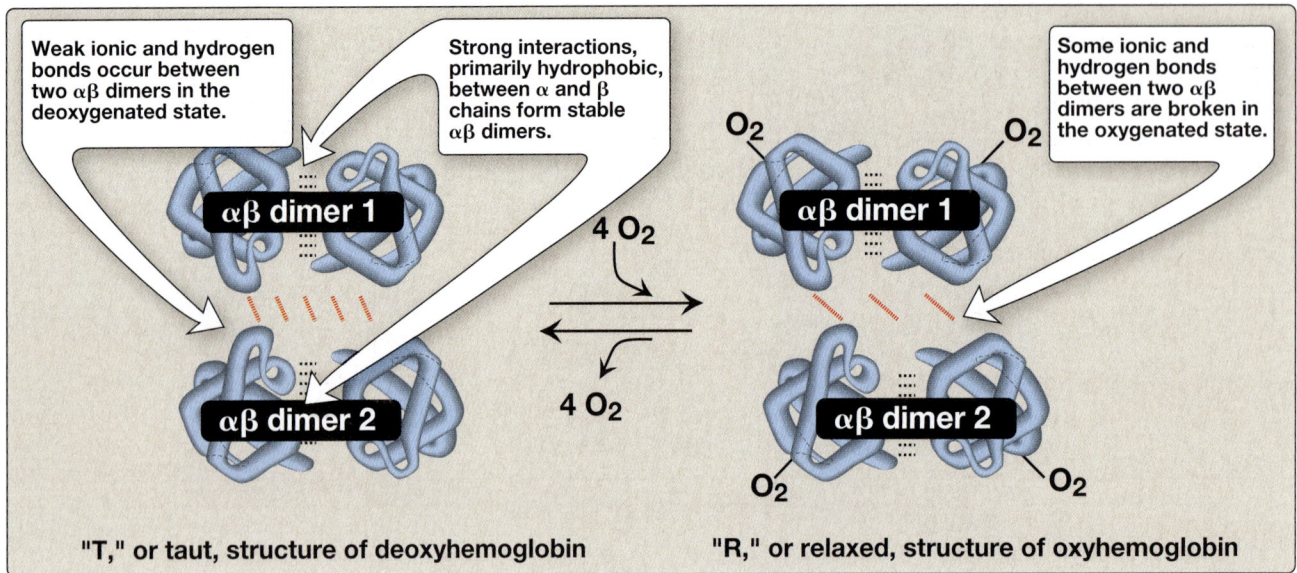

Figure 3.4
Structural changes resulting from oxygenation and deoxygenation of hemoglobin.

in the two dimers occupying different relative positions in deoxy-hemoglobin as compared with oxyhemoglobin (see Fig. 3.4).

a. **T form:** The deoxy form of hemoglobin is called the "T" or taut (tense) form. In the T form, the two $\alpha\beta$ dimers interact through a network of ionic and hydrogen bonds that constrain the movement of the polypeptide chains. The Fe^{2+} iron is pulled out of the heme planar structure. The T conformation is the low–oxygen-affinity form of hemoglobin (Fig. 3.5A).

Obtaining O_2 from the atmosphere solely by diffusion greatly limits the size of organisms. Circulatory systems overcome this, but transport molecules such as hemoglobin are also required because O_2 is only slightly soluble in aqueous solutions such as blood.

b. **R form:** The binding of O_2 to hemoglobin causes the rupture of some of the polar bonds between the two $\alpha\beta$ dimers, allowing movement of the Fe^{2+} with respect to the planar heme structure. Specifically, the binding of O_2 to the heme Fe^{2+} pulls the iron more directly into the plane of the heme ring structure (see Fig. 3.5B). Because the iron is also linked to the proximal histidine (F8), the resulting movement of the globin chains alters the interface between the $\alpha\beta$ dimers, leading to a structure called the "R" or relaxed form (see Fig. 3.4). The R conformation is the high–oxygen-affinity form of hemoglobin.

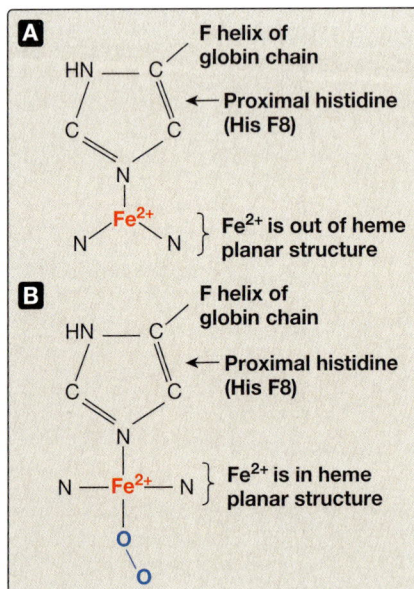

Figure 3.5
Movement of heme iron (Fe^{2+}). **A.** Out of the plane of the heme when oxygen (O_2) is not bound. **B.** Into the plane of the heme on O_2 binding.

D. Oxygen binding to myoglobin and hemoglobin

Myoglobin can bind only one molecule of O_2, because it contains only one heme group. In contrast, hemoglobin can bind four molecules of O_2, one at each of its four heme groups (one heme group/globin chain). The degree of saturation (Y) of these oxygen-binding sites on all myoglobin or hemoglobin molecules can vary between zero (all sites are empty) and 100% (all sites are full), as shown in Figure 3.6. [Note: Pulse oximetry is a noninvasive, indirect method of measuring the oxygen saturation of arterial blood based on differences in light absorption by oxyhemoglobin and deoxyhemoglobin.]

1. **Oxygen-dissociation curve:** A plot of Y measured at different partial pressures of oxygen (pO_2) is called the oxygen-dissociation curve. [Note: pO_2 may also be represented as PO_2.] The curves for myoglobin and hemoglobin show important differences (see Fig. 3.6). This graph illustrates that myoglobin has a higher oxygen affinity at all pO_2 values than does hemoglobin. The partial pressure of oxygen needed to achieve half saturation of the binding sites (P_{50}) is ~1 mm Hg for myoglobin and ~26 mm Hg for hemoglobin. The higher the oxygen affinity (i.e., the more tightly O_2 binds), the lower the P_{50}.

   a. **Myoglobin:** The oxygen-dissociation curve for myoglobin has a hyperbolic shape (see Fig. 3.6). This reflects the fact that myoglobin reversibly binds a single molecule of O_2. Thus, oxygenated (MbO_2) and deoxygenated (Mb) myoglobin exists in a simple equilibrium:

$$Mb + O_2 \rightleftarrows MbO_2$$

   The equilibrium is shifted to the right or to the left as O_2 is added to or removed from the system. [Note: Myoglobin is designed to bind O_2 released by hemoglobin at the low pO_2 found in muscle. Myoglobin, in turn, releases O_2 within the muscle cell in response to oxygen demand.]

   b. **Hemoglobin:** The oxygen-dissociation curve for hemoglobin is sigmoidal in shape (see Fig. 3.6), indicating that the subunits cooperate in binding O_2. Cooperative binding of O_2 by the four subunits of hemoglobin means that the binding of an oxygen molecule at one subunit increases the oxygen affinity of the remaining subunits in the same hemoglobin tetramer (Fig. 3.7). Although it is more difficult for the first oxygen molecule to bind to hemoglobin, the subsequent binding of oxygen molecules occurs with high affinity, as shown by the steep upward curve in the region near 20 to 30 mm Hg (see Fig. 3.6).

E. Allosteric effectors

The ability of hemoglobin to reversibly bind O_2 is affected by the pO_2, the pH of the environment, the partial pressure of carbon dioxide (pCO_2), and the concentration of 2,3-bisphosphoglycerate (2,3-BPG). These are collectively called allosteric ("other site") effectors because their interaction at one site on the tetrameric hemoglobin molecule causes structural changes that affect the binding of O_2 to the heme iron at other sites on the molecule. [Note: The binding of O_2 to monomeric myoglobin is not influenced by allosteric effectors.]

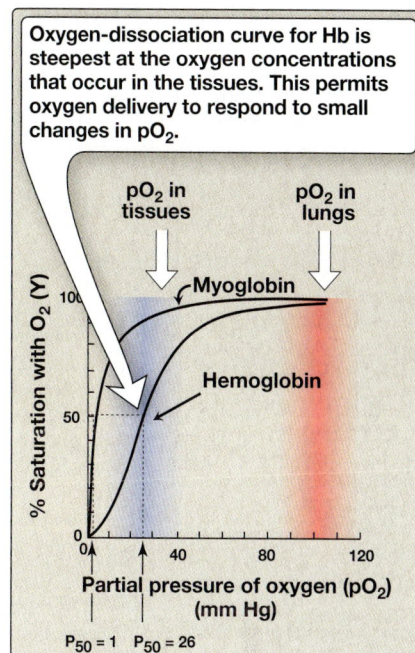

Figure 3.6
Oxygen-dissociation curves for myoglobin and hemoglobin (Hb).

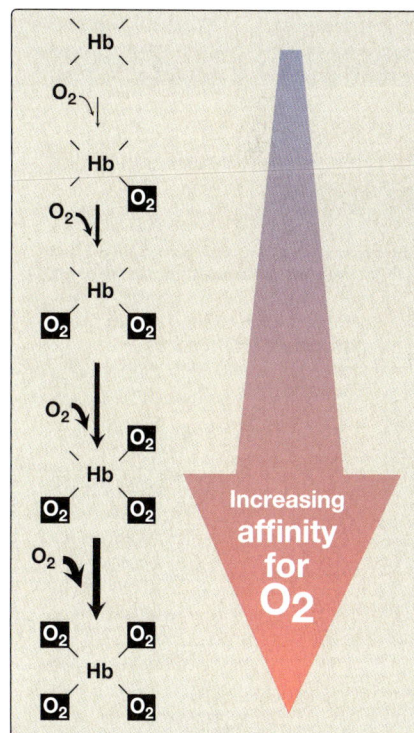

Figure 3.7
Hemoglobin (Hb) binds successive molecules of oxygen (O_2) with increasing affinity.

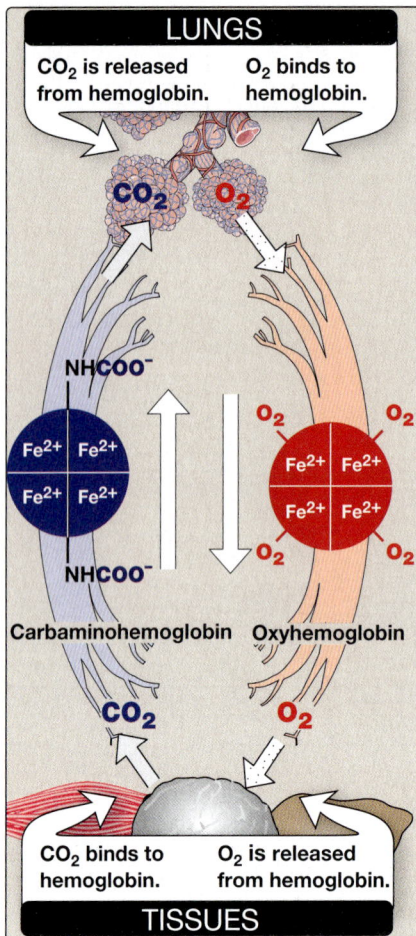

Figure 3.8
Transport of oxygen and carbon dioxide by hemoglobin. Fe, iron.

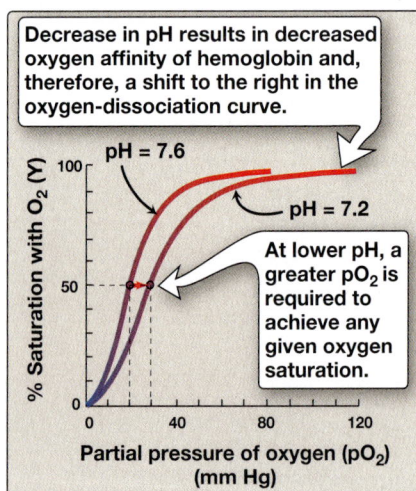

Figure 3.9
Effect of pH on the oxygen affinity of hemoglobin. Protons are allosteric effectors of hemoglobin.

1. **Oxygen:** The sigmoidal oxygen-dissociation curve reflects specific structural changes that are initiated at one subunit and transmitted to other subunits in the hemoglobin tetramer. The net effect of this cooperativity is that the affinity of hemoglobin for the last oxygen molecule bound is ~300 times greater than its affinity for the first oxygen molecule bound. Oxygen, then, is an allosteric effector of hemoglobin. It stabilizes the R form.

   a. **Loading and unloading oxygen:** The cooperative binding of O_2 allows hemoglobin to deliver more O_2 to the tissues in response to relatively small changes in pO_2. This can be seen in Figure 3.6, which indicates pO_2 in the alveoli of the lung and the capillaries of the tissues. For example, in the lungs, oxygen concentration is high, and hemoglobin becomes virtually saturated (or "loaded") with O_2. In contrast, in the peripheral tissues where the pO_2 is much lower than in the lungs, oxyhemoglobin releases (or "unloads") much of its O_2 for use in the oxidative metabolism of the tissues (Fig. 3.8).

   b. **Significance of the sigmoidal oxygen-dissociation curve:** The steep slope of the oxygen-dissociation curve over the range of oxygen concentrations that occur between the lungs and the tissues permits hemoglobin to carry and deliver O_2 efficiently from sites of high to sites of low pO_2. A molecule with a hyperbolic oxygen-dissociation curve, such as myoglobin, could not achieve the same degree of O_2 release within this range of pO_2. Instead, it would have maximum affinity for O_2 throughout this oxygen pressure range and, therefore, would deliver no O_2 to the tissues.

2. **Bohr effect:** The release of O_2 from hemoglobin is enhanced when the pH is lowered (proton concentration [H^+] is increased) or when the hemoglobin is in the presence of an increased pCO_2. Both result in decreased oxygen affinity of hemoglobin, a shift to the right in the oxygen-dissociation curve, and stabilization of the T (deoxy) form (Fig. 3.9). This change in oxygen binding is called the Bohr effect. Conversely, raising the pH or lowering the concentration of CO_2 results in a greater oxygen affinity, a shift to the left in the oxygen-dissociation curve, and stabilization of the R (oxy) form.

   a. **Source of the protons that lower pH:** The concentration of both H^+ and CO_2 in the capillaries of metabolically active tissues is higher than that observed in alveolar capillaries of the lungs (where CO_2 is released into the expired air). In the tissues, zinc-containing carbonic anhydrase converts CO_2 to carbonic acid:

   $$CO_2 + H_2O \rightleftarrows H_2CO_3$$

   which spontaneously ionizes to bicarbonate (the major blood buffer) and H^+:

   $$H_2CO_3 \rightleftarrows HCO_3^- + H^+$$

   The H^+ produced by this pair of reactions contributes to the lowering of pH. This differential pH gradient (i.e., lungs having a higher pH and tissues having a lower pH) favors the loading of O_2 in the lung and the unloading of O_2 in the peripheral tissues, respectively. Thus, the oxygen affinity of the hemoglobin molecule responds to small shifts in pH between the lungs and oxygen-consuming tissues, making hemoglobin a more efficient transporter of O_2.

b. **Mechanism of the Bohr effect:** The Bohr effect reflects the fact that deoxyhemoglobin has a greater affinity for H^+ than does oxyhemoglobin. This is caused by ionizable functional groups such as specific histidine side chains that have a higher pK_a in deoxyhemoglobin than in oxyhemoglobin. Therefore, an increase in the concentration of H^+ (resulting in a decrease in pH) causes these groups to become protonated (charged) and able to form ionic bonds (salt bridges). These bonds preferentially stabilize deoxyhemoglobin, producing a decrease in oxygen affinity. [Note: Hemoglobin, then, is an important blood buffer.]

The Bohr effect can be represented schematically as:

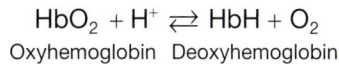

$$HbO_2 + H^+ \rightleftarrows HbH + O_2$$

Oxyhemoglobin   Deoxyhemoglobin

where an increase in H^+ concentration (or a lower pO_2) shifts the equilibrium to the right (favoring deoxyhemoglobin), whereas an increase in pO_2 (or a decrease in H^+ concentration) shifts the equilibrium to the left.

3. **2,3-BPG effect on oxygen affinity:** 2,3-BPG is an important regulator of the binding of O_2 to hemoglobin. It is the most abundant organic phosphate in RBCs, where its concentration is approximately that of hemoglobin. 2,3-BPG is synthesized from an intermediate of the glycolytic pathway (Fig. 3.10).

a. **2,3-BPG binding to deoxyhemoglobin:** 2,3-BPG decreases the oxygen affinity of hemoglobin by binding to deoxyhemoglobin but not to oxyhemoglobin. This preferential binding stabilizes the T conformation of hemoglobin. The effect of binding 2,3-BPG can be represented schematically as:

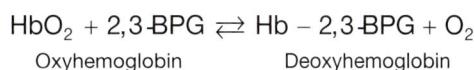

$$HbO_2 + 2,3\text{-}BPG \rightleftarrows Hb - 2,3\text{-}BPG + O_2$$

Oxyhemoglobin   Deoxyhemoglobin

b. **2,3-BPG–binding site:** One molecule of 2,3-BPG binds to a pocket, formed by the two β-globin chains, in the center of the deoxyhemoglobin tetramer (Fig. 3.11). This pocket contains several positively charged amino acids that form ionic bonds with the negatively charged phosphate groups of 2,3-BPG. [Note: Replacement of one of these amino acids can result in hemoglobin variants with abnormally high oxygen affinity that may be compensated for by increased RBC production, or erythrocytosis.] Oxygenation of hemoglobin narrows the pocket and causes 2,3-BPG to be released.

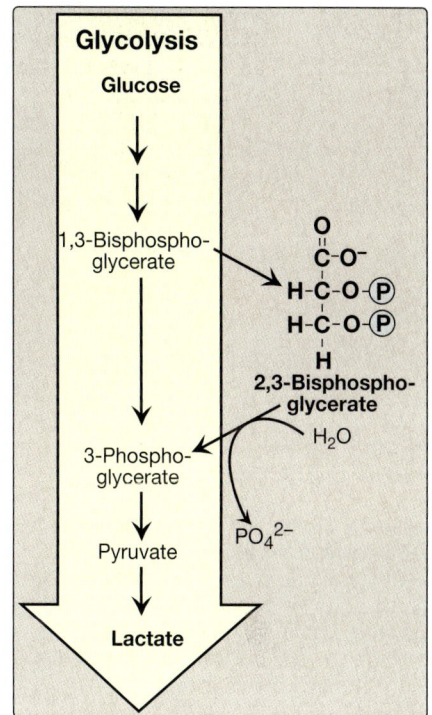

Figure 3.10
Synthesis of 2,3-bisphosphoglycerate (2,3-BPG). [Note: Ⓟ is a phosphoryl group, PO_3^{2-}.] In older literature, 2,3-BPG may be referred to as 2,3-diphosphoglycerate (2,3-DPG).

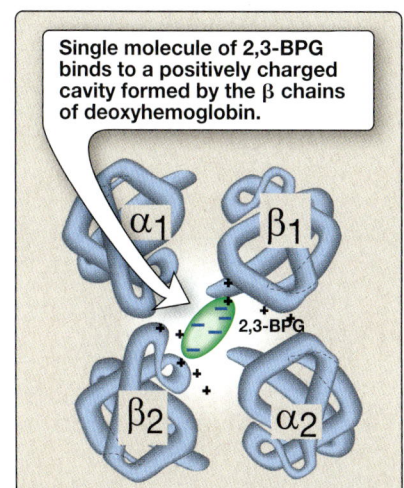

Figure 3.11
Binding of 2,3-bisphosphoglycerate (2,3-BPG) by deoxyhemoglobin.

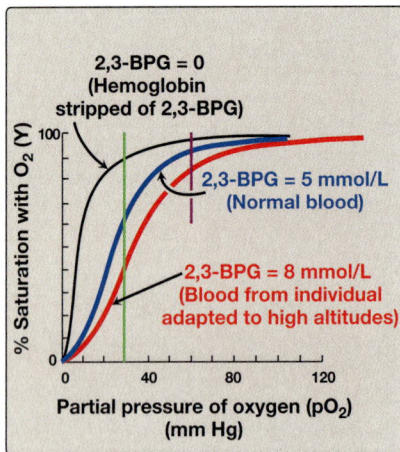

Figure 3.12
Allosteric effect of 2,3-bisphospho-glycerate (2,3-BPG) on the oxygen affinity of hemoglobin. Partial pressure of oxygen in the tissues is indicated by the *green line*. Partial pressure of oxygen in the lungs at high altitude is indicated by the *purple line*.

c. **Oxygen-dissociation curve shift:** Hemoglobin from which 2,3-BPG has been removed has high oxygen affinity. However, the presence of 2,3-BPG significantly reduces the oxygen affinity of hemoglobin, shifting the oxygen-dissociation curve to the right (Fig. 3.12). This reduced affinity enables hemoglobin to release O_2 efficiently at the partial pressures found in the tissues.

d. **2,3-BPG levels in chronic hypoxia or anemia:** The concentration of 2,3-BPG in the RBCs increases in response to chronic hypoxia, such as that observed in chronic obstructive pulmonary disease (COPD) like emphysema, or at high altitudes, where pO_2 is lower, and circulating hemoglobin may have difficulty receiving sufficient O_2. Intracellular levels of 2,3-BPG are also elevated in chronic anemia, in which fewer than normal RBCs are available to supply the body's oxygen needs. Elevated 2,3-BPG levels lower the oxygen affinity of hemoglobin, permitting greater unloading of O_2 in the capillaries of tissues (Fig. 3.12).

e. **2,3-BPG in transfused blood:** 2,3-BPG is essential for the normal oxygen transport function of hemoglobin. However, blood bank–stored blood gradually becomes depleted in 2,3-BPG. Consequently, stored blood displays an abnormally high oxygen affinity and fails to unload its bound O_2 properly in the tissues. Thus, hemoglobin deficient in 2,3-BPG would act as an oxygen "trap" rather than as an oxygen delivery system. Transfused RBCs are able to restore their depleted supplies of 2,3-BPG in 6 to 24 hours. However, severely ill patients may be compromised if transfused with large quantities of such

---

Clinical Application 3.1: 2,3-BPG Offloads Oxygen to the Tissues

To illustrate the use of 2,3-BPG to offload oxygen to the tissues, consider two conditions: one individual living at sea level with 5 mmol/L 2,3-BPG (see Fig. 3.12, *blue curve*), who travels to a high altitude where the pO_2 is lower, and another individual who lives at a high altitude and compensates by elevating their 2,3-BPG levels to 8 mmol/L (see Fig. 3.12, *red curve*). Hemoglobin in the lungs of the individual from sea level with 5 mmol/L 2,3-BPG will be fully saturated at sea level. In the tissues (indicated by the *green line* in Fig. 3.12), their hemoglobin is ~60% saturated, delivering ~40% of the bound oxygen to their tissues. At high altitudes with 5 mmol/L of 2,3-BPG, this same individual's hemoglobin will be only 90% saturated in the lungs (indicated by the *purple line* in Fig. 3.12), so oxygen delivery to their tissues is only 30%. This is why a person living at sea level may experience problems at high altitudes. However, the individual living at a high altitude has adapted to have hemoglobin with 8 mmol/L of 2,3-BPG. The oxygen-binding curve shifts to the right. Oxygen saturation in the lungs is now only ~80% (indicated by the *purple line* in Fig. 3.12) and oxygen saturation in the tissues is ~40% (indicated by the *green line* in Fig. 3.12), providing a similar 40% delivery of the bound oxygen to the tissues by the increase in 2,3-BPG levels. The shift in O_2-binding affinity allowed a comparable 40% oxygen delivery to the tissues.

2,3-BPG–depleted blood. Stored blood, therefore, is treated with a "rejuvenation" solution that rapidly restores 2,3-BPG. [Note: Rejuvenation also restores ATP lost during storage.]

4. **Carbon dioxide binding:** Most of the CO_2 produced in metabolism is converted to carbonic acid, and transported in the blood as bicarbonate ion (see Fig. 1.12). However, some CO_2 is carried as carbamate bound to the terminal amino groups of hemoglobin (forming carbaminohemoglobin as shown in Fig. 3.8), which can be represented schematically as follows:

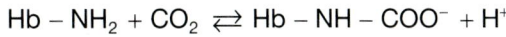

$$Hb - NH_2 + CO_2 \rightleftarrows Hb - NH - COO^- + H^+$$

The binding of CO_2 stabilizes the T, or deoxy, form of hemoglobin, resulting in a decrease in its oxygen affinity and a right shift in the oxygen-dissociation curve. In the lungs, CO_2 dissociates from the hemoglobin and is released in the exhaled breath.

5. **Carbon monoxide binding:** Carbon monoxide (CO) binds tightly (but reversibly) to the hemoglobin iron, forming carboxyhemoglobin. When CO binds to one or more of the four heme sites, hemoglobin shifts to the R conformation, causing the remaining heme sites to bind O_2 with very high affinity. This shifts the oxygen-dissociation curve further to the left and changes the normal sigmoidal shape toward a hyperbola. As a result, the affected hemoglobin is unable to release O_2 to the tissues (Fig. 3.13). [Note: The affinity of hemoglobin for CO is 220 times greater than for O_2. Consequently, even minute concentrations of CO in the environment can produce toxic concentrations of carboxyhemoglobin in the blood. For example, increased levels of CO are found in the blood of tobacco smokers. CO toxicity appears to result from a combination of tissue hypoxia and direct CO-mediated damage at the cellular level.] CO poisoning is treated with 100% O_2 at high pressure (hyperbaric oxygen therapy), which facilitates the dissociation of CO from hemoglobin. [Note: CO also inhibits complex IV of the electron transport chain.] Nitric oxide gas (NO) also is carried by hemoglobin. NO is a potent vasodilator. It can be taken up (salvaged) or released from RBCs, thereby modulating NO availability and influencing blood vessel diameter.

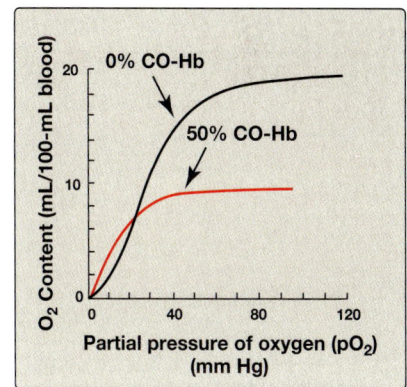

F. Minor hemoglobins

Importantly, the adult form of human HbA is just one member of a functionally and structurally related family of hemoglobin proteins (Fig. 3.14). Each of these oxygen-carrying proteins is a tetramer, composed of two α-globin (or α-like) polypeptides and two β-globin (or β-like) polypeptides. HbF is synthesized during fetal development but is less than 2% of the hemoglobin in adult blood. HbF is concentrated in RBCs known as F cells. HbA_2 is also synthesized in the adult, although at low levels compared with HbA. HbA can become modified by the covalent addition of a hexose (HbA1c).

1. **Fetal hemoglobin:** HbF is a tetramer consisting of two α-globin chains identical to those found in HbA, plus two γ-globin chains ([$\alpha_2\gamma_2$] see Fig. 3.14). The γ chains are members of the β-globin gene family.

Figure 3.13
Effect of carbon monoxide (CO) on the oxygen affinity of hemoglobin. CO competes with O_2 for binding the heme iron. CO-Hb, carboxyhemoglobin (carbon monoxyhemoglobin).

Form	Chain composition	Fraction of total hemoglobin
HbA	$\alpha_2\beta_2$	90%
HbA_2	$\alpha_2\delta_2$	2%–3%
HbF	$\alpha_2\gamma_2$	<2%
HbA_{1c}	$\alpha_2\beta_2$-glucose	4%–6%

Figure 3.14
Human hemoglobins found in adult blood. HbA_{1c} is a subtype of HbA (or HbA_1). [Note: The α chains in these hemoglobins are identical.] Hb, hemoglobin.

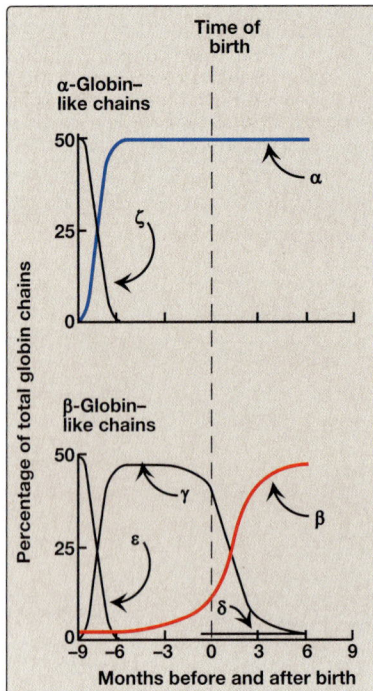

Figure 3.15
Developmental changes in globin production.

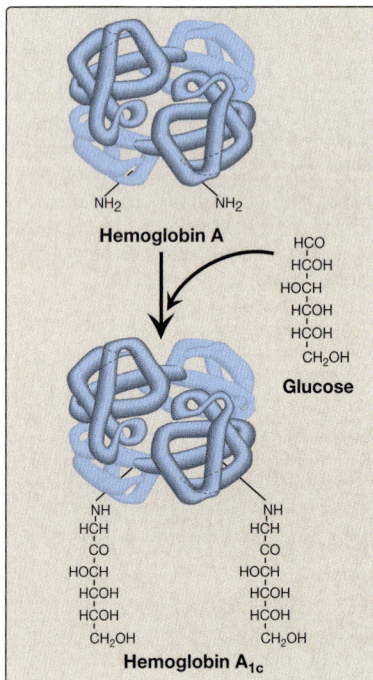

Figure 3.16
Nonenzymatic addition of glucose to hemoglobin. The nonenzymatic addition of a sugar to a protein is referred to as glycation.

a. **Fetal hemoglobin synthesis during development:** In the first month after conception, embryonic hemoglobins such as Hb Gower 1, composed of two α-like zeta (ζ) chains and two β-like epsilon (ϵ) chains ($\zeta_2\epsilon_2$), are synthesized by the embryonic yolk sac. In week 5 of gestation, the site of globin synthesis shifts, first to the liver and then to the marrow, and the primary product is HbF. HbF is the major hemoglobin found in the fetus and newborn, accounting for ~60% of the total hemoglobin in the RBCs during the last months of fetal life (Fig. 3.15). HbA synthesis starts in the bone marrow around the eighth month of pregnancy and gradually replaces HbF. Figure 3.15 shows the relative production of each type of hemoglobin chain during fetal and postnatal life.

b. **2,3-BPG binding to fetal hemoglobin:** Under physiologic conditions, HbF has a higher oxygen affinity than does HbA, mainly because HbF weakly binds 2,3-BPG. [Note: The γ-globin chains of HbF lack some of the positively charged amino acids that are responsible for binding 2,3-BPG in the β-globin chains.] Because 2,3-BPG serves to reduce the oxygen affinity of hemoglobin, the weaker interaction between 2,3-BPG and HbF results in a higher oxygen affinity for HbF relative to HbA. In contrast, if both HbA and HbF are stripped of their 2,3-BPG, they then have a similar oxygen affinity. The higher oxygen affinity of HbF facilitates the transfer of O_2 from the maternal circulation across the placenta to the RBCs of the developing fetus.

2. **Hemoglobin A$_2$:** HbA$_2$ is a minor component of normal adult hemoglobin, first appearing shortly before birth and, ultimately, constituting ~2% of the total hemoglobin. It is composed of two α-globin chains and two δ-globin chains ([$\alpha_2\delta_2$] see Fig. 3.14). Like the γ chains, the δ chains are also members of the β-globin gene family.

3. **Hemoglobin A$_{1c}$:** Under physiologic conditions, sugar molecules, predominantly glucose, are added nonenzymatically to HbA in a process referred to as glycation. The extent of glycation depends on the plasma concentration of the hexose. The most abundant and stable form of glycated hemoglobin is HbA$_{1c}$. In HbA$_{1c}$, glucose residues are attached to the amino groups of the N-terminal valines of the β-globin chains (Fig. 3.16). Increased amounts of HbA$_{1c}$ are found in RBCs of patients with diabetes mellitus, because their HbA has contact with higher glucose concentrations during the 120-day lifetime of these cells (see Chapter 25, II. A. for a discussion of the use of HbA$_{1c}$ levels in assessing average blood glucose levels in patients with diabetes). In a healthy individual, HbA$_{1c}$ comprises approximately 5% of total HbA.

III. GLOBIN GENE ORGANIZATION

To understand diseases resulting from genetic alterations in the structure or synthesis of hemoglobin, it is necessary to grasp how the hemoglobin genes are structurally organized into gene families, and how they

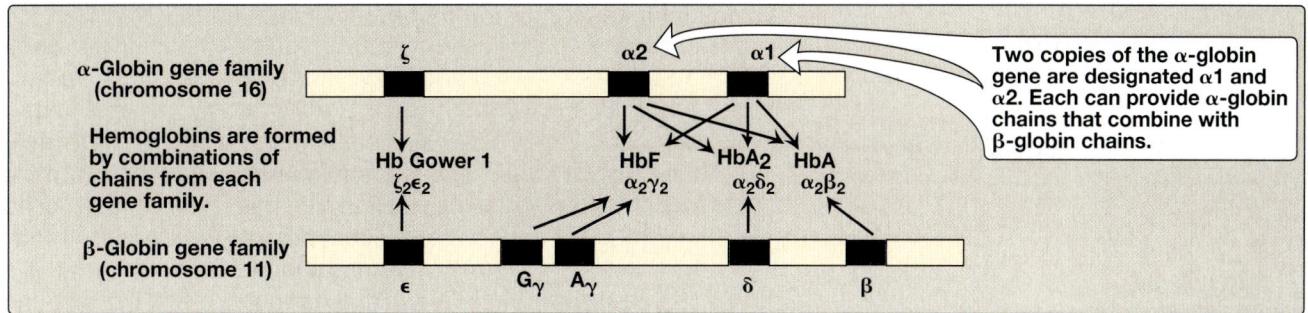

Figure 3.17
Organization of the globin gene families. Hb, hemoglobin.

are expressed. Expression of a globin gene begins in RBC precursors, where the DNA sequence encoding the gene is transcribed. Two introns are spliced out to join together three exons into the mature mRNA for translation. A more detailed description of gene expression is presented in Unit VII, Chapters 30, 31, and 32.

A. α-Gene family

The genes coding for the α-globin and β-globin subunits of hemoglobin occur in two separate gene clusters (or families) located on two different chromosomes (Fig. 3.17). The α-gene cluster on chromosome 16 contains two genes for the α-globin chains. It also contains the ζ gene that is expressed early in development as an α-globin–like component of embryonic hemoglobin. [Note: Globin gene families also contain globin-like genes that are not expressed (i.e., their genetic information is not used to produce globin chains). These are called pseudogenes.]

B. β-Gene family

The β-gene cluster contains a single gene for the β-globin chain, located on chromosome 11 (Fig. 3.17). There are an additional four β-globin–like genes within the gene cluster: the ε gene (which, like the ζ gene, is expressed only during early embryonic development), two γ genes (G_γ and A_γ that are expressed in HbF), and the δ gene that codes for the globin chain found in the minor adult hemoglobin, HbA_2.

IV. HEMOGLOBINOPATHIES

Hemoglobinopathies are defined as a group of genetic disorders caused by production of a structurally abnormal hemoglobin molecule; synthesis of insufficient quantities of normal hemoglobin; or, rarely, both. Sickle cell anemia (HbS), hemoglobin C disease (HbC), hemoglobin SC disease (HbS + HbC = HbSC), and thalassemias are representative hemoglobinopathies that can have severe clinical consequences. The first three conditions result from production of hemoglobin with an altered amino acid sequence (qualitative hemoglobinopathy), whereas the thalassemias are caused by decreased production of normal hemoglobin (quantitative hemoglobinopathy).

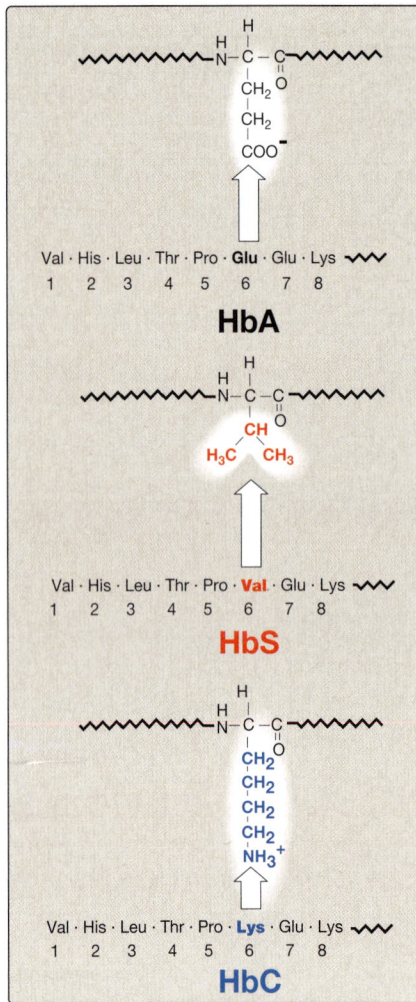

Figure 3.18
Amino acid substitutions in hemoglobin S (HbS) and hemoglobin C (HbC).

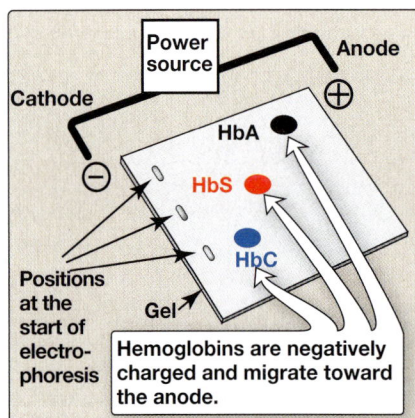

Figure 3.19
Diagram of hemoglobins HbA, HbS, and HbC after electrophoresis.

A. Sickle cell anemia (hemoglobin S disease)

Sickle cell anemia is a genetic disorder caused by a single nucleotide substitution (point mutation) in the gene for β-globin. The alteration in the amino acid sequence of HbS causes RBC morphology to form into sickle or crescent shapes, rather than the round biconcave shape of a normal RBC expressing normal HbA. This abnormal cell morphology is referred to as sickling. Sickle cell anemia is the most common inherited blood disorder in the United States, affecting 50,000 Americans. It occurs primarily in the African American population, affecting 1 in 500 African Americans. Sickle cell anemia is an autosomal-recessive disorder. It occurs in individuals who have inherited two mutant alleles (one from each parent) that code for synthesis of only the mutant β chains. [Note: The mutant β-globin chain is designated β^S, and the resulting hemoglobin, $\alpha_2\beta_2^S$, is referred to as HbS.] An infant does not begin showing symptoms of the disease until sufficient HbF has been replaced by HbS such that sickling can occur. Sickle cell anemia is characterized by lifelong episodes of pain ("crises"), chronic hemolytic anemia with associated hyperbilirubinemia, and increased susceptibility to infections, usually beginning in infancy. [Note: The lifespan of RBCs in sickle cell anemia is <20 days, compared with 120 days for normal RBCs, hence, the anemia.] Other symptoms include acute chest syndrome, stroke, splenic and renal dysfunction, and bone changes due to marrow hyperplasia. Life expectancy is reduced (mid-40s median age). Heterozygotes, representing 1 in 12 African Americans, have one normal β allele and one sickle cell β^S allele. The blood cells of such heterozygotes contain both HbS and HbA, and these individuals have sickle cell trait, not sickle cell disease. They usually do not show clinical signs or symptoms (but may be under conditions of extreme physical exertion with dehydration), can have a normal life span, and are genetic carriers of sickle cell anemia.

1. **Amino acid substitution in HbS β chains:** In a patient with sickle cell anemia, a molecule of HbS contains two normal α-globin chains and two mutant β-globin chains (β^S), in which glutamate at position six has been replaced with valine (Fig. 3.18). The resulting exchange of negatively charged polar glutamate residues for neutral nonpolar valine residues in the two β chains renders HbS less negatively charged than HbA. Therefore, during electrophoresis at alkaline pH, HbS migrates more slowly toward the anode (positive electrode) than does HbA (Fig. 3.19). Electrophoresis of hemoglobin obtained from lysed RBCs is routinely used in the diagnosis of sickle cell trait and sickle cell anemia (or sickle cell disease). DNA analysis is also used to diagnose sickle cell anemia.

2. **Sickling and tissue anoxia:** The replacement of the charged glutamate with the nonpolar valine forms a hydrophobic protrusion on the β chain that fits into a complementary hydrophobic site on the β chain of another HbS molecule in the cell (Fig. 3.20). At low oxygen tension, deoxygenated HbS polymerizes inside the RBC, forming a network of insoluble fibrous polymers that stiffen and distort the cell shape, producing rigid, sickle-shaped RBCs. Such sickled cells frequently block the flow of blood in the narrow

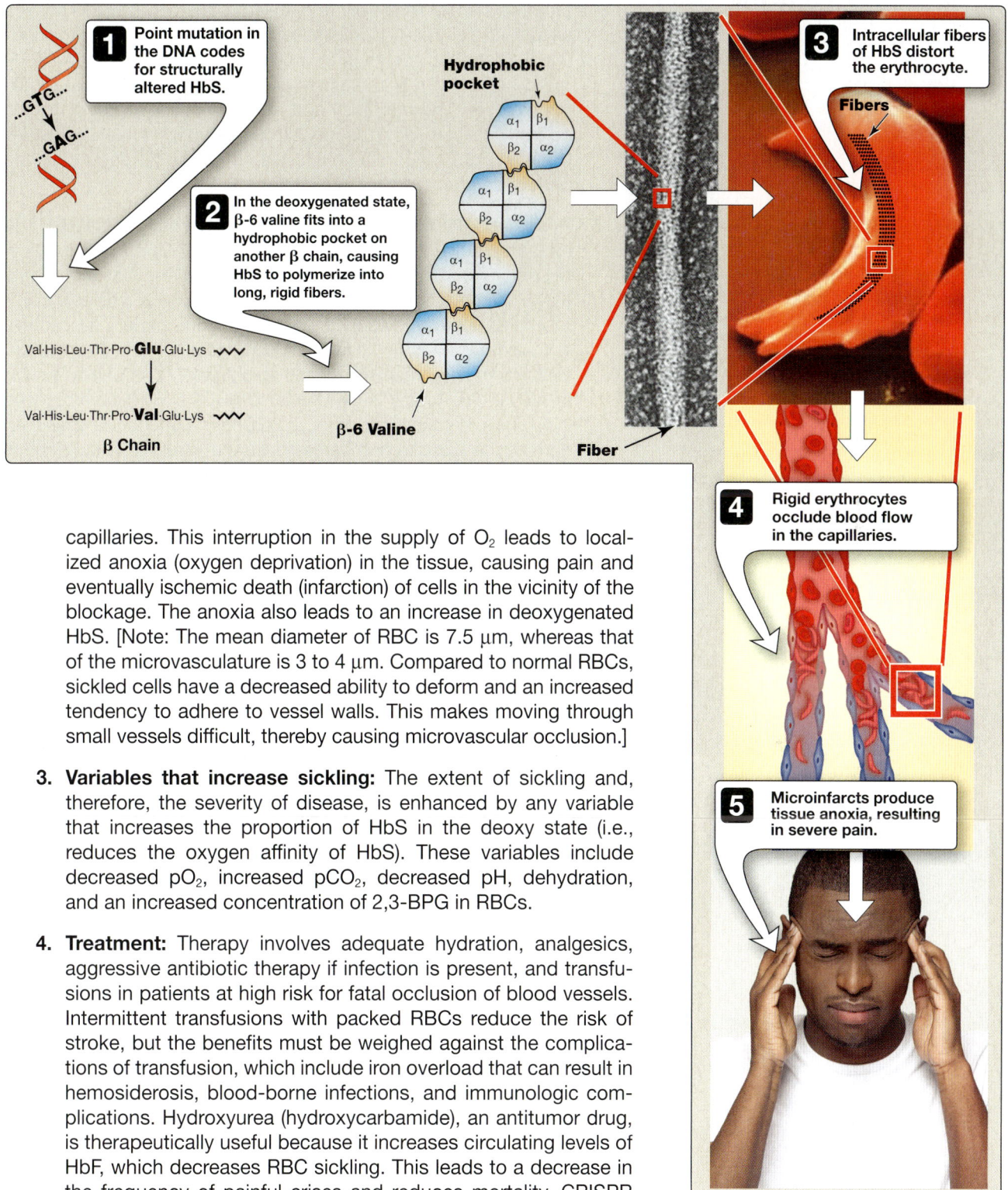

1 Point mutation in the DNA codes for structurally altered HbS.

...GTG...
...GAG...

2 In the deoxygenated state, β-6 valine fits into a hydrophobic pocket on another β chain, causing HbS to polymerize into long, rigid fibers.

Hydrophobic pocket

Val·His·Leu·Thr·Pro·**Glu**·Glu·Lys

Val·His·Leu·Thr·Pro·**Val**·Glu·Lys

β Chain

β-6 Valine

Fiber

3 Intracellular fibers of HbS distort the erythrocyte.

Fibers

4 Rigid erythrocytes occlude blood flow in the capillaries.

5 Microinfarcts produce tissue anoxia, resulting in severe pain.

capillaries. This interruption in the supply of O_2 leads to localized anoxia (oxygen deprivation) in the tissue, causing pain and eventually ischemic death (infarction) of cells in the vicinity of the blockage. The anoxia also leads to an increase in deoxygenated HbS. [Note: The mean diameter of RBC is 7.5 μm, whereas that of the microvasculature is 3 to 4 μm. Compared to normal RBCs, sickled cells have a decreased ability to deform and an increased tendency to adhere to vessel walls. This makes moving through small vessels difficult, thereby causing microvascular occlusion.]

3. **Variables that increase sickling:** The extent of sickling and, therefore, the severity of disease, is enhanced by any variable that increases the proportion of HbS in the deoxy state (i.e., reduces the oxygen affinity of HbS). These variables include decreased pO_2, increased pCO_2, decreased pH, dehydration, and an increased concentration of 2,3-BPG in RBCs.

4. **Treatment:** Therapy involves adequate hydration, analgesics, aggressive antibiotic therapy if infection is present, and transfusions in patients at high risk for fatal occlusion of blood vessels. Intermittent transfusions with packed RBCs reduce the risk of stroke, but the benefits must be weighed against the complications of transfusion, which include iron overload that can result in hemosiderosis, blood-borne infections, and immunologic complications. Hydroxyurea (hydroxycarbamide), an antitumor drug, is therapeutically useful because it increases circulating levels of HbF, which decreases RBC sickling. This leads to a decrease in the frequency of painful crises and reduces mortality. CRISPR gene editing in hematopoietic stem cells has recently been used successfully to reduce sickling in a small number of HbS patients. This gene editing introduces a benign genetic mutation, referred to as hereditary persistence of fetal hemoglobin (HPFH). HPFH inhibits the normal switch from γ-globin to β-globin after birth. As much as 30% of the total circulating Hb will be HbF

Figure 3.20
Molecular and cellular events leading to sickle cell crisis. HbS, hemoglobin S.

after the stem cells are reintroduced into the patient. This mimics what hydroxyurea accomplishes but without the need for continual treatment. It should be noted that coinheritance of HPFH in individuals with β^S mutations alleviates the HbS clinical symptoms. [Note: The morbidity and mortality associated with sickle cell anemia have led to its inclusion in newborn screening panels to allow prophylactic antibiotic therapy to begin soon after the birth of an affected child.]

5. **Possible selective advantage of the heterozygous state:** The high frequency of the β^S mutation among African Americans, despite its damaging effects in the homozygous state, suggests that a selective advantage exists for heterozygous individuals. For example, heterozygotes for the sickle cell gene are less susceptible to the severe effects of malaria caused by the parasite *Plasmodium falciparum.* This organism spends an obligatory part of its life cycle in RBCs. One theory is that because RBCs in HbS heterozygous individuals, like those in homozygotes, have a shorter life span than normal, the parasite cannot complete the intracellular stage of its development. This may provide a selective advantage to heterozygotes living in regions where malaria is a major cause of death. For example, in Africa, the geographic distribution of sickle cell anemia is similar to that of malaria.

B. Hemoglobin C disease

Like HbS, HbC is a hemoglobin variant that has a single amino acid substitution in the sixth position of the β-globin chain (see Fig. 3.18). In HbC, however, a positively charged lysine is substituted for the negatively charged glutamate (as compared with a nonpolar hydrophobic valine substitution in HbS). [Note: This substitution causes HbC to move more slowly toward the anode than either HbA or HbS does (see Fig. 3.19)]. Rare patients homozygous for HbC generally have a relatively mild, chronic hemolytic anemia. They do not experience infarctive crises, and no specific therapy is required.

C. Hemoglobin SC disease

HbSC disease is another of the RBC sickling diseases. In this disease, one β-globin allele (and chain) has the sickle cell mutation, whereas the other allele (and chain) carries the mutation found in HbC disease. [Note: Patients with HbSC disease are doubly heterozygous. They are called compound heterozygotes because both of their β-globin genes are abnormal, although different from each other.] Hemoglobin levels tend to be higher in HbSC disease than in sickle cell anemia and may even be at the low end of the normal range. The clinical course of adults with HbSC anemia differs from that of sickle cell anemia in that symptoms such as painful crises are less frequent and less severe. However, there is significant clinical variability.

D. Methemoglobinemias

Oxidation of the heme iron in hemoglobin from Fe^{2+} to Fe^{3+} produces methemoglobin (HbM), which cannot bind O_2. This oxidation may be acquired and caused by the action of certain drugs, such as nitrates, or endogenous products such as reactive oxygen species.

The oxidation may also result from congenital defects. For example, a deficiency of NADH-cytochrome b_5 reductase (also called NADH-methemoglobin reductase), the enzyme responsible for the conversion of methemoglobin (Fe^{3+}) to hemoglobin (Fe^{2+}), leads to the accumulation of methemoglobin (Fig. 3.21). [Note: The RBCs of newborns have approximately half the capacity of those of adults to reduce methemoglobin.] Additionally, rare mutations in the α- or β-globin chain can cause the production of HbM, an abnormal hemoglobin that is resistant to the reductase. The methemoglobinemias are characterized by "chocolate cyanosis" (blue coloration of the skin and mucous membranes and brown-colored blood) as a result of the dark-colored methemoglobin. Symptoms are related to the degree of tissue hypoxia and include anxiety, headache, and dyspnea (labored breathing). In rare cases, coma and death can occur. Treatment is with methylene blue, which is oxidized, as Fe^{3+} is reduced to Fe^{2+}.

E. Thalassemias

The thalassemias are hereditary hemolytic diseases in which an imbalance occurs in the synthesis of globin chains. As a group, they are the most common single-gene disorders in humans. Normally, synthesis of the α- and β-globin chains is coordinated, so that each α-globin chain has a β-globin chain partner. This leads to the formation of $\alpha_2\beta_2$ (HbA). In the thalassemias, the synthesis of either the α- or the β-globin chain is defective, creating a quantitative imbalance between α- and β-globin chain synthesis. As a result, hemoglobin concentration is also reduced. A thalassemia can be caused by a variety of mutations, including entire gene deletions, or substitutions or deletions of one of many nucleotides in the DNA. [Note: Each thalassemia can be classified as either a disorder in which essentially no functional globin chains are produced (α^0- or β^0-thalassemia), or one in which some chains are synthesized but at a reduced level (α^+- or β^+-thalassemia).]

1. **β-Thalassemias:** In β-thalassemias, the synthesis of β-globin chains is decreased or absent. This is usually the result of point mutations, either in promoter regions or slice junctions that affect the normal production of functional beta-globin mRNA. However, the synthesis of the α-globin chain is normal. The excess α-globin chains cannot form stable tetramers, and so precipitate, causing the premature death of cells initially destined to become mature RBCs. There is also an increase in $\alpha_2\delta_2$ (HbA$_2$) and $\alpha_2\gamma_2$ (HbF) as a percentage of total hemoglobin in the resulting RBCs. There are only two copies of the β-globin gene in each cell (one on each chromosome 11). Therefore, individuals with β-globin gene defects have either β-thalassemia trait (β-thalassemia minor) if they have only one defective β-globin gene (genotypically β^+/β or β^0/β), or β-thalassemia major (Cooley anemia) if both genes are defective (genotypically β^0/β^+ or β^0/β^0; Fig. 3.22). Because the β-globin gene is not expressed until late in prenatal development, the physical manifestations of β-thalassemias appear only several months after birth. Those individuals with β-thalassemia minor make some β chains and usually do not require specific treatment. However, infants born with β-thalassemia major are seemingly healthy at birth but become severely anemic due to ineffective erythropoiesis, usually during the first or second year

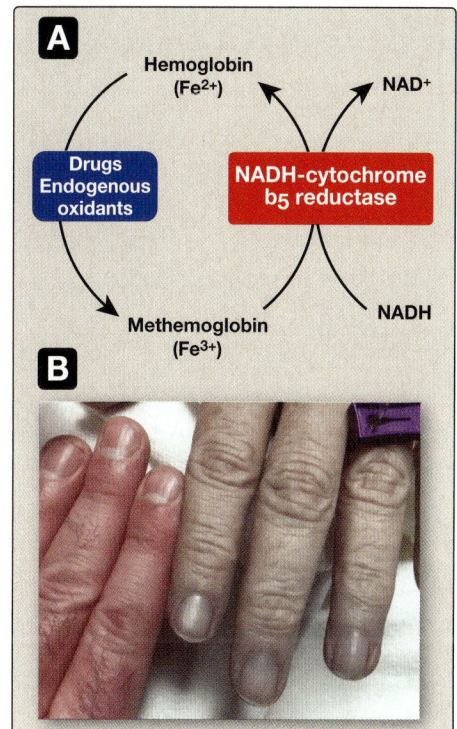

Figure 3.21
A. Formation of methemoglobin and its reduction to hemoglobin by NADH-cytochrome b_5 reductase. **B.** Methemoglobinemia.

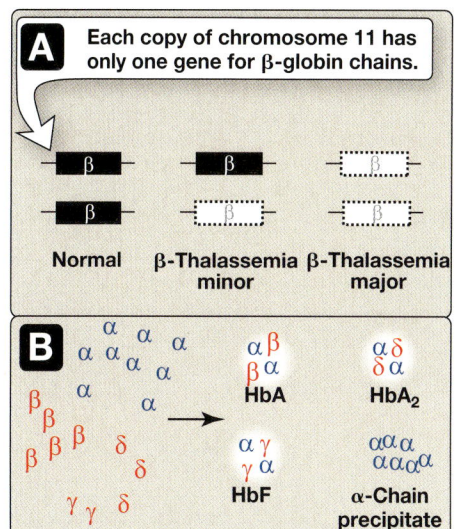

Figure 3.22
A. β-Globin gene mutations in the β-thalassemias. **B.** Hemoglobin (Hb) tetramers formed in β-thalassemias.

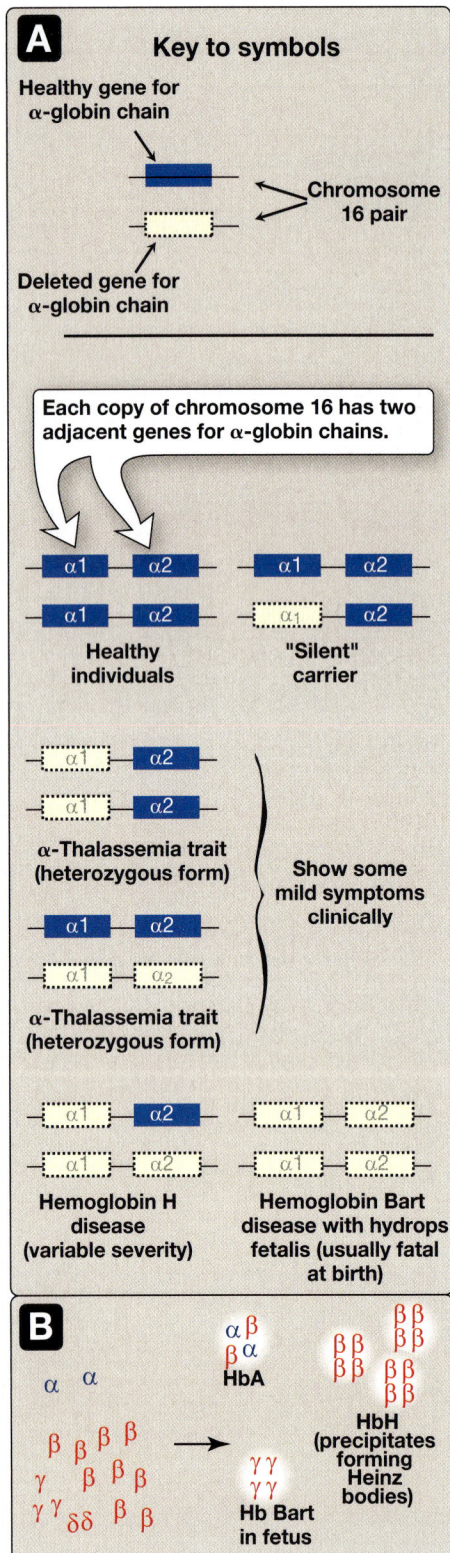

A

Key to symbols

Healthy gene for
α-globin chain

Chromosome
16 pair

Deleted gene for
α-globin chain

Each copy of chromosome 16 has two
adjacent genes for α-globin chains.

| α1 | α2 |
| α1 | α2 |

**Healthy
individuals**

| α1 | α2 |
| α1 | α2 |

**"Silent"
carrier**

| α1 | α2 |
| α1 | α2 |

**α-Thalassemia trait
(heterozygous form)**

| α1 | α2 |
| α1 | α2 |

**α-Thalassemia trait
(heterozygous form)**

Show some
mild symptoms
clinically

| α1 | α2 |
| α1 | α2 |

**Hemoglobin H
disease
(variable severity)**

| α1 | α2 |
| α1 | α2 |

**Hemoglobin Bart
disease with hydrops
fetalis (usually fatal
at birth)**

B

α   α
HbA

αβ
βα

ββ
βα
ββ

ββ
ββ
ββ
ββ
ββ
ββ

HbH
(precipitates
forming
Heinz
bodies)

β β β β
γ   β β
γ γ δδ   β β

γγ
γγ

Hb Bart
in fetus

Figure 3.23
A. α-Globin gene deletions in the
α-thalassemias. **B.** Hemoglobin (Hb)
tetramers formed in α-thalassemias.

of life. Skeletal changes as a result of extramedullary hematopoiesis are also seen. These patients require regular transfusions of blood. [Note: Although this treatment is lifesaving, the cumulative effect of the transfusions is iron overload. Use of iron chelation therapy has improved morbidity and mortality.] The only curative option available is hematopoietic stem cell transplantation.

2. **α-Thalassemias:** In α-thalassemias, synthesis of the α-globin chain is decreased or absent, typically as a result of deletional mutations. Because each individual's genome contains four copies of the α-globin gene (two on each chromosome 16), there are several levels of α-globin chain deficiencies (Fig. 3.23). If one of the four alleles encodes for a defective globin protein, the individual is termed a "silent" carrier of α-thalassemia, because no physical manifestations of the disease occur. If two α-globin alleles encode for defective globin proteins, the individual is designated as having α-thalassemia trait. If three α-globin alleles encode for defective globin proteins, the individual has hemoglobin H (β_4) disease, a hemolytic anemia of variable severity. If all four α-globin alleles encode for defective globin proteins, hemoglobin Bart (γ_4) disease with hydrops fetalis and fetal death results, because α-globin chains are required for the synthesis of HbF. [Note: Heterozygote advantage against malaria is seen in both α- and β-thalassemias.]

V. CHAPTER SUMMARY

- **Hemoglobin A (HbA)**, the major hemoglobin in adults, is composed of four polypeptide chains (two α chains and two β chains, $\alpha_2\beta_2$) held together by noncovalent interactions (Fig. 3.24).

- The subunits occupy different relative positions in deoxyhemoglobin compared with oxyhemoglobin. The **deoxy form** of Hb is called the "**T**," or **taut (tense)**, **conformation**. It has a constrained structure that limits the movement of the polypeptide chains. The T form is the **low–oxygen-affinity form** of Hb.

- The binding of oxygen (O_2) to the heme iron causes rupture of some of the ionic and hydrogen bonds and movement of the dimers. This leads to a structure called the "**R**" or **relaxed conformation**. The R form is the **high–oxygen-affinity form** of Hb.

- The **oxygen-dissociation curve** for Hb is **sigmoidal** in shape (in contrast to that of **myoglobin**, which is **hyperbolic**), indicating that the subunits cooperate in binding O_2. The binding of an oxygen molecule at one heme group increases the oxygen affinity of the remaining heme groups in the same Hb molecule (**cooperativity**).

- Hb's ability to bind O_2 reversibly is affected by the partial pressure of oxygen (**pO_2**), the **pH** of the environment, the partial pressure of carbon dioxide (**pCO_2**), and the availability of **2,3-bisphosphoglycerate (2,3-BPG)**. For example, the release of O_2 from Hb is enhanced when the pH is lowered or the pCO_2 is increased (the **Bohr effect**), such as in **exercising muscle**, and the oxygen-dissociation curve of Hb is shifted to the right.

- To cope long term with the effects of **chronic hypoxia** or **anemia**, the concentration of **2,3-BPG** in **RBCs** increases. **2,3-BPG** binds to the Hb

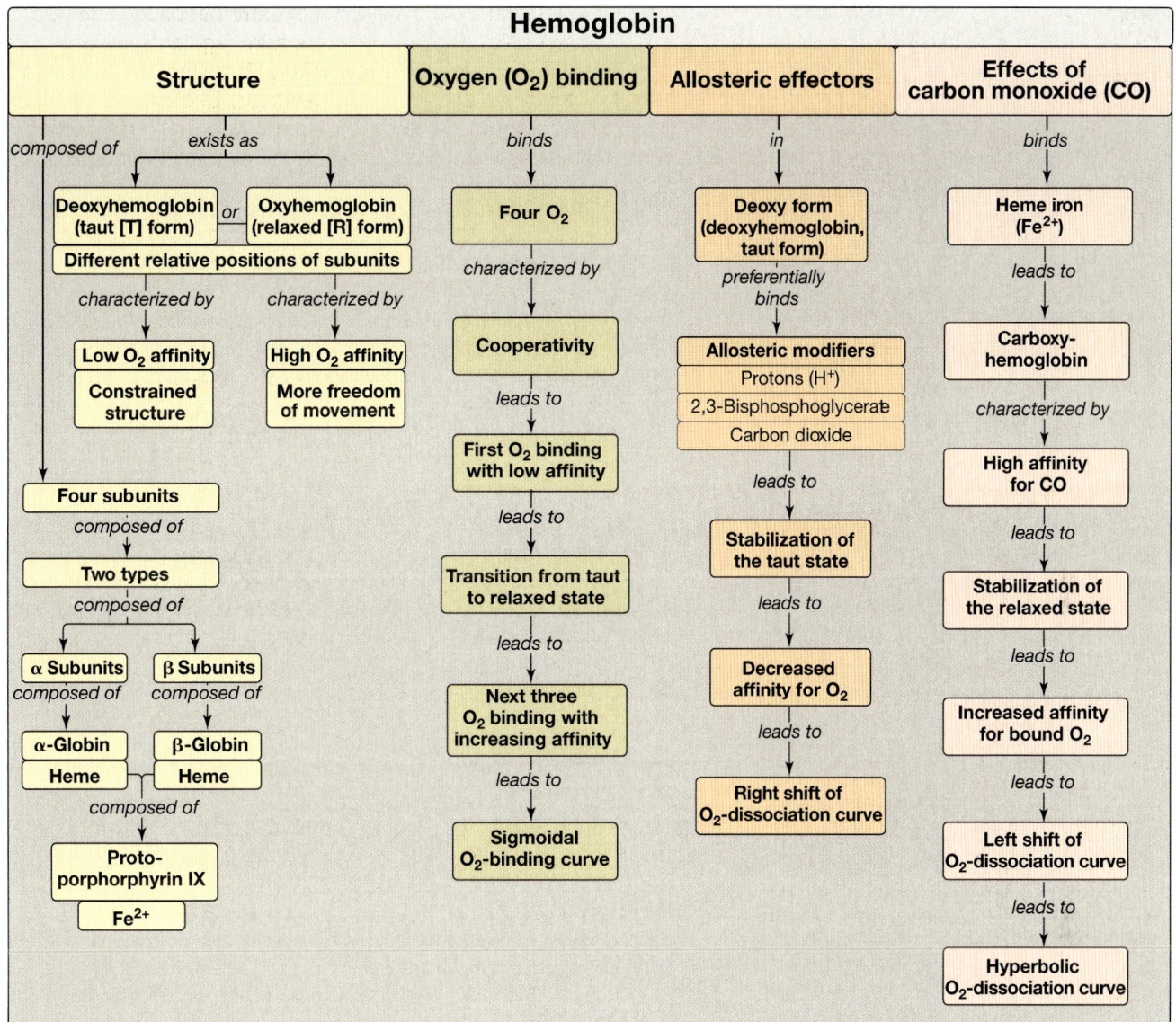

Hemoglobin

Structure	Oxygen (O₂) binding	Allosteric effectors	Effects of carbon monoxide (CO)

Structure

composed of / exists as

Deoxyhemoglobin (taut [T] form) *or* Oxyhemoglobin (relaxed [R] form)

Different relative positions of subunits

characterized by → Low O_2 affinity → Constrained structure

characterized by → High O_2 affinity → More freedom of movement

Four subunits
composed of
Two types
composed of
α Subunits | β Subunits
composed of | *composed of*
α-Globin | β-Globin
Heme | Heme
composed of
Proto-porphorphyrin IX
Fe^{2+}

Oxygen (O₂) binding

binds → Four O_2

characterized by → Cooperativity

leads to → First O_2 binding with low affinity

leads to → Transition from taut to relaxed state

leads to → Next three O_2 binding with increasing affinity

leads to → Sigmoidal O_2-binding curve

Allosteric effectors

in → Deoxy form (deoxyhemoglobin, taut form)

preferentially binds → Allosteric modifiers
- Protons (H^+)
- 2,3-Bisphosphoglycerate
- Carbon dioxide

leads to → Stabilization of the taut state

leads to → Decreased affinity for O_2

leads to → Right shift of O_2-dissociation curve

Effects of carbon monoxide (CO)

binds → Heme iron (Fe^{2+})

leads to → Carboxy-hemoglobin

characterized by → High affinity for CO

leads to → Stabilization of the relaxed state

leads to → Increased affinity for bound O_2

leads to → Left shift of O_2-dissociation curve

leads to → Hyperbolic O_2-dissociation curve

Figure 3.24
Key concept map for hemoglobin structure and function. Fe^{2+}, ferrous iron.

and decreases its oxygen affinity. This shifts the oxygen-dissociation curve to the right, offloading more oxygen at any given pO_2.

- **Fetal hemoglobin (HbF)** binds 2,3-BPG less tightly than does HbA and has a higher oxygen affinity. This facilitates the transfer of oxygen across the placenta from maternal to fetal blood during development.
- **Carbon monoxide (CO)** binds tightly (but reversibly) to the Hb iron, forming **carboxyhemoglobin**. This interaction shifts the oxygen-dissociation curve to the left, preventing oxygen from offloading to the tissues. This is why CO is so toxic.
- **Hemoglobinopathies** are disorders primarily caused either by production of a structurally abnormal Hb molecule as in **sickle cell anemia** or synthesis of insufficient quantities of normal Hb subunits as in the **thalassemias** (Fig. 3.25).

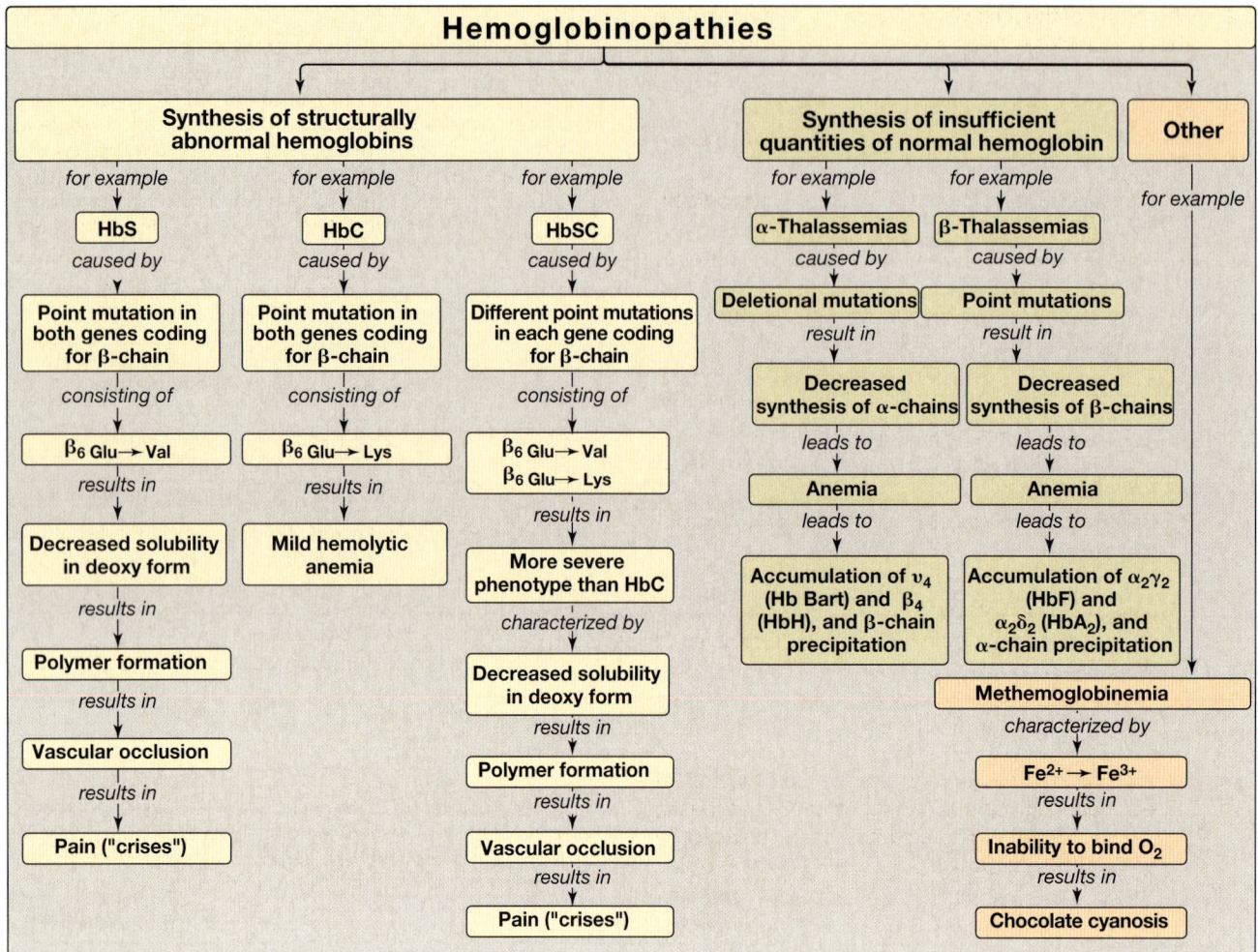

Figure 3.25
Key concept map for hemoglobinopathies. Hb, hemoglobin; Fe, iron; O$_2$, oxygen.

Study Questions

Choose the ONE best answer.

3.1. Which one of the following statements concerning different forms of hemoglobin is correct?

    A. HbA is the most abundant hemoglobin in normal adults.

    B. Fetal blood has a lower affinity for oxygen than does adult blood because HbF has an increased affinity for 2,3-bisphosphoglycerate.

    C. The globin chain composition of HbF is $\alpha_2\delta_2$.

    D. HbA1c differs from HbA by a single, genetically determined amino acid substitution.

    E. HbA$_2$ appears early in fetal life.

Correct answer = A. HbA accounts for >90% of the hemoglobin in a normal adult. If HbA$_{1c}$ is included, the percentage rises to ~97%. Because 2,3-bisphosphoglycerate (2,3-BPG) reduces the affinity of hemoglobin for oxygen, the weaker interaction between 2,3-BPG and HbF results in a higher oxygen affinity for HbF relative to HbA. HbF consists of $\alpha_2\gamma_2$. HbA1c is the most common and stable form of glycated HbA, formed nonenzymatically in red blood cells proportionate to blood glucose levels. Normal blood glucose levels should reflect 4% to 6% HbA$_{1c}$. HbA$_2$ ($\alpha_2\delta_2$) is a minor component of normal adult hemoglobin, first appearing shortly before birth and rising to adult levels (~2% of the total hemoglobin) by age 6 months.

3.2. Which one of the following statements concerning the ability of acidosis to precipitate a crisis in sickle cell anemia is correct?

    A. Acidosis decreases the solubility of HbS.

    B. Acidosis increases the oxygen affinity of hemoglobin.

    C. Acidosis favors the conversion of hemoglobin from the taut to the relaxed conformation.

    D. Acidosis shifts the oxygen-dissociation curve to the left.

    E. Acidosis decreases the ability of 2,3-bisphosphoglycerate to bind to hemoglobin.

> Correct answer = A. HbS is significantly less soluble in the deoxygenated form, compared with oxyhemoglobin S. Decreased pH (acidosis) causes the oxygen-dissociation curve to shift to the right, indicating decreased oxygen affinity (increased delivery). This favors the formation of the deoxy, or taut, form of hemoglobin and can precipitate a sickle cell crisis. The binding of 2,3-bisphosphoglycerate is increased, because it binds only to the deoxy form of hemoglobin.

3.3. Which one of the following statements concerning the binding of oxygen by hemoglobin is correct?

    A. The Bohr effect results in a lower oxygen affinity at higher pH values.

    B. Carbon dioxide increases oxygen affinity of hemoglobin.

    C. The oxygen affinity of hemoglobin increases as the percentage saturation increases.

    D. The hemoglobin tetramer binds four molecules of 2,3-bisphosphoglycerate.

    E. Oxyhemoglobin and deoxyhemoglobin have the same affinity for protons.

> Correct answer = C. The binding of oxygen at one heme group increases the oxygen affinity of the remaining heme groups in the same molecule. A rise in pH results in increased oxygen affinity. Carbon dioxide decreases oxygen affinity because it lowers the pH. Moreover, binding of carbon dioxide to the N-termini stabilizes the taut, deoxy form. Hemoglobin binds one molecule of 2,3-bisphosphoglycerate. Deoxyhemoglobin has a greater affinity for protons than does oxyhemoglobin.

3.4. β-Lysine 82 in HbA is important for the binding of 2,3-bisphosphoglycerate. In Hb Helsinki, this basic, positively charged amino acid has been replaced by the noncharged amino acid methionine. Which of the following should be true concerning Hb Helsinki?

    A. It should be stabilized in the taut, rather than the relaxed, form.

    B. It should decrease oxygen delivery to tissues.

    C. The Hb Helsinki oxygen-dissociation curve should be shifted to the right relative to HbA.

    D. It results in anemia.

    E. It should decrease hemoglobin affinity to oxygen.

> Correct answer = B. Substitution of positively charged lysine by neutral methionine decreases the ability of negatively charged phosphate groups in 2,3-bisphosphoglycerate (2,3-BPG) to bind the β subunits of hemoglobin. Because 2,3-BPG decreases the affinity of hemoglobin for oxygen, a reduction in 2,3-BPG should result in increased oxygen affinity and decreased oxygen (O_2) delivery to tissues. The relaxed form is the high–oxygen-affinity form of hemoglobin. Increased oxygen affinity (decreased delivery) results in a left shift in the oxygen-dissociation curve. Decreased delivery of O_2 is compensated for by increased RBC production.

3.5. A 67-year-old male presented to the emergency department with a 1-week history of angina and shortness of breath. He reported that his face and extremities had taken on a blue color. His medical history included chronic stable angina treated with isosorbide dinitrate and nitroglycerin. Blood obtained for analysis was brown. Which one of the following is the most likely diagnosis?

    A. Carboxyhemoglobinemia

    B. Hemoglobin SC disease

    C. Methemoglobinemia

    D. Sickle cell anemia

    E. β-Thalassemia

> Correct answer = C. Oxidation of the ferrous (Fe^{2+}) iron to the ferric (Fe^{3+}) state in the heme prosthetic group of hemoglobin forms methemoglobin. This may be caused by the action of certain drugs such as nitrates. The methemoglobinemias are characterized by chocolate cyanosis (blue coloration of the skin and mucous membranes and brown blood) as a result of the dark-colored methemoglobin. Symptoms are related to tissue hypoxia and include anxiety, headache, and dyspnea. In rare cases, coma and death can occur. [Note: Benzocaine, an aromatic amine used as a topical anesthetic, is a cause of acquired methemoglobinemia.]

3.6. In a patient with an NADH-cytochrome b_5 reductase deficiency, which one of the following would most likely be increased?

A. 2,3-BPG

B. Fe^{3+}

C. HbA2

D. HbF

E. HbS

Correct answer = B. NADH-cytochrome b_5 reductase catalyzes the reduction of the ferric (Fe^{3+}) state to ferrous (Fe^{2+}) iron state in the heme prosthetic group of hemoglobin, converting methemoglobin to hemoglobin. A deficiency in this enzyme would increase the levels of ferric (Fe^{3+}) state (methemoglobin).

3.7. A 1-year-old becomes anemic and is unable to synthesize β-globin protein. This patient has which one of the following disease states?

A. β-Thalassemia minor

B. Cooley anemia

C. Hemoglobin H disease

D. Methemoglobinemia

E. Sickle cell disease

Correct answer = B. Individuals with β-globin gene defects have either β-thalassemia trait (β-thalassemia minor) if they have only one defective β-globin gene (genotypically $β^+/β$ or $β^0/β$), or β-thalassemia major (Cooley anemia) if both genes are defective (genotypically $β^0/β^+$ or $β^0/β^0$). With both alleles affected, patients with β-thalassemia major synthesize little to no β-globin. Because the β-globin gene is not expressed until late in prenatal development, the physical manifestations of β-thalassemias appear only several months after birth. Those individuals with β-thalassemia minor make some β chains and usually do not require specific treatment. However, infants born with β-thalassemia major are seemingly healthy at birth but become severely anemic due to ineffective erythropoiesis, usually during the first or second year of life.

3.8. A patient with a defect in three of the four α-globin genes has which one of the following diseases?

A. Hemoglobin Bart disease

B. Hemoglobin C disease

C. Hemoglobin H disease

D. Methemoglobinemia

E. Sickle cell disease

Correct answer = C. In α-thalassemias, synthesis of α-globin chains is decreased or absent, typically as a result of deletional mutations. Because each individual's genome contains four copies of the α-globin gene (two on each chromosome 16), there are several levels of α-globin chain deficiencies. If one of the four alleles encodes for a defective globin protein, the individual is termed a "silent" carrier of α-thalassemia, because no physical manifestations of the disease occur. If two α-globin alleles encode for defective globin proteins, the individual is designated as having α-thalassemia trait. If three α-globin alleles encode for defective globin proteins, the individual has hemoglobin H (β4) disease, a hemolytic anemia of variable severity. If all four α-globin alleles encode for defective globin proteins, hemoglobin Bart (γ4) disease with hydrops fetalis and fetal death results, because α-globin chains are required for the synthesis of HbF.

3.9. Why is hemoglobin C disease a nonsickling disease?

In hemoglobin C, the polar glutamate is replaced by polar lysine rather than by nonpolar valine as in hemoglobin S. No nonpolar pocket is created in to generate sickling, as seen between $β^S$ chains.

3.10. What would be true about the extent of RBC sickling in individuals with both HbS and hereditary persistence of HbF?

The increased levels of fetal hemoglobin would reduce overall HbS concentration and inhibit sickling polymerization of deoxy HbS.

Fibrous Proteins

4

I. OVERVIEW

Fibrous proteins are usually folded into either extended filaments or sheet-like structures and possess repeated amino acid sequences. They are relatively insoluble and provide structural or protective function in our tissues, such as in connective tissues, tendons, bone, and muscle fibers. Collagen and elastin are examples of commonly occurring, well-characterized fibrous proteins of the extracellular matrix (ECM). Collagen and elastin serve structural functions in the body and are components of the skin, connective tissue, blood vessel walls, and the sclera and cornea of the eye. Each fibrous protein exhibits special mechanical properties, resulting from its unique structure, which is obtained by combining specific amino acids into repeated, secondary structural elements. This is in contrast to globular proteins (discussed in Chapter 3), whose shapes are the result of complex interactions between secondary; tertiary; and, sometimes, quaternary structural elements.

II. COLLAGEN

Collagen is the most abundant protein in the human body. A typical collagen molecule is a long, rigid structure in which three polypeptides (referred to as α chains) are wound around one another in a rope-like triple helix (Fig. 4.1). Although this triple helix is found in all collagen molecules throughout the body, the many subtypes of collagen are further organized and dictated by the structural role collagen plays in a particular organ. In some tissues, collagen may be dispersed as a gel that gives support to the structure, such as in the ECM or the vitreous humor of the eye. In other tissues, collagen may be bundled in tight, parallel fibers that provide great strength, as in tendons. In the cornea, collagen is stacked to transmit light with a minimum of scattering. Collagen of bone occurs as fibers arranged at an angle to each other so as to resist mechanical shear from any direction.

A. Types

The collagen superfamily of proteins includes more than 25 collagen types as well as additional proteins that have collagen-like domains. The three polypeptide α chains are held together by interchain hydrogen bonds. Variations in the amino acid sequence of the α chains result in structural components that are about the same size (~1,000 amino acids long) but with slightly different properties. These α chains are combined to form the various types of collagen found in the tissues. For example, the most common collagen, type I, contains two chains called $\alpha 1$ and one chain called $\alpha 2$ ($\alpha 1_2\alpha 2$),

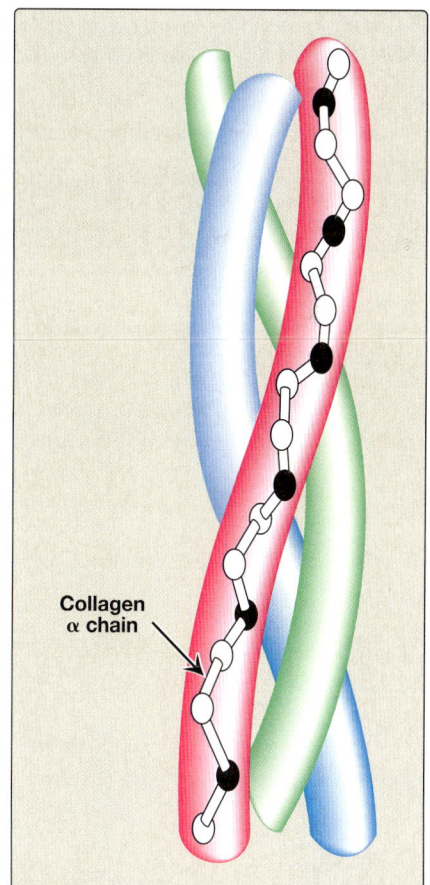

Figure 4.1
Triple-stranded helix of collagen formed from three α chains. [Note: The α chains themselves are helical in structure.]

Collagen
α chain

TYPE	TISSUE DISTRIBUTION
Fibril forming	
I	Skin, bone, tendon, blood vessels, cornea
II	Cartilage, intervertebral disk, vitreous body
III	Blood vessels, skin, muscle
Network forming	
IV	Basement membrane
VIII	Corneal and vascular endothelium
Fibril associated*	
IX	Cartilage
XII	Tendon, ligaments, some other tissues

Figure 4.2
The most abundant types of collagen.
[Note: *Fibril-associated collagens with interrupted triple helices are known as FACIT.]

whereas type II collagen contains three $\alpha 1$ chains ($\alpha 1_3$). The collagens can be organized into three major groups, based on their location and functions in the body (Fig. 4.2).

1. **Fibril-forming collagens:** Types I, II, and III are the fibrillar collagens, with the rope-like structure described for a typical collagen molecule. In the electron microscope, these linear polymers of fibrils have characteristic banding patterns, reflecting the regular staggered packing of the individual collagen molecules in the fibril (Fig. 4.3). Type I collagen fibers (composed of collagen fibrils) are found in supporting elements of high tensile strength (e.g., tendons and corneas), whereas fibers formed from type II collagen molecules are restricted to cartilaginous structures. The fibers derived from type III collagen are prevalent in more distensible tissues such as blood vessels.

2. **Network-forming collagens:** Types IV and VIII form a three-dimensional mesh, rather than distinct fibrils (Fig. 4.4). For

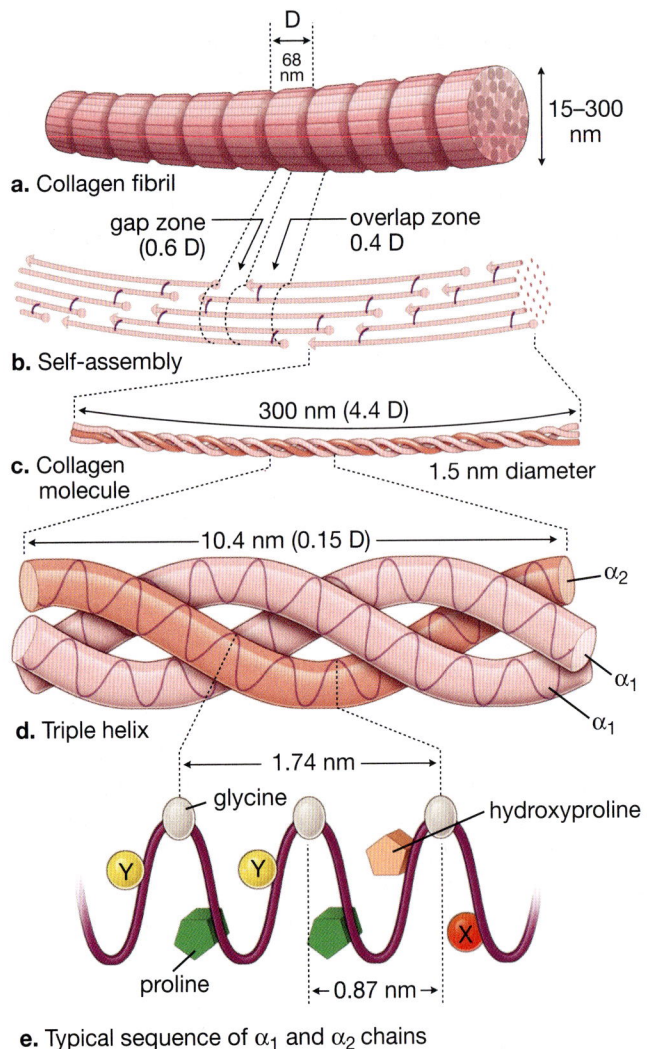

Figure 4.3
Diagram showing the molecular character of a type I collagen fibril in increasing order of structure. (From Pawlina W. *Histology: A Text and Atlas.* 9th ed. Wolters Kluwer; 2024.)

example, type IV molecules assemble into a sheet or meshwork that constitutes a major part of basement membranes.

> || Basement membranes are thin, sheet-like structures that provide mechanical support for adjacent cells and function as a semipermeable filtration barrier to macromolecules in organs such as the kidney and the lung.

3. **Fibril-associated collagens:** Types IX and XII bind to the surface of collagen fibrils, linking these fibrils to one another and to other components in the ECM (see Fig. 4.2).

B. Structure

Unlike most globular proteins that are folded into compact structures, collagen, a fibrous protein, has an elongated, triple-helix structure that is stabilized by interchain hydrogen bonds.

1. **Amino acid sequence:** Collagen is rich in proline and glycine, both of which are important in the formation of the triple-stranded helix. Proline facilitates the formation of the helical conformation of each α chain because its ring structure causes "kinks" in the peptide chain. [Note: The presence of proline dictates that the helical conformation of the α chain cannot be an α helix.] Glycine, the smallest amino acid, is found in every third position of each polypeptide chain. Glycine fits into the restricted spaces where the three chains of the helix come together. The glycine residues are part of a repeating sequence, –Gly–X–Y–, where X is frequently proline, and Y is often hydroxyproline (but can be hydroxylysine; Fig. 4.5). Thus, most of the α chain can be regarded as a polytripeptide whose sequence can be represented as (–Gly–Pro–Hyp–)$_{333}$.

2. **Hydroxyproline and hydroxylysine:** Collagen contains hydroxyproline and hydroxylysine, which are nonstandard amino acids not present in most other proteins. These unique amino acids result from the hydroxylation of some of the proline and lysine residues after their incorporation into polypeptide chains (Fig. 4.6). Therefore, hydroxylation is a posttranslational modification. [Note: The presence of hydroxyproline maximizes formation of interchain hydrogen bonds that stabilize the triple-helical structure.]

3. **Glycosylation:** The hydroxyl group of the hydroxylysine residues of collagen may be enzymatically glycosylated. Most commonly, glucose and galactose are sequentially attached to the polypeptide chain prior to triple-helix formation (Fig. 4.7). The hydroxylysine residues of collagen type IV are more frequently glycosylated than in fibril-forming type collagens.

C. Biosynthesis

The polypeptide precursors of the collagen molecule are synthesized in fibroblasts (or in the related osteoblasts of bone and chondroblasts of cartilage). They are enzymatically modified and form the triple helix, which gets secreted into the ECM. After additional enzymic modification, the mature extracellular collagen fibrils aggregate and become cross-linked to form collagen fibers.

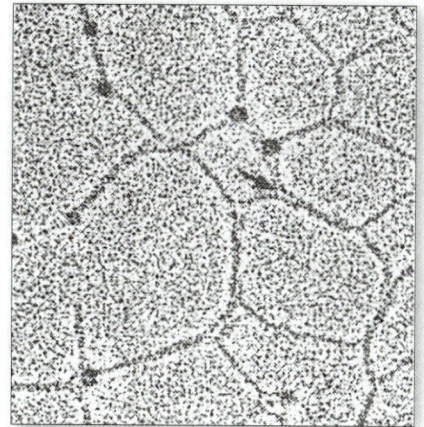

Figure 4.4
Electron micrograph of a polygonal network formed by association of collagen type IV monomers.

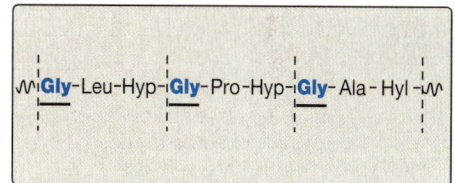

Figure 4.5
Amino acid sequence of a portion of the α1 chain of collagen. Hyp, hydroxyproline; Hyl, hydroxylysine.

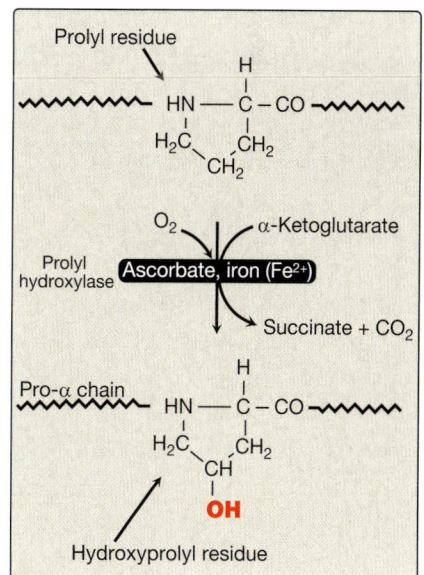

Figure 4.6
Hydroxylation of proline residues in pro-α chains of collagen by *prolyl hydroxylase*. [Note: Fe^{2+} (hydroxylase cofactor) is protected from oxidation to Fe^{3+} by ascorbate (vitamin C).]

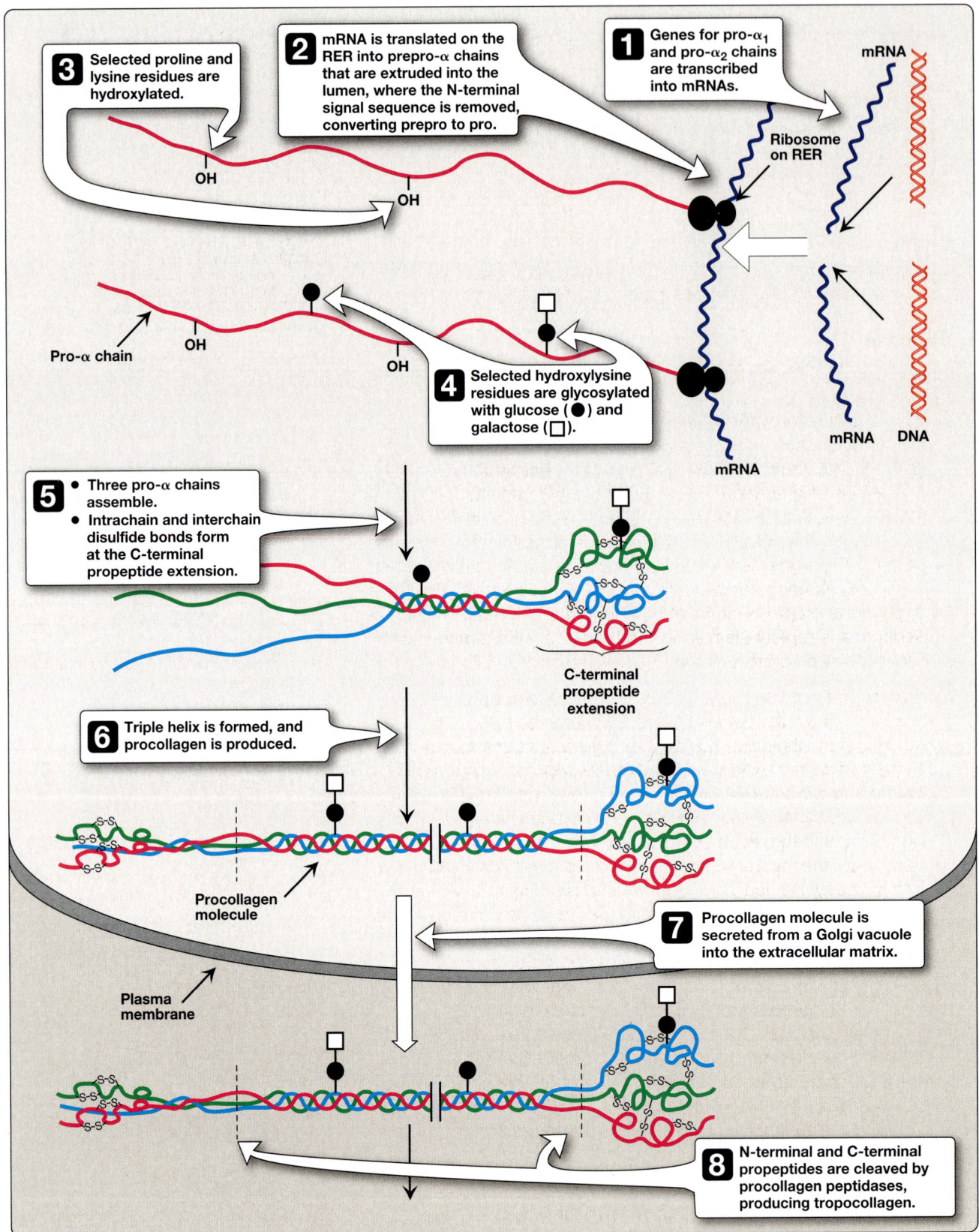

Figure 4.7
Synthesis of collagen. RER, rough endoplasmic reticulum; mRNA, messenger RNA. (continued on the next page)

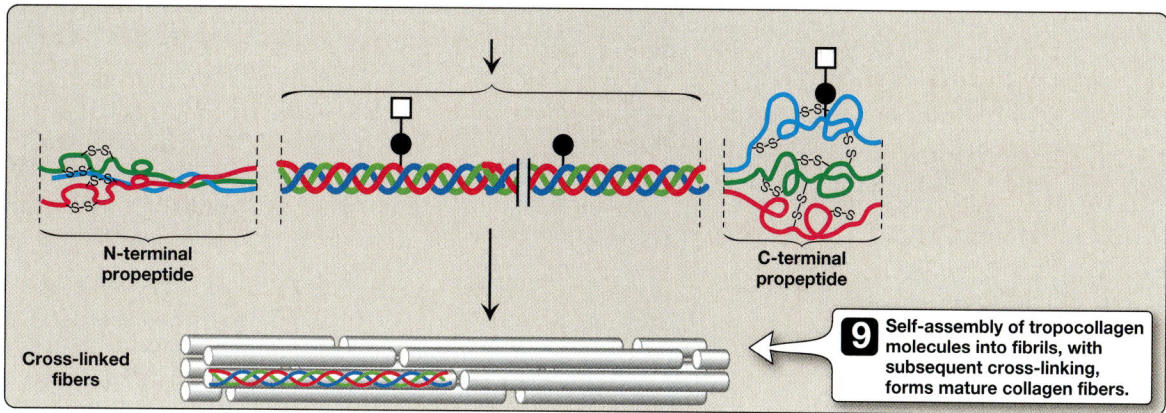

Figure 4.7
Synthesis of collagen. (continued from the previous page)

1. **Pro-α chain formation:** Collagen is one of many proteins that normally function outside of cells. The newly synthesized polypeptide precursors of α chains (prepro-α chains) contain a special amino acid sequence at their amino (N)-terminal ends. This sequence acts as a signal that, in the absence of additional signals, targets the polypeptide being synthesized for secretion from the cell. The signal sequence facilitates the binding of ribosomes to the rough endoplasmic reticulum (RER) and directs the passage of the prepro-α chain into the lumen of the RER. The signal sequence is rapidly cleaved in the lumen to yield a precursor of collagen called a pro-α chain (Fig. 4.7, steps 1 and 2).

2. **Hydroxylation:** The pro-α chains are processed by a number of enzymes within the lumen of the RER while the polypeptides are still being synthesized. Proline and lysine residues found in the Y-position of the –Gly–X–Y– sequence can be hydroxylated to form hydroxyproline and hydroxylysine residues (Fig. 4.7, step 3). These hydroxylation reactions require molecular oxygen, ferrous iron (Fe^{2+}), and the reducing agent vitamin C (ascorbic acid), without which the hydroxylating enzymes, prolyl hydroxylase, and lysyl hydroxylase, are unable to function (see Fig. 4.6). In the case of ascorbic acid deficiency (and, therefore, a lack of proline and lysine hydroxylation), the formation of interchain H-bonds and the formation of a stable triple helix are impaired. Additionally, collagen fibrils cannot be cross-linked (see Section 7), greatly decreasing the tensile strength of the assembled fiber. The resulting ascorbic acid deficiency disease is known as scurvy. Patients with scurvy often show ecchymoses (bruise-like discolorations) and petechiae on the limbs because of subcutaneous extravasation (leakage) of blood due to capillary fragility (Fig. 4.8). Other symptoms also include gum disease, loosening of the teeth, and poor wound healing.

3. **Glycosylation:** Some hydroxylysine residues are modified by glycosylation with glucose or glucosyl-galactose (Fig. 4.7, step 4).

4. **Assembly and secretion:** After hydroxylation and glycosylation, three pro-α chains form procollagen, a precursor of collagen that has a central region of triple helix flanked by the nonhelical N- and

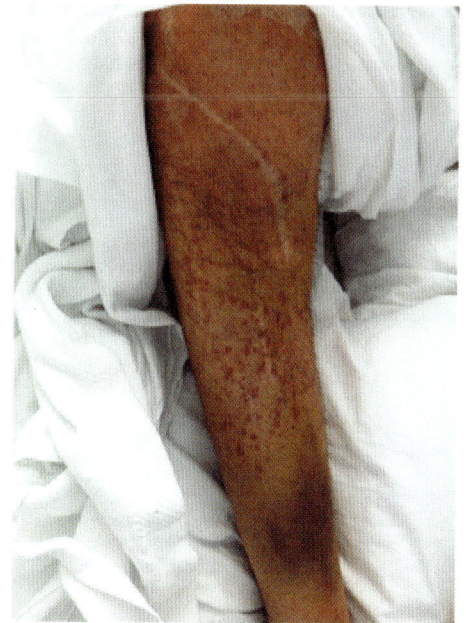

Figure 4.8
Legs of a 46-year-old man with scurvy. (Council ML, Sheinbein D, Cornelius LA; *The Washington Manual of Dermatology Diagnostics*. Wolters Kluwer Health, 2016.)

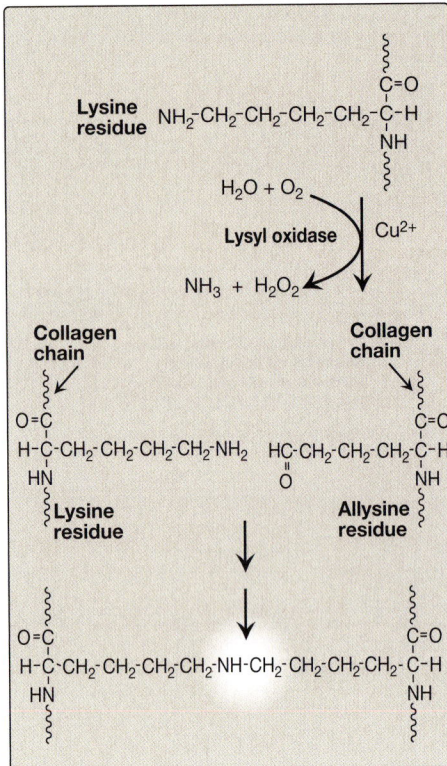

Figure 4.9
Formation of cross-links in collagen. [Note: Lysyl oxidase is irreversibly inhibited by a toxin present in seeds from *Lathyrus odoratus* (sweet pea), leading to a condition known as lathyrism that is characterized by skeletal and vascular problems.] Cu^{2+}, copper; NH_3, ammonia; H_2O_2, hydrogen peroxide.

carboxyl (C)-terminal extensions called propeptides (Fig. 4.7, step 5). The formation of procollagen begins with formation of interchain disulfide bonds between the C-terminal extensions of the pro-α chains. This brings the three α chains into an alignment favorable for triple helix formation. The procollagen molecules move through the Golgi apparatus, where they are packaged in secretory vesicles. The vesicles fuse with the cell membrane, causing the release of procollagen molecules into the extracellular space (Fig. 4.7, steps 6 and 7).

5. **Extracellular cleavage of procollagen molecules:** After their release, the triple-helical procollagen molecules are cleaved by *N-* and *C-procollagen peptidases*, which remove the terminal propeptides, producing tropocollagen molecules (Fig. 4.7, step 8).

6. **Collagen fibril formation:** Tropocollagen molecules spontaneously associate to form collagen fibrils. The fibrils form an ordered, parallel array, with adjacent collagen molecules arranged in a staggered pattern formed by approximately three quarters of each molecule overlapping the neighboring molecule (Fig. 4.7, step 9).

7. **Cross-link formation:** The array of collagen fibril molecules serves as a substrate for lysyl oxidase. This copper-containing extracellular enzyme oxidatively deaminates some of the lysine and hydroxylysine residues in collagen. The reactive aldehydes (allysine and hydroxyallysine) that result from the deamination reactions can spontaneously condense with lysine or hydroxylysine residues in neighboring collagen molecules to form covalent cross-links and, thus, mature collagen fibers (Fig. 4.7, step 9 and Fig. 4.9). [Note: Cross-links can also form between two allysine residues.]

> Lysyl oxidase is one of several copper-containing enzymes. Others include ceruloplasmin, cytochrome c oxidase, dopamine hydroxylase, superoxide dismutase, and tyrosinase. Disruption in copper homeostasis causes copper deficiency (X-linked Menkes syndrome) or overload (Wilson disease).

D. Degradation

Normal collagen fibers are highly stable molecules, having half-lives as long as several years. However, connective tissue is dynamic and is constantly being remodeled, often in response to growth or injury of the tissue. Breakdown of collagen fibers depends on the proteolytic action of collagenases, which are part of a large family of matrix metalloproteinases. For type I collagen, the cleavage site is specific, generating three-quarter and one-quarter length fragments. These fragments are further degraded by other matrix proteinases.

E. Collagenopathies

Defects in any one of the many steps in collagen fiber synthesis can result in a genetic disease involving an inability of collagen to form fibers properly and, therefore, an inability to provide tissues with the needed tensile strength normally provided by collagen. More than 1,000 mutations have been identified in 23 genes coding for 13 of the collagen types. The following are examples of diseases (collagenopathies) that are the result of defective collagen synthesis.

1. **Ehlers–Danlos syndrome:** Ehlers–Danlos syndrome (EDS) is a heterogeneous group of connective tissue disorders that result from heritable defects in the metabolism of fibrillar collagen molecules. EDS can be caused by a deficiency of collagen-processing enzymes (e.g., lysyl hydroxylase or N-procollagen peptidase) or from mutations in the amino acid sequences of collagen types I, III, and V. The classic form of EDS, caused by defects in type V collagen, is characterized by skin extensibility and fragility and joint hypermobility (Fig. 4.10). The vascular form, due to defects in type III collagen, is the most serious form of EDS because it is associated with potentially lethal arterial rupture. [Note: The classic and vascular forms show autosomal-dominant inheritance.] Collagen that contains mutant chains may have altered structure, secretion, or distribution, and it frequently is degraded. [Note: Incorporation of just one mutant chain may result in degradation of the triple helix. This is known as a "dominant-negative" effect.]

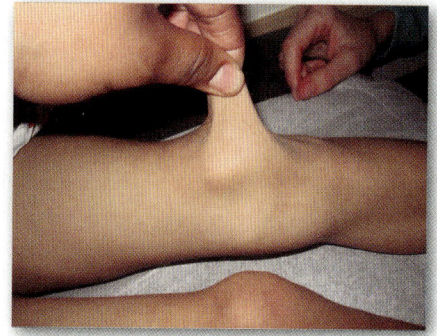

Figure 4.10
Stretchy skin of classic Ehlers–Danlos syndrome.

2. **Osteogenesis imperfecta:** This syndrome, known as "brittle-bone" disease, is a genetic disorder of bone fragility characterized by bones that fracture easily, with minor or no trauma (Fig. 4.11). More than 80% of cases of osteogenesis imperfecta (OI) are caused by dominant mutations to the genes that encode the $\alpha 1$ or $\alpha 2$ chains in type I collagen. The most common mutations cause the replacement of glycine (in –Gly–X–Y–) by amino acids with bulky side chains. The resultant structurally abnormal α chains prevent the formation of the required triple-helical conformation. Phenotypic severity ranges from mild to lethal. Type I OI, the most common form, is characterized by mild bone fragility, hearing loss, and blue sclerae. Type II, the most severe form, is typically lethal in the perinatal period because of pulmonary complications. *In utero* fractures are seen (Fig. 4.11 A). Type III is also a severe form and is characterized by multiple fractures at birth, short stature, spinal curvature leading to a humped-back (kyphotic) appearance, and blue sclerae. Dentinogenesis imperfecta, a disorder of tooth development, may be seen in OI. OI is treated with bisphosphonates, which function by inactivating osteoclasts, the cells that break down bone tissue (Fig. 4.11 B). Bisphosphonates also increase apoptosis (cell death) of osteoclasts and, therefore, inhibit the resorption of bone material. Bisphosphonates also decrease apoptosis of osteoblasts, the cells that lay down new bone matrix.

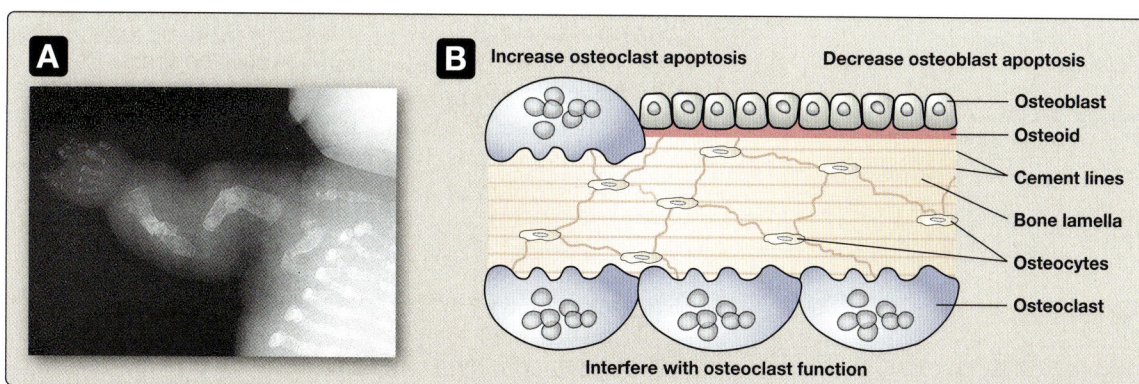

Figure 4.11
A. Lethal form (type II) of osteogenesis imperfecta in which the fractures appear *in utero*, as revealed by this radiograph of a stillborn fetus. **B.** Mechanism of action of bisphosphonates to treat osteogenesis imperfecta (OI).

3. **Alport syndrome:** This is a group of heterogeneous inherited disorders of basement membranes of the kidney and, frequently, the cochlea and the eye characterized by glomerulonephritis, hematuria, proteinuria, hypertension, and progression to end-stage renal disease (ESRD) and hearing loss during the second to fourth decades of life. This disorder is the result of mutations in type IV collagen genes, with a genetic frequency of ~1 case in 5,000. The most common form inherits as X-linked autosomal dominant. The pattern of inheritance and symptoms differ, depending on which type IV collagen gene is involved.

III. ELASTIN

In contrast to collagen, which forms fibers that are tough and have high tensile strength, elastin is a fibrous protein with rubber-like properties found in connective tissue. Elastic fibers composed of elastin and glycoprotein microfibrils are found in the lungs, walls of large arteries, and elastic ligaments. They can be stretched to several times their normal length but recoil to their original shape when the stretching force is relaxed.

A. Structure

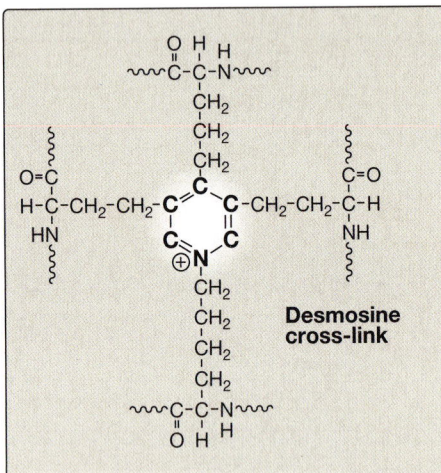

Figure 4.12
Desmosine cross-link unique to elastin.

Elastin is an insoluble protein polymer generated from a precursor, tropoelastin, which is a soluble polypeptide composed of ~700 amino acids that are primarily small and nonpolar (e.g., glycine, alanine, and valine). Elastin is also rich in proline and lysine but contains few hydroxyproline and hydroxylysine. Tropoelastin is secreted by the cell into the ECM. There, it interacts with specific glycoprotein microfibrils, such as fibrillin, which function as a scaffold onto which tropoelastin is deposited. Some of the lysyl side chains of the tropoelastin polypeptides are oxidatively deaminated by lysyl oxidase, forming allysine residues. Three of the allysyl side chains plus one unaltered lysyl side chain from the same or neighboring polypeptides form a desmosine cross-link (Fig. 4.12). This produces elastin, an extensively interconnected, rubbery network that can stretch and bend in any direction when stressed, giving connective tissue elasticity (Fig. 4.13).

B. Marfan syndrome

Mutations in the fibrillin-1 protein are responsible for Marfan syndrome, a connective tissue disorder characterized by impaired structural integrity in the skeleton, eye, and cardiovascular system. With this disease, abnormal fibrillin protein is incorporated into microfibrils along with normal fibrillin, inhibiting the formation of functional microfibrils. Patients with Marfan syndrome are frequently tall, with long slender arms, legs, fingers, and toes. They have flexible joints and may have scoliosis. The heart and aorta are often affected as well, and they have an increased risk for mitral valve prolapse or aortic aneurysm. [Note: Patients with Marfan syndrome, OI, or EDS may have blue sclerae due to tissue thinning that allows underlying pigment to show through.]

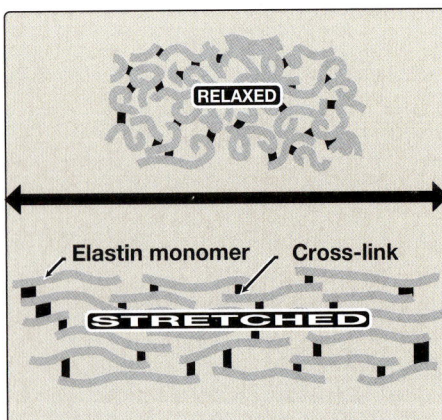

Figure 4.13
Elastin fibers in relaxed and stretched conformations.

C. α_1-Antitrypsin in elastin degradation

Blood and other body fluids contain the protein α_1-antitrypsin (AAT), which inhibits a few proteolytic enzymes (called peptidases, proteases,

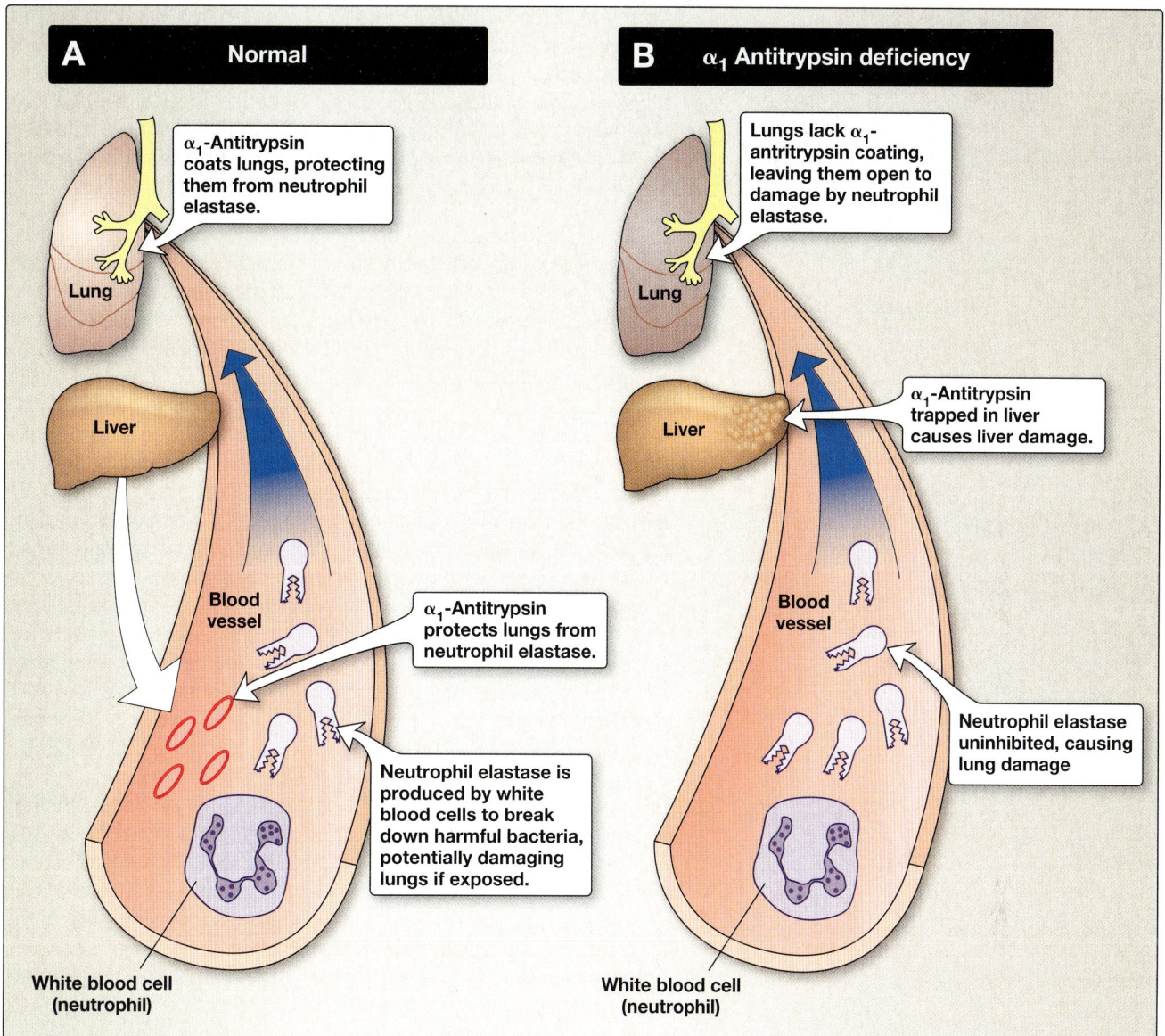

Figure 4.14
α_1-Antitrypsin (AAT) deficiency. **A.** AAT produced and secreted from hepatocytes inhibits neutrophil elastase from damaging extracellular matrix of the lung. **B.** In AAT deficiency, misfolding of AAT in the endoplasmic reticulum prevents its secretion from hepatocytes, causing liver damage. Elastase is more active and can lead to destruction of alveolar tissue and emphysema.

or proteinases) that hydrolyze and degrade proteins. [Note: The inhibitor was originally named AAT because it inhibits the activity of trypsin, a proteolytic enzyme synthesized as trypsinogen by the pancreas.] AAT has the important physiologic role of inhibiting neutrophil elastase, a powerful protease that is released into the extracellular space and degrades elastin of alveolar walls as well as other structural proteins in a variety of tissues (Fig. 4.14). Most of the AAT found in plasma is synthesized and secreted by the liver. Extrahepatic synthesis also occurs.

1. **α_1-Antitrypsin in the lungs:** In the normal lung, the alveoli are chronically exposed to low levels of neutrophil elastase released

from activated and degenerating neutrophils. The proteolytic activity of elastase can destroy the elastin in alveolar walls if unopposed by the action of AAT, the most important inhibitor of neutrophil elastase (Fig. 4.14). Because lung tissue cannot regenerate, the destruction of the connective tissue of alveolar walls caused by an imbalance between the elastase and its inhibitor results in pulmonary disease.

2. **α_1-Antitrypsin deficiency and emphysema:** In the United States, ~2% to 5% of patients with emphysema are predisposed to the disease by inherited defects in AAT. Several different mutations in the AAT gene are known to cause a deficiency of the protein, but one single purine base mutation (GAG to AAG, resulting in the substitution of lysine for glutamic acid at position 342 of the protein) is clinically the most widespread and severe. [Note: The mutated protein is termed the "Z variant."] The mutation causes the normally monomeric AAT protein to misfold, polymerize, and aggregate within the RER of hepatocytes, resulting in decreased secretion of AAT by the liver. AAT deficiency is, therefore, a misfolded protein disease. [Note: The polymer that accumulates in hepatocytes may cause hepatocellular damage and result in cirrhosis. Such hepatic damage is a leading cause for pediatric end-stage liver failure, which requires liver transplantation.] Because less AAT is secreted by the liver, blood levels of AAT are reduced, as is the amount of AAT that is available to lung tissues. In the United States, the AAT mutation is most common in Caucasians of Northern European ancestry. An individual must inherit two abnormal AAT alleles to be at risk for the development of emphysema. In a heterozygote, with one normal and one defective allele, the levels of AAT are sufficient to protect the alveoli from damage. [Note: Methionine 358 in AAT is required for the binding of the inhibitor to its target proteases. Smoking causes the oxidation and subsequent inactivation of the methionine, thereby rendering the inhibitor powerless to neutralize elastase. Smokers with AAT deficiency, therefore, have a considerably elevated rate of lung destruction and a poorer survival rate than nonsmokers with the deficiency.] The deficiency of elastase inhibitor can be treated by weekly augmentation therapy, that is, intravenous administration of AAT. The AAT diffuses from the blood into the lung, where it reaches therapeutic levels in the fluid surrounding the lung epithelial cells.

IV. CHAPTER SUMMARY

- Collagen and elastin are structural fibrous proteins of the ECM (Fig. 4.15).
- **Collagen** contains an abundance of **proline, lysine,** and **glycine,** the latter occurring at every third position in the primary structure. It also contains **hydroxyproline, hydroxylysine,** and **glycosylated hydroxylysine,** each formed by posttranslational modification.
- Fibrillar collagen has a long, rigid structure, in which three collagen polypeptide α chains are wound around one another in a rope-like **triple helix** stabilized by **interchain hydrogen bonds.** Diseases of fibrillar collagen synthesis affect bones, joints, skin, and blood vessels.
- **Elastin** is a connective tissue protein with rubber-like properties in tissues such as the lung. **AAT,** produced primarily by the liver, inhibits **elastase**-catalyzed degradation of elastin in the alveolar walls. A deficiency of AAT increases elastin degradation and can cause **emphysema** and, in some cases, **cirrhosis** of the liver.

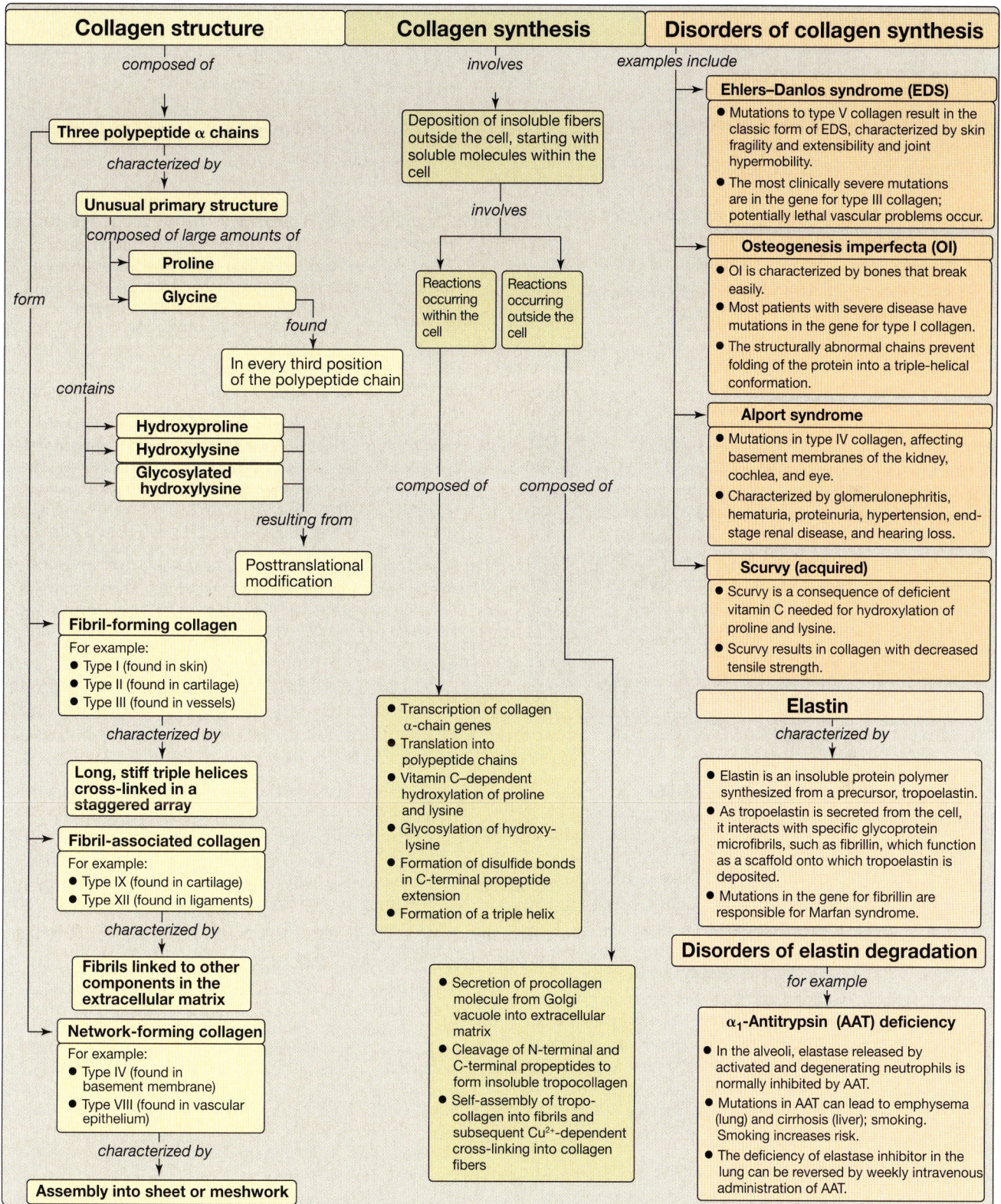

Collagen structure

composed of

Three polypeptide α chains

characterized by

Unusual primary structure

composed of large amounts of

Proline

Glycine

found

In every third position of the polypeptide chain

contains

Hydroxyproline

Hydroxylysine

Glycosylated hydroxylysine

resulting from

Posttranslational modification

form

Fibril-forming collagen

For example:
- Type I (found in skin)
- Type II (found in cartilage)
- Type III (found in vessels)

characterized by

Long, stiff triple helices cross-linked in a staggered array

Fibril-associated collagen

For example:
- Type IX (found in cartilage)
- Type XII (found in ligaments)

characterized by

Fibrils linked to other components in the extracellular matrix

Network-forming collagen

For example:
- Type IV (found in basement membrane)
- Type VIII (found in vascular epithelium)

characterized by

Assembly into sheet or meshwork

Collagen synthesis

involves

Deposition of insoluble fibers outside the cell, starting with soluble molecules within the cell

involves

Reactions occurring within the cell

Reactions occurring outside the cell

composed of

- Transcription of collagen α-chain genes
- Translation into polypeptide chains
- Vitamin C–dependent hydroxylation of proline and lysine
- Glycosylation of hydroxylysine
- Formation of disulfide bonds in C-terminal propeptide extension
- Formation of a triple helix

composed of

- Secretion of procollagen molecule from Golgi vacuole into extracellular matrix
- Cleavage of N-terminal and C-terminal propeptides to form insoluble tropocollagen
- Self-assembly of tropocollagen into fibrils and subsequent Cu^{2+}-dependent cross-linking into collagen fibers

Disorders of collagen synthesis

examples include

Ehlers–Danlos syndrome (EDS)

- Mutations to type V collagen result in the classic form of EDS, characterized by skin fragility and extensibility and joint hypermobility.
- The most clinically severe mutations are in the gene for type III collagen; potentially lethal vascular problems occur.

Osteogenesis imperfecta (OI)

- OI is characterized by bones that break easily.
- Most patients with severe disease have mutations in the gene for type I collagen.
- The structurally abnormal chains prevent folding of the protein into a triple-helical conformation.

Alport syndrome

- Mutations in type IV collagen, affecting basement membranes of the kidney, cochlea, and eye.
- Characterized by glomerulonephritis, hematuria, proteinuria, hypertension, end-stage renal disease, and hearing loss.

Scurvy (acquired)

- Scurvy is a consequence of deficient vitamin C needed for hydroxylation of proline and lysine.
- Scurvy results in collagen with decreased tensile strength.

Elastin

characterized by

- Elastin is an insoluble protein polymer synthesized from a precursor, tropoelastin.
- As tropoelastin is secreted from the cell, it interacts with specific glycoprotein microfibrils, such as fibrillin, which function as a scaffold onto which tropoelastin is deposited.
- Mutations in the gene for fibrillin are responsible for Marfan syndrome.

Disorders of elastin degradation

for example

$α_1$-Antitrypsin (AAT) deficiency

- In the alveoli, elastase released by activated and degenerating neutrophils is normally inhibited by AAT.
- Mutations in AAT can lead to emphysema (lung) and cirrhosis (liver); smoking. Smoking increases risk.
- The deficiency of elastase inhibitor in the lung can be reversed by weekly intravenous administration of AAT.

Figure 4.15
Key concept map for the fibrous proteins collagen and elastin. Cu^{2+}, copper.

Study Questions

Choose the ONE best answer.

4.1. A 30-year-old female of Northern European ancestry presents with progressive dyspnea (shortness of breath). She denies the use of cigarettes. Family history reveals that her sister also has lung problems. Which of the following etiologies most likely explains this patient's pulmonary symptoms?

A. Deficiency in dietary vitamin C
B. Deficiency of α_1-antitrypsin
C. Deficiency of prolyl hydroxylase
D. Decreased elastase activity
E. Increased collagenase activity

Correct answer = B. α_1-Antitrypsin (AAT) deficiency is a genetic disorder that can cause pulmonary damage and emphysema even in the absence of cigarette use. AAT deficiency permits increased elastase activity to destroy elastin in the alveolar walls. It should be suspected when chronic obstructive pulmonary disease develops in a patient younger than age 45 years who does not have a history of chronic bronchitis or tobacco use, or when multiple family members develop obstructive lung disease at an early age. Choices A, C, and E refer to collagen, not elastin.

4.2. A 7-month-old child "fell over" while crawling and now presents with a swollen leg. Imaging reveals a fracture of a bowed femur, secondary to minor trauma, and thin bones (see x-ray). Blue sclerae are also noted. At age 1 month, the infant had multiple fractures in various states of healing (right clavicle, right humerus, and right radius). A careful family history has ruled out nonaccidental trauma (child abuse) as a cause of the bone fractures. Which pairing of a defective (or deficient) molecule and the resulting pathology best fits this clinical description?

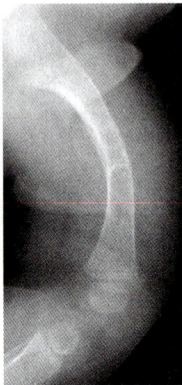

A. Elastin and emphysema
B. Fibrillin and Marfan disease
C. Type I collagen and osteogenesis imperfecta
D. Type V collagen and Ehlers–Danlos syndrome
E. Vitamin C and scurvy

Correct answer = C. The child most likely has osteogenesis imperfecta. Most cases arise from a defect in the genes encoding type I collagen. Bones in affected patients are thin, osteoporotic, often bowed, and extremely prone to fracture. Pulmonary problems are not seen in this child. Individuals with Marfan syndrome have impaired structural integrity of the skeleton, eyes, and cardiovascular system. Defects in type V collagen cause the classic form of Ehlers–Danlos syndrome characterized by skin extensibility and fragility and joint hypermobility. Scurvy caused by vitamin C deficiency is characterized by capillary fragility.

4.3. A 60-year-old homeless male presents to the emergency room complaining of progressive fatigue, leg pain, and generalized weakness. He has bloody stools, shortness of breath, easy bruising, leg swelling, and a red rash on his arms and legs. He is taking no medications. On further questioning, he reveals that his diet consists entirely of bread, canned meat, and beer. Closer examination of the rashes on his legs reveals corkscrew hairs and subepidermal red blood cell extravasation surrounding the hair follicles. What is the possible underlying problem in this patient?

A. Mutation of type V collagen
B. Mutation of type I collagen
C. Decreased prolyl hydroxylase and lysyl hydroxylase activity
D. Decreased circulating AAT levels
E. Mutation of fibrillin

Correct answer = C. The patient has scurvy, caused by a vitamin C deficiency. Vitamin C is required for prolyl hydroxylase and lysyl hydroxylase activity. Hydroxylation of proline and lysine residues in the –Gly–X–Y– sequence of collagen is essential for interchain H-bond formation and a stable collagen triple helix. A mutation in type V collagen is characteristic of Ehlers–Danlos syndrome. A mutation in type I collagen is characteristic of osteogenesis imperfecta. Decreased circulating AAT levels are the basis of AAT deficiency, which results in possible pulmonary damage and emphysema symptoms or pediatric end-stage liver failure. A mutation in fibrillin is characteristic of Marfan syndrome.

4.4. A 30-year-old patient presents with hearing loss and hypertension. Analysis of his urine indicates proteinuria. He is found to have a defect in type IV collagen. Which of the following is the most likely diagnosis in this patient?

A. Alport syndrome
B. Ehlers–Danlos syndrome
C. Marfan syndrome
D. Menkes syndrome
E. Osteogenesis imperfecta

Correct answer = A. The patient has Alport syndrome, a group of inherited disorders of the basement membrane of the kidney caused by mutations in type IV collagen, resulting in hereditary nephritis, proteinuria, and hypertension. This frequently can involve the eye or cochlea as well. Patients progress to end-stage renal disease by ages 30 to 40 years. Ehlers–Danlos syndrome is a deficiency in procollagen peptidase (*ADAMTS2* gene) affecting procollagen processing in the extracellular matrix. Collagen fibrils do not form. Marfan syndrome is caused by a defect in fibrillin, an elastin-binding protein. Menkes syndrome is caused by a defect in copper metabolism, affecting lysyl oxidase activity, preventing cross-linking of collagen into stable tropocollagen fibrils. Osteogenesis imperfecta is caused by a mutation in type I collagen.

4.5. A 30-year-old patient presents in the emergency department with sharp pain in his jaw and neck and difficulty swallowing and breathing; he is hypotensive. He also has scoliosis, a sunken breastbone, and long slender limbs and fingers. Which of the following is the most likely underlying problem in this patient?

A. Decreased circulating AAT levels
B. Decreased lysyl oxidase activity
C. Decreased procollagen peptidase activity
D. Mutation of fibrillin
E. Mutation of type I collagen

Correct answer = D. The patient has Marfan syndrome, caused by a mutation in fibrillin, an elastase-binding protein. Patients often have long slender limbs and fingers, a sunken breastbone, scoliosis, and bilateral lens dislocation. The most serious complications of this disease include aortic aneurysm and aortic dissection. A decrease in circulating AAT levels causes AAT deficiency, with damage to lung and liver tissues. A decrease in lysyl oxidase activity results in Menkes syndrome, a defect in copper metabolism that prevents the proper cross-linking of tropocollagen into stable fibrils. A decrease in procollagen peptidase results in Ehlers–Danlos syndrome, preventing the cleavage of procollagen into tropocollagen in the extracellular matrix. A mutation in type I collagen results in osteogenesis imperfecta, or "brittle-bone" disease.

4.6. A 20-year-old presents for his annual checkup. He has overly flexible joints and stretchy, fragile skin, which bruises easily. Which of the following is the most likely diagnosis for this patient?

A. Alport syndrome
B. Ehlers–Danlos syndrome
C. Marfan syndrome
D. Menkes syndrome
E. Scurvy

Correct answer = B. The patient has Ehlers–Danlos syndrome, a deficiency in procollagen peptidase (*ADAMTS2* gene) affecting procollagen processing to tropocollagen in the extracellular matrix. Alport syndrome is a group of inherited disorders of the basement membrane of the kidney caused by mutations in type IV collagen, resulting in hereditary nephritis, proteinuria, and hypertension. Marfan syndrome is caused by a defect in fibrillin, an elastin-binding protein. Menkes syndrome is caused by a defect in copper metabolism, affecting lysyl oxidase activity, preventing cross-linking of collagen into stable tropocollagen fibrils. Scurvy is a vitamin C deficiency, affecting the activity of prolyl hydroxylase and decreasing the presence of hydroxyproline in the third position of the –Gly–X–Y– sequence of collagen. A vitamin C deficiency can also affect the activity of lysyl hydroxylase, decreasing hydroxylysine (also found in the third position). Hydroxyproline and hydroxylysine are required for interstrand hydrogen bonding stabilization of the procollagen triple helix.

4.7. What is the differential basis of the liver and lung pathology seen in α_1-antitrypsin deficiency?

> With α_1-antitrypsin (AAT) deficiency, hepatic damage (including possible cirrhosis) is caused by retention/buildup of AAT in the rough endoplasmic reticulum in hepatocytes and lack of AAT secretion from these cells. Circulating AAT (a serine protease inhibitor) normally inhibits elastase (a serine protease), secreted by neutrophils. Lung damage (including emphysema) is due to the lack of functional circulating AAT, causing unopposed and excessive elastase activity and degradation of elastin in the alveolar extracellular matrix.

4.8. How and why is proline hydroxylated in collagen?

> Proline is hydroxylated by prolyl hydroxylase, an enzyme of the endoplasmic reticulum that requires oxygen, ferrous iron, and vitamin C. Hydroxylation increases interchain hydrogen bond formation, strengthening the triple helix of procollagen. Vitamin C deficiency impairs hydroxylation, causing scurvy. Symptoms include generalized weakness, bleeding gums, subcutaneous hemorrhage and bruising, anemia, and the presence of corkscrew hairs.

Enzymes

5

I. OVERVIEW

Essentially, all reactions in the body are mediated by enzymes, which are protein catalysts, usually within cells, that increase the rate of reactions without being changed in the overall process. Among the many biologic reactions that are energetically possible, enzymes selectively channel substrates into useful pathways and direct all metabolic events. This chapter examines the nature of these catalytic molecules and their mechanisms of action.

II. NOMENCLATURE

Each enzyme is assigned two names. The first is its short, recommended name, convenient for everyday use. The second is the more complete systematic name, which is used when an enzyme must be identified without ambiguity.

A. Recommended name

Most commonly used enzyme names have the suffix "-ase" attached to the substrate of the reaction, such as glucosidase and urease. Names of other enzymes include a description of the action performed, for example, lactate dehydrogenase (LDH) and adenylyl cyclase. Some enzymes retain their original general names, which give no hint of the associated enzymatic reaction, for example, trypsin and pepsin.

B. Systematic name

In the systematic naming system, enzymes are divided into six major classes (Fig. 5.1), each with numerous subgroups. For a given enzyme, the suffix -ase is attached to a fairly complete description of the chemical reaction catalyzed, including the names of all the substrates, for example, lactate: nicotinamide adenine dinucleotide (NAD$^+$) oxidoreductase. [Note: Each enzyme is also assigned a classification number. Lactate: NAD$^+$ oxidoreductase is 1.1.1.27.] The systematic names are unambiguous and informative but are frequently too cumbersome to be of general use.

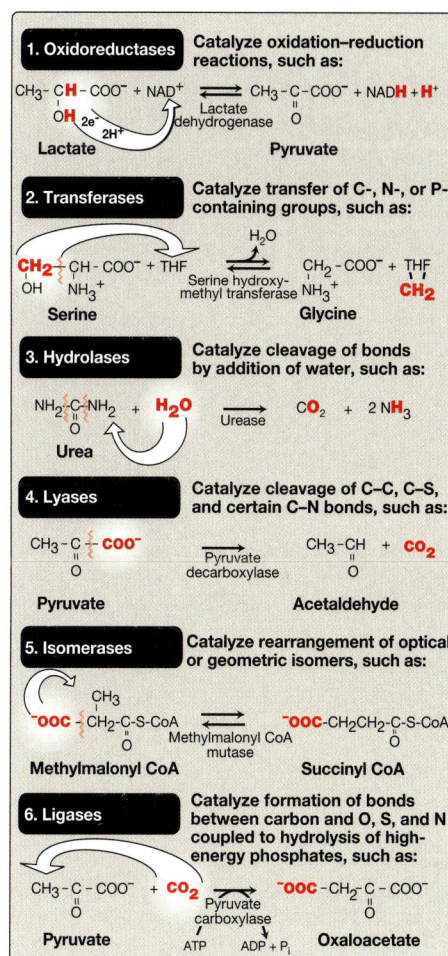

Figure 5.1
Six major classes of enzymes with examples. NAD(H), nicotinamide adenine dinucleotide; THF, tetrahydrofolate; CoA, coenzyme A; CO_2, carbon dioxide; NH_3, ammonia; ADP, adenosine diphosphate; P_i, inorganic phosphate.

Potentially confusing enzyme nomenclature includes enzymes with similar names but different functions or mechanisms. For example, synthetases require adenosine triphosphate (ATP), while synthases do not require ATP. Phosphatases use water to remove a phosphate group, while phosphorylases use inorganic phosphate to break a bond and generate a phosphorylated product. Dehydrogenases (using NAD^+ or flavin adenine dinucleotide [FAD]) accept electrons in a redox reaction. Oxidases use oxygen as the acceptor, with no oxygen atoms incorporated into the substrate, while oxygenases do incorporate oxygen atoms into their substrates.

III. PROPERTIES

An enzyme is an efficient, specific protein catalyst that combines with a substrate at the enzyme active site and performs chemistry on that substrate to convert it to a product. Without enzymes, most biochemical reactions would not occur quickly enough to have physiologic importance in the human body. Enzymes increase the velocity of a chemical reaction but are not consumed or altered during the reaction. [Note: Some ribonucleic acids [RNAs] can catalyze reactions that affect phosphodiesterase and peptide bonds. RNAs with catalytic activity are called ribozymes and are much less common than protein catalysts.]

A. Active site

Enzyme molecules contain a special pocket or cleft called the active site, formed by folding of the protein. The active site contains amino acid residues whose side chains participate in substrate binding and catalysis (Fig. 5.2). The substrate first binds the enzyme, forming an enzyme–substrate (ES) complex. Binding causes a conformational change in the enzyme (induced fit model) that allows a rapid conversion of the ES to enzyme–product (EP) complex that subsequently dissociates to free enzyme and product.

B. Efficiency

Enzyme-catalyzed reactions are highly efficient, proceeding from 10^3 to 10^8 times faster than uncatalyzed reactions. The number of substrate molecules converted to product per enzyme molecule per second is the turnover number, or k_{cat}, and typically is 10^2 to 10^4 second^{-1}. [Note: k_{cat} is the rate constant for the conversion of ES to E + P.]

C. Specificity

Enzymes are capable of interacting with one or a very few substrates and can catalyze only one type of chemical reaction. The set of enzymes synthesized within a cell determines which reactions occur in that cell.

D. Holoenzymes, apoenzymes, cofactors, and coenzymes

Some enzymes require nonprotein components to have enzymatic activity. **Holoenzyme** refers to the complete functional enzyme

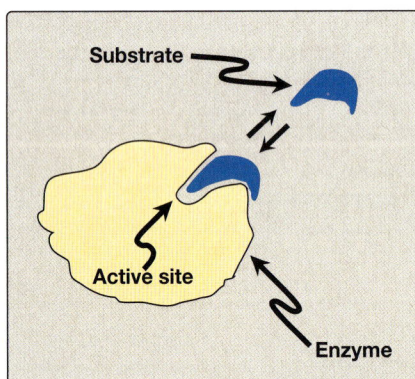

Figure 5.2
Schematic representation of an enzyme with one active site binding a substrate molecule.

consisting of the protein component of the enzyme along with its nonprotein component, whereas the enzyme without its nonprotein moiety is termed an **apoenzyme** and is inactive. For enzymes that require nonprotein components, those components must be present for the enzyme to function in catalysis.

If the nonprotein moiety is a metal ion, such as zinc (Zn^{2+}) or iron (Fe^{2+}), it is called a **cofactor**. If the nonprotein component is instead a small organic molecule, it is termed a **coenzyme or cosubstrate**. Coenzymes are commonly derived from vitamins and only transiently associated with the enzyme. For example, NADH produced in glycolysis is a derivative of niacin (vitamin B3), and $FADH_2$ produced in the Krebs cycle contains riboflavin (vitamin B2). Coenzymes dissociate from the enzyme in an altered state (e.g., NAD^+), hence considered cosubstrates. If the coenzyme is permanently associated with the enzyme and returned to its original form, it is called a **prosthetic group** (e.g., FAD).

E. Regulation

Enzyme activity can often be increased or decreased, so that the rate of product formation responds to the present cellular needs.

F. Location within the cell

Most enzymes function inside cells, within the confines of plasma membranes. Many enzymes are localized in specific organelles within the cell (Fig. 5.3). Such compartmentalization serves to isolate the reaction substrate or product from other competing reactions. This provides a favorable environment for the reaction and organizes the thousands of enzymes present in the cell into purposeful pathways.

IV. MECHANISM OF ENZYME ACTION

Mechanisms of enzyme action can be viewed in terms of energy changes and from how the active site chemically modifies the substrate. In terms of energy changes that occur during the reaction, enzymes provide alternate, energetically favorable reaction pathways different from those followed in the uncatalyzed reaction. The second perspective describes how the active site chemically facilitates catalysis.

A. Energy changes occurring during the reaction

Chemical reactions have an energy barrier separating the reactants and the products, called the activation energy (E_a). This is the energy difference between the reactants and a high-energy intermediate, the transition state (T^*), formed during conversion of reactant to product. Figure 5.4 shows the energy changes during conversion of a molecule of reactant A to product B as it proceeds through the transition state.

$$A \rightleftarrows T^* \rightleftarrows B$$

1. **Activation energy:** The peak of energy in Figure 5.4 is the difference in free energy between the reactant and T^*, in which the high-energy, short-lived intermediate is formed during the conversion of the reactant to the product. Because of the high E_a, the rates of uncatalyzed chemical reactions are often slow.

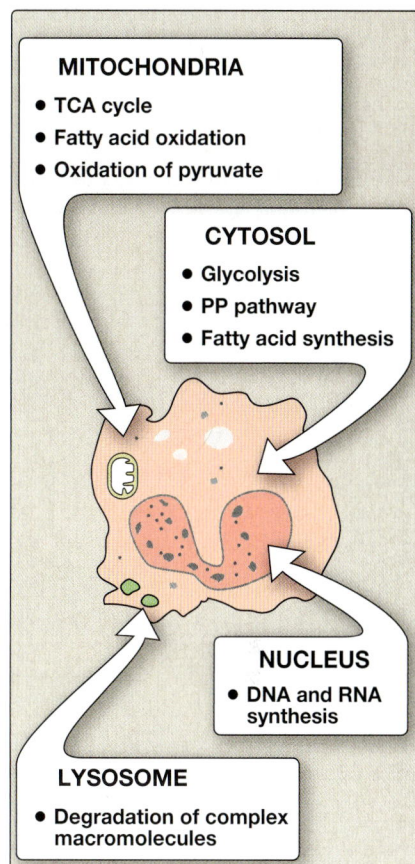

MITOCHONDRIA
- TCA cycle
- Fatty acid oxidation
- Oxidation of pyruvate

CYTOSOL
- Glycolysis
- PP pathway
- Fatty acid synthesis

NUCLEUS
- DNA and RNA synthesis

LYSOSOME
- Degradation of complex macromolecules

Figure 5.3
Intracellular location of some important biochemical pathways. TCA, tricarboxylic acid; PP, pentose phosphate.

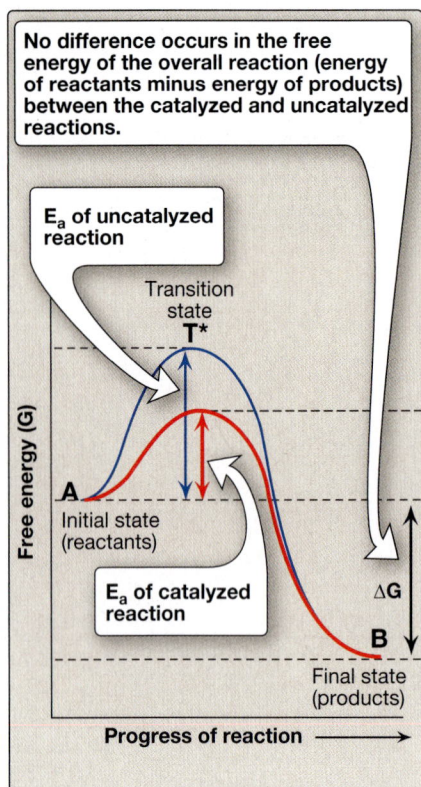

Figure 5.4
Effect of an enzyme on the activation energy (E_a) of a reaction. ΔG, change in free energy.

2. **Rate of reaction:** For molecules to react, they must contain sufficient energy to overcome the energy barrier of the transition state. In the absence of an enzyme, only a small proportion of a population of molecules may possess enough energy to achieve the transition state between reactant and product. The rate of reaction is determined by the number of such energized molecules. In general, the lower the E_a, the more molecules have sufficient energy to pass through the transition state and, therefore, the faster the rate of the reaction.

3. **Alternate reaction pathway:** An enzyme allows a reaction to proceed rapidly under conditions prevailing in the cell by providing this alternate reaction pathway with a lower E_a (see Fig. 5.4). The enzyme does not change the free energies of the reactants (substrates) or products and, therefore, does not change the equilibrium of the reaction. It does, however, accelerate the rate by which equilibrium is reached.

B. Active site chemistry

The active site is not a passive receptacle for binding the substrate but, rather, is a complex molecular machine that employs diverse chemical mechanisms to facilitate the conversion of substrate to product. A number of factors are responsible for the catalytic efficiency of enzymes, including the following examples.

1. **Transition-state stabilization:** The active site often acts as a flexible molecular template that binds the substrate and initiates its conversion to the transition state, a structure in which the bonds are not like those in the substrate or the product (see T* at the top of the curve in Fig. 5.4). By stabilizing the transition state, the enzyme greatly increases the concentration of the reactive intermediate that can be converted to product and, thus, accelerates the reaction. [Note: The transition state cannot be isolated.]

2. **Catalysis:** The active site can provide catalytic groups that enhance the probability that the transition state is formed. In some enzymes, these groups can participate in general acid–base catalysis in which amino acid residues provide or accept protons. In other enzymes, catalysis may involve the transient formation of a covalent ES complex.

> The mechanism of action of chymotrypsin, an enzyme of protein digestion in the intestine, includes general base, general acid, and covalent catalysis. A histidine at the active site of the enzyme gains (general base) and loses (general acid) protons, mediated by the pK of histidine in proteins being close to physiologic pH. Serine at the active site forms a transient covalent bond with the substrate.

3. **Transition-state visualization:** The enzyme-catalyzed conversion of substrate to product can be depicted as being similar to removing a sweater (chemical group) from an uncooperative

infant (substrate) (Fig. 5.5). The process has a high E_a because the only reasonable strategy for removing the garment requires that both arms being fully extended over the head, an unlikely posture to be adopted without a catalyst. We can envision a parent acting as an enzyme, first coming in contact with the infant (forming ES) and then guiding the infant's arms into an extended, vertical position, analogous to the transition state. This posture (conformation) facilitates the removal of the sweater, forming the disrobed infant, which represents the product. [Note: The substrate bound to the enzyme (ES) is at a slightly lower energy than unbound substrate (S) and explains the small dip in the curve at ES. Imagine it being likened to a brief period of time when the child is gently restrained by the parent and temporarily stops wiggling.] Once the transition state has been achieved, removal of the sweater is easy to accomplish, and substrate is converted to product, the infant without the sweater.

V. FACTORS AFFECTING REACTION VELOCITY

Enzymes can be isolated from cells and their properties studied in a test tube, that is, *in vitro*. Different enzymes show different responses to changes in substrate concentration, temperature, and pH. This section describes factors that influence the reaction velocity of enzymes. Enzymatic responses to these factors provide valuable clues as to how enzymes function in living cells, that is, *in vivo*.

A. Substrate concentration

Because enzymes modify their substrates to form reaction products, the concentrations of available substrates will influence the rate of product formation.

1. **Maximal velocity:** The rate or velocity of a reaction (v) is the number of substrate molecules converted to product per unit time. Velocity is ordinarily expressed as μmol of product formed per second. The rate of an enzyme-catalyzed reaction increases with substrate concentration until a maximal velocity (V_{max}) is reached (Fig. 5.6). Leveling off of the reaction rate at high substrate concentrations reflects that all available active sites on the enzyme are saturated with substrate. Addition of more substrate molecules will not increase the reaction rate since they have no place to bind.

2. **Shape of the enzyme kinetics curve:** Most enzymes follow Michaelis–Menten kinetics (see Section VI), in which a plot of initial reaction velocity (v_o) against substrate concentration is hyperbolic (similar in shape to that of the oxygen-dissociation curve of myoglobin; see Chapter 3). In contrast, allosteric enzymes do not follow Michaelis–Menten kinetics and instead show a sigmoidal curve (see Fig. 5.6) that is similar in shape to the oxygen-dissociation curve of hemoglobin.

B. Temperature

The temperature at which an enzyme is combined with its substrate will influence the rate of conversion of substrate to product.

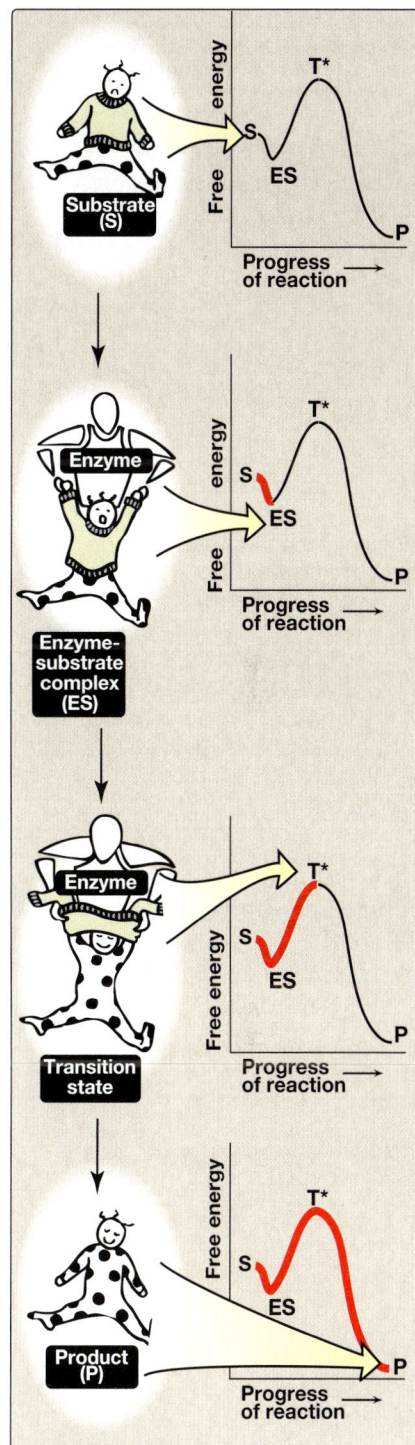

Figure 5.5
Schematic representation of energy changes accompanying formation of an enzyme–substrate complex and subsequent formation of a transition state.

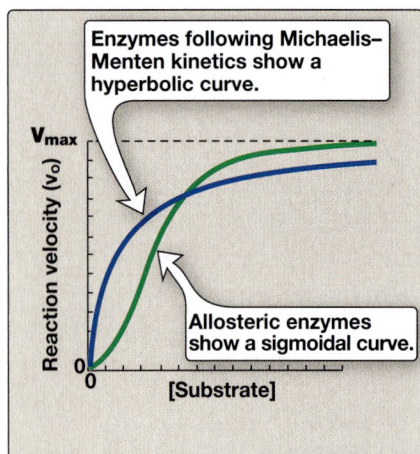

Figure 5.6
Effect of substrate concentration on reaction velocity.

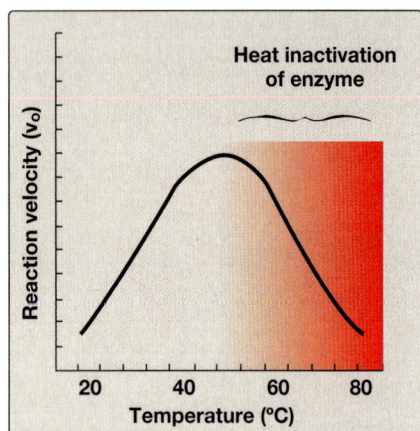

Figure 5.7
Effect of temperature on an enzyme-catalyzed reaction.

1. **Velocity increases with temperature:** The reaction velocity of an enzyme-catalyzed reaction increases with temperature until a peak velocity is reached (Fig. 5.7). This increase is the result of an increased number of substrate molecules with sufficient energy to pass over the energy barrier of the transition state when the reaction temperature is increased.

2. **Velocity decreases with higher temperature:** However if the temperature becomes too high, the structure of the enzyme can be affected, rendering the enzyme incapable of catalyzing a reaction. Therefore, too high a temperature will cause a decrease in reaction velocity as a result of temperature-induced denaturation of the enzyme (see Fig. 5.7).

Normal human body temperature has been reported to be 37 °C (98.6 °F) since 1851. The optimum temperature for most human enzymes is close to this normal body temperature. Basal metabolic rate and levels of inflammation appear to impact normal body temperature, and, as both have decreased in the past several hundred years, so has average normal human temperature, which is now closer to 36.7 °C (98.0 °F). Human enzymes function best at 35 °C–37 °C (95 °F–98.6 °F) and start to denature at temperatures >40 °C (104 °F), meaning very high fevers impact enzyme structure and function. Other species have higher normal body temperatures, and their enzymes function at temperatures that could cause human enzymes to denature. For example, normal body temperature of dogs and cats is 38.3 °C–39.20 °C (101.0 °F–102.5 °F). Adult chicken body temperature ranges from 40.6 °C–41.7 °C (105 °F–107 °F). Thermophilic bacteria found in hot springs have optimum temperatures of 70 °C (158 °F), and their enzymes are adapted to function well at that extreme temperature.

C. pH

The pH of the solution in which an enzyme interacts with its substrate can influence the catalytic activity of the enzyme, with extremes in pH damaging the protein structure of the enzyme. Each enzyme has an optimum pH at which it has maximal activity.

1. **pH effect on active site ionization:** The concentration of protons ($[H^+]$) affects reaction velocity in several ways. First, the catalytic process usually requires that the enzyme and substrate have specific chemical groups in either an ionized or unionized state in order to interact. For example, catalytic activity may require that an amino group of the enzyme be in the protonated form ($-NH_3^+$). Because this group is deprotonated at alkaline pH, the rate of the reaction declines.

2. **pH effect on enzyme denaturation:** Extremes of pH can also lead to denaturation of the enzyme, because the structure of the catalytically active protein molecule depends on the ionic character of the amino acid side chains.

3. **Variable pH optimum:** The pH at which maximal enzyme activity is achieved is different for different enzymes and often reflects the [H^+] at which the enzyme functions in the body. For example, pepsin, a digestive enzyme in the stomach, is maximally active at pH 2, whereas other enzymes, designed to work at neutral pH, are denatured by such an acidic environment (Fig. 5.8).

VI. MICHAELIS–MENTEN KINETICS

In a paper published in 1913, Leonor Michaelis and Maud Menten mathematically expressed the relationship between substrate concentration and the rate at which an enzyme converts its substrate to product. This model accounts for most features of many enzyme-catalyzed reactions. In it, the enzyme reversibly combines with its substrate to form an ES complex that subsequently yields product, regenerating the free enzyme. The reaction model, involving one substrate molecule, is represented below:

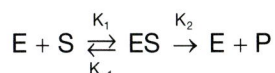

$$E + S \underset{K_{-1}}{\overset{K_1}{\rightleftharpoons}} ES \xrightarrow{K_2} E + P$$

Where   S is the substrate;
E is the enzyme;
ES is the enzyme–substrate complex;
P is the product;
k_1, k_{-1}, and k_2 (or, k_{cat}) are rate constants.

Figure 5.8
Effect of pH on enzyme-catalyzed reactions.

A. Michaelis–Menten equation

The Michaelis–Menten equation describes how reaction velocity varies with substrate concentration:

$$V_O = \frac{V_{max}[S]}{K_m + [S]}$$

Where v_o = initial reaction velocity;
V_{max} = maximal velocity = $k_{cat}[E]_{Total}$;
K_m = Michaelis constant = $(k_{-1} + k_2)/k_1$;
[S] = substrate concentration.

The following assumptions are made in deriving the Michaelis–Menten rate equation.

1. **Enzyme and substrate relative concentrations:** The substrate concentration ([S]) is much greater than the concentration of enzyme so that the percentage of total substrate bound by the enzyme at any one time is small.

2. **Steady-state assumption:** The concentration of the ES complex does not change with time (steady-state assumption), that is, the rate of formation of ES is equal to that of the breakdown of ES (to E + S and to E + P). In general, an intermediate in a series of reactions is said to be in a steady state when its rate of synthesis is equal to its rate of degradation.

3. **Initial velocity:** Initial reaction velocities (v_o) are used in the analysis of enzyme reactions. This means that the rate of the reaction

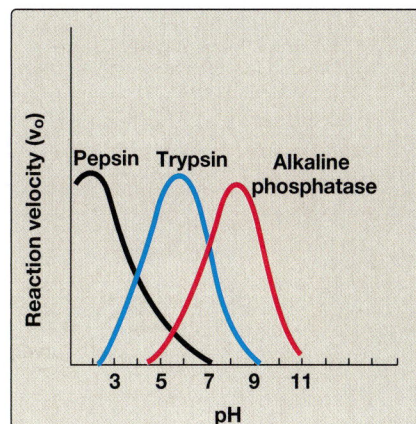

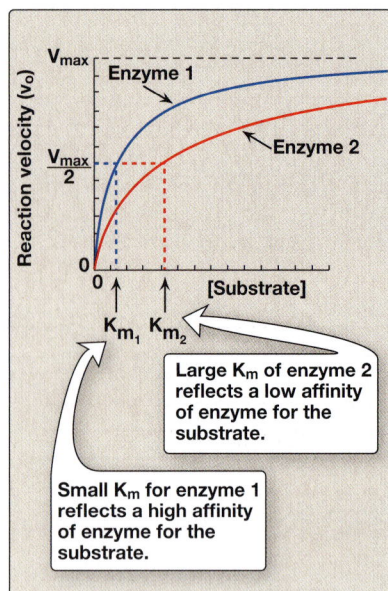

Figure 5.9
Effect of substrate concentration on reaction velocities for two enzymes: enzyme 1 with a small Michaelis constant (K_m) and enzyme 2 with a large K_m. V_{max}, maximal velocity.

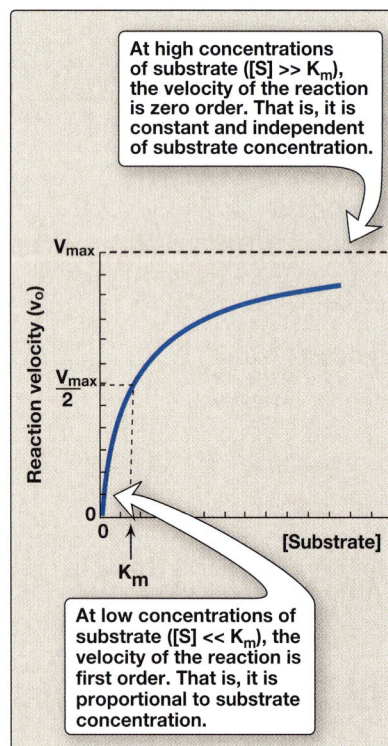

Figure 5.10
Effect of substrate concentration on reaction velocity for an enzyme-catalyzed reaction. V_{max}, maximal velocity; K_m, Michaelis constant.

is measured as soon as the enzyme and substrate are mixed. At that time, the concentration of product is very small, and, therefore, the rate of the reverse reaction from product to substrate can be ignored.

B. Important conclusions

We can draw the following conclusions from these principles.

1. **K_m characteristics:** K_m, the Michaelis constant, is characteristic of an enzyme and its particular substrate and reflects the **affinity** of the enzyme for that substrate. K_m is numerically equal to the substrate concentration at which the reaction velocity is equal to one-half V_{max}. K_m does not vary with enzyme concentration.

    a. **Small K_m:** A numerically small (low) K_m reflects a high affinity of the enzyme for substrate, because a low concentration of substrate is needed to half-saturate the enzyme—that is, to reach a velocity that is one-half V_{max} (Fig. 5.9).

    b. **Large K_m:** A numerically large (high) K_m reflects a low affinity of the enzyme for substrate because a high concentration of substrate is needed to half saturate the enzyme.

2. **Velocity relationship to enzyme concentration:** The rate of the reaction is directly proportional to the enzyme concentration because [S] is not limiting. For example, if the enzyme concentration is halved, the initial rates of the reaction (v_o) and that of V_{max} are reduced to half that of the original.

3. **Reaction order:** When [S] is much less ($<<$) than K_m, the velocity of the reaction is approximately proportional to the substrate concentration (Fig. 5.10). The rate of reaction is then said to be first order with respect to substrate. When [S] is much greater ($>>$) than K_m, the velocity is constant and equal to V_{max}. The rate of reaction is then independent of substrate concentration because the enzyme is saturated with substrate and is said to be zero order with respect to substrate concentration (see Fig. 5.10).

C. Lineweaver–Burk plot

When v_o is plotted against [S], it is not always possible to determine when V_{max} has been achieved because of the gradual upward slope of the hyperbolic curve at high substrate concentrations. However, as Hans Lineweaver and Dean Burk first described in 1934, if $1/v_o$ is plotted versus $1/[S]$, then a straight line is obtained (Fig. 5.11). This Lineweaver–Burk plot, also called a double-reciprocal plot, can be used to calculate K_m and V_{max} as well as to determine the mechanism of action of enzyme inhibitors.

The equation describing the Lineweaver–Burk plot is:

$$\frac{1}{V_o} = \frac{K_m}{V_{max}[S]} + \frac{1}{V_{max}}$$

where the intercept on the x-axis is equal to $-1/K_m$, and the intercept on the y-axis is equal to $1/V_{max}$. [Note: The slope = K_m/V_{max}.]

VII. ENZYME INHIBITION

Any substance that can decrease the velocity of an enzyme-catalyzed reaction is considered to be an **inhibitor**. Inhibitors can be reversible or irreversible. Irreversible inhibitors bind to enzymes through covalent bonds. Lead, for example, can act as an irreversible inhibitor of some enzymes. It forms covalent bonds with the sulfhydryl side chain of cysteine in proteins. Ferrochelatase, an enzyme involved in heme synthesis, is irreversibly inhibited by lead. Reversible inhibitors bind to enzymes through noncovalent bonds forming an enzyme–inhibitor complex. Dilution of the enzyme–inhibitor complex results in dissociation of the reversibly bound inhibitor and recovery of enzyme activity. While four types of enzyme inhibition are often described, the two most commonly encountered types of reversible inhibition are competitive and noncompetitive. In uncompetitive inhibition, the inhibitor binds to the ES complex only, and, in mixed inhibition, inhibitor can bind E or ES; these are the other two types of inhibitors described.

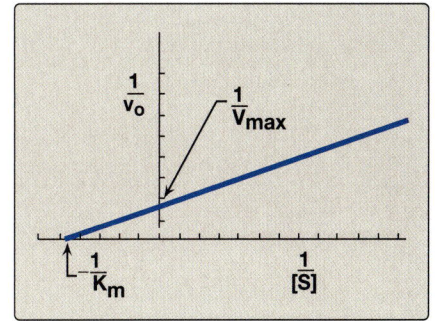

Figure 5.11
Lineweaver–Burk plot. v_o, initial reaction velocity; V_{max}, maximal velocity; K_m, Michaelis constant; [S], substrate concentration.

A. Competitive inhibition

This type of inhibition occurs when the inhibitor binds reversibly to the same site that the substrate would normally occupy and, therefore, competes with the substrate for binding to the enzyme active site.

1. **Effect on V_{max}:** The effect of a competitive inhibitor is reversed by increasing the concentration of substrate. At a sufficiently high [S], the reaction velocity reaches the V_{max} observed in the absence of inhibitor, that is, V_{max} is unchanged in the presence of a competitive inhibitor (Fig. 5.12).

2. **Effect on K_m:** A competitive inhibitor increases the apparent K_m for a given substrate. This means that, in the presence of a competitive inhibitor, more substrate is needed to achieve half V_{max}.

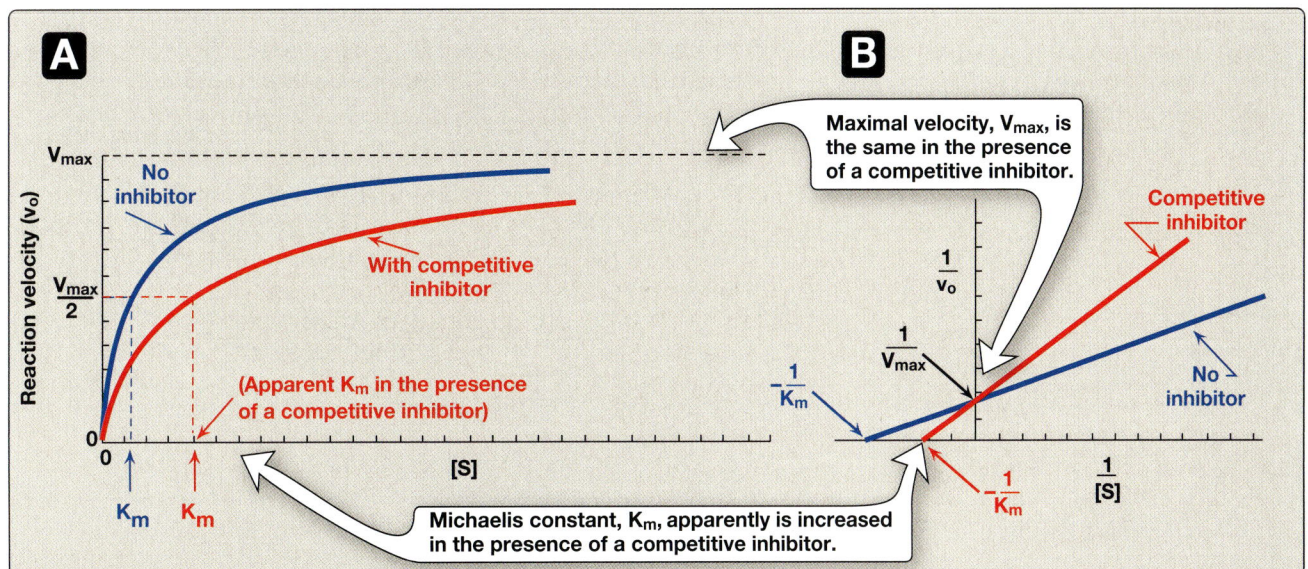

Figure 5.12
A. Effect of a competitive inhibitor on the reaction velocity versus substrate concentration ([S]) plot. **B.** Lineweaver–Burk plot of competitive inhibition of an enzyme. [Note: The slope increases if inhibitor concentration increases.]

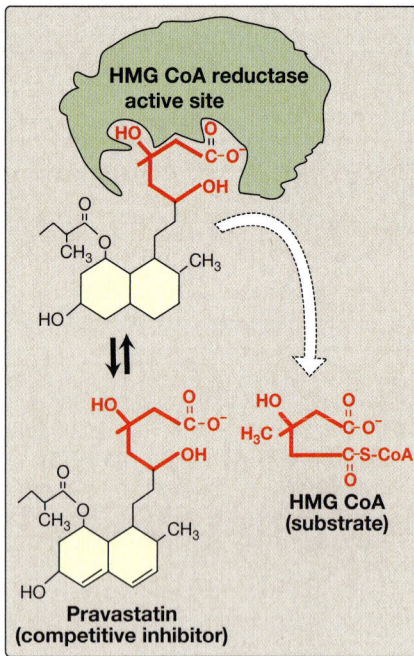

Figure 5.13
Pravastatin competes with hydroxymethylglutaryl coenzyme A (HMG CoA) for the active site of HMG CoA reductase.

3. **Effect on the Lineweaver–Burk plot:** Competitive inhibition shows a characteristic Lineweaver–Burk plot in which the plots of the inhibited and uninhibited reactions intersect on the y-axis at $1/V_{max}$ (V_{max} is unchanged). The inhibited and uninhibited reactions show different x-axis intercepts, indicating that the apparent K_m is increased in the presence of the competitive inhibitor because $-1/K_m$ moves closer to zero from a negative value (see Fig. 5.12). [Note: An important group of competitive inhibitors are the transition state analogs, stable molecules that approximate the structure of the transition state, and, therefore, bind the enzyme more tightly than does the substrate.]

4. **Statin drugs as examples of competitive inhibitors:** Statin drugs are cholesterol-lowering agents that competitively inhibit the rate-limiting (slowest) step in cholesterol biosynthesis. This reaction is catalyzed by hydroxymethylglutaryl coenzyme A reductase ([HMG CoA reductase] see Chapter 18). Statins, such as atorvastatin and pravastatin, are structural analogs of the natural substrate for this enzyme and compete effectively to inhibit HMG CoA reductase. By doing so, they inhibit *de novo* cholesterol synthesis (Fig. 5.13).

B. Noncompetitive inhibition

Noncompetitive inhibitors can bind to enzyme with equal affinity whether or not substrate has bound to the active site. This type of inhibition is recognized by its characteristic effect, causing a decrease in V_{max} (Fig. 5.14). Noncompetitive inhibition occurs when the inhibitor and substrate bind at different sites on the enzyme. The noncompetitive inhibitor prevents conversion of

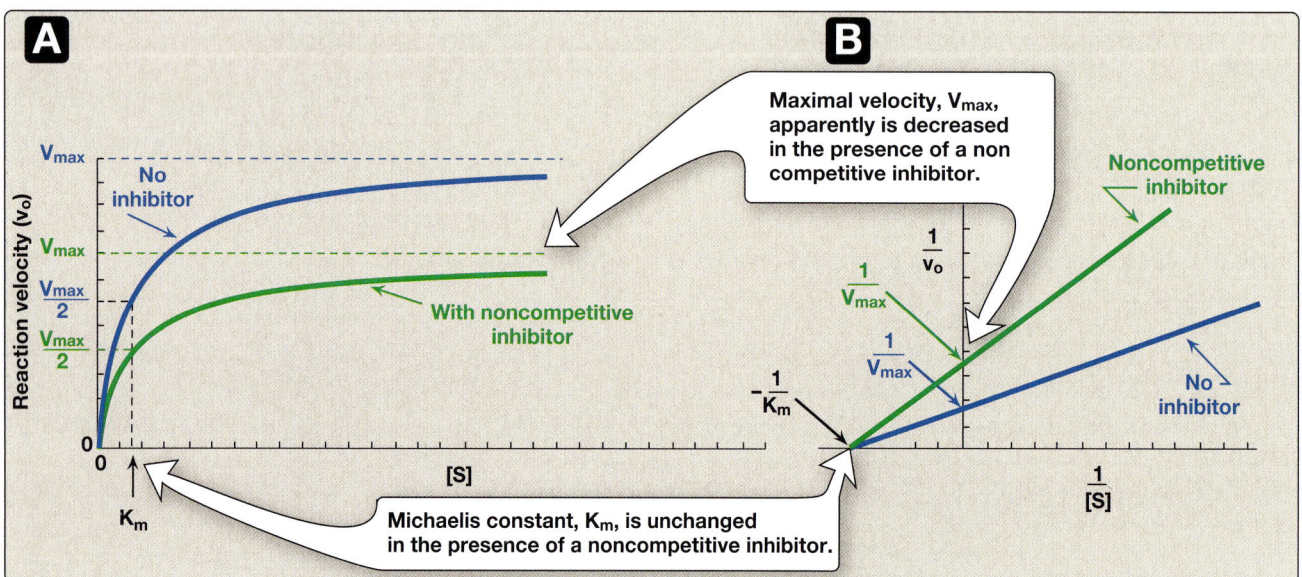

Figure 5.14
A. Effect of a noncompetitive inhibitor on the reaction velocity versus substrate concentration ([S]) plot. **B.** Lineweaver–Burk plot of noncompetitive inhibition of an enzyme. [Note: The slope increases if inhibitor concentration increases.]

substrate to product whether it binds to free enzyme or the ES complex (Fig. 5.15).

1. **Effect on V_{max}:** Effects of a noncompetitive inhibitor cannot be overcome by increasing the concentration of substrate. Therefore, noncompetitive inhibitors decrease the apparent V_{max} of the reaction.

2. **Effect on K_m:** Noncompetitive inhibitors do not interfere with the binding of substrate to enzyme. Therefore, the enzyme shows the same K_m in the presence or absence of the noncompetitive inhibitor, that is, K_m is unchanged in the presence of a noncompetitive inhibitor.

3. **Effect on Lineweaver–Burk plot:** Noncompetitive inhibition is readily differentiated from competitive inhibition by plotting $1/v_o$ versus $1/[S]$ and noting that the apparent V_{max} decreases in the presence of a noncompetitive inhibitor, whereas K_m is unchanged (see Fig. 5.14).

C. Enzyme inhibitors as drugs

Enzymes are important targets of drug therapies because an immediate effect can be observed when a critical enzyme is inhibited. It is estimated that close to half (47%) of drugs currently used function as enzyme inhibitors. The widely prescribed β-lactam antibiotics, such as penicillin and amoxicillin, act by inhibiting enzymes involved in bacterial cell wall synthesis. Drugs may also act by inhibiting extracellular reactions. This is illustrated by angiotensin-converting enzyme (ACE) inhibitors. They lower blood pressure by blocking plasma ACE that cleaves angiotensin I to form the potent vasoconstrictor, angiotensin II. These drugs, which include captopril, enalapril, and lisinopril, cause vasodilation and, therefore, a reduction in blood pressure. Aspirin, a nonprescription drug, irreversibly inhibits prostaglandin and thromboxane synthesis by inhibiting the enzyme cyclooxygenase.

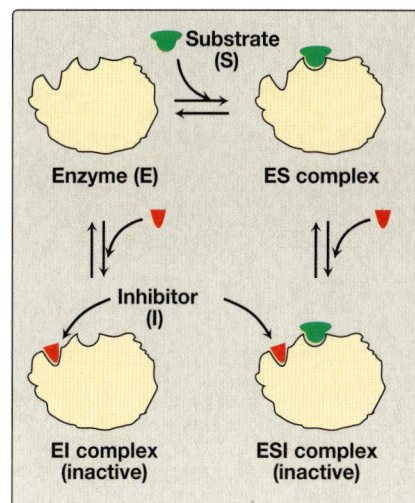

Figure 5.15
Noncompetitive inhibitor binding to both free enzyme and enzyme–substrate (ES) complex. EI, enzyme–inhibitor complex; ESI, enzyme–substrate–inhibitor complex.

VIII. ENZYME REGULATION

Regulation of the reaction velocities of enzymes is essential for an organism to coordinate its numerous metabolic processes. The rates of most enzymes are responsive to changes in substrate concentration, because the intracellular level of many substrates is in the range of the K_m. Thus, an increase in substrate concentration prompts an increase in reaction rate, which tends to return the concentration of substrate towards the normal. In addition, some enzymes with specialized regulatory functions respond to allosteric effectors and/or covalent modification, or they show altered rates of enzyme synthesis (or degradation) when physiologic conditions are changed.

A. Allosteric enzymes

Allosteric enzymes do not follow Michaelis–Menten kinetics but are regulated by molecules called **effectors** that bind to them noncovalently at a site other than the active site. In a plot of their v_o versus

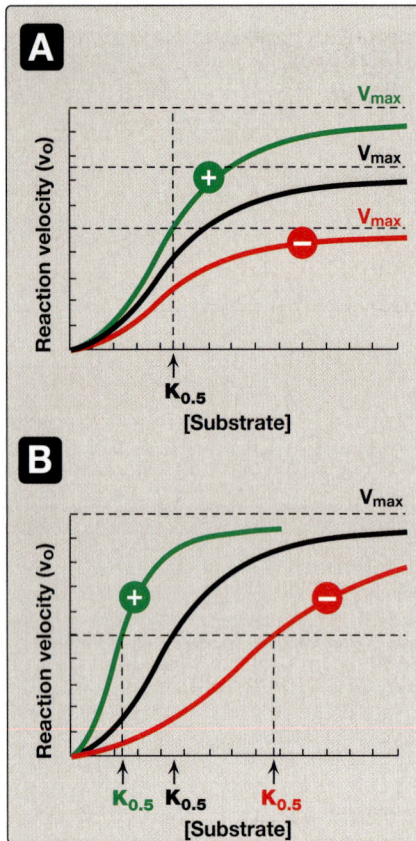

Figure 5.16
Effects of negative or positive effectors on an allosteric enzyme. **A.** Maximal velocity (V_{max}) is altered. **B.** The substrate concentration that gives half maximal velocity ($K_{0.5}$) is altered.

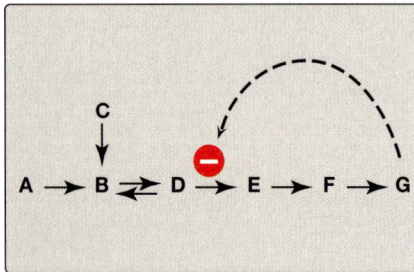

Figure 5.17
Feedback inhibition of a metabolic pathway.

substrate concentration, a sigmoidal-shaped curve results. These enzymes are almost always composed of multiple subunits, and the regulatory (allosteric) site that binds the effector is different from the substrate-binding site and may be located on a subunit that is not itself catalytic.

Effectors that inhibit enzyme activity are negative effectors, whereas those that increase enzyme activity are positive effectors. Positive and negative effectors can affect the affinity of the enzyme for its substrate ($K_{0.5}$), modify the maximal catalytic activity of the enzyme (V_{max}), or both (Fig. 5.16). [Note: Allosteric enzymes frequently catalyze the committed step, often the rate-limiting step, early in a pathway.]

1. **Homotropic effectors:** When the substrate itself serves as an effector by binding to a site other than the active site, the effect is said to be homotropic, or same as the substrate. In such a case, the presence of a substrate molecule at one site on the enzyme enhances the catalytic properties of the other substrate-binding sites. That is, their binding sites cooperate with each other for substrate binding and are said to exhibit **cooperativity**. Allosteric enzymes show a sigmoidal curve when v_o is plotted against substrate concentration, as shown in Figure 5.16. This contrasts with the hyperbolic curve characteristic of enzymes following Michaelis–Menten kinetics, as previously discussed. [Note: The concept of cooperativity of substrate binding is analogous to the binding of oxygen to hemoglobin; see Chapter 3.]

2. **Heterotropic effectors:** When the effector is a different molecule than the substrate, it is considered a heterotropic effector. For example, consider the feedback inhibition shown in Figure 5.17. The enzyme that converts D to E has an allosteric site that binds the end product, G. If the concentration of G increases (e.g., because it is not used as rapidly as it is synthesized), the first irreversible step unique to the pathway is typically inhibited. Feedback inhibition provides the cell with the appropriate amounts of a product it needs by regulating the flow of substrate molecules through the pathway that synthesizes that product. Heterotropic effectors are commonly encountered. For example, the glycolytic enzyme phosphofructokinase-1 is allosterically inhibited by citrate, which is not a substrate for the enzyme.

B. Covalent modification

Many enzymes are regulated by covalent modification, most often by the addition or removal of phosphate groups from specific serine, threonine, or tyrosine residues of the enzyme. Protein phosphorylation is recognized as one of the primary ways in which cellular processes are regulated.

1. **Phosphorylation and dephosphorylation:** Phosphorylation reactions are catalyzed by enzymes called protein **kinases** that catalyze the addition of a phosphate group to its protein or enzyme substrate, using ATP as the phosphate donor. **Phosphorylases**

Table 5.1: Mechanisms for Regulating Enzyme Activity

Regulator Event	Typical Effector	Results	Time Required for Change
Substrate availability	Substrate	Change in velocity (V_o)	Immediate
Product inhibition	Reaction product	Change in V_{max} and/or K_m	Immediate
Allosteric control	Pathway end product	Change in V_{max} and/or $K_{0.5}$	Immediate
Covalent modification	Another enzyme	Change in V_{max} and/or K_m	Immediate to minutes
Synthesis or degradation of enzyme	Hormone or metabolite	Change in the amount of enzyme	Hours to days

[Note: Inhibition by pathway end product is also referred to as feedback inhibition.]

also enzymatically add phosphate to substrates but do so without the need for ATP. **Phosphatases** cleave phosphate groups from phosphorylated substrates including phosphorylated enzymes (Fig. 5.18).

2. **Enzyme response to phosphorylation:** Depending on the specific enzyme, its phosphorylated form may be more or less active than the unphosphorylated enzyme. For example, hormone-mediated phosphorylation of glycogen phosphorylase, an enzyme that degrades glycogen, increases its activity, whereas phosphorylation of glycogen synthase, an enzyme that synthesizes glycogen, decreases its activity (see Chapter 11).

C. Enzyme synthesis

The regulatory mechanisms described can modify the activity of existing enzyme molecules. Cells can also regulate the amount of enzyme present by altering the rate of enzyme degradation or, more typically, the rate of enzyme synthesis. The increase (induction) or decrease (repression) of enzyme synthesis leads to an alteration in the total population of active sites. Enzymes subject to the regulation of synthesis are often those needed under selected physiologic conditions. For example, elevated levels of insulin as a result of high blood glucose levels cause an increase in the synthesis of key enzymes involved in glucose metabolism (see Chapter 23).

In contrast, enzymes that are in constant use are usually not regulated by altering the rate of enzyme synthesis. Alterations in enzyme levels as a result of induction or repression of protein synthesis are slow (hours to days), compared with allosterically or covalently regulated changes in enzyme activity, which occur in seconds to minutes. Table 5.1 summarizes the common ways that enzyme activity is regulated.

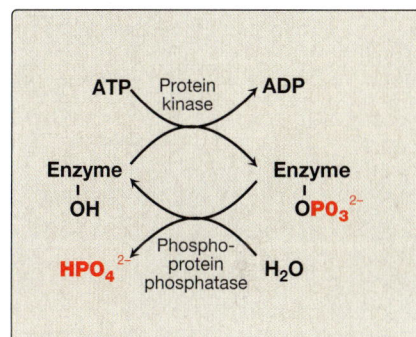

Figure 5.18
Covalent modification by the addition and removal of phosphate groups. [Note: HPO_4^{2-} may be represented as P_i and PO_3^{2-} as P.] ADP, adenosine diphosphate.

IX. ENZYMES IN HUMAN BLOOD

While most enzymes function intracellularly, enzymes can be found outside cells in fluids including blood plasma, the fluid portion of blood. Enzymes that appear in blood plasma of healthy persons can be classified into two major groups. First, a relatively small group of enzymes are

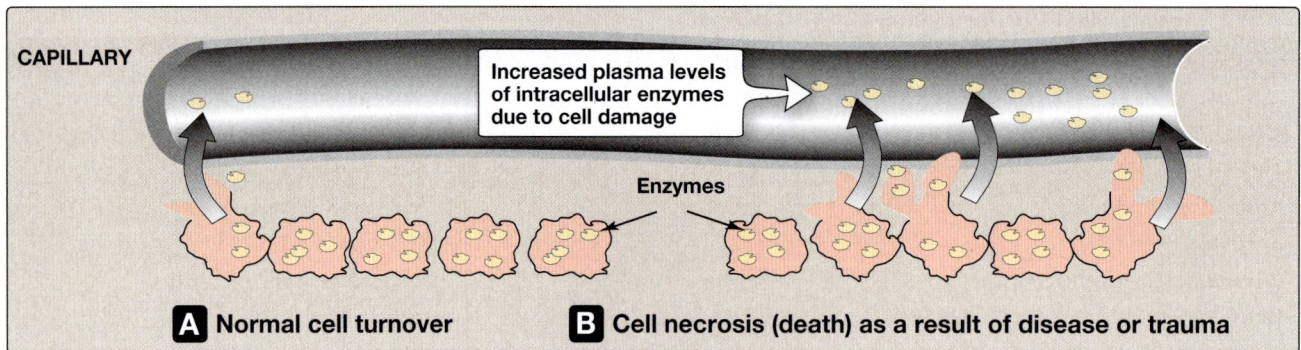

Figure 5.19
Release of enzymes from normal (**A**) and diseased or traumatized (**B**) cells.

actively secreted into the blood by certain cell types. For example, the liver secretes zymogens (inactive precursors) of the protease enzymes involved in blood coagulation (see Chapter 35), and the complement cascade involves a series of liver-produced zymogen proteases activated in turn to form functional enzymes that are an integral component of our innate immune responses.[1] Zymogen forms of proteases can be activated by other enzymes and then have enzymatic functions in the blood. Second, enzymes are released from cells during normal cell turnover. These enzymes almost always function only intracellularly and have no ability to catalyze reactions in blood plasma. In healthy individuals, the levels of these enzymes are fairly constant and represent a steady state in which the rate of release from damaged cells into the plasma is balanced by an equal rate of removal from the plasma. Increased blood plasma levels of these enzymes may indicate tissue damage and cell death that is greater than cell death from normal turnover (Fig. 5.19).

> Blood plasma is the fluid, noncellular fraction of blood. Laboratory assays of enzyme activity most often use serum, which is the fluid obtained by centrifugation of whole blood after it has been allowed to coagulate. Plasma is a physiologic fluid, whereas serum is a fluid prepared in the laboratory from a patient's whole blood sample.

A. Blood plasma enzyme levels in disease states

Many diseases cause tissue damage that includes the rupture of plasma membranes and lysis of cells in the tissue. As a result, the damaged cells release their contents into fluids, including the blood plasma, causing an increased concentration of enzymes in the plasma. These enzymes are normally intracellular and cannot catalyze reactions when present outside their normal cellular location. However, they are routinely measured

[1]For further discussion of apoptosis, see *LIR Cell and Molecular Biology*, 2nd Edition, Chapter 23.

Table 5.2: Some Clinically Useful Enzymes

Enzyme	Abbreviation	Main Tissue Source(s)	Useful to Assess
Alanine aminotransferase	ALT	Liver	Liver damage or disease
Alkaline phosphatase	ALP	Liver, bone	Liver, hepatobiliary diseases, and bone diseases
Amylase	Amylase	Pancreas	Pancreatic diseases
Aspartate aminotransferase	AST	Liver, muscle	Liver and muscle diseases
Creatine kinase	CK	Muscle	Muscle damage or disease
Gamma glutamyl transferase	GGT	Liver, bile duct	Hepatobiliary disease (obstructive jaundice)
Lipase	Lipase	Pancreas	Pancreatic diseases
Lactate dehydrogenase	LDH	Red blood cells, liver, muscle—most cells	General marker of cell death; particularly in hemolysis, hepatic or muscle diseases
5′ Nucleotidase	5′NT	Liver	Hepatobiliary disease (obstructive jaundice)

[Note: Appearance of these enzymes in blood can indicate damage to cells in the tissue where the enzyme normally functions.]

in patients' blood samples for diagnostic purposes. The level of specific enzyme activity in the plasma frequently correlates with the extent of tissue damage. Therefore, determining the extent of elevation of a particular enzyme activity in the blood plasma is often useful in evaluating the extent of tissue damage, response to therapies, and patient prognosis.

B. Plasma enzymes as diagnostic tools

Some enzymes show relatively high activity in only one or a few tissues (Table 5.2). Therefore, the presence of increased levels of these enzymes in blood plasma reflects damage to the corresponding tissue. For example, the enzyme alanine aminotransferase (ALT) is one of many enzymes that are abundant in the liver. The appearance of elevated levels of ALT in plasma signals possible damage to hepatic tissue. Measurement of ALT released into a patient's blood from dying cells is part of the liver function test panel. Increases in plasma levels of enzymes with a wide tissue distribution provide a less specific indication of the site of cellular injury, which limits their diagnostic value.

C. Isoenzymes

Isoenzymes are variant forms of a particular enzyme that all catalyze the same reaction but have slightly different physical properties because of genetically determined differences in amino acid sequence. For this reason, isoenzymes may contain different numbers of charged amino acids, which allows them to be separated from each other by electrophoresis (the movement of charged particles in an electric field) (Fig. 5.20).

Different organs commonly contain characteristic proportions of different isoenzymes. LDH is found in relatively high concentration in most tissues; five isoenzyme forms of LDH exist, LD 1–5, with LD5 prevalent in liver and skeletal muscle, LD2 in red blood cells, and LD1 in myocardial muscle, for example. The pattern of isoenzymes found in the blood plasma may, therefore, serve as a means of identifying the site of tissue damage. The plasma levels

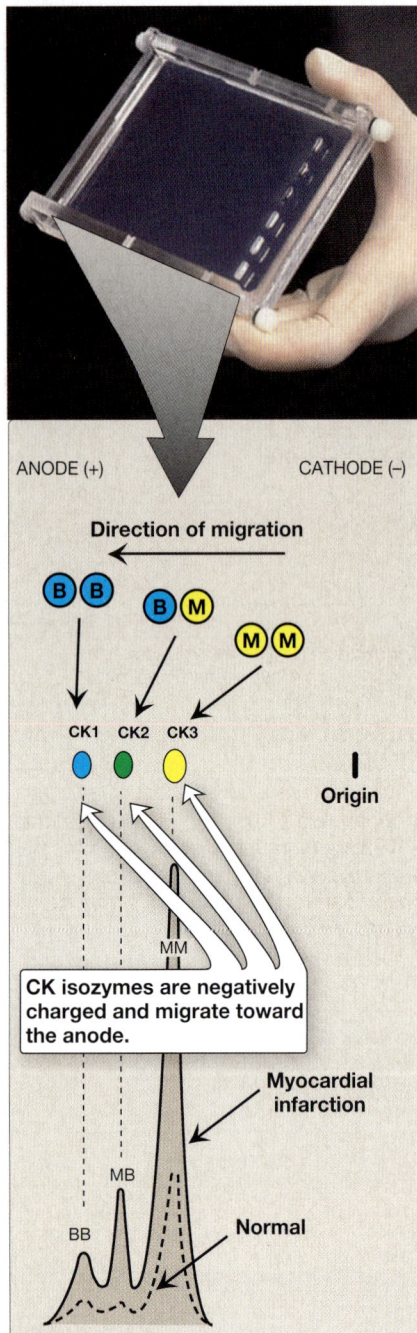

Figure 5.20
Subunit composition, electrophoretic mobility, and enzyme activity of creatine kinase (CK) isoenzymes.

of various isoenzyme forms of LDH and of creatine kinase (CK) vary under different disease states.

1. **Isoenzyme quaternary structure:** Isoenzymes of a given enzyme often contain different subunits in various combinations. For example, LDH occurs as five isoenzymes and each exists as a tetramer, containing four subunits (combinations of subunits called H and M for heart and skeletal muscle where they were first discovered) such that LD1 = HHHH, LD2 = HHHM, LD3 = HHMM, LD4 = HMMM, and LD5 = MMMM. CK occurs as three isoenzymes. Each CK isoenzyme is a dimer composed of two polypeptide subunits (called B and M subunits for brain and skeletal muscle) associated in one of three combinations: CK1 = BB, CK2 = MB, and CK3 = MM. Each CK isoenzyme shows a characteristic electrophoretic mobility (see Fig. 5.20). [Note: Essentially all CK in brain is BB isoform, whereas it is MM in skeletal muscle. In cardiac muscle, a majority of CK is MM, but the presence of CK-MB is unique to myocardium.]

2. **Historical use in diagnosis of myocardial infarction (MI):** Measurement of blood levels of isoenzymes with cardiac specificity (biomarkers) had an important use in diagnosis of MI or heart attack prior to the advent of testing for cardiac proteins known as troponins (see Clinical Application 5.1). Because myocardial muscle is the only tissue that contains more than 5% of the total CK activity as the CK-MB (CK2) isoenzyme, its appearance in blood plasma is virtually specific for damage to myocardial muscle and is seen after an acute MI. CK-MB appears in a patient's blood plasma within 4 to 8 hours following the onset of chest pain in an acute MI, reaches a peak of activity ~24 hours later, and returns to baseline after 48 to 72 hours (Fig. 5.21).

Clinical Application 5.1: Diagnostic Use of Troponins

Troponins T (TnT) and I (TnI) are regulatory proteins involved in muscle contractility. Cardiac-specific isoforms (cTn) of troponins are released into the plasma in response to cardiac damage, and provide a highly sensitive and specific indication of damage to cardiac tissue. cTn appears in plasma within 4–6 hours after an MI, peaks in 24–36 hours, and remains elevated for 3–10 days. Elevated cTn, in combination with the clinical presentation and characteristic changes in the ECG, are currently considered the "gold standard" in the diagnosis of an MI. While the appearance characteristics of cTN in blood plasma after an acute MI are similar to those of CK-MB, the change from baseline to peak values is much greater for cTN (see Fig. 5.21).

X. CHAPTER SUMMARY

- Enzymes are **protein catalysts** that increase the velocities of reactions by providing an alternate reaction pathway with a lower activation energy (Fig. 5.22).

- Enzymes contain a specialized cleft called the **active site**, which binds substrate, forming an **enzyme–substrate (ES) complex**, with subsequent conversion to product (P): ES → EP → E + P.

- Most enzymes follow **Michaelis–Menten kinetics**, and a plot of **initial reaction velocity (v_o)** against **[S]** has a **hyperbolic** shape. **Allosteric** enzymes show a **sigmoidal** curve when v_o is plotted against [S].

- A **Lineweaver–Burk** or double-reciprocal plot of $1/v_0$ and $1/[S]$ transforms the hyperbolic shaped curve to a straight line and allows easier determination of **V_{max}** (maximal velocity) and **K_m** (Michaelis constant, which reflects affinity for substrate).

- **Inhibitors** decrease the velocities of enzyme-catalyzed reactions.

- The two most common types of enzyme inhibition are **competitive**, which **increases** the apparent **K_m**, and **noncompetitive**, which **decreases** the apparent **V_{max}**.

- **Allosteric enzymes** are composed of subunits and are regulated by **effectors** that bind noncovalently at a site other than the active site.

- **Positive** allosteric effectors increase enzyme activity, and **negative** effectors decrease it.

- Enzymes can be regulated by **covalent modification,** most often via phosphorylation catalyzed by protein kinases and dephosphorylation catalyzed by phosphatases.

- Regulation can also occur by changes in the rate of synthesis or degradation.

- Since most enzymes function intracellularly, their appearance in blood plasma can indicate damage to a corresponding tissue, giving enzymes a diagnostic value in medicine.

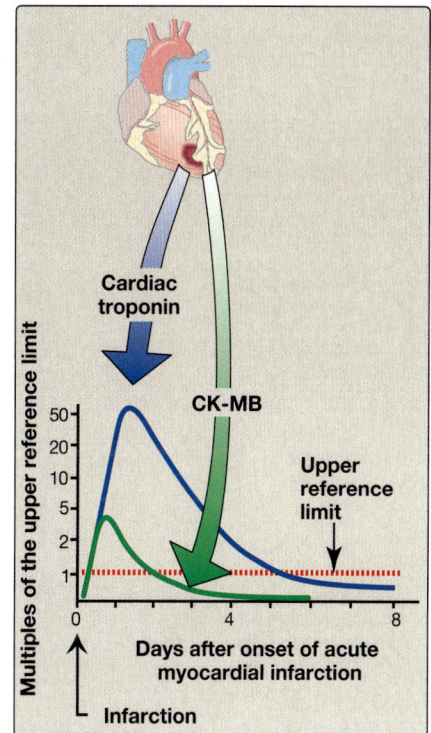

Figure 5.21
Appearance of creatine kinase isozyme CK-MB and cardiac troponin in plasma after myocardial infarction. [Note: Either cardiac troponin T or I may be measured.]

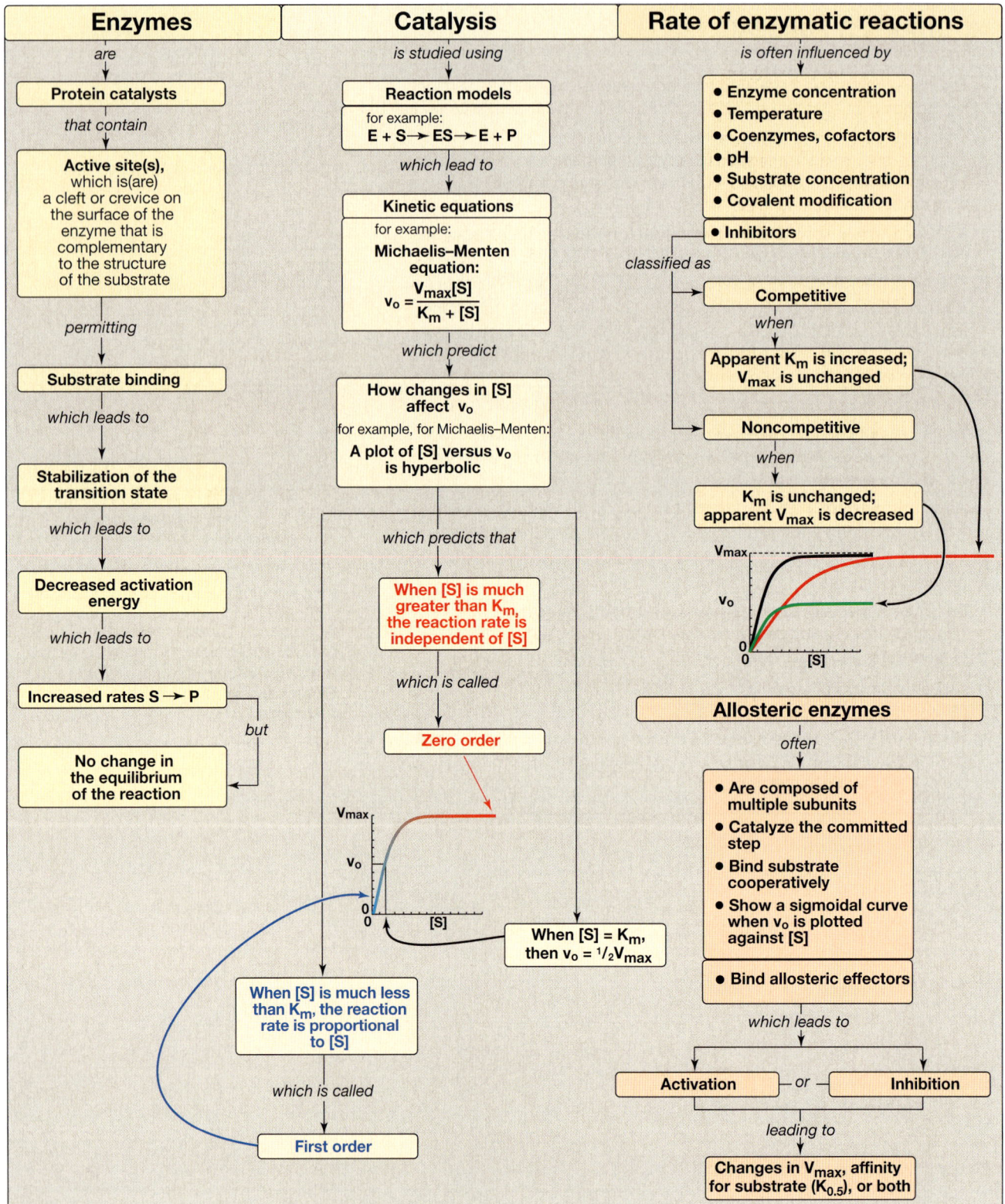

Figure 5.22
Key concept map for the enzymes. S, substrate; [S], substrate concentration; P, product; E, enzyme; v_o, initial velocity; V_{max}, maximal velocity; K_m, Michaelis constant; $K_{0.5}$, substrate concentration that gives half maximal velocity.

Study Questions

Choose the ONE best answer.

5.1. In cases of ethylene glycol poisoning and its characteristic metabolic acidosis, treatment involves correction of the acidosis, removal of any remaining ethylene glycol, and administration of an inhibitor of alcohol dehydrogenase (ADH), the enzyme that oxidizes ethylene glycol to the organic acids that cause the acidosis. Ethanol (grain alcohol) frequently is the inhibitor given to treat ethylene glycol poisoning. Results of experiments using ADH with and without ethanol are shown. Based on these data, what type of inhibition is caused by the ethanol?

A. Competitive
B. Feedback
C. Irreversible
D. Noncompetitive

Substrate Concentration with Ethanol (mM)	Rate of Reaction (mol/L/s)	Substrate Concentration without Ethanol	Rate of Reaction (mol/L/s)
5	3.0×10^{-7}	5	8.0×10^{-7}
10	5.0×10^{-7}	10	1.2×10^{-6}
20	1.0×10^{-6}	20	1.8×10^{-6}
40	1.6×10^{-6}	40	1.9×10^{-6}
80	2.0×10^{-6}	80	2.0×10^{-6}

Correct answer = A. At a sufficiently high [S], the reaction velocity reaches the V_{max} observed in the absence of inhibitor, a characteristic of competitive inhibition. A competitive inhibitor also increases the apparent K_m for a given substrate. This means that, in the presence of a competitive inhibitor, more substrate is needed to achieve one half V_{max}. But, the effect of a competitive inhibitor is reversed by increasing substrate concentration ([S]).

5.2. Alcohol dehydrogenase (ADH) requires oxidized nicotinamide adenine dinucleotide (NAD^+) for catalytic activity. In the reaction catalyzed by ADH, an alcohol is oxidized to an aldehyde as NAD^+ is reduced to NADH and dissociates from the enzyme. The NAD^+ is functioning as a/an:

A. apoenzyme.
B. coenzyme–cosubstrate.
C. coenzyme–prosthetic group.
D. cofactor.
E. heterotropic effector.

Correct answer = B. Coenzymes–cosubstrates are small organic molecules that associate transiently with an enzyme and leave the enzyme in a changed form. Coenzyme–prosthetic groups are small organic molecules that associate permanently with an enzyme and are returned to their original form on the enzyme. Cofactors are metal ions. Heterotropic effectors are not substrates.

For Questions 5.3 and 5.4, use the graph below that shows the changes in free energy when a reactant is converted to a product in the presence and absence of an enzyme. Select the letter that best represents:

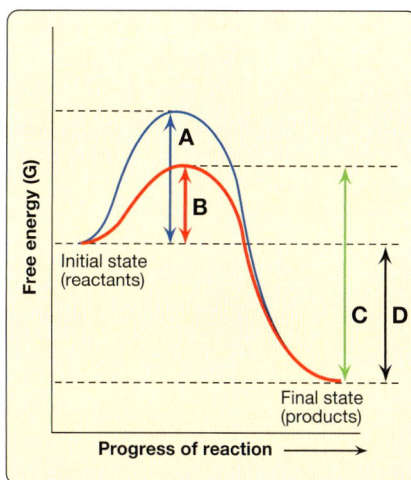

5.3. Activation energy of the catalyzed forward reaction

5.4. Free energy of the reaction

Correct answers = B, D. Enzymes (protein catalysts) provide an alternate reaction pathway with a lower activation energy. However, they do not change the free energy of the reactant or product. A is the activation energy of the uncatalyzed reaction. C is the activation energy of the catalyzed reverse reaction.

5.5. If a noncompetitive inhibitor is included in the reaction of an enzyme with its substrate, then:

A. addition of sufficient concentrations of substrate will overcome the inhibition.
B. K_m will be decreased because of reduced enzyme–substrate affinity.
C. inhibitor and substrate will bind to different sites on the enzyme.
D. the curve when plotting velocity versus (substrate) will become sigmoidal.
E. V_{max} will remain the same as for the uninhibited reaction.

Correct answer = C. Noncompetitive inhibitors do not bind to the enzyme active site but to other binding sites on the enzyme. Substrates bind to enzyme active sites. Because binding is to a different site, the addition of substrate will not overcome the inhibition. K_m will remain the same since substrate will continue to bind to the active site; however, it will be an inactive complex when a noncompetitive inhibitor is also bound. For enzymes that follow Michaelis–Menten kinetics, the hyperbolic shape curve will be shifted to a lower V_{max} in the presence of a noncompetitive inhibitor.

5.6. In an enzyme-catalyzed reaction, protein Q is the substrate for enzyme X, a kinase. The product of this reaction will be:

A. ATP.
B. dephosphorylated enzyme X.
C. phosphorylated enzyme X.
D. dephosphorylated protein Q.
E. phosphorylated protein Q.

Correct answer = E. The substrate of a kinase will be converted to its phosphorylated form as a result of the reaction. Adenosine triphosphate (ATP) is used as a phosphate donor and will be hydrolyzed to adenosine diphosphate (ADP) during the course of the reaction. The kinase enzyme itself will not be altered by the reaction.

5.7. Sulfa drugs can be effective in limiting bacterial infections in humans while not producing toxic effects within human cells. To account for these characteristics, sulfa drugs most likely act as a(n):

A. allosteric effector increasing catalytic action of a bacterial enzyme.
B. antimetabolite that interrupts replication in all types of dividing cells.
C. competitive inhibitor of an enzyme required by bacteria but not human cells.
D. noncompetitive inhibitors of several regulatory steps in glycolysis.
E. positive allosteric effector of an enzyme in bacterial cell wall synthesis.

Correct answer = C. Acting as a competitive inhibitor of an enzyme required only by bacteria, sulfa drugs and other antibiotics can halt bacterial growth without damaging human cells. Agents such as antimetabolites that interrupt all dividing cells would harm host human cells as well as bacteria. Most drugs that act as enzyme inhibitors are competitive and designed to resemble the substrate structurally to be able to bind to the active site. Both human and bacterial cells undergo glycolysis. Allosteric effectors that increase an enzyme's catalytic function are positive effectors. Positive allosteric effectors enhance enzyme activity and do not inhibit it.

Bioenergetics and Oxidative Phosphorylation

6

I. OVERVIEW

Bioenergetics describes the transfer and utilization of energy in biologic systems and concerns the initial and final energy states of the reaction components. Bioenergetics makes use of a few basic ideas from the field of thermodynamics, particularly the concept of free energy. Because changes in free energy provide a measure of the energetic feasibility of a chemical reaction, they allow for an estimation of the likelihood that a reaction or process will take place. Bioenergetics then predicts if a process is possible, whereas kinetics measures the reaction rate.

II. FREE ENERGY

The direction and extent to which a chemical reaction proceeds are determined by the degree of change of enthalpy and entropy during the reaction. Enthalpy (ΔH) is a measure of the change [Δ] in heat content of the reactants and products, whereas entropy (ΔS) is a measure of the change in randomness or disorder of the reactants and products, as shown in Figure 6.1. Neither of these thermodynamic quantities by itself is sufficient to determine whether a chemical reaction will proceed spontaneously in the direction it is written. However, when combined mathematically (see Fig. 6.1), enthalpy and entropy can be used to define a third quantity, free energy (G), which predicts the direction in which a reaction will spontaneously proceed.

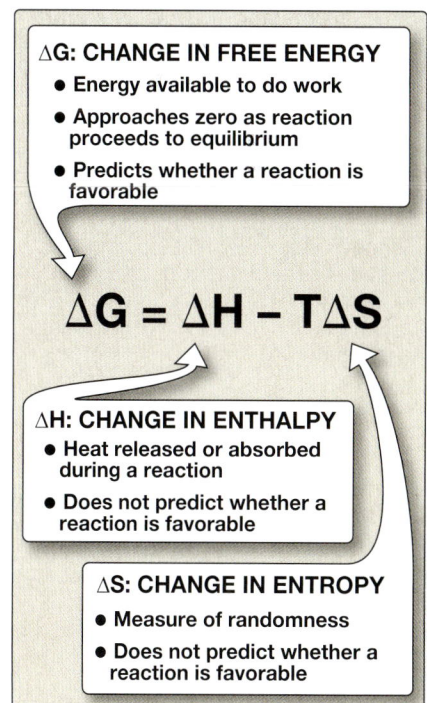

ΔG: CHANGE IN FREE ENERGY
- Energy available to do work
- Approaches zero as reaction proceeds to equilibrium
- Predicts whether a reaction is favorable

$$\Delta G = \Delta H - T\Delta S$$

ΔH: CHANGE IN ENTHALPY
- Heat released or absorbed during a reaction
- Does not predict whether a reaction is favorable

ΔS: CHANGE IN ENTROPY
- Measure of randomness
- Does not predict whether a reaction is favorable

Figure 6.1
Relationship between changes in free energy (G), enthalpy (H), and entropy (S). T is the absolute temperature in Kelvin (K), where K = °C + 273.

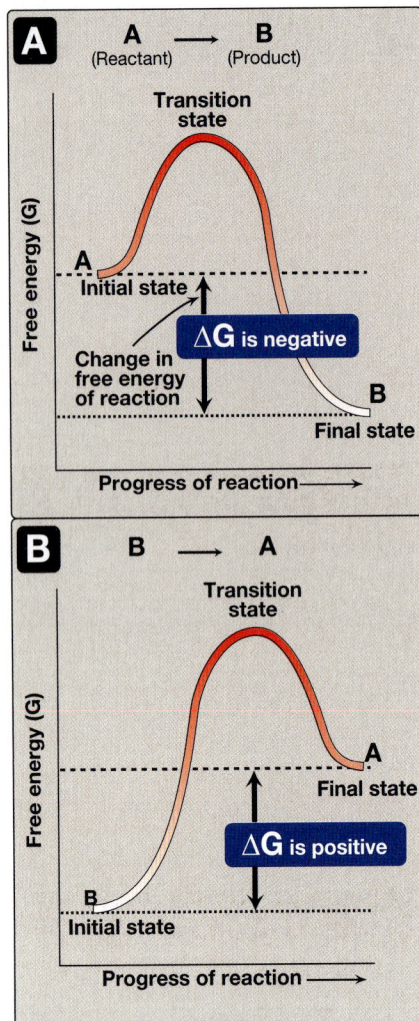

Figure 6.2
Change in free energy (ΔG) during a reaction. **A.** The product has a lower free energy (G) than the reactant. **B.** The product has a higher free energy than the reactant.

III. FREE ENERGY CHANGE

The **change in free energy** is represented in two ways, ΔG and ΔG^0. ΔG represents the change in free energy and, thus, the direction of a reaction at any specified concentration of products and reactants. ΔG, then, is a variable. This contrasts with the standard free energy change, ΔG^0, which is the energy change when reactants and products are at a concentration of 1 mol/L. [Note: The concentration of protons [H^+] is assumed to be 10^{-7} mol/L (i.e., pH = 7). This may be shown by a prime sign, as in $\Delta G^{0\prime}$]. Although ΔG^0, a constant, represents energy changes at these nonphysiologic concentrations of reactants and products, it is nonetheless useful in comparing the energy changes of different reactions. Furthermore, ΔG^0 can readily be determined from the measurement of the equilibrium constant.

A. ΔG and reaction direction

The negative or positive sign preceding ΔG can be used to predict the direction of a reaction at constant temperature and pressure. Consider the reaction:

$$A \rightleftarrows B$$

If ΔG is negative, the reaction is considered exergonic with a net loss of energy. In this case, the reaction proceeds spontaneously as written, with A converted to B (Fig. 6.2A). If ΔG is positive, the reaction is endergonic with a net gain of energy. Energy must be added to the system for the reaction from B to A to take place (Fig. 6.2B). In cases when $\Delta G = 0$, the reaction is in equilibrium. Note that when a reaction is proceeding spontaneously (ΔG is negative), the reaction will continue until ΔG reaches zero, and equilibrium is established.

B. ΔG of the forward and reverse reactions

The free energy of the forward reaction (A → B) is equal in magnitude but opposite in sign to that of the reverse reaction (B → A). For example, if ΔG of the forward reaction is –5 kcal/mol, then ΔG of the back reaction is +5 kcal/mol. [Note: ΔG can also be expressed in kilojoules per mole, or kJ/mol (1 kcal = 4.2 kJ).]

C. ΔG and reactant and product concentrations

The ΔG of the reaction A → B depends on the concentrations of the reactant and of the product. At constant temperature and pressure, the following relationship can be derived:

$$\Delta G = \Delta G^0 + RT \ln \frac{[B]}{[A]}$$

where ΔG^0 is the standard free energy change (see D. below)
R is the gas constant (1.987 cal/mol K)
T is the absolute temperature (K)
[A] and [B] are the actual concentrations of the reactant and product
ln represents the natural logarithm.

A reaction with a positive ΔG^0 can proceed in the forward direction if the ratio of products to reactants ([B]/[A]) is sufficiently small (i.e., the ratio of reactants to products is large) to make ΔG negative. For example, consider the reaction:

Glucose 6-phosphate $\rightleftarrows$ fructose 6-phosphate

Figure 6.3A shows reaction conditions in which the concentration of the reactant, glucose 6-phosphate, is high compared with the concentration of product, fructose 6-phosphate. This means that the ratio of the product to reactant is small, and the ratio of fructose 6-phosphate/glucose 6-phosphate is large and negative, causing ΔG to be negative despite ΔG^0 being positive. Thus, the reaction can proceed in the forward direction.

D. Standard free energy change

The standard free energy change, ΔG^0, is equal to the free energy change, ΔG, under standard conditions, when reactants and products are at 1 mol/L concentrations (Fig. 6.3B). Under these conditions, the natural logarithm of the ratio of products to reactants is zero ($\ln 1 = 0$), and, therefore, the equation shown at the bottom of the previous page becomes:

$$\Delta G = \Delta G^0 + 0$$

1. **ΔG^0 and reaction direction:** Under standard conditions, ΔG^0 can be used to predict the direction a reaction proceeds because, under these conditions, ΔG^0 is equal to ΔG. However, ΔG^0 cannot predict the direction of a reaction under physiologic conditions because it is composed solely of constants (R, T, and K_{eq} [see Section 2.]) and is not, therefore, altered by changes in product or substrate concentrations.

2. **Relationship between ΔG^0 and K_{eq}:** In a reaction A $\rightleftarrows$ B, a point of equilibrium is reached at which no further net chemical change takes place. In this state, the ratio of [B] to [A] is constant, regardless of the actual concentrations of the two compounds:

$$K_{eq} = \frac{[B]_{eq}}{[A]_{eq}}$$

where K_{eq} is the equilibrium constant, and $[A]_{eq}$ and $[B]_{eq}$ are the concentrations of A and B at equilibrium. If the reaction A $\rightleftarrows$ B is allowed to reach equilibrium at constant temperature and pressure, then, at equilibrium, the overall ΔG is zero (Fig. 6.3C). Therefore,

$$\Delta G = 0 = \Delta G^0 + RT \ln \frac{[B]_{eq}}{[A]_{eq}}$$

where the actual concentrations of A and B are equal to the equilibrium concentrations of reactant and product ($[A]_{eq}$ and $[B]_{eq}$), and their ratio is equal to the K_{eq}. Thus,

$$\Delta G^0 = -RT \ln K_{eq}$$

This equation allows some simple predictions:

If $K_{eq} = 1$, then $\Delta G^0 = 0$

If $K_{eq} > 1$, then $\Delta G^0 < 0$

If $K_{eq} < 1$, then $\Delta G^0 > 0$

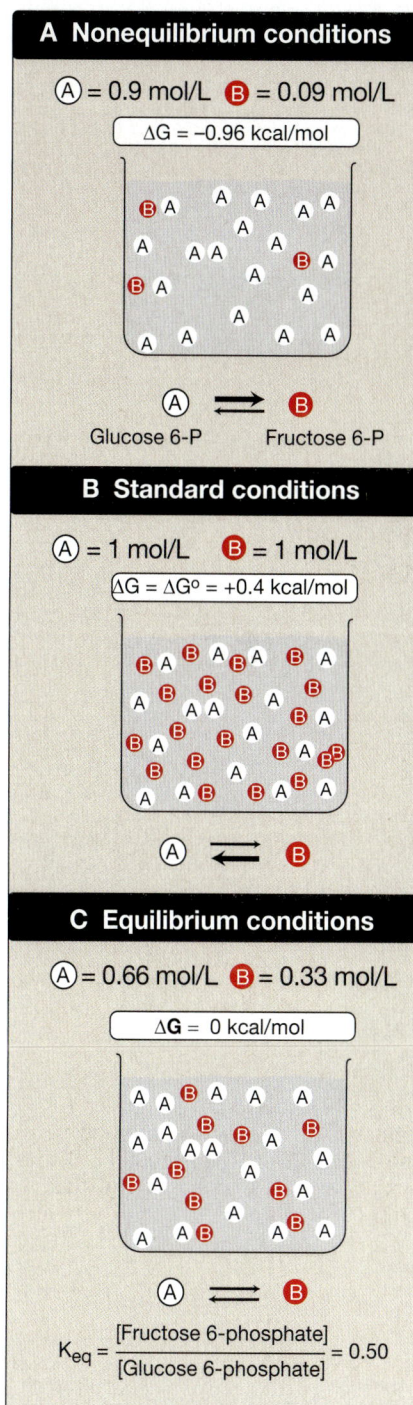

A Nonequilibrium conditions

(A) = 0.9 mol/L (B) = 0.09 mol/L

$\Delta G = -0.96$ kcal/mol

(A) $\rightleftarrows$ (B)

Glucose 6-P   Fructose 6-P

B Standard conditions

(A) = 1 mol/L   (B) = 1 mol/L

$\Delta G = \Delta G^0 = +0.4$ kcal/mol

(A) $\rightleftarrows$ (B)

C Equilibrium conditions

(A) = 0.66 mol/L (B) = 0.33 mol/L

$\Delta G = 0$ kcal/mol

(A) $\rightleftarrows$ (B)

$$K_{eq} = \frac{[\text{Fructose 6-phosphate}]}{[\text{Glucose 6-phosphate}]} = 0.50$$

Figure 6.3
Free energy change (ΔG) of a reaction depends on the concentration of reactant and product. For the conversion of glucose 6-phosphate to fructose 6-phosphate, ΔG is negative when the ratio of reactant to product is large (top, **panel A**), is positive under standard conditions (middle, **panel B**), and is zero at equilibrium (bottom, **panel C**). ΔG^0, standard free energy change.

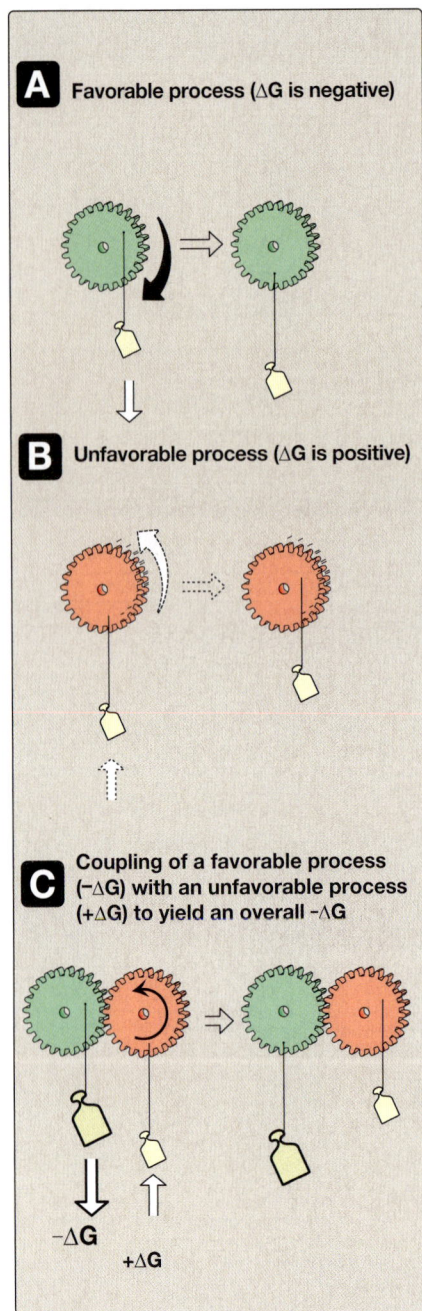

A Favorable process (ΔG is negative)

B Unfavorable process (ΔG is positive)

C Coupling of a favorable process (−ΔG) with an unfavorable process (+ΔG) to yield an overall −ΔG

−ΔG

+ΔG

Figure 6.4
Mechanical model of the coupling of favorable and unfavorable processes.
A. Gear with weight attached spontaneously turns in the direction that achieves the lowest energy state.
B. The reverse movement is energetically unfavorable (not spontaneous).
C. The energetically favorable movement can drive the unfavorable one. ΔG, change in free energy.

3. **ΔG^0s of two consecutive reactions:** The ΔG^0s are additive in any sequence of consecutive reactions, as are the ΔGs. For example:

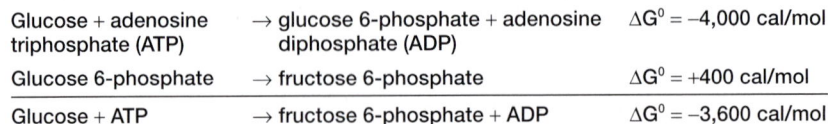

Glucose + adenosine triphosphate (ATP)	→ glucose 6-phosphate + adenosine diphosphate (ADP)	$\Delta G^0 = -4{,}000$ cal/mol
Glucose 6-phosphate	→ fructose 6-phosphate	$\Delta G^0 = +400$ cal/mol
Glucose + ATP	→ fructose 6-phosphate + ADP	$\Delta G^0 = -3{,}600$ cal/mol

4. **ΔGs of a pathway:** The additive property of ΔG is very important in biochemical pathways through which substrates must pass in a particular direction (e.g., $A \rightarrow B \rightarrow C \rightarrow D \rightarrow \cdots$). As long as the sum of the ΔGs of the individual reactions is negative, the pathway can proceed as written, even if some of the individual reactions of the pathway have a positive ΔG. However, the actual rates of the reactions depend on the lowering of activation energies (E_a) by the enzymes that catalyze the reactions.

IV. ADENOSINE TRIPHOSPHATE: AN ENERGY CARRIER

Reactions with a large, positive ΔG are made possible by coupling the endergonic movement of ions with a second, spontaneous process with a large, negative ΔG, such as the exergonic hydrolysis of ATP. Figure 6.4 shows a mechanical model of energy coupling. The simplest example of energy coupling in biologic reactions occurs when the energy-requiring and the energy-yielding reactions share a common intermediate.

A. Common intermediates

Two chemical reactions have a common intermediate when they occur sequentially, in that the product of the first reaction is a substrate for the second. For example, given the reactions

$$A + B \rightarrow C + D$$
$$D + X \rightarrow Y + Z$$

D is the common intermediate and can serve as a carrier of chemical energy between the two reactions. [Note: The intermediate may be linked to an enzyme.] Many coupled reactions use ATP to generate a common intermediate. These reactions may involve the transfer of a phosphate group from ATP to another molecule. Other reactions involve the transfer of phosphate from an energy-rich intermediate to ADP, forming ATP.

B. Energy carried by ATP

ATP consists of a molecule of adenosine to which three phosphate groups are attached (Fig. 6.5). Removal of one phosphate produces ADP, and removal of two phosphates produces adenosine monophosphate (AMP). For ATP, the ΔG^0 of hydrolysis is approximately −7.3 kcal/mol for each of the two terminal phosphate groups. Because of this large negative ΔG^0 of hydrolysis, ATP is called a high-energy phosphate compound. [Note: Adenine nucleotides are interconverted (2 ADP $\rightleftarrows$ ATP + AMP) by adenylate kinase.]

@ V. ELECTRON TRANSPORT CHAIN

Energy-rich molecules, such as glucose, are metabolized by a series of oxidation reactions, ultimately yielding carbon dioxide and water (H_2O) (Fig. 6.6). The metabolic intermediates of these reactions donate electrons to specific coenzymes, nicotinamide adenine dinucleotide (NAD^+) and flavin adenine dinucleotide (FAD), to form the energy-rich reduced forms, NADH and flavin adenine dinucleotide ($FADH_2$). These reduced coenzymes can, in turn, each donate a pair of electrons to a specialized set of electron carriers, collectively called the electron transport chain (ETC). As electrons are passed down the ETC, they lose much of their free energy. This energy is used to move H^+ across the inner mitochondrial membrane, creating an H^+ gradient that drives the production of ATP from ADP and inorganic phosphate (P_i). The coupling of electron transport with ATP synthesis is called oxidative phosphorylation (OXPHOS). It proceeds continuously in all tissues that contain mitochondria. Note that the free energy not trapped as ATP is used to drive ancillary reactions such as the transport of calcium ions into mitochondria and to generate heat.

A. Mitochondrial electron transport chain

The ETC (except for cytochrome c) is located in the inner mitochondrial membrane and is the final common pathway by which electrons derived from different fuels of the body flow to oxygen (O_2), reducing it to H_2O (see Fig. 6.6).

1. **Mitochondrial membranes:** Mitochondria have outer and inner membranes separated by the intermembrane space. The outer membrane contains specialized channels formed by the protein porin, making it freely permeable to most ions and small molecules. The inner membrane is a specialized structure impermeable to most small ions, including H^+, and small molecules such as ATP, ADP, pyruvate, and other metabolites important to mitochondrial function (Fig. 6.7). Transport proteins are required to move ions or molecules across this membrane. The inner mitochondrial membrane is unusually rich in proteins, more than half of which are directly involved in OXPHOS. It also contains convolutions, called cristae, which greatly increase its surface area.

2. **Mitochondrial matrix:** The gel-like solution of the interior of the mitochondria, the matrix, is also rich in proteins. These include the enzymes responsible for the oxidation of pyruvate, amino acids, and fatty acids (by β-oxidation) as well as those of the tricarboxylic acid (TCA) cycle. The synthesis of glucose, urea, and heme occurs partially in the matrix of mitochondria. In addition, the matrix contains NAD^+ and FAD (the oxidized forms of the two coenzymes that are required as electron acceptors), and ADP and P_i, which are used to produce ATP. [Note: The matrix also contains mitochondrial deoxyribonucleic acid (mtDNA), ribonucleic acid (mtRNA), and ribosomes.]

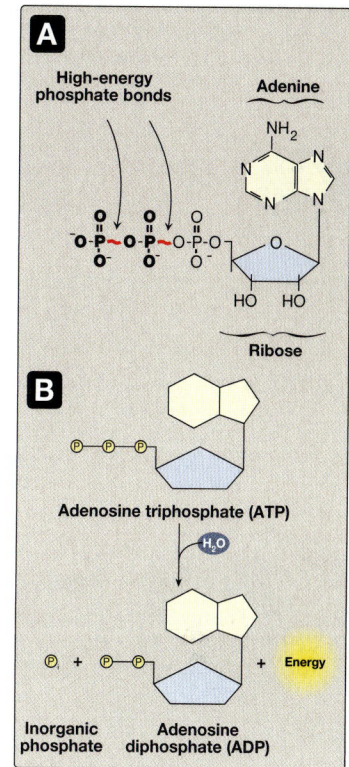

Figure 6.5
A. Adenosine triphosphate (ATP).
B. Hydrolysis of ATP.

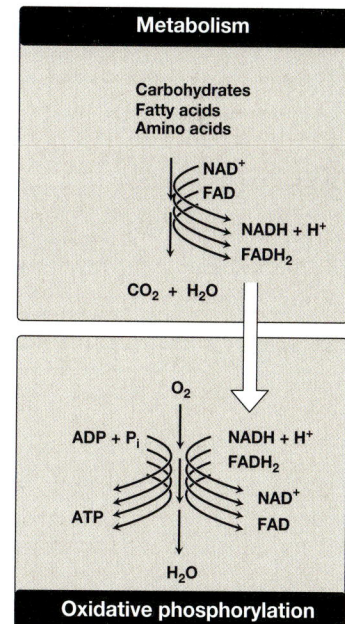

Figure 6.6
Metabolic breakdown of energy-yielding molecules. NAD(H), nicotinamide adenine dinucleotide; FAD(H_2), flavin adenine dinucleotide; ADP, adenosine diphosphate; P_i, inorganic phosphate; CO_2, carbon dioxide.

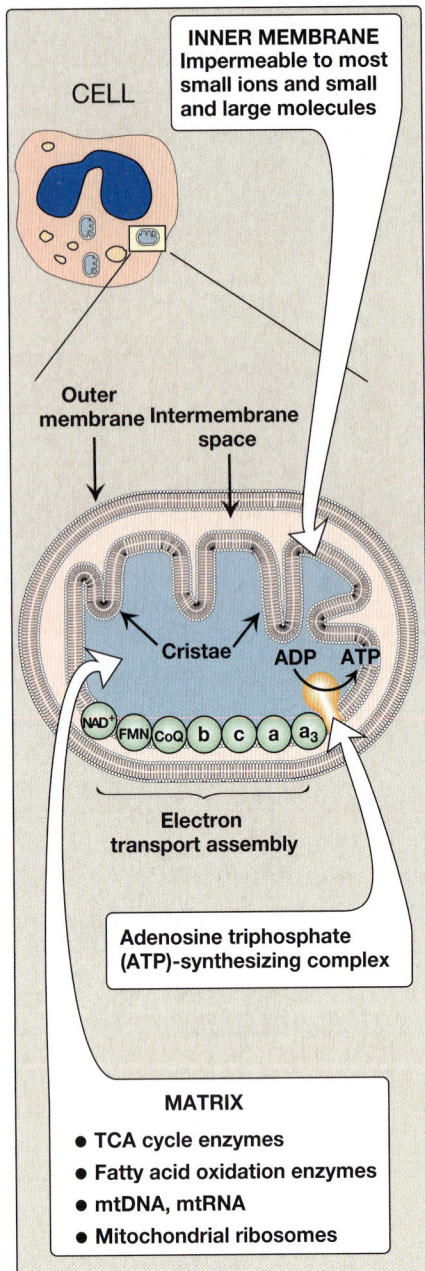

Figure 6.7
Structure of a mitochondrion showing schematic representation of the electron transport chain and the ATP synthesizing complex on the inner membrane. [Note: Unlike the inner membrane, the outer membrane is highly permeable, and the milieu of the intermembrane space is like that of the cytosol.] mt, mitochondrial; RNA, ribonucleic acid; ADP, adenosine diphosphate; TCA, tricarboxylic acid.

B. Organization

The inner mitochondrial membrane contains four separate protein complexes, I, II, III, and IV, that each contains part of the ETC (Fig. 6.8). These complexes accept or donate electrons to the relatively mobile electron carrier coenzyme Q (CoQ) and **cytochrome c**. Each carrier in the ETC can receive electrons from an electron donor and can subsequently donate electrons to the next acceptor in the chain. The electrons ultimately combine with O_2 and H^+ to form H_2O. This requirement for O_2 makes the electron transport process the respiratory chain, which accounts for the greatest portion of the body's use of O_2.

C. Reactions

Except for CoQ, which is a lipid-soluble quinone, all members of the ETC are proteins. These may function as enzymes, as is the case with the flavin-containing dehydrogenases, or may contain iron as part of an iron–sulfur (Fe–S) center, iron as part of the porphyrin prosthetic group of heme as in the cytochromes, or copper (Cu) as does the cytochrome $a + a_3$ complex.

1. **NADH formation:** NAD^+ is reduced to NADH by dehydrogenases that remove two hydrogen atoms from their substrate. [Note: For examples of these reactions, see the discussion of the dehydrogenases of the TCA cycle.] Both electrons but only one H^+ (i.e., a hydride ion [:H^-]) are transferred to the NAD^+, forming NADH plus a free H^+.

2. **NADH dehydrogenase:** The free H^+ plus the hydride ion carried by NADH are transferred to NADH dehydrogenase, a protein complex (complex I) embedded in the inner mitochondrial membrane. Complex I has a tightly bound molecule of flavin mononucleotide (FMN), a coenzyme structurally related to FAD that accepts the two hydrogen atoms (two electrons + two H^+), becoming $FMNH_2$. NADH dehydrogenase also contains peptide subunits with Fe–S centers (Fig. 6.9). At complex I, electrons move from NADH to FMN to the iron of the Fe–S centers and then to CoQ. As electrons flow, they lose energy. This energy is used to pump four H^+ across the inner mitochondrial membrane, from the matrix to the intermembrane space.

3. **Succinate dehydrogenase:** At complex II, electrons from the succinate dehydrogenase–catalyzed oxidation of succinate to fumarate move from the coenzyme, $FADH_2$, to an Fe–S protein, and then to CoQ. [Note: Because no energy is lost in this process, no H^+ are pumped at complex II.]

4. **Coenzyme Q:** CoQ, also known as ubiquinone, is a quinone derivative with a long, hydrophobic isoprenoid tail, made from an intermediate of cholesterol synthesis (see Chapter 18). CoQ is a mobile electron carrier and can accept electrons from NADH dehydrogenase (complex I); from succinate dehydrogenase (complex II); and from other mitochondrial dehydrogenases, such as glycerol 3-phosphate dehydrogenase and acyl CoA dehydrogenases. CoQ transfers electrons to complex III (cytochrome

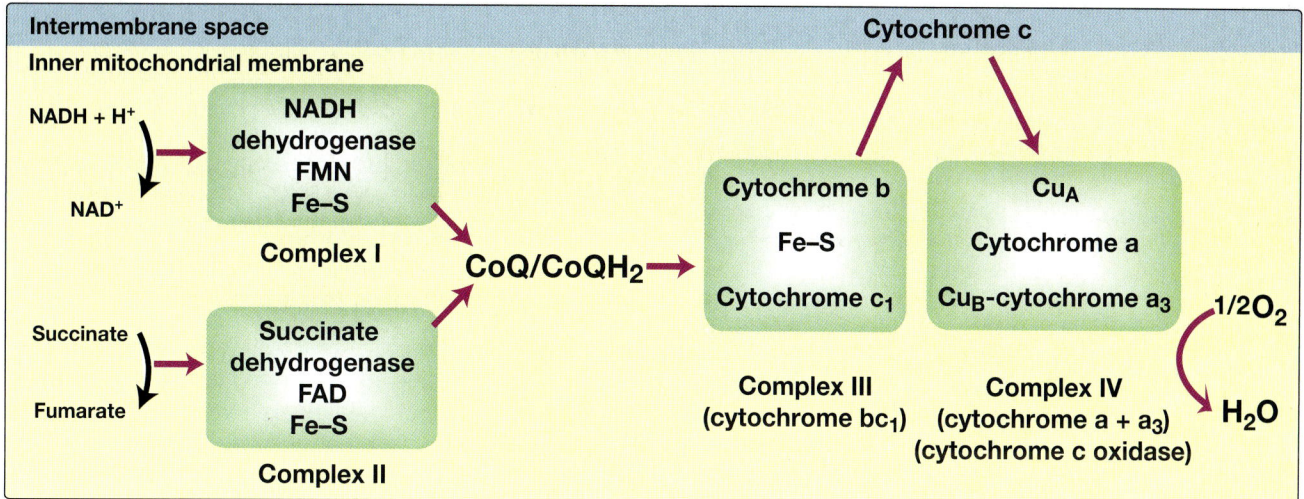

Figure 6.8
Electron transport chain. Electron flow is shown by *magenta arrows*. NAD(H), nicotinamide adenine dinucleotide; FMN, flavin mononucleotide; FAD, flavin adenine dinucleotide; Fe–S, iron–sulfur; CoQ, coenzyme Q; Cu, copper.

bc_1). Thus, a function of CoQ is to link the flavoprotein dehydrogenases to the cytochromes.

5. **Cytochromes:** The remaining members of the ETC are cytochrome proteins. Each contains a heme group, which is a porphyrin ring plus iron. Unlike the heme groups of hemoglobin, the cytochrome iron is reversibly converted from its ferric (Fe^{3+}) to its ferrous (Fe^{2+}) form as a normal part of its function as an acceptor and donor of electrons. Electrons are passed along the chain from cytochrome bc_1 (complex III), to cytochrome c, and then to cytochromes a + a_3 ([complex IV] see Fig. 6.8). As electrons flow, four H^+ are pumped across the inner mitochondrial membrane at complex III and two at complex IV. [Note: Cytochrome c is located in the intermembrane space, loosely associated with the outer face of the inner membrane. As seen with CoQ, cytochrome c is a mobile electron carrier.]

6. **Cytochrome a + a_3:** Because complex IV is the only electron carrier in which the heme iron has an available coordination site that can react directly with O_2, it also is called cytochrome c oxidase. At complex IV, the transported electrons, O_2, and free H^+ are brought together, and O_2 is reduced to H_2O (see Fig. 6.8). [Note: Four electrons are required to reduce one molecule of O_2 to two molecules of H_2O.] Cytochrome c oxidase contains Cu atoms that are required for this complicated reaction to occur. Electrons move from Cu_A to cytochrome a to cytochrome a_3 (in association with Cu_B) to O_2.

7. **Site-specific inhibitors:** Inhibitors of specific sites in the ETC are illustrated in Figure 6.10. These respiratory inhibitors prevent the passage of electrons by binding to a component of the chain, blocking the oxidation–reduction reaction. Therefore, all electron carriers before the block are fully reduced, whereas those located after the block are oxidized. [Note: Inhibition of the ETC inhibits ATP synthesis because these processes are tightly coupled.]

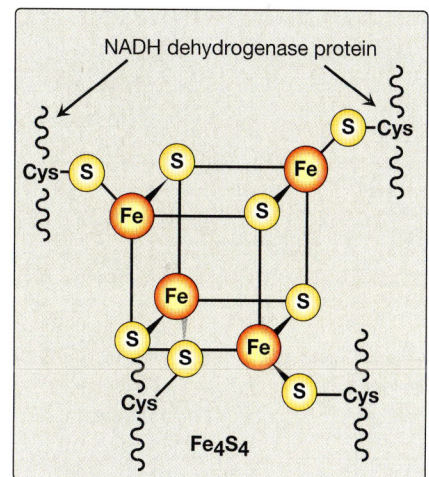

Figure 6.9
Iron–sulfur (Fe–S) center of complex I. [Note: Complexes II and III also contain Fe–S centers.] Fe_4S_4, iron–sulfur center; NADH, nicotinamide adenine dinucleotide; Cys, cysteine.

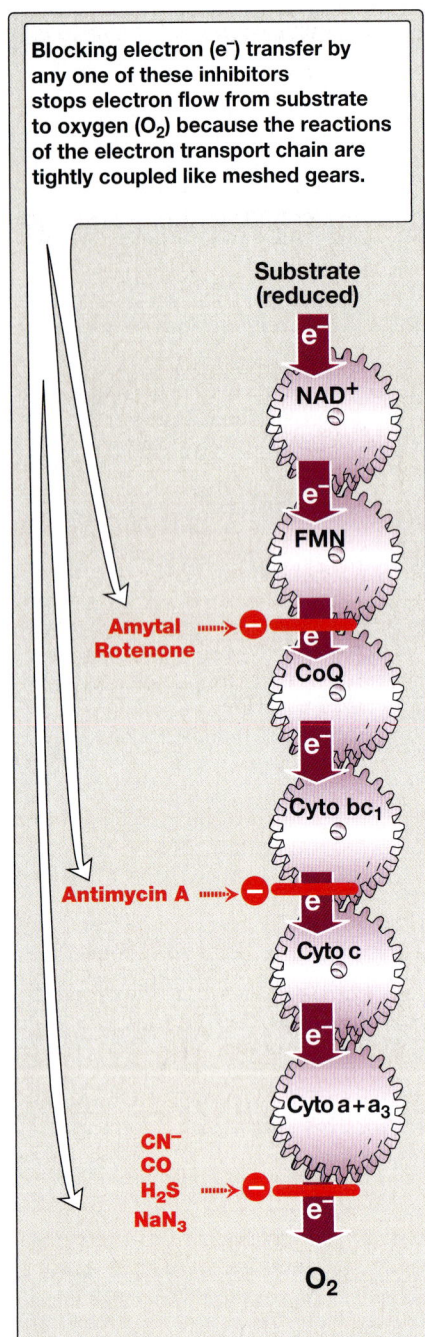

Blocking electron (e⁻) transfer by any one of these inhibitors stops electron flow from substrate to oxygen (O₂) because the reactions of the electron transport chain are tightly coupled like meshed gears.

Figure 6.10
Site-specific inhibitors of electron transport shown using a mechanical model for the coupling of oxidation–reduction reactions. [Note: Normal direction of electron flow is illustrated.] NAD⁺, nicotinamide adenine dinucleotide; FMN, flavin mononucleotide; CoQ, coenzyme Q; Cyto, cytochrome; CN⁻, cyanide; CO, carbon monoxide; H₂S, hydrogen sulfide; NaN₃, sodium azide.

Leakage of electrons from the ETC produces reactive oxygen species (ROS), such as superoxide (O_2), hydrogen peroxide (H_2O_2), and hydroxyl radicals (OH). ROS damage DNA and proteins and cause lipid peroxidation. Enzymes such as superoxide dismutase (SOD), catalase, and glutathione peroxidase are cellular defenses against ROS.

D. Free energy release during electron transport

The free energy released as electrons is transferred along the ETC from an electron donor (reducing agent or reductant) to an electron acceptor (oxidizing agent or oxidant) is used to pump H⁺ at complexes I, III, and IV. [Note: The electrons can be transferred as hydride ions to NAD⁺; as hydrogen atoms to FMN, CoQ, and FAD; or as electrons to cytochromes.]

1. **Redox pairs:** Oxidation (loss of electrons) of one substance is always accompanied by reduction (gain of electrons) of a second. For example, Figure 6.11 shows the oxidation of NADH to NAD⁺ by NADH dehydrogenase at complex I, accompanied by the reduction of FMN, the prosthetic group, to $FMNH_2$. Such redox reactions can be written as the sum of two separate half reactions, one an oxidation and the other a reduction (see Fig. 6.11). NAD⁺ and NADH form a redox pair, as do FMN and $FMNH_2$. Redox pairs differ in their tendency to lose electrons. This tendency is a characteristic of a particular redox pair and can be quantitatively specified by a constant, E_0 (the standard reduction potential), with units in volts.

2. **Standard reduction potential:** The E_0 of various redox pairs can be ordered from the most negative E_0 to the most positive. The more negative the E_0 of a redox pair, the greater the tendency of the reductant member of that pair to lose electrons. The more positive the E_0, the greater the tendency of the oxidant member of that pair to accept electrons. Therefore, electrons flow from the pair with the more negative E_0 to that with the more positive E_0. The E_0 values for some members of the ETC are shown in Figure 6.12. [Note: The components of the chain are arranged in order of increasingly positive E_0 values.]

3. **Relationship of ΔG^0 to ΔE_0:** The ΔG^0 is related directly to the magnitude of the change in E_0:

$$\Delta G^0 = -nF \Delta E_0,$$

where   n = number of electrons transferred (1 for a cytochrome, 2 for NADH, FADH₂, and CoQ)
F = Faraday constant (23.1 kcal/volt mol)
ΔE_0 = E_0 of the electron-accepting pair minus the E_0 of the electron-donating pair
ΔG^0 = change in the standard free energy

4. **ΔG^0 of ATP:** The ΔG^0 for the phosphorylation of ADP to ATP is +7.3 kcal/mol. The transport of a pair of electrons from NADH to O₂ through the ETC releases 52.6 kcal. Therefore, more than

sufficient energy is available to produce three ATP from three ADP and three P_i ($3 \times 7.3 = 21.9$ kcal/mol), sometimes expressed as a P/O ratio (ATP made per O atom reduced) of 3:1. The remaining calories are used for ancillary reactions or released as heat. [Note: The P:O for $FADH_2$ is 2:1 because complex I is bypassed.]

@ VI. PHOSPHORYLATION OF ADP TO ATP

The transfer of electrons down the ETC is energetically favored because NADH is a strong electron donor and O_2 is an avid electron acceptor. However, the flow of electrons does not directly result in ATP synthesis.

A. Chemiosmotic hypothesis

The chemiosmotic hypothesis (also known as the Mitchell hypothesis) explains how the free energy generated by the transport of electrons by the ETC is used to produce ATP from ADP + P_i.

1. **Proton pump: Electron transport** is coupled to ADP phosphorylation by the pumping of H^+ across the inner mitochondrial membrane, from the matrix to the intermembrane space, at complexes I, III, and IV. For each pair of electrons transferred from NADH to O_2, 10 H^+ are pumped. This creates an electrical gradient (with more positive charges on the cytosolic side of the membrane than on the matrix side) and a pH (chemical) gradient (the cytosolic side of the membrane is at a lower pH than the matrix side), as shown in Figure 6.13. The energy (proton-motive force) generated by these gradients is sufficient to drive ATP synthesis. Thus, the H^+ gradient serves as the common intermediate that couples oxidation to phosphorylation.

2. **ATP synthase:** The multisubunit enzyme ATP synthase ([complex V]; Fig. 6.14) synthesizes ATP using the energy of the H^+ gradient. It contains a membrane domain (F_0) that spans the inner mitochondrial membrane and an extramembranous domain (F_1) that appears as a sphere that protrudes into the mitochondrial matrix (see Fig. 6.13). The chemiosmotic hypothesis proposes that after H^+ have been pumped to the cytosolic side of the inner mitochondrial membrane, they reenter the matrix by passing through an H^+ channel in the F_0 domain, driving rotation of the c ring of F_0 and, at the same time, dissipating the pH and electrical gradients. Rotation in F_0 causes conformational changes in the three β subunits of F_1 that allow them to bind ADP + P_i, phosphorylate ADP to ATP, and release ATP. One complete rotation of the c ring produces three ATP. [Note: ATP synthase is also called F_1/F_0-ATPase because the enzyme can also catalyze the hydrolysis of ATP to ADP and P_i.]

   a. **Coupling in OXPHOS:** In normal mitochondria, ATP synthesis is coupled to electron transport through the H^+ gradient. Increasing (or decreasing) one process has the same effect on the other. For example, hydrolysis of ATP to ADP and P_i in energy-requiring reactions increases the availability of substrates for ATP synthase and, thus, increases H^+ flow through

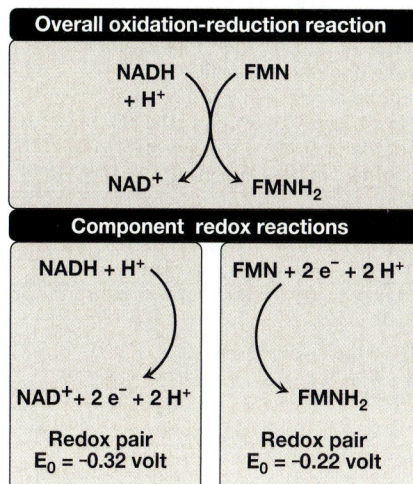

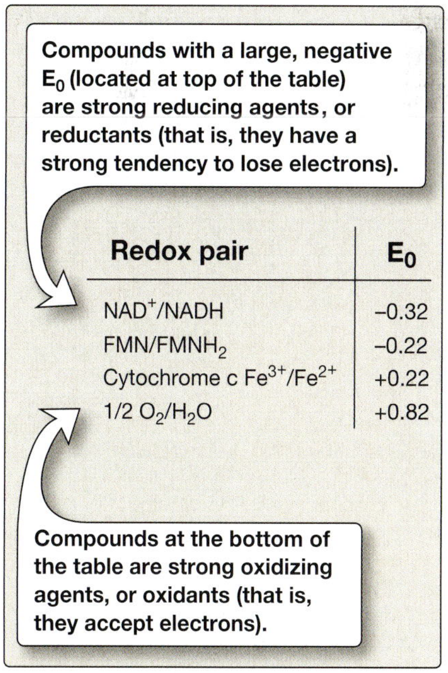

Figure 6.11
Oxidation of NADH by FMN, separated into two component half reactions. NAD(H), nicotinamide adenine dinucleotide; FMN(H$_2$), flavin mononucleotide; e$^-$, electron; H$^+$, proton; E$_0$, standard reduction potential.

Figure 6.12
Standard reduction potentials (E$_0$) of some reactions. NAD(H), nicotinamide adenine dinucleotide; FMN(H$_2$), flavin mononucleotide; Fe, iron.

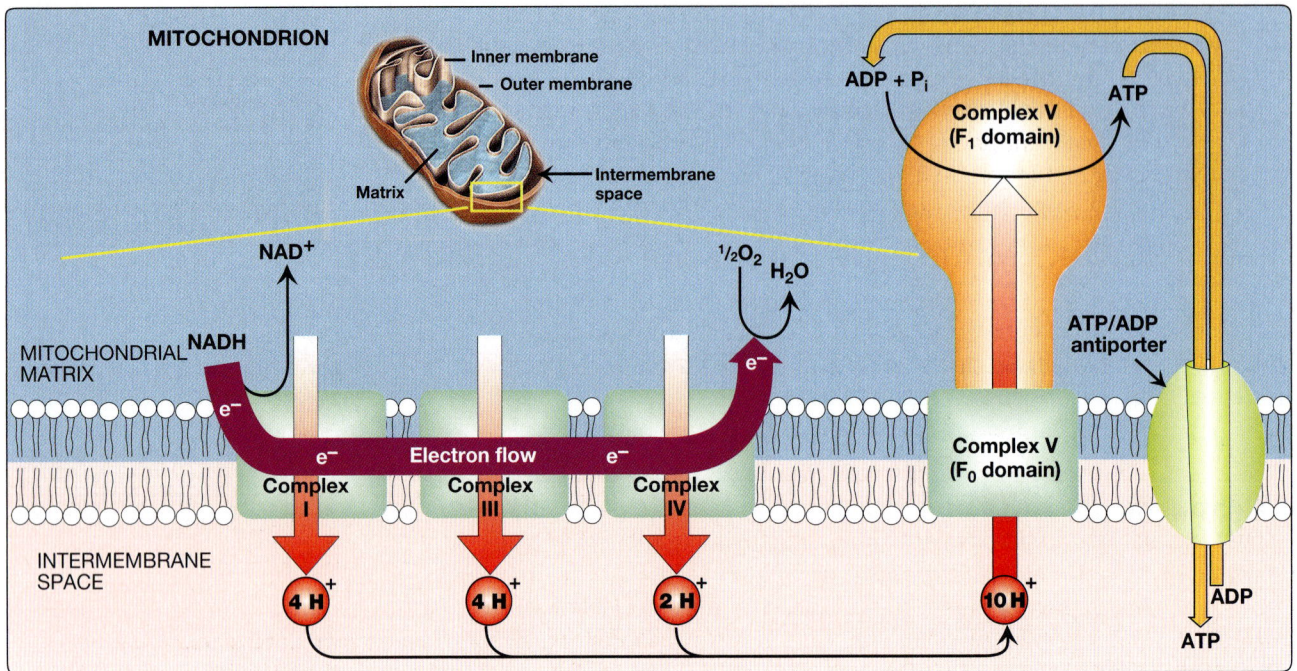

Figure 6.13
Electron transport chain shown in association with proton (H^+) pumping; 10 H^+ are pumped for each nicotinamide adenine dinucleotide (NADH) oxidized. [Note: H^+ are not pumped at complex II.] e^-, electron; complex V, ATP synthase.

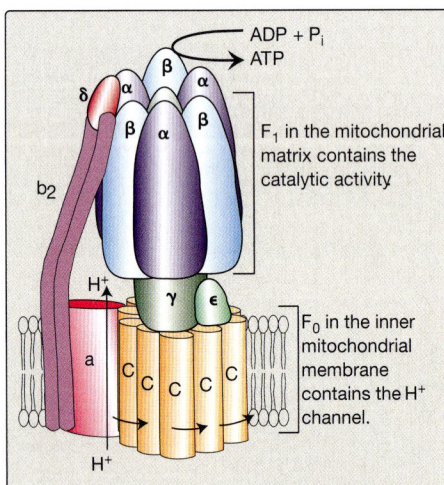

Figure 6.14
ATP synthase (F_1/F_0-ATPase). [Note: The C-ring of vertebrates contains eight subunits. One complete turn of the ring is driven by eight H^+ (protons) moving through the F_0 domain. The resulting conformational changes in the three β subunits of the F_1 domain allow phosphorylation of three adenosine diphosphate (ADP) to three ATP.] P_i, inorganic phosphate.

the enzyme. Electron transport and H^+ pumping by the ETC increase to maintain the H^+ gradient and allow ATP synthesis.

b. **Oligomycin:** This drug binds to the F_0 (hence the letter "o") domain of ATP synthase, closing the H^+ channel and preventing reentry of H^+ into the matrix, thereby inhibiting phosphorylation of ADP to ATP. Because the pH and electrical gradients cannot be dissipated in the presence of this phosphorylation inhibitor, electron transport stops because of the difficulty of pumping any more H^+ against the steep concentration gradient. This dependency of cellular respiration on the ability to phosphorylate ADP to ATP is known as respiratory control and is the consequence of the tight coupling of these processes.

c. **Uncoupling proteins:** Uncoupling proteins (UCPs) occur in the inner mitochondrial membrane of mammals, including humans. These proteins form channels that allow H^+ to reenter the mitochondrial matrix without energy being captured as ATP (Fig. 6.15). The energy is released as heat, a process called nonshivering thermogenesis. UCP1, also called thermogenin, is responsible for heat production in the mitochondria-rich brown adipocytes of mammals. [Note: Cold causes catecholamine-dependent activation of UCP1 expression.] In brown fat, unlike the more abundant white fat, ~90% of its respiratory energy is used for thermogenesis in infants in response to cold. Thus, brown fat is involved in energy expenditure, whereas white fat is involved in energy storage. [Note: Brown fat deposits have recently been found in adults.]

d. **Synthetic uncouplers:** Electron transport and phosphorylation of ADP can also be uncoupled by compounds that shuttle H^+ across the inner mitochondrial membrane, dissipating the gradient. The classic example is 2,4-dinitrophenol, a lipophilic H^+ carrier (ionophore) that readily diffuses through the mitochondrial membrane (Fig. 6.16). This uncoupler causes electron transport to proceed at a rapid rate without establishing an H^+ gradient, much as do the UCP. Again, energy is released as heat rather than being used to synthesize ATP. [Note: In high doses, aspirin and other salicylates uncouple OXPHOS, explaining the hyperthermia (abnormally high body temperature) that accompanies toxic overdoses of these drugs.]

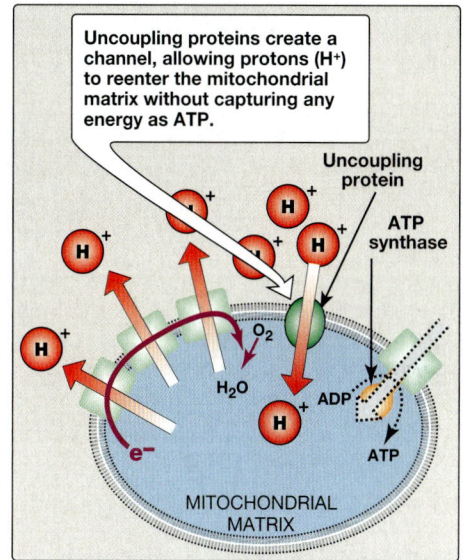

Figure 6.15
Transport of protons across the mitochondrial membrane by an uncoupling protein. ADP, adenosine diphosphate; e^-, electrons.

Clinical Application 6.1: UCP-1 as Possible Obesity Treatment

Thermogenin (UCP-1) is activated in brown adipose tissue in response to cold temperatures and causes subsequent release of free fatty acids. UCP-1 uncouples electron transfer from ATP synthesis, increasing the ETC, and generating heat. Overexpression of UCP-1 could be useful in combating obesity and, possibly, help control diabetes mellitus.

B. **Membrane transport systems**

The inner mitochondrial membrane is impermeable to most charged or hydrophilic substances. However, it contains numerous transport proteins that permit the passage of certain molecules from the cytosol to the mitochondrial matrix.

1. **ATP and ADP transport:** The inner membrane contains specialized carriers to transport ADP and P_i from the cytosol (where ATP is hydrolyzed to ADP in many energy-requiring reactions) into mitochondria, where ATP can be resynthesized. An adenine nucleotide antiporter imports one ADP from the cytosol into the matrix, while exporting one ATP from the matrix into the cytosol (see Fig. 6.13). A symporter cotransports P_i and H^+ from the cytosol into the matrix.

2. **Reducing equivalent transport:** The inner mitochondrial membrane lacks an NADH transporter, and NADH produced in the cytosol (e.g., in glycolysis) cannot directly enter the mitochondrial matrix. However, reducing equivalents of NADH are transported from the cytosol into the matrix using substrate shuttles. In the glycerol 3-phosphate shuttle, two electrons are transferred from NADH to dihydroxyacetone phosphate by cytosolic glycerol 3-phosphate dehydrogenase (Fig. 6.17A). The glycerol 3-phosphate produced is oxidized by the mitochondrial isozyme as FAD is reduced to $FADH_2$. CoQ of the ETC oxidizes the $FADH_2$. Therefore, the glycerol 3-phosphate shuttle results in the synthesis of the equivalent of 1.5 ATP for each cytosolic NADH oxidized. This contrasts with the malate–aspartate shuttle which produces NADH (rather than $FADH_2$) in the mitochondrial matrix, thereby yielding the equivalent of 2.5 ATP for each

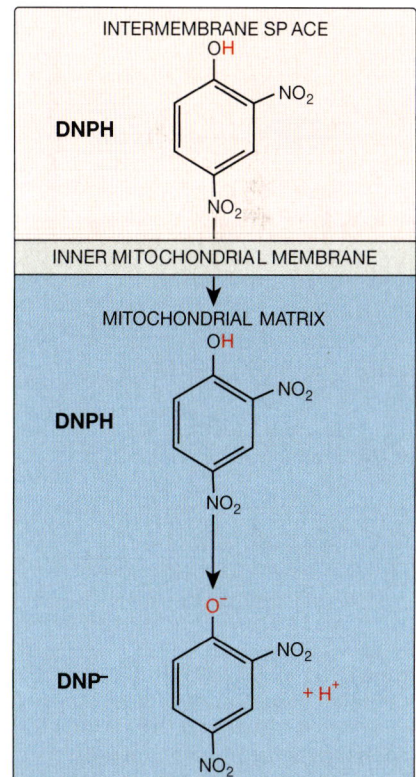

Figure 6.16
2,4-Dinitrophenol (DNP), a proton (H^+) carrier, shown in its reduced (DNPH) and oxidized (DNP^-) forms.

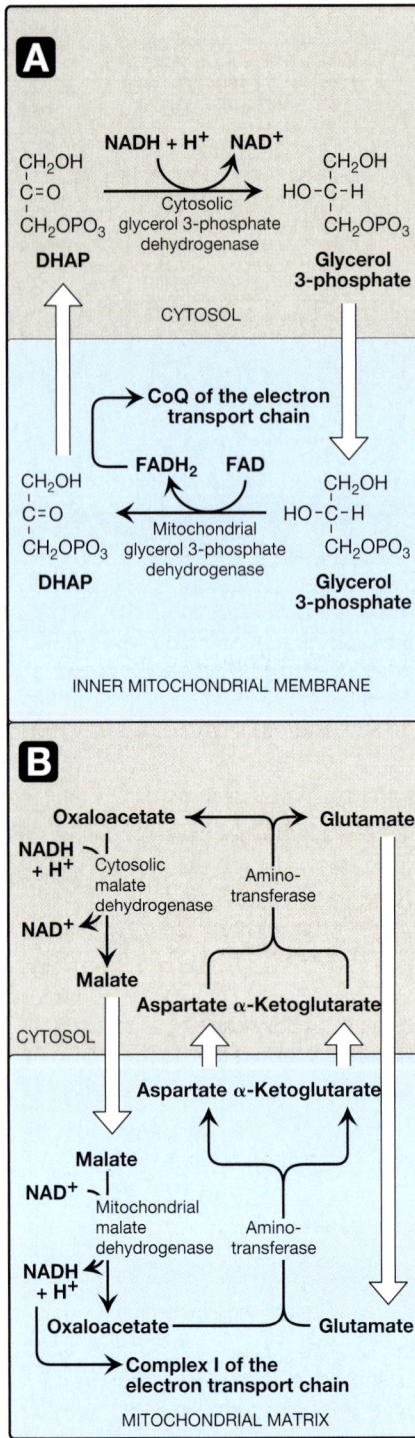

cytosolic NADH oxidized by malate dehydrogenase as oxalo-acetate is reduced to malate (Fig. 6.17B). A transport protein moves malate into the mitochondrial matrix.

C. Inherited defects in oxidative phosphorylation

Of the ~90 polypeptides required for OXPHOS, 13 are encoded by mtDNA and the proteins are synthesized in mitochondria. The remaining ~77 proteins are encoded by nuclear DNA, synthesized in the cytosol, and then transported into mitochondria. Defects in OXPHOS are more likely a result of alterations in mtDNA, which has a mutation rate about 10 times greater than that of nuclear DNA. Cells in tissues with high ATP requirements, including those in brain, nerves, retina, skeletal and heart muscle, and the liver, are particularly vulnerable. Impairments in OXPHOS usually cause lactic acidosis, particularly in the muscles, central nervous system, and retina. Clinical manifestations of OXPHOS disorders include seizures, ophthalmoplegia, muscle weakness, and cardiomyopathy (Table 6.1). Some medications known to affect mitochondrial function should be avoided in persons with mitochondrial disorders.

Table 6.1: Disorders of Mitochondrial Oxidative Phosphorylation

Disease	Characteristics
Kearns–Sayre syndrome	• Weakness or paralysis of eye muscles with drooping eyelids (ptosis), vision loss, cardiac conduction defects, unsteadiness when walking (ataxia), muscle weakness in limbs, kidney problems, deterioration of cognitive function (dementia), and short stature • Features appear before age 20 • Caused by mutation in mtDNA
Leber hereditary optic neuropathy (LHON)	• Bilateral central vision loss caused by retinal detachment • Onset usually in the 20s or 30s • Caused by mitochondrial inheritance along maternal line; however, four times more men are affected than women
Leigh disease	• Severe neurologic disorder that manifests in the first year of life; progressive swallowing problems, poor weight gain, hypotonia, weakness, ataxia, nystagmus, and optic atrophy accompany lactic acidosis • Death usually occurs between ages 2 and 3 years from respiratory failure • Caused by mutations in nuclear or mtDNA
Mitochondrial encephalomyopathy, lactic acidosis, and stroke-like episodes (MELAS)	• Progressive neurodegeneration • Repeated episodes of lactic acidosis and myopathy • Cells often contain mutant and wild-type mtDNA; expression is variable
Myoclonic epilepsy with ragged-red fibers (MERRF)	• Progressive condition • Uncontrolled muscle contractions, dementia, ataxia, and myopathy • Caused by mutation in mtDNA; expression is variable
Neuropathy, ataxia, and retinitis pigmentosa (NARP) syndrome	• Progressive condition • Sensory neuropathy with numbness or tingling in the extremities, muscle weakness, ataxia and vision loss, cognitive decline, and seizures • Caused by mutation in mtDNA altering ATP synthase and reducing ability to make ATP

Figure 6.17
Substrate shuttles for the transport of reducing equivalents across the inner mitochondrial membrane. **A.** Glycerol 3-phosphate shuttle. **B.** Malate–aspartate shuttle. DHAP, dihydroxyacetone phosphate; NAD(H), nicotinamide adenine dinucleotide; H^+, proton; FAD(H_2), flavin adenine dinucleotide; CoQ, coenzyme Q.

[Note: mtDNA is maternally inherited because mitochondria from the sperm do not survive the fertilization process and only those from the oocyte survive in the developing embryo and into the adult individual.]

D. Mitochondria and apoptosis

The process of apoptosis (programmed cell death) may be initiated through the intrinsic or mitochondrial-mediated pathway in response to irreparable damage within the cell. During this process, channel proteins including Bax and Bak are inserted in the outer mitochondrial membrane and allow cytochrome c to leave the intermembrane space and enter the cytosol. There, cytochrome c associates with proapoptotic factors to form a structure called the apoptosome, which then activates a family of proteolytic enzymes (caspases). These proteases in turn cause cleavage of key intracellular proteins and results in the morphologic and biochemical changes characteristic of cell death via apoptosis.[1]

---

[1]For further discussion of apopotosis, see *LIR Cell and Molecular Biology*, 3e Chapter 23.

VII. CHAPTER SUMMARY

- The **change** in **free energy** (ΔG) occurring during a reaction predicts the **direction** in which that reaction will spontaneously proceed (Fig. 6.18).

- If ΔG is **negative**, the reaction is **spontaneous** as written. If ΔG is **positive**, the reaction is **not spontaneous**. If $\Delta G = 0$, the reaction is in **equilibrium**.

- The ΔG of the forward reaction is equal in magnitude but opposite in sign to that of the reverse reaction.

- Reactions with a large, positive ΔG are made possible by **coupling** with those that have a large, negative ΔG such as **adenosine triphosphate (ATP) hydrolysis**.

- The reduced coenzymes **nicotinamide adenine dinucleotide (NADH)** and **flavin adenine dinucleotide (FADH$_2$)** each donate a pair of electrons to a specialized set of **electron carriers**, consisting of **flavin mononucleotide (FMN)**; Fe–S centers, coenzyme Q (**CoQ**); and a series of heme-containing **cytochromes**, collectively called the **electron transport chain (ETC)**.

- This pathway is present in the **inner mitochondrial membrane** and is the final common pathway by which electrons derived from different fuels of the body flow to O_2, which has a large, positive **reduction potential (E$_0$)**, reducing it to water.

- The terminal cytochrome, **cytochrome c oxidase**, is the only cytochrome able to bind O_2.

- Electron transport results in the **pumping of protons (H$^+$)** across the inner mitochondrial membrane from the matrix to the intermembrane space, 10 H$^+$ per NADH oxidized.

- This process creates **electrical** and **pH gradients** across the inner mitochondrial membrane. After H$^+$ have been transferred to the cytosolic side of the membrane, they reenter the matrix by passing through the **F$_0$** H$^+$ channel in **ATP synthase (complex V)**, dissipating the pH and electrical gradients and causing conformational changes in the **F$_1$** β subunits of the synthase that result in the synthesis of ATP from adenosine diphosphate (ADP) + inorganic phosphate.

- **Electron transport** and **phosphorylation** are tightly coupled in **oxidative phosphorylation**. These processes can be uncoupled by uncoupling proteins (UCPs), such as **UCP1 (thermogenin)** of the inner mitochondrial membrane of brown adipocytes and by synthetic compounds such as **2,4-dinitrophenol** and **aspirin**, all of which dissipate the H$^+$ gradient.

- In uncoupled mitochondria, the energy produced by electron transport is released as **heat** rather than being used to synthesize ATP.

- Defects in oxidative phosphorylation are usually a result of alterations in mitochondrial (mt)DNA. Impairments in oxidative phosphorylation usually cause **lactic acidosis**, particularly in the muscles, central nervous system, and retina. Clinical manifestations of oxidative phosphorylation disorders include seizures, ophthalmoplegia, muscle weakness, and cardiomyopathy.

- The release of **cytochrome c** from mitochondria into the cytosol stimulates the generation of the apoptosome and subsequent activation of proteolytic caspases that results in **apoptotic cell death**.

Oxidative phosphorylation

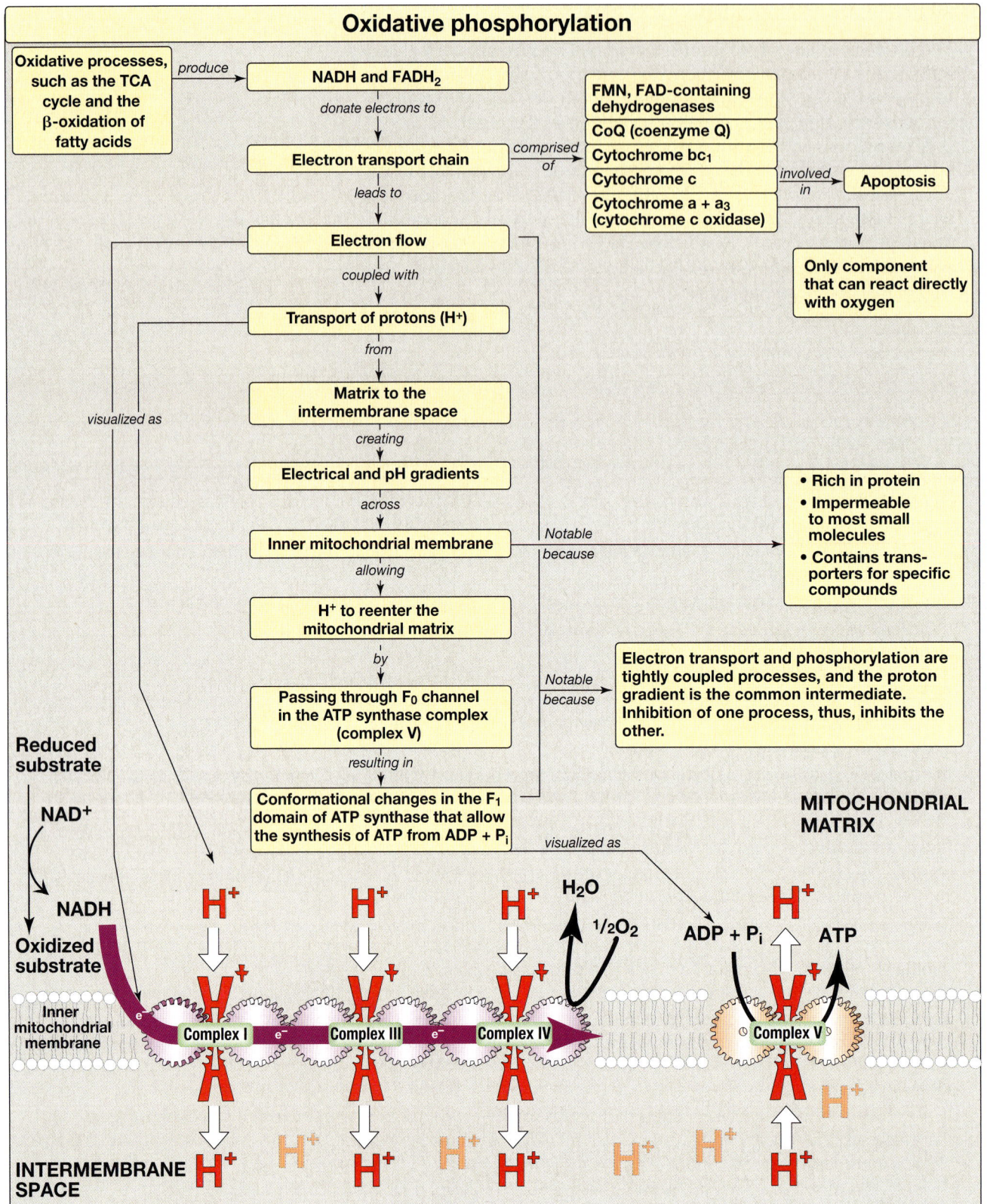

Oxidative processes, such as the TCA cycle and the β-oxidation of fatty acids → *produce* → **NADH and FADH$_2$**

donate electrons to

Electron transport chain — *comprised of* →

- **FMN, FAD-containing dehydrogenases**
- **CoQ (coenzyme Q)**
- **Cytochrome bc$_1$**
- **Cytochrome c** — *involved in* → **Apoptosis**
- **Cytochrome a + a$_3$ (cytochrome c oxidase)**

Only component that can react directly with oxygen

leads to

Electron flow

coupled with

Transport of protons (H$^+$)

from

Matrix to the intermembrane space

creating

Electrical and pH gradients

across

Inner mitochondrial membrane — *Notable because* →
- Rich in protein
- Impermeable to most small molecules
- Contains trans-porters for specific compounds

allowing

H$^+$ to reenter the mitochondrial matrix

by

Passing through F$_0$ channel in the ATP synthase complex (complex V) — *Notable because* → Electron transport and phosphorylation are tightly coupled processes, and the proton gradient is the common intermediate. Inhibition of one process, thus, inhibits the other.

resulting in

Conformational changes in the F$_1$ domain of ATP synthase that allow the synthesis of ATP from ADP + P$_i$

visualized as

MITOCHONDRIAL MATRIX

visualized as

Reduced substrate

NAD$^+$

NADH

Oxidized substrate

Inner mitochondrial membrane

H$^+$ H$^+$ H$^+$ H$_2$O $1/2$ O$_2$ ADP + P$_i$ ATP   H$^+$

Complex I   e$^-$ Complex III   e$^-$ Complex IV   Complex V

INTERMEMBRANE SPACE   H$^+$ H$^+$ H$^+$ H$^+$ H$^+$ H$^+$ H$^+$

Figure 6.18
Key concept map for oxidative phosphorylation (OXPHOS). [Note: Electron [e$^-$] flow and ATP synthesis are shown as sets of interlocking gears to emphasize coupling.] TCA, tricarboxylic acid; NAD(H), nicotinamide adenine dinucleotide; FAD(H$_2$), flavin adenine dinucleotide; FMN, flavin mononucleotide; ADP, adenosine diphosphate.

Study Questions

Choose the ONE best answer.

6.1. 2,4-Dinitrophenol (DNP), an uncoupler of oxidative phosphorylation, was used as a weight-loss agent in the 1930s, but reports of fatal overdoses led to its discontinuation in 1939. Which of the following would most likely contribute to the adverse effects experienced by individuals taking 2,4-DNP?

A. ATP levels in the mitochondria were greater than normal.

B. Body temperature became elevated as a result of hypermetabolism.

C. Delivery of electrons to O_2 was blocked.

D. The H+ gradient across the inner mitochondrial membrane was greater than normal.

E. The rate of electron transport became abnormally low.

Correct answer = B. When phosphorylation is uncoupled from electron flow, a decrease in the proton gradient across the inner mitochondrial membrane and impaired ATP synthesis are expected. In an attempt to compensate for this defect in energy capture, metabolism and electron flow to oxygen are increased. This hypermetabolism will be accompanied by elevated body temperature because the energy in fuels is largely wasted, appearing as heat. Cyanide, not DNP, blocks the electron transport chain.

6.2. Which of the following has the strongest tendency to gain electrons?

A. Coenzyme Q

B. Cytochrome c

C. Flavin adenine dinucleotide

D. Nicotinamide adenine dinucleotide

E. Oxygen

Correct answer = E. Oxygen is the terminal acceptor of electrons in the electron transport chain (ETC). Electrons flow down the ETC to oxygen because it has the highest (most positive) reduction potential (E_0). The other choices precede oxygen in the ETC and have lower E_0 values.

6.3. Why and how does the malate–aspartate shuttle move nicotinamide adenine dinucleotide (NADH)-reducing equivalents from the cytosol to the mitochondrial matrix?

There is no transporter for NADH in the inner mitochondrial membrane. However, cytoplasmic NADH can be oxidized to NAD+ by malate dehydrogenase as oxaloacetate (OAA) is reduced to malate. The malate is transported across the inner membrane to the matrix, where the mitochondrial isozyme of malate dehydrogenase oxidizes it to OAA as mitochondrial NAD+ is reduced to NADH. This NADH can be oxidized by complex I of the electron transport chain, generating three ATP through the coupled processes of oxidative phosphorylation.

6.4. Carbon monoxide (CO) binds to and inhibits complex IV of the electron transport chain. What effect, if any, should this respiratory inhibitor have on phosphorylation of adenosine diphosphate (ADP) to ATP?

Inhibition of electron transport by respiratory inhibitors such as CO results in an inability to maintain the proton (H+) gradient. Therefore, phosphorylation of ADP to ATP is inhibited, as are ancillary reactions such as calcium uptake by mitochondria, because they also require the H+ gradient.

6.5. Defects in oxidative phosphorylation most often develop from:

A. acquired damage to autosomal genes.

B. inheritance of a mutation on mtDNA.

C. mutations inherited from their father.

D. X-linked inheritance from their mother.

Correct answer = B. Defects in oxidative phosphorylation are more likely a result of alterations in mtDNA, which has a mutation rate about 10 times greater than that of nuclear DNA. Mitochondria and mtDNA are inherited exclusively from the mother. X-linked inheritance is of nuclear DNA, not mtDNA.

6.6. A 25-year-old man presents to the emergency department with increasing weakness and confusion, followed by dizziness and a seizure. He is diagnosed with cyanide poisoning. His signs and symptoms resulted from his inability to synthesize cellular ATP and instead using oxygen, owing to cyanide binding to and impairing the function of:

A. ATP synthase.

B. complex III.

C. complex IV.

D. NADH dehydrogenase.

E. ubiquinone.

Correct answer = C. Complex IV, also called cytochrome c oxidase, is the only electron carrier in which the heme iron has an available coordination site that can react directly with O_2. When cyanide binds to complex IV, it prevents the normal function of complex IV and halts the electron transport chain. None of the other electron carriers interacts directly with O_2.

6.7. An 18-year-old woman presents for evaluation of recent-onset ptosis (drooping eyelids) and vision loss. She also appears unsteady when walking. Examination reveals muscle weakness in her limbs, and testing shows early stages of renal impairment. A mutation in mtDNA is suspected. Assuming the diagnosis is supported by additional test findings, this patient may be expected to also develop:

A. deterioration of cognitive function.

B. lactic acidosis.

C. progressive swallowing problems.

D. retinal detachment.

E. sensory neuropathy.

Correct answer = A. This patient has signs and symptoms of Kearns–Sayre syndrome, and dementia (deterioration of cognitive function) occurs. Features appear before age 20. Lactic acidosis occurs in Leigh disease, a severe neurologic disorder that manifests in the first year of life and includes progressive swallowing problems. Bilateral vision loss and retinal detachment are seen in Leber hereditary optic neuropathy, and neuropathy, ataxia, and retinitis pigmentosa syndrome is a progressive condition that includes sensory neuropathy.

6.8. Drugs that can cause overexpression of uncoupling protein ([UCP-1] thermogenin) might cause:

A. a decrease in the electron transport chain.

B. development of diabetes mellitus type 2.

C. increased coupling of electron transfer from ATP synthesis.

D. less heat to be generated in brown adipose tissue.

E. weight loss.

Correct answer = A. Thermogenin (UCP-1) is normally activated in brown adipose tissue exposed to cold temperatures and causes the release of free fatty acids. UCP-1 uncouples electron transfer from ATP synthesis, increasing the electron transport chain, and generating heat. Overexpression of UCP-1 could be useful in combating obesity by promoting weight loss and, possibly, help control diabetes mellitus.

Introduction to Carbohydrates

7

I. OVERVIEW

Carbohydrates are the most abundant organic molecules in nature. They have a wide range of functions, including providing a significant fraction of the dietary calories for most organisms and acting as a storage form of energy in the body and as cell membrane components that mediate some forms of intercellular communication. Carbohydrates also serve as structural components of many organisms, including the cell walls of bacteria, the exoskeleton of insects, and the fibrous cellulose of plants. The empiric formula for many of the simpler carbohydrates is $(CH_2O)_n$, where *n* is at least 3, hence the name "hydrate of carbon."

II. CLASSIFICATION AND STRUCTURE

Monosaccharides or simple sugars are classified according to the number of carbon atoms they contain. Examples of monosaccharides commonly found in humans are listed in Figure 7.1. They can also be classified by the type of carbonyl group they contain. Carbohydrates with an aldehyde as their carbonyl group are called "aldoses," such as glyceraldehyde, and those with a keto as their carbonyl group are called "ketoses," such as dihydroxyacetone (Fig. 7.2). Carbohydrates that have a free carbonyl group have the suffix "-ose." [Note: Ketoses have an additional "ul" in their suffix such as xylulose. There are exceptions, such as fructose, to this rule.] Monosaccharides can be linked by glycosidic bonds to create larger structures (Fig. 7.3). Disaccharides contain two monosaccharide units, oligosaccharides contain 3 to 10 monosaccharide units, and polysaccharides contain more than 10 monosaccharide units and can be hundreds of sugar units in length.

A. Isomers and epimers

Compounds with the same chemical formula but different structures are isomers of each other. For example, fructose, glucose, mannose, and galactose all have the same chemical formula, $C_6H_{12}O_6$, with different structures. Carbohydrate isomers that differ in configuration around only one specific carbon atom (with the exception of the carbonyl carbon, see Section C. 1.) are defined as epimers of each other. For example, glucose and galactose are C-4 epimers because their structures differ only in the position of the –OH (hydroxyl) group at carbon 4. [Note: The carbons in sugars are numbered beginning

Figure 7.1
Examples of monosaccharides found in humans, classified according to the number of carbons they contain.

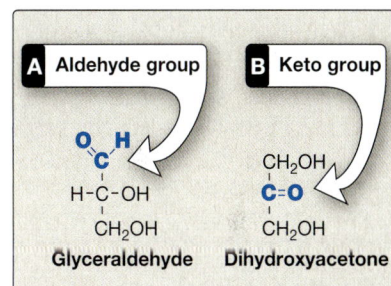

GENERIC NAMES	EXAMPLES
3 Carbons: trioses	Glyceraldehyde
4 Carbons: tetroses	Erythrose
5 Carbons: pentoses	Ribose
6 Carbons: hexoses	Glucose
7 Carbons: heptoses	Sedoheptulose
9 Carbons: nonoses	Neuraminic acid

Figure 7.2
Examples of an aldose (**A**) and a ketose (**B**) sugar.

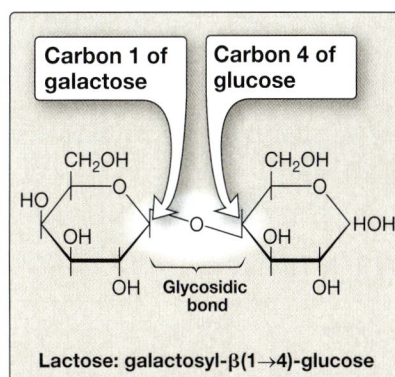

Figure 7.3
Glycosidic bond between two hexoses producing a disaccharide.

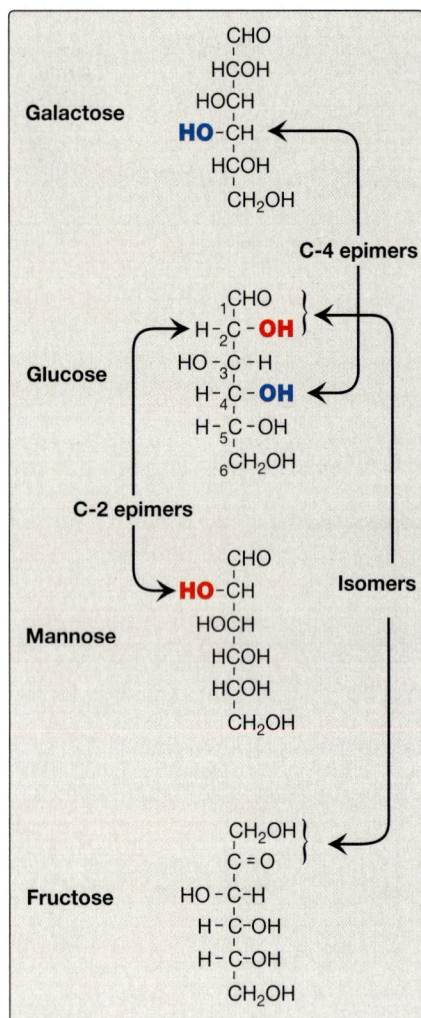

Figure 7.4
Carbon-2 (C-2) and C-4 epimers and an isomer of glucose.

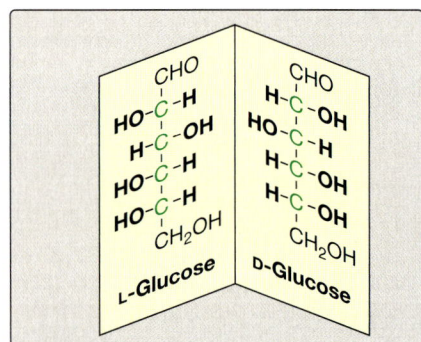

Figure 7.5
Enantiomers (mirror images) of glucose. Designation of D- and L- is by comparison to a triose, glyceraldehyde. [Note: The asymmetric carbons are shown in *green*.]

at the end that contains the carbonyl carbon (i.e., the aldehyde or keto group), as shown in Fig. 7.4.] Glucose and mannose are C-2 epimers. However, because galactose and mannose differ in the position of –OH groups at two carbons (carbons 2 and 4), they are isomers rather than epimers (see Fig. 7.4).

B. Enantiomers

A special type of isomerism is found in the pairs of structures that are mirror images of each other. These mirror images are called enantiomers, and the two members of the pair are designated as a D- and an L-sugar (Fig. 7.5). Most sugars in humans are D-isomers. In the D-isomeric form, the –OH group on the asymmetric carbon (a carbon linked to four different atoms or groups) farthest from the carbonyl carbon is on the right, whereas, in the L-isomer, it is on the left. Enzymes are specific for either the D- or the L-form, and enzymes known as isomerases interconvert D- and L-isomers.

C. Monosaccharide cyclization

Less than 1% of each of the monosaccharides with five or more carbons exists in the open-chain (acyclic) form in solution. Rather, they are predominantly found in a ring or cyclic form, in which the aldehyde (or keto) group has reacted with a hydroxyl group on the same sugar, making the carbonyl carbon (carbon 1 for an aldose, carbon 2 for a ketose) asymmetric. This asymmetric carbon is referred to as the anomeric carbon.

1. **Anomers:** Creation of an anomeric carbon (the former carbonyl carbon) generates a new pair of isomers, the α and β configurations of the sugar (e.g., α-D–glucopyranose and β-D–glucopyranose), as shown in Figure 7.6, that are anomers of each other. [Note: In the α configuration, the –OH group on the anomeric carbon projects to the same side as the ring in a modified Fischer projection formula (see Fig. 7.6A) and is trans to the CH_2OH group in a Haworth projection formula (see Fig. 7.6B). The α and β forms are not mirror images, and they are referred to as diastereomers.] Enzymes can distinguish between these two structures and use one or the other preferentially. For example, glycogen is synthesized from α-D–glucopyranose, whereas cellulose is synthesized from β-D–glucopyranose. The cyclic α and β anomers of a sugar in solution spontaneously (but slowly) form an equilibrium mixture, a process known as mutarotation (see Fig. 7.6). [Note: For glucose, the α form makes up 36% of the mixture.]

2. **Reducing sugars:** If the hydroxyl group on the anomeric carbon of a cyclized sugar is not linked to another compound by a glycosidic bond (see Section E.), the ring can open. The sugar can act as a reducing agent and is termed a reducing sugar. Such sugars can react with chromogenic agents (e.g., the Benedict reagent) causing the reagent to be reduced and take on a color, as the aldehyde group of the acyclic sugar is oxidized to a carboxyl group. All monosaccharides, but not all disaccharides, are reducing sugars. [Note: Fructose, a ketose, is a reducing sugar because it can be isomerized to an aldose.]

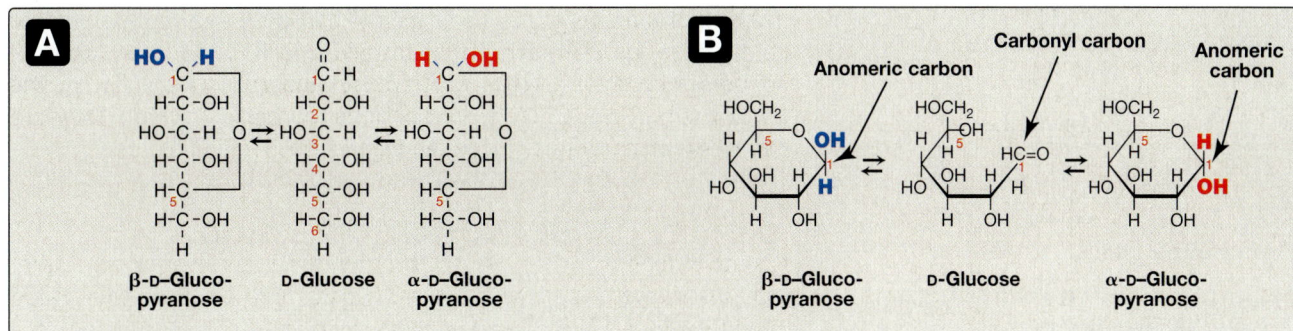

Figure 7.6
A. Interconversion (mutarotation) of the α and β anomeric forms of glucose shown as modified Fischer projection formulas. **B.** The interconversion shown as Haworth projection formulas. [Note: A sugar with a six-membered ring (5 C + 1 O) is termed a pyranose, whereas one with a five-membered ring (4 C + 1 O) is a furanose. Virtually all glucose in solution is in the pyranose form.]

A colorimetric or color change test on a urine specimen can detect the presence of reducing sugars, which are not normally present in urine. A positive test result can indicate an underlying pathology. Glucose is the sugar most commonly tested for in urine, and human urine can also contain small amounts of galactose, lactose, fructose, xylose, and other pentoses.

D. Monosaccharide joining

Monosaccharides can be joined together to form disaccharides, oligosaccharides, and polysaccharides. Important disaccharides include lactose (galactose + glucose), sucrose (glucose + fructose), and maltose (glucose + glucose). Important polysaccharides include branched glycogen (from animal sources) and starch (plant sources) and unbranched cellulose (plant sources). Each is a polymer of glucose.

E. Glycosidic bonds

The bonds that link sugars are called glycosidic bonds. They are formed by enzymes known as glycosyltransferases that use nucleotide sugars (activated sugars) such as uridine diphosphate glucose as substrates. Glycosidic bonds between sugars are named according to the numbers of the connected carbons and with regard to the position of the anomeric hydroxyl group of the first sugar involved in the bond. If this anomeric hydroxyl is in the α configuration, the linkage is an α-bond. If it is in the β configuration, the linkage is a β-bond. Lactose, for example, is synthesized by forming a glycosidic bond between carbon 1 of β-galactose and carbon 4 of glucose. Therefore, the linkage is a β(1→4) glycosidic bond (see Fig. 7.3). [Note: Because the anomeric end of the glucose residue is not involved in the glycosidic linkage, it (and, therefore, lactose) remains a reducing sugar.]

F. Carbohydrate linkage to noncarbohydrates

Carbohydrates can be attached by glycosidic bonds to noncarbohydrate structures, including purine and pyrimidine bases in nucleic

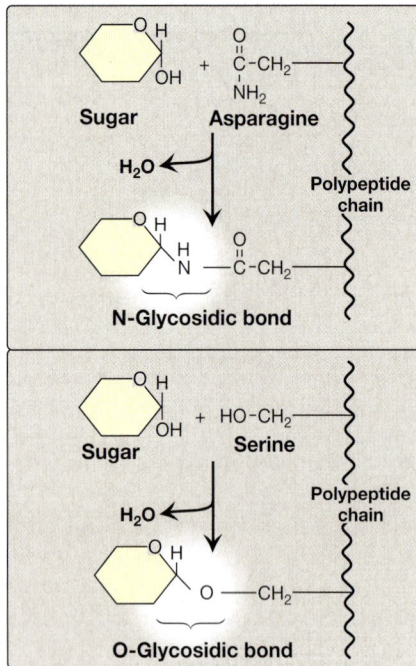

Figure 7.7
Examples of N- and O-glycosidic bonds in glycoproteins.

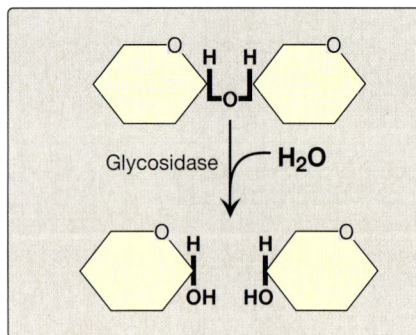

Figure 7.8
Hydrolysis of a glycosidic bond.

acids, aromatic rings such as those found in steroids, proteins, and lipids. If the group on the noncarbohydrate molecule to which the sugar is attached is an $-NH_2$ group, the bond is called an N-glycosidic link. If the group is an $-OH$, the bond is an O-glycosidic link (Fig. 7.7). [Note: All sugar–sugar glycosidic bonds are O-type linkages.]

III. DIETARY CARBOHYDRATE DIGESTION

The principal sites of dietary carbohydrate digestion are the mouth and intestinal lumen. This digestion is rapid and is catalyzed by enzymes known as glycoside hydrolases (glycosidases) that hydrolyze glycosidic bonds (Fig. 7.8). Because little monosaccharide is present in diets of mixed animal and plant origin, the enzymes are primarily endoglycosidases that hydrolyze polysaccharides and oligosaccharides and disaccharidases that hydrolyze tri- and disaccharides into their reducing sugar components. Glycosidases are usually specific for the structure and configuration of the glycosyl residue to be removed as well as for the type of bond to be broken. The final products of carbohydrate digestion are the monosaccharides glucose, galactose, and fructose that are absorbed by cells (enterocytes) of the small intestine.

A. Salivary α-amylase

The major dietary polysaccharides consumed by humans are of plant (starch, composed of amylose and amylopectin) and animal (glycogen) origin. Salivary α-amylase is secreted into saliva by the salivary glands, and, during mastication or chewing, it acts briefly on dietary starch and glycogen, hydrolyzing random $\alpha(1 \rightarrow 4)$ bonds. [Note: There are both $\alpha(1 \rightarrow 4)$- and $\beta(1 \rightarrow 4)$-endoglucosidases in nature, but humans do not produce the latter. Therefore, we are unable to digest cellulose, a carbohydrate of plant origin containing $\beta(1 \rightarrow 4)$ glycosidic bonds between glucose residues.] Because branched amylopectin and glycogen also contain $\alpha(1 \rightarrow 6)$ bonds, which α-amylase cannot hydrolyze, the digest resulting from its action contains a mixture of short, branched, and unbranched oligosaccharides known as dextrins (Fig. 7.9). [Note: Disaccharides are also present as they, too, are resistant to amylase.] Carbohydrate digestion halts temporarily in the stomach, because the high acidity inactivates salivary α-amylase.

B. Pancreatic α-amylase

When the acidic stomach contents reach the small intestine, they are neutralized by bicarbonate secreted by the pancreas, and pancreatic α-amylase is secreted by pancreatic exocrine glands and continues the process of starch digestion.

C. Intestinal disaccharidases

The final digestive processes occur primarily at the mucosal lining of the duodenum and upper jejunum and include the action of several disaccharidases (see Fig. 7.9). For example, isomaltase cleaves the $\alpha(1 \rightarrow 6)$ bond in isomaltose, and maltase cleaves the $\alpha(1 \rightarrow 4)$ bond in maltose and maltotriose, each producing glucose. Sucrase cleaves the α(1→2) bond in sucrose, producing glucose and fructose, and lactase (β-galactosidase) cleaves the β(1→4) bond in lactose, producing galactose and glucose. [Note: The substrates for isomaltase are broader than its name suggests, and it hydrolyzes the majority of maltose.] Trehalose, an α(1→1) disaccharide of glucose found in mushrooms and other fungi, is cleaved by trehalase. These enzymes are transmembrane proteins of the brush border on the luminal (apical) surface of the enterocytes.

> Sucrase and isomaltase are both enzymatic activities of a single protein that is cleaved into two functional subunits that remain associated in the cell membrane and form the sucrase–isomaltase (SI) complex. In contrast, maltase is one of two enzymic activities of the single membrane protein maltase–glucoamylase (MGA) that is not cleaved. Its second enzymatic activity, glucoamylase, cleaves α(1→4) glycosidic bonds in dextrins.

D. Intestinal absorption of monosaccharides

The upper jejunum absorbs the bulk of the monosaccharide products of digestion. However, different sugars have different mechanisms of absorption (Fig. 7.10). For example, galactose and glucose are taken into enterocytes by secondary active transport that requires a concurrent uptake (symport) of sodium (Na^+) ions. The transport protein is the sodium-dependent glucose cotransporter 1 (SGLT-1), that transports sodium with its concentration gradient and glucose against its gradient. Glucose transport via SGLT-1 is driven by the Na^+ gradient created by the Na^+/potassium [K^+]-ATPase that moves Na^+ out of enterocyte and K^+ in (see Fig. 7.10).[1] Fructose absorption utilizes an energy- and Na^+-independent monosaccharide transporter (GLUT-5). All three monosaccharides (fructose, galactose, and glucose) are subsequently transported from enterocytes into the portal circulation by yet another transporter, GLUT-2. [Note: See also Chapter 8 for a more detailed discussion of these transporters.]

E. Abnormal degradation of disaccharides

The overall process of carbohydrate digestion and absorption is so efficient in healthy persons that, ordinarily, all digestible dietary carbohydrate is absorbed by the time the ingested material reaches the lower jejunum. However, because only monosaccharides are absorbed, any deficiency (genetic or acquired) that a person has in a specific disaccharidase activity of the intestinal mucosa will cause the passage of undigested carbohydrate into the large intestine. The presence of this osmotically active material causes water to

[1]See LIR Cell and Molecular Biology, 3e Chapters 14 and 15.

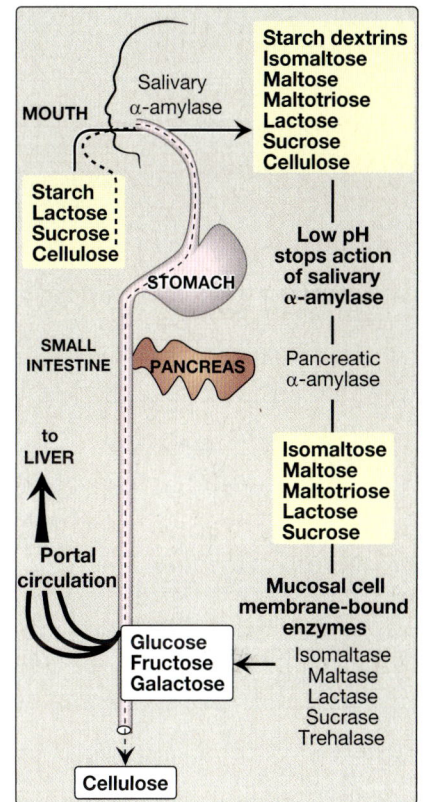

Figure 7.9
Digestion of carbohydrates. [Note: Indigestible cellulose enters the colon and is excreted.]

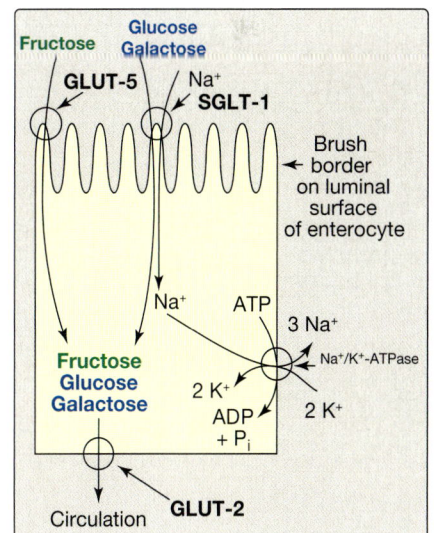

Figure 7.10
Absorption by enterocytes of the monosaccharide products of carbohydrate digestion. GLUT, glucose transporter; K^+, potassium; SGLT-1, sodium (Na^+)-dependent glucose cotransporter.

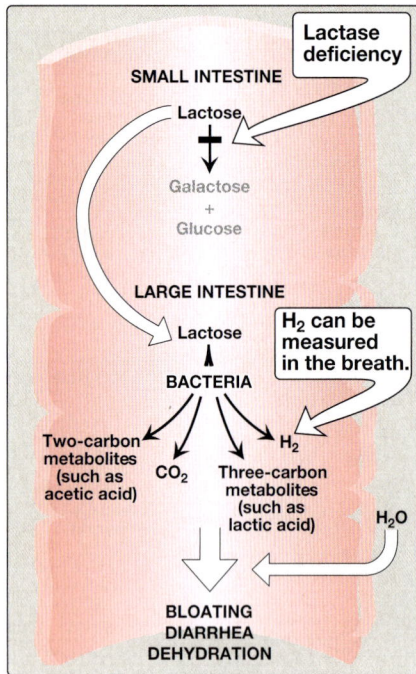

Figure 7.11
Abnormal lactose metabolism. CO_2, carbon dioxide; H_2, hydrogen gas.

be drawn from the mucosa into the large intestine, with resulting osmotic diarrhea. This is reinforced by the bacterial fermentation of the remaining carbohydrate to two- and three-carbon compounds, which are also osmotically active, plus large volumes of carbon dioxide and hydrogen gas (H_2), causing abdominal cramps, diarrhea, and flatulence.

1. **Digestive enzyme deficiencies:** Genetic deficiencies of the individual disaccharidases result in disaccharide intolerance. Alterations in disaccharide degradation can also be caused by a variety of intestinal diseases, malnutrition, and drugs that injure the mucosa of the small intestine. For example, brush border enzymes are rapidly lost in normal individuals with severe diarrhea, causing a temporary, acquired enzyme deficiency. Therefore, patients experiencing or recovering from such a disorder cannot drink or eat significant amounts of dairy products or sucrose without exacerbating the diarrhea.

2. **Lactose intolerance:** More than 60% of the world's adults experience lactose malabsorption because they lack the enzyme lactase (Fig. 7.11). Those of Northern European heritage are most likely to maintain the ability to digest lactose into adulthood. Up to 90% of adults of African or Asian descent are lactase deficient. Consequently, they are less able to metabolize lactose than individuals of Northern European origin. The age-dependent loss of lactase activity starting at approximately 2 years of age results from a reduction in the amount of enzyme produced. It is thought to be caused by small variations in the DNA sequence of a region on chromosome 2 that controls expression of the gene for lactase, also on chromosome 2. Treatment for this disorder is to reduce the consumption of most dairy products. To ensure adequate calcium intake, affected persons may be encouraged to use dietary supplements and to eat yogurt and some cheeses, in which bacterial action and the cheese aging process has decreased lactose content, as well as calcium-containing green vegetables, such as broccoli. The use of lactase-treated products and lactase in pill form prior to eating dairy can also be helpful. Rare cases of congenital lactase deficiency are known.

3. **SI deficiency:** SI deficiency results in intolerance of ingested sucrose. This condition was previously considered quite rare, but now it is recognized that up to 9% of Americans of European descent are estimated to be affected by a form of SI deficiency.

   Initially considered to be exclusively an autosomal-recessive disorder, those with one mutation (heterozygous carriers) sometimes express disease manifestations. Twenty-five different mutations in the human sucrose gene are known. Individuals homozygous for mutations express congenital SI deficiency and experience osmotic diarrhea, mild steatorrhea, irritability, and vomiting after consuming sucrose. Heterozygous carriers often have symptoms including chronic diarrhea, abdominal pain, and bloating. Treatment includes the dietary restriction of sucrose and enzyme replacement therapy.

4. **Diagnosis of enzyme deficiencies:** Identification of a specific enzyme deficiency can be obtained by performing oral tolerance tests with the individual disaccharides. Measurement of H_2 in the breath is a reliable test for determining the amount of ingested carbohydrate not absorbed by the body, but which is metabolized instead by the intestinal flora (see Fig. 7.11).

IV. CHAPTER SUMMARY

- **Monosaccharides** containing an aldehyde group are called **aldoses**, and those with a keto group are called **ketoses** (Fig. 7.12).

- **Disaccharides, oligosaccharides,** and **polysaccharides** consist of monosaccharides linked by **glycosidic bonds**.

- Compounds with the same chemical formula but different structures are called **isomers**.

- Two monosaccharide isomers differing in configuration around one specific carbon atom (not the carbonyl carbon) are defined as **epimers**.

- In **enantiomers** (mirror images), the members of the sugar pair are designated as D- and L-**isomers**. When the aldehyde group on an acyclic sugar that is oxidized as a chromogenic agent is reduced, that sugar is a **reducing sugar**.

- When a sugar cyclizes, an **anomeric carbon** is created from the carbonyl carbon of the aldehyde or keto group. The sugar can have two configurations, forming α or β **anomers**.

- A sugar can have its anomeric carbon linked to an $-NH_2$ or an $-OH$ group on another structure through **N-** and **O-glycosidic bonds**, respectively.

- **Salivary α-amylase** initiates the digestion of **dietary polysaccharides** (e.g., starch or glycogen), producing oligosaccharides. **Pancreatic α-amylase** continues the process. The final digestive processes occur at the **mucosal lining** of the **small intestine**.

- Several disaccharidases (e.g., **lactase [β-galactosidase], sucrase, isomaltase,** and **maltase**) produce monosaccharides (glucose, galactose, and fructose). These enzymes are **transmembrane proteins** of the luminal **brush border** of **intestinal mucosal cells (enterocytes)**.

- Absorption of the monosaccharides requires specific **transporters**. If carbohydrate degradation is deficient (as a result of heredity, disease, or drugs that injure the intestinal mucosa), undigested carbohydrate will pass into the large intestine, where it can cause **osmotic diarrhea**.

- Bacterial fermentation of the material produces large volumes of carbon dioxide and hydrogen gas, causing abdominal cramps, diarrhea, and flatulence. **Lactose intolerance**, primarily caused by the age-dependent loss of **lactase (adult-type hypolactasia)**, is by far the most common of these deficiencies.

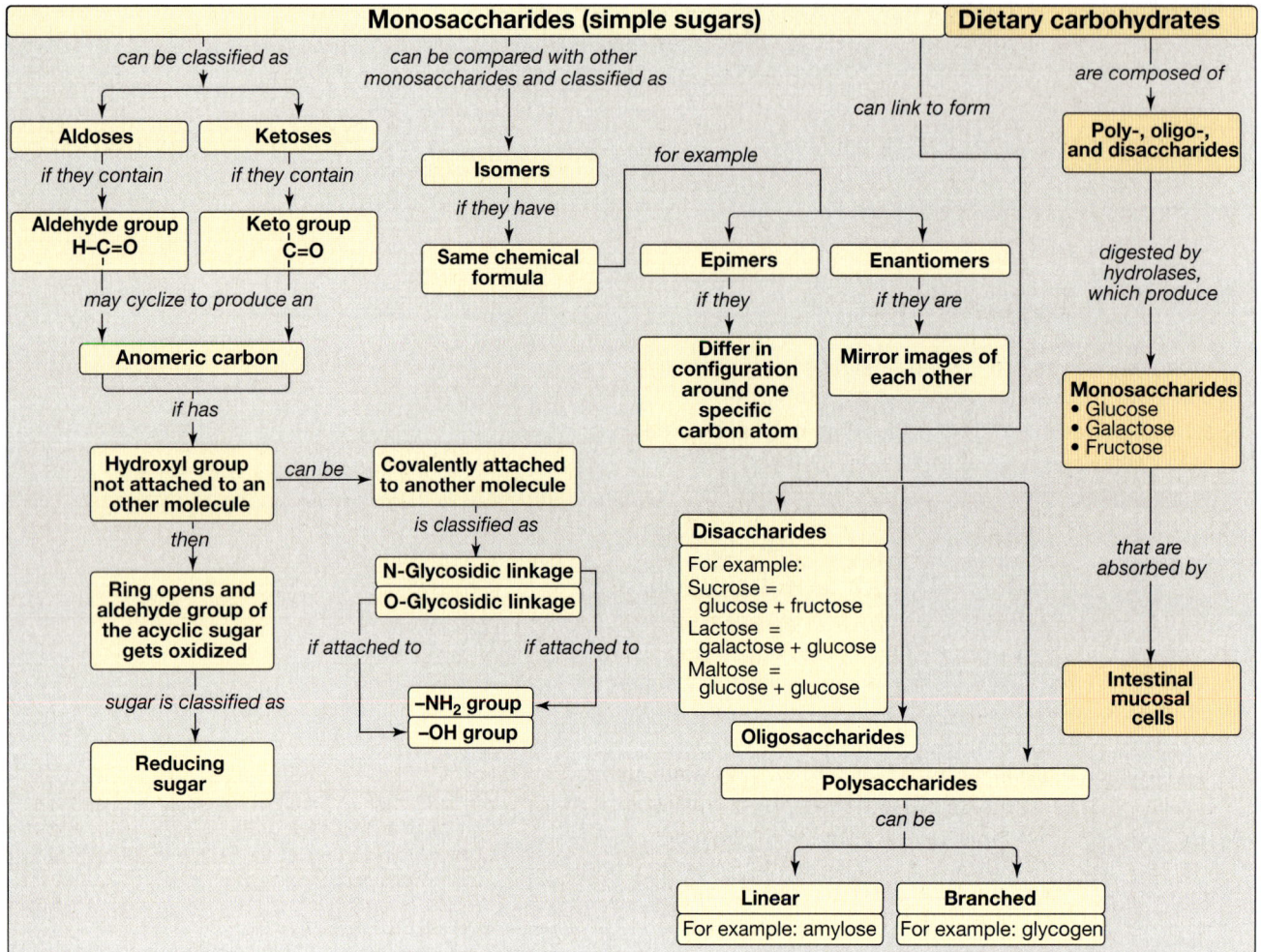

Figure 7.12
Key concept map for the classification and structure of monosaccharides and the digestion of dietary carbohydrates.

Study Questions

Choose the ONE best answer.

7.1. Glucose is:

A. a C-4 epimer of galactose.

B. a ketose and usually exists as a furanose ring in solution.

C. produced from dietary starch by the action of α-amylase.

D. utilized in biologic systems only in the L–isomeric form.

Correct answer = A. Because glucose and galactose differ only in configuration around carbon 4, they are C-4 epimers that are interconvertible by the action of an epimerase. Glucose is an aldose sugar that typically exists as a pyranose ring in solution. Fructose, however, is a ketose with a furanose ring. α-Amylase does not produce monosaccharides. The D-isomeric form of carbohydrates is the form typically found in biologic systems, in contrast to amino acids that typically are found in the L–isomeric form.

7.2. A 28-year-old man presents in the office with a chief concern of recurrent bloating and diarrhea. His eyes appear sunken, and the physician notes additional signs of dehydration. The patient's temperature is normal. He explains that the most recent episode occurred last night soon after he had ice cream for dessert. This clinical picture is most likely caused by a deficiency in the activity of:

A. isomaltase.
B. lactase.
C. pancreatic α-amylase.
D. salivary α-amylase.
E. sucrase.

> Correct answer = B. The physical symptoms suggest a deficiency in an enzyme responsible for carbohydrate degradation. The symptoms observed following the ingestion of dairy products suggest that the patient is deficient in lactase as a result of the age-dependent reduction in expression of the enzyme.

7.3. A 6-month-old girl is evaluated for diarrhea and steatorrhea. Her mother reports that she recently began giving her daughter fruit juice, but she vomits and cries after drinking it from her bottle. Inability to digest which of the following is most likely causing her signs and symptoms?

A Cellulose
B. Glycogen
C. Lactose
D. Starch
E. Sucrose

> Correct answer = E. Sucrase–isomaltase (SI) deficiency results in intolerance of ingested sucrose. Those with congenital SI deficiency experience osmotic diarrhea, mild steatorrhea, irritability, and vomiting after consuming sucrose. Treatment includes restricting dietary sucrose and enzyme replacement therapy.

7.4. Routine examination of the urine of an asymptomatic pediatric patient showed a positive reaction for a test to detect reducing sugars, but a negative reaction with the glucose oxidase test specifically for detecting glucose. Which of the sugars could (YES) or could not (NO) be present in the urine of this individual?

Sugar	Yes	No
Fructose		
Galactose		
Glucose		
Lactose		
Sucrose		
Xylulose		

> Each of the listed sugars, except for sucrose and glucose, could be present in the urine of this individual. The first test is a more general test for all reducing sugars (fructose, galactose, glucose, lactose, and xylulose). Because sucrose is not a reducing sugar, it is not detected by this test. The glucose oxidase test will detect only glucose, and it cannot detect other sugars. The negative glucose oxidase test coupled with a positive reducing sugar test means that glucose cannot be the reducing sugar in the patient's urine.

7.5. Why can α-glucosidase inhibitors such as acarbose and miglitol, which are taken with meals, be used in the treatment of some patients with diabetes mellitus? What effect should these drugs have on the digestion of lactose?

> α-Glucosidase inhibitors slow the production of glucose from dietary carbohydrates, thereby reducing the postprandial rise in blood glucose and facilitating better blood glucose control in patients with diabetes. These drugs have no effect on lactose digestion because the disaccharide lactose contains a β-glycosidic bond, not an α-glycosidic bond.

7.6. For intestinal absorption to occur, glucose must first be transported into enterocytes using which mechanism?

    A. ATP-dependent primary active transport

    B. catalyzed diffusion via insulin-independent glucose transporter (GLUT-2)

    C. cotransport with sodium via symport using sodium-dependent glucose cotransporter 1 (SGLT-1)

    D. facilitated diffusion via insulin-dependent GLUT-4

Correct answer = C. Glucose is transported against its concentration into enterocytes using sodium-dependent glucose cotransporter (SGLT), a symporter that cotransports sodium with its concentration gradient. Catalyzed diffusion and facilitated diffusion both describe the same type of transport, where the substrate is moved with its concentration gradient. Glucose must be transported against its gradient into enterocytes that normally already contain higher glucose concentrations than outside the cell. The sodium gradient drives transport of glucose against its gradient via SGLT-1. This process is not directly dependent on ATP and is secondary active transport, not primary active transport.

Introduction to Metabolism and Glycolysis

I. METABOLISM OVERVIEW

Within cells, enzymatic reactions rarely occur in isolation from each other. Instead, they are organized into multistep sequences called pathways, such as that of glycolysis (Fig. 8.1). In a pathway, the product of one reaction often serves as the substrate (see Chapter 5) of the subsequent reaction. Most pathways can be classified as either **catabolic** (degradative) or **anabolic** (synthetic). Catabolic pathways break down complex molecules, including proteins, polysaccharides, and lipids, to a few simple molecules (e.g., carbon dioxide, ammonia, and water). Anabolic pathways form complex end products from simple precursors. For example, the synthesis of the polysaccharide glycogen from glucose is an anabolic process. Different pathways can also intersect and form an integrated and purposeful network of chemical reactions. Metabolism is the sum of all the chemical changes occurring in a cell, a tissue, or the body. Metabolites are intermediate products of metabolism. The next several chapters focus on the central metabolic pathways that are involved in synthesizing and degrading carbohydrates, lipids, and amino acids.

A. Metabolic map

Metabolism is best understood by examining its component pathways. Each pathway is composed of sequences of enzyme-catalyzed reactions, and each enzyme, in turn, may exhibit important catalytic or regulatory features. A metabolic map containing the important central pathways of energy metabolism is presented in Figure 8.2. This "big picture" view of metabolism can be useful in tracing connections between pathways, visualizing the purposeful movement of metabolites, and depicting the effect on the flow of intermediates if a pathway is inhibited or blocked, for example, by a drug or an inherited deficiency of an enzyme. Throughout the next three units of this book, each pathway under discussion will be repeatedly featured as part of the major metabolic map shown in Figure 8.2.

B. Catabolic pathways

Catabolic reactions are exergonic; they generate energy.

They capture chemical energy in the form of adenosine triphosphate (ATP) from the degradation of energy-rich fuel molecules.

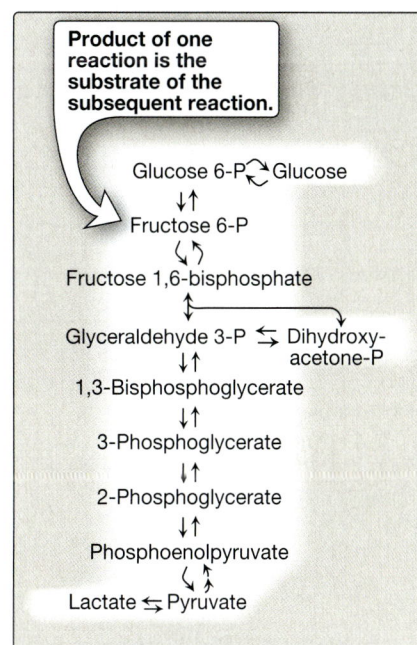

Figure 8.1
Glycolysis, an example of a metabolic pathway. [Note: Pyruvate to phosphoenolpyruvate requires two reactions.] *Curved reaction arrows* (⤸) indicate forward and reverse reactions that are catalyzed by different enzymes. P, phosphate.

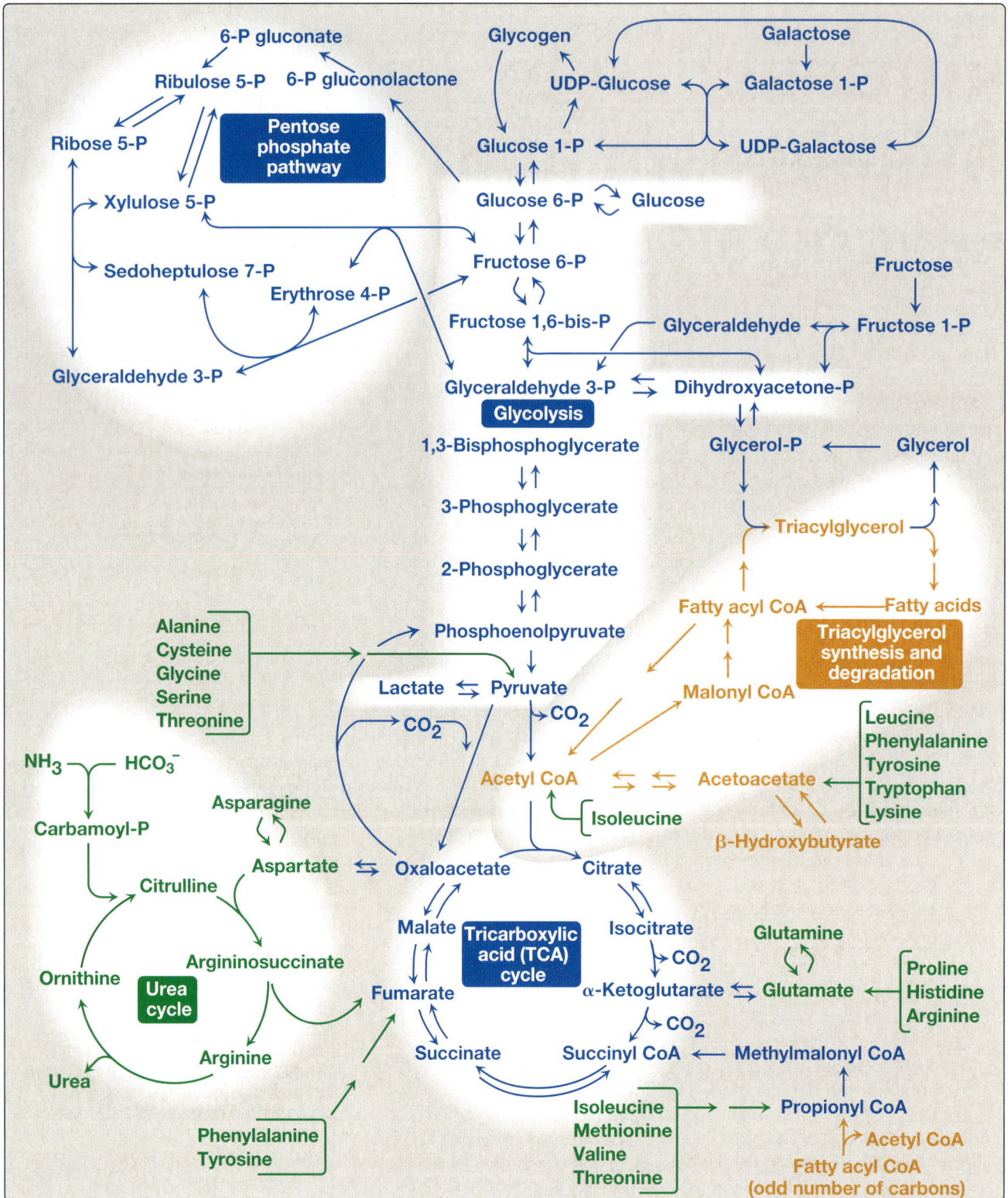

Figure 8.2
Important reactions of intermediary metabolism. Several important pathways to be discussed in later chapters are highlighted. *Curved reaction arrows* (↻) indicate forward and reverse reactions that are catalyzed by different enzymes. The *straight arrows* (⇄) indicate forward and reverse reactions that are catalyzed by the same enzyme. **Blue text** = intermediates of carbohydrate metabolism; **brown text** = intermediates of lipid metabolism; **green text** = intermediates of protein metabolism. UDP, uridine diphosphate; P, phosphate; CoA, coenzyme A; CO_2, carbon dioxide; HCO_3^- bicarbonate; NH_3, ammonia.

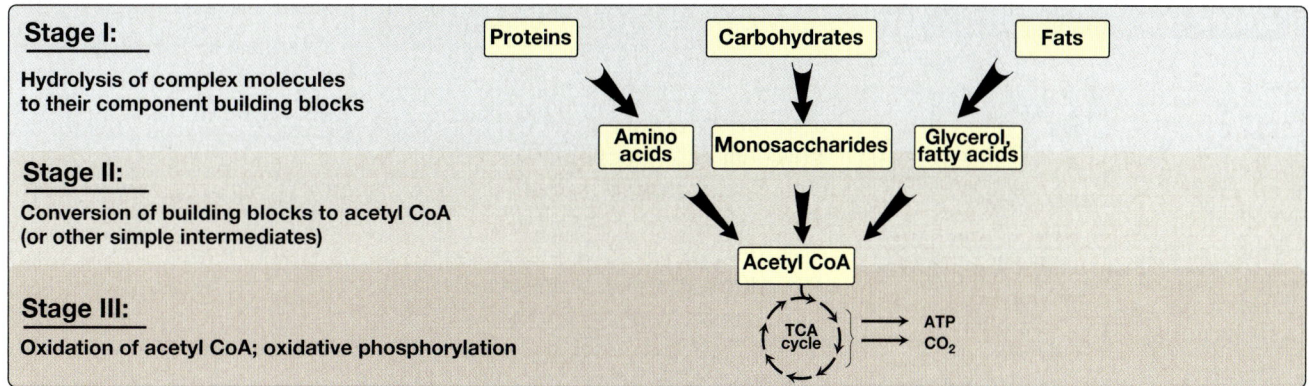

Figure 8.3
Three stages of catabolism. CoA, coenzyme A; TCA, tricarboxylic acid; CO_2, carbon dioxide; ATP, adenosine triphosphate.

ATP generation by degradation of complex molecules occurs in three stages, as shown in Figure 8.3. [Note: Catabolic pathways are typically oxidative and require oxidized coenzymes such as nicotinamide adenine dinucleotide (NAD^+).] Catabolism also allows molecules in the diet or nutrient molecules stored in cells, to be converted into basic building blocks needed for the synthesis of complex molecules. Catabolism, then, is a convergent process in which a wide variety of molecules are transformed into a few common end products.

1. **Hydrolysis of complex molecules:** In the first stage, complex molecules are broken down into their component building blocks. For example, proteins are degraded to amino acids, polysaccharides to monosaccharides, and fats (triacylglycerols) to free fatty acids and glycerol.

2. **Conversion of building blocks to simple intermediates:** In the second stage, diverse building blocks are further degraded to acetyl coenzyme A (CoA) and a few other simple molecules. Some energy is captured as ATP, but the amount is small compared with the energy produced during the third stage of catabolism.

3. **Oxidation of acetyl CoA:** The tricarboxylic acid (TCA) cycle (see Chapter 9) is the final common pathway in the oxidation of fuel molecules that produce acetyl CoA. Oxidation of acetyl CoA generates large amounts of ATP via oxidative phosphorylation as electrons flow from nicotinamide adenine dinucleotide (NADH) and flavin adenine dinucleotide ($FADH_2$) to oxygen ($[O_2]$, see Chapter 6).

C. Anabolic pathways

Anabolic reactions are endergonic. They require energy, which is generally provided by the hydrolysis of ATP to adenosine diphosphate (ADP) and inorganic phosphate (P_i). Anabolism is a divergent process in which a few biosynthetic precursors such as amino acids form a wide variety of complex products such as proteins (Fig. 8.4). Anabolic reactions often involve chemical reductions in which the reducing power is most frequently provided by the electron donor NADPH (phosphorylated NADH; see Chapter 13).

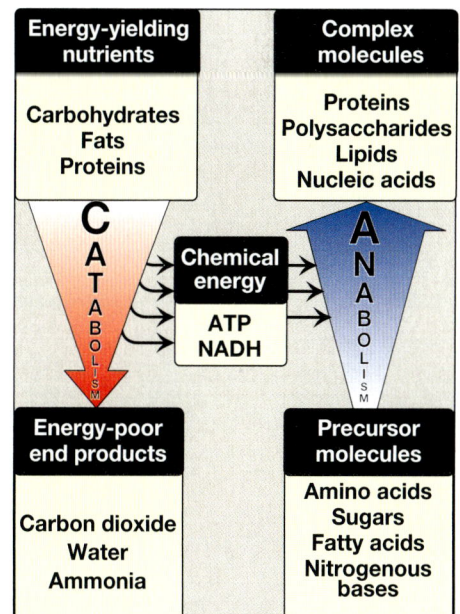

Figure 8.4
Comparison of catabolic and anabolic pathways. NADH, nicotinamide adenine dinucleotide; ATP, adenosine triphosphate.

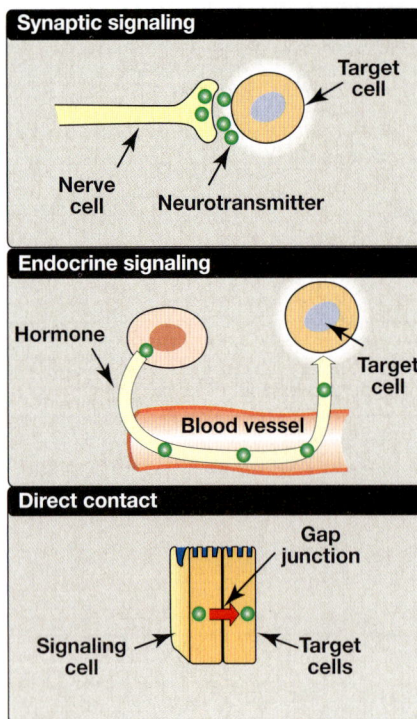

Figure 8.5
Some commonly used mechanisms for transmission of regulatory signals between cells.

II. METABOLISM REGULATION

Metabolic pathways must be coordinated so that the production of energy or the synthesis of end products meets the needs of the cell. Furthermore, individual cells function as part of tissues, and tissues interact with each other. Metabolism does not occur in isolation. Thus, a sophisticated communication system has evolved to coordinate the functions of the body by acting initially on individual cells. Regulatory signals inform individual cells about the metabolic state of the body as a whole and include hormones, neurotransmitters, and the availability of nutrients. These regulatory molecules, in turn, influence signals generated within cells (Fig. 8.5).

A. Intracellular communication

The rate of a metabolic pathway can change in response to regulatory signals that arise from within the cell. For example, the rate may be influenced by the availability of substrates, product inhibition, or alterations in the levels of allosteric activators or inhibitors. These intracellular signals typically elicit rapid responses and are important for the moment-to-moment regulation of metabolism.

B. Intercellular communication

The ability to respond to intercellular signals is essential for the development and survival of organisms. Signaling between cells provides for long-range integration of metabolism and usually results in a response, such as a change in gene expression that is slower than is seen with intracellular signals. Communication between cells can be mediated, for example, by surface-to-surface contact and, in some tissues, by formation of gap junctions, allowing direct communication between the cytoplasms of adjacent cells. However, for energy metabolism, the most important route of communication is chemical signaling between cells by blood-borne hormones and by neurotransmitters.

C. G protein–linked receptors and second-messenger systems

Hormones and neurotransmitters act as signals to cells that have specific receptors for them. The signals are ligands that bind to receptor proteins often found embedded in the plasma membranes of their target cells. The target cells respond to ligands bound to their receptors by initiating a series of reactions that ultimately result in specific intracellular responses. Many receptors that regulate metabolism are linked to intracellular guanosine triphosphate (GTP)-binding proteins called G proteins and are known as **G protein–coupled receptors** (**GPCRs**). This type of receptor regulates the production of molecules known as second messengers, which are so named because they intervene between the original extracellular messenger (the neurotransmitter or hormone) and the ultimate intracellular effect. **Second messengers** are part of the cascade of events that converts (transduces) ligand binding into a response by the cell.

Two of the most widely recognized second messenger systems regulated by G proteins are the **phospholipase C** system that involves calcium and phosphatidylinositol hydrolysis and the **adenylyl cyclase** (AC) (adenylate cyclase) system, which is particularly important in regulating the pathways of intermediary metabolism. Both these second-messenger systems are initiated by the binding of hormone ligands, such as epinephrine or glucagon, to specific GPCRs embedded with the plasma membrane of the target cell that will respond to the hormone.

GPCRs are characterized by an extracellular ligand-binding domain; seven transmembrane α helices; and an intracellular domain that interacts with heterotrimeric G proteins composed of α, β, and γ subunits (Fig. 8.6). [Note: Insulin, another key regulator of metabolism, does not signal via GPCRs but instead acts via a receptor with tyrosine kinase activity (see Chapter 23).[1]]

D. Adenylyl cyclase

Binding of the hormone ligand by some GPCRs, including the β- and α_2-adrenergic receptors, triggers either an increase or a decrease in the activity of AC. This is a membrane-bound enzyme that converts ATP to 3',5'-adenosine monophosphate (cyclic AMP [cAMP]) when active.

1. **GTP–dependent regulatory proteins or G proteins:** The effect of the activated, occupied GPCR on second-messenger formation is mediated by specialized heterotrimeric G proteins (α, β, and γ subunits) found on the inner face of the plasma membrane. G proteins are named because their α subunit binds GTP when activated. In the inactive form of a G protein, the α subunit is bound to GDP (Fig. 8.7). Ligand binding causes a conformational change in the receptor, triggering the replacement of this GDP with GTP. The GTP-bound form of the α subunit dissociates from the $\beta\gamma$ subunits and moves to the membrane-bound AC enzyme, affecting its enzyme activity. Many molecules of active Gα protein are formed by one activated receptor. [Note: The ability of a hormone or neurotransmitter to stimulate or inhibit AC depends on the type of Gα protein that is linked to the receptor. One type, designated G$_s$, stimulates AC (see Fig. 8.7), whereas G$_i$ inhibits AC (not shown).]

Activated AC converts ATP to the second messenger **cAMP**, which then activates the serine/threonine protein kinase known as **protein kinase A** (**PKA**). The actions of the Gα–GTP complex are short-lived because Gα has inherent GTPase activity, resulting in the rapid hydrolysis of GTP to guanosine diphosphate (GDP). This causes inactivation of Gα, its dissociation from AC, and its reassociation with the $\beta\gamma$ dimer.

> Toxins from *Vibrio cholerae* (cholera) and *Bordetella pertussis* (whooping cough) cause inappropriate activation of adenylyl cyclase (AC) through covalent modification (ADP-ribosylation) of different G proteins that interact with AC. With cholera toxin, the GTPase activity of Gα_s is inhibited in intestinal cells. With whooping cough, the pertussis toxin inactivates Gα_i in respiratory tract cells. The result in both situations is increased AC activity and excess production of the second messenger cAMP.

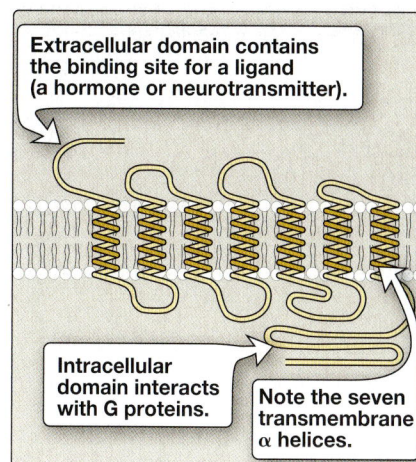

Figure 8.6
Structure of a typical G protein–coupled receptor of the plasma membrane.

---

[1]For more information on GPCR signaling and second messengers, see *LIR Cell and Molecular Biology*, 3e Chapter 17.

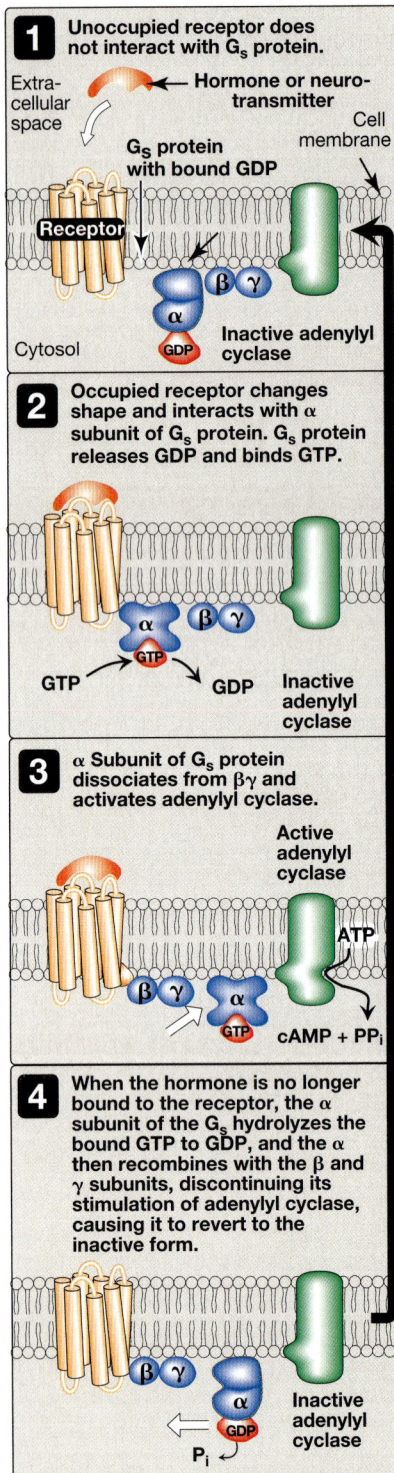

1 Unoccupied receptor does not interact with G_S protein.

Extracellular space — Hormone or neurotransmitter

G_S protein with bound GDP

Cell membrane

Receptor

β γ

α GDP

Inactive adenylyl cyclase

Cytosol

2 Occupied receptor changes shape and interacts with α subunit of G_S protein. G_S protein releases GDP and binds GTP.

α GTP β γ

GTP → GDP   Inactive adenylyl cyclase

3 α Subunit of G_S protein dissociates from βγ and activates adenylyl cyclase.

Active adenylyl cyclase

β γ

α GTP

ATP

cAMP + PP$_i$

4 When the hormone is no longer bound to the receptor, the α subunit of the G_S hydrolyzes the bound GTP to GDP, and the α then recombines with the β and γ subunits, discontinuing its stimulation of adenylyl cyclase, causing it to revert to the inactive form.

β γ

α GDP

Inactive adenylyl cyclase

P$_i$

Figure 8.7
Recognition of chemical signals by certain membrane receptors triggers an increase (or, less often, a decrease) in the activity of adenylyl cyclase. GDP and GTP, guanosine di- and triphosphates; cAMP, cyclic adenosine monophosphate.

2. **Protein kinases:** The next step in the cAMP second-messenger system is the activation of a family of enzymes called cAMP-dependent protein kinases, including PKA, as shown in Figure 8.8. cAMP activates PKA by binding to its two regulatory subunits, causing the release of its two catalytically active subunits. Active PKA is a serine/threonine kinase because it functions to transfer phosphate from ATP to specific serine or threonine residues of its specific protein substrates. The phosphorylated proteins may act directly on the cell's ion channels or, if enzymes, may become activated or inhibited.

3. **Protein phosphatases:** The phosphate groups added to proteins by protein kinases are removed by phosphoprotein phosphatases, enzymes that hydrolytically cleave phosphate esters (see Fig. 8.8). Actions of phosphatases ensure that changes in protein activity induced by phosphorylation are not permanent.

4. **cAMP hydrolysis:** cAMP is rapidly hydrolyzed to 5'-AMP by cAMP phosphodiesterase that cleaves the cyclic 3',5'-phosphodiester bond. 5'-AMP is not an intracellular signaling molecule. Therefore, the effects of neurotransmitter- or hormone-mediated increases of cAMP are rapidly terminated if the extracellular signal is removed. [Note: cAMP phosphodiesterase is inhibited by caffeine, a methylxanthine derivative.]

III. GLYCOLYSIS OVERVIEW

The glycolytic pathway is used by all tissues for the oxidation of glucose to provide energy as ATP and intermediates for other metabolic pathways. Glycolysis is at the hub of carbohydrate metabolism because virtually all sugars, whether arising from the diet or from catabolic reactions in the body, can ultimately be converted to glucose (Fig. 8.9A). **Pyruvate** is the end product of glycolysis in cells with mitochondria and an adequate supply of O_2. This series of 10 reactions is called aerobic glycolysis because O_2 is required to reoxidize the NADH formed during the oxidation of glyceraldehyde 3-phosphate (Fig. 8.9B). Aerobic glycolysis sets the stage for the oxidative decarboxylation of pyruvate to acetyl CoA, a major fuel of the TCA cycle. Alternatively, pyruvate is reduced to **lactate** as NADH is oxidized to NAD^+ (Fig. 8.9C). This conversion of glucose to lactate is called anaerobic glycolysis because it can occur without the participation of O_2. Anaerobic glycolysis allows the production of ATP in tissues that lack mitochondria (e.g., red blood cells [RBCs] and parts of the eye) or in cells deprived of sufficient O_2 (as in hypoxia).

IV. GLUCOSE TRANSPORT INTO CELLS

Glucose enters cells by one of two transport systems: passive transport or a Na^+- and ATP-dependent cotransport system.

A. Passive transport of glucose

This passive system also known as catalyzed diffusion, permits one molecule of glucose to enter a cell at a time and is known as **uniport**. It is mediated by a family of 14 glucose transporter (GLUT) isoforms found in cell membranes designated GLUT-1 to GLUT-14. These monomeric protein transporters exist in the membrane in two conformational states (Fig. 8.10). Extracellular glucose binds to the transporter, which then alters its conformation, transporting glucose across the cell membrane via facilitated or catalyzed diffusion.[2]

1. **Tissue specificity:** GLUT displays a tissue-specific pattern of expression (Table 8.1). For example, GLUT-1 is abundant in most tissues, GLUT-4 is abundant in muscle and adipose tissue, and GLUT-5 transports fructose. [Note: The number of GLUT-4 transporters active in these tissues is increased by insulin (see Chapter 23 for a discussion of insulin and glucose transport).] GLUT-2 is abundant in the liver, kidneys, and pancreatic β cells. The other GLUT isoforms also have tissue-specific distributions.

2. **Specialized functions:** In facilitated diffusion, transporter-mediated glucose movement is down a concentration gradient (i.e., from a high concentration to a lower one, therefore requiring no energy). For example, GLUT-1, GLUT-3, and GLUT-4 are primarily involved in glucose uptake from the blood. In contrast, GLUT-2 in the liver and kidneys can either transport glucose into these cells when blood glucose levels are high or transport glucose from these cells when blood glucose levels are low (e.g., during fasting). GLUT-5 is unusual in that it is the primary transporter for fructose (not glucose) in the small intestine and the testes.

B. Sodium- and ATP-dependent cotransport of glucose

This type of glucose cotransport with sodium occurs in the epithelial cells of the intestine, the renal tubules, and the choroid plexus.

---

[2]For more information on glucose transport, see *LIR Cell and Molecular Biology*, 3e Chapter 15.

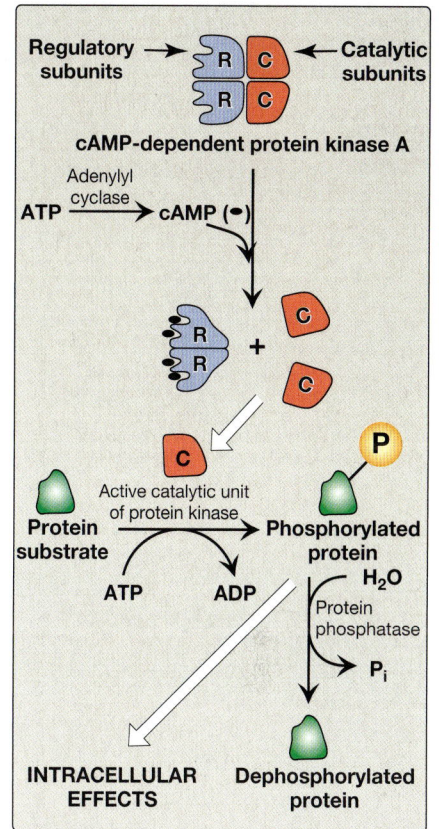

Figure 8.8
Actions of cyclic adenosine monophosphate (cAMP). (P), phosphate; ADP, adenosine diphosphate; P$_i$, inorganic phosphate.

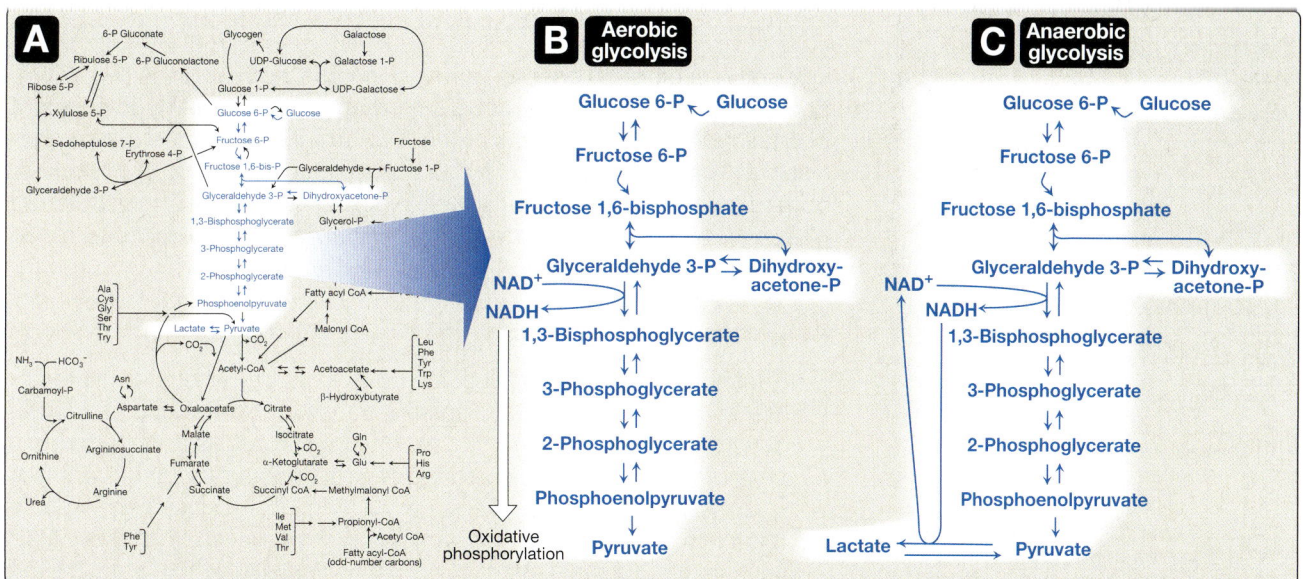

Figure 8.9
A. Glycolysis shown as one of the essential pathways of energy metabolism. **B.** Reactions of aerobic glycolysis. **C.** Reactions of anaerobic glycolysis. NAD(H), nicotinamide adenine dinucleotide; P, phosphate.

Table 8.1: Tissue Distribution of Selected Glucose Transporters

	Location	Function	K_m (mM)
GLUT-1	Most tissues	Basal glucose uptake	1
GLUT-2	Liver, kidneys, pancreas	Removes excess glucose from blood	15–20
GLUT-3	Most tissues	Basal glucose uptake	1
GLUT-4	Muscle and fat	Removes excess glucose from blood	5
GLUT-5	Small intestine, testes	Transport of fructose	10

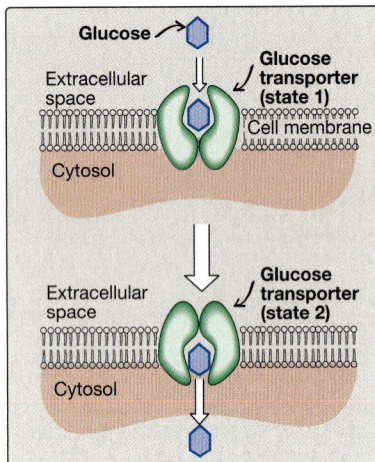

Figure 8.10
Schematic representation of the facilitated transport of glucose through a cell membrane. [Note: Glucose transporter proteins are monomeric and contain 12 transmembrane α helices.]

This is an energy-requiring process that transports glucose against or up its concentration gradient, from low extracellular concentrations to higher intracellular concentrations while Na^+ is transported down its electrochemical gradient. The extracellular concentration of Na^+ is much higher than the intracellular concentration, which is the result of the Na^+/K^+-ATPase. The Na^+ concentration gradient powers the transport of glucose against its concentration gradient; ATP hydrolysis is an indirect energy source because it is necessary to establish the Na^+ gradient (see Fig. 7.10). Because this secondary active transport of glucose requires the concurrent uptake (symport) of Na^+, the transporter is a sodium-dependent glucose cotransporter (SGLT). [Note: The choroid plexus, part of the blood–brain barrier, also contains GLUT-1.[3]]

> SGLT-2 functions in the kidneys and is the major transporter for glucose reabsorption back into the blood. Gliflozins are SGLT-2 inhibitors, which reduce reabsorption of glucose in the kidney and, therefore, lower blood sugar. SGLT-2 inhibitors are used to treat hyperglycemia in people with type 2 diabetes.

V. GLYCOLYSIS REACTIONS

The conversion of glucose to pyruvate occurs in two stages (Fig. 8.11). The first five reactions of glycolysis correspond to an energy-investment phase in which the phosphorylated forms of intermediates are synthesized at the expense of ATP. The subsequent reactions of glycolysis constitute an energy-generation phase in which net two molecules of ATP are formed by substrate-level phosphorylation per glucose molecule metabolized.

A. Glucose phosphorylation

Phosphorylated sugar molecules do not readily penetrate cell membranes because there are no specific transmembrane carriers for these compounds and because they are too polar to diffuse through the lipid core of membranes. Therefore, the irreversible phosphorylation of glucose effectively traps the sugar as cytosolic glucose 6-phosphate and commits it to further metabolism in the cell (Fig. 8.12). Mammals have four isozymes (I–IV) of the enzyme hexokinase that catalyze the phosphorylation of glucose to glucose 6-phosphate.

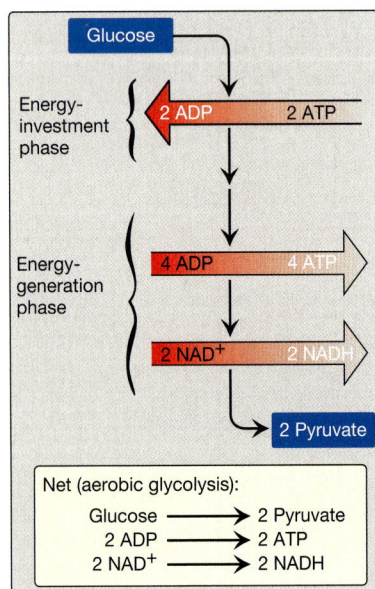

Figure 8.11
Two phases of aerobic glycolysis. NAD(H), nicotinamide adenine dinucleotide; ADP, adenosine diphosphate.

---

[3]For further information, see *LIR Cell and Molecular Biology*, 3e Chapters 14 and 15.

1. **Hexokinases I–III:** In most tissues, glucose phosphorylation is catalyzed by one of these isozymes of hexokinase, which is one of three regulatory enzymes of glycolysis (along with phosphofructokinase [PFK] and pyruvate kinase [PK]). They are inhibited by the reaction product glucose 6-phosphate, which accumulates when further metabolism of this hexose phosphate is reduced. Hexokinases I–III have a low Michaelis constant (K_m) and, therefore, a high affinity for glucose. This permits the efficient phosphorylation and subsequent metabolism of glucose even when tissue concentrations of glucose are low (Fig. 8.13). However, because these isozymes have a low maximal velocity ([V_{max}],) for glucose, they do not sequester (trap) cellular phosphate in the form of phosphorylated glucose or phosphorylate more glucose than the cell can use. [Note: These isozymes have broad substrate specificity and are able to phosphorylate several hexoses in addition to glucose.]

2. **Hexokinase IV:** In liver parenchymal cells and pancreatic β cells, **glucokinase** (the hexokinase IV isozyme) is the predominant enzyme responsible for glucose phosphorylation. In β cells, glucokinase functions as a glucose sensor, determining the threshold for insulin secretion. [Note: Hexokinase IV also serves as a glucose sensor in hypothalamic neurons, playing a key role in the adrenergic response to hypoglycemia.] In the liver, the glucokinase enzyme facilitates glucose phosphorylation during hyperglycemia. Despite the popular but misleading name glucokinase, the sugar specificity of the enzyme is similar to that of other hexokinase isoenzymes, and it can phosphorylate other hexoses along with glucose.

   a. **Kinetics:** Glucokinase differs from hexokinases I–III in several important properties. For example, it has a much higher K_m, requiring a higher glucose concentration for half-saturation (see Fig. 8.13). Thus, glucokinase functions only when the intracellular concentration of glucose in the hepatocyte is elevated such as during the brief period following consumption of a carbohydrate-rich meal, when high levels of glucose are delivered to the liver via the portal vein. Glucokinase has a high V_{max}, allowing the liver to effectively remove the flood of glucose delivered by the portal blood. This prevents large amounts of glucose from entering the systemic circulation following such a meal, thereby minimizing hyperglycemia during the absorptive period. [Note: GLUT-2 ensures that blood glucose equilibrates rapidly across the hepatocyte membrane.]

   b. **Regulation:** Glucokinase activity is not directly inhibited by glucose 6-phosphate as are the other hexokinases. Instead, it is indirectly inhibited by fructose 6-phosphate (which is in equilibrium with glucose 6-phosphate, a product of glucokinase) and is indirectly stimulated by glucose (a substrate of glucokinase). Regulation is achieved by reversible binding to the hepatic protein glucokinase regulatory protein (GKRP). In the presence of fructose 6-phosphate, glucokinase binds tightly to GKRP and is translocated to the nucleus, thereby rendering the enzyme inactive (Fig. 8.14). When glucose

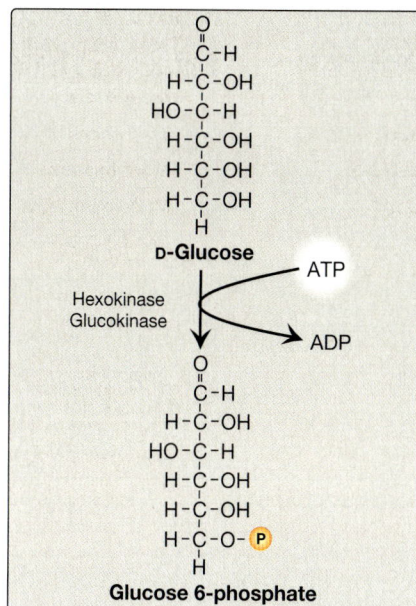

Figure 8.12
Energy-investment phase: phosphorylation of glucose. [Note: Kinases utilize ATP complexed with a divalent metal ion, most typically magnesium.] ADP, adenosine diphosphate; ℗ phosphate.

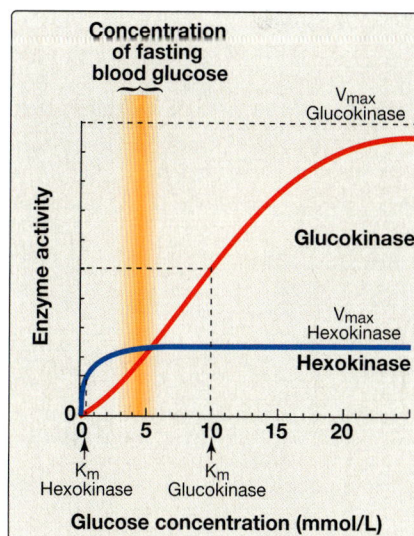

Figure 8.13
Effect of glucose concentration on the rate of phosphorylation catalyzed by hexokinase and glucokinase. K_m, Michaelis constant; V_{max}, maximal velocity.

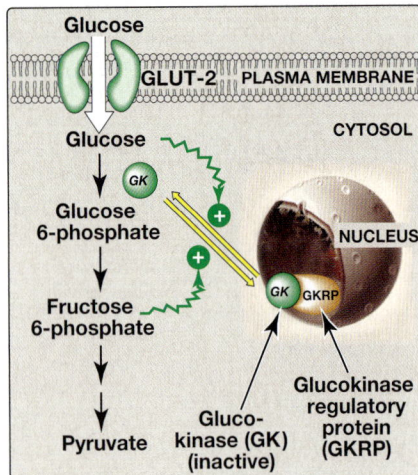

Figure 8.14
Regulation of glucokinase activity by glucokinase regulatory protein. GLUT, glucose transporter.

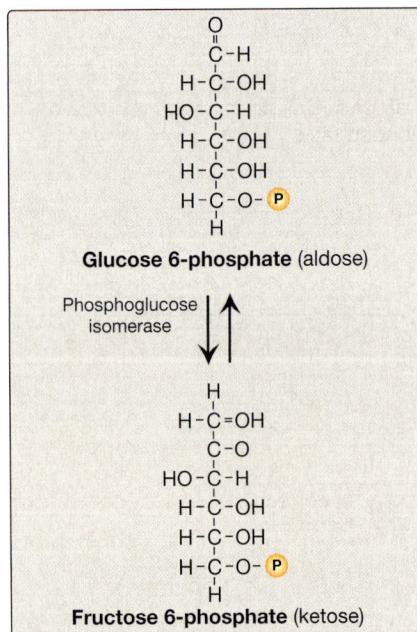

Figure 8.15
Aldose-ketose isomerization of glucose 6-phosphate to fructose 6-phosphate. ℗, phosphate.

levels in the blood (and also in the hepatocyte, as a result of GLUT-2) increase, glucokinase is released from GKRP, and the enzyme reenters the cytosol where it phosphorylates glucose to glucose 6-phosphate. [Note: GKRP is a competitive inhibitor of glucose use by glucokinase.]

> Glucokinase functions as a glucose sensor in blood glucose homeostasis. Inactivating mutations of glucokinase are the cause of a rare form of diabetes, maturity-onset diabetes of the young type 2 (MODY2) that is characterized by impaired insulin secretion and hyperglycemia.

B. Glucose 6-phosphate isomerization

The isomerization of glucose 6-phosphate to fructose 6-phosphate is catalyzed by phosphoglucose isomerase (Fig. 8.15). The reaction is readily reversible and is not a rate-limiting or regulated step.

C. Fructose 6-phosphate phosphorylation

The irreversible phosphorylation reaction catalyzed by PFK-1 is the most important control point and the rate-limiting and committed step of glycolysis (Fig. 8.16). PFK-1 is controlled by the available concentrations of the substrates ATP and fructose 6-phosphate as well as by other regulatory molecules.

1. **Regulation by intracellular energy levels:** PFK-1 is inhibited allosterically by elevated levels of ATP, which act as an energy-rich signal indicating an abundance of high-energy compounds. Elevated levels of citrate, an intermediate in the TCA cycle, also inhibit PFK-1. [Note: Inhibition by citrate favors the use of glucose for glycogen synthesis.] Conversely, PFK-1 is activated allosterically by high concentrations of AMP, which signal that the cell's energy stores are depleted.

2. **Regulation by fructose 2,6-bisphosphate:** Fructose 2,6-bisphosphate is the most potent activator of PFK-1 (see Fig. 8.16) and is able to activate the enzyme even when ATP levels are high. It is formed from fructose 6-phosphate by PFK-2. Unlike PFK-1, PFK-2 is a bifunctional protein that has both the kinase activity that produces fructose 2,6-bisphosphate and the phosphatase activity that dephosphorylates fructose 2,6-bisphosphate to fructose 6-phosphate. In the liver isozyme, phosphorylation of PFK-2 inactivates the kinase domain and activates the phosphatase domain (Fig. 8.17). The opposite is seen in the cardiac isozyme. Skeletal PFK-2 is not covalently regulated. [Note: Fructose 2,6-bisphosphate is an inhibitor of fructose 1,6-bisphosphatase, an enzyme of gluconeogenesis. The reciprocal actions of fructose 2,6-bisphosphate on glycolysis (activation) and gluconeogenesis (inhibition) ensure that both pathways are not fully active at the same time, preventing a futile cycle of glucose oxidation to pyruvate followed by glucose resynthesis from pyruvate.]

a. **During the well-fed state:** Decreased levels of glucagon and elevated levels of insulin (such as occur following a carbohydrate-rich meal) cause an increase in hepatic fructose 2,6-bisphosphate (PFK-2 is dephosphorylated) and, thus, in the rate of glycolysis (see Fig. 8.17). Therefore, fructose 2,6-bisphosphate acts as an intracellular signal of glucose abundance.

b. **During fasting:** By contrast, the elevated levels of glucagon and low levels of insulin that occur during fasting cause a decrease in hepatic fructose 2,6-bisphosphate (PFK-2 is phosphorylated). This results in inhibition of glycolysis and activation of gluconeogenesis.

D. Fructose 1,6-bisphosphate cleavage

Aldolase cleaves fructose 1,6-bisphosphate to dihydroxyacetone phosphate (DHAP) and glyceraldehyde 3-phosphate (see Fig. 8.16). The reaction is reversible and not regulated. [Note: Aldolase B, the hepatic isoform, also cleaves fructose 1-phosphate and functions in dietary fructose metabolism.]

E. Dihydroxyacetone phosphate isomerization

Triose phosphate isomerase interconverts DHAP and **glyceraldehyde 3-phosphate** (see Fig. 8.16). DHAP must be isomerized to glyceraldehyde 3-phosphate for further metabolism by the glycolytic pathway. This isomerization results in the net production of two molecules of glyceraldehyde 3-phosphate from the cleavage products of fructose 1,6-bisphosphate. [Note: DHAP is utilized in triacylglycerol synthesis.]

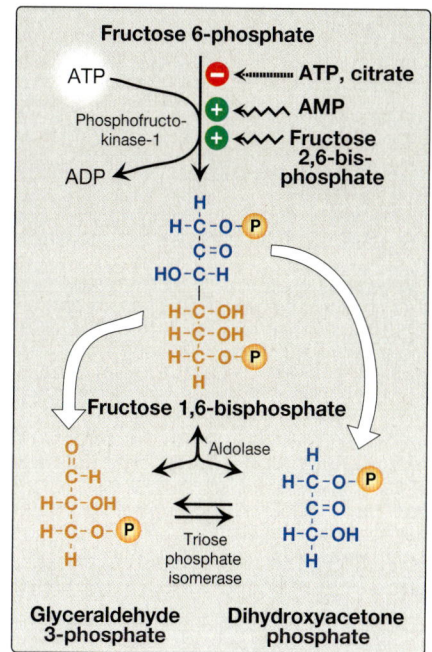

Figure 8.16
Energy-investment phase (continued): conversion of fructose 6-phosphate to triose phosphates. ℗, phosphate; AMP and ADP, adenosine mono- and diphosphates.

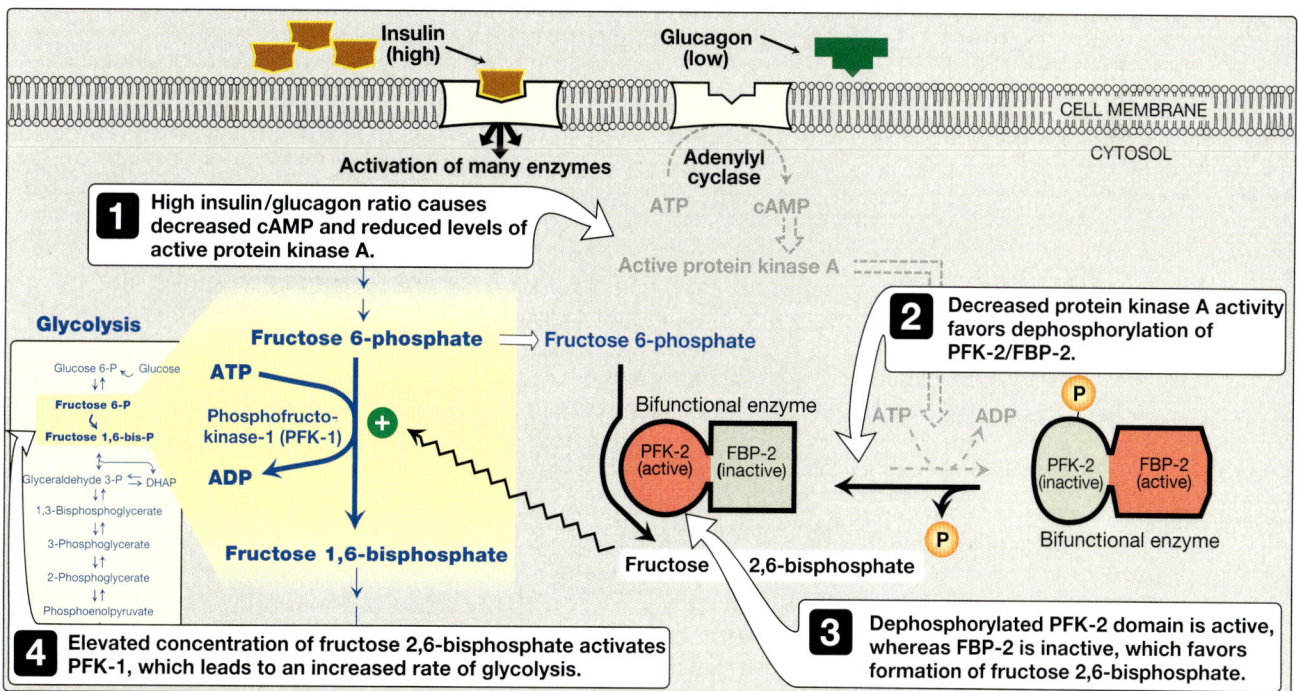

Figure 8.17
Effect of elevated insulin concentration on the intracellular concentration of fructose 2,6-bisphosphate in the liver. PFK-2, phosphofructokinase-2; FBP-2, fructose 2,6-bisphosphatase; AMP and ADP, adenosine mono- and diphosphates; cAMP, cyclic AMP; ℗, phosphate.

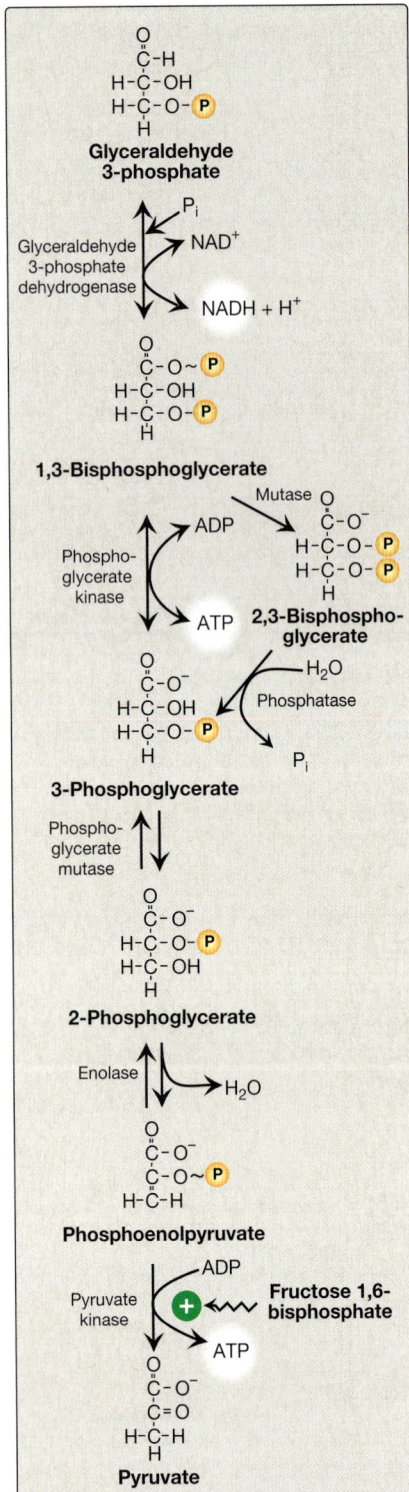

Figure 8.18
Energy-generating phase: conversion of glyceraldehyde 3-phosphate to pyruvate. NAD(H) = nicotinamide adenine dinucleotide; **P** = phosphate; P_i = inorganic phosphate; ~ = high-energy bond; ADP = adenosine diphosphate.

Clinical Application 8.1: Triose Phosphate Isomerase Deficiency

Triose phosphate isomerase (TPI) deficiency results in the accumulation of dihydroxyacetone phosphate, especially in RBCs. This is a severe autosomal recessive disorder characterized by hemolytic anemia and progressive neurodegeneration. TPI deficiency often results in death during childhood. TPI normally functions to regulate the equilibrium between dihydroxyacetone phosphate and glyceraldehyde-3-phosphate, both of which are connected also with the pentose phosphate pathway and to lipid metabolism.

F. Glyceraldehyde 3-phosphate oxidation

The conversion of glyceraldehyde 3-phosphate to 1,3-bisphosphoglycerate (1,3-BPG) by glyceraldehyde 3-phosphate dehydrogenase is the first oxidation–reduction reaction of glycolysis (Fig. 8.18). [Note: Because of the limited amount of NAD+ in the cell, the NADH formed by the dehydrogenase reaction must be oxidized for glycolysis to continue. Two major mechanisms for oxidizing NADH to NAD+ are the reduction of pyruvate to lactate by lactate dehydrogenase (LD) anaerobic, and the electron transport chain ([ETC] aerobic)]. Because NADH cannot cross the inner mitochondrial membrane, the ETC requires the malate–aspartate and glycerol 3-phosphate substrate shuttles to move NADH-reducing equivalents into the mitochondrial matrix.

1. **1,3-BPG synthesis:** The oxidation of the aldehyde group of glyceraldehyde 3-phosphate to a carboxyl group is coupled to the attachment of P_i to the carboxyl group. This phosphate group, linked to carbon 1 of the 1,3-BPG product by a high-energy bond, conserves much of the free energy produced by the oxidation of glyceraldehyde 3-phosphate. This high-energy phosphate drives ATP synthesis in the next reaction of glycolysis.

2. **2,3-Bisphosphoglycerate synthesis in red blood cells:** Some of the 1,3-BPG is converted to 2,3-BPG by the action of bisphosphoglycerate mutase (Fig. 8.18). 2,3-BPG, which is found in only trace amounts in most cells, is present at high concentrations in

Clinical Application 8.2: Arsenic Poisoning

The toxicity of arsenic is due primarily to the inhibition of enzymes such as the pyruvate dehydrogenase complex (PDHC), which require lipoic acid as a coenzyme. However, pentavalent arsenic (arsenate) can prevent net ATP and NADH production by glycolysis without inhibiting the pathway itself. It does so by competing with P_i as a substrate for glyceraldehyde 3-phosphate dehydrogenase, forming a complex that spontaneously hydrolyzes to form 3-phosphoglycerate (see Fig. 8.18). By bypassing the synthesis of and phosphate transfer from 1,3-BPG, the cell is deprived of energy usually obtained from the glycolytic pathway. [Note: Arsenate also competes with P_i binding to the F_1 domain of ATP synthase resulting in formation of ADP-arsenate that is rapidly hydrolyzed.]

RBCs and serves to increase O_2 delivery. 2,3-BPG is hydrolyzed by a phosphatase to 3-phosphoglycerate, which is also an intermediate in glycolysis (see Fig. 8.18). In the RBC, glycolysis is modified by the inclusion of these shunt reactions.

G. 3-Phosphoglycerate synthesis and ATP production

When 1,3-BPG is converted to 3-phosphoglycerate, the high-energy phosphate group of 1,3-BPG is used to synthesize ATP from ADP (see Fig. 8.18). This reaction is catalyzed by phosphoglycerate kinase, which, unlike most other kinases, is physiologically reversible. Because two molecules of 1,3-BPG are formed from each glucose molecule, this kinase reaction replaces the two ATP molecules consumed by the earlier formation of glucose 6-phosphate and fructose 1,6-bisphosphate. [Note: This reaction is an example of substrate-level phosphorylation, in which the energy needed for the production of a high-energy phosphate comes from a substrate rather than from the ETC (see Section J. for other examples).]

H. Phosphate group shift

The shift of the phosphate group from carbon 3 to carbon 2 of phosphoglycerate by phosphoglycerate mutase is freely reversible.

I. 2-Phosphoglycerate dehydration

The dehydration of 2-phosphoglycerate by enolase redistributes the energy within the substrate, forming phosphoenolpyruvate (PEP), which contains a high-energy enol phosphate (see Fig. 8.18). The reaction is reversible, despite the high-energy nature of the product. [Note: Fluoride inhibits enolase, and water fluoridation reduces lactate production by mouth bacteria, decreasing dental caries.]

J. Pyruvate synthesis and ATP production

The conversion of PEP to pyruvate, catalyzed by PK, is the third irreversible reaction of glycolysis. The high-energy enol phosphate in PEP is used to synthesize ATP from ADP and is another example of substrate-level phosphorylation (see Fig. 8.18).

1. **Feedforward regulation:** PK is activated by fructose 1,6-bisphosphate, the product of the PFK-1 reaction. This feedforward (instead of the more usual feedback) regulation has the effect of linking the two kinase activities: increased PFK-1 activity results in elevated levels of fructose 1,6-bisphosphate, which activates PK. [Note: PK is inhibited by ATP.]

2. **Covalent regulation in the liver:** Phosphorylation by cAMP-dependent PKA leads to inactivation of the hepatic isozyme of PK (Fig. 8.19). When blood glucose levels are low, elevated glucagon increases the intracellular level of cAMP, which causes the phosphorylation and inactivation of PK in the liver only. Therefore, PEP is unable to continue in glycolysis and, instead, enters the gluconeogenesis pathway. This partly explains the observed inhibition of hepatic glycolysis and stimulation of gluconeogenesis by glucagon. Dephosphorylation of PK by a phosphatase results in the reactivation of the enzyme.

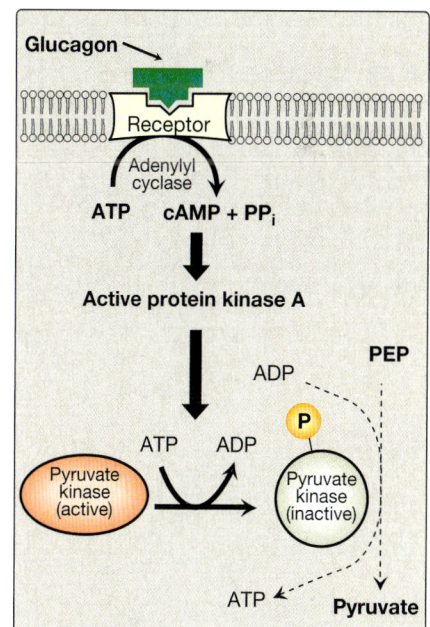

Figure 8.19
Covalent modification of hepatic pyruvate kinase results in inactivation of the enzyme. cAMP, cyclic adenosine monophosphate; PEP, phosphoenolpyruvate; Ⓟ, phosphate; PP_i, pyrophosphate; ADP, adenosine diphosphate.

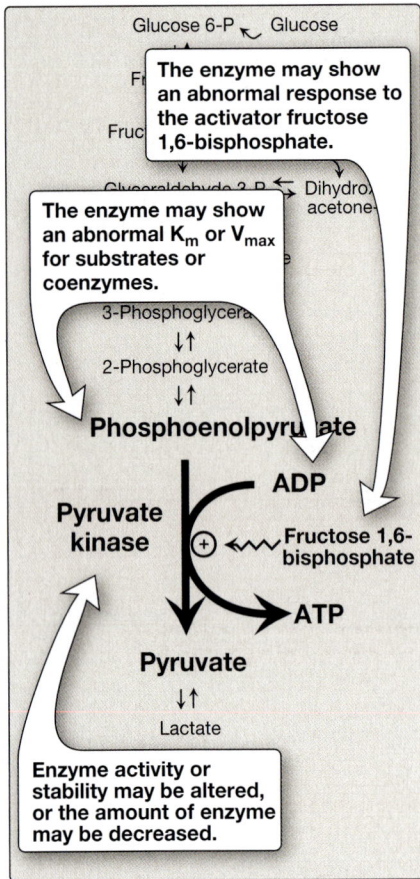

Figure 8.20
Alterations observed with various mutant forms of pyruvate kinase. K_m, Michaelis constant; V_{max}, maximal velocity; ADP, adenosine diphosphate.

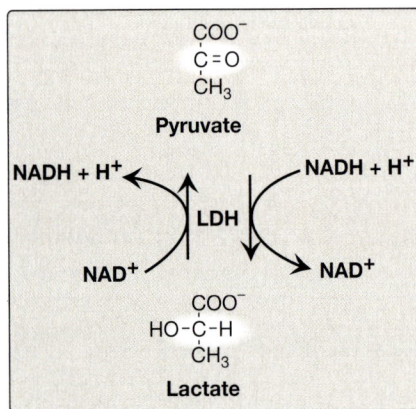

Figure 8.21
Interconversion of pyruvate and lactate by lactate dehydrogenase (LDH). NAD(H), nicotinamide adenine dinucleotide.

3. **PK deficiency:** Because mature RBCs lack mitochondria, they are completely dependent on glycolysis for ATP production. ATP is required to meet the metabolic needs of RBCs and to fuel the ion pumps necessary for the maintenance of the flexible, biconcave shape that allows them to squeeze through narrow capillaries. The anemia observed in glycolytic enzyme deficiencies is a consequence of the reduced rate of glycolysis, leading to decreased ATP production by substrate-level phosphorylation. The resulting alterations in the RBC membrane lead to changes in cell shape and, ultimately, to phagocytosis by cells of the mononuclear phagocyte system, particularly splenic macrophages. The premature death and lysis of RBC result in mild-to-severe hemolytic anemia, with the severe form requiring regular transfusions. Most patients with rare genetic defects of glycolytic enzymes have a deficiency in PK. [Note: Liver PK is encoded by the same gene as the RBC isozyme. However, liver cells show no effect because they can synthesize more PK and can also generate ATP by oxidative phosphorylation.] Severity depends both on the degree of enzyme deficiency (generally 5% to 35% of normal levels) and on the extent to which RBC compensate by synthesizing increased levels of 2,3-BPG. Almost all individuals with PK deficiency have a mutant enzyme that shows altered kinetics or decreased stability (Fig. 8.20). Individuals heterozygous for PK deficiency have resistance to the most severe forms of malaria.

> The tissue-specific expression of PK in RBC and the liver results from the use of different start sites in transcription (see Chapter 30) of the gene that encodes the enzyme.

K. Pyruvate reduction to lactate

Lactate, formed from pyruvate by LDH, is the final product of anaerobic glycolysis in eukaryotic cells (Fig. 8.21). Reduction to lactate is the major fate for pyruvate in tissues that are poorly vascularized (e.g., the lens and cornea of the eye and the kidney medulla) or in RBCs that lack mitochondria.

1. **Lactate formation in muscle:** In exercising skeletal muscle, NADH production (by glyceraldehyde 3-phosphate dehydrogenase and by the three NAD+-linked dehydrogenases of the TCA cycle; see also Chapter 9) exceeds the oxidative capacity of the ETC. This results in an elevated NADH/NAD+ ratio, favoring the reduction of pyruvate to lactate by LDH. Therefore, during intense exercise, lactate accumulates in muscle, causing a drop in the intracellular pH, potentially resulting in cramps. Much of this lactate eventually diffuses into the bloodstream and can be used by the liver to make glucose.

2. **Lactate utilization:** The direction of the LDH reaction depends on the relative intracellular concentrations of pyruvate and lactate and on the ratio of NADH/NAD+. For example, in the liver and heart, this ratio is lower than in exercising muscle. Consequently,

the liver and heart oxidize lactate (obtained from the blood) to pyruvate. In the liver, pyruvate is either converted to glucose by gluconeogenesis or converted to acetyl CoA that is oxidized in the TCA cycle. Heart muscle exclusively oxidizes lactate to carbon dioxide and water via the TCA cycle.

3. **Lactic acidosis:** Elevated concentrations of lactate in the plasma, termed lactic acidosis (type of metabolic acidosis), occur when the circulatory system collapses, such as with myocardial infarction, pulmonary embolism, and uncontrolled hemorrhage, or when an individual is in shock. The failure to bring adequate amounts of O_2 to the tissues results in impaired oxidative phosphorylation and decreased ATP synthesis. To survive, the cells rely on anaerobic glycolysis for generating ATP, producing lactic acid as the end product. [Note: Production of even meager amounts of ATP may be lifesaving during the period required to reestablish adequate blood flow to the tissues.] The additional O_2 required to recover from a period when O_2 availability has been inadequate is termed the "O_2 debt." [Note: The O_2 debt is often related to patient morbidity or mortality. In many clinical situations, measuring the blood levels of lactic acid allows the rapid, early detection of O_2 debt in patients and the monitoring of their recovery.]

L. Energy yield from glycolysis

Despite the production of some ATP by substrate-level phosphorylation during glycolysis, the end product, pyruvate or lactate, still contains most of the energy originally contained in glucose. The TCA cycle is required to release that energy completely.

1. **Anaerobic glycolysis:** A net two molecules of ATP are generated for each molecule of glucose converted to two molecules of lactate (Fig. 8.22). There is no net production or consumption of NADH.

2. **Aerobic glycolysis:** The generation of ATP is the same as in anaerobic glycolysis (i.e., a net gain of two ATP per molecule of glucose). Two molecules of NADH are also produced per molecule of glucose. Ongoing aerobic glycolysis requires the oxidation of most of this NADH by the ETC, producing the equivalent of 2.5 ATP for each NADH molecule entering the chain. [Note: NADH cannot cross the inner mitochondrial membrane, and substrate shuttles are required.]

VI. HORMONAL REGULATION

Regulation of the activity of the irreversible glycolytic enzymes by allosteric activation/inhibition or covalent phosphorylation/dephosphorylation is short term (i.e., the effects occur over minutes or hours). Superimposed on these effects on the activity of preexisting enzyme molecules are the long-term hormonal effects on the number of new enzyme molecules. These hormonal effects can result in 10- to 20-fold increases in enzyme synthesis that typically occur over hours to days.

Regular consumption of meals rich in carbohydrate or administration of insulin initiates an increase in the amount of glucokinase, PFK-1, and

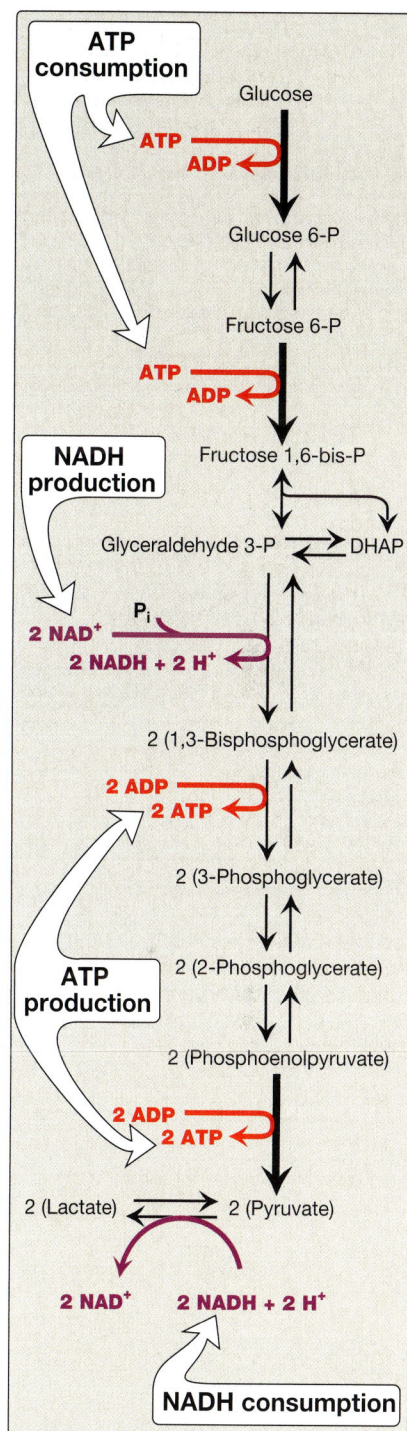

Figure 8.22
Summary of anaerobic glycolysis. Reactions involving the production or consumption of ATP or nicotinamide adenine dinucleotide (NADH) are indicated. The three irreversible reactions of glycolysis are shown with thick arrows. DHAP, dihydroxyacetone phosphate; ADP, adenosine diphosphate; P, phosphate.

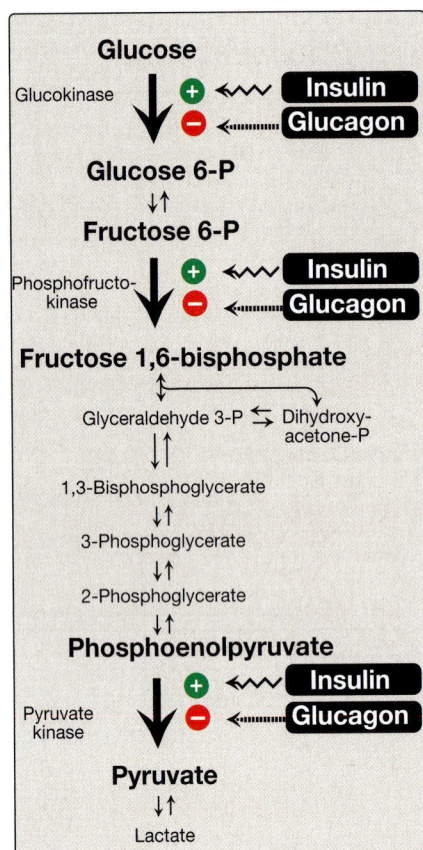

Figure 8.23
Effect of insulin and glucagon on the expression of key enzymes of glycolysis in the liver. P, phosphate.

PK in the liver (Fig. 8.23). The change reflects an increase in gene transcription, resulting in increased enzyme synthesis. Increased availability of these three enzymes favors the conversion of glucose to pyruvate, a characteristic of the absorptive state. [Note: The transcriptional effects of insulin and carbohydrate (specifically glucose) are mediated by the transcription factors sterol regulatory element–binding protein-1c and carbohydrate response element–binding protein, respectively. These factors also regulate the transcription of genes involved in fatty acid synthesis).] Conversely, gene expression of the three enzymes is decreased when plasma glucagon is high and insulin is low (e.g., as seen in fasting or diabetes).

VII. ALTERNATE FATES OF PYRUVATE

Pyruvate can be metabolized to products other than lactate.

A. Oxidative decarboxylation to acetyl CoA

Oxidative decarboxylation of pyruvate by the PDHC is an important pathway in tissues with a high oxidative capacity such as cardiac muscle (Fig. 8.24). PDHC irreversibly converts pyruvate, the end product of aerobic glycolysis, into acetyl CoA, a TCA cycle substrate and the carbon source for fatty acid synthesis.

B. Carboxylation to oxaloacetate

Carboxylation of pyruvate to oxaloacetate by pyruvate carboxylase is a biotin-dependent reaction (Fig. 8.24). This irreversible reaction is important because it replenishes the TCA cycle intermediate and provides substrate for gluconeogenesis.

C. Reduction to ethanol (microorganisms)

The reduction of pyruvate to ethanol occurs by the two reactions summarized in Figure 8.24. The decarboxylation of pyruvate to acetaldehyde by thiamine-requiring pyruvate decarboxylase occurs in yeast and certain other microorganisms but not in humans.

VIII. CHAPTER SUMMARY

- Most **pathways** can be classified as either **catabolic** (degrade complex molecules to a few simple products with **adenosine triphosphate [ATP] production**) or **anabolic** (synthesize complex end products from simple precursors with **ATP hydrolysis**).
- **Intercellular** signaling provides for the integration of metabolism. The primary route of this communication is chemical signaling by **hormones** or **neurotransmitters**.
- **Second-messenger molecules** are regulated in response to signaling via G protein–coupled receptors (GPCRs) and amplify the signal sent to the cell by the hormone.

- **Adenylyl cyclase** (AC) **is** a cell membrane enzyme regulated by GPCR that catalyzes the synthesis of **cyclic adenosine monophosphate** (**cAMP**) in response to hormones **glucagon** and **epinephrine**, as well as other hormones
- The cAMP produced activates **protein kinase A** (**PKA**), which phosphorylates a variety of enzymes, on serine/threonine residues, causing their activation or deactivation.
- Phosphorylation is reversed by **phosphoprotein phosphatases**.
- **Aerobic glycolysis**, in which **pyruvate** is the end product, occurs in cells with mitochondria and an adequate supply of O_2 (Fig. 8.25).
- **Anaerobic glycolysis**, in which **lactic acid** is the end product, occurs in cells that lack mitochondria and in cells deprived of sufficient O_2.
- Glucose is passively transported across membranes by **glucose transporters** (**GLUTs**) that have tissue-specific distributions.
- The oxidation of glucose to pyruvate (**glycolysis**; see Fig. 8.25) occurs through an **energy-investment** phase, in which phosphorylated intermediates are synthesized at the expense of ATP, and an **energy-generation** phase in which ATP is produced by **substrate-level phosphorylation**.
- **Hexokinase** has a **high affinity** (**low K_m**) and a **low maximal velocity** (**V_{max}**) for glucose and is inhibited by glucose 6-phosphate. **Glucokinase** has a high K_m and a high V_{max} for glucose. It is regulated indirectly by fructose 6-phosphate (inhibits) and glucose (activates) via **glucokinase regulatory protein** (**GKRP**).
- Glucose 6-phosphate is isomerized to **fructose 6-phosphate**, which is phosphorylated to **fructose 1,6-bisphosphate** by **phosphofructokinase-1** (**PFK-1**). **Two ATPs** are used during this phase of glycolysis.
- Fructose 1,6-bisphosphate is cleaved to form two trioses that are further metabolized by the glycolytic pathway, forming pyruvate. During this phase, **four ATP** and **two reduced nicotinamide adenine dinucleotides** (**NADHs**) are produced per glucose molecule.
- The final step in pyruvate synthesis from **phosphoenolpyruvate** (**PEP**) is catalyzed by **pyruvate kinase** (**PK**). **PK deficiency** accounts for the majority of all inherited defects in glycolytic enzymes. Effects are restricted to **red blood cells** and present as mild-to-severe **chronic, hemolytic anemia**.
- Glycolytic gene **transcription** is enhanced by insulin and glucose.

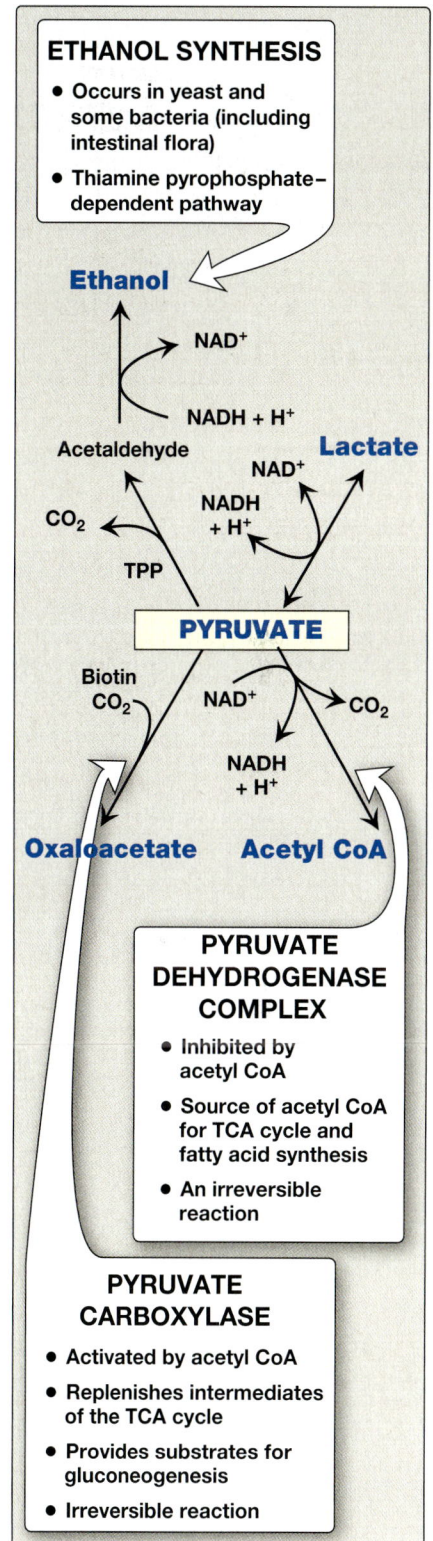

ETHANOL SYNTHESIS
- Occurs in yeast and some bacteria (including intestinal flora)
- Thiamine pyrophosphate–dependent pathway

PYRUVATE DEHYDROGENASE COMPLEX
- Inhibited by acetyl CoA
- Source of acetyl CoA for TCA cycle and fatty acid synthesis
- An irreversible reaction

PYRUVATE CARBOXYLASE
- Activated by acetyl CoA
- Replenishes intermediates of the TCA cycle
- Provides substrates for gluconeogenesis
- Irreversible reaction

Figure 8.24
Summary of the metabolic fates of pyruvate. TPP, thiamine pyrophosphate; TCA, tricarboxylic acid; NAD(H), nicotinamide adenine dinucleotide; CoA, coenzyme A; CO_2, carbon dioxide.

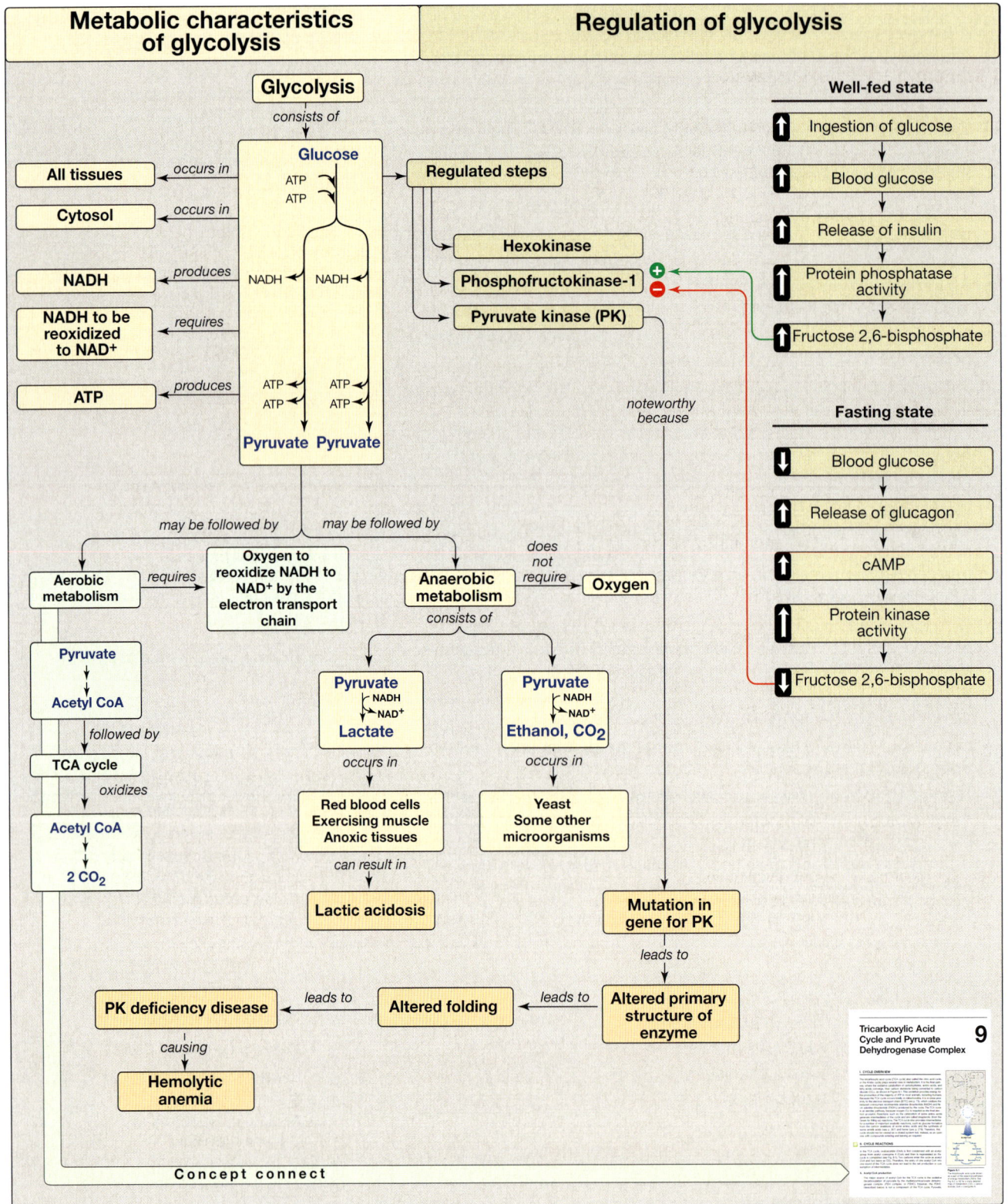

Metabolic characteristics of glycolysis

Regulation of glycolysis

Glycolysis

consists of

Glucose

ATP
ATP

NADH   NADH

ATP   ATP
ATP   ATP

Pyruvate   Pyruvate

occurs in → **All tissues**

occurs in → **Cytosol**

produces → **NADH**

requires → **NADH to be reoxidized to NAD⁺**

produces → **ATP**

Regulated steps

Hexokinase

Phosphofructokinase-1 ⊕ ⊖

Pyruvate kinase (PK)

noteworthy because

Well-fed state

↑ Ingestion of glucose

↑ Blood glucose

↑ Release of insulin

↑ Protein phosphatase activity

↑ Fructose 2,6-bisphosphate

Fasting state

↓ Blood glucose

↑ Release of glucagon

↑ cAMP

↑ Protein kinase activity

↓ Fructose 2,6-bisphosphate

may be followed by   *may be followed by*

Aerobic metabolism —*requires*→ **Oxygen to reoxidize NADH to NAD⁺ by the electron transport chain**

Anaerobic metabolism —*does not require*→ **Oxygen**

consists of

Pyruvate
NADH
NAD⁺
Lactate

Pyruvate
NADH
NAD⁺
Ethanol, CO₂

occurs in

**Red blood cells
Exercising muscle
Anoxic tissues**

**Yeast
Some other microorganisms**

can result in

Lactic acidosis

Pyruvate
↓
Acetyl CoA

followed by

TCA cycle

oxidizes

Acetyl CoA
↓
2 CO₂

Mutation in gene for PK

leads to

PK deficiency disease ←*leads to*— **Altered folding** ←*leads to*— **Altered primary structure of enzyme**

causing

Hemolytic anemia

Concept connect

Figure 8.25

Key concept map for glycolysis. NAD(H), nicotinamide adenine dinucleotide; cAMP, cyclic adenosine monophosphate; CoA, coenzyme A; TCA, tricarboxylic acid; CO₂, carbon dioxide.

Study Questions

Choose the ONE best answer.

8.1. Which of the following best describes the activity level and phosphorylation state of the listed hepatic enzymes in an individual who consumed a carbohydrate-rich meal about an hour ago? PFK-1, phosphofructokinase-1; PFK-2, phosphofructokinase-2; (P), phosphorylated.

Choice	PFK-1		PFK-2		Pyruvate Kinase	
	Activity	(P)	Activity	(P)	Activity	(P)
A.	Low	No	Low	No	Low	No
B.	High	Yes	Low	Yes	Low	Yes
C.	High	No	High	No	High	No
D.	High	Yes	High	Yes	High	Yes

Correct answer = C. Immediately following a meal, blood glucose levels and hepatic uptake of glucose increase. The glucose is phosphorylated to glucose 6-phosphate and used in glycolysis. In response to the rise in blood glucose, the insulin/glucagon ratio increases. As a result, the kinase domain of PFK-2 is dephosphorylated and active. Its product, fructose 2,6-bisphosphate, allosterically activates PFK-1. (PFK-1 is not covalently regulated). Active PFK-1 produces fructose 1,6-bisphosphate that is a feedforward activator of pyruvate kinase. Hepatic pyruvate kinase is covalently regulated, and the rise in insulin favors dephosphorylation and activation.

8.2. Which of the following statements is true for anabolic pathways only?

- A. Their irreversible (nonequilibrium) reactions are regulated.
- B. They are called cycles if they regenerate an intermediate.
- C. They are convergent and generate a few simple products.
- D. They are synthetic and require energy.
- E. They typically require oxidized coenzymes.

Correct answer = D. Anabolic processes are synthetic and energy requiring (endergonic). Statements A and B apply to both anabolic and catabolic processes, whereas C and E apply only to catabolic processes.

8.3. Compared with the resting state, vigorously contracting skeletal muscle shows:

- A. decreased AMP/ATP ratio.
- B. decreased levels of fructose 2,6-bisphosphate.
- C. decreased NADH/NAD⁺ ratio.
- D. increased oxygen availability.
- E. increased reduction of pyruvate to lactate.

Correct answer = E. Vigorously contracting skeletal muscle shows an increase in the reduction of pyruvate to lactate compared with resting muscle. The levels of reduced nicotinamide adenine dinucleotide (NADH) increase and exceed the oxidative capacity of the electron transport chain. Consequently, the levels of adenosine monophosphate (AMP) increase. The concentration of fructose 2,6-bisphosphate is not a key regulatory factor in skeletal muscle.

8.4. Glucose transport into:

- A. brain cells is through active transport.
- B. intestinal mucosal cells requires insulin.
- C. liver cells involves a glucose transporter.
- D. most cells is through simple diffusion.

Correct answer = C. Glucose uptake in the liver, brain, muscle, and adipose tissue is down a concentration gradient, and the transport is facilitated by tissue-specific glucose transporters (GLUTs). In adipose and muscle tissues, insulin is required for glucose uptake. Moving glucose against a concentration gradient requires energy and is seen with the sodium-dependent glucose cotransporter 1 (SGLT-1) of intestinal mucosal cells. Except for some gasses, membrane transport into cells does not occur via simple diffusion. All glucose transport utilizes GLUT or SGLT transport proteins.

8.5. Given that the K_m of glucokinase for glucose is 10 mM, whereas that of hexokinase is 0.1 mM, which isozyme will more closely approach V_{max} at the normal blood glucose concentration of 5 mM?

> Correct answer = Hexokinase. K_m (Michaelis constant) is the substrate concentration that gives one half V_{max} (maximal velocity). When blood glucose concentration is 5 mM, hexokinase ($K_m = 0.1$ mM) will be saturated, but glucokinase ($K_m = 10$ mM) will not.

8.6. In patients with pertussis infection and symptoms of whooping cough, $G\alpha_i$ is inhibited. How does this inhibition lead to a rise in cyclic adenosine monophosphate (cAMP)?

> Correct answer = G proteins of the $G\alpha_i$ type inhibit adenylyl cyclase (AC) when their associated G protein–coupled receptor is bound by ligand. If $G\alpha_i$ is inhibited by pertussis toxin, AC production of cAMP is inappropriately activated.

8.7. Which enzyme catalyzes conversion of 1,3-bisphosphglycerate to 3-phosphoglycerate in glycolysis?

    A. Enolase
    B. Glyceraldehyde-3-phosphate dehydrogenase
    C. Hexokinase
    D. Phosphoglycerate kinase
    E. Pyruvate kinase

> Correct answer = D. Phosphoglycerate kinase facilitates the transfer of a phosphate group from 1,3-bisphosphoglycerate to ADP, forming ATP and 3-phosphoglycerate. Hexokinase initiates glycolysis, and the other enzymes listed catalyze other reactions and not the conversion mentioned.

8.8. A 19-year-old man presents with unexplained weight loss, fatigue, and muscle weakness. Laboratory studies show elevated blood glucose and increased lactate. A partial deficiency in a regulatory enzyme of glycolysis is suspected. Deficiency of which of the following enzymes is most likely?

    A. Aldolase
    B. Hexokinase
    C. Phosphofructokinase-1 (PFK-1)
    D. Pyruvate kinase
    E. Triose phosphate isomerase (TPI)

> Correct answer = D. Pyruvate kinase deficiency can lead to impaired ATP production and to the unexplained weight loss, fatigue, and muscle weakness along with elevated blood glucose and lactate. The unique combination of symptoms, also including muscle weakness, makes pyruvate kinase the most likely deficient enzyme. Aldolase deficiency is associated with fructose intolerance. Hexokinase deficiency would impair the entire glycolytic pathway. PFK-1 deficiency is rare and could cause decreased glycolytic flux, with exercise intolerance and myopathy most likely to result. TPI deficiency is also very rare and most likely to cause hemolytic anemia and neurologic symptoms.

Tricarboxylic Acid Cycle and Pyruvate Dehydrogenase Complex

9

I. CYCLE OVERVIEW

The **tricarboxylic acid (TCA) cycle** can also be referred to as the citric acid cycle or the Krebs cycle and plays several roles in metabolism. It is the final pathway where the oxidative catabolism of carbohydrates, amino acids, and fatty acids converge, their carbon skeletons being converted to carbon dioxide (CO_2), as shown in Figure 9.1. This oxidation provides the energy to produce the majority of adenosine triphosphate (ATP) in most animals, including humans. Because the TCA cycle occurs totally in mitochondria, it is in close proximity to the electron transport chain (ETC), which oxidizes the reduced coenzymes nicotinamide adenine dinucleotide (NADH) and flavin adenine dinucleotide ($FADH_2$) produced by the cycle. The TCA cycle is an aerobic pathway because oxygen (O_2) is required as the final electron acceptor. Reactions such as the catabolism of some amino acids generate intermediates of the cycle and are called anaplerotic (from the Greek for "filling up") reactions. The TCA cycle also provides intermediates for important anabolic reactions, such as glucose formation from the carbon skeletons of some amino acids and the synthesis of some amino acids (see Chapter 20, V.) and heme (see Chapter 21, II. B.). This cycle should not be viewed as a closed system, but, instead, as an open one with compounds entering and leaving as required.

@ II. CYCLE REACTIONS

In the TCA cycle, oxaloacetate (OAA) is first condensed with an acetyl group from **acetyl coenzyme A (CoA)** and then is regenerated as the cycle is completed (see Fig. 9.1). Two carbons enter the cycle as acetyl CoA and two leave as CO_2. Therefore, the entry of one acetyl CoA into one round of the TCA cycle does not lead to the net production or consumption of intermediates.

A. Acetyl CoA production

The major source of acetyl CoA for the TCA cycle is the oxidative decarboxylation of **pyruvate** by the multienzyme **pyruvate dehydrogenase (PDH) complex (PDHC)**. However, the PDHC is not a component of

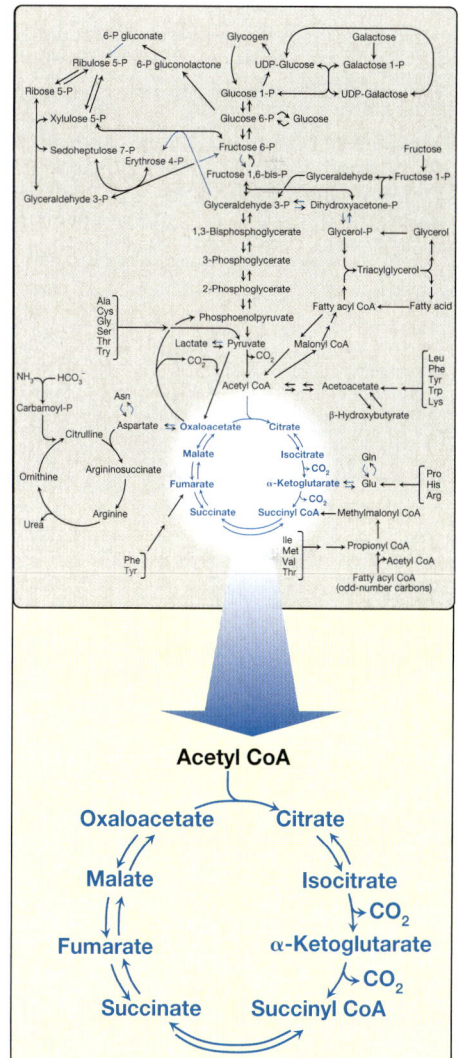

Figure 9.1
Tricarboxylic acid cycle shown as a part of the essential pathways of energy metabolism. [Note: See Fig. 8.2 for a more detailed map of metabolism.]
CO_2, carbon dioxide; CoA, coenzyme A.

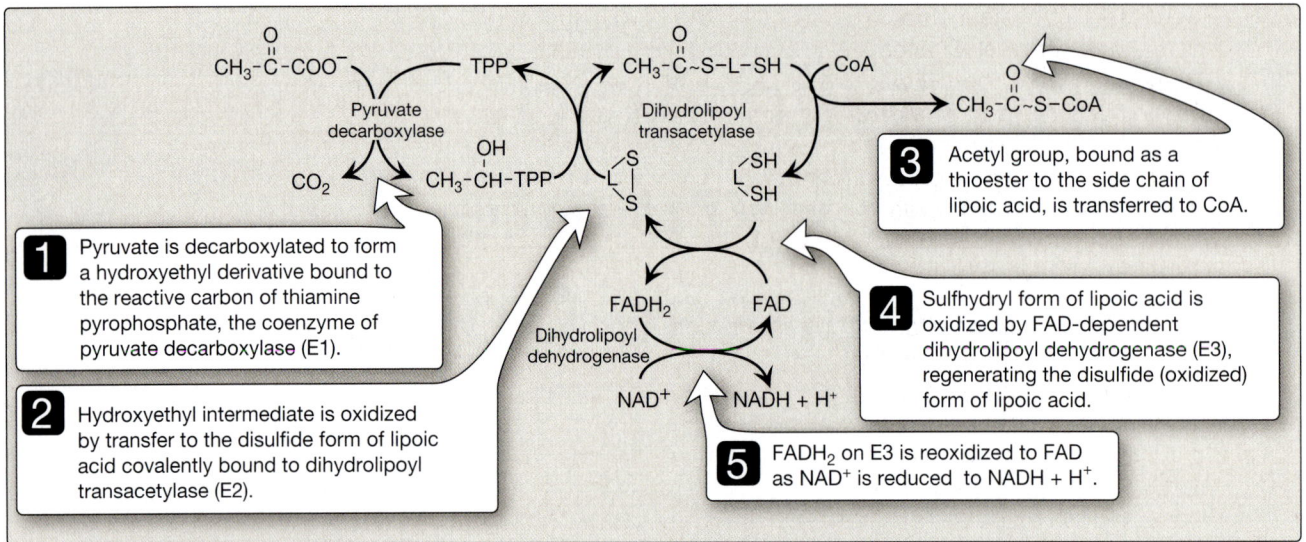

Figure 9.2
Mechanism of action of the enzymes (E) of the pyruvate dehydrogenase complex. [Note: All the coenzymes of the complex, except for lipoic acid, are derived from vitamins. TPP is from thiamine, FAD from riboflavin, NAD from niacin, and CoA from pantothenic acid.] CO_2, carbon dioxide; TPP, thiamine pyrophosphate; L, lipoic acid; CoA, coenzyme A; FAD(H_2) and NAD(H), flavin and nicotinamide adenine dinucleotides; ~, high-energy bond.

the TCA cycle. Pyruvate, the end product of glycolysis, is transported from the cytosol into the mitochondrial matrix by the pyruvate mitochondrial carrier of the inner mitochondrial membrane. In the matrix, the PDHC converts pyruvate to acetyl CoA. [Note: Fatty acid oxidation is another source of acetyl CoA (see Chapter 16, IV.).]

1. **PDHC component enzymes:** The PDHC is a protein aggregate of multiple copies of three enzymes, pyruvate decarboxylase ([E1] sometimes called PDH), dihydrolipoyl transacetylase (E2), and dihydrolipoyl dehydrogenase (E3). Each catalyzes a part of the overall reaction (Fig. 9.2). Their physical association links the reactions in proper sequence without the release of intermediates. In addition to the enzymes participating in the conversion of pyruvate to acetyl CoA, the PDHC also contains two regulatory enzymes, **PDH kinase** and **PDH phosphatase**.

2. **Coenzymes:** The PDHC contains five coenzymes that act as carriers or oxidants for the intermediates of the reactions shown in Figure 9.2. E1 requires **thiamine pyrophosphate (TPP)**, E2 requires **lipoic acid** and CoA, and E3 requires **FAD** and **NAD$^+$**. [Note: TPP, lipoic acid, and FAD are tightly bound to the enzymes and function as coenzymes–prosthetic groups.]

Deficiencies of thiamine or niacin can cause serious central nervous system problems because brain cells are unable to produce sufficient ATP via the TCA cycle if the PDHC is inactive. Wernicke–Korsakoff, an encephalopathy–psychosis syndrome caused by thiamine deficiency, may be seen in persons with alcohol use disorder.

3. **Regulation:** Covalent modifications by the two regulatory enzymes of the PDHC alternately activate and inactivate E1. **PDH kinase** phosphorylates and inactivates E1, whereas PDH phosphatase dephosphorylates and activates E1 (Fig. 9.3). The kinase itself is allosterically activated by ATP, acetyl CoA, and NADH. Therefore, in the presence of these high-energy products, the PDHC is turned off. [Note: It is the rise in the ATP/adenosine diphosphate (ADH), NADH/NAD$^+$, or acetyl CoA/CoA ratios that affects enzymic activity.]

Pyruvate is a potent inhibitor of PDH kinase, and, if pyruvate concentrations are elevated, E1 will be maximally active. Calcium (Ca^{2+}) is a strong activator of PDH phosphatase, stimulating E1 activity. This is particularly important in skeletal muscle, where Ca^{2+} release during contraction stimulates the PDHC and, thus, energy production. [Note: Although covalent regulation by the kinase and phosphatase is primary, the PDHC is also subject to product (NADH and acetyl CoA) inhibition.]

4. **Deficiency:** A deficiency of the α subunits of the tetrameric E1 component of the PDHC, although very rare, is the most common biochemical cause of **congenital lactic acidosis**. The deficiency results in a decreased ability to convert pyruvate to acetyl CoA, causing pyruvate to be shunted to lactate via lactate dehydrogenase. Because the brain relies on the TCA cycle for most of its energy and is particularly sensitive to acidosis, the central nervous system is impacted by this deficiency. Symptoms are variable and include neurodegeneration; abnormal muscle tightness due to prolonged contraction (muscle spasticity); and, in the neonatal-onset form, early death. The gene for the α-subunit is located on the X chromosome. The inheritance pattern is X-linked dominant in which inheritance of one X chromosome with the mutant allele results in disease. Both XY males and XX females are affected. Although there is no proven treatment for PDHC deficiency, dietary restriction of carbohydrate and supplementation with thiamine may reduce symptoms in select patients.

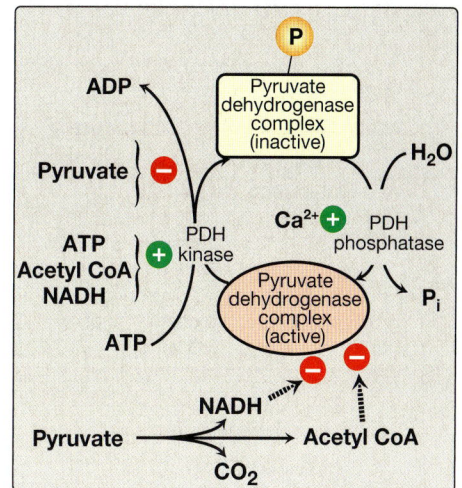

Figure 9.3
Regulation of pyruvate dehydrogenase (PDH) complex. (P) = phosphate (⤏ denotes product inhibition).

> Leigh syndrome (subacute necrotizing encephalomyelopathy) is a rare, progressive, neurodegenerative disorder caused by defects in mitochondrial ATP production, primarily caused by mutations in genes that encode proteins of the PDHC, the ETC, or ATP synthase. Both nuclear and mitochondrial DNA can be affected.

5. **Arsenic poisoning:** As described in Chapter 8, pentavalent arsenic (arsenate) can interfere with glycolysis at the glyceraldehyde 3-phosphate step, thereby decreasing ATP production. However, **arsenic poisoning** is due primarily to inhibition of enzyme complexes that require lipoic acid as a coenzyme, including PDH, α-ketoglutarate dehydrogenase (see Section E.), and branched-chain α-keto acid dehydrogenase (see Chapter 20, III.). Arsenite (the trivalent form of arsenic) forms a stable complex with the

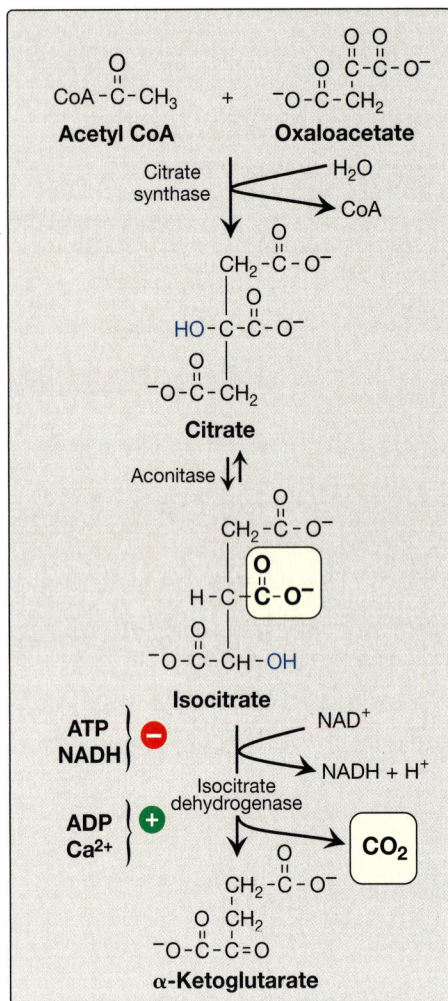

Figure 9.4
Formation of α-ketoglutarate from acetyl coenzyme A (CoA) and oxaloacetate. NAD(H), nicotinamide adenine dinucleotide; CO_2, carbon dioxide.

thiol ($-SH$) groups of lipoic acid, making that compound unavailable to serve as a coenzyme. When it binds to lipoic acid in the PDHC, pyruvate (and, consequently, lactate) accumulates. As with PDHC deficiency, this particularly affects the brain, causing neurologic disturbances and death.

B. Citrate synthesis

The irreversible condensation of acetyl CoA and OAA to form citrate (a TCA) is catalyzed by citrate synthase, the initiating enzyme of the TCA cycle (Fig. 9.4). This aldol condensation has a highly negative change in standard free energy (ΔG^0), which strongly favors citrate formation. The enzyme is inhibited by citrate (product inhibition). Substrate availability is another means of regulation for citrate synthase. The binding of OAA greatly increases the enzyme's affinity for acetyl CoA. [Note: Citrate, in addition to being an intermediate in the TCA cycle, is a source of acetyl CoA for the cytosolic synthesis of fatty acids and cholesterol. Citrate also inhibits phosphofructokinase-1 (PFK-1), the rate-limiting enzyme of glycolysis, and activates acetyl CoA carboxylase (the rate-limiting enzyme of fatty acid synthesis; see Chapter 16, III.).]

C. Citrate isomerization

Citrate is isomerized to **isocitrate** through hydroxyl group migration catalyzed by **aconitase** (aconitate hydratase), an iron-sulfur protein (see Fig. 9.4). [Note: Aconitase is inhibited by fluoroacetate, a plant toxin that is used as a pesticide. Fluoroacetate is converted to fluoroacetyl CoA that condenses with OAA to form fluorocitrate, a potent inhibitor of aconitase.]

D. Oxidative decarboxylation of isocitrate

Isocitrate dehydrogenase catalyzes the irreversible oxidative decarboxylation of isocitrate to α-**ketoglutarate**, yielding the first of three NADH molecules produced by the cycle and the first release of CO_2 (see Fig. 9.4). This is one of the rate-limiting steps of the TCA cycle. The enzyme is allosterically activated by ADP (a low-energy signal) and Ca^{2+} and is inhibited by ATP and NADH, levels of which are elevated when the cell has abundant energy stores.

E. Oxidative decarboxylation of α-ketoglutarate

The irreversible conversion of α-**ketoglutarate** to **succinyl CoA** is catalyzed by the α-ketoglutarate dehydrogenase complex, a protein aggregate of multiple copies of three enzymes (Fig. 9.5). The mechanism of this oxidative decarboxylation is very similar to that used for the conversion of pyruvate to acetyl CoA by the PDHC. The reaction releases the second CO_2 and produces the second NADH of the cycle. The coenzymes required are TPP, lipoic acid, FAD, NAD^+, and CoA. Each functions as part of the catalytic mechanism in a way analogous to that described for the PDHC. The large negative ΔG^0 of the reaction favors the formation of succinyl CoA, a high-energy thioester similar to acetyl CoA. The α-ketoglutarate dehydrogenase complex is inhibited by its products, NADH and succinyl CoA, and activated by

Ca^{2+}. However, it is not regulated by phosphorylation/dephosphorylation reactions as described for the PDHC. [Note: α-Ketoglutarate is also produced by the oxidative deamination and transamination of the amino acid glutamate.]

F. Succinyl CoA cleavage

Succinate thiokinase (also called succinyl CoA synthetase, named for the reverse reaction) cleaves the high-energy thioester bond of succinyl CoA (see Fig. 9.5). This reaction is coupled to phosphorylation of guanosine diphosphate (GDP) to guanosine triphosphate (GTP). GTP and ATP are energetically interconvertible by the nucleoside diphosphate kinase reaction:

$$GTP + ADP \rightleftarrows GDP + ATP$$

The generation of GTP by succinate thiokinase is another example of **substrate-level** phosphorylation (see Chapter 8). [Note: Succinyl CoA is also produced from propionyl CoA derived from the metabolism of fatty acids with an odd number of carbon atoms and from the metabolism of several amino acids. It can be converted to pyruvate for gluconeogenesis (see Chapter 10) or used in heme synthesis (see Chapter 21).]

G. Succinate oxidation

Succinate is oxidized to **fumarate** by succinate dehydrogenase, as its coenzyme FAD is reduced to $FADH_2$ (see Fig. 9.5). Succinate dehydrogenase is the only enzyme of the TCA cycle that is embedded in the inner mitochondrial membrane. As such, it functions as complex II of the ETC. [Note: FAD, rather than NAD^+, is the electron acceptor because the reducing power of succinate is not sufficient to reduce NAD^+.]

H. Fumarate hydration

Fumarate is hydrated to **malate** in a freely reversible reaction catalyzed by **fumarase** (fumarate hydratase; see Fig. 9.5). [Note: Fumarate is also produced by the urea cycle, in purine synthesis (see Fig. 22.7), and during catabolism of the amino acids phenylalanine and tyrosine.]

I. Malate oxidation

Malate is oxidized to OAA by malate dehydrogenase (Fig. 9.6). This reaction produces the third and final **NADH** of the cycle. The ΔG^0 of the reaction is positive, but the reaction is driven in the direction of OAA by the highly exergonic citrate synthase reaction. [Note: OAA is also produced by the transamination of the amino acid aspartic acid.]

III. ENERGY PRODUCED BY THE CYCLE

Four pairs of electrons are transferred during one turn of the TCA cycle: three pairs reducing three NAD^+ to NADH and one pair reducing FAD to $FADH_2$. Oxidation of one NADH by the ETC leads to the

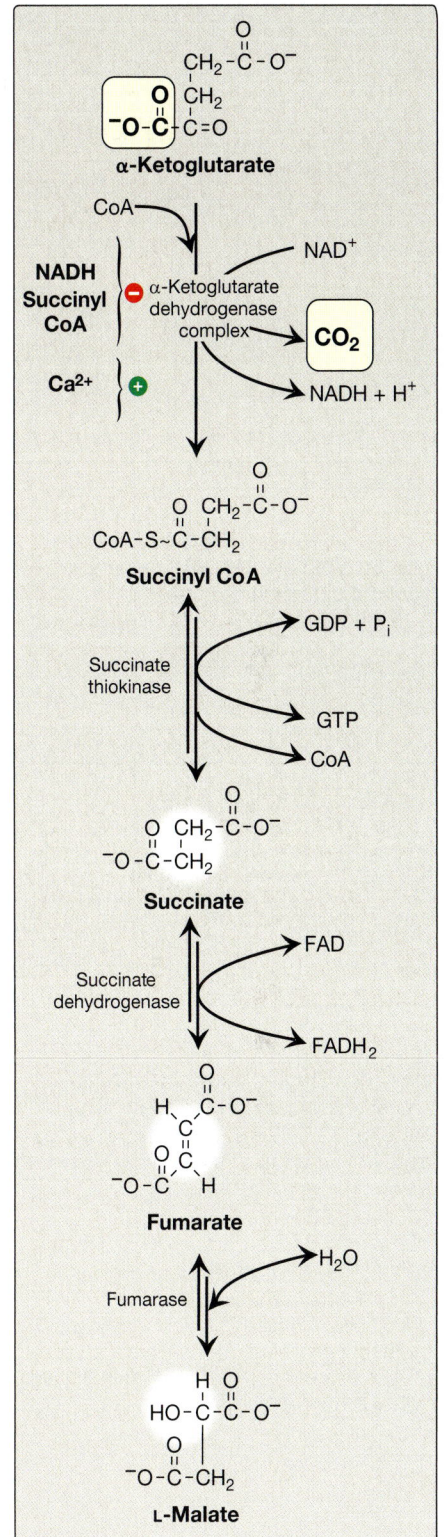

Figure 9.5
Formation of malate from α-ketoglutarate. FAD(H₂) and NAD(H), flavin and nicotinamide adenine dinucleotides; GDP and GTP, guanosine di- and triphosphates; ~, high-energy bond; CoA, coenzyme A.

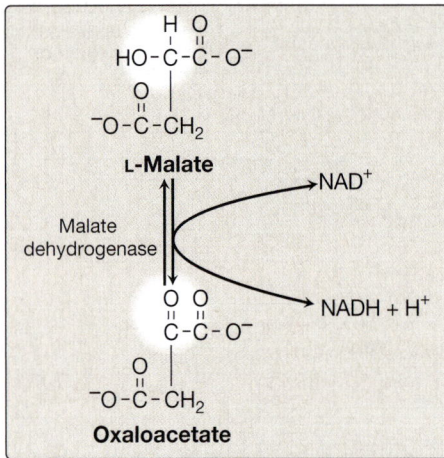

Figure 9.6
Formation (regeneration) of oxaloacetate from malate. NAD(H), nicotinamide adenine dinucleotide.

Figure 9.7
Number of ATP molecules produced from the oxidation of one molecule of acetyl coenzyme A (CoA) using both substrate-level and oxidative phosphorylation. NAD(H) and FAD(H$_2$), nicotinamide and flavin adenine dinucleotides; GDP and GTP, guanosine di- and triphosphates; P$_i$, inorganic phosphate. [Note: These numbers were previously reported as 3 ATP per NADH and 2 ATP per FADH$_2$ giving 12 ATP per acetyl CoA oxidized; however, recent recalculations of ATP yields are 2.5 ATP per NADH and 1.5 ATP per FADH$_2$.]

formation of 2.5 ATP, whereas oxidation of FADH$_2$ produces 1.5 ATP. The total yield of ATP from the oxidation of one acetyl CoA is shown in Figure 9.7. Figure 9.8 summarizes the reactions of the TCA cycle. [Note: The cycle does not involve the net consumption or production of intermediates. Two carbons entering as acetyl CoA are balanced by two CO$_2$ exiting.]

IV. CYCLE REGULATION

In contrast to glycolysis, which is regulated primarily by PFK-1, the TCA cycle is controlled by the regulation of several enzymes (see Fig. 9.8). The most important of these regulated enzymes are those that catalyze reactions with highly negative ΔG^0: citrate synthase, isocitrate dehydrogenase, and the α-ketoglutarate dehydrogenase complex. Reducing equivalents needed for oxidative phosphorylation are generated by the PDHC and the TCA cycle, and both processes are upregulated in response to a decrease in the ATP/ADP ratio.

V. CHAPTER SUMMARY

- In the **tricarboxylic acid (TCA) cycle**, also called the **Krebs cycle, pyruvate** is oxidatively decarboxylated by the **pyruvate dehydrogenase (PDH) complex (PDHC)**, producing **acetyl CoA** (Fig. 9.9).

- The multienzyme PDHC requires five coenzymes: **thiamine pyrophosphate (TPP), lipoic acid, flavin adenine dinucleotide (FAD), nicotinamide adenine dinucleotide (NAD$^+$)**, and **coenzyme A (CoA)**.

- The PDHC is regulated by covalent modification of **E1**, by **PDH kinase** and **PDH phosphatase**: phosphorylation inhibits E1.

- PDH kinase is allosterically activated by ATP, acetyl CoA, and NADH and inhibited by pyruvate. The phosphatase is activated by calcium (Ca^{2+}).

- **Pyruvate decarboxylase deficiency** is the most common biochemical cause of **congenital lactic acidosis**. The brain is particularly affected in this **X-linked dominant** disorder.

- **Arsenic poisoning** causes inactivation of the PDHC by binding to lipoic acid. In the TCA cycle, **citrate** is synthesized from **oxaloacetate (OAA)** and **acetyl CoA** by **citrate synthase**, which is inhibited by product.

- **Citrate** is isomerized to **isocitrate** by **aconitase (aconitate hydratase)**. Isocitrate is oxidatively decarboxylated by **isocitrate dehydrogenase** to **α-ketoglutarate**, producing **CO$_2$** and **NADH**. The enzyme is inhibited by ATP and NADH and activated by ADP and Ca^{2+}.

- α-Ketoglutarate is oxidatively decarboxylated to **succinyl CoA** by the **α-ketoglutarate dehydrogenase complex**, producing CO_2 and NADH. The enzyme is very similar to the PDHC and uses the same coenzymes.

- The α-ketoglutarate dehydrogenase complex is activated by Ca^{2+} and inhibited by NADH and succinyl CoA but is not covalently regulated. Succinyl CoA is cleaved by **succinate thiokinase** producing **succinate** and **guanosine triphosphate (GTP)**. This is an example of **substrate-level phosphorylation**.

- Succinate is oxidized to **fumarate** by **succinate dehydrogenase**, producing **$FADH_2$**. Fumarate is hydrated to **malate** by **fumarase (fumarate hydratase)**, and malate is oxidized to OAA by **malate dehydrogenase**, producing **NADH**.

- **Three NADH** and **one $FADH_2$** are produced by one round of the TCA cycle.

- Generation of acetyl CoA by oxidation of pyruvate via the PDHC also produces an NADH. Oxidation of the NADH and $FADH_2$ by the ETC yields 9 ATP per pyruvate, and 18 ATP from the 2 pyruvate per glucose. The terminal phosphate of the GTP produced by substrate-level phosphorylation in the TCA cycle can be transferred to ADP by nucleoside diphosphate kinase, yielding another ATP per pyruvate. Overall 20 ATP are generated via the citric acid cycle.

- Therefore, in aerobic metabolism, a total of 30–32 ATP are produced per glucose, (depending on the shuttle used [see Chapter 6]): 5–7 ATP from glucose to pyruvate, 5 ATP from the 2 NADH of the PDH complex, and 20 ATP from the citric acid cycle.

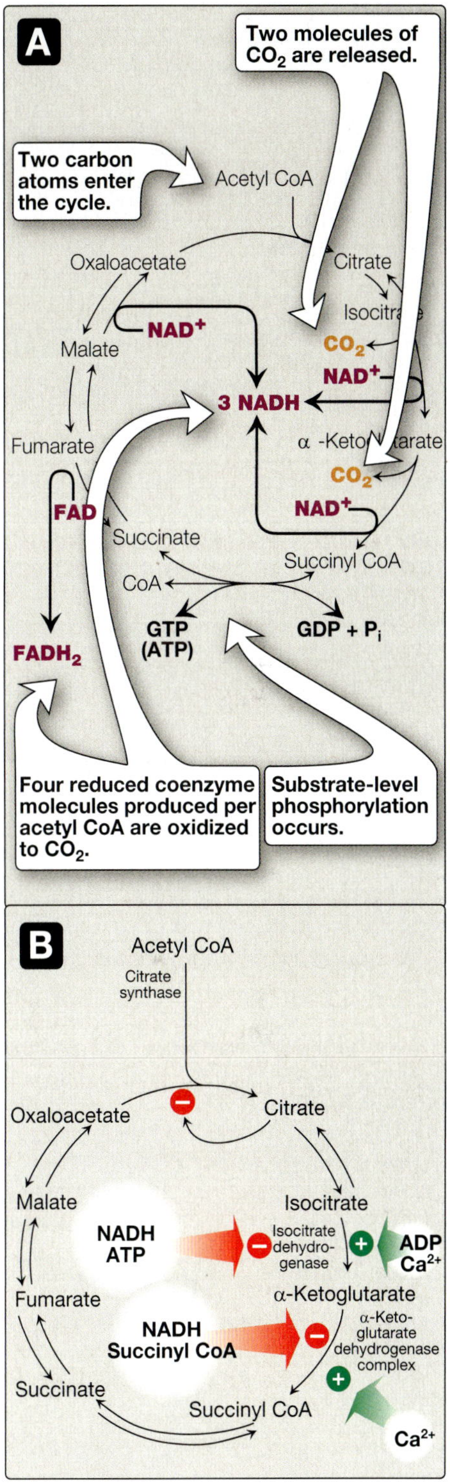

Figure 9.8
A. Production of reduced coenzymes, ATP, and carbon dioxide (CO_2) in the tricarboxylic acid cycle. [Note: Guanosine triphosphate (GTP) and ATP are interconverted by nucleoside diphosphate kinase.] **B.** Inhibitors and activators of the cycle.

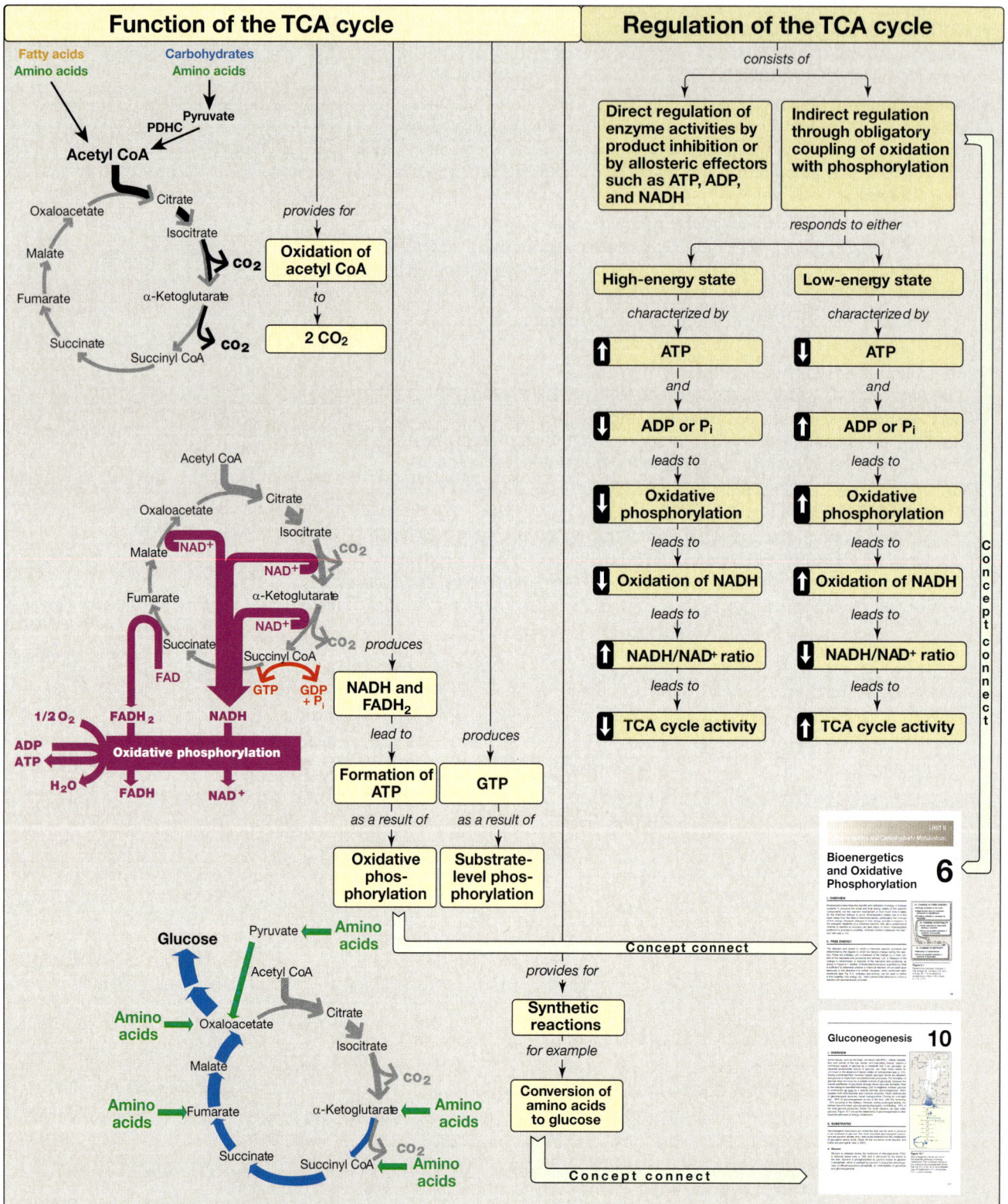

Figure 9.9
Key concept map for the tricarboxylic acid (TCA) cycle. PDHC, pyruvate dehydrogenase complex; CoA, coenzyme A; CO_2, carbon dioxide; NAD(H), nicotinamide adenine dinucleotide; $FAD(H_2)$, flavin adenine dinucleotide; GDP and GTP, guanosine di- and triphosphates; ADP, adenosine diphosphate; P_i, inorganic phosphate.

Study Questions

Choose the ONE best answer.

9.1. The conversion of pyruvate to acetyl coenzyme A and carbon dioxide:

    A. involves the participation of lipoic acid.

    B. is activated when pyruvate decarboxylase of the PDHC is phosphorylated by PDH kinase in the presence of ATP.

    C. is reversible.

    D. occurs in the cytosol.

    E. requires the coenzyme biotin.

Correct answer = A. Lipoic acid is an intermediate acceptor of the acetyl group formed in the reaction. [Note: Lipoic acid linked to a lysine residue in E2 functions as a "swinging arm" that allows interaction with E1 and E3.] The PDHC catalyzes an irreversible reaction that is inhibited when the decarboxylase component (E1) is phosphorylated. The PDHC is located in the mitochondrial matrix. Biotin is utilized by carboxylases, not decarboxylases.

9.2. Which one of the following conditions decreases the oxidation of acetyl coenzyme A by the citric acid cycle?

    A. High availability of calcium

    B. High acetyl CoA/CoA ratio

    C. Low ATP/ADP ratio

    D. Low NAD^+/NADH ratio

Correct answer = D. A low NAD^+/NADH (oxidized to reduced nicotinamide adenine dinucleotide) ratio limits the rates of the NAD^+-requiring dehydrogenases. High availability of calcium and substrate (acetyl coenzyme A) and a low ATP/ADP (adenosine tri- to diphosphate) ratio stimulate the cycle.

9.3. The following is the sum of three steps in the citric acid cycle.

$$A + B + FAD + H_2O \rightarrow C + FADH_2 + NADH.$$

Choose the lettered answer that corresponds to the missing "A," "B," and "C" in the equation.

Reactant A	Reactant B	Product C
A. Succinyl CoA	GDP	Succinate
B. Succinate	NAD^+	Oxaloacetate
C. Fumarate	NAD^+	Oxaloacetate
D. Succinate	NAD^+	Malate
E. Fumarate	GTP	Malate

Correct answer = B. Succinate + NAD^+ + FAD + H_2O → oxaloacetate + NADH + $FADH_2$.

9.4. A 1-month-old child shows neurologic problems and lactic acidosis. An enzyme activity assay for pyruvate dehydrogenase complex (PDHC) performed on extracts of cultured skin fibroblasts showed 5% of normal activity with a low concentration of thiamine pyrophosphate (TPP) but 80% of normal activity when the assay contained a 1,000-fold higher concentration of TPP. Which one of the following statements concerning this patient is correct?

    A. Administration of thiamine is expected to reduce his serum lactate level and improve his clinical symptoms.

    B. A high-carbohydrate diet would be expected to be beneficial for this patient.

    C. Citrate production from aerobic glycolysis is expected to be increased.

    D. PDH kinase, a regulatory enzyme of the PDHC, is expected to be active.

Correct answer = A. The patient appears to have a thiamine-responsive PDHC deficiency. The pyruvate decarboxylase (E1) component of the PDHC fails to bind thiamine pyrophosphate at low concentrations but shows significant activity at a high concentration of the coenzyme. This mutation, which affects the K_m (Michaelis constant) of the enzyme for the coenzyme, is present in some, but not all, cases of PDHC deficiency. Because the PDHC is an integral part of carbohydrate metabolism, a diet low in carbohydrates would be expected to blunt the effects of the enzyme deficiency. Aerobic glycolysis generates pyruvate, the substrate of the PDHC. Decreased activity of the complex decreases production of acetyl coenzyme A, a substrate for citrate synthase. Because PDH kinase is allosterically inhibited by pyruvate, it is inactive.

9.5. Which coenzyme–cosubstrate is used by dehydrogenases in both glycolysis and the tricarboxylic acid cycle?

Oxidized nicotinamide adenine dinucleotide (NAD^+) is used by glyceraldehyde 3-phosphate dehydrogenase of glycolysis and by isocitrate dehydrogenase, α-ketoglutarate dehydrogenase, and malate dehydrogenase of the tricarboxylic acid cycle. [Note: E3 of the pyruvate dehydrogenase complex requires oxidized flavin adenine dinucleotide (FAD) and NAD^+.]

10 Gluconeogenesis

I. OVERVIEW

Some tissues, including the brain, erythrocytes, kidney, medulla, lens and cornea of the eye, testes, and exercising skeletal muscle, require a continuous supply of glucose as a metabolic fuel. Liver glycogen is an essential source of stored glucose that can meet these needs for less than 24 hours in the absence of dietary intake of carbohydrate. During a prolonged fast, however, hepatic glycogen stores are depleted, and glucose must be synthesized from noncarbohydrate precursors. The formation of glucose does not occur by a simple reversal of glycolysis, because the overall equilibrium of glycolysis strongly favors pyruvate formation. Instead, glucose is synthesized *de novo* by a special pathway, called gluconeogenesis, which requires both mitochondrial and cytosolic enzymes. Deficiencies of enzymes involved in gluconeogenesis cause hypoglycemia. During an overnight fast, ~90% of gluconeogenesis occurs in the liver, with the remaining ~10% occurring in the kidneys. However, during prolonged fasting of 48 hours or longer, the kidneys contribute ~40% of the total glucose production. The small intestine can also make glucose. Figure 10.1 shows the relationship of gluconeogenesis to other essential pathways of energy metabolism.

II. SUBSTRATES

Gluconeogenic precursors are molecules that can be used to produce a net synthesis of glucose. The most important of these are glycerol, lactate, and α-**keto acids** obtained from the metabolism of glucogenic amino acids; 18 of the 20 amino acids are gluconeogenic, all except leucine and lysine.

A. Glycerol

Glycerol is released during the hydrolysis of triacylglycerols (TAGs) in adipose tissue and is delivered by the blood to the liver. Glycerol is phosphorylated by glycerol kinase to glycerol 3-phosphate, which is oxidized by glycerol 3-phosphate dehydrogenase to dihydroxyacetone phosphate, an intermediate of glycolysis and gluconeogenesis.

B. Lactate

Lactate from anaerobic glycolysis is released into the blood by exercising skeletal muscle and by erythrocytes that lack mitochondria.

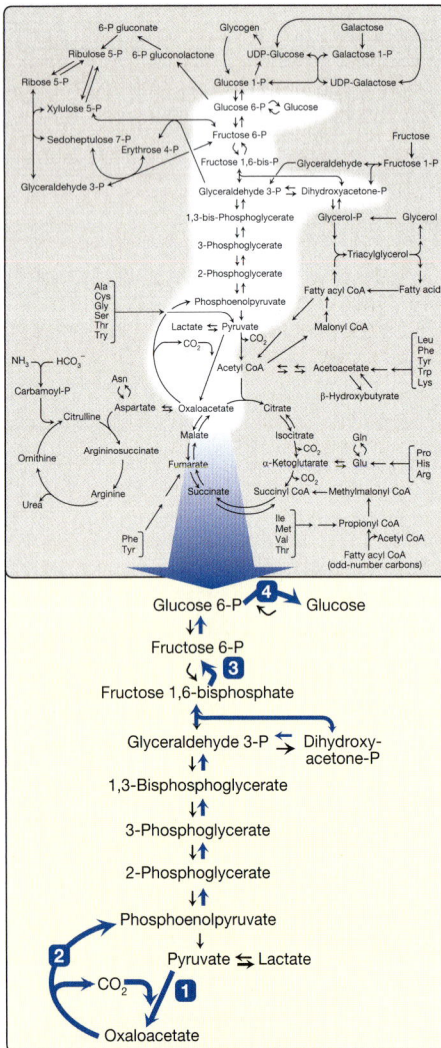

Figure 10.1
Gluconeogenesis shown as one of the essential pathways of energy metabolism. The numbered reactions are unique to gluconeogenesis. [Note: See Fig. 8.2, for a more detailed map of metabolism.] P, phosphate; CO_2, carbon dioxide.

In the Cori cycle, this lactate is taken up by the liver and oxidized to pyruvate that is converted to glucose, which is released back into the circulation (Fig. 10.2).

C. Amino acids

Amino acids produced by hydrolysis of tissue proteins are the major sources of glucose during a fast. Their metabolism generates α-keto acids, such as **pyruvate** that is converted to glucose, or **α-keto-glutarate** that can enter the tricarboxylic acid (TCA) cycle and form oxaloacetate (OAA), a direct precursor of **phosphoenolpyruvate (PEP)**. [Note: **Acetyl coenzyme A (CoA)** and compounds that give rise only to acetyl CoA (e.g., acetoacetate, lysine, and leucine) cannot give rise to a net synthesis of **glucose**. This is because of the irreversible nature of the pyruvate dehydrogenase complex (PDHC), which converts pyruvate to acetyl CoA. These compounds give rise instead to ketone bodies and are termed ketogenic.]

III. REACTIONS

Seven glycolytic reactions are reversible and are used in the synthesis of glucose from lactate or pyruvate. Three reactions in glycolysis are irreversible and must be circumvented by four alternate reactions that energetically favor the synthesis of glucose. These irreversible reactions, which together are unique to gluconeogenesis, are described below.

A. Pyruvate carboxylation

The first roadblock to overcome in the synthesis of glucose from pyruvate is the irreversible conversion in glycolysis of PEP to pyruvate in a reaction catalyzed by **pyruvate kinase** (**PK**). In gluconeogenesis, pyruvate is carboxylated by pyruvate carboxylase (PC) to OAA, which is converted to PEP by PEP-carboxykinase (PEPCK) (Fig. 10.3).

1. **Biotin:** PC requires the coenzyme biotin covalently bound to the ε-amino group of a lysine residue in the enzyme (see Fig. 10.3). ATP hydrolysis drives the formation of an enzyme–biotin–carbon dioxide (CO_2) intermediate, which then carboxylates pyruvate to form OAA. [Note: HCO_3^- provides the CO_2.] The PC reaction occurs in the mitochondria of liver and kidney cells and has two purposes: to allow the production of PEP, an important substrate for gluconeogenesis, and to provide OAA that can replenish the TCA cycle intermediates that may become depleted. Muscle cells also contain PC but use the OAA product only for the replenishment (anaplerotic) purpose and do not synthesize glucose. [Note: Pyruvate carrier protein moves pyruvate from the cytosol into mitochondria.]

> PC is one of several carboxylases that require biotin. Others include acetyl CoA carboxylase (p. 214), propionyl CoA carboxylase (p. 225), and methylcrotonyl CoA carboxylase (p. 225).

2. **Allosteric regulation:** PC is allosterically activated by acetyl CoA. Elevated levels of acetyl CoA in mitochondria signal

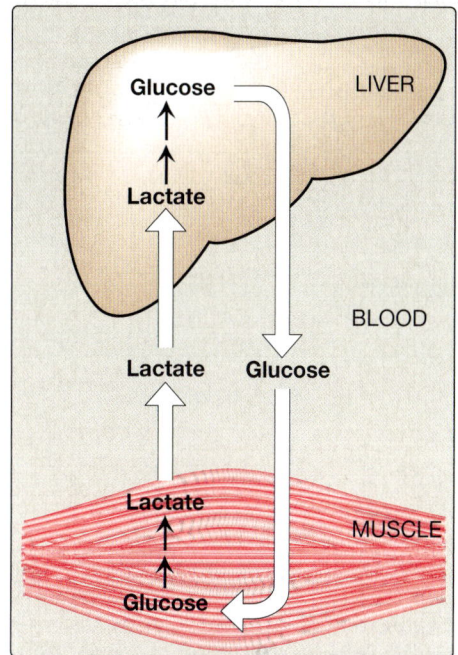

Figure 10.2
Intertissue Cori cycle links gluconeogenesis with glycolysis. [Note: Diffusion of lactate and glucose across membranes is facilitated by transport proteins.]

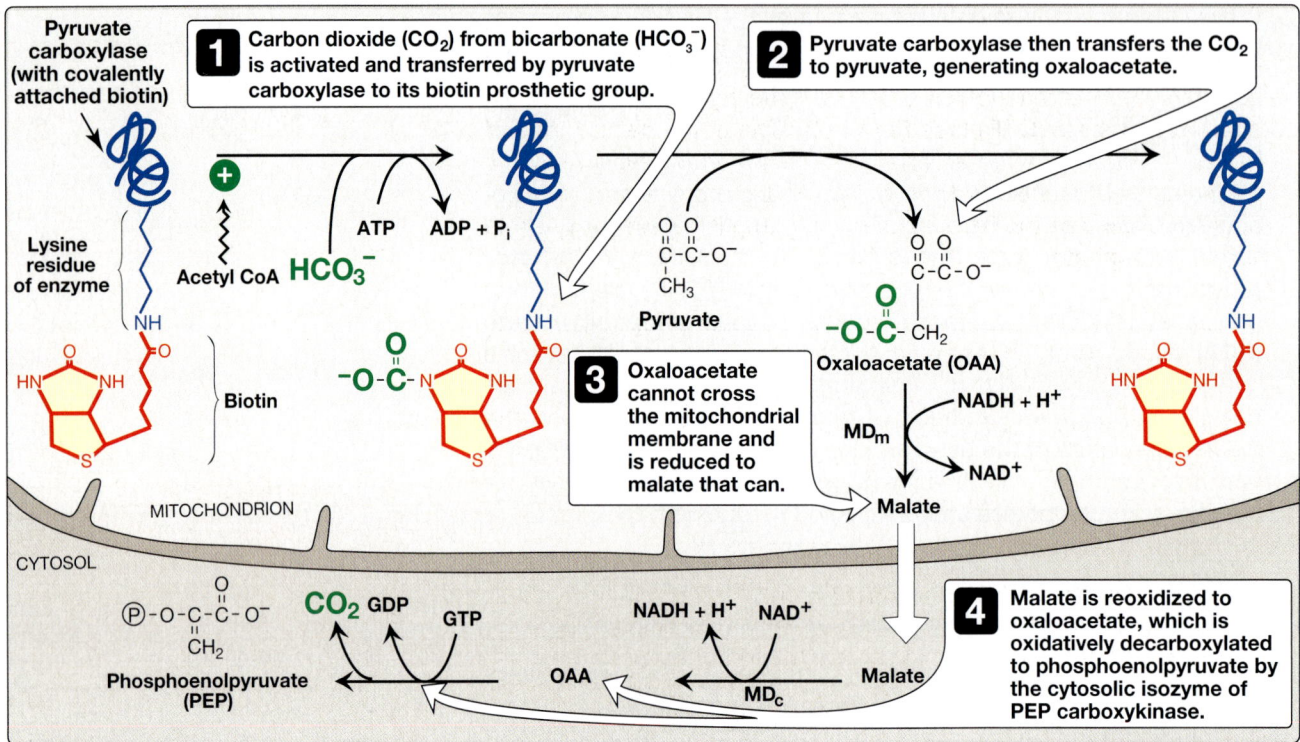

Figure 10.3
PEP synthesis in the cytosol. [Note: The process moves nicotinamide adenine dinucleotide (NADH)-reducing equivalents required for gluconeogenesis out of mitochondria into the cytosol.] MD_m and MD_c, mitochondrial and cytosolic isozymes of malate dehydrogenase; GTP and GDP, guanosine tri- and diphosphates; ADP, adenosine diphosphate.

a metabolic state that requires increased OAA synthesis. For example, this occurs during fasting, when OAA is used for gluconeogenesis in the liver and kidneys. Conversely, at low levels of acetyl CoA, PC is largely inactive, and pyruvate is primarily oxidized by the PDHC to acetyl CoA that can be further oxidized by the TCA cycle.

B. Oxaloacetate transport to the cytosol

For gluconeogenesis to continue, OAA must be converted to PEP by PEPCK. PEP production in the cytosol requires the transport of OAA out of mitochondria. However, the inner mitochondrial membrane lacks an OAA transporter, and OAA is first reduced to malate by mitochondrial malate dehydrogenase (MD). Malate is transported into the cytosol and reoxidized to OAA by cytosolic MD as nicotinamide adenine dinucleotide (NAD^+) is reduced to NADH (see Fig. 10.3). The NADH is used in the reduction of 1,3-bisphosphoglycerate to glyceraldehyde 3-phosphate by glyceraldehyde 3-phosphate dehydrogenase, a reaction common to glycolysis and gluconeogenesis. [Note: When abundant, lactate is oxidized to pyruvate as NAD^+ is reduced. The pyruvate is transported into mitochondria and carboxylated by PC to OAA, which can be converted to PEP by the mitochondrial isozyme of PEPCK. PEP is transported to the cytosol. OAA can also be converted to aspartate that is transported into the cytosol.]

C. Cytosolic oxaloacetate decarboxylation

OAA is decarboxylated and phosphorylated to PEP in the cytosol by PEPCK. The reaction is driven by the hydrolysis of guanosine triphosphate (GTP) (see Fig. 10.3). The combined actions of PC and PEPCK provide an energetically favorable pathway from pyruvate to PEP. PEP is then acted on by the reactions of glycolysis running in the reverse direction until it becomes fructose 1,6-bisphosphate.

> The pairing of carboxylation with decarboxylation drives reactions that would otherwise be energetically unfavorable. This strategy is also used in fatty acid (FA) synthesis.

D. Fructose 1,6-bisphosphate dephosphorylation

Hydrolysis of fructose 1,6-bisphosphate by fructose 1,6-bisphosphatase, found in the liver and kidneys, bypasses the irreversible **phosphofructokinase-1** (**PFK-1**) reaction of glycolysis and provides an energetically favorable pathway for the formation of fructose 6-phosphate (Fig. 10.4). This reaction is an important regulatory site of gluconeogenesis.

1. **Regulation by intracellular energy levels:** Fructose 1,6-bisphosphatase is inhibited by a rise in the ratio of adenosine monophosphate (AMP) to ATP, called the AMP-to-ATP ratio, which signals a low-energy state in the cell. Conversely, low AMP and high ATP levels stimulate gluconeogenesis, an energy-requiring pathway.

2. **Regulation by fructose 2,6-bisphosphate:** Fructose 1,6-bisphosphatase is inhibited by fructose 2,6-bisphosphate, an allosteric effector whose concentration is influenced by the insulin/glucagon ratio. When glucagon is high, the effector is not made by hepatic PFK-2, and thus, the phosphatase is active (Fig. 10.5). [Note: The signals that inhibit (low-energy, high fructose 2,6-bisphosphate) or activate (high-energy, low fructose 2,6-bisphosphate) gluconeogenesis have the opposite effect on glycolysis, providing reciprocal control of the pathways that synthesize and oxidize glucose.]

E. Glucose 6-phosphate dephosphorylation

Glucose 6-phosphate hydrolysis by **glucose 6-phosphatase** bypasses the irreversible hexokinase/glucokinase reaction and provides an energetically favorable pathway for the formation of free glucose (Fig. 10.6). The liver is the primary organ that produces free glucose from glucose 6-phosphate. This process requires a complex of two proteins found only in gluconeogenic tissue: glucose 6-phosphate translocase, which transports glucose 6-phosphate across the endoplasmic reticular (ER) membrane, and glucose 6-phosphatase, which removes the phosphate, producing free glucose (see Fig. 10.6). These ER membrane proteins are also required for the final step of glycogen degradation.

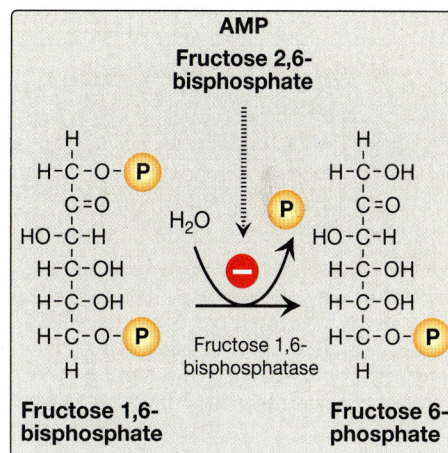

Figure 10.4
Dephosphorylation of fructose 1,6-bisphosphate. AMP, adenosine monophosphate; (P), phosphate.

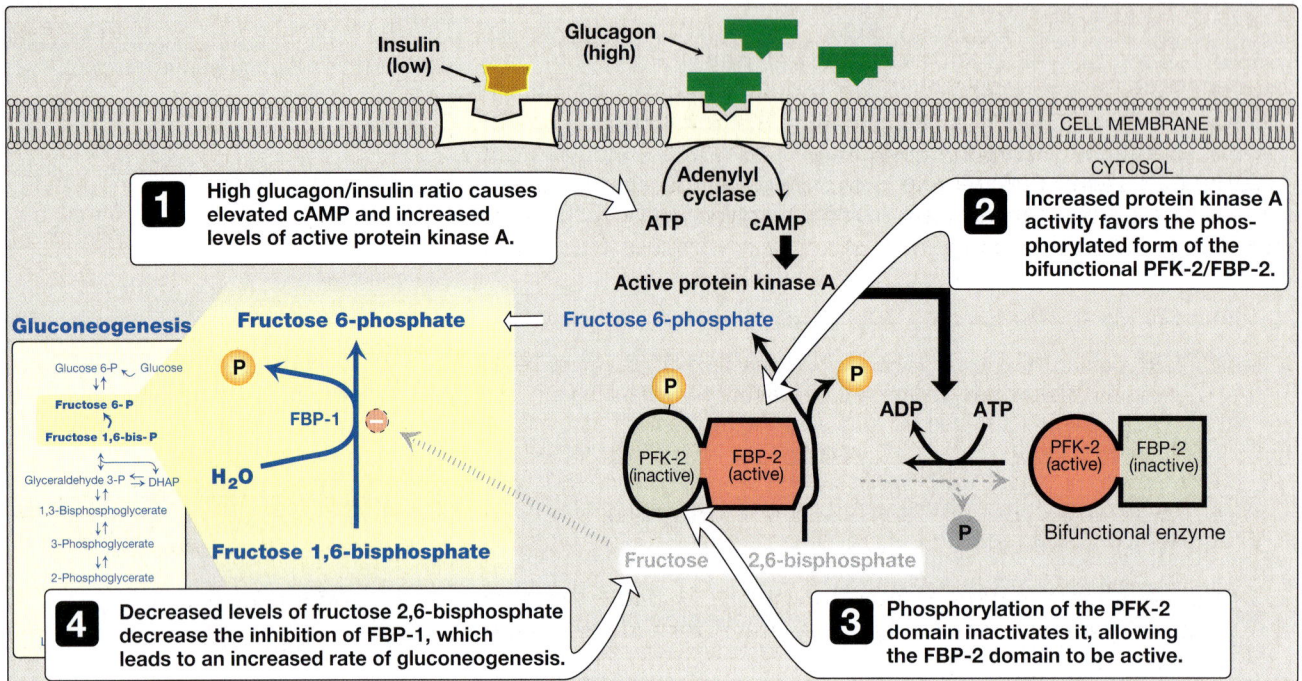

Figure 10.5
Effect of elevated glucagon on the intracellular concentration of fructose 2,6-bisphosphate in the liver. AMP and ADP, adenosine mono- and diphosphates; cAMP, cyclic AMP; PFK-2, phosphofructokinase-2; FBP-2, fructose 2,6-bisphosphatase; FBP-1, fructose 1,6-bisphosphatase; (P) and (P), phosphate.

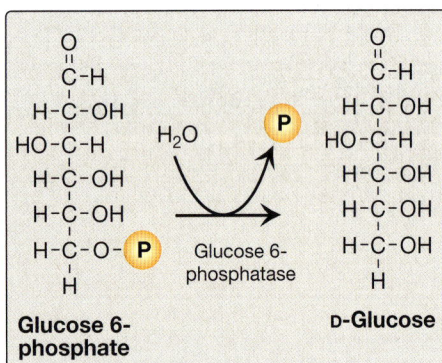

Figure 10.6
Dephosphorylation of glucose 6-phosphate allows release of free glucose from gluconeogenic tissues (primarily the liver) into blood. (P), phosphate.

Glycogen storage diseases Ia and Ib, caused by deficiencies in the phosphatase and the translocase, respectively, are characterized by severe fasting hypoglycemia, because free glucose is unable to be produced from either gluconeogenesis or glycogenolysis. Specific transporters are responsible for moving the free glucose into the cytosol and then into blood.

F. Summary of the reactions of glycolysis and gluconeogenesis

Of the 11 reactions required to convert pyruvate to free glucose, 7 are catalyzed by reversible glycolytic enzymes (Fig. 10.7). The 3 irreversible reactions, catalyzed by hexokinase/glucokinase, PFK-1, and PK, are circumvented by reactions catalyzed by glucose 6-phosphatase, fructose 1,6-bisphosphatase, PC, and PEPCK. In gluconeogenesis, the equilibria of the reversible glycolytic reactions are pushed toward glucose synthesis as a result of the essentially irreversible formation of PEP, fructose 6-phosphate, and glucose by the gluconeogenic enzymes. [Note: The stoichiometry of gluconeogenesis from two pyruvate molecules couples the cleavage of six high-energy phosphate bonds and the oxidation of 2 NADH with the formation of one glucose molecule (see Fig. 10.7).]

IV. REGULATION

The moment-to-moment regulation of gluconeogenesis is determined primarily by the circulating level of glucagon and by the availability of gluconeogenic substrates. In addition, slow adaptive changes in enzyme amount result from an alteration in the rate of enzyme synthesis or degradation or both. [Note: The hormone insulin normally inhibits gluconeogenesis. Hormonal control of the glucoregulatory system is presented in Chapter 23.]

A. Glucagon

Glucagon is a peptide hormone secreted by pancreatic islet α-cells, and it stimulates gluconeogenesis by three mechanisms.

1. **Changes in allosteric effectors:** Glucagon lowers hepatic fructose 2,6-bisphosphate, resulting in fructose 1,6-bisphosphatase activation and PFK-1 inhibition, thereby favoring gluconeogenesis over glycolysis (see Fig. 10.5).

2. **Covalent modification of enzyme activity:** Glucagon binds its G protein–coupled receptor and, via an elevation in cyclic AMP (cAMP) level and cAMP-dependent protein kinase A activity, stimulates the conversion of hepatic PK to its inactive (phosphorylated) form. This decreases PEP conversion to pyruvate, which has the effect of diverting PEP to gluconeogenesis (Fig. 10.8).

3. **Induction of enzyme synthesis:** Glucagon increases transcription of the gene for PEPCK via the transcription factor cAMP response element–binding (CREB) protein, thereby increasing the availability of this enzyme as levels of its substrate rise during fasting. Cortisol, a glucocorticoid, also increases expression of the gene, whereas insulin decreases expression.

B. Substrate availability

The availability of gluconeogenic precursors, particularly glucogenic amino acids, significantly influences the rate of glucose synthesis. Decreased insulin levels favor mobilization of amino acids from muscle protein to provide the carbon skeletons for gluconeogenesis. The ATP and NADH coenzymes required for gluconeogenesis are primarily provided by FA oxidation.

C. Allosteric activation by acetyl CoA

Allosteric activation of hepatic PC by acetyl CoA occurs during fasting. As a result of increased TAG hydrolysis in adipose tissue, the liver is flooded with FA. The rate of formation of acetyl CoA by β-oxidation of these FA exceeds the capacity of the liver to oxidize it to CO_2 and water. As a result, acetyl CoA accumulates and activates PC. [Note: Acetyl CoA inhibits the PDHC (by activating PDH kinase). Thus, this single compound can divert pyruvate toward gluconeogenesis and away from the TCA cycle (Fig. 10.9).]

D. Allosteric inhibition by AMP

Fructose 1,6-bisphosphatase is inhibited by AMP, a compound that activates PFK-1. This results in reciprocal regulation of glycolysis

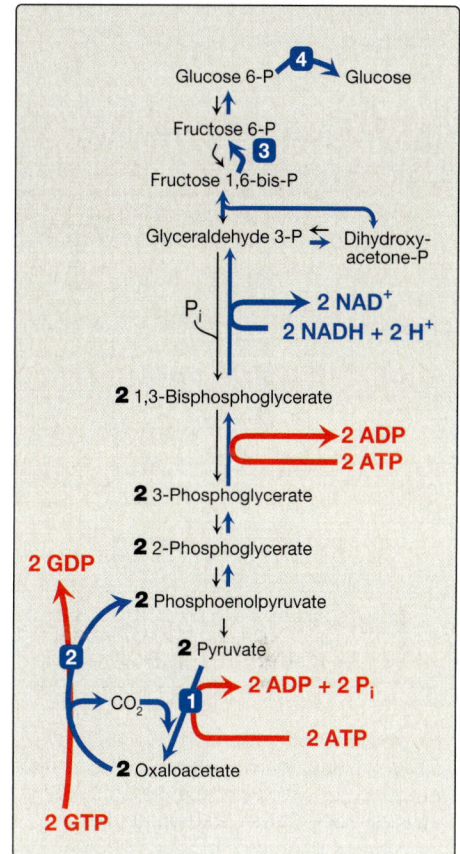

Figure 10.7
Summary of the reactions of glycolysis and gluconeogenesis, showing the energy requirements of gluconeogenesis. The numbered reactions are unique to gluconeogenesis. P, phosphate; GDP and GTP, guanosine di- and triphosphates; NAD(H), nicotinamide adenine dinucleotide; ADP, adenosine diphosphate.

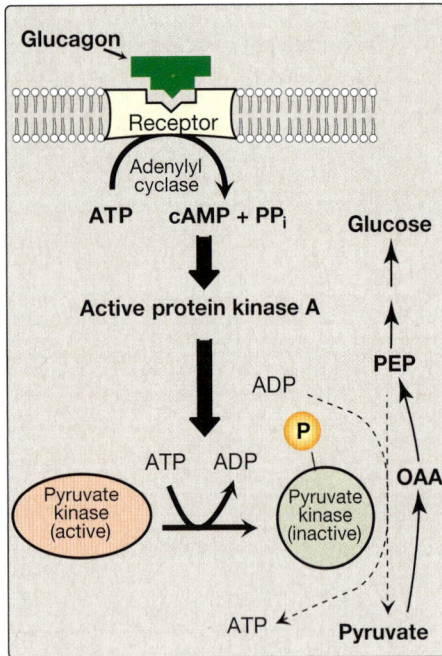

Figure 10.8
Covalent modification of pyruvate kinase results in inactivation of the enzyme. [Note: Only the hepatic isozyme is subject to covalent regulation.] OAA, oxaloacetate; PEP, phosphoenolpyruvate; PP$_i$, pyrophosphate; P, phosphate; AMP and ADP, adenosine mono- and diphosphates; cAMP, cyclic AMP.

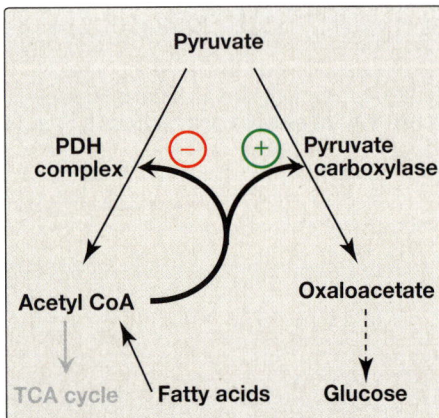

Figure 10.9
Acetyl coenzyme A (CoA) diverts pyruvate away from oxidation and toward gluconeogenesis. PDH, pyruvate dehydrogenase; TCA, tricarboxylic acid.

and gluconeogenesis seen previously with **fructose 2,6-bisphosphate**. Thus, elevated AMP stimulates energy-producing pathways and inhibits energy-requiring ones.

V. CHAPTER SUMMARY

- **Gluconeogenic precursors** include **glycerol** released during triacylglycerol hydrolysis in adipose tissue, **lactate** released by cells that lack mitochondria and by exercising skeletal muscle, and **α-keto acids** (e.g., **α-ketoglutarate** and **pyruvate**) derived from glucogenic amino acid metabolism (Fig. 10.10).

- Seven of the reactions of glycolysis are reversible and are used for gluconeogenesis in the liver and kidneys.

- Three reactions, catalyzed by **pyruvate kinase**, **phosphofructokinase-1**, and glucokinase/hexokinase, are physiologically irreversible and must be circumvented.

- **Pyruvate** is converted to **oxaloacetate** and then to **phosphoenolpyruvate** (**PEP**) by **pyruvate carboxylase** and **PEP-carboxykinase** (**PEPCK**).

- PC requires **biotin** and adenosine triphosphate (**ATP**) and is allosterically activated by **acetyl CoA**, and PEPCK requires **guanosine triphosphate** (**GTP**).

- Fructose 1,6-bisphosphate is converted to fructose 6-phosphate by **fructose 1,6-bisphosphatase**. This enzyme is inhibited by a high adenosine monophosphate (AMP)/ATP ratio and by **fructose 2,6-bisphosphate**, the primary allosteric activator of glycolysis.

- **Glucose 6-phosphate** is dephosphorylated to **glucose** by **glucose 6-phosphatase**. This enzyme of the ER membrane catalyzes the final step in gluconeogenesis and in glycogen degradation. Its deficiency results in severe, fasting hypoglycemia.

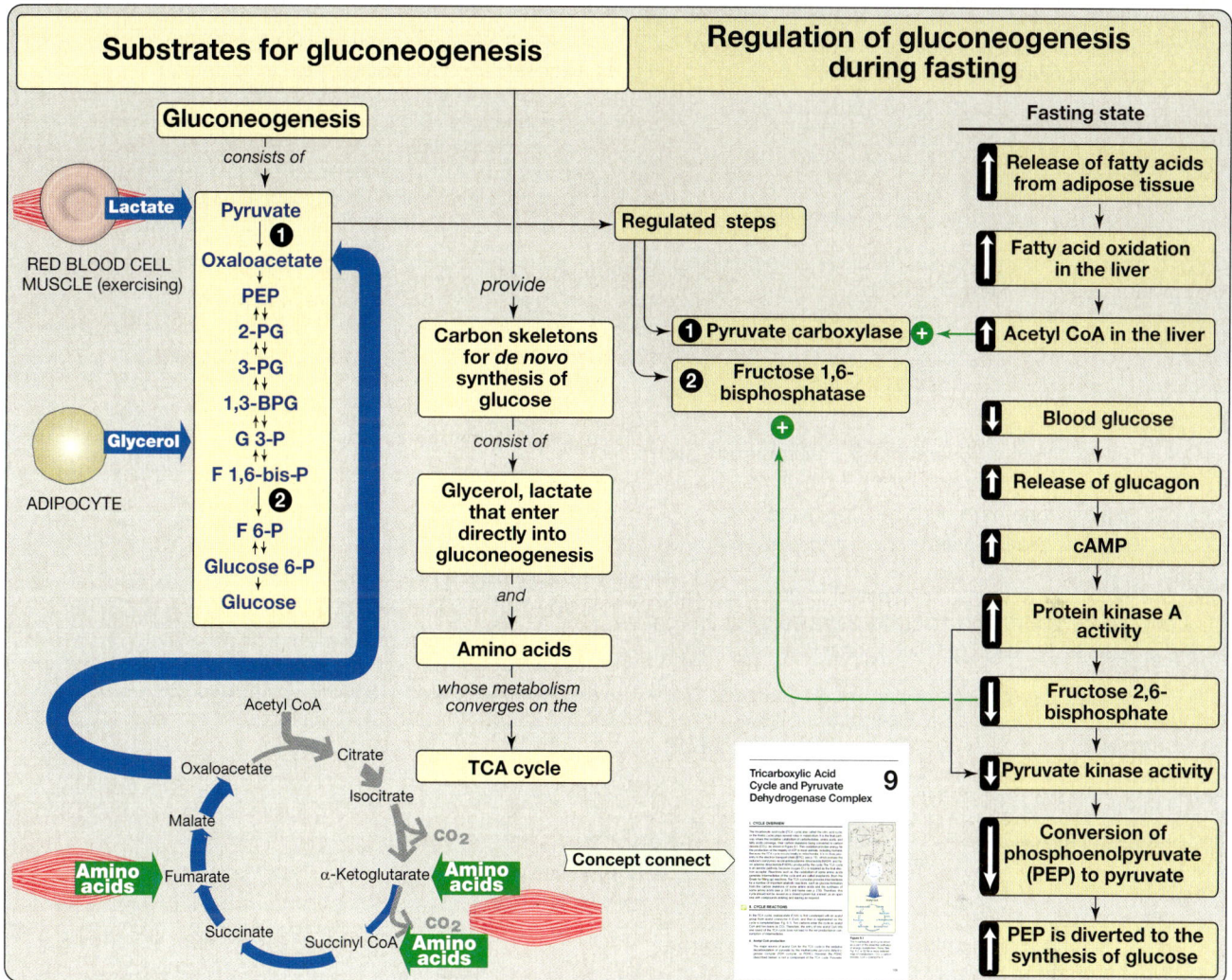

Substrates for gluconeogenesis

Gluconeogenesis

consists of

Lactate → Pyruvate ❶
RED BLOOD CELL
MUSCLE (exercising)

Pyruvate ❶
Oxaloacetate
PEP
2-PG
3-PG
1,3-BPG
G 3-P
F 1,6-bis-P ❷
F 6-P
Glucose 6-P
Glucose

ADIPOCYTE → Glycerol

provide

Carbon skeletons for *de novo* synthesis of glucose

consist of

Glycerol, lactate that enter directly into gluconeogenesis

and

Amino acids

whose metabolism converges on the

TCA cycle

Acetyl CoA
Oxaloacetate
Citrate
Isocitrate
Malate
CO_2
Fumarate
α-Ketoglutarate → Amino acids
Succinate
Succinyl CoA → Amino acids
CO_2
Amino acids

Concept connect

Tricarboxylic Acid Cycle and Pyruvate Dehydrogenase Complex 9

Regulation of gluconeogenesis during fasting

Fasting state

↑ Release of fatty acids from adipose tissue

↑ Fatty acid oxidation in the liver

↑ Acetyl CoA in the liver

Regulated steps

❶ Pyruvate carboxylase ⊕

❷ Fructose 1,6-bisphosphatase ⊕

↓ Blood glucose

↑ Release of glucagon

↑ cAMP

↑ Protein kinase A activity

↓ Fructose 2,6-bisphosphate

↓ Pyruvate kinase activity

↓ Conversion of phosphoenolpyruvate (PEP) to pyruvate

↑ PEP is diverted to the synthesis of glucose

Figure 10.10
Key concept map for gluconeogenesis. TCA, tricarboxylic acid; CoA, coenzyme A; cAMP, cyclic adenosine monophosphate; P, phosphate; (B)PG, (bis)phosphoglycerate; G, glyceraldehyde; F, fructose; CO_2, carbon dioxide.

Study Questions

Choose the ONE best answer.

10.1. Which of the following statements concerning gluconeogenesis is correct?

A. It is an energy-producing (exergonic) process.

B. It is important in maintaining blood glucose during a 2-day fast.

C. It is inhibited by a fall in the insulin/glucagon ratio.

D. It occurs in the cytosol of muscle cells.

E. It uses carbon skeletons provided by fatty acid degradation.

Correct answer = B. During a 2-day fast, glycogen stores are depleted, and gluconeogenesis maintains blood glucose. This is an energy-requiring (endergonic) pathway (both ATP and GTP get hydrolyzed) that occurs primarily in the liver, with the kidneys becoming major glucose producers in prolonged fasting. Gluconeogenesis uses both mitochondrial and cytosolic enzymes and is stimulated by a fall in the insulin/glucagon ratio. Fatty acid degradation yields acetyl coenzyme A (CoA), which cannot be converted to glucose. This is because the tricarboxylic acid cycle yields no net gain of carbons from acetyl CoA, and the pyruvate dehydrogenase complex is physiologically irreversible. It is the carbon skeletons of most amino acids that are glucogenic.

10.2. Which reaction in the diagram would be inhibited in the presence of large amounts of avidin, an egg white protein that binds and sequesters biotin?

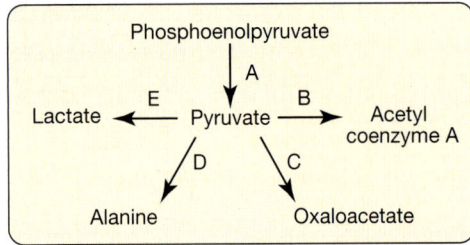

Correct answer = C. Pyruvate is carboxylated to oxaloacetate by pyruvate carboxylase, a biotin-requiring enzyme. B (pyruvate dehydrogenase complex) requires thiamine pyrophosphate, lipoic acid, flavin, and nicotinamide adenine dinucleotides (FAD and NAD^+, respectively), and coenzyme A; D (transaminase) requires pyridoxal phosphate; E (lactate dehydrogenase) requires NADH.

10.3. Which of the following reactions is unique to gluconeogenesis?

A. 1,3-Bisphosphoglycerate → 3-phosphoglycerate
B. Lactate → pyruvate
C. Oxaloacetate → phosphoenolpyruvate
D. Phosphoenolpyruvate → pyruvate

Correct answer = C. The other reactions are common to both gluconeogenesis and glycolysis.

10.4. Use the chart to show the effect of adenosine monophosphate (AMP) and fructose 2,6-bisphosphate on the listed enzymes of gluconeogenesis and glycolysis.

Enzyme	Fructose 2,6-bisphosphate	AMP
Fructose 1,6-bisphosphatase		
Phosphofructokinase-1		

Both fructose 2,6-bisphosphate and adenosine monophosphate inhibit fructose 1,6-bisphosphatase of gluconeogenesis and activate phosphofructokinase-1 of glycolysis. This results in reciprocal regulation of the two pathways.

10.5. The metabolism of ethanol by alcohol dehydrogenase produces reduced nicotinamide adenine dinucleotide (NADH) from the oxidized (NAD^+) form. What effect is the fall in the NAD^+/NADH ratio expected to have on gluconeogenesis?

The increase in NADH as ethanol is oxidized decreases the availability of oxaloacetate (OAA) because the reversible oxidation of malate to OAA by malate dehydrogenase of the tricarboxylic acid cycle is driven in the reverse direction by NADH. Additionally, the reversible reduction of pyruvate to lactate by lactate dehydrogenase is driven to lactate by NADH. Thus, two important gluconeogenic substrates, OAA and pyruvate, decrease as a result of the increase in NADH during ethanol metabolism. Consequently, gluconeogenesis decreases.

10.6. Given that acetyl coenzyme A cannot be a substrate for gluconeogenesis, why is its production in fatty acid oxidation essential for gluconeogenesis?

Acetyl coenzyme A inhibits the pyruvate dehydrogenase complex and activates pyruvate carboxylase, pushing pyruvate to gluconeogenesis and away from oxidation.

Glycogen Metabolism

11

I. OVERVIEW

Having a constant source of blood glucose is an absolute requirement for human life. Glucose is the greatly preferred energy source for the brain and the required energy source for cells with few or no mitochondria such as mature red blood cells. Glucose is also essential as an energy source for exercising muscle, where it is the substrate for anaerobic glycolysis. Blood glucose can be obtained from three primary sources: diet, glycogen degradation, and gluconeogenesis. Dietary intake of glucose and glucose precursors, such as starch (polysaccharide), disaccharides, and monosaccharides, is sporadic and, depending on the diet, is not always a reliable source of blood glucose. In contrast, gluconeogenesis can provide a sustained synthesis of glucose, but it is somewhat slow in responding to a falling blood glucose level. Therefore, the body has developed mechanisms for storing a supply of glucose in a rapidly mobilized form, namely, glycogen. In the absence of a dietary source of glucose, this sugar is rapidly released into the blood from liver glycogen. Similarly, muscle glycogen is extensively degraded in exercising muscle to provide that tissue with an important energy source. When glycogen stores are depleted, specific tissues synthesize glucose *de novo*, using glycerol, lactate, pyruvate, and amino acids as carbon sources for gluconeogenesis (see Chapter 10). Figure 11.1 shows the reactions of glycogen synthesis and degradation as part of the essential pathways of energy metabolism.

II. STRUCTURE AND FUNCTION

The main stores of glycogen are found in skeletal muscle and liver, although most other cells store small amounts of glycogen for their own use. The function of muscle glycogen is to serve as a fuel reserve for the synthesis of ATP during muscle contraction, while the purpose of liver glycogen is to maintain the blood glucose concentration, particularly during the early stages of a fast (Fig. 11.2). [Note: Liver glycogen can maintain blood glucose for <24 hours.]

A. Amounts in liver and muscle

Approximately 400 g of glycogen make up 1% to 2% of the fresh weight of resting muscle, and ~100 g of glycogen make up to 10% of the fresh weight of a well-fed adult liver. What limits the production of glycogen at these levels is not clear. However, in some glycogen

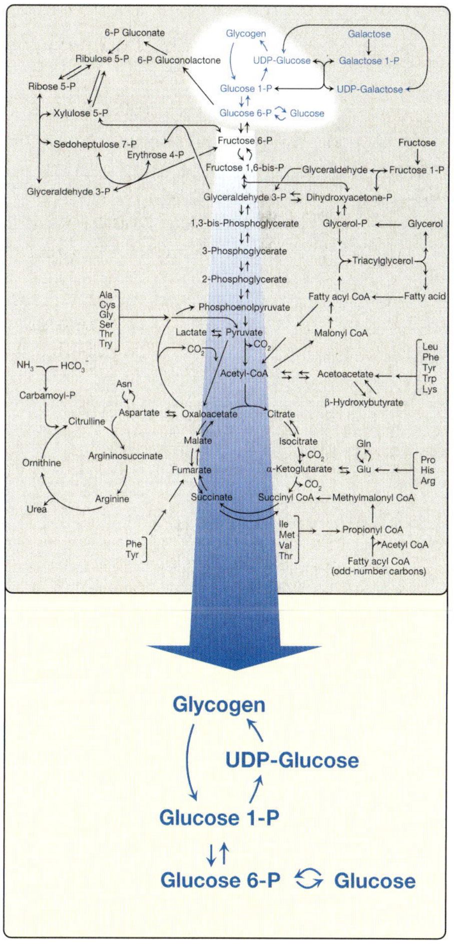

Figure 11.1
Glycogen synthesis and degradation shown as a part of the essential pathways of energy metabolism. [Note: See Fig. 8.2, for a more detailed map of metabolism.] P, phosphate; UDP, uridine diphosphate.

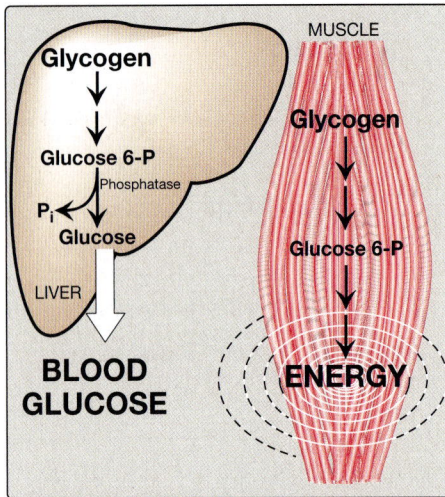

Figure 11.2
Functions of muscle and liver glycogen. [Note: The presence of glucose 6-phosphatase in liver allows release of glucose into blood.] P, phosphate; P_i, inorganic phosphate.

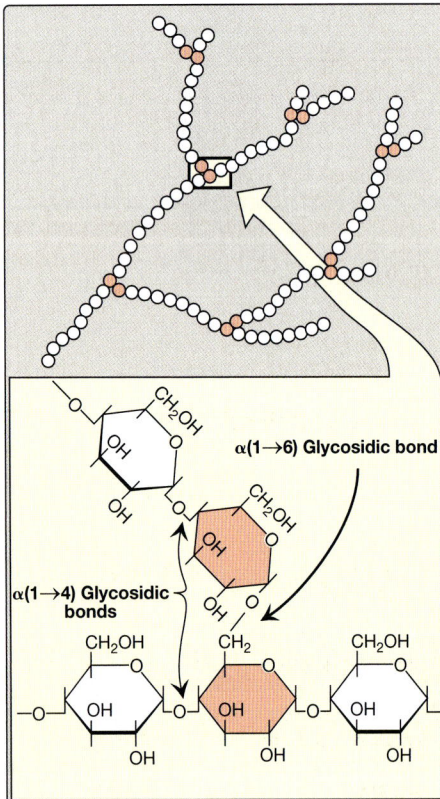

Figure 11.3
Branched structure of glycogen, showing $\alpha(1\rightarrow4)$ and $\alpha(1\rightarrow6)$ glycosidic bonds.

storage diseases (GSDs) (see Fig. 11.8), the amount of glycogen in the liver and/or muscle can be significantly higher. [Note: In the body, muscle mass is greater than liver mass. Consequently, most of the body's glycogen is found in skeletal muscle.]

B. Structure

Glycogen is a branched-chain polysaccharide made exclusively from α-D–glucose. The primary glycosidic bond is an $\alpha(1\rightarrow4)$ linkage. After an average of 8 to 14 glucosyl residues, there is a branch containing an $\alpha(1\rightarrow6)$ linkage (Fig. 11.3). A single glycogen molecule can contain up to 55,000 glucosyl residues. These polymers of glucose exist as large, spherical, cytoplasmic granules (particles) that also contain most of the enzymes necessary for glycogen synthesis and degradation.

C. Glycogen store fluctuation

Liver glycogen stores increase during the well-fed state and are depleted during a fast. Muscle glycogen is not affected by short periods of fasting (a few days) and is only moderately decreased in prolonged fasting (weeks). Muscle glycogen is synthesized to replenish muscle stores after they have been depleted following strenuous exercise. Glycogen synthesis and degradation go on continuously. The difference between the rates of these two processes determines the levels of stored glycogen during specific physiologic states.

III. SYNTHESIS (GLYCOGENESIS)

Glycogen is synthesized from molecules of α-D–glucose. The process occurs in the cytosol and requires energy supplied by adenosine triphosphate (ATP), for the phosphorylation of glucose, and uridine triphosphate (UTP).

A. Uridine diphosphate glucose synthesis

α-D–Glucose attached to uridine diphosphate (UDP) is the source of all the glucosyl residues that are added to the growing glycogen molecule (Fig. 11.4). UDP-glucose is synthesized from glucose 1-phosphate and UTP by UDP–glucose pyrophosphorylase (Fig. 11.5). Pyrophosphate (PP_i), the second product of the reaction, is hydrolyzed to two inorganic phosphates (P_i) by pyrophosphatase. The hydrolysis is exergonic, which ensures that the UDP–glucose pyrophosphorylase reaction proceeds in the direction of UDP-glucose production. [Note: Glucose 1-phosphate is generated from glucose 6-phosphate by phosphoglucomutase. Glucose 1,6-bisphosphate is an obligatory intermediate in this reversible reaction (Fig. 11.6).]

B. Primer requirement and synthesis

Glycogen synthase catalyzes the $\alpha(1\rightarrow4)$ linkages in glycogen. This enzyme cannot initiate chain synthesis using free glucose as an acceptor of a molecule of glucose from UDP-glucose. Instead, it can only elongate already existing chains of glucose and, therefore, requires a primer. A fragment of glycogen can serve as a primer. In the absence of a fragment, the homodimeric protein glycogenin can serve as an acceptor of glucose from UDP-glucose (see Fig. 11.5). The side-chain hydroxyl group of tyrosine-194 in the protein is the site at which the initial glucosyl unit is attached. Because the reaction is catalyzed by glycogenin itself via autoglucosylation, glycogenin is an enzyme. Glycogenin then catalyzes the transfer of at least four molecules of glucose from UDP-glucose, producing a short, $\alpha(1\rightarrow4)$-linked glucosyl chain. This short chain serves as a primer that is able to be elongated by glycogen synthase, which is recruited by glycogenin, as described next. [Note: Glycogenin stays associated with and forms the core of a glycogen granule.]

C. Elongation by glycogen synthase

Elongation of a glycogen chain involves the transfer of glucose from UDP-glucose to the nonreducing end of the growing chain, forming a new glycosidic bond between the anomeric hydroxyl group of carbon 1 of the activated glucose and carbon 4 of the accepting glucosyl residue (see Fig. 11.5). [Note: The nonreducing end of a carbohydrate chain is one in which the anomeric carbon of the terminal sugar is linked by a glycosidic bond to another molecule, making the terminal sugar nonreducing.] The enzyme responsible for making the $\alpha(1\rightarrow4)$ linkages in glycogen is glycogen synthase. [Note: The UDP released when the new $\alpha(1\rightarrow4\_)$ glycosidic bond is made can be phosphorylated to UTP by nucleoside diphosphate kinase (UDP + ATP $\rightleftarrows$ UTP + ADP).]

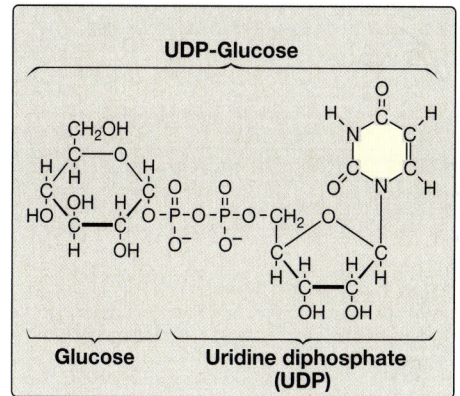

Figure 11.4
Structure of UDP-glucose, a nucleotide sugar.

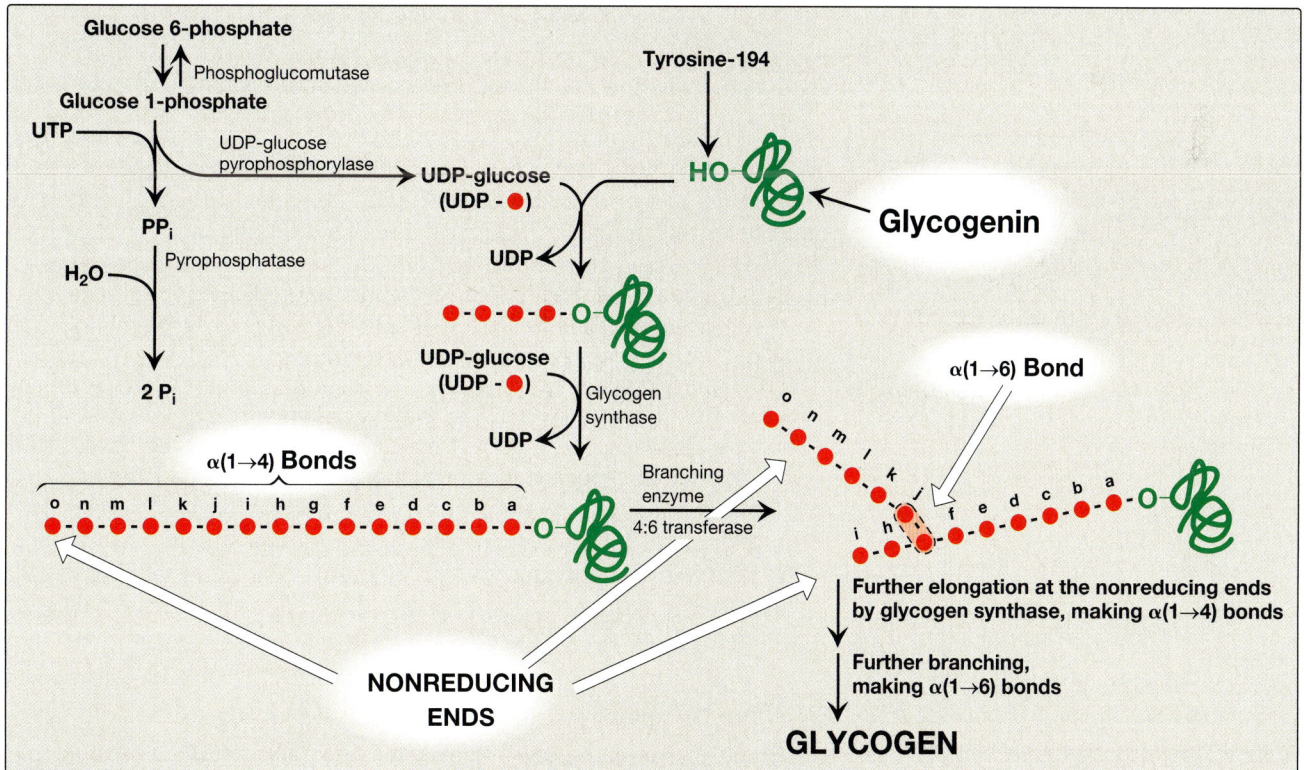

Figure 11.5
Glycogen synthesis. UDP and UTP, uridine di- and triphosphates; PP_i, pyrophosphate; P_i, inorganic phosphate.

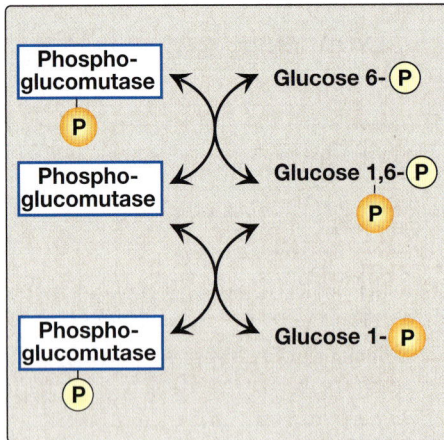

Figure 11.6
Interconversion of glucose 6-phosphate and glucose 1-phosphate by phospho-glucomutase. (P) and (P), phosphate.

D. Branch formation

If no other synthetic enzyme acted on the chain, the resulting structure would be a linear (unbranched) chain of glucosyl residues attached by α(1→4) linkages. Such a compound is found in plant tissues and is called amylose. In contrast, glycogen has branches located, on average, eight glucosyl residues apart, resulting in a highly branched, tree-like structure (see Fig. 11.3) that is far more soluble than the unbranched amylose. Branching also increases the number of nonreducing ends to which new glucosyl residues can be added (and also, as described in Section IV, from which these residues can be removed), thereby greatly accelerating the rate at which glycogen synthesis can occur and dramatically increasing the size of the glycogen molecule.

1. **Branch synthesis:** Branches are made by the action of the branching enzyme, amylo-α(1→4)→α(1→6)-transglycosylase. This enzyme removes a set of six to eight glucosyl residues from the nonreducing end of the glycogen chain, breaking an α(1→4) bond to another residue on the chain, and attaches it to a nonterminal glucosyl residue by an α(1→6) linkage, thus functioning as a 4:6 transferase. The resulting new, nonreducing end (see "i" in Fig. 11.5), as well as the old nonreducing end from which the six to eight residues were removed (see "o" in Fig. 11.5), can now be further elongated by glycogen synthase.

2. **Additional branch synthesis:** After elongation of these two ends has been accomplished, their terminal six to eight glucosyl residues can be removed and used to make additional branches.

IV. DEGRADATION (GLYCOGENOLYSIS)

The degradative pathway that mobilizes stored glycogen in the liver and skeletal muscle is not a reversal of the synthetic reactions. Instead, a separate set of cytosolic enzymes is required. When glycogen is degraded, the primary product is glucose 1-phosphate, obtained by breaking α(1→4) glycosidic bonds. In addition, free glucose is released from each α(1→6)–linked glucosyl residue (branch point).

A. Chain shortening

Glycogen phosphorylase sequentially cleaves the α(1→4) glycosidic bonds between the glucosyl residues at the nonreducing ends of the glycogen chains by simple phosphorolysis (producing glucose 1-phosphate) until four glucosyl units remain on each chain at a branch point (Fig. 11.7). The resulting structure is called a limit dextrin, and phosphorylase cannot degrade it any further (Fig. 11.8). [Note: Phosphorylase requires pyridoxal phosphate (derivative of vitamin B_6) as a coenzyme.]

B. Branch removal

Branches are removed by the two enzymic activities of a single bifunctional protein, the debranching enzyme (see Fig. 11.8). First, oligo-α(1→4)→α(1→4)-glucantransferase activity removes the outer three of the four glucosyl residues remaining at a branch. It next

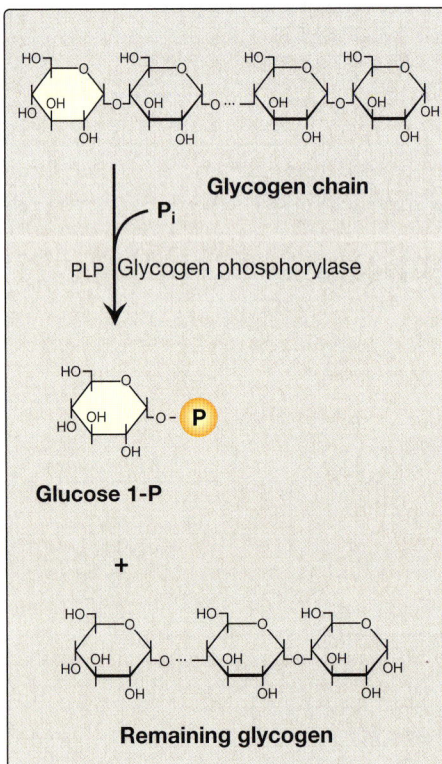

Figure 11.7
Cleavage of an α(1→4)-glycosidic bond. PLP, pyridoxal phosphate; P_i, inorganic phosphate; (P), phosphate.

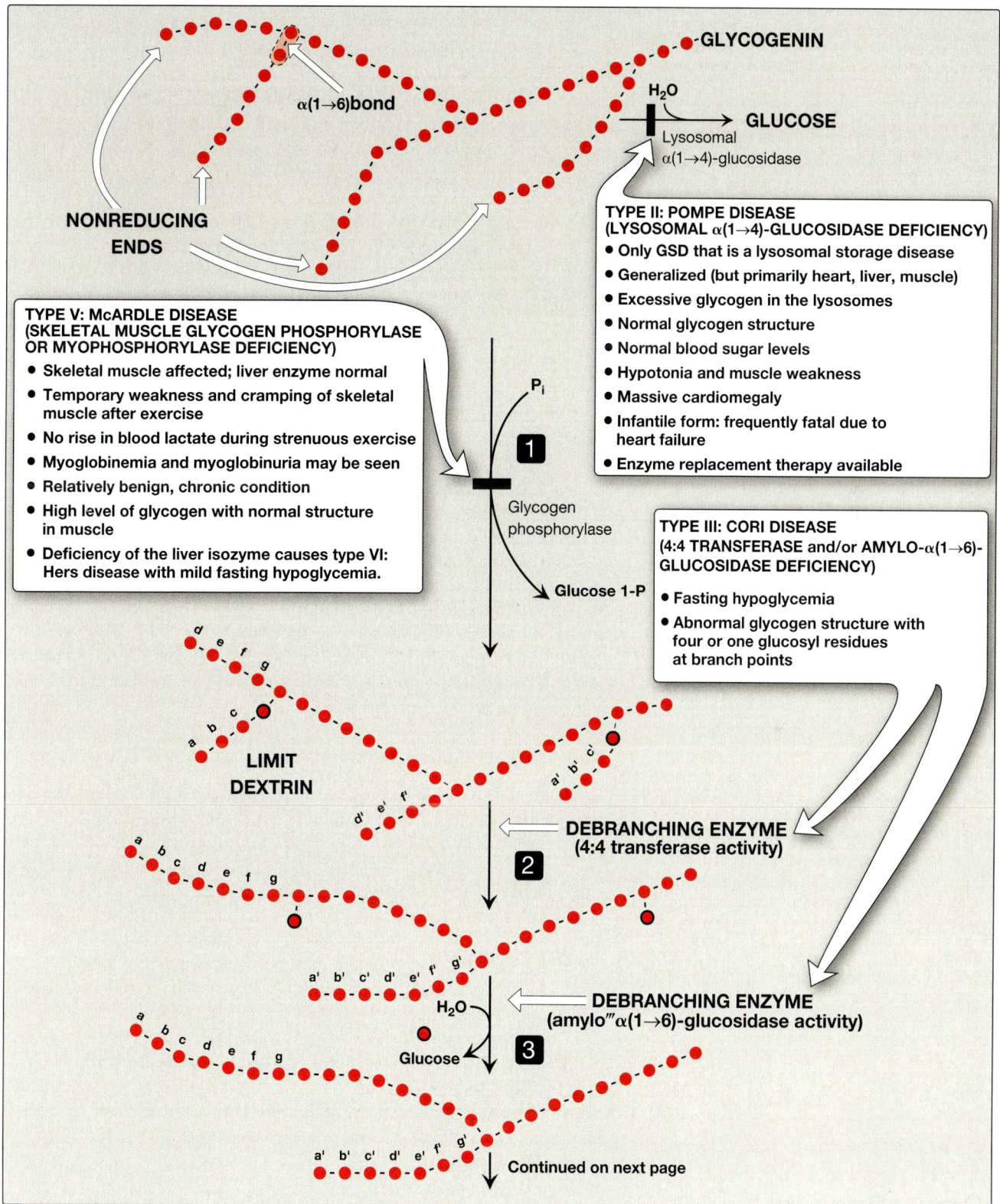

GLYCOGENIN

α(1→6)bond

H_2O

GLUCOSE

Lysosomal α(1→4)-glucosidase

NONREDUCING ENDS

TYPE II: POMPE DISEASE (LYSOSOMAL α(1→4)-GLUCOSIDASE DEFICIENCY)
- Only GSD that is a lysosomal storage disease
- Generalized (but primarily heart, liver, muscle)
- Excessive glycogen in the lysosomes
- Normal glycogen structure
- Normal blood sugar levels
- Hypotonia and muscle weakness
- Massive cardiomegaly
- Infantile form: frequently fatal due to heart failure
- Enzyme replacement therapy available

TYPE V: McARDLE DISEASE (SKELETAL MUSCLE GLYCOGEN PHOSPHORYLASE OR MYOPHOSPHORYLASE DEFICIENCY)
- Skeletal muscle affected; liver enzyme normal
- Temporary weakness and cramping of skeletal muscle after exercise
- No rise in blood lactate during strenuous exercise
- Myoglobinemia and myoglobinuria may be seen
- Relatively benign, chronic condition
- High level of glycogen with normal structure in muscle
- Deficiency of the liver isozyme causes type VI: Hers disease with mild fasting hypoglycemia.

P_i

1

Glycogen phosphorylase

Glucose 1-P

TYPE III: CORI DISEASE (4:4 TRANSFERASE and/or AMYLO-α(1→6)-GLUCOSIDASE DEFICIENCY)
- Fasting hypoglycemia
- Abnormal glycogen structure with four or one glucosyl residues at branch points

d e f g
a b c
LIMIT DEXTRIN
d' e' f'

a' b' c'

a b c d e f g

DEBRANCHING ENZYME (4:4 transferase activity)

2

a' b' c' d' e' f' g'

H_2O

DEBRANCHING ENZYME (amylo‴α(1→6)-glucosidase activity)

3

Glucose

a b c d e f g

a' b' c' d' e' f' g'

Continued on next page

Figure 11.8
Glycogen degradation, showing some of the glycogen storage diseases (GSDs). [Note: GSD type IV is caused by defects in branching enzyme, an enzyme of synthesis, resulting in liver cirrhosis that can be fatal in early childhood.] P_i, inorganic phosphate; P, phosphate. (continued on next page)

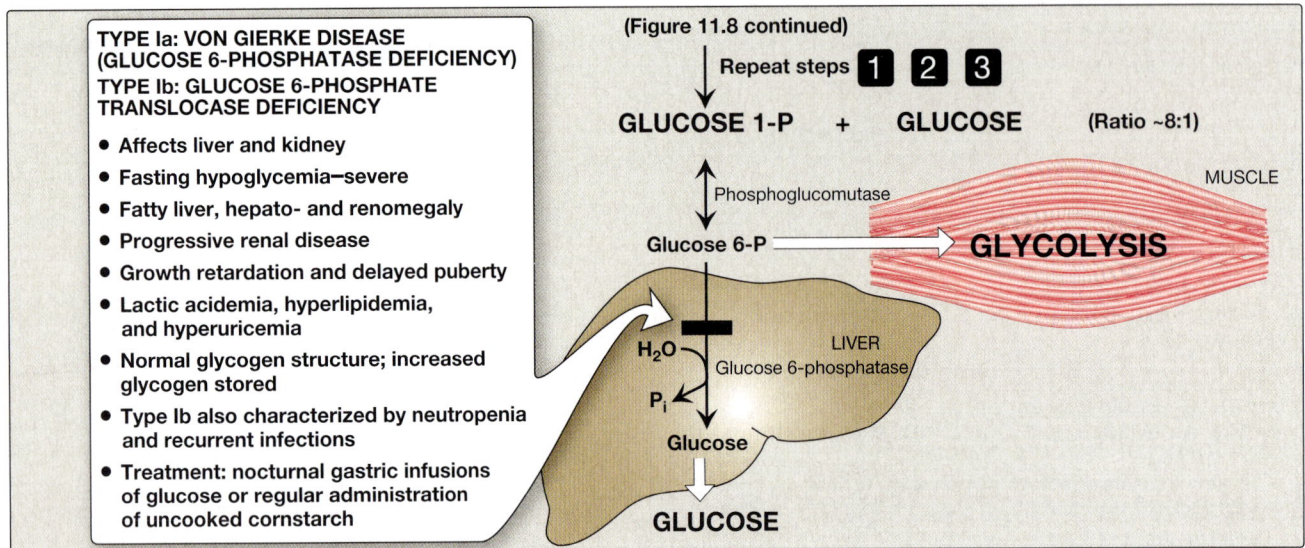

TYPE Ia: VON GIERKE DISEASE (GLUCOSE 6-PHOSPHATASE DEFICIENCY)
TYPE Ib: GLUCOSE 6-PHOSPHATE TRANSLOCASE DEFICIENCY

- Affects liver and kidney
- Fasting hypoglycemia–severe
- Fatty liver, hepato- and renomegaly
- Progressive renal disease
- Growth retardation and delayed puberty
- Lactic acidemia, hyperlipidemia, and hyperuricemia
- Normal glycogen structure; increased glycogen stored
- Type Ib also characterized by neutropenia and recurrent infections
- Treatment: nocturnal gastric infusions of glucose or regular administration of uncooked cornstarch

Figure 11.8
(continued from the previous page)

transfers them to the nonreducing end of another chain, lengthening it accordingly. Thus, an $\alpha(1\rightarrow4)$ bond is broken and an $\alpha(1\rightarrow4)$ bond is made, and the enzyme functions as a 4:4 transferase. Next, the remaining glucose residue attached in an $\alpha(1\rightarrow6)$ linkage is removed hydrolytically by amylo-$\alpha(1\rightarrow6)$-glucosidase activity, releasing free (nonphosphorylated) glucose. The glucosyl chain is now available again for degradation by glycogen phosphorylase until four glucosyl units in the next branch are reached.

C. Glucose 1-phosphate isomerization to glucose 6-phosphate

Glucose 1-phosphate, produced by glycogen phosphorylase, is isomerized in the cytosol to glucose 6-phosphate by phosphoglucomutase (see Fig. 11.6). In the liver, glucose 6-phosphate is transported into the endoplasmic reticulum (ER) by glucose 6-phosphate translocase. There, it is dephosphorylated to glucose by glucose 6-phosphatase (the same enzyme used in the last step of gluconeogenesis). The glucose is then transported from the ER to the cytosol. Hepatocytes release glycogen-derived glucose into the blood to help maintain blood glucose levels until the gluconeogenic

---

Clinical Application 11.1: Glycogen Storage Disease V (McCardle Disease)

Glycogen storage disease V (GSD V), McArdle disease, is an autosomal recessive disorder that mainly affects skeletal muscles and is a deficiency of muscle glycogen phosphorylase. Glycogen in muscles cannot be degraded effectively to produce glucose. Tiredness and muscle pain and cramping during exercise are key features. The muscle problems are relieved by rest, and the person can resume exercise without difficulty. With vigorous exercise, affected persons may experience rhabdomyolysis, a serious disorder where muscles break down and release myoglobin that can damage the kidneys.

Table 11.1: Descriptions of Glycogen Storage Diseases

Type[a]	Deficient Enzyme	Main Signs/Symptoms
I: Von Gierke disease	Glucose-6-phosphatase	Lactic acidosis, hypoglycemia, hyperuricemia, impaired growth, bone thinning
II: Pompe disease[a]	Acid α-glucosidase (acid maltase)	Excess glycogen in lysosomes, normal blood sugar. Enlarged liver and heart; muscle weakness and heart problems in severe forms
III: Cori disease[a]	Glycogen debranching enzyme (4:4 transferase)	Enlarged liver, growth delay, fasting hypoglycemia, abnormal glycogen structure, elevated fat in blood, possible muscle weakness
IV: Andersen disease	Glycogen branching enzyme (4:6 transferase)	Growth delay, enlarged liver, myopathy; death by age 5 usually
V: McArdle disease[a]	Muscle glycogen phosphorylase (myophosphorylase)	Muscle weakness and cramping after exercise; usually a relatively benign, chronic condition
VI: Hers disease	Liver glycogen phosphorylase	Liver enlargement, hypoglycemia, developmental delay
VII: Tarui disease	Muscle phosphofructokinase	Exercise-induced muscle cramps, developmental delay, hemolytic anemia in some

[a]This describes 7 of the 15 types of GSDs. See also Figure 11.3.

pathway is actively producing glucose. [Note: Muscle lacks glucose 6-phosphatase. Consequently, glucose 6-phosphate cannot be dephosphorylated and sent into the blood. Instead, it enters glycolysis, providing energy needed for muscle contraction.]

D. Lysosomal degradation

A small amount (1% to 3%) of glycogen is degraded by the lysosomal enzyme, acid α(1→4)-glucosidase (acid maltase). The purpose of this autophagic pathway is unknown. However, a deficiency of this enzyme causes accumulation of glycogen in vacuoles in the lysosomes, resulting in the serious GSD type II, Pompe disease (see Table 11.1 and Fig. 11.8). [Note: Pompe disease, caused by acid maltase deficiency, is the only GSD that is a lysosomal storage disease.]

> Lysosomal storage diseases are genetic disorders characterized by the accumulation of abnormal amounts of carbohydrates or lipids primarily due to their decreased lysosomal degradation resulting from the absence or decreased activity or amount of the specific lysosomal acid hydrolase that is normally responsible for its degradation.

V. GLYCOGENESIS AND GLYCOGENOLYSIS REGULATION

Because of the importance of maintaining blood glucose levels, the synthesis and degradation of its glycogen storage form are tightly regulated. In the liver, glycogenesis accelerates during periods when the body has been well fed, whereas glycogenolysis accelerates during periods of fasting. In skeletal muscle, glycogenolysis occurs during active exercise, and glycogenesis begins as soon as the muscle is again at rest.

Regulation of synthesis and degradation is accomplished on two levels. First, glycogen synthase and glycogen phosphorylase are hormonally regulated (by covalent phosphorylation/dephosphorylation) to meet the needs of the body as a whole. Second, these same enzymes are allosterically regulated (by effector molecules) to meet the needs of a particular tissue.

A. Covalent activation of glycogenolysis

The binding of hormones, such as glucagon or epinephrine, to plasma membrane G protein–coupled receptors (GPCRs) signals the need for glycogen to be degraded, either to elevate blood glucose levels or to provide energy for exercising muscle.

1. **Protein kinase A activation:** Binding of glucagon or epinephrine to their specific hepatocyte GPCR, or of epinephrine to a specific myocyte GPCR, results in the G protein–mediated activation of adenylyl cyclase. This enzyme catalyzes the synthesis of cyclic adenosine monophosphate (cAMP), which activates cAMP-dependent protein kinase A (PKA). cAMP binds the two regulatory subunits of tetrameric PKA, releasing two individual catalytic subunits that are active (Fig. 11.9). PKA then phosphorylates several enzymes of glycogen metabolism, affecting their activity. [Note: When cAMP is removed, the inactive PKA tetramer reforms.]

2. **Phosphorylase kinase activation:** Phosphorylase kinase exists in two forms: an inactive "b" form and an active "a" form. Active PKA phosphorylates the inactive "b" form of phosphorylase kinase, producing the active "a" form (see Fig. 11.9).

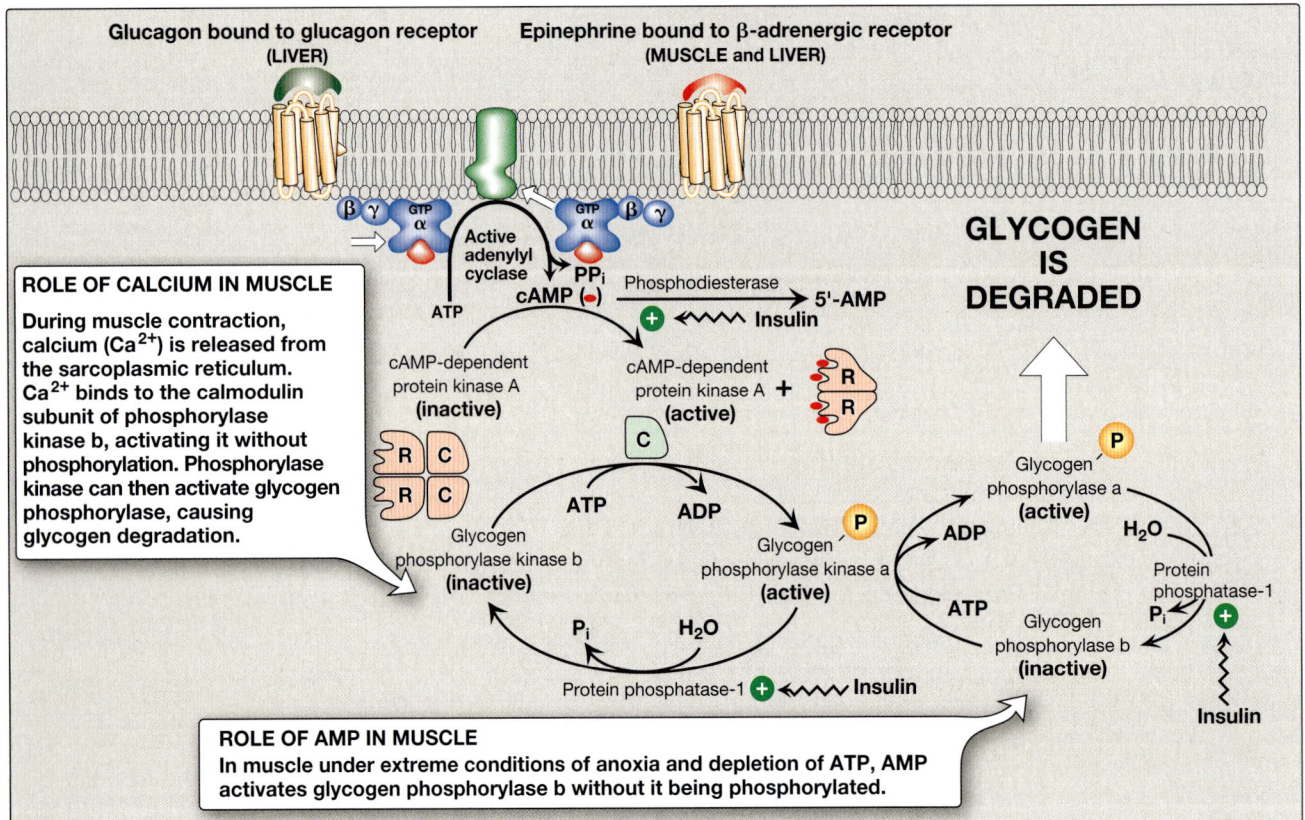

Figure 11.9
Stimulation and inhibition of glycogen degradation. AMP, adenosine monophosphate; cAMP, cyclic AMP; GTP, guanosine triphosphate; (P), phosphate; PP$_i$, pyrophosphate; R, regulatory subunit; C, catalytic subunit.

3. **Glycogen phosphorylase activation:** Glycogen phosphorylase also exists in a dephosphorylated, inactive "b" form and a phosphorylated, active "a" form. Phosphorylase kinase a is the only enzyme that phosphorylates glycogen phosphorylase b to its active "a" form, which then begins glycogenolysis (see Fig. 11.9).

4. **Signal amplification:** The cascade of reactions just described activates glycogenolysis. The large number of sequential steps serves to amplify the effect of the hormonal signal, that is, a few hormone molecules binding to their GPCR result in a number of PKA molecules being activated that can each activate many phosphorylase kinase molecules. This causes the production of many molecules of active glycogen phosphorylase a that can degrade glycogen.

5. **Phosphorylated state maintenance:** The phosphate groups added to phosphorylase kinase and phosphorylase in response to cAMP are maintained because the enzyme that hydrolytically removes the phosphate, protein phosphatase-1 (PP1), is inactivated by inhibitor proteins that are also phosphorylated and activated in response to cAMP (see Fig. 11.9). Because insulin also activates the phosphodiesterase that degrades cAMP, it opposes the effects of glucagon and epinephrine.

B. Covalent inhibition of glycogenesis

The regulated enzyme in glycogenesis, glycogen synthase, also exists in two forms, the active "a" form and the inactive "b" form. However, in contrast to phosphorylase kinase and phosphorylase, the active form of glycogen synthase is dephosphorylated, whereas the inactive form is phosphorylated at several sites on the enzyme, with the level of inactivation proportional to the degree of phosphorylation (Fig. 11.10). Phosphorylation is catalyzed by several different protein kinases in response to cAMP (e.g., PKA and phosphorylase kinase) or other signaling mechanisms (described next). Glycogen synthase b can be reconverted to the "a" form by PP1. Figure 11.11 summarizes the covalent regulation of glycogen metabolism.

C. Allosteric regulation of glycogenesis and glycogenolysis

In addition to hormonal signals, glycogen synthase and glycogen phosphorylase respond to the levels of metabolites and energy needs of the cell. Glycogenesis is stimulated when glucose and energy levels are high, whereas glycogenolysis is increased when glucose and energy levels are low. This allosteric regulation allows a rapid response to the needs of a cell and can override the effects of hormone-mediated covalent regulation. [Note: The "a" and "b" forms of the allosteric enzymes of glycogen metabolism are each in an equilibrium between the R (relaxed, more active) and T (tense, less active) conformations. The binding of effectors shifts the equilibrium and alters enzymic activity without directly altering the covalent modification.]

1. **Regulation in the well-fed state:** In the well-fed state, glycogen synthase b in both liver and muscle is allosterically activated by glucose 6-phosphate, which is present in elevated concentrations (Fig. 11.12). In contrast, glycogen phosphorylase a is allosterically

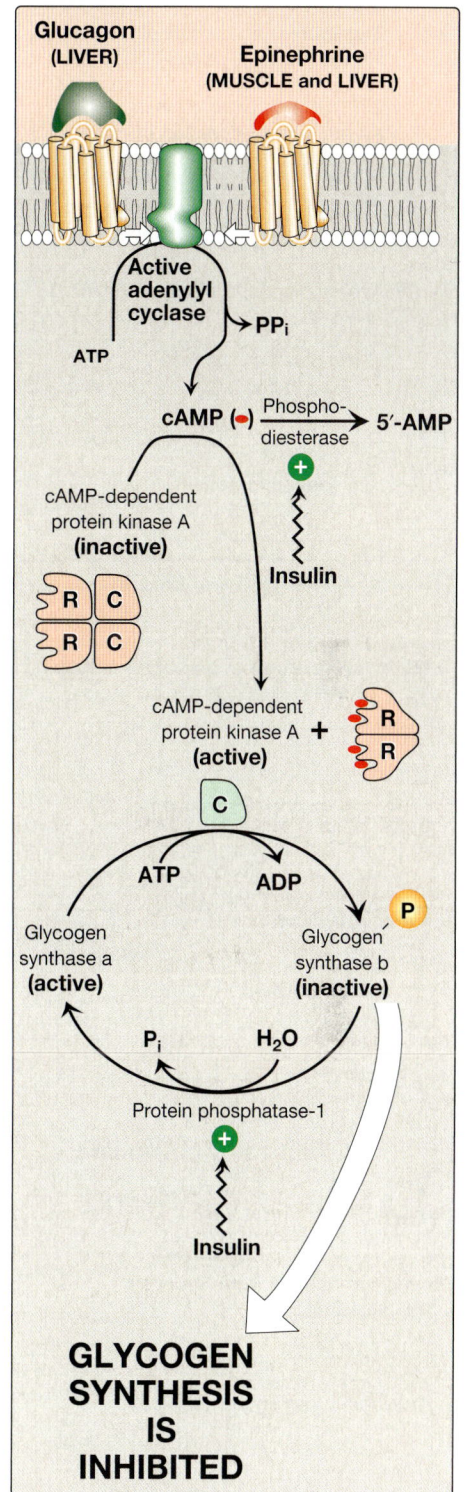

Figure 11.10
Hormonal regulation of glycogen synthesis. [Note: In contrast to glycogen phosphorylase, glycogen synthase is inactivated by phosphorylation.] cAMP, cyclic adenosine monophosphate; Ⓟ, phosphate; PPi, pyrophosphate; R, regulatory subunit; C, catalytic subunit; ADP, adenosine diphosphate.

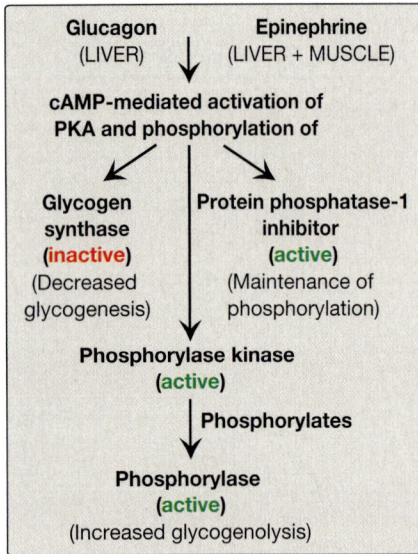

Figure 11.11
Summary of the hormone-mediated covalent regulation of glycogen metabolism. cAMP, cyclic adenosine monophosphate; PKA, protein kinase A.

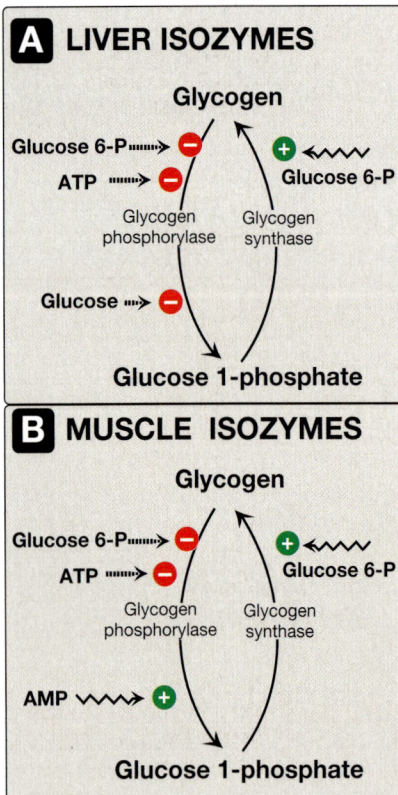

Figure 11.12
Allosteric regulation of glycogenesis and glycogenolysis in liver (**A**) and muscle (**B**). P, phosphate; AMP, adenosine monophosphate.

inhibited by glucose 6-phosphate, as well as by ATP, a high-energy signal. [Note: In liver, but not muscle, free glucose is also an allosteric inhibitor of glycogen phosphorylase a.]

2. **Glycogenolysis activation by AMP:** Muscle glycogen phosphorylase (myophosphorylase), but not the liver isozyme, is active in the presence of the high AMP concentrations that occur under extreme conditions of anoxia and ATP depletion. AMP binds to glycogen phosphorylase b, causing its activation without phosphorylation (see Fig. 11.9). Recall that AMP also activates phosphofructokinase-1 of glycolysis, allowing glucose from glycogenolysis to be oxidized.

3. **Glycogenolysis activation by calcium:** Calcium (Ca^{2+}) is released into the sarcoplasm in muscle cells (myocytes) in response to neural stimulation and in the liver in response to epinephrine binding to α_1-adrenergic receptors. The Ca^{2+} binds to calmodulin (CaM), the most widely distributed member of a family of small, Ca^{2+}-binding proteins. The binding of four molecules of Ca^{2+} to CaM triggers a conformational change such that the activated Ca^{2+}–CaM complex binds to and activates protein molecules, often enzymes that are inactive in the absence of this complex (Fig. 11.13). Thus, CaM functions as an essential subunit of many complex proteins. One such protein is the tetrameric phosphorylase kinase, whose "b" form is activated by the binding of Ca^{2+} to its δ subunit (CaM) without the need for the kinase to be phosphorylated by PKA. [Note: Epinephrine at β-adrenergic receptors signals through a rise in cAMP, not via a rise in Ca^{2+}.]

   a. **Muscle phosphorylase kinase activation:** During muscle contraction, ATP is rapidly and urgently needed. It is supplied by the degradation of muscle glycogen to glucose 6-phosphate, which enters glycolysis. Nerve impulses cause membrane depolarization, which promotes Ca^{2+} release from the sarcoplasmic reticulum into the sarcoplasm of myocytes. The Ca^{2+} binds the CaM subunit, and the complex activates muscle phosphorylase kinase b (see Fig. 11.9).

   b. **Liver phosphorylase kinase activation:** During physiologic stress, epinephrine is released from the adrenal medulla and signals the need for blood glucose. This glucose initially comes from hepatic glycogenolysis. Binding of epinephrine to hepatocyte α_1-adrenergic GPCR activates a phospholipid-dependent cascade that results in movement of Ca^{2+} from the ER into the cytoplasm. A Ca^{2+}–CaM complex forms and activates hepatic phosphorylase kinase b. [Note: The released Ca^{2+} also helps to activate protein kinase C that can phosphorylate and inactivate glycogen synthase a.]

VI. GLYCOGEN STORAGE DISEASES

GSDs are a group of genetic diseases caused by defects in enzymes required for glycogen degradation or, more rarely, glycogen synthesis. The most common symptoms are hypoglycemia (low blood glucose), enlarged liver, slow growth, and muscle weakness or cramping.

These disorders result either in formation of glycogen that has an abnormal structure or in the accumulation of excessive amounts of normal glycogen in specific tissues as a result of impaired degradation. A particular enzyme may be defective in a single tissue, such as the liver (resulting in hypoglycemia) or muscle (causing muscle weakness), or the defect may be more generalized, affecting a variety of tissues, such as the heart and kidneys. Severity ranges from fatal in early childhood to mild disorders that are not life threatening. Overall, 15 types of GSDs are recognized, some quite rare. The more prevalent types are described in Table 11.1, and three of the most common GSDs are illustrated in Figure 11.8.

VII. CHAPTER SUMMARY

- The main stores of **glycogen** in the body are found in **skeletal muscle**, where they serve as a fuel reserve for the synthesis of **adenosine triphosphate (ATP)** during muscle **contraction**, and in the **liver**, where they are used to maintain the **blood glucose** concentration, particularly during the early stages of a **fast**.

- Glycogen is a highly **branched polymer** of α-D–glucose.

- **Uridine diphosphate (UDP)-glucose**, the building block of glycogen, is synthesized from **glucose 1-phosphate** and **uridine triphosphate (UTP)** by **UDP–glucose pyrophosphorylase** (Fig. 11.14).

- **Glucose** from UDP-glucose is transferred to the **nonreducing ends** of glycogen chains by primer-requiring **glycogen synthase**, which makes the α(1→4) linkages. The **primer** is made by **glycogenin**. Branches are formed by **amylo-α(1→4)→α(1→6)-transglycosylase (a 4:6 transferase)**, which transfers a set of six to eight glucosyl residues from the nonreducing end of the glycogen chain (breaking an α[1→4] linkage), and making an α(1→6) linkage to another residue in the chain.

- **Glycogen phosphorylase** cleaves the α(1→4) bonds between glucosyl residues at the nonreducing ends of the glycogen chains, producing **glucose 1-phosphate**.

- Glucose 1-phosphate is converted to **glucose 6-phosphate** by **phosphoglucomutase**.

- In **muscle**, glucose 6-phosphate enters glycolysis. In **liver**, the phosphate is removed by **glucose 6-phosphatase**, releasing free glucose that can be used to maintain blood glucose levels at the beginning of a fast.

- Glucose 6-phosphatase deficiency causes **von Gierke disease** and results in an inability of the liver to provide free glucose to the body during a fast. It affects both glycogen degradation and gluconeogenesis.

- Glycogen synthesis and degradation are **reciprocally regulated** to meet whole-body needs by the same hormonal signals, namely, an **elevated insulin** level results in overall **increased glycogenesis** and **decreased glycogenolysis**, whereas an **elevated glucagon**, or **epinephrine**, level causes the opposite effects.

- Key enzymes are phosphorylated by **protein kinases**, some of which are dependent on **cyclic adenosine monophosphate (cAMP)**, a compound increased by glucagon and epinephrine. Phosphate groups are removed by **protein phosphatase-1 (PP1)**.

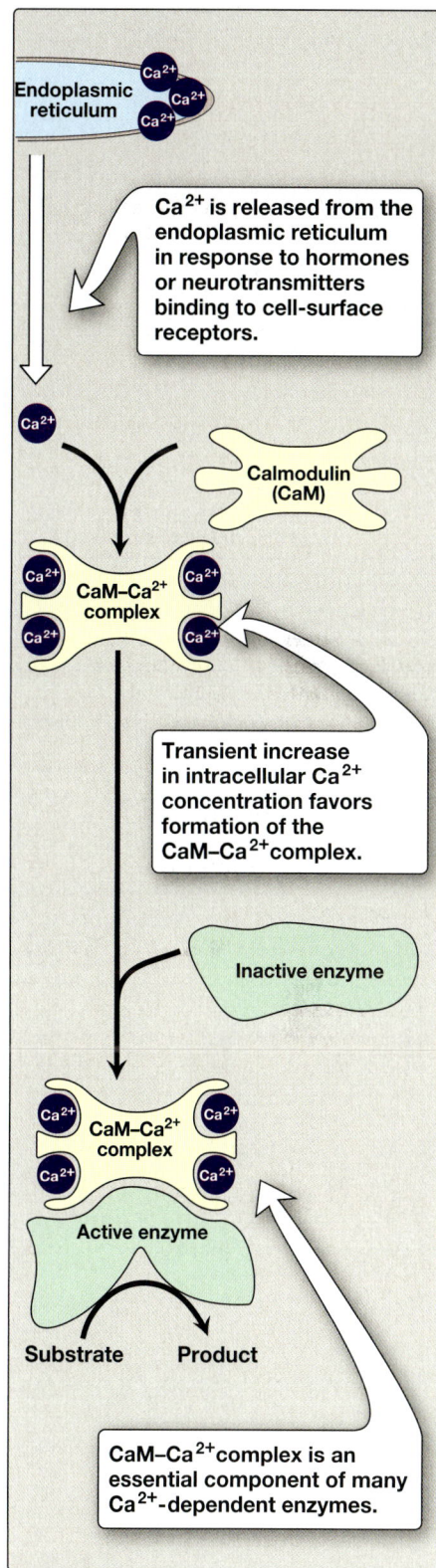

Ca^{2+} is released from the endoplasmic reticulum in response to hormones or neurotransmitters binding to cell-surface receptors.

Calmodulin (CaM)

CaM–Ca^{2+} complex

Transient increase in intracellular Ca^{2+} concentration favors formation of the CaM–Ca^{2+} complex.

Inactive enzyme

CaM–Ca^{2+} complex

Active enzyme

Substrate   Product

CaM–Ca^{2+} complex is an essential component of many Ca^{2+}-dependent enzymes.

Figure 11.13
Calmodulin mediates many effects of intracellular calcium (Ca^{2+}). [Note: Ca^{2+} activates phosphorylase kinase in the liver and muscles.]

- In addition to this **covalent regulation**, **glycogen synthase**, **phosphorylase kinase**, and **phosphorylase** are **allosterically regulated** to meet tissues' needs.
- In the well-fed state, glycogen synthase is activated by glucose 6-phosphate, but glycogen phosphorylase is inhibited by glucose 6-phosphate as well as by ATP.
- In the liver, free glucose also serves as an allosteric inhibitor of glycogen phosphorylase.
- The rise in **calcium** in muscle during exercise and in the liver in response to epinephrine activates phosphorylase kinase by binding to the enzyme's **calmodulin** subunit. This allows the enzyme to activate glycogen phosphorylase, thereby causing glycogen degradation.

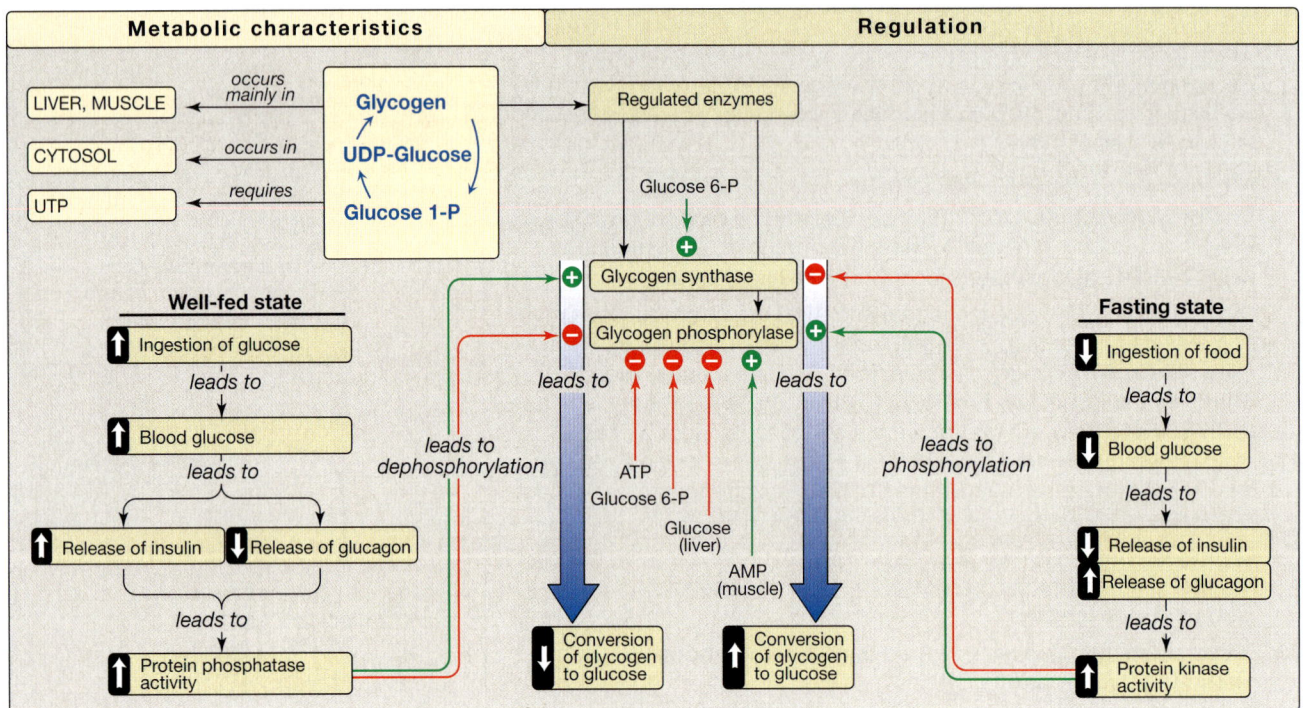

Figure 11.14
Key concept map for glycogen metabolism in the liver. [Note: Glycogen phosphorylase is phosphorylated by phosphorylase kinase, the "b" form of which can be activated by calcium.] UDP and UTP, uridine di- and triphosphates; P, phosphate; AMP, adenosine monophosphate.

Study Questions

Choose the ONE best answer.

For Questions 11.1 to 11.4, match the deficient enzyme to the clinical finding in selected glycogen storage diseases (GSDs).

Choice	GSD	Deficient Enzyme
A	Von Gierke disease type Ia	Glucose 6-phosphatase
B	Pompe disease type II	Acid maltase
C	Cori disease type III	4:4 Transferase
D	Andersen disease type IV	4:6 Transferase
E	McArdle disease type V	Myophosphorylase
F	Hers disease type VI	Liver phosphorylase

11.1. Exercise intolerance, with no rise in blood lactate during exercise

Correct answer = E. Myophosphorylase (the muscle isozyme of glycogen phosphorylase) deficiency (or, McArdle disease) prevents glycogen degradation in muscle, depriving muscle of glycogen-derived glucose, resulting in decreased glycolysis and its anaerobic product, lactate.

11.2. Fatal, progressive cirrhosis and glycogen with longer-than-normal outer chains

Correct answer = D. 4:6 Transferase (branching enzyme) deficiency (Andersen disease), a defect in glycogen synthesis, results in glycogen with fewer branches and decreased solubility.

11.3. Generalized accumulation of glycogen, severe hypotonia, and death from heart failure

Correct answer = B. Acid maltase (acid $\alpha[1\rightarrow4]$-glucosidase) deficiency (or, Pompe disease) prevents degradation of any glycogen brought into lysosomes. A variety of tissues are affected, with the most severe pathology resulting from heart damage.

11.4. Severe fasting hypoglycemia, lactic acidemia, hyperuricemia, and hyperlipidemia

Correct answer = A. Glucose 6-phosphatase deficiency (von Gierke disease) prevents the liver from releasing free glucose into the blood, causing severe fasting hypoglycemia, lactic acidemia, hyperuricemia, and hyperlipidemia.

11.5. Both epinephrine and glucagon have which effect on hepatic glycogen metabolism?

A. Phosphorylate and activate glycogen phosphorylase and glycogen synthase

B. Phosphorylate and inactivate glycogen phosphorylase and glycogen synthase

C. Cause increased glycogen degradation and decreased synthesis in the liver

D. Cause the synthesis of glycogen to have a net increase

Correct answer = C. Epinephrine and glucagon both cause increased glycogen degradation and decreased synthesis in the liver through covalent modification (phosphorylation) of key enzymes of glycogen metabolism. Glycogen phosphorylase is phosphorylated and active ("a" form), whereas glycogen synthase is phosphorylated and inactive ("b" form). Glucagon does not cause a rise in calcium.

11.6. In contracting skeletal muscle, a sudden elevation of the sarcoplasmic calcium concentration will result in:

  A. activation of cAMP-dependent protein kinase A.
  B. conversion of cAMP to AMP by phosphodiesterase.
  C. direct activation of glycogen synthase b.
  D. direct activation of phosphorylase kinase b.
  E. inactivation of phosphorylase kinase a by the action of protein phosphatase-1.

Correct answer = D. Calcium (Ca^{2+}) released from the sarcoplasmic reticulum during exercise binds to the calmodulin subunit of phosphorylase kinase, thereby allosterically activating the dephosphorylated "b" form of this enzyme. The other choices are not caused by an elevation of cytosolic Ca^{2+}. [Note: Ca^{2+} also activates hepatic phosphorylase kinase b.]

11.7. Why is the hypoglycemia seen with type 1 von Gierke disease (glucose 6-phosphatase deficiency) severe, whereas that seen with type VI Hers disease (liver phosphorylase deficiency) is mild?

With von Gierke disease, the liver is unable to generate free glucose either from glycogenolysis or gluconeogenesis because both processes produce glucose 6-phosphate. With Hers disease, the liver is still able to produce free glucose from gluconeogenesis, but glycogenolysis is inhibited.

11.8. A 2-year-old boy presents for evaluation of growth delay. Myopathy and hepatomegaly are noted. Liver biopsy reveals accumulation of abnormal glycogen with short outer branches. Deficiency of which enzyme is most likely responsible for the signs and symptoms?

  A. Glycogen synthase
  B. Glycogen phosphorylase
  C. Branching enzyme
  D. Debranching enzyme

Correct answer = C. Branching enzyme catalyzes the last step in glycogen biosynthesis. Its deficiency would lead to accumulation of abnormal glycogen with short outer branches. Glycogen storage disease (GSD) type IV, Anderson disease, is most likely in this situation. Deficiencies in glycogen synthase and phosphorylase would not cause the glycogen structure described. Deficiency of debranching enzyme, GSD type II, Cori disease, would cause accumulation of glycogen with long, not short outer branches, but could have similar presentation.

11.9. A 9-year-old girl experiences muscle cramps and weakness after beginning to play soccer on a youth team. Her symptoms improve with rest. When her blood is tested after a period of exercise, her blood glucose is low. What enzyme is most likely deficient?

  A. Glycogen synthase
  B. Glycogen phosphorylase
  C. Glycogenin
  D. Branching enzyme
  E. Debranching enzyme

Correct answer = B. This child exhibits signs and symptoms of McCardle disease, a deficiency in glycogen phosphorylase. Impaired glycogen breakdown leads to low blood glucose during exercise. Glycogen synthase is involved in glycogen synthesis, not its breakdown; debranching and branching enzyme play roles in glycogen structure but not in glycogenolysis. Glycogenin is the primer for glycogen synthesis.

Monosaccharide and Disaccharide Metabolism

12

I. OVERVIEW

Glucose is the most common monosaccharide consumed by humans but two other monosaccharides, fructose and galactose, also occur in significant amounts in the diet, primarily in disaccharides, and make important contributions to energy metabolism. In addition, galactose is an important component of glycosylated proteins. Figure 12.1 shows the metabolism of fructose and galactose as part of the essential pathways of energy metabolism.

II. FRUCTOSE METABOLISM

About 10% of the calories in the typical Western diet are supplied by fructose (~55 g/day). The major source of fructose is the disaccharide sucrose, which, when cleaved in the intestine, releases equimolar amounts of fructose and glucose. Fructose is also found as a free monosaccharide in many fruits, honey, and high-fructose corn syrup (typically, 55% fructose and 45% glucose), which is used to sweeten soft drinks and many foods. Fructose transport into cells is not insulin dependent (unlike that of glucose into certain tissues), and, in contrast to glucose, fructose does not promote the secretion of insulin.

A. Phosphorylation

For fructose to enter the pathways of intermediary metabolism, it must first be phosphorylated (Fig. 12.2). This can be accomplished by actions of either hexokinase or fructokinase. Hexokinase phosphorylates glucose in most cells of the body, and several additional hexoses can serve as substrates for this enzyme. However, it has a low affinity (high K_m) for fructose.

Therefore, unless the intracellular concentration of fructose becomes unusually high, the normal presence of saturating concentrations of glucose means that little fructose is phosphorylated by hexokinase. Fructokinase provides the primary mechanism for fructose phosphorylation (see Fig. 12.2). The enzyme has a low K_m for fructose and a high V_{max} (maximal velocity). It is found in the liver (which processes most of the dietary fructose), kidneys, and the small intestine and

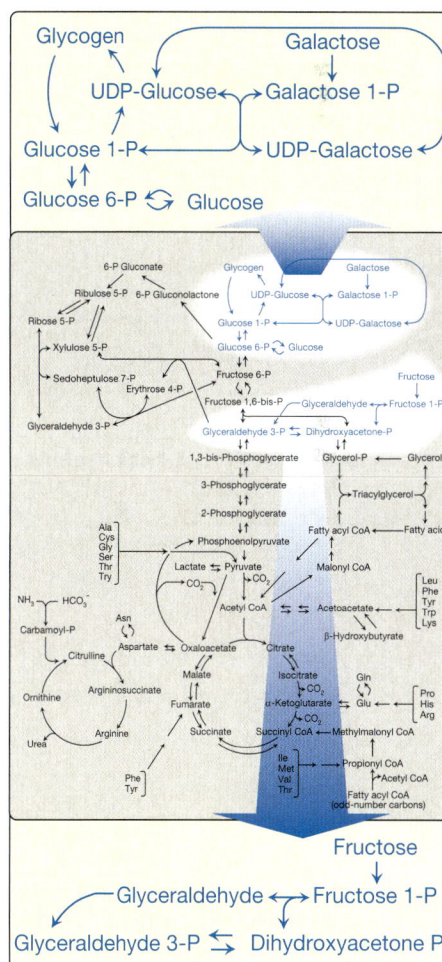

Figure 12.1
Galactose and fructose metabolism as part of the essential pathways of energy metabolism. [Note: See Fig. 8.2, for a more detailed map of metabolism.]
P, phosphate; *UDP*, uridine diphosphate.

159

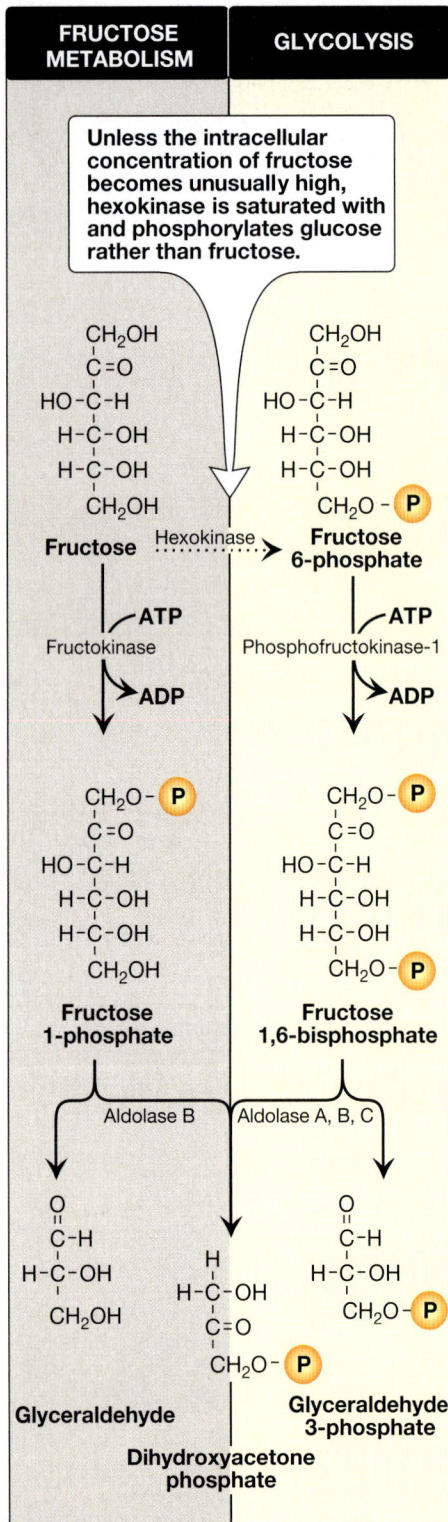

Figure 12.2
Fructose phosphorylation products and their cleavage. P, phosphate; *ADP*, adenosine diphosphate.

converts fructose to fructose 1-phosphate, using adenosine triphosphate (ATP) as the phosphate donor. [Note: These three tissues also contain aldolase B, discussed next.]

B. Fructose 1-phosphate cleavage

Fructose 1-phosphate is not phosphorylated to fructose 1,6-bisphosphate as is fructose 6-phosphate but is cleaved by aldolase B (also called fructose 1-phosphate aldolase) to two trioses, dihydroxyacetone phosphate (DHAP) and glyceraldehyde. [Note: Humans express three distinct aldolase isoenzymes, the products of three different genes: aldolase A in most tissues; aldolase B in the liver, kidneys, and small intestine; and aldolase C in the brain. All cleave fructose 1,6-bisphosphate produced during glycolysis to DHAP and glyceraldehyde 3-phosphate, but only aldolase B cleaves fructose 1-phosphate.] DHAP can be used in glycolysis or gluconeogenesis, whereas glyceraldehyde can be metabolized by a number of pathways, as illustrated in Figure 12.3.

C. Kinetics

The rate of fructose metabolism is more rapid than that of glucose because triose production from fructose 1-phosphate bypasses phosphofructokinase-1, the major rate-limiting step in glycolysis.

D. Disorders

A deficiency of one of the key enzymes required for the entry of fructose into metabolic pathways can result in either a benign condition because of fructokinase deficiency (essential fructosuria) or a severe disturbance of liver and kidney metabolism because of aldolase B deficiency, or hereditary fructose intolerance (HFI), which occurs in ~1:20,000 live births (see Fig. 12.3).

The first symptoms of HFI appear when an infant is weaned from lactose-containing milk and begins ingesting food containing sucrose or fructose. Fructose 1-phosphate accumulates, resulting in a drop in the level of inorganic phosphate (P_i) and, therefore, of ATP production. As ATP falls, adenosine monophosphate (AMP) rises. The AMP is degraded, causing hyperuricemia and lactic acidemia. The decreased availability of hepatic ATP decreases gluconeogenesis (causing hypoglycemia with vomiting) and protein synthesis (causing a decrease in blood-clotting factors and other essential proteins). Renal reabsorption of P_i is also decreased. [Note: The drop in P_i also inhibits glycogenolysis.]

Diagnosis of HFI can be made on the basis of fructose in the urine, enzyme assay using liver cells, or by DNA-based testing (see Chapter 34). With HFI, sucrose, as well as fructose, must be removed from the diet to prevent liver failure and possible death. Persons with HFI tend to display a life-long aversion to sweets.

E. Mannose conversion to fructose 6-phosphate

Mannose, the C-2 epimer of glucose, is an important component of glycoproteins. Hexokinase phosphorylates mannose, producing mannose 6-phosphate, which, in turn, is reversibly isomerized to

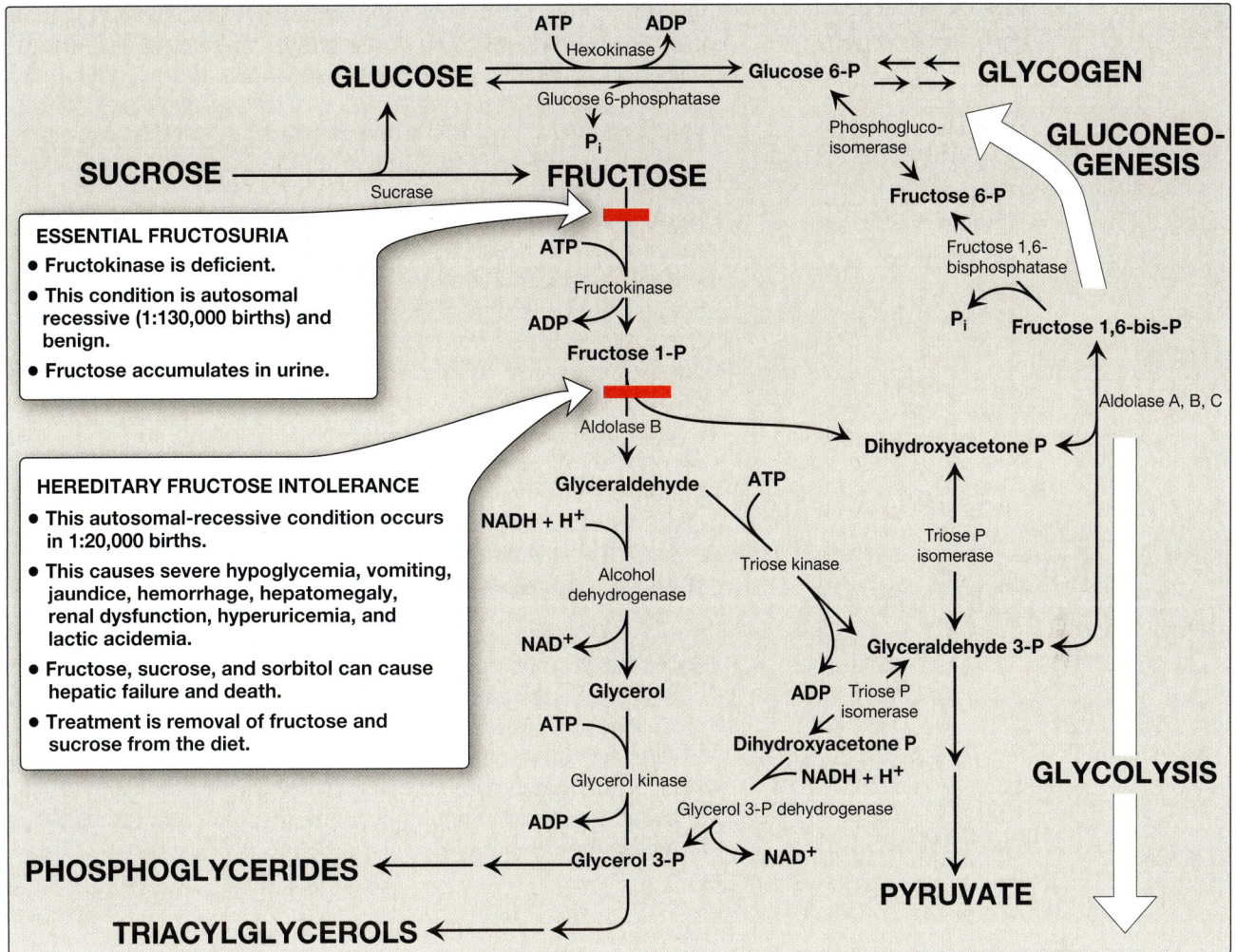

Figure 12.3
Summary of fructose metabolism. *ADP*, adenosine diphosphate; *NAD(H)*, nicotinamide adenine dinucleotide; *P*, phosphate; P_i, inorganic phosphate.

fructose 6-phosphate by phosphomannose isomerase. [Note: Most intracellular mannose is synthesized from fructose or is pre-existing mannose produced by the degradation of glycoproteins and salvaged by hexokinase. Dietary carbohydrates contain little mannose.]

F. Glucose conversion to fructose via sorbitol

Most sugars are rapidly phosphorylated following their entry into cells. Therefore, they are trapped within the cells, because organic phosphates cannot freely cross membranes without specific transporters. An alternate mechanism for metabolizing a monosaccharide is to convert it to a polyol (sugar alcohol) by the reduction of an aldehyde group, thereby producing an additional hydroxyl group.

1. **Sorbitol synthesis:** Aldose reductase reduces glucose, producing sorbitol (or, glucitol; Fig. 12.4), but the K_m is high. This enzyme is found in many tissues, including the retina, lens, kidneys, peripheral nerves, ovaries, and seminal vesicles. A second enzyme, sorbitol dehydrogenase, can oxidize sorbitol to fructose

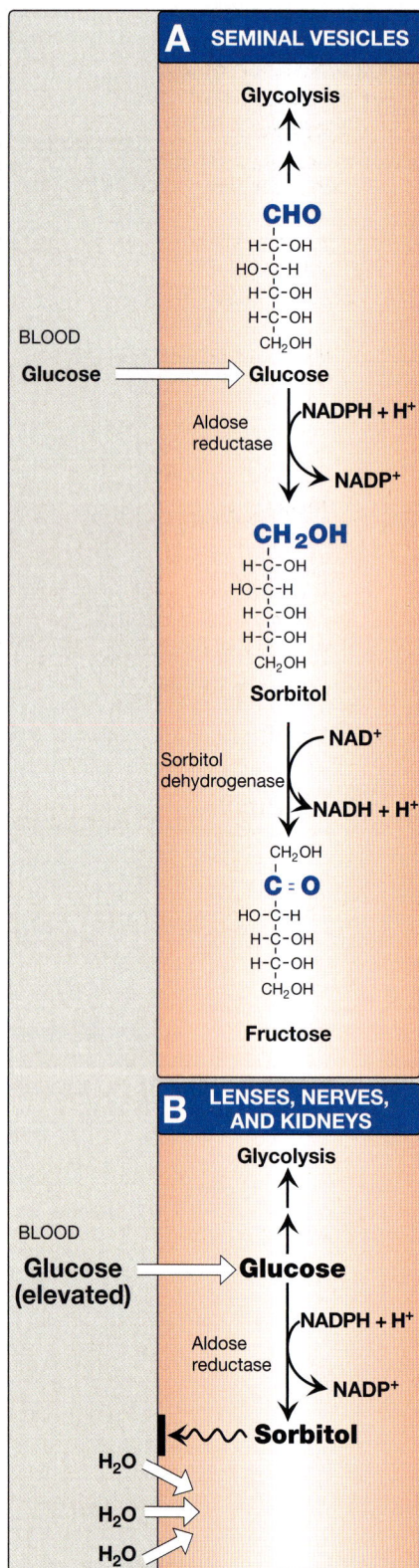

Figure 12.4
Sorbitol metabolism. *NAD(H)*, nicotin-amide adenine dinucleotide; *NADP(H)*, nicotinamide adenine dinucleotide phosphate.

in cells of the liver, ovaries, and seminal vesicles (see Fig. 12.4). The two-reaction pathway from glucose to fructose in the seminal vesicles benefits sperm cells, which use fructose as a major carbohydrate energy source. The pathway from sorbitol to fructose in the liver provides a mechanism by which any available sorbitol is converted into a substrate that can enter glycolysis.

2. **Hyperglycemia and sorbitol metabolism:** Because insulin is not required for the entry of glucose into cells of the retina, lens, kidneys, and peripheral nerves, large amounts of glucose may enter these cells during times of hyperglycemia (e.g., in poorly controlled diabetes mellitus). An elevated intracellular glucose concentration and an adequate supply of reduced nicotinamide adenine dinucleotide phosphate (NADPH) cause aldose reductase to produce a significant increase in sorbitol within the cell, which cannot pass efficiently through cell membranes and, therefore, remains trapped inside the cell (see Fig. 12.4). This is exacerbated when sorbitol dehydrogenase production is low or absent. As a result, sorbitol accumulates in these cells, causing strong osmotic effects and cell swelling due to water influx and retention.

Some of the pathologic alterations associated with diabetes mellitus can be partly attributed to this osmotic stress, including cataract formation, peripheral neuropathy, and microvascular problems leading to nephropathy and retinopathy. Use of NADPH in the aldose reductase reaction decreases the generation of reduced glutathione, an important antioxidant, and may also be related to complications of diabetes.

III. GALACTOSE METABOLISM

The major dietary source of galactose is lactose (galactosyl β-1,4-glucose) obtained from milk and milk products. [Note: The digestion of lactose by β-galactosidase, also called lactase, was discussed.] Some galactose can also be obtained by lysosomal degradation of glycoproteins and glycolipids. Like fructose (and mannose), the transport of galactose into cells is not insulin dependent.

A. Phosphorylation

Like fructose, galactose must be phosphorylated before it can be further metabolized. Most tissues have a specific enzyme for this purpose, galactokinase, which produces galactose 1-phosphate (Fig. 12.5). As with other kinases, ATP is the phosphate donor.

B. Uridine diphosphate–galactose formation

Galactose 1-phosphate cannot enter the glycolytic pathway unless it is first converted to uridine diphosphate (UDP)-galactose (Fig. 12.6). This occurs in an exchange reaction, in which UDP-glucose reacts with galactose 1-phosphate, producing UDP-galactose and glucose 1-phosphate (see Fig. 12.5). The reaction is catalyzed by galactose 1-phosphate uridylyltransferase (GALT). [Note: The glucose 1-phosphate product can

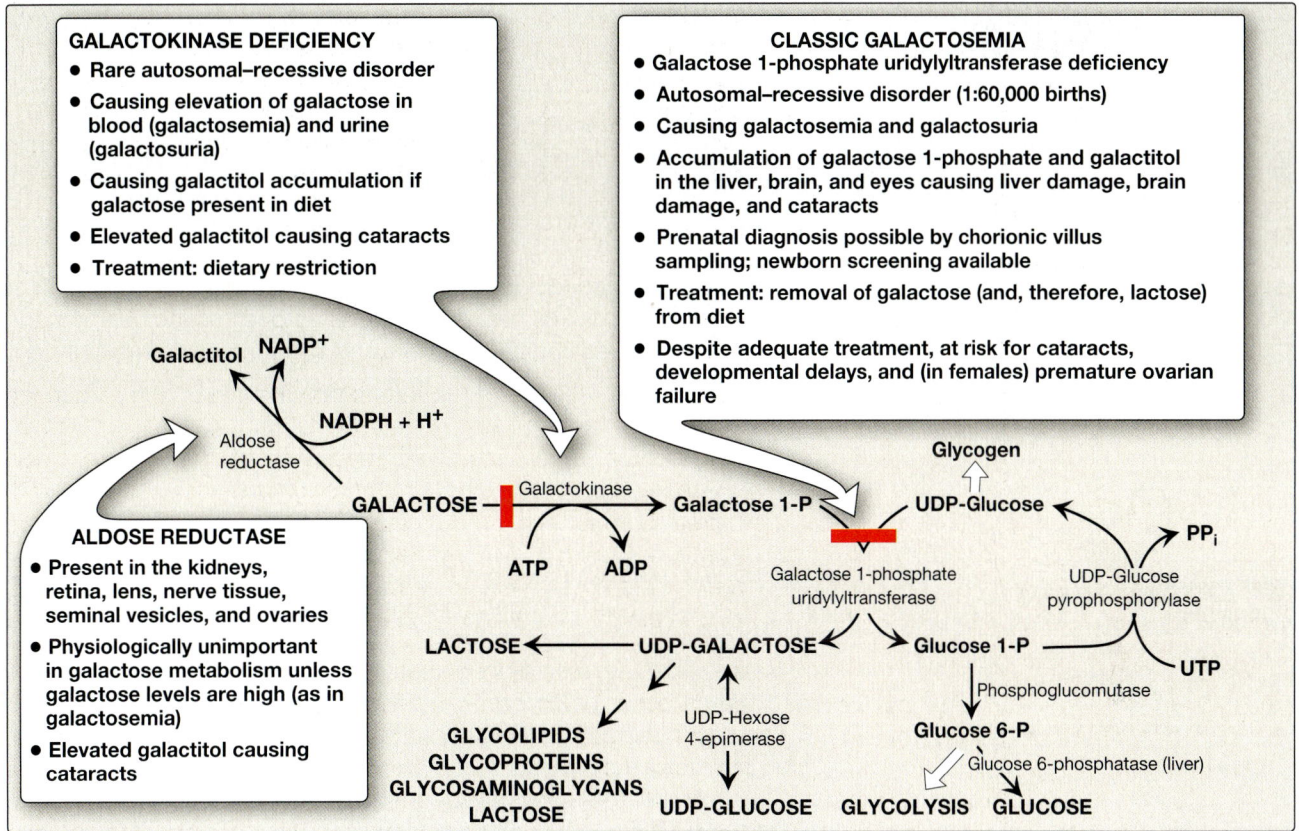

Figure 12.5
Metabolism of galactose. *ADP*, adenosine diphosphate; *NADP(H)*, nicotinamide adenine dinucleotide phosphate; *P*, phosphate; *PP$_i$*, pyrophosphate; *UDP* and *UTP*, uridine di- and triphosphates.

be isomerized to glucose 6-phosphate, which can enter glycolysis or be dephosphorylated.]

C. UDP-galactose conversion to UDP-glucose

For UDP-galactose to enter the mainstream of glucose metabolism, it must first be isomerized to its C-4 epimer, UDP-glucose, by UDP-hexose 4-epimerase. This "new" UDP-glucose (produced from the original UDP-galactose) can participate in biosynthetic reactions (e.g., glycogenesis) as well as in the GALT reaction. [Note: See Fig. 12.5 for a summary of the interconversions.]

D. UDP-galactose in biosynthetic reactions

UDP-galactose can serve as the donor of galactose units in a number of synthetic pathways, including synthesis of lactose (see Section IV), glycoproteins, glycolipids, and glycosaminoglycans. [Note: If galactose is not provided by the diet (e.g., when it cannot be released from lactose owing to a lack of β-galactosidase in people who are lactose intolerant), all tissue requirements for UDP-galactose can be met by the action of UDP-hexose 4-epimerase on UDP-glucose, which is efficiently produced from glucose 1-phosphate and uridine triphosphate (see Fig. 12.5).]

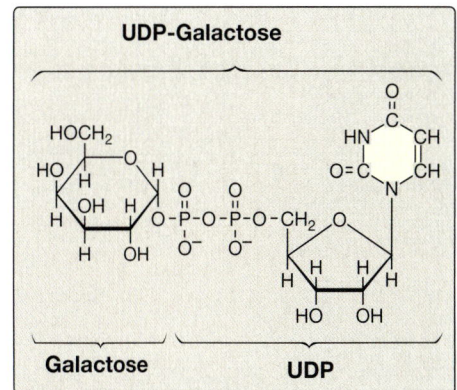

Figure 12.6
Structure of UDP-galactose. *UDP*, uridine diphosphate.

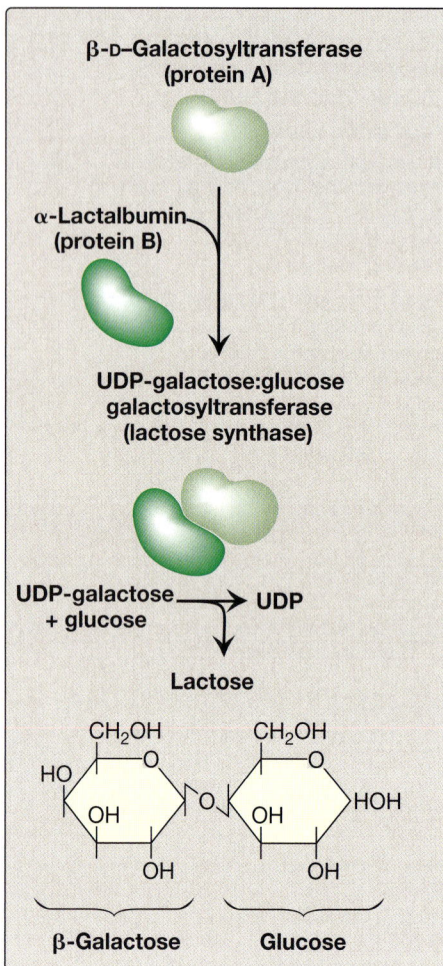

Figure 12.7
Lactose synthesis. *UDP*, uridine diphosphate.

E. Disorders

GALT is severely deficient in individuals with classic galactosemia (see Fig. 12.5). In this disorder, galactose 1-phosphate and, therefore, galactose accumulate. Physiologic consequences are similar to those found in HFI, but a broader spectrum of tissues is affected. The accumulated galactose is shunted into side pathways such as that of galactitol production. This reaction is catalyzed by aldose reductase, the same enzyme that reduces glucose to sorbitol. GALT deficiency is part of the newborn screening panel. Treatment of galactosemia requires the removal of galactose and lactose from the diet. Deficiencies in galactokinase and the epimerase result in less severe disorders of galactose metabolism, although cataracts are common (see Fig. 12.5).

IV. LACTOSE SYNTHESIS

Lactose is a disaccharide that consists of a molecule of β-galactose attached by a β(1→4) linkage to glucose. Therefore, lactose is galactosyl β(1→4)-glucose. Because lactose, the sugar in milk, is made by lactating (milk-producing) mammary glands, milk and other dairy products are the dietary sources of lactose.

Lactose synthase (UDP-galactose: glucose galactosyltransferase) catalyzes lactose synthesis in the Golgi. This enzyme, composed of A and B proteins, transfers galactose from UDP-galactose to glucose, releasing UDP (Fig. 12.7). Protein A is a β-D–galactosyltransferase and is found in a number of body tissues. In tissues other than the lactating mammary gland, this enzyme transfers galactose from UDP-galactose to N-acetyl-D–glucosamine, forming the same β(1→4) linkage found in lactose, and producing N-acetyllactosamine, a component of the structurally important N-linked glycoproteins. In contrast, protein B is found only in lactating mammary glands. It is α-lactalbumin, and its synthesis is stimulated by the peptide hormone prolactin. Protein B forms a complex with the enzyme, protein A, changing the specificity of that transferase (by decreasing the K_m for glucose) so that lactose, rather than N-acetyllactosamine, is produced (see Fig. 12.7).

Clinical Application 12-1: Lactose Intolerance

Lactose intolerance, also called lactose malabsorption, affects up to 60% of adults with ancestries other than Northern European. It results from deficiency of β-galactosidase or lactase in the small intestine. Recall from Chapter 7 that insufficient lactase causes an inability to fully digest dairy products. After consuming dairy, persons with lactose intolerance can experience cramping, diarrhea, and bloating. Lactase supplements and avoidance of dairy products can be effective in treating the condition.

V. CHAPTER SUMMARY

- The major source of fructose is the disaccharide **sucrose**, which, when cleaved, releases equimolar amounts of **fructose** and **glucose** (Fig. 12.8).

- Transport of fructose into cells is **insulin independent**.

- Fructose is first phosphorylated to **fructose 1-phosphate** by **fructokinase** and then cleaved by **aldolase B** to dihydroxyacetone phosphate and **glyceraldehyde**. These enzymes are found in the **liver, kidneys**, and **small intestine**.

- A deficiency of fructokinase causes a benign condition, **essential fructosuria**, whereas a deficiency of aldolase B causes **hereditary fructose intolerance (HFI)**, in which **severe hypoglycemia** and **liver failure** lead to **death** if fructose (and sucrose) is not removed from the diet.

- **Mannose**, an important component of **glycoproteins**, is phosphorylated by **hexokinase** to **mannose 6-phosphate**, which is reversibly isomerized to **fructose 6-phosphate** by **phosphomannose isomerase**.

- Glucose can be reduced to **sorbitol (glucitol)** by **aldose reductase** in many tissues, including the **lens, retina, peripheral nerves, kidneys, ovaries**, and **seminal vesicles**. In the liver, ovaries, and seminal vesicles, a second enzyme, **sorbitol dehydrogenase**, can oxidize sorbitol to produce **fructose**.

- **Hyperglycemia** results in the accumulation of sorbitol in those cells lacking sorbitol dehydrogenase. The resulting **osmotic events** cause cell swelling and may contribute to the **cataract formation, peripheral neuropathy, nephropathy**, and **retinopathy** seen in **diabetes**.

- The major dietary source of **galactose** is **lactose**. The transport of galactose into cells is insulin independent. Galactose is first phosphorylated to galactose 1-phosphate by **galactokinase**, a deficiency of which results in cataracts.

- Galactose 1-phosphate is converted to **uridine diphosphate (UDP)-galactose** by **galactose 1-phosphate uridylyltransferase**, with the nucleotide supplied by UDP-glucose. A deficiency of this enzyme causes **classic galactosemia**. Galactose 1-phosphate accumulates, and excess galactose is converted to **galactitol** by **aldose reductase**. This causes **liver damage, brain damage**, and **cataracts**. Treatment requires removal of galactose (and lactose) from the diet.

- For UDP-galactose to enter the mainstream of glucose metabolism, it must first be isomerized to UDP-glucose by **UDP-hexose 4-epimerase**. This enzyme can also be used to produce UDP-galactose from UDP-glucose when the former is required for glycoprotein and glycolipid synthesis.

- **Lactose** is a disaccharide of **galactose** and **glucose**. **Dairy products** are the dietary sources of lactose. Lactose is synthesized by **lactose synthase** from **UDP-galactose** and **glucose** in the **lactating mammary gland**. The enzyme has two subunits, **protein A** (which is a **galactosyltransferase** found in most cells where it synthesizes **N-acetyllactosamine**) and **protein B** (α-lactalbumin, which is found only in lactating mammary glands, and whose synthesis is stimulated by the peptide hormone **prolactin**). When both subunits are present, the transferase produces lactose.

Important dietary monosaccharides

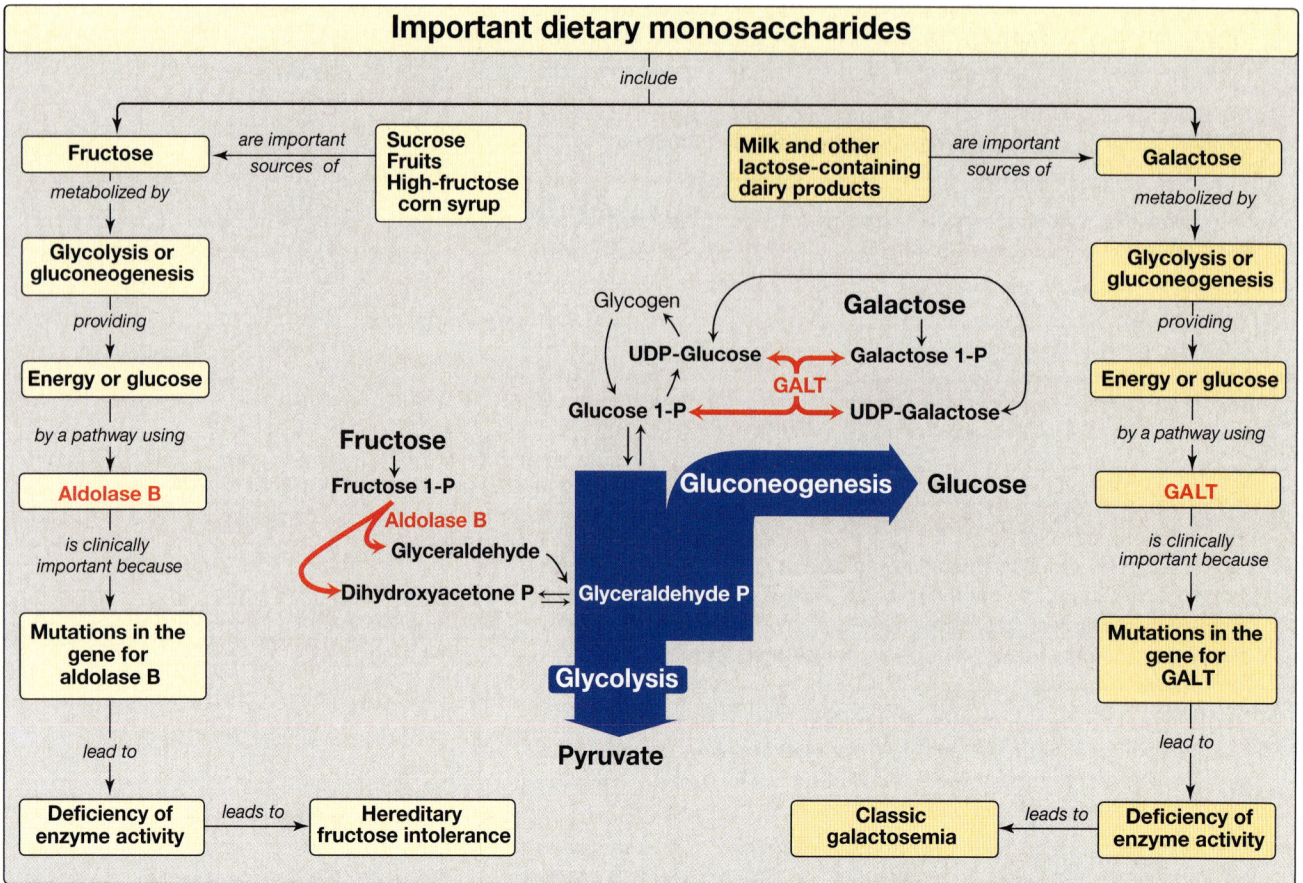

Figure 12.8
Key concept map for metabolism of fructose and galactose. *GALT*, galactose 1-phosphate uridylyltransferase; *UDP*, uridine diphosphate; *P*, phosphate.

Study Questions

Choose the ONE best answer.

12.1. A woman with classic galactosemia who is on a galactose-free diet delivers a full-term infant and successfully breastfeeds the child. She can produce lactose in her breast milk because:

   A. galactose can be produced from fructose by isomerization.

   B. galactose can be produced from a glucose metabolite by epimerization.

   C. hexokinase can efficiently phosphorylate galactose to galactose 1-phosphate.

   D. the enzyme affected in galactosemia is activated by a mammary gland hormone.

Correct answer = B. Uridine diphosphate (UDP)-glucose is converted to UDP-galactose by UDP-hexose 4-epimerase, thereby providing the appropriate form of galactose for lactose synthesis. Isomerization of fructose to galactose does not occur in the human body. Galactose is not converted to galactose 1-phosphate by hexokinase. A galactose-free diet provides no galactose. Galactosemia is the result of an enzyme (galactose 1-phosphate uridylyltransferase) deficiency.

12.2. A 7-month-old boy is brought to his pediatrician because of vomiting, night sweats, and tremors. History reveals that these symptoms began after fruit juices were introduced to his diet as he was being weaned off breast milk. The physical examination was remarkable for hepatomegaly. Tests on his urine were positive for reducing sugar but negative for glucose. The infant most likely has a deficiency of:

A. aldolase B.

B. fructokinase.

C. galactokinase.

D. β-galactosidase.

Correct answer = A. The symptoms suggest hereditary fructose intolerance, a deficiency in aldolase B. Deficiencies in fructokinase or galactokinase result in relatively benign conditions characterized by elevated levels of fructose or galactose in the blood and urine. Deficiency in β-galactosidase (lactase) results in a decreased ability to degrade lactose. Congenital lactase deficiency is quite rare and would have presented much earlier in this child (and with different symptoms). Typical lactase deficiency (adult lactose intolerance) presents at a later age.

12.3. In lactose synthesis:

A. α-lactalbumin decreases the affinity of protein A for glucose.

B. α-lactalbumin expression is decreased by the hormone prolactin.

C. galactosyltransferase catalyzes transfer of galactose from galactose 1-phosphate to glucose.

D. protein A is used exclusively.

E. protein B expression is stimulated by prolactin.

Correct answer = A. α-Lactalbumin (protein B) decreases affinity of protein A for glucose, and its expression is increased by the hormone prolactin. Uridine diphosphate–galactose is the form used by the galactosyltransferase (protein A). Protein A is also involved in the synthesis of the amino sugar N-acetyllactosamine. Protein B decreases the Michaelis constant (K_m) and, so, increases the affinity of protein A for glucose.

12.4. A 3-month-old child is evaluated for cloudiness of her eyes. Her physical examination reveals cataracts. Other than not having a social smile or being able to track objects visually, all other aspects of her examination are normal. Tests on her urine are positive for reducing sugar but negative for glucose. Which enzyme is most likely deficient in this child?

A. Aldolase B

B. Fructokinase

C. Galactokinase

D. Galactose 1-phosphate uridylyltransferase

Correct answer = C. The child is deficient in galactokinase and is unable to appropriately phosphorylate galactose. Galactose accumulates in the blood (and urine). In the lens of the eye, galactose is reduced by aldose reductase to galactitol, a sugar alcohol, which causes osmotic effects that result in cataract formation. Deficiency of galactose 1-phosphate uridylyltransferase also results in cataracts but is characterized by liver damage and neurologic effects. Fructokinase deficiency is a benign condition. Aldolase B deficiency is severe, with effects on several tissues but cataracts are not typically seen.

12.5. In a person with elevated blood glucose and an adequate supply of nicotinamide adenine dinucleotide phosphate (NADPH), which of the following will be produced in high concentration and then remain trapped in the cell?

A. Fructose

B. Galactose

C. Lactose

D. Sorbitol

E. Sucrose

Correct answer = D. Sorbitol will be elevated in this situation. An elevated intracellular glucose concentration and an adequate supply of reduced NADPH cause aldose reductase to produce a significant increase in sorbitol, which cannot pass efficiently through cell membranes and, therefore, remains trapped inside the cell. Sorbitol trapped in the cells then contributes to complications of diabetes mellitus including cataract formation, peripheral neuropathy, and microvascular problems.

12.6. A 3-month-old is evaluated for feeding difficulties, frequent vomiting, and failure to thrive. Hepatomegaly is noted during an examination. Elevated galactose is found in her blood. Which enzyme is most likely to be deficient?

A. Aldolase

B. Lactase

C. Galactose 1-phosphate uridylyltransferase

D. Glucose 6-phosphatase

E. Phosphoglucomutase

Correct answer = C. This presentation is of classic galactosemia. Galactose 1-phosphate uridylyltransferase converts galactose 1-phosphate to glucose 1-phosphate. Deficiency would lead to accumulation of galactose 1-phosphate, hepatomegaly, feeding difficulties, and failure to thrive. Aldolase deficiency causes intolerance to fructose, and lactase deficiency impairs lactose digestion, both with symptoms different from those in this case. Glucose 6-phosphatase is involved in gluconeogenesis and glycogenolysis, not in galactose metabolism. Deficiency of phosphoglucomutase is associated with a glycogen storage disorder.

13 Pentose Phosphate Pathway and Nicotinamide Adenine Dinucleotide Phosphate

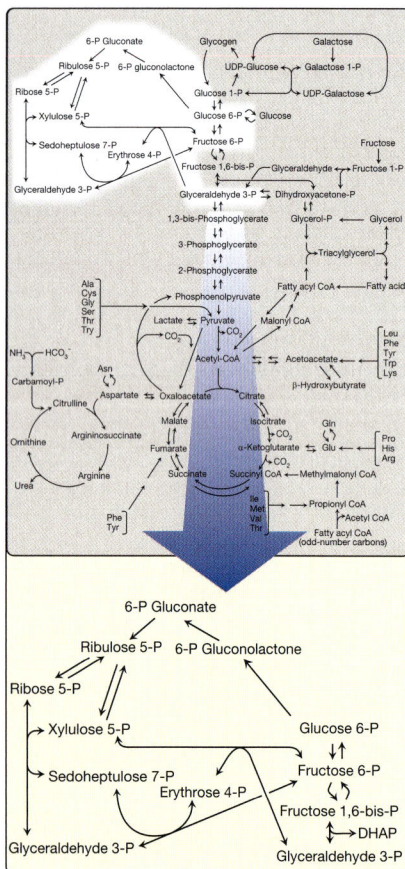

Figure 13.1
Pentose phosphate pathway shown as a component of the metabolic map. [Note: See Fig. 8.2 for a more detailed map of metabolism.] P, phosphate; DHAP, dihydroxyacetone phosphate.

I. OVERVIEW

The pentose phosphate pathway, also known as the hexose monophosphate shunt, does not use or consume adenosine triphosphate (ATP) but is important as it provides the body's main source of **nicotinamide adenine dinucleotide phosphate** (**NADPH**), a biochemical reductant. The reactions of the pathway occur in the cytosol and include an irreversible oxidative phase, followed by a series of reversible sugar–phosphate interconversions (Fig. 13.1). In the oxidative phase, carbon 1 of a glucose 6-phosphate molecule is released as carbon dioxide (CO_2), and one pentose sugar-phosphate plus two reduced NADPHs are produced. The rate and direction of the reversible reactions are determined by the supply of and demand for intermediates of the pathway. The pentose phosphate pathway also produces **ribose 5-phosphate**, required for nucleotide biosynthesis (see also Chapter 22, III.), and provides a mechanism for the conversion of pentose sugars to triose and hexose intermediates of glycolysis.

II. IRREVERSIBLE OXIDATIVE REACTIONS

The oxidative portion of the pentose phosphate pathway consists of three irreversible reactions that lead to the formation of **ribulose 5-phosphate**, CO_2, and two molecules of NADPH for each molecule of glucose 6-phosphate oxidized (Fig. 13.2). This portion of the pathway is particularly important in the liver, lactating mammary glands, and adipose tissue for the NADPH-dependent biosynthesis of fatty acids (see also Chapter 15, III. D.); in the testes, ovaries, placenta, and adrenal cortex for the NADPH-dependent biosynthesis of steroid hormones (see also Chapter 18, VII.); and in red blood cells (RBCs) for the NADPH-dependent reduction of glutathione.

A. Glucose 6-phosphate dehydrogenation

Glucose 6-phosphate dehydrogenase (**G6PD**) catalyzes the oxidation of glucose 6-phosphate to 6-phosphogluconolactone as the

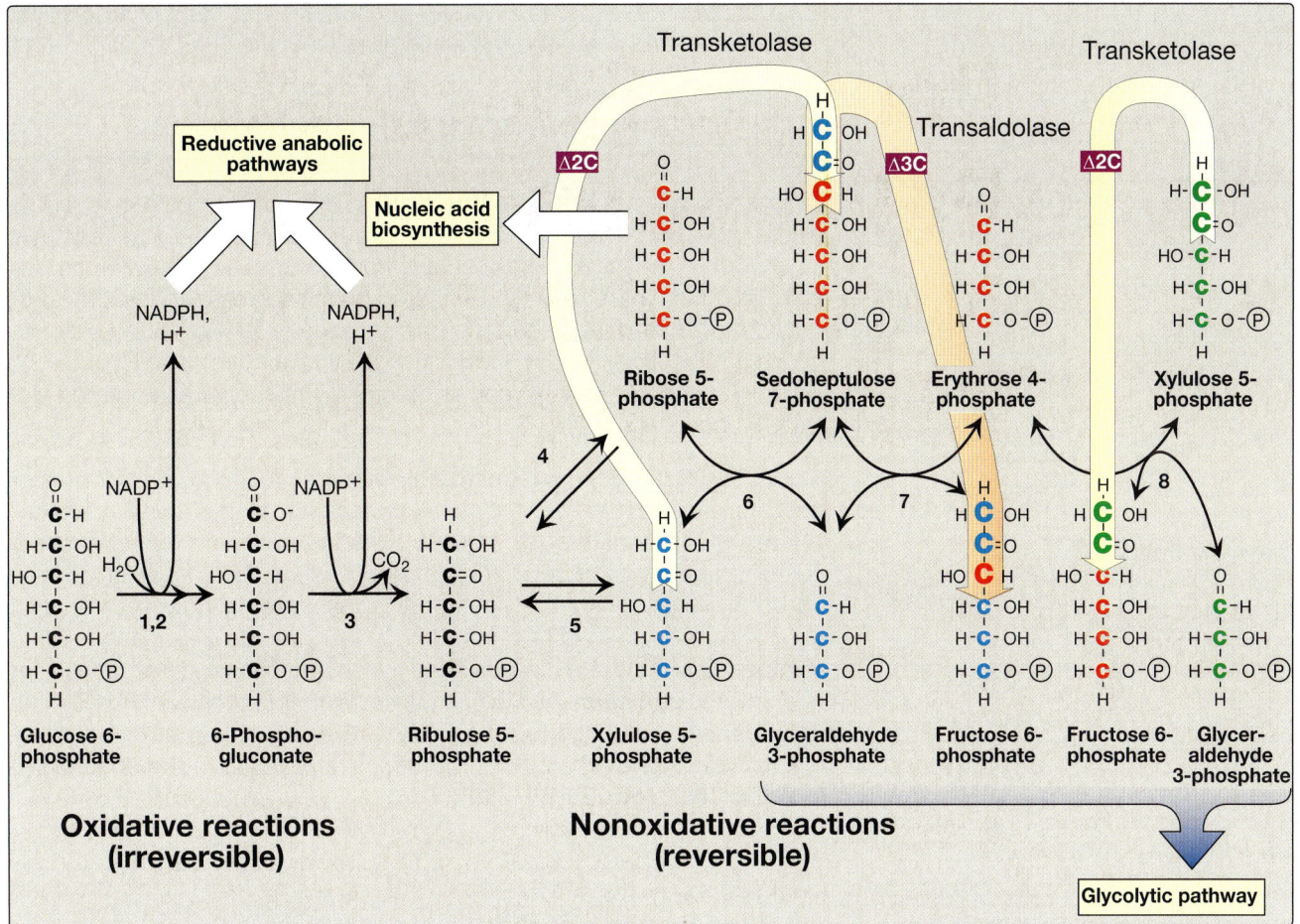

Figure 13.2
Reactions of the pentose phosphate pathway. Enzymes are: (1, 2) glucose 6-phosphate dehydrogenase and 6-phospho-gluconolactone hydrolase, (3) 6-phosphogluconate dehydrogenase, (4) ribose 5-phosphate isomerase, (5) phosphopentose epimerase, (6, 8) transketolase (coenzyme: thiamine pyrophosphate), and (7) transaldolase. Δ2C = two carbons are transferred from a ketose donor to an aldose acceptor in transketolase reactions; Δ3C = three carbons are transferred in the transaldolase reaction. This can be represented as: 5C sugar + 5C sugar → 7C sugar + 3C sugar → 4C sugar + 6C sugar. NADP(H), nicotinamide adenine dinucleotide phosphate; (P), phosphate; CO_2, carbon dioxide.

coenzyme $NADP^+$ is reduced to NADPH. This initial reaction is the committed, rate-limiting, and regulated step of the pathway. NADPH is a potent competitive inhibitor of G6PD, and the ratio of NADPH/$NADP^+$ is sufficiently high to substantially inhibit the enzyme under most metabolic conditions. However, with increased demand for NADPH, the ratio of NADPH/$NADP^+$ decreases, and flux through the pathway increases in response to the enhanced activity of G6PD. It should be noted that insulin upregulates expression of the *G6PD* gene, and flux through the pathway increases in the absorptive state (see also Chapter 24, III. A. 3.).

B. Ribulose 5-phosphate formation

6-Phosphogluconolactone is hydrolyzed by 6-phosphogluconolactone hydrolase in the second step. The oxidative decarboxylation of the product, 6-phosphogluconate, is catalyzed by 6-phosphogluconate dehydrogenase. This third irreversible step produces ribulose

5-phosphate, a pentose sugar–phosphate, CO_2 (from carbon 1 of glucose), and a second molecule of NADPH (see Fig. 13.2).

III. REVERSIBLE NONOXIDATIVE REACTIONS

The nonoxidative reactions of the pentose phosphate pathway occur in all cell types synthesizing nucleotides and nucleic acids. These reactions catalyze the interconversion of sugars containing three to seven carbons (see Fig. 13.2). These reversible reactions permit ribulose 5-phosphate produced by the oxidative portion of the pathway to be converted either to ribose 5-phosphate needed for nucleotide synthesis (see also Chapter 22, III.) or to intermediates of glycolysis, fructose 6-phosphate and glyceraldehyde 3-phosphate.

Many cells that carry out reductive biosynthetic reactions have a greater need for NADPH than for ribose 5-phosphate. In this case, **transketolase**, which transfers two-carbon units in a thiamine pyrophosphate (TPP)-requiring reaction, and **transaldolase,** which transfers three-carbon units, convert the ribulose 5-phosphate produced as an end product of the oxidative phase to glyceraldehyde 3-phosphate and fructose 6-phosphate. In contrast, when the demand for ribose for nucleotides and nucleic acids is greater than the need for NADPH, the nonoxidative reactions can provide the ribose 5-phosphate from glyceraldehyde 3-phosphate and fructose 6-phosphate in the absence of the oxidative steps (Fig. 13.3).

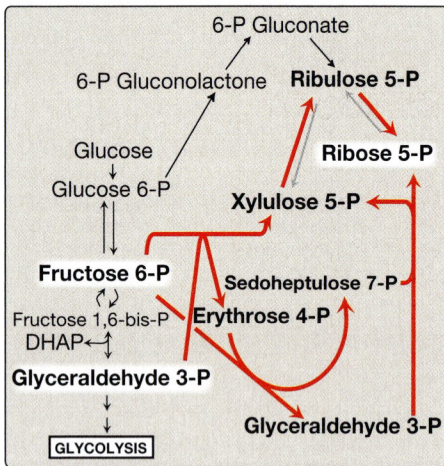

Figure 13.3
Formation of ribose 5-phosphate from intermediates of glycolysis. P, phosphate; DHAP, dihydroxyacetone phosphate.

In addition to transketolase, TPP is required by the multienzyme complexes pyruvate dehydrogenase (see also Chapter 9, II. A. 2.), α-ketoglutarate dehydrogenase of the tricarboxylic acid cycle (see also Chapter 9, II. E.), and branched-chain α-keto acid dehydrogenase of branched-chain amino acid catabolism (see also Chapter 20, III. H.).

IV. NICOTINAMIDE ADENINE DINUCLEOTIDE PHOSPHATE USES

The coenzyme NADPH differs from nicotinamide adenine dinucleotide (NADH) only by the presence of a phosphate group on one of the ribose units (Fig. 13.4). This seemingly small change in structure allows NADPH to interact with NADPH-specific enzymes that have unique roles in the cell. For example, in the cytosol of hepatocytes, the steady-state $NADP^+$/NADPH ratio is ~0.1, which favors the use of NADPH in reductive biosynthetic reactions. This contrasts with the high NAD^+/NADH ratio (~1,000), which favors an oxidative role for NAD^+. This summarizes some important NADPH-specific functions in reductive biosynthesis and detoxification reactions.

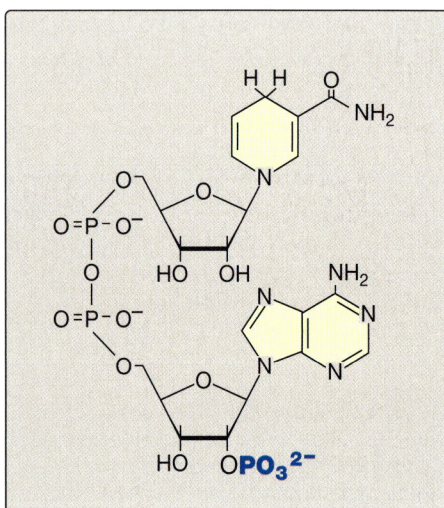

Figure 13.4
Structure of reduced nicotinamide adenine dinucleotide phosphate (NADPH).

A. Reductive biosynthesis

Like NADH, NADPH can be thought of as a high-energy molecule. However, the electrons of NADPH are used for reductive biosynthesis,

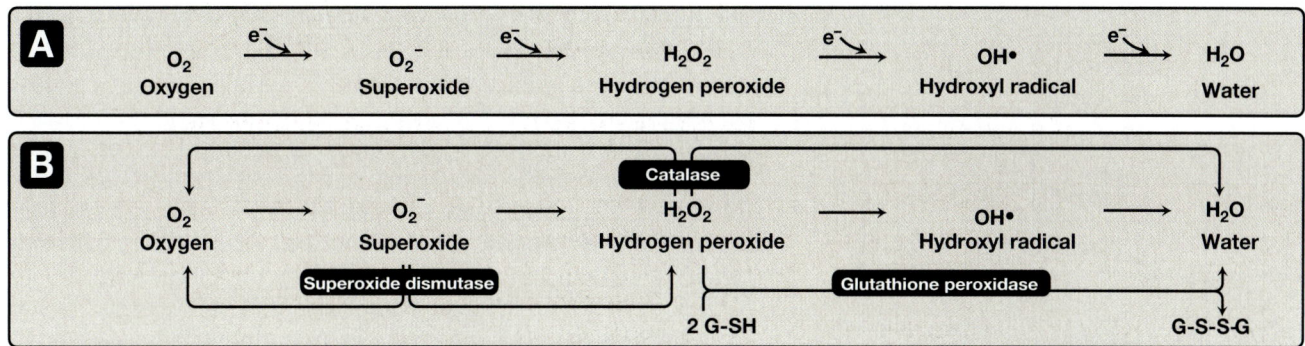

Figure 13.5
A. Formation of reactive intermediates from oxygen. e⁻, electrons. **B.** Actions of antioxidant enzymes. G-SH, reduced glutathione; G-S-S-G, oxidized glutathione. [Note: See Fig. 13.6B for the regeneration of G-SH.]

rather than for transfer to the electron transport chain as is seen with NADH (see Chapter 6, V. C.). In the metabolic transformations of the pentose phosphate pathway, part of the energy of glucose 6-phosphate is conserved in NADPH, a molecule with a negative reduction potential (see Chapter 6, V. D.), that, therefore, can be used in reactions requiring an electron donor, such as fatty acid (see Chapter 16, III.), cholesterol, and steroid hormone synthesis (see also Chapter 18, III. C. and VII.).

B. Hydrogen peroxide reduction

Hydrogen peroxide (H_2O_2) is one of a family of reactive oxygen species (ROS) that are formed from the partial reduction of molecular oxygen, O_2 (Fig. 13.5A). These compounds are generated continuously as byproducts of aerobic metabolism, through reactions with drugs and environmental toxins, or when the level of antioxidants is diminished, all creating the condition of oxidative stress. These highly reactive oxygen intermediates can cause serious chemical damage to DNA, proteins, and unsaturated lipids and can lead to cell death. ROS have been implicated in a number of pathologic processes, including reperfusion injury, cancer, inflammatory disease, and aging. The cell has several protective mechanisms that minimize the toxic potential of these compounds. ROS can also be generated in the killing of microbes by white blood cells (see Section D.).

1. **Enzymes that catalyze antioxidant reactions: Reduced glutathione (G-SH)**, a tripeptide-thiol (γ-glutamyl cysteinylglycine) present in most cells, can chemically detoxify H_2O_2 (Fig. 13.5B). This reaction, catalyzed by glutathione peroxidase, forms oxidized glutathione (G-S-S-G), which no longer has protective properties. The cell regenerates G-SH in a reaction catalyzed by glutathione reductase, using NADPH as a source of reducing equivalents. Thus, NADPH indirectly provides electrons for the reduction of H_2O_2 (Fig. 13.6). Additional enzymes, such as superoxide dismutase and catalase, catalyze the conversion of other ROS to harmless products (see Fig. 13.5B). As a group, these enzymes serve as a defense system to guard against the toxic effects of ROS.

2. **Antioxidant chemicals:** A number of intracellular reducing agents, such as ascorbate or vitamin C, vitamin E, and β-carotene,

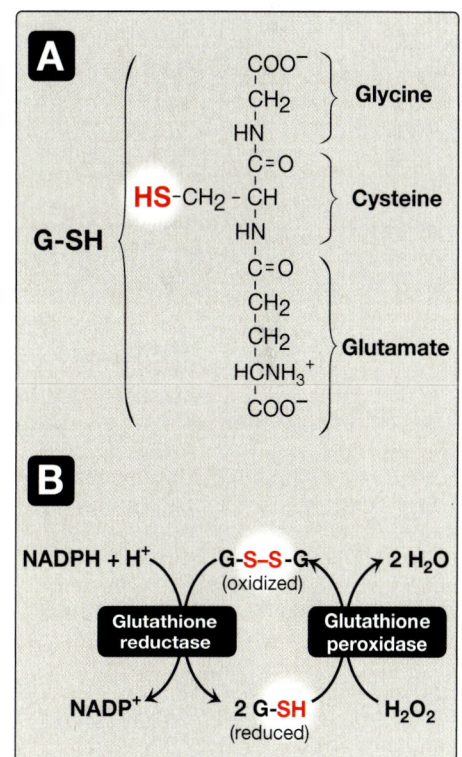

Figure 13.6
A. Structure of reduced glutathione (G-SH). [Note: Glutamate is linked to cysteine through a γ-carboxyl, rather than an α-carboxyl.] **B.** The roles of G-SH and reduced nicotinamide adenine dinucleotide phosphate (NADPH) in the reduction of hydrogen peroxide (H_2O_2) to water. G-S-S-G, oxidized glutathione.

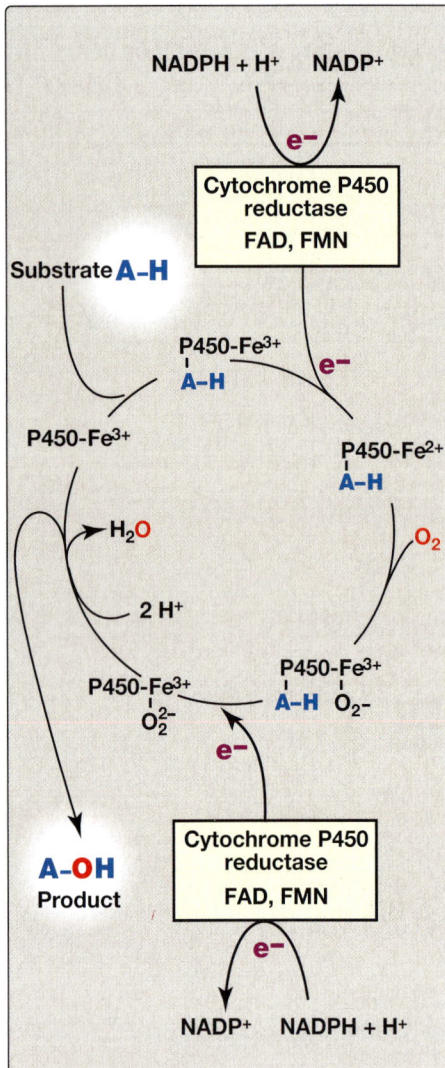

are able to reduce and, thereby, detoxify ROS in the laboratory. Consumption of foods rich in these antioxidant compounds has been correlated with a reduced risk for certain types of cancers as well as decreased frequency of certain other chronic health problems. Therefore, it is tempting to speculate that the effects of these compounds are, in part, an expression of their ability to quench the toxic effect of ROS. However, clinical trials with antioxidants as dietary supplements have failed to show clear beneficial effects. In the case of dietary supplementation with β-carotene, the rate of lung cancer in smokers increased rather than decreased. Thus, the health-promoting effects of dietary fruits and vegetables likely reflect a complex interaction among many naturally occurring compounds, which has not been duplicated by the consumption of isolated antioxidant compounds. (See also Chapter 28, IV., XI., and XIV.)

C. Cytochrome P450 monooxygenase system

Monooxygenases (mixed-function oxidases) incorporate one atom from O_2 into a substrate (creating a hydroxyl group), with the other atom being reduced to water (H_2O). In the cytochrome P450 (CYP) monooxygenase system, NADPH provides the reducing equivalents required by this series of reactions (Fig. 13.7). This system performs different functions in two separate locations in cells. The overall reaction catalyzed by a CYP enzyme is:

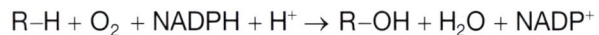

$$R–H + O_2 + NADPH + H^+ \rightarrow R–OH + H_2O + NADP^+$$

where R may be a steroid, drug, or other chemical. *CYP* enzymes are actually a superfamily of related, heme-containing monooxygenases that participate in a broad variety of reactions. The P450 in the name reflects the absorbance at 450 nm by the protein.

1. **Mitochondrial system:** An important function of the CYP monooxygenase system found associated with the inner mitochondrial membrane is the biosynthesis of steroid hormones. In steroidogenic tissues, such as the placenta, ovaries, testes, and adrenal cortex, it is used to hydroxylate intermediates in the conversion of cholesterol to steroid hormones, a process that makes these hydrophobic compounds more water soluble (see Chapter 18, VII.). The liver uses this same system in bile acid synthesis (see Chapter 18, V. B.) and the hydroxylation of cholecalciferol to 25-hydroxycholecalciferol ([vitamin D_3] see Chapter 28, XII.), and the kidney uses it to hydroxylate vitamin D_3 to its biologically active 1,25-dihydroxylated form.

2. **Microsomal system:** The microsomal CYP monooxygenase system found associated with the membrane of the smooth endoplasmic reticulum, particularly in the liver, functions primarily in the detoxification of foreign compounds or xenobiotics. These include numerous drugs and such varied pollutants as petroleum products and pesticides. CYP enzymes of the microsomal system, for example, CYP3A4, can be used to hydroxylate these toxins (phase I). The purpose of these modifications is twofold. First, it may itself activate or inactivate a drug and second, make a toxic compound more soluble, thereby facilitating its excretion

Figure 13.7
Cytochrome P450 (CYP) monooxygenase catalytic cycle (simplified). Electrons (e^-) move from nicotinamide adenine dinucleotide phosphate (NADPH) to flavin adenine dinucleotide (FAD) to flavin adenine mononucleotide (FMN) of the reductase and then to the heme iron (Fe) of the microsomal CYP enzyme. [Note: In the mitochondrial system, e^- move from FAD to an iron-sulfur protein and then to the CYP enzyme.]

in the urine or feces. Frequently, however, the new hydroxyl group will serve as a site for conjugation with a polar molecule, such as glucuronic acid (see Chapter 14, III. B. 1.), which will significantly increase the compound's solubility (phase II). It should be noted that polymorphisms (see Chapter 34) in the genes for CYP enzymes can lead to differences in drug metabolism.

D. White blood cell phagocytosis and microbe killing

Phagocytosis is the ingestion by receptor-mediated endocytosis of microorganisms, foreign particles, and cellular debris by leukocytes such as neutrophils and macrophages (monocytes). It is an important defense mechanism, particularly in bacterial infections. Neutrophils and monocytes are armed with both oxygen-independent and oxygen-dependent mechanisms for killing bacteria.

1. **Oxygen-independent:** Oxygen-independent mechanisms use pH changes in phagolysosomes and lysosomal enzymes to destroy pathogens.

2. **Oxygen-dependent:** Oxygen-dependent mechanisms include the enzymes NADPH oxidase and **myeloperoxidase** (**MPO**) that work together in killing bacteria (Fig. 13.8). Overall, the MPO system is the most potent of the bactericidal mechanisms. An invading bacterium is recognized by the immune system and attacked by antibodies that bind it to a receptor on a phagocytic cell. After internalization of the microorganism has occurred, NADPH oxidase, located in the leukocyte cell membrane, is activated and reduces O_2 from the surrounding tissue to superoxide (O_2^-), a free radical ROS, as NADPH is oxidized. The rapid consumption of O_2 that accompanies its formation is referred to as the respiratory burst. [Note: Active NADPH oxidase is a membrane-associated complex containing a flavocytochrome plus additional peptides that translocate from the cytoplasm upon activation of the leukocyte. Electrons move from NADPH to O_2 via flavin adenine nucleotide (FAD) and heme, generating O_2^-.]

Rare genetic deficiencies in NADPH oxidase cause chronic granulomatous disease (CGD) characterized by severe, persistent infections and the formation of granulomas (nodular areas of inflammation) that sequester the bacteria that were not destroyed. Next, O_2^- is converted to H_2O_2 (also a ROS), either spontaneously or catalyzed by superoxide dismutase. In the presence of MPO, a heme-containing lysosomal enzyme present within the phagolysosome, peroxide plus chloride ions are converted to hypochlorous acid, HOCl, the major component of household bleach, which kills the bacteria. The peroxide can also be partially reduced to the hydroxyl radical (OH•), a ROS, or be fully reduced to H_2O by catalase or glutathione peroxidase. Deficiencies in MPO do not confer increased susceptibility to infection because peroxide from NADPH oxidase is bactericidal.

E. Nitric oxide synthesis

Nitric oxide (NO) is recognized as a mediator in a broad array of biologic systems. NO is the endothelium-derived relaxing factor that

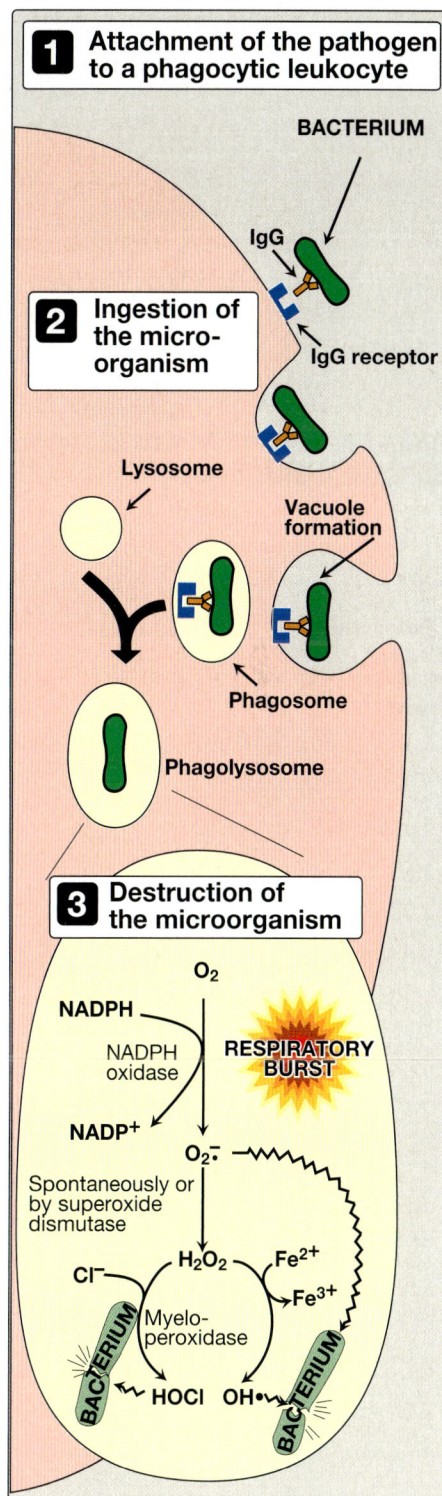

Figure 13.8
Phagocytosis and the oxygen (O_2)-dependent pathway of microbial killing. IgG, immunoglobulin G; NADP(H), nicotinamide adenine dinucleotide phosphate; O_2^-, superoxide; H_2O_2, hydrogen peroxide; HOCl, hypochlorous acid; OH•, hydroxyl radical.

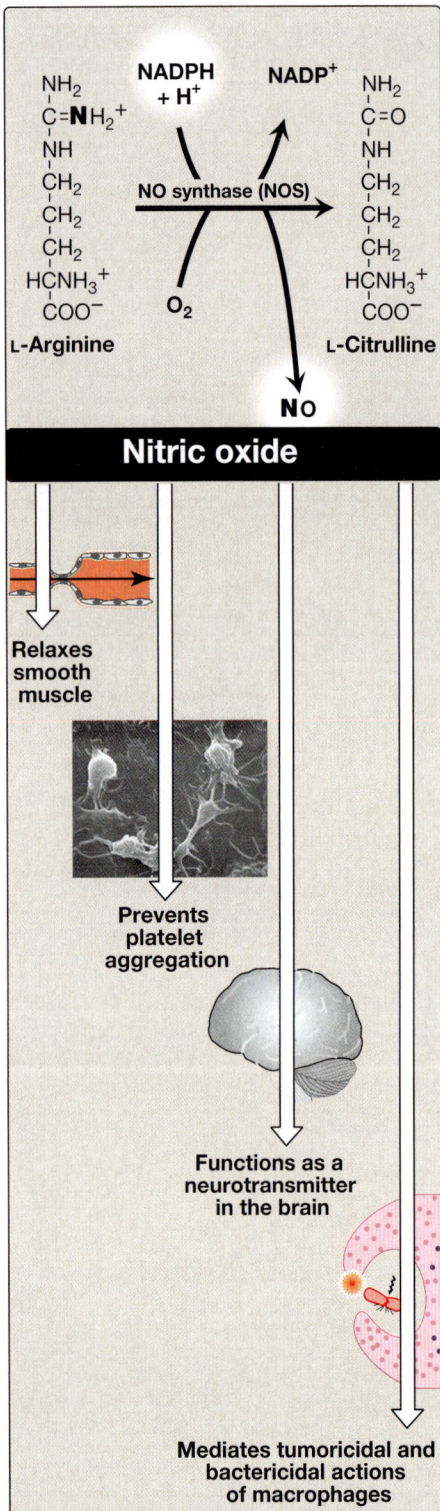

Figure 13.9
Synthesis and some actions of nitric oxide (NO). [Note: Flavin mononucleotide, flavin adenine dinucleotide, heme, and tetrahydrobiopterin are additional coenzymes required by NOS.] NADP(H), nicotinamide adenine dinucleotide phosphate.

causes vasodilation by relaxing vascular smooth muscle. It also acts as a neurotransmitter, prevents platelet aggregation, and plays an essential role in macrophage function. It has a very short half-life in tissues (3 to 10 seconds) because it reacts with O_2 and is converted into nitrates and nitrites including peroxynitrite (O = NOO⁻), a reactive nitrogen species (RNS). Note that NO is a free radical gas that is often confused with nitrous oxide (N_2O), the "laughing gas" that is used as an anesthetic and is chemically stable.

1. **Nitric oxide synthase:** Arginine, O_2, and NADPH are substrates for cytosolic NO synthase ([NOS] Fig. 13.9). Flavin mononucleotide (FMN), FAD, heme, and tetrahydrobiopterin (see Chapter 20, V. E.) are coenzymes, and NO and citrulline are products of the reaction. Three NOS isozymes, each the product of a different gene, have been identified. Two are constitutive (synthesized at a constant rate), calcium (Ca^{2+})–calmodulin (CaM)-dependent enzymes (see Chapter 11, V. C.). They are found primarily in endothelium (eNOS) and neural tissue (nNOS) and constantly produce very low levels of NO for vasodilation and neurotransmission. An inducible, Ca^{2+}-independent enzyme (iNOS) can be expressed in many cells, including macrophages and neutrophils, as an early defense against pathogens. The specific inducers for iNOS vary with cell type and include proinflammatory cytokines, such as tumor necrosis factor-α (TNF-α) and interferon-γ (IFN-γ), and bacterial endotoxins such as lipopolysaccharide (LPS). These compounds promote the synthesis of iNOS, which can result in large amounts of NO being produced over hours or even days.

2. **Nitric oxide and vascular endothelium:** NO is an important mediator in the control of vascular smooth muscle tone. NO is synthesized by eNOS in endothelial cells and diffuses to vascular smooth muscle, where it activates the cytosolic form of guanylyl cyclase (or, guanylate cyclase) to form cyclic guanosine monophosphate (cGMP). This reaction is analogous to the formation of cyclic adenosine monophosphate (cAMP) by adenylyl cyclase (see Chapter 8, II. D.). The resultant rise in cGMP causes activation of protein kinase G, which phosphorylates Ca^{2+} channels, causing decreased entry of Ca^{2+} into smooth muscle cells. This decreases the Ca^{2+}–CaM activation of myosin light-chain kinase, thereby decreasing smooth muscle contraction and favoring relaxation.

Vasodilator nitrates, such as nitroglycerin, are metabolized to NO, which causes relaxation of vascular smooth muscle and, therefore, lowers blood pressure. Thus, NO can be envisioned as an endogenous nitrovasodilator. Note that under hypoxic conditions, nitrite (NO_2^-) can be reduced to NO, which binds to deoxyhemoglobin. The NO is released into the blood, causing vasodilation and increasing blood flow.

3. **Nitric oxide and macrophage bactericidal activity:** In macrophages, iNOS activity is normally low, but synthesis of the enzyme is significantly stimulated by bacterial LPS and by the release of IFN-γ and TNF-α in response to infection. Activated macrophages form radicals that combine with NO to form intermediates that decompose, producing the highly bactericidal OH• radical.

4. **Additional functions:** NO is a potent inhibitor of platelet adhesion and aggregation (by activating the cGMP pathway). It is also characterized as a neurotransmitter in the central and peripheral nervous systems.

V. GLUCOSE 6-PHOSPHATE DEHYDROGENASE DEFICIENCY

Deficiency of G6PD deficiency is a hereditary condition characterized by hemolytic anemia and caused by the inability of RBCs (erythrocytes) to detoxify oxidizing agents. Inherited as an **X-linked trait**, G6PD deficiency affects mostly persons with one X and one Y sex chromosome and is the most common disease-producing enzyme abnormality in humans. More than 400 million individuals are affected worldwide. This enzyme deficiency has the highest prevalence in persons whose ancestries include the Middle East, tropical Africa and Asia, and parts of the Mediterranean. G6PD deficiency is a family of deficiencies caused by several different mutations in the G6PD gene. Only some of the resulting protein variants cause clinical symptoms.

In addition to periodic bouts of hemolytic anemia in response to oxidant stress, a common clinical manifestation of G6PD deficiency is neonatal jaundice appearing 1 to 4 days after birth. The jaundice, which may be severe, typically results from increased production of unconjugated bilirubin (see Chapter 21, II. E.). The life span of individuals with a severe form of G6PD deficiency may be somewhat shortened as a result of complications arising from chronic hemolysis. This negative effect of G6PD deficiency has been balanced in evolution by an increased resistance to malaria caused by *Plasmodium falciparum*.

A. G6PD role in erythrocytes

Diminished G6PD activity impairs the ability of the cell to form NADPH that is essential for the maintenance of the G-SH pool. This results in a decrease in the detoxification of free radicals and peroxides formed within the cell (Fig. 13.10). G-SH also helps maintain the reduced states of sulfhydryl groups in proteins, including hemoglobin. Oxidation of

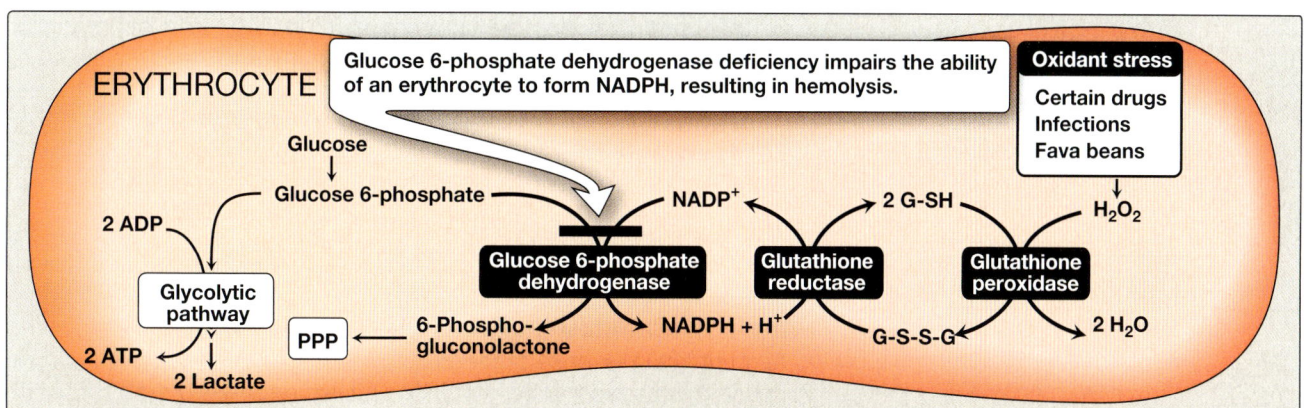

Figure 13.10
Pathways of glucose 6-phosphate metabolism in the erythrocyte. NADP(H), nicotinamide adenine dinucleotide phosphate; G-SH, reduced glutathione; G-S-S-G, oxidized glutathione; H_2O_2, hydrogen peroxide; PPP, pentose phosphate pathway.

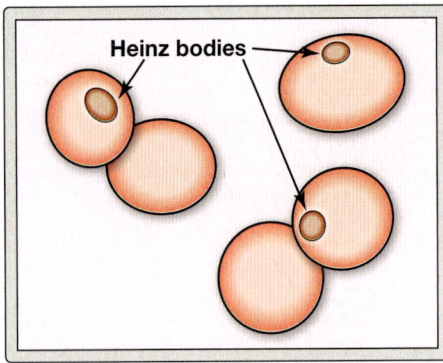

Figure 13.11
Heinz bodies in the erythrocytes of a patient with glucose 6-phosphate dehydrogenase deficiency.

those sulfhydryl groups leads to the formation of denatured proteins that form insoluble masses called Heinz bodies, that attach to RBC membranes (Fig. 13.11). Additional oxidation of membrane proteins causes RBC membranes to be rigid (less deformable), and they are removed from the circulation by macrophages in the spleen and liver.

Although G6PD deficiency occurs in all cells of the affected individual, it is most severe in RBCs, where the pentose phosphate pathway provides the only means of generating NADPH. Additionally, since RBCs have no nucleus or ribosomes, they cannot renew their supply of the enzyme, thus are particularly vulnerable to enzyme variants with diminished stability. Other tissues have an alternative pathway to produce NADPH (via $NADP^+$-dependent malate dehydrogenase [malic enzyme]; see Chapter 16, III. D.).

B. Precipitating factors in G6PD deficiency

Persons with only one X chromosome who inherit a *G6PD* mutation on that lone X chromosome are considered to be hemizygous for the G6PD deficiency trait. Affected individuals will normally remain asymptomatic until treated with an oxidant drug, ingest fava beans, or contract a severe infection. Hemolytic anemia results in response to these oxidant-stress inducing agents.

1. **Oxidant drugs:** Drugs that can cause oxidant stress and produce hemolytic anemia in patients with G6PD deficiency are often in categories that begin with the letter A: some **antibiotics** (particularly sulfa drugs), some **antimalarials**, some **analgesics**, and some **antipyretics**. Only certain drugs in each category are implicated. Drug lists are available for prescribers that include usually safe agents and those best avoided by G6PD-deficient individuals.

2. **Favism:** Persons with some forms of G6PD deficiency, especially the Mediterranean variant, are particularly susceptible to the hemolytic effect of the fava (broad) bean, a dietary staple in the Mediterranean region. Favism, the hemolytic effect of ingesting fava beans, is not observed in all individuals with G6PD deficiency, but all patients with favism do have G6PD deficiency.

3. **Infection:** Infection is a common precipitating factor of hemolysis in persons with G6PD deficiency. The inflammatory response to infection results in the generation of free radicals in macrophages. The radicals can diffuse into RBCs and cause oxidative damage.

C. *G6PD* gene variants

The cloning and sequencing of the *G6PD* gene (see Chapter 34) have led to identification of more than 400 *G6PD* variants that result in G6PD enzyme deficiency. Some mutations do not affect enzymatic activity. Most mutations that do result in low G6PD enzyme function are missense point mutations (see Chapter 32, II. C.); some cause decreased catalytic activity, others decreased stability, while other *G6PD* mutations alter the binding affinity for $NADP^+$ or glucose 6-phosphate. Active G6PD enzyme exists as a homodimer or tetramer. Mutations at the interface between subunits can affect enzyme stability.

The severity of hemolytic anemia in those with G6PD deficiency usually correlates with the amount of residual enzyme activity in the patient's RBCs. G6PD variants can be classified as shown in Figure 13.12. G6PD A⁻ is the prototype of the moderate (class III) form of the disease. RBCs contain an unstable but kinetically normal G6PD, with most of the enzyme activity present in the reticulocytes and younger RBCs (Fig. 13.13). The oldest RBCs have the lowest level of G6PD activity and are preferentially removed in a hemolytic episode. Because hemolysis does not affect younger cells, the episodes are self-limiting. G6PD Mediterranean is the prototype of a more severe (class II) deficiency. Class I mutations (rare) are the most severe and are associated with chronic nonspherocytic hemolytic anemia, even in the absence of oxidative stress.

Both G6PD A⁻ and G6PD Mediterranean proteins represent mutant enzymes that differ from the respective normal variants by a single amino acid. Large deletions or frameshift mutations have not been identified, suggesting that complete absence of G6PD enzyme activity is most likely lethal.

Class	Clinical symptoms	Residual enzyme activity
I	Very severe (chronic, nonspherocytic hemolytic anemia)	<10%
*II	Severe (acute hemolytic anemia)	<10%
*III	Moderate	10%-60%
IV	None	>60%

Figure 13.12
Classification of glucose 6-phosphate dehydrogenase (G6PD) deficiency variants. [Note: Class V variants (not shown) result in overproduction of G6PD.] *, most common.

VI. CHAPTER SUMMARY

- The **pentose phosphate pathway** is the main producer of **nicotinamide adenine dinucleotide phosphate** (**NADPH**) in the body.

- **No adenosine triphosphate** (**ATP**) is used or consumed in the pathway.

- The pathway includes an irreversible oxidative phase followed by a series of reversible sugar–phosphate interconversions (Fig. 13.14).

- **Reversible nonoxidative reactions** interconvert sugars. Ribulose 5-phosphate is converted to **ribose 5-phosphate**, required for nucleotide and nucleic acid synthesis, or to **fructose 6-phosphate** and **glyceraldehyde 3-phosphate** (glycolytic intermediates).

- The NADPH-producing **oxidative portion** of the pathway provides reducing equivalents for reductive biosynthesis and detoxification reactions.

- In this part of the pathway, **glucose 6-phosphate** is irreversibly converted to **ribulose 5-phosphate**, and **two NADPH** are produced. The regulated step is catalyzed by **glucose 6-phosphate dehydrogenase** (**G6PD**), which is strongly inhibited by a rise in the **NADPH/NADP⁺ ratio**.

- NADPH is a source of **reducing equivalents** in **reductive biosynthesis**, such as the production of fatty acids in liver, adipose tissue, and the mammary gland; cholesterol in the liver; and steroid hormones in the placenta, ovaries, testes, and adrenal cortex.

- NADPH is also required by erythrocytes for the reduction of **hydrogen peroxide** produced as a consequence of aerobic metabolism.

- **Reduced glutathione** (**G-SH**) is used by **glutathione peroxidase** to reduce the peroxide to water. The **oxidized glutathione** (**G-S-S-G**) produced is reduced by **glutathione reductase**, using NADPH as the source of electrons.

- NADPH provides reducing equivalents for the **mitochondrial cytochrome P450 monooxygenase system**, which is used in **steroid hormone synthesis** in steroidogenic tissue, **bile acid synthesis** in the liver, and **vitamin D activation** in the liver and kidneys.

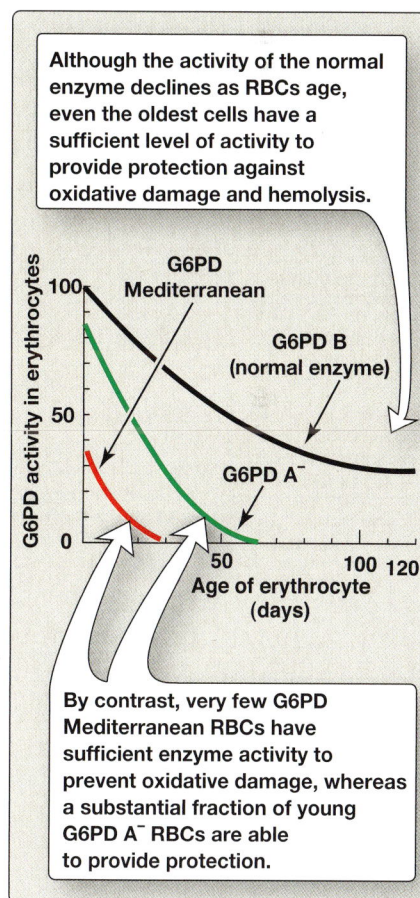

Although the activity of the normal enzyme declines as RBCs age, even the oldest cells have a sufficient level of activity to provide protection against oxidative damage and hemolysis.

By contrast, very few G6PD Mediterranean RBCs have sufficient enzyme activity to prevent oxidative damage, whereas a substantial fraction of young G6PD A⁻ RBCs are able to provide protection.

Figure 13.13
Decline of red blood cell (RBC) glucose 6-phosphate dehydrogenase (G6PD) activity with cell age for the three most commonly encountered forms of the enzyme.

- The **microsomal system** uses NADPH to **detoxify** xenobiotics, such as drugs and a variety of pollutants. NADPH provides the reducing equivalents for phagocytes involved in eliminating invading microorganisms. **NADPH oxidase** uses molecular oxygen (O_2) and electrons from NADPH to produce **superoxide radicals**, which, in turn, can be converted to peroxide by **superoxide dismutase**.
- **G6PD deficiency, an X-linked disease that affects mostly males,** impairs **erythrocyte** ability to form NADPH essential for maintaining the G-SH pool. Erythrocytes are most affected because they do not have additional sources of NADPH. Characterized by **hemolytic anemia** caused by the production of free radicals and peroxides after exposure to oxidant stress include severe infection, **oxidant drugs, or fava beans**. The extent of the anemia depends on the amount of residual enzyme. Neonates with G6PD deficiency may experience prolonged **neonatal jaundice**.

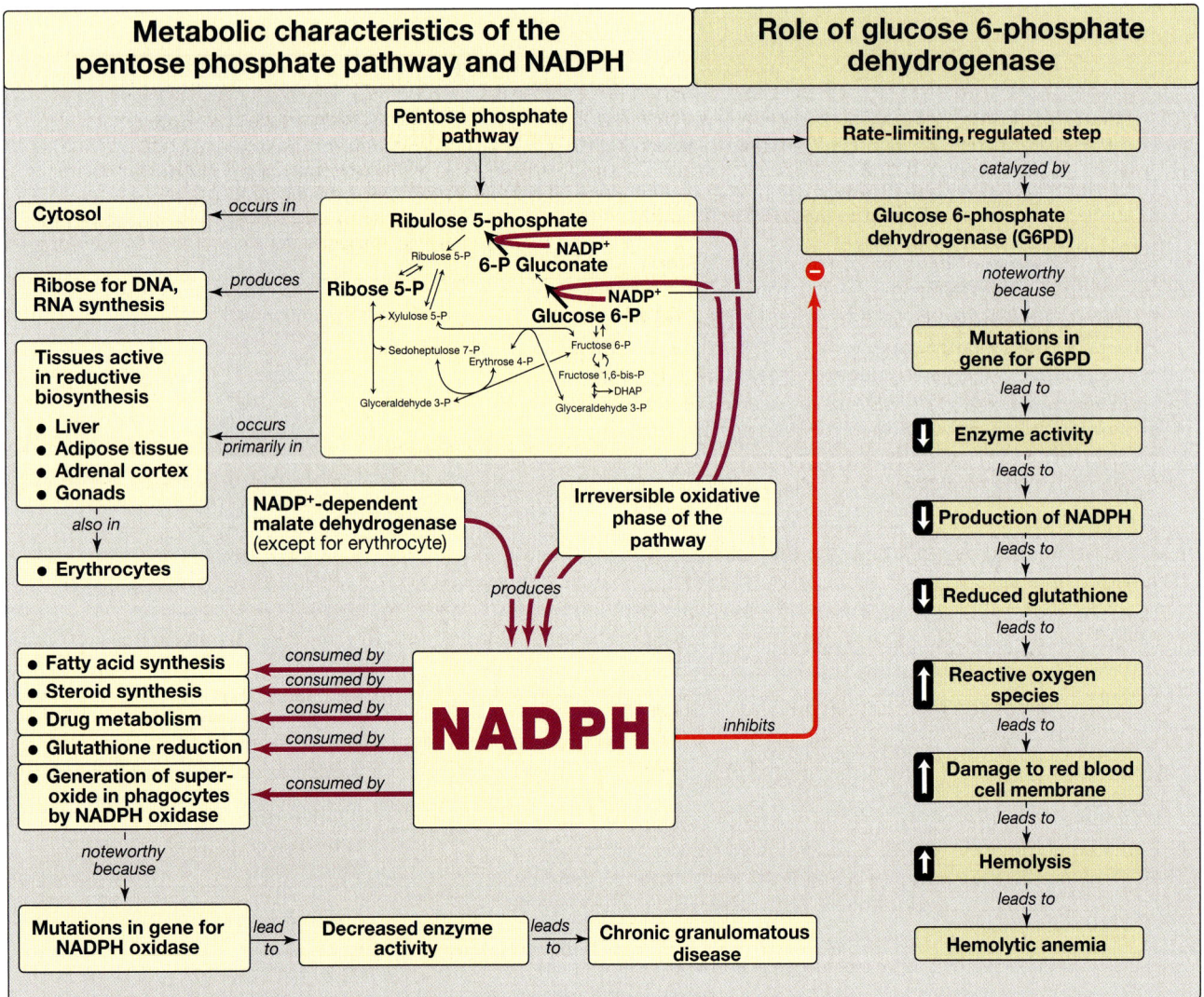

Figure 13.14
Key concept map for the pentose phosphate pathway and nicotinamide adenine dinucleotide phosphate (NADPH).

Study Questions

Choose the ONE best answer.

13.1. In preparation for a trip to a place where malaria is prevalent, a 21-year-old man is given an antimalarial drug prophylactically. Several days after initiation of this therapy, he develops jaundice and is diagnosed with anemia. A less-than-normal level of which of the following is a consequence of the enzyme deficiency that is the most likely underlying cause his signs and symptoms?

A. Glucose 6-phosphate
B. Oxidized form of nicotinamide adenine dinucleotide
C. Reduced form of glutathione
D. Ribose 5-phosphate

Correct answer = C. Glutathione (G-SH) is essential for RBC integrity and is maintained in this reduced (functional) form by nicotinamide adenine dinucleotide phosphate (NADPH)-dependent glutathione reductase. The NADPH is from the oxidative portion of the pentose phosphate pathway. Individuals with a deficiency of the regulated enzyme of this pathway, glucose 6-phosphate dehydrogenase (G6PD), have a decreased ability to generate NADPH and, therefore, a decreased ability to keep G-SH reduced. When treated with an oxidant drug such as primaquine, some patients with G6PD deficiency develop hemolytic anemia. Primaquine does not affect glucose 6-phosphate levels. Nicotinamide adenine dinucleotide (NAD[H]) is neither produced by the pathway nor used as a coenzyme by G-SH reductase. A decrease in ribose 5-phosphate does not cause hemolysis.

13.2. Low blood pressure (hypotension) is a sign of septic shock, resulting from a severe inflammatory response to a bacterial infection. Which of the following is a likely cause of this hypotension?

A. Activation of endothelial nitric oxide synthase causing a decrease in nitric oxide
B. Long half-life of nitric oxide promotes long-term, excess vasoconstriction
C. Lysine, the nitrogen source for nitric oxide synthesis, is deaminated by bacteria
D. Bacterial endotoxin promoting iNOS synthesis causing increased NO production

Correct answer = D. Overproduction of short-lived nitric oxide (NO) by calcium-independent, inducible nitric oxide synthase (iNOS) results in excessive vasodilation leading to hypotension. The endothelial enzyme (eNOS) is constitutive and produces low levels of NO at a consistent rate. NOS use arginine, not lysine, as the source of the nitrogen.

13.3. An individual who has recently been prescribed a statin drug to lower cholesterol levels is advised to limit consumption of grapefruit juice, because high intake of the juice reportedly results in an increased level of the drug in the blood, increasing the risk of side effects. This drug is a substrate for the cytochrome P450 enzyme CYP3A4, and grapefruit juice inhibits the enzyme. Based on this information, CYP enzymes most likely:

A. accept electrons from reduced nicotinamide adenine dinucleotide.
B. catalyze the hydroxylation of hydrophobic molecules.
C. differ from nitric oxide synthase in that they contain heme.
D. function in association with an oxidase.

Correct answer = B. The CYP enzymes hydroxylate hydrophobic compounds, making them more water soluble. Reduced nicotinamide adenine dinucleotide phosphate (NADPH) from the pentose phosphate pathway is the electron donor. Both the CYP enzymes and the nitric oxide synthase isozymes contain heme.

13.4. In persons who are hemizygous for glucose 6-phosphate dehydrogenase deficiency, pathophysiologic consequences are more apparent in red blood cells (RBCs) than in other cells such as in the liver. The best explanation for these findings is that:

    A. excess glucose 6-phosphate in the liver, but not in RBCs, can be channeled to glycogen, thereby averting cellular damage

    B. liver cells, in contrast to RBCs, have alternative mechanisms for supplying the reduced nicotinamide adenine dinucleotide phosphate required for maintaining cell integrity

    C. RBC production of ATP required to maintain cell integrity depends exclusively on the shunting of glucose 6-phosphate to the pentose phosphate pathway

    D. in contrast to liver cells, RBC glucose 6-phosphatase activity decreases the level of glucose 6-phosphate, resulting in cell damage

Correct answer = B. Cellular damage is directly related to the decreased ability of the cell to regenerate reduced glutathione, for which large amounts of reduced nicotinamide adenine dinucleotide phosphate (NADPH) are needed, and RBCs have no means other than the pentose phosphate pathway of generating NADPH. It is decreased product (NADPH), not increased substrate (glucose 6-phosphate), that is the problem. RBCs do not have glucose 6-phosphatase. The pentose phosphate pathway does not generate ATP.

13.5. An essential coenzyme for several enzymes of metabolism is derived from the vitamin thiamine. Measurement of the activity of which enzyme in RBCs could be used to determine thiamine status in the body?

    A. Glucose-6-phosphate dehydrogenase

    B. Glutathione peroxidase

    C. Pyruvate dehydrogenase

    D. Transketolase

Correct answer = D. Red blood cells do not have mitochondria and, so, do not contain mitochondrial enzymes such as pyruvate dehydrogenase that require the thiamine-derived coenzyme thiamine pyrophosphate (TPP). However, they do contain the cytosolic TPP-requiring transketolase, whose activity is used clinically to assess thiamine status.

13.6. The oxidative branch of the pentose phosphate pathway is inhibited by a high ratio of which of the following?

    A. ATP/ADP

    B. $NADP^+$/NADPH

    C. NADPH/NADP+

    D. Ribose-5-phosphate/ribulose-5-phosphate

    E. Ribulose-5-phosphate/glyceraldehyde-3-phosphate

Correct answer = C. High levels of NADPH inhibit G6PD, the rate-limiting enzyme in the pentose phosphate pathway, and the first enzyme in the oxidative branch. The ratio of NADPH to NADP+ is sufficiently high to inhibit the enzyme under most metabolic conditions. None of the other answer choices would have an inhibitory impact on the pathway.

13.7. In a liver cell carrying out reductive biosynthesis to produce fatty acids, what will be the fate of most of the ribulose-5-phosphate generated in the oxidative branch of the pentose phosphate pathway?

    A. Catabolism to yield 2 NADPH

    B. Conversion to glyceraldehyde-3-phosphate and fructose-6-phosphate

    C Metabolism to ribose-5-phosphate for use in nucleotide synthesis

    D. Acting as an inhibitor of transaldolase and transketolase

    E. Use as a substrate for cytosolic nitric oxide synthase

Correct answer = B. When a cell is undergoing reductive biosynthesis, its need for NADPH predominates. It will convert the ribulose-5-phosphate to intermediates that can rejoin glycolysis. Glyceraldehyde-3-phosphate and fructose-6-phosphate are generated through the actions of enzymes including transaldolase and transketolase. The oxidative branch is composed of irreversible reactions, so it is not possible to take the ribulose-5-phosphate and move in the reverse direction of the pathway to produce more NADPH.

Glycosaminoglycans, Proteoglycans, and Glycoproteins

I. GLYCOSAMINOGLYCAN OVERVIEW

Glycosaminoglycans (GAGs) are large complexes of negatively charged heteropolysaccharide chains. They are generally associated with a small amount of protein, forming structures known as **proteoglycans**, which typically consist of up to 95% carbohydrate. GAGs have the special ability to bind large amounts of water, producing the gel-like matrix that forms the basis of the body's ground substance, which, along with fibrous structural proteins such as collagen, elastin, and fibrillin-1, and adhesive proteins such as fibronectin, makes up the extracellular matrix (ECM).[1] Hydrated GAGs serve as a flexible support for the ECM, interacting with the structural and adhesive proteins, and as a molecular sieve, influencing movement of materials through the ECM. The viscous, lubricating properties of mucous secretions also result from the presence of GAGs, which led to the original naming of these compounds as mucopolysaccharides.

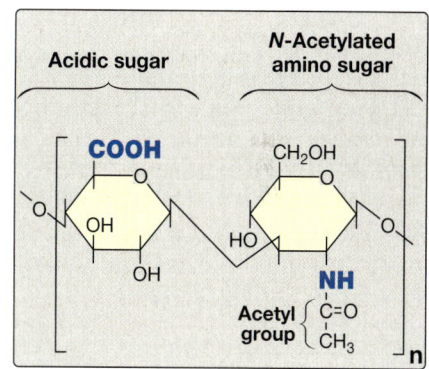

Figure 14.1
Repeating disaccharide unit of glycosaminoglycans.

II. STRUCTURE

GAGs are long, unbranched, heteropolysaccharides composed of repeating disaccharide chains in which one of the sugars is an N-acetylated amino sugar, either N-acetylglucosamine or N-acetylgalactosamine (Fig. 14.1), and the other is an acidic sugar. A single exception is keratan sulfate, which contains galactose rather than an acidic sugar. The amino sugar is either D-glucosamine or D-galactosamine, in which the amino group is usually acetylated, eliminating its positive charge. The amino sugar may also be sulfated on carbon 4 or 6 or on a nonacetylated nitrogen. The acidic sugar is either D-glucuronic acid or its C-5 epimer L-iduronic acid (Fig. 14.2). These uronic sugars contain carboxyl groups that are negatively charged at physiologic pH and, together with the sulfate groups ($-SO_4^{2-}$), give GAGs their strongly negative nature.

A. Structure–function relationship

Because of the high concentration of negative charges, these repeating disaccharide chains tend to be extended in solution. Because

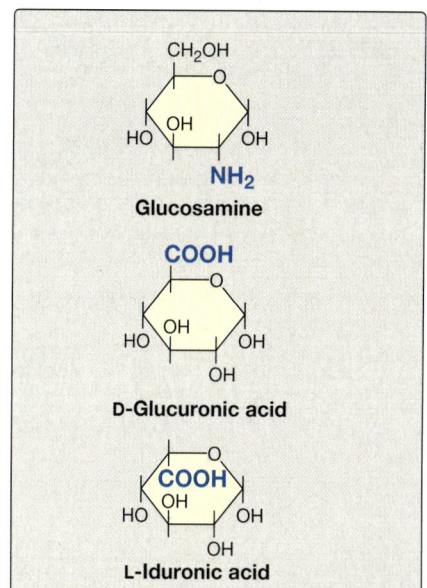

Figure 14.2
Some monosaccharide units found in glycosaminoglycans.

---

[1]For more information on ECM, see *LIR Cell and Molecular Biology*, 3rd ed., Chapter 2.

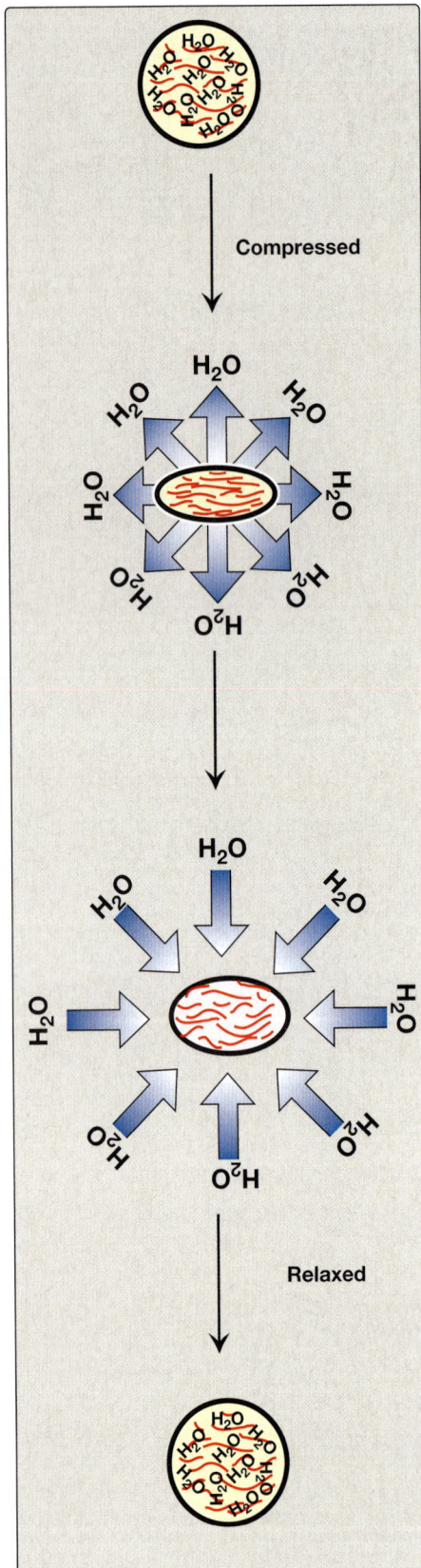

Figure 14.3
Resilience of glycosaminoglycans.

of their negative charges, repel each other and are surrounded by a shell of water molecules. In solution, they slide past each other, much as two magnets with the same polarity seem to slide past each other. This charge repulsion in solutions containing GAGs produces the slippery consistency of mucous secretions and synovial fluid. When a structure containing GAGs is compressed, the water is squeezed out, and the GAGs are forced to occupy a smaller volume. When the compression is released, the GAGs spring back to their original, hydrated volume because of the repulsion of their negative charges. This property contributes to the resilience of cartilage, synovial fluid, and the vitreous humor of the eye (Fig. 14.3).

B. Classification

The six major types of GAGs are divided according to monomeric composition, type of glycosidic linkages, and degree and location of sulfate units. The structure of the GAGs and their distribution in the body is illustrated in Figure 14.4. All GAGs, except for hyaluronic acid, are sulfated and are found covalently attached to protein, forming proteoglycan monomers.

C. Proteoglycans

Proteoglycans are found in the ECM and on the outer surface of cells.

1. **Monomer structure:** A proteoglycan monomer found in cartilage consists of a core protein to which up to 100 linear chains of GAGs are covalently attached. These chains, which may each be composed of up to 200 disaccharide units, extend out from the core protein and remain separated from each other because of charge repulsion. The resulting structure is described as resembling a wire bottle brush (Fig. 14.5). In proteoglycans of cartilage, chondroitin sulfate and keratan sulfate are the main types of GAGs. Proteoglycans are grouped into gene families that encode core proteins with common structural features. The aggrecan family (aggrecan, versican, neurocan, and brevican), abundant in cartilage, is an example.

2. **GAG–protein linkage:** GAGs attached to core protein via covalent linkage is most commonly through a trihexoside (galactose-galactose-xylose) and a serine residue in the protein. An O-glycosidic

---

Clinical Application 14.1: Proteoglycans, Cartilage, and Osteoarthritis

Osteoarthritis affects millions of individuals worldwide. In this disease, joint cartilage is degraded and proteoglycans that normally help provide a cushion for the joint are lost. Without the resilience of the cartilage protecting the joint, pain, stiffness, and swelling result, with progressive worsening of signs and symptoms. Glucosamine and chondroitin have been reported both to relieve pain and to stop progression of osteoarthritis. These compounds are readily available as over-the-counter dietary supplements in the United States. Based on several well-controlled clinical studies, it appears that glucosamine sulfate (but not glucosamine hydrochloride) and chondroitin sulfate may have a small-to-moderate effect in relieving symptoms of osteoarthritis.

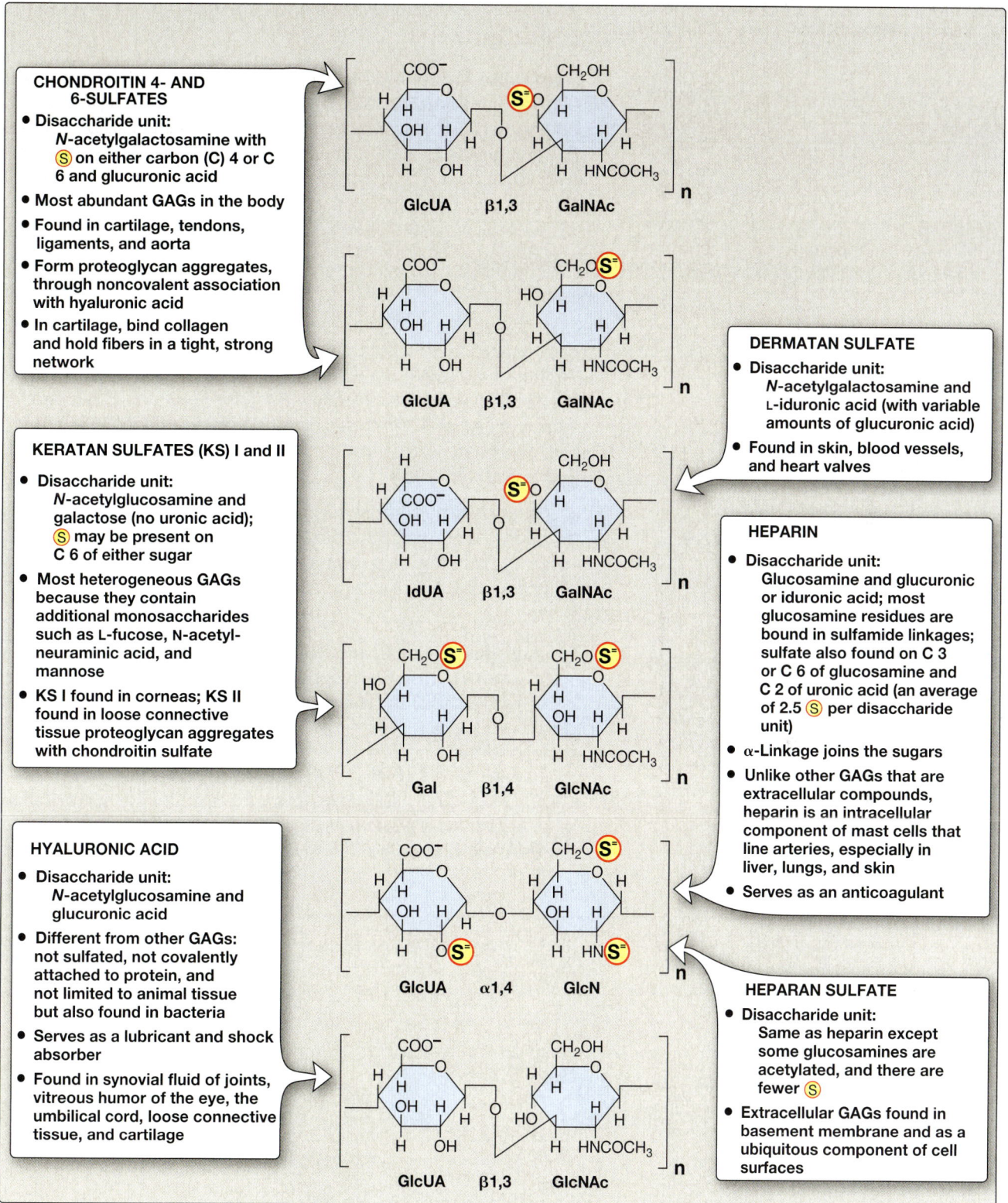

Figure 14.4
Structure of repeating units in and distribution of glycosaminoglycans (GAGs). Sulfate groups (S) are shown in all possible positions. GlcUA and IdUA, glucuronic and iduronic acids; GalNAc, N-acetylgalactosamine; GlcNAc, N-acetylglucosamine; GlcN, glucosamine; Gal, galactose.

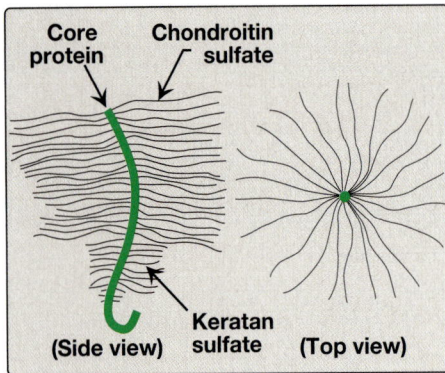

Figure 14.5
Bottle brush model of a cartilage proteo-glycan monomer.

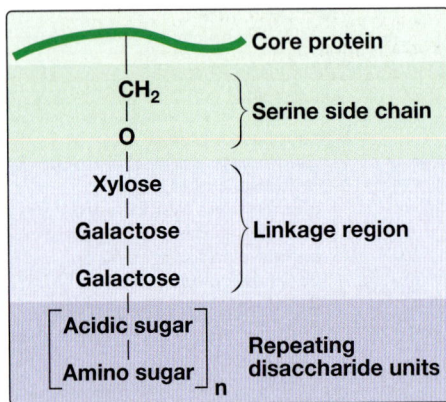

Figure 14.6
Glycosaminoglycan linkage regions.

bond is formed between the xylose and the hydroxyl group of the serine (Fig. 14.6).

3. **Aggregate formation:** Many proteoglycan monomers can asso-ciate with one molecule of hyaluronic acid to form proteoglycan aggregates. The association is not covalent and occurs primar-ily through ionic interactions between the core protein and the hyaluronic acid. The association is stabilized by additional small proteins called link proteins (Fig. 14.7).

III. SYNTHESIS

The heteropolysaccharide chains are elongated by the sequential addi-tion of alternating acidic and amino sugars donated primarily by their uridine diphosphate (UDP) derivatives. The reactions are catalyzed by a family of specific glycosyltransferases. Because GAGs are produced for export from the cell, their synthesis occurs primarily in the Golgi.

A. Amino sugar synthesis

Amino sugars are essential components of glycoconjugates such as proteoglycans, glycoproteins, and glycolipids. The synthetic pathway of amino sugars (hexosamines) is very active in connec-tive tissues, where as much as 20% of glucose flows through this pathway.

1. **N-Acetylglucosamine and N-acetylgalactosamine:** The mono-saccharide fructose 6-phosphate is the precursor of N-acetylglu-cosamine (GlcNAc) and N-acetylgalactosamine (GalNAc). A hydroxyl group on the fructose is replaced by the amide nitrogen of a gluta-mine, and the glucosamine 6-phosphate product gets acetylated, isomerized, and activated, producing the nucleotide sugar UDP-GlcNAc (Fig. 14.8). UDP-GalNAc is generated by the epimerization of UDP-GlcNAc. These nucleotide sugar forms of the amino sugars are used to elongate the carbohydrate chains.

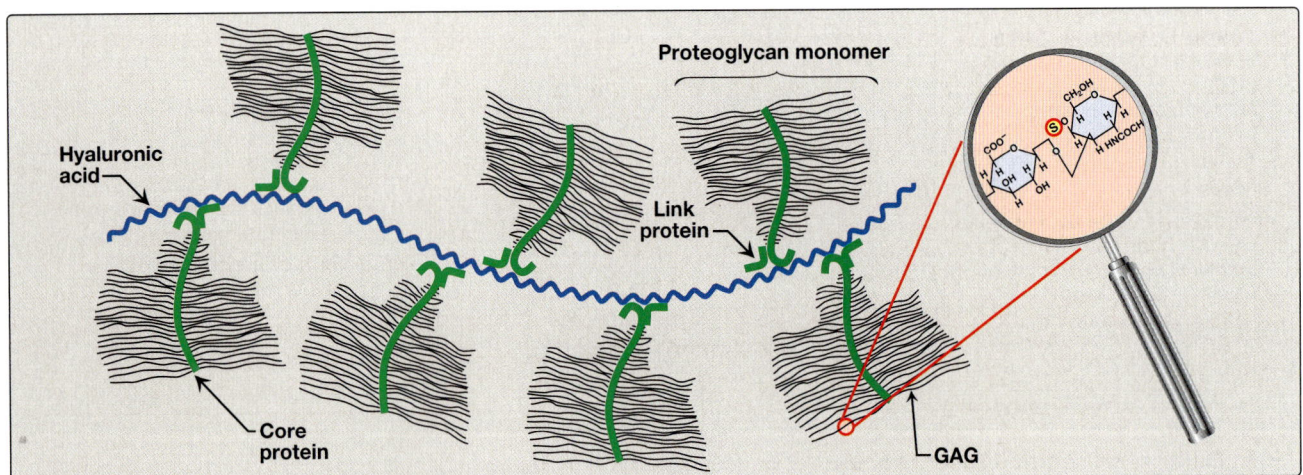

Figure 14.7
Proteoglycan aggregate. GAGs, glycosaminoglycan.

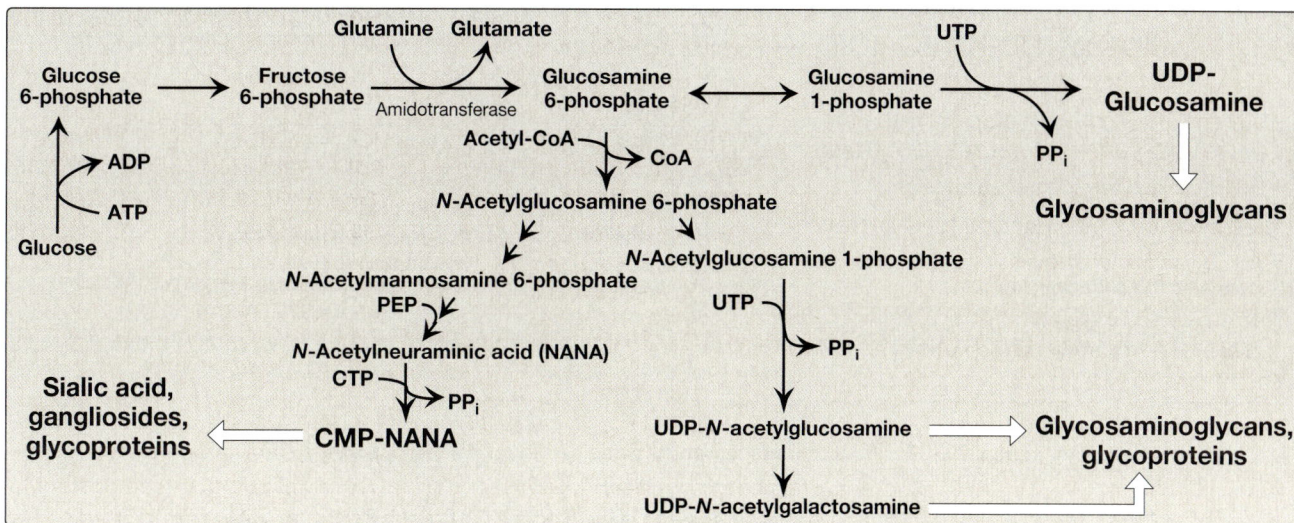

Figure 14.8
Synthesis of the amino sugars. ADP, adenosine diphosphate; UTP and UDP, uridine tri- and diphosphates; CoA, coenzyme A; PEP, phosphoenolpyruvate; CTP and CMP, cytidine tri- and monophosphates; PP$_i$, pyrophosphate.

2. **N-Acetylneuraminic acid:** N-Acetylneuraminic acid (NANA), a nine-carbon, acidic monosaccharide (see Fig. 17.15), is a member of the family of sialic acids, each of which is acylated at a different site. These compounds are usually found as terminal carbohydrate residues of oligosaccharide side chains of glycoproteins, glycolipids, or (less frequently) GAGs. N-Acetylmannosamine 6-phosphate (derived from fructose 6-phosphate) and phosphoenolpyruvate (an intermediate in glycolysis) are the immediate sources of the carbons and nitrogens for NANA synthesis (see Fig. 14.8). Before NANA can be added to a growing oligosaccharide, it must be activated to cytidine monophosphate (CMP)-NANA by reacting with cytidine triphosphate (CTP). CMP-NANA synthetase catalyzes the reaction. CMP-NANA is the only nucleotide sugar in human metabolism in which the carrier nucleotide is a monophosphate rather than a diphosphate.

B. Acidic sugar synthesis

D-Glucuronic acid, whose structure is that of glucose with an oxidized carbon 6 ($-CH_2OH \rightarrow -COOH$), and its C-5 epimer, L-iduronic acid, are essential components of GAGs. Glucuronic acid is also required for the detoxification of lipophilic compounds, such as bilirubin; steroids; and many drugs, including the statins, because conjugation with glucuronate (glucuronidation) increases water solubility. In plants and mammals (other than guinea pigs and primates, including humans), glucuronic acid is a precursor of ascorbic acid (vitamin C) as shown in Figure 14.9. This uronic acid pathway also provides a mechanism by which dietary D-xylulose can enter the central metabolic pathways.

1. **Glucuronic acid:** Glucuronic acid can be obtained in small amounts from the diet and from the lysosomal degradation of GAGs. It also can be synthesized by the uronic acid pathway, in which glucose 1-phosphate reacts with uridine triphosphate (UTP)

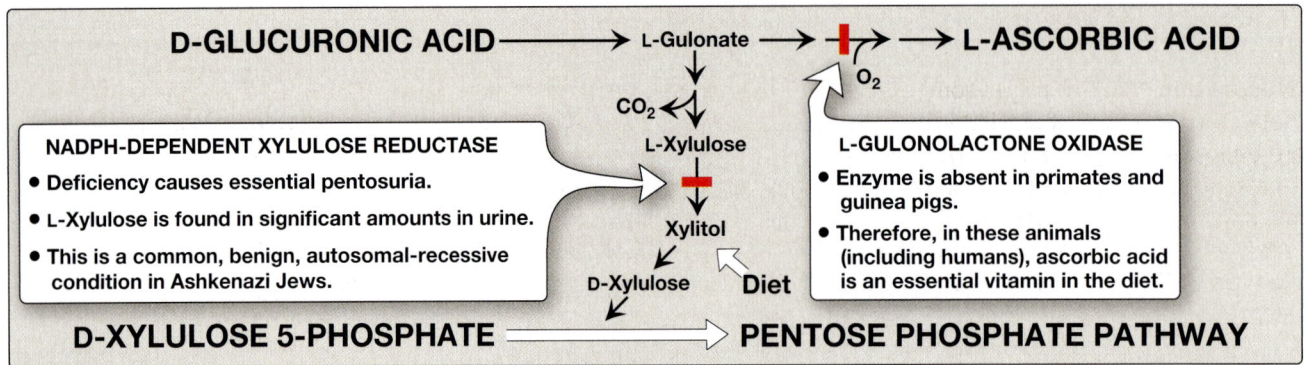

Figure 14.9
Metabolism of glucuronic acid. NADPH, reduced nicotinamide adenine dinucleotide phosphate; CO_2, carbon dioxide.

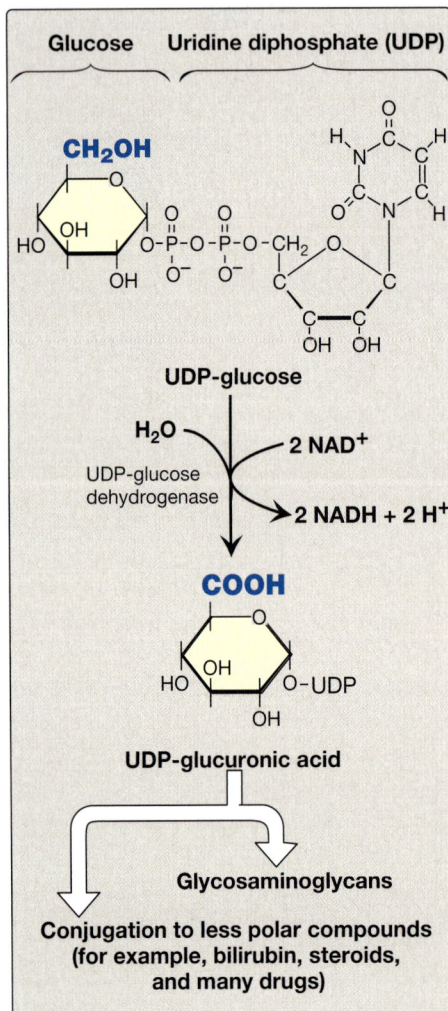

Figure 14.10
Oxidation of UDP-glucose to UDP-glucuronic acid. NAD(H), nicotinamide adenine dinucleotide.

and is converted to UDP-glucose. Oxidation of UDP-glucose produces UDP-glucuronic acid, the form that supplies glucuronic acid for GAGs synthesis and glucuronidation (Fig. 14.10). The end product of glucuronic acid metabolism in humans is D-xylulose 5-phosphate, which can enter the pentose phosphate pathway and produce the glycolytic intermediates glyceraldehyde 3-phosphate and fructose 6-phosphate (see Fig. 14.9; see also Fig. 13.2).

2. **L-Iduronic acid:** Synthesis of L-iduronic acid occurs after D-glucuronic acid has been incorporated into the carbohydrate chain. Uronosyl 5-epimerase causes epimerization of the D- to the L-sugar.

C. Core protein synthesis

The core protein is made by ribosomes on the rough endoplasmic reticulum (rER), enters the rER lumen, and then moves to the Golgi, where it is glycosylated by membrane-bound glycosyltransferases.

D. Carbohydrate chain synthesis

Carbohydrate chain formation is initiated by synthesis of a short linker on the core protein on which carbohydrate chain synthesis will occur. The most common linker is a trihexoside formed by the transfer of xylose from UDP-xylose to the hydroxyl group of a serine (or threonine) catalyzed by xylosyltransferase. Two galactose molecules are then added, completing the trihexoside. This is followed by sequential addition of alternating acidic and amino sugars (Fig. 14.11) and epimerization of some D-glucuronyl to L-iduronyl residues.

E. Sulfate group addition

Sulfation of a GAG occurs after the monosaccharide to be sulfated has been incorporated into the growing carbohydrate chain. The source of the sulfate is 3′-phosphoadenosyl-5′-phosphosulfate ([PAPS] a molecule of adenosine monophosphate with a sulfate group attached to the 5′-phosphate; see also Fig. 17.16). The sulfation reaction is catalyzed by sulfotransferases. Synthesis of the sulfated GAG chondroitin sulfate is shown in Figure 14.11. Note that PAPS is also the sulfur donor in glycosphingolipid synthesis.

IV. DEGRADATION

GAGs are degraded in lysosomes, which contain hydrolytic enzymes that are most active at a pH of ~5. Therefore, as a group, these enzymes are called **acid hydrolases**. The low pH optimum within lysosomes is a protective mechanism that prevents the enzymes from destroying the cell should leakage occur into the cytosol where the pH is neutral.[2] The half-lives of GAGs vary from minutes to months and are influenced by the type of GAG and its location in the body.

A. Glycosaminoglycans and phagocytosis

Because GAGs are extracellular or cell-surface compounds, they must first be engulfed by invagination of the cell membrane (phagocytosis), forming a vesicle inside of which are the GAGs to be degraded. This vesicle then fuses with a lysosome, forming a single digestive vesicle in which the GAGs are efficiently degraded for a discussion of phagocytosis.

B. Lysosomal degradation

The lysosomal degradation of GAGs requires many different acid hydrolases for complete digestion. First, the polysaccharide chains are cleaved by endoglycosidases, producing oligosaccharides. Further degradation of the oligosaccharides occurs sequentially from the nonreducing end of each chain, the last group (sulfate or sugar) added during synthesis being the first group removed, by action of sulfatases or exoglycosidases. Examples of some of these enzymes and the bonds they hydrolyze are shown in Figure 14.12. [Note: Endo- and exoglycosidases are also involved in the lysosomal degradation of glycoproteins and glycolipids. Deficiencies in these enzymes result in the accumulation of partially degraded carbohydrates, causing tissue damage.]

> Multiple sulfatase deficiency (Austin disease) is a rare lysosomal storage disease in which all sulfatases are nonfunctional because of a defect in the formation of formylglycine, an amino acid derivative required at the active site for enzymatic activity to occur.

V. MUCOPOLYSACCHARIDOSES

Mucopolysaccharidoses are hereditary diseases (~1:25,000 live births) caused by a deficiency of any one of the lysosomal hydrolases normally involved in the degradation of heparan sulfate, dermatan sulfate and/or keratin sulfate (summarized in Fig. 14.12). They are progressive disorders characterized by lysosomal accumulation of GAGs in various tissues, causing a range of symptoms, such as skeletal and ECM deformities and intellectual disability. All are autosomal-recessive disorders except Hunter syndrome, which is X-linked.

[2]For a discussion of lysosomes, see *LIR Cell and Molecular Biology*, 3rd ed., Chapter 5.

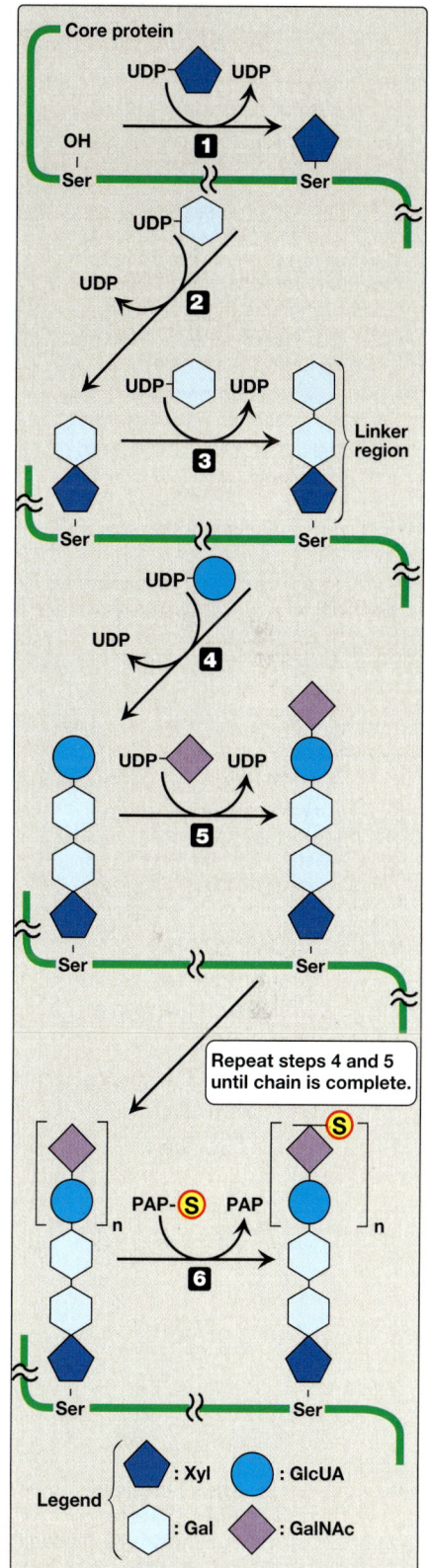

Figure 14.11
Synthesis of chondroitin sulfate. PAP-Ⓢ, 3′-phosphoadenosyl-5′-phosphosulfate; Ser, serine.

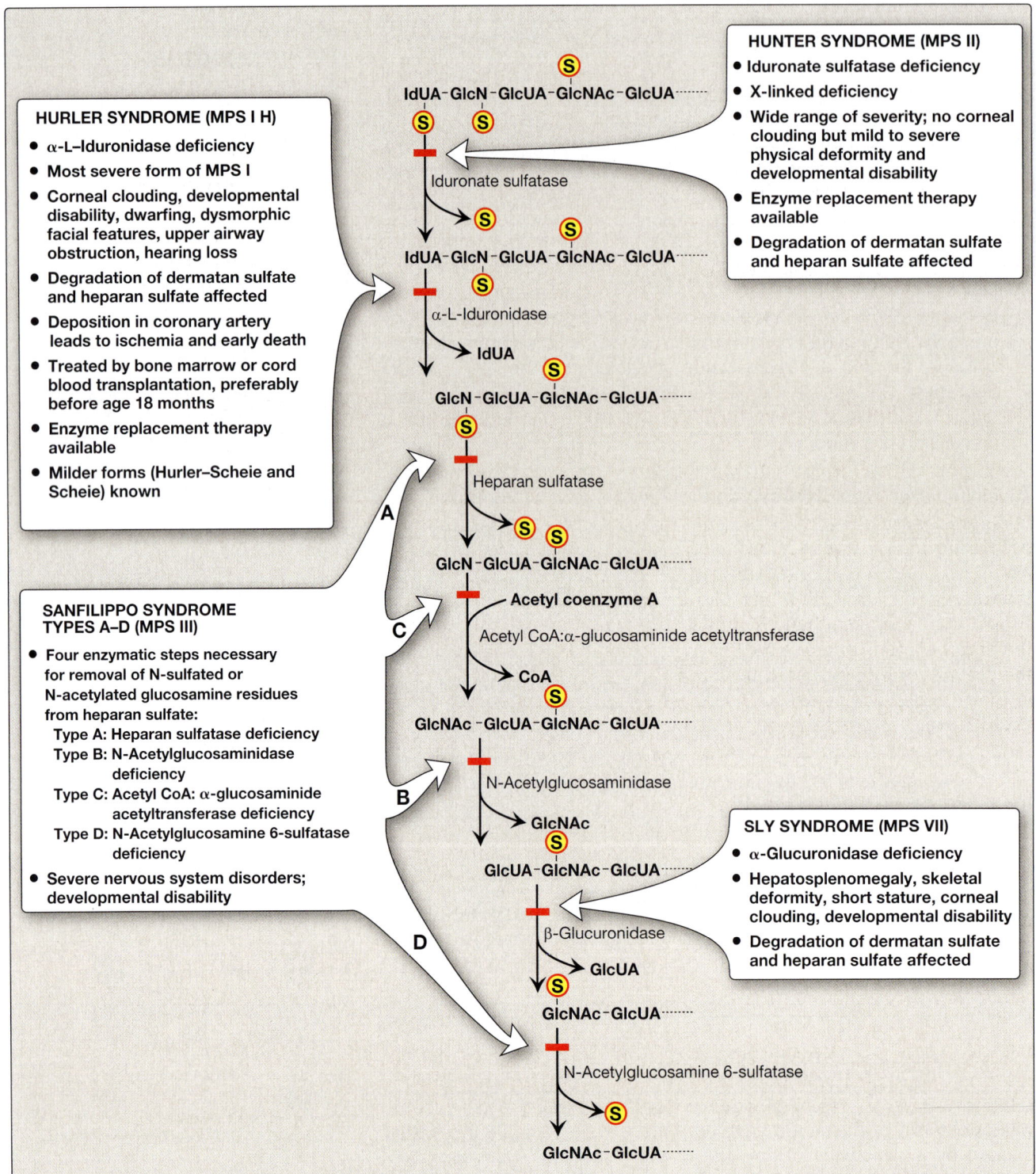

HUNTER SYNDROME (MPS II)
- Iduronate sulfatase deficiency
- X-linked deficiency
- Wide range of severity; no corneal clouding but mild to severe physical deformity and developmental disability
- Enzyme replacement therapy available
- Degradation of dermatan sulfate and heparan sulfate affected

HURLER SYNDROME (MPS I H)
- α-L–Iduronidase deficiency
- Most severe form of MPS I
- Corneal clouding, developmental disability, dwarfing, dysmorphic facial features, upper airway obstruction, hearing loss
- Degradation of dermatan sulfate and heparan sulfate affected
- Deposition in coronary artery leads to ischemia and early death
- Treated by bone marrow or cord blood transplantation, preferably before age 18 months
- Enzyme replacement therapy available
- Milder forms (Hurler–Scheie and Scheie) known

SANFILIPPO SYNDROME TYPES A–D (MPS III)
- Four enzymatic steps necessary for removal of N-sulfated or N-acetylated glucosamine residues from heparan sulfate:
  Type A: Heparan sulfatase deficiency
  Type B: N-Acetylglucosaminidase deficiency
  Type C: Acetyl CoA: α-glucosaminide acetyltransferase deficiency
  Type D: N-Acetylglucosamine 6-sulfatase deficiency
- Severe nervous system disorders; developmental disability

IdUA-GlcN-GlcUA-GlcNAc-GlcUA ⋯⋯

Iduronate sulfatase

IdUA-GlcN-GlcUA-GlcNAc-GlcUA ⋯⋯

α-L-Iduronidase

→ IdUA

GlcN-GlcUA-GlcNAc-GlcUA ⋯⋯

Heparan sulfatase

GlcN-GlcUA-GlcNAc-GlcUA ⋯⋯

Acetyl coenzyme A

Acetyl CoA:α-glucosaminide acetyltransferase

→ CoA

GlcNAc-GlcUA-GlcNAc-GlcUA ⋯⋯

N-Acetylglucosaminidase

→ GlcNAc

GlcUA-GlcNAc-GlcUA ⋯⋯

β-Glucuronidase

→ GlcUA

GlcNAc-GlcUA ⋯⋯

N-Acetylglucosamine 6-sulfatase

GlcNAc-GlcUA ⋯⋯

SLY SYNDROME (MPS VII)
- α-Glucuronidase deficiency
- Hepatosplenomegaly, skeletal deformity, short stature, corneal clouding, developmental disability
- Degradation of dermatan sulfate and heparan sulfate affected

Figure 14.12
Degradation of the glycosaminoglycan heparan sulfate by lysosomal enzymes, indicating sites of enzyme deficiencies in some representative mucopolysaccharidoses (MPS). [Note: Deficiencies in *galactosamine 6-sulfatase* and *β-galactosidase* that degrade keratan sulfate result in Morquio syndrome (MPS IV), A and B, respectively. Deficiencies in arylsulfatase B that degrades dermatan sulfate result in Maroteaux–Lamy syndrome (MPS VI).] GlcUA and IdUA, glucuronic and iduronic acids; GalNAc, N-acetylgalactosamine; GlcNAc, N-acetylglucosamine; GlcN, glucosamine; Ⓢ, sulfate.

Children homozygous for any one of these diseases have a normal phenotype at birth and then gradually deteriorate. In severe deficiencies, death occurs in childhood. Currently, there is no cure. Incomplete lysosomal degradation of GAGs results in the presence of oligosaccharides in the urine. These fragments can be used to diagnose the specific mucopolysaccharidosis by identifying the structure present on the nonreducing end of the oligosaccharide, because that residue would have been the substrate for the missing enzyme. Diagnosis is confirmed by measuring the patient's cellular level of the lysosomal hydrolases. Bone marrow and cord blood transplants, in which transplanted macrophages produce the enzymes that degrade GAGs, have been used to treat Hurler and Hunter syndromes, with limited success. Enzyme replacement therapy is available for both syndromes but does not prevent neurologic damage.

VI. GLYCOPROTEIN OVERVIEW

Glycoproteins are proteins to which oligosaccharides (glycans) are covalently attached; **glycosylation** is the most common posttranslational modification of proteins. [Note: Nonenzymatic addition of carbohydrate to proteins is known as glycation.] Glycoproteins contain highly variable amounts of carbohydrate but typically much less than that of proteoglycans. For example, the glycoprotein immunoglobulin G (IgG) contains less than 4% of its mass as carbohydrate, whereas the proteoglycan aggrecan contains more than 80%. In glycoproteins, the glycan is relatively short, usually 2 to 10 sugar residues in length; is often branched instead of linear; and may or may not be negatively charged. Membrane-bound glycoproteins participate in a broad range of cellular phenomena, including cell-surface recognition by other cells, hormones, and viruses; cell-surface antigenicity (such as the blood group antigens); and as components of the ECM and of the mucins of the gastrointestinal and urogenital tracts, where they act as protective biologic lubricants. In addition, almost all of the globular proteins present in human plasma are glycoproteins, although albumin is an exception. Figure 14.13 summarizes some glycoprotein functions.

VII. OLIGOSACCHARIDE STRUCTURE

The oligosaccharide (glycan) components of glycoproteins are generally branched heteropolymers composed primarily of D-hexoses, with the addition in some cases of neuraminic acid (a nonose) and of L-fucose, a 6-deoxyhexose.

A. Carbohydrate–protein linkage

The glycan may be attached to the protein through an N- or an O-glycosidic link. In the former case, the sugar chain is attached to the amide group of an asparagine side chain and, in the latter case, to the hydroxyl group of either a serine or threonine side chain. In the case of collagen, there is an O-glycosidic linkage between galactose or glucose and the hydroxyl group of hydroxylysine.

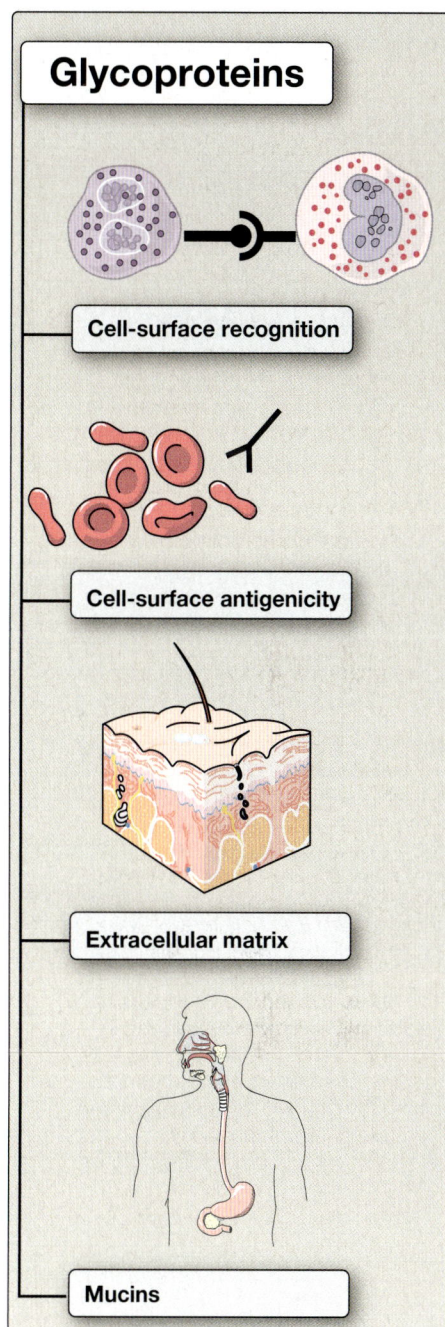

Glycoproteins

Cell-surface recognition

Cell-surface antigenicity

Extracellular matrix

Mucins

Figure 14.13
Functions of glycoproteins.

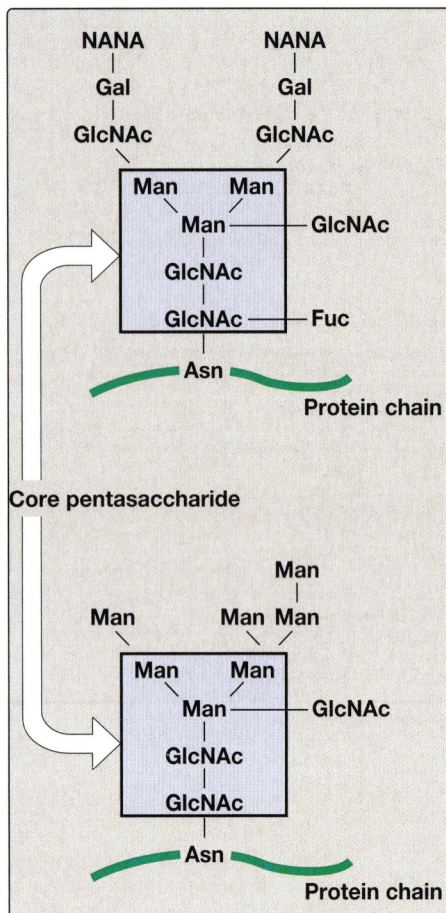

Figure 14.14
Complex (**top**) and high-mannose (**bottom**) N-linked oligosaccharides. [Note: Members of each class contain the same pentasaccharide core (shown inside the box).] NANA, N-acetylneuraminic acid; Gal, galactose; GlcNAc, N-acetylglucosamine; Man, mannose; Fuc, fucose; Asn, asparagine.

B. N- and O-Linked oligosaccharides

A glycoprotein may contain only one type of glycosidic linkage (N- or O-linked) or may have both types within the same molecule.

1. **O-Linked:** The O-linked glycans may have one or more of a wide variety of sugars arranged in either a linear or a branched pattern. Many are found in extracellular glycoproteins or as membrane glycoprotein components. For example, O-linked oligosaccharides on the surface of red blood cells help provide the ABO blood group determinants. If the terminal sugar on the glycan is GalNAc, the blood group is A. If it is galactose, the blood group is B. If neither GalNAc nor galactose is present, the blood group is O.

2. **N-Linked:** The N-linked glycans fall into two broad classes: complex oligosaccharides and high-mannose oligosaccharides. Both contain the same pentasaccharide core shown in Figure 14.14, but the complex oligosaccharides contain a diverse group of additional sugars, for example, GlcNAc, GalNAc, L-fucose, and NANA, whereas the high-mannose oligosaccharides contain primarily mannose.

VIII. GLYCOPROTEIN SYNTHESIS

Proteins destined to function in the cytoplasm are synthesized on free cytosolic ribosomes. However, proteins, including glycoproteins, that are destined for cellular membranes, lysosomes, or to be exported from the cell, are synthesized on ribosomes attached to the endoplasmic reticulum. These proteins contain specific signal sequences that act as molecular addresses, targeting the proteins to their proper destinations. An N-terminal hydrophobic sequence initially directs these proteins to the ER, allowing the growing polypeptide to be extruded into the lumen. The proteins are then transported via secretory vesicles to the Golgi, which acts as a sorting center (Fig. 14.15). In the Golgi, those glycoproteins that are to be secreted from the cell or targeted for lysosomes are packaged into vesicles that fuse with the plasma or lysosomal membrane and release their contents. Those that are destined to become components of the cell membrane are integrated into the Golgi membrane, which buds off, forming vesicles that add their membrane-bound glycoproteins to the cell membrane and are oriented with the carbohydrate portion facing toward the outside of the cell (see Fig. 14.15).

A. Carbohydrate components

The precursors of the carbohydrate components of glycoproteins are nucleotide sugars, which include UDP-glucose, UDP-galactose, UDP-GlcNAc, and UDP-GalNAc. In addition, guanosine diphosphate (GDP)-mannose, GDP-L-fucose (synthesized from GDP-mannose), and CMP-NANA may donate sugars to the growing chain. When the acidic NANA is present, the oligosaccharide has a negative charge at physiologic pH. The oligosaccharides are covalently attached to the side chains of specific amino acids in the protein, where the three-dimensional structure of the protein determines whether or not a specific amino acid is glycosylated.

B. O-Linked glycoprotein synthesis

Synthesis of the O-linked glycoproteins is very similar to that of the GAGs. First, the protein to which sugars are to be attached is synthesized on the RER and extruded into its lumen. Glycosylation begins with the transfer of GalNAc (from UDP-GalNAc) to the hydroxyl group of a specific serine or threonine residues. The glycosyltransferases responsible for the stepwise synthesis (from individual sugars) of the oligosaccharides are bound to the membranes of the Golgi. They act in a specific order, without using a template as is required for DNA, ribonucleic acid (RNA), and protein synthesis (see Unit VII), but instead by recognizing the actual structure of the growing oligosaccharide as the appropriate substrate.

C. N-Linked glycoprotein synthesis

Synthesis of N-linked glycoproteins occurs in the lumen of the rER and requires the participation of the phosphorylated form of dolichol (dolichol pyrophosphate), a lipid of the RER membrane (Fig. 14.16). The initial product is processed in the rER and Golgi.

1. **Dolichol-linked oligosaccharide synthesis:** As with the O-linked glycoproteins, the protein is synthesized on the rER and enters its lumen. However, it does not become glycosylated with individual sugars. Instead, a lipid-linked oligosaccharide is first constructed. This consists of dolichol, an rER membrane lipid made

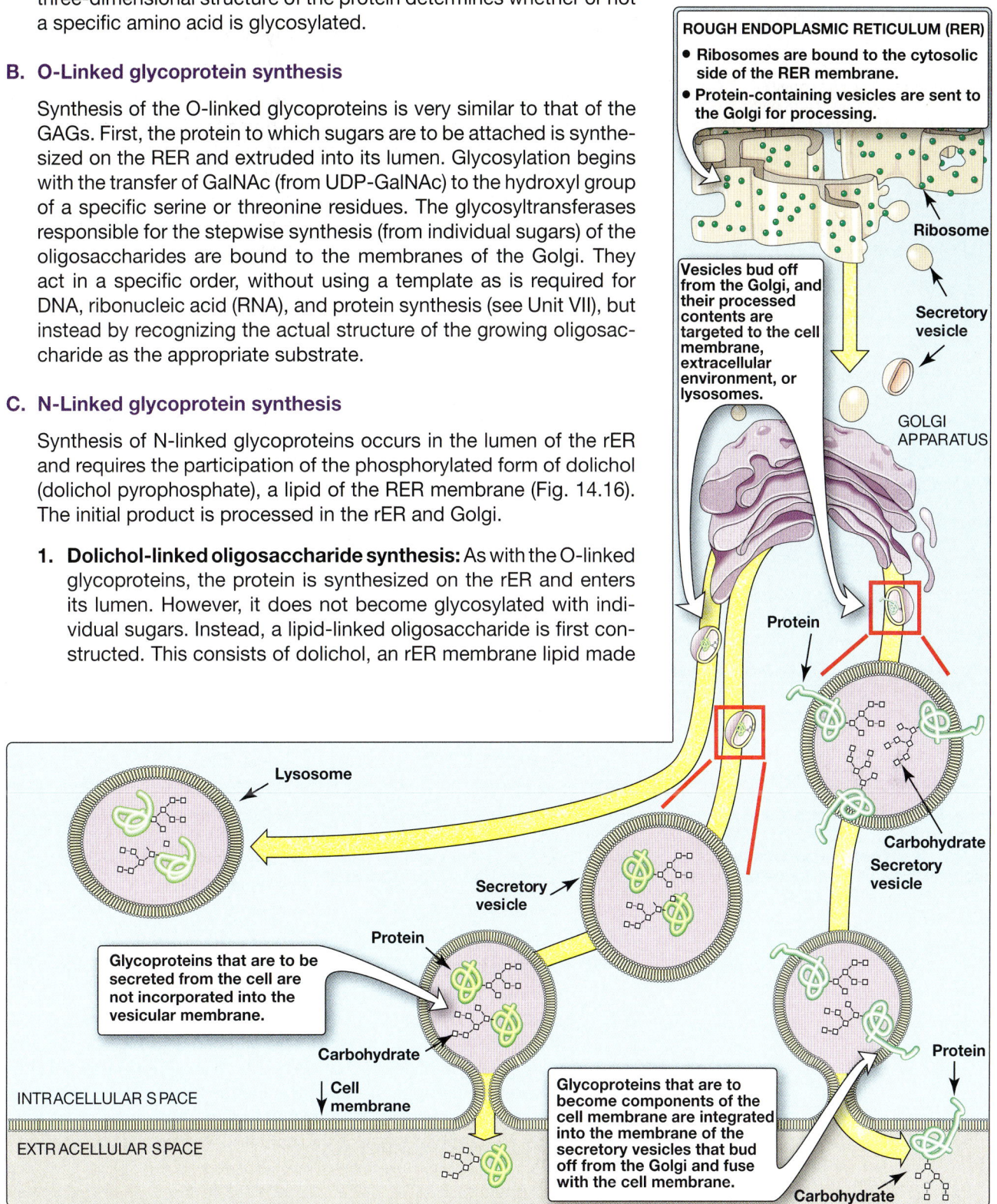

ROUGH ENDOPLASMIC RETICULUM (RER)
- Ribosomes are bound to the cytosolic side of the RER membrane.
- Protein-containing vesicles are sent to the Golgi for processing.

Ribosome

Secretory vesicle

GOLGI APPARATUS

Vesicles bud off from the Golgi, and their processed contents are targeted to the cell membrane, extracellular environment, or lysosomes.

Protein

Carbohydrate

Secretory vesicle

Lysosome

Secretory vesicle

Protein

Glycoproteins that are to be secreted from the cell are not incorporated into the vesicular membrane.

Carbohydrate

Cell membrane

Protein

Carbohydrate

INTRACELLULAR SPACE

EXTRACELLULAR SPACE

Glycoproteins that are to become components of the cell membrane are integrated into the membrane of the secretory vesicles that bud off from the Golgi and fuse with the cell membrane.

Figure 14.15
Transport of glycoproteins to and through the Golgi and their subsequent secretion or incorporation into a lysosome or the cell membrane.

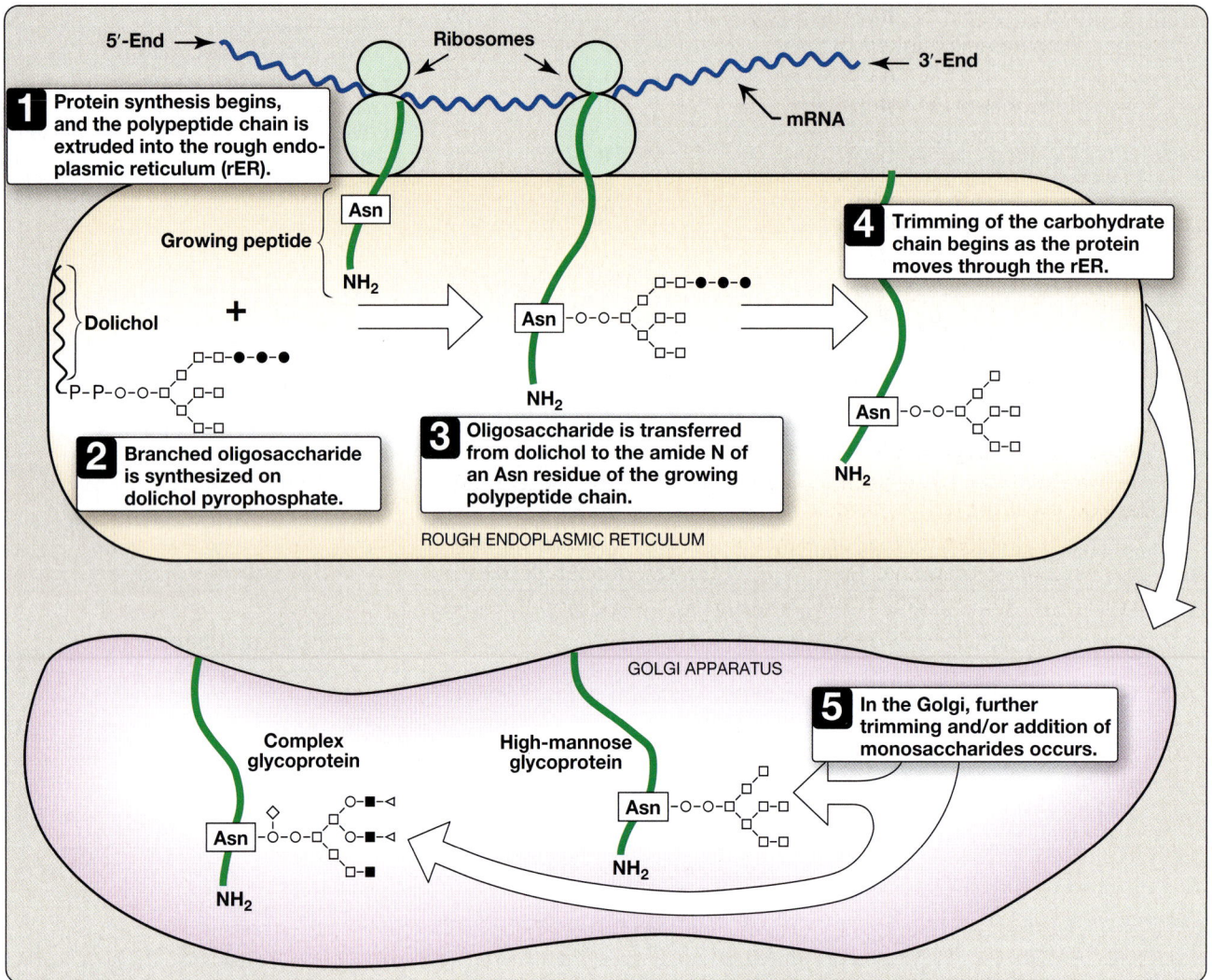

Figure 14.16
Synthesis of N-linked glycoproteins. ○, N-acetylglucosamine; □, mannose; ●, glucose; ■, galactose; ◇ or ◁, terminal group (fucose or N-acetylneuraminic acid); mRNA, messenger RNA; Asn, asparagine.

from an intermediate of cholesterol synthesis attached through a pyrophosphate linkage to an oligosaccharide containing GlcNAc, mannose, and glucose. The sugars to be added sequentially to the dolichol by membrane-bound glycosyltransferases are first GlcNAc, followed by mannose and glucose (see Fig. 14.16). The entire 14-sugar oligosaccharide is then transferred from dolichol to the amide nitrogen of an asparagine residue in the protein to be glycosylated by a protein–oligosaccharide transferase in the rER. [Note: The antibiotic tunicamycin inhibits N-linked glycosylation.]

Congenital disorders of glycosylation (CDG) are syndromes caused primarily by defects in the N-linked glycosylation of proteins, either oligosaccharide assembly (type I) or processing (type II).

2. **N-Linked oligosaccharide processing:** After addition to the protein, the N-linked oligosaccharide is processed by the removal of specific mannosyl and glucosyl residues as the glycoprotein moves through the rER. Finally, the oligosaccharide chains are completed in the Golgi by addition of a variety of sugars (e.g., GlcNAc, GalNAc, and additional mannoses and then fucose or NANA as terminal groups) to produce a complex glycoprotein. Alternatively, they are not processed further, leaving branched, mannose-containing chains in a high-mannose glycoprotein (see Fig. 14.16). The ultimate fate of N-linked glycoproteins is the same as that of the O-linked glycoproteins (e.g., they can be released by the cell or become part of a cell membrane). In addition, N-linked glycoproteins can be targeted to the lysosomes.

3. **Lysosomal enzymes:** N-Linked glycoproteins being processed in the Golgi can be phosphorylated on carbon 6 of one or more mannosyl residues. UDP-GlcNAc provides the phosphate in a reaction catalyzed by a phosphotransferase. Receptors, located in the Golgi membrane, bind the mannose 6-phosphate (M6P) residues of these proteins, which are then packaged into vesicles and sent to the lysosomes (Fig. 14.17).

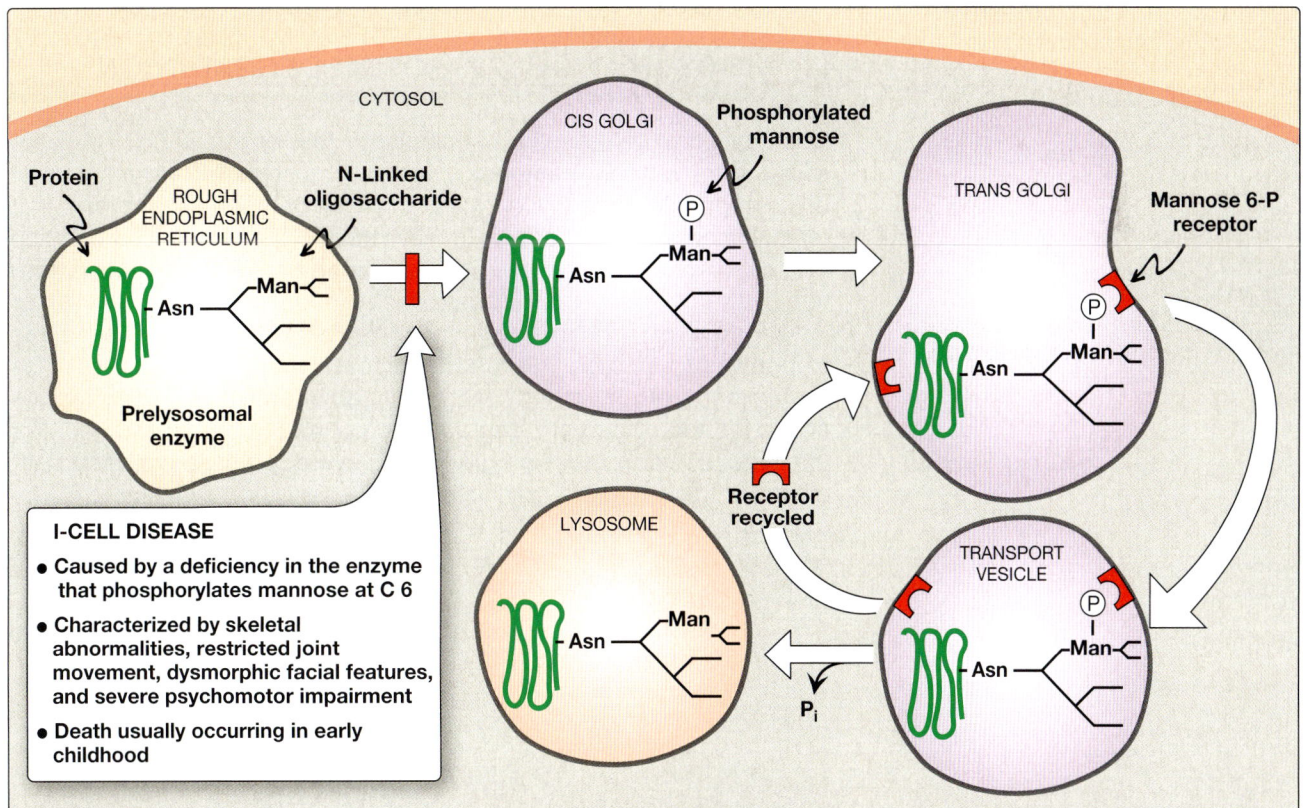

Figure 14.17
Mechanism for transport of N-linked glycoproteins to the lysosomes. Asn, asparagine; Man, mannose; (P), phosphate; P$_i$, inorganic phosphate.

Clinical Application 14.2: I-Cell Disease

I-Cell disease is a rare lysosomal storage disease named for large inclusion bodies seen in cells of patients with the disease. N-Acetylglucosamine phosphotransferase is deficient, and mannose 6-phosphate is not generated on proteins destined for lysosomes. Lack of M6P on amino acid residues causes the mistargeting of precursor acid hydrolases to the plasma membrane for secretion, instead of trafficking to lysosomes. Consequently, the acid hydrolases are absent from lysosomes, and the macromolecule substrates for these digestive enzymes accumulate within the lysosomes, generating the inclusion bodies that define the disorder. In addition, high concentrations of lysosomal enzymes are found in the patient's plasma and urine, indicating that the targeting process to lysosomes is deficient.

I-Cell disease is characterized by skeletal abnormalities, restricted joint movement, dysmorphic facial features, and severe psychomotor impairment. Because I-cell disease has features in common with the mucopolysaccharidoses and sphingolipidoses, it is considered a mucolipidosis (ML II). Currently, there is no cure, and death from cardiopulmonary complications usually occurs in early childhood. Pseudo-Hurler polydystrophy (ML III) is a less-severe mucolipidosis form of I-cell disease, in which the phosphotransferase maintains some residual enzymatic activity, and symptomatically resembles a mild form of Hurler syndrome.

IX. LYSOSOMAL GLYCOPROTEIN DEGRADATION

Degradation of glycoproteins is similar to that of the GAGs. The lysosomal acid hydrolases are each generally specific for the removal of one component of the glycoprotein. They are primarily exoenzymes that remove their respective groups in the reverse order of their incorporation (last on, first off). If any of the degradative enzyme is missing, degradation by the other exoenzymes cannot continue.

A group of very rare autosomal-recessive diseases called the glycoprotein storage diseases (oligosaccharidoses), caused by a deficiency of any one of the degradative enzymes, results in the accumulation of partially degraded structures in the lysosomes. For example, α-mannosidosis type 3 is a severe, progressive, fatal deficiency of the enzyme α-mannosidase. Presentation is similar to Hurler syndrome, but immune deficiency is also seen. Mannose-rich oligosaccharide fragments appear in the urine. Diagnosis is by enzyme activity assay.

X. CHAPTER SUMMARY

- **Glycosaminoglycans** (**GAGs**) are synthesized in the Golgi as **long, negatively charged, unbranched, heteropolysaccharide chains** generally composed of a **repeating disaccharide unit** (acidic sugar–amino sugar)$_n$ (Fig. 14.18).

- The **amino sugar** is either D-**glucosamine** or D-**galactosamine**, and the **acidic sugar** is either D-**glucuronic acid** or its C-5 epimer **l-iduronic acid**.

- GAGs bind water, thereby producing the gel-like matrix that forms the basis of the body's **ground substance** and the lubricating properties of mucous secretions.

- There are six major types of GAGs: **chondroitin 4-** and **6-sulfates, keratan sulfate, dermatan sulfate, heparin, heparan sulfate**, and **hyaluronic acid**.

- All GAGs, except hyaluronic acid, are found covalently attached to a **core protein**, forming **proteoglycan monomers**. Many proteoglycan monomers associate with a molecule of **hyaluronic acid** to form **proteoglycan aggregates**.

- The completed proteoglycans are secreted into the **extracellular matrix** (**ECM**) or remain associated with the outer surface of cells.

- GAGs are degraded by **lysosomal acid hydrolases**. A deficiency of any one of the hydrolases results in a **mucopolysaccharidosis** in which GAGs accumulate in tissues, causing symptoms such as **skeletal** and **ECM deformities** and **intellectual disability**. Examples of these genetic diseases include **Hunter** (X-linked) and **Hurler syndromes**.

- **Glycoproteins** are synthesized in the rough endoplasmic reticulum (rER) and Golgi and are proteins to which **oligosaccharides** (**glycans**) are covalently attached.

- **Membrane-bound** glycoproteins participate in **cell-surface recognition**; **cell-surface antigenicity**; and as components of the ECM and of the **mucins** of the gastrointestinal and urogenital tracts, where they act as protective biologic lubricants. Almost all of the globular proteins present in human plasma are glycoproteins.

- Precursors of carbohydrate components of glycoproteins are **nucleotide sugars**. **O-Linked glycoproteins** are produced in the Golgi by the sequential transfer of sugars from their nucleotide carriers to the hydroxyl group of a Ser or Thr residue in the protein. **N-Linked glycoproteins** are created by the transfer of a preformed oligosaccharide from its rER membrane lipid carrier, **dolichol pyrophosphate**, to the amide N of an Asn residue in the protein. They contain varying amounts of **mannose**.

- A deficiency in **N-acetylglucosamine phosphotransferase** that phosphorylates mannose residues at carbon 6 in N-linked glycoprotein enzymes destined for the lysosomes results in **I-cell disease**.

- Glycoproteins are normally degraded in lysosomes by **acid hydrolases**. A deficiency of any one of these enzymes results in a **lysosomal glycoprotein storage disease**, resulting in the accumulation of partially degraded glycoproteins in the lysosome and causing a range of symptoms including skeletal deformity and intellectual disability.

Glycosaminoglycans

composed of

Proteins → Heteropoly-saccharide chains

including

Core proteins | Link proteins

physical/chemical properties include
- Bind large amounts of water
- Viscous
- Lubricating

classified as
- Chondroitin sulfates
- Keratan sulfates
- Dermatan sulfate
- Hyaluronic acid
- Heparin
- Heparan sulfate

degraded by

Lysosomal hydrolases

of interest because

Hereditary enzyme deficiencies

lead to

Mucopoly-saccharidoses
- Hurler syndrome
- Hunter syndrome
- Sanfilippo syndrome
- Sly syndrome

characterized as
- Long
- Unbranched
- Negatively charged
- Composed of repeating disaccharide units

function as
- Flexible support for the proteins of the ECM
- Molecular sieve

Glycoproteins

composed of

Protein | Oligosaccharides

contain

N-terminal signal sequences

which target

The growing poly-peptide chain

to the

Lumen of the endoplasmic reticulum → **for** N-Glycosylation

and subsequently to the

Golgi apparatus → **for** O-Glycosylation

where

Proteins are sorted and sent to their proper destinations

for example

Secretion from the cell | Incorporation into the cell membrane | Targeted to lysosomes

defect results in

I-Cell disease

composed of

Heterooligo-saccharide chains
- Short
- Branched
- May or may not be negatively charged
- No repeating disaccharide units

function in
- Cell-surface recognition
- Cell-surface antigenicity
- Components of extracellular matrix and mucins
- Globular proteins in plasma

degraded by

Lysosomal hydrolases

of interest because

Hereditary enzyme deficiencies

lead to

Oligo-saccharidoses

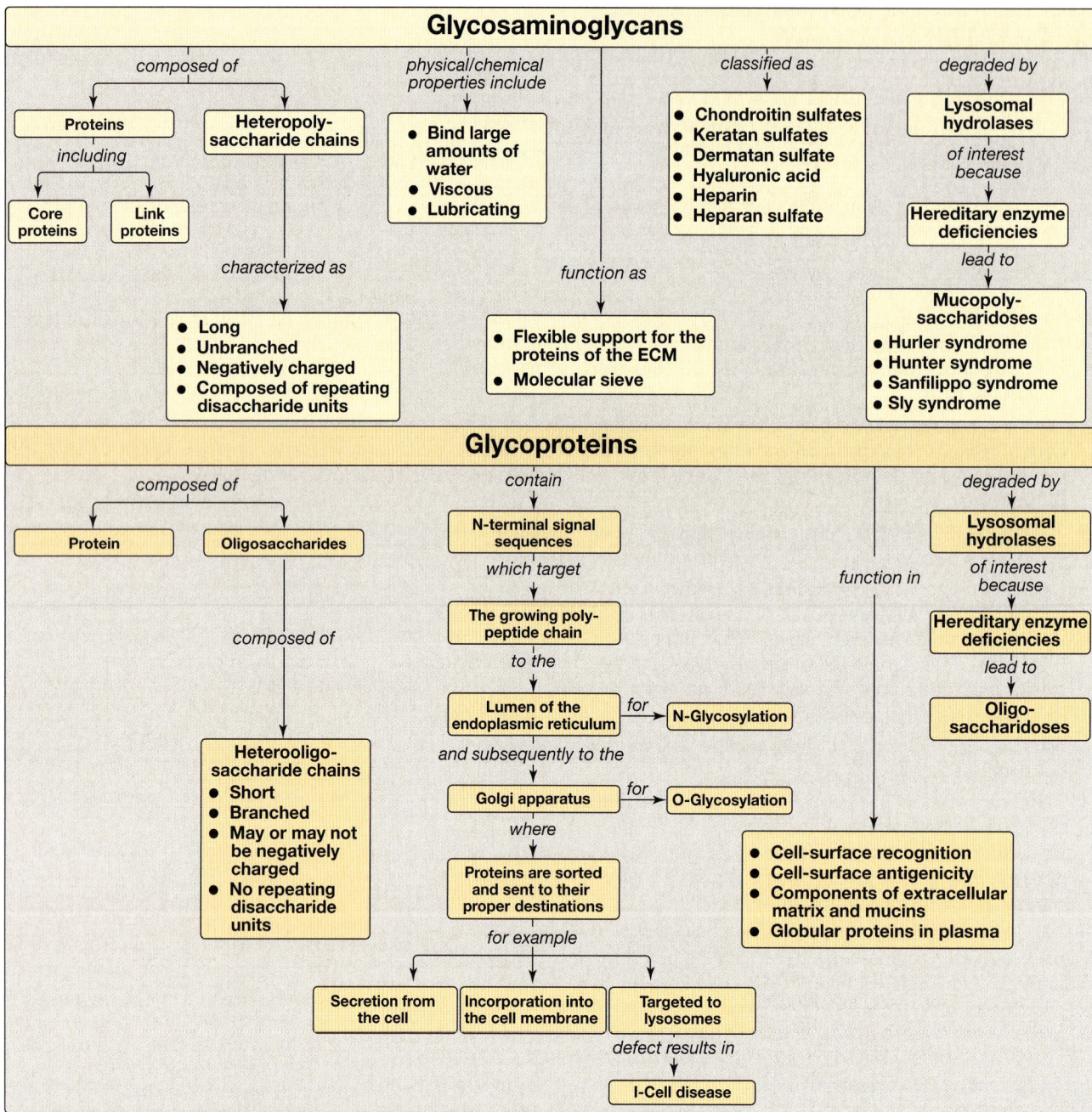

Figure 14.18
Key concept map for glycosaminoglycans and glycoproteins. ECM, extracellular matrix.

Study Questions

Choose the ONE best answer.

14.1. Mucopolysaccharidoses are hereditary lysosomal storage diseases caused by:

    A. abnormal targeting of acid hydrolase enzymes to lysosomes.

    B. increased rate of synthesis of the carbohydrate component of proteoglycans.

    C. insufficient rate of synthesis of proteolytic enzymes.

    D. defects in the degradation of glycosaminoglycans.

    E. synthesis of abnormally small amounts of core proteins.

Correct answer = D. The mucopolysaccharidoses are caused by deficiencies in any one of the lysosomal acid hydrolases responsible for the degradation of glycosaminoglycans (not proteins). The enzyme is correctly targeted to the lysosome, so blood levels of the enzyme do not increase, but it is nonfunctional. In these diseases, synthesis of the protein and carbohydrate components of proteoglycans is unaffected, in terms of both structure and amount.

14.2. Which of the following are the main structural units found within glycosaminoglycans?

    A. Compact globular glycoproteins

    B. Disaccharide of an acid sugar and an N-acetylated amino sugar

    C. GalNAc and GlcNAc

    D. UDP glucose and UDP GlcNAc

Correct answer = B. Repeating disaccharides are the main structural units found in GAGs.

14.3. The presence of the compound shown in the urine of a patient suggests a deficiency in which of the following enzymes?

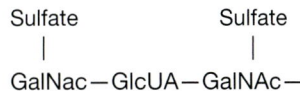

$$\begin{array}{ccc} \text{Sulfate} & & \text{Sulfate} \\ | & & | \\ \text{GalNac} - \text{GlcUA} - \text{GalNAc} - \end{array}$$

    A. Galactosidase

    B. Glucuronidase

    C. Iduronidase

    D. Mannosidase

    E. Sulfatase

Correct answer = E. Degradation of glycoproteins follows the rule: last on, first off. Because sulfation is the last step in the synthesis of this sequence, a sulfatase is required for the next step in the degradation of the compound shown.

14.4. An 8-month-old boy with dysmorphic facial features is evaluated for skeletal abnormalities and delays in both growth and development. I-Cell disease is suspected. Which of the following will be observed in this patient if that diagnosis is correct?

    A. Decreased production of cell surface O-linked glycoproteins

    B. Elevated levels of acid hydrolases in the blood

    C. Inability to N-glycosylate proteins

    D. Increased synthesis of proteoglycans

    E. Oligosaccharides in the urine

Correct answer = B. I-Cell disease is a lysosomal storage disease caused by deficiency of the phosphotransferase needed for synthesis of the mannose 6-phosphate signal that targets acid hydrolases to the lysosomal matrix. This results in the secretion of these enzymes from the cell and accumulation of materials within the lysosome because of impaired degradation. None of the other choices relates to I-cell disease or lysosomal function. Oligosaccharides in the urine are characteristic of the muco- and polysaccharidoses but not I-cell disease (a type II mucolipidosis).

14.5. An infant with corneal clouding has dermatan sulfate and heparan sulfate in his urine. Decreased activity of which of the enzymes listed below would confirm the suspected diagnosis of Hurler syndrome?

    A. α-L-Iduronidase

    B. α-Glucuronidase

    C. Glycosyltransferase

    D. Iduronate sulfatase

Correct answer = A. Hurler syndrome, a defect in the lysosomal degradation of glycosaminoglycans (GAGs) with corneal clouding, is due to a deficiency in α-L-iduronidase. β-Glucuronidase is deficient in Sly syndrome, and iduronate sulfatase is deficient in Hunter syndrome. Glycosyltransferases are enzymes of GAGs synthesis.

14.6. A 67-year-old man presents for evaluation of pain and stiffness in his left knee. Osteoarthritis is diagnosed. Decreases in which of the following contribute to his symptoms?

    A. Cartilage proteoglycans

    B. Cell-surface O-linked glycoproteins

    C. Golgi phosphotransferase

    D. Lysosomal acid hydrolases

Correct answer = A. Proteoglycans contribute to the resilience of cartilage. In osteoarthritis, cartilage has degraded and the protection normally provided by proteoglycans is lost. The disease is not caused by lysosomal defects including trafficking or function of acid hydrolases.

14.7. Lysosomal acid hydrolase enzymes are tagged with which of the following during the processing in order to be sent to lysosomes where they will function?

    A. Dolichol pyrophosphate

    B. GalNAc

    C. Mannose 6-phosphate (M6P)

    D. UDP-GlcNAc

Correct answer = C. Acid hydrolase precursors are tagged with M6P in a reaction catalyzed by a phosphotransferase. The tag binds to M6P receptors on the Golgi membrane, and the acid hydrolase precursors are packaged into vesicles and sent to lysosomes, where they become functional at acidic pH.

Dietary Lipid Metabolism

15

I. OVERVIEW

Lipids are a heterogeneous group of water-insoluble (hydrophobic) organic molecules (Fig. 15.1). Because of their insolubility in aqueous solutions, body lipids are generally found compartmentalized, as in the case of membrane-associated lipids or droplets of triacylglycerol (TAG) in adipocytes, or transported in blood in association with protein, as in lipoprotein particles or on albumin. Lipids are a major source of energy for the body, and they also provide the hydrophobic barrier that permits partitioning of the aqueous contents of cells and subcellular structures. Lipids serve additional functions in the body (e.g., some fat-soluble vitamins have regulatory or coenzyme functions, and the prostaglandins and steroid hormones play major roles in the control of the body's homeostasis). Deficiencies or imbalances of lipid metabolism can lead to some of the major clinical problems encountered by physicians, such as atherosclerosis, diabetes, and obesity.

II. DIGESTION, ABSORPTION, SECRETION, AND UTILIZATION

The 2020–2025 Dietary Guidelines for Americans recommends 22 to 44 g/day lipids for adults. The adult average daily intake is ~78 g, of which more than 90% is TAG, also known as triglycerides (TGs) that consist of three fatty acids (FAs) esterified to a glycerol backbone (see Fig. 15.1).

The remainder of the dietary lipids consists primarily of cholesterol, cholesteryl esters, phospholipids, and nonesterified (free) FAs (FFAs). The digestion of dietary lipids begins in the stomach and is completed in the small intestine. The process is summarized in Figure 15.2.

A. Digestion in the stomach

Lipid digestion in the stomach is limited. It is catalyzed by lingual lipase that originates from serous glands located on both sides of

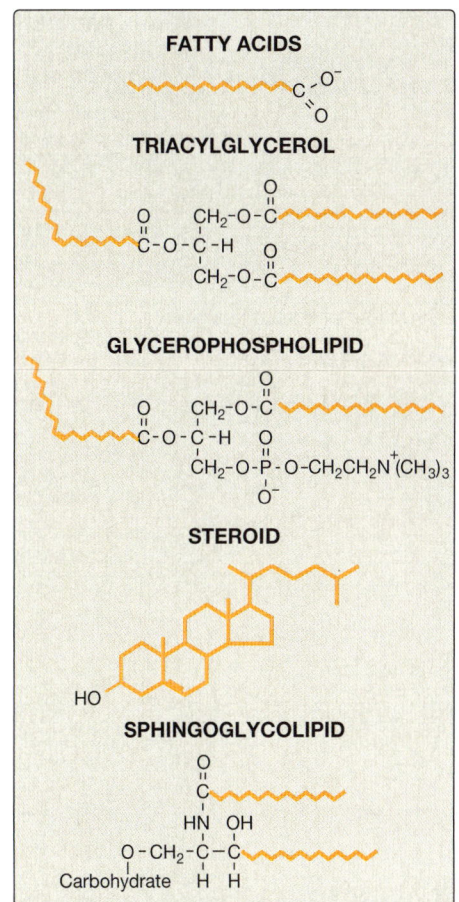

Figure 15.1
Structures of some common classes of lipids. Hydrophobic portions of the molecules are shown in orange.

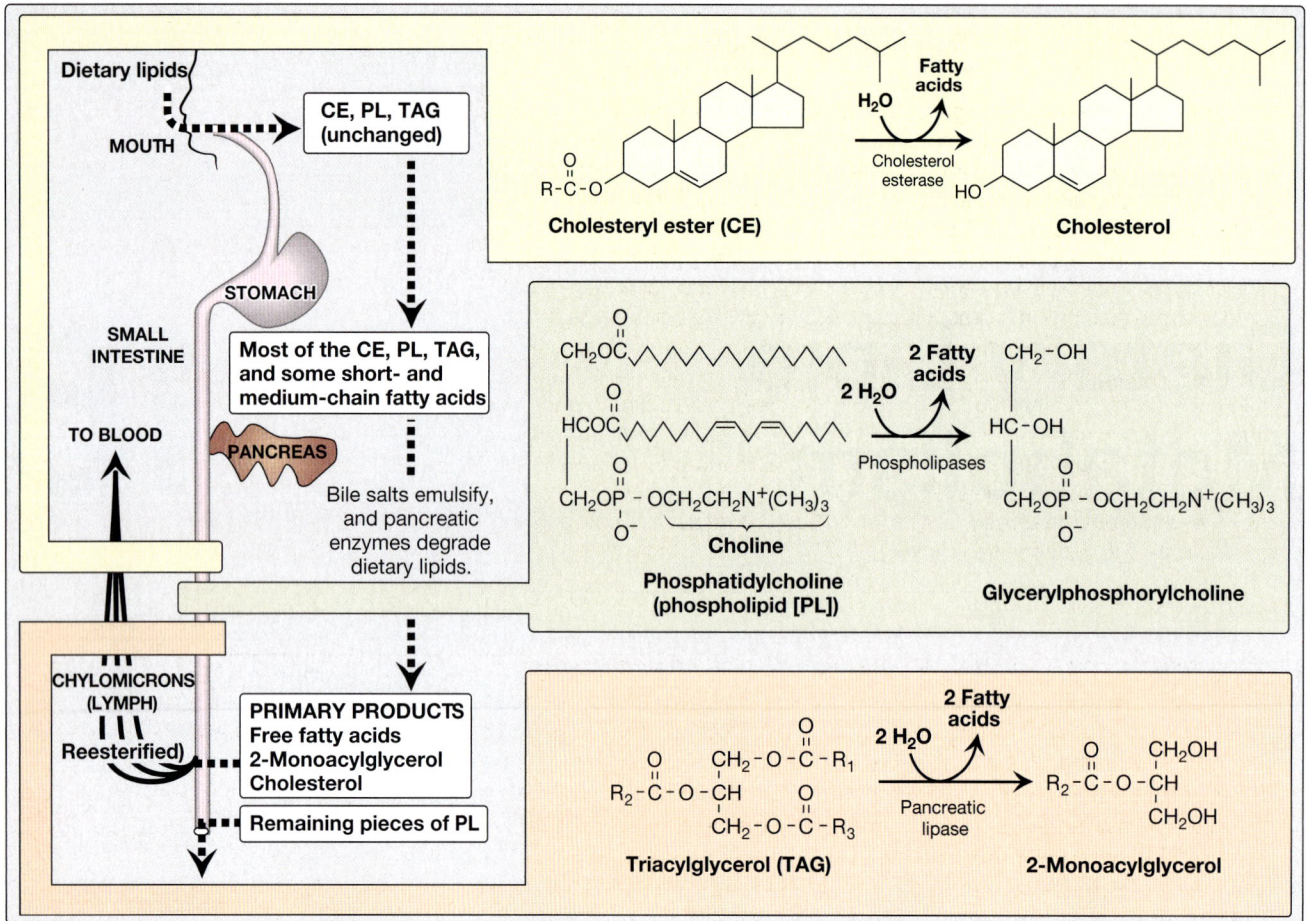

Figure 15.2
Overview of lipid digestion.

the oral cavity toward the back of the tongue and by gastric lipase that is secreted by the chief cells of gastric mucosa. Both enzymes are relatively acid stable, with optimal pH values of 4 to 6. These acid lipases hydrolyze FAs from TAG molecules, particularly those containing short- or medium–chain-length (≤12 carbons) FAs such as are found in milk fat. Consequently, these lipases play a particularly important role in lipid digestion in infants for whom milk fat is the primary source of calories. They also become important digestive enzymes in individuals with pancreatic insufficiency such as those with cystic fibrosis (CF). Lingual and gastric lipases aid these patients in degrading TAG molecules (especially those with short- to medium-chain FAs) despite a near or complete absence of pancreatic lipase (see Section D. 1.).

B. Emulsification in the small intestine

The critical process of dietary lipid emulsification occurs in the duodenum. Emulsification increases the surface area of the hydrophobic lipid droplets so that the digestive enzymes, which work at the interface of the droplet and the surrounding aqueous solution, can act effectively. Emulsification is accomplished by two complementary

Clinical Application 15.1: Cystic fibrosis

CF is the most common autosomal recessive disorder in the United States, where it has an incidence of more than 1 in 6,000 births among Whites whose ancestors are from northern Europe, although it affects all ethnic groups. CF is caused by mutations in the gene for the CF transmembrane conductance regulator (CFTR) protein that functions as a chloride channel on epithelium in the pancreas, lungs, testes, and sweat glands. Defective CFTR results in decreased secretion of chloride and increased uptake of sodium and water that thickens the mucus layer on the epithelial surfaces of these tissues. In the pancreas, the thickened mucus clogs the pancreatic ducts, preventing pancreatic enzymes from reaching the intestine, thereby leading to pancreatic insufficiency. CF also causes chronic lung infections with progressive pulmonary disease and male infertility. Patients receive pancreatic enzymatic replacement therapy (PERT) that includes pancreatic lipase, protease, and amylase to replace the endogenous pancreatic enzymes and support digestion. The fat-soluble vitamins (A, D, E, and K) are also provided, as patients are often deficient in these nutrients.

mechanisms, namely, use of the detergent properties of the conjugated bile salts and mechanical mixing due to peristalsis. Bile salts, made in the liver and stored in the gallbladder, are amphipathic derivatives of cholesterol. Conjugated bile salts consist of a hydroxylated sterol ring structure with a side chain to which a molecule of glycine or taurine is covalently attached by an amide linkage (Fig. 15.3). These emulsifying agents interact with the dietary lipid droplets and the aqueous duodenal contents, thereby stabilizing the droplets as they become smaller from peristalsis and preventing them from coalescing. [Note: See Chapter 18 for a more complete discussion of bile salt metabolism.]

C. Degradation by pancreatic enzymes

The dietary TAG, cholesteryl esters, and phospholipids are enzymatically degraded (digested) in the small intestine by pancreatic enzymes, whose secretion is hormonally controlled.

1. **Triacylglycerol degradation:** TAG molecules are too large to be taken up efficiently by the mucosal cells (enterocytes) of the intestinal villi. Therefore, they are hydrolyzed by an esterase, pancreatic lipase, which preferentially removes the FAs at carbons 1 and 3. The primary products of hydrolysis are, thus, a mixture of 2-monoacylglycerol (2-MAG) and FFAs (see Fig. 15.2 and Table 15.1). [Note: Pancreatic lipase is found in high concentrations in pancreatic secretions (2% to 3% of the total protein present), and it is highly efficient catalytically, thus ensuring that only severe pancreatic deficiency, such as that seen in CF, results in significant malabsorption of fat and fat soluble vitamins.] A second coenzyme protein, colipase, which activates pancreatic lipase is also secreted by the pancreas, binds the lipase at a ratio of 1:1 and anchors it at the lipid–aqueous interface. Colipase restores activity to lipase in the presence of inhibitory substances like bile salts that bind the micelles. [Note: Colipase is secreted as the zymogen, procolipase, which is activated in the intestine by trypsin.]

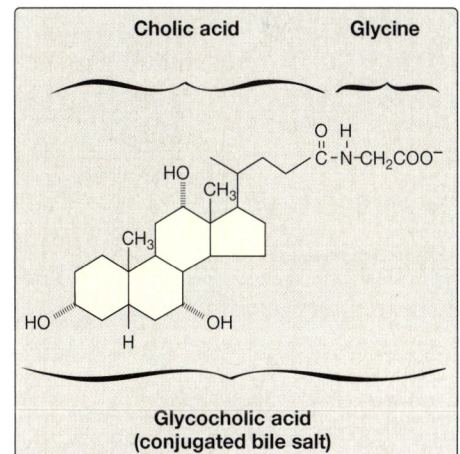

Figure 15.3
Structure of glycocholic acid.

Table 15.1: **Lipases**

	Source	Function
Lingual lipase	Ebner glands (lingual serous glands)	Ancillary: Hydrolyze short- or medium–chain-length triacylglycerol (TAG) molecules Important role in lipid digestion in premature infants
Gastric lipase	Chief cells of gastric mucosa	Ancillary: Hydrolyze short- or medium–chain-length TAG molecules Important role in lipid digestion in premature infants
Pancreatic lipase	Exocrine pancreas	Hydrolyze all TAG molecules: Primary products of hydrolysis are, thus, a mixture of 2-monoacylglycerol (2-MAG) and free fatty acids
Colipase	Exocrine pancreas	Zymogen: Digested by trypsin Activates pancreatic lipase
Phospholipase A2	Exocrine pancreas	Zymogen: Digested by trypsin Hydrolyze phospholipids

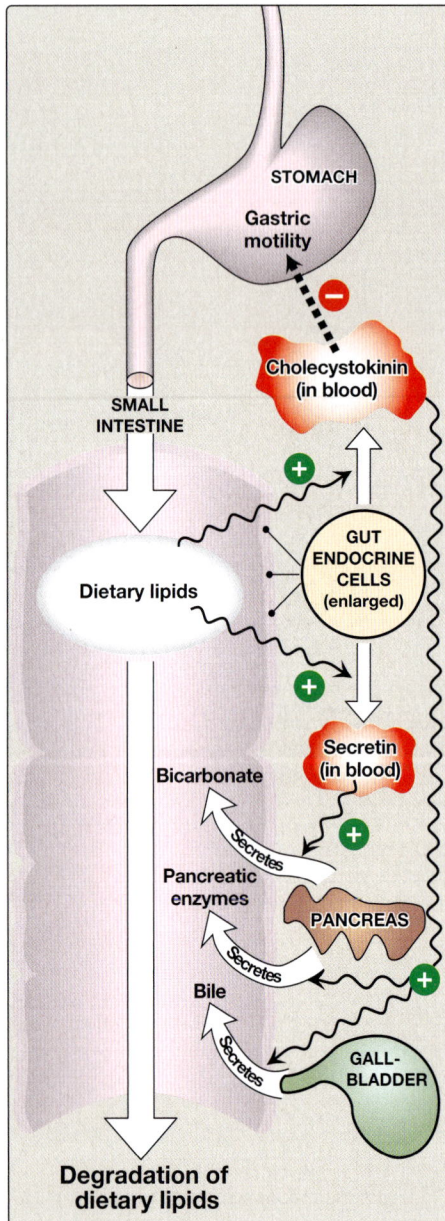

Figure 15.4
Hormonal control of lipid digestion in the small intestine. [Note: The small intestine is divided into three parts: duodenum (upper 5%), jejunum, and ileum (lower 55%).]

The antiobesity drug Orlistat, which reversibly inhibits gastric and pancreatic lipases decreases fat absorption, resulting in weight loss. It is the only U.S. Food and Drug Administration–approved, over-the-counter weight loss pill.

2. **Cholesteryl ester degradation:** Most dietary cholesterol is present in the free (nonesterified) form, with 10% to 15% present in the esterified form. Cholesteryl esters are hydrolyzed by pancreatic cholesteryl ester hydrolase (cholesterol esterase), which produces cholesterol plus FFAs (see Fig. 15.2). Activity of this enzyme is greatly increased in the presence of bile salts.

3. **Phospholipid degradation:** Pancreatic juice is rich in the proenzyme of phospholipase A_2 that, like procolipase, is activated by trypsin and, like cholesteryl ester hydrolase, requires bile salts for optimum activity. Phospholipase A_2 removes one FA from carbon 2 of a phospholipid, leaving a lysophospholipid. For example, phosphatidylcholine (predominant phospholipid of digestion) becomes lysophosphatidylcholine. The remaining FA at carbon 1 can be removed by lysophospholipase, leaving a glycerylphosphoryl base (e.g., glycerylphosphorylcholine, see Fig. 15.2) that may be excreted in the feces, further degraded, or absorbed.

4. **Control:** Pancreatic secretion of the hydrolytic enzymes that degrade dietary lipids in the small intestine is hormonally controlled (Fig. 15.4). Enteroendocrine cells found throughout the small intestine secrete several hormones such as cholecystokinin (CCK) and secretin. Enteroendocrine I cells in the mucosa of the lower duodenum and jejunum produce the peptide hormone CCK, in response to the presence of lipids and partially digested proteins entering these regions of the upper small intestine. CCK acts on the gallbladder (causing it to contract and release bile,

Clinical Application 15.2: Somatostatinoma

Somatostatin, a peptide hormone, exerts inhibitory effects on the secretion of CCK, secretin, glucagon, insulin, and gastrin. Somatostatinomas are rare tumors found in pancreatic islet cells, resulting in impaired absorption of fats and glucose. Common symptoms are reduced appetite, nausea and vomiting, abdominal pain, diarrhea, and unintended weight loss.

a mixture of bile salts, phospholipids, and free cholesterol) and on the exocrine cells of the pancreas (causing them to release digestive enzymes). It also decreases gastric motility, resulting in a slower release of gastric contents into the small intestine. Enteroendocrine S cells produce another peptide hormone, secretin, in response to the low pH of the chyme entering the intestine from the stomach. Secretin causes the pancreas to release a solution rich in bicarbonate that helps neutralize the pH of the intestinal contents, bringing them to the appropriate pH for digestive activity by pancreatic enzymes.

D. Absorption by enterocytes

FFAs, free cholesterol, and 2-MAG are the primary products of lipid digestion in the jejunum, which, plus bile salts and fat-soluble vitamins (A, D, E, and K), form mixed micelles (i.e., disc-shaped clusters of a mixture of amphipathic lipids that coalesce with their hydrophobic groups on the inside and their hydrophilic groups on the outside). Therefore, mixed micelles are soluble in the aqueous environment of the intestinal lumen (Fig. 15.5). These particles approach the primary site of lipid absorption, the brush border membrane of the enterocytes. This microvilli-rich apical membrane is separated from the liquid contents of the intestinal lumen by an unstirred water layer that mixes poorly with the bulk fluid. The hydrophilic surface of the micelles facilitates the transport of the hydrophobic lipids through the unstirred water layer to the brush border membrane where they are absorbed. Bile salts are absorbed in the terminal ileum, with less than 5% being lost in the feces. Cholesterol and plant sterols are taken up by the enterocytes through the Niemann–Pick C1-like 1 (NPC1L1) protein in the brush border cells. Because short- and medium-chain FAs are water soluble, they do not require the assistance of mixed micelles for absorption by the intestinal mucosa.

Ezetimibe, a cholesterol-lowering drug, inhibits NPC1L1, thus reducing cholesterol absorption in the small intestine. It can be used along with statins, drugs which inhibit hydroxy β-methylglutaryl-CoA (HMG-CoA) reductase, to better control cholesterol levels in patients with resistant hypercholesterolemia.

E. Triacylglycerol and cholesteryl ester resynthesis

The mixture of lipids absorbed by the enterocytes migrates to the smooth endoplasmic reticulum (sER) where biosynthesis of complex

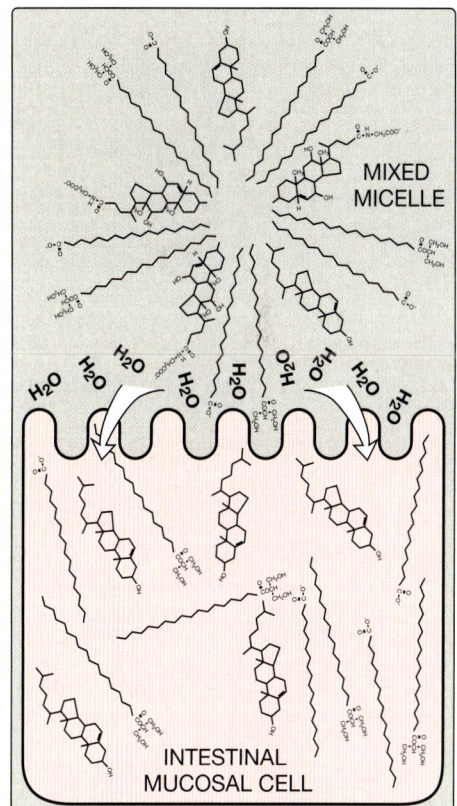

Figure 15.5
Absorption of lipids contained in a mixed micelle by an intestinal mucosal cell. The micelle itself is not absorbed. [Note: Short- and medium–chain-length fatty acids do not require incorporation into micelles.]

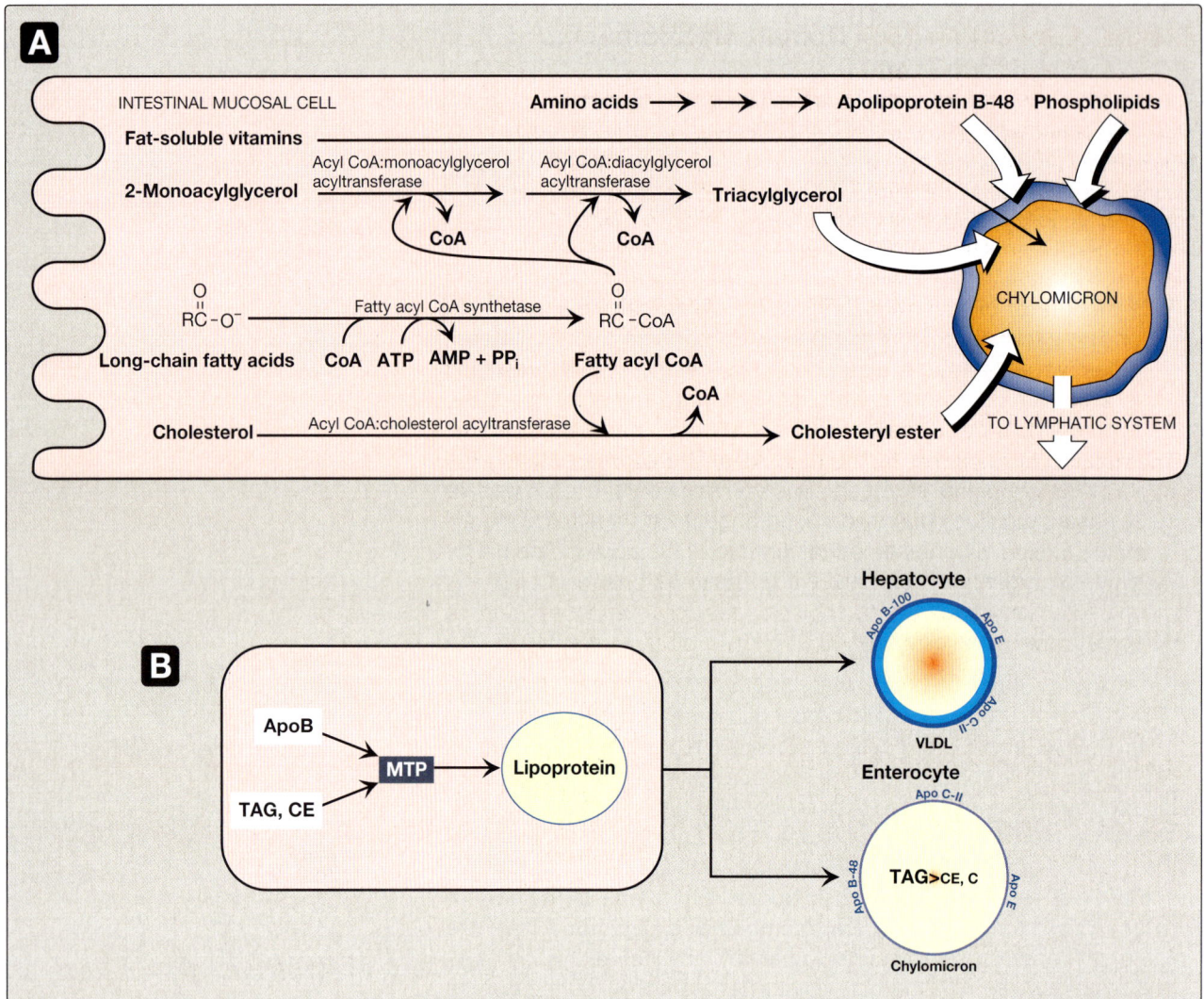

Figure 15.6
A. Assembly and secretion of chylomicrons by intestinal mucosal cells. [Note: Short- and medium-chain-length fatty acids do not require incorporation into chylomicrons and directly enter into the blood.] CoA, coenzyme A; AMP, adenosine monophosphate; PPi, pyrophosphate. **B.** Intracellular formation of VLDL and chylomicrons within liver and intestinal cells via the microsomal TG transfer protein (MTP) pathway.

lipids takes place. Long-chain FAs are first converted into their activated form by fatty acyl-coenzyme A (CoA) synthetase (thiokinase), as shown in Figure 15.6. Using the fatty acyl CoA derivatives, the 2-MAG absorbed by the enterocytes is converted to TAG through sequential reacylations by two acyltransferases, acyl CoA: MAG acyltransferase and acyl CoA: diacylglycerol acyltransferase. Lysophospholipids are reacylated to form phospholipids by a family of acyltransferases, and cholesterol is acylated primarily by acyl CoA: cholesterol acyltransferase.

Virtually all long-chain FAs entering the enterocytes are used in this fashion to form TAG, phospholipids, and cholesteryl esters. Short- and medium-chain FAs are not converted to their CoA derivatives and are not re-esterified to 2-MAG. Instead, they are released into the portal circulation, where they are carried by serum albumin to the liver.

Clinical Application 15.3: Abetalipoproteinemia/Hypobetalipoproteinemia

Mutations in the *MTP* gene lead to reduced or absent levels of cholesterol; triglycerides; and vitamins A, E, and K in the bloodstream, resulting in two distinct conditions: familial hypobetalipoproteinemia (FHBL) and abetalipoproteinemia (ABL). Patients with these conditions cannot form apo B–containing lipoproteins, such as chylomicrons. Symptoms manifest as gastrointestinal issues, which stem from inadequate fat absorption. These symptoms include pale, bulky, and malodorous stools (steatorrhea), diarrhea, vomiting, abdominal distension, and neurologic complications.

F. Secretion from enterocytes

The newly resynthesized TAG and cholesteryl esters are very hydrophobic and aggregate in an aqueous environment. Therefore, they must be packaged as particles of lipid droplets surrounded by a thin layer composed of phospholipids, nonesterified cholesterol, and a molecule of the protein apolipoprotein (apo) B-48. This layer stabilizes the particle and increases its solubility, thereby preventing multiple particles from coalescing. Microsomal TG transfer protein (MTP) is essential for the assembly of all TAG-rich apo B–containing particles in the ER. The lipoprotein particles are released by exocytosis from enterocytes into the lacteals (lymphatic vessels in the villi of the small intestine). The presence of these particles in the lymph after a lipid-rich meal gives it a milky appearance. This lymph is called chyle (as opposed to chyme, the name given to the semifluid mass of partially digested food that passes from the stomach to the duodenum), and the particles are named chylomicrons. Chylomicrons follow the lymphatic system to the thoracic duct and are then carried to the left subclavian vein, where they enter the blood. The steps in the production of chylomicrons are summarized in Figure 15.6. Once released into the blood, the nascent (immature) chylomicrons pick up apolipoproteins E and C-II from high-density lipoproteins and mature. [Note: For a more detailed description of chylomicron structure and metabolism, see Chapter 18.]

G. Lipid malabsorption

Lipid malabsorption, resulting in increased lipid (including the fat-soluble vitamins and essential FAs; see Chapter 16) in the feces, a condition known as steatorrhea, can be caused by disturbances in lipid digestion and/or absorption (Fig. 15.7). Such disturbances can result from several conditions, including CF (causing poor digestion), celiac disease, short-bowel syndrome (causing decreased absorption), and bariatric surgery (insufficient secretion of pancreatic enzymes).

The ability of short- and medium-chain FAs to be taken up by enterocytes without the aid of mixed micelles has made them important in medical nutrition therapy for individuals with malabsorption disorders. Coconut oil and palm kernel oil are rich in medium-chain TAGs, but these oils also contain long-chain FAs. Commercial formulations are now available for therapeutic use that contain 100% medium-chain TAGs.

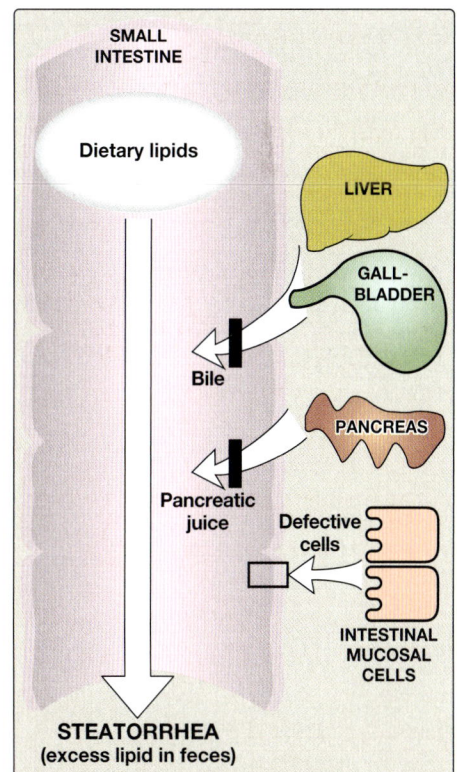

Figure 15.7
Possible causes of steatorrhea.

H. Use by the tissues

Most of the TAG contained in chylomicrons is broken down in the capillary beds of skeletal and cardiac muscle and adipose tissue. The TAG is degraded to FFAs and glycerol by lipoprotein lipase (LPL). This enzyme is synthesized and secreted primarily by adipocytes and muscle cells. Secreted LPL is anchored to the luminal surface of endothelial cells in the capillaries of muscle and adipose tissues. LPL is activated when bound to the cofactor, apo C-II, which resides on the circulating lipoprotein particles.

> Familial chylomicronemia (type I hyperlipoproteinemia) is a rare, autosomal-recessive disorder caused by a deficiency of LPL or its coenzyme apo C-II. The result is fasting chylomicronemia and severe hypertriacylglycerolemia, which can cause pancreatitis.

1. **Fate of free fatty acids:** The FFAs derived from the hydrolysis of TAG may either directly enter adjacent muscle cells and adipocytes or be transported in the blood in association with serum albumin until they are taken up by cells.

   Human serum albumin is a large protein secreted by the liver. It transports a number of primarily hydrophobic compounds in the circulation, including FFA and some drugs. Most cells can oxidize FA to produce energy. Adipocytes can also re-esterify FFAs to produce TAG molecules, which are stored until the FAs are needed by the body.

2. **Fate of glycerol:** Glycerol released from TAG is taken up from the blood and phosphorylated by hepatic glycerol kinase to produce glycerol 3-phosphate, which can enter either glycolysis or gluconeogenesis by oxidation to dihydroxyacetone phosphate or be used in TAG synthesis (see Chapter 16).

3. **Fate of chylomicron remnants:** After most of the TAG has been removed, the chylomicron remnants (which contain cholesteryl esters, phospholipids, apolipoproteins, fat-soluble vitamins, and a small amount of TAG) bind to receptors on the liver (apo E is the ligand; see Chapter 18) and are endocytosed. The intracellular remnants are hydrolyzed to their component parts. Cholesterol and the nitrogenous bases of phospholipids (e.g., choline) can be recycled by the body.

> A genetic mutation in apo E leads to familial dysbetalipoproteinemia, also called familial dysbetalipoproteinemia or broad-beta disease. When chylomicron remnants lack apo E, they cannot bind their receptor on the liver and accumulate in the plasma, leading to the development of multiple small, yellow skin growths (xanthomas) and atherosclerosis.

III. CHAPTER SUMMARY

- **Dietary lipid digestion** begins in the stomach and continues in the small intestine (Fig. 15.8).

- **Cholesteryl esters, phospholipids,** and triacylglycerols (**TAGs**) containing **long–chain-length fatty acids** (**FAs**) are degraded in the **small intestine** by **pancreatic enzymes**. The most important of these enzymes are **cholesterol esterase, phospholipase A$_2$,** and **pancreatic lipase**. In **cystic fibrosis** (**CF**), thickened mucus prevents these enzymes from reaching the intestine.

- TAGs in **milk fat** contain **short-** to **medium–chain-length FAs** and are degraded in the **stomach** by **acid lipases** (**lingual lipase** and **gastric lipase**).

- The **hydrophobic** nature of lipids requires that dietary lipids be **emulsified** for efficient degradation. Emulsification occurs in the small intestine using **peristaltic action** (mechanical mixing) and **bile salts** (detergents).

- The primary products of dietary lipid degradation are **2-monoacylglycerol** (**2-MAG**), nonesterified (free) **cholesterol,** and **free FAs.** These compounds, plus the **fat-soluble vitamins,** form **mixed micelles** that facilitate dietary lipid absorption by **intestinal mucosal cells** (**enterocytes**). These cells use activated long-chain FAs to regenerate TAG and cholesteryl esters and also synthesize protein **apolipoprotein (apo) B-48,** all of which are then assembled with the fat-soluble vitamins into **lipoprotein particles** called **chylomicrons**. Short- and medium-chain FAs enter blood directly.

- Chylomicrons are first released into the **lymph** and then enter the **blood,** where their lipid core is degraded by **lipoprotein lipase** (**LPL**) (with **apo C-II** as the coenzyme) in the **capillaries** of **muscle** and **adipose** tissues. Thus, dietary lipids are made available to the peripheral tissues.

- A deficiency in the ability to degrade chylomicron components, or remove chylomicron remnants after TAG has been degraded, results in accumulation of these particles in blood.

- **Fat maldigestion** or malabsorption causes **steatorrhea** (lipid in the feces).

Hydrophobic substrates:
Triacylglycerol
Cholesterol

present chal- *lenges for the*

Digestion of dietary lipids

are also substrates for

Cholesteryl esters
Phospholipids

is initiated in the

LIVER

produces

BILE

stored in

GALLBLADDER

releases bile containing bile salts

STOMACH

has

Acid-stable lipases (lingual and gastric lipases)

aid

Digestion of TAG (found in milk) in neonate

and continues in the

SMALL INTESTINE

into

Bicarbonate
Pancreatic lipase
Cholesterol esterase
Phospholipase A$_2$

secretes

PANCREAS

stimulates

Cholecystokinin

aided by

Enzymes
Bile salts
Peristalsis (for emulsification of the lipids)

stimulates secretion of bicarbonate

stimulates secretion of enzymes

Secretin

Cholecystokinin

Hormonal regulation

Hormonal regulation

produces

Digestive products

Free fatty acids (medium and short chain)

2-Monoacylglycerol
Cholesterol
Free fatty acids (long chain)

absorbed directly into

produces

Bile salts
Fat-soluble vitamins

Mixed micelles

facilitate absorption of dietary lipids by

INTESTINAL MUCOSAL CELLS (ENTEROCYTES)

assemble

Lipids (with characteristic protein apolipoprotein B-48)

to form

Chylomicrons

Concept connect

Cholesterol, Lipoprotein, and Steroid Metabolism 18

secreted into

LYMPHATIC SYSTEM

which transports them to

BLOOD

which transports TAG-rich chylomicrons to

MUSCLE, ADIPOSE

which transports free fatty acids to

where lipoprotein lipase degrades TAG to

TISSUES SUCH AS MUSCLE, ADIPOSE, AND LIVER

Fatty acids   **Glycerol**

metabolized by   *taken up by*

MOST TISSUES   **LIVER**

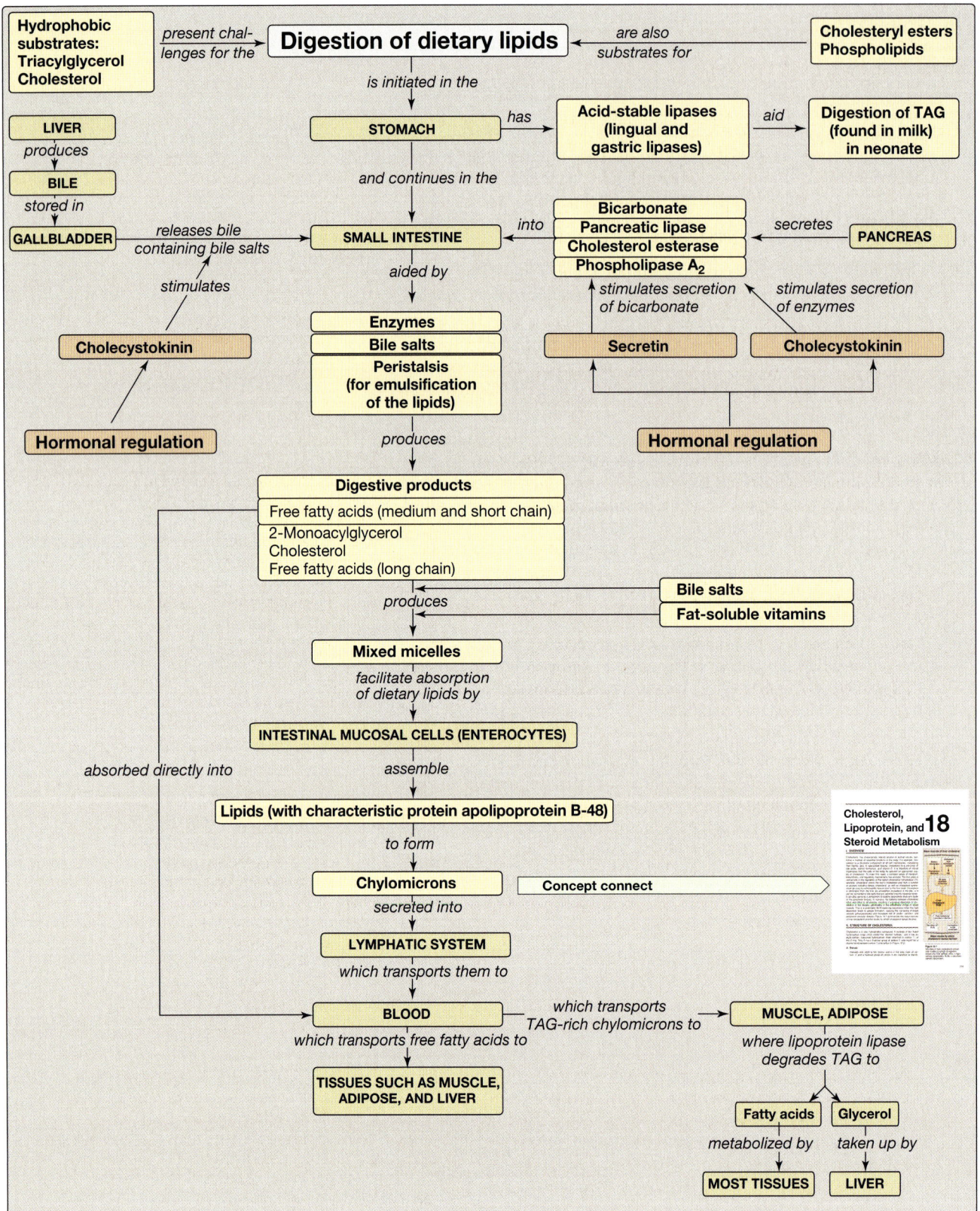

Figure 15.8
Key concept map for metabolism of dietary lipids. TAGs, triacylglycerols.

Study Questions

Choose the ONE best answer.

15.1. Which one of the following statements about lipid digestion is correct?

A. Large lipid droplets are emulsified in the mouth through the act of chewing.

B. Colipase facilitates the binding of bile salts to mixed micelles, maximizing pancreatic lipase activity.

C. Secretin causes the gallbladder to contract and release bile.

D. Patients with cystic fibrosis have difficulties with digestion because their pancreatic secretions are less able to reach the small intestine.

E. Formation of triacylglycerol-rich chylomicrons is independent of protein synthesis in the intestinal mucosa.

Correct answer = D. Patients with cystic fibrosis, a genetic disease resulting in a deficiency of a functional chloride transporter, have thickened mucus that impedes the flow of pancreatic enzymes into the duodenum. Emulsification occurs through peristalsis, which provides mechanical mixing, and bile salts that function as detergents. Colipase restores activity to pancreatic lipase in the presence of inhibitory bile salts that bind the micelles. Cholecystokinin is the hormone that causes the contraction of the gallbladder and release of stored bile, and secretin causes release of bicarbonate. Chylomicron formation requires synthesis of apolipoprotein B-48.

15.2. Which of the following statements about lipid absorption from the intestine is correct?

A. Dietary triacylglycerol must be completely hydrolyzed to free fatty acids and glycerol before absorption.

B. The triacylglycerol carried by chylomicrons is degraded by lipoprotein lipase, producing fatty acids that are taken up by muscle and adipose tissues and glycerol that is taken up by the liver.

C. Fatty acids that contain ≤12 carbon atoms are absorbed and enter the circulation primarily via the lymphatic system.

D. Deficiencies in the ability to absorb fat result in excessive amounts of chylomicrons in the blood.

Correct answer = B. The triacylglycerols (TAGs) in chylomicrons are degraded to fatty acids (FAs) and glycerol by lipoprotein lipase on capillary endothelial surfaces in muscle and adipose tissue, thus providing a source of FA to these tissues for degradation or storage and providing glycerol for hepatic metabolism. In the duodenum, TAGs are degraded to one 2-monoacylglycerol + two free FAs that get absorbed. Medium- and short-chain FAs enter directly into blood (not lymph), and they neither require micelles nor get packaged into chylomicrons. Because chylomicrons contain dietary lipids that were digested and absorbed, a defect in fat absorption would result in decreased production of chylomicrons.

15.3. A 2-year-old female is brought to the physician because of recurrent respiratory tract infections, weight loss, and foul-smelling diarrhea. This patient most likely has defective secretion in which of the following?

A. Cholecystokinin

B. Pancreatic enzymes

C. Chylomicron

D. Secretin

Correct answer: B. This patient most likely has cystic fibrosis (CF), which causes defective secretion of pancreatic enzymes such as lipase and colipase due to mutations in the cystic fibrosis transmembrane conductance receptor (CFTR). These enzymes are important for the digestion and absorption of lipids. Cholecystokinin and secretin are released from enteroendocrine cells. Although they are important for lipid digestion and absorption, they are not defective in CF. Chylomicron formation and release into lymphatic system is not affected in CF.

15.4. A 45-year-old female is brought to the emergency department due to acute pain, nausea, and vomiting. Computed tomography indicates acute pancreatitis that leads to an increased activation of trypsin. Which of the following is most likely activated in this condition?

A. Gastric lipase

B. Pancreatic lipase

C. Lysophospholipase

D. Colipase

Correct answer: D. Colipase is secreted as the zymogen, procolipase, which is activated in the intestine by trypsin. Colipase is important for pancreatic lipase for hydrolyzing triacylglycerols. Gastric lipase hydrolyzes short- and medium-chain fatty acids in milk, especially important for infants and patients with pancreatic insufficiency. Lysophospholipase is important for the digestion of phospholipids.

15.5. A 22-month-old child is brought to the physician by her parents because of refusal to feed, chronic diarrhea, abdominal distension, and weight loss. She is diagnosed with chylomicron retention disease, which prevents the release chylomicrons into the lymphatics. This patient most likely has a deficiency in which of the following vitamins?

- A. Ascorbic acid
- B. Beta-carotene
- C. Folate
- D. Pyridoxine

Correct answer: B. Chylomicrons are important for the absorption of fat-soluble vitamins A, D, E, and K. Beta-carotene is a provitamin A packaged into chylomicrons before its release into the lymphatics. Ascorbic acid is vitamin C, folate is vitamin B9, and pyridoxine is vitamin B6. These three vitamins are water soluble.

15.6. A 45-year-old male presents with fatigue, chronic diarrhea, steatorrhea, and unintentional weight loss. A qualitative assay of stool with Sudan III stain shows the presence of unabsorbed fat. A computed tomography scan demonstrates a mass in the pancreas. Biopsy of the mass reveals a somatostatinoma. Which of the following mechanisms best explains the fat malabsorption in this patient?

- A. Decreased release of pancreatic enzymes
- B. Inhibition of bile acid synthesis
- C. Decreased production of cholecystokinin
- D. Increased production of secretin

Correct answer: C. Somatostatinomas are neuroendocrine tumors that secrete somatostatin, a hormone that inhibits the release of various hormones, including cholecystokinin (CCK), secretin, insulin, and glucagon. Somatostatinomas can inhibit the secretion of CCK from the duodenal mucosa. CCK is a hormone responsible for stimulating the release of bile from the gallbladder and the secretion of pancreatic enzymes, both of which are crucial for fat digestion. Decreased CCK production leads to reduced bile flow and diminished pancreatic enzyme secretion, resulting in fat malabsorption. However, it does not directly affect the release of pancreatic enzymes, so it is not the primary cause of fat malabsorption. Also, somatostatin also does not inhibit bile acid synthesis.

Fatty Acid, Triacylglycerol, and Ketone Body Metabolism

16

I. OVERVIEW

Fatty acids (FAs) exist free in the body (i.e., they are nonesterified) and as fatty acyl esters in more complex molecules such as triacylglycerols (TAGs). Low levels of free fatty acids (FFAs) occur in all tissues, but substantial amounts can sometimes be found in the plasma, particularly during fasting. Plasma FFAs (transported on serum albumin) are en route from their point of origin (TAGs of adipose tissue or circulating lipoproteins) to their site of consumption (most tissues). FFAs can be oxidized by many tissues, particularly the liver to provide the substrate for ketone body synthesis and muscle to provide energy. FAs are also structural components of membrane lipids, such as phospholipids and glycolipids (see Chapter 17). FAs attached to certain proteins enhance the ability of those proteins to associate with membranes. FAs are also precursors of the hormone-like prostaglandins (see Chapter 17). Esterified FAs, in the form of TAG stored in white adipose tissue (WAT), serve as the major energy reserve of the body. Alterations in FA metabolism are associated with obesity and diabetes. Figure 16.1 illustrates the metabolic pathways of FA synthesis and degradation and their relationship to carbohydrate metabolism.

II. FATTY ACID STRUCTURE

An FA consists of a hydrophobic hydrocarbon chain with a terminal carboxyl group that has a pK_a of ~4.8 (Fig. 16.2). At physiologic pH, the terminal carboxyl group (–COOH) ionizes, becoming $-COO^-$. [Note: When the pH is above the pK, the deprotonated form predominates.] This anionic group has an affinity for water, giving the FA its amphipathic nature (having both a hydrophilic and a hydrophobic component). However, for long–chain-length FAs (LCFAs), the hydrophobic portion is predominant. These molecules are highly water insoluble and must be transported in the circulation in association with protein. More than 90% of the FAs found in plasma are in the form of FA esters (primarily TAGs, cholesteryl esters, and phospholipids) contained in circulating lipoprotein particles (see Chapter 18).

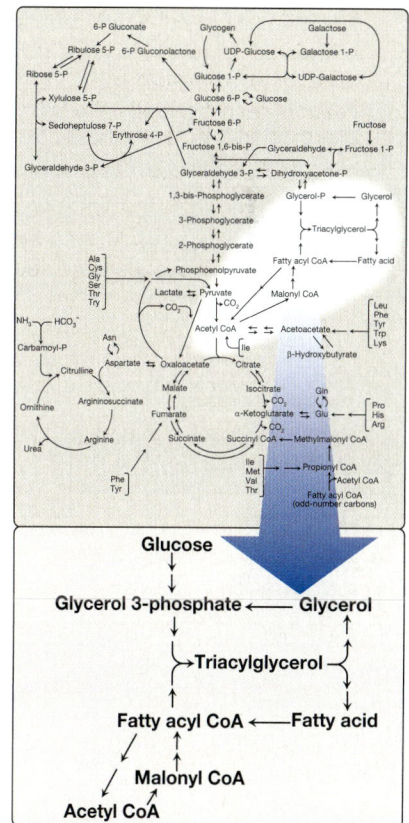

Figure 16.1
Triacylglycerol synthesis and degradation. CoA, coenzyme A.

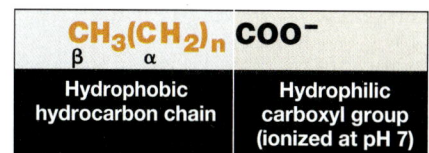

$$\underset{\beta}{CH_3}\underset{\alpha}{(CH_2)_n}\ COO^-$$

Hydrophobic hydrocarbon chain	Hydrophilic carboxyl group (ionized at pH 7)

Figure 16.2
Structure of a fatty acid. The carbon next to carbonyl group is designated as alpha (α). The next carbon is the beta carbon (β). When the chain is longer, the last carbon in the chain is designated as the ω carbon.

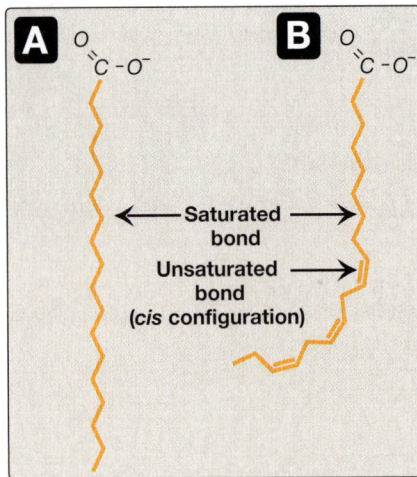

Figure 16.3
A saturated (**A**) and an unsaturated (**B**) fatty acid. Orange denotes hydrophobic portions of the molecules. [Note: *Cis* double bonds cause a fatty acid to kink.]

Fatty acids with chain lengths of 4 to 10 carbons are found in significant quantities in milk.

Structural lipids and triacylglycerols contain primarily fatty acids of at least 16 carbons.

COMMON NAME	STRUCTURE
Formic acid	1
Acetic acid	2:0
Propionic acid	3:0
Butyric acid	4:0
Capric acid	10:0
Palmitic acid	16:0
Palmitoleic acid	16:1(9)
Stearic acid	18:0
Oleic acid	18:1(9)
Linoleic acid	18:2(9,12)
α-Linolenic acid	18:3(9,12,15)
Arachidonic acid	20:4(5, 8,11,14)
Lignoceric acid	24:0
Nervonic acid	24:1(15)

Precursor of prostaglandins

Essential fatty acids

Figure 16.4
Some fatty acids of physiologic importance. [Note: A fatty acid containing 2 to 4 carbons is considered short; 6 to 12, medium; 14 to 20, long; and ≥22, very long.]

FFAs are transported in the circulation in association with albumin, the most abundant protein in serum.

A. Fatty acid saturation

FA chains may contain no double bonds (i.e., saturated) or one or more double bonds (i.e., mono- or polyunsaturated, respectively). In humans, the majority are saturated or monounsaturated. When double bonds are present, they are nearly always in the *cis* rather than in the *trans* configuration. The introduction of a *cis* double bond causes the FA to bend or kink at that position (Fig. 16.3). If the FA has two or more double bonds, they are always spaced at three-carbon intervals. [Note: In general, addition of double bonds decreases the melting temperature (T_m) of a FA, whereas increasing the chain length increases the T_m. Because membrane lipids typically contain LCFAs, the presence of double bonds in some FAs helps maintain the fluid nature of those lipids. [See Chapter 27, V. B. 3. for a discussion of the dietary occurrence of cis and trans unsaturated FAs.]

B. Fatty acid chain length and double-bond positions

The common names and structures of some FAs of physiologic importance are listed in Figure 16.4. In humans, FAs with an even number of carbon atoms (16, 18, or 20) predominate, with longer FAs (>22 carbons) being found in the brain. The carbon atoms are numbered, beginning with the carbonyl carbon as carbon 1. The number before the colon indicates the number of carbons in the chain, and those after the colon indicate the numbers and positions (relative to the carboxyl end) of double bonds. For example, as denoted in Figure 16.4, arachidonic acid, 20:4(5,8,11,14), is 20 carbons long and has 4 double bonds (between carbons 5–6, 8–9, 11–12, and 14–15). [Note: Carbon 2, the carbon to which the carboxyl group is attached, is also called the α-carbon, carbon 3 is the β-carbon, and carbon 4 is the γ-carbon. The carbon of the terminal methyl group is called the ω-carbon regardless of the chain length.] The double bonds in an FA can also be referenced relative to the ω (methyl) end of the chain. Arachidonic acid is referred to as an ω-6 FA because the terminal double bond is six bonds from the ω end (Fig. 16.5A). [Note: The equivalent designation of n-6 may also be used (Fig. 16.5B).] Another ω-6 FA is the essential linoleic acid 18:2(9,12). In contrast, α-linolenic acid, 18:3(9,12,15), is an essential ω-3 FA.

C. Essential fatty acids

Linoleic acid and α-linolenic acid are essential FAs, meaning that we need to obtain them through our diet, because we lack the enzymes that can form carbon–carbon double bonds after carbon 9 from the methyl (ω) end of a fatty acid. Linoleic acid is the precursor of ω-6 arachidonic acid, which is the substrate for prostaglandin synthesis, and α-linolenic acid, which is the precursor of ω-3 FAs that are important for growth and development. Plants provide us with these essential FAs. [Note: Arachidonic acid becomes essential if linoleic acid is deficient in the diet. See Chapter 27 for a discussion of the nutritional significance of ω-3 and ω-6 FAs.]

While ω-6 FAs as part of sphingolipids are important for structural integrity and barrier function of the skin, both ω-6 and ω-3 FAs are essential for giving rise to different types of eicosanoids, with ω-6 FAs often leading to proinflammatory eicosanoids and ω-3 FAs often being precursors to anti-inflammatory ones (see Chapter 17). Although a deficiency in essential FAs is rare, its consequences can be significant. Such a deficiency can manifest as dry, scaly dermatitis due to an inability to synthesize molecules that provide the water barrier in the skin, impaired immunity leading to increased susceptibility to infection, and poor wound healing.

A Double bonds relative to the ω end

$$HOOC(CH_2)_3C=C-CH_2-C=C-CH_2-C=C-CH_2-C=C(CH_2)_4CH_3$$

B Double bonds relative to the carboxy end

(5–6)
(8–9)
(11–12)
(14–15)

$$HOOC(CH_2)_3C=C-CH_2-C=C-CH_2-C=C-CH_2-C=C(CH_2)_4CH_3$$

Figure 16.5
Arachidonic acid, 20:4(5,8,11,14), illustrating the position of the double bonds. **A.** Arachidonic acid is an ω-6 fatty acid because the first double bond from the ω end is 6 carbons from that end. **B.** It is also referred to as an n-6 fatty acid because the last double bond from the carboxyl end is 14 carbons from that end: 20 – 14 = 6 = n. Thus, the "ω" and "n" designations are equivalent (see*).

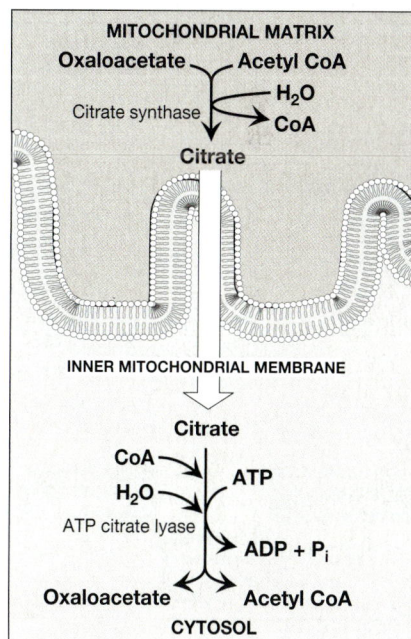

III. FATTY ACID DE NOVO SYNTHESIS

Carbohydrates and proteins obtained from the diet in excess of the body's needs for these nutrients can be converted to FAs. In adults, *de novo* FA synthesis occurs primarily in the liver and lactating mammary glands and, to a lesser extent, in adipose tissue. This cytosolic process is endergonic and reductive. It incorporates carbons from acetyl coenzyme A (CoA) into the growing FA chain, using adenosine triphosphate (ATP) and reduced nicotinamide adenine dinucleotide phosphate (NADPH). [Note: Dietary TAGs also supply FAs. See Chapter 24 for a discussion of the metabolism of dietary nutrients in the well-fed state.]

A. Cytosolic acetyl CoA production

The first step in FA synthesis is the transfer of acetate units from mitochondrial acetyl CoA to the cytosol. Mitochondrial acetyl CoA is produced by the oxidation of pyruvate (see Chapter 9) and by the catabolism of certain amino acids (see Chapter 20). However, the CoA portion of acetyl CoA cannot cross the inner mitochondrial membrane, and only the acetyl portion enters the cytosol. It does so as part of citrate produced by the condensation of acetyl CoA with oxaloacetate (OAA) by citrate synthase (Fig. 16.6). [Note: The transport of citrate to the cytosol occurs when the mitochondrial citrate concentration is high. This is observed when isocitrate dehydrogenase of the tricarboxylic acid (TCA) cycle is inhibited by the presence of large amounts of ATP, causing citrate and isocitrate to accumulate. Therefore, cytosolic citrate may be viewed as a high-energy signal. Because a large amount of ATP is needed for FA synthesis, the increase in both ATP and citrate enhances this pathway.] In the cytosol, citrate is cleaved to OAA and acetyl CoA by ATP citrate lyase.

B. Acetyl CoA carboxylation to malonyl CoA

The energy for the carbon-to-carbon condensations in FA synthesis is supplied by the carboxylation and then decarboxylation of acyl groups in the cytosol. The carboxylation of acetyl CoA to malonyl CoA is catalyzed by acetyl CoA carboxylase (ACC) (Fig. 16.7). ACC transfers carbon dioxide (CO_2) from bicarbonate (HCO_3^-) in an ATP-requiring reaction. The coenzyme is biotin (vitamin B_7), which is covalently bound to a lysyl residue of the carboxylase (see Chapter 28).

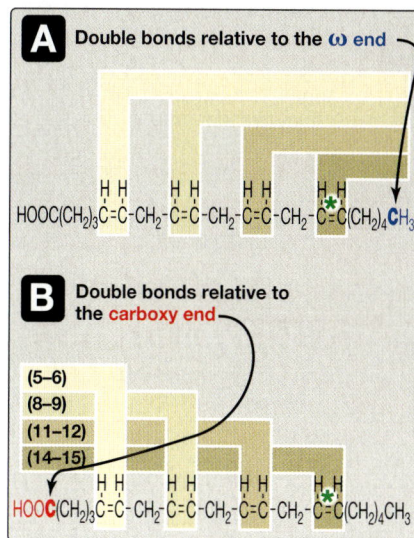

Figure 16.6
Production of cytosolic acetyl coenzyme A (CoA). [Note: Citrate is transported by the tricarboxylate transporter system.] ADP, adenosine monophosphate; P_i, inorganic phosphate.

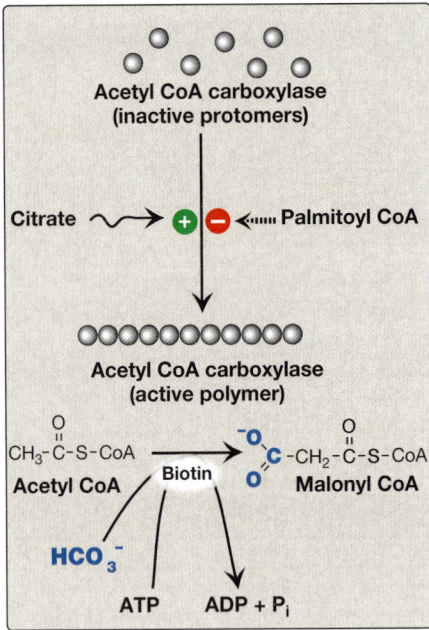

Figure 16.7
Allosteric regulation of malonyl coenzyme A (CoA) synthesis by acetyl CoA carboxylase. The carboxyl group contributed by bicarbonate (HCO_3^-) is shown in *blue*. P_i, inorganic phosphate; ADP, adenosine diphosphate.

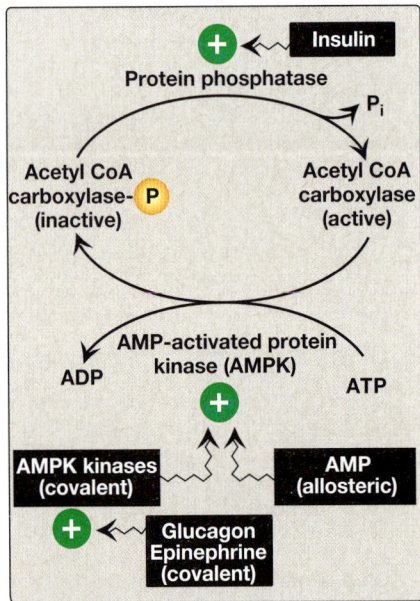

Figure 16.8
Covalent regulation of acetyl CoA carboxylase by AMPK, which itself is regulated both covalently and allosterically. CoA, coenzyme A; ADP and AMP, adenosine di- and monophosphates; ⓟ, phosphate; P_i, inorganic phosphate.

ACC carboxylates the bound biotin, which transfers the activated carboxyl group to acetyl CoA.

1. **Acetyl CoA carboxylase short-term regulation:** This carboxylation is both the rate-limiting and the regulated step in FA synthesis (see Fig. 16.7). The inactive form of ACC is a protomer (complex of ≥2 polypeptides). The enzyme is allosterically activated by citrate, which causes protomers to polymerize, and allosterically inactivated by palmitoyl CoA (the end product of the pathway), which causes depolymerization. A second mechanism of short-term regulation is by reversible phosphorylation. Adenosine monophosphate–activated protein kinase (AMPK) phosphorylates and inactivates ACC. AMPK itself is activated allosterically by AMP and covalently by phosphorylation via several kinases. At least one of these AMPK kinases is activated by cyclic AMP (cAMP)-dependent protein kinase A (PKA). Thus, in the presence of counterregulatory hormones, such as epinephrine and glucagon, ACC is phosphorylated and inactive (Fig. 16.8). In the presence of insulin, ACC is dephosphorylated and active. [Note: This is analogous to the regulation of glycogen synthase (see Chapter 11).]

2. **Acetyl CoA carboxylase long-term regulation:** Prolonged consumption of a diet containing excess calories (particularly high-carbohydrate, low-fat diets) causes an increase in ACC synthesis, thereby increasing FA synthesis. A low-calorie or a high-fat, low-carbohydrate diet has the opposite effect. [Note: ACC synthesis is upregulated by carbohydrate (specifically glucose) via the transcription factor carbohydrate response element–binding protein (ChREBP) and by insulin via the transcription factor sterol regulatory element–binding protein-1c (SREBP-1c). FA synthase (FAS; see Section C.) is similarly regulated. The function and regulation of SREBP are described in Chapter 18.]

> Metformin stimulates AMPK activation, leading to the phosphorylation-induced inhibition of ACC activity and a reduction in FAS expression via the downregulation of SREBP-1c. Consequently, this medication reduces plasma TAG levels. Additionally, it lowers blood glucose by enhancing muscle glucose uptake through the AMPK-mediated pathway.

C. Eukaryotic fatty acid synthase

The remaining series of reactions of FA synthesis in eukaryotes is catalyzed by the multifunctional, homodimeric enzyme FAS. The process involves the addition of two carbons from malonyl CoA to the carboxyl end of a series of acyl acceptors. Each FAS monomer is a multicatalytic polypeptide with six different enzymic domains plus a 4′-phosphopantetheine-containing acyl carrier protein (ACP) domain. 4′-Phosphopantetheine, a derivative of pantothenic acid (vitamin B_5; see Chapter 28), carries acyl units on its terminal thiol (–SH) group and presents them to the catalytic domains of FAS during FA synthesis. It also is a component of CoA. The reaction numbers in brackets below refer to Figure 16.9.

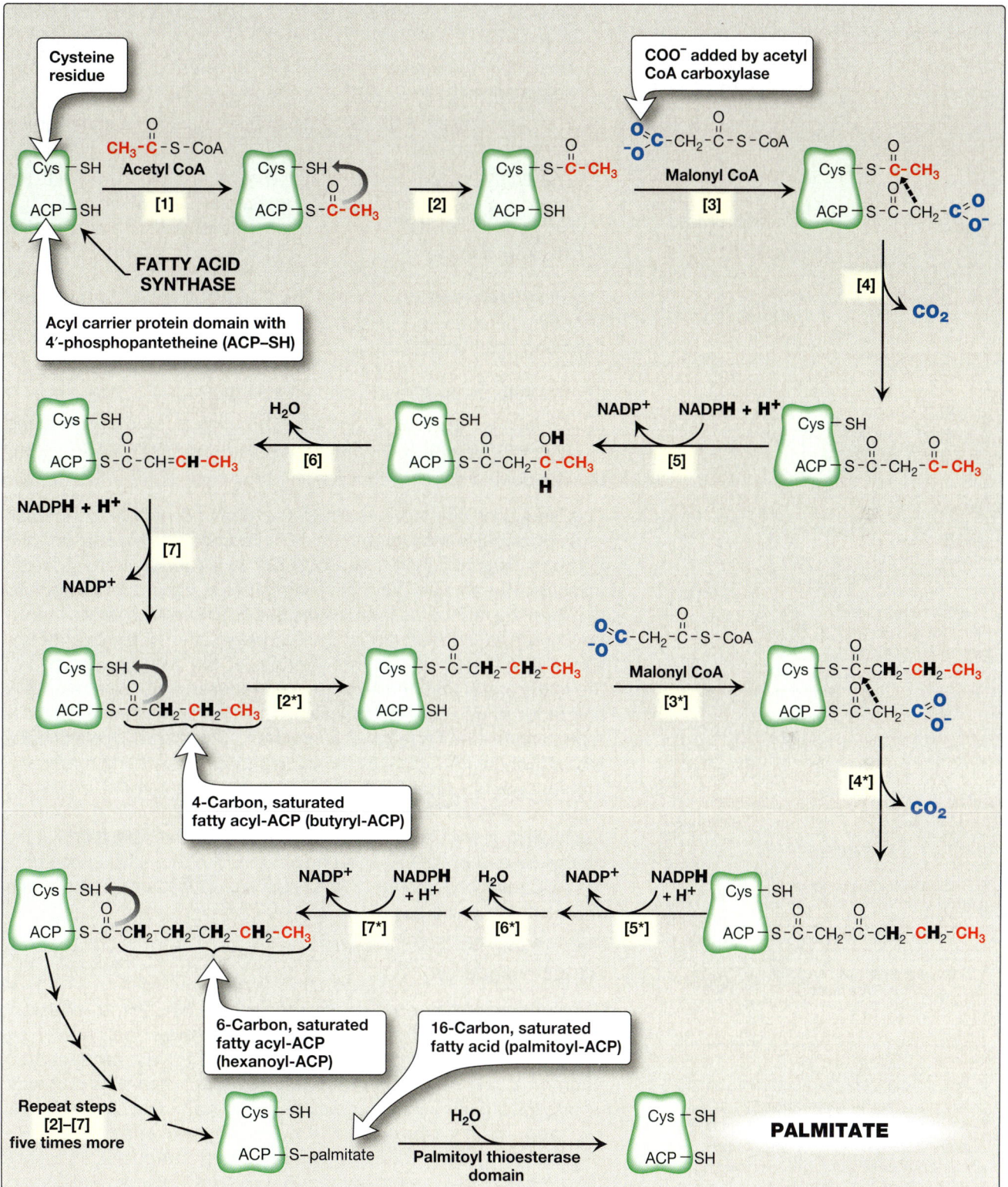

Figure 16.9
Synthesis of palmitate (16:0) by multifunctional fatty acid synthase. [Note: Numbers in brackets correspond to bracketed numbers in the text. A second repetition of the steps is indicated by numbers with an *asterisk* [*]. Carbons provided directly by acetyl coenzyme A (CoA) are shown in *red*.] ACP, acyl carrier protein domain; CO_2, carbon dioxide; NADP(H), nicotinamide adenine dinucleotide phosphate.

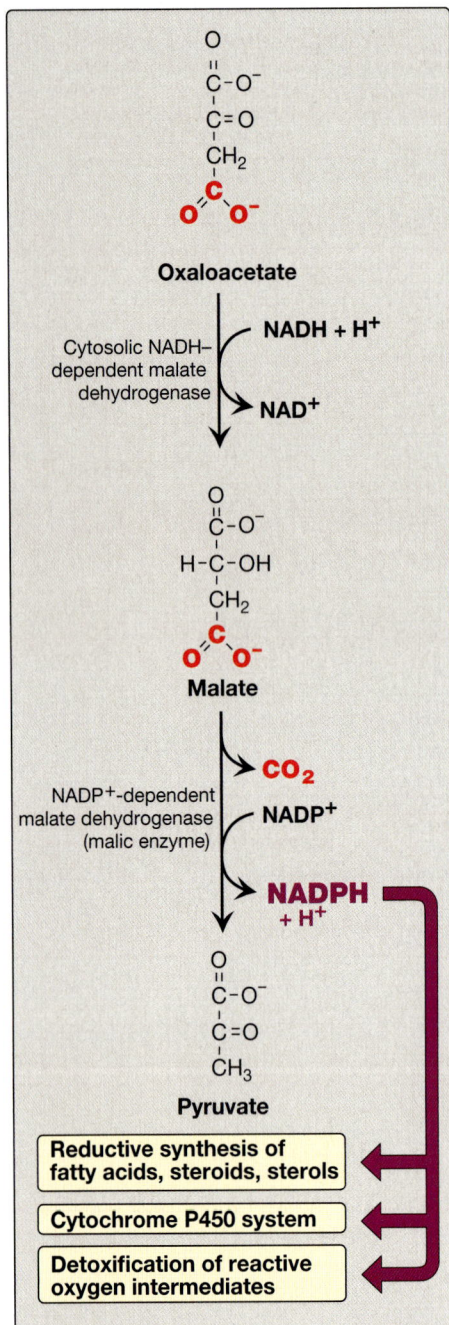

Figure 16.10
Cytosolic conversion of oxaloacetate to pyruvate with the generation of nicotinamide adenine dinucleotide phosphate (NADPH). [Note: The pentose phosphate pathway is also a source of NADPH.] NAD(H), nicotinamide adenine dinucleotide; CO_2, carbon dioxide.

(1) An acetyl group is transferred from acetyl CoA to the –SH group of the ACP.

(2) Next, this two-carbon fragment is transferred to a temporary holding site.

(3) The now-vacant ACP accepts a three-carbon malonyl group from malonyl CoA.

(4) The acetyl group on the cysteine residue condenses with the malonyl group on ACP with the release of CO_2, which was originally added by ACC. The result is a four-carbon unit attached to the ACP domain.

The next three reactions convert the 3-ketoacyl group to the corresponding saturated acyl group by a pair of NADPH-requiring reductions and a dehydration step.

(5) The keto group is reduced to an alcohol.

(6) A molecule of water is removed, creating a trans double bond between carbons 2 and 3 (the α- and β-carbons).

(7) The double bond is reduced.

This sequence of steps results in the production of a four-carbon group (butyryl) whose three terminal carbons are fully saturated and which remains attached to the ACP domain. The steps are repeated (indicated by an asterisk), beginning with the transfer of the butyryl unit from the ACP to the cysteine residue (2*), the attachment of a malonyl group to the ACP (3*), and the condensation of the two groups liberating CO_2 (4*). The carbonyl group at the β-carbon (carbon 3, the third carbon from the sulfur) is then reduced (5*), dehydrated (6*), and reduced (7*), generating hexanoyl-ACP. This cycle of reactions is repeated until the FA reaches a length of 16 carbons. The final catalytic activity of FAS, cleaves the thioester bond, releasing a fully saturated molecule of palmitate (16:0). [Note: All the carbons in palmitic acid have passed through malonyl CoA except the two donated by the original acetyl CoA (the first acyl acceptor), which are found at the methyl (ω) end of the FA. This underscores the rate-limiting nature of the ACC reaction.] Shorter-length FAs are produced only in the lactating mammary gland.

D. Reductant sources

The synthesis of one palmitate requires 14 NADPH, a reductant (reducing agent). The pentose phosphate pathway (see Chapter 13) is a major supplier of the NADPH. Two NADPH are produced for each molecule of glucose 6-phosphate that enters this pathway. The cytosolic conversion of malate to pyruvate, in which malate is oxidized and decarboxylated by cytosolic malic enzyme (NADP+-dependent malate dehydrogenase), also produces cytosolic NADPH (and CO_2), as shown in Figure 16.10. [Note: Malate can arise from the reduction of OAA by cytosolic NADH-dependent malate dehydrogenase (see Fig. 16.10). One source of the cytosolic NADH required for this reaction is glycolysis. OAA, in turn, can arise from citrate cleavage by ATP citrate lyase.] A summary of the interrelationship between glucose metabolism and palmitate synthesis is shown in Figure 16.11.

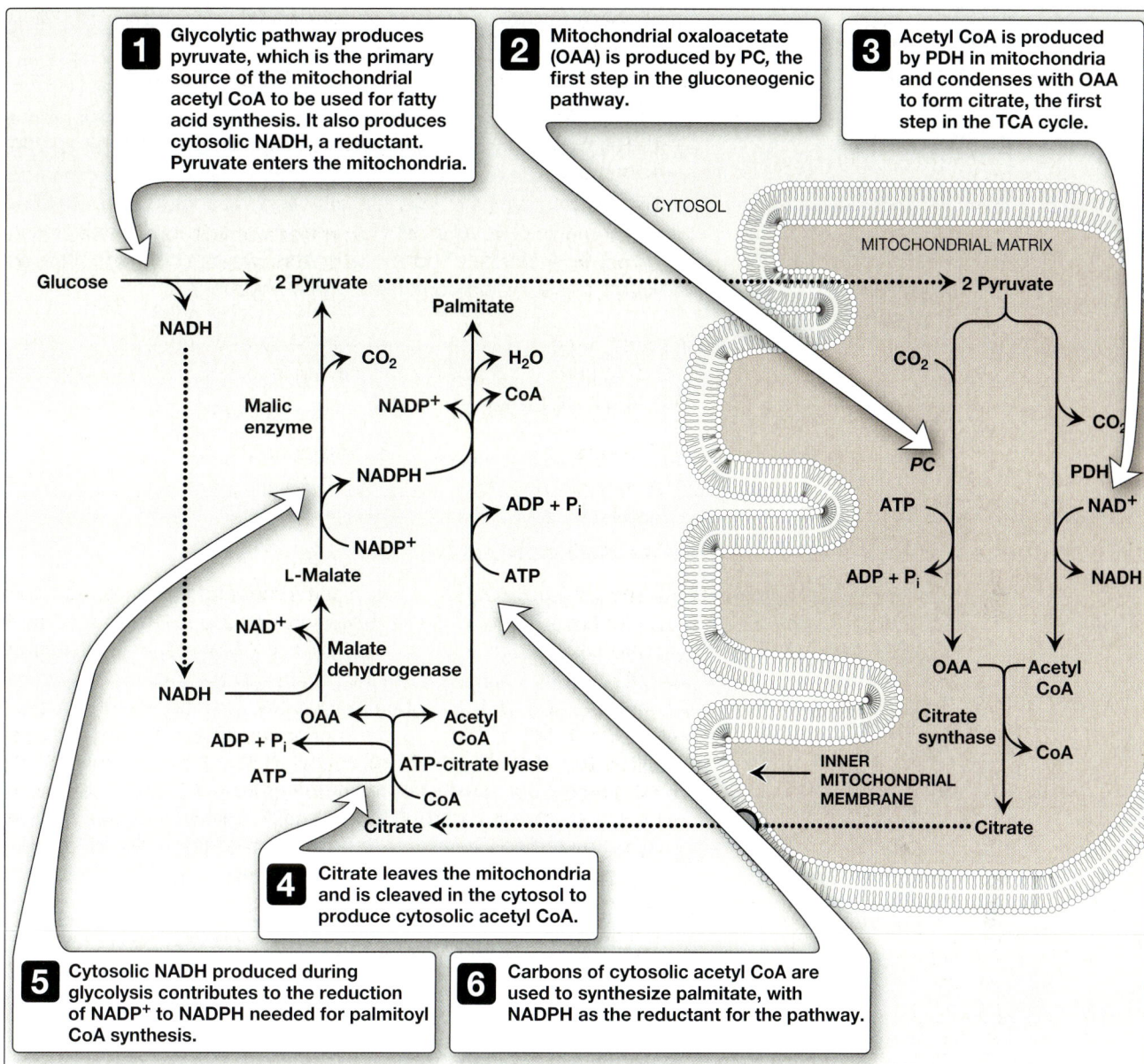

Figure 16.11
Interrelationship between glucose metabolism and palmitate synthesis. CoA, coenzyme A; NAD(H), nicotinamide adenine nucleotide; NADP(H), nicotinamide adenine dinucleotide phosphate; ADP, adenosine diphosphate; P_i, inorganic phosphate; CO_2, carbon dioxide; TCA, tricarboxylic acid; PC, pyruvate carboxylase; PDH, pyruvate dehydrogenase.

E. Further elongation

Although palmitate, a 16-carbon, fully saturated LCFA (16:0), is the primary end product of FAS activity, it can be further elongated by the addition of two-carbon units to the carboxylate end primarily in the smooth endoplasmic reticulum (sER). Elongation requires a system of separate enzymes rather than a multifunctional enzyme. Malonyl CoA is the two-carbon donor, and NADPH supplies the electrons. The brain has additional elongation capabilities, allowing it to produce the very–long-chain FAs ([VLCFAs] >22 carbons) that are required for synthesis of brain lipids.

F. Chain desaturation

Enzymes (fatty acyl CoA desaturases) also present in the sER are responsible for desaturating LCFA (i.e., adding *cis* double bonds). The desaturation reactions require oxygen (O_2), NADH, cytochrome b_5, and its flavin adenine dinucleotide (FAD)-linked reductase. The FA and the NADH get oxidized as the O_2 gets reduced to H_2O. The first double bond is typically inserted between carbons 9 and 10, producing primarily oleic acid, 18:1(9), and small amounts of palmitoleic acid, 16:1(9). A variety of polyunsaturated FAs can be made through additional desaturation combined with elongation.

> Humans have carbon 9, 6, 5, and 4 desaturases but lack the ability to introduce double bonds from carbon 10 to the ω end of the chain. This is the basis for considering ω-6 linoleic acid and ω-3 linolenic acid as essential nutrients.

G. Storage as triacylglycerol components

Mono-, di-, and triacylglycerols consist of one, two, or three molecules of FA esterified to a molecule of glycerol. FAs are esterified through their carboxyl groups, resulting in a loss of negative charge and formation of neutral fat. [Note: An acylglycerol that is solid at room temperature is called a fat. If liquid, it is an oil.]

1. **Arrangement:** The three FAs esterified to a glycerol molecule to form a TAG are usually not the same type. The FA on carbon 1 is typically saturated, carbon 2 is typically unsaturated, and carbon 3 can be either. Recall that the presence of the unsaturated FAs decreases the T_m of the lipid. An example of a TAG molecule is shown in Figure 16.12.

2. **Triacylglycerol storage and function:** Because TAGs are only slightly soluble in water and cannot form stable micelles by themselves, they coalesce within white adipocytes to form large oily droplets that are nearly anhydrous. These cytosolic lipid droplets are the major energy reserve of the body. [Note: TAGs stored in brown adipocytes serve as a source of heat through nonshivering thermogenesis (see Chapter 6).]

3. **Glycerol 3-phosphate synthesis:** Glycerol 3-phosphate is the initial acceptor of FAs during TAG synthesis. There are two major pathways for its production (Fig. 16.13). [Note: A third process (glyceroneogenesis) is described in Section IV. A.3.] In both liver (primary site of TAG synthesis) and adipose tissue, glycerol 3-phosphate can be produced from glucose, first using the reactions of the glycolytic pathway to produce dihydroxyacetone phosphate (DHAP). DHAP is reduced by glycerol 3-phosphate dehydrogenase to glycerol 3-phosphate. A second pathway found in the liver, but not in adipose tissue, uses glycerol kinase to convert free glycerol to glycerol 3-phosphate (see Fig. 16.13). [Note: The glucose transporter in adipocytes (GLUT-4) is insulin dependent (see Chapter 23). Thus, when plasma glucose levels are low, adipocytes have only a limited

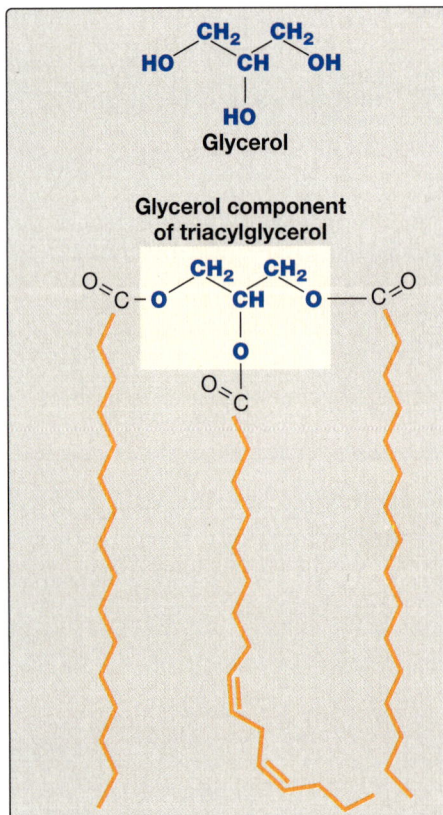

Figure 16.12
Triacylglycerol with an unsaturated fatty acid on carbon 2. *Orange* denotes the hydrophobic portions of the molecule.

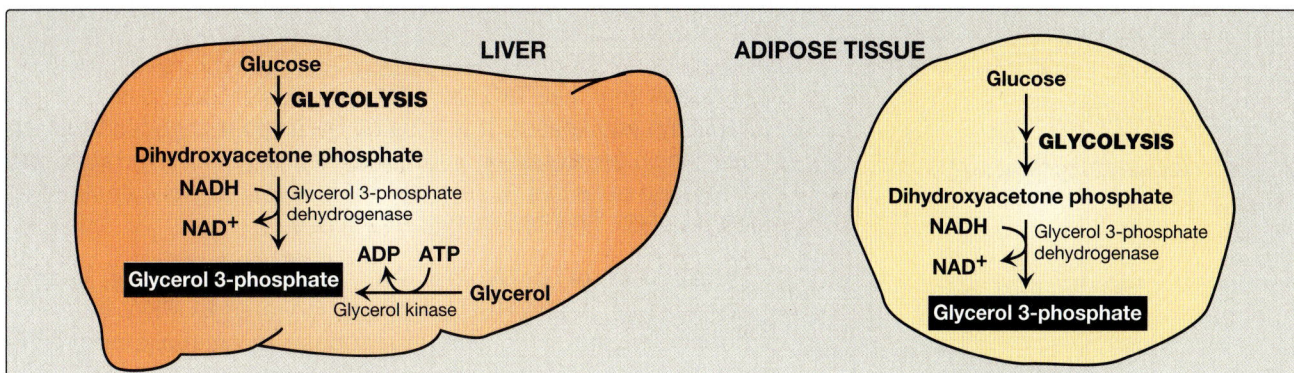

Figure 16.13
Pathways for production of glycerol 3-phosphate in liver and adipose tissue. [Note: Glycerol 3-phosphate can also be generated by glyceroneogenesis.] NAD(H), nicotinamide adenine dinucleotide; ADP, adenosine diphosphate.

ability to synthesize glycerol phosphate and cannot produce TAGs *de novo*.]

4. **Fatty acid activation:** An FFA must be converted to its activated form (bound to CoA through a thioester link) before it can participate in metabolic processes such as TAG synthesis. This reaction, illustrated in Figure 15.6A, is catalyzed by a family of fatty acyl CoA synthetases (thiokinases).

5. **Triacylglycerol synthesis:** This pathway from glycerol 3-phosphate involves four reactions, shown in Figure 16.14. These include the sequential addition of two FAs from fatty acyl CoA, the removal of phosphate, and the addition of the third FA.

H. Triacylglycerol fate in liver and adipose tissue

In WAT, TAG is stored in a nearly anhydrous form as fat droplets in the cytosol of the cells. The fat droplets are coated with a family of proteins known as perilipins that sequester and protect TAGs from lipolysis until the body requires FAs for fuel. (Perilipins may play a role in pathologic conditions such as type 2 diabetes, atherosclerosis, and cardiovascular disease.) Scant TAG is stored in a healthy liver. Instead, most are exported, packaged with other lipids and apolipoproteins to form very–low-density lipoproteins (VLDLs). Nascent VLDLs are secreted directly into the blood where they mature and function to deliver the endogenously derived lipids to the peripheral tissues. [Note: Recall from Chapter 15 that chylomicrons carry dietary (exogenously derived) lipids. Plasma lipoproteins are discussed in Chapter 18.]

IV. FAT MOBILIZATION AND FATTY ACID OXIDATION

FAs stored in WAT, in the form of neutral TAG, serve as the body's major fuel storage reserve. TAGs provide concentrated stores of metabolic energy because they are highly reduced and largely anhydrous. The yield from the complete oxidation of FAs to CO_2 and H_2O is 9 kcal/g fat (as compared to 4 kcal/g protein or carbohydrate; see Chapter 27).

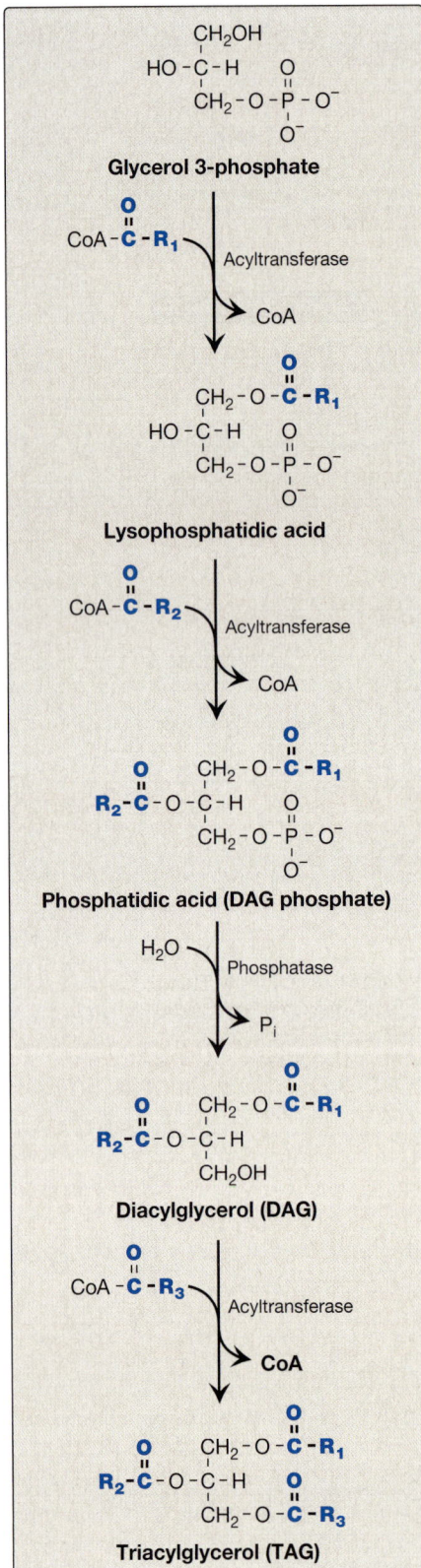

Figure 16.14
Synthesis of TAG. R1–R3, activated fatty acids. CoA, coenzyme A; P_i, inorganic phosphate.

A. Fatty acid release from fat

The mobilization of stored fat requires the hydrolytic release of FFAs and glycerol from their TAG form. This process of lipolysis is achieved by perilipins and lipases. It is initiated by adipose triglyceride lipase (ATGL), which generates a diacylglycerol that is the preferred substrate for hormone-sensitive lipase (HSL). The monoacylglycerol (MAG) product of HSL is acted on by MAG lipase.

1. **Regulation of perilipins and hormone-sensitive lipase:** Both perilipins and HSL are phosphorylated by PKA, a cAMP-dependent protein kinase, in response to catabolic hormones such as epinephrine, norepinephrine, and glucagon. cAMP is produced in adipocytes when catecholamines (such as epinephrine) bind to cell-membrane β-adrenergic receptors and activate adenylyl cyclase (Fig. 16.15). The process is similar to that of the activation of glycogen phosphorylase (see Chapter 11). Phosphorylation of perilipin by PKA allows the translocation and binding of phosphorylated HSL (active HSL) to the droplet. [Note: Because ACC is inhibited by hormone-directed phosphorylation, when the cAMP-mediated cascade is activated (see Fig. 16.8), FA synthesis is turned off, and TAG degradation is turned on.] In the presence of high plasma levels of insulin, HSL is dephosphorylated and inactivated. Insulin also suppresses expression of ATGL.

2. **Fate of glycerol:** The glycerol released during TAG degradation cannot be metabolized by adipocytes because they lack glycerol kinase. Rather, glycerol is transported through the blood to the liver, which has the kinase. The resulting glycerol 3-phosphate can be used to form TAG in the liver or can be converted to DHAP by reversal of the glycerol 3-phosphate dehydrogenase reaction illustrated in Figure 16.13. DHAP can participate in glycolysis or gluconeogenesis.

3. **Fate of fatty acids:** FFAs move through the cell membrane of the adipocyte and bind to serum albumin in the blood. They are transported to tissues such as muscle, enter cells, get activated to their CoA derivatives, and are oxidized for energy in mitochondria. Regardless of their levels, plasma FFA cannot be used for fuel by red blood cells (RBCs), which have no mitochondria. The brain does not use FAs for energy to any appreciable extent to prevent β-oxidation–associated reactive oxygen species, which are neurotoxic and have low antioxidative defense mechanisms in the brain. [Note: More than 50% of the FAs released from adipose TAG are re-esterified to glycerol 3-phosphate. WAT does not express glycerol kinase, and the glycerol 3-phosphate is produced by glyceroneogenesis, an incomplete version of gluconeogenesis: pyruvate to OAA via pyruvate carboxylase (PC) and OAA to phosphoenolpyruvate (PEP) via phosphoenolpyruvate carboxykinase. The PEP is converted (by reactions common to glycolysis and gluconeogenesis) to DHAP, which is reduced to glycerol 3-phosphate. The process decreases plasma FFAs, which are associated with insulin resistance in type 2 diabetes and obesity (see Chapter 25).]

B. Fatty acid β-oxidation

The major pathway for catabolism of FAs is a mitochondrial pathway called β-oxidation, in which two-carbon fragments are successively removed from the carboxyl end of the fatty acyl CoA, producing acetyl CoA, NADH, and $FADH_2$.

1. **Long-chain fatty acid transport into cytosol and mitochondria:** The uptake of FAs may occur using several different mechanisms. It may involve passive diffusion or one of several lipid transportation proteins, such as FA translocase (FAT), FA-binding protein (FABP) and FA transport protein (FATP). LCFAs are taken up by FATP. After an LCFA enters a cell, it is converted in the cytosol to its CoA derivative by long-chain fatty acyl CoA synthetase (thiokinase), an enzyme of the outer mitochondrial membrane. Because β-oxidation occurs in the mitochondrial matrix, the FA must be transported across the inner mitochondrial membrane that is impermeable to CoA. Therefore, a specialized carrier transports the long-chain acyl group from the cytosol into the mitochondrial matrix. This carrier is carnitine, and this rate-limiting transport process is called the carnitine shuttle (Fig. 16.16).

   a. **Translocation steps:** First, the acyl group is transferred from CoA to carnitine by carnitine palmitoyltransferase I (CPT-I), an enzyme of the outer mitochondrial membrane. [Note: CPT-I is also known as CAT-I, for carnitine acyltransferase I.] This reaction forms an acylcarnitine and regenerates free CoA. Second, the acylcarnitine is transported into the mitochondrial matrix in exchange for free carnitine by carnitine–acylcarnitine translocase. Carnitine palmitoyltransferase 2 (CPT-II, or CAT-II), an enzyme of the inner mitochondrial membrane, catalyzes the transfer of the acyl group from carnitine to CoA in the mitochondrial matrix, thus regenerating free carnitine.

   b. **Carnitine shuttle inhibitor:** Malonyl CoA inhibits CPT-I, thus preventing the entry of long-chain acyl groups into the mitochondrial matrix. Therefore, when FA synthesis is occurring in the cytosol (as indicated by the presence of malonyl CoA),

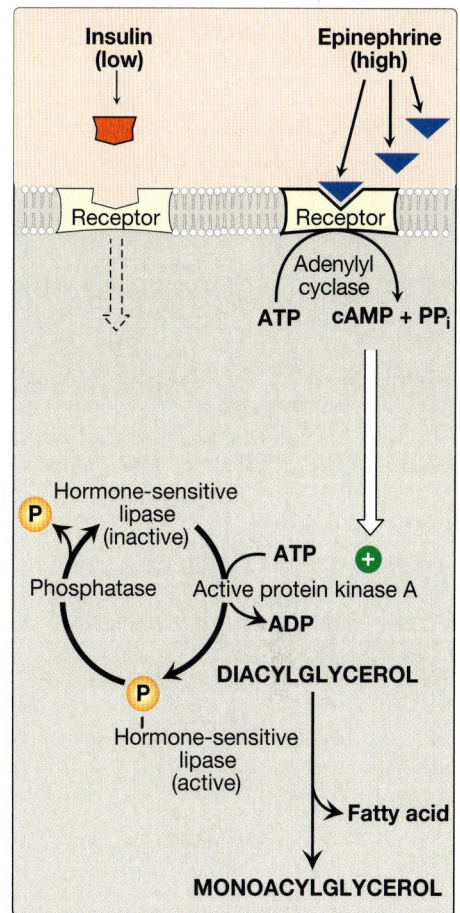

Figure 16.15
Hormonal regulation of diacylglycerol degradation in the adipocyte. [Note: Triacylglycerol is degraded to diacylglycerol by adipose triglyceride lipase.] cAMP, cyclic adenosine monophosphate; PP_i, pyrophosphate; ADP, adenosine diphosphate; P, phosphate.

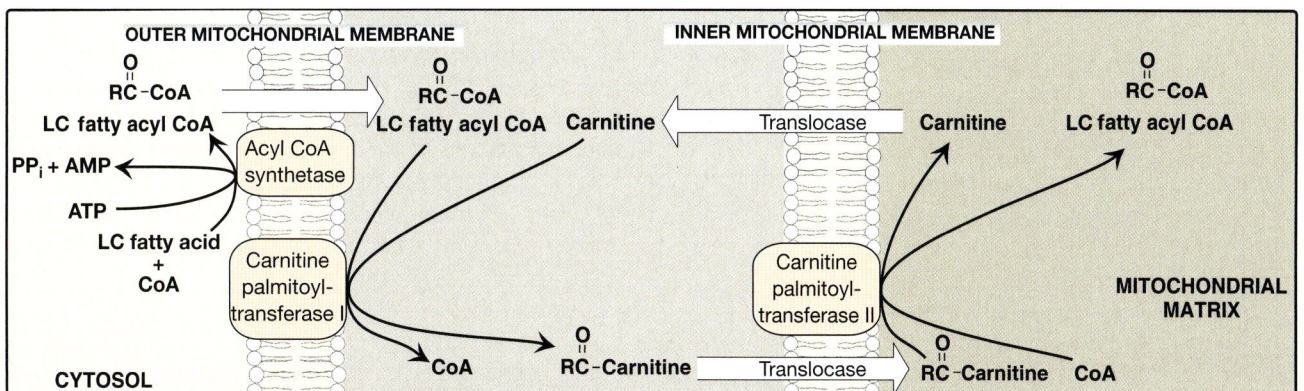

Figure 16.16
Carnitine shuttle. The net effect is that a long-chain (LC) fatty acyl coenzyme A (CoA) is transported from the outside to the inside of mitochondria. AMP, adenosine monophosphate; PP_i, pyrophosphate.

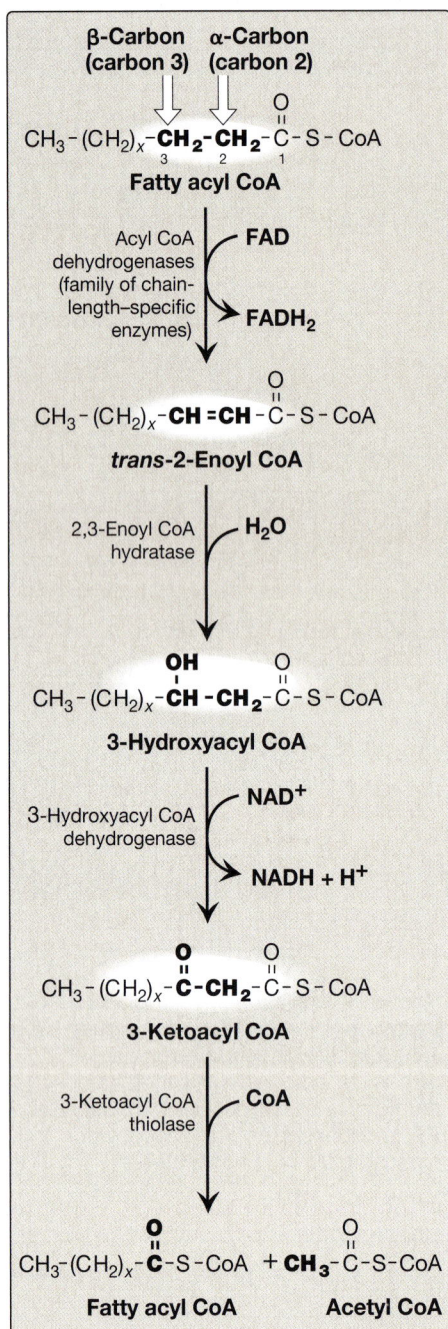

Figure 16.17
Enzymes involved in the β-oxidation of fatty acyl coenzyme A (CoA). [Note: 2,3-Enoyl CoA hydratase requires a trans double bond between carbon 2 and carbon 3.] FAD(H$_2$), flavin adenine dinucleotide; NAD(H), nicotinamide adenine dinucleotide.

the newly made palmitate cannot be transferred into mitochondria and degraded. [Note: Muscle tissue, although it does not synthesize FAs, contains the mitochondrial isozyme of ACC (ACC2), allowing regulation of β-oxidation. The liver contains both isozymes.] FA oxidation is also regulated by the acetyl CoA/CoA ratio: As the ratio increases, the CoA-requiring thiolase reaction decreases (Fig. 16.17).

c. **Carnitine sources:** Carnitine can be obtained from the diet, where it is found primarily in meat products. It can also be synthesized from the amino acids lysine and methionine by an enzymatic pathway found in the liver and kidneys but not in skeletal or cardiac muscle. Therefore, these latter tissues are totally dependent on uptake of carnitine provided by endogenous synthesis or the diet and distributed by the blood. [Note: Skeletal muscle contains ~97% of all carnitine in the body.]

Carnitine enters into cells through carnitine transporters. In heart, muscle, and kidney, the high-affinity transporter is organic cation transporter novel 2 (OCTN2). The liver has a different, low-affinity, high-capacity carnitine transporter.

2. **Shorter-chain fatty acid entry into mitochondria:** FAs 12 carbons or not more can cross the inner mitochondrial membrane without the aid of carnitine or the CPT system. Once inside the mitochondria, they are activated to their CoA derivatives by matrix enzymes and are oxidized. [Note: Medium-chain FAs are plentiful in human milk. Because their oxidation is not dependent on CPT-I, malonyl CoA is not inhibitory.]

3. **β-Oxidation reactions:** The first cycle of β-oxidation is shown in Figure 16.17. It consists of a sequence of four reactions involving the β-carbon (carbon 3) that results in shortening the FA by two carbons at the carboxylate end. The steps include an oxidation

---

Clinical Application 16.1: Carnitine Inborn Errors of Metabolism

Primary carnitine deficiency develops due to a defect in OCTN2, resulting in urinary loss of carnitine and low levels of both serum and cellular carnitine. Such deficiencies result in decreased ability of tissues to use LCFA as a fuel. Treatment includes carnitine supplementation.

Secondary carnitine deficiency primarily occurs as a result of defects in FA oxidation, leading to the accumulation of acylcarnitines that are excreted in the urine, decreasing carnitine availability. Acquired secondary carnitine deficiency can be seen, for example, in patients with liver disease (decreased carnitine synthesis) or those taking the antiseizure drug valproic acid (decreased renal reabsorption).

Defects in mitochondrial oxidation can also be caused by deficiencies in CPT-I and CPT-II. CPT-I deficiency affects the liver, where an inability to use LCFAs for fuel greatly impairs that tissue's ability to synthesize glucose (an endergonic process) during a fast. This can lead to severe hypoglycemia, coma, and death. CPT-II deficiency can affect the liver and cardiac and skeletal muscle. The most common (and least severe) form affects skeletal muscle. It presents as muscle weakness with myoglobinemia following prolonged exercise. Treatment includes avoidance of fasting and adopting a diet high in carbohydrates and low in fat but supplemented with medium-chain TAGs.

that produces FADH$_2$, a hydration, a second oxidation that produces NADH, and a CoA-dependent thiolytic cleavage that releases a molecule of acetyl CoA. Each step is catalyzed by enzymes with chain-length specificity. [Note: For LCFAs, the last three steps are catalyzed by a trifunctional protein.] These four steps are repeated for saturated FAs of even-numbered carbon chains (n/2) − 1 time (where n is the number of carbons), each cycle producing one acetyl CoA plus one NADH and one FADH$_2$. The final cycle produces two acetyl CoA. The acetyl CoA can be oxidized or used in hepatic ketogenesis (see Section V.). The reduced coenzymes are oxidized by the electron transport chain, NADH by complex I, and FADH$_2$ by coenzyme Q. [Note: Acetyl CoA is a positive allosteric effector of PC (see Chapter 10), thus linking FA oxidation and gluconeogenesis.]

4. **β-Oxidation energy yield:** The energy yield from FA β-oxidation is high. For example, the oxidation of a molecule of palmitoyl (C16) CoA to CO_2 and H_2O produces 8 acetyl CoA, 7 NADH, and 7 FADH$_2$, from which 131 ATP can be generated. However, activation of the FA requires two ATP. Therefore, the net yield from palmitate is 129 ATP (Fig. 16.18). A comparison of the processes

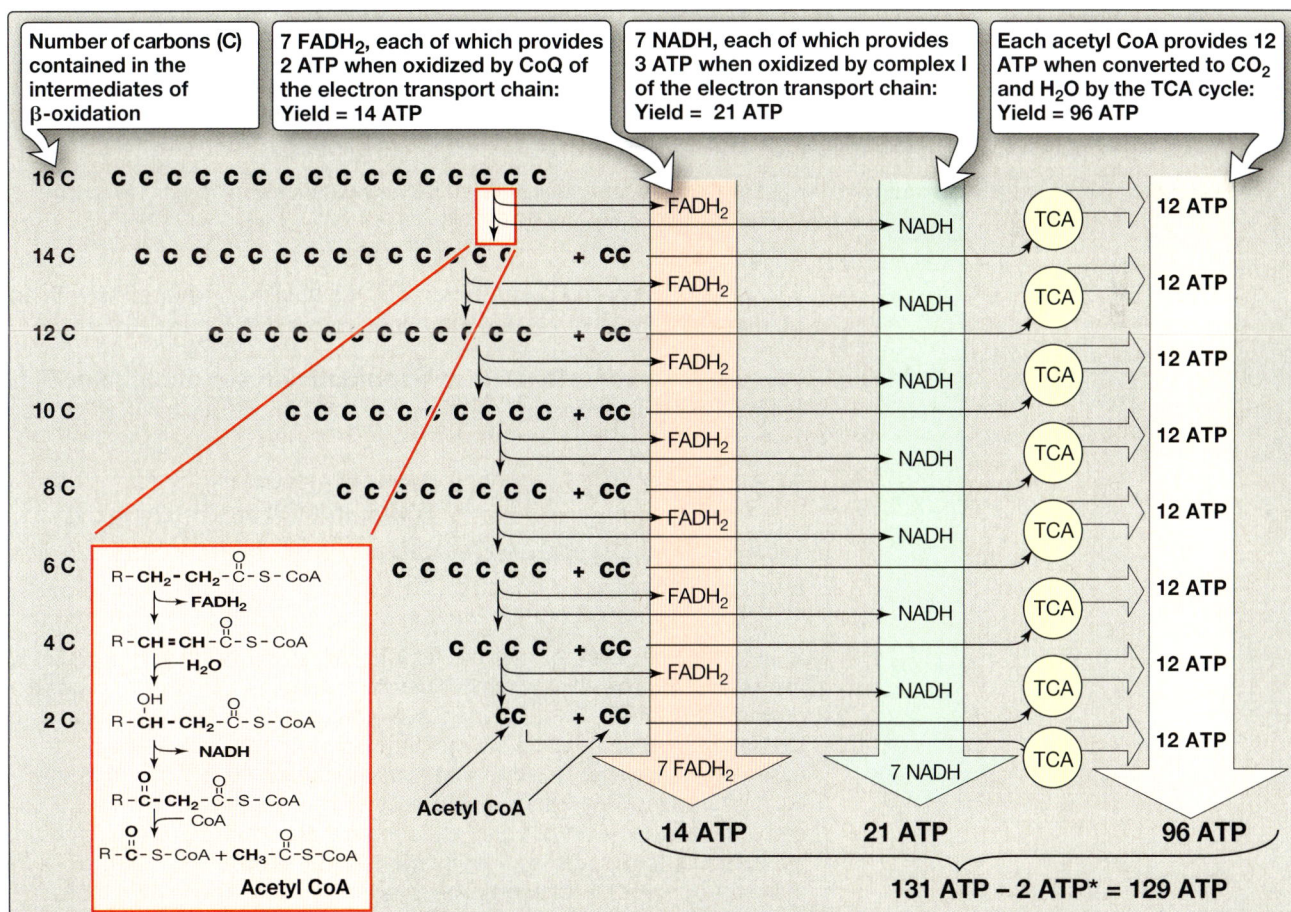

Figure 16.18
Summary of the energy yield from the oxidation of palmitoyl coenzyme A (CoA) (16 carbons). [Note: *Activation of palmitate to palmitoyl CoA requires the equivalent of 2 ATP (ATP → AMP + PP$_i$).] FADH$_2$, flavin adenine dinucleotide; NADH, nicotinamide adenine dinucleotide; TCA, tricarboxylic acid; CoQ, coenzyme Q; CO_2, carbon dioxide.

VARIABLE	SYNTHESIS	DEGRADATION
Greatest flux through pathway	After carbohydrate-rich meal	In starvation
Hormonal state favoring pathway	High insulin/glucagon ratio	Low insulin/glucagon ratio
Major tissue site	Primarily liver	Muscle, liver
Subcellular location	Cytosol	Primarily mitochondria
Carriers of acyl/acetyl groups between mitochondria and cytosol	Citrate (mitochondria to cytosol)	Carnitine (cytosol to mitochondria)
Phosphopantetheine-containing active carriers	Acyl carrier protein domain, coenzyme A	Coenzyme A
Oxidation/reduction coenzymes	NADPH (reduction)	NAD^+, FAD (oxidation)
Two-carbon donor/product	Malonyl CoA: donor of one acetyl group	Acetyl CoA: product of β-oxidation
Activator	Citrate	—
Inhibitor	Palmitoyl CoA (inhibits acetyl CoA carboxylase)	Malonyl CoA (inhibits carnitine palmitoyltransferase-I)
Product of pathway	Palmitate	Acetyl CoA
Repetitive four-step process	Condensation, reduction dehydration, reduction	Dehydrogenation, hydration dehydrogenation, thiolysis

Figure 16.19
Comparison of the synthesis and degradation of long-chain, even-numbered, saturated fatty acids. NADPH, nicotinamide adenine dinucleotide phosphate; NAD^+, nicotinamide adenine dinucleotide; FAD, flavin adenine dinucleotide; CoA, coenzyme A.

of synthesis and degradation of long-chain saturated FAs with an even number of carbon atoms is provided in Figure 16.19.

5. **Oxidation of fatty acids with an odd number of carbons:** This process proceeds by the same reaction steps as that of FAs

---

Clinical Application 16.2: Medium-Chain Fatty Acyl CoA Dehydrogenase Deficiency and Hypoketotic Hypoglycemia

In mitochondria, each of the four fatty acyl CoA dehydrogenase species has distinct but overlapping specificity for either short–, medium–, long-, or very–long-chain FAs. Deficiency in each of these dehydrogenases has been observed, however medium-chain fatty acyl CoA dehydrogenase (MCAD) deficiency is the most common inborn error of β-oxidation. MCAD deficiency, an autosomal-recessive disorder, is found in 1:14,000 births worldwide, with a higher incidence in Northern European ancestry. It results in a decreased ability to oxidize FAs with 6 to 10 carbons, decreased production of acetyl CoA, and increased reliance on glucose for energy, which causes hypoketotic hypoglycemia. Laboratory urine studies show an accumulation of medium-chain acyl carnitines and medium-chain dicarboxylic acids. Treatment includes avoidance of fasting.

with an even number of carbons, until the final three carbons are reached. This product, propionyl CoA, is metabolized by a three-step pathway (Fig. 16.20). [Note: Propionyl CoA is also produced during the metabolism of certain amino acids (see Chapter 20).]

a. **D–Methylmalonyl CoA synthesis:** First, propionyl CoA is carboxylated, forming D–methylmalonyl CoA. The enzyme propionyl CoA carboxylase has an absolute requirement for the coenzymes biotin and ATP, as do ACC and most other carboxylases.

b. **L–Methylmalonyl CoA formation:** Next, the D-isomer is converted to the L-form by the enzyme methylmalonyl CoA racemase.

c. **Succinyl CoA synthesis:** Finally, the carbons of L–methylmalonyl CoA are rearranged, forming succinyl CoA, which can enter the TCA cycle. [Note: This is the only example of a glucogenic precursor generated from FA oxidation.] The enzyme methylmalonyl CoA mutase requires a coenzyme form of vitamin B_{12} (deoxyadenosylcobalamin). The mutase reaction is one of only two reactions in the body that require vitamin B_{12} (see Chapter 28). The other vitamin B_{12}–dependent enzyme is methionine synthase, which catalyzes the synthesis of methionine from homocysteine. This reaction is essential for folate recycling and for the conversion of B_{12} into its coenzyme form.

6. **Unsaturated fatty acid β-oxidation:** The oxidation of unsaturated FAs generates intermediates that cannot serve as substrates for 2,3-enoyl CoA hydratase (see Fig. 16.17). Consequently, additional enzymes are required. Oxidation of a double bond at an odd-numbered carbon, such as 18:1(9) (oleic acid), requires one additional enzyme, 3,2-enoyl CoA isomerase, which converts the 3-*cis* derivative obtained after three rounds of β-oxidation to the 2-*trans* derivative required by the hydratase. Oxidation of a double bond at an even-numbered carbon, such as 18:2(9,12) (linoleic acid), requires an NADPH-dependent 2,4-dienoyl CoA

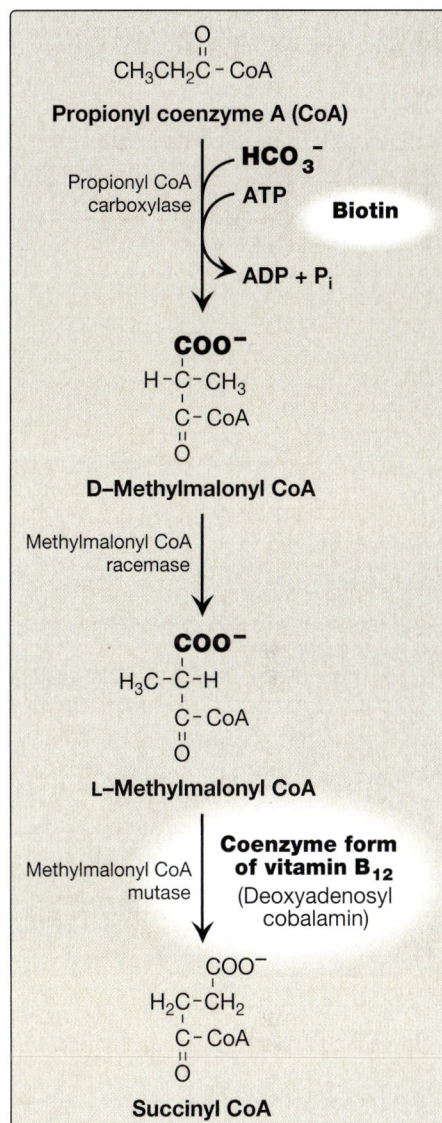

Figure 16.20
Metabolism of propionyl CoA. ADP, adenosine diphosphate; HCO_3^-, bicarbonate; P_i, inorganic phosphate.

Clinical Application 16.3: Distinguishing Between Folate and Vitamin B_{12} Deficiencies

Distinguishing between folate (vitamin B_9) and vitamin B_{12} (cobalamin) deficiencies can be challenging, because the early sign of both deficiencies is megaloblastic anemia. At later stages of vitamin B_{12} deficiency, neurologic symptoms arise, such as paresthesias, numbness, and ataxia. However, key diagnostic tests can differentiate between the deficiencies. In patients with vitamin B_{12} deficiency, both propionic and methylmalonic acid (MMA) are excreted in the urine. Elevated serum MMA is a useful diagnostic measurement to distinguish vitamin B_{12} deficiency from folate deficiency. Vitamin supplementation should correct the symptoms in either case. A heritable defect in methylmalonyl CoA mutase or methionine synthase may also produce methylmalonic acidemia and aciduria, but the toxic accumulation of these metabolites would likely occur in infancy causing symptoms such as excessive tiredness, vomiting, weak muscle tone, and acid–base imbalance.

reductase in addition to the isomerase. [Note: Because unsaturated FAs are less reduced than saturated FAs, fewer reducing equivalents are produced by their oxidation.]

7. **Peroxisomal β-oxidation:** Peroxisome is the primary site for the β-oxidation of VLCFAs with a length of 22 carbons or more. Unlike medium and long-chain FAs, which are oxidized in mitochondria, VLCFAs are exclusively oxidized in peroxisomes. After oxidation, the resulting shortened FA, linked to carnitine, diffuses to a mitochondrion for further oxidation. In contrast to mitochondrial β-oxidation, the initial dehydrogenation in peroxisomes is catalyzed by a FAD-containing acyl CoA oxidase. The $FADH_2$ produced is oxidized by O_2, resulting in the reduction of oxygen to hydrogen peroxide (H_2O_2). Importantly, no ATP is generated during this step. Subsequently, catalase (as discussed in Chapter 13) reduces H_2O_2 to H_2O.

Note that very–long-chain fatty acyl dehydrogenase (VLCAD) is distinct from acyl-CoA oxidase, which has a broader substrate specificity, oxidizing VLCFAs, branched-chain FAs, and some other lipid molecules. VLCAD primarily acts on long-chain FAs, whether they are derived from the diet or synthesized in the cytoplasm. However, VLCAD does not directly oxidize very–long-chain FAs synthesized within peroxisomes.

C. Peroxisomal α-oxidation

Branched-chain phytanic acid, a product of chlorophyll metabolism, is not a substrate for acyl CoA dehydrogenase because of the methyl group on its β-carbon (Fig. 16.21). Instead, it is hydroxylated at the α-carbon by phytanoyl CoA α-hydroxylase (PhyH); carbon 1 is released as CO_2; and the product, 15-carbon-long pristanal, is oxidized to pristanic acid, which is activated to its CoA derivative and undergoes β-oxidation. [Note: ω-Oxidation (at the methyl terminus)

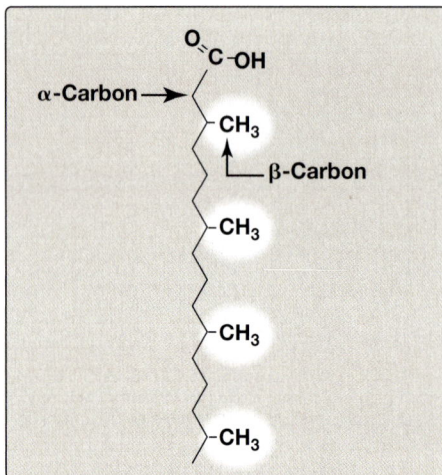

Figure 16.21
Phytanic acid, a branched-chain fatty acid 16 carbons in length.

Clinical Application 16.4: Mitochondrial Very–Long-Chain Fatty Acyl Dehydrogenase versus Peroxisomal Acyl-CoA Oxidase

Zellweger syndrome and VLCAD deficiency are rare metabolic disorders inherited in an autosomal recessive pattern that impairs the ability to break down VLCFAs, leading to their accumulation and potential health problems. Zellweger syndrome is a peroxisomal biogenesis disorder leading to accumulation of VLCFAs and other peroxisomal metabolites, which differs from VLCAD deficiency that specifically affects mitochondrial β-oxidation of VLCFAs. Both disorders cause widespread dysfunction in various organ systems, with especially severe impacts on the brain and liver. All cases of LCAD deficiencies are actually VLCAD deficiencies. Management of VLCAD deficiency involves avoiding fasting and adopting a diet that limits long-chain fats and incorporating medium-chain triacylglycerols (MCTs).

Deficiencies in peroxisomal acyl-CoA oxidase are associated with various peroxisomal disorders, such as X-linked adrenoleukodystrophy (X-ALD), which results in the accumulation of VLCFAs in tissues and can have serious neurologic and adrenal gland–related consequences.

also is known and generates dicarboxylic acids. Normally a minor pathway of the sER, its upregulation is seen with conditions such as MCAD deficiency that limit FA β-oxidation.]

> Refsum disease is a rare, autosomal-recessive disorder caused by a deficiency of peroxisomal PhyH. This results in the accumulation of phytanic acid in the plasma and tissues. The symptoms are primarily neurologic, and the treatment involves dietary restriction to halt disease progression.

V. KETONE BODIES: ALTERNATIVE FUEL FOR CELLS

Liver mitochondria have the capacity to convert acetyl CoA derived from FA oxidation into ketone bodies. The molecules categorized as ketone bodies are acetoacetate, 3-hydroxybutyrate (also called β-hydroxybutyrate), and acetone (a nonmetabolized side product; Fig. 16.22). [Note: The two functional ketone bodies are organic acids.] Acetoacetate and 3-hydroxybutyrate are transported in the blood to the peripheral tissues. There they can be reconverted to acetyl CoA, which can be oxidized by the TCA cycle. Ketone bodies are important sources of energy for the peripheral tissues because they (1) are soluble in aqueous solution and, therefore, do not need to be incorporated into lipoproteins or carried by albumin as do the other lipids; (2) are produced in the liver during periods when the amount of acetyl CoA present exceeds the oxidative capacity of the liver; and (3) are used in proportion to their concentration in the blood by extrahepatic tissues, such as skeletal and cardiac muscle, the intestinal mucosa, and the renal cortex. Even the brain can use ketone bodies to help meet its energy needs if the blood levels rise sufficiently. Thus, ketone bodies spare glucose, which is particularly important during prolonged periods of fasting (see Chapter 24).

Clinical Application 16.5: Fatty-Acid Oxidation Disorders and Hypoketotic Hypoglycemia

Fatty-acid oxidation disorders (FODs) are a group of rare genetic metabolic disorders, each caused by a specific enzyme or transport protein deficiency in the FA oxidation pathway. Some of the specific diseases within the FOD category include medium-chain acyl-CoA dehydrogenase deficiency (MCADD), very–long-chain acyl-CoA dehydrogenase deficiency (VLCADD or LCADD), short-chain acyl-CoA dehydrogenase deficiency (SCADD), and carnitine palmitoyltransferase deficiency (CPT) I and II.

Each of these specific FODs has its own unique biochemical basis and clinical presentation, but they all share the common feature of impairing the body's ability to efficiently utilize FAs for energy production, resulting in various metabolic disturbances and symptoms, including hypoketotic hypoglycemia. Treatment often involves dietary management; avoidance of fasting; and, in some cases, specific medications or supplements to support energy metabolism. Early diagnosis and intervention are crucial for managing these conditions effectively.

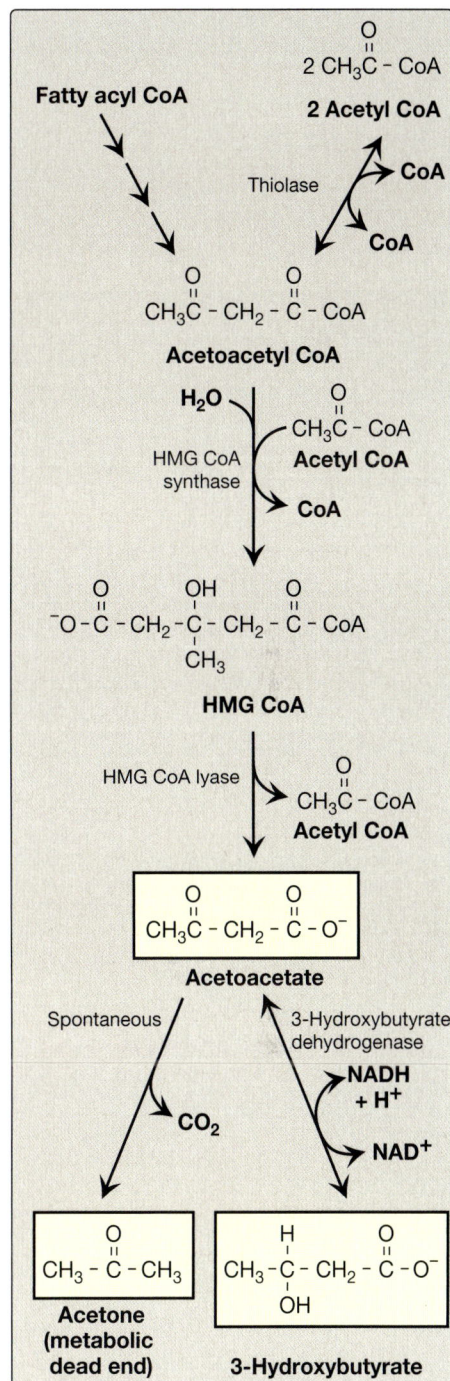

Figure 16.22
Synthesis of ketone bodies. [Note: The release of CoA in ketogenesis supports continued fatty acid oxidation.] CoA, coenzyme A; HMG, hydroxymethylglutarate; NAD(H), nicotinamide adenine dinucleotide; CO_2, carbon dioxide.

A. Ketone body synthesis by the liver: ketogenesis

During a fast, the liver is flooded with FAs mobilized from adipose tissue. The resulting elevated hepatic acetyl CoA produced by FA oxidation inhibits pyruvate dehydrogenase and activates PC. The OAA produced by PC is used by the liver for gluconeogenesis rather than for the TCA cycle. Additionally, FA oxidation decreases the NAD^+/NADH ratio, and the rise in NADH shifts OAA to malate. The decreased availability of OAA for condensation with acetyl CoA results in the increased use of acetyl CoA for ketone body synthesis. [Note: Acetyl CoA for ketogenesis is also generated by the catabolism of ketogenic amino acids.]

1. **3-Hydroxy-3-methylglutaryl CoA synthesis:** The first step, formation of acetoacetyl CoA, occurs by reversal of the final thiolase reaction of FA oxidation (see Fig. 16.17). Mitochondrial 3-hydroxy-3-methylglutaryl (HMG) CoA synthase combines a third molecule of acetyl CoA with acetoacetyl CoA to produce HMG CoA. HMG CoA synthase is the rate-limiting step in the synthesis of ketone bodies and is present in significant quantities only in the liver. [Note: HMG CoA is also an intermediate in cytosolic cholesterol synthesis. The two pathways are separated by location in, and conditions of, the cell.]

2. **Ketone body synthesis:** HMG CoA is cleaved by HMG CoA lyase to produce acetoacetate and acetyl CoA, as shown in Figure 16.22. Acetoacetate can be reduced to form 3-hydroxybutyrate with NADH as the electron donor. [Note: Because ketone bodies are not linked to CoA, they can cross the inner mitochondrial membrane.] Acetoacetate can also spontaneously decarboxylate in the blood to form acetone, a volatile, biologically nonmetabolized compound that can be detected in the breath. The equilibrium between acetoacetate and 3-hydroxybutyrate is determined by the NAD^+/NADH ratio. Because this ratio is low during FA oxidation, 3-hydroxybutyrate synthesis is favored.

B. Ketone body use by the peripheral tissues: ketolysis

Although the liver constantly synthesizes low levels of ketone bodies, their production increases during fasting when ketone bodies are needed to provide energy to the peripheral tissues. 3-Hydroxybutyrate is oxidized to acetoacetate by 3-hydroxybutyrate dehydrogenase, producing NADH (Fig. 16.23). Acetoacetate is then provided with a CoA molecule taken from succinyl CoA by succinyl CoA:acetoacetate CoA transferase (thiophorase). This reaction is reversible, but the product, acetoacetyl CoA, is actively removed by its cleavage to two acetyl CoA by thiolase. This pulls the reaction forward. Extrahepatic tissues, including the brain but excluding cells lacking mitochondria (e.g., RBCs), efficiently oxidize acetoacetate and 3-hydroxybutyrate in this manner. In contrast, although the liver actively produces ketone bodies, it lacks thiophorase (also called succinyl-CoA-3-ketoacid-CoA transferase) and, therefore, is unable to use ketone bodies as fuel.

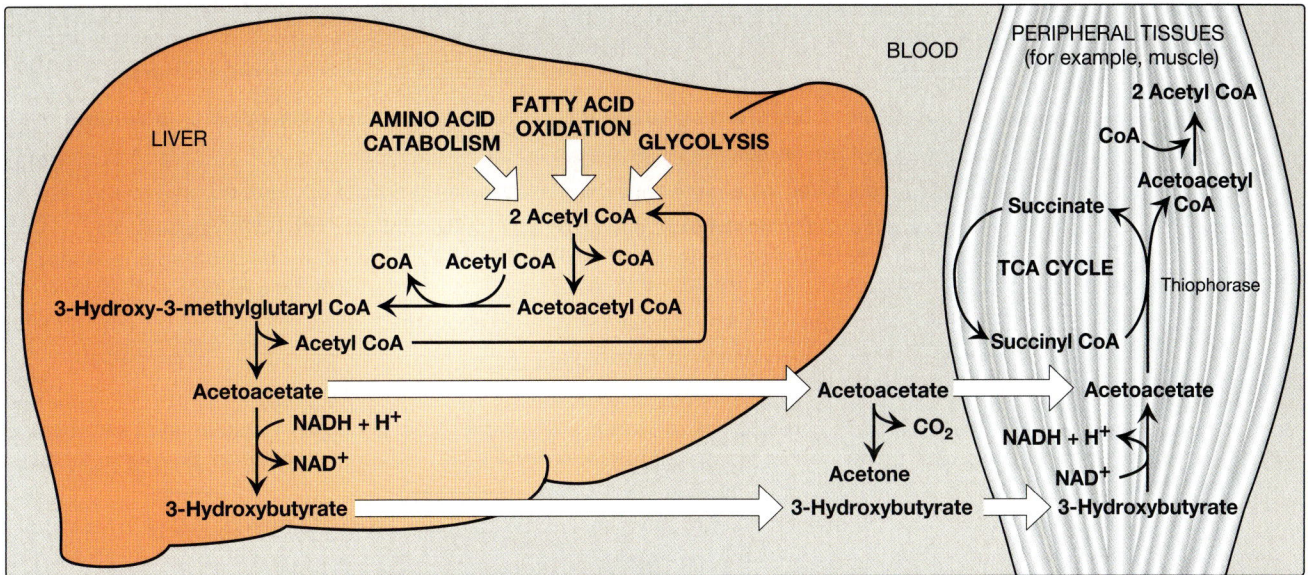

Figure 16.23
Ketone body synthesis in the liver and use in peripheral tissues. The liver and red blood cells cannot use ketone bodies. [Note: Thiophorase is also known as succinyl CoA:acetoacetate CoA transferase.] CoA, coenzyme A; NAD(H), nicotinamide adenine dinucleotide; TCA, tricarboxylic acid; CO_2, carbon dioxide.

Clinical Application 16.6: Diabetic Ketoacidosis

When the rate of formation of ketone bodies is greater than the rate of their use, their levels begin to rise in the blood (ketonemia) and, eventually, in the urine (ketonuria). This is seen most often in cases of uncontrolled type 1 diabetes mellitus (T1D), in which the blood concentration of ketone bodies may reach 90 mg/dL (vs. <3 mg/dL in healthy individuals), and urinary excretion of ketone bodies may be as high as 5,000 mg/24 h. The elevation of the ketone body concentration in the blood can result in acidemia. [Note: The carboxyl group of a ketone body has a pKa of ~4. Therefore, each ketone body loses a proton [H^+] as it circulates in the blood, which lowers the pH. Also, in uncontrolled T1D, urinary loss of glucose and ketone bodies results in dehydration. Therefore, the increased number of H^+ circulating in a decreased volume of plasma can cause severe acidosis (ketoacidosis; Fig. 16.24) known as diabetic ketoacidosis (DKA).] A frequent symptom of DKA is a fruity breath odor, which results from increased production of acetone. Ketoacidosis may also be seen in cases of prolonged fasting and excessive ethanol consumption.

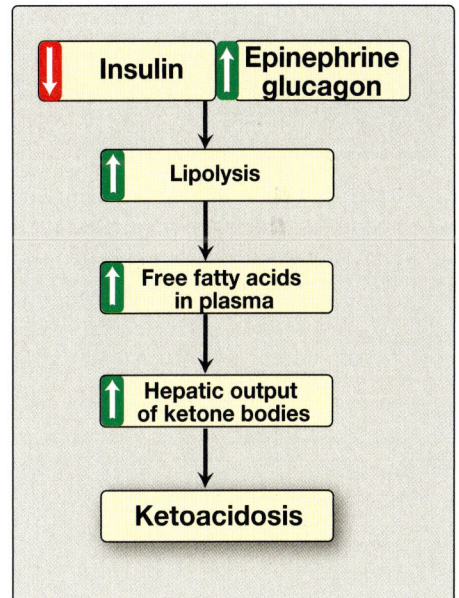

Figure 16.24
Mechanism of diabetic ketoacidosis seen in uncontrolled type 1 diabetes.

VI. CHAPTER SUMMARY

- A **fatty acid** (**FA**), generally a linear hydrocarbon chain with a terminal carboxyl group, can be **saturated** or **unsaturated**.

- Two unsaturated FAs are dietary essentials: **linoleic** and **α-linolenic acids**.

- FAs are synthesized in the **liver cytosol** following a meal containing excess carbohydrate and protein.

- Carbons used to synthesize FAs are provided by **acetyl CoA**, energy by **adenosine triphosphate** (**ATP**), and reducing equivalents by **nicotinamide adenine dinucleotide phosphate** (**NADPH**) (Fig. 16.25) provided by the **pentose phosphate pathway** and **malic enzyme**.

- **Citrate** carries two-carbon acetyl units from the mitochondrial matrix to the cytosol.

- The regulated step in FA synthesis is the carboxylation of acetyl CoA to **malonyl CoA** by **biotin**- and **ATP**-requiring **acetyl CoA carboxylase** (**ACC**).

- **Citrate** allosterically activates ACC, and **palmitoyl CoA** inhibits it. **ACC** can also be activated by **insulin** and inactivated by **adenosine monophosphate** (**AMP**)-**activated protein kinase** (**AMPK**) in response to **epinephrine**, **glucagon**, or a rise in **AMP**.

- The remaining steps in FA synthesis are catalyzed **fatty acid synthase** (**FAS**), which produces **palmitoyl CoA** by adding two-carbon units from malonyl CoA to a series of acyl acceptors.

- FAs can be **elongated** and **desaturated** in the **smooth endoplasmic reticulum** (**sER**).

- When FAs are required for energy, hormone-sensitive lipase (**activated by epinephrine**, and **inhibited by insulin**), along with other lipases, degrades **triacylglycerols** (**TAGs**) stored in **adipocytes**.

- The FAs are carried by **serum albumin** to the liver and peripheral tissues, where their oxidation provides energy. The **glycerol** backbone of the degraded TAG is carried by the blood to the **liver**, where it serves as a **gluconeogenic precursor**.

- FA degradation (**β-oxidation**) occurs in **mitochondria**.

- The **carnitine shuttle** is required to transport long-chain FAs from the cytosol to the mitochondrial matrix. Carnitine palmitoyltransferase-I is inhibited by **malonyl CoA**, thereby preventing simultaneous synthesis and degradation of FAs.

- Mitochondrial FA β-oxidation produces **acetyl CoA**, **NADH**, and **flavin adenine dinucleotide** (**FADH$_2$**).

- The first step in β-oxidation is catalyzed by one of four acyl CoA dehydrogenases, each with chain-length specificity.

- **Medium-chain fatty acyl coenzyme A dehydrogenase deficiency** causes a decrease in FA oxidation resulting in **hypoketonemia** and severe **hypoglycemia**.

- Oxidation of FAs with an **odd number** of carbons produces **propionyl CoA**, which is carboxylated to **methylmalonyl CoA** (by **biotin**- and **ATP-requiring propionyl CoA carboxylase**), which is then converted to **succinyl CoA** (gluconeogenic precursor) by **vitamin B$_{12}$–requiring methylmalonyl CoA mutase**.

- A genetic error in the mutase or vitamin B$_{12}$ deficiency causes **methylmalonic acidemia** and **aciduria**. β-Oxidation of **unsaturated** FAs requires additional enzymes.

- **ω-Oxidation**, normally a minor pathway, occurs in the sER.

- Liver mitochondria can convert acetyl CoA derived from FA oxidation into **acetoacetate** and **3-hydroxybutyrate** (**ketone bodies**).

- Peripheral tissues possessing mitochondria can oxidize 3-hydroxybutyrate to acetoacetate, which can be cleaved to two acetyl CoA, thereby producing energy for the cell.

- Unlike FAs, ketone bodies are utilized by the **brain** and, therefore, are important fuels during a fast.

- Because the liver lacks **thiophorase** required to degrade ketone bodies, it synthesizes them specifically for the peripheral tissues.

- **Ketoacidosis** occurs when the rate of ketone body formation is greater than the rate of use, as is seen in cases of uncontrolled **type 1 diabetes mellitus**.

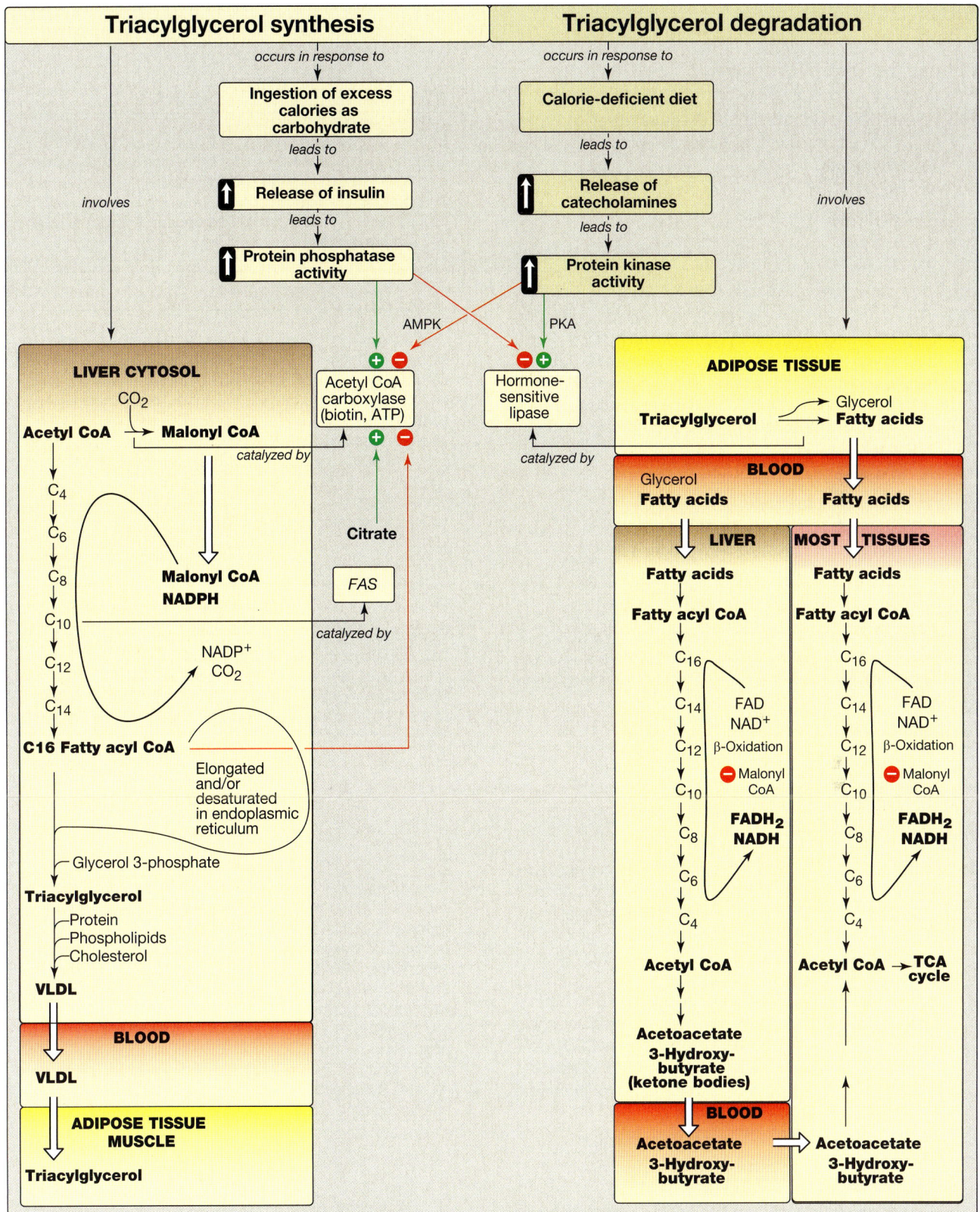

Figure 16.25
Key concept map for fatty acid and triacylglycerol metabolism. AMPK, adenosine monophosphate–activated protein kinase; PKA, protein kinase A; CoA, coenzyme A; NADP(H), nicotinamide adenine dinucleotide phosphate; FAD(H$_2$), flavin adenine dinucleotide; FAS, fatty acid synthase; CO$_2$, carbon dioxide; NAD(H), nicotinamide adenine dinucleotide; TCA, tricarboxylic acid; VLDL, very–low-density lipoprotein.

Study Questions

Choose the ONE best answer.

16.1. When oleic acid, 18:1(9), is desaturated at carbon 6 and then elongated, what is the correct representation of the product?

    A. 19:2(7,9)
    B. 20:2(ω-6)
    C. 20:2(6,9)
    D. 20:2(8,11)

> Correct answer = D. Fatty acids are elongated in the smooth endoplasmic reticulum by adding two carbons at a time to the carboxylate end (carbon 1) of the molecule. This pushes the double bonds at carbon 6 and carbon 9 farther away from carbon 1. The 20:2(8,11) product is an ω-9 (n-9) fatty acid.

16.2. A 4-month-old child is being evaluated for fasting hypoglycemia. Laboratory tests at admission reveal low levels of ketone bodies (hypoketonemia), free carnitine, and long-chain acylcarnitines in the blood. Free fatty acid levels in the blood were elevated. Deficiency of which of the following would best explain these findings?

    A. Adipose triglyceride lipase
    B. Carnitine transporter
    C. Carnitine palmitoyltransferase-I
    D. Long-chain fatty acid dehydrogenase

> Correct answer = B. A defect in the carnitine transporter (primary carnitine deficiency) would result in low levels of carnitine in the blood (as a result of increased urinary loss) and low levels in the tissues. In the liver, this decreases fatty acid (FA) oxidation and ketogenesis. Consequently, blood levels of free FAs rise. Deficiencies of adipose triglyceride lipase would decrease FA availability. Deficiency of carnitine palmitoyltransferase I would result in elevated blood carnitine. Defects in any of the enzymes of β-oxidation would result in secondary carnitine deficiency, with a rise in acylcarnitines.

16.3. A teenager, concerned about his weight, attempts to maintain a fat-free diet for a period of several weeks. If his ability to synthesize various lipids are examined, which of the following is most likely to be most deficient in his ability to synthesize?

    A. Cholesterol
    B. Glycolipids
    C. Phospholipids
    D. Prostaglandins
    E. Triacylglycerol

> Correct answer = D. Prostaglandins are synthesized from arachidonic acid. Arachidonic acid is synthesized from linoleic acid, an essential fatty acid obtained by humans from dietary lipids. The teenager would be able to synthesize all other compounds but, presumably, in somewhat decreased amounts.

16.4. A 6-month-old male was hospitalized following a seizure. History revealed that for several days prior, his appetite was decreased owing to a stomach virus. At admission, his blood glucose was 24 mg/dL (age-referenced normal is 60 to 100). His urine was negative for ketone bodies and positive for a variety of dicarboxylic acids. Blood carnitine levels (free and acyl bound) were normal. A tentative diagnosis of medium-chain fatty acyl coenzyme A dehydrogenase (MCAD) deficiency is made. In patients with MCAD deficiency, which of the following is most likely explanation for the fasting hypoglycemia?

    A. Decreased acetyl coenzyme A production
    B. Decreased ability to convert acetyl coenzyme A to glucose
    C. Increased conversion of acetyl coenzyme A to acetoacetate
    D. Increased production of ATP and nicotinamide adenine dinucleotide

> Correct answer = A. Impaired oxidation of fatty acids <12 carbons in length results in decreased production of acetyl-coenzyme A (CoA), which is the allosteric activator of pyruvate carboxylase, a gluconeogenic enzyme; thus, glucose levels fall. Acetyl CoA can never be used for the net synthesis of glucose. Acetoacetate is a ketone body, and with medium-chain fatty acyl CoA dehydrogenase deficiency, ketogenesis is decreased as a result of decreased production of the substrate, acetyl CoA. Impaired fatty acid oxidation means that less ATP and nicotinamide adenine dinucleotide are made, and both are needed for gluconeogenesis.

16.5. A 6-week-old female is brought to the emergency room due to hypotonia and failure to thrive. Physical examination shows dysmorphic facial features and hepatomegaly. Laboratory studies show high levels of very–long-chain fatty acids and phytanic acid. Which of the following is the most likely diagnosis?

A. X-linked adrenoleukodystrophy
B. Refsum disease
C. Zellweger syndrome
D. Very–long-chain fatty acid (VLCFA) deficiency

> Correct answer = C. Zellweger syndrome is caused by an inability to target proteins to the peroxisome. Therefore, all peroxisomal activities are affected because functional peroxisomes are not formed. As a result, laboratory studies show elevation in both VLCFAs and phytanic acid in serum. Refsum disease caused by a deficiency of peroxisomal PhyH. This results in the accumulation of phytanic acid in the plasma and tissue. In X-linked adrenoleukodystrophy, the defect is an inability to transport VLCFAs into the peroxisome, but other peroxisomal functions, such as α-oxidation, are normal.

16.6. A 30-year-old patient presents with fatigue, weakness, and paresthesias. A comprehensive metabolic panel reveals an elevated methylmalonic acid level. Further investigation shows the patient has been following a strict vegan diet for the past year. Which of the following is the most likely explanation for the elevated methylmalonic acid level in this patient?

A. Deficiency of even-numbered fatty acids
B. Deficiency of odd-numbered fatty acids
C. Deficiency of vitamin B12
D. Deficiency of vitamin B9

> Correct answer = C. Elevated methylmalonic acid levels are indicative of a deficiency in vitamin B_{12}, which is essential for the conversion of methylmalonyl-CoA to succinyl-CoA. In a vegan diet, which excludes animal-derived products, the risk for vitamin B_{12} deficiency is higher due to the absence of dietary sources of this vitamin. Answer choice B is incorrect. Odd-numbered fatty acid metabolism produces methylmalonic acid, which accumulates in the absence of vitamin B_{12}. So, the elevation of methylmalonic acid levels is primarily associated with vitamin B_{12} deficiency. Answer choice D is incorrect since green leafy vegetables, a source of folic acid (vitamin B_9), are likely plentiful in a vegan diet.

16.7. A 25-year-old patient with a history of type 1 diabetes presents to the emergency department with excessive thirst, frequent urination, and abdominal pain for the past 24 hours. The patient reports missing his last insulin injection. Which of the following mechanisms is most accurate?

A. Increased ketone production in the liver
B. Decreased lipolysis at the adipose tissue
C. Increased plasma pH
D. Decreased gluconeogenesis

> Correct answer = A. This patient's presentation is indicative of diabetic ketoacidosis (DKA), a severe complication associated with type 1 diabetes. Insufficient supply of insulin leads to an increase in the breakdown of fatty acids (lipolysis) within adipose tissue. Subsequently, the fatty acids are oxidized in the liver and produce ketone bodies, including β-hydroxybutyrate and acetoacetate. The production of these ketones reduces the serum pH and causes metabolic acidosis, which manifests as symptoms such as abdominal pain. The excessive thirst and frequent urination are symptoms of the high blood glucose due to insufficient insulin. Insulin normally inhibits gluconeogenesis in the liver.

17 Phospholipid, Glycosphingolipid, and Eicosanoid Metabolism

I. PHOSPHOLIPID OVERVIEW

Membrane lipids are composed of four major types: phospholipids, sphingolipids, glycolipids, and cholesterol. In this chapter, only the polar membrane lipids are discussed (Fig. 17.1A). Phospholipids are ionic compounds composed of an alcohol attached by a phosphodiester bond to either diacylglycerol (DAG) or sphingosine. Like fatty acids (FAs), phospholipids are amphipathic in nature. That is, each has a hydrophilic head, which is the phosphate group plus whatever alcohol is attached to it (e.g., serine, ethanolamine, and choline; highlighted in *blue* in Fig. 17.1B), and a long, hydrophobic tail containing FAs or FA-derived hydrocarbons (shown in *orange* in Fig. 17.1B). Phospholipids are the predominant lipids of cell membranes. In membranes, the hydrophobic portion of a phospholipid molecule is associated with the nonpolar portions of other membrane constituents, such as glycolipids, proteins, and cholesterol. The hydrophilic (polar) head of the phospholipid extends outward, interacting with the intracellular or extracellular aqueous environment (see Fig. 17.1B). Membrane phospholipids also function as a reservoir for intracellular messengers, and, for some proteins, phospholipids serve as anchors to cell membranes. Nonmembrane phospholipids serve additional functions in the body, for example, as components of lung surfactant and essential components of bile, where their detergent properties aid cholesterol solubilization.

II. PHOSPHOLIPID STRUCTURE

There are two classes of phospholipids: those that have glycerol (from glucose) as a backbone and those that have sphingosine (from serine and palmitate). Both classes are found as structural components of membranes, and both play a role in the generation of lipid-signaling molecules.

A. Glycerophospholipids

Phospholipids that contain glycerol are called glycerophospholipids (or phosphoglycerides). Glycerophospholipids constitute the major class of phospholipids and are the predominant lipids in membranes.

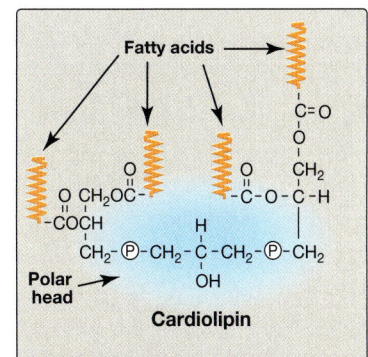

Figure 17.1
A: Polar membrane lipids. B: Structures of some glycerophospholipids. C: Phosphatidic acid. $\text{\textcircled{P}}$, phosphate (an anion).

All contain (or are derivatives of) phosphatidic acid (PA), which is DAG with a phosphate group on carbon 3 (Fig. 17.1C). Despite the apparent symmetry of the three-carbon glycerol backbone, phospholipids are directionally dependent, and C-1 is not interchangeable with C-3 of the glycerol backbone. PA is the simplest phosphoglyceride and is the precursor of the other members of this group.

1. **From phosphatidic acid and alcohol:** The phosphate group on PA can be esterified to a compound containing an alcohol group (see Fig. 17.1B). For example:

Serine	+ PA →	phosphatidylserine (PS)
Ethanolamine	+ PA →	phosphatidylethanolamine (PE)
Choline	+ PA →	phosphatidylcholine (PC) (lecithin)
Inositol	+ PA →	phosphatidylinositol (PI)
Glycerol	+ PA →	phosphatidylglycerol (PG)

2. **Cardiolipin:** Two molecules of PA esterified through their phosphate groups to an additional molecule of glycerol form cardiolipin, or diphosphatidylglycerol (Fig. 17.2). Cardiolipin is found in membranes in prokaryotes and eukaryotes. In eukaryotes, cardiolipin is virtually exclusive to the inner mitochondrial membrane, where it maintains the structure and function of certain respiratory complexes of the electron transport chain. The fatty acyl groups of cardiolipin undergo remodeling in muscle tissue so that the majority of the cardiolipin contains four linoleic acid (18:2) groups. Barth syndrome, a rare X-linked disorder caused by a defect in cardiolipin remodeling, results in cardiomyopathy, skeletal muscle weakness, growth retardation, and neutropenia during infancy, and eventually in heart failure.

3. **Plasmalogens:** When the FA at carbon 1 of a glycerophospholipid is replaced by an unsaturated alkyl group attached by an ether (rather than by an ester) linkage to the core glycerol molecule, an

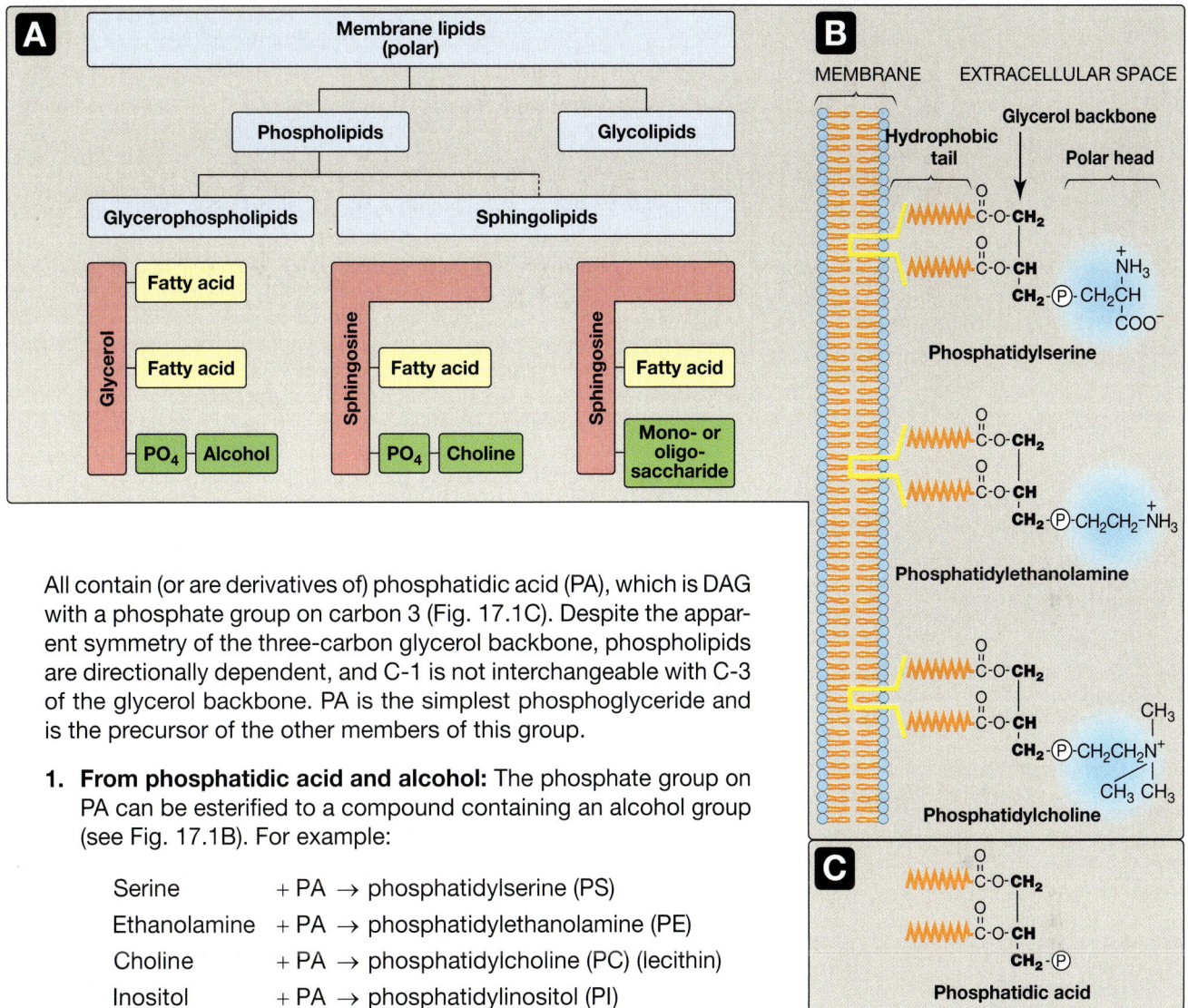

Figure 17.2
Structure of cardiolipin (diphosphatidylglycerol). $\text{\textcircled{P}}$, phosphate.

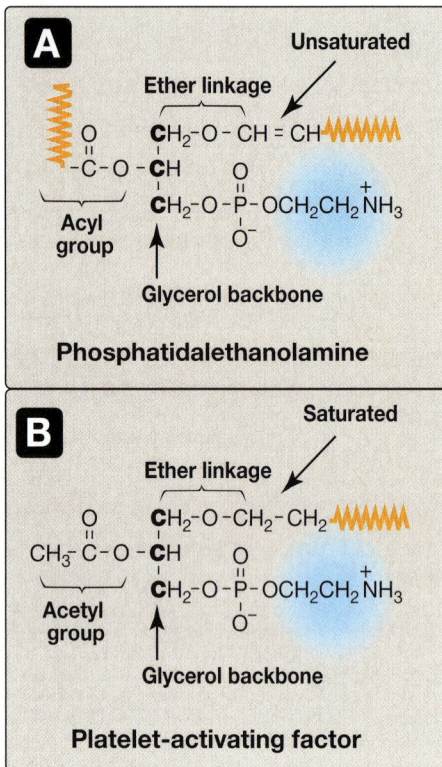

Figure 17.3
The ether glycerophospholipids. **A:** The plasmalogen phosphatidylethanolamine. **B:** Platelet-activating factor. (〰〰〰 is a long, hydrophobic hydrocarbon chain.)

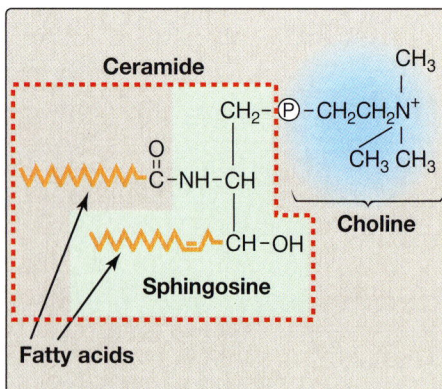

Figure 17.4
Structure of sphingomyelin, showing sphingosine (*green box*) and ceramide components (*dashed box*). Ⓟ, phosphate.

Clinical Application 17.1: Anticardiolipin Antibodies

Antibodies against cardiolipin may be generated by the immune system in a normal defense response to infection or as a consequence of an abnormal response to cardiolipin within the body. Cardiolipin antibodies are produced by the immune system in response to infection with *Treponema pallidum*, the bacterium responsible for syphilis. These antibodies are directed against cardiolipin molecules present in the outer membrane of *T. pallidum*. The Wasserman test, used to diagnose syphilis, detects anticardiolipin antibodies in a patient's serum. Normally, the human immune system is trained during early development to recognize molecules in the body as "self" and to not react to them as it would foreign antigens. However, patients with autoimmune disorders, such as systemic lupus erythematosus (SLE), develop antibodies against self-molecules, which are called autoantibodies. The presence of cardiolipin autoantibodies is a diagnostic criterion for antiphospholipid syndrome, a condition associated with a risk of blood clots that can cause deep vein thrombosis and stroke.

ether phosphoglyceride known as a plasmalogen is produced. For example, phosphatidylethanolamine (PE), which is abundant in nerve tissue (Fig. 17.3A), is the plasmalogen similar in structure to PE. Phosphatidylcholine (PC) abundant in heart muscle is the other quantitatively significant ether lipid in mammals. [Note: Plasmalogens have "al" rather than "yl" in their names.]

4. **Platelet-activating factor (PAF):** A second example of an ether glycerophospholipid is PAF, which has a saturated alkyl group in an ether link to carbon 1 and an acetyl residue (rather than a FA) at carbon 2 of the glycerol backbone (Fig. 17.3B). PAF is synthesized and released by a variety of cell types. It binds to surface receptors, triggering potent thrombotic and acute inflammatory events. For example, PAF activates inflammatory cells and mediates hypersensitivity, acute inflammatory, and anaphylactic reactions. It causes platelets to aggregate and activate and neutrophils and alveolar macrophages to generate superoxide radicals to kill bacteria (see Chapter 13). It also lowers blood pressure. PAF is one of the most potent bioactive molecules known, causing effects at concentrations as low as 10^{-11} mol/L.

B. Sphingophospholipids: sphingomyelin

The backbone of sphingomyelin is the amino alcohol sphingosine, rather than glycerol (Fig. 17.4). A long-chain FA (LCFA) is attached to the amino group of sphingosine through an amide linkage, producing a ceramide, which can also serve as a precursor of glycolipids. The alcohol group at carbon 1 of sphingosine is esterified to phosphorylcholine, producing sphingomyelin, the only significant sphingophospholipid in humans. Sphingomyelin is an important constituent of the myelin sheath of nerve fibers, and it is essential for myelin integrity and function. [Note: The myelin sheath is a layered, membranous structure that insulates and protects neuronal axons of the central nervous system (CNS). It also allows rapid neuronal conduction along axons.]

III. PHOSPHOLIPID SYNTHESIS

Glycerophospholipid synthesis involves either the donation of PA from cytidine diphosphate (CDP)-DAG to an alcohol or the donation of the phosphomonoester of the alcohol from CDP-alcohol to DAG (Fig. 17.5). In both cases, the CDP-bound structure is considered an activated intermediate, and cytidine monophosphate (CMP) is released as a side product. Therefore, a key concept in glycerophospholipid synthesis is activation, of either DAG or the alcohol to be added, by linkage with CDP. [Note: This is similar in principle to the activation of sugars by their attachment to uridine diphosphate (UDP; see Chapter 11).] The FA esterified to the glycerol alcohol groups can vary widely, contributing to the heterogeneity of this group of compounds, with saturated FA typically found at carbon 1 and unsaturated ones at carbon 2. Most phospholipids are synthesized in the smooth endoplasmic reticulum (sER). From there, they are transported to the Golgi and then to membranes of organelles or the plasma membrane or are secreted from the cell by exocytosis. [Note: Ether lipid synthesis from dihydroxyacetone phosphate begins in peroxisomes.]

A. Phosphatidic acid

PA is the precursor of other glycerophospholipids. The steps in its synthesis from glycerol 3-phosphate and two fatty acyl coenzyme A (CoA) molecules are illustrated in Figure 16.14, in which PA is shown as a precursor of triacylglycerol (TAG).

> Essentially all cells except mature erythrocytes can synthesize phospholipids, whereas TAG synthesis occurs essentially only in the liver, adipose tissue, lactating mammary glands, and intestinal mucosal cells.

B. Phosphatidylcholine and phosphatidylethanolamine

The neutral phospholipids PC and PE are the most abundant phospholipids in most eukaryotic cells. The primary route of their synthesis uses choline and ethanolamine obtained either from the diet or from the turnover of the body's phospholipids. [Note: In the liver, PC also can be synthesized from PS and PE (see Section III. B. 2.).]

1. **Synthesis from pre-existing choline and ethanolamine:** These synthetic pathways involve the phosphorylation of choline or ethanolamine by kinases, followed by conversion to the activated form, CDP-choline or CDP-ethanolamine. Finally, choline phosphate or ethanolamine phosphate is transferred from the nucleotide (leaving CMP) to a molecule of DAG (see Fig. 17.5).

   a. **Significance of choline reutilization:** The reutilization of choline is important because, although humans can synthesize choline *de novo*, the amount made is insufficient for our needs. Thus, choline is an essential dietary nutrient with an adequate intake of 550 mg for men and 425 mg for women. [Note: Choline is also used for the synthesis of acetylcholine, a neurotransmitter.] Although choline deficiency is rare, it may lead to muscle damage and nonalcoholic fatty liver disease.

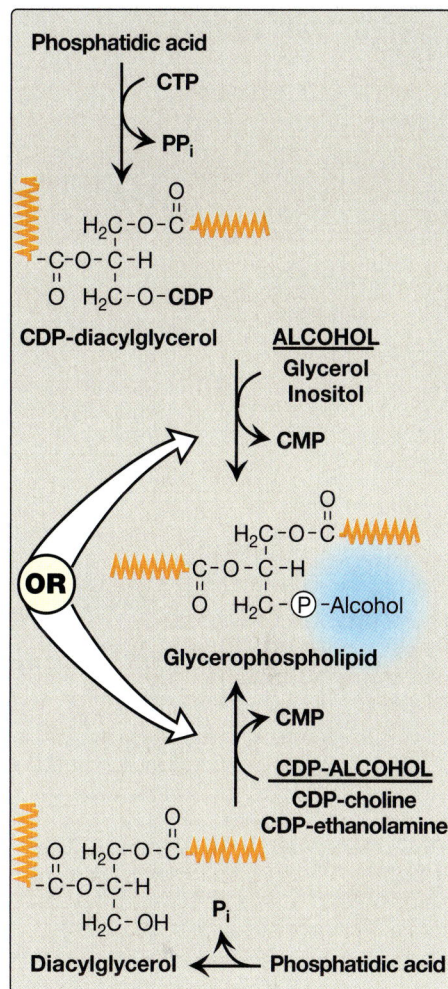

Figure 17.5
Glycerophospholipid synthesis requires activation of either diacylglycerol or an alcohol by linkage to cytidine diphosphate (CDP). CMP and CTP, cytidine mono- and triphosphates; P_i, inorganic phosphate; PP_i, pyrophosphate. (〰〰 is a fatty acid hydrocarbon chain.)

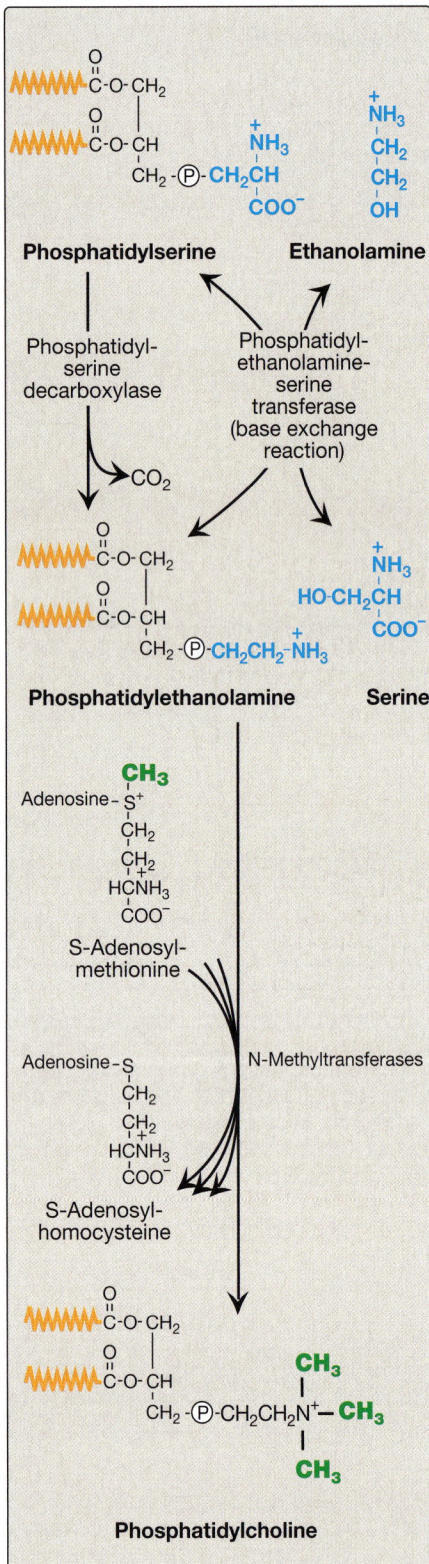

Figure 17.6
Synthesis of phosphatidylcholine from phosphatidylserine in the liver. ($\wedge\wedge\wedge\wedge\wedge$ is a fatty acid hydrocarbon chain.) (P), phosphate; CO_2, carbon dioxide.

Clinical Application 17.2: Fetal Lung Maturity and Respiratory Distress Syndrome

Fetal lung maturity is crucial for a newborn baby to breathe air. Before birth, lung surfactant must reach a sufficient concentration of DPPC (also called dipalmitoyl lecithin) to reduce surface tension within the alveoli, facilitating efficient lung function. Fetal lung maturity can be gauged by determining the lecithin/sphingomyelin (L/S) ratio in amniotic fluid. A value ≥2 is evidence of maturity because it reflects the shift from sphingomyelin to DPPC synthesis that occurs in pneumocytes at ~32 weeks' gestation. Premature infants, especially those born before 32 weeks gestation, often do not have enough surfactant production and may experience respiratory distress syndrome (RDS), a condition in which lung alveoli collapse and the infant struggles to breathe. To promote fetal lung maturation, the mother may be given glucocorticoids shortly before delivery to promote the synthesis of surfactant. Following birth, the infant may receive exogenous surfactant treatments to improve lung function. Acute RDS may occur in all age groups as the result of alveolar damage (due to infection, injury, or aspiration) that causes fluid to accumulate in the alveoli, impeding the exchange of oxygen (O_2) and carbon dioxide (CO_2).

   b. **PC in lung surfactant:** The pathway just described is the principal pathway for the synthesis of dipalmitoylphosphatidylcholine ([DPPC] or, dipalmitoyl lecithin). In DPPC, positions 1 and 2 on the glycerol are occupied by palmitate, a saturated LCFA. DPPC, made and secreted by type II pneumocytes, is a major lipid component of lung surfactant, which is the extracellular fluid layer lining the alveoli. Surfactant serves to decrease the surface tension of this fluid layer, reducing the pressure needed to reinflate alveoli, thereby preventing alveolar collapse (atelectasis). [Note: Surfactant is a complex mixture of lipids (90%) and proteins (10%), with DPPC being the major component for reducing surface tension.]

2. **PC synthesis from phosphatidylserine (PS):** The liver requires a mechanism for producing PC, even when free choline levels are low, because it exports significant amounts of PC in the bile and as a component of plasma lipoproteins. To provide the needed PC, PS is decarboxylated to PE by PS decarboxylase. PE then undergoes three methylation steps to produce PC, as illustrated in Figure 17.6. S-Adenosylmethionine is the methyl group donor (see Chapter 20).

C. Phosphatidylserine

PS synthesis in mammalian tissues is provided by the base exchange reaction, in which the ethanolamine of PE is exchanged for free serine (see Fig. 17.6). This reaction, although reversible, is used primarily to produce the PS required for membrane synthesis. PS has a net negative charge. (See Chapter 35 for the role of PS in clotting.)

D. Phosphatidylinositol

Phosphatidylinositol (PI) is synthesized from free inositol and CDP-DAG, as shown in Figure 17.5. PI is an unusual phospholipid in that it most frequently contains stearic acid on carbon 1 and arachidonic acid on carbon 2 of the glycerol. Therefore, PI serves as a reservoir of arachidonic acid in membranes and, thus, provides the substrate for prostaglandin (PG) synthesis when required. Like PS, PI has a net negative charge. [Note: There is asymmetry in the phospholipid composition of the cell membrane. PS and PI, for example, are found primarily on the inner leaflet. Asymmetry is achieved by ATP-dependent enzymes known as "flippases" and "floppases."]

@ 1. **Role in signal transduction across membranes:** The phosphorylation of membrane-bound PI produces polyphosphoinositides such as phosphatidylinositol 4,5-bisphosphate ([PIP_2] Fig. 17.7). The cleavage of PIP_2 by phospholipase C occurs in response to the binding of various neurotransmitters, hormones, and growth factors to G protein–coupled receptors (GPCRs), such as the α_1-adrenergic receptor, on the cell membrane and activation of the G_q α-subunit (Fig. 17.8). The products of this cleavage, inositol 1,4,5-trisphosphate (IP_3) and DAG, mediate the mobilization of intracellular calcium and the activation of protein kinase C, which act synergistically to evoke specific cellular responses. Signal transduction across the membrane is, thus, accomplished.

2. **Role in membrane protein anchoring:** Specific proteins can be covalently attached through a carbohydrate bridge to

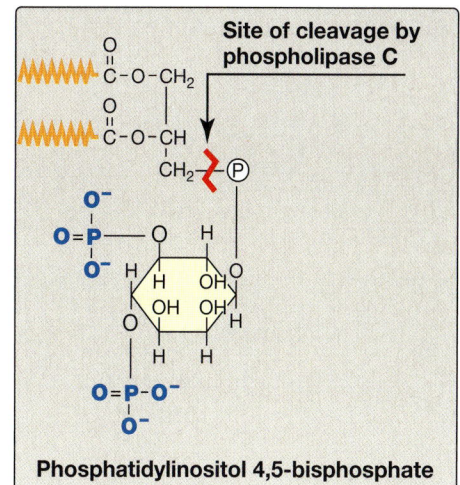

Figure 17.7
Structure of phosphatidylinositol 4,5-bisphosphate (PIP_2). Cleavage by *phospholipase C* produces inositol 1,4,5-trisphosphate (IP_3) and diacylglycerol. (〰 is a fatty acid hydrocarbon chain.) Ⓟ, phosphate.

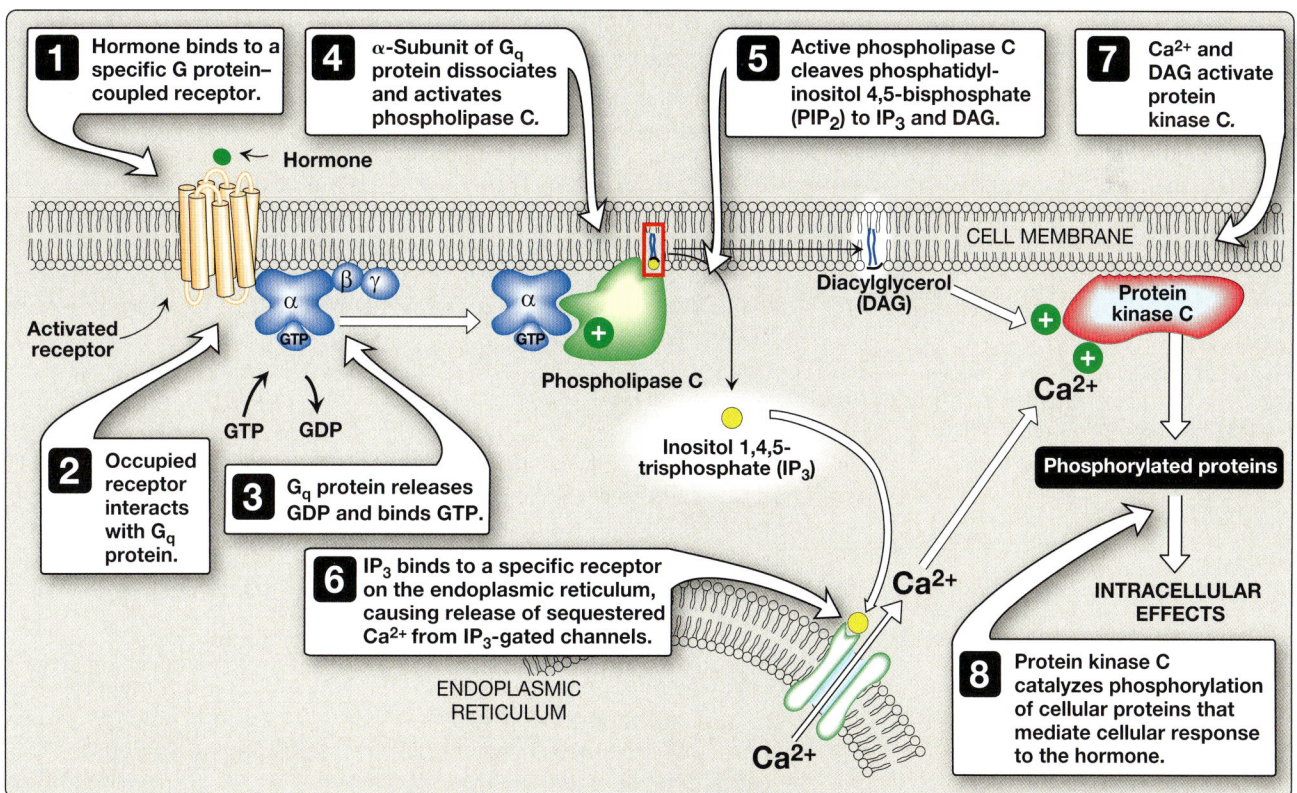

Figure 17.8
Role of inositol triphosphate and diacylglycerol in cell signaling. GDP and GTP, guanosine di- and triphosphates; Ca^{2+}, calcium.

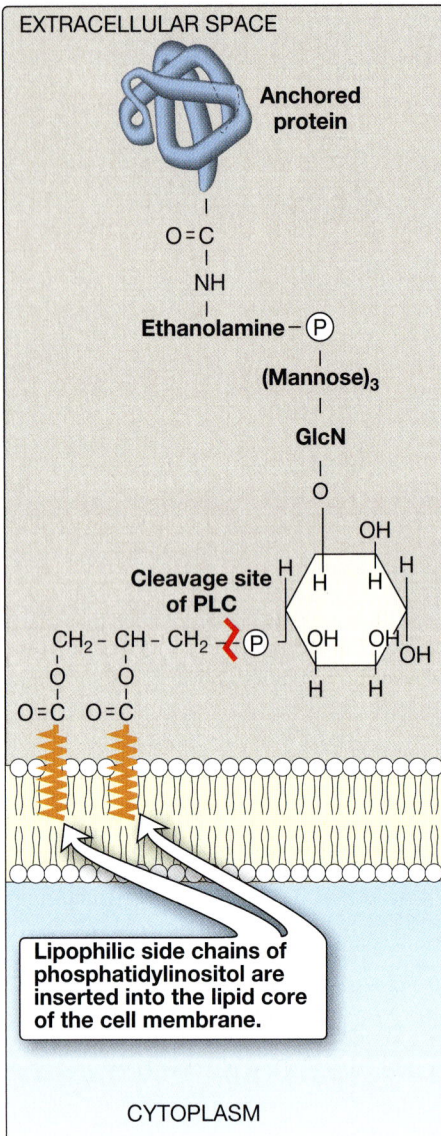

EXTRACELLULAR SPACE

Anchored protein

$O=C$

NH

Ethanolamine – (P)

(Mannose)$_3$

GlcN

O

OH

Cleavage site of PLC

$CH_2 - CH - CH_2$ (P)

O O

$O=C$ $O=C$

Lipophilic side chains of phosphatidylinositol are inserted into the lipid core of the cell membrane.

CYTOPLASM

Figure 17.9
Example of a glycosylphosphatidyli-nositol (GPI) membrane protein anchor. GlcN, glucosamine; (P), phosphate; PLC, phospholipase C.

membrane-bound PI (Fig. 17.9). For example, lipoprotein lipase, an enzyme that degrades TAG in lipoprotein particles, is attached to capillary endothelial cells by a glycosylphosphatidylinositol (GPI) anchor. [Note: GPI-linked proteins are also found in a variety of parasitic protozoans, such as trypanosomes and leishmania.] Being attached to a membrane lipid (rather than being an integral part of the membrane) allows GPI-anchored proteins to increase lateral mobility on the extracellular surface of the plasma membrane. The protein can be cleaved from its anchor by the action of phospholipase C (see Fig. 17.9).

E. Phosphatidylglycerol and cardiolipin

Phosphatidylglycerol is found in relatively large concentrations in mitochondrial membranes and is a precursor of cardiolipin (diphos-phatidylglycerol). It is synthesized from CDP-DAG and glycerol 3-phosphate. Cardiolipin (see Fig. 17.2) is synthesized by the transfer of DAG 3-phosphate from CDP-DAG to a pre-existing molecule of phosphatidylglycerol.

F. Sphingomyelin

Sphingomyelin, a sphingosine-based phospholipid, is found in cell membranes and in the myelin sheath. The synthesis of sphingo-myelin is shown in Figure 17.10. Briefly, palmitoyl CoA condenses with serine, as CoA and the carboxyl group (as CO_2) of serine are lost. [Note: This reaction, like the decarboxylation reactions involved in the synthesis of PE from PS and of regulators from amino acids (e.g., the catecholamines from tyrosine; see Chapter 21), requires pyridoxal phosphate (vitamin B_6 derivative) as a coenzyme.] The product is reduced in a nicotinamide adenine dinucleotide phosphate (NADPH)-requiring reaction to sphinganine (dihydrosphin-gosine). The sphinganine is acylated at the amino group with one of a variety of LCFAs and then desaturated to produce a ceramide, the immediate precursor of sphingomyelin (and other sphingolipids, as described in Section V.).

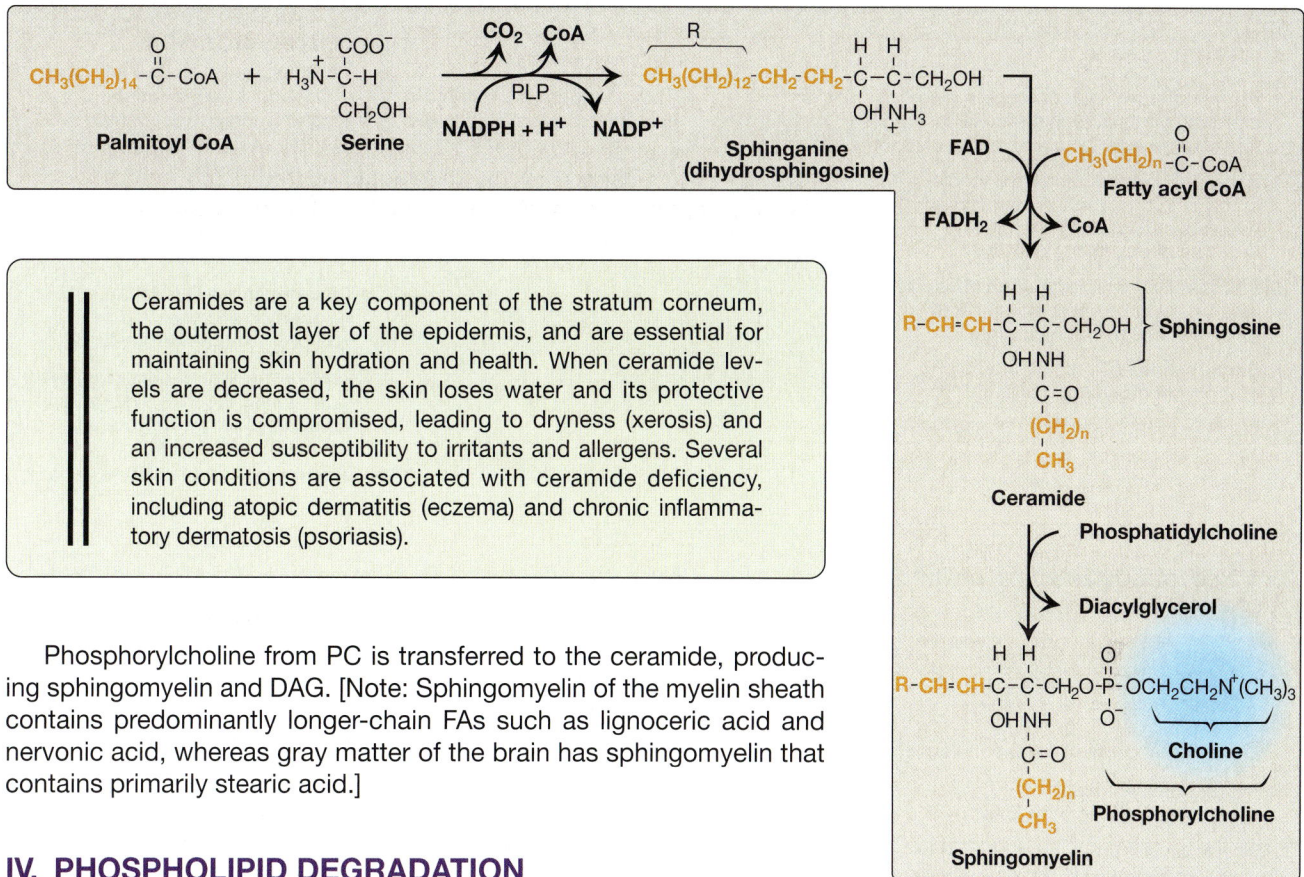

Ceramides are a key component of the stratum corneum, the outermost layer of the epidermis, and are essential for maintaining skin hydration and health. When ceramide levels are decreased, the skin loses water and its protective function is compromised, leading to dryness (xerosis) and an increased susceptibility to irritants and allergens. Several skin conditions are associated with ceramide deficiency, including atopic dermatitis (eczema) and chronic inflammatory dermatosis (psoriasis).

Phosphorylcholine from PC is transferred to the ceramide, producing sphingomyelin and DAG. [Note: Sphingomyelin of the myelin sheath contains predominantly longer-chain FAs such as lignoceric acid and nervonic acid, whereas gray matter of the brain has sphingomyelin that contains primarily stearic acid.]

IV. PHOSPHOLIPID DEGRADATION

The degradation of phosphoglycerides is performed by phospholipases found in all tissues and pancreatic juice. [Note: For a discussion of phospholipid digestion, see Chapter 15, II. C. 3] A number of toxins and venoms have phospholipase activity, and several pathogenic bacteria produce phospholipases that dissolve cell membranes and allow the spread of infection. Sphingomyelin is degraded by the lysosomal phospholipase, sphingomyelinase (see Section B.).

A. Phosphoglycerides

Phospholipases hydrolyze the phosphodiester bonds of phosphoglycerides, with each enzyme cleaving the phospholipid at a specific site. The major phospholipases are shown in Figure 17.11. [Note: Removal of the FA from carbon 1 or 2 of a phosphoglyceride produces a lysophosphoglyceride, which is the substrate for lysophospholipases.] Phospholipases release molecules that can serve as second messengers (e.g., DAG and IP_3) or that are the substrates for synthesis of messengers (e.g., arachidonic acid). Phospholipases are responsible not only for degrading phospholipids but also for remodeling them. For example, phospholipases A_1 and A_2 remove specific FAs from membrane-bound phospholipids, which can be replaced with different FAs using fatty acyl CoA transferase. This mechanism is one way to create the unique lung surfactant DPCC and to ensure that carbon 2 of PI (and sometimes of PC) is bound to arachidonic acid. A similar remodeling of FAs occurs in the cardiolipin of muscle mitochondria.

Figure 17.10
Synthesis of sphingomyelin. PLP, pyridoxal phosphate; NADP(H), nicotinamide adenine dinucleotide phosphate; FAD(H_2), flavin adenine dinucleotide; CoA, coenzyme A.

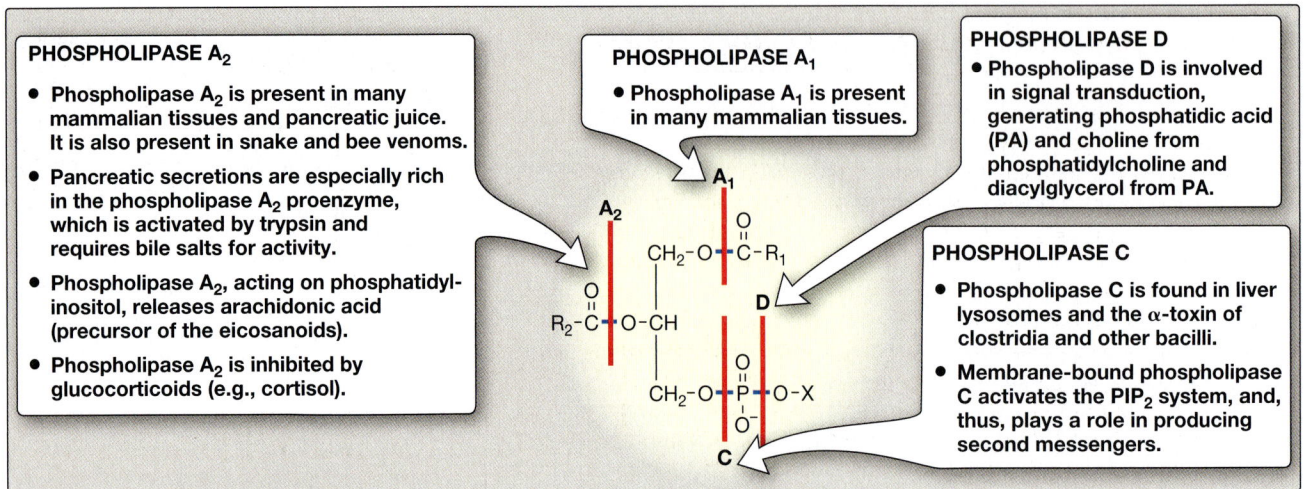

Figure 17.11
Degradation of glycerophospholipids by phospholipases. PIP_2, phosphatidylinositol 4,5-bisphosphate; R_1 and R_2, fatty acids; X, an alcohol.

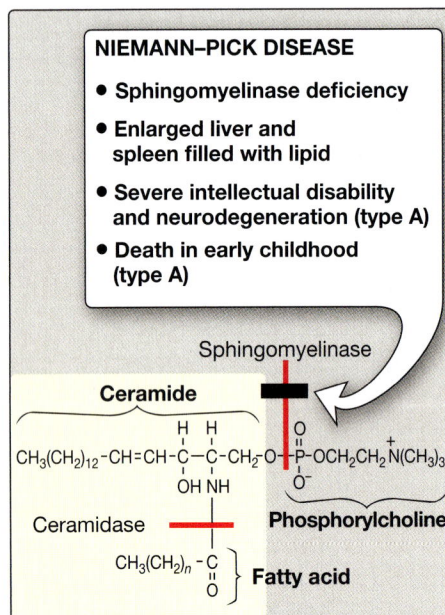

Figure 17.12
Degradation of sphingomyelin. [Note: Type B is the non-neuropathic form. It has a later age of onset and a longer survival time than type A.]

B. Sphingomyelin

Sphingomyelin is degraded by sphingomyelinase, a lysosomal enzyme that removes phosphorylcholine, leaving a ceramide. The ceramide, in turn, is cleaved by ceramidase into sphingosine and a free FA (Fig. 17.12). The released ceramide and sphingosine regulate signal transduction pathways, in part by influencing the activity of protein kinase C and, thus, the phosphorylation of its protein substrates. They also promote apoptosis. Niemann–Pick disease (types A and B) is an autosomal-recessive disorder caused by the inability to degrade sphingomyelin due to a deficiency of sphingomyelinase, a type of phospholipase C.

V. GLYCOLIPID OVERVIEW

Glycolipids are molecules that contain both carbohydrate and lipid components. Like the phospholipid sphingomyelin, glycolipids are derivatives of ceramides in which an LCFA is attached to the amino alcohol sphingosine. Therefore, they are more precisely called glycosphingolipids. [Note: Thus, ceramides are the precursors of both phosphorylated and glycosylated sphingolipids.] Like the phospholipids, glycosphingolipids are essential components of all membranes in the body, but they are found in greatest amounts in nerve tissue. They are located in the outer leaflet of the plasma membrane, where they interact with the extracellular environment. As such, they play a role in the regulation of cellular interactions (e.g., adhesion and recognition), growth, and development.

Membrane glycosphingolipids associate with cholesterol and GPI-anchored proteins to form lipid rafts, laterally mobile microdomains of the plasma membrane that function to organize and regulate membrane signaling and trafficking functions.

Clinical Application 17.4: Niemann–Pick Disease

Niemann–Pick disease is a group of autosomal-recessive disorders that fall under the category of lysosomal storage diseases.

Type A and B (Niemann–Pick disease type A/B): In the severe infantile form (type A, which shows <1% of normal enzymic activity), the liver and spleen are the primary sites of lipid deposits; therefore, hepatospleno-megaly develops. The lipid consists primarily of the sphingomyelin that cannot be degraded (Fig. 17.13). Macrophages of the reticuloendothelial system become engorged with sphingomyelin, which gives them a foamy histologic appearance. Infants with this lysosomal storage disease experience rapid and progressive neurodegeneration as a result of deposition of sphingomyelin in the CNS. A cherry-red spot in the macula of the eye develops due to lipid deposition and edema in the retinal ganglion cells. These infants die in early childhood. A less severe variant (type B, which shows up to 10% of normal activity) with a later age of onset and a longer survival time causes little to no damage to neural tissue, but lungs, spleen, liver, and bone marrow are affected, resulting in a chronic form of the disease. Although Niemann–Pick disease occurs in all ethnic groups, type A occurs with greater frequency in the Ashkenazi Jewish population.

Niemann–Pick type C: It is a distinct condition characterized by the abnormal accumulation of cholesterol and glycosphingolipids within lysosomes due to mutations in either the *NPC1* or *NPC2* gene disrupting transport of cholesterol and glycosphingolipids, such as glycosphingolipids and sphingosine from lysosomes to other cellular compartments. It is characterized by progressive neurologic symptoms including ataxia, dystonia, and cognitive decline, with onset from infancy to adulthood.

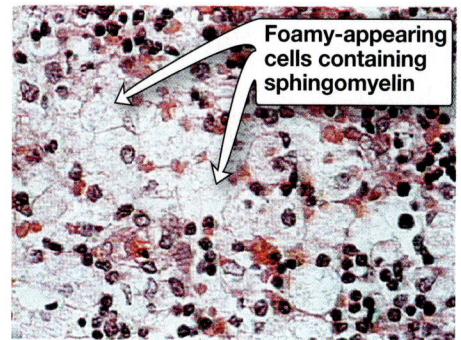

Figure 17.13
Accumulation of lipids in spleen cells from a patient with Niemann–Pick disease.

Glycosphingolipids are antigenic and are the source of ABO blood group antigens (see Chapter 14, VII. B.), various embryonic antigens specific for particular stages of fetal development, and some tumor antigens. [Note: The carbohydrate portion of a glycolipid is the antigenic determinant, and the lipid portion serves as the membrane anchor.] They have been co-opted for use as cell surface receptors for cholera and tetanus toxins as well as for certain viruses and microbes. Genetic disorders associated with an inability to properly degrade the glycosphingolipids result in lysosomal accumulation of these compounds. [Note: Changes in the carbohydrate portion of glycosphingolipids (and glycoproteins) are characteristic of transformed cells (cells with dysregulated growth).]

VI. GLYCOSPHINGOLIPID STRUCTURE

The glycosphingolipids differ from sphingomyelin in that they do not contain phosphate, and the polar head function is provided by a monosaccharide or oligosaccharide attached directly to the ceramide by an O-glycosidic bond (Fig. 17.14). The number and type of carbohydrate moieties present determine the type of glycosphingolipid.

A. Neutral glycosphingolipids

The simplest neutral glycosphingolipids are the cerebrosides. These are ceramide monosaccharides that contain either a molecule of galactose (forming ceramide-galactose or galactocerebroside, the

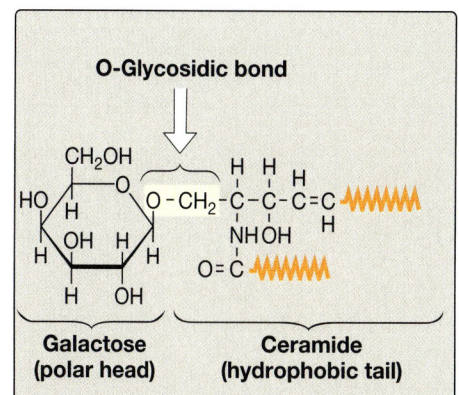

Figure 17.14
Structure of a neutral glycosphingolipid, galactocerebroside. (〰〰 is a hydrophobic hydrocarbon chain.)

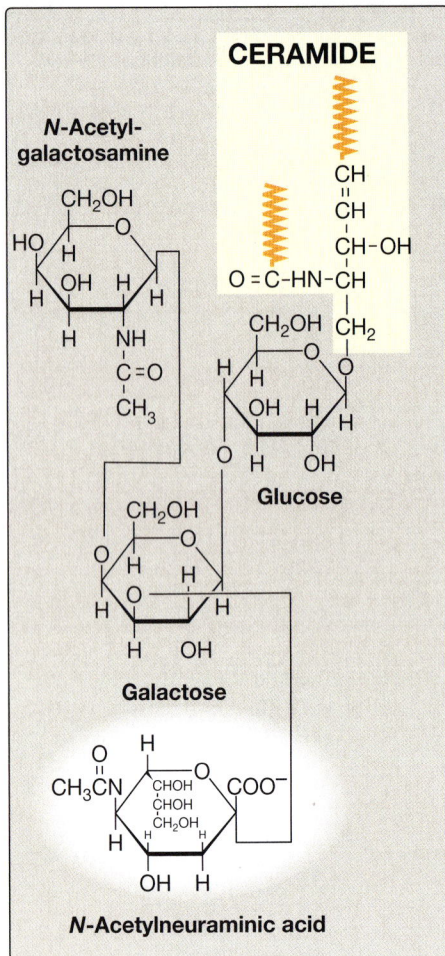

Figure 17.15
Structure of the ganglioside G_{M2}.
(〰〰〰 is a hydrophobic hydrocarbon chain.)

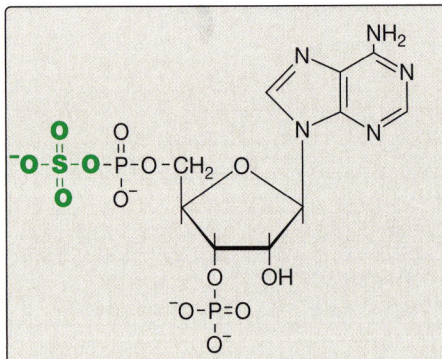

Figure 17.16
Structure of 3′-phosphoadenosine-5′-phosphosulfate.

most common cerebroside found in myelin, as shown in Fig. 17.14) or glucose (forming ceramide-glucose or glucocerebroside, an intermediate in the synthesis and degradation of the more complex glycosphingolipids). [Note: Members of a group of galacto- or glucocerebrosides may also differ from each other in the type of FA attached to the sphingosine.] As their name implies, cerebrosides are found predominantly in the brain and peripheral nerves, with high concentrations in the myelin sheath. Ceramide oligosaccharides (or globosides) are produced by attaching additional monosaccharides to a glucocerebroside, for example, ceramide–glucose–galactose (also known as lactosylceramide). The additional monosaccharides can include substituted sugars such as N-acetylgalactosamine.

B. Acidic glycosphingolipids

Acidic glycosphingolipids are negatively charged at physiologic pH. The negative charge is provided by N-acetylneuraminic acid ([NANA] a sialic acid, as shown in Fig. 17.15) in gangliosides or by sulfate groups in sulfatides.

1. **Gangliosides:** These are the most complex glycosphingolipids and are found primarily in the ganglion cells of the CNS, particularly at the nerve endings. They are derivatives of ceramide oligosaccharides and contain one or more molecules of NANA (from CMP-NANA). The notation for these compounds is G (for ganglioside) plus a subscript M, D, T, or Q to indicate whether the ganglioside contains one (mono), two (di), three (tri), or four (quatro) molecules of NANA, respectively. Additional numbers and letters in the subscript designate the monomeric sequence of the carbohydrate attached to the ceramide. (See Fig. 17.15 for the structure of G_{M2}.) Gangliosides are of medical interest because several lipid storage disorders involve the accumulation of NANA-containing glycosphingolipids in cells (see Fig. 17.19).

2. **Sulfatides:** These sulfoglycosphingolipids are sulfated galactocerebrosides that are negatively charged at physiologic pH. Sulfatides are found predominantly in the brain and kidneys.

VII. GLYCOSPHINGOLIPID SYNTHESIS AND DEGRADATION

Synthesis of glycosphingolipids occurs primarily in the Golgi by sequential addition of glycosyl monomers transferred from UDP-sugar donors to the acceptor molecule. The mechanism is similar to that used in glycoprotein synthesis (see Chapter 14).

A. Enzymes involved in synthesis

The enzymes involved in the synthesis of glycosphingolipids are glycosyltransferases that are specific for the type and location of the glycosidic bond formed. [Note: These enzymes can recognize both glycosphingolipids and glycoproteins as substrates.]

B. Sulfate group addition

A sulfate group from the sulfate carrier 3′-phosphoadenosine-5′-phosphosulfate ([PAPS] Fig. 17.16) is added by a sulfotransferase to the 3′-hydroxyl group of the galactose in a galactocerebroside, forming the sulfatide galactocerebroside 3-sulfate (Fig. 17.17). [Note: PAPS is also the sulfur donor in glycosaminoglycan synthesis and steroid hormone catabolism (see Chapter 18).] An overview of the synthesis of sphingolipids is shown in Figure 17.18.

C. Glycosphingolipid degradation

Glycosphingolipids are internalized by phagocytosis as described for the glycosaminoglycans (see Chapter 14). All of the enzymes required for the degradative process are present in lysosomes, which fuse with the phagosomes. The lysosomal enzymes hydrolytically and irreversibly cleave specific bonds in the glycosphingolipid. As seen with the glycosaminoglycans and glycoproteins, degradation is a sequential process following the rule "last on, first off," in which the last group added during synthesis is the first group removed in degradation. Therefore, defects in the degradation of the polysaccharide chains in these three glycoconjugates result in lysosomal storage diseases.

D. Sphingolipidoses

In a healthy individual, synthesis and degradation of glycosphingolipids are balanced, so that the amount of these compounds present in membranes is constant. If a specific lysosomal acid hydrolase required for degradation is partially or totally missing, a sphingolipid accumulates. Lysosomal lipid storage diseases caused by these deficiencies are called sphingolipidoses. The result of a specific acid hydrolase deficiency may be seen dramatically in nerve tissue, where neurologic deterioration can lead to early death. Figure 17.19 provides an outline of the pathway of sphingolipid degradation and descriptions of some sphingolipidoses. [Note: Some sphingolipidoses can also result from defects in lysosomal activator proteins (e.g., the saposins) that facilitate access of the hydrolases to short carbohydrate chains as degradation proceeds.] A specific lysosomal hydrolytic enzyme is deficient in the classic form of each disorder. Therefore, usually, only a single sphingolipid (the substrate for the deficient enzyme) accumulates in the involved organs in each disease. [Note: The rate of biosynthesis

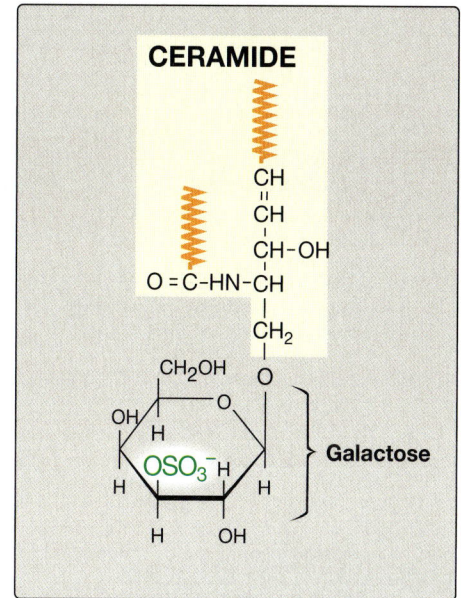

Figure 17.17
Structure of galactocerebroside 3-sulfate. (〰〰 is a hydrophobic hydrocarbon chain.)

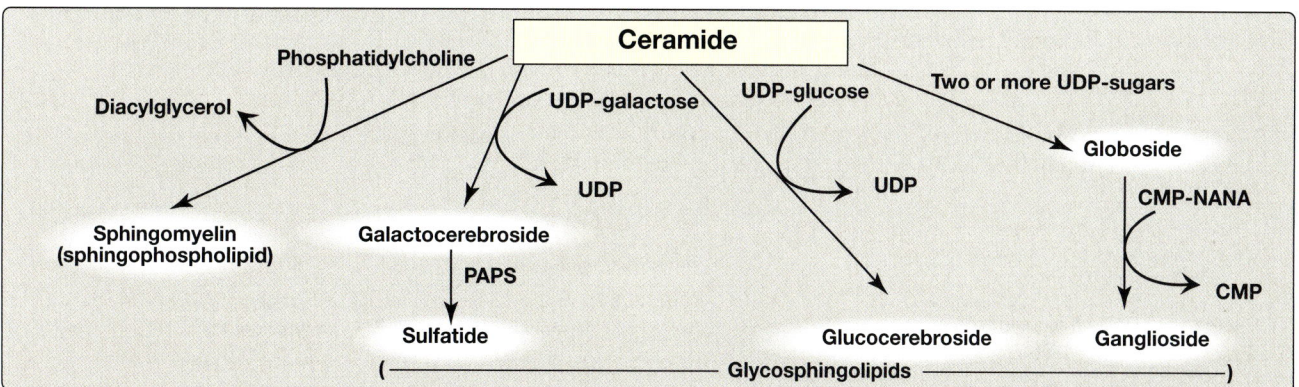

Figure 17.18
Overview of sphingolipid synthesis. UDP, uridine diphosphate; CMP, cytidine monophosphate; NANA, N-acetylneuraminic acid; PAPS, 3′-phosphoadenosine-5′-phosphosulfate.

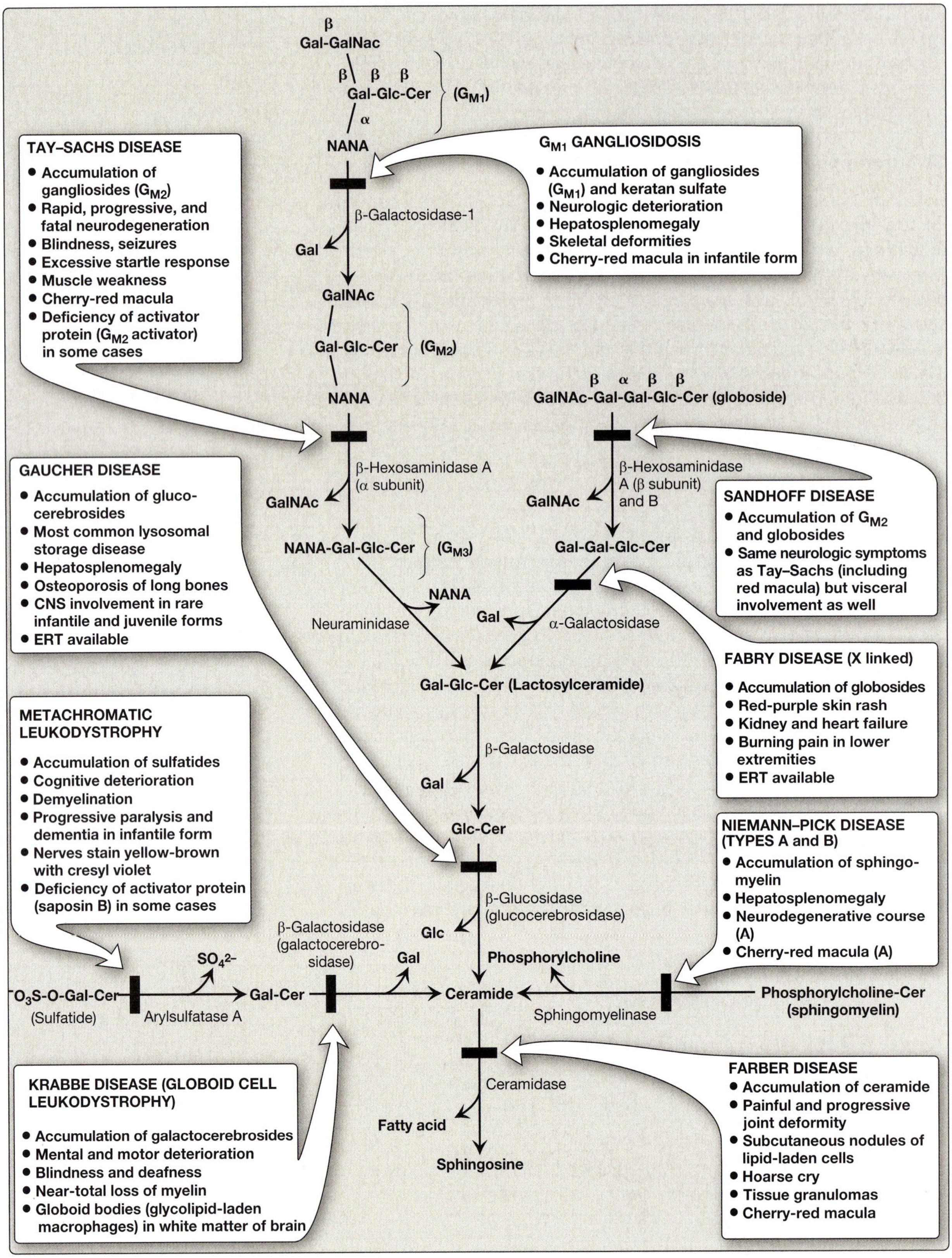

Figure 17.19
Degradation of sphingolipids showing the lysosomal enzymes affected in related genetic diseases, the sphingolipidoses. All are autosomal-recessive diseases except Fabry, which is X-linked, and all can be fatal in early life. Cer, ceramide; Gal, galactose; Glc, glucose; GalNAc, N-acetylgalactosamine; NANA, N-acetylneuraminic acid; CNS, central nervous system. SO_4^{2-}, sulfate; ERT, enzyme replacement therapy.

Clinical Application 17.5: Diagnosis and Treatment of Spingolipidoses

Spingolipidoses are diagnosed through a combination of clinical evaluation, biochemical tests, enzymatic activity assays, and genetic testing. A specific sphingolipidosis can be diagnosed by measuring enzyme activity in patient-cultured fibroblasts or peripheral leukocytes or by analyzing patient DNA. Histologic examination of the affected tissue is also useful. For example, shell-like inclusion bodies are seen in Tay–Sachs disease, and a crumpled tissue paper appearance of the cytosol is seen in Gaucher disease (Fig. 17.20). Prenatal diagnosis, using cultured amniocytes or chorionic villi, is also possible. In general, the treatments for lysosomal lipid storage diseases are largely supportive, focusing on improving quality of life and managing complications. Only a few disease-specific treatments are available, and these may be expensive. Enzyme replacement therapy (ERT) involves intravenous infusion of a recombinant form of the defective enzyme. Lysosomal storage diseases with approved ERT include Gaucher disease, Fabry disease, and Niemann–Pick disease type B. Substrate reduction therapy (SRT) for sphingolipidoses is focused on reducing the production of sphingolipids, thus alleviating the cellular burden for their degradation. For example, Gaucher disease is treated pharmacologically with miglustat, which reduces glucosylceramide, the substrate for the defective enzyme.

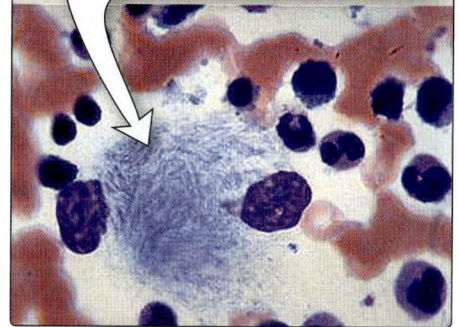

Crumpled tissue–paper appearance of the cytoplasm of Gaucher cells is caused by enlarged, elongated lysosomes filled with glucocerebroside.

Figure 17.20
Aspirated bone marrow cells from a patient with Gaucher disease.

of the accumulating lipid is normal.] The disorders are progressive and, although many are fatal in childhood, extensive phenotypic variability is seen leading to the designation of different clinical types, such as types A and B in Niemann–Pick disease. Genetic variability is also seen because a given disorder can be caused by any one of a variety of mutations within a single gene. The sphingolipidoses are autosomal-recessive disorders, except for Fabry disease, which is X-linked. The incidence of the sphingolipidoses is low in most populations, except for Gaucher and Tay–Sachs diseases, which, like Niemann–Pick disease, show a high frequency in the Ashkenazi Jewish population. [Note: Tay–Sachs disease also has a high frequency in Irish American, French Canadian, and Louisiana Cajun populations.] Metachromatic leukodystrophy, a demyelinating condition due to sulfatide accumulation, can arise from mutations in one of several genes, including arylsulfatase A (ARSA), which degrades sulfatide, or saposin B, which facilitates ARSA activity. This is an example of locus heterogeneity in the development of a disorder.

VIII. EICOSANOIDS: PROSTAGLANDINS, THROMBOXANES, AND LEUKOTRIENES

PGs, thromboxanes (TXs), and leukotrienes (LTs) are collectively known as eicosanoids to reflect their origin from ω-3 and ω-6 polyunsaturated FAs with 20 carbons (eicosa = 20). They are extremely potent compounds that elicit a wide range of responses, both physiologic (inflammatory response) and pathologic (hypersensitivity). They ensure gastric integrity and renal function, regulate smooth muscle contraction

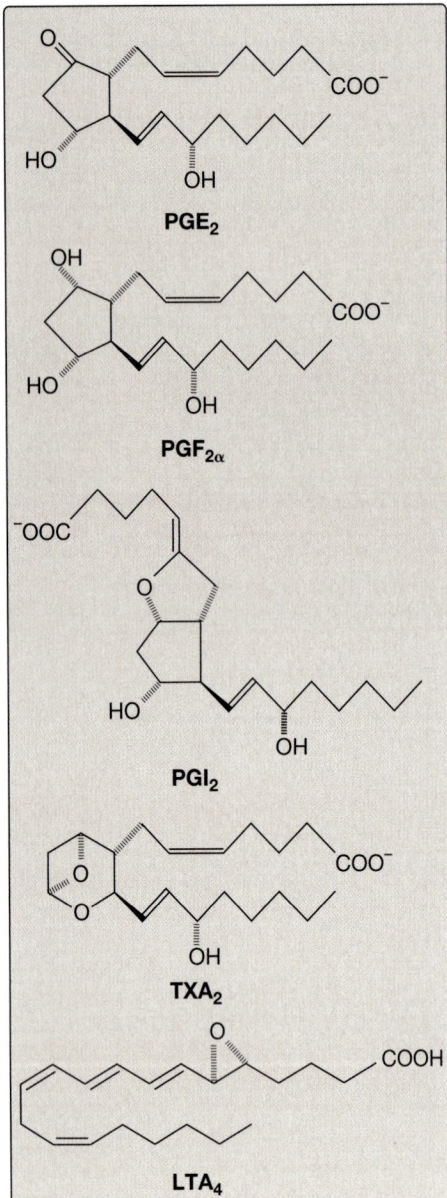

Figure 17.21
Examples of eicosanoid structures.
[Note: Prostaglandins are named as
follows: PG plus a third letter (e.g., E),
which designates the type and arrangement of functional groups in the molecule. The subscript number indicates
the number of double bonds in the
molecule. PGI_2 is also known as prostacyclin. Thromboxanes are designated
by TXs and leukotrienes by LTs.]

(intestine and uterus are key sites) and blood vessel diameter, and
maintain platelet homeostasis. Although they have been compared to
hormones in terms of their actions, eicosanoids differ from endocrine
hormones in that they are produced in very small amounts in almost
all tissues rather than in specialized glands and act locally rather than
after transport in the blood to distant sites. Eicosanoids are not stored,
and they have an extremely short half-life, being rapidly metabolized
to inactive products. Their biologic actions are mediated by plasma
membrane GPCRs (see Chapter 8, II. C.), which are different in different organ systems and typically result in changes in cyclic adenosine
monophosphate production. Examples of eicosanoid structures are
shown in Figure 17.21.

A. Prostaglandin and thromboxane synthesis

Arachidonic acid, an ω-6 FA containing 20 carbons and four double bonds (eicosatetraenoic FA), is the immediate precursor of the
predominant type of human PG (series 2 or those with two double bonds, as shown in Fig. 17.22). It is derived by the elongation
and desaturation of the essential FA linoleic acid, also an ω-6 FA.
Arachidonic acid is incorporated into membrane phospholipids (typically PI) at carbon 2, from which it is released by phospholipase A_2
(Fig. 17.23) in response to a variety of signals, such as a rise in calcium. [Note: Series 1 PGs contain one double bond and are derived
from an ω-6 eicosatrienoic FA, dihomo-γ-linolenic acid, whereas
series 3 PGs contain three double bonds and are derived from eicosapentaenoic acid (EPA), an ω-3 FA.]

1. **Prostaglandin H_2 synthase:** The first step in PG and TX synthesis is the oxidative cyclization of free arachidonic acid to yield
   PGH_2 by PGH_2 synthase (or, PG endoperoxide synthase). This
   enzyme is an ER membrane-bound protein that has two catalytic activities: FA cyclooxygenase (COX), which requires two
   molecules of O_2, and peroxidase, which requires reduced glutathione (see Chapter 13). PGH_2 is converted to a variety of PGs
   and TXs, as shown in Figure 17.23, by cell-specific synthases.
   [Note: PGs contain a five-carbon ring, whereas TXs contain a
   heterocyclic six-membered oxane ring (see Fig. 17.21).] Two
   isozymes of PGH_2 synthase, usually denoted as COX-1 and
   COX-2, are known. COX-1 is made constitutively in most tissues and is required for maintenance of healthy gastric tissue,
   renal homeostasis, and platelet aggregation. COX-2 is inducible
   in a limited number of tissues in response to products of activated immune and inflammatory cells. [Note: The increase in
   PG synthesis subsequent to the induction of COX-2 mediates
   the pain, heat, redness, and swelling of inflammation and the
   fever of infection.]

2. **Synthesis inhibition:** The synthesis of PG and TX can be inhibited by unrelated compounds. For example, cortisol (steroidal
   anti-inflammatory agent) leads to the inhibition of phospholipase
   A_2 activity (see Fig. 17.23) and, therefore, arachidonic acid, the
   substrate for PG and TX synthesis, is not released from membrane phospholipids. Aspirin, indomethacin, and phenylbuta-
   zone (all nonsteroidal anti-inflammatory drugs [NSAIDs]) inhibit

both COX-1 and COX-2 and, thus, prevent the synthesis of the parent molecule, PGH_2. [Note: Systemic inhibition of COX-1, with subsequent damage to the stomach and the kidneys and impaired clotting of blood, is the basis of aspirin's toxicity.] Aspirin (but not other NSAIDs) also induces synthesis of lipoxins (anti-inflammatory mediators made from arachidonic acid) and resolvins and protectins (inflammation-resolving mediators made from EPA). Inhibitors specific for COX-2 ("coxibs") are designed to reduce the pathologic inflammatory processes mediated by COX-2 while maintaining the physiologic functions of COX-1. Currently, celecoxib is the only U.S. Food and Drug Administration–approved coxib.

B. Thromboxanes and prostaglandins in platelet homeostasis

Thromboxane A_2 (TXA_2) is produced by COX-1 in activated platelets. It promotes platelet adhesion and aggregation and contraction of vascular smooth muscle, thereby promoting formation of blood clots (thrombi). (See Chapter 35.) Prostacyclin (PGI_2), produced by COX-2 in vascular endothelial cells, inhibits platelet aggregation and stimulates vasodilation and, so, impedes thrombogenesis. The opposing effects of TXA_2 and PGI_2 limit thrombi formation to sites of vascular injury.

C. Leukotriene synthesis

Arachidonic acid is converted to a variety of linear hydroperoxy (–OOH) acids by a separate pathway involving a family of lipoxygenases (LOXs). For example, 5-LOX converts arachidonic acid to 5-hydroperoxy-6,8,11,14 eicosatetraenoic acid ([5-HPETE] see Fig. 17.23). 5-HPETE is converted to a series of LTs containing four double bonds, the nature of the final products varying according to the tissue. LTs are mediators of allergic response and inflammation. Inhibitors of 5-LOX and LT-receptor antagonists are used in the treatment of asthma. [Note: LT synthesis is inhibited by cortisol and not by NSAIDs. Aspirin-exacerbated respiratory disease is a response to LT overproduction with NSAID use in ~10% of individuals with asthma.]

Clinical Application 17.6: Aspirin for Prevention of Cardiovascular Disease

Aspirin has an antithrombogenic effect. It inhibits TXA_2 synthesis by COX-1 in platelets and PGI_2 synthesis by COX-2 in endothelial cells through irreversible acetylation of these isozymes (Fig. 17.24). COX-1 inhibition cannot be overcome in platelets, which lack nuclei. However, COX-2 inhibition can be overcome in endothelial cells because they have a nucleus and, therefore, can generate more of the enzyme. This difference is the basis of low-dose aspirin therapy used for prevention of cardiovascular disease (CVD).

Aspirin plays a vital role in secondary prevention for patients with CVD because of its antiplatelet properties.

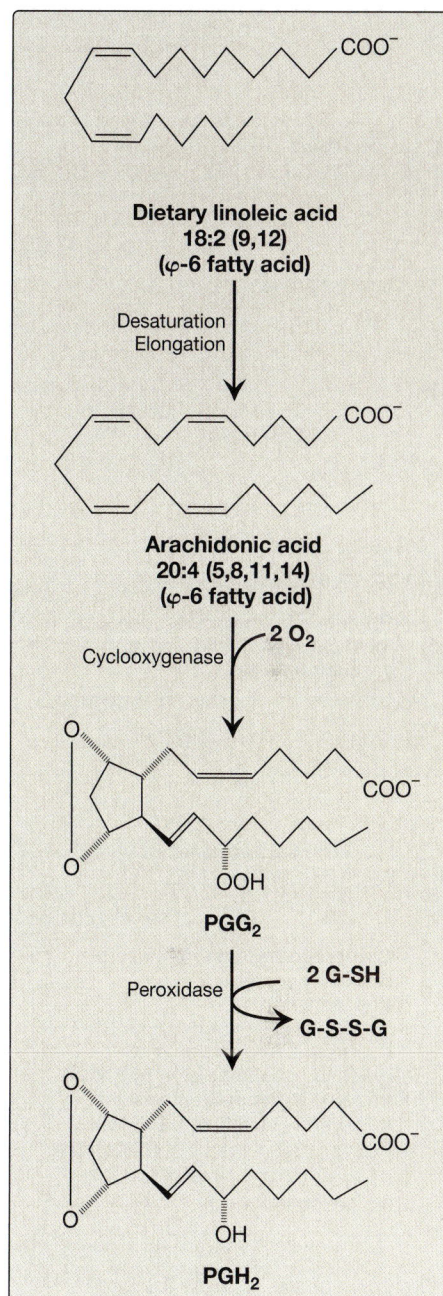

Figure 17.22
Oxidation and cyclization of arachidonic acid by the two catalytic activities (cyclooxygenase and peroxidase) of PGH_2 synthase (prostaglandin-endoperoxide synthase). G-SH, reduced glutathione; G-S-S-G, oxidized glutathione; PG, prostaglandin.

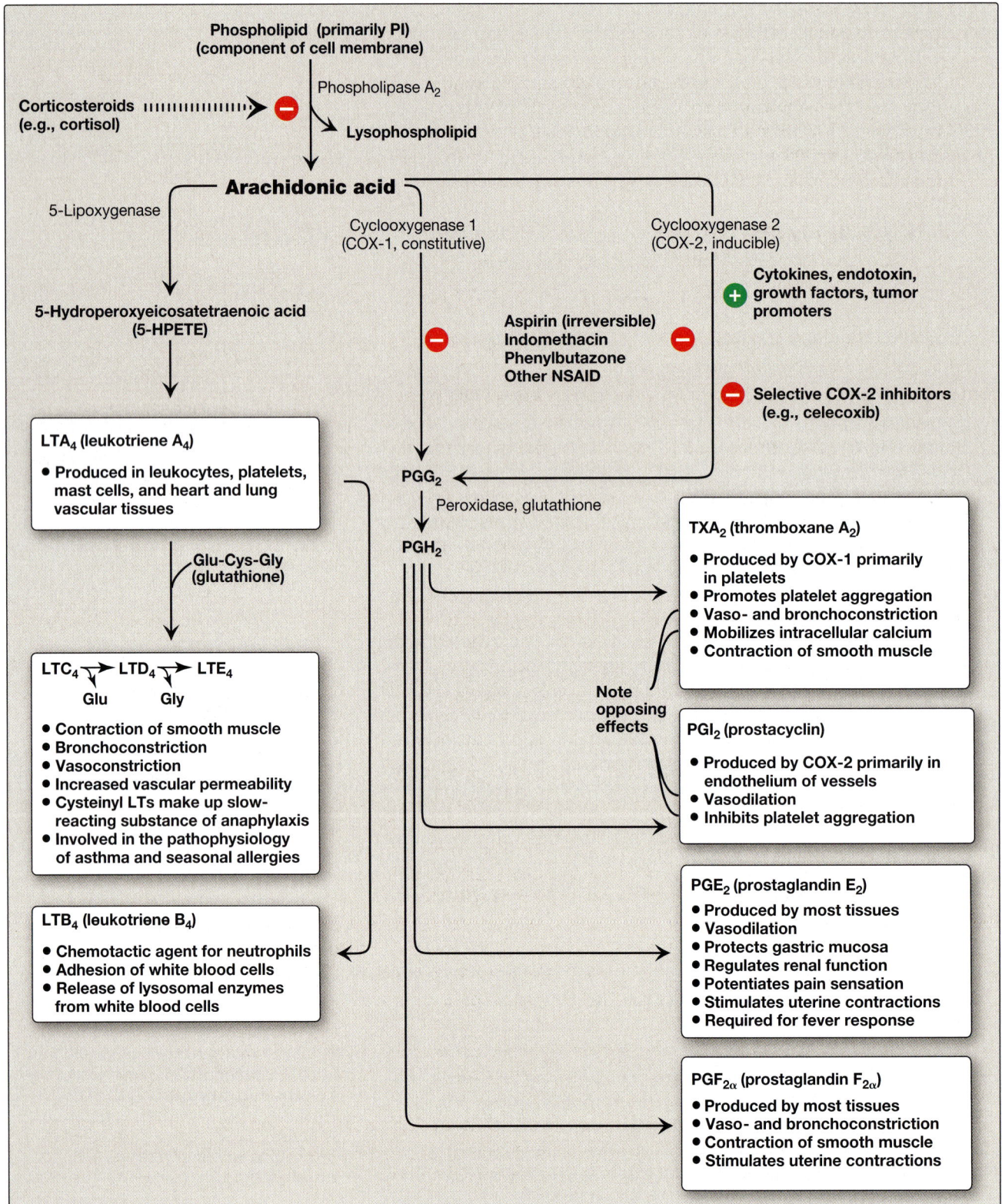

Figure 17.23
Overview of the biosynthesis and function of some important prostaglandins (PGs), leukotrienes (LT), and a thromboxane (TX) from arachidonic acid. [Note: The arachidonic acid in the membrane phospholipid was derived from the ω-6 essential fatty acid (FA), linoleic, also an ω-6 FA.] PI, phosphatidylinositol; NSAID, nonsteroidal anti-inflammatory drugs; Glu, glutamate; Cys, cysteine; Gly, glycine.

IX. CHAPTER SUMMARY

- **Phospholipids** are polar, ionic compounds composed of an **alcohol** (e.g., **choline** or **ethanolamine**) attached by a phosphodiester bond either to **diacylglycerol (DAG)**, producing **phosphatidylcholine** or **phosphatidylethanolamine**, or to the amino alcohol **sphingosine** (Fig. 17.25).

- Addition of a long-chain fatty acid (LCFA) to sphingosine produces a **ceramide**.

- Addition of **phosphorylcholine** produces the phospholipid **sphingomyelin**.

- Phospholipids are the predominant lipids of **cell membranes**.

- Nonmembrane phospholipids serve as components of **lung surfactant** and **bile**.

- **Dipalmitoylphosphatidylcholine**, also called **dipalmitoyl lecithin**, is the major lipid component of **lung surfactant**.

- Insufficient surfactant production causes **respiratory distress syndrome (RDS)**.

- **Phosphatidylinositol (PI)** serves as a reservoir for **arachidonic acid** in membranes.

- The phosphorylation of membrane-bound PI produces **phosphatidylinositol 4,5-bisphosphate (PIP$_2$)**. This compound is degraded by **phospholipase C** in response to the binding of various neurotransmitters, hormones, and growth factors to membrane **G protein–coupled receptors (GPCRs)**.

- The products of **phospholipase C**, inositol 1,4,5-trisphosphate (**IP$_3$**), and **DAG,** mediate the mobilization of intracellular **calcium** and the activation of **protein kinase C**, which act synergistically to evoke cellular responses.

- Specific proteins can be covalently attached via a carbohydrate bridge to membrane-bound PI, forming a **glycosyl phosphatidylinositol (GPI) anchor**. A deficiency in GPI synthesis in hematopoietic cells results in the hemolytic disease **paroxysmal nocturnal hemoglobinuria (PNH)**.

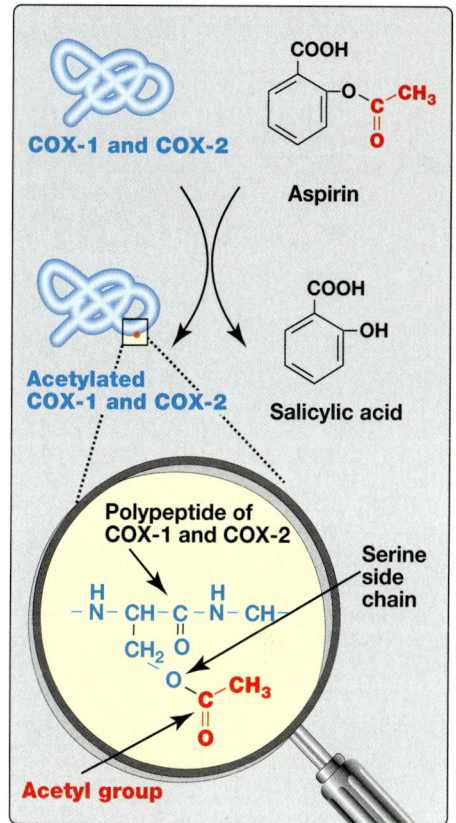

Figure 17.24
Irreversible acetylation of cyclooxygenase (COX)-1 and COX-2 by aspirin.

- The degradation of phosphoglycerides is performed by **phospholipases** found in all tissues and pancreatic juice.

- **Sphingomyelin** is degraded to a ceramide plus phosphorylcholine by the lysosomal enzyme **sphingomyelinase**, a deficiency of which causes **Niemann–Pick (A and B) disease**.

- **Glycosphingolipids** are derivatives of **ceramides** to which carbohydrates have been attached. Adding one sugar molecule to the ceramide produces a **cerebroside**, adding an oligosaccharide produces a **globoside**, and adding an acidic **N-acetylneuraminic acid (NANA)** molecule produces a **ganglioside**.

- Glycosphingolipids are found predominantly in cell membranes of the **brain** and **peripheral nervous tissue**, with high concentrations in the **myelin sheath**. They are **antigenic**. Glycolipids are degraded in the **lysosomes** by **acid hydrolases**. A deficiency of any one of these enzymes causes a **sphingolipidosis**, in which a characteristic sphingolipid accumulates.

- **Prostaglandins (PGs)**, **thromboxanes (TXs)**, and **leukotrienes (LTs)**, the **eicosanoids**, are produced in very small amounts in almost all tissues, act locally, and have an extremely short half-life.

- **Eicosanoids** serve as mediators of the **inflammatory response**. **Arachidonic acid** is the immediate precursor of the predominant class of human PGs (those with two double bonds). It is derived by the elongation and desaturation of the essential FA **linoleic acid** and is stored in the membrane as a component of a phospholipid, generally PI.

- Arachidonic acid is released from the phospholipid by **phospholipase A$_2$** (inhibited by **cortisol**).

- Synthesis of the **PG** and **TX** begins with the oxidative cyclization of free arachidonic acid to yield PGH$_2$ by **PGH$_2$ synthase** (or, **prostaglandin endoperoxide synthase**), an endoplasmic reticular membrane protein that has two catalytic activities: **FA cyclooxygenase (COX)** and **peroxidase**.

- There are two isozymes of PGH$_2$ synthase: **COX-1** (constitutive) and **COX-2** (inducible).

- **Aspirin** irreversibly inhibits **COX-1** at low dose and both **COX-1** and **COX-2** at high dose. Opposing effects of PGI$_2$ and TXA$_2$ limit clot formation. Low-dose aspirin is recommended for the prevention of clot formation.

- **LTs** are linear molecules produced from arachidonic acid by the **5-lipoxygenase** pathway. They mediate allergic response. Their synthesis is inhibited by cortisol and not by aspirin.

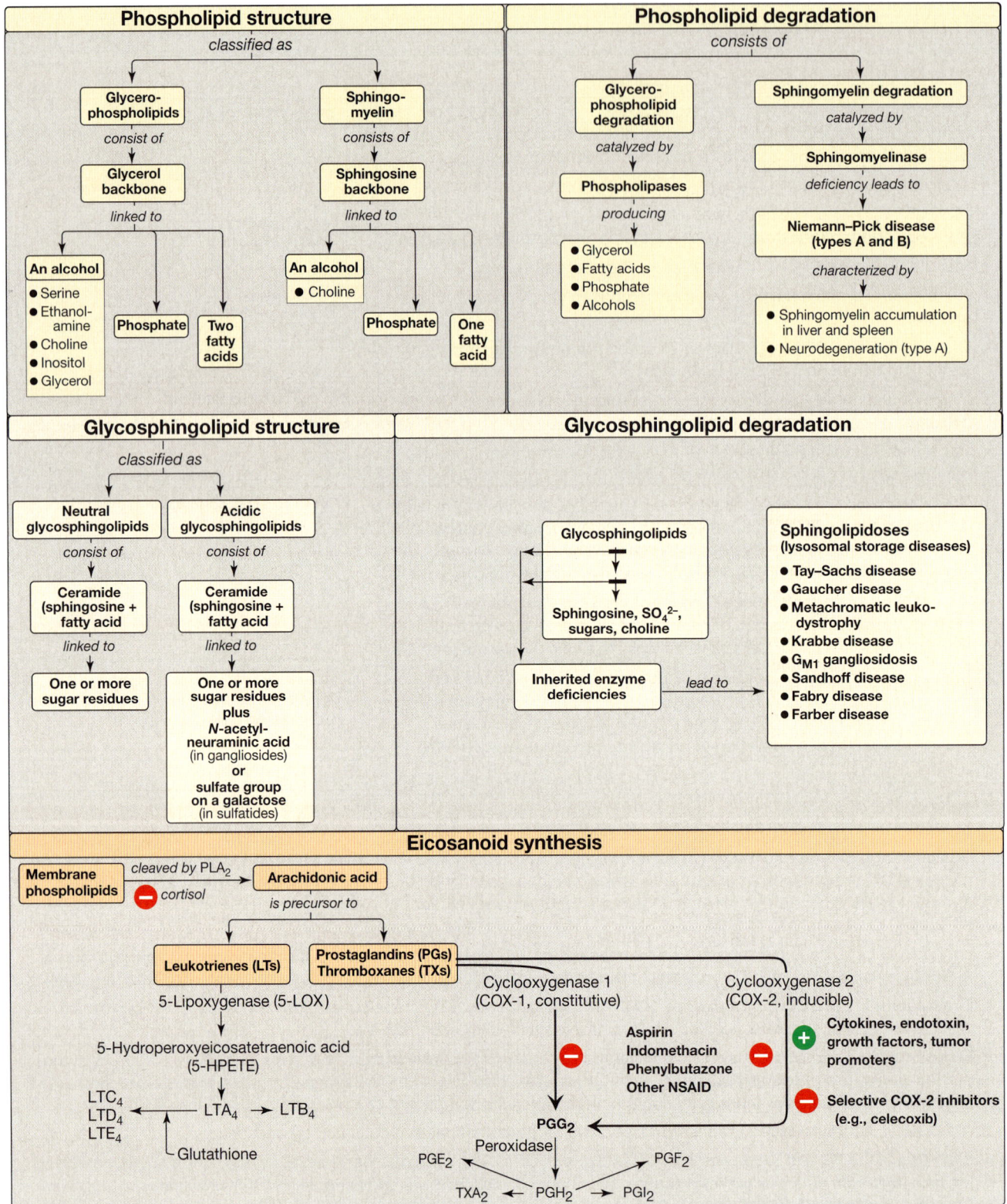

Figure 17.25
Key concept map for phospholipids, glycosphingolipids, and eicosanoids. PLA$_2$, phospholipase A$_2$; SO$_4^{2-}$, sulfate ion; NSAID, nonsteroidal anti-inflammatory drug.

Study Questions

Choose the ONE best answer.

17.1. A 28-year-old female complains of severe menstrual cramps and heavy menstrual bleeding. Her medical history reveals a diagnosis of endometriosis. Which group of compounds plays a critical role in mediating the inflammatory response associated with endometriosis?

   A. Phospholipids

   B. Eicosanoids

   C. Glycosphingolipids

   D. Steroids

Correct Answer = B. Eicosanoids, including prostaglandins and leukotrienes, are derived from arachidonic acid and are involved in inflammation and pain pathways. In endometriosis, excessive eicosanoid production contributes to the associated symptoms.

17.2. Aspirin-exacerbated respiratory disease (AERD) is a severe reaction to nonsteroidal anti-inflammatory drugs (NSAIDs) characterized by bronchoconstriction 30 minutes to several hours after ingestion. Which of the following statements about NSAIDs best explains the symptoms seen in patients with AERD?

   A. Inhibition of the activity of the cystic fibrosis transmembrane conductance regulator protein, resulting in thickened mucus that block airways.

   B. Inhibition of cyclooxygenase but not lipoxygenase, resulting in the flow of arachidonic acid to leukotriene synthesis.

   C. Activation of the cyclooxygenase activity of prostaglandin H_2 synthase, resulting in increased synthesis of prostaglandins that promote vasodilation.

   D. Activation phospholipases, resulting in decreased amounts of dipalmitoylphosphatidylcholine and alveolar collapse (atelectasis).

Correct answer = B. NSAIDs inhibit cyclooxygenase but not lipoxygenase, so any arachidonic acid available is used for the synthesis of bronchoconstricting leukotrienes. NSAIDs have no effect on the cystic fibrosis (CF) transmembrane conductance regulator protein, defects which are the cause of CF. Steroids, not NSAIDs, inhibit phospholipase A_2. Cyclooxygenase is inhibited by NSAIDs, not activated. NSAIDs have no effect on phospholipases.

17.3. An infant, born at 28 weeks gestation, rapidly gave evidence of respiratory distress. Clinical laboratory and imaging results supported the diagnosis of infant respiratory distress syndrome (RDS). Which of the following is the most accurate statement about RDS?

   A. It is unrelated to the baby's premature birth.

   B. It is a consequence of too few type II pneumocytes.

   C. The lecithin/sphingomyelin ratio in the amniotic fluid is likely to be high (>2).

   D. The concentration of dipalmitoylphosphatidylcholine in the amniotic fluid would be expected to be lower than that of a full-term infant.

   E. It is an easily treated disorder with low mortality.

   F. It is treated by administering surfactant to the mother just before giving birth.

Correct answer = D. Dipalmitoylphosphatidylcholine ([DPPC] or, dipalmitoyl lecithin) is the lung surfactant found in mature, healthy lungs. RDS can occur in lungs that make too little of this compound. If the lecithin/sphingomyelin (L/S) ratio in amniotic fluid is ≥2, a newborn's lungs are considered to be sufficiently mature (premature lungs would be expected to have a ratio <2). The RDS would not be due to too few type II pneumocytes, which would simply be secreting sphingomyelin rather than DPPC at 28 weeks gestation. The mother is given a glucocorticoid, not surfactant, prior to giving birth (antenatally). Surfactant would be administered to the infant postnatally to reduce surface tension.

17.4. A 10-year-old male was evaluated for burning sensations in his feet and clusters of small, red-purple spots on his skin. Laboratory studies revealed protein in his urine. Enzymatic analysis revealed a deficiency of α-galactosidase, and enzyme replacement therapy was recommended. Which of the following is the most likely working diagnosis?

A. Fabry disease
B. Farber disease
C. Gaucher disease
D. Krabbe disease
E. Niemann–Pick disease

Correct answer = A. Fabry disease, a deficiency of α-galactosidase, is the only X-linked sphingolipidosis. It is characterized by pain in the extremities, a red-purple skin rash (generalized angiokeratomas), and kidney and cardiac complications. Protein in his urine indicates kidney damage. Enzyme replacement therapy is available.

17.5. A 5-year-old child is brought to the pediatrician by his mother due to abdominal distention and pain in his leg. The mother states that her son started having difficulty walking and began to fall repeatedly. Physical examination shows developmental delay and hepatosplenomegaly. Fundoscopic examination shows cherry-red spots in the macula. Which of the following histologic finding of the affected tissue is most likely to confirm the diagnosis?

A. Shell-like inclusion bodies in neuronal cells
B. Crumpled tissue paper appearance of the cytosol
C. Foamy macrophages in the bone marrow
D. Globoid bodies in macrophages

Correct answer = C. Niemann–Pick disease type B is the most likely diagnosis due to the presence of hepatomegaly, neurologic defects leading to falls, and cherry-red areas in the macula. The histologic finding is the foamy appearance of macrophages of the reticuloendothelial system because of sphingomyelin accumulation.

17.6. A 65-year-old male patient with a history of cardiovascular disease is advised by his physician to take a daily low dose of aspirin as a preventive measure. Which of the following explains the rationale behind it?

A. Aspirin increases the synthesis of thromboxane A2, promoting platelet aggregation.
B. Aspirin enhances the production of PGI2 to cause vasodilation and inhibit platelet aggregation.
C. Aspirin irreversibly inhibits COX-1 in the platelets, reducing the synthesis of thromboxane A2 and decreasing platelet aggregation.
D. Aspirin selectively and irreversibly inhibits COX-2 in endothelial cells, leading to an anti-inflammatory effect.

Correct Answer = C. The primary rationale behind recommending a daily low dose of aspirin in patients with a history of cardiovascular disease is its antithrombotic effect. Aspirin irreversibly inhibits the cyclooxygenase enzyme COX-1 in platelets, which in turn reduces the synthesis of thromboxane A_2 (TXA_2). TXA_2 is a potent vasoconstrictor and promotes platelet aggregation. By reducing TXA_2 synthesis, aspirin decreases platelet aggregation and provides a protective antithrombotic effect, reducing the risk of clot formation in blood vessels. While aspirin does affect PGI_2 synthesis, it does not enhance its production. Instead, aspirin reduces the synthesis of PGI_2 along with TXA_2, with a more pronounced and clinically relevant effect on reducing TXA_2.

17.7. A 6-month-old infant is brought to the pediatrician's office by concerned parents. The child has shown a progressive decline in motor skills, exaggerated startle response to loud noises, and a cherry-red spot on macular examination. Laboratory analysis reveals deficient activity of the enzyme hexosaminidase A. Which of the following compounds accumulates in the affected individuals due to the enzyme deficiency?

A. Cerebroside
B. Ceramide
C. Sphingomyelin
D. Ganglioside G_{M2}

Correct Answer = D. Tay–Sachs disease is characterized by a deficiency of hexosaminidase A, leading to the accumulation of ganglioside G_{M2}, primarily in neuronal cells. The clinical features, including neurodegeneration and the cherry-red spot on macula, result from this accumulation.

17.8. A 35-year-old woman presents with a history of recurrent miscarriages and a recent episode of deep vein thrombosis. Laboratory tests reveal the presence of cardiolipin antibodies. Which of the following conditions or infections is most strongly associated with the presence of cardiolipin antibodies?

A. Systemic lupus erythematosus
B. Rheumatoid arthritis
C. Syphilis
D. Chronic obstructive pulmonary disease

Correct Answer = A. Cardiolipin antibodies are strongly associated with autoimmune disorders, particularly systemic lupus erythematosus (SLE). Patients with SLE often have various autoantibodies, including cardiolipin antibodies, which can lead to hypercoagulability and an increased risk of thrombotic events like deep vein thrombosis and recurrent miscarriages. While cardiolipin antibodies can also be found in individuals with syphilis, the strongest association is with SLE.

18 Cholesterol, Lipoprotein, and Steroid Metabolism

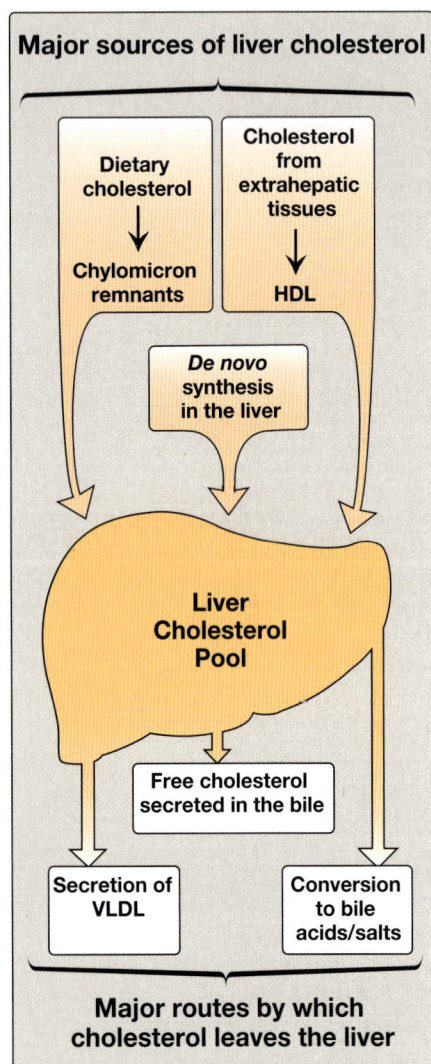

I. OVERVIEW

Cholesterol, the major steroid alcohol in animals, performs a number of essential functions in the body. For example, cholesterol is a structural component of all cell membranes, modulating their fluidity, and, in specialized tissues, cholesterol is a precursor of bile acids, steroid hormones, and vitamin D. Therefore, it is critically important that the cells of the body be assured an appropriate supply of cholesterol. To meet this need, a complex series of transport, biosynthetic, and regulatory mechanisms has evolved. The liver plays a central role in the regulation of the body's cholesterol homeostasis. For example, cholesterol enters the hepatic cholesterol pool from a number of sources including dietary cholesterol as well as cholesterol synthesized *de novo* by extrahepatic tissues and by the liver itself. Cholesterol is eliminated from the liver as unmodified cholesterol in the bile, or it can be converted to bile salts that are secreted into the intestinal lumen. It can also serve as a component of plasma lipoproteins that carry lipids to the peripheral tissues. In humans, the balance between cholesterol influx and efflux is not precise, resulting in a gradual deposition of cholesterol in the tissues, particularly in the endothelial linings of blood vessels. This is a potentially life-threatening occurrence when the lipid deposition leads to plaque formation, causing the narrowing of blood vessels (atherosclerosis) and increased risk of cardio-, cerebro-, and peripheral vascular disease. Figure 18.1 summarizes the major sources of liver cholesterol and the routes by which cholesterol leaves the liver.

II. CHOLESTEROL STRUCTURE

Cholesterol is a very hydrophobic compound. It consists of four fused hydrocarbon rings (A–D) called the steroid nucleus, and it has an eight-carbon, branched hydrocarbon chain attached to carbon 17 of the D ring. Ring A has a hydroxyl group at carbon 3, and ring B has a double bond between carbon 5 and carbon 6 (Fig. 18.2).

A. Sterols

Steroids with 8 to 10 carbon atoms in the side chain at carbon 17 and a hydroxyl group at carbon 3 are classified as sterols. Cholesterol

Figure 18.1
Sources of liver cholesterol (influx) and routes by which cholesterol leaves the liver (efflux). HDLs and VLDLs, high- and very–low-density lipoproteins.

Major sources of liver cholesterol

Dietary cholesterol → Chylomicron remnants

Cholesterol from extrahepatic tissues → HDL

De novo synthesis in the liver

Liver Cholesterol Pool

Free cholesterol secreted in the bile

Secretion of VLDL

Conversion to bile acids/salts

Major routes by which cholesterol leaves the liver

is the major sterol in animal tissues. It arises from *de novo* synthesis and absorption of dietary cholesterol. Intestinal uptake of cholesterol is mediated by the Niemann–Pick C1-like 1 protein (NPC1L1). This transporter is the target of the drug ezetimibe that reduces absorption of dietary cholesterol (see Chapter 15). Plant sterols (phytosterols), such as β-sitosterol, are poorly absorbed by NPC1L1 compared to cholesterol. After entering the enterocytes, sterols are actively transported back into the intestinal lumen by the ABCG5/8 efflux transporter. When phytosterols are ingested, they can reduce cholesterol absorption by competing with cholesterol for incorporation into micelles in the intestinal lumen (see Chapter 15). Therefore, less cholesterol gets into the enterocytes and some of this amount is returned to the lumen and excreted from the body. Sitosterolemia (also called phytosterolemia) is an autosomal recessive disorder associated with defects in the ABCG5/8 transporter. In this disorder, there is excessive absorption of cholesterol and plant sterols from the intestine resulting in their accumulation in the blood. Sitosterolemia can reduce blood flow and increase the risk of a heart attack or stroke that may cause sudden death.

B. Cholesteryl esters

Most plasma cholesterol is in an esterified form (with a fatty acid [FA] attached at carbon 3, as shown in Fig. 18.2), which makes the structure even more hydrophobic than free (nonesterified) cholesterol. Cholesteryl esters are not found in membranes and are normally present only at low levels in most cells. Because of their hydrophobicity, cholesterol and its esters must be transported in association with protein as a component of a lipoprotein particle or be solubilized by phospholipids and bile salts in the bile.

III. CHOLESTEROL SYNTHESIS

Cholesterol is synthesized by virtually all tissues in humans, although liver, intestine, adrenal cortex, and reproductive tissues, including ovaries, testes, and placenta, make the largest contributions to the cholesterol pool. As with FA, all the carbon atoms in cholesterol are provided by acetyl coenzyme A (CoA), and nicotinamide adenine dinucleotide phosphate (NADPH) provides the reducing equivalents. The pathway is endergonic, being driven by hydrolysis of the high-energy thioester bond of acetyl CoA and the terminal phosphate bond of adenosine triphosphate (ATP). Synthesis requires enzymes in the cytosol, the membrane of the smooth endoplasmic reticulum (sER), and the peroxisome. The pathway is responsive to changes in cholesterol concentration, and regulatory mechanisms exist to balance the rate of cholesterol synthesis against the rate of cholesterol excretion. An imbalance in this regulation can lead to an elevation in circulating levels of plasma cholesterol, with the potential for vascular disease.

A. 3-Hydroxy-3-methylglutaryl coenzyme A synthesis

The first two reactions in the cholesterol biosynthetic pathway are similar to those in the pathway that produces ketone bodies (see Fig. 16.22). They result in the production of 3-hydroxy-3-methylglutaryl CoA ([HMG CoA] Fig. 18.3). First, two acetyl CoA molecules

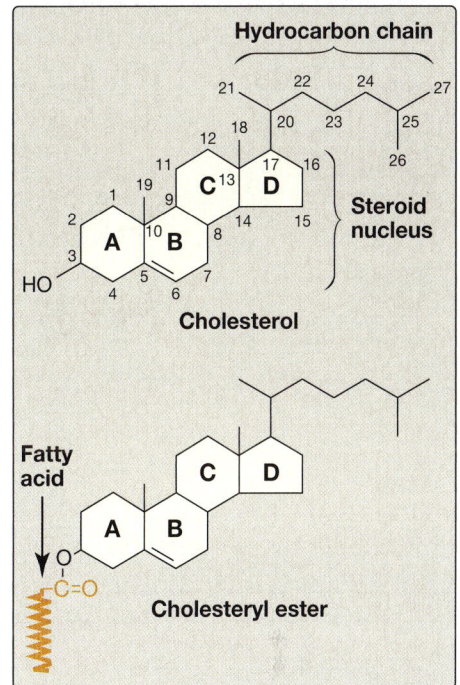

Figure 18.2
Structure of cholesterol and its ester.

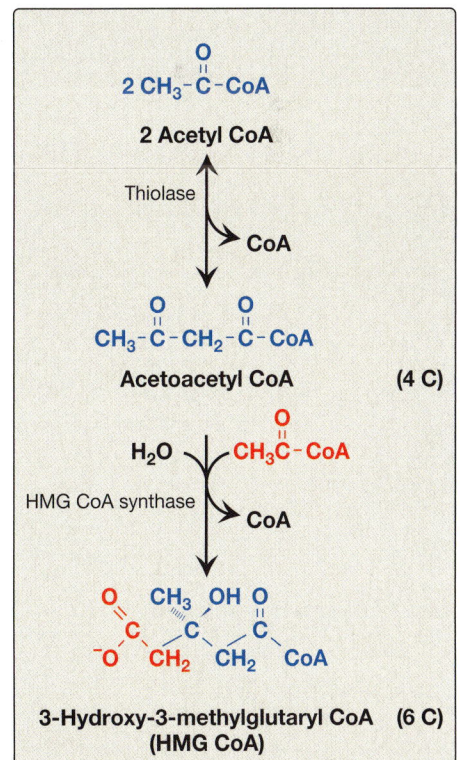

Figure 18.3
Synthesis of HMG CoA. CoA, coenzyme A.

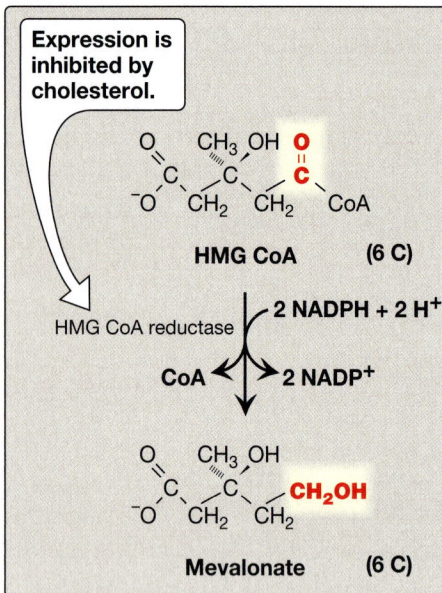

Figure 18.4
Synthesis of mevalonate. HMG CoA, hydroxymethylglutaryl coenzyme A; NADP(H), nicotinamide adenine dinucleotide phosphate.

condense to form acetoacetyl CoA. Next, a third molecule of acetyl CoA is added by HMG CoA synthase, producing HMG CoA, a six-carbon compound. [Note: Liver parenchymal cells contain two isoenzymes of the synthase. The cytosolic enzyme participates in cholesterol synthesis, whereas the mitochondrial enzyme functions in the pathway for ketone body synthesis.]

B. Mevalonate synthesis

HMG CoA is reduced to mevalonate by HMG CoA reductase. This is the rate-limiting and key regulated step in cholesterol synthesis. It occurs in the cytosol, uses two molecules of NADPH as the reducing agent, and releases CoA, making the reaction irreversible (Fig. 18.4). [Note: HMG CoA reductase is an integral membrane protein of the sER, with its catalytic domain projecting into the cytosol. Regulation of reductase activity is discussed in Section D.]

C. Cholesterol synthesis from mevalonate

The reactions and enzymes involved in the synthesis of cholesterol from mevalonate are illustrated in Figure 18.5. [Note: The numbers shown in brackets below correspond to numbered reactions shown in this figure.]

(1) Mevalonate is converted to 5-pyrophosphomevalonate in two steps, each of which transfers a phosphate group from ATP.

(2) A five-carbon isoprene unit, isopentenyl pyrophosphate (IPP), is formed by the decarboxylation of 5-pyrophosphomevalonate. The reaction requires ATP. [Note: IPP is the precursor of a family of molecules with diverse functions, the isoprenoids. Cholesterol is a sterol isoprenoid. Nonsterol isoprenoids include dolichol and ubiquinone (or, coenzyme Q).]

(3) IPP is isomerized to 3,3-dimethylallyl pyrophosphate (DPP).

(4) IPP and DPP condense to form 10-carbon geranyl pyrophosphate (GPP).

(5) A second molecule of IPP then condenses with GPP to form 15-carbon farnesyl pyrophosphate (FPP). [Note: Covalent attachment of farnesyl to proteins, a process known as prenylation, is one mechanism for anchoring proteins (e.g., Ras)] to the inner face of plasma membranes.]

(6) Two molecules of FPP combine, releasing pyrophosphate, and are reduced, forming the 30-carbon compound squalene. [Note: Squalene is formed from six isoprenoid units. Because 3 ATP are hydrolyzed per mevalonate residue converted to IPP, a total of 18 ATP are required to make the polyisoprenoid squalene.]

(7) Squalene is converted to the sterol lanosterol by a sequence of two reactions catalyzed by SER-associated enzymes that use molecular oxygen (O_2) and NADPH. The hydroxylation of linear squalene triggers the cyclization of the structure to lanosterol.

(8) The conversion of lanosterol to cholesterol is a multistep process involving shortening of the side chain, oxidative removal of methyl groups, reduction of double bonds, and migration of a double bond. Smith–Lemli–Opitz syndrome (SLOS), an autosomal recessive disorder of cholesterol biosynthesis, is caused

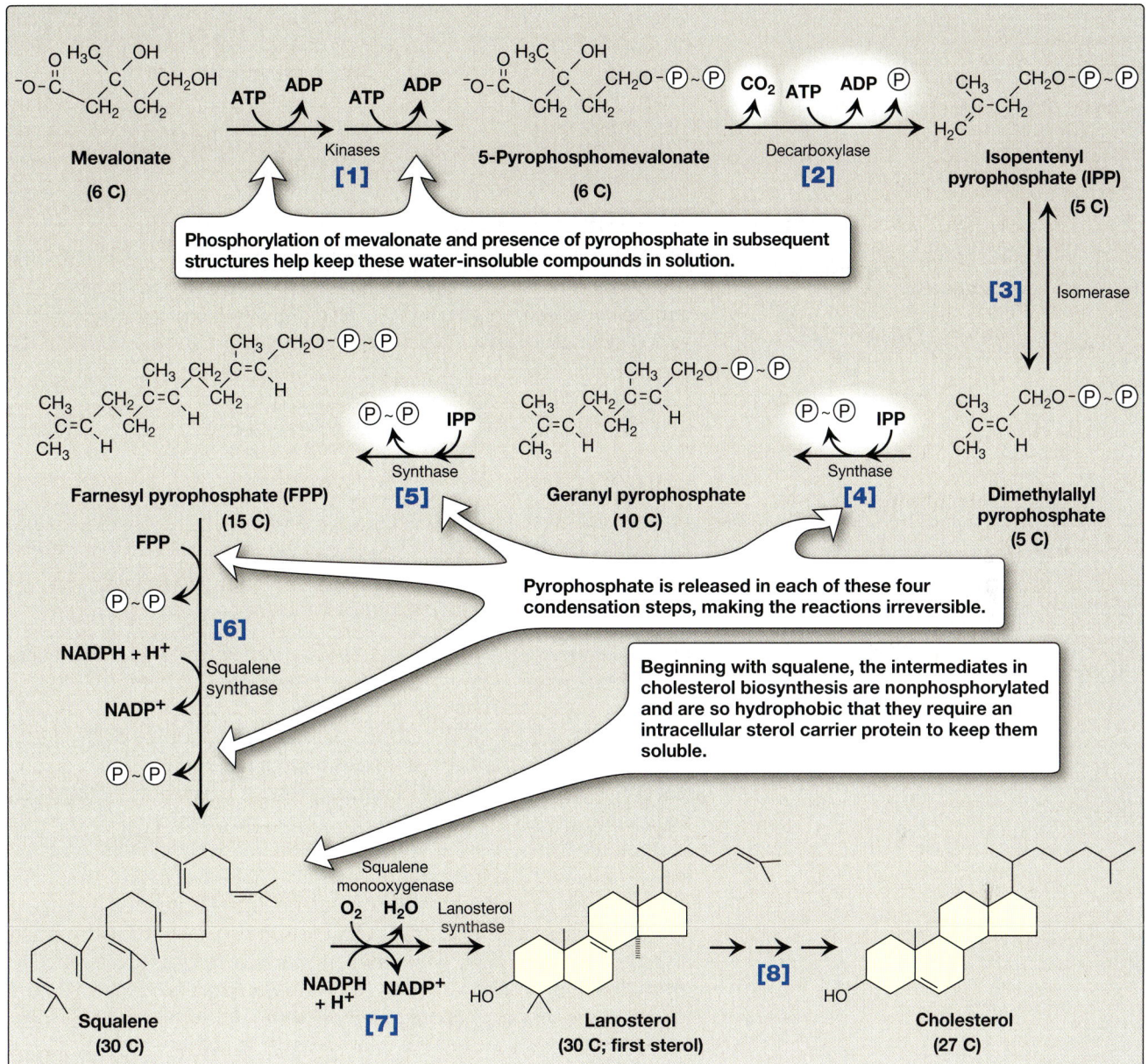

Figure 18.5
Synthesis of cholesterol from mevalonate. ADP, adenosine diphosphate; (P), phosphate; (P) ~ (P), pyrophosphate; NADP(H), nicotinamide adenine dinucleotide phosphate.

by a partial deficiency in 7-dehydrocholesterol-7-reductase, the enzyme that reduces the double bond in 7-dehydrocholesterol (7-DHC), thereby converting it to cholesterol. SLOS is one of several multisystem, embryonic malformation syndromes associated with impaired cholesterol synthesis. [Note: 7-DHC is converted to vitamin D_3 in the skin (see Chapter 28).]

D. Branch-point reactions in the biosynthesis of cholesterol

The intermediates of cholesterol synthesis are shunted for modification of other molecules. The first branch point starts with step 2 above, the

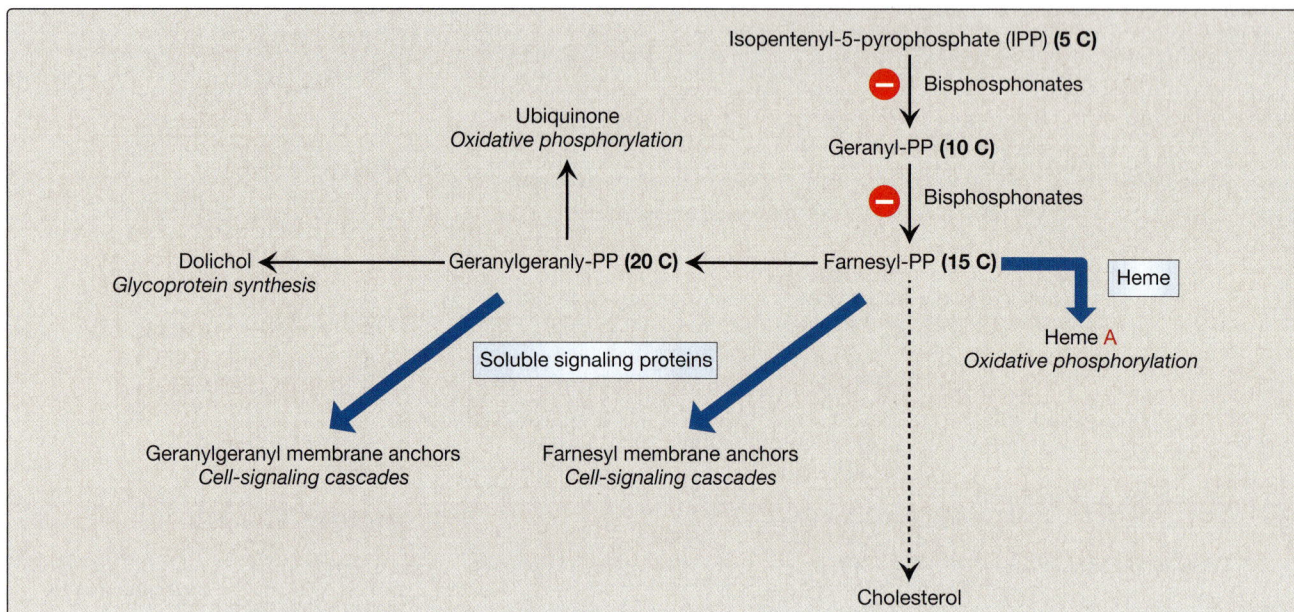

Figure 18.6
Branch-point reactions in the biosynthesis of cholesterol. The *thin arrows* indicate an enzymatic or chemical conversion to a product. *Thick arrows* indicate protein modifications. The italics below the protein modifications indicate the processes that the products are involved in. PP, pyrophosphate.

synthesis of IPP **(5 C)** (Fig. 18.6). Subsequent addition of 5-carbon isoprene units results in the synthesis of geranyl-PP **(10 C)**, farnesyl-PP **(15 C)**, and geranylgeranyl-PP **(20 C)**, respectively. These molecules can modify proteins so that they can be anchored into the membrane lipids. Farnesylation of heme creates heme A, a specialized heme in cytochrome *a* of the electron transport chain. Farnesylation and geranylgeranylation of proteins such as ras oncogene can lead to activation of cellular signaling pathways for proliferation. Geranylgeranylation also produces dolichol, which is important for sugar transfer during glycoprotein synthesis, and ubiquinone, a lipid-soluble electron carrier in oxidative phosphorylation.

Bisphosphonates are used to inhibit bone resorption in osteoporosis and Paget disease. The new generation of bisphosphonates has been shown to kill cancer cells by inhibiting the synthesis of farnesyl-PP and geranylgeranyl-PP.

E. Cholesterol synthesis regulation

HMG CoA reductase is the major control point for cholesterol biosynthesis and is subject to different kinds of metabolic control.

1. **Sterol-dependent regulation of gene expression:** Expression of the gene for HMG CoA reductase is controlled by the trans-acting factor, sterol regulatory element–binding protein-2 (SREBP-2), which binds DNA at the *cis*-acting sterol regulatory element (SRE) upstream of the *reductase* gene. Inactive SREBP-2 is an integral protein of the SER membrane and associates with a second sER membrane protein, SREBP cleavage–activating protein (SCAP). When sterol levels in the sER are low, the SREBP-2–SCAP

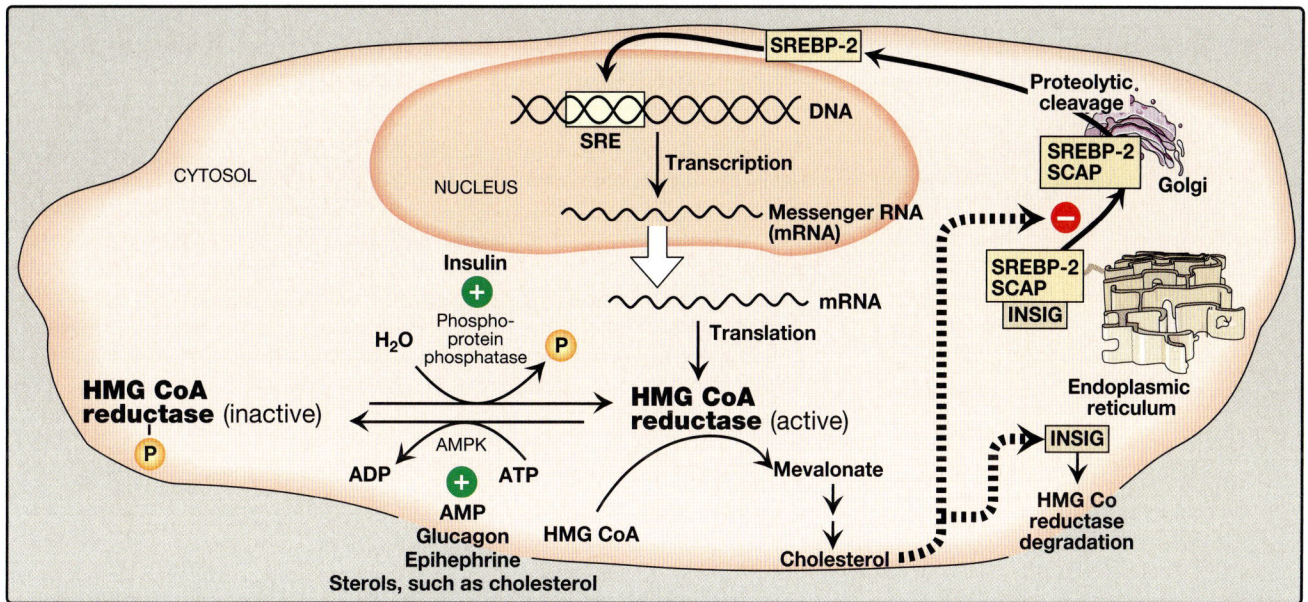

Figure 18.7
Regulation of hydroxymethylglutaryl coenzyme A (HMG CoA) reductase. SRE, sterol regulatory element; SREBP, SRE-binding protein; SCAP, SREBP cleavage–activating protein; AMPK, adenosine monophosphate–activated protein kinase; ADP, adenosine diphosphate ⓟ, phosphate; INSIG, insulin-induced gene protein.

complex moves from the endoplasmic reticulum (ER) to the Golgi. In the Golgi membrane, SREBP-2 is sequentially acted upon by two proteases, which generate a soluble fragment that enters the nucleus, binds the SRE, and functions as a transcription factor. This results in increased synthesis of HMG CoA reductase and, therefore, increased cholesterol synthesis (Fig. 18.7). However, if sterols are abundant, they bind SCAP at its sterol-sensing domain and induce the binding of SCAP to yet other ER membrane proteins, the insulin-induced gene proteins (INSIGs). This results in the retention of the SCAP–SREBP complex in the sER, thereby preventing the activation of SREBP-2 and leading to downregulation of cholesterol synthesis. [Note: SREBP-1c upregulates expression of enzymes involved in FA synthesis in response to insulin.]

2. **Sterol-accelerated enzyme degradation:** The reductase itself is a sterol-sensing integral protein of the SER membrane. When sterol levels in the SER are high, the enzyme binds to INSIG proteins (see Fig. 18.7). Binding leads to cytosolic transfer, ubiquitination, and proteasomal degradation of the reductase (see Chapter 19).

3. **Sterol-independent phosphorylation/dephosphorylation:** HMG CoA reductase activity is controlled covalently through the actions of adenosine monophosphate (AMP)–activated protein kinase (AMPK) and a phosphoprotein phosphatase (see Fig. 18.7). The phosphorylated form of the enzyme is inactive, whereas the dephosphorylated form is active. [Note: Because AMPK is activated by AMP, cholesterol synthesis, like FA synthesis, is decreased when ATP availability is decreased.]

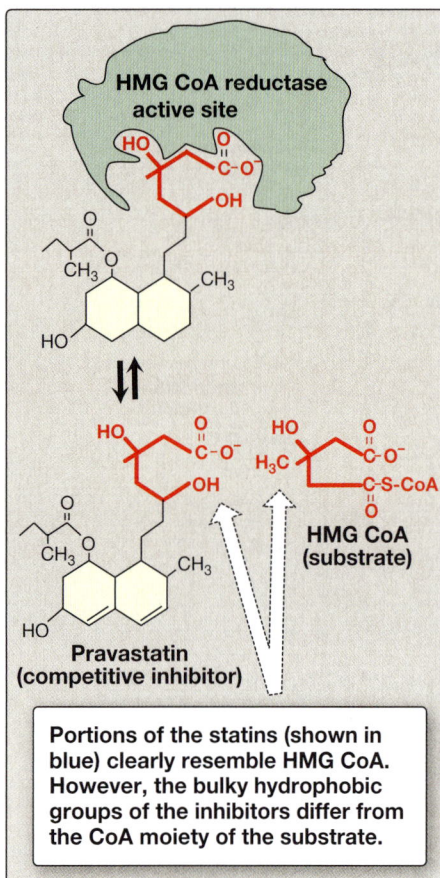

Figure 18.8
Structural similarity of hydroxymethyl-glutaric acid (HMG) and pravastatin, a clinically useful cholesterol-lowering drug of the statin family. CoA, coenzyme A.

Clinical Application 18.1: Statin Drugs for Regulating Cholesterol Synthesis

The statin drugs (atorvastatin, fluvastatin, lovastatin, pravastatin, rosuvastatin, and simvastatin) are structural analogs of HMG CoA. They are (or are metabolized to) reversible, competitive inhibitors of HMG CoA reductase (Fig. 18.8) and are used to decrease plasma cholesterol levels in patients with hypercholesterolemia. The most commonly reported side effects of statins include muscle pain, fatigue and weakness (myopathy), and even muscle damage (rhabdomyolysis). These effects may be due to the inhibition of heme A and ubiquinone synthesis, which are both essential for oxidative phosphorylation for energy (Fig. 18.6). Genetic polymorphisms also influence the physiologic response to statins. For example, organic anion–transporting polypeptide ([OATP1B1] also known as SLCO1B1) has a polymorphism at nucleotide 521 T>C that is used as a biomarker for simvastatin myopathy.

4. **Hormonal regulation:** The activity of HMG CoA reductase is controlled hormonally. An increase in insulin favors dephosphorylation (activation) of the reductase, whereas an increase in glucagon and epinephrine and elevated levels of cholesterol has the opposite effect (see Fig. 18.7).

IV. CHOLESTEROL DEGRADATION

Humans cannot metabolize the cholesterol ring structure to carbon dioxide and water. Rather, the intact steroid nucleus is eliminated from the body by conversion to bile acids and bile salts, a small percentage of which is excreted in the feces, and by secretion of cholesterol into the bile, which transports it to the intestine for elimination. Some of the cholesterol in the intestine is modified by bacteria before excretion. The primary compounds made are the isomers coprostanol and cholestanol, which are reduced derivatives of cholesterol. Together with cholesterol, these compounds make up the bulk of neutral fecal sterols.

V. BILE ACIDS AND BILE SALTS

Bile consists of a watery mixture of organic and inorganic compounds. Phosphatidylcholine (PC), or lecithin (see Chapter 17), and conjugated bile salts are quantitatively the most important organic components of bile. Bile can either pass directly from the liver, where it is synthesized, into the duodenum through the common bile duct, or be stored in the gallbladder when not immediately needed for digestion.

A. Structure

The bile acids contain 24 carbons, with two or three hydroxyl groups and a side chain that terminates in a carboxyl group (Fig. 18.9A). The carboxyl group has a pK_a of ~6. In the duodenum (pH~6), this group will be protonated in half of the molecules (the bile acids) and deprotonated in the rest (the bile salts). The terms bile acid and bile salt are frequently used interchangeably, however. Both forms have

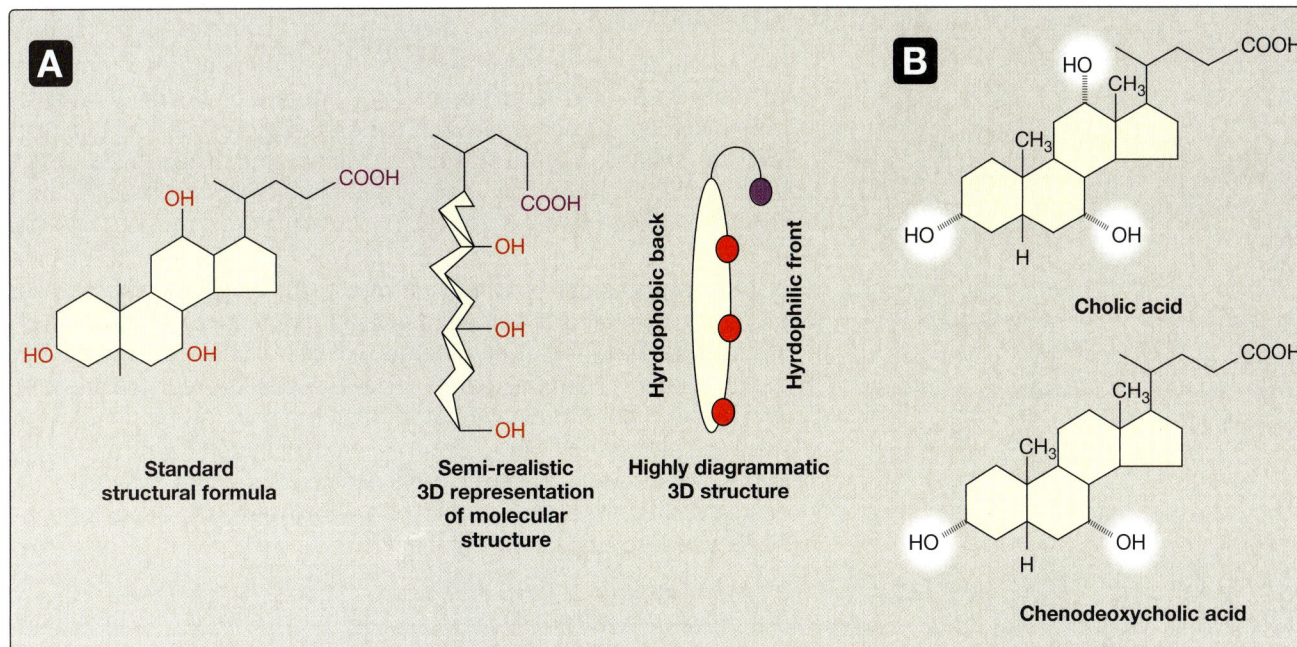

Figure 18.9
A. Structure of bile acids. The hydrophobic (nonpolar) and hydrophilic (polar) surfaces help emulsifying and the digestion of fats in the intestine surfaces. The *oval shape* indicates the cholesterol backbone. The *red balls* indicate the hydroxyl groups. The *purple ball* indicates the carboxyl group. **B.** The most abundant bile acids in human bile: cholic acid and chenodeoxycholic acid.

hydroxyl groups that are α in orientation (they lie below the plane of the rings) and methyl groups that are β (they lie above the plane of the rings). Therefore, the molecules have both a polar and a nonpolar surface and can act as emulsifying agents in the intestine, helping prepare dietary fat (triacylglycerol [TAG]) and other complex lipids for degradation by pancreatic digestive enzymes (Fig. 18.9B).

B. Synthesis

Bile acids are synthesized in the liver by a multistep, multiorganelle pathway in which hydroxyl groups are inserted at specific positions on the steroid structure; the double bond of the cholesterol B ring is reduced; and the hydrocarbon chain is shortened by three carbons, introducing a carboxyl group at the end of the chain. The most common resulting compounds, cholic acid (a triol) and chenodeoxycholic acid (a diol), as shown in Figure 18.9B, are called primary bile acids. The rate-limiting step in bile acid synthesis is the introduction of a hydroxyl group at carbon 7 of the steroid nucleus by 7-α-hydroxylase, a sER-associated cytochrome P450 (CYP) monooxygenase found only in liver. Expression of the enzyme is downregulated by bile acids and cholesterol (Fig. 18.10). Expression of cholesterol-7-α hydroxylase is upregulated by cholesterol and downregulated by bile acids. Elevated levels of cholesterol in the liver stimulate the nuclear receptor liver X factor (LXR), which increases the transcription of cholesterol-7-alpha hydroxylase. Elevated levels of bile acids activate another nuclear receptor bile acid receptor ([BAR] also known as farnesoid X receptor [FXR]), which downregulates the transcription of cholesterol-7-α hydroxylase.

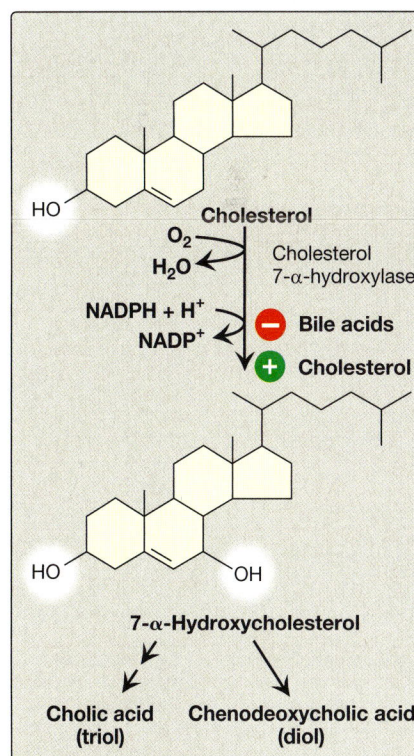

Figure 18.10
Synthesis and regulation of the bile acids, cholic acid and chenodeoxycholic acid from cholesterol.

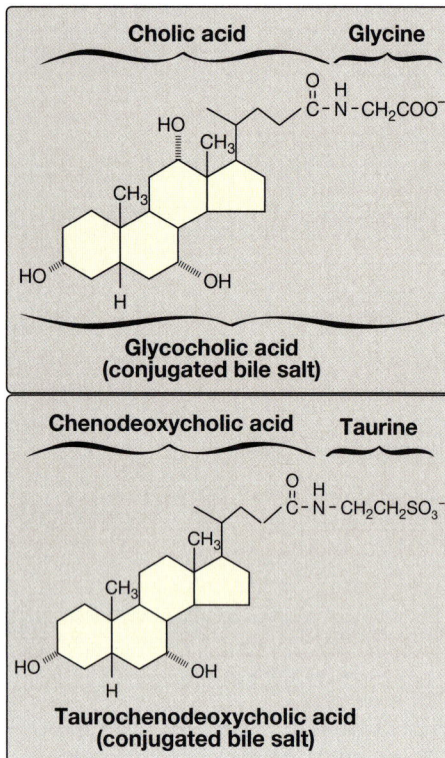

Figure 18.11
Conjugated bile salts. Note "cholic" in the names.

C. Conjugation

Before the bile acids leave the liver, they are conjugated to a molecule of either glycine or taurine (end product of cysteine metabolism) by an amide bond between the carboxyl group of the bile acid and the amino group of the added compound. These new structures include glycocholic and glycochenodeoxycholic acids and taurocholic and taurochenodeoxycholic acids (Fig. 18.11). The ratio of glycine to taurine forms in the bile is ~3/1. Addition of glycine or taurine results in the presence of a carboxyl group with a lower pK_a (from glycine) or a sulfonate group (from taurine), both of which are fully ionized (negatively charged) at the alkaline pH of bile and the duodenum. The conjugated, ionized bile salts are more effective detergents than the unconjugated ones because of their enhanced amphipathic nature. Therefore, only the conjugated forms are found in the bile. Individuals with genetic deficiencies in the conversion of cholesterol to bile acids are treated with exogenously supplied chenodeoxycholic acid.

> Bile salts provide the only significant mechanism for cholesterol excretion, both as a metabolic product of cholesterol and as a solubilizer of cholesterol in bile.

D. Enterohepatic circulation

Bile salts secreted into the intestine are efficiently reabsorbed (>95%) and reused. The liver actively secretes bile salts via the bile salt export pump. In the intestine, they are reabsorbed in the terminal ileum via the apical sodium (Na^+)-bile salt cotransporter and returned to the blood via a separate transport system. [Note: Lithocholic acid is only poorly absorbed.] They are efficiently taken up from the blood by the hepatocytes via an isoform of the cotransporter. Albumin binds bile salts and transports them through the blood as was seen with FA. The continuous cycle of bile salt secretion into the bile, passage through the duodenum (where some are deconjugated then dehydroxylated to secondary bile salts), uptake in the ileum, and subsequent return to the liver (as a mixture of primary and secondary forms) is termed the enterohepatic circulation (Fig. 18.12). The total amount of bile salt in circulation is 2 to 4 g. This amount recycles 6 to 10 times daily, occurring multiple times with each meal. So, the total daily secretion of bile salts from the liver into the duodenum is 15 to 30 g. Despite this, only about 0.5 g (<3%) is excreted from the body in the feces each day. To compensate for this loss, ~0.5 g is synthesized daily from cholesterol in the liver. Bile acid sequestrants, such as cholestyramine, bind bile salts in the gut and prevent their reabsorption, thereby promoting their excretion. The increased excretion of bile salts relieves the inhibition of bile acid synthesis in the liver so that cholesterol is used in this pathway. This mechanism of action makes bile acid sequestrants useful in the treatment of hypercholesterolemia. Dietary fiber acts in a similar manner by binding bile salts and increasing their excretion (see Chapter 27).

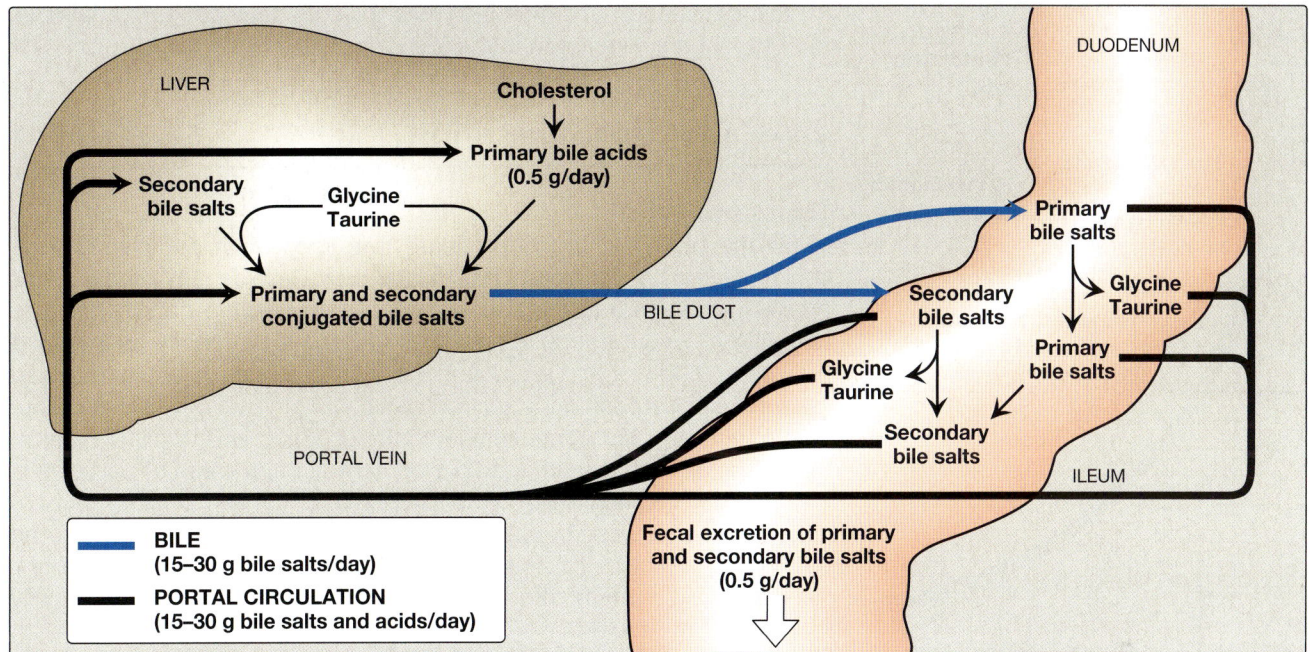

Figure 18.12
Enterohepatic circulation of bile salts. [Note: Ionized bile acids are called bile salts.]

E. Bacterial action on bile salts

After enterohepatic circulation, a small amount of secreted bile salts reaches the colon, where they are exposed to bacterial modification by the gut microbiome. Bacteria of the intestinal microbiota can deconjugate (remove glycine and taurine) bile salts. They can also dehydroxylate carbon 7, producing secondary bile acids such as deoxycholic acid from cholic acid and lithocholic acid from chenodeoxycholic acid. A small proportion of these secondary bile acids are absorbed by the colonic epithelium and may be conjugated and

---

Clinical Application 18.2: Gallstone Disease (Cholelithiasis) Due to Bile Salt Deficiency

The movement of cholesterol from the liver into the bile must be accompanied by the simultaneous secretion of phospholipid and bile salts. If this dual process is disrupted and more cholesterol is present than can be solubilized by the bile salts and PC present, the cholesterol may precipitate in the gallbladder, leading to cholesterol gallstone disease or cholelithiasis (Fig. 18.13). This disorder is typically caused by a decrease of bile acids in the bile. Cholelithiasis may also result from increased secretion of cholesterol into bile, as seen with the use of fibrates (e.g., gemfibrozil) to reduce cholesterol (and TAG) in the blood. Laparoscopic cholecystectomy (surgical removal of the gallbladder through a small incision) is currently the treatment of choice. However, for patients who are unable to undergo surgery, oral administration of ursodeoxycholic acid to supplement the body's supply of bile acids results in a gradual (months to years) dissolution of the gallstones. Cholesterol stones account for >85% of cases of cholelithiasis, with bilirubin and mixed stones accounting for the rest.

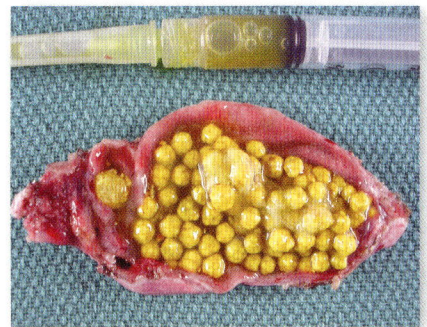

Figure 18.13
Gallbladder with gallstones.

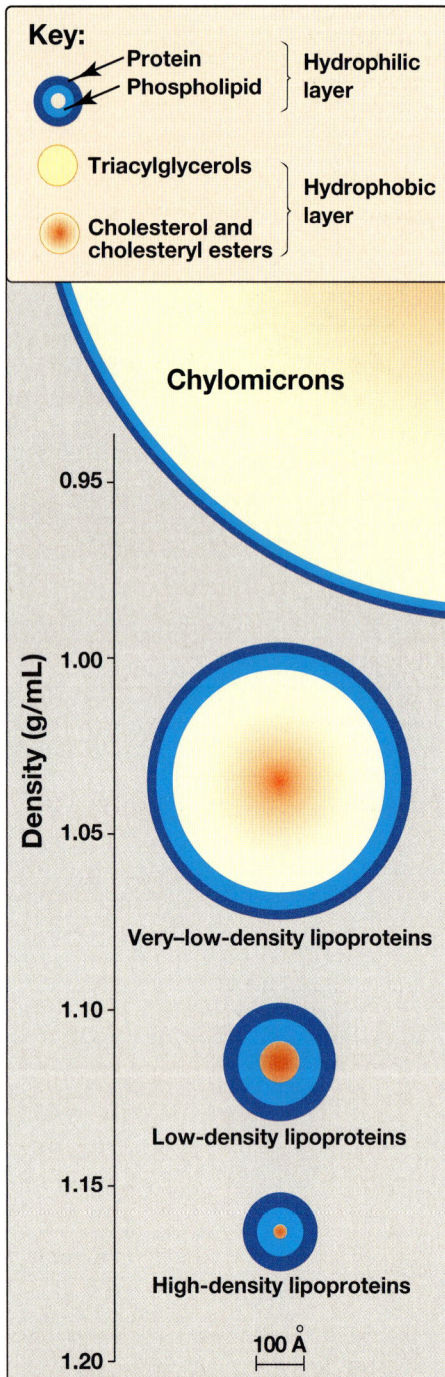

Key:

Protein — Phospholipid — Hydrophilic layer

Triacylglycerols — Hydrophobic layer

Cholesterol and cholesteryl esters

Chylomicrons

Density (g/mL): 0.95, 1.00, 1.05, 1.10, 1.15, 1.20

Very–low-density lipoproteins

Low-density lipoproteins

High-density lipoproteins

100 Å

Figure 18.14
Plasma lipoprotein particles exhibit a range of sizes and densities, and typical values are shown. Ring widths approximate the amount of each component. [Note: Although cholesterol and its esters are shown as one component in the center of each particle, physically, cholesterol is on the surface, whereas cholesteryl esters are in the interior.]

hydroxylated by the liver enzymes to produce secondary bile salts. The rest are eliminated in the feces.

VI. PLASMA LIPOPROTEINS

The plasma lipoproteins are spherical macromolecular complexes of lipids and proteins (apolipoproteins). The lipoprotein particles include chylomicrons, chylomicron remnants, very–low-density lipoproteins (VLDLs), VLDL remnants (also known as intermediate-density lipoproteins [IDLs]), low-density lipoproteins (LDLs), high-density lipoproteins (HDLs), and lipoprotein (a) (Lp[a]). They differ in site of origin and lipid and protein composition, size, and density (Fig. 18.14). [Note: Because lipoprotein particles constantly interchange lipids and apolipoproteins, the actual apolipoprotein and lipid content of each class of particles is somewhat variable.] Lipoproteins function both to keep their component lipids soluble as they transport them in the plasma and to provide an efficient mechanism for transporting their lipid contents to (and from) the tissues. In humans, there is a gradual deposition of lipid (especially cholesterol) in tissues.

A. Composition

Lipoproteins are composed of a neutral lipid core (containing TAG and cholesteryl esters) surrounded by a shell of amphipathic apolipoproteins, phospholipid, and nonesterified (free) cholesterol (Fig. 18.15). These amphipathic compounds are oriented such that their polar portions are exposed on the surface of the lipoprotein, thereby rendering the particle soluble in aqueous solution. The TAG and cholesterol carried by the lipoproteins are obtained either from the diet (exogenous source) or from *de novo* synthesis (endogenous source). [Note: The cholesterol (C) content of plasma lipoproteins is now routinely measured in fasting blood. Friedewald equation (LDL-C = Total C – HDL-C – TAG/5) is used to calculate LDL-C once the total C, HDL, and TAG are measured in serum. This formula assumes the TAG/cholesterol ratio in VLDL is 5:1. The goal value for total cholesterol is <200 mg/dL.]

1. **Size and density:** Chylomicrons are the lipoprotein particles lowest in density and largest in size and that contain the highest percentage of lipid (as TAG) and the lowest percentage of protein. VLDLs and LDLs are successively denser, having higher ratios of protein to lipid. HDL particles are the smallest and densest. Plasma lipoproteins can be separated on the basis of their electrophoretic mobility, as shown in Figure 18.16, or on the basis of their density by ultracentrifugation.

2. **Apolipoproteins:** The apolipoproteins associated with lipoprotein particles have a number of diverse functions, such as providing recognition sites for cell-surface receptors and serving as activators or coenzymes for enzymes involved in lipoprotein metabolism. Some of the apolipoproteins are required as essential structural components of the particles and cannot be removed (in fact, the particles cannot be produced without them), whereas others are transferred freely between lipoproteins. Apolipoproteins are divided by structure and function into several major classes,

denoted by letters, with each class having subclasses (e.g., apolipoprotein [apo] C-I, apo C-II, and apo C-III). [Note: The functions of all the apolipoproteins are not yet known.]

B. Chylomicron metabolism

Chylomicrons are assembled in intestinal mucosal cells and carry dietary (exogenous) TAGs, cholesterol, fat-soluble vitamins, and cholesteryl esters to the peripheral tissues (Fig. 18.17). [Note: TAGs account for close to 90% of the lipids in a chylomicron.]

1. **Apolipoprotein synthesis:** Apo B-48 is unique to chylomicrons. Its synthesis begins on the rough ER (rER), and it is glycosylated as it moves through the rER and Golgi. [Note: Apo B-48 is so named because it constitutes the N-terminal 48% of the protein encoded by the gene for apo B. Apo B-100, which is synthesized by the liver and found in VLDL and LDL, represents the entire protein encoded by this gene. Posttranscriptional editing (see Chapter 33) of a cytosine to a uracil in intestinal apo B-100 messenger RNA (mRNA) creates a nonsense (stop) codon (see Chapter 33), allowing translation of only 48% of the mRNA.]

2. **Chylomicron assembly:** Many enzymes involved in TAG, cholesterol, and phospholipid synthesis are located in the sER. Assembly of the apolipoprotein and lipid into chylomicrons requires microsomal triglyceride transfer protein (MTP), which loads apo B-48 with lipid. [Note: MTP is also expressed in the liver where it loads lipids, mainly TAGs, onto apo B-100 to form nascent VLDL.] This occurs before transition from the ER to the Golgi, where the particles are packaged in secretory vesicles. These fuse with the plasma membrane releasing the lipoproteins, which then enter the lymphatic system and, ultimately, the blood. [Note: Chylomicrons leave the lymphatic system via the thoracic duct that empties into the left subclavian vein.]

3. **Nascent chylomicron modification:** The particle released by the intestinal mucosal cell is called a nascent chylomicron because it is functionally incomplete. When it reaches the plasma, the particle is rapidly modified, receiving apo E (which is recognized by hepatic receptors) and apo C. The latter includes apo C-II, which is necessary for the activation of lipoprotein lipase (LPL), the enzyme that degrades the TAG contained in the chylomicron. The source of these apolipoproteins is circulating HDL (see Fig. 18.17). [Note: Apo C-III on TAG-rich lipoproteins inhibits LPL.]

4. **TAG degradation by LPL:** LPL is an extracellular enzyme that is anchored to the capillary walls of most tissues but predominantly those of adipose tissue and cardiac and skeletal muscle. The adult liver does not express LPL. Instead, hepatic lipase, a distinct enzyme found on the surface of endothelial cells of the liver, plays a role in TAG degradation in chylomicrons and VLDL and is important in HDL metabolism. LPL, activated by apo C-II on circulating chylomicrons, hydrolyzes the TAG in these particles to FA and glycerol. The FA are stored (in adipose) or used for energy (in muscle). The glycerol is taken up by the liver, converted to dihydroxyacetone phosphate (an intermediate of glycolysis),

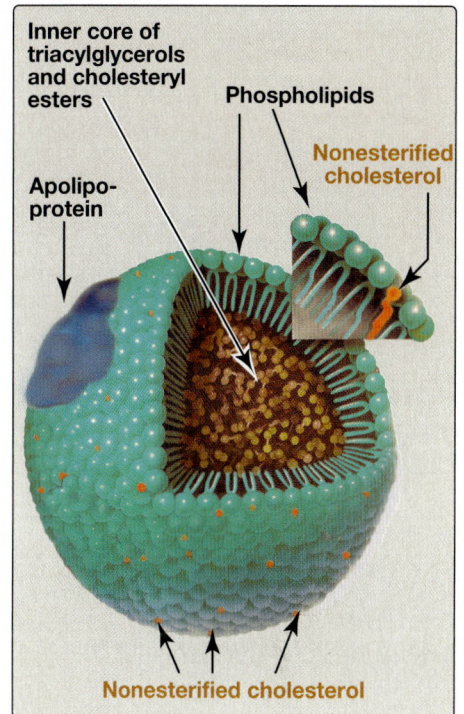

Figure 18.15
Structure of a typical lipoprotein particle.

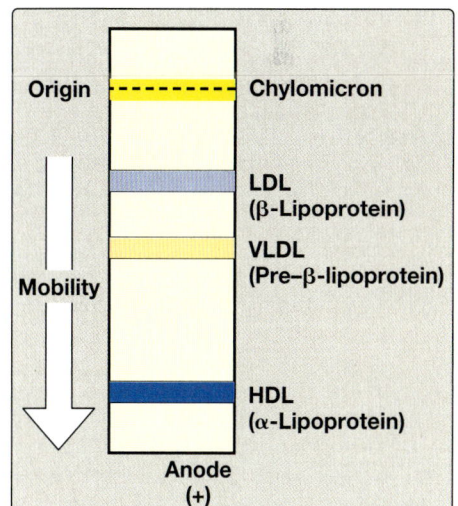

Figure 18.16
Electrophoretic mobility of plasma lipoprotein particles. [Note: The order of low-density lipoprotein [LDL] and very–low-density lipoprotein [VLDL] is reversed if ultracentrifugation is used as the separation technique.] HDL, high-density lipoprotein.

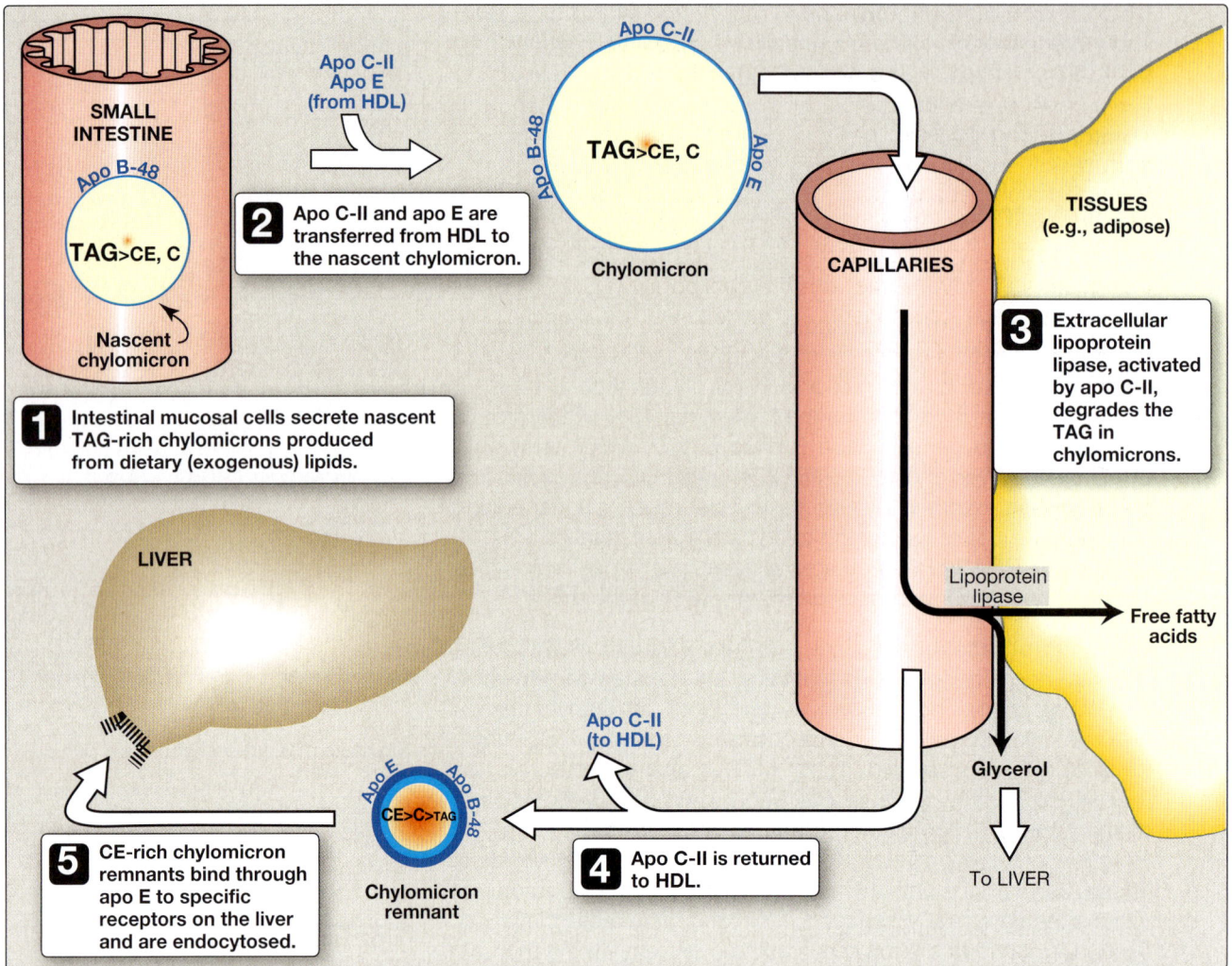

Figure 18.17
Metabolism of chylomicrons. Apo B-48, apo C-II, and apo E are apolipoproteins found as components of plasma lipoprotein particles. The particles are not drawn to scale (see Fig. 18.14 for details of their size and density). TAG, triacylglycerol; C, cholesterol; CE, cholesteryl ester; HDL, high-density lipoprotein.

and used in lipid synthesis or gluconeogenesis. Patients with a deficiency of LPL (type I hyperlipoproteinemia or familial chylomicronemia) show a dramatic accumulation ($\geq$1,000 mg/dL) of chylomicron-TAG in the plasma (hypertriacylglycerolemia) even in the fasted state. They are at increased risk for acute pancreatitis. Treatment is the reduction of dietary fat. [Note: A similar condition, type IB hyperlipoproteinemia, results from a deficiency in apo C-II.]

5. **LPL expression:** LPL is synthesized by adipose tissue and by cardiac and skeletal muscle. Expression of the tissue-specific isozymes is regulated by nutritional state and hormonal level. For example, in the fed state (elevated insulin levels), LPL synthesis is increased in adipose but decreased in muscle tissue. Fasting (decreased insulin) favors LPL synthesis in muscle. [Note: The highest concentration of LPL is in cardiac muscle, reflecting the use of FA to provide much of the energy needed for cardiac function.]

6. **Chylomicron remnant formation:** As the chylomicron circulates, and more than 90% of the TAG in its core is degraded by LPL, the particle decreases in size and increases in density. In addition, the C apolipoproteins (but not apo B or E) are returned to HDL. The remaining particle, called a remnant, is rapidly removed from the circulation by the liver, whose cell membranes contain lipoprotein receptors that recognize apo E (see Fig. 18.17). Chylomicron remnants bind to these receptors and are taken into the hepatocytes by endocytosis. The endocytosed vesicle then fuses with a lysosome, and the apolipoproteins, cholesteryl esters, and other components of the remnant are hydrolytically degraded, releasing amino acids, free cholesterol, and FAs. The receptor is recycled. [Note: The mechanism of receptor-mediated endocytosis is illustrated for LDL in Fig. 18.21.]

C. Very–low-density lipoprotein metabolism

VLDLs are produced in the liver (Fig. 18.18). They are composed predominantly of endogenous TAGs (~60%), and their function is to carry this lipid from the liver (site of synthesis) to the peripheral

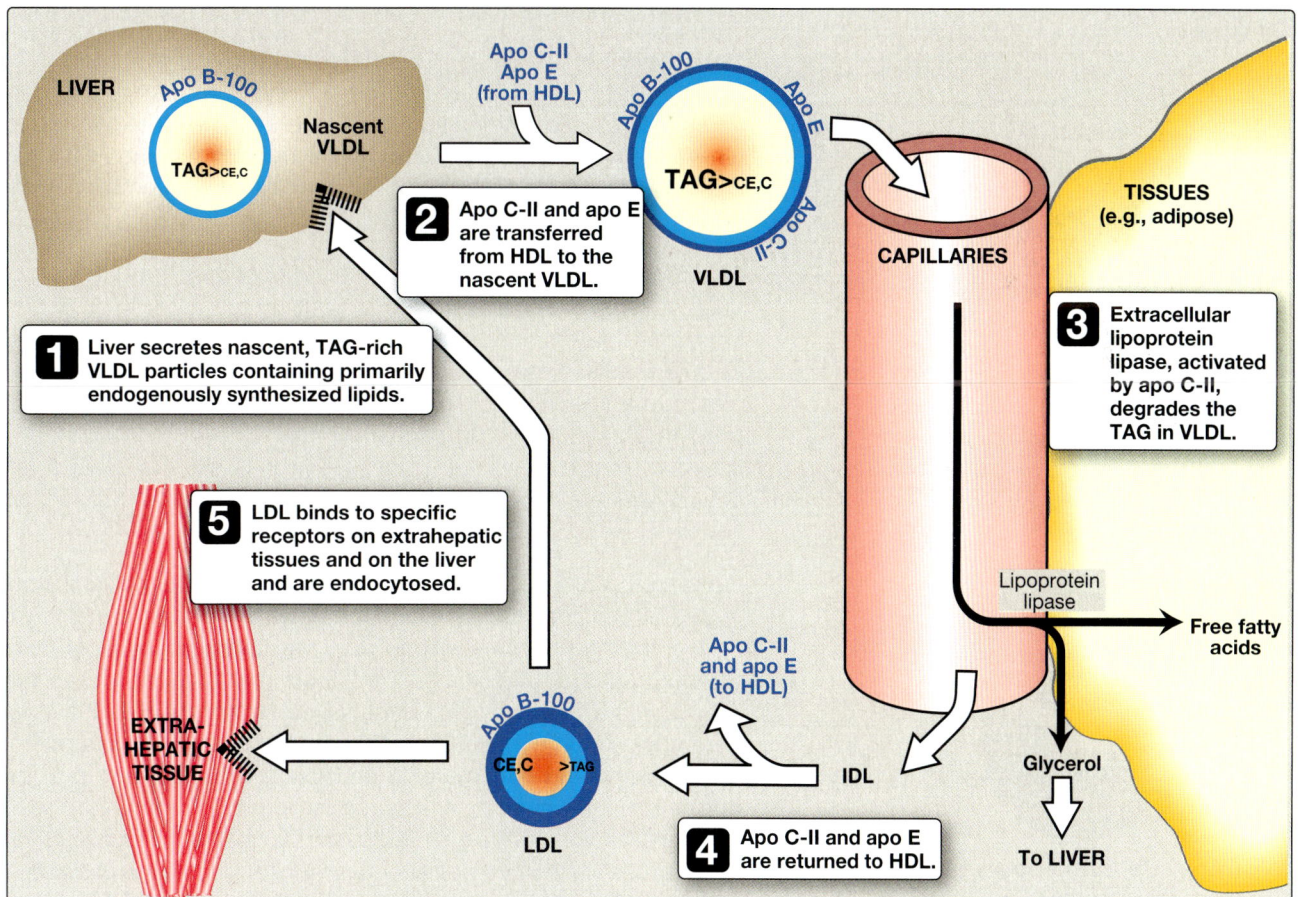

Figure 18.18
Metabolism of very–low-density lipoprotein (VLDL) and low-density lipoprotein (LDL) particles. Apo B-100, C-II, and E are apolipoproteins found as components of plasma lipoprotein particles. The particles are not drawn to scale (see Fig. 18.14 for details of their size and density). [Note: IDL can also be taken up by liver.] TAG, triacylglycerol; HDLs and IDLs, high- and intermediate-density lipoproteins; C, cholesterol; CE, cholesteryl ester.

Clinical Application 18.3: Abetalipoproteinemia

Abetalipoproteinemia is a rare autosomal recessive genetic disorder caused by a defect in the *MTTP* gene that codes for MTP. This defect impairs lipid loading on apo B proteins in the intestine and liver. In the intestine, the failure to load lipids onto apo B-48 prevents the formation of chylomicrons and reduces the absorption of lipid and fat-soluble vitamins. In the liver, nascent VLDL formation is disrupted because TAGs cannot be loaded on apo B-100. Patients have very low LDL-C levels (hypolipoproteinemia), present with signs and symptoms related to vitamin A and E deficiencies (see Chapter 28), and may develop hepatic steatosis. A similar disorder, familial hypobetalipoproteinemia, results from mutations in the APOB gene that decrease the production of functional apolipoprotein B proteins.

---

tissues. There, the TAG is degraded by LPL, as discussed for chylomicrons. [Note: Nonalcoholic fatty liver (hepatic steatosis) occurs in conditions in which there is an imbalance between hepatic TAG synthesis and the secretion of VLDL. Such conditions include obesity and type 2 diabetes mellitus (see Chapter 25).]

1. **Release from the liver:** VLDLs are produced in the liver when lipids are added to apo B-100 by MTP. The nascent VLDL particles, composed primarily of TAG, are secreted into the blood where they obtain apo C-II and apo E from circulating HDL (see Fig. 18.18). As with chylomicrons, apo C-II is required for activation of LPL.

2. **Modification in the circulation:** As VLDL passes through the circulation, TAG is degraded by LPL, causing the VLDL to decrease in size and become denser. Surface components, including the C and E apolipoproteins, are returned to HDL, but the particles retain apo B-100. Additionally, some TAGs are transferred from VLDL to HDL in an exchange reaction that concomitantly transfers cholesteryl esters from HDL to VLDL. This exchange is accomplished by cholesteryl ester transfer protein (CETP), as shown in Figure 18.19.

3. **Conversion to LDLs:** With these modifications, the VLDL is converted in the plasma to LDL. IDLs of varying sizes are formed during this transition. IDL can also be taken up by liver cells through receptor-mediated endocytosis that uses apo E as the ligand. Apo E is normally present in three isoforms, E-2 (the least common), E-3 (the most common), and E-4. Apo E-2 binds poorly to receptors, and patients who are homozygotic for apo E-2 are deficient in the clearance of IDL and chylomicron remnants. These individuals have familial type III hyperlipoproteinemia (familial dysbetalipoproteinemia or broad beta disease), with hypercholesterolemia and premature atherosclerosis. These patients have a risk for coronary artery disease that is 5 to 10 times greater than that of the general population. The apo E-4 isoform confers increased susceptibility to an earlier age of onset of the late-onset form of Alzheimer disease. The effect is dose dependent, with homozygotes being at greatest risk. Estimates of the risk vary.

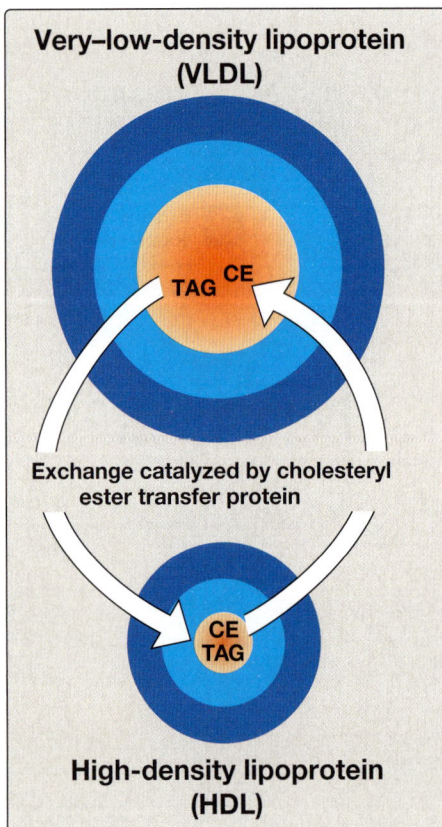

Figure 18.19
Transfer of cholesteryl ester (CE) from HDL to VLDL in exchange for triacylglycerol (TAG).

D. Low-density lipoprotein metabolism

LDL particles contain much less TAG than their VLDL predecessors and have a high concentration of cholesterol and cholesteryl esters (Fig. 18.20). About 70% of plasma cholesterol is in LDL.

1. **Receptor-mediated endocytosis:** The primary function of LDL particles is to provide cholesterol to the peripheral tissues (or return it to the liver). They do so by binding to plasma membrane LDL receptors that recognize apo B-100 (but not apo B-48). Because these LDL receptors can also bind apo E, they are known as apo B-100/apo E receptors. A summary of the uptake and degradation of LDL particles is presented in Figure 18.21. [Note: The numbers in brackets below refer to corresponding numbers on that figure.] A similar mechanism of receptor-mediated endocytosis is used for the uptake and degradation of chylomicron remnants and IDLs by the liver.

   [1] LDL receptors are negatively charged glycoproteins that are clustered in pits on cell membranes. The cytosolic side of the pit is coated with the protein clathrin, which stabilizes the pit.

   [2] After binding, the LDL–receptor complex is endocytosed. Defects in the LDL receptor or in apo B-100 that disable this step cause a significant elevation in plasma LDL-C, coronary artery disease, and lipid deposits in tissues. Mutations in the gene for the LDL receptor can cause either autosomal dominant or autosomal recessive forms of familial hypercholesterolemia 1 (FHCL1). Autosomal dominant familial hypercholesterolemia 2 (FHCL2) is caused by defects in apo B-100 that reduce its binding to the LDL receptor. Autosomal dominant familial hypercholesterolemia 3 (FHCL3) is associated with a gain of function in the proprotein convertase subtilisin/kexin type 9 (PCSK9) protease, which promotes endocytosis and lysosomal degradation of the LDL receptor. Current therapeutic approaches for FHCL involve the administration of statins, PCSK9 inhibitors, ezetimibe, and bempedoic acid. Bempedoic acid acts by inhibiting ATP-citrate lyase (Chapter 16, Figure 16.11), which is involved in the early steps of the cholesterol synthesis pathway, converting citrate to acetyl-CoA, which is a precursor for the synthesis of cholesterol and FAs. By inhibiting ATP-citrate lyase, bempedoic acid reduces the production of cholesterol in the liver.

   [3] The vesicle containing LDL loses its clathrin coat and fuses with other similar vesicles, forming larger vesicles called endosomes.

   [4] The pH of the endosome falls (due to the proton-pumping activity of endosomal ATPase), which allows separation of the LDL from its receptor. The receptors then migrate to one side of the endosome, whereas the LDL stay free within the lumen of the vesicle.

   [5] The receptors can be recycled, whereas the lipoprotein remnants in the vesicle are transferred to lysosomes and degraded by lysosomal acid hydrolases, releasing free cholesterol, amino acids, FA, and phospholipids. These compounds can be reutilized by the cell. [Note: A few of the lysosomal storage

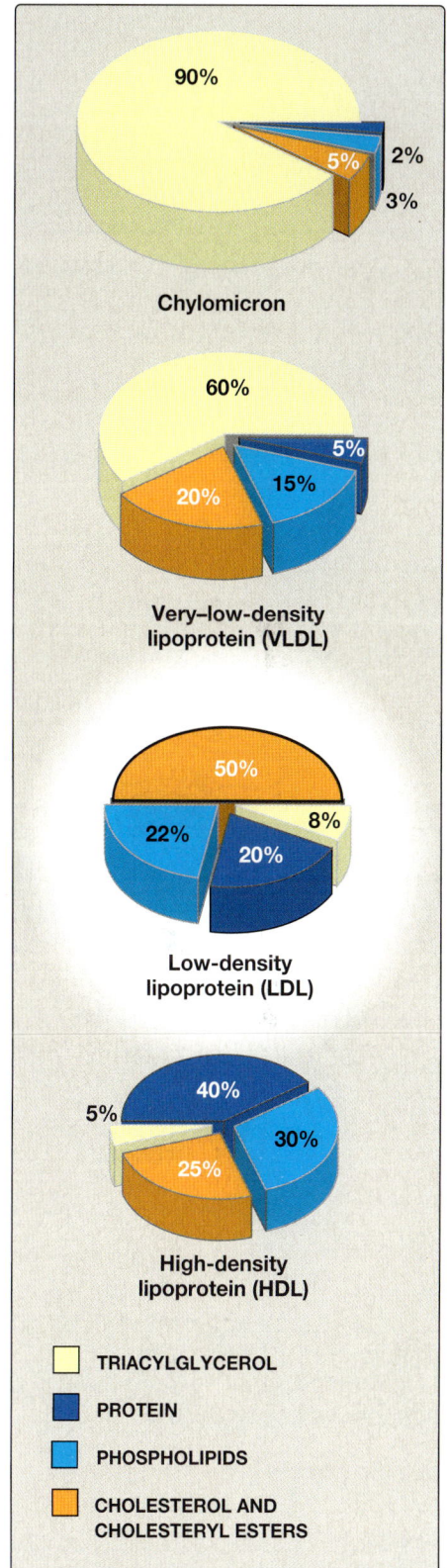

TRIACYLGLYCEROL

PROTEIN

PHOSPHOLIPIDS

CHOLESTEROL AND CHOLESTERYL ESTERS

Figure 18.20
Composition of the plasma lipoprotein particles. Note the high concentration of cholesterol and cholesteryl esters in LDL.

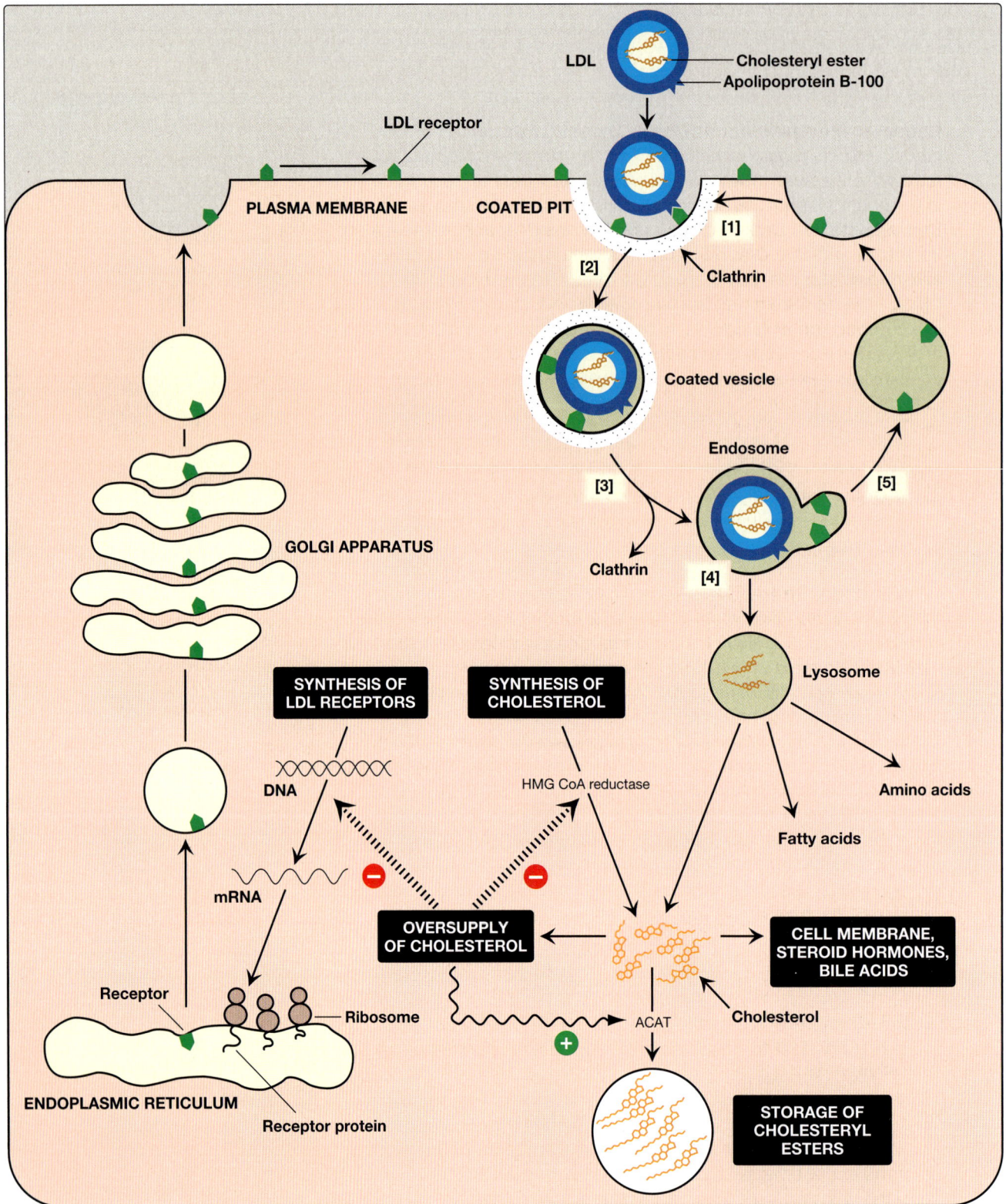

Figure 18.21
Cellular uptake and degradation of low-density lipoprotein (LDL) particles. [Note: Oversupply of cholesterol accelerates the degradation of HMG CoA reductase. It also decreases transcription of its gene as seen with the LDL receptor.] ACAT, acyl CoA:cholesterol acyltransferase; HMG CoA, hydroxymethylglutaryl coenzyme A; mRNA, messenger RNA.

diseases result from rare autosomal recessive deficiencies in the ability to hydrolyze lysosomal cholesteryl esters (Wolman disease) or to transport free cholesterol out of the lysosome (Niemann–Pick disease, type C).]

2. **Endocytosed cholesterol and cholesterol homeostasis:** The chylomicron remnant-, IDL-, and LDL-derived cholesterol affects cellular cholesterol content in several ways (Fig. 18.21). First, expression of the gene for HMG CoA reductase is inhibited by high cholesterol, and *de novo* cholesterol synthesis decreases as a result. Additionally, degradation of the reductase is accelerated. Second, synthesis of new LDL receptor protein is reduced by decreasing the expression of the LDL receptor gene, thus limiting further entry of LDL-C into cells. [Note: As was seen with the *reductase* gene, transcriptional regulation of the LDL receptor gene involves an SRE and SREBP-2. This allows coordinate regulation of the expression of these proteins.] Third, if the cholesterol is not required immediately for some structural or synthetic purpose, it is esterified by acyl CoA:cholesterol acyltransferase (ACAT). ACAT transfers a FA from a fatty acyl CoA to cholesterol, producing a cholesteryl ester that can be stored in the cell (Fig. 18.22). The activity of ACAT is enhanced in the presence of increased intracellular cholesterol.

3. **Uptake by macrophage scavenger receptors (SRs):** In addition to the highly specific and regulated receptor-mediated pathway for LDL uptake described above, macrophages possess high levels of SR activity. These receptors, known as scavenger receptor class A (SR-A), can bind a broad range of ligands and mediate the endocytosis of chemically modified LDL in which the lipid or apo B component has been oxidized. Unlike the LDL receptor, the SR is not downregulated in response to increased intracellular cholesterol. Cholesteryl esters accumulate in macrophages and cause their transformation into "foam" cells, which participate in the formation of atherosclerotic plaque (Fig. 18.23). LDL-C is the primary cause of atherosclerosis.

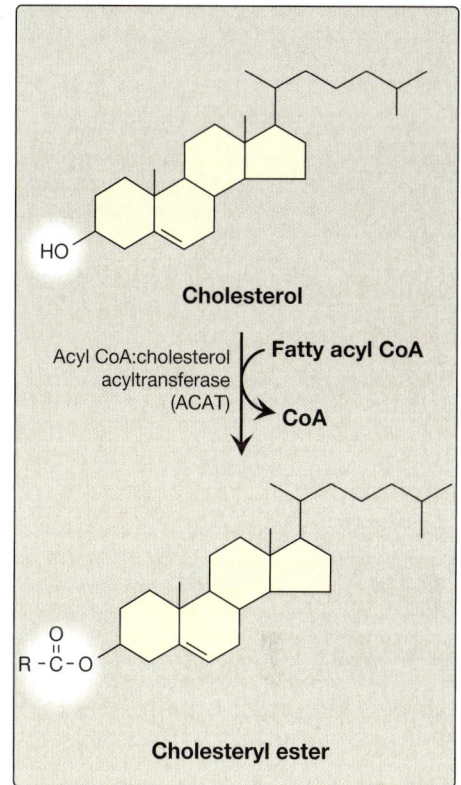

Figure 18.22
Synthesis of intracellular cholesteryl ester by ACAT. [Note: Lecithin:cholesterol acyl transferase (LCAT) is the extracellular enzyme that esterifies cholesterol using phosphatidylcholine (lecithin) as the source of the fatty acid.] CoA, coenzyme A.

E. High-density lipoprotein metabolism

HDLs comprise a heterogeneous family of lipoproteins with a complex metabolism that is not yet completely understood. HDL particles are formed in the blood by the addition of lipid to apo A-1, an apolipoprotein made and secreted by the liver and intestine. Apo A-1 accounts for ~70% of the apolipoproteins in HDL. HDLs perform a number of important functions, including the following.

1. **Apolipoprotein supply:** HDL particles serve as a circulating reservoir of apo C-II (the apolipoprotein that is transferred to VLDL and chylomicrons and is an activator of LPL) and apo E (the apolipoprotein required for the receptor-mediated endocytosis of IDLs and chylomicron remnants).

2. **Nonesterified cholesterol uptake:** Nascent HDLs are disc-shaped particles containing primarily phospholipid (largely PC) and apo A, C, and E. They take up cholesterol from nonhepatic (peripheral) tissues and return it to the liver as cholesteryl esters

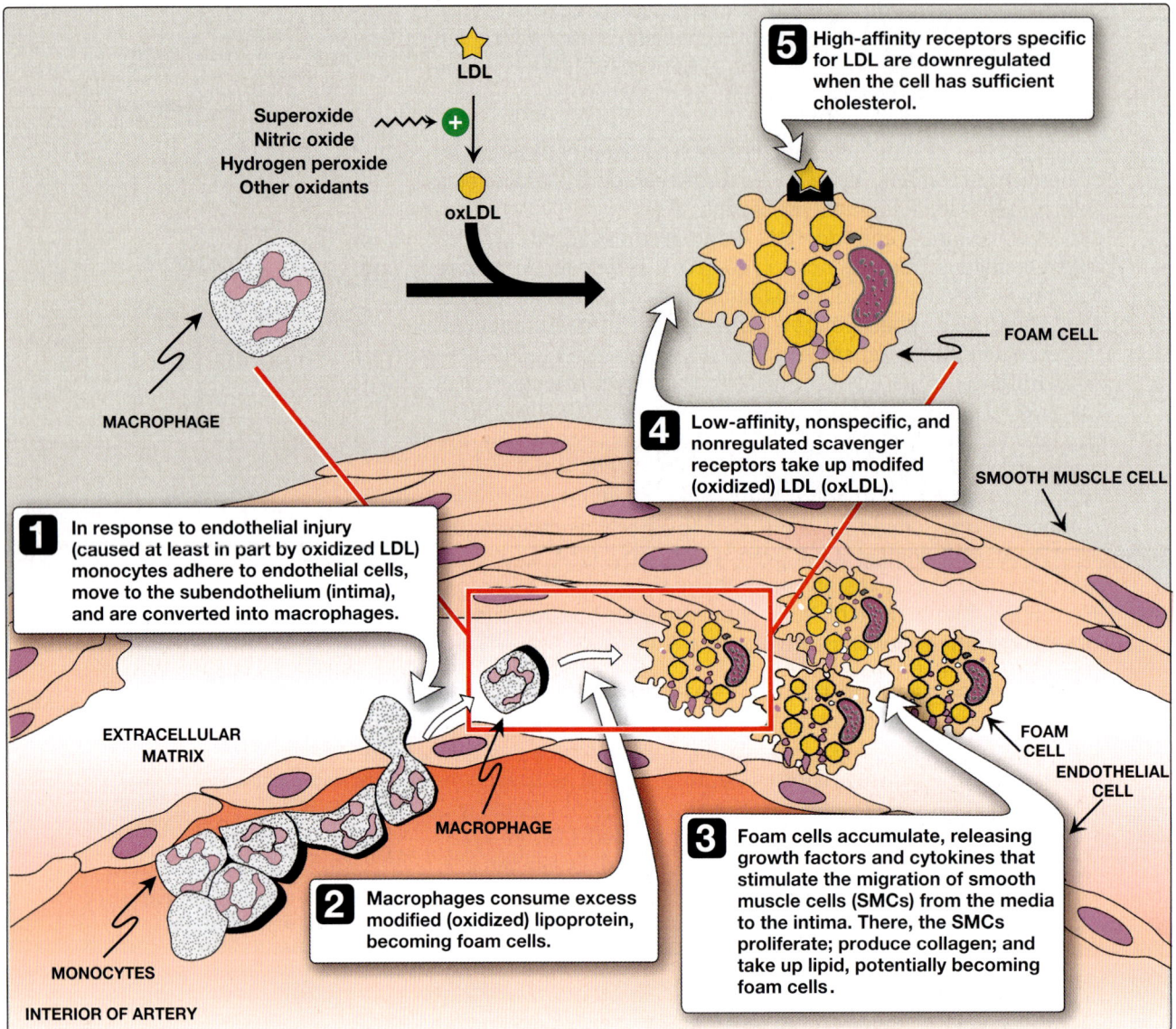

Figure 18.23
Role of oxidized low-density lipoprotein (LDL) particles in plaque formation in an arterial wall.

(Fig. 18.24). [Note: HDL particles are excellent acceptors of non-esterified cholesterol as a result of their high concentration of phospholipids, which are important solubilizers of cholesterol.]

3. **Cholesterol esterification:** The cholesterol taken up by HDL is immediately esterified by the plasma enzyme lecithin:cholesterol acyltransferase ([LCAT] also known as PCAT, in which P stands for PC, the source of the FA). This enzyme is synthesized and secreted by the liver. LCAT binds to nascent HDL and is activated by apo A-I. LCAT transfers the FA from carbon 2 of PC to cholesterol. This produces a hydrophobic cholesteryl ester, which is sequestered in the core of the HDL, and lysophosphatidylcholine, which binds to albumin. [Note: Esterification maintains the cholesterol

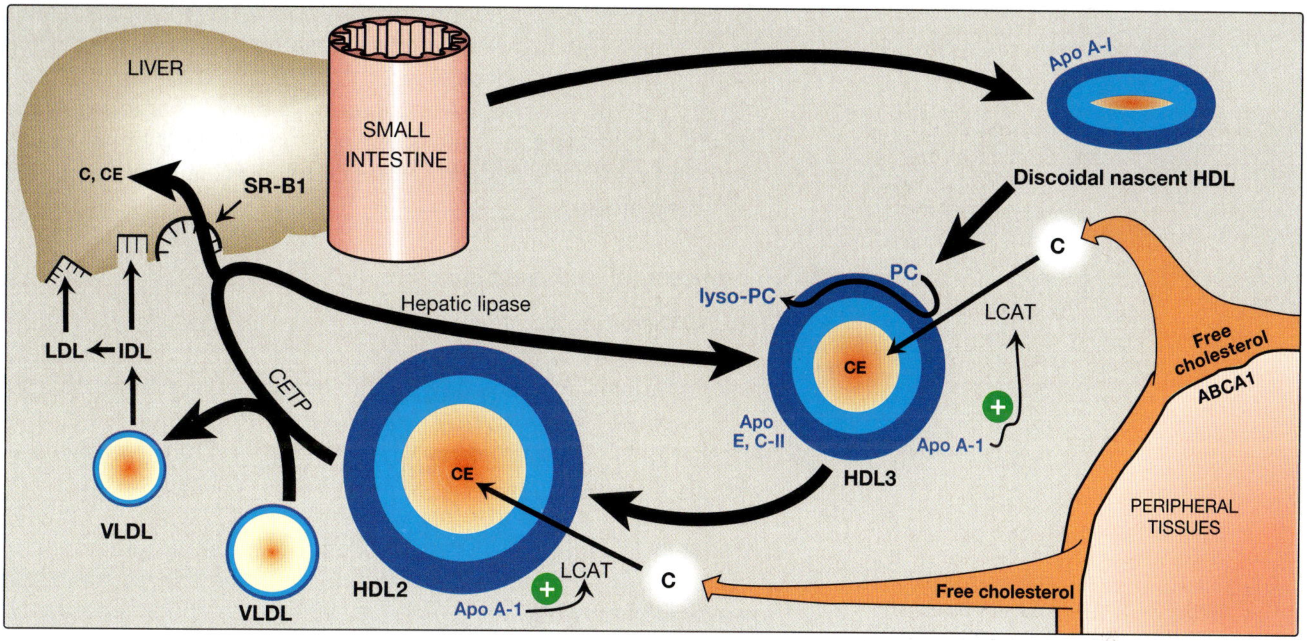

Figure 18.24
Metabolism of high-density lipoprotein (HDL) particles. Apo, apolipoprotein; ABCA1, ATP-binding cassette transport protein; C, cholesterol; CE, cholesteryl ester; LCAT, lecithin:cholesterol acyltransferase; VLDLs, IDLs, and LDLs, very–low-, intermediate-, and low-density lipoproteins; CETP, cholesteryl ester transfer protein; SR-B1, scavenger receptor B1.

concentration gradient, allowing continued efflux of cholesterol to HDL.] As the discoidal nascent HDL accumulates cholesteryl esters, it first becomes a spherical, relatively cholesteryl ester–poor HDL3 and, eventually, a cholesteryl ester–rich HDL2 particle that carries these esters to the liver. Hepatic lipase, which degrades TAGs and phospholipids, participates in the conversion of HDL2 to HDL3 (see Fig. 18.24). CETP transfers some of the cholesteryl esters from HDL to VLDL in exchange for TAG, relieving product inhibition of LCAT. Because VLDLs are catabolized to LDLs, the cholesteryl esters transferred by CETP are ultimately taken up by the liver.

4. **Reverse cholesterol transport (RCT):** The selective transfer of cholesterol from peripheral cells to HDL and from HDL to the liver for bile acid synthesis or disposal via the bile is a key component of cholesterol homeostasis. This process of RCT is, in part, the basis for the inverse relationship seen between plasma HDL concentration and atherosclerosis and for the designation of HDL as the "good" cholesterol carrier. [Note: Exercise and estrogen raise HDL levels.] RCT involves efflux of cholesterol from peripheral cells to HDL, esterification of the cholesterol by LCAT, binding of the cholesteryl ester–rich HDL (HDL2) to liver (and, perhaps, steroidogenic cells), selective transfer of the cholesteryl esters into these cells, and release of lipid-depleted HDL (HDL3). The efflux of cholesterol from peripheral cells is mediated primarily by the transport protein ATP-binding cassette A1 (ABCA1) protein. [Note: Tangier disease is a very rare deficiency of ABCA1 and is characterized by the virtual absence of HDL

particles due to degradation of lipid-poor apo A-1.] Cholesteryl ester uptake by the liver is mediated by the cell-surface receptor scavenger receptor class B type 1 (SR-B1) that binds HDL (see Section VI. D. 3. for SR-A receptors). The HDL particle itself is not taken up. Instead, there is selective uptake of the cholesteryl ester from the HDL particle. [Note: Low HDL-C is a risk factor for atherosclerosis.]

F. Lipoprotein (a) and heart disease

Lp(a), is nearly identical in structure to an LDL particle. Its distinguishing feature is the presence of an additional apolipoprotein molecule, apo(a), which is covalently linked at a single site to apo B-100. Apo(a) is structurally homologous to plasminogen, the precursor of a blood protease whose target is fibrin. Fibrin is the main protein component of blood clots (see Chapter 35). Lp(a) is an independent risk factor for coronary heart disease. The apo(a) component of Lp(a) particles is indicated to promote atherogenesis. Circulating levels of Lp(a) are determined primarily by genetics. However, diet may play some role, as trans FA has been reported to increase Lp(a). On the other hand, niacin reduces Lp(a), as well as LDL-C and TAG, but raises HDL-C.

VII. STEROID HORMONES

Cholesterol is the precursor of all classes of steroid hormones: glucocorticoids (e.g., cortisol), mineralocorticoids (e.g., aldosterone), and the sex hormones (i.e., androgens, estrogens, and progestins), as shown in Figure 18.25. [Note: Glucocorticoids and mineralocorticoids are collectively called corticosteroids.] Synthesis and secretion occur in the adrenal cortex (cortisol, aldosterone, and androgens), ovaries, and placenta (estrogens and progestins), and testes (testosterone). Steroid hormones are transported by the blood from their sites of synthesis to their target organs. Because of their hydrophobicity, they must be complexed with a plasma protein. Albumin can act as a nonspecific carrier and does carry aldosterone. However, specific steroid-carrier plasma proteins bind the steroid hormones more tightly than does albumin (e.g., corticosteroid-binding globulin, or transcortin, is responsible for transporting cortisol). A number of genetic diseases are caused by deficiencies in specific steps in the biosynthesis of steroid hormones. Some representative diseases are described in Figure 18.26.

A. Synthesis

Synthesis involves shortening the hydrocarbon chain of cholesterol and hydroxylating the steroid nucleus. The initial and rate-limiting reaction converts cholesterol to the 21-carbon pregnenolone. It is catalyzed by the cholesterol side-chain cleavage enzyme, a cytochrome P450 (CYP) mixed function oxidase of the inner mitochondrial membrane that is also known as $P450_{scc}$ and desmolase. NADPH and O_2 are required for the reaction. The cholesterol substrate can be newly synthesized, taken up from lipoproteins, or released by an esterase from cholesteryl esters stored in the cytosol of steroidogenic tissues. The cholesterol moves to the outer mitochondrial membrane.

An important control point is the subsequent movement from the outer to the inner mitochondrial membrane. This process is mediated by steroidogenic acute regulatory (StAR) protein. Pregnenolone is the parent compound for all steroid hormones (see Fig. 18.26). It is oxidized and then isomerized to progesterone, which is further modified to the other steroid hormones by CYP protein–catalyzed hydroxylation reactions in the SER and mitochondria. A defect in the activity or amount of an enzyme in this pathway can lead to a deficiency in the synthesis of hormones beyond the affected step and to an excess in the hormones or metabolites before that step. Because all members of the pathway have potent biologic activity, serious metabolic imbalances occur with enzyme deficiencies (see Fig. 18.26). Collectively, these disorders are known as congenital adrenal hyperplasia (CAH), because they result in enlarged adrenals. [Note: Addison disease, due to autoimmune destruction of the adrenal cortex, is characterized by adrenocortical insufficiency.]

B. Adrenal cortical steroid hormones

Steroid hormones are synthesized and secreted in response to hormonal signals. The corticosteroids and androgens are made in different regions of the adrenal cortex and are secreted into blood in response to different signals. [Note: The adrenal medulla makes catecholamines (see Chapter 21).]

1. **Cortisol:** Its production in the middle layer (zona fasciculata) of the adrenal cortex is controlled by the hypothalamus, to which the pituitary gland is attached (Fig. 18.27). In response to severe stress (e.g., infection), corticotropin-releasing hormone (CRH), produced by the hypothalamus, travels through capillaries to the anterior lobe of the pituitary, where it induces the production and secretion of adrenocorticotropic hormone (ACTH), a peptide. ACTH stimulates the adrenal cortex to synthesize and secrete the glucocorticoid cortisol, the stress hormone. [Note: ACTH binds to a membrane G protein–coupled receptor, resulting in cyclic AMP (cAMP) production and activation of protein kinase A (PKA). PKA phosphorylates and activates both the esterase that converts cholesteryl ester to free cholesterol and StAR protein.] Cortisol allows the body to respond to stress through its effects on intermediary metabolism (e.g., increased gluconeogenesis) and the inflammatory and immune responses (which are decreased). As cortisol levels rise, the release of CRH and ACTH is inhibited. [Note: The reduction of cortisol in CAH results in a rise in ACTH that causes adrenal hyperplasia.]

2. **Aldosterone:** Its production in the outer layer (zona glomerulosa) of the adrenal cortex is induced by a decrease in the plasma

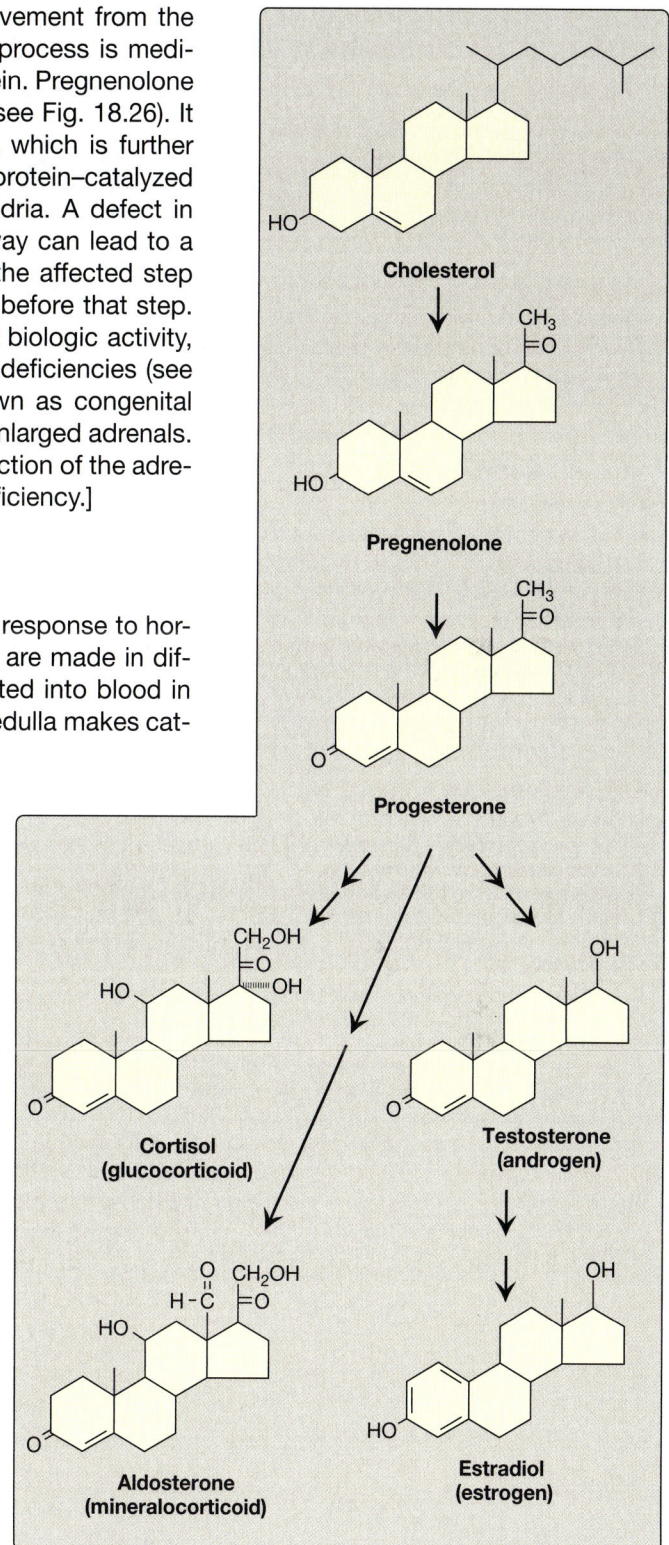

Figure 18.25
Key steroid hormones.

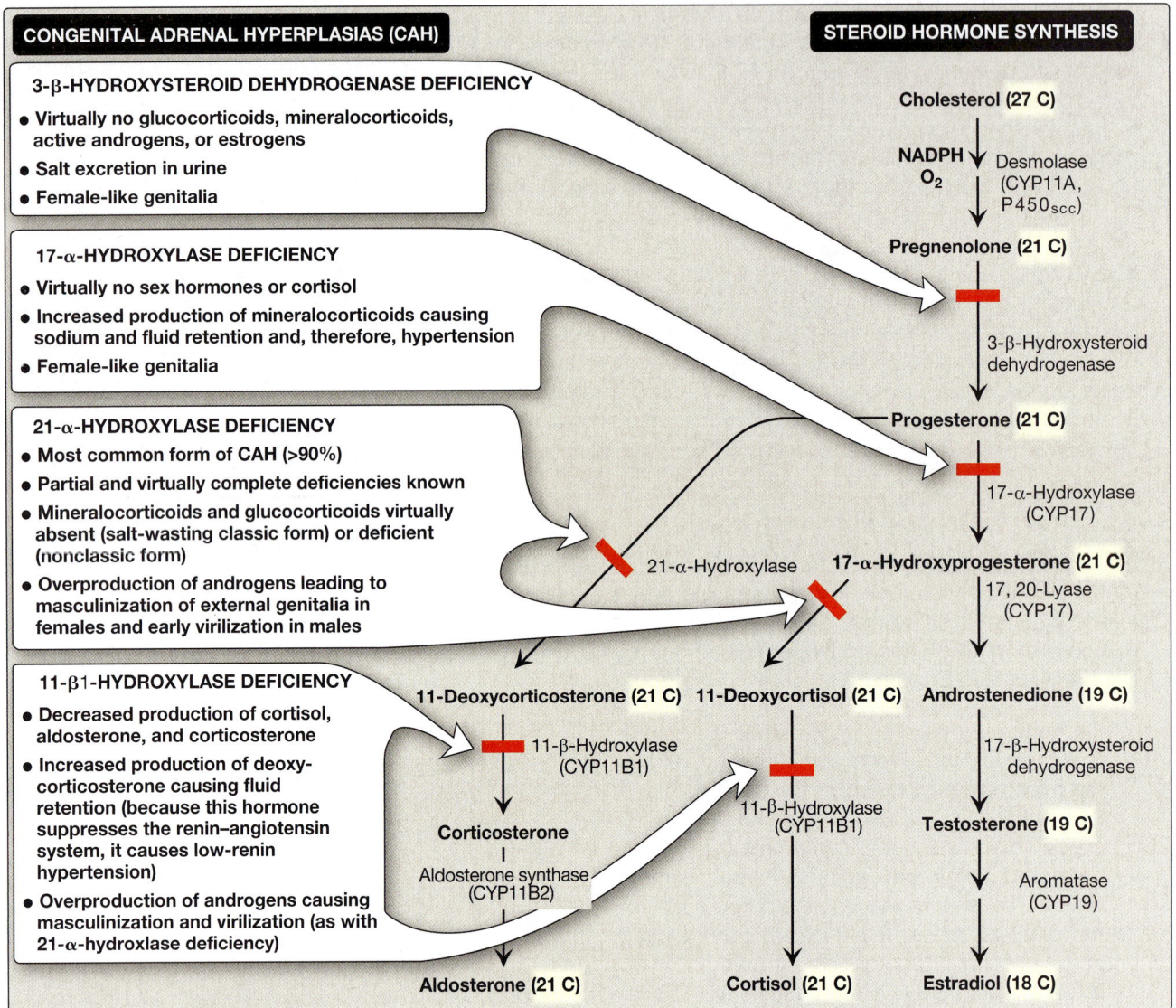

Figure 18.26
Steroid hormone synthesis and associated diseases. [Note: 3-β-Hydroxysteroid dehydrogenase, CYP17, and CYP11B2 are multifunctional enzymes. Synthesis of testosterone and the estrogens occurs primarily outside of the adrenal gland.] NADPH, nicotinamide adenine dinucleotide phosphate; CYP, cytochrome P450.

Na^+/potassium (K^+) ratio and by the hormone angiotensin II (Ang-II). Ang-II (an octapeptide) is produced from angiotensin I ([Ang-I] a decapeptide) by angiotensin-converting enzyme (ACE), an enzyme found predominantly in the lungs but also distributed widely in the body. [Note: Ang-I is produced in the blood by cleavage of an inactive precursor, angiotensinogen, secreted by the liver. Cleavage is catalyzed by renin and made and secreted by the kidneys.] Ang-II binds to cell surface receptors. However, in contrast to ACTH, its effects are mediated through the phosphatidylinositol 4,5-bisphosphate pathway and not by cAMP. Aldosterone's primary effect is on the kidney tubules, where it stimulates Na^+ and water uptake and K^+ excretion (Fig. 18.28). [Note: An effect of aldosterone is an increase in blood pressure.

Competitive inhibitors of ACE are used to treat renin-dependent hypertension.]

3. **Androgens:** Both the inner (zona reticularis) and middle layers of the adrenal cortex produce androgens, primarily dehydroepiandrosterone and androstenedione. Although adrenal androgens themselves are weak, they are converted by aromatase (CYP19) to testosterone, a stronger androgen, in the testes and to estrogens in the ovaries (primarily) of premenopausal women. [Note: Postmenopausal women produce estrogen at extragonadal sites such as the breast. Aromatase inhibitors are used in the treatment of estrogen-responsive breast cancer in these women.]

C. Gonadal steroid hormones

The testes and ovaries (gonads) synthesize hormones necessary for sexual differentiation and reproduction. A single hypothalamic-releasing factor, gonadotropin-releasing hormone, stimulates the anterior pituitary to release the glycoproteins luteinizing hormone (LH) and follicle-stimulating hormone (FSH). Like ACTH, LH and FSH bind to surface receptors and cause an increase in cAMP. LH stimulates the testes to produce testosterone and the ovaries to produce estrogens and progesterone (see Fig. 18.27). FSH regulates the growth of ovarian follicles and stimulates testicular spermatogenesis.

D. Mechanism

Each steroid hormone diffuses across the plasma membrane of its target cell and binds to a specific cytosolic or nuclear receptor. These receptor–ligand complexes accumulate in the nucleus, dimerize, and bind to specific regulatory DNA sequences (hormone-response elements [HREs]) in association with coactivator proteins, thereby causing increased transcription of targeted genes (Fig. 18.29). An HRE is found in the promoter or an enhancer element (see Chapter 33) for genes that respond to a specific steroid hormone, thus ensuring coordinated regulation of these genes. Hormone–receptor complexes can also inhibit transcription in association with corepressors. [Note: The binding of a hormone to its receptor causes a conformational change in the receptor that uncovers its DNA-binding domain, allowing the complex to interact through a zinc finger motif with the appropriate DNA sequence. Receptors for the steroid hormones, plus those for thyroid hormone, retinoic acid, and 1,25-dihydroxycholecalciferol (vitamin D), are members of a superfamily of structurally related gene regulators that function in a similar way.]

E. Further metabolism

Steroid hormones are generally converted into inactive metabolic excretion products in the liver. Reactions include the reduction of unsaturated bonds and the introduction of additional hydroxyl groups. The resulting structures are made more soluble by conjugation with glucuronic acid or sulfate (from 3′-phosphoadenosyl-5′-phosphosulfate). These conjugated metabolites are fairly water soluble and do not need protein carriers. They are eliminated in feces and urine.

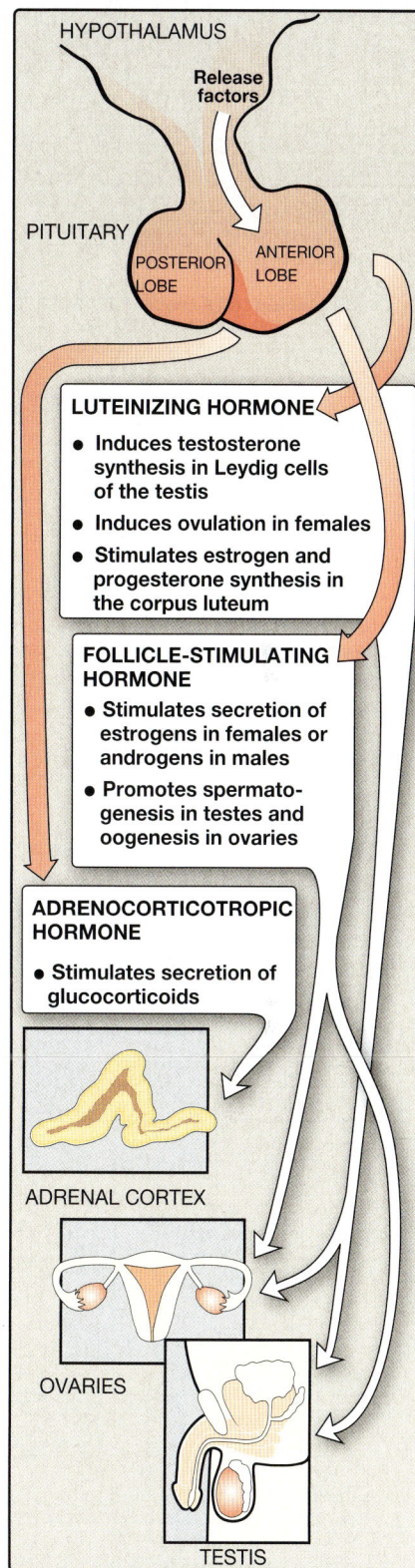

Figure 18.27
Pituitary hormone stimulation of steroid hormone synthesis and secretion.

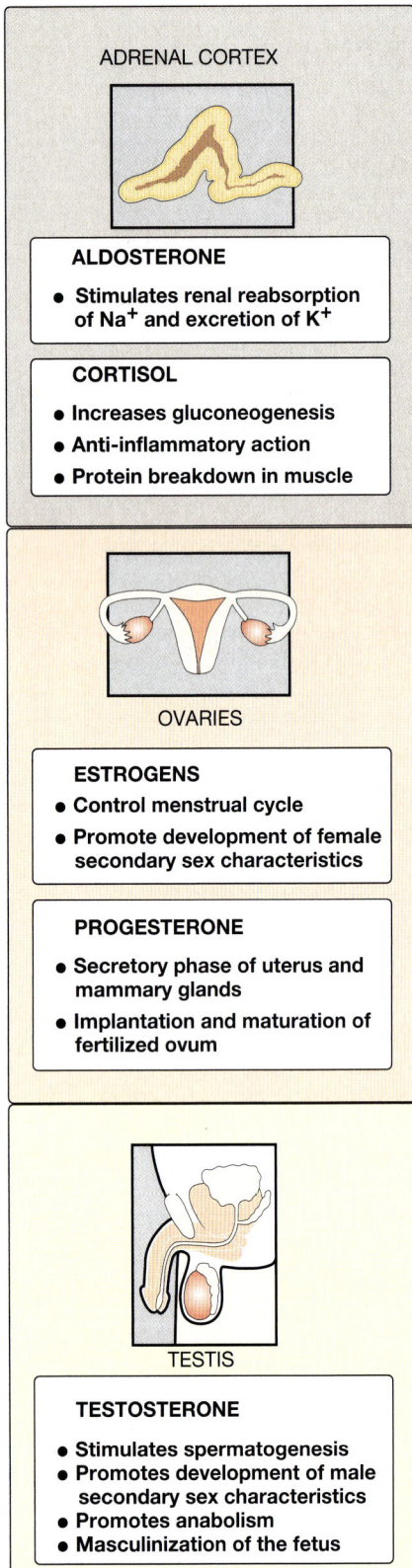

ADRENAL CORTEX

ALDOSTERONE

- Stimulates renal reabsorption of Na^+ and excretion of K^+

CORTISOL

- Increases gluconeogenesis
- Anti-inflammatory action
- Protein breakdown in muscle

OVARIES

ESTROGENS

- Control menstrual cycle
- Promote development of female secondary sex characteristics

PROGESTERONE

- Secretory phase of uterus and mammary glands
- Implantation and maturation of fertilized ovum

TESTIS

TESTOSTERONE

- Stimulates spermatogenesis
- Promotes development of male secondary sex characteristics
- Promotes anabolism
- Masculinization of the fetus

Figure 18.28
Actions of steroid hormones. Na^+, sodium; K^+, potassium.

VIII. CHAPTER SUMMARY

- **Cholesterol** is a hydrophobic compound, with a single hydroxyl group to which a fatty acid (FA) can be attached, producing an even more hydrophobic **cholesteryl ester**.

- Cholesterol is synthesized by virtually all human tissues, although primarily by the **liver**, **intestine**, **adrenal cortex**, and **reproductive tissues.**

- Synthesis requires enzymes of the **cytosol**, **smooth endoplasmic reticulum (sER)**, and **peroxisomes**.

- The rate-limiting and regulated step in cholesterol synthesis is catalyzed **HMG CoA reductase**, which produces **mevalonate** from HMG CoA.

- **HMG CoA reductase** is highly regulated by a number of mechanisms: (1) via the transcription factor **sterol regulatory element–binding protein-2 (SREBP-2)**, (2) accelerated **degradation** of the **protein** when cholesterol levels are high, (3) **phosphorylation** causing **inactivation** of the enzyme by **adenosine monophosphate (AMP) kinase (AMPK)**, and (4) hormonal regulation by **insulin** and **glucagon**.

- **Statins** are **competitive inhibitors** of HMG CoA reductase. These drugs are used to decrease plasma cholesterol in patients with **hyper-cholesterolemia**.

- The ring structure of cholesterol cannot be degraded in humans. It is eliminated from the body either by conversion to bile salts or by secretion into the **bile**.

- The rate-limiting step in bile acid synthesis is catalyzed by **cholesterol-7-α-hydroxylase**, which is inhibited by **bile acids**.

- Before the bile acids leave the liver, they are conjugated. Conjugated bile acids are known as bile salts which are **more** ionized and **more** water **soluble** than bile acids at the alkaline pH of the bile.

- Intestinal **bacteria** modify bile salts producing the **secondary bile salts.**

- Bile salts are efficiently reabsorbed (>95%) and return to the liver by **enterohepatic circulation.**

- Enterohepatic circulation of bile salts is reduced by **bile acid sequestrants.**

- If more cholesterol enters the bile than can be solubilized by the available bile salts and PC, **cholesterol gallstone disease (cholelithiasis)** can occur.

- The plasma lipoproteins (see Fig. 18.30) include **chylomicrons, very–low-density lipoproteins (VLDLs), intermediate-density lipoproteins (IDLs), low-density lipoproteins (LDLs)**, and **high-density lipoproteins (HDLs)**. They function to keep lipids soluble as they transport them between tissues.

- Lipoproteins are composed of **triacylglycerol (TAG)** and **cholesteryl esters** in the core surrounded by a shell of **amphipathic apolipoproteins, phospholipid**, and **nonesterified cholesterol**.

- **Chylomicrons** are assembled in **intestinal mucosal cells** from **dietary lipids**. Each nascent chylomicron particle has one molecule of apolipoprotein **(apo) B-48**.

- Due to their large size, **chylomicrons** are released from the cells into the lymphatic system and travel to the blood. Apo C-II activates endothelial **lipoprotein lipase (LPL)**, which degrades the TAGs in chylomicrons to FAs and glycerol. The **FAs** that are released are stored in **adipose tissue** or used for energy in **muscle**. The **glycerol** is metabolized by the **liver**.

- After most of the TAG is removed, the **chylomicron remnant**, carrying most of the **dietary cholesterol**, binds to a liver receptor that recognizes apo E.

- Patients with a **deficiency** of LPL or apo C-II show a dramatic accumulation of chylomicrons in the plasma (**type I hyperlipoproteinemia** or **familial chylomicronemia**) even if fasted.

- Nascent VLDLs are produced in the liver and are composed predominantly of TAGs. They contain a single molecule of **apo B-100**. VLDLs carry hepatic TAG to the peripheral tissues where LPL degrades the lipid.

- The VLDL particle receives **cholesteryl esters** from HDL in exchange for TAG. This process is accomplished by **cholesteryl ester transfer protein** (**CETP**).

- VLDL in the plasma is first converted to IDL and then to LDL.

- Apo B-100 on LDL is recognized by the **LDL receptor** which results in the **receptor-mediated endocytosis of LDL. The contents of LDL are** degraded in the **lysosomes**, and the **LDL receptor is recycled**. The protease **PCSK9** promotes degradation of LDL receptors in lysosomes. PCSK9 inhibitors keep LDL receptors on the cell surface to remove cholesterol from circulation.

- Defective uptake of these chylomicron remnants and IDL causes **type III hyperlipoproteinemia** or **dysbetalipoproteinemia**.

- Defects in the synthesis of functional LDL receptors cause **type IIa hyperlipoproteinemia**.

- HDLs are created by **lipidation** of **apo A-1** synthesized in the liver and intestine. They have a number of functions, including (1) serving as a circulating **reservoir** of apo C-II and apo E for chylomicrons and VLDL; (2) removing **cholesterol** from peripheral tissues via ATP-binding cassette transport protein (ABCA)-1 and esterifying it using **lecithin:cholesterol acyl transferase** (**LCAT**), a liver-synthesized plasma enzyme that is activated by **apo A-1**; and (3) delivering these cholesteryl esters to the liver (reverse cholesterol transport [**RCT**]) for uptake via **scavenger receptor B1** (**SR-B1**).

- Cholesterol is the precursor of all classes of **steroid hormones**, which include **glucocorticoids**, **mineralocorticoids**, and the **sex hormones**. Synthesis occurs in the **adrenal cortex (glucocorticoids, mineralocorticoids, androgens)**, **gonads**, and **placenta**.

- The initial and rate-limiting step is the conversion of cholesterol to **pregnenolone** by the side-chain cleavage enzyme $P450_{scc}$. Deficiencies in synthesis lead to **congenital adrenal hyperplasia** (**CAH**).

- Each steroid hormone binds to a specific intracellular receptor in its target cell. These **receptor–hormone complexes** bind to specific regulatory DNA sequences (**hormone-response elements** [**HREs**]) in association with coactivator proteins/corepressors, thereby regulating **transcription** of targeted genes.

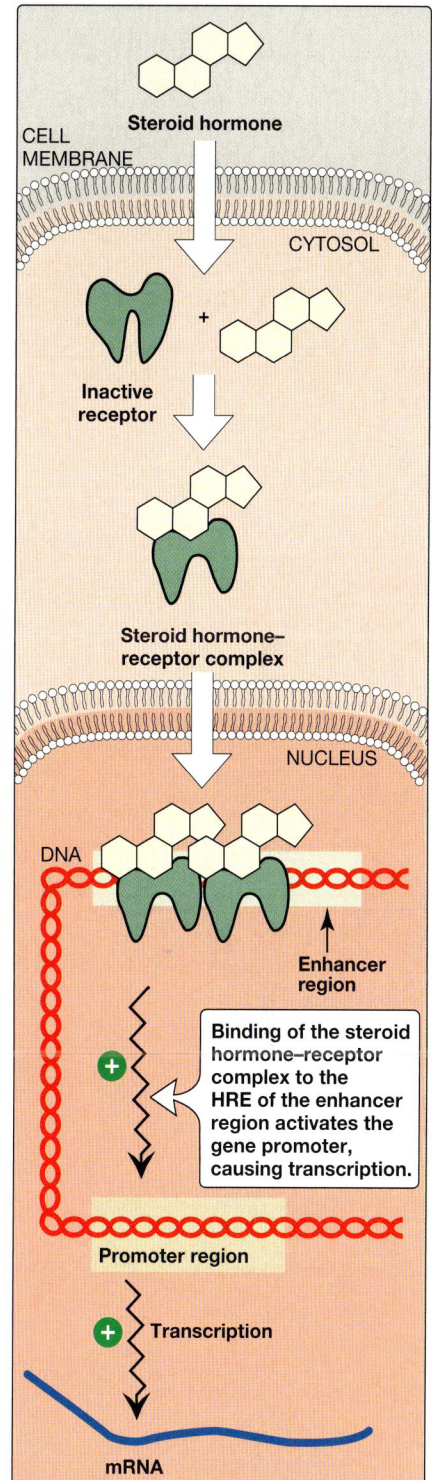

Figure 18.29
Activation of transcription by interaction of steroid hormone–receptor complex with hormone response element (HRE). The receptor contains domains that bind the hormone, DNA, and coactivating proteins. mRNA, messenger RNA.

Cholesterol synthesis

Rate-limiting step ← *catalyzes* — 3 Acetyl CoA

Drug target in therapy for hyper-cholesterolemia by the statin drugs — *is* ← HMG CoA reductase

HMG CoA reductase ← *catalyzed by* — HMG CoA ⊖ ⊕

All tissues, particularly in the liver ← *occurs in* — Mevalonic acid

Cytosol ← *occurs in*

ATP ← *utilizes*

NADPH ← *utilizes*

Acetate as sole source of carbon atoms ← *utilizes*

5 C
10 C
15 C
30 C

Cholesterol (27 C)

Major regulatory mechanism: sterol-dependent regulation of gene expression

responds to

↑ Intracelluar cholesterol — *leads to* — ↓ SREBP-2 activity — *leads to* — ↓ Gene transcription — *leads to* — ↓ Amount of enzyme

↓ Intracelluar cholesterol — *leads to* — ↑ SREBP-2 activity — *leads to* — ↑ Gene transcription — *leads to* — ↑ Amount of enzyme

Lipoproteins

Intestine — *generates* — **Chylomicrons** — *composed of* — **Highest TAG** / **Lowest cholesterol** — *functions to* — Deliver dietary (exogenous) TAG to peripheral tissues

Liver — *generates* — **VLDL** — *composed of* — **High TAG** / **Low cholesterol** — *functions to* — Deliver endogenous TAG to peripheral tissues

VLDL — *generates* — **LDL** — *composed of* — **Low TAG** / **Highest Cholesterol** — *functions to* — Deliver cholesterol to peripheral tissues and back to liver

Liver and intestine — *generate* — **HDL** — *composed of* — **Lowest TAG** / **High cholesterol** — *functions to* — Deliver cholesterol from peripheral tissues to the liver for elimination

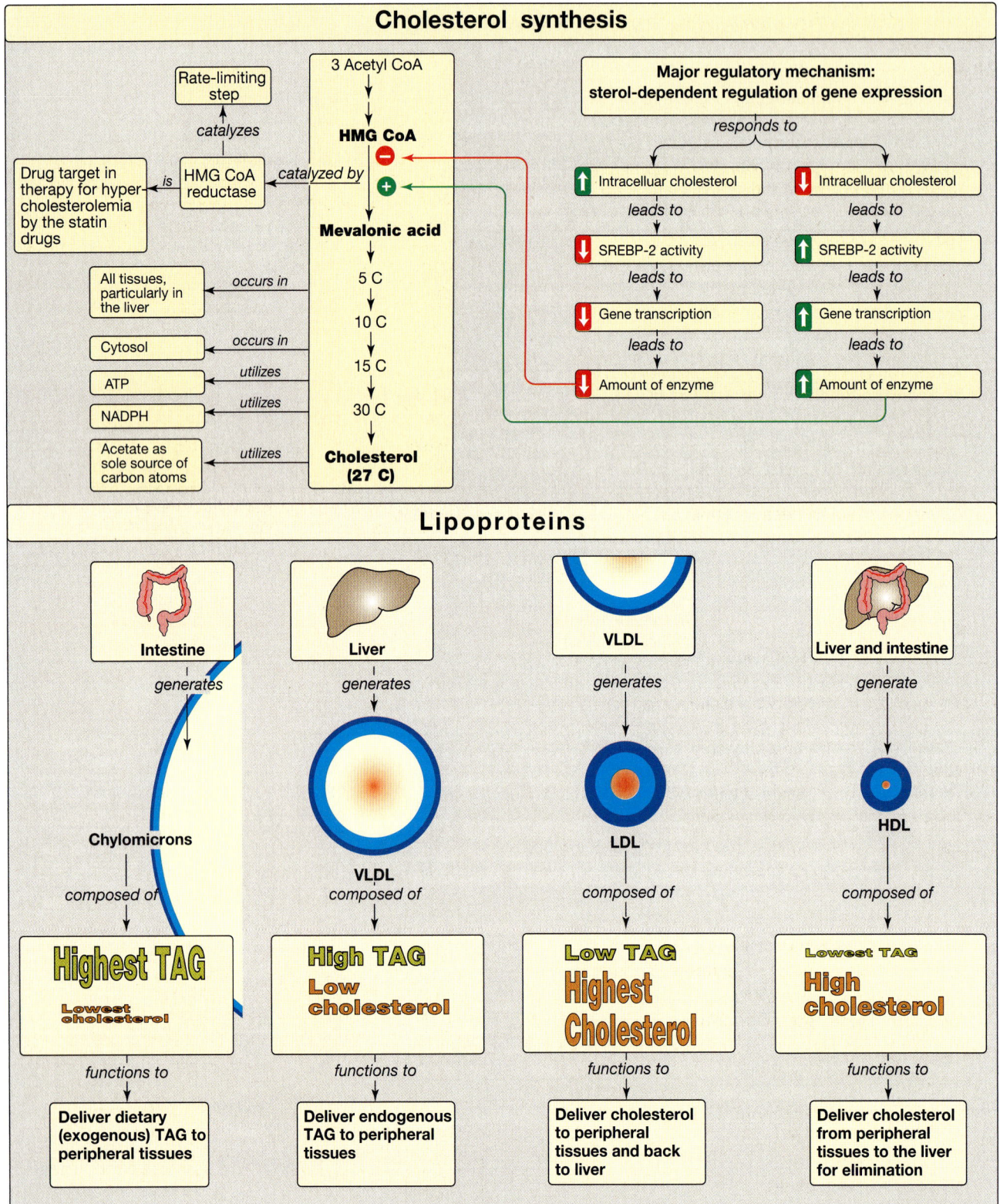

Figure 18.30
Concept map for cholesterol and the lipoproteins. HMG CoA, hydroxymethylglutaryl coenzyme A; SREBP, sterol regulatory element–binding protein; HDLs, LDLs, and VLDLs, high-, low-, and very–low-density lipoproteins; TAG, triacylglycerol; NADPH, nicotinamide adenine dinucleotide phosphate; C, carbon.

Study Questions

Choose the ONE best answer.

18.1. Mice were genetically engineered to contain hydroxymethylglutaryl coenzyme A reductase in which serine 871, a phosphorylation site, was replaced by alanine. Which of the following statements concerning the modified form of the enzyme is most likely to be correct?

A. The enzyme is nonresponsive to ATP depletion.

B. The enzyme is nonresponsive to statin drugs.

C. The enzyme is nonresponsive to the sterol response element–sterol response element–binding protein system.

D. The enzyme is unable to be degraded by the ubiquitin–proteasome system.

Correct answer = A. The reductase is regulated by covalent phosphorylation and dephosphorylation. Depletion of ATP results in a rise in adenosine monophosphate (AMP), which activates AMP kinase (AMPK), thereby phosphorylating and inactivating the reductase. In the absence of the serine, a common phosphorylation site, the enzyme cannot be phosphorylated by AMPK. The enzyme is also regulated physiologically through changes in transcription and degradation and pharmacologically by statin drugs (competitive inhibitors), but none of these depends on serine phosphorylation.

18.2. What is the amount of cholesterol in the low-density lipoproteins in an individual whose fasting blood gave the following lipid-panel test results: total cholesterol = 300 mg/dL, high-density lipoprotein cholesterol = 25 mg/dL, triglycerides = 150 mg/dL?

A. 55 mg/dL

B. 95 mg/dL

C. 125 mg/dL

D. 245 mg/dL

Correct answer = D. The total cholesterol in the blood of a fasted individual is equal to the sum of the cholesterol in low-density lipoproteins plus the cholesterol in high-density lipoproteins plus the cholesterol in very–low-density lipoproteins (VLDLs). This last term is calculated by dividing the triacylglycerol value by 5 because cholesterol accounts for about a fifth of the volume of VLDL in fasted blood.

For Questions 18.3 and 18.4, use the following scenario.

A young female with a history of severe abdominal pain was taken to her local hospital at 5 AM in severe distress. Blood was drawn, and the plasma appeared milky, with the triacylglycerol level >2,000 mg/dL (normal = 4 to 150 mg/dL). The patient was placed on a diet extremely limited in fat but supplemented with medium-chain triglycerides.

18.3. Which of the following lipoprotein particles are most likely responsible for the appearance of the patient's plasma?

A. Chylomicrons

B. High-density lipoproteins

C. Intermediate-density lipoproteins

D. Low-density lipoproteins

E. Very–low-density lipoproteins

Correct answer = A. The milky appearance of her plasma was a result of triacylglycerol-rich chylomicrons. Because 5 AM is presumably several hours after her evening meal, the patient must have difficulty degrading these lipoprotein particles. Intermediate-, low-, and high-density lipoproteins contain primarily cholesteryl esters, and, if one or more of these particles were elevated, it would cause hypercholesterolemia. Very–low-density lipoproteins do not cause the described milky appearance of plasma.

18.4. Which one of the following proteins is most likely to be deficient in this patient?

A. Apolipoprotein A-I

B. Apolipoprotein B-48

C. Apolipoprotein C-II

D. Cholesteryl ester transfer protein

E. Microsomal triglyceride transfer protein

Correct answer = C. The triacylglycerol (TAG) in chylomicrons is degraded by endothelial lipoprotein lipase (LPL), which requires apolipoprotein (apo) C-II as a coenzyme. Deficiency of LPL or apo C-II results in decreased ability to degrade chylomicrons to their remnants, which get cleared (via apo E) by liver receptors. Apo A-I is the coenzyme for lecithin:cholesterol acyltransferase; apo B-48 is the characteristic structural protein of chylomicrons; cholesteryl ester transfer protein catalyzes the cholesteryl ester–TAG exchange between high-density and very–low-density lipoproteins (VLDLs); and microsomal triglyceride transfer protein is involved in the formation, not degradation, of chylomicrons (and VLDLs).

18.5. Complete the table below for an individual with classic 21-α-hydroxylase deficiency relative to a normal individual.

Variable	Increased	Decreased
Aldosterone		
Androstenedione		
Cortisol		
Blood glucose		
Adrenocorticotropic hormone		
Blood pressure		

How might the results be changed if this individual were deficient in 17-α-hydroxylase, rather than 21-α-hydroxylase?

Classic 21-α-hydroxylase deficiency causes mineralocorticoids (aldosterone) and glucocorticoids (cortisol) to be virtually absent. Because aldosterone increases blood pressure, and cortisol increases blood glucose, their deficiencies result in a decrease in blood pressure and blood glucose, respectively. Cortisol normally feeds back to inhibit adrenocorticotropic hormone (ACTH) release by the pituitary, and, so, its absence results in an elevation in ACTH. The loss of 21-α-hydroxylase pushes progesterone and pregnenolone to androgen synthesis and, therefore, causes androstenedione levels to rise. With 17-α-hydroxylase deficiency, sex hormone synthesis would be decreased. Mineralocorticoid production would be increased, leading to hypertension.

Amino Acids: Nitrogen Disposal

19

I. OVERVIEW

Unlike fats and carbohydrates, amino acids are not stored by the body. That is, no protein exists whose sole function is to maintain a storage form of excess amino acids for future use. Therefore, amino acids must be obtained from the diet, synthesized *de novo*, or produced from the degradation of previous body proteins. Any amino acids in excess of the biosynthetic needs of the cell are rapidly degraded. The first phase of catabolism involves the removal of the α-amino groups (usually by transamination and subsequent oxidative deamination), forming ammonia and the corresponding α-keto acids, the carbon skeletons of amino acids. A portion of the free ammonia is excreted in the urine, but most is used in the synthesis of urea (Fig. 19.1), which is quantitatively the most important route for disposing of nitrogen from the body. In the second phase of amino acid catabolism, described in Chapter 20, the carbon skeletons of the α-keto acids are converted to common intermediates of energy-producing metabolic pathways. These compounds can be metabolized to carbon dioxide (CO_2) and water (H_2O), glucose, fatty acids, or ketone bodies by the central pathways of metabolism described in Chapters 8 through 13 and 16.

II. OVERALL NITROGEN METABOLISM

Amino acid catabolism is part of the larger process of the metabolism of nitrogen-containing molecules. Nitrogen enters the body in a variety of compounds present in food, the most important being amino acids contained in dietary protein. Nitrogen leaves the body as urea, ammonia, and other products derived from amino acid metabolism (such as creatinine). The role of body proteins in these transformations involves two important concepts: the amino acid pool and protein turnover.

A. Amino acid pool

Free amino acids are present throughout the body, such as in cells, blood, and extracellular fluids. For this discussion, envision all of

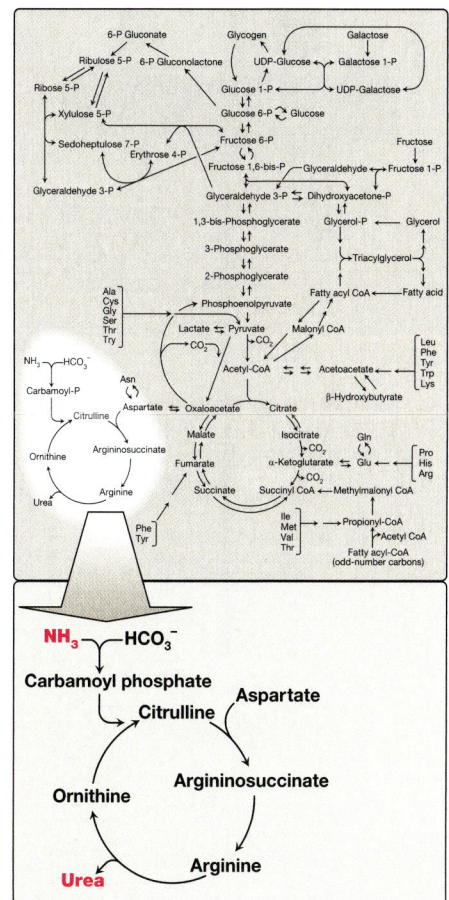

Figure 19.1
Urea cycle shown as part of the essential pathways of energy metabolism. [Note: see Fig. 8.2 for a more detailed map of metabolism.] NH_3, ammonia; CO_2, carbon dioxide.

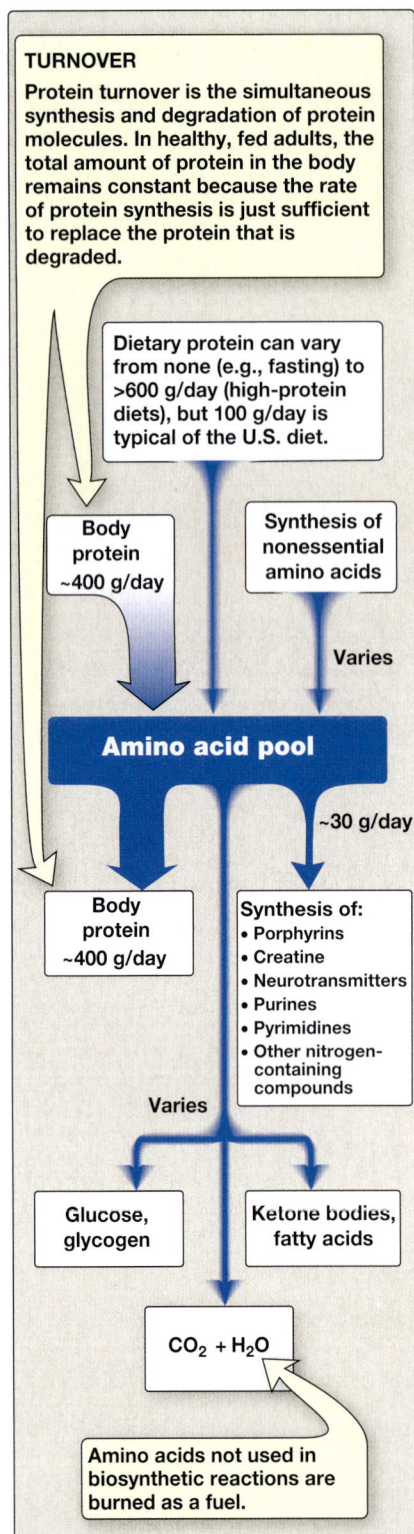

Dietary protein can vary from none (e.g., fasting) to >600 g/day (high-protein diets), but 100 g/day is typical of the U.S. diet.

Body protein ~400 g/day

Synthesis of nonessential amino acids

Varies

Amino acid pool

~30 g/day

Body protein ~400 g/day

Synthesis of:
• Porphyrins
• Creatine
• Neurotransmitters
• Purines
• Pyrimidines
• Other nitrogen-containing compounds

Varies

Glucose, glycogen

Ketone bodies, fatty acids

$CO_2 + H_2O$

Amino acids not used in biosynthetic reactions are burned as a fuel.

Figure 19.2
Sources and fates of amino acids. [Note: Nitrogen from amino acid degradation is released as ammonia, which is converted to urea and excreted.] CO_2, carbon dioxide.

these amino acids as if they belonged to a single entity, called the amino acid pool. This pool is supplied by three sources: (1) amino acids provided by the degradation of endogenous (body) proteins, most of which are reutilized; (2) amino acids derived from exogenous (dietary) protein; and (3) nonessential amino acids synthesized from simple intermediates of metabolism (Fig. 19.2). Conversely, the amino acid pool is depleted by three routes: (1) synthesis of body protein; (2) consumption of amino acids as precursors of essential nitrogen-containing small molecules; and (3) conversion of amino acids to glucose, glycogen, fatty acids, and ketone bodies or oxidation to $CO_2 + H_2O$ (see Fig. 19.2). Although the amino acid pool is small (comprising ~90 to 100 g of amino acids) in comparison with the amount of protein in the body (~12 kg in a 70-kg man), it is conceptually at the center of whole-body nitrogen metabolism.

> In healthy, well-fed individuals, input to the amino acid pool is balanced by output. That is, the amount of amino acids contained in the pool is constant. The amino acid pool is said to be in a steady state, and the individual is said to be in nitrogen balance.

B. Protein turnover

Most proteins in the body are constantly being synthesized and then degraded (turned over), permitting the removal of abnormal or unneeded proteins. For many proteins, regulation of synthesis determines the concentration of protein in the cell, with protein degradation assuming a minor role. For other proteins, the rate of synthesis is constitutive (i.e., essentially constant), and cellular levels of the protein are controlled by selective degradation.

1. **Rate:** In healthy adults, the total amount of protein in the body remains constant because the rate of protein synthesis is just sufficient to replace the protein that is degraded. This process, called protein turnover, leads to the hydrolysis and resynthesis of 300 to 400 g of body protein each day. The rate of protein turnover varies widely for individual proteins. Short-lived proteins (e.g., many regulatory proteins and misfolded proteins) are rapidly degraded, having half-lives measured in minutes or hours. Long-lived proteins, with half-lives of days to weeks, constitute the majority of proteins in the cell. Structural proteins, such as collagen, are metabolically stable and have half-lives measured in months or years.

2. **Protein degradation:** There are two major enzyme systems responsible for degrading proteins: the adenosine triphosphate (ATP)-dependent ubiquitin (Ub)–proteasome system of the cytosol and the ATP-independent degradative enzyme system of the lysosomes. Proteasomes selectively degrade damaged or short-lived proteins. Lysosomes use acid hydrolases to nonselectively degrade intracellular proteins (autophagy) and extracellular proteins (heterophagy), such as plasma proteins that are taken into the cell by endocytosis.

a. **Ubiquitin–proteasome system:** Proteins selected for degradation by the cytosolic Ub–proteasome system are first modified by the covalent attachment of Ub, a small, globular, nonenzymic protein that is highly conserved across eukaryotic species. Ubiquitination of the target substrate occurs through isopeptide linkage of the α-carboxyl group of the C-terminal glycine of Ub to the ϵ-amino group of a lysine in the protein substrate by a three-step, enzyme-catalyzed, ATP-dependent process. [Note: Enzyme 1 (E1, an activating enzyme) activates Ub, which is then transferred to E2 (a conjugating enzyme). E3 (a ligase) identifies the protein to be degraded and interacts with E2–Ub. There are many more E3 proteins than E1 and E2 proteins.] The consecutive addition of four or more Ub molecules to the target protein generates a polyubiquitin chain. Proteins tagged with Ub chains are recognized by a large, barrel-shaped, macromolecular, proteolytic complex called a proteasome (Fig. 19.3). The proteasome unfolds, deubiquitinates, and cuts the target protein into fragments that are then further degraded by cytosolic proteases to amino acids, which enter the amino acid pool. The Ub is recycled. Notably, the selective degradation of proteins by the Ub–proteosome complex (unlike simple hydrolysis by proteolytic enzymes) requires ATP hydrolysis.

b. **Degradation signals:** Because proteins have different half-lives, it is clear that protein degradation cannot be random but, rather, is influenced by some structural aspect of the protein that serves as a degradation signal, which is recognized and bound by an E3. The half-life of a protein is also influenced by the amino (N)-terminal residue, the so-called "N-end rule," and ranges from minutes to hours. Destabilizing N-terminal amino acids include arginine and posttranslationally modified amino acids such as acetylated alanine. In contrast, serine is a stabilizing amino acid. Additionally, proteins rich in sequences containing proline, glutamate, serine, and threonine (called PEST sequences after the one-letter designations for these amino acids) are rapidly ubiquitinated and degraded and, therefore, have short half-lives.

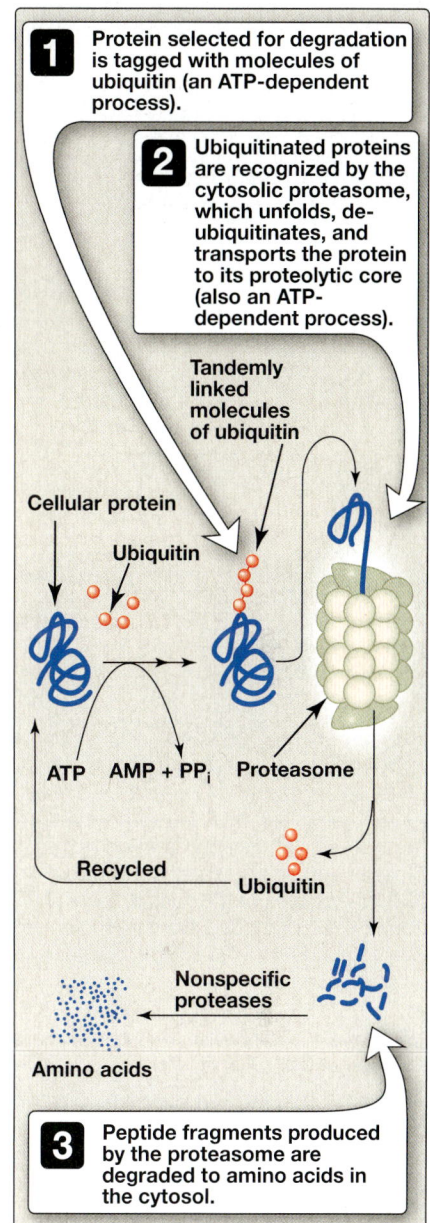

Figure 19.3
The ubiquitin–proteasome degradation pathway of proteins. AMP, adenosine monophosphate; PP$_i$, pyrophosphate.

III. DIETARY PROTEIN DIGESTION

Most of the nitrogen in the diet is consumed in the form of protein, typically amounting to 70 to 100 g/day in the American diet (see Fig. 19.2). Proteins are generally too large to be absorbed by the intestine. [Note: An example of an exception to this rule is that newborns can take up maternal antibodies in breast milk.] Therefore, proteins must be hydrolyzed to yield di- and tripeptides as well as individual amino acids, which can be absorbed. Proteolytic enzymes responsible for degrading proteins are produced by three different organs: the stomach, pancreas, and small intestine (Fig. 19.4).

A. Digestion by gastric secretion

The digestion of proteins begins in the stomach, which secretes gastric juice, a unique solution containing hydrochloric acid (HCl) and the proenzyme pepsinogen.

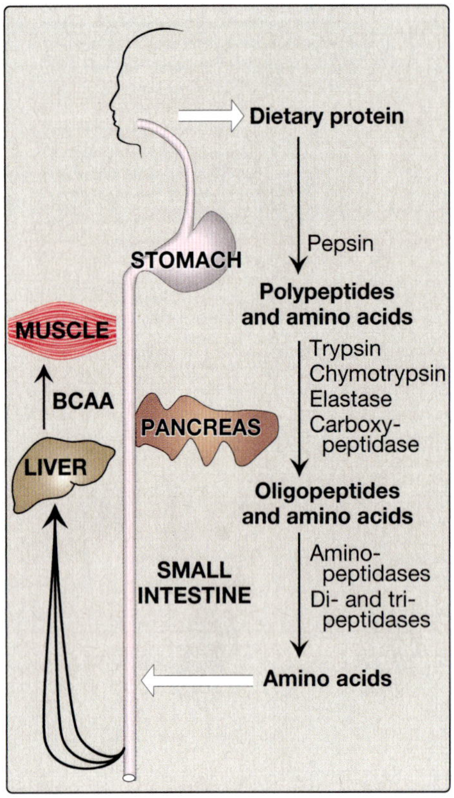

Figure 19.4
Digestion of dietary proteins by the proteolytic enzymes of the gastrointestinal tract. BCAA, branched chain amino acids.

1. **Hydrochloric acid:** Stomach HCl is too dilute (pH 2 to 3) to hydrolyze proteins. The acid, secreted by the parietal cells of the stomach, functions instead to kill some bacteria and to denature proteins, thereby making them more susceptible to subsequent hydrolysis by proteases.

2. **Pepsin:** This acid-stable endopeptidase is secreted by the chief cells of the stomach as an inactive zymogen (or proenzyme), pepsinogen. [Note: In general, zymogens contain extra amino acids in their sequences that prevent them from being catalytically active. Removal of these amino acids permits the proper folding required for an active enzyme.] In the presence of HCl, pepsinogen undergoes a conformational change that allows it to cleave itself (autocatalysis) to the active form, pepsin, which releases polypeptides and a few free amino acids from dietary proteins.

B. Digestion by pancreatic enzymes

On entering the small intestine, the polypeptides produced in the stomach by the action of pepsin are further cleaved to oligopeptides and amino acids by a group of pancreatic proteases that include both endopeptidases (that cleave within) and exopeptidases (that cut at an end). [Note: Bicarbonate (HCO_3^-), secreted by the pancreas in response to the intestinal hormone secretin, raises the intestinal pH.]

1. **Specificity:** Each of these enzymes has a different specificity for the amino acid R-groups adjacent to the susceptible peptide bond (Fig. 19.5). For example, trypsin cleaves only when the carbonyl group of the peptide bond is contributed by arginine or lysine. These enzymes, like pepsin, are synthesized and secreted as inactive zymogens.

2. **Zymogen release:** The release and activation of the pancreatic zymogens are mediated by the secretion of cholecystokinin, a polypeptide hormone of the small intestine.

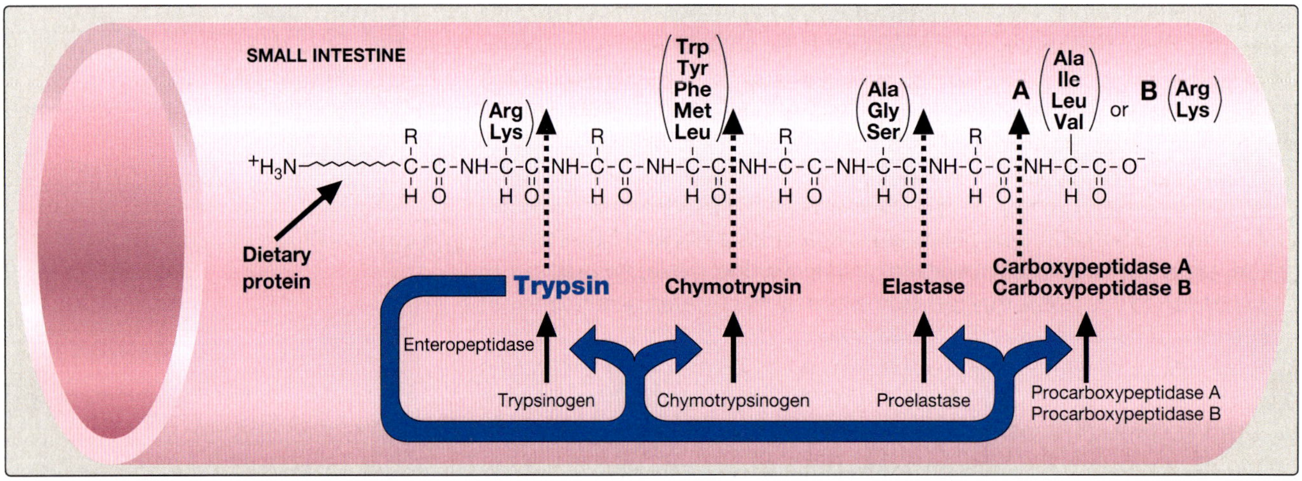

Figure 19.5
Cleavage of dietary protein in the small intestine by pancreatic proteases. The peptide bonds susceptible to hydrolysis are shown for each of the five major pancreatic proteases. [Note: The first three are serine endopeptidases, whereas the last two are exopeptidases. Each is produced from an inactive zymogen.]

3. **Zymogen activation:** Enteropeptidase (also called enterokinase), a serine protease synthesized by and present on the luminal (apical) surface of intestinal mucosal cells (enterocytes) of the brush border, converts the pancreatic zymogen trypsinogen to trypsin by removal of a hexapeptide from the N-terminus of trypsinogen. Trypsin subsequently converts other trypsinogen molecules to trypsin by cleaving a limited number of specific peptide bonds in the zymogen. Thus, enteropeptidase unleashes a cascade of proteolytic activity because trypsin is the common activator of all the pancreatic zymogens (Fig. 19.5).

4. **Digestion abnormalities:** In individuals with a deficiency in pancreatic secretion (e.g., because of chronic pancreatitis, cystic fibrosis, or surgical removal of the pancreas), the digestion and absorption of fat and protein are incomplete. This results in the abnormal appearance of lipids in the feces (steatorrhea) as well as undigested protein.

Celiac disease (celiac sprue) is a disease of malabsorption resulting from immune-mediated damage to the small intestine in response to ingestion of gluten (or gliadin produced from gluten), a protein found in wheat, barley, and rye.

C. Digestion of oligopeptides by small intestine enzymes

The luminal surface of the enterocytes contains aminopeptidase, an exopeptidase that repeatedly cleaves the N-terminal residue from oligopeptides to produce even smaller peptides and free amino acids.

D. Amino acid and small peptide intestinal absorption

Most free amino acids are taken into enterocytes via sodium-dependent secondary active transport by solute carrier (SLC) proteins of the apical membrane. At least seven different transport systems with overlapping amino acid specificities are known. Di- and tripeptides, however, are taken up by a proton-linked peptide transporter (PepT1). The peptides are then hydrolyzed to free amino acids. Regardless of their source, free amino acids are released from enterocytes into the portal system by sodium-independent transporters of the basolateral membrane. Therefore, only free amino acids are found in the portal vein after a meal containing protein. These amino acids are either metabolized by the liver or released into the general circulation. [Note: Branched-chain amino acids (BCAAs) are not metabolized by the liver but, instead, are sent from the liver to muscle via the blood (see Fig. 19.4).]

E. Absorption abnormalities

The small intestine and the proximal tubules of the kidneys have common transport systems for amino acid uptake. Consequently, a defect in any one of these systems results in an inability to absorb particular amino acids into the intestine and into the kidney tubules. For example, one system is responsible for the uptake of cystine and the dibasic amino acids ornithine, arginine, and lysine (represented

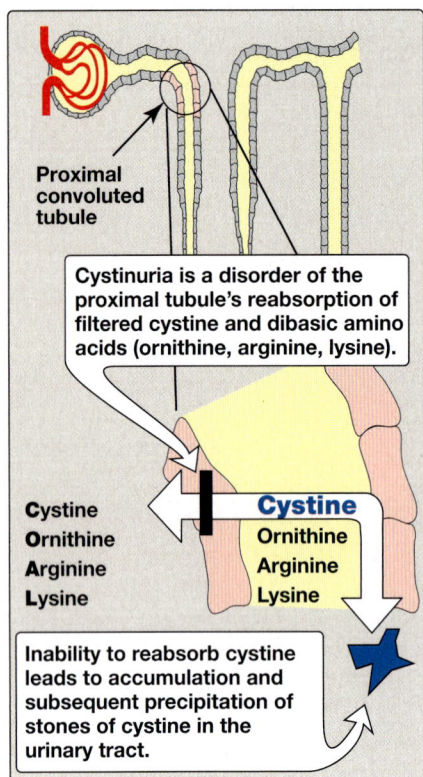

Figure 19.6
Genetic defect seen in cystinuria. [Note: Cystinuria is distinct from cystinosis, a rare defect in the transport of cystine out of lysosomes that results in the formation of cystine crystals within the lysosome and widespread tissue damage.]

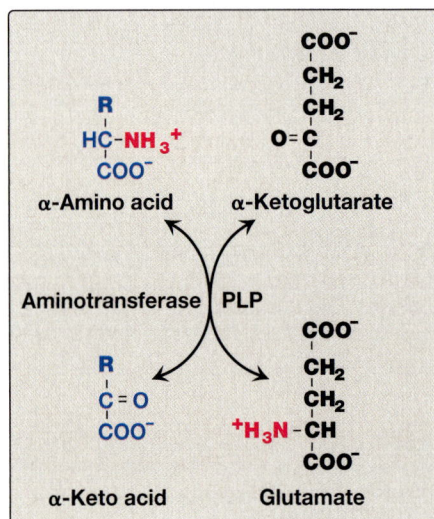

Figure 19.7
Aminotransferase reaction using α-ketoglutarate as the amino group acceptor. PLP, pyridoxal phosphate.

as COAL). In the inherited disorder cystinuria, this carrier system is defective, and all four amino acids appear in the urine (Fig. 19.6). Cystinuria occurs at a frequency of 1 in 7,000 individuals, making it one of the most common inherited diseases and the most common genetic error of amino acid transport. The disease expresses itself clinically by the precipitation of cystine to form kidney stones (calculi), which can block the urinary tract. Oral hydration is an important part of treatment for this disorder. [Note: Defects in the uptake of tryptophan by a neutral amino acid transporter can result in Hartnup disorder and pellagra-like dermatologic and neurologic symptoms.]

IV. NITROGEN REMOVAL FROM AMINO ACIDS

The presence of the α-amino group keeps amino acids safely locked away from oxidative breakdown. Removing the α-amino group is essential for producing energy from any amino acid and is an obligatory step in the catabolism of all amino acids. Once removed, this nitrogen can be incorporated into other compounds or excreted as urea, with the carbon skeletons being metabolized. This section describes transamination and oxidative deamination, reactions that ultimately provide ammonia and aspartate, the two sources of urea nitrogen.

A. Transamination: funneling amino groups to form glutamate

The first step in the catabolism of most amino acids is the transfer of their α-amino group to α-ketoglutarate (Fig. 19.7), producing an α-keto acid (derived from the original amino acid) and glutamate (derived from α-ketoglutarate). The citric acid cycle keto acid intermediate α-ketoglutarate plays a pivotal role in amino acid metabolism by accepting the amino groups from most amino acids, thereby becoming its structurally related amino acid, glutamate. Glutamate produced by transamination can be oxidatively deaminated (see Section B.) or used as an amino group donor in the synthesis of nonessential amino acids. This transfer of amino groups from one carbon skeleton to another is catalyzed by a family of readily reversible enzymes called aminotransferases (also called transaminases). These enzymes are found in the cytosol and mitochondria of cells throughout the body. All amino acids, with the exception of lysine and threonine, participate in transamination at some point in their catabolism. [Note: These two amino acids lose their α-amino groups by deamination.]

1. **Substrate specificity:** Each aminotransferase is specific for one or, at most, a few amino group donors. Aminotransferases are named after the specific amino group donor, because the acceptor of the amino group is almost always α-ketoglutarate. Two important aminotransferase reactions are catalyzed by alanine aminotransferase (ALT) and aspartate aminotransferase (AST), as shown in Figure 19.8. All aminotransferases require the coenzyme pyridoxal phosphate (a derivative of vitamin B_6), which is covalently linked to the ε-amino group of a specific lysine residue at the active site of the enzyme.

   a. **ALT:** ALT is present in many tissues. The enzyme catalyzes the transfer of the amino group of alanine to α-ketoglutarate, resulting in the formation of pyruvate and glutamate, respectively. The reaction is readily reversible. However, during amino acid catabolism, ALT (like most aminotransferases) functions mainly in the direction of glutamate synthesis. In effect, α-ketoglutarate acts as a nitrogen acceptor from most amino acids, forming glutamate.

b. **AST:** AST is an exception to the rule that aminotransferases funnel amino groups to form glutamate. During amino acid catabolism, AST primarily transfers amino groups from glutamate to oxaloacetate, forming α-ketoglutarate and aspartate, respectively. Aspartate is used as a source of nitrogen in the urea cycle. Like other transaminations, the AST reaction is reversible.

2. **Mechanism:** Figure 19.9 shows the mechanistic reactions for the transamination catalyzed by AST. Aminotransferases act by transferring the amino group of an amino acid substrate (glutamate) to the pyridoxal part of the coenzyme to generate pyridoxamine phosphate. The amino acid substrate glutamate is thus converted to an α-keto acid product (α-ketoglutarate). The pyridoxamine form of the coenzyme then reacts with an α-keto acid substrate (oxaloacetate) to form an amino acid product (aspartate), at the same time regenerating the original aldehyde form of the coenzyme.

3. **Equilibrium:** For most transamination reactions, the equilibrium constant is near 1. This allows the reaction to function in both amino acid degradation through removal of α-amino groups (e.g., after consumption of a protein-rich meal) and biosynthesis of nonessential amino acids through addition of amino groups to the carbon skeletons of α-keto acids (e.g., when the supply of amino acids from the diet is not adequate to meet the synthetic needs of cells).

4. **Diagnostic value:** Aminotransferases are normally intracellular enzymes, with the low levels found in the plasma representing the release of cellular contents during normal cell turnover. Elevated plasma levels of aminotransferases indicate damage to cells rich in these enzymes. For example, physical trauma or a disease process can cause cell lysis, resulting in release of intracellular enzymes into the blood. AST and ALT are of particular diagnostic value when they are found in the plasma.

a. **Hepatic disease:** Plasma AST and ALT are elevated in nearly all hepatic diseases but are particularly high in conditions that cause extensive cell necrosis, such as severe viral hepatitis, toxic injury, and prolonged circulatory collapse. ALT is more specific than AST for liver disease, but the latter is more sensitive because the liver contains larger amounts of AST. Serial measurements of AST and ALT (liver function tests) are often useful in determining the course of liver damage. Figure 19.10 shows the early release of ALT into the blood, following ingestion of a liver toxin. [Note: The elevation in bilirubin results from hepatocellular damage that decreases the hepatic conjugation and excretion of bilirubin.]

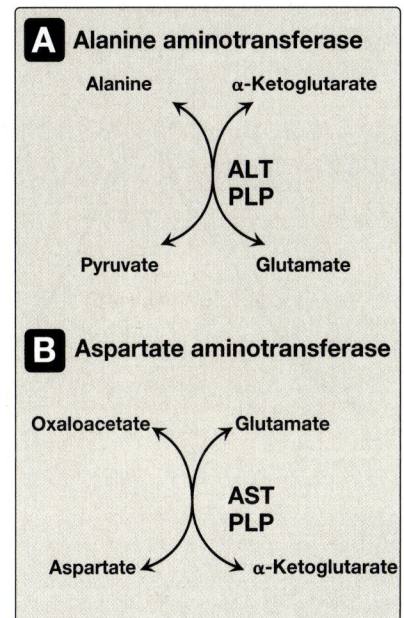

Figure 19.8
Reactions catalyzed during amino acid catabolism. **A.** Alanine aminotransferase (ALT). **B.** Aspartate aminotransferase (AST). PLP, pyridoxal phosphate.

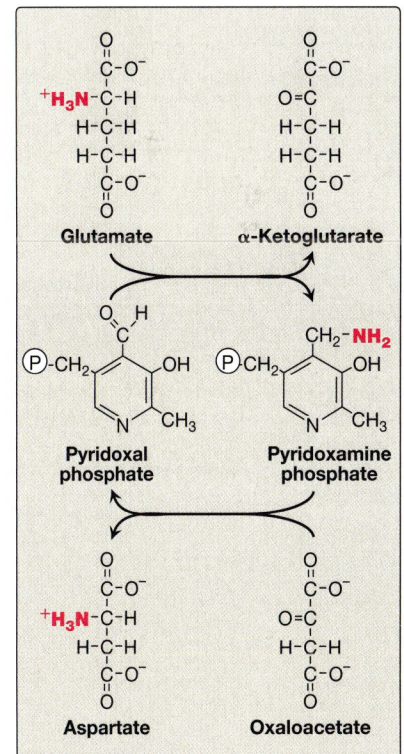

Figure 19.9
Cyclic interconversion of pyridoxal phosphate and pyridoxamine phosphate during the aspartate aminotransferase reaction. Ⓟ, phosphate group.

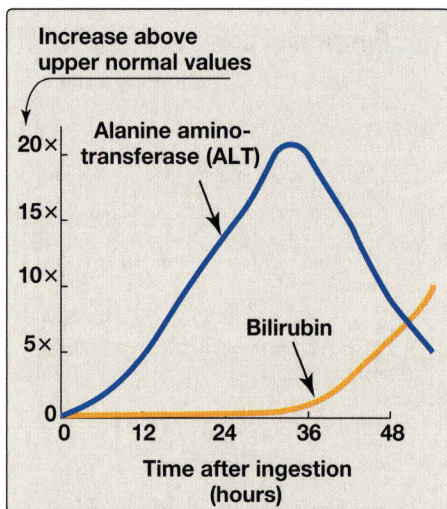

Figure 19.10
Pattern of ALT and bilirubin in the plasma, following poisoning by ingestion of the toxic mushroom *Amanita phalloides*.

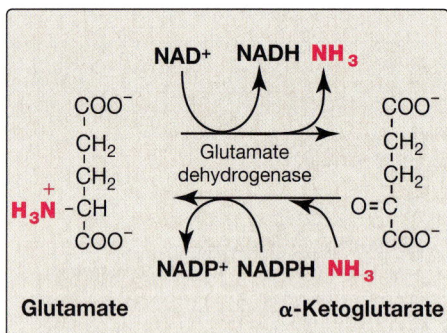

Figure 19.11
Oxidative deamination by glutamate dehydrogenase. [Note: The enzyme is unusual in that it uses both nicotinamide adenine dinucleotide (NAD⁺) and nicotinamide adenine dinucleotide phosphate (NADPH).] NH₃, ammonia.

b. **Nonhepatic disease:** Aminotransferases may be elevated in nonhepatic diseases such as those that cause damage to cardiac or skeletal muscle. However, these disorders can usually be distinguished clinically from liver disease using additional lab tests. When muscle damage is suspected, creatine kinase, lactate dehydrogenase, and myoglobin plasma levels, in addition to AST and ALT levels, may also be increased. Blood urea nitrogen, bilirubin, γ-glutamyl transferase (GGT), and alkaline phosphatase (ALP) levels would be in the normal range. If bone disease is suspected, ALP levels will be disproportionately higher than the AST, ALT, and GGT levels.

B. Oxidative deamination: amino group removal

In contrast to transamination reactions that transfer amino groups, oxidative deamination reactions result in the liberation of the amino group as free ammonia (Fig. 19.11). These reactions occur primarily in the liver and kidney. They provide α-keto acids that can enter the central pathways of energy metabolism and ammonia, which is a source of nitrogen in hepatic urea synthesis. [Note: Ammonia exists primarily as ammonium (NH_4^+) in aqueous solution, but it is the unionized form (NH_3) that crosses membranes.]

1. **Glutamate dehydrogenase (GDH):** As described, the amino groups of most amino acids are ultimately funneled to α-ketoglutarate, forming glutamate by means of a transamination reaction. Glutamate is unique in that it is the only amino acid that undergoes rapid oxidative deamination, a reaction catalyzed by GDH (Fig. 19.11). Therefore, the sequential action of transamination (resulting in the transfer of amino groups from most amino acids to α-ketoglutarate to produce glutamate) and the oxidative deamination of that glutamate (regenerating α-ketoglutarate) provide a pathway whereby the amino groups of most amino acids can be released as ammonia (Fig. 19.12).

   a. **Coenzymes:** GDH, a mitochondrial enzyme, is unusual in that it can use either oxidized nicotinamide adenine dinucleotide (NAD⁺) or reduced nicotinamide adenine dinucleotide phosphate form (NADPH) as coenzymes (Fig. 19.11). NAD⁺ is used primarily in oxidative deamination (the simultaneous loss of ammonia coupled with the oxidation of the carbon skeleton, as shown in Fig. 19.12A), whereas NADPH is used in reductive amination (the simultaneous gain of ammonia coupled with the reduction of the carbon skeleton, as shown in Fig. 19.12B).

   b. **Reaction direction:** The direction of the reaction depends on the relative concentrations of glutamate, α-ketoglutarate, and ammonia and the ratio of oxidized to reduced coenzymes. For example, after ingestion of a meal containing protein, glutamate levels in the liver are elevated, and the reaction proceeds in the direction of amino acid degradation and the formation of ammonia (Fig. 19.12A). High ammonia levels are required to drive the reaction to glutamate synthesis.

   c. **Allosteric regulators:** Guanosine triphosphate is an allosteric inhibitor of GDH, whereas adenosine diphosphate is an activator.

Therefore, when energy levels are low in the cell, amino acid degradation by GDH is high, facilitating energy production from the carbon skeletons derived from amino acids.

2. **D–Amino acid oxidase (DAO):** D–Amino acids are supplied by the diet but are not used in the synthesis of mammalian proteins. They are, however, efficiently metabolized to α-keto acids, ammonia, and hydrogen peroxide in the peroxisomes of liver and kidney cells by flavin adenine dinucleotide–dependent DAO. The α-keto acids can enter the general pathways of amino acid metabolism and be reaminated to L-isomers or catabolized for energy. [Note: DAO degrades D-serine, the isomeric form of serine that modulates N-methyl-D-aspartate (NMDA)-type glutamate receptors. Increased DAO activity has been linked to increased susceptibility to schizophrenia. DAO also converts glycine to glyoxylate.] L–Amino acid oxidases are found in snake venom.

C. Ammonia transport to the liver

Two mechanisms are available in humans for the transport of ammonia from peripheral tissues to the liver for conversion to urea. Both are important in, but not exclusive to, skeletal muscle. The first uses glutamine synthetase to combine ammonia with glutamate to form glutamine, a nontoxic transport form of ammonia (Fig. 19.13). The glutamine is transported in the blood to the liver, where it is cleaved by glutaminase to glutamate and ammonia. The glutamate is oxidatively deaminated to ammonia and α-ketoglutarate by GDH. The ammonia is converted to urea. The second transport mechanism involves the formation of alanine by the transamination of pyruvate produced from both aerobic glycolysis and metabolism of the succinyl coenzyme A (CoA) generated by the catabolism of the BCAA isoleucine and valine. Alanine is transported in the blood to the liver, where it is transaminated by ALT to pyruvate. The pyruvate is used to synthesize glucose, which can enter the blood and be used by muscle, a pathway called the glucose–alanine cycle. The glutamate product of ALT can be deaminated by GDH, generating ammonia. Thus, both alanine and glutamine carry ammonia to the liver.

V. UREA CYCLE

Urea (H_2NCNH_2, with O double-bonded) is the major disposal form of amino groups derived from amino acids and accounts for ~90% of the nitrogen-containing components of urine. One nitrogen of the urea molecule is supplied by free ammonia and the other nitrogen by aspartate. [Note: Glutamate is the immediate precursor of both ammonia (through oxidative deamination by GDH) and aspartate nitrogen (through transamination of oxaloacetate by AST).] The carbon and oxygen of urea are derived from CO_2 (as HCO_3^-). Urea is produced by the liver and then is transported in the blood (blood urea nitrogen) to the kidneys for excretion in the urine.

A. Reactions

The first two reactions leading to the synthesis of urea occur in the mitochondrial matrix, whereas the remaining urea cycle enzymes

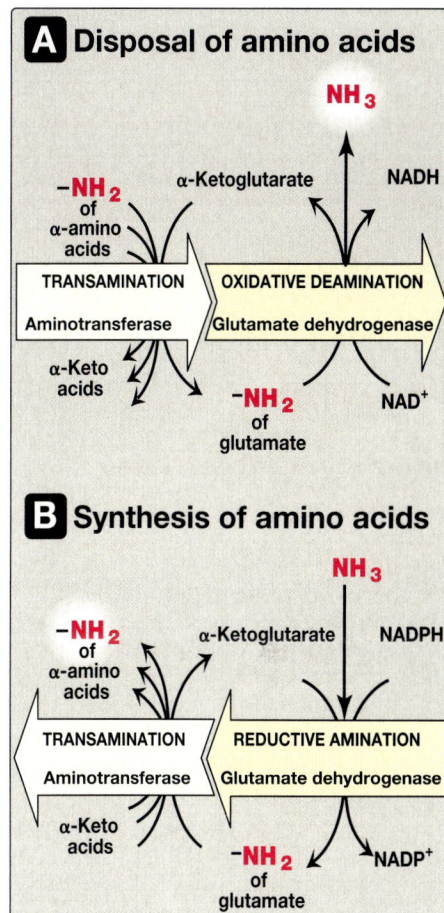

A Disposal of amino acids

B Synthesis of amino acids

Figure 19.12
A, B. Combined actions of aminotransferase and glutamate dehydrogenase reactions. [Note: Reductive amination occurs only when ammonia (NH_3) level is high.] NAD(H), nicotinamide adenine dinucleotide; NADP(H), nicotinamide adenine dinucleotide phosphate.

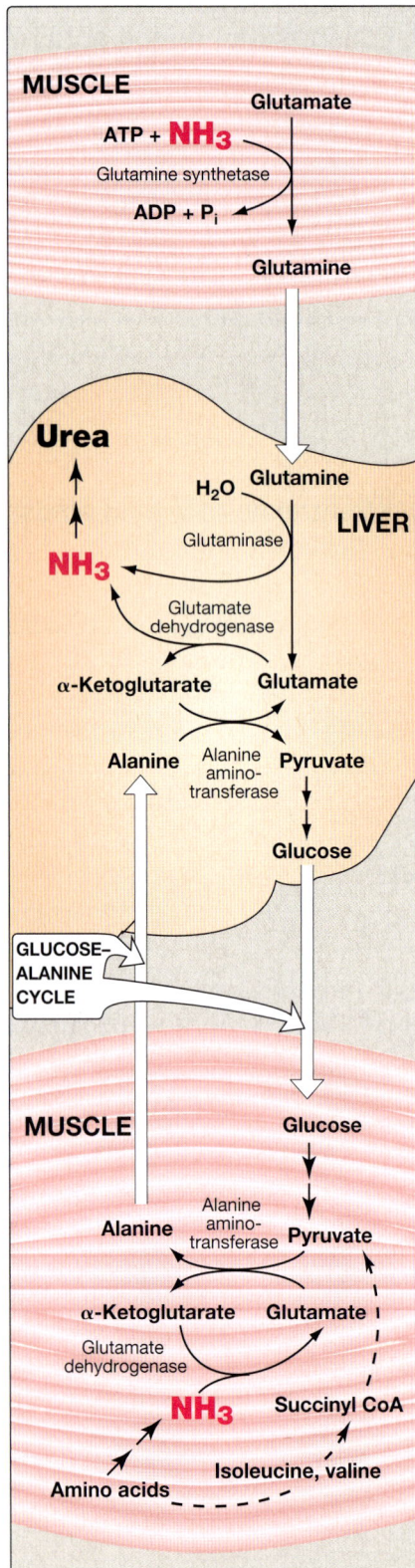

Figure 19.13
Transport of ammonia (NH_3) from muscle to the liver. ADP, adenosine diphosphate; P_i, inorganic phosphate; CoA, coenzyme A.

are in the cytosol (Fig. 19.14). [Note: Gluconeogenesis and heme synthesis also involve both mitochondrial matrix and the cytosolic enzymes.]

1. **Carbamoyl phosphate formation:** Formation of carbamoyl phosphate by carbamoyl phosphate synthetase I (CPS I) is driven by cleavage of two molecules of ATP. Ammonia incorporated into carbamoyl phosphate is provided primarily by the oxidative deamination of glutamate by mitochondrial GDH (Fig. 19.11). Ultimately, the nitrogen atom derived from this ammonia becomes one of the nitrogens of urea. CPS I requires N-acetylglutamate (NAG) as a positive allosteric activator (Fig. 19.14). [Note: CPS II participates in the biosynthesis of pyrimidines. It does not require NAG, uses glutamine as the nitrogen source, and occurs in the cytosol.]

2. **Citrulline formation:** The carbamoyl portion of carbamoyl phosphate is transferred to ornithine by ornithine transcarbamylase (OTC), while the phosphate is released as inorganic phosphate. The reaction product, citrulline, is transported to the cytosol. [Note: Ornithine and citrulline move across the inner mitochondrial membrane via an antiporter. These basic amino acids are not incorporated into cellular proteins because there are no codons for them.] Ornithine is regenerated with each turn of the urea cycle, much in the same way that oxaloacetate is regenerated by the reactions of the citric acid cycle.

3. **Argininosuccinate formation:** Argininosuccinate synthetase combines citrulline with aspartate to form argininosuccinate. The α-amino group of aspartate provides the second nitrogen that is ultimately incorporated into urea. The formation of argininosuccinate is driven by the cleavage of ATP to adenosine monophosphate and pyrophosphate. This is the third and final molecule of ATP consumed in the formation of urea.

4. **Argininosuccinate cleavage:** Argininosuccinate is cleaved by argininosuccinate lyase to yield arginine and fumarate. The arginine serves as the immediate precursor of urea. The fumarate is hydrated to malate, providing a link with several metabolic pathways. Malate can be oxidized by malate dehydrogenase to oxaloacetate, which can be transaminated to aspartate (see Fig. 19.8) and enter the urea cycle (Fig. 19.14). Alternatively, malate can be transported into the mitochondrial matrix via the malate–aspartate shuttle, reenter the citric acid cycle, and get oxidized to oxaloacetate, which can be used for gluconeogenesis. [Note: Malate oxidation generates NADH for oxidative phosphorylation, thereby reducing the energy cost of the urea cycle.]

5. **Arginine cleavage to ornithine and urea:** Arginase-I hydrolyzes arginine to ornithine and urea and is virtually exclusive to the liver. Therefore, only the liver can cleave arginine, thereby synthesizing urea, whereas other tissues, such as the kidney, can synthesize arginine from citrulline. [Note: Arginase-II in kidneys controls arginine availability for nitric oxide synthesis.]

6. **Fate of urea:** Urea diffuses from the liver and is transported in the blood to the kidneys, where it is filtered and excreted in the urine.

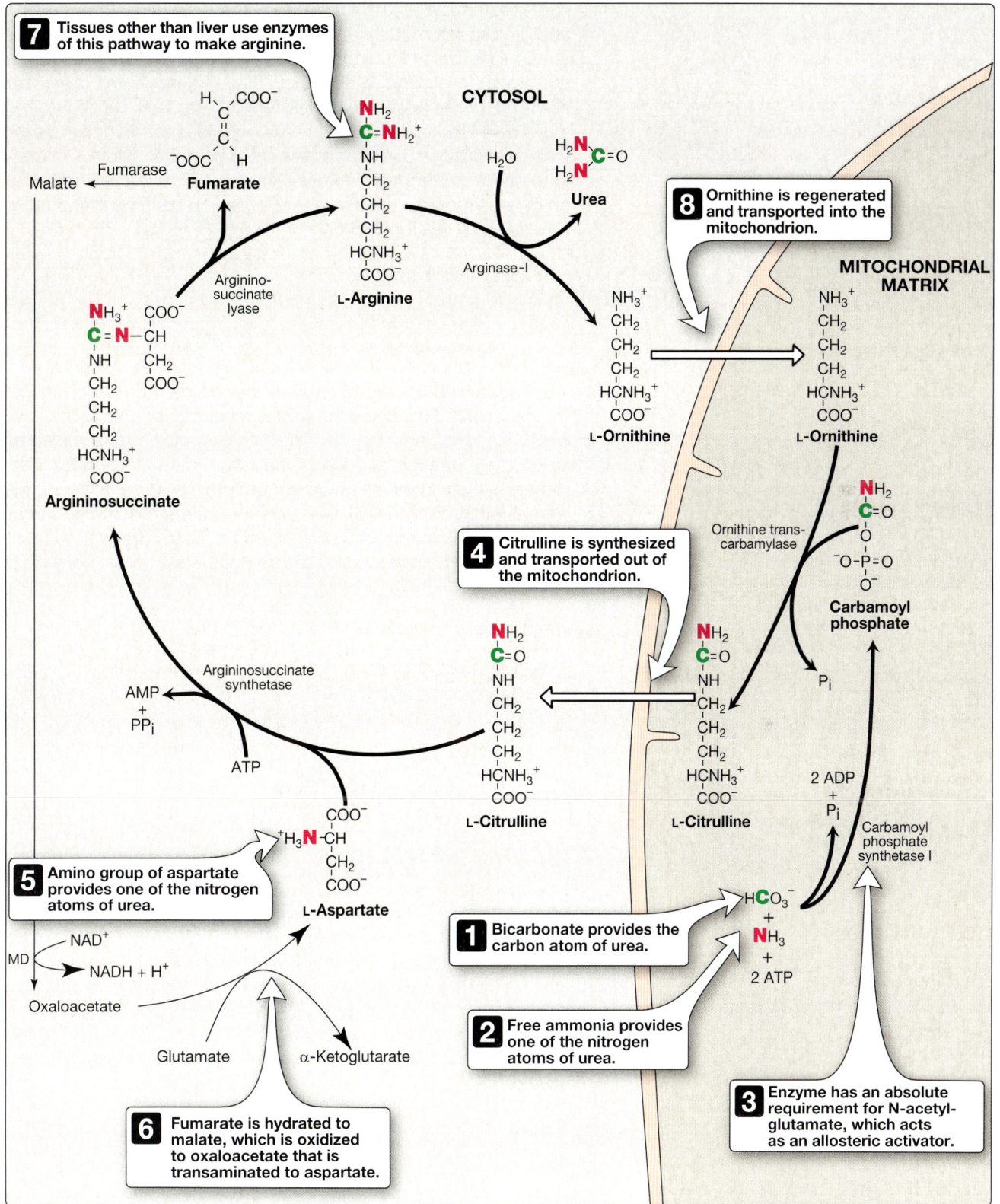

Figure 19.14
Reactions of the urea cycle. [Note: An antiporter transports citrulline and ornithine across the inner mitochondrial membrane.] ADP, adenosine diphosphate; AMP, adenosine monophosphate; PP$_i$, pyrophosphate; P$_i$, inorganic phosphate; NAD(H), nicotinamide adenine dinucleotide; MD, malate dehydrogenase.

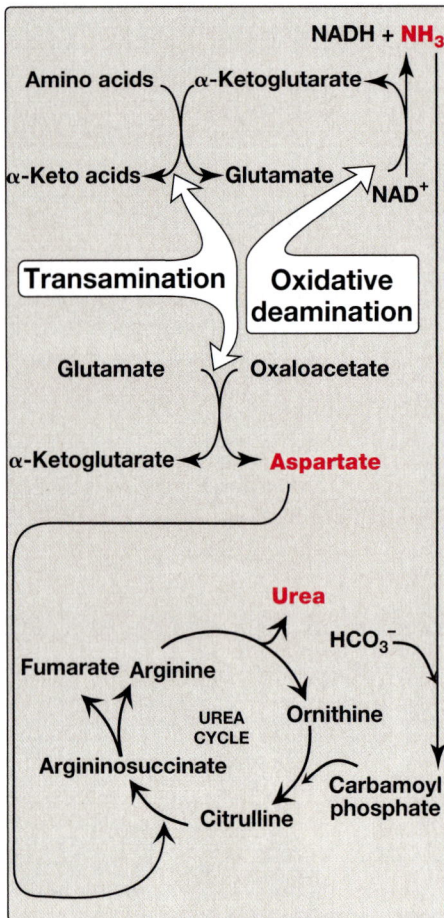

Figure 19.15
Flow of nitrogen from amino acids to urea. Amino groups for urea synthesis are collected in the form of ammonia (NH_3) and aspartate. NAD(H), nicotinamide adenine dinucleotide; HCO_3^-, bicarbonate.

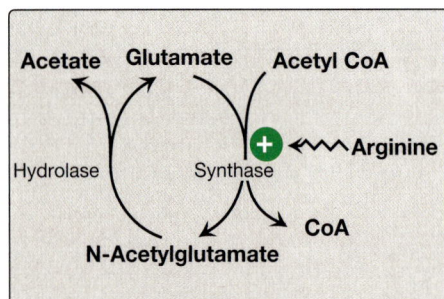

Figure 19.16
Formation and degradation of N-acetylglutamate, an allosteric activator of carbamoyl phosphate synthetase I. CoA, coenzyme A.

A portion of the urea diffuses from the blood into the intestine and is cleaved to CO_2 and ammonia by bacterial urease. The ammonia is partly lost in the feces and is partly reabsorbed into the blood. In patients with kidney failure, plasma urea levels are elevated, promoting a greater transfer of urea from blood into the gut. The intestinal action of urease on this urea becomes a clinically important source of ammonia, contributing to the hyperammonemia often seen in these patients. Oral administration of antibiotics reduces the number of intestinal bacteria responsible for this ammonia production.

B. Overall stoichiometry

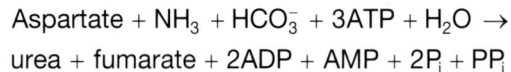

$$Aspartate + NH_3 + HCO_3^- + 3ATP + H_2O \rightarrow$$
$$urea + fumarate + 2ADP + AMP + 2P_i + PP_i$$

Because four high-energy phosphate bonds are consumed in the synthesis of each molecule of urea, the synthesis of urea is irreversible, with a large, negative ΔG. One nitrogen of the urea molecule is supplied by free ammonia and the other nitrogen by aspartate. Glutamate is the immediate precursor of both ammonia (through oxidative deamination by GDH) and aspartate nitrogen (through transamination of oxaloacetate by AST). In effect, both nitrogen atoms of urea arise from glutamate, which, in turn, gathers nitrogen from other amino acids (Fig. 19.15).

C. Regulation

NAG is an essential activator for CPS I, the rate-limiting step in the urea cycle. It increases the affinity of CPS I for ATP. NAG is synthesized from acetyl CoA and glutamate by N-acetylglutamate synthase (NAGS), as shown in Figure 19.16, in a reaction for which arginine is an activator. The cycle is also regulated by substrate availability (short-term regulation) and enzyme induction (long term).

VI. AMMONIA METABOLISM

Ammonia is produced by all tissues during the metabolism of a variety of compounds, and it is disposed of primarily by the formation of urea in the liver. However, the blood ammonia level must be kept very low, because even slightly elevated concentrations (hyperammonemia) are toxic to the central nervous system (CNS). Therefore, a mechanism is required for the transport of nitrogen from the peripheral tissues to the liver for ultimate disposal as urea while keeping circulating levels of free ammonia low.

A. Sources

Amino acids are quantitatively the most important source of ammonia because most Western diets are high in protein and provide excess amino acids, which travel to the liver and undergo transdeamination (i.e., the linking of the aminotransferase and GDH reactions), producing ammonia. [Note: The liver catabolizes straight-chain amino acids, primarily.] However, substantial amounts of ammonia can be obtained from other sources.

1. **Glutamine:** An important source of plasma glutamine is from the catabolism of BCAA in skeletal muscle. This glutamine is taken up by cells of the intestine, liver, and kidneys. The liver and kidneys generate ammonia from glutamine by the actions of glutaminase (Fig. 19.17) and GDH. In the kidneys, most of this ammonia is excreted into the urine as NH_4^+, which provides an important mechanism for maintaining the body's acid–base balance through the excretion of protons. In the liver, the ammonia is detoxified to urea and excreted. [Note: α-Ketoglutarate, the second product of GDH, is a glucogenic precursor in the liver and kidneys.] Ammonia is also generated by intestinal glutaminase. Enterocytes obtain glutamine either from the blood or from digestion of dietary protein. [Note: Intestinal glutamine metabolism also produces alanine, which is used by the liver for gluconeogenesis, and citrulline, which is used by the kidneys to synthesize arginine.]

2. **Intestinal bacteria:** Ammonia is formed from urea by the action of bacterial urease in the lumen of the intestine. This ammonia is absorbed from the intestine by way of the portal vein, and virtually all is removed by the liver via conversion to urea.

3. **Amines:** Amines obtained from the diet and monoamines that serve as hormones or neurotransmitters give rise to ammonia by the action of monoamine oxidase.

4. **Purines and pyrimidines:** In the catabolism of purines and pyrimidines, amino groups attached to the ring atoms are released as ammonia (see Fig. 22.15).

B. Transport in the circulation

Although ammonia is constantly produced in the tissues, it is present at very low levels in blood. This is due both to the rapid removal of blood ammonia by the liver and to the fact that several tissues, particularly muscle, release amino acid nitrogen in the form of glutamine and alanine, rather than as free ammonia (see Fig. 19.13).

1. **Urea:** Formation of urea in the liver is quantitatively the most important disposal route for ammonia. Urea travels in the blood from the liver to the kidneys, where it passes into the glomerular filtrate.

2. **Glutamine:** This amide of glutamate provides a nontoxic storage and transport form of ammonia (Fig. 19.18). The ATP-requiring formation of glutamine from glutamate and ammonia by glutamine synthetase occurs primarily in skeletal muscle and the liver but is also important in the CNS, where it is the major mechanism for the removal of ammonia in the brain. Glutamine is found in plasma at concentrations higher than other amino acids, a finding consistent with its transport function. [Note: The liver keeps blood ammonia levels low through glutaminase, GDH, and the urea cycle in periportal (close to inflow of blood) hepatocytes and through glutamine synthetase as an ammonia scavenger in the perivenous hepatocytes.] Ammonia metabolism is summarized in Figure 19.19.

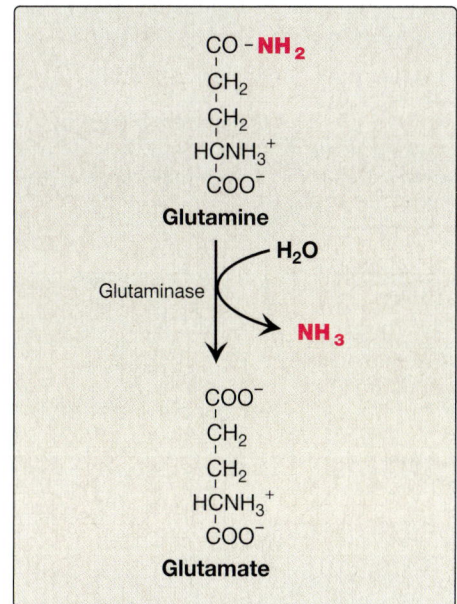

Figure 19.17
Hydrolysis of glutamine to form ammonia (NH_3).

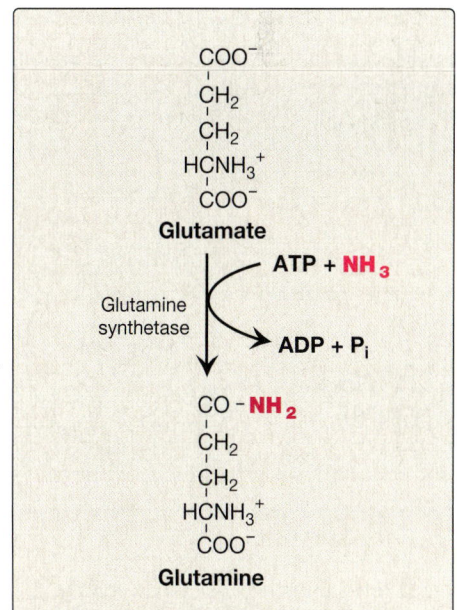

Figure 19.18
Synthesis of glutamine. ADP, adenosine diphosphate; P_i, inorganic phosphate; NH_3, ammonia.

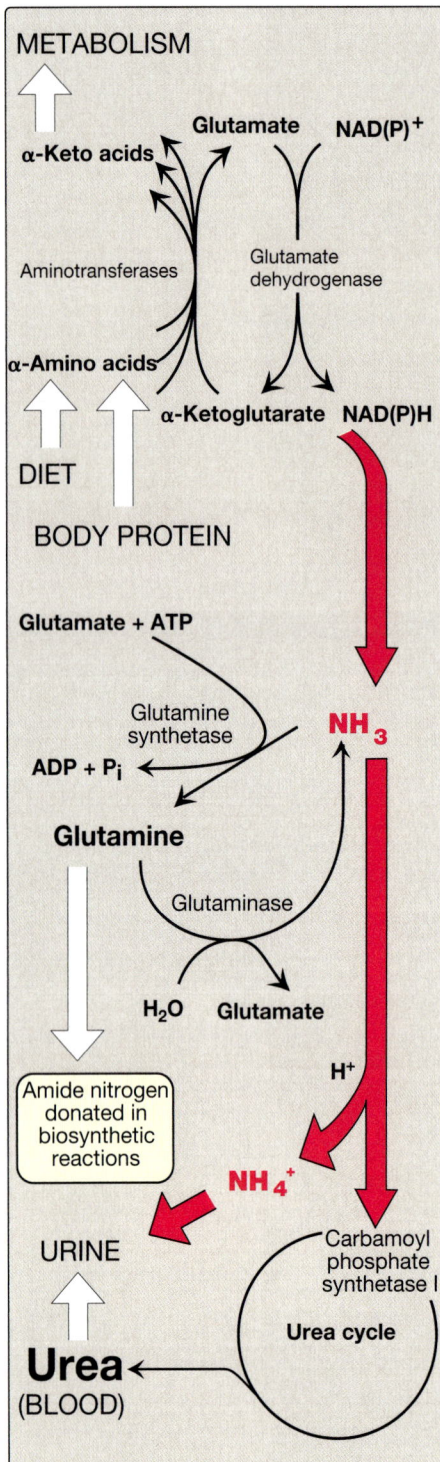

METABOLISM

Glutamate NAD(P)⁺
α-Keto acids
Aminotransferases   Glutamate dehydrogenase
α-Amino acids
α-Ketoglutarate NAD(P)H

DIET

BODY PROTEIN

Glutamate + ATP

Glutamine synthetase
ADP + Pᵢ

Glutamine

Glutaminase

H_2O Glutamate

NH_3

H^+

Amide nitrogen donated in biosynthetic reactions

NH_4^+

URINE

Carbamoyl phosphate synthetase I

Urea cycle

Urea
(BLOOD)

Figure 19.19
Ammonia (NH_3) metabolism. Urea content in the urine is reported as urinary urea nitrogen, or UUN. Urea in blood is reported as BUN (blood urea nitrogen). [Note: The enzymes glutamate dehydrogenase, glutamine synthetase, and carbamoyl phosphate synthetase I fix NH_3 into organic molecules.]

C. Hyperammonemia

The capacity of the hepatic urea cycle exceeds the normal rates of ammonia generation, and the levels of blood ammonia are normally low (5 to 35 µmol/L). However, when liver function is compromised, due to either genetic defects of the urea cycle or liver disease, blood ammonia levels can be more than 1,000 µmol/L. Such hyperammonemia is a medical emergency because ammonia has a direct neurotoxic effect on the CNS. For example, elevated concentrations of ammonia in the blood cause the symptoms of ammonia intoxication, which include tremors, slurred speech, somnolence (drowsiness), vomiting, cerebral edema, and blurred vision. At high concentrations, ammonia can cause coma and death. There are two major types of hyperammonemia.

1. **Acquired:** Liver disease is a common cause of acquired hyperammonemia in adults and may be due to viral hepatitis or chronic exposure to hepatotoxins such as alcohol. Cirrhosis of the liver may result in the formation of collateral circulation around the liver, where portal blood is shunted directly into the systemic circulation and has less access to the liver. Therefore, the conversion of ammonia to urea is severely impaired, leading to elevated levels of ammonia.

2. **Congenital:** Genetic deficiencies of each of the five enzymes of the urea cycle (and NAGS) have been described, with an overall incidence of ~1:25,000 live births. OTC deficiency is X linked, predominantly affecting males, although female carriers may become symptomatic. All the other urea cycle disorders follow an autosomal-recessive inheritance pattern. In each case, the failure to synthesize urea leads to hyperammonemia during the first weeks following birth. Combinations of other symptoms common to hyperammonemia (tremors; slurred speech; drowsiness; vomiting; cerebral edema; blurred vision; intellectual and developmental disability; and, in severe hyperammonemia, even coma and death) can also be seen in different urea cycle deficiencies. Diagnosis is based upon symptoms, laboratory testing, and genetic testing. Historically, congenital urea cycle defects have a high morbidity (neurologic manifestations) and mortality. Additional information for specific urea cycle deficiencies is summarized in the following sections.

   a. **OTC deficiency:** OTC deficiency is the most common urea cycle disorder. Specific laboratory test results include a decrease in the reaction downstream products citrulline and arginine. Interestingly, detectable serum and urinary orotic acid levels also increase. Carbamoyl phosphate, one of the OTC substrates, instead becomes a substrate for pyrimidine biosynthesis, entering the pyrimidine *de novo* pathway downstream of the regulatory reaction (see Fig. 22.2). As a result, orotic acid is an overproduced pyrimidine biosynthesis pathway intermediate. [Note: Elevated orotic acid is also seen in hereditary orotic aciduria, due to a pyrimidine biosynthesis enzyme deficiency in uridine monophosphate (UMP) synthase (UMPS). Along with genetic testing, OTC deficiency can be differentially diagnosed from UMPS deficiency based on other symptoms. Hyperammonemia is a symptom of OTC deficiency, but not of UMPS deficiency; instead, megaloblastic anemia may be a symptom of UMPS deficiency.]

b. **Argininosuccinate synthetase deficiency:** This deficiency is also referred to as citrullinemia type 1, as the substrate for the reaction, citrulline, accumulates in blood and urine. There may be a neonatal acute (classic) form, a milder late-onset form, a form that begins during or after pregnancy, and an asymptomatic form. In the neonatal acute form, citrulline can be detected as part of newborn screening. This detection is critical to prevent hyperammonemia and brain damage.

c. **Argininosuccinate lyase deficiency:** In argininosuccinate lyase deficiency, argininosuccinate, the substrate for the reaction, accumulates in the urine, resulting in argininosuccinic aciduria. This is diagnostic and part of the newborn screening. In more severe- and late-onset forms of the deficiency, the aciduria may be associated with neurologic abnormalities, developmental delays, and cognitive impairment.

d. **Arginase-I deficiency:** In arginase-I deficiency, arginine, the substrate for the reaction, accumulates in the blood and urine, and is often referred to as argininemia or hyperargininemia. The hyperammonemia seen with arginase deficiency is often less severe because arginine contains two waste nitrogens and can be excreted in the urine. As such, patients with this deficiency may appear to be healthy at birth and have normal development during the first 1 to 3 years. After this, the first symptoms of arginase deficiency may appear with apparent developmental delays, loss of developmental milestones, and intellectual disability. Hyperammonemia may be episodic, associated with high-protein meals or periods of stress, such as illness or fasting.

e. **NAGS deficiency:** Like in arginase-I deficiency, a deficiency in NAGS can result in developmental delays and intellectual disability. Less severe forms may be episodic later in life, associated with periods of high-protein meals, stress, or fasting. Carglumic acid is a U.S. Food and Drug Administration–approved therapy for NAGS deficiency. It is a synthetic form of NAG, the positive allosteric activator of CPS I.

f. **Treatment for hyperammonemia:** Treatment for urea cycle enzyme deficiencies involves both limiting protein intake in the diet in the presence of sufficient calories to prevent protein catabolism and the removal of excess ammonia in the blood. This can vary depending on the enzyme deficiency and the defect severity. Patients adhere to a low-protein diet, with minimal protein levels needed to maintain good health. This can vary depending on patient age and weight. Drinks with special formulas and/or medical foods are available in which protein levels are tailored to the patient's needs. Nitrogen-scavenging medications, including the aromatic acids benzoate and phenylbutyrate, can reduce ammonia levels in the blood. Benzoate combines with glycine to form hippurate.

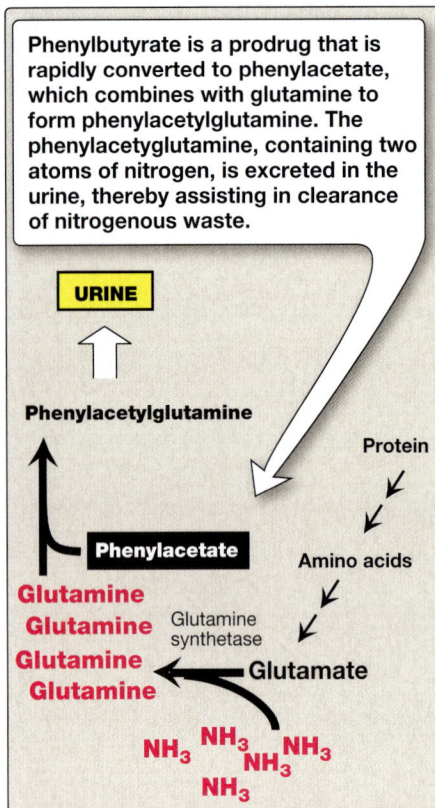

Phenylbutyrate is a prodrug that is rapidly converted to phenylacetate, which combines with glutamine to form phenylacetylglutamine. The phenylacetylglutamine, containing two atoms of nitrogen, is excreted in the urine, thereby assisting in clearance of nitrogenous waste.

URINE

Phenylacetylglutamine

Protein

Phenylacetate

Amino acids

Glutamine
Glutamine   Glutamine
Glutamine   synthetase
Glutamine   ← Glutamate

NH_3 NH_3 NH_3
        NH_3
    NH_3

Figure 19.20
Treatment of patients with urea cycle defects by administration of phenylbutyrate to aid in excretion of ammonia (NH_3).

Phenylbutyrate is converted to phenylacetate and combines with glutamine to form phenylacetylglutamine (Fig. 19.20). Both end products, hippurate and phenylacetylglutamine, are readily excreted in the urine. The combined excretion of glycine and glutamine, and their subsequent biosynthesis, effectively lowers ammonia levels and the potential for hyperammonemia. During severe hyperammonemia, patients may also require dialysis, intravenous fluids, or other treatments to quickly reduce blood ammonia levels and prevent permanent brain damage.

VII. CHAPTER SUMMARY

- **Nitrogen enters** the body in a variety of compounds present in food, the most important being **amino acids** contained in **dietary protein**.

- **Nitrogen leaves** the body as **urea**, **ammonia**, and other products derived from amino acid metabolism (Fig. 19.21).

- Free amino acids in the body are produced by hydrolysis of dietary protein by **proteases** activated from their **zymogen** form in the stomach and intestine, degradation of tissue proteins, and de novo synthesis. This **amino acid pool** is consumed in the synthesis of body protein, metabolized for energy, or its members used as precursors for other nitrogen-containing compounds.

- Free amino acids from digestion are taken up by intestinal **enterocytes** via **sodium-dependent secondary active transport**. Small peptides are taken up via **proton-linked transport**.

- Body protein is simultaneously degraded and resynthesized, a process known as **protein turnover**. The concentration of a cellular protein may be determined by regulation of its synthesis or degradation. The adenosine triphosphate (ATP)-dependent, cytosolic, selective **ubiquitin (Ub)–proteasome** and ATP-independent, relatively nonselective **lysosomal acid hydrolases** are the two major enzyme systems that are responsible for **degrading proteins**.

- Nitrogen cannot be stored, and amino acids in excess of the biosynthetic needs of the cell are quickly degraded. The first phase of **catabolism** involves the transfer of the α-amino groups through **transamination** by **pyridoxal phosphate**–dependent **aminotransferases** (**transaminases**), followed by **oxidative deamination of glutamate** by **glutamate dehydrogenase**, forming **ammonia** and the corresponding α-keto acids.

- A portion of the **free ammonia** is excreted in the **urine**. Some ammonia is used in converting glutamate to **glutamine** for safe transport, but most is used in the hepatic synthesis of **urea**, which is quantitatively the most important route for disposing of nitrogen from the body. **Alanine** also carries nitrogen to the liver for disposal as urea.

- The two major causes of **hyperammonemia** (with its neurologic effects) are acquired liver disease and congenital deficiencies of urea cycle enzymes such as X-linked **ornithine transcarbamylase**.

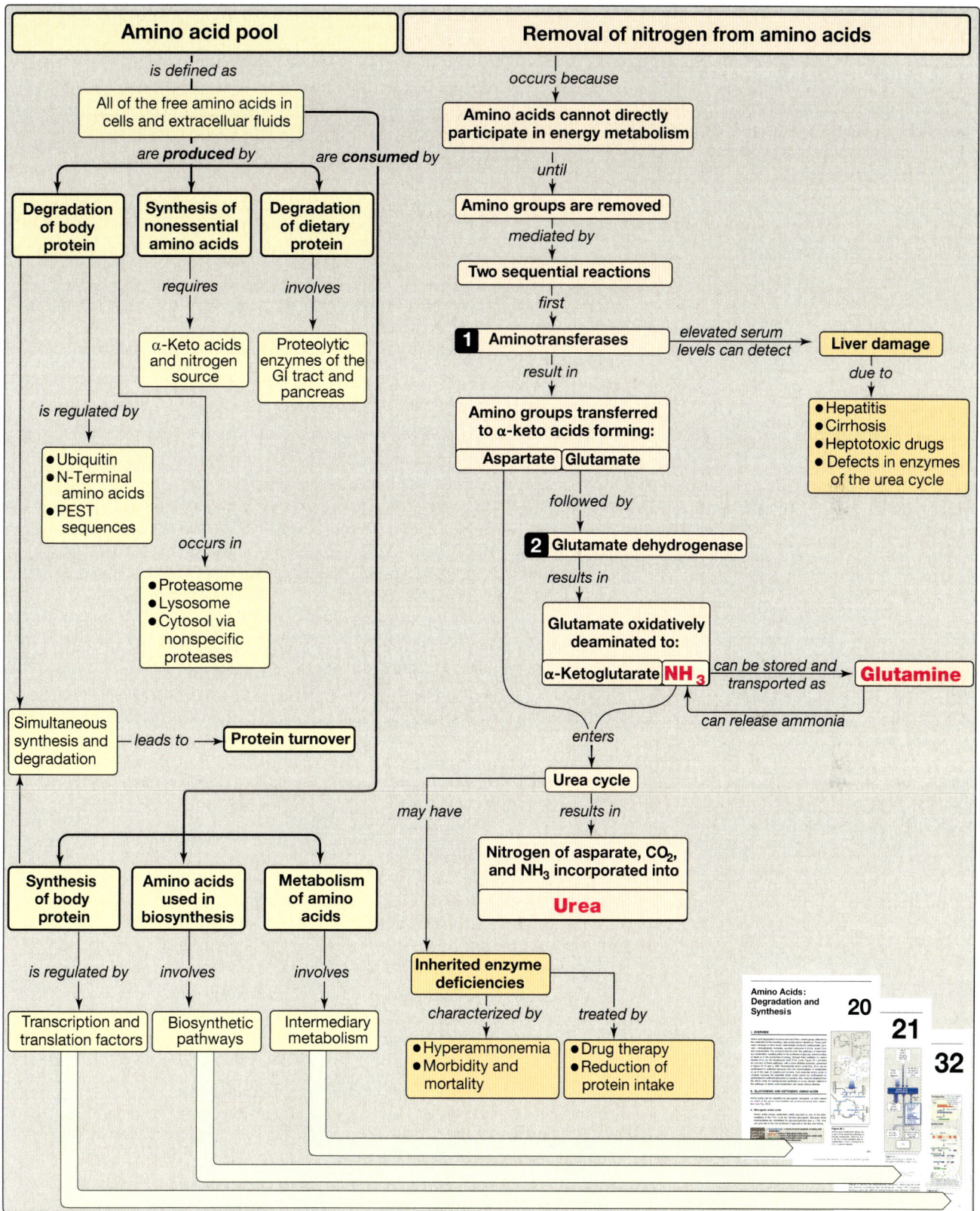

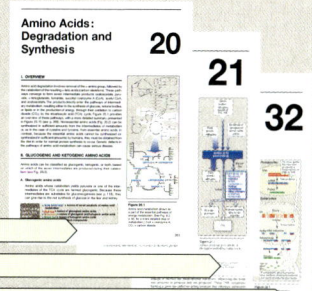

Amino acid pool

is defined as

All of the free amino acids in cells and extracelluar fluids

*are **produced** by*   *are **consumed** by*

Degradation of body protein

Synthesis of nonessential amino acids

Degradation of dietary protein

requires   *involves*

α-Keto acids and nitrogen source

Proteolytic enzymes of the GI tract and pancreas

is regulated by

- Ubiquitin
- N-Terminal amino acids
- PEST sequences

occurs in

- Proteasome
- Lysosome
- Cytosol via nonspecific proteases

Simultaneous synthesis and degradation   —*leads to*→   **Protein turnover**

Synthesis of body protein

Amino acids used in biosynthesis

Metabolism of amino acids

is regulated by   *involves*   *involves*

Transcription and translation factors

Biosynthetic pathways

Intermediary metabolism

Removal of nitrogen from amino acids

occurs because

Amino acids cannot directly participate in energy metabolism

until

Amino groups are removed

mediated by

Two sequential reactions

first

1 **Aminotransferases**   *elevated serum levels can detect* →   **Liver damage**

result in   *due to*

Amino groups transferred to α-keto acids forming:

| Aspartate | Glutamate |

- Hepatitis
- Cirrhosis
- Heptotoxic drugs
- Defects in enzymes of the urea cycle

followed by

2 **Glutamate dehydrogenase**

results in

Glutamate oxidatively deaminated to:

α-Ketoglutarate | **NH$_3$**   *can be stored and transported as* → **Glutamine**

can release ammonia

enters

Urea cycle

may have   *results in*

Nitrogen of asparate, CO$_2$, and NH$_3$ incorporated into

Urea

Inherited enzyme deficiencies

characterized by   *treated by*

- Hyperammonemia
- Morbidity and mortality

- Drug therapy
- Reduction of protein intake

Amino Acids: Degradation and Synthesis **20** **21** **32**

Figure 19.21
Key concept map for nitrogen metabolism. GI, gastrointestinal; PEST, proline, glutamate, serine, threonine; NH$_3$, ammonia; CO$_2$, carbon dioxide.

Study Questions

Choose the ONE best answer.

19.1. In this transamination reaction, which of the following are the products X and Y?

Oxaloacetate → X

Glutamate → Y

A. Alanine, α-ketoglutarate
B. Aspartate, α-ketoglutarate
C. Glutamate, alanine
D. Pyruvate, aspartate
E. Alanine, pyruvate

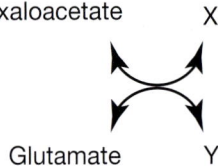

Correct answer = B. Transamination reactions always have an amino acid and an α-keto acid as substrates. The products of the reaction are also an amino acid (corresponding to the α-keto substrate) and an α-keto acid (corresponding to the amino acid substrate). Three amino acid/α-keto acid pairs commonly encountered in metabolism are alanine/pyruvate, aspartate/oxaloacetate, and glutamate/α-ketoglutarate. In this question, glutamate is deaminated to form α-ketoglutarate, and oxaloacetate is aminated to form aspartate.

19.2. Which of the following statements about amino acids and their metabolism is correct?

A. Free amino acids are taken into the enterocytes by a single proton-linked transport system.
B. In healthy, well-fed individuals, input to the amino acid pool exceeds output.
C. The liver uses ammonia to buffer protons.
D. Muscle-derived glutamine is deaminated in liver and kidney tissue to ammonia + a gluconeogenic precursor.
E. The first step in the catabolism of most amino acids is their oxidative deamination.
F. The toxic ammonia generated from the amide nitrogen of amino acids is transported through blood as arginine.

Correct answer = D. Glutamine, produced by the catabolism of branched-chain amino acids in muscle, is deaminated by glutaminase to ammonia + glutamate. The glutamate is deaminated by glutamate dehydrogenase to ammonia + α-ketoglutarate, which can be used for gluconeogenesis. Free amino acids are taken into enterocytes by several different sodium-linked transport systems. Healthy, well-fed individuals are in nitrogen balance, in which nitrogen input equals output. The liver converts ammonia to urea, and the kidneys use ammonia to buffer protons. Amino acid catabolism begins with transamination that generates glutamate. The glutamate undergoes oxidative deamination. Toxic ammonia is transported as glutamine and alanine. Arginine is synthesized and hydrolyzed in the hepatic urea cycle.

For Questions 19.3 to 19.5, use the following clinical scenario.

A female neonate appeared healthy until age ~24 hours, when she became lethargic. A sepsis workup proved negative. At 56 hours, she started showing focal seizure activity. The plasma ammonia level was found to be 887 µmol/L (normal 5 to 35 µmol/L). Quantitative plasma amino acid levels revealed a marked elevation of citrulline but not argininosuccinate.

19.3. Which of the following enzymic activities is most likely to be deficient in this patient?

A. Arginase
B. Argininosuccinate lyase
C. Argininosuccinate synthetase
D. Carbamoyl phosphate synthetase I
E. Ornithine transcarbamylase

Correct answer = C. Genetic deficiencies of each of the five enzymes of the urea cycle, as well as deficiencies in N-acetylglutamate synthase, have been described. The accumulation of citrulline (but not argininosuccinate) in the plasma of this patient means that the enzyme required for the conversion of citrulline to argininosuccinate (argininosuccinate synthetase) is defective, whereas the enzyme that cleaves argininosuccinate (argininosuccinate lyase) is functional.

19.4. Which of the following would also be elevated in this patient's blood?

A. Asparagine
B. Glutamine
C. Lysine
D. Urea
E. Arginine

Correct answer = B. Deficiencies of the enzymes of the urea cycle result in the failure to synthesize urea and lead to hyperammonemia in the first few weeks after birth. Glutamine will also be elevated because it acts as a nontoxic storage and transport form of ammonia. Therefore, elevated glutamine accompanies hyperammonemia. Asparagine, lysine, and arginine do not serve this sequestering role. Urea would be decreased because of impaired activity of the urea cycle. [Note: Alanine would also be elevated in this patient.]

19.5. Why might supplementation with arginine benefit this patient?

The arginine will be cleaved by arginase to urea and ornithine. Ornithine will be combined with carbamoyl phosphate by ornithine transcarbamylase to form citrulline. Citrulline, containing one waste nitrogen, will be excreted.

19.6. A patient is found comatose and presents with hyperammonemia and orotic acidemia. Which of the following enzymes is most likely to be deficient in this patient?

A. Arginase
B. Argininosuccinate lyase
C. Argininosuccinate synthetase
D. Carbamoyl phosphate synthetase I
E. N-Acetylglutamate synthase
F. Ornithine transcarbamylase

Correct answer = F. A patient with a defect in ornithine transcarbamylase would prevent the condensation of carbamoyl phosphate and ornithine to form citrulline. Carbamoyl phosphate can instead enter the cytosol and contribute to pyrimidine synthesis, resulting in orotic acid buildup (orotic acid is an intermediate in the pyrimidine synthesis pathway). Elevated orotic acid is also seen in a UMP synthase deficiency (in the pyrimidine synthesis pathway). Hyperammonemia confirms the diagnosis as a urea cycle enzyme deficiency, not a pyrimidine synthesis deficiency. A UMP synthase deficiency would be accompanied with a megaloblastic anemia, not hyperammonemia.

19.7. Parents bring their 1-week-old infant to the emergency room. The infant is lethargic and feeding poorly. Lab tests show mild hyperammonemia. How might the infant best be treated?

A. Dialysis
B. Glucose solution
C. Insulin therapy
D. Low-protein diet
E. Phenylbutyrate

Correct answer = E. Nitrogen-scavenging medications, including the aromatic acids benzoate and phenylbutyrate, can reduce ammonia levels in the blood. Benzoate combines with glycine to form hippurate. Phenylbutyrate is converted to phenylacetate, which combines with glutamine to form phenylacetylglutamine. Both end products, hippurate and phenylacetylglutamine, are readily excreted in the urine. The combined excretion of glycine and glutamine, and their subsequent biosynthesis, effectively lowers ammonia levels and the potential for hyperammonemia. During more severe hyperammonemia, patients may also require dialysis, intravenous fluids, or other treatments to quickly reduce blood ammonia levels and to prevent permanent brain damage.

19.8. A patient with viral hepatitis is seen in the emergency department with hyperammonemia, jaundice, and abdominal swelling. Which of the following might be elevated in the blood of this patient?

A. Alanine aminotransferase
B. Citrulline
C. Glucose
D. Ketoacids
E. Orotic acid

Correct answer = A. Plasma aspartate aminotransferase (AST) and ALT are elevated in nearly all hepatic diseases but are particularly high in acute conditions that cause extensive cell necrosis, such as severe viral hepatitis, toxic injury, and prolonged circulatory collapse. ALT is more specific than AST for liver disease, but the latter is more sensitive because the liver contains larger amounts of AST. Serial measurements of AST and ALT (liver function tests) are often useful in determining the course of liver damage. Hyperammonemia results from a decrease in urea cycle hepatic function. Jaundice results from a decrease in the hepatic conjugation and excretion of bilirubin.

20 Amino Acids: Degradation and Synthesis

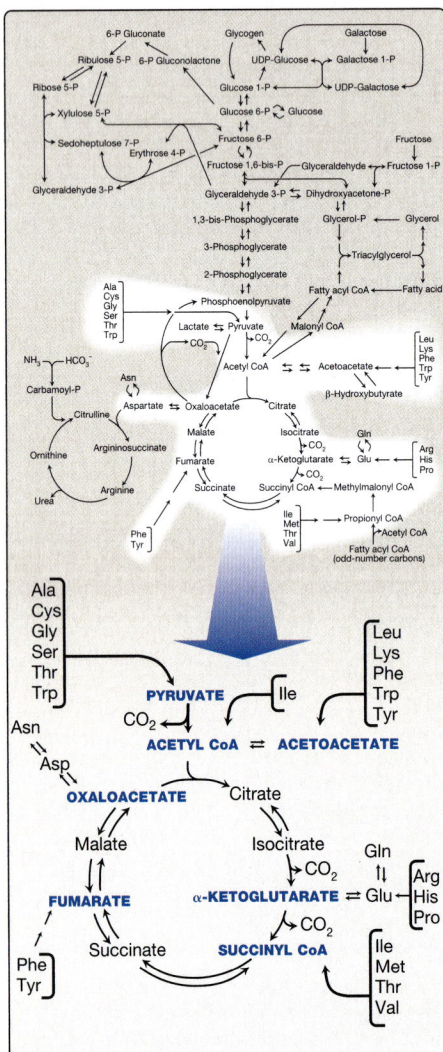

Figure 20.1
Amino acid metabolism shown as a part of the essential pathways of energy metabolism. (See Fig. 8.2 for a more detailed map of metabolism.) CoA, coenzyme A; CO_2, carbon dioxide.

I. OVERVIEW

Amino acid degradation involves the removal of the α-amino group, followed by the catabolism of the resulting α-keto acids (carbon skeletons). The degradation pathways of the various amino acids converge to form seven intermediate products: oxaloacetate, pyruvate, α-ketoglutarate, fumarate, succinyl coenzyme A (CoA), acetyl CoA, and acetoacetate. The products directly enter the pathways of intermediary metabolism, resulting either in the synthesis of glucose, ketone bodies, or lipids or in the production of energy through their oxidation to carbon dioxide (CO_2) by the citric acid cycle. Figure 20.1 provides an overview of these pathways, with a more detailed summary presented in Figure 20.15. Nonessential amino acids can be synthesized in sufficient amounts from the intermediates of metabolism or, as in the case of cysteine and tyrosine, from essential amino acids (Fig. 20.2). In contrast, because the essential amino acids cannot be synthesized (or synthesized in sufficient amounts) by humans, they must be obtained from the diet for normal protein synthesis to occur. Genetic defects in the pathways of amino acid metabolism can cause serious disease.

II. GLUCOGENIC AND KETOGENIC AMINO ACIDS

Amino acids can be classified as glucogenic, ketogenic, or both, based on which of the seven intermediates are produced during their catabolism (see Fig. 20.2).

A. Glucogenic amino acids

Amino acids whose catabolism yields pyruvate or one of the intermediates of the citric acid cycle are termed glucogenic. Because these intermediates can ultimately form oxaloacetate within the citric acid cycle, they are potential substrates for gluconeogenesis (see Chapter 10, II. C.) and can give rise to the net synthesis of glucose in the liver and kidney.

Color-coding used in this chapter:	• **BLUE CAPS TEXT** = names of seven products of amino acid metabolism • **Red text** = names of glucogenic amino acids • **Brown text** = names of glucogenic and ketogenic amino acids • **Green text** = names of ketogenic amino acids • **Cyan text** = one-carbon compounds

B. Ketogenic amino acids

Amino acids whose catabolism yields either acetyl CoA (directly, without pyruvate serving as an intermediate) or acetoacetate (or its precursor acetoacetyl CoA) are termed ketogenic (see Fig. 20.2). Acetoacetate is one of the ketone bodies, which also include 3-hydroxybutyrate and acetone. Leucine and lysine are the only exclusively ketogenic amino acids found in proteins. Their carbon skeletons are not substrates for gluconeogenesis and cannot give rise to the synthesis of glucose.

III. AMINO ACID CARBON SKELETON CATABOLISM

The pathways by which amino acids are catabolized are conveniently organized according to which one (or more) of the seven intermediates listed above is produced from a particular amino acid.

A. Amino acids that form oxaloacetate

Asparagine is hydrolyzed by asparaginase, liberating ammonia and aspartate (Fig. 20.3). Aspartate is converted to its corresponding ketoacid by transamination to form oxaloacetate (Fig. 20.3).

> Some rapidly dividing leukemic cells are unable to synthesize sufficient asparagine to support their growth. This makes asparagine an essential amino acid for these cells, which, therefore, require asparagine from the blood. Asparaginase, which hydrolyzes asparagine to aspartate, can be administered systemically to treat leukemia. Asparaginase lowers the level of asparagine in the plasma, thereby depriving cancer cells of a required nutrient.

B. Amino acids that form α-ketoglutarate via glutamate

1. **Glutamine:** This amino acid is hydrolyzed to glutamate and ammonia by the enzyme glutaminase (see Chapter 19, VI. A.). Glutamate is converted to α-ketoglutarate by transamination or through oxidative deamination by glutamate dehydrogenase (see Chapter 19, IV. B.).

2. **Proline:** This amino acid is oxidized to glutamate. Glutamate is transaminated or oxidatively deaminated to form α-ketoglutarate.

3. **Arginine:** This amino acid is hydrolyzed by arginase to produce ornithine and urea. The reaction occurs primarily in the liver as part of the urea cycle (see Chapter 19, V. A.). Ornithine is subsequently converted to α-ketoglutarate, with glutamate semialdehyde as an intermediate.

4. **Histidine:** Histidine is oxidatively deaminated by histidase to urocanic acid, which subsequently forms N-formiminoglutamate (FIGlu) (Fig. 20.4). FIGlu donates its formimino group to tetrahydrofolate (THF), leaving glutamate, which is degraded as described. A deficiency in histidase results in the relatively benign inborn error of metabolism histidinemia, characterized by elevated levels of histidine in blood and urine.

Figure 20.2
Classification of amino acids. [Note: Some amino acids can become conditionally essential; e.g., supplementation with glutamine and arginine has been shown to improve outcomes in patients with trauma, postoperative infections, and immunosuppression.]

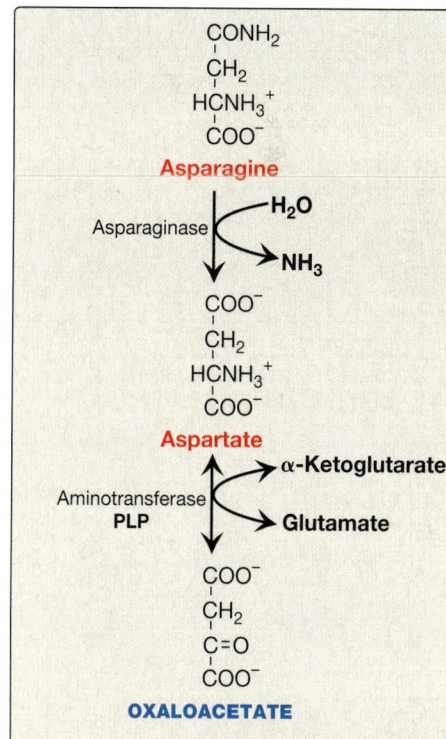

	Glucogenic	Glucogenic and Ketogenic	Ketogenic
Nonessential	Alanine Arginine Asparagine Aspartate Cysteine Glutamate Glutamine Glycine Proline Serine	Tyrosine	
Essential	Histidine Methionine Threonine Valine	Isoleucine Phenyl-alanine Tryptophan	Leucine Lysine

Figure 20.3
Metabolism of asparagine and aspartate. PLP, pyridoxal phosphate; NH_3, ammonia.

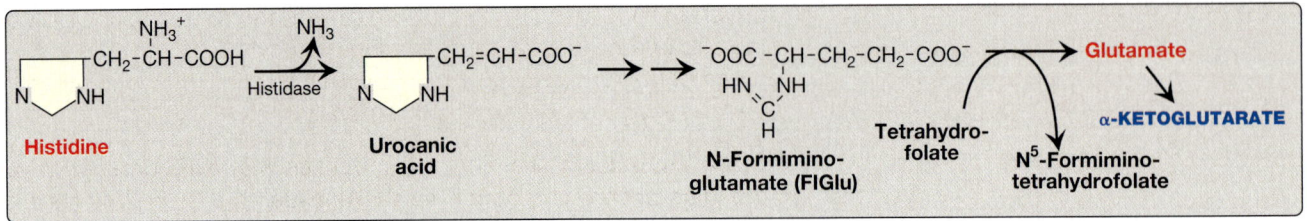

Figure 20.4
Degradation of histidine. NH_3, ammonia.

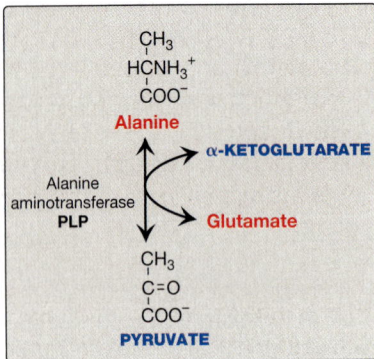

Figure 20.5
Transamination of alanine to pyruvate. PLP, pyridoxal phosphate.

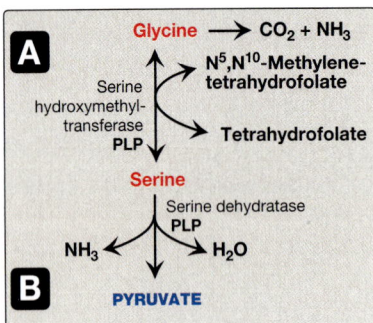

Figure 20.6
A. Interconversion of serine and glycine and oxidation of glycine. **B.** Dehydration of serine to pyruvate. PLP, pyridoxal phosphate; NH_3, ammonia.

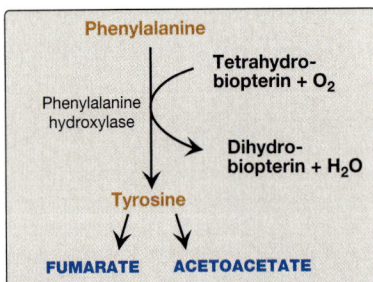

Figure 20.7
Degradation of phenylalanine.

Individuals deficient in folic acid excrete increased amounts of FIGlu in the urine, particularly after ingestion of a large dose of histidine. The FIGlu excretion test has been used in diagnosing a deficiency of folic acid.

C. Amino acids that form pyruvate

1. **Alanine:** This amino acid loses its amino group by transamination to form pyruvate (Fig. 20.5).

2. **Serine:** This amino acid can be converted to glycine as THF becomes N^5,N^{10}-methylenetetrahydrofolate (N^5,N^{10}-MTHF) (Fig. 20.6A). Serine can also be converted to pyruvate (Fig. 20.6B).

3. **Glycine:** This amino acid can be converted to serine by the reversible addition of a methylene group from N^5,N^{10}-MTHF (see Fig. 20.6A) or oxidized to CO_2 and ammonia by the glycine cleavage system. Glycine can be deaminated to glyoxylate by a D–amino acid oxidase (see Chapter 19, IV. B.), which can be oxidized to oxalate or transaminated to glycine. Deficiency of the transaminase in liver peroxisomes causes overproduction of oxalate, the formation of oxalate stones, and kidney damage (primary oxaluria type 1).

4. **Cysteine:** This sulfur-containing amino acid undergoes desulfurization to yield pyruvate. [Note: The sulfate released can be used to synthesize 3′-phosphoadenosine-5′-phosphosulfate (PAPS), an activated sulfate donor to a variety of acceptors.] Cysteine can also be oxidized to its disulfide derivative, cystine.

5. **Threonine:** This amino acid is converted to pyruvate in most organisms but is (at best) a minor pathway in humans.

D. Amino acids that form fumarate

Hydroxylation of phenylalanine produces tyrosine (Fig. 20.7). This irreversible reaction, catalyzed by tetrahydrobiopterin (BH_4)-requiring phenylalanine hydroxylase (PAH), initiates the catabolism of phenylalanine. Thus, phenylalanine metabolism and tyrosine metabolism merge, ultimately forming fumarate and acetoacetate. Therefore, phenylalanine and tyrosine are both glucogenic and ketogenic amino acids.

E. Amino acids that form succinyl CoA: methionine

Methionine is one of four amino acids that form succinyl CoA. This sulfur-containing amino acid deserves special attention because it is

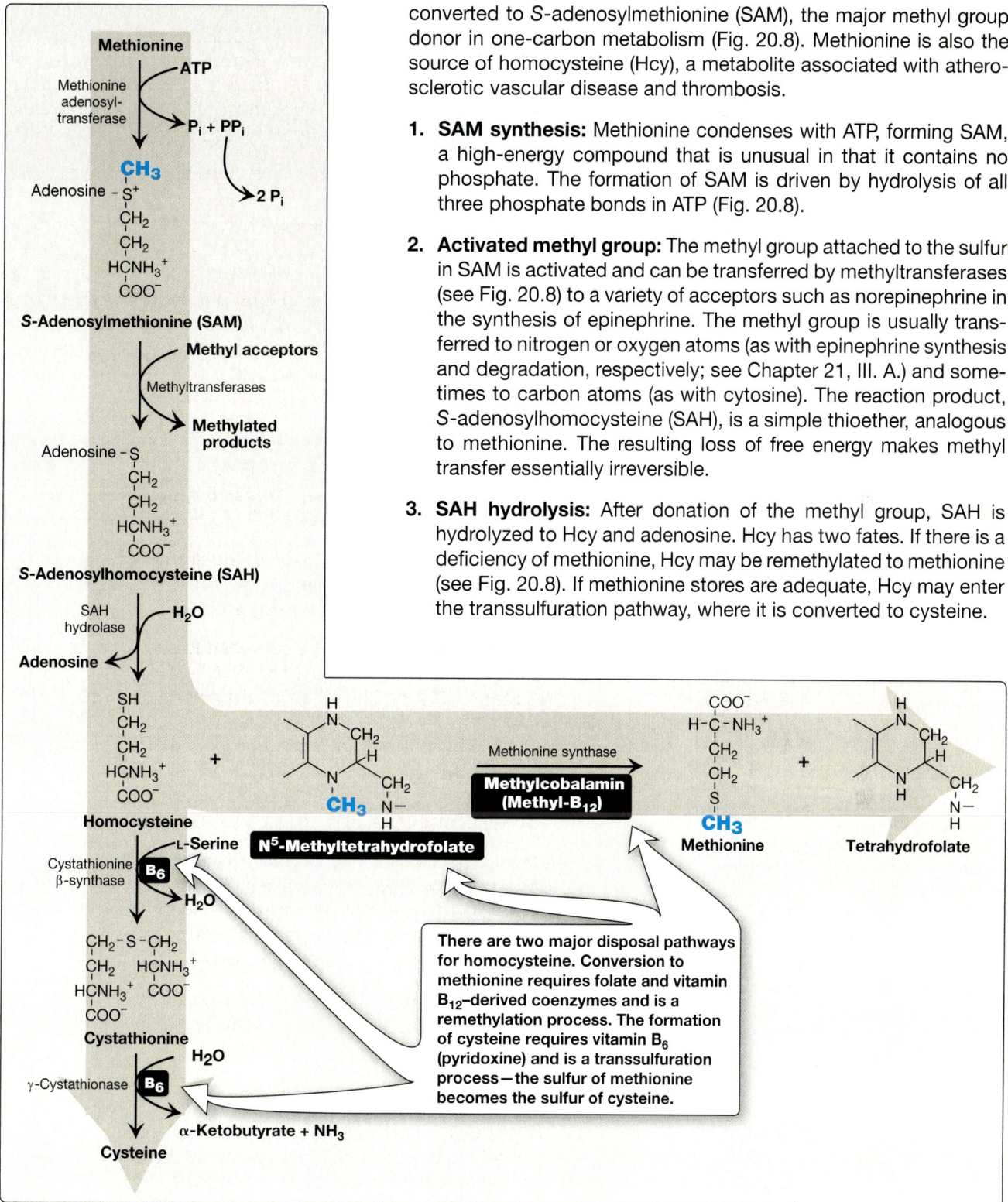

converted to S-adenosylmethionine (SAM), the major methyl group donor in one-carbon metabolism (Fig. 20.8). Methionine is also the source of homocysteine (Hcy), a metabolite associated with atherosclerotic vascular disease and thrombosis.

1. **SAM synthesis:** Methionine condenses with ATP, forming SAM, a high-energy compound that is unusual in that it contains no phosphate. The formation of SAM is driven by hydrolysis of all three phosphate bonds in ATP (Fig. 20.8).

2. **Activated methyl group:** The methyl group attached to the sulfur in SAM is activated and can be transferred by methyltransferases (see Fig. 20.8) to a variety of acceptors such as norepinephrine in the synthesis of epinephrine. The methyl group is usually transferred to nitrogen or oxygen atoms (as with epinephrine synthesis and degradation, respectively; see Chapter 21, III. A.) and sometimes to carbon atoms (as with cytosine). The reaction product, S-adenosylhomocysteine (SAH), is a simple thioether, analogous to methionine. The resulting loss of free energy makes methyl transfer essentially irreversible.

3. **SAH hydrolysis:** After donation of the methyl group, SAH is hydrolyzed to Hcy and adenosine. Hcy has two fates. If there is a deficiency of methionine, Hcy may be remethylated to methionine (see Fig. 20.8). If methionine stores are adequate, Hcy may enter the transsulfuration pathway, where it is converted to cysteine.

Figure 20.8
Degradation and resynthesis of methionine. [Note: The resynthesis of methionine from homocysteine is the only reaction in which tetrahydrofolate both carries and donates a methyl (–CH_3) group. In all other reactions, SAM is the methyl group carrier and donor.] PP_i, pyrophosphate; P_i, inorganic phosphate; NH_3, ammonia.

a. **Methionine resynthesis:** Hcy accepts a methyl group from N^5-methyltetrahydrofolate (N^5-methyl-THF) in a reaction requiring methylcobalamin, a coenzyme derived from vitamin B_{12} (see Chapter 28, III.). The methyl group is transferred by methionine synthase from the B_{12} derivative to Hcy, regenerating methionine. Cobalamin is remethylated from N^5-methyl-THF.

b. **Cysteine synthesis:** Catalyzed by cystathionine β-synthase, Hcy condenses with serine, forming cystathionine, which is hydrolyzed to α-ketobutyrate and cysteine (see Fig. 20.8). This vitamin B_6–requiring sequence has the net effect of converting serine to cysteine and Hcy to α-ketobutyrate, which is oxidatively decarboxylated to form propionyl CoA. Propionyl CoA is converted to succinyl CoA (Fig. 16.20). Because Hcy is synthesized from the essential amino acid methionine, cysteine is not an essential amino acid if sufficient methionine is available.

4. **Relationship of Hcy to vascular disease:** Elevation in plasma Hcy levels promotes oxidative damage, inflammation, and endothelial dysfunction and is an independent risk factor for occlusive vascular diseases such as cardiovascular disease (CVD) and stroke (Fig. 20.9). Mild elevations (hyperhomocysteinemia) are seen in ~7% of the population. Epidemiologic studies have shown that plasma Hcy levels are inversely related to plasma levels of folate, B_{12}, and B_6, the three vitamins involved in the conversion of Hcy to methionine and cysteine. Supplementation with these vitamins has been shown to reduce circulating levels of Hcy. However, in patients with established CVD, vitamin therapy does not decrease cardiovascular events or death. This raises the question as to whether Hcy is a cause of the vascular damage or merely a marker of such damage. Large elevations in plasma Hcy as a result of rare deficiencies in cystathionine β-synthase (see Section VI. D.) of the transsulfuration pathway are seen in patients with classic homocystinuria, resulting from severe hyperhomocysteinemia (>100 μmol/L). Deficiencies in the remethylation reaction also result in a rise in Hcy.

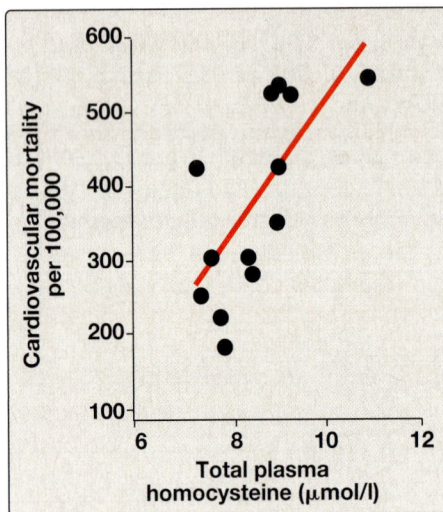

Figure 20.9
Association between cardiovascular disease mortality and total plasma homocysteine.

In pregnant women, elevated Hcy levels usually indicate a deficiency in folic acid, which is associated with an increased incidence of neural tube defects (improper closure, as in spina bifida) in the fetus (see Chapter 28, II. B.). Periconceptual supplementation with folate reduces the risk of such defects.

F. **Other amino acids that form succinyl CoA**

Degradation of valine, isoleucine, and threonine also results in the production of succinyl CoA, a citric acid cycle intermediate and gluconeogenic compound.

1. **Valine and isoleucine:** These amino acids are branched-chain amino acids (BCAAs) that generate propionyl CoA, which is converted to methylmalonyl CoA and then succinyl CoA by biotin- and vitamin B_{12}–requiring reactions (see Fig. 16.20).

2. **Threonine:** This amino acid is dehydrated to α-ketobutyrate, which is converted to propionyl CoA and then to succinyl CoA. Propionyl CoA, then, is generated by the catabolism of the amino acids methionine, valine, isoleucine, and threonine. Propionyl CoA also is generated by the oxidation of odd-numbered fatty acids.

G. Amino acids that form acetyl CoA or acetoacetyl CoA

Tryptophan, leucine, isoleucine, and lysine form acetyl CoA or acetoacetyl CoA directly, without pyruvate serving as an intermediate. As noted earlier, phenylalanine and tyrosine also give rise to acetoacetate during their catabolism (see Fig. 20.7). Therefore, there are a total of six partly or wholly ketogenic amino acids.

1. **Tryptophan:** This amino acid is both glucogenic and ketogenic, because its catabolism yields alanine and acetoacetyl CoA (Fig. 20.10). Quinolinate from tryptophan catabolism is used in the synthesis of nicotinamide adenine dinucleotide (NAD) (see Chapter 28, VII.).

2. **Leucine:** This amino acid is exclusively ketogenic, because its catabolism yields acetyl CoA and acetoacetate (Fig. 20.11). The first two reactions in the catabolism of leucine and the other BCAAs, isoleucine and valine, are catalyzed by enzymes that use all three BCAAs (or their derivatives) as substrates (see Section H.).

3. **Isoleucine:** This amino acid is both ketogenic and glucogenic, because its metabolism yields acetyl CoA and propionyl CoA.

4. **Lysine:** This amino acid is exclusively ketogenic and is unusual in that neither of its amino groups undergoes transamination as the first step in catabolism. Lysine is ultimately converted to acetoacetyl CoA.

H. Branched-chain amino acid degradation

The BCAAs, isoleucine, leucine, and valine, are essential amino acids. In contrast to other amino acids, they are catabolized primarily by the peripheral tissues (particularly muscle), rather than by the liver. Because these three amino acids have a similar route of degradation, it is convenient to describe them as a group (see Fig. 20.11).

1. **Transamination:** Transfer of the amino groups of all three BCAAs to α-ketoglutarate is catalyzed by a single, vitamin B_6–requiring enzyme, BCAA aminotransferase that is expressed primarily in skeletal muscle.

2. **Oxidative decarboxylation:** Removal of the carboxyl group of the α-keto acids derived from leucine, valine, and isoleucine is catalyzed by a single multienzyme complex, branched-chain α-keto acid dehydrogenase (BCKD) complex. An enzymatic deficiency in this complex results in maple syrup urine disease (MSUD) (see Fig. 20.11 and Section VI. B.). This complex uses thiamine pyrophosphate, lipoic acid, oxidized flavin adenine dinucleotide (FAD), NAD^+, and CoA as its coenzymes and produces NADH. This reaction is similar to the conversion of pyruvate to acetyl CoA by the pyruvate dehydrogenase (PDH) complex

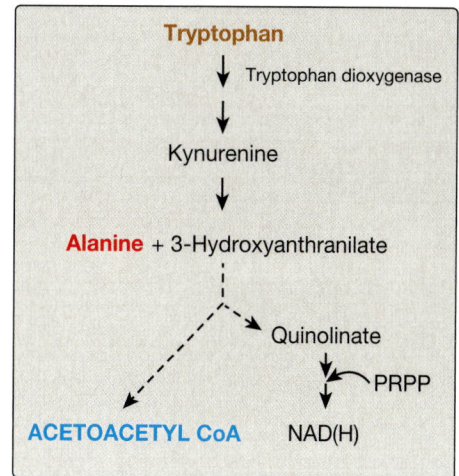

Figure 20.10
Metabolism of tryptophan by the kynurenine pathway (abbreviated). CoA, coenzyme A; PRPP, phosphoribosyl pyrophosphate; NAD(H), nicotinamide adenine dinucleotide.

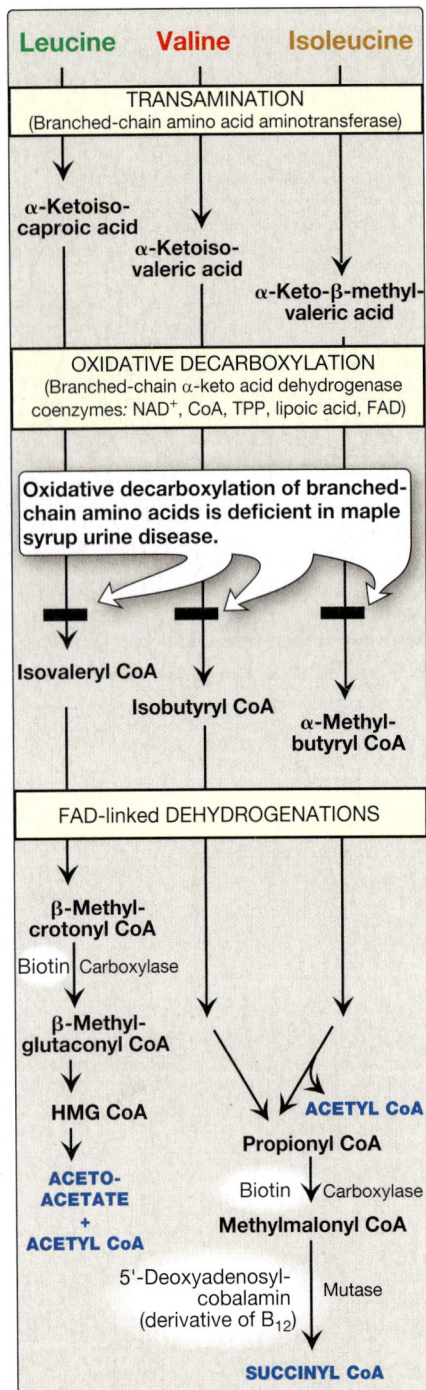

Figure 20.11
Degradation of leucine, valine, and isoleucine. [Note: β-Methylcrotonyl CoA carboxylase is one of four biotin-requiring carboxylases discussed in this book. The other three are pyruvate carboxylase, acetyl CoA carboxylase, and propionyl CoA carboxylase.] TPP, thiamine pyrophosphate; FAD, flavin adenine dinucleotide; CoA, coenzyme A; NAD, nicotinamide adenine dinucleotide; HMG, hydroxymethylglutarate.

(see Chapter 9, II. A.) and α-ketoglutarate to succinyl CoA by the α-ketoglutarate dehydrogenase complex (see Chapter 9, II. E.). The dihydrolipoyl dehydrogenase (enzyme 3, or E3) component is identical in all three complexes.

3. **Dehydrogenations:** Oxidation of the products formed in the BCKD reaction produces α-β-unsaturated acyl CoA derivatives and $FADH_2$. These reactions are analogous to the FAD-linked dehydrogenation in the β-oxidation of fatty acids (see Chapter 16, IV. B.). A deficiency in the dehydrogenase specific for isovaleryl CoA causes neurologic problems and is associated with a "sweaty feet" odor in body fluids.

4. **End products:** The catabolism of isoleucine ultimately yields acetyl CoA and succinyl CoA, rendering it both ketogenic and glucogenic. Valine yields succinyl CoA and is glucogenic. Leucine is ketogenic, being metabolized to acetoacetate and acetyl CoA. In addition, NADH and $FADH_2$ are produced in the decarboxylation and dehydrogenation reactions, respectively. BCAA catabolism also results in glutamine and alanine being synthesized and sent out into the blood from muscle (see Chapter 19, IV. C.).

IV. FOLIC ACID AND AMINO ACID METABOLISM

Some synthetic pathways require the addition of single-carbon groups that exist in a variety of oxidation states, including formyl, methenyl, methylene, and methyl. These single-carbon groups can be transferred from carrier compounds such as THF and SAM to specific structures that are being synthesized or modified. The "one-carbon pool" refers to the single-carbon units attached to this group of carriers. CO_2, coming from bicarbonate (HCO_3^-), is carried by the vitamin biotin (see Chapter 28, IX.), which is a prosthetic group for most carboxylation reactions but is not considered a member of the one-carbon pool. Defects in the ability to add or remove biotin from carboxylases result in multiple carboxylase deficiency. Treatment is supplementation with biotin.

A. Folic acid and one-carbon metabolism

The active form of folic acid, THF, is produced from folate by dihydrofolate reductase in a two-step reaction requiring two reduced nicotinamide adenine dinucleotide phosphate (NADPH). The one-carbon unit carried by THF is bound to N^5 or N^{10} or to both N^5 and N^{10}. Figure 20.12 shows the structures of the various members of the THF family and their interconversions and indicates the sources of the one-carbon units and the synthetic reactions in which the specific members participate. Folate deficiency presents as a megaloblastic anemia because of decreased availability of the purines and of the thymidine monophosphate needed for DNA synthesis (see Chapter 22, VI. E.).

V. BIOSYNTHESIS OF NONESSENTIAL AMINO ACIDS

Nonessential amino acids are synthesized from intermediates of metabolism or, as in the case of tyrosine and cysteine, from the essential amino acids phenylalanine and methionine, respectively. The synthetic reactions for the nonessential amino acids are described below. Some amino acids found in proteins, such as hydroxyproline and hydroxylysine (see Chapter 4, II. B.), are produced by posttranslational modification (after incorporation into a protein) of their precursor (parent) amino acids.

A. Synthesis from α-keto acids

Alanine, aspartate, and glutamate are synthesized by transfer of an amino group to the α-keto acids pyruvate, oxaloacetate, and α-ketoglutarate, respectively. These transamination reactions (Fig. 20.13) are the most direct of the biosynthetic pathways. Glutamate is unusual in that it can also be synthesized by reversal of oxidative deamination, catalyzed by glutamate dehydrogenase, when ammonia levels are high (see Fig. 19.11).

B. Synthesis by amidation

Amidation is the process of creating a compound containing an acyl group (R–C = O) linked to a nitrogen atom, most commonly derived from ammonia. The amide group forms the peptide bond in the main chain of a protein and an isopeptide bond when it is part of the side chain, as in the amino acids asparagine and glutamine. In these examples, the carboxylic acid in the R group of either glutamic acid or aspartic acid is converted into an amide group, generating glutamine or asparagine, respectively.

1. **Glutamine:** This amino acid, which contains an amide linkage with ammonia at the γ-carboxyl, is formed from glutamate and ammonia by glutamine synthetase (see Fig. 19.18). The reaction is driven by the hydrolysis of ATP. In addition to producing glutamine for protein synthesis, the reaction also serves as a major mechanism for the transport of ammonia in a nontoxic form.

2. **Asparagine:** This amino acid, which contains an amide linkage with ammonia at the β-carboxyl, is formed from aspartate by asparagine synthetase, using glutamine as the amide donor. Like the synthesis of glutamine, the reaction requires ATP and has an equilibrium far in the direction of amide synthesis.

C. Proline

Glutamate via glutamate semialdehyde is converted to proline by cyclization and reduction reactions. [Note: The semialdehyde can also be transaminated to ornithine.]

D. Serine, glycine, and cysteine

The pathways of synthesis for these amino acids are interconnected.

1. **Serine:** This amino acid arises from 3-phosphoglycerate, a glycolytic intermediate, which is first oxidized to 3-phosphopyruvate and then transaminated to 3-phosphoserine. Serine is formed by hydrolysis of the phosphate ester. Serine can also be formed from glycine through transfer of a hydroxymethyl group by serine hydroxymethyltransferase using N^5,N^{10}-MTHF as the one-carbon donor (see Fig. 20.6A).

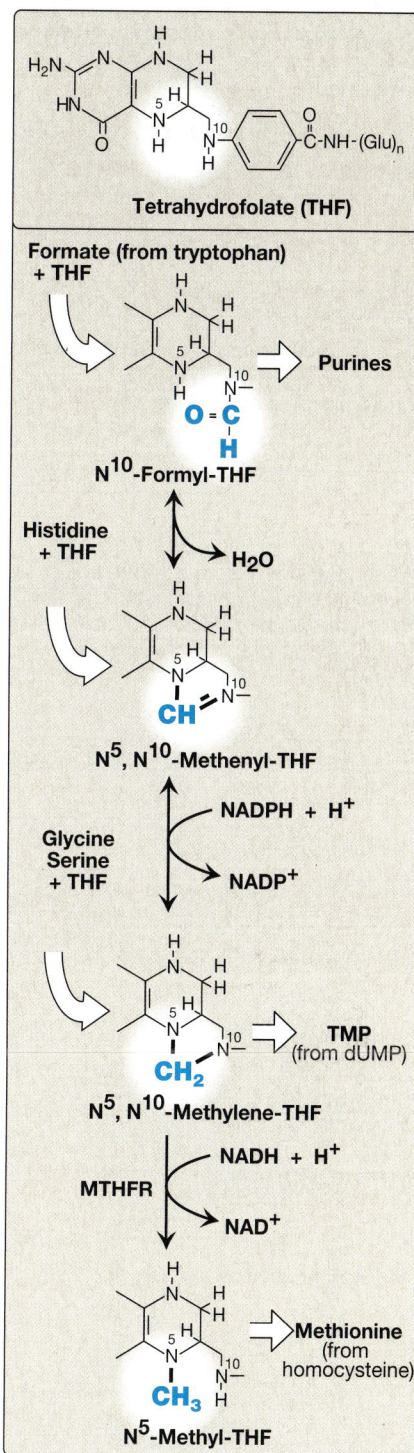

Figure 20.12
Summary of the interconversions and uses of tetrahydrofolate (THF). [Note: N^5, N^{10}-Methenyl-THF also arises from N^5-formimino-THF (see Fig. 20.4).] NADP(H), nicotinamide adenine dinucleotide phosphate; NAD(H), nicotinamide adenine dinucleotide; TMP, thymidine monophosphate; dUMP, deoxyuridine monophosphate; MTHFR, N^5,N^{10}-methylene-THF reductase.

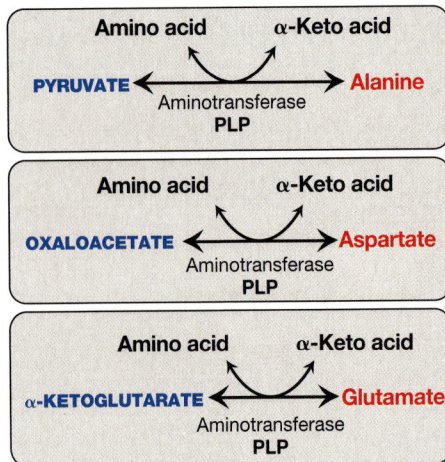

Figure 20.13
Formation of alanine, aspartate, and glutamate from the corresponding α-keto acids by transamination. PLP, pyridoxal phosphate.

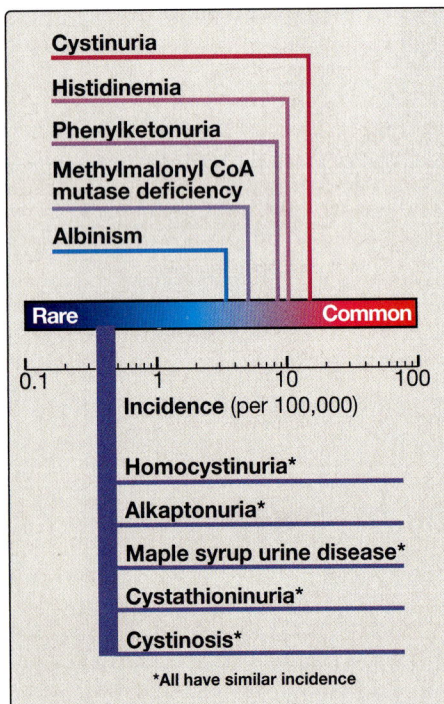

Figure 20.14
Incidence of inherited diseases of amino acid metabolism. [Note: Cystinuria is the most common inborn error of amino acid transport.]

2. **Glycine:** This amino acid is synthesized from serine by removal of a hydroxymethyl group, also by serine hydroxymethyltransferase (see Fig. 20.6A). THF is the one-carbon acceptor.

3. **Cysteine:** This amino acid is synthesized by two consecutive reactions in which Hcy combines with serine, forming cystathionine, which, in turn, is hydrolyzed to α-ketobutyrate and cysteine (see Fig. 20.8). Hcy is derived from methionine. Because methionine is an essential amino acid, cysteine synthesis requires adequate dietary intake of methionine.

E. Tyrosine

Tyrosine is formed from phenylalanine by PAH (see Section III. D.). The reaction requires molecular oxygen and the coenzyme BH_4, which is synthesized from guanosine triphosphate. One atom of molecular oxygen becomes incorporated into the hydroxyl group of tyrosine, and the other atom is reduced to water. During the reaction, BH_4 is oxidized to dihydrobiopterin (BH_2). BH_4 is regenerated from BH_2 by NADH-requiring dihydropteridine reductase. Tyrosine, like cysteine, is formed from an essential amino acid and is nonessential in the presence of adequate dietary phenylalanine.

VI. AMINO ACID METABOLISM DISORDERS

These single-gene disorders, a subset of the inborn errors of metabolism, are generally caused by loss-of-function mutations in enzymes involved in amino acid metabolism. The inherited defects may be expressed as a total loss of enzyme activity or, more frequently, as a partial deficiency in catalytic activity. Without treatment, the amino acid disorders almost invariably result in intellectual disability or other developmental abnormalities, as a consequence of harmful accumulation of metabolites. Although more than 50 of these disorders have been described, many are rare, occurring in fewer than 1 per 250,000 in most populations (Fig. 20.14). Collectively, they constitute a very significant portion of pediatric genetic diseases, which are summarized in their amino acid pathways in Figure 20.15.

A. Phenylketonuria

Phenylketonuria (PKU) is the most common clinically encountered inborn error of amino acid metabolism (incidence 1:15,000). Classical PKU is an autosomal-recessive disorder resulting from loss of function mutations in the gene coding for PAH (Fig. 20.16; also see Fig. 20.15). Biochemically, PKU is characterized by hyperphenylalaninemia. Phenylalanine is present in high concentrations (10 times normal) not only in plasma but also in urine and body tissues. Tyrosine, which is formed from phenylalanine by PAH, is deficient. Treatment includes dietary restriction of phenylalanine and supplementation with tyrosine.

1. **Additional characteristics:** As the name suggests, PKU is also characterized by elevated levels of a phenylketone in the urine.

   a. **Elevated phenylalanine metabolites:** Phenylpyruvate (a phenylketone), phenylacetate, and phenyllactate, which are

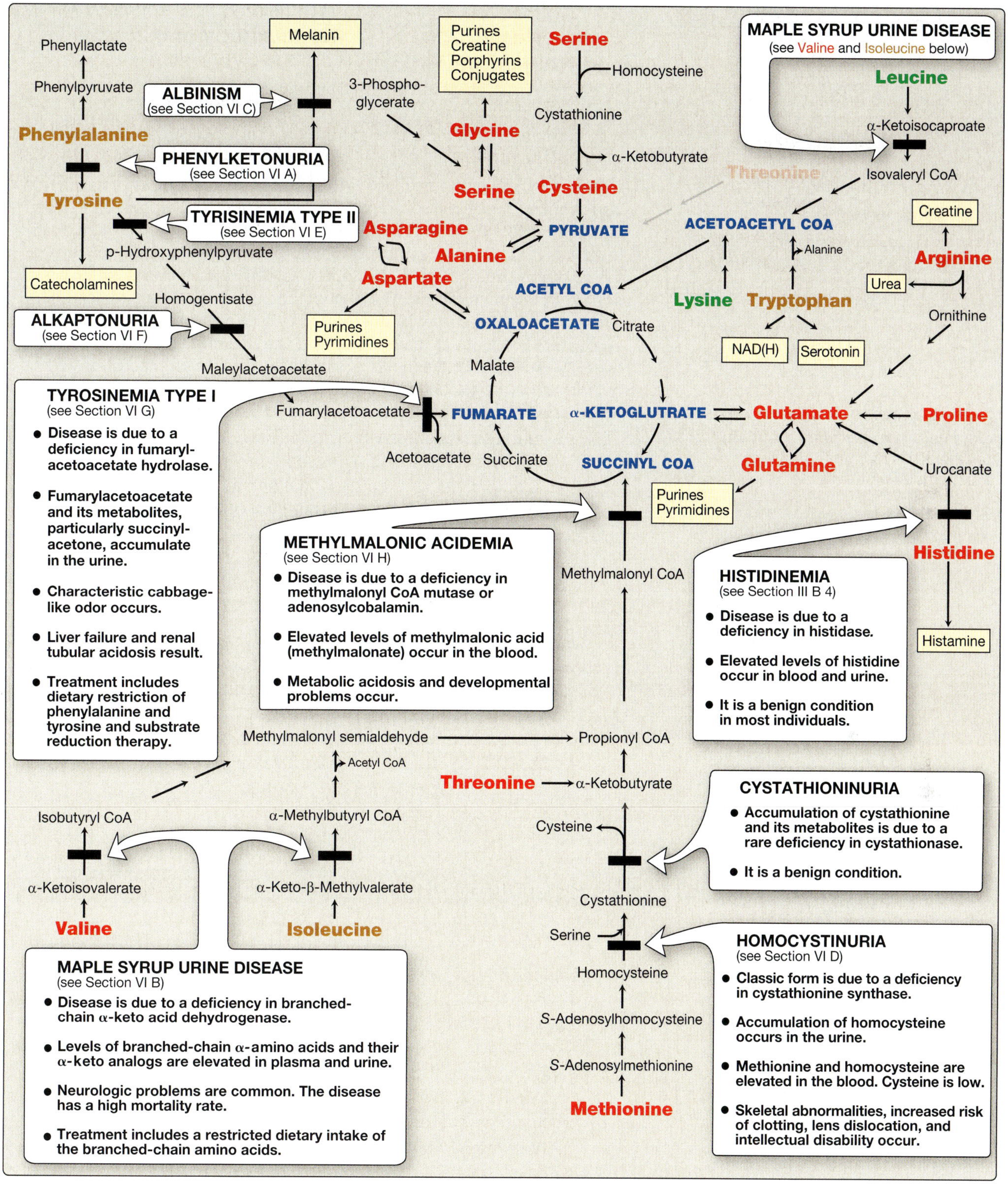

Figure 20.15
Summary of the metabolism of amino acids in humans. Genetically determined enzyme deficiencies are summarized in white boxes. More information can be found at the indicated section headings. Nitrogen-containing compounds derived from amino acids are shown in small, yellow boxes. Classification of amino acids is color coded: Red, glucogenic; brown, glucogenic and ketogenic; green, ketogenic. Compounds in BLUE ALL CAPS are the seven metabolites to which all amino acid metabolism converges. CoA, coenzyme A; NAD(H), nicotinamide adenine dinucleotide.

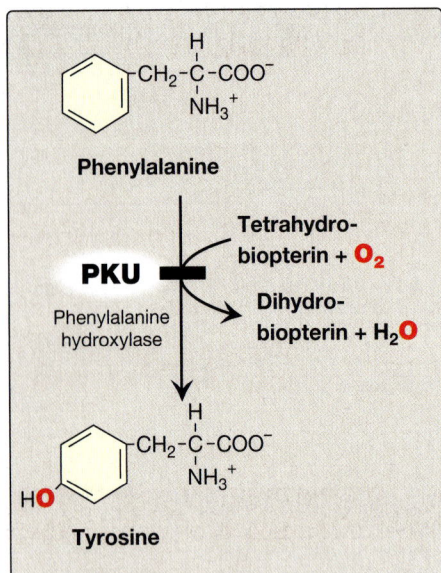

Figure 20.16
Phenylalanine hydroxylase deficiency results in the disease phenylketonuria (PKU).

Clinical Application 20.1: Hyperphenylalaninemia and Neurotransmitters

Hyperphenylalaninemia may also be caused by deficiencies in enzymes required to synthesize BH_4 or in dihydropteridine reductase, which regenerates BH_4 from BH_2 (Fig. 20.17). Such deficiencies indirectly raise phenylalanine concentrations because PAH requires BH_4 as a coenzyme. BH_4 is also required for tyrosine hydroxylase and tryptophan hydroxylase, which catalyze reactions for the synthesis of catecholamine and serotonin neurotransmitters. Simply restricting dietary phenylalanine does not reverse the central nervous system effects due to deficiencies in the synthesis of certain neurotransmitters. Supplementation with BH_4 and replacement therapy with L–3,4-dihydroxyphenylalanine (L-DOPA) and 5-hydroxytryptophan, both products of the affected tyrosine hydroxylase– and tryptophan hydroxylase–catalyzed reactions, improves the clinical neurotransmitter outcome in these variant forms of hyperphenylalaninemia (see Chapter 21, III. A. and III. C., respectively).

not normally produced in significant amounts in the presence of functional PAH, are also elevated in PKU, in addition to phenylalanine (Fig. 20.18). These metabolites give urine a characteristic musty ("mousy") odor.

b. **Central nervous system effects:** Severe intellectual disability, developmental delay, microcephaly, and seizures are characteristic findings in untreated PKU. The affected individual typically shows symptoms of intellectual disability by age 1 year and rarely achieves an intelligence quotient (IQ) more than 50 (Fig. 20.19). Fortunately, these clinical manifestations are now rarely seen as a result of newborn screening programs, which allow early diagnosis and treatment.

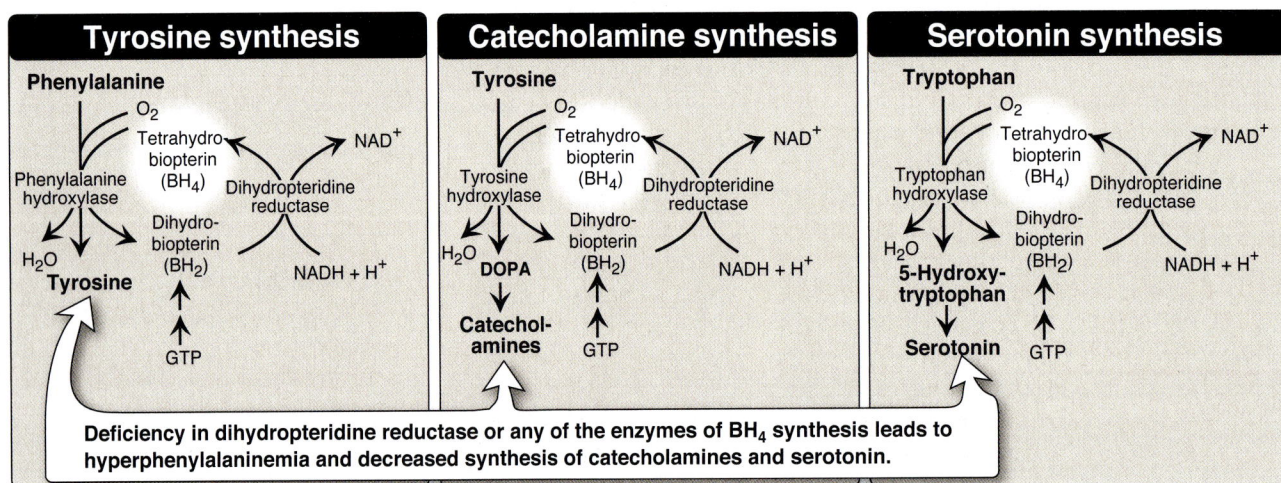

Figure 20.17
Biosynthetic reactions involving amino acids and tetrahydrobiopterin. [Note: Aromatic amino acid hydroxylases use BH_4 and not pyridoxal phosphate (PLP).] NAD(H), nicotinamide adenine dinucleotide; GTP, guanosine triphosphate; DOPA, L–3,4-dihydroxyphenylalanine.

c. **Hypopigmentation:** Patients with untreated PKU may show a deficiency of pigmentation (fair hair, light skin color, and blue eyes). The hydroxylation of tyrosine by copper-requiring tyrosinase, which is the first step in the formation of the pigment melanin, is decreased in PKU because tyrosine levels are decreased.

2. **Newborn screening and diagnosis:** Early diagnosis of PKU is important because the disease is treatable by dietary means. Because of the lack of neonatal symptoms, laboratory testing for elevated blood levels of phenylalanine is mandatory for detection. However, the infant with PKU frequently has normal blood levels of phenylalanine at birth because the mother clears increased blood phenylalanine in her affected fetus through the placenta. Normal levels of phenylalanine may persist until the newborn is exposed to 24 to 48 hours of protein feeding. Thus, screening tests are typically done after this time to avoid false negatives. For newborns with a positive screening test, diagnosis is confirmed through quantitative determination of phenylalanine levels.

> Screening of newborns for a number of treatable disorders, including inborn errors of amino acid metabolism, is done by tandem mass spectrometry of blood obtained from a heel prick. By law, all states must screen for >20 disorders, with some screening for >50. All states screen for PKU.

3. **Prenatal diagnosis:** Classic PKU is caused by any of 100 or more different mutations in the gene that encodes PAH. The frequency of any given mutation varies among populations, and the disease is often doubly heterozygous (i.e., the *PAH* gene has a different mutation in each allele). Despite this complexity, prenatal diagnosis is possible (see Chapter 34, VI. C.).

4. **Treatment:** Because most natural protein contains phenylalanine, an essential amino acid, it is impossible to satisfy the body's protein requirement without exceeding the phenylalanine limit when ingesting a normal diet. Therefore, in PKU, blood phenylalanine level is maintained close to the normal range by feeding synthetic amino acid preparations free of phenylalanine, supplemented with some natural foods (such as fruits, vegetables, and certain cereals) selected for their low phenylalanine content. The amount is adjusted according to the tolerance of the individual as measured by blood phenylalanine levels. The earlier treatment is started, the more completely neurologic damage can be prevented. Individuals who are appropriately treated can have normal intelligence. Treatment must begin during the first 7 to 10 days of life to prevent cognitive impairment. Because phenylalanine is an essential amino acid, overzealous treatment that results in blood phenylalanine levels below normal is avoided. In patients with PKU, tyrosine cannot be synthesized from phenylalanine, and, therefore, it becomes an essential amino acid and must be supplied in the diet. Discontinuance of the phenylalanine-restricted diet in early childhood is associated with poor performance on

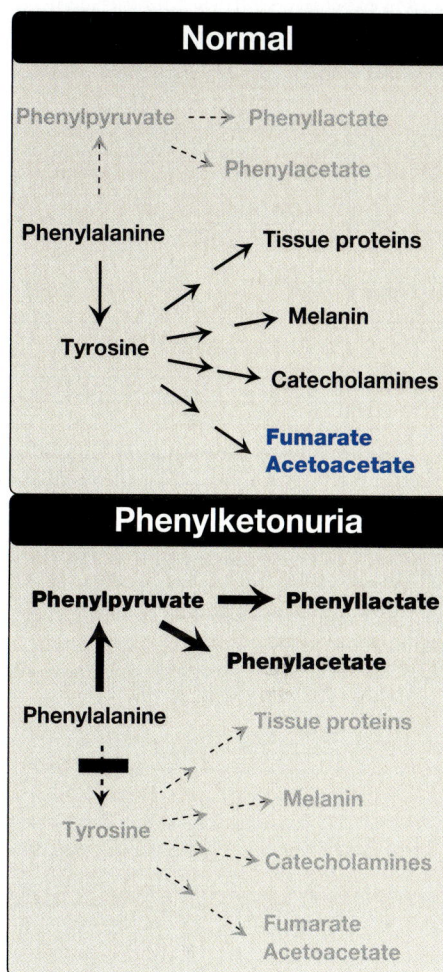

Figure 20.18
Pathways of phenylalanine metabolism in normal individuals and in patients with phenylketonuria.

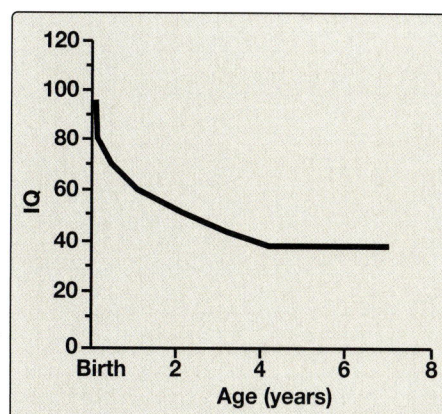

Figure 20.19
Typical intellectual ability in untreated patients of different ages with phenylketonuria. IQ, intelligence quotient.

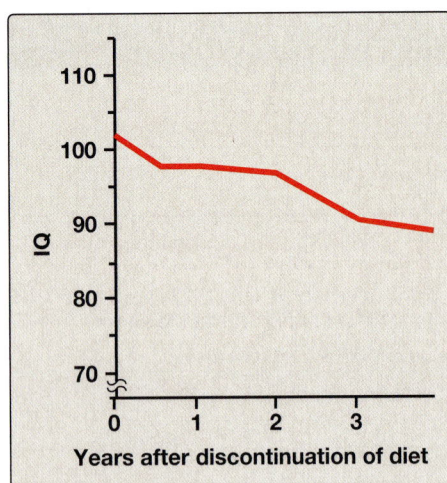

Figure 20.20
Changes in intelligence quotient (IQ) scores after discontinuation of low-phenylalanine diet in patients with phenylketonuria.

IQ tests. Adults with PKU show deterioration of IQ scores after discontinuation of the diet (Fig. 20.20). Therefore, lifelong restriction of dietary phenylalanine is recommended. Individuals with PKU are advised to avoid aspartame, an artificial sweetener that contains phenylalanine.

5. **Maternal PKU:** If women with PKU who are not on a low-phenylalanine diet become pregnant, the offspring can still be affected with maternal PKU syndrome. Even if the fetus has not inherited the disease (i.e., the fetus is heterozygous for the *PAH* mutation), high blood phenylalanine in the mother has a teratogenic effect, causing microcephaly and congenital heart abnormalities in the fetus. Because these developmental responses to high phenylalanine occur during the first months of pregnancy, dietary control of blood phenylalanine must begin prior to conception and be maintained throughout the pregnancy.

B. Maple syrup urine disease

MSUD is a rare (1:185,000), autosomal-recessive disorder in which there is a partial or complete deficiency in BCKD, the mitochondrial enzyme complex that oxidatively decarboxylates the ketoacid metabolites of leucine, isoleucine, and valine (see Figs. 20.11 and 20.15). These BCAAs and their corresponding α-keto acids accumulate in the blood, causing a toxic effect that interferes with brain functions. The disease is characterized by feeding problems, vomiting, ketoacidosis, changes in muscle tone, neurologic problems that can result in coma (primarily because of the rise in leucine), and a characteristic maple syrup–like odor of the urine because of the rise in isoleucine. If untreated, the disease is fatal. If treatment is delayed, intellectual disability results.

1. **Classification:** MSUD includes a classic type and several variant forms. The classic, neonatal-onset form is the most common type of MSUD. Leukocytes or cultured skin fibroblasts from these patients show little or no BCKD activity. Infants with classic MSUD show symptoms within the first several days of life. If not diagnosed and treated, classic MSUD is lethal in the first weeks of life. Patients with intermediate forms have a higher level of enzyme activity (up to 30% of normal). The symptoms are milder and show an onset from infancy to adolescence. Patients with the rare thiamine-dependent variant of MSUD respond to large doses of this vitamin.

2. **Screening and diagnosis:** As with PKU, prenatal diagnosis and newborn screening are available, and most affected individuals are compound heterozygotes.

3. **Treatment:** MSUD is treated with a synthetic formula that is free of BCAA, supplemented with limited amounts of leucine, isoleucine, and valine to allow for normal growth and development without producing toxic levels. [Note: Elevated leucine is the cause of the neurologic damage in MSUD, and its level is carefully monitored.] Early diagnosis and lifelong dietary treatment are essential if the child with MSUD is to develop normally. BCAAs are an important

energy source in times of metabolic need, and individuals with MSUD are at risk of decompensation during periods of increased protein catabolism.

C. Albinism

Albinism refers to a group of conditions in which a defect in tyrosine metabolism results in a deficiency in the production of melanin. These defects result in the partial or full absence of pigment from the skin, hair, and eyes. Albinism appears in different forms, and it may be inherited by one of several modes: autosomal recessive (primary mode), autosomal dominant, or X-linked. Total absence of pigment from the hair, eyes, and skin (Fig. 20.21; also see Fig. 20.15), tyrosinase-negative oculocutaneous albinism (type 1 albinism, 1:40,000), results from an absent or defective copper-requiring tyrosinase, which catalyzes the first two steps in the synthesis of melanin from tyrosine. It is the most severe form of the condition. In addition to hypopigmentation, affected individuals have vision defects and photophobia (sunlight hurts their eyes). They also are at increased risk for skin cancer.

Figure 20.21
Patient with oculocutaneous albinism, showing white eyebrows and lashes. (Prostock-studio/Shutterstock.)

D. Homocystinuria

The homocystinurias are a group of disorders (1:100,000) involving defects in the metabolism of Hcy. These autosomal-recessive diseases are characterized by high urinary levels of Hcy, high plasma levels of Hcy and methionine, and low plasma levels of cysteine. The most common cause of homocystinuria is a defect in the enzyme cystathionine β-synthase, which converts Hcy to cystathionine (Fig. 20.22; also see Fig. 20.15). Individuals homozygous for cystathionine β-synthase deficiency exhibit dislocation of the lens (ectopia lentis), skeletal anomalies (long limbs and fingers), intellectual disability, and an increased risk for developing thrombi (blood clots). Thrombosis is the major cause of early death in these individuals. Treatment includes restriction of methionine and supplementation with vitamin B_{12} and folate. Cysteine becomes an essential amino acid and must be supplemented. As glutathione is synthesized from cysteine (see Fig. 13.6), adding cysteine to the diet is also helpful to reduce oxidative stress. Additionally, some patients are responsive to oral administration of pyridoxine (vitamin B_6), which is converted to pyridoxal phosphate, the coenzyme of cystathionine β-synthase. These patients usually have a milder and later onset of clinical symptoms compared with B_6-nonresponsive patients. Deficiencies in methylcobalamin (see Fig. 20.8) or N^5,N^{10}-MTHF reductase (MTHFR) (see Fig. 20.12) can also result in elevated Hcy.

E. Tyrosinemia type II

Tyrosinemia type II is a rare (1:100,000) autosomal-recessive disorder caused by a deficiency in tyrosine aminotransferase. This reaction, which converts tyrosine into p-hydroxyphenylpyruvate, occurs early in the degradative pathway of tyrosine to fumarate and acetoacetate (see Figs. 20.7 and 20.15). The deficiency can be diagnosed by the accumulation of tyrosine in the blood. Symptoms can include eye pain, light sensitivity (photophobia), painful callus buildup on the hands and feet

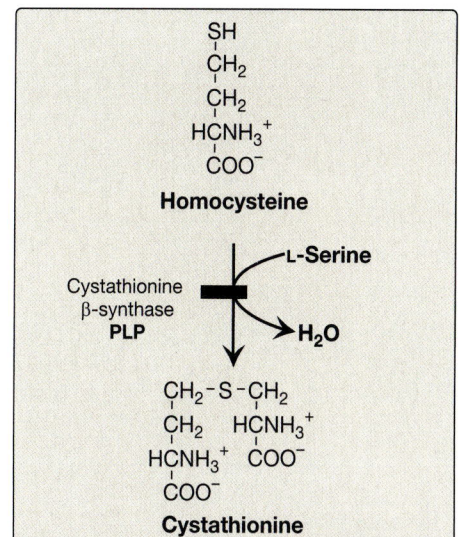

Figure 20.22
Enzyme deficiency in homocystinuria. PLP, pyridoxal phosphate.

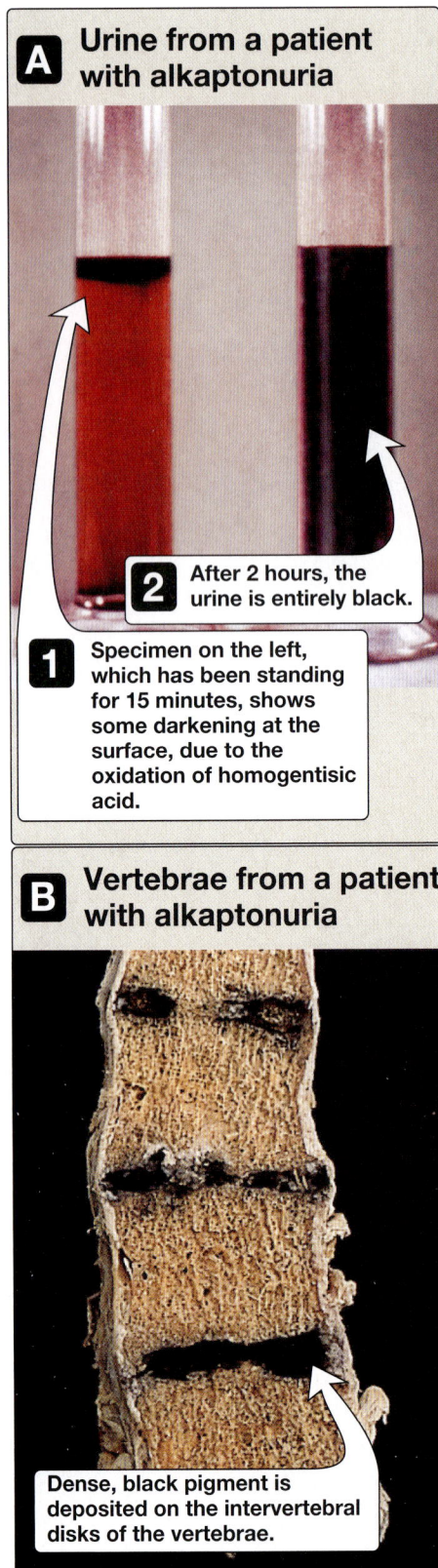

A Urine from a patient with alkaptonuria

2 After 2 hours, the urine is entirely black.

1 Specimen on the left, which has been standing for 15 minutes, shows some darkening at the surface, due to the oxidation of homogentisic acid.

B Vertebrae from a patient with alkaptonuria

Dense, black pigment is deposited on the intervertebral disks of the vertebrae.

Figure 20.23
Specimens from a patient with alkaptonuria. **A.** Urine. **B.** Vertebrae.

(palmoplantar hyperkeratosis), and intellectual disability. Treatment includes dietary restriction of phenylalanine and tyrosine.

F. Alkaptonuria

Alkaptonuria is a rare (1:250,000) organic aciduria involving a deficiency in homogentisic acid (HA) oxidase, resulting in the accumulation of HA, an intermediate in the degradative pathway of tyrosine (see Figs. 20.7 and 20.15). The condition has three characteristic symptoms: homogentisic aciduria (the urine contains elevated levels of HA, which is oxidized to a dark pigment on standing, as shown in Fig. 20.23A), early onset of arthritis in the large joints, and deposition of black pigment (ochronosis) in cartilage and collagenous tissue (Fig. 20.23B). Dark staining of diapers can indicate the disease in infants, but usually no symptoms are present until about age 40 years. Treatment includes dietary restriction of phenylalanine and tyrosine to reduce HA levels. Although alkaptonuria is not life threatening, the associated arthritis may be severely crippling.

G. Tyrosinemia type I

Tyrosinemia type I is a rare (1:250,000) autosomal-recessive disorder caused by a deficiency in fumarylacetoacetate hydrolase, which is the final enzyme in the degradative pathway converting tyrosine to fumarate and acetoacetate (see Figs. 20.7 and 20.15). This reaction occurs primarily in the liver and kidney. The deficiency can be diagnosed by the accumulation of fumarylacetoacetate and its metabolite succinylacetone in the urine, resulting in a characteristic cabbage-like odor. Symptoms can include poor growth and hepatomegaly, but if untreated can lead to chronic liver damage, liver failure, as well as renal tubular acidosis. Treatment includes dietary restriction of phenylalanine and tyrosine and substrate reduction therapy.

H. Methylmalonic acidemia

Methylmalonic acidemia (MMA) is a rare (1:100,000) autosomal-recessive disorder caused by a deficiency in methylmalonyl CoA mutase, which converts L-methylmalonyl CoA to succinyl CoA (Fig. 20.15). Since the mutase requires vitamin B_{12}, the disease can also result from a severe B_{12} deficiency. The breakdown of odd-chain length fatty acids, valine, isoleucine, methionine, and threonine can all result in MMA, due to this enzyme deficiency. Elevation of methylmalonate in the blood and urine can result in a metabolic acidosis. There may also be an increase in propionyl-CoA, exacerbating the aciduria with an accumulation of additional propionic acid. Symptoms appear in early infancy, varying due to the degree of the enzyme deficiency, including failure to thrive, vomiting, dehydration, hypotonia, developmental delay, seizures, hepatomegaly, hyperammonemia, and a progressive encephalopathy. If severe and left untreated, it can lead to intellectual disability, chronic renal or hepatic damage, pancreatitis, and coma or death. Treatment includes a low-protein, high-calorie diet, and vitamin B_{12} supplementation. The diet limits the intake of isoleucine, threonine, methionine, and valine, as these amino acids can lead to the buildup of methylmalonic acid by the mutase deficiency.

VII. CHAPTER SUMMARY

- **Amino acids** whose catabolism yields **pyruvate** or an **intermediate** of the **citric acid cycle** are termed **glucogenic** (Fig. 20.24). They can give rise to the net formation of **glucose** in the **liver** and **kidneys**. The solely glucogenic amino acids are glutamine, glutamate, proline, arginine, histidine, alanine, serine, glycine, cysteine, methionine, valine, threonine, aspartate, and asparagine.

- Amino acids whose catabolism yields either acetyl CoA (directly, without pyruvate serving as an intermediate) or acetoacetate (or its precursor acetoacetyl CoA) are termed **ketogenic**. Leucine and lysine are solely ketogenic.

- Tyrosine, phenylalanine, tryptophan, and isoleucine are both ketogenic and glucogenic.

- **Nonessential amino acids** can be synthesized from metabolic intermediates or from the carbon skeletons of essential amino acids.

- **Essential amino acids** need to be obtained from the **diet**. They include histidine, methionine, threonine, valine, isoleucine, phenylalanine, tryptophan, leucine, and lysine.

- **Phenylketonuria** (**PKU**) is caused by a **deficiency** of **phenylalanine hydroxylase** (**PAH**), which converts phenylalanine to tyrosine. **Hyperphenylalaninemia** may also be caused by deficiencies in the enzymes that synthesize or regenerate the coenzyme for PAH, **BH_4**. Untreated individuals with PKU suffer from severe intellectual disability, developmental delay, microcephaly, seizures, and a characteristic musty (mousy) smell of the urine. Treatment involves controlling dietary phenylalanine. **Tyrosine** becomes an essential dietary component for people with PKU.

- **Maple syrup urine disease** is caused by a partial or complete deficiency in **branched-chain α-keto acid dehydrogenase**, the enzyme that decarboxylates the branched-chain amino acids (**BCAAs**), **leucine**, **isoleucine**, and **valine**. Symptoms include feeding problems, vomiting, ketoacidosis, changes in muscle tone, and a characteristic sweet smell of the urine. If untreated, the disease leads to neurologic problems that result in death. Treatment involves controlling BCAA intake.

- Other important genetic diseases associated with amino acid metabolism include **albinism**, **homocystinuria**, **tyrosinemia** (**types I** and **II**), **alkaptonuria**, **methylmalonic acidemia**, **histidinemia**, and **cystathioninuria**.

Metabolism of amino acids	Some clinically important amino acids

Catabolism of amino acids

involves

Removal of α-amino group → **Metabolism of carbon skeletons**

converges to produce

Seven products

consisting of

ACETYL CoA	**PYRUVATE**
ACETOACETYL CoA	**OXALOACETATE**
	FUMARATE
	α-KETOGLUTARATE
	SUCCINYL CoA

classified as   *classified as*

Ketogenic   **Glucogenic**

provide   *provide*

Lipids	Lipids
Energy	Energy
	Glucose

Synthesis of amino acids

involves

Transamination of α-keto acids, for example, pyruvate → alanine

Amidation, for example, asparate → asparagine

Synthesis from other amino acids, for example, phenylalanine → tyrosine

Some clinically important amino acids

Methionine
- Source of methyl groups in metabolism
- Precursor of cysteine

Arginine
- Component of urea cycle
- Precursor of nitric oxide

Glutamine
- Storage and transport form of ammonia
- Precursor of purines and pyrimidines

Phenylalanine
- Precursor of tyrosine
- Elevated in phenylketonuria

Histidine
- Precursor of histamine
- Elevated in histidinemia

Tryptophan
- Precursor of serotonin

Alanine
- Transport form of ammonia from muscle
- Key glucogenic amino acid

Metabolic defects in amino metabolism

characterized by

Family of defects in enzymes of amino acid metabolism —*can be*→ **Screened for in newborns**

caused by

Point mutations, deletions, splicing errors —*usually*→ **Inheritance is recessive; heterozygotes usually do not show symptoms**

which can lead to   *treated by*

Partially or completely inactive enzyme   **Dietary restriction**

which leads to

Accumulation of substrate and a deficiency of the defective enzyme's product —*can result in*→ **Characteristic smell of the urine**

which leads to

Disturbances in metabolism, particularly the central nervous system (CNS)

which leads to

Seizures, intellectual disability, other CNS effects

Concept connect

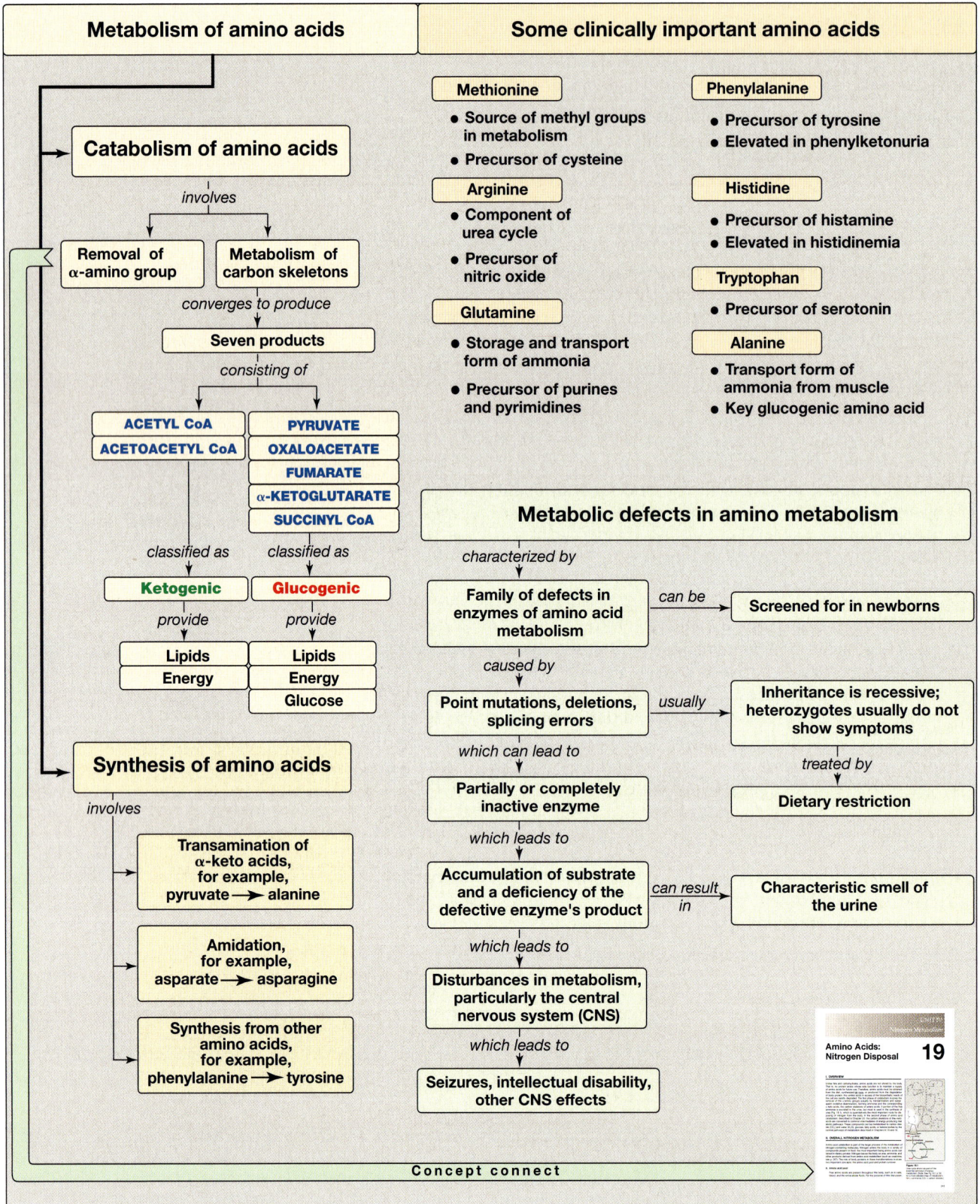

Figure 20.24
Key concept map for amino acid metabolism. CoA, coenzyme A.

Study Questions

Choose the ONE best answer.

For Questions 20.1 to 20.5, match the deficient enzyme with the associated clinical sign or laboratory finding in urine.

A. Black pigmentation of cartilage
B. Propionic acidemia
C. Cystine crystals in urine
D. White hair, red eye color
E. Increased branched-chain amino acids
F. Increased homocysteine
G. Liver failure and renal tubular acidosis
H. Increased phenylalanine in the urine

20.1. Cystathionine β-synthase

20.2. Homogentisic acid oxidase

20.3. Tyrosinase

20.4. Fumarylacetoacetate hydroxylase

20.5. Methylmalonyl CoA mutase

Correct answers = F, A, D, G, B, respectively. A deficiency in cystathionine β-synthase of methionine degradation results in a rise in homocysteine (homocystinuria). A deficiency in homogentisic acid oxidase of tyrosine degradation results in a rise in homogentisic acid, which forms a black pigment that is deposited in connective tissue (alkaptonuria). A deficiency in tyrosinase results in decreased formation of melanin from tyrosine in the skin, hair, and eyes (albinism). A deficiency in fumarylacetoacetate hydroxylase can result in liver failure and renal tubular acidosis (tyrosinemia type I). A defect in methylmalonyl coenzyme A (CoA) mutase results in acidosis from methylmalonic acidosis. Propionic acidosis can also result since propionic acid can feed into the same pathway. Cystine crystals in urine are seen with cystinuria, a defect in intestinal and renal cystine absorption. A defect in the branched-chain amino acid (BCAA) dehydrogenase complex results in an increase in BCAAs in the urine (maple syrup urine disease). A defect in phenylalanine hydroxylase results in an increase in phenylalanine and phenylketones in the urine (phenylketonuria).

20.6. A 1-week-old infant, who was born at home in a rural, medically underserved area, has undetected classic phenylketonuria. Which statement about this infant and/or the proper treatment is correct?

A. Diet devoid of all phenylalanine should be initiated immediately.
B. Dietary treatment can be discontinued in adulthood.
C. Supplementation with vitamin B_6 is required.
D. Tyrosine becomes an essential amino acid.
E. Folic acid supplementation may increase PAH activity.

Correct answer = D. In patients with phenylketonuria, tyrosine cannot be synthesized from phenylalanine and, hence, becomes essential and must be supplied in the diet. Phenylalanine in the diet must be controlled but cannot be eliminated entirely because it is an essential amino acid. Dietary treatment must begin during the first 7 to 10 days of life to prevent intellectual disability, and lifelong restriction of phenylalanine is recommended to prevent cognitive decline. Additionally, elevated levels of phenylalanine are teratogenic to a developing fetus. The cofactor for PAH is tetrahydrobiopterin (BH_4). BH_4 supplementation may help reduced phenylalanine levels if the enzyme defect is in BH_4 production or its reduction from dihydrobiopterin.

20.7. Which of the following statements concerning amino acids is correct?

A. Alanine is ketogenic.
B. Amino acids that are catabolized directly to acetyl CoA (without forming pyruvate as an intermediate) are glucogenic.
C. Branched-chain amino acids are catabolized primarily in the liver.
D. Cysteine is essential for individuals consuming a diet severely limited in methionine.
E. Alanine is an essential amino acid.

Correct answer = D. Methionine is the precursor of cysteine, which becomes essential if methionine is severely restricted. Alanine is a key glucogenic amino acid. Acetyl coenzyme A (CoA) cannot be used for the net synthesis of glucose. Amino acids catabolized to acetyl CoA, acetoacetate, and acetoacetyl CoA are ketogenic. Branched-chain amino acids are catabolized primarily in skeletal muscle. Alanine is a nonessential amino acid, synthesized from pyruvate by a transaminase.

20.8. In an individual with the dihydrolipoyl dehydrogenase (E3)-deficient form of maple syrup urine disease, why would lactic acidosis be an expected finding?

The three α-ketoacid dehydrogenase complexes (pyruvate dehydrogenase [PDH], α-ketoglutarate dehydrogenase, and branched-chain α-keto acid dehydrogenase [BCKD]) have the identical enzyme 3, or E3 in common. In E3-deficient maple syrup urine disease, in addition to the branched-chain amino acids and their α-keto acid derivatives accumulating as a result of decreased activity of BCKD, lactate will also be increased because of decreased activity of PDH.

20.9. In contrast to the vitamin B$_6$–derived pyridoxal phosphate required in most enzymic reactions involving amino acids, what coenzyme is required by the aromatic amino acid hydroxylases?

Tetrahydrobiopterin, made from guanosine triphosphate, is the required coenzyme.

20.10. A 4-day-old newborn is brought into the emergency department by his parents. The parents report that the infant is lethargic, having difficulty feeding, and had a seizure. His urine in reported to be brownish in color and has a sickly sweet smell. Which one of the following enzymes would likely be defective in the newborn screening?

A. Branched-chain α-keto acid dehydrogenase
B. Cystathionine β-synthase
C. Homogentisate dioxygenase
D. Methylmalonyl CoA mutase
E. Phenylalanine hydroxylase

Correct answer = A. The newborn has maple syrup urine disease, characterized by a defect in Branched-chain α-keto acid dehydrogenase and the inability to break down branched-chain amino acids leucine, lysine, and valine. A defect in cystathionine β-synthase results in homocystinuria. A defect in homogentisate oxidase results in alkaptonuria. A defect in methylmalonyl CoA mutase results in methylmalonic acidemia. A defect in phenylalanine hydroxylase results in phenylketonuria.

Amino Acids: Conversion to Specialized Products

21

I. OVERVIEW

In addition to serving as building blocks for proteins, amino acids are precursors of many nitrogen (N)-containing compounds that have both important physiologic functions and medical relevance (Fig. 21.1). These molecules include porphyrins, creatine, neurotransmitters, hormones, purines, and pyrimidines.

II. PORPHYRIN METABOLISM

Porphyrins are cyclic compounds that readily bind metal ions, usually ferrous (Fe^{2+}) or ferric (Fe^{3+}) iron. The most prevalent metalloporphyrin in humans is heme, which consists of one Fe^{2+} coordinated in the center of the tetrapyrrole ring of protoporphyrin IX (see Section II. B.). Heme is the prosthetic group for hemoglobin (Hb), myoglobin, the cytochromes, including the cytochrome P450 (CYP) monooxygenase system, catalase, nitric oxide synthase, and peroxidase. These heme proteins are rapidly synthesized and degraded. For example, 6 to 7 g of Hb is synthesized each day to replace heme lost through the normal turnover of erythrocytes. The synthesis and degradation of the associated porphyrins and recycling of the iron are coordinated with the turnover of heme proteins.

A. Structure

Porphyrins are cyclic planar molecules formed by the linkage of four pyrrole rings through methenyl bridges (Fig. 21.2). Three structural features of these molecules are relevant to understanding their medical significance.

1. **Side chains:** Different porphyrins vary in the nature of the side chains attached to each of the four pyrrole rings. Uroporphyrin contains acetate ($-CH_2-COO-$) and propionate ($-CH_2-CH_2-COO-$) side chains; coproporphyrin contains methyl ($-CH_3$) and propionate groups; and protoporphyrin IX (and heme b, the most common heme) contains vinyl ($-CH = CH_2$), methyl, and propionate groups. [Note: The methyl and vinyl groups are produced by decarboxylation of acetate and propionate side chains, respectively.]

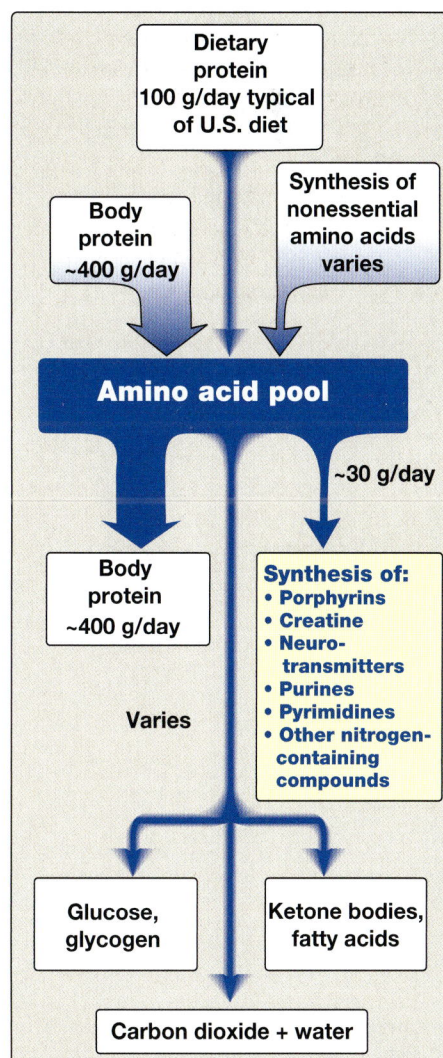

Figure 21.1
Amino acids as precursors of nitrogen-containing compounds.

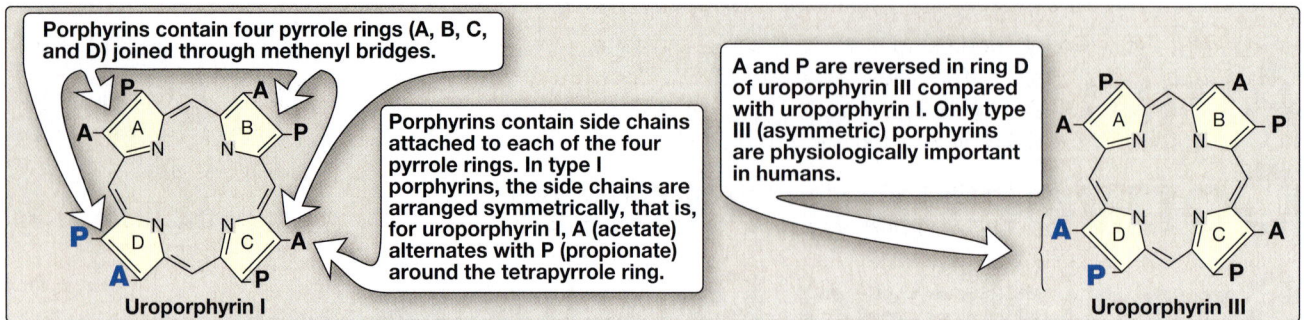

Figure 21.2
Structures of uroporphyrin I and uroporphyrin III.

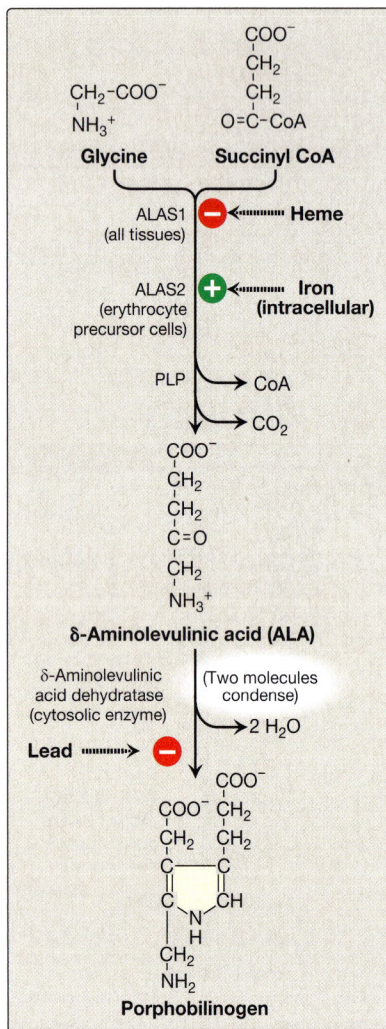

Figure 21.3
Pathway of porphyrin synthesis: Formation of porphobilinogen. [Note: ALAS1 is regulated by heme; ALAS2 is regulated by iron.] ALAS, δ-aminolevulinic acid synthase; CoA, coenzyme A; CO_2, carbon dioxide; PLP, pyridoxal phosphate. (Continued in Figs. 21.4 and 21.5.)

2. **Side-chain distribution:** The side chains of porphyrins can be ordered around the tetrapyrrole nucleus in four different ways, designated by Roman numerals I to IV. Only type III porphyrins, which contain an asymmetric substitution on ring D (see Fig. 21.2), are physiologically important in humans. [Note: Protoporphyrin IX is a member of the type III series.]

3. **Porphyrinogens:** These porphyrin precursors (e.g., uroporphyrinogen) exist in a chemically reduced, colorless form and serve as intermediates between porphobilinogen (PBG) and the oxidized, colored protoporphyrins in heme biosynthesis.

B. Heme biosynthesis

The major sites of heme biosynthesis are the liver and the erythrocyte-producing cells of the bone marrow. In the liver, which synthesizes a number of heme proteins (particularly the CYP proteins), the rate of heme synthesis is highly variable, responding to alterations in the cellular heme pool caused by fluctuating demands for heme proteins. In contrast, heme synthesis in erythroid cells, which are active in Hb synthesis, is relatively constant and is matched to the rate of globin synthesis. [Note: More than 85% of all heme synthesis occurs in erythroid tissue. Mature red blood cells (RBCs) lack mitochondria and are unable to synthesize heme.] The initial reaction and the last three steps in the formation of porphyrins occur in mitochondria, whereas the intermediate steps of the biosynthetic pathway occur in the cytosol.

1. **δ-Aminolevulinic acid (ALA) formation:** All the carbon and nitrogen atoms of the porphyrin molecule are provided by glycine (a nonessential amino acid) and succinyl coenzyme A (a citric acid cycle intermediate) that condense to form ALA in a reaction catalyzed by ALA synthase ([ALAS] Fig. 21.3). This reaction requires pyridoxal phosphate ([PLP] see Fig. 28.10) as a coenzyme and is the committed and rate-limiting step in porphyrin biosynthesis. There are two ALAS isoforms, each produced by different genes and controlled by different mechanisms. ALAS1 is found in all tissues, whereas ALAS2 occurs in the bone marrow and is erythroid specific.

   a. **Heme (hemin) effects:** When porphyrin production exceeds the availability of the apoproteins that require it, heme accumulates and is converted to hemin by the oxidation of Fe^{2+}

to Fe^{3+}. Hemin decreases the amount (and, thus, the activity) of ALAS1 by repressing transcription of its gene, increasing degradation of its messenger RNA, and decreasing the import of the enzyme into mitochondria. ALAS2 is controlled by the availability of intracellular iron (see Fig. 21.3).

b. **Drug effects:** Administration of any of a large number of drugs (and various environmental xenobiotic chemicals, present in certain foods, cosmetics, and commercial products) results in a significant increase in hepatic ALAS1 activity. These molecules are metabolized by the microsomal CYP monooxygenase system, a heme protein oxidase system found in the liver (see Chapter 13, IV. C.). In response to these drugs, the synthesis of CYP proteins increases, leading to enhanced consumption of heme, a component of these proteins. This, in turn, causes a decrease in the concentration of free or unbound heme in liver cells. The lower intracellular concentration of unbound heme leads to an increase in the synthesis of ALAS1 and prompts a corresponding increase in the synthesis of ALA.

2. **Porphobilinogen formation:** The cytosolic condensation of two ALA to form PBG by zinc-containing ALA dehydratase (PBG synthase) is extremely sensitive to inhibition by heavy metal ions (e.g., lead) that replace the zinc (see Fig. 21.3). This inhibition is, in part, responsible for the elevation in ALA and the anemia caused by lead poisoning.

3. **Uroporphyrinogen formation:** Condensation of four PBG molecules, catalyzed by hydroxymethylbilane synthase, produces the linear tetrapyrrole hydroxymethylbilane. A deficiency in this enzyme results in acute intermittent porphyria ([AIP] Fig. 21.4). Uroporphyrinogen III synthase cyclizes and isomerizes hydroxymethylbilane to produce the asymmetric uroporphyrinogen III. A deficiency in this enzyme results in congenital erythropoietic porphyria (CEP). Uroporphyrinogen III

Clinical Application 21.1: X-linked Sideroblastic Anemia

Since the majority of heme synthesized in the body is used in hemoglobin, a loss-of-function *ALAS2* mutation results in X-linked sideroblastic anemia. With lower levels of heme being produced, iron is not incorporated into heme efficiently, thus resulting in an intracellular iron overload. Sideroblasts are immature red blood cell precursors containing both mitochondria and the nucleus, commonly found in bone marrow (mature RBCs in circulation lack both mitochondria and the nucleus). In X-linked sideroblastic anemia, the excess intracellular iron accumulates in the mitochondria. When five or more iron-engorged mitochondria form a ring around at least a third of the nucleus (perinuclear), the sideroblast is referred to as a ringed sideroblast. Ringed sideroblasts can comprise 40% of the erythroid cells in a bone marrow aspirate of patients with sideroblastic anemia. Sideroblastic anemia patients also have elevated plasma iron and ferritin levels, contributing to iron toxicity damage to the liver, heart, and kidney (in addition to typical anemia symptoms such as fatigue, paleness, and dizziness). X-linked (congenital sideroblastic anemia) is the most common form of sideroblastic anemia. Approximately one-third of patients may be female, due to X chromosome inactivation.

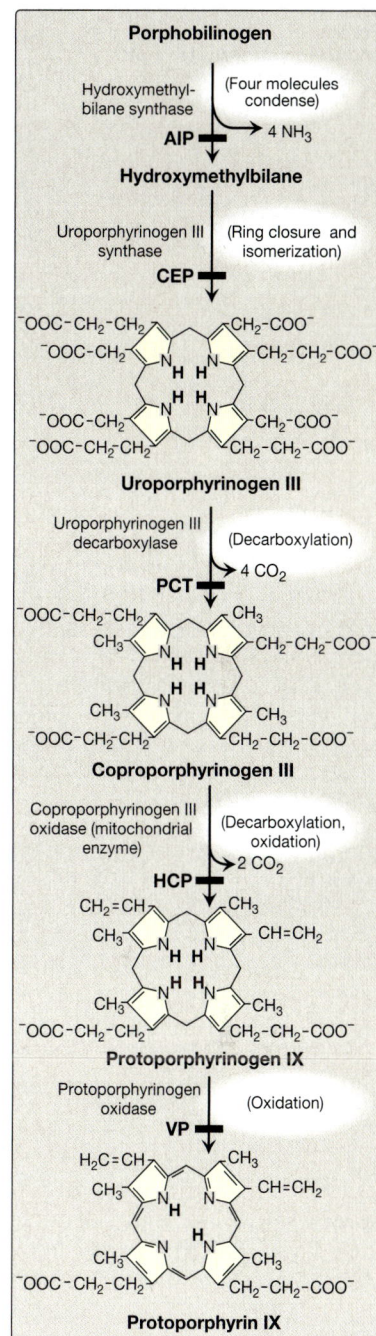

Figure 21.4
Pathway of porphyrin synthesis: Formation of protoporphyrin IX. (Continued from Fig. 21.3.) The prefixes uro- (urine) and copro- (feces) reflect initial sites of discovery. Enzyme deficiencies in porphyrias are indicated with black bars. AIP, acute intermittent porphyria; CEP, congenital erythropoietic porphyria; PCT, porphyria cutanea tarda; HCP, hereditary coproporphyria; VP, variegate porphyria. [Note: Deficiency in uroporphyrinogen III synthase prevents isomerization but not ring closure, resulting in production of type I porphyrins.]

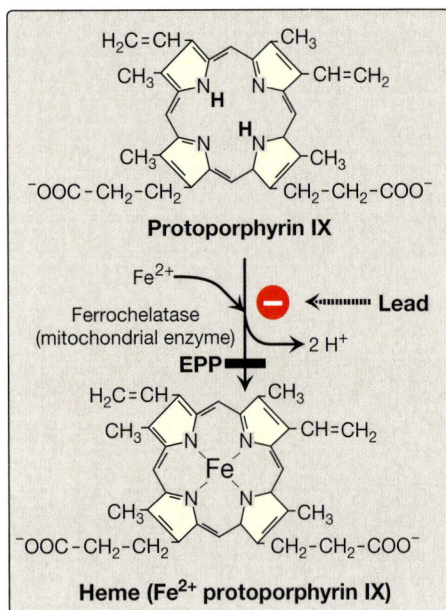

Figure 21.5
Pathway of porphyrin synthesis: Formation of heme b. (Continued from Figs. 21.3 and 21.4.) Enzyme deficiency in porphyria is indicated with a *black bar*. EPP, erythropoietic protoporphyria; Fe^{2+}, ferrous iron.

Clinical Application 21.2: Lead Poisoning

Lead poisoning is a buildup of lead in the body over a period of months to years. Common sources for lead include exposure to lead-based paints and paint dust or flakes common in older buildings; lead in household plumbing pipes may also contaminate drinking water. Exposure can occur through inhalation, contact with the skin or mucous membranes, or ingestion. Lead has a sweet taste, and ingestion exposure is of special concern for infants and toddlers. Symptoms of lead poisoning may include developmental delays, learning disabilities and low IQ, abdominal pain, constipation, neurologic changes, and irritability. Very high lead levels can be fatal. Lead inhibits δ-Aminolevulinic acid (ALA) dehydratase and ferrochelatase, both enzymes involved in the synthesis of heme, and therefore causes a decrease in heme synthesis. Further, high levels of lead impair iron utilization. This results in increased use of zinc (instead of iron) as substrate for chelation to protoporphyrin IX by ferrochelatase. Consequently, patients with lead poisoning may present with anemia and elevated levels of zinc protoporphyrin. The increase in ALA can be toxic to neurons. Lead can also cross the blood–brain barrier and is neurotoxic. The usual treatment is to remove the source of exposure to the lead contaminant, but in cases of severe lead poisoning (greater than 45 μg/dL measured in the serum), divalent chelators such as succimer (DMSA, 2,3-dimercaptosuccinic acid), calcium disodium ethylenediaminetetraacetic acid (EDTA) or others may be used to remove excess lead ions from the blood.

undergoes decarboxylation of its acetate groups by uroporphyrinogen III decarboxylase (UROD), generating coproporphyrinogen III. A deficiency in this enzyme results in porphyria cutanea tarda (PCT). These three reactions occur in the cytosol.

4. **Heme formation:** Coproporphyrinogen III enters the mitochondrion, and two propionate side chains are decarboxylated to vinyl groups by coproporphyrinogen III oxidase, generating protoporphyrinogen IX. A deficiency in this enzyme results in hereditary coproporphyria (HCP). Protoporphyrinogen IX is oxidized to protoporphyrin IX by protoporphyrinogen oxidase. A deficiency in this enzyme results in variegate porphyria (VP). The introduction of iron (as Fe^{2+}) into protoporphyrin IX produces heme. This step can occur spontaneously, but the rate is enhanced by ferrochelatase, an enzyme that, like ALA dehydratase, is inhibited by lead (Fig. 21.5). A deficiency in this enzyme results in erythropoietic protoporphyria (EPP).

C. Porphyrias

Porphyrias are rare, inherited (or sometimes acquired) defects in heme synthesis, resulting in the accumulation and increased excretion of porphyrins or porphyrin precursors. They may be inherited as autosomal-dominant (AD) or autosomal-recessive (AR) disorders. Each porphyria results in the accumulation of a unique pattern of intermediates caused by the deficiency of an enzyme in the heme synthetic pathway. Porphyria, derived from the Greek for "purple," refers to the red-blue color caused by pigment-like porphyrins in the urine of some patients with defects in heme synthesis.

1. **Clinical manifestations:** The porphyrias are classified as erythropoietic or hepatic, depending on whether the enzyme deficiency occurs in the erythropoietic cells of the bone marrow or in the liver. Hepatic porphyrias can be further classified as chronic or acute. In general, individuals with an enzyme defect prior to the synthesis of the tetrapyrroles manifest abdominal and neuropsychiatric symptoms (accumulation of ALA, PBG, neurotoxic porphyrin precursors), whereas those with enzyme defects leading to the accumulation of tetrapyrrole intermediates show photosensitivity (i.e., their skin itches and burns [pruritus] when exposed to sunlight). Cutaneous photosensitivity is a result of the oxidation of colorless porphyrinogens to colored porphyrins, photosensitizing molecules participating in the formation of superoxide radicals from oxygen. These radicals oxidatively damage membranes and cause the release of destructive enzymes from lysosomes. Immediate painful photosensitivity is often due to the accumulation of lipid-soluble protoporphyrin IX. Delayed blistering photosensitivity is often due to the accumulation of water-soluble porphyrins, uroporphyrin and coproporphyrin.

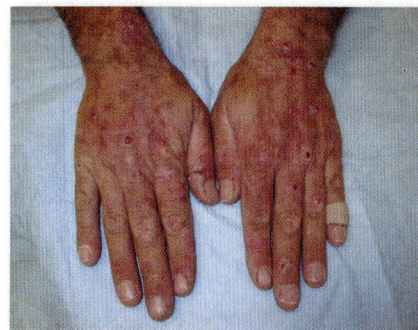

Figure 21.6
Skin eruptions in a patient with porphyria cutanea tarda.

   a. **Chronic hepatic porphyria:** PCT, the most common porphyria, is a chronic disease of the liver. The disease is associated with severe deficiency of UROD, with buildup of uroporphyrin in the urine. Clinical expression of the deficiency is influenced by various factors, including hepatic iron overload, exposure to sunlight, alcohol ingestion, estrogen therapy, and the presence of hepatitis B or C or HIV infections. Mutations to UROD are found in only 20% of affected individuals. Inheritance is AD. Clinical onset is typically during the fourth or fifth decade of life. Porphyrin accumulation leads to cutaneous symptoms, as shown in Figure 21.6, as well as urine that is red to brown in natural light, as shown in Figure 21.7, and pink to red in fluorescent light. PCT and other porphyrias are summarized in Figure 21.8 as they relate to the heme synthesis pathway and its regulation.

Figure 21.7
Urine from a patient with porphyria cutanea tarda (**right**) and from a patient with normal porphyrin excretion (**left**).

   b. **Acute hepatic porphyrias:** Acute hepatic porphyrias (ALA dehydratase–deficiency porphyria, AIP, HCP, and VP; see Fig. 21.8) are characterized by acute attacks of gastrointestinal (GI), neuropsychiatric, and motor symptoms that may be accompanied by photosensitivity (HCP and VP). Porphyrias leading to accumulation of ALA and PBG, such as AIP, cause abdominal pain and neuropsychiatric disturbances, ranging from anxiety to delirium. Symptoms of the acute hepatic porphyrias are often precipitated by use of drugs, such as barbiturates and ethanol, which induce the synthesis of the heme-containing CYP microsomal drug-oxidation system. This further decreases the amount of available heme, which, in turn, promotes increased synthesis of ALAS1.

   c. **Erythropoietic porphyrias:** The chronic erythropoietic porphyrias (CEP and EPP) cause photosensitivity characterized by skin rashes and blisters that appear in early childhood (see Fig. 21.8).

2. **Increased ALAS activity:** One common feature of the hepatic porphyrias is decreased synthesis of heme. In the liver, heme normally

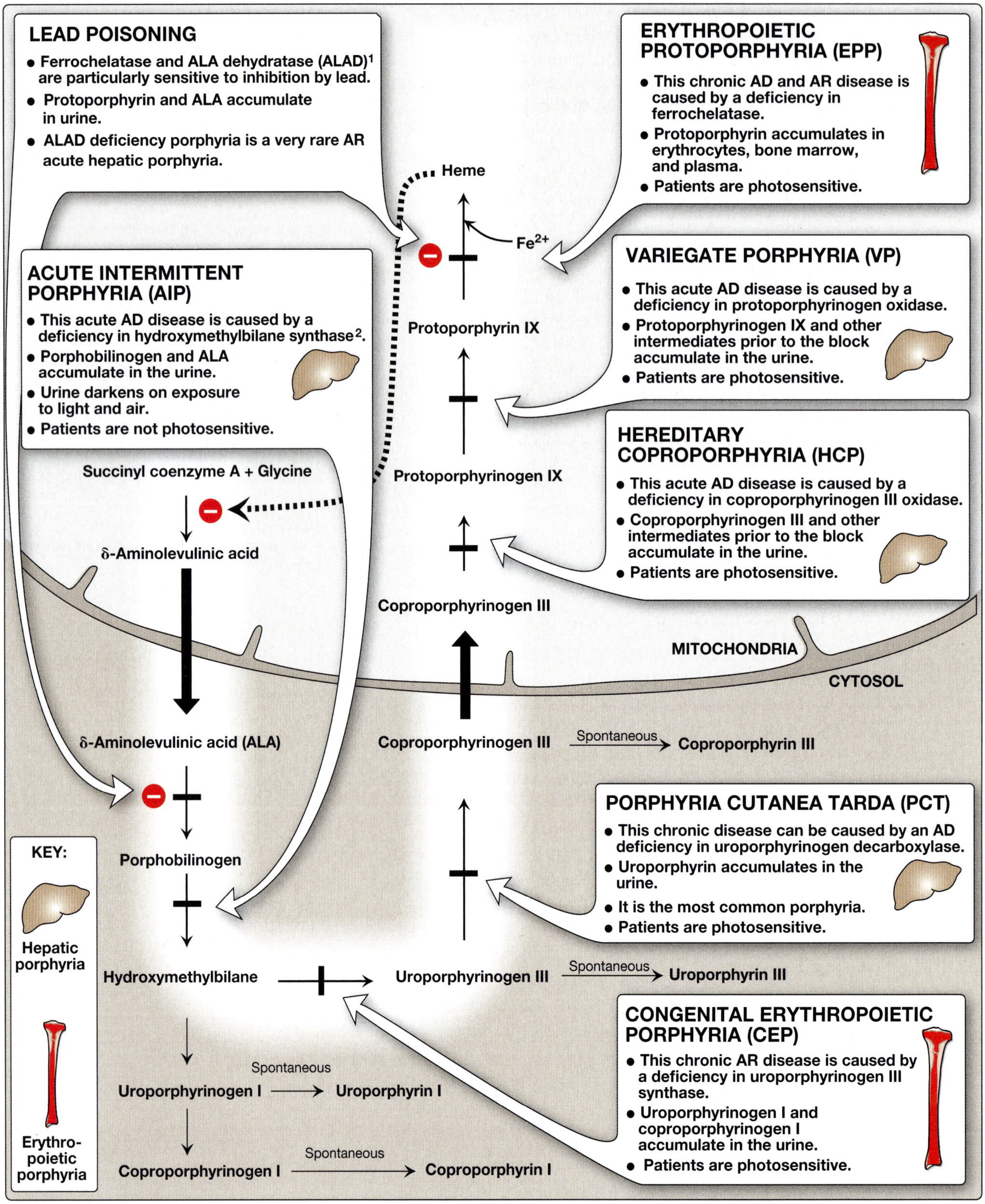

LEAD POISONING
- Ferrochelatase and ALA dehydratase (ALAD)[1] are particularly sensitive to inhibition by lead.
- Protoporphyrin and ALA accumulate in urine.
- ALAD deficiency porphyria is a very rare AR acute hepatic porphyria.

ERYTHROPOIETIC PROTOPORPHYRIA (EPP)
- This chronic AD and AR disease is caused by a deficiency in ferrochelatase.
- Protoporphyrin accumulates in erythrocytes, bone marrow, and plasma.
- Patients are photosensitive.

ACUTE INTERMITTENT PORPHYRIA (AIP)
- This acute AD disease is caused by a deficiency in hydroxymethylbilane synthase[2].
- Porphobilinogen and ALA accumulate in the urine.
- Urine darkens on exposure to light and air.
- Patients are not photosensitive.

VARIEGATE PORPHYRIA (VP)
- This acute AD disease is caused by a deficiency in protoporphyrinogen oxidase.
- Protoporphyrinogen IX and other intermediates prior to the block accumulate in the urine.
- Patients are photosensitive.

HEREDITARY COPROPORPHYRIA (HCP)
- This acute AD disease is caused by a deficiency in coproporphyrinogen III oxidase.
- Coproporphyrinogen III and other intermediates prior to the block accumulate in the urine.
- Patients are photosensitive.

Heme

Fe^{2+}

Protoporphyrin IX

Protoporphyrinogen IX

Succinyl coenzyme A + Glycine

δ-Aminolevulinic acid

Coproporphyrinogen III

MITOCHONDRIA

CYTOSOL

δ-Aminolevulinic acid (ALA)

Coproporphyrinogen III —Spontaneous→ Coproporphyrin III

KEY:

Hepatic porphyria

Porphobilinogen

PORPHYRIA CUTANEA TARDA (PCT)
- This chronic disease can be caused by an AD deficiency in uroporphyrinogen decarboxylase.
- Uroporphyrin accumulates in the urine.
- It is the most common porphyria.
- Patients are photosensitive.

Hydroxymethylbilane ——┼—— Uroporphyrinogen III —Spontaneous→ Uroporphyrin III

Erythro-poietic porphyria

Uroporphyrinogen I ——→ Uroporphyrin I

Spontaneous

Coproporphyrinogen I —Spontaneous→ Coproporphyrin I

CONGENITAL ERYTHROPOIETIC PORPHYRIA (CEP)
- This chronic AR disease is caused by a deficiency in uroporphyrinogen III synthase.
- Uroporphyrinogen I and coproporphyrinogen I accumulate in the urine.
- Patients are photosensitive.

Figure 21.8
Summary of heme synthesis. [1]Also referred to as porphobilinogen synthase. [2]Also referred to as porphobilinogen deaminase. [Note: Symptomatic deficiencies in ALA synthase-1 (ALAS1) are unknown. Deficiencies in X-linked ALAS2 result in an anemia.] ALA, δ-aminolevulinic acid; AD, autosomal dominant; AR, autosomal recessive; Fe, iron.

functions as a repressor of the *ALAS1* gene (see Fig. 21.3). Therefore, the absence of heme results in an increase in ALAS1 activity (derepression/activation). This causes an increased synthesis of intermediates that occur prior to the genetic block. The accumulation of these toxic intermediates is the major pathophysiology of the porphyrias.

3. **Treatment:** During acute porphyria attacks, patients require medical support, particularly treatment for pain and vomiting. The severity of acute symptoms of the porphyrias can be diminished by intravenous injection of hemin and glucose. Hemin consists of a protoporphyrin structure with a ferric iron (Fe^{3+}) coordinated with a chloride ion. Hemin administration reduces the deficit of porphyrins. This in turn decreases synthesis of ALAS1 and minimizes the production of toxic porphyrin intermediates. High doses of glucose can also decrease porphyrin biosynthesis in the liver. These treatments are particularly effective to treat AIP and other acute porphyrias. Protection from sunlight, ingestion of β-carotene (provitamin A; see Chapter 28, XI.) to scavenge free radicals, and phlebotomy (removes porphyrins) are helpful in porphyrias with photosensitivity.

D. Heme degradation

After ~120 days in the circulation, RBCs are taken up and degraded by the mononuclear phagocyte system (MPS), particularly in the liver and spleen (Fig. 21.9). Approximately 85% of heme destined for degradation comes from senescent RBCs (Fig. 21.10). The remainder is from the degradation of heme proteins other than Hb.

1. **Bilirubin formation:** The first step in the degradation of heme is catalyzed by microsomal heme oxygenase in macrophages of the MPS. In the presence of reduced nicotinamide adenine dinucleotide phosphate and oxygen, the enzyme catalyzes three successive oxygenation reactions that result in opening of the porphyrin ring (converting cyclic heme to linear biliverdin), production of carbon monoxide (CO), and release of Fe^{2+} (Fig. 21.9). The CO has a biologic function, acting as a signaling molecule and anti-inflammatory. Biliverdin, a green pigment, is then reduced by biliverdin reductase, forming the red-orange bilirubin. Bilirubin and its derivatives are collectively termed bile pigments. The changing colors of a bruise reflect the varying pattern of intermediates that occurs during heme degradation.

2. **Bilirubin uptake by the liver:** Because bilirubin is only slightly soluble in plasma, it is transported through blood to the liver by binding noncovalently to albumin (Fig. 21.10). Bilirubin dissociates from the carrier albumin molecule, enters a hepatocyte via facilitated diffusion, and binds to intracellular proteins, particularly the protein ligandin.

Certain anionic drugs, such as salicylates and sulfonamides, can displace bilirubin from albumin, permitting bilirubin to enter the central nervous system (CNS). This causes the potential for neural damage in infants.

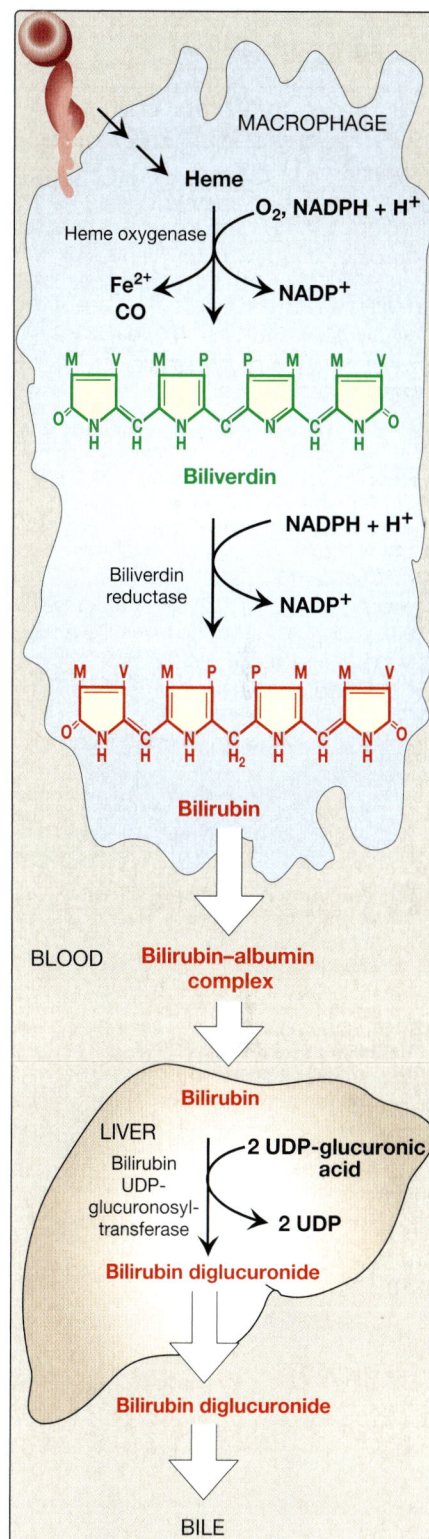

Figure 21.9
Formation of bilirubin from heme and its conversion to bilirubin diglucuronide. UDP, uridine diphosphate; Fe, iron; CO, carbon monoxide; NADP(H), nicotinamide adenine dinucleotide phosphate.

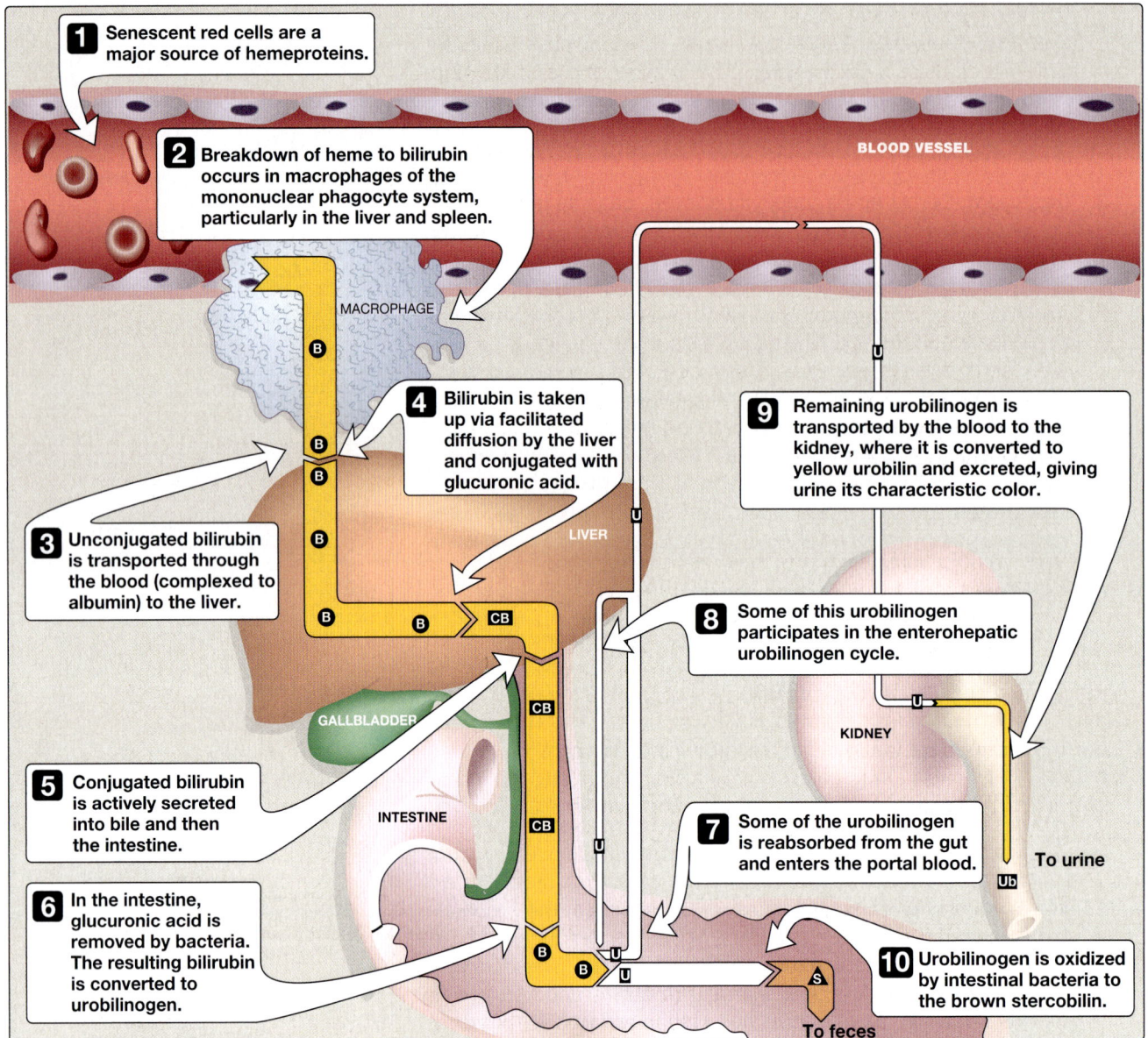

Figure 21.10
Catabolism of heme. ❽, bilirubin; 🄲🄱, conjugated bilirubin; 🅄, urobilinogen; 🅄🄱, urobilin; 🄰, stercobilin.

3. **Bilirubin diglucuronide formation:** In the hepatocyte, bilirubin solubility is increased by the sequential addition of two molecules of glucuronic acid in a process called conjugation. The reactions are catalyzed by microsomal bilirubin uridine diphosphate (UDP)-glucuronosyltransferase (bilirubin UGT) using UDP-glucuronic acid as the glucuronate donor. The bilirubin diglucuronide product is referred to as conjugated bilirubin ([CB] see Fig. 21.10). Varying degrees of deficiency of bilirubin UGT result in Crigler–Najjar I and II and Gilbert syndromes, with Crigler–Najjar I syndrome being the most severe.

4. **Bilirubin secretion into bile:** CB is actively transported against a concentration gradient into the bile canaliculi and then into the bile.

This energy-dependent, rate-limiting step is susceptible to impairment in liver disease. A rare deficiency in the protein required for transport of CB out of the liver results in Dubin–Johnson syndrome. Unconjugated bilirubin (UCB) is normally not secreted into bile.

5. **Urobilinogen, urobilin, and stercobilin formation:** CB is hydrolyzed and reduced by gut bacteria to yield urobilinogen (see Fig. 21.10), a colorless compound. Most of the urobilinogen is further oxidized by bacteria to stercobilin, which gives feces its characteristic brown color. However, some is reabsorbed from the gut and enters the portal blood. A portion of this urobilinogen participates in the enterohepatic urobilinogen cycle in which it is taken up by the liver and then resecreted into the bile. The remainder of the urobilinogen is transported by the blood to the kidney, where it is converted to yellow urobilin and excreted, giving urine its characteristic color. The metabolism of bilirubin is summarized in Figure 21.10.

E. Jaundice

Jaundice (or, icterus) refers to the yellow color of skin, nail beds, and sclerae (whites of the eyes) caused by bilirubin deposition, secondary to increased bilirubin levels in the blood (hyperbilirubinemia) as shown in Figure 21.11. Although not a disease, jaundice is usually a symptom of an underlying disorder. Blood bilirubin levels are normally no more than 1 mg/dL. Jaundice is seen at 2 to 3 mg/dL.

1. **Types:** Jaundice can be classified into three major types described below. However, in clinical practice, jaundice is often more complex than indicated in this simple classification. For example, the accumulation of bilirubin may be a result of defects at more than one step in its metabolism.

    a. **Hemolytic (prehepatic):** The liver has the capacity to conjugate and excrete more than 3,000 mg of bilirubin/day, whereas the normal production of bilirubin is only 300 mg/day, primarily from Hb breakdown of senescent RBCs. This excess capacity allows the liver to respond to increased heme degradation with a corresponding increase in conjugation and secretion of CB. However, extensive RBC hemolysis due to a hemolytic anemia (e.g., in patients with sickle cell anemia or deficiency of pyruvate kinase or glucose 6-phosphate dehydrogenase) may produce bilirubin faster than it can be conjugated in the liver. UCB levels in the blood become elevated (unconjugated hyperbilirubinemia), causing jaundice (Fig. 21.12A). Due to hemolysis, CB levels may be greatly elevated to the uppermost range of normal hepatic capacity and excreted into the bile. The amount of urobilinogen entering the enterohepatic circulation is also increased, as well as urinary urobilin. Still, CB, urobilinogen, stercobilin, and urobilin levels are all seen at the higher side of their normal ranges and are not abnormally elevated. In hemolytic jaundice, only UCB levels are abnormally high in the blood.

    b. **Hepatocellular (hepatic):** Damage to liver cells (e.g., in patients with cirrhosis or hepatitis) can compromise the ability

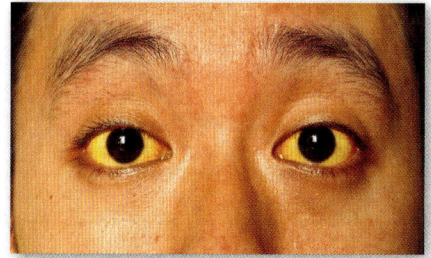

Figure 21.11
Jaundiced patient with the sclerae of his eyes appearing yellow. (From Zay Nyi Nyi/Shutterstock.com)

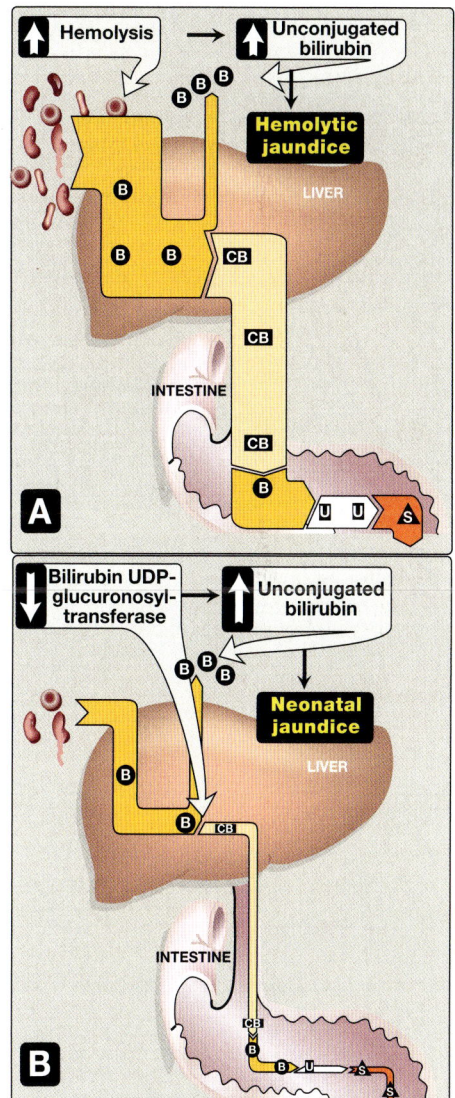

Figure 21.12
Alterations in the metabolism of heme. **A.** Hemolytic jaundice. **B.** Neonatal jaundice. CB, conjugated bilirubin; B, bilirubin; U, urobilinogen; S, stercobilin; UDP, uridine diphosphate.

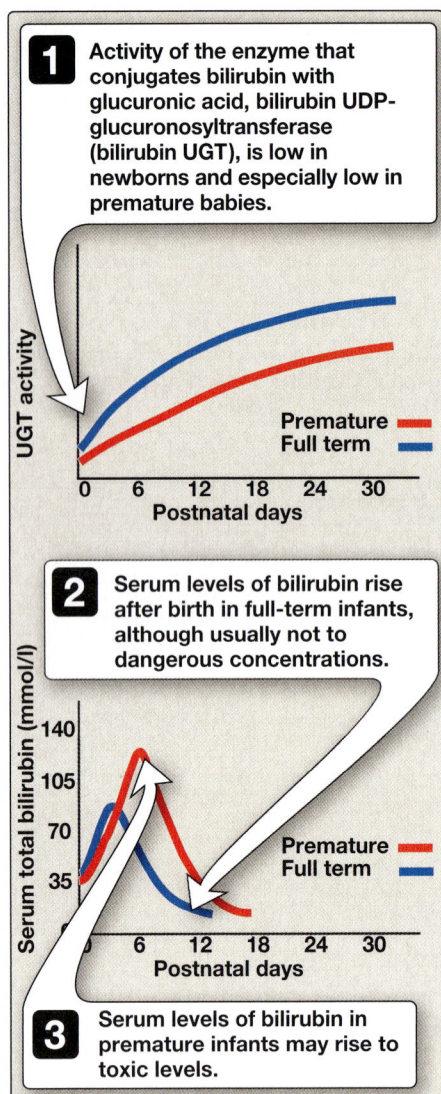

Figure 21.13
Neonatal jaundice. UDP, uridine diphosphate.

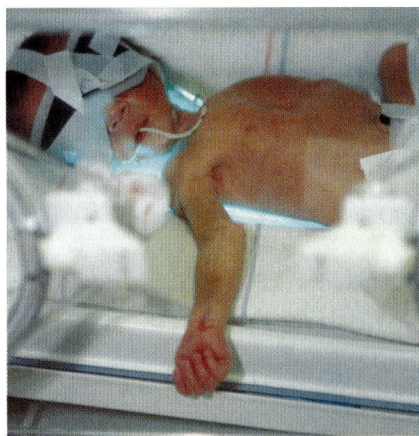

Figure 21.14
Phototherapy in neonatal jaundice.

to conjugate even normal levels of bilirubin, also resulting in unconjugated hyperbilirubinemia (a result of decreased conjugation). Urobilinogen is increased in the urine because hepatic damage decreases the enterohepatic circulation of this compound, allowing more to enter the blood, from which it is filtered into the urine. The urine consequently darkens, whereas stools may be a pale, clay color. Plasma levels of alanine and aspartate transaminases (ALT and AST, respectively; see Chapter 19, IV. A.) are elevated, suggesting hepatic damage. If CB is made but is not efficiently secreted from the liver into bile (intrahepatic cholestasis), it can also leak into the blood (regurgitation), causing conjugated hyperbilirubinemia. In hepatic jaundice, both UCB and CB levels are abnormally elevated in the blood (UCB and CB hyperbilirubinemia).

   c. **Obstructive (posthepatic):** In this instance, jaundice is not caused by overproduction of bilirubin or decreased conjugation but, instead, results from obstruction of the common bile duct (extrahepatic cholestasis). For example, the presence of a tumor or bile stones may block the duct, preventing passage of CB into the intestine. Patients with obstructive jaundice experience GI pain and nausea and produce stools that are a pale, clay color. The CB regurgitates into the blood. The CB is eventually excreted in the urine (which darkens over time) and is referred to as urinary bilirubin. Urinary urobilinogen is absent. In obstructive jaundice, only CB hyperbilirubinemia results.

2. **Jaundice in newborns:** Most newborn infants (60% of full term and 80% of preterm) show a rise in UCB in the first postnatal week (a transient, physiologic jaundice) because the activity of hepatic bilirubin UGT is low at birth (it reaches adult levels in ~4 weeks), as shown in Figure 21.12B and Figure 21.13. Elevated UCB, in excess of the binding capacity of albumin (20 to 25 mg/dL), can diffuse into the basal ganglia, causing toxic encephalopathy (kernicterus) and pathologic jaundice. Therefore, newborns with significantly elevated bilirubin levels are treated with blue fluorescent light (phototherapy), as shown in Figure 21.14, which converts bilirubin to more polar and, therefore, water-soluble isomers. These photoisomers can be excreted into the bile without conjugation to glucuronic acid. Because of solubility differences, only UCB crosses the blood–brain barrier, and only CB appears in urine.

3. **Bilirubin measurement:** Bilirubin is commonly measured by the van den Bergh reaction, in which diazotized sulfanilic acid reacts with bilirubin to form red azodipyrroles that are measured colorimetrically. In aqueous solution, the water-soluble CB reacts rapidly with the reagent (within 1 minute) and is said to be direct reacting (or direct bilirubin). The UCB, which is much less soluble in aqueous solution, reacts more slowly. When the reaction is carried out in methanol, both CB and UCB are soluble and react with the reagent, providing the total bilirubin value. The indirect reacting bilirubin, which corresponds to the UCB, is obtained by subtracting the direct reacting bilirubin from the total bilirubin. In normal plasma, only ~4% of the total bilirubin is conjugated, or direct reacting, because most is secreted into bile.

III. OTHER NITROGEN-CONTAINING COMPOUNDS

Many nitrogen-containing molecules are derived from amino acids, including certain neurotransmitters, muscle creatine phosphate, and melanin. Following is a discussion of their function, synthesis, degradation, and relevant medical implications.

A. Catecholamines

Dopamine, norepinephrine (NE), and epinephrine (or, adrenaline) are biologically active (biogenic) amines that are collectively termed catecholamines. Dopamine and NE are synthesized in the brain and function as neurotransmitters. Epinephrine is synthesized from NE in the adrenal medulla.

1. **Function:** Outside the central nervous system (CNS), NE and its methylated derivative, epinephrine, are hormone regulators of carbohydrate and lipid metabolism. NE and epinephrine are released from storage vesicles in the adrenal medulla in response to fright, exercise, cold, and low levels of blood glucose. They increase the degradation of glycogen and triacylglycerol as well as increase blood pressure and the output of the heart. These effects are part of a coordinated response to prepare the individual for stress and are often called the "fight-or-flight" reactions.

2. **Synthesis:** The catecholamines are synthesized from tyrosine, as shown in Figure 21.15. Tyrosine is first hydroxylated by tyrosine hydroxylase to form L-3,4-dihydroxyphenylalanine ([DOPA] a catechol) in a reaction analogous to that described for the hydroxylation of phenylalanine (see Chapter 20, III. D. and Fig. 20.7). The tetrahydrobiopterin (BH_4)-requiring enzyme is abundant in the CNS, the sympathetic ganglia, and the adrenal medulla, and it catalyzes the rate-limiting step of the pathway. DOPA is then decarboxylated in a reaction catalyzed by DOPA decarboxylase (DDC) and requiring PLP to form dopamine (the first catecholamine in the pathway). Dopamine is next hydroxylated by dopamine β-hydroxylase to yield NE in a reaction that requires ascorbic

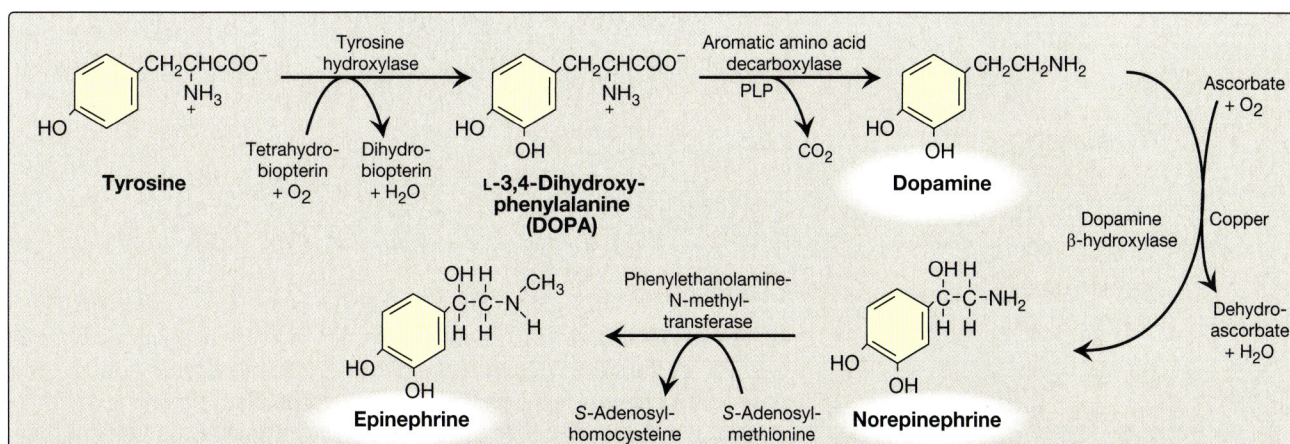

Figure 21.15
Synthesis of catecholamines. [Note: Catechols have two adjacent hydroxyl groups.] PLP, pyridoxal phosphate.

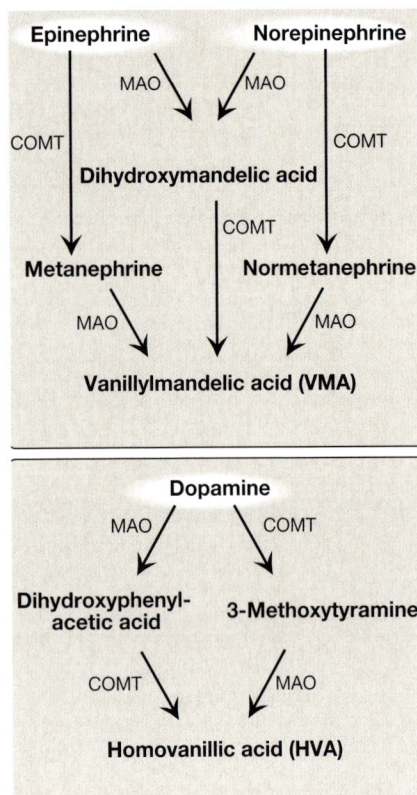

Figure 21.16
Metabolism of the catecholamines by catechol-O-methyltranferase (COMT) and monoamine oxidase (MAO). [Note: COMT requires *S*-adenosylmethionine.]

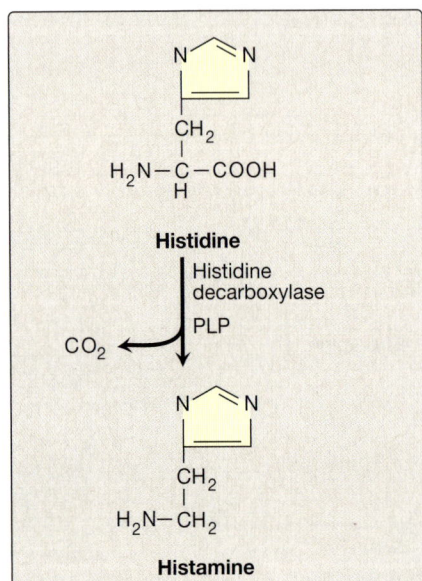

Figure 21.17
Biosynthesis of histamine. PLP, pyridoxal phosphate.

Clinical Application 21.3: Parkinson Disease

Parkinson disease, a neurodegenerative movement disorder, involves insufficient dopamine production because of the idiopathic loss of dopamine-producing cells in the substantia nigra, a region of the brain that controls movement. Administration of levodopa (L-DOPA) is the most common treatment because dopamine cannot cross the blood–brain barrier. Carbidopa is a drug that inhibits DDC activity, preventing the conversion of L-DOPA to dopamine in the peripheral nervous system. Since carbidopa cannot cross the blood–brain barrier, when used in tandem with L-DOPA, this allows more peripheral L-DOPA to cross the blood–brain barrier, so dopamine can reach a more therapeutic range in the central nervous system. In the case of a BH₄-deficiency (used in the tyrosine hydroxylase reaction), L-DOPA may be given as a neurotransmitter supplement to produce dopamine, NE, and epinephrine.

acid (vitamin C) and copper. Epinephrine is formed from NE by an N-methylation reaction using *S*-adenosylmethionine (SAM) as the methyl donor (see Chapter 20, section IIIE, Fig. 20.8).

3. **Degradation:** Catecholamines are inactivated by oxidative deamination catalyzed by monoamine oxidase (MAO) and by O-methylation catalyzed by catechol-O-methyltransferase (COMT) using SAM as the methyl donor (Fig. 21.16). The reactions can occur in either order. The aldehyde products of the MAO reaction are oxidized to the corresponding acids. The products of these reactions are excreted in the urine as vanillylmandelic acid (VMA) from epinephrine and NE and homovanillic acid (HVA) from dopamine. VMA and the metanephrines are increased with pheochromocytomas, rare tumors of the adrenal gland characterized by excessive production of catecholamines.

4. **MAO inhibitors:** MAO is found in neural and other tissues, such as the intestine and liver. In the neuron, this enzyme oxidatively deaminates and inactivates any excess neurotransmitter molecules (NE, dopamine, or serotonin) that may leak out of synaptic vesicles when the neuron is at rest. MAO inhibitors (MAOIs) may irreversibly or reversibly inactivate the enzyme, permitting neurotransmitter molecules to escape degradation and, therefore, both to accumulate within the presynaptic neuron and to leak into the synaptic space. This increases activation of NE and serotonin receptors and may be responsible for the antidepressant action of MAOIs. MAOIs can also be used for panic and social anxiety disorders, as well as for Parkinson disease.

B. Histamine

Histamine is a chemical messenger that mediates a wide range of cellular responses, including allergic and inflammatory reactions and gastric acid secretion. A powerful vasodilator, histamine is formed by decarboxylation of histidine in a reaction catalyzed by histidine decarboxylase and requiring PLP as a cofactor (Fig. 21.17). It is secreted by mast cells as a result of allergic reactions or trauma. Histamine has no clinical applications, but antihistamines that interfere with the action of histamine have important therapeutic applications. Antihistamines are generally histamine analogs that block histamine binding to its receptors to reduce histamine responses.

C. Serotonin

Serotonin, also called 5-hydroxytryptamine (5-HT), is synthesized and/or stored at several sites in the body (Fig. 21.18). The largest amount by far is found in the intestinal mucosa. Smaller amounts occur in the CNS, where it functions as a neurotransmitter, and in platelets (see online Chapter 35). Serotonin is synthesized from tryptophan, which is hydroxylated in a BH_4-requiring reaction analogous to reactions catalyzed by phenylalanine or tyrosine hydroxylases. The product, 5-hydroxytryptophan, is decarboxylated to 5-HT. In the case of a BH_4-deficiency, 5-hydroxytryptophan may be given as a supplement to produce serotonin. Serotonin has multiple physiologic roles including pain perception and regulation of sleep, appetite, temperature, blood pressure, cognitive functions, and mood (causes a feeling of well-being). Selective serotonin reuptake inhibitors (SSRIs) maintain serotonin levels, thereby functioning as antidepressants. Serotonin is degraded by MAO to 5-hydroxy-3-indoleacetic acid (5-HIAA).

D. Creatine

Creatine phosphate (also called phosphocreatine), the phosphorylated derivative of creatine found in muscle, is a high-energy compound that provides a small but rapidly mobilized reserve of high-energy phosphates that can be reversibly transferred to adenosine diphosphate to maintain the intracellular level of ATP by substrate-level phosphorylation during the first few minutes of intense muscular contraction (Fig. 21.19). The amount of creatine phosphate in the body is proportional to the muscle mass.

1. **Synthesis:** Creatine is synthesized in the liver and kidneys from glycine and the guanidino group of arginine, plus a methyl group from SAM (see Fig. 21.19). Animal products are dietary sources. Creatine is reversibly phosphorylated to creatine phosphate by creatine kinase (CK), using ATP as the phosphate donor. The presence of CK (cardiac-specific MB isozyme) in the plasma is indicative of heart damage and historically was used in the diagnosis of myocardial infarction. Cardiac troponin has replaced cardiac CK isoenzymes as the test of choice for the diagnosis of heart attacks.

2. **Degradation:** Creatine and creatine phosphate spontaneously cyclize at a slow but constant rate to form creatinine, which is excreted in the urine. The amount excreted is proportional to the total creatine phosphate content of the body and, therefore, can be used to estimate muscle mass. When muscle mass decreases for any reason (e.g., muscle atrophy from paralysis or muscular dystrophy), the creatinine content of the urine falls. In addition, a rise in blood creatinine is a sensitive indicator of renal dysfunction, because creatinine normally is rapidly cleared from the blood and excreted. A typical adult male excretes ~1 to 2 g of creatinine/day.

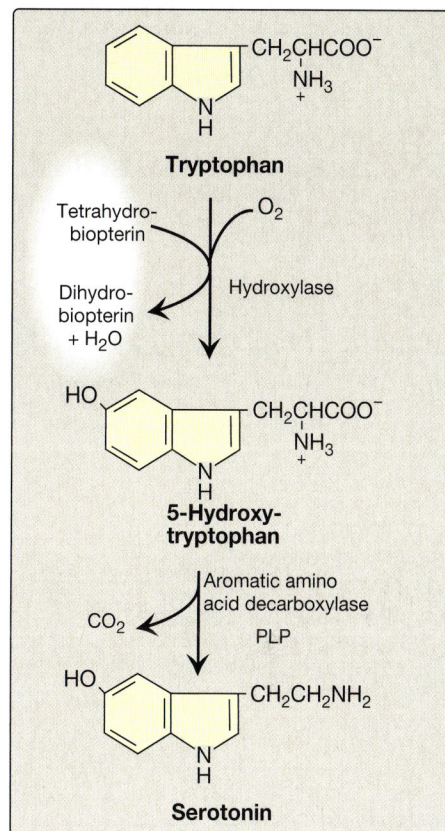

Figure 21.18
Synthesis of serotonin. [Note: Serotonin is converted to melatonin, a regulator of circadian rhythm, in the pineal gland.] PLP, pyridoxal phosphate; CO_2, carbon dioxide.

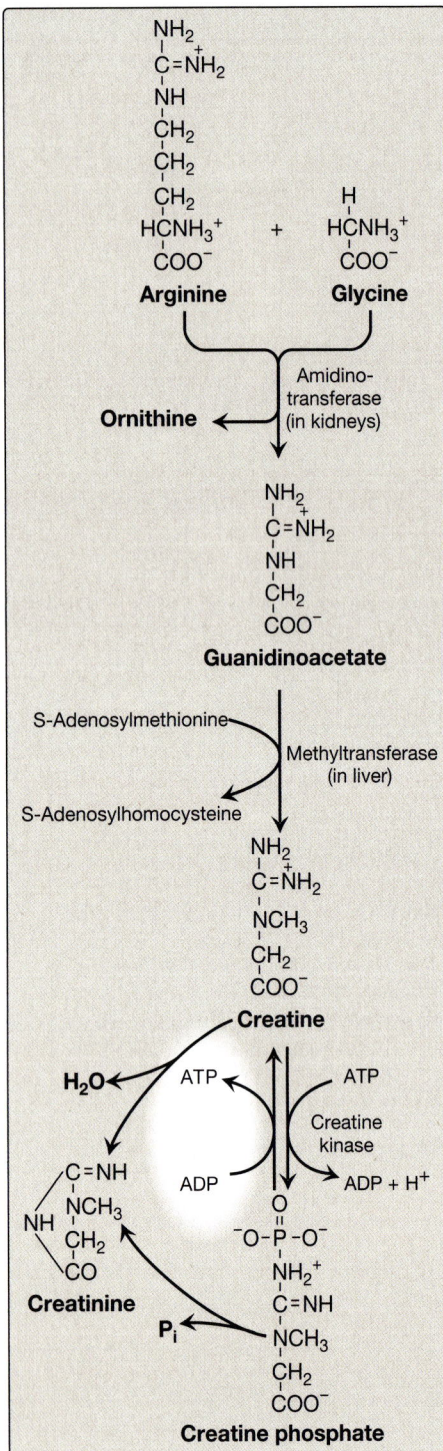

Figure 21.19
Synthesis of creatine. ADP, adenosine diphosphate; P_i, inorganic phosphate.

E. Melanin

Melanin is a pigment that occurs in several tissues, particularly the eye, hair, and skin. It is synthesized from tyrosine in melanocytes (pigment-forming cells) of the epidermis. It functions to protect underlying cells from the harmful effects of sunlight. A defect in melanin production results in oculocutaneous albinism, the most common type being due to defects in copper-containing tyrosinase (see Chapter 20, VI. C. and Fig. 20.15).

IV. CHAPTER SUMMARY

- **Amino acids** are **precursors** of many N-containing compounds including **porphyrins**, which, in combination with Fe^{2+} **iron**, form **heme** (Fig. 21.20).

- The major sites of **heme biosynthesis** are the **liver** and the **erythrocyte-producing cells** of the bone marrow. In the liver, the rate of heme synthesis is highly variable, responding to alterations in the cellular heme pool caused by fluctuating demands for heme proteins (particularly **CYP enzymes**). In contrast, heme synthesis in erythroid cells is relatively constant and is matched to the rate of Hb synthesis.

- Heme synthesis starts with **glycine** and **succinyl coenzyme A**. The **committed step** is the formation of **δ-aminolevulinic acid (ALA)**. This mitochondrial reaction is catalyzed by ALA synthase 1 (**ALAS1**) in the liver (inhibited by **hemin**, the oxidized form of heme that accumulates when heme is being underutilized) and **ALAS2** in erythroid tissues (regulated by iron).

- **Porphyrias** are caused by inherited or acquired (**lead poisoning**) defects in heme synthesis, resulting in the accumulation and increased excretion of porphyrins or porphyrin precursors. Enzyme defects early in the pathway cause **abdominal pain** and **neuropsychiatric symptoms**, whereas later defects cause **photosensitivity**.

- **Degradation** of heme occurs in the **mononuclear phagocyte system (MPS)**, particularly in the **liver** and **spleen**. The first step is the production by **heme oxygenase** of **biliverdin**, which is subsequently reduced to **bilirubin**. Bilirubin is transported by **albumin** to the liver, where its solubility is increased by the addition of two molecules of **glucuronic acid** by **bilirubin UGT** (uridine diphosphate [UDP]-glucuronosyltransferase). **Bilirubin diglucuronide** (**conjugated bilirubin [CB]**) is transported into the **bile canaliculi**, where it is first hydrolyzed and reduced by gut bacteria to yield **urobilinogen**, which is further oxidized by bacteria to **stercobilin**.

- **Jaundice (icterus)** refers to the yellow color of the skin and sclerae that is caused by deposition of bilirubin, secondary to increased bilirubin levels in the blood. Three commonly encountered types of jaundice are **hemolytic (prehepatic)**, **obstructive (posthepatic)**, and **hepatocellular (hepatic)** (see Fig. 21.20).

- Other important N-containing compounds derived from amino acids include the **catecholamines (dopamine, norepinephrine, and epinephrine), creatine, histamine, serotonin, melanin,** and **nitric oxide**.

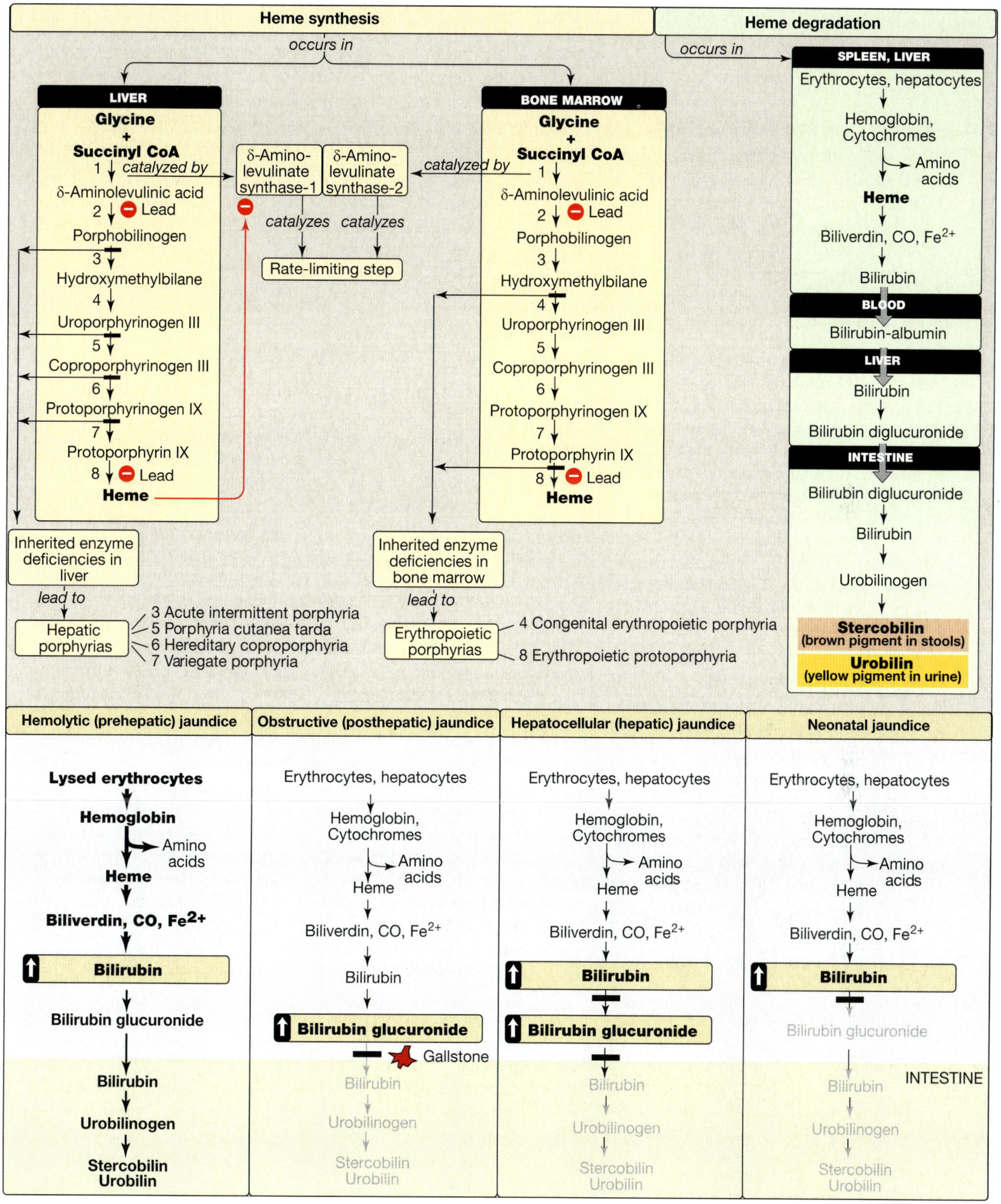

Figure 21.20
Key concept map for heme metabolism. [Note: Hepatocellular jaundice can be caused by decreased conjugation of bilirubin or decreased secretion of conjugated bilirubin from the liver into bile.] ▬▬, Block in the pathway. CoA, coenzyme A; CO, carbon monoxide; Fe, iron.

Study Questions

Choose the ONE best answer.

21.1. δ-Aminolevulinic acid synthase activity:

    A. catalyzes the committed step in porphyrin biosynthesis.

    B. is decreased by iron in erythrocytes.

    C. is decreased in the liver in individuals treated with certain drugs such as the barbiturate phenobarbital.

    D. occurs in the cytosol.

    E. requires tetrahydrobiopterin as a coenzyme.

> Correct answer = A. δ-Aminolevulinic acid synthase is mitochondrial and catalyzes the rate-limiting and regulated step of porphyrin synthesis. It requires pyridoxal phosphate as a coenzyme. Iron increases the production of the erythroid isozyme. The hepatic isozyme is increased in patients treated with certain drugs.

21.2. A 50-year-old male presented with painful blisters on the backs of his hands. He was a golf instructor and indicated that the blisters had erupted shortly after the golfing season began. He did not have recent exposure to common skin irritants. He had partial complex seizure disorder that had begun ~3 years earlier after a head injury. The patient had been taking phenytoin (his only medication) since the onset of the seizure disorder. He reported an average weekly ethanol intake of ~18 12-oz cans of beer. The patient's urine was reddish orange. Cultures obtained from skin lesions failed to grow organisms. A 24-hour urine collection showed elevated uroporphyrin (1,000 mg; normal, <27 mg). What is the most likely diagnosis?

    A. Acute intermittent porphyria

    B. Congenital erythropoietic porphyria

    C. Erythropoietic protoporphyria

    D. Hereditary coproporphyria

    E. Porphyria cutanea tarda

> Correct answer = E. The disease is associated with a deficiency in uroporphyrinogen III decarboxylase (UROD), but clinical expression of the enzyme deficiency is influenced by hepatic injury caused by environmental (e.g., ethanol) and infectious (e.g., hepatitis B virus) agents. Exposure to sunlight can also be a precipitating factor. Clinical onset is typically during the fourth or fifth decade of life. Porphyrin accumulation leads to cutaneous symptoms and urine that is red to brown. Treatment of the patient's seizure disorder with phenytoin caused increased synthesis of δ-aminolevulinic acid synthase and, therefore, of uroporphyrinogen, the substrate of the deficient UROD. The laboratory and clinical findings are inconsistent with other porphyrias.

21.3. A patient presents with jaundice, abdominal pain, and nausea. Clinical laboratory results indicate conjugated hyperbilirubinemia and the presence of urinary bilirubin. What is the most likely cause of the jaundice?

    A. Decreased hepatic conjugation of bilirubin

    B. Decreased hepatic uptake of bilirubin

    C. Decreased secretion of bile into the intestine

    D. Increased hemolysis

> Correct answer = C. The data are consistent with an obstructive jaundice, in which a block in the common bile duct decreases the secretion of bile-containing conjugated bilirubin (CB) into the intestine (stool will be pale in color). The CB regurgitates into the blood (conjugated hyperbilirubinemia). The CB is excreted in the urine (which darkens) and is referred to as urinary bilirubin. Urinary urobilinogen is not present because its normal source is intestinal urobilinogen, which is low.

21.4. A patient with hemolytic jaundice would be expected to have which of the following profiles?

	Unconjugated Bilirubin	Conjugated Bilirubin	Urobilinogen
A	Increased	Increased	Decreased
B	Increased	Normal	Normal
C	Increased	Decreased	Decreased
D	Normal	Increased	Decreased

Correct answer = B. In hemolytic jaundice, there is more bilirubin than the liver can conjugate, resulting in unconjugated hyperbilirubinemia. The liver conjugates and excretes bilirubin at the upper level of normal, producing normal urobilinogen in the intestines. Profile A would be consistent with hepatocellular jaundice, with both conjugated and unconjugated hyperbilirubinemia and a decrease in urobilinogen. Profile C would be consistent with neonatal jaundice, with unconjugated hyperbilirubinemia and a decrease in both conjugated bilirubin and urobilinogen. Profile D would be consistent with obstructive jaundice, with conjugated hyperbilirubinemia and a decrease in urobilinogen. Unconjugated bilirubin is normal.

21.5. A 2-year-old child was brought to his pediatrician for evaluation of gastrointestinal problems. The parents report that the boy has been listless for the last few weeks. Lab tests reveal microcytic, hypochromic anemia. Blood lead levels are elevated. Which enzyme is most likely to have higher-than-normal activity in the patient's liver?

A. δ-Aminolevulinic acid synthase
B. Bilirubin UDP-glucuronosyltransferase
C. Ferrochelatase
D. Heme oxygenase
E. Porphobilinogen synthase

Correct answer = A. This child has the acquired porphyria of lead poisoning. Lead inhibits both δ-aminolevulinic acid dehydratase and ferrochelatase and, consequently, heme synthesis. The decrease in heme derepresses δ-aminolevulinic acid synthase-1 (the hepatic isozyme), resulting in an increase in its activity. The decrease in heme also results in decreased hemoglobin synthesis, and anemia is seen. Ferrochelatase is directly inhibited by lead. The other choices are enzymes of heme degradation.

21.6. A 50-year-old male presents with hand tremors, a slow unsteady gait, and stiffness. After neurologic scans and additional testing, the patient is diagnosed with Parkinson disease. Which of the following treatments would be most effective in this patient?

A. Biopterin
B. β-Carotene
C. Hemin
D. Levodopa–carbidopa
E. Serotonin reuptake inhibitors

Correct answer = D. Levodopa (L-DOPA) can cross the blood–brain barrier to be used as a substrate for DOPA decarboxylase to increase dopamine levels in the central nervous system. Carbidopa cannot cross the blood–brain barrier and inhibits peripheral DOPA decarboxylase. This provides higher therapeutic L-DOPA levels for the central nervous system. Biopterin can be provided as a useful therapeutic agent for aromatic amino acid hydroxylase reactions when the cofactor is deficient. β-Carotene is an antioxidant, which can scavenge free radicals. Along with phlebotomy, it can help with photosensitivity in acute porphyria cases. Hemin reduces the deficit of porphyrins. This in turn decreases the synthesis of ALAS1 and minimizes the production of toxic porphyrin intermediates. Serotonin reuptake inhibitors help maintain serotonin levels, and function as antidepressants.

21.7. Kidney malfunction in a patient may be indicated by which of the following lab tests?

A. Increased blood creatine kinase MB isoenzyme levels
B. Increased urine vanillylmandelic acid and metanephrine levels
C. Increased blood bilirubin diglucuronide levels
D. Decreased urine creatinine levels
E. Increased blood creatinine levels

Correct answer = E. Creatinine is normally very rapidly cleared from the blood by the kidneys and excreted in the urine. An increase in blood creatinine concentration levels indicates renal malfunction. Increased blood creatine kinase MB isoenzyme levels would be indicative of heart damage and/or myocardial infarction. Increased urine vanillylmandelic acid and metanephrine levels would be indicative of tumors of the adrenal gland, characterized by increased production of catecholamines. Increased blood bilirubin diglucuronide levels would be indicative of obstructive jaundice. Decreased urine creatinine levels would be indicative of decreased muscle mass, such as muscle atrophy from paralysis or muscular dystrophy.

21.8. Which of the following catecholamines is generated from tyrosine?

 A. Creatinine

 B. Histamine

 C. Epinephrine

 D. Melanin

 E. Serotonin

Correct answer = C. Hydroxylation of tyrosine by tyrosine hydroxylase generates DOPA, which is a catechol. Decarboxylation of DOPA forms dopamine, the first catecholamine. Norepinephrine is formed from dopamine. Epinephrine is formed from norepinephrine. Creatinine is not a catecholamine or a neurotransmitter. It is formed spontaneously from creatine and creatine phosphate in muscle tissue and is readily excreted in the urine proportionate to muscle mass. Histamine is formed from histidine by histidine decarboxylase. Melanin is formed from tyrosine by tyrosinase in melanocytes. Serotonin is formed from tryptophan by hydroxylation followed by decarboxylation.

21.9. A patient has a defect in dihydropteridine reductase. What supplement can be used to allow the synthesis of serotonin?

 A. Carbidopa

 B. 5-Hydroxytryptophan

 C. L-DOPA

 D. Monoamine oxidase inhibitor

 E. Tyrosine

Correct answer = B. Tryptophan is first hydroxylated by tryptophan hydroxylase, forming 5-hydroxtryptophan, followed by decarboxylation to form serotonin. Tryptophan hydroxylase is a tetrahydrobiopterin (BH_4)-requiring enzyme. Dihydropteridine reductase regenerates BH_4 from dihydrobiopterin (BH_2). To allow synthesis in the absence of the BH_4 cofactor, the downstream product 5-hydroxytryptophan can be supplemented. Carbidopa is an inhibitor of aromatic amino acid decarboxylase, functioning in the peripheral nervous system. It is used in conjunction with levodopa (L-DOPA) to increase dopamine production in the central nervous system in Parkinson disease. MAOIs inhibit monoamine oxidase, which breaks down catecholamines. Supplementation of MAOIs would increase catecholamine levels. Tyrosine could be supplemented in patients with phenylketonuria, since tyrosine would become an essential amino acid. Tyrosine levels would then allow the downstream synthesis of catecholamines.

Nucleotide Metabolism

22

I. OVERVIEW

Ribonucleoside and deoxyribonucleoside phosphates (nucleotides) are essential for all cells. Without them, neither ribonucleic acid (RNA) nor deoxyribonucleic acid (DNA) can be produced, and, therefore, proteins cannot be synthesized, nor can cells proliferate. Nucleotides also serve as carriers of activated intermediates in the synthesis of some carbohydrates, lipids, and conjugated proteins (e.g., uridine diphosphate [UDP]-glucose and cytidine diphosphate [CDP]-choline) and are structural components of several essential coenzymes, such as coenzyme A, flavin adenine dinucleotide (FAD[H_2]), nicotinamide adenine dinucleotide (NAD[H]), and nicotinamide adenine dinucleotide phosphate (NADP[H]). Nucleotides, such as cyclic adenosine monophosphate (cAMP) and cyclic guanosine monophosphate (cGMP), serve as second messengers in signal transduction pathways. In addition, nucleotides play an important role as energy sources in the cell. Finally, nucleotides are important regulatory compounds for many of the pathways of intermediary metabolism, inhibiting or activating key enzymes. The purine and pyrimidine bases found in nucleotides can be synthesized *de novo* or can be obtained through salvage pathways that allow the reuse of the preformed bases resulting from normal cell turnover. Few of the purines and pyrimidines supplied by diet are utilized; instead, nearly all of the nucleic acids that enter the gastrointestinal (GI) tract are degraded.

II. STRUCTURE

Nucleotides are composed of a nitrogenous base; a pentose monosaccharide; and one, two, or three phosphate groups. The nitrogen-containing bases belong to two families of compounds: the purines and the pyrimidines.

A. Purine and pyrimidine bases

Purines are double-ringed structures, whereas pyrimidines have a single ring. Both DNA and RNA contain the same purine bases: adenine (A) and guanine (G). Both DNA and RNA contain the pyrimidine cytosine (C), but they differ in their second pyrimidine base: DNA contains thymine (T), whereas RNA contains uracil (U). T and U differ in that only T has a methyl group (Fig. 22.1). Unusual (modified) bases are occasionally found in some species of DNA (e.g., in some viral DNA) and RNA (e.g., in transfer RNA [tRNA]). Base modifications

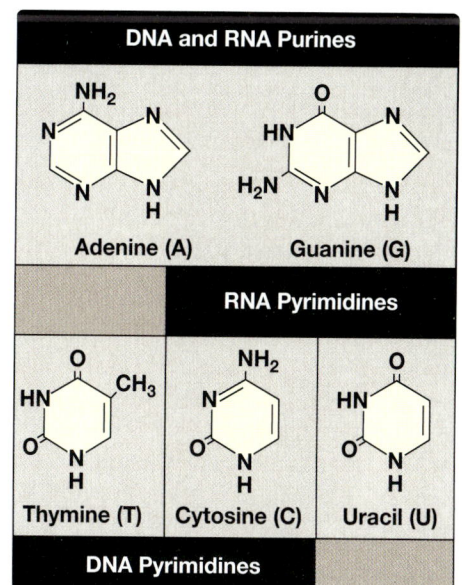

Figure 22.1
Purines and pyrimidines commonly found in DNA and RNA.

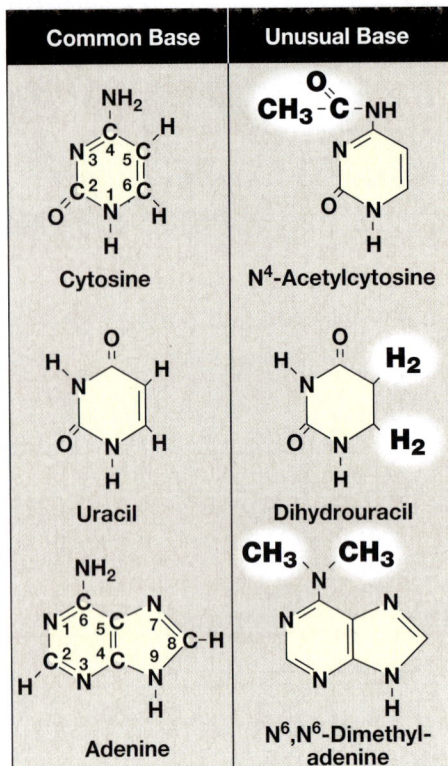

Figure 22.2
Examples of unusual bases.

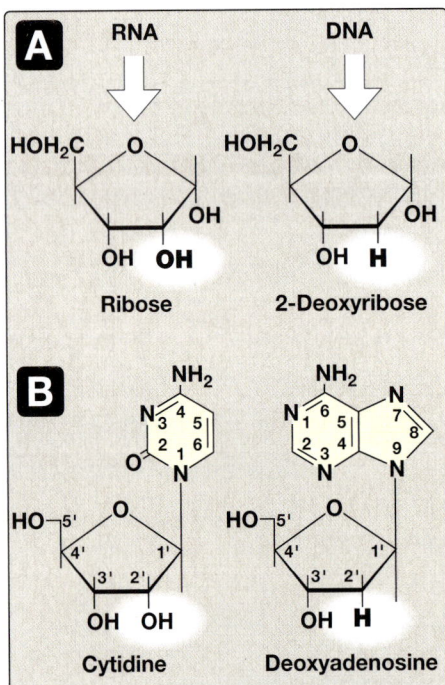

Figure 22.3
A. Pentoses found in nucleic acids.
B. Examples of the numbering systems for purine- and pyrimidine-containing nucleosides.

include methylation, glycosylation, acetylation, and reduction. Some examples of unusual bases are shown in Figure 22.2. The presence of an unusual base in a nucleotide sequence may aid in its recognition by specific enzymes or protect it from being degraded by nucleases.

B. Nucleosides

The addition of a pentose sugar to a base through an N-glycosidic bond (see Chapter 7, II. E.) produces a nucleoside. If the sugar is ribose, a ribonucleoside is produced, and if the sugar is 2-deoxyribose, a deoxyribonucleoside is produced (Fig. 22.3A). The ribonucleosides of A, G, C, and U are named adenosine, guanosine, cytidine, and uridine, respectively. The deoxyribonucleosides of A, G, C, and T have the added prefix deoxy- (e.g., deoxyadenosine). [Note: The compound deoxythymidine is often simply called thymidine, with the deoxy- prefix being understood, because it is incorporated into DNA only.] The carbon and nitrogen atoms in the rings of the base and the sugar are numbered separately (see Fig. 22.3B). Carbons in the pentose are numbered 1' to 5'. Thus, when the 5'-carbon of a nucleoside (or nucleotide) is referred to, a carbon atom in the pentose, rather than an atom in the base, is being specified.

C. Nucleotides

The addition of one or more phosphate groups to a nucleoside produces a nucleotide. The first phosphate group is attached by an ester linkage to the 5'-OH of the pentose, forming a nucleoside 5'-phosphate or a 5'-nucleotide. The type of pentose is denoted by the prefix in the names 5'-ribonucleotide and 5'-deoxyribonucleotide. If one phosphate group is attached to the 5'-carbon of the pentose, the structure is a nucleoside monophosphate, like adenosine monophosphate ([AMP] or adenylate). If a second or third phosphate is added to the nucleoside, a nucleoside diphosphate (e.g., adenosine diphosphate [ADP]) or triphosphate (e.g., adenosine triphosphate [ATP]) results (Fig. 22.4). The second and third phosphates are each connected in tandem to the nucleotide by a "high-energy bond" (a bond with a large, negative change in free energy [$-\Delta G$; see Chapter 6, III. A.] of hydrolysis). The phosphate groups are responsible for the negative charges associated with nucleotides and cause DNA and RNA to be referred to as nucleic acids.

III. PURINE NUCLEOTIDE SYNTHESIS

The atoms of the purine ring are contributed by a number of compounds, including amino acids (aspartate, glycine, and glutamine), carbon dioxide (CO_2), and N^{10}-formyltetrahydrofolate (N^{10}-formyl-THF), as shown in Figure 22.5. The purine ring is constructed primarily in the liver by a series of reactions that add the donated carbons and nitrogens to a preformed ribose 5-phosphate. Synthesis of ribose 5-phosphate from glucose 6-phosphate by the pentose phosphate pathway is discussed in Chapter 13, III.

A. 5-Phosphoribosyl-1-pyrophosphate synthesis

5-Phosphoribosyl-1-pyrophosphate (PRPP) is an activated pentose that participates in the synthesis and salvage of purines and pyrimidines. Synthesis of PRPP from ATP and ribose 5-phosphate is catalyzed by PRPP synthetase (Fig. 22.6). This X-linked enzyme is activated by inorganic phosphate and inhibited by purine nucleotides (end-product inhibition). Because the sugar moiety of PRPP is ribose, ribonucleotides are the end products of *de novo* purine synthesis. When deoxyribonucleotides are required for DNA synthesis, the ribose sugar moiety is reduced (see Section IV.).

B. 5-Phosphoribosylamine synthesis

Synthesis of 5-phosphoribosylamine from PRPP and glutamine is the first step of purine *de novo* synthesis (Fig. 22.7). The amide group of glutamine replaces the pyrophosphate group attached to carbon 1 of PRPP. This is the committed step in purine nucleotide biosynthesis. The enzyme that catalyzes the reaction, glutamine:phosphoribosylpyrophosphate amidotransferase (GPAT), is inhibited by the purine 5′-nucleotides AMP and guanosine monophosphate ([GMP] or guanylate), the end products of the pathway. The rate of the reaction is also controlled by the intracellular concentration of PRPP. The concentration of PRPP is normally far below the Michaelis constant (K_m) for the GPAT. Therefore, any small change in the PRPP concentration causes a proportional change in rate of the reaction (see Chapter 5, VI.).

C. Inosine monophosphate synthesis

The next nine steps in purine nucleotide biosynthesis lead to the synthesis of inosine monophosphate (IMP), whose base is hypoxanthine (see Fig. 22.7). IMP is the precursor purine nucleotide for the synthesis of AMP and GMP. Four steps in this pathway require ATP as an energy source, and two steps in the pathway require N^{10}-formyl-THF as a one-carbon donor (see Chapter 20, IV.).

D. Synthetic inhibitors

Certain synthetic inhibitors of purine synthesis are analogs of para-aminobenzoic acid (e.g., sulfonamides), which inhibit bacterial synthesis of folic acid and serve as antimicrobial agents (see Fig. 22.7). Since humans cannot synthesize folic acid and must take in folic acid

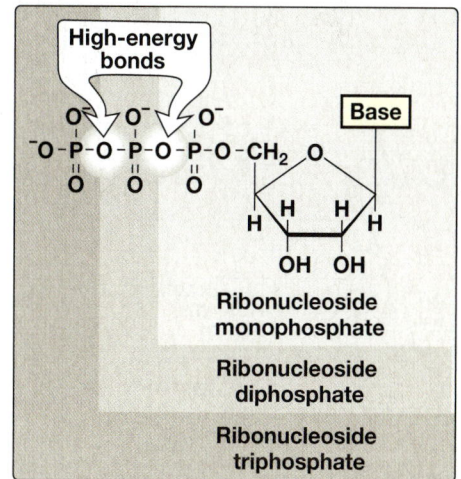

Figure 22.4
Ribonucleoside monophosphate, diphosphate, and triphosphate.

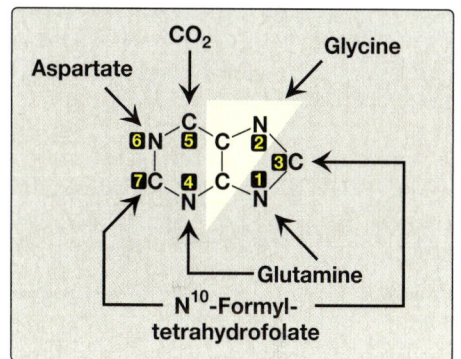

Figure 22.5
Sources of the individual atoms in the purine ring. The order in which the atoms are added is shown by the numbers in the black boxes (see Fig. 22.7). CO_2, carbon dioxide.

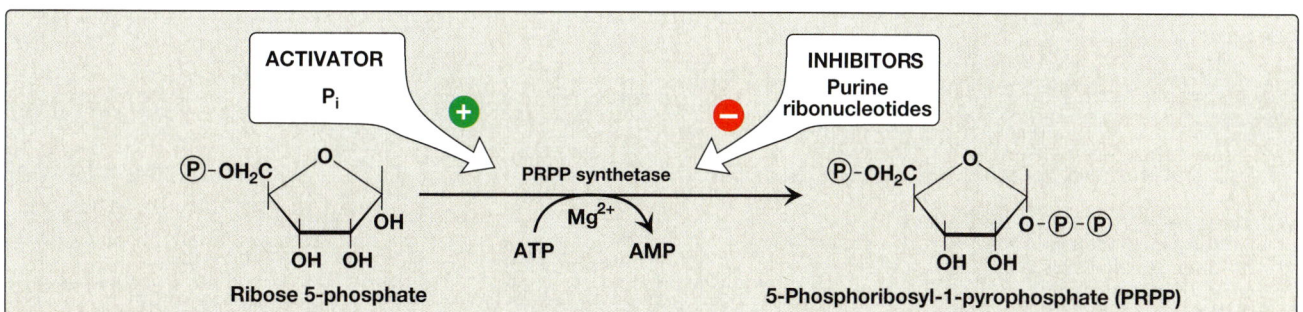

Figure 22.6
Synthesis of 5-phosphoribosyl-1-pyrophosphate (PRPP), showing the activator and inhibitors of the reaction. [Note: This is not the committed step of purine synthesis because PRPP is used in other pathways such as salvage.] Ⓟ, phosphate; P_i, inorganic phosphate; AMP, adenosine monophosphate; ATP, adenosine triphosphate; Mg, magnesium.

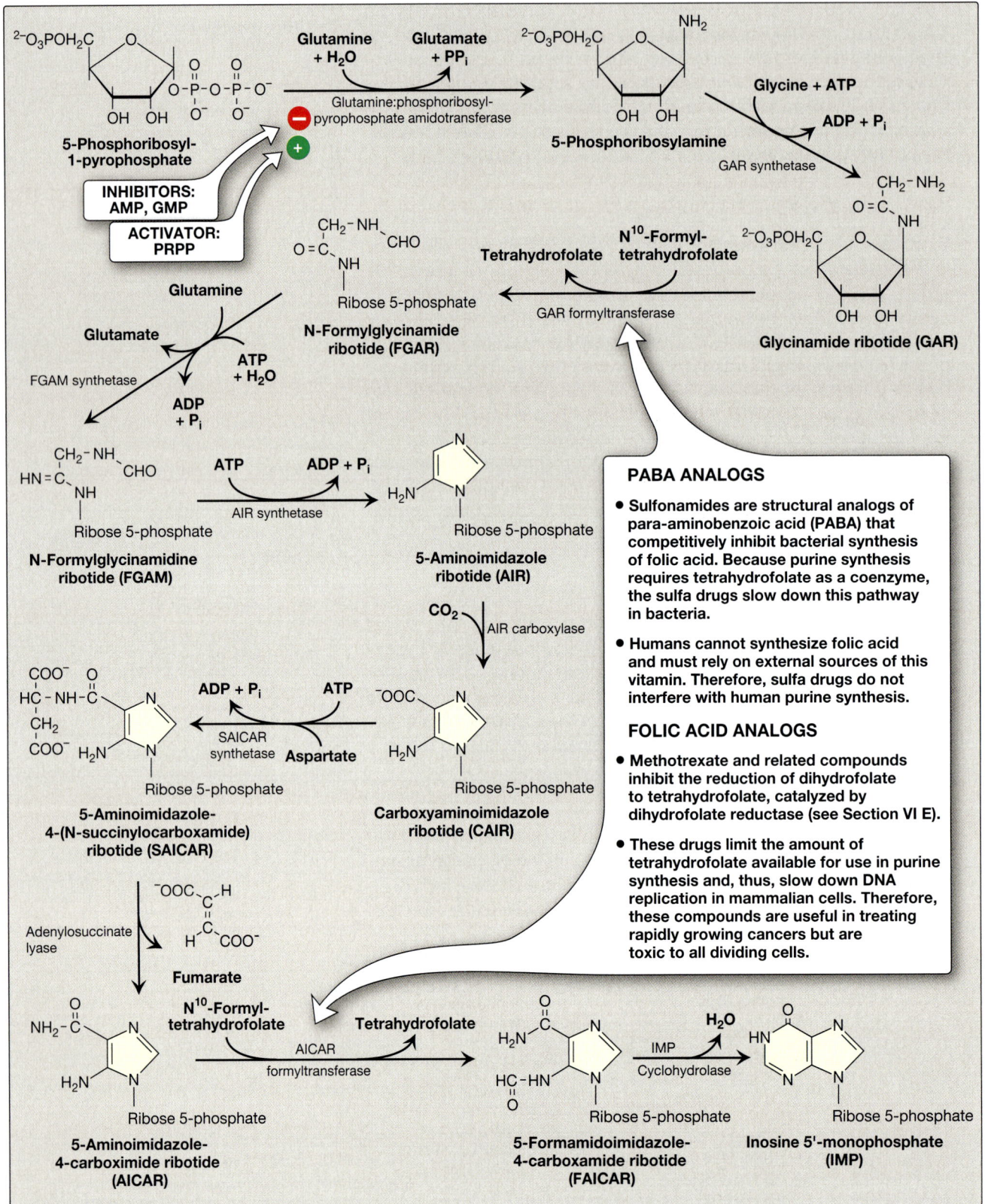

PABA ANALOGS

- Sulfonamides are structural analogs of para-aminobenzoic acid (PABA) that competitively inhibit bacterial synthesis of folic acid. Because purine synthesis requires tetrahydrofolate as a coenzyme, the sulfa drugs slow down this pathway in bacteria.

- Humans cannot synthesize folic acid and must rely on external sources of this vitamin. Therefore, sulfa drugs do not interfere with human purine synthesis.

FOLIC ACID ANALOGS

- Methotrexate and related compounds inhibit the reduction of dihydrofolate to tetrahydrofolate, catalyzed by dihydrofolate reductase (see Section VI E).

- These drugs limit the amount of tetrahydrofolate available for use in purine synthesis and, thus, slow down DNA replication in mammalian cells. Therefore, these compounds are useful in treating rapidly growing cancers but are toxic to all dividing cells.

Figure 22.7

De novo synthesis of purine nucleotides, showing the inhibitory effect of some structural analogs. AMP and ADP, adenosine mono- and diphosphates; GMP, guanosine monophosphate; PRPP, 5-phosphoribosyl-1-pyrophosphate; P_i, inorganic phosphate; PP_i, pyrophosphate; CO_2, carbon dioxide.

from our diet, sulfa drugs do not interfere with purine synthesis in our cells. Other purine synthesis inhibitors are structural analogs of folic acid (e.g., methotrexate and trimethoprim), which inhibit the reduction of dihydrofolate (DHF) to tetrahydrofolate by DHF reductase (see Fig. 22.7). As such, these drugs limit the active form of folic acid in reactions in which tetrahydrofolate is used as a cofactor. Methotrexate is used pharmacologically to control the spread of cancer by interfering with the synthesis of nucleotides and, therefore, of DNA and RNA. In bacteria, trimethoprim functions similarly as an antimicrobial agent.

> Inhibitors of human purine synthesis are extremely toxic to tissues, especially to developing structures such as those in a fetus, or to cell types that normally replicate rapidly, including those of bone marrow, skin, GI tract, and the immune system and hair follicles. As a result, individuals taking such anticancer drugs can experience adverse effects, including anemia, scaly skin, GI tract disturbance, immunodeficiency, and hair loss.

E. Adenosine and guanosine monophosphate synthesis

The conversion of IMP to either AMP or GMP uses a two-step, energy- and nitrogen-requiring pathway (Fig. 22.8). AMP synthesis requires

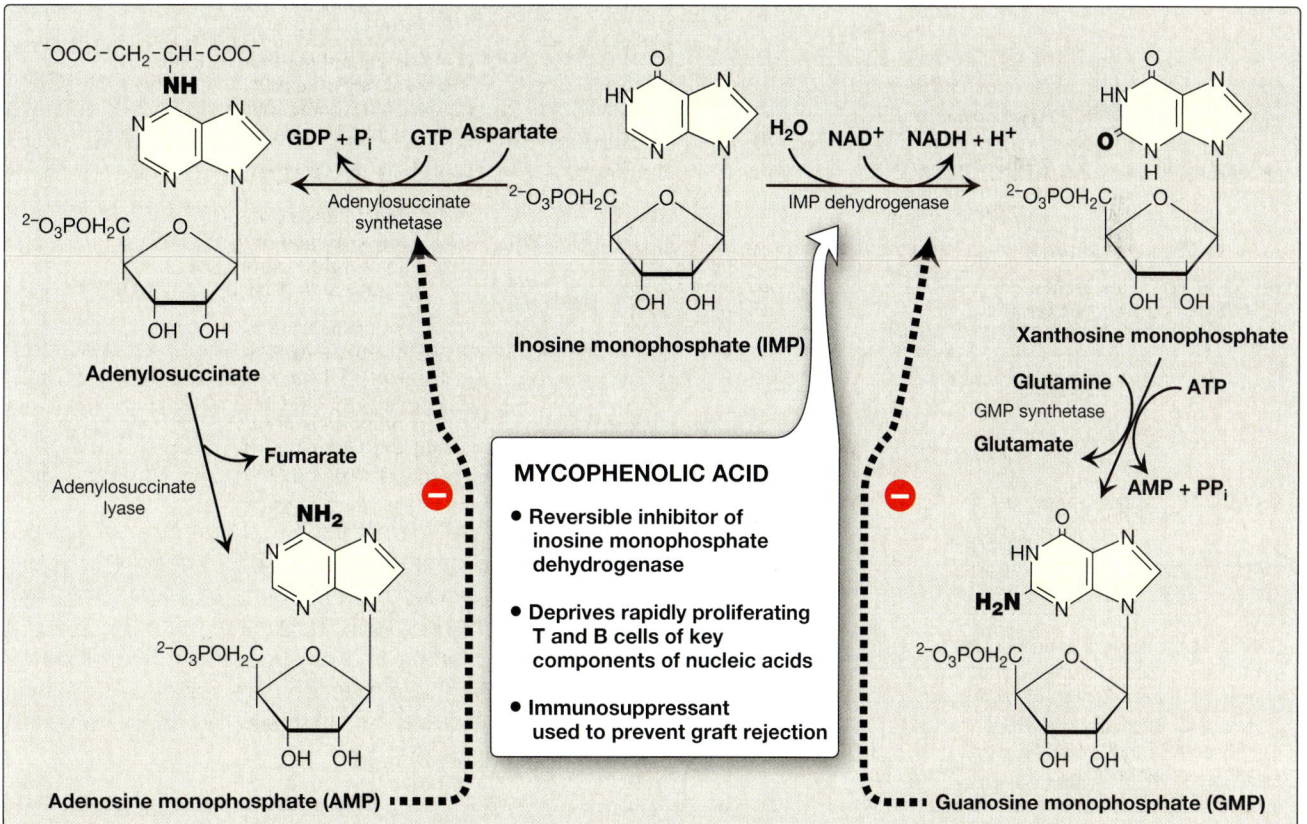

Figure 22.8
Conversion of IMP to AMP (or, adenylate) and GMP (or, guanylate) showing feedback inhibition. NAD(H), nicotinamide adenine dinucleotide; GDP and GTP, guanosine di- and triphosphates; P_i, inorganic phosphate; PP_i, pyrophosphate.

guanosine triphosphate (GTP) as an energy source and aspartate as a nitrogen source, whereas GMP synthesis requires ATP and glutamine. The first reaction in each pathway is inhibited by the end product of that pathway. This provides a mechanism for diverting IMP to the synthesis of the purine present in lesser amounts. If both AMP and GMP are present in adequate amounts, the *de novo* pathway of purine nucleotide synthesis is inhibited at the GPAT step.

Mycophenolic acid is a reversible inhibitor of IMP dehydrogenase, the first enzyme used to generate GMP from IMP. Proliferating T and B lymphocytes are highly susceptible to low levels of this key purine nucleotide, so mycophenolic acid is an effective immunosuppressant agent to prevent organ transplant rejection (kidney, heart, and liver) as well as to treat certain immune disorders such as systemic lupus erythematosus and Crohn disease.

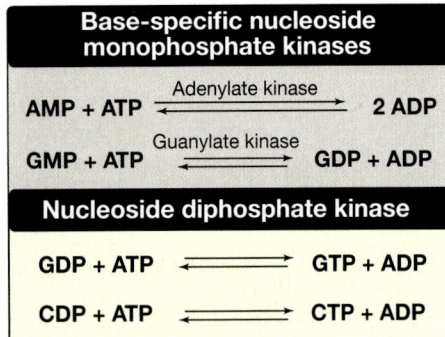

Figure 22.9
Conversion of nucleoside monophosphates to di- and triphosphates. AMP, ADP, and ATP, adenosine mono-, di-, and triphosphates; GMP, GDP, and GTP, guanosine mono-, di-, and triphosphates; CDP and CTP, cytidine di- and triphosphates.

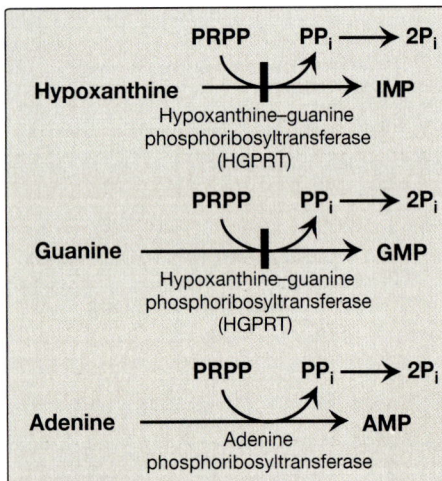

Figure 22.10
Salvage pathways of purine nucleotide synthesis. [Note: Virtually complete deficiency of hypoxanthine-guanine phosphoribosyltransferase (HGPRT) results in Lesch–Nyhan syndrome. Partial deficiencies of HGPRT are known. As the amount of functional enzyme increases, the severity of the symptoms decreases.] IMP, GMP, and AMP, inosine, guanosine, and adenosine monophosphates; PRPP, 5-phosphoribosyl-1-pyrophosphate; PP$_i$, pyrophosphate

F. Nucleoside di- and triphosphate synthesis

Nucleoside diphosphates are synthesized from the corresponding nucleoside monophosphates by base-specific nucleoside monophosphate kinases (Fig. 22.9). These kinases do not discriminate between ribose and deoxyribose in the substrate. ATP is generally the source of the transferred phosphate because it is present in higher concentrations than the other nucleoside triphosphates. Adenylate kinase is particularly active in the liver and in muscle, where the turnover of energy from ATP is high. Its function is to maintain equilibrium among the adenine nucleotides (AMP, ADP, and ATP). Nucleoside diphosphates and triphosphates are interconverted by nucleoside diphosphate kinase, an enzyme that, unlike the monophosphate kinases, has broad substrate specificity.

G. Purine salvage pathway

Purines that result from the normal turnover of cellular nucleic acids, or the small amount that is obtained from the diet and not degraded, can be converted to nucleoside triphosphates and used by the body. This is referred to as the salvage pathway for purines. Salvage is particularly important in the brain.

1. **Purine base salvage to nucleotides:** Two enzymes are involved: adenine phosphoribosyltransferase (APRT) and X-linked hypoxanthine–guanine phosphoribosyltransferase (HGPRT). Both use PRPP as the source of the ribose 5-phosphate group (Fig. 22.10). The release of pyrophosphate and its subsequent hydrolysis by pyrophosphatase makes these reactions irreversible. Adenosine is the only purine nucleoside to be salvaged. It is phosphorylated to AMP by adenosine kinase.

2. **Lesch–Nyhan syndrome:** This is a rare, X-linked recessive disorder associated with a virtually complete deficiency of HGPRT. The deficiency results in an inability to salvage hypoxanthine or G. The lack of this salvage pathway causes increased PRPP levels and decreased IMP and GMP levels. As a result, GPAT

(the regulated step in purine synthesis) has excess substrate and decreased inhibitors available, and *de novo* purine synthesis is increased. The combination of decreased purine reutilization and increased purine synthesis results in increased degradation of purines and the production of large amounts of uric acid (see Section V.), making HGPRT deficiency an inherited cause of hyperuricemia. In patients with Lesch–Nyhan syndrome, the hyperuricemia frequently results in the formation of uric acid stones in the kidneys (urolithiasis) and the deposition of urate crystals in the joints (gouty arthritis) and soft tissues. In addition, the syndrome is characterized by motor dysfunction, cognitive deficits, and behavioral disturbances that include self-mutilation (e.g., biting of lips and fingers; Fig. 22.11).

IV. DEOXYRIBONUCLEOTIDE SYNTHESIS

The nucleotides described thus far all contain ribose (ribonucleotides). DNA synthesis, however, requires 2′-deoxyribonucleotides, which are produced from ribonucleoside diphosphates by the enzyme ribonucleotide reductase during the S-phase of the cell cycle (see Chapter 30, IV.). The same enzyme acts on pyrimidine ribonucleotides.

A. Ribonucleotide reductase

Ribonucleotide reductase (ribonucleoside diphosphate reductase) is a dimer composed of two nonidentical subunits, R1 (or, α) and the smaller R2 (or, β), and is specific for the reduction of purine nucleoside diphosphates (ADP and GDP) and pyrimidine nucleoside diphosphates (CDP and UDP) to their deoxy forms (dADP, dGDP, dCDP, and dUDP; Fig. 22.12). The immediate donors of the hydrogen atoms needed for the reduction of the 2′-hydroxyl group are two sulfhydryl (–SH) groups on the enzyme itself (R1 subunit), which form a disulfide bond during the reaction (see Chapter 2, IV. B.). R2 contains the stable tyrosyl radical required for catalysis at R1.

1. **Reduced enzyme regeneration:** In order for ribonucleotide reductase to continue to produce deoxyribonucleotides at R1, the disulfide bond created during the production of the 2′-deoxy carbon must be reduced. The source of the reducing equivalents is thioredoxin, a protein coenzyme of ribonucleotide reductase. Thioredoxin contains two cysteine residues separated by two amino acids in the peptide chain. The two –SH groups of thioredoxin donate their hydrogen atoms to ribonucleotide reductase, forming a disulfide bond in the process (see Fig. 22.12).

2. **Reduced thioredoxin regeneration:** Thioredoxin must be converted back to its reduced form in order to continue performing its function. The reducing equivalents are provided by NADPH + H$^+$, and the reaction is catalyzed by thioredoxin reductase, a selenoprotein (see Chapter 20, V. D.).

B. Deoxyribonucleotide synthesis regulation

Ribonucleotide reductase is responsible for maintaining a balanced supply of the deoxyribonucleotides required for DNA synthesis.

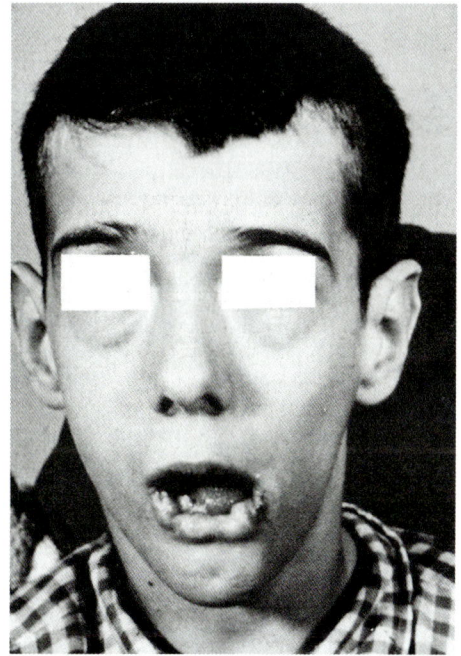

Figure 22.11
Lesions on the lips of a patient with Lesch–Nyhan syndrome.

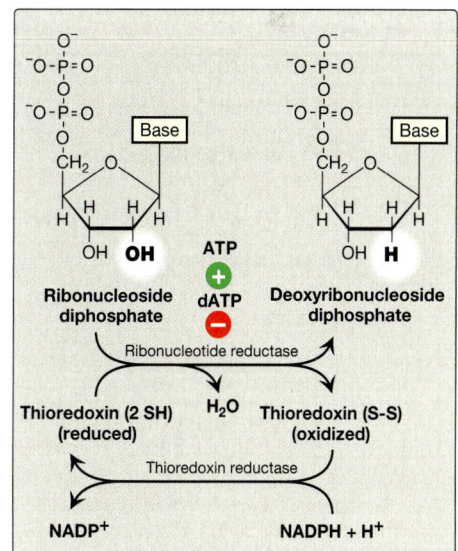

Figure 22.12
Conversion of ribonucleotides to deoxyribonucleotides. NADPH, nicotinamide adenine dinucleotide phosphate; dATP. deoxyadenosine triphosphate.

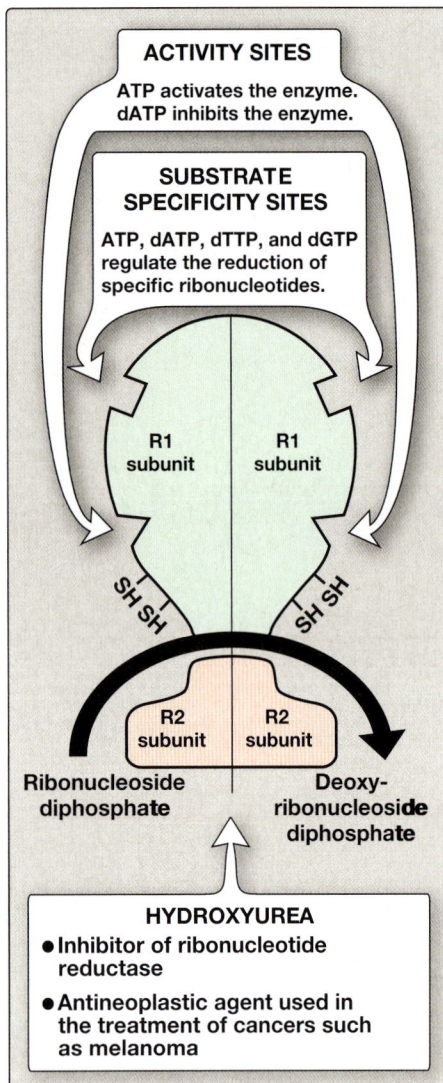

Figure 22.13
Regulation of ribonucleotide reductase. [Note: The R1 subunit is also referred to as α and the R2 as β.] dATP, dTTP, and dGTP, deoxyadenosine, deoxythymidine, and deoxyguanosine triphosphates.

Clinical Application 22.1: Hydroxyurea

The drug hydroxyurea (hydroxycarbamide) inhibits ribonucleotide reductase, thereby inhibiting the generation of substrates for DNA synthesis. The drug is an antineoplastic agent and is used in the treatment of cancers such as melanoma. Hydroxyurea is also used in the treatment of sickle cell anemia (see Chapter 3, IV. A.). However, the increase in fetal hemoglobin seen with hydroxyurea treatment is because of changes in gene expression and not due to ribonucleotide reductase inhibition.

Consequently, the regulation of the enzyme is complex. In addition to the catalytic site, R1 contains two distinct allosteric sites involved in regulating enzymic activity (Fig. 22.13).

1. **Activity sites:** The binding of deoxyadenosine triphosphate (dATP) to allosteric sites (known as activity sites) on R1 inhibits the overall catalytic activity of the enzyme and, therefore, prevents the reduction of any of the four nucleoside diphosphates. This effectively prevents DNA synthesis and explains the toxicity of increased levels of dATP seen in conditions such as adenosine deaminase (ADA) deficiency (see Section V. C.). In contrast, ATP bound to these sites activates the enzyme.

2. **Substrate specificity sites:** The binding of nucleoside triphosphates to additional allosteric sites (known as substrate specificity sites) on R1 regulates substrate specificity, causing an increase in the conversion of different species of ribonucleotides to deoxyribonucleotides as they are required for DNA synthesis. For example, deoxythymidine triphosphate binding at the specificity site causes a conformational change that allows reduction of GDP to dGDP at the catalytic site when ATP is at the activity site.

V. PURINE NUCLEOTIDE DEGRADATION

Degradation of dietary nucleic acids occurs in the small intestine, where pancreatic nucleases hydrolyze them to nucleotides. The nucleotides are sequentially degraded by intestinal enzymes to nucleosides, phosphorylated sugars, and free bases. Uric acid is the end product of intestinal purine degradation. Purine nucleotides from *de novo* synthesis are degraded in the liver primarily. The free bases are sent out from the liver and salvaged by peripheral tissues.

A. Degradation in the small intestine

Ribonucleases and deoxyribonucleases, secreted by the pancreas, hydrolyze dietary RNA and DNA to oligonucleotides that are further hydrolyzed by pancreatic phosphodiesterases, producing a mixture of 3'- and 5'-mononucleotides. At the intestinal mucosal surface, nucleotidases remove the phosphate groups hydrolytically, releasing nucleosides that are taken into enterocytes by sodium-dependent transporters and degraded by nucleosidases (nucleoside phosphorylases) to free bases plus (deoxy)ribose 1-phosphate. Dietary purine bases are not used to any appreciable extent for the synthesis of tissue nucleic acids. Instead, they are degraded to uric acid in the enterocytes. Most of the uric acid enters the blood and is eventually excreted in the urine. A summary of this pathway is shown in Figure 22.14. Mammals other than primates express urate oxidase (uricase), which cleaves the purine ring, generating allantoin. Modified recombinant urate oxidase is now used clinically to lower urate levels.

B. Uric acid formation

A summary of the steps in the production of uric acid and the genetic diseases associated with deficiencies of specific degradative enzymes are shown in Figure 22.15. The bracketed numbers refer to specific reactions in the figure.

1. An amino group is removed from AMP to produce IMP by AMP (adenylate) deaminase or from adenosine to produce inosine (hypoxanthine-ribose) by ADA.

2. IMP and GMP are converted into their respective nucleoside forms, inosine and guanosine, by the action of 5′-nucleotidase.

3. Purine nucleoside phosphorylase converts inosine and guanosine into ribose 1-phosphate and the respective purine bases hypoxanthine and G. A mutase interconverts ribose 1- and ribose 5-phosphate.

4. G is deaminated to form xanthine.

5. Hypoxanthine is oxidized by molybdenum-containing xanthine oxidase (XO) to xanthine, which is further oxidized by XO to uric acid, the final product of human purine degradation. Uric acid is excreted primarily in the urine.

C. Diseases associated with purine degradation

Three disease states are associated with the degradation of purine nucleotides: gout, ADA deficiency, and purine nucleoside phosphorylase deficiency (two immunodeficiency disorders).

1. **Gout:** Gout is a disorder initiated by high levels of uric acid (the end product of purine catabolism) in blood (hyperuricemia), as a result of either the overproduction or underexcretion of uric acid (Fig. 22.15). The hyperuricemia can lead to the deposition of monosodium urate (MSU) crystals in the joints and an inflammatory response to the crystals, causing first acute and then progressing to chronic gouty arthritis. Nodular masses of MSU crystals (tophi) may be deposited in the soft tissues, resulting in chronic tophaceous gout (Fig. 22.16). Formation of uric acid stones in the kidney (urolithiasis) may also be seen. Hyperuricemia is not sufficient to cause gout, but gout is always preceded by hyperuricemia. Hyperuricemia is typically asymptomatic but may be indicative of comorbid conditions such as hypertension. The definitive diagnosis of gout requires aspiration and examination of synovial fluid (Fig. 22.17) from an affected joint (or material from a tophus) using polarized light microscopy to confirm the presence of needle-shaped MSU crystals (Fig. 22.18).

   a. **Uric acid underexcretion:** In more than 90% of individuals with hyperuricemia, the cause is underexcretion of uric acid. Underexcretion can be primary, because of as-yet-unidentified

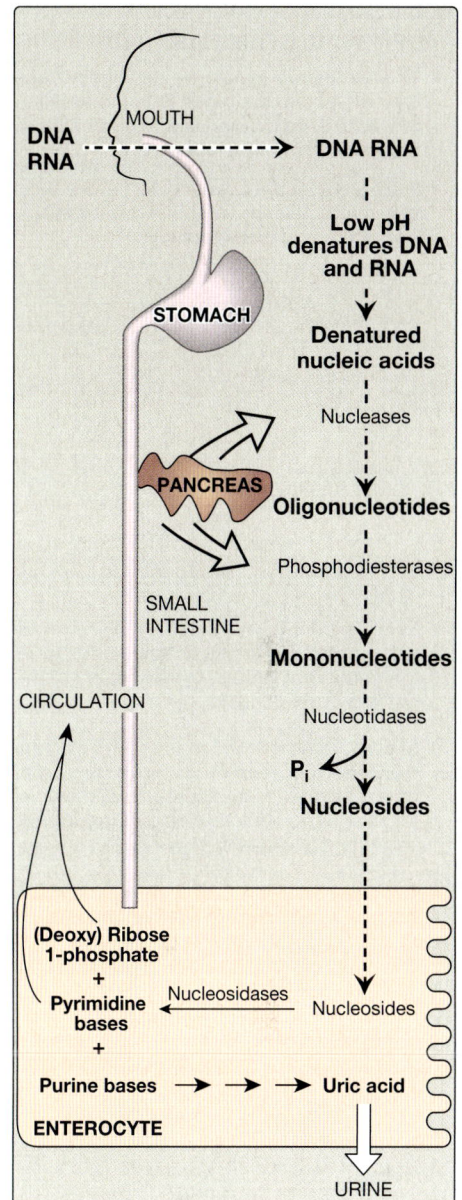

Figure 22.14
Digestion of dietary nucleic acids. P_i, inorganic phosphate.

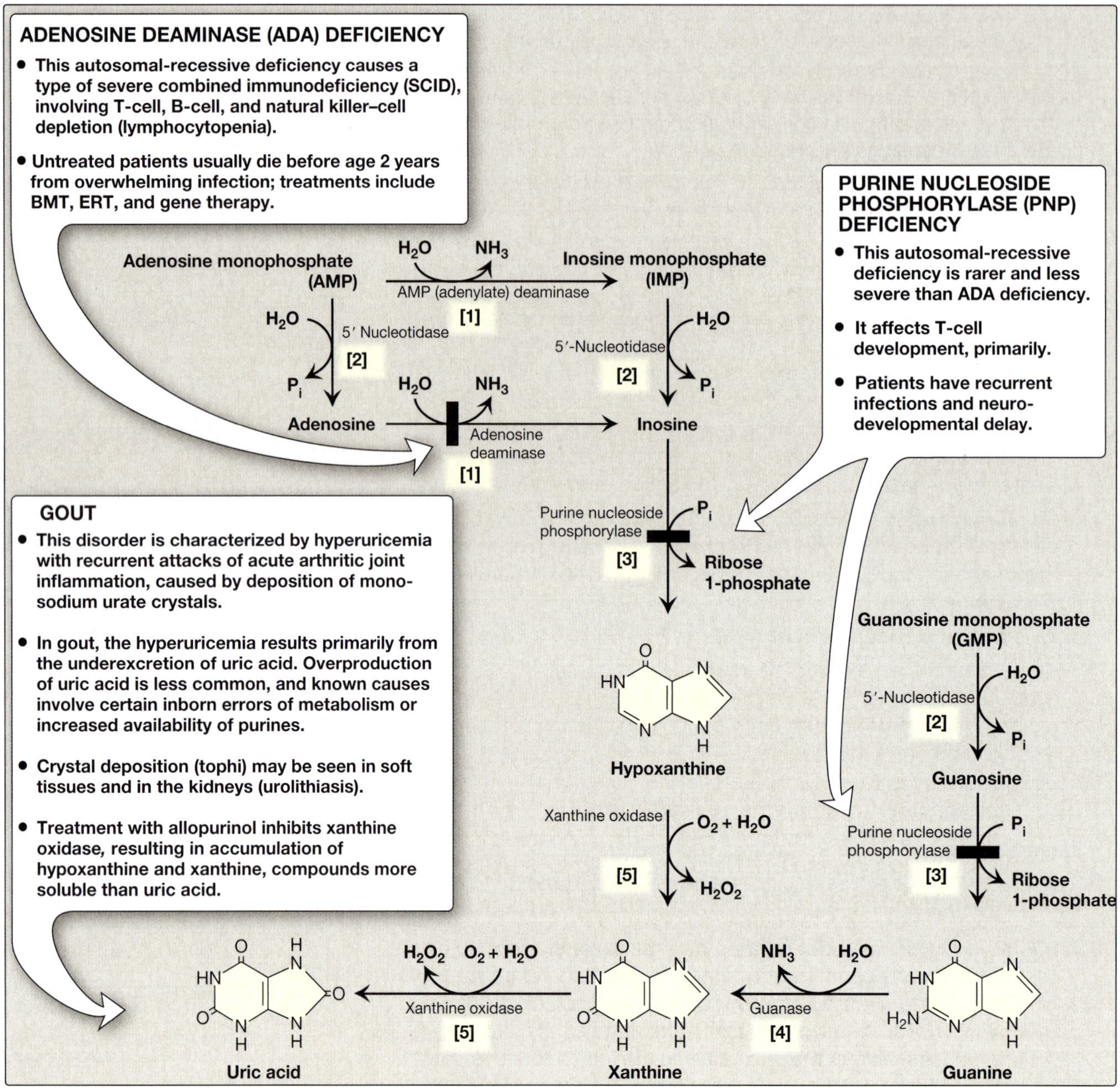

ADENOSINE DEAMINASE (ADA) DEFICIENCY
- This autosomal-recessive deficiency causes a type of severe combined immunodeficiency (SCID), involving T-cell, B-cell, and natural killer–cell depletion (lymphocytopenia).
- Untreated patients usually die before age 2 years from overwhelming infection; treatments include BMT, ERT, and gene therapy.

PURINE NUCLEOSIDE PHOSPHORYLASE (PNP) DEFICIENCY
- This autosomal-recessive deficiency is rarer and less severe than ADA deficiency.
- It affects T-cell development, primarily.
- Patients have recurrent infections and neuro-developmental delay.

GOUT
- This disorder is characterized by hyperuricemia with recurrent attacks of acute arthritic joint inflammation, caused by deposition of mono-sodium urate crystals.
- In gout, the hyperuricemia results primarily from the underexcretion of uric acid. Overproduction of uric acid is less common, and known causes involve certain inborn errors of metabolism or increased availability of purines.
- Crystal deposition (tophi) may be seen in soft tissues and in the kidneys (urolithiasis).
- Treatment with allopurinol inhibits xanthine oxidase, resulting in accumulation of hypoxanthine and xanthine, compounds more soluble than uric acid.

Figure 22.15
Degradation of purine nucleotides to uric acid, illustrating some of the genetic diseases associated with this pathway. [Note: The numbers in brackets refer to the corresponding numbered citations in the text.] BMT, bone marrow transplantation; ERT, enzyme replacement therapy; P_i, inorganic phosphate; H_2O_2, hydrogen peroxide; NH_3, ammonia.

inherent excretory defects, or secondary to known disease processes that affect how the kidney handles urate (e.g., in lactic acidosis, lactate increases renal urate reabsorption, thereby decreasing its excretion) and to environmental factors such as the use of drugs (e.g., thiazide diuretics) or exposure to lead (saturnine gout).

b. **Uric acid overproduction:** A less common cause of hyperuricemia is from the overproduction of uric acid. Primary hyperuricemia is, for the most part, idiopathic (having no known cause). However, several identified mutations in the gene for X-linked PRPP synthetase result in the enzyme having an increased maximal velocity ($[V_{max}]$ see Chapter 5, V. A.) for the production of PRPP, a lower K_m (see Chapter 5, VI. B.) for ribose 5-phosphate, or a decreased sensitivity to purine nucleotides, its allosteric inhibitors (see Chapter 5, VIII. A.). In each case, increased availability of PRPP increases purine production, followed by an increase in degradation, resulting in elevated levels of plasma uric acid. Lesch–Nyhan syndrome (see Section III. G. and Fig. 22.11) also causes hyperuricemia as a result of the decreased salvage of hypoxanthine and G and the subsequent increased availability of PRPP. Secondary hyperuricemia is typically the consequence of increased availability of purines (e.g., in patients with myeloproliferative disorders or who are undergoing chemotherapy and so have a high rate of cell turnover). Hyperuricemia can also be the result of seemingly unrelated metabolic diseases, such as von Gierke disease (see Fig. 11.8) or hereditary fructose intolerance (see Chapter 12, II. D.).

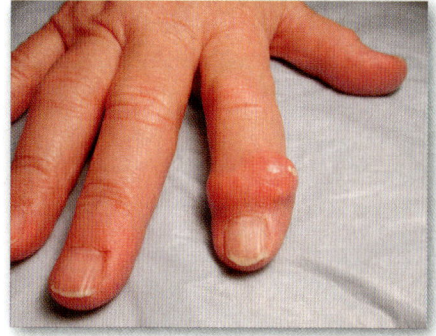

Figure 22.16
Tophaceous gout.

A diet rich in meat, seafood (particularly shellfish), and ethanol is associated with increased risk of gout, whereas a diet rich in low-fat dairy products is associated with a decreased risk.

c. **Treatment:** Acute attacks of gout are treated with anti-inflammatory agents. Colchicine, steroidal drugs such as prednisone, and nonsteroidal drugs such as indomethacin are used. Colchicine prevents formation of microtubules, thereby decreasing the movement of neutrophils into the affected area. Like the other anti-inflammatory drugs, it has no effect on uric acid levels. Long-term therapeutic strategies for gout involve lowering the uric acid level below its saturation point (6.5 mg/dL), thereby preventing the deposition of MSU crystals. Uricosuric agents, such as probenecid or sulfinpyrazone, that increase renal excretion of uric acid, are used in patients who are underexcretors of uric acid. Allopurinol, a structural analog of hypoxanthine, inhibits uric acid synthesis and is used in patients who are overproducers of uric acid. Allopurinol is oxidized to oxypurinol, a long-lived inhibitor of XO. This results in an accumulation of hypoxanthine and xanthine (see Fig. 22.15), compounds more soluble than uric acid and, therefore, less likely to initiate an inflammatory response. In patients with normal levels of HGPRT, the hypoxanthine can be salvaged, reducing the levels of PRPP and, therefore, *de novo* purine synthesis. Febuxostat, a nonpurine inhibitor of XO, is also available. Uric acid levels in the blood normally are close to the saturation point. One reason for this may be the strong antioxidant effects of uric acid.

2. **Adenosine deaminase deficiency:** ADA is expressed in a variety of tissues, but, in humans, lymphocytes have the highest activity

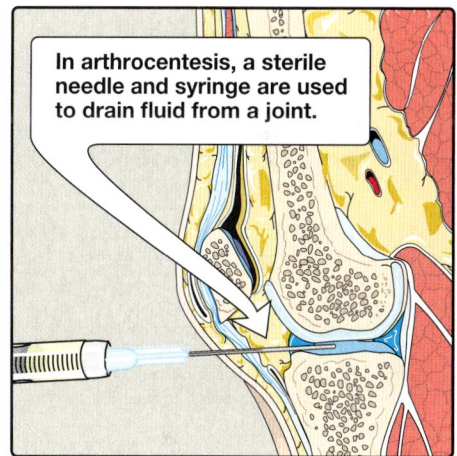

Figure 22.17
Analysis of joint fluid can help to define causes of joint swelling and arthritis, such as infection, gout, and rheumatoid disease.

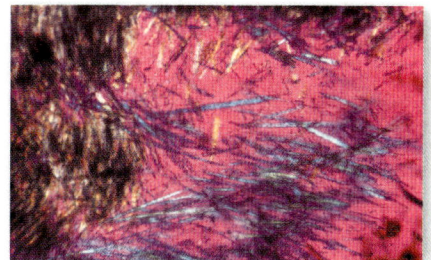

Figure 22.18
Gout can be diagnosed by the presence of negatively birefringent monosodium urate crystals in aspirated synovial fluid examined by polarized light microscopy. Here, crystals are seen within polymorphonuclear leukocytes.

of this cytoplasmic enzyme. A deficiency of ADA results in an accumulation of adenosine, which is converted to its ribonucleotide or deoxyribonucleotide forms by cellular kinases. As dATP levels rise, ribonucleotide reductase is inhibited, thereby preventing the production of all deoxyribose-containing nucleotides (see Section IV. B.). Consequently, cells cannot make DNA and divide. The dATP and adenosine that accumulate in ADA deficiency lead to developmental arrest and apoptosis of lymphocytes. In its most severe form, this autosomal-recessive disorder causes a type of severe combined immunodeficiency disease (SCID), involving a decrease in T cells, B cells, and natural killer cells. ADA deficiency accounts for ~14% of cases of SCID in the United States. Treatments include bone marrow transplantation, enzyme replacement therapy, and gene therapy (see Chapter 34, IX.). Without appropriate treatment, children with this disorder usually die from infection by age 2 years. Purine nucleoside phosphorylase deficiency results in a less severe immunodeficiency primarily involving T cells.

VI. PYRIMIDINE SYNTHESIS AND DEGRADATION

Unlike the synthesis of the purine ring, which is constructed on a preexisting ribose 5-phosphate, the pyrimidine ring is fully synthesized before being attached to ribose 5-phosphate, which is donated by PRPP. The sources of the atoms in the pyrimidine ring are glutamine, CO_2, and aspartate (Fig. 22.19).

A. Carbamoyl phosphate synthesis

The regulated step of this pathway in mammalian cells is the synthesis of carbamoyl phosphate from glutamine and CO_2, catalyzed by carbamoyl phosphate synthetase (CPS) II. CPS II is inhibited by uridine triphosphate ([UTP] the end product of this pathway, which can be converted into the other pyrimidine nucleotides) and is activated by PRPP. Carbamoyl phosphate, synthesized by CPS I, is also a precursor of urea (see Chapter 19, V. A.). Defects in ornithine transcarbamylase (OTC) of the urea cycle promote pyrimidine synthesis because of the increased availability of carbamoyl phosphate. A comparison of CPS I and CPS II enzymes is presented in Figure 22.20.

B. Orotic acid synthesis

The second step in pyrimidine synthesis is the formation of carbamoylaspartate, catalyzed by aspartate transcarbamoylase. The pyrimidine ring is then closed by dihydroorotase. The resulting dihydroorotate is oxidized to produce orotic acid (orotate), as shown in Figure 22.21. Flavin mononucleotide (FMN) is reduced in this reaction.

C. Pyrimidine nucleotide synthesis

The completed pyrimidine ring is converted to the nucleotide orotidine monophosphate (OMP) in the second stage of pyrimidine nucleotide synthesis (see Fig. 22.21). As seen with the purines, PRPP is

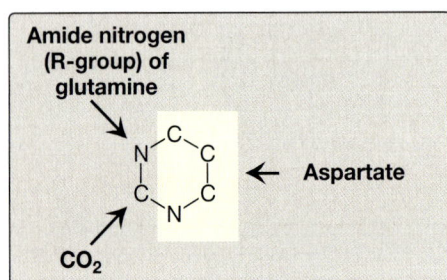

Figure 22.19
Sources of the individual atoms in the pyrimidine ring. CO_2, carbon dioxide.

Variable	CPS I	CPS II
Cellular location	Mitochondria	Cytosol
Pathway involved	Urea cycle	Pyrimidine synthesis
Source of nitrogen	Ammonia	γ-Amide group of glutamine
Regulators	Activator: N-acetyl-glutamate	Activator: PRPP \n Inhibitor: UTP

Figure 22.20
Summary of the differences between carbamoyl phosphate synthetase (CPS) I and II. PRPP, 5-phosphoribosyl-1-pyrophosphate; UTP, uridine triphosphate.

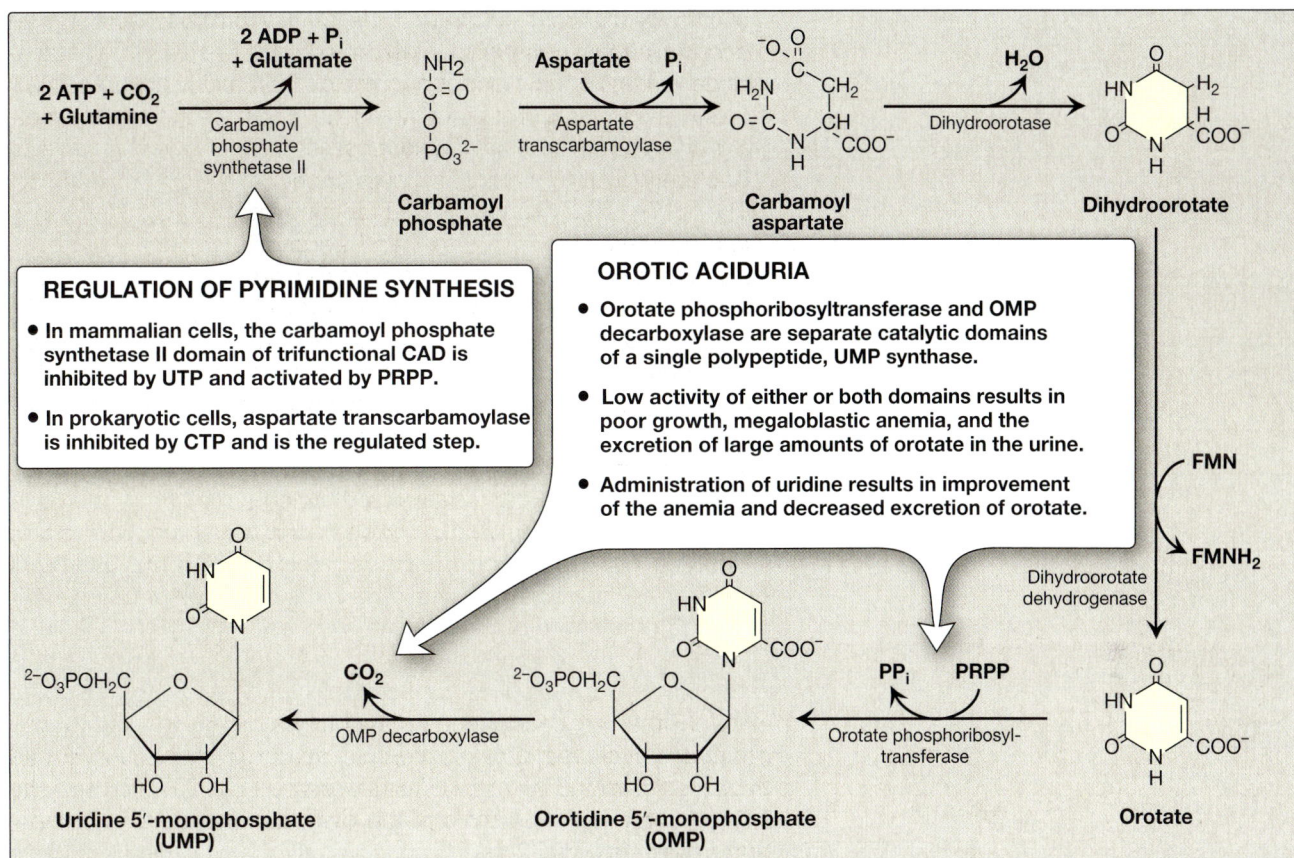

Figure 22.21
De novo pyrimidine synthesis. ADP, adenosine diphosphate; P_i, inorganic phosphate; FMN(H_2), flavin mononucleotide; CTP, cytidine triphosphate; PRPP, 5-phosphoribosyl-1-pyrophosphate; PP_i, pyrophosphate.

the ribose 5-phosphate donor. The enzyme orotate phosphoribos-yltransferase produces OMP and releases pyrophosphate, thereby making the reaction biologically irreversible. OMP (orotidylate) is decarboxylated to uridine monophosphate (UMP) by orotidylate decarboxylase. The phosphoribosyltransferase and decarboxy-lase activities are separate catalytic domains of a single polypep-tide called UMP synthase. Hereditary orotic aciduria (a very rare disorder) may be caused by a deficiency of one or both activities of this bifunctional enzyme, resulting in orotic acid in the urine (see Fig. 22.21). Since the first reaction of pyrimidine biosynthesis is feed-back inhibited by UTP, hereditary orotic aciduria and its associated megaloblastic anemia is treated with uridine. Recall that a deficiency of OTC in the urea cycle would present with elevated urinary levels of orotate (see Chapter 19, VI. C.). This is because the carbamoyl phos-phate substrate of OTC is funneled instead into pyrimidine synthesis after CPS II. Uridine would not be effective in treating OTC defi-ciency. UMP is sequentially phosphorylated to UDP and UTP. The UDP is a substrate for ribonucleotide reductase, which generates dUDP. The dUDP is phosphorylated to dUTP, which is rapidly hydro-lyzed to deoxyuridine monophosphate (dUMP) by UTP diphospha-tase (dUTPase). Thus, dUTPase plays an important role in reducing

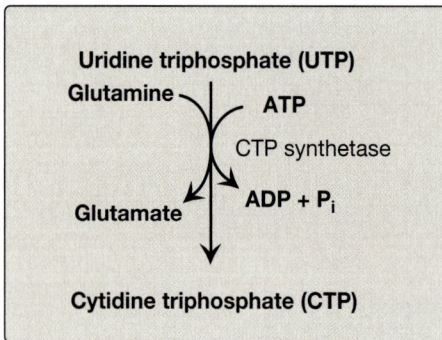

Figure 22.22
Synthesis of cytidine triphosphate (CTP) from UTP. [Note: CTP, required for RNA synthesis, is converted to dCTP for DNA synthesis.] ADP, adenosine diphosphate; P_i, inorganic phosphate.

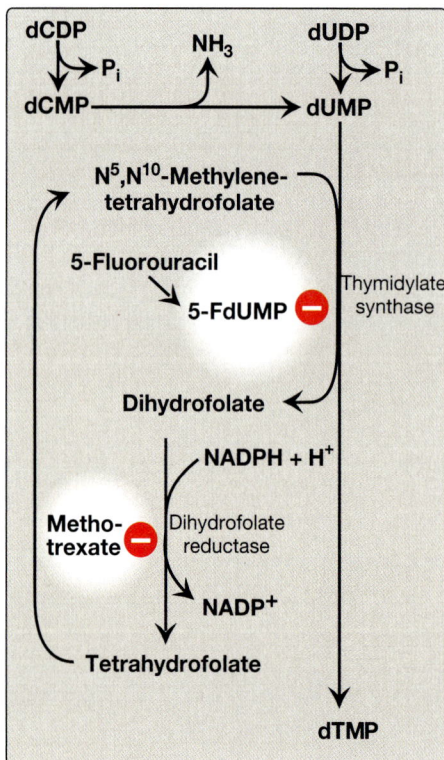

Figure 22.23
Synthesis of dTMP from dUMP, illustrating sites of action of antineoplastic drugs.

availability of dUTP as a substrate for DNA synthesis, thereby preventing erroneous incorporation of U into DNA. The dUMP is used to make deoxythymidine monophosphate ([dTMP] see Section VI. E.).

D. Cytidine triphosphate synthesis

Cytidine triphosphate (CTP) is produced by amination of UTP by CTP synthetase (Fig. 22.22), with glutamine providing the nitrogen. Some of this CTP is dephosphorylated to CDP, which is a substrate for ribonucleotide reductase. The dCDP product can be phosphorylated to dCTP for DNA synthesis or dephosphorylated to dCMP that is deaminated to dUMP.

E. Deoxythymidine monophosphate synthesis

dUMP is converted to dTMP by thymidylate synthase, which uses N^5,N^{10}-methylene-THF as the source of the methyl group (see Chapter 20, IV. and Fig. 20.12). This is an unusual reaction in that THF contributes not only a one-carbon unit but also two hydrogen atoms from the pteridine ring, resulting in the oxidation of THF to DHF (Fig. 22.23). Inhibitors of thymidylate synthase include the T analog 5-fluorouracil, which serve as an antitumor agent. 5-Fluorouracil is metabolically converted to 5-fluorodeoxyuridine monophosphate (5-FdUMP), which becomes permanently bound to the inactivated thymidylate synthase, making the drug a suicide inhibitor. DHF can be reduced back to THF by DHF reductase (see Fig. 28.2), an enzyme that is inhibited by folate analogs such as methotrexate. By decreasing the supply of THF, these drugs not only inhibit purine synthesis (see Fig. 22.7), but, by preventing methylation of dUMP to dTMP, they also decrease dTMP availability. DNA synthesis is inhibited and cell growth slowed. Thus, these drugs are used to treat cancer.

Acyclovir (a purine analog) and 3′-azido-3′-deoxythymidine ([AZT] a pyrimidine analog) are used to treat infections of herpes simplex virus and human immunodeficiency virus, respectively. Each inhibits the specific viral DNA polymerase.

F. Pyrimidine salvage and degradation

Unlike the purine ring, which is not cleaved in humans and is excreted as poorly soluble uric acid, the pyrimidine ring is opened and degraded to highly soluble products, β-alanine (from the degradation of CMP and UMP) and β-aminoisobutyrate (from TMP degradation), with the production of ammonia and CO_2. Pyrimidine bases can be salvaged to nucleosides, which are phosphorylated to nucleotides. However, their high solubility makes pyrimidine salvage less significant clinically than purine salvage. The salvage of pyrimidine nucleosides is the basis for using uridine in the treatment of hereditary orotic aciduria (see Fig. 22.21).

VII. CHAPTER SUMMARY

- **Nucleotides** are composed of a **nitrogenous base** (**A**, **G**, **C**, **U**, and **T**); a **pentose sugar**; and one, two, or three **phosphate groups** (Fig. 22.24).

- A and G are **purines**, and C, U, and T are **pyrimidines**.

- If the sugar is **ribose**, the nucleotide is a **ribonucleoside phosphate** (e.g., adenosine monophosphate [AMP]), and it can have several functions in the cell, including being a component of **RNA**. If the sugar is **deoxyribose**, the nucleotide is a **deoxyribonucleoside phosphate** (e.g., deoxyAMP) and will be found almost exclusively as a component of **DNA**.

- The **committed step** in **purine synthesis** uses **5-phosphoribosyl-1-pyrophosphate** (**PRPP**) (an activated pentose that provides the **ribose 5-phosphate** for *de novo* purine and pyrimidine synthesis and salvage) and nitrogen from **glutamine** to produce phosphoribosylamine. The enzyme is **glutamine:phosphoribosylpyrophosphate amidotransferase** (**GPAT**) and is inhibited by AMP and GMP (the end products of the pathway) and is activated by PRPP.

- Purine nucleotides can also be produced from preformed purine bases by using **salvage reactions** catalyzed by **adenine phosphoribosyltransferase** (**APRT**) and X-linked **hypoxanthine–guanine phosphoribosyltransferase** (**HGPRT**). A near-total deficiency of HGPRT causes **Lesch–Nyhan syndrome**, a severe, inherited form of hyperuricemia accompanied by compulsive self-mutilation.

- All deoxyribonucleotides are synthesized from ribonucleotides by the enzyme **ribonucleotide reductase**. This enzyme is highly regulated (e.g., it is strongly inhibited by **deoxyadenosine triphosphate (dATP)**, a compound that is overproduced in bone marrow cells in individuals with **adenosine deaminase [ADA] deficiency**). ADA deficiency causes **severe combined immunodeficiency disease (SCID)**.

- The end product of purine degradation is **uric acid**, a compound of low solubility whose overproduction or undersecretion causes **hyperuricemia** that, if accompanied by the deposition of **monosodium urate** (**MSU**) **crystals** in joints and soft tissues and an inflammatory response to those crystals, results in **gout**.

- The first step in **pyrimidine synthesis**, the production of carbamoyl phosphate by **carbamoyl phosphate synthetase** (**CPS**) **II**, is the **regulated** step in this pathway (it is inhibited by **uridine triphosphate [UTP]** and activated by PRPP). The UTP produced by this pathway can be converted to cytidine triphosphate (CTP).

- **Deoxyuridine monophosphate** can be converted to dTMP by **thymidylate synthase**, an enzyme targeted by anticancer drugs such as **5-fluorouracil**.

- The regeneration of **tetrahydrofolate** from dihydrofolate produced in the thymidylate synthase reaction requires **dihydrofolate reductase**, an enzyme targeted by the drug **methotrexate**.

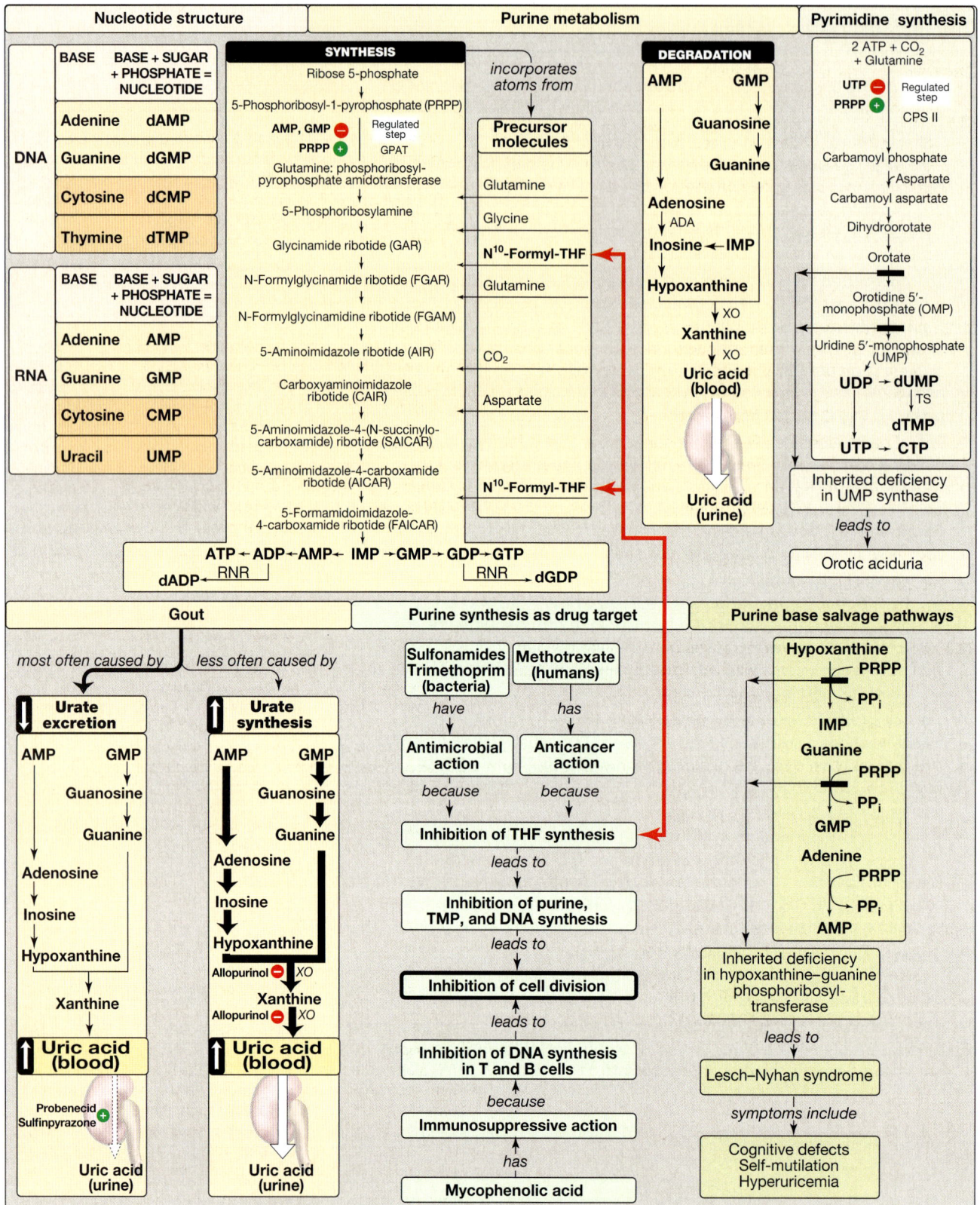

Nucleotide structure

DNA

BASE	BASE + SUGAR + PHOSPHATE = NUCLEOTIDE
Adenine	dAMP
Guanine	dGMP
Cytosine	dCMP
Thymine	dTMP

RNA

BASE	BASE + SUGAR + PHOSPHATE = NUCLEOTIDE
Adenine	AMP
Guanine	GMP
Cytosine	CMP
Uracil	UMP

Purine metabolism

SYNTHESIS

Ribose 5-phosphate
↓
5-Phosphoribosyl-1-pyrophosphate (PRPP)

AMP, GMP ⊖ / PRPP ⊕ — Regulated step — GPAT
↓
Glutamine: phosphoribosyl-pyrophosphate amidotransferase
↓
5-Phosphoribosylamine
↓
Glycinamide ribotide (GAR)
↓
N-Formylglycinamide ribotide (FGAR)
↓
N-Formylglycinamidine ribotide (FGAM)
↓
5-Aminoimidazole ribotide (AIR)
↓
Carboxyaminoimidazole ribotide (CAIR)
↓
5-Aminoimidazole-4-(N-succinylo-carboxamide) ribotide (SAICAR)
↓
5-Aminoimidazole-4-carboxamide ribotide (AICAR)
↓
5-Formamidoimidazole-4-carboxamide ribotide (FAICAR)

incorporates atoms from
↓

Precursor molecules

- Glutamine
- Glycine
- N^{10}-Formyl-THF
- Glutamine
- CO_2
- Aspartate
- N^{10}-Formyl-THF

ATP ← ADP ← AMP ← IMP → GMP → GDP → GTP
dADP ← RNR / RNR → dGDP

DEGRADATION

AMP   GMP
↓   ↓
            Guanosine
              ↓
            Guanine
Adenosine   ↓
↓ ADA
Inosine ← IMP
↓
Hypoxanthine
↓ XO
Xanthine
↓ XO
Uric acid (blood)
↓
Uric acid (urine)

Pyrimidine synthesis

2 ATP + CO_2 + Glutamine
↓
UTP ⊖ / PRPP ⊕ — Regulated step — CPS II
↓
Carbamoyl phosphate
↓ ← Aspartate
Carbamoyl aspartate
↓
Dihydroorotate
↓
Orotate
↓
Orotidine 5'-monophosphate (OMP)
↓
Uridine 5'-monophosphate (UMP)
↓
UDP → dUMP
↓ TS
UTP → CTP   dTMP

Inherited deficiency in UMP synthase
↓ leads to
Orotic aciduria

Gout

most often caused by / less often caused by

↓ Urate excretion

AMP   GMP
↓   ↓
            Guanosine
              ↓
            Guanine
Adenosine
↓
Inosine
↓
Hypoxanthine
↓
Xanthine
↓
↑ Uric acid (blood)

Probenecid Sulfinpyrazone ⊕
↓
Uric acid (urine)

↑ Urate synthesis

AMP   GMP
↓   ↓
            Guanosine
              ↓
            Guanine
Adenosine
↓
Inosine
↓
Hypoxanthine
Allopurinol ⊖ XO
↓
Xanthine
Allopurinol ⊖ XO
↓
↑ Uric acid (blood)
↓
Uric acid (urine)

Purine synthesis as drug target

Sulfonamides Trimethoprim (bacteria)	Methotrexate (humans)
have	has
Antimicrobial action	Anticancer action
because	because

↓
Inhibition of THF synthesis
↓ leads to
Inhibition of purine, TMP, and DNA synthesis
↓ leads to
Inhibition of cell division
↑ leads to
Inhibition of DNA synthesis in T and B cells
↑ because
Immunosuppressive action
↑ has
Mycophenolic acid

Purine base salvage pathways

Hypoxanthine
PRPP → ↓ → PP_i
IMP
Guanine
PRPP → ↓ → PP_i
GMP
Adenine
PRPP → ↓ → PP_i
AMP

Inherited deficiency in hypoxanthine–guanine phosphoribosyl-transferase
↓ leads to
Lesch–Nyhan syndrome
↓ symptoms include
Cognitive defects
Self-mutilation
Hyperuricemia

Figure 22.24

Key concept map for nucleotide metabolism. THF, tetrahydrofolate; GPAT, glutamine:phosphoribosylpyrophosphate amidotransferase; ADA, adenosine deaminase; XO, xanthine oxidase; TS, thymidylate synthase; RNR, ribonucleotide reductase; CPS II, carbamoyl phosphate synthetase II; AMP, GMP, CMP, TMP, and IMP, adenosine, guanosine, cytidine, thymidine, and inosine monophosphates; d, deoxy; PP_i, pyrophosphate; PRPP, 5-phosphoribosyl-1-pyrophosphate.

Study Questions

Choose the ONE best answer.

22.1. Azaserine, a drug with research applications, inhibits glutamine-dependent enzymes. Incorporation of which of the ring nitrogens (N) in the generic purine structure shown would most likely be affected by azaserine?

   A. 1
   B. 3
   C. 7
   D. 9

Correct answer = D. The N at position 9 is supplied by glutamine in the first step of purine *de novo* synthesis, and its incorporation would be affected by azaserine. The N at position 1 is supplied by aspartate and at position 7 by glycine. The N at position 3 is also supplied by glutamine, but azaserine would have inhibited purine synthesis prior to this step.

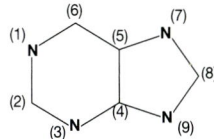

22.2. A 42-year-old male undergoing radiation therapy for prostate cancer develops severe pain in the metatarsal phalangeal joint of his right big toe. Monosodium urate crystals are detected by polarized light microscopy in fluid obtained from this joint by arthrocentesis. This patient's pain is directly caused by the overproduction of the end product of which of the following metabolic pathways?

   A. *De novo* pyrimidine biosynthesis
   B. Pyrimidine degradation
   C. *De novo* purine biosynthesis
   D. Purine salvage
   E. Purine degradation

Correct answer = E. The patient's pain is caused by gout, resulting from an inflammatory response to the crystallization of excess urate (as monosodium urate) in his joints. Radiation therapy caused cell death, with degradation of nucleic acids and their constituent purines. Uric acid, the end product of purine degradation, is a relatively insoluble compound that can cause gout (and kidney stones). Pyrimidine metabolism is not associated with uric acid production. Overproduction of purines can indirectly result in hyperuricemia. Purine salvage decreases uric acid production.

22.3. Which of the following enzymes of nucleotide metabolism is correctly paired with its pharmacologic inhibitor?

   A. Dihydrofolate reductase—methotrexate
   B. Inosine monophosphate dehydrogenase—hydroxyurea
   C. Ribonucleotide reductase—5-fluorouracil
   D. Thymidylate synthase—allopurinol
   E. Xanthine oxidase—probenecid

Correct answer = A. Methotrexate interferes with folate metabolism by acting as a competitive inhibitor of the enzyme dihydrofolate reductase. This starves cells for tetrahydrofolate and makes them unable to synthesize purines and thymidine monophosphate. Inosine monophosphate dehydrogenase is inhibited by mycophenolic acid. Ribonucleotide reductase is inhibited by hydroxyurea. Thymidylate synthase is inhibited by 5-fluorouracil. Xanthine oxidase is inhibited by allopurinol. Probenecid increases renal excretion of urate but does not inhibit its production.

22.4. A 1-year-old patient is lethargic, weak, and anemic. Her height and weight are low for her age. Her urine contains an elevated level of orotic acid. Activity of uridine monophosphate (UMP) synthase is low. Administration of which of the following is most likely to alleviate her symptoms?

   A. Adenine
   B. Guanine
   C. Hypoxanthine
   D. Thymidine
   E. Uridine

Correct answer = E. The elevated excretion of orotic acid and low activity of UMP synthase indicate that the patient has orotic aciduria, a very rare genetic disorder affecting *de novo* pyrimidine synthesis. Deficiencies in one or both catalytic domains of UMP synthase leave the patient unable to synthesize pyrimidines. Uridine, a pyrimidine nucleoside, is a useful treatment because it bypasses the missing activities and can be salvaged to UMP, which can be converted to all the other pyrimidines. Although thymidine is a pyrimidine nucleoside, it cannot be converted to other pyrimidines. Hypoxanthine, guanine, and adenine are all purine bases and cannot be converted to pyrimidines.

22.5. What laboratory results would help in distinguishing an orotic aciduria caused by ornithine transcarbamylase (OTC) deficiency from that caused by uridine monophosphate (UMP) synthase deficiency?

Both OTC and UMP synthase deficiencies result in elevated urinary orotate levels. Hyperammonemia would also be expected to be elevated in OTC deficiency that affects the urea cycle. Megaloblastic anemia may also be associated with orotic aciduria due to a UMP synthase deficiency in the pyrimidine synthesis pathway.

22.6. A buildup of deoxyadenosine triphosphate (dATP) and inhibition of ribonucleotide reductase is part of the mechanism of which of the following inborn errors of nucleotide metabolism disease states?
   A. Adenosine deaminase deficiency
   B. Gout
   C. Lesch–Nyhan syndrome
   D. Orotic aciduria
   E. Megaloblastic anemia

Correct answer = A. In adenosine deaminase (ADA) deficiency, dATP builds up, which inhibits ribonucleotide reductase. The decrease in deoxyribonucleotide production leads to developmental arrest and apoptosis of B and T lymphocytes (lymphocytopenia), causing severe combined immunodeficiency. Gout is caused by hyperuricemia, forming urate crystals. This is more commonly due to underexcretion of uric acid in the kidneys but can also be caused by overproduction of uric acid due to too much purine degradation. Lesch–Nyhan is due to the inability to salvage guanine and hypoxanthine purine bases. The increase in PRPP, leads to excessive purine synthesis and subsequent degradation. Orotic aciduria is caused by uridine monophosphate (UMP) synthase deficiency, blocking the addition of orotate to PRPP in pyrimidine synthesis pathway. Megaloblastic anemia is often associated with orotic aciduria and UMP synthase deficiency. Other causes for megaloblastic anemia include folic acid deficiency.

22.7. A patient with a urinary tract infection is prescribed which of the following antimicrobial folic acid analogs?
   A. Allopurinol
   B. Methotrexate
   C. Mycophenolic acid
   D. Sulfonamides
   E. Trimethoprim

Correct answer = E. Trimethoprim is an antimicrobial folic acid analog. It inhibits bacterial dihydrofolate reductase, preventing the regeneration of tetrahydrofolate from dihydrofolate. Methotrexate functions in a similar mechanism in humans, not bacteria. It is used as a cancer therapeutic agent. Allopurinol inhibits xanthine oxidase and the overproduction of uric acid from hypoxanthine and xanthine in purine degradation to treat hyperuricemia. Mycophenolic acid inhibits inosine monophosphate dehydrogenase and the production of GMP. It is an effective immunosuppressant agent to prevent organ transplant rejection. Sulfonamides inhibit the synthesis of folic acid in bacteria and are antimicrobial in function. Sulfonamides have no effect in humans since humans cannot synthesize folic acid. All folic acid must be obtained from our diet.

22.8. A patient presents with uric acid stones in the kidney, gouty arthritis, and cognitive defects and is self-mutilating (biting their lips and fingers). Which of the following enzymes is defective?
   A. Adenosine deaminase
   B. Hypoxanthine-guanine phosphoribosyltransferase
   C. Thymidylate synthase
   D. Ornithine transcarbamylase
   E. Uridine monophosphate (UMP) synthase

Correct answer = B. Hypoxanthine-guanine phosphoribosyltransferase (HGPRT) is defective in Lesch–Nyhan syndrome. Adenosine deaminase deficiency causes severe combined immunodeficiency involving B- and T-lymphocytes. No disease is associated with a deficiency in thymidylate synthase, but it can be inhibited by 5-fluorouracil to treat cancers. Ornithine transcarbamylase is deficient in the urea cycle, causing hyperammonemia and an increase in urinary orotic acid. Uridine monophosphate (UMP) synthase deficiency results in orotic aciduria (and megaloblastic anemia) but is due to a defect in the pyrimidine synthesis pathway, not the urea cycle.

Metabolic Effects of Insulin and Glucagon

23

I. OVERVIEW

Four major tissues play a dominant role in fuel metabolism: liver, adipose, muscle, and brain. These tissues contain unique sets of enzymes, such that each tissue is specialized for the storage, use, or generation of specific fuels. These tissues do not function in isolation but, rather, form part of a network in which one tissue may provide substrates to another or process compounds produced by other tissues. Communication between tissues is mediated by the nervous system, by the availability of circulating substrates, and by variation in the levels of plasma hormones (Fig. 23.1). The integration of energy metabolism is controlled primarily by the actions of two peptide hormones, insulin, and glucagon (secreted in response to changing substrate levels in the blood), with the catecholamines epinephrine and norepinephrine (secreted in response to neural signals) playing a supporting role. Changes in the circulating levels of these hormones allow the body to store energy when food is abundant (well-fed state) or to make stored energy available such as during the fasting state or for survival crises (e.g., famine, severe injury, and "fight-or-flight" situations). This chapter describes the structure, secretion, and metabolic effects of the two hormones that most profoundly affect energy metabolism in the well-fed and fasting states, insulin and glucagon.

II. INSULIN

Insulin is a peptide hormone produced by the β cells of the islets of Langerhans, which are clusters of cells embedded in the endocrine portion of the pancreas (Fig. 23.2). "Insulin" is from the Latin for island. The islets make up only about 1% to 2% of the total cells of the pancreas. Insulin is the most important hormone coordinating the use of fuels by tissues in a well-fed state. Its metabolic effects are anabolic, favoring, for example, the synthesis of glycogen, triacylglycerol (TAG), and proteins.

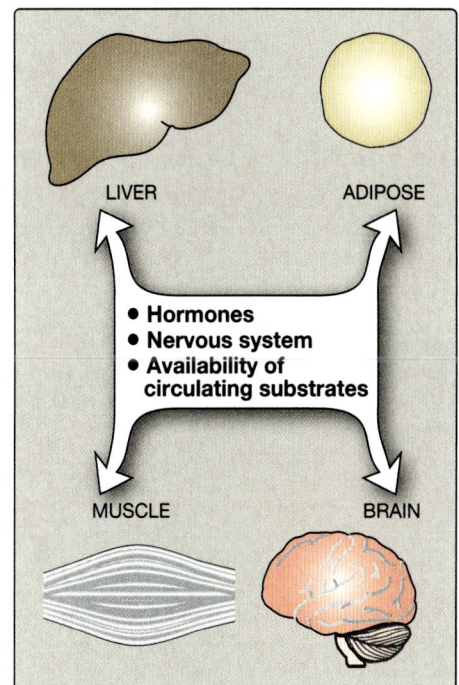

Figure 23.1
Mechanisms of communication between four major tissues.

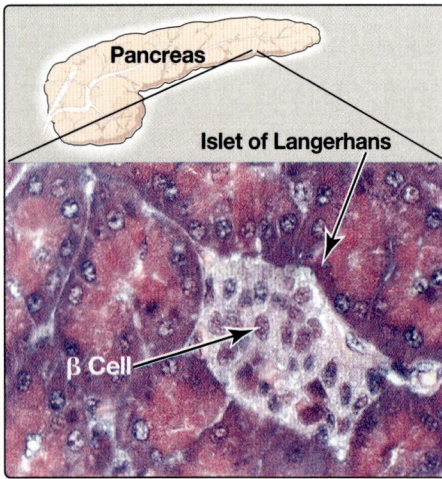

Figure 23.2
Islet of Langerhans.

A. Structure

Insulin is composed of 51 amino acids arranged in two polypeptide chains, designated A (21 amino acids) and B (30 amino acids), which are linked together by two disulfide bonds (Fig. 23.3A). The insulin molecule also contains an intramolecular disulfide bond between cysteine residues of the A chain. Insulin was the first peptide for which the primary structure was determined, and the first therapeutic molecule made by recombinant DNA technology (see Chapter 34, III. B.).

B. Synthesis and degradation

The processing and transport of intermediates that occur during the synthesis of insulin are shown in Figures 23.3B and 23.4. Biosynthesis involves the production of two inactive precursors, preproinsulin and proinsulin, which contain the A and B chains as a single peptide. These precursors are sequentially cleaved to form the active two-chain hormone plus a connecting or C-peptide in a 1:1 ratio. Prior to cleavage, the C-peptide is essential for proper insulin folding. Also, because its half-life in plasma is longer than that of insulin, the C-peptide level is a good indicator of insulin production and secretion. Insulin is stored in cytosolic granules that, given the proper stimulus, are released by exocytosis. Insulin is degraded by insulin-degrading enzyme, which is present in the liver and, to a lesser extent, in the kidneys. Insulin has a plasma half-life of ~6 minutes. This short duration of action permits rapid changes in circulating levels of the hormone.

C. Secretion regulation

Secretion of insulin is regulated by blood-borne fuels and hormones.

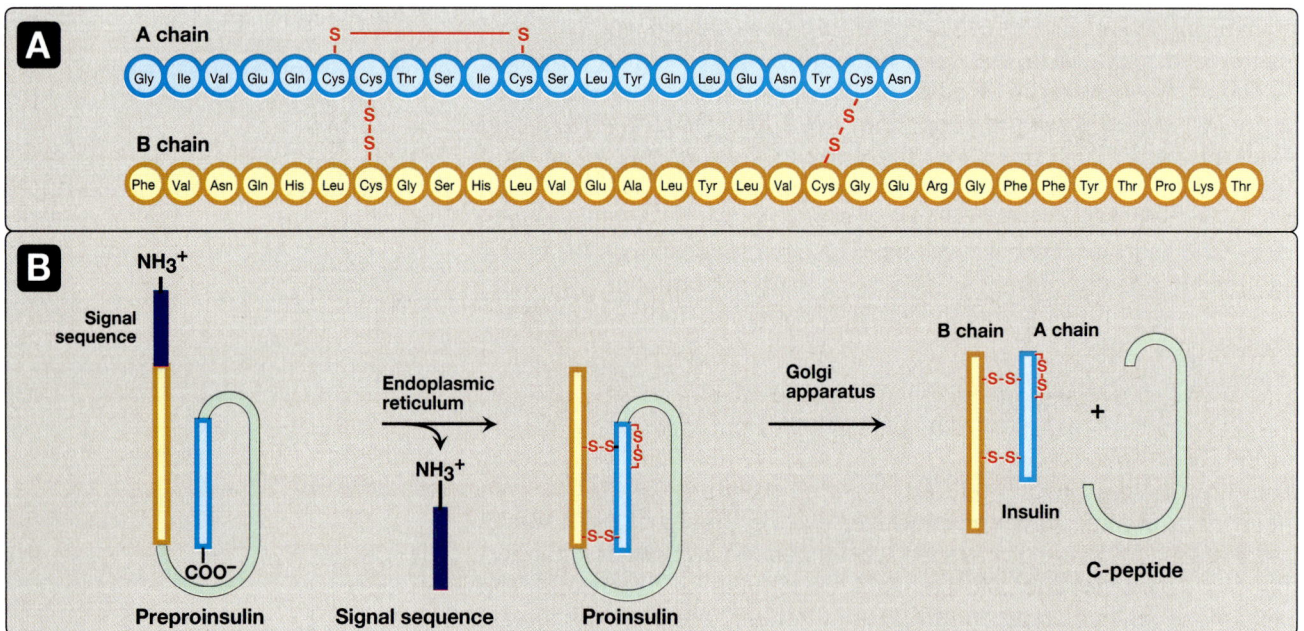

Figure 23.3
A. Structure of insulin. **B**. Formation of human insulin from preproinsulin. S-S, disulfide bond.

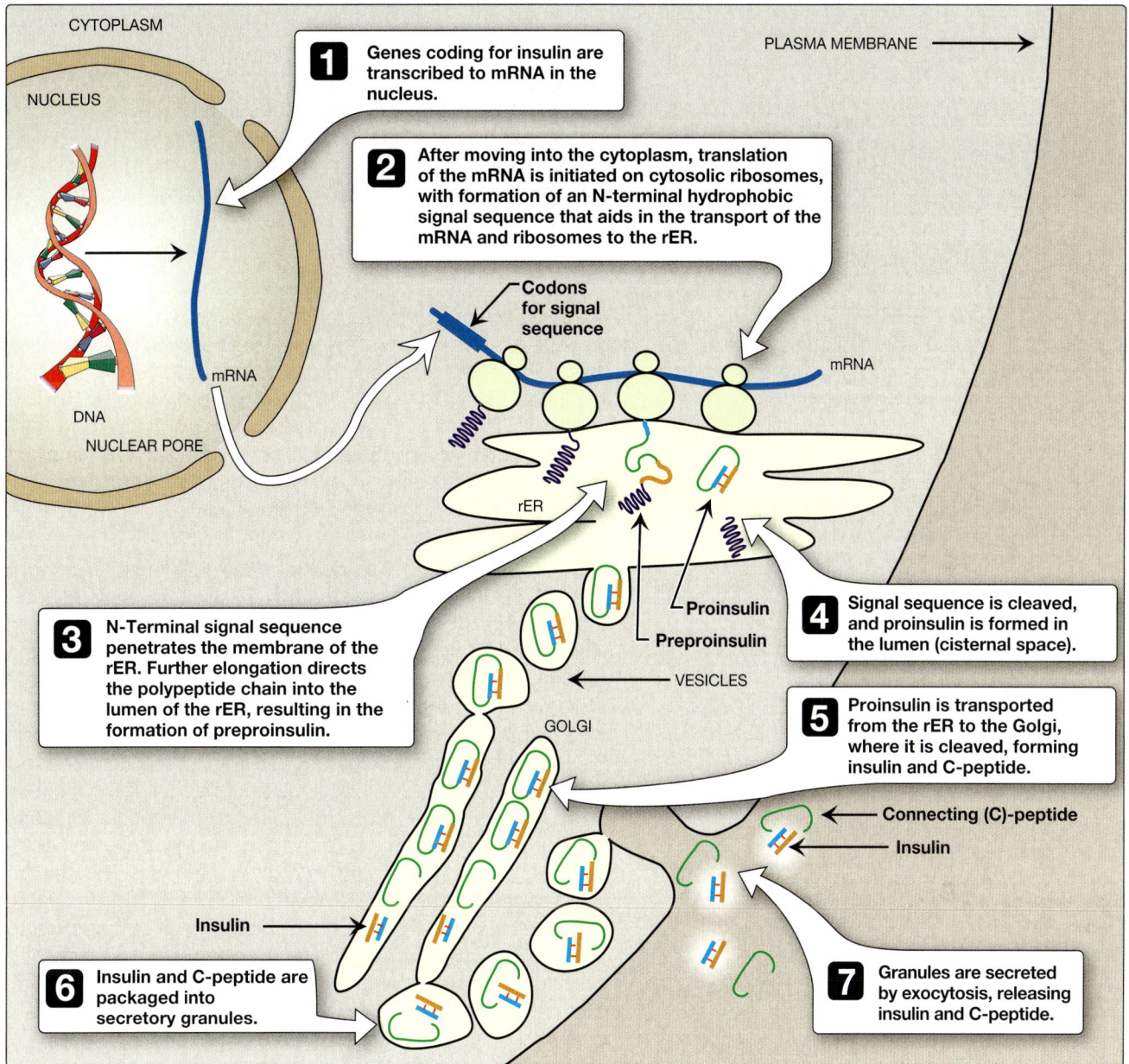

Figure 23.4
Intracellular movements of insulin and its precursors. mRNA, messenger RNA; rER, rough endoplasmic reticulum.

1. **Increased secretion:** Insulin secretion by the pancreatic β cells is closely coordinated with the secretion of glucagon by pancreatic α cells (Fig. 23.5). The relative amounts of glucagon and insulin released are normally regulated such that the rate of hepatic glucose production is kept equal to the use of glucose by peripheral tissues. This maintains blood glucose between 70 and 140 mg/dL (3.9 to 7.8 mmol/L). In view of its coordinating role, it is not surprising that the β cell responds to a variety of stimuli. In particular, insulin secretion is increased by glucose, amino acids, and gastrointestinal peptide hormones.

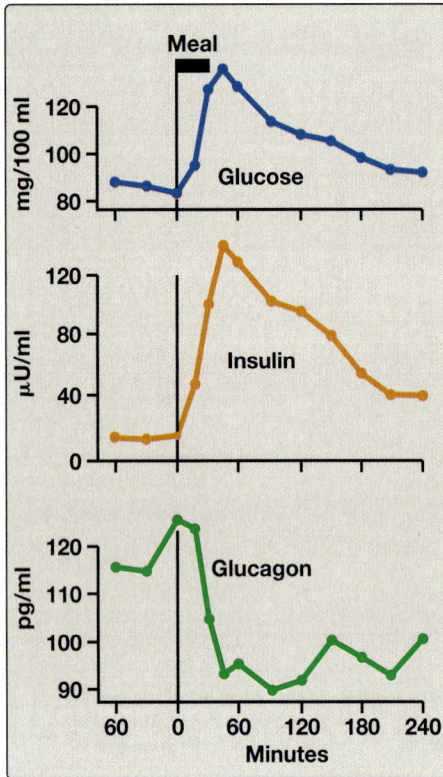

Figure 23.5
Changes in blood levels of glucose, insulin, and glucagon after ingestion of a carbohydrate-rich meal.

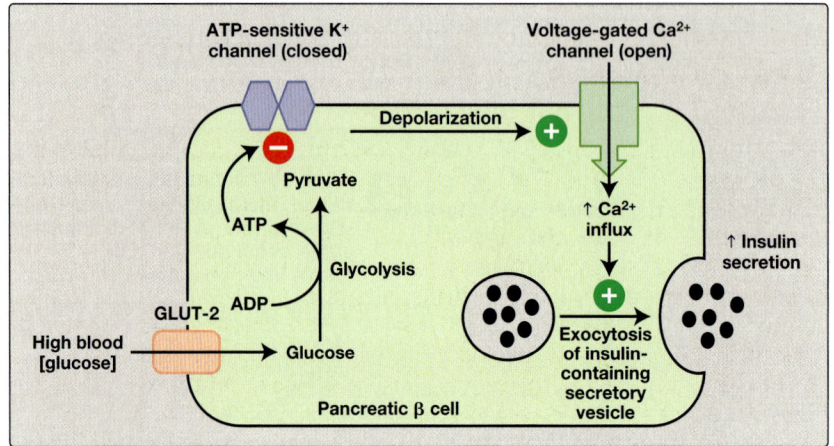

Figure 23.6
Glucose-dependent pancreatic secretion of insulin. ADP, adenosine diphosphate; ATP, adenosine triphosphate; Ca²⁺, calcium; K⁺, potassium.

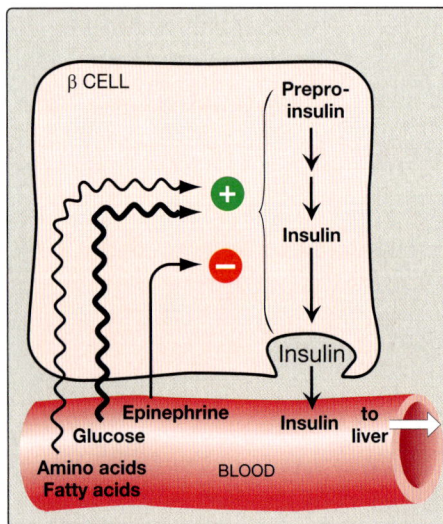

Figure 23.7
Regulation of insulin release from pancreatic β cells. Gastrointestinal peptide hormones also stimulate insulin release.

a. **Glucose:** Ingestion of a carbohydrate-rich meal leads to a rise in blood glucose, the primary stimulus for insulin secretion (see Fig. 23.5). At the same time that glucose and insulin levels rise, glucagon levels drop. This high insulin/glucagon ratio represents the well-fed state. The β cells are the most important glucose-sensing cells in the body. Like the liver, β cells contain GLUT-2 transporters and express glucokinase (hexokinase IV; see Chapter 8, V.). At blood glucose levels higher than 45 mg/dL, glucokinase phosphorylates glucose proportionate to the blood glucose concentration. Glucose taken into β cells by GLUT-2 and phosphorylated by glucokinase is metabolized by the glycolytic pathway. This results in a rise in intracellular ATP levels. ATP-sensitive potassium (K⁺) channels in the plasma membrane close in response to the rise in intracellular ATP levels, causing depolarization across the plasma membrane. Depolarization is detected by and mediates the opening of voltage-gated calcium (Ca²⁺) channels in the plasma membrane, causing an influx of Ca²⁺. An increase in cytosolic Ca²⁺ signals exocytosis of insulin-containing vesicles from the β cells. Therefore, secretion of insulin from the β cells into blood is signaled through an intracellular rise in Ca²⁺ concentration, depending on and proportionate to the concentration of blood glucose. This relationship is summarized in Figure 23.6.

b. **Amino acids:** Ingestion of protein causes a transient rise in plasma amino acid levels (e.g., arginine) that enhances the glucose-stimulated secretion of insulin from endocrine pancreatic β cells. Fatty acids (FAs) have a similar effect (Fig. 23.7).

c. **Gastrointestinal peptide hormones:** The intestinal peptides glucagon-like peptide-1 (GLP-1) and gastric inhibitory polypeptide ([GIP] also called glucose-dependent insulinotropic peptide) increase the sensitivity of β cells to glucose. They are released from the small intestine after the ingestion of food,

> ## Clinical Application 23.1: Therapeutic Applications Affecting Insulin Secretion
>
> Sulfonylureas and meglitinides are antihyperglycemic drugs used to treat hyperglycemia in type 2 diabetics. Both classes of drugs function to increase insulin secretion by closing the ATP-sensitive K^+ channels, leading to more Ca^{2+} influx and more insulin secretion (see Fig. 23.6). As such, both are referred to as "insulin secretagogues." Meglitinides have a weaker binding affinity and higher dissociation constant for the ATP-sensitive K^+ channels (as compared to sulfonylureas) and, therefore, are shorter-acting insulin secretagogues. Sulfonylureas are longer-acting and may therefore increase the risk of hypoglycemia as a negative side effect. Conversely, diazoxides open the ATP-sensitive K^+ channel, leading to decreased insulin secretion. Diazoxides are used to treat hypoglycemia caused by congenital hyperinsulinemia or in insulinomas (insulin-producing tumors).

causing an anticipatory rise in insulin levels and, thus, are referred to as incretins. Their action may account for the fact that the same amount of glucose given orally induces a much greater secretion of insulin than if given intravenously (IV).

2. **Decreased secretion:** The synthesis and release of insulin are decreased when dietary fuels are scarce and also during periods of physiologic stress (e.g., infection, hypoxia, and vigorous exercise), thereby preventing hypoglycemia. These effects are mediated primarily by the catecholamines norepinephrine and epinephrine, which are made from tyrosine in the sympathetic nervous system (SNS) and the adrenal medulla and then secreted. Secretion is largely controlled by neural signals. The catecholamines (primarily epinephrine) have a direct effect on energy metabolism, causing rapid mobilization of energy-yielding fuels, including glucose from the liver (produced by glycogenolysis or gluconeogenesis; see Chapters 10 and 11) and FAs from adipose tissue (produced by TAG lipolysis; see Chapter 16). In addition, these biogenic amines can override the normal glucose-stimulated release of insulin. Thus, in emergency situations, the SNS largely replaces the plasma glucose concentration as the controlling influence over β-cell secretion. The regulation of insulin secretion is summarized in Figures 23.6 and 23.7.

D. Metabolic effects

Insulin promotes the cellular uptake of nutrients into cells (primarily glucose). It also promotes anabolic synthesis and storage of nutrients, including glycogen, TAG, and protein, and inhibits their mobilization.

1. **Effects on carbohydrate metabolism:** The effects of insulin on glucose promote its metabolism and storage most prominently in three tissues: liver, muscle, and adipose. In liver and muscle, insulin increases glycogen synthesis. In muscle and adipose, insulin increases glucose uptake by increasing the number of glucose transporters (GLUT-4; see Chapter 8, IV.) in the cell membrane. Thus, the IV administration of insulin causes an immediate decrease in blood glucose level. In the liver, insulin decreases

the production of glucose through the inhibition of glycogeno-lysis and gluconeogenesis. Insulin can also increase the utilization of glucose in the glycolytic pathway (see Chapter 8, VI.). The effects of insulin are due not just to changes in enzyme activity but also in enzyme amount insofar as insulin signaling affects gene transcription.

2. **Effects on lipid metabolism:** A rise in insulin rapidly causes a significant reduction in the release of FAs from adipose tissue by inhibiting the activity of hormone-sensitive lipase, a key enzyme of TAG degradation in adipocytes. Insulin acts by promoting the dephosphorylation and, hence, inactivation of the enzyme (see Chapter 16, IV.). Insulin also increases the transport and metabolism of glucose into adipocytes, providing the glycerol 3-phosphate substrate for TAG synthesis (see Chapter 16, III.). Expression of lipoprotein lipase (LL), which degrades TAG in circulating chylomicrons and very–low-density lipoprotein (VLDL) complexes (see Chapter 18, VI.), is increased by insulin in adipose, thereby providing FAs for esterification to glycerol 3-phosphate. Insulin also promotes FA synthesis (see Chapter 16, III.) and the storage of the FAs to TAGs in the liver. These TAGs are secreted in VLDL complexes.

3. **Effects on protein synthesis:** In most tissues, insulin stimulates both the entry of amino acids into cells and protein synthesis (translation). Insulin stimulates protein synthesis through covalent activation of factors required for translation initiation.

E. Mechanism

Insulin binds to specific, high-affinity receptors in the cell membrane of most tissues, including liver, muscle, and adipose. This is the first step in a cascade of reactions ultimately leading to a diverse array of biologic actions, summarized in Figure 23.8.

1. **Insulin receptor:** The insulin receptor is synthesized as a single polypeptide that is glycosylated and cleaved into α and β subunits, which are then assembled into a tetramer linked by disulfide bonds (Fig. 23.8). The extracellular α subunits contain the insulin-binding site. A hydrophobic domain in each β subunit spans the plasma membrane. The cytosolic domain of the β subunit is a tyrosine kinase, which is activated by insulin. As a result, the insulin receptor is classified as a receptor tyrosine kinase.

2. **Signal transduction:** The binding of insulin to the α subunits of the insulin receptor induces conformational changes that are transmitted to the β subunits. This promotes a rapid autophosphorylation of specific tyrosine residues on each β subunit (see Fig. 23.8). Autophosphorylation initiates a cascade of cell-signaling responses, including phosphorylation of a family of proteins called insulin receptor substrates (IRSs). At least four IRSs have been identified that show similar structures but different tissue distributions. Phosphorylated IRS proteins interact with other signaling molecules through specific domains (known as SH2), activating a number of pathways that affect gene expression, cell metabolism, and growth. The actions of insulin are terminated by dephosphorylation of the receptor.

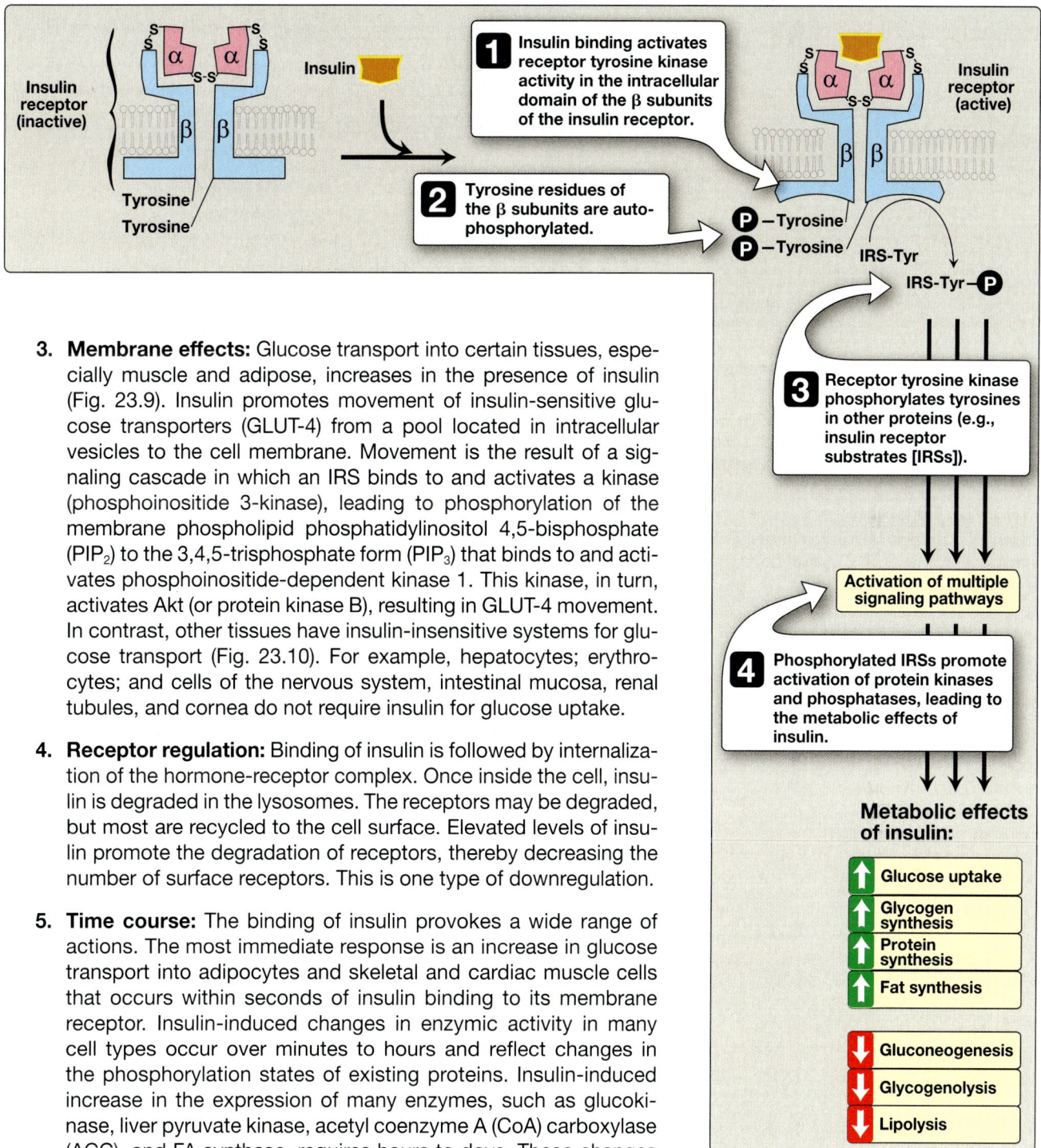

Figure 23.8
Mechanism of action of insulin.
P, phosphate; Tyr, tyrosine; S-S, disulfide bond.

3. **Membrane effects:** Glucose transport into certain tissues, especially muscle and adipose, increases in the presence of insulin (Fig. 23.9). Insulin promotes movement of insulin-sensitive glucose transporters (GLUT-4) from a pool located in intracellular vesicles to the cell membrane. Movement is the result of a signaling cascade in which an IRS binds to and activates a kinase (phosphoinositide 3-kinase), leading to phosphorylation of the membrane phospholipid phosphatidylinositol 4,5-bisphosphate (PIP_2) to the 3,4,5-trisphosphate form (PIP_3) that binds to and activates phosphoinositide-dependent kinase 1. This kinase, in turn, activates Akt (or protein kinase B), resulting in GLUT-4 movement. In contrast, other tissues have insulin-insensitive systems for glucose transport (Fig. 23.10). For example, hepatocytes; erythrocytes; and cells of the nervous system, intestinal mucosa, renal tubules, and cornea do not require insulin for glucose uptake.

4. **Receptor regulation:** Binding of insulin is followed by internalization of the hormone-receptor complex. Once inside the cell, insulin is degraded in the lysosomes. The receptors may be degraded, but most are recycled to the cell surface. Elevated levels of insulin promote the degradation of receptors, thereby decreasing the number of surface receptors. This is one type of downregulation.

5. **Time course:** The binding of insulin provokes a wide range of actions. The most immediate response is an increase in glucose transport into adipocytes and skeletal and cardiac muscle cells that occurs within seconds of insulin binding to its membrane receptor. Insulin-induced changes in enzymic activity in many cell types occur over minutes to hours and reflect changes in the phosphorylation states of existing proteins. Insulin-induced increase in the expression of many enzymes, such as glucokinase, liver pyruvate kinase, acetyl coenzyme A (CoA) carboxylase (ACC), and FA synthase, requires hours to days. These changes reflect an increase in gene expression through increased transcription and translation.

III. GLUCAGON

Glucagon is a peptide hormone secreted by the α cells of the pancreatic islets of Langerhans. Glucagon, along with epinephrine, norepinephrine, cortisol, and growth hormone ([GH], counterregulatory hormones),

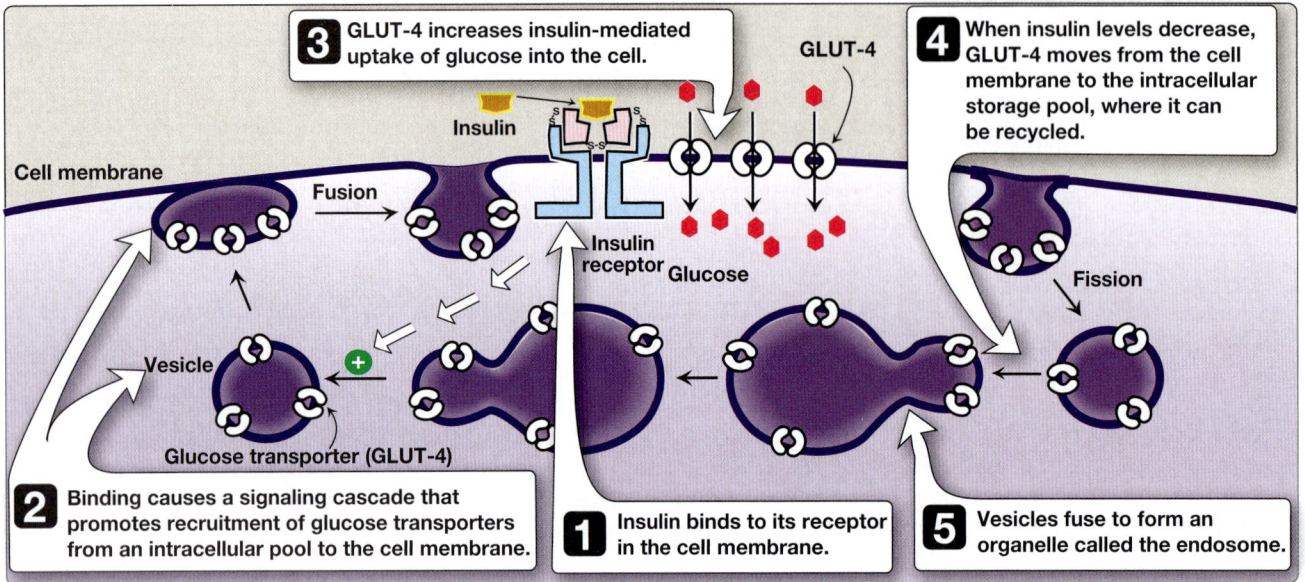

Figure 23.9
Insulin-mediated recruitment of GLUT-4 from intracellular stores to the cell membrane in skeletal and cardiac muscle and adipose tissue. S-S, disulfide bond.

opposes many of the actions of insulin (Fig. 23.11). Most importantly, glucagon acts to maintain blood glucose levels by activation of hepatic glycogenolysis and gluconeogenesis. Glucagon is composed of 29 amino acids arranged in a single polypeptide chain. Unlike insulin, the amino acid sequence of glucagon is the same in all mammalian species examined to date. Glucagon is synthesized as a large precursor molecule (preproglucagon) that is converted to glucagon through a series of selective proteolytic cleavages, similar to those described for insulin biosynthesis (see Fig. 23.3). In contrast to insulin, preproglucagon is processed to different products in different tissues, for example, GLP-1 in intestinal L cells. Like insulin, glucagon has a short half-life.

A. Increased secretion

The α cell is responsive to a variety of stimuli that signal actual or potential hypoglycemia (Fig. 23.12). Specifically, glucagon secretion is increased by low blood glucose, amino acids, and catecholamines, including epinephrine.

1. **Low blood glucose:** A decrease in plasma glucose concentration is the primary stimulus for glucagon release. During an overnight or prolonged fast, elevated glucagon levels prevent hypoglycemia (see Section IV. for a discussion of hypoglycemia).

2. **Amino acids:** Amino acids (e.g., arginine) derived from a meal containing protein stimulate the release of glucagon. Glucagon effectively prevents hypoglycemia that would otherwise occur as a result of the increased insulin secretion that also occurs after a protein meal.

3. **Catecholamines:** Elevated levels of circulating epinephrine (from the adrenal medulla), norepinephrine (from sympathetic innervation

	Active transport	Facilitated transport
Insulin sensitive		Skeletal and cardiac muscle and adipose tissue (together account for largest tissue mass)
Insulin insensitive	Intestinal epithelia Renal tubules Choroid plexus	Erythrocytes Leukocytes Lens of eyes Corneas Liver Brain

Figure 23.10
Characteristics of glucose transport in various tissues.

of the pancreas), or both stimulate the release of glucagon. Thus, during periods of physiologic stress, the elevated catecholamine levels can override the effect on the α cell of circulating substrates. In these situations, regardless of the concentration of blood glucose, glucagon levels are elevated in anticipation of increased glucose use. In contrast, insulin levels are depressed.

B. Decreased secretion

Glucagon secretion is significantly decreased by elevated blood glucose and by insulin. Both substances are increased following ingestion of glucose or a carbohydrate-rich meal (see Fig. 23.5). The regulation of glucagon secretion is summarized in Figure 23.12.

C. Metabolic effects

Glucagon is a catabolic hormone that promotes the maintenance of blood glucose levels. Its primary target is the liver. Glucagon also has an effect on the mobilization and utilization of FAs in adipose and muscle tissues.

1. **Effects on carbohydrate metabolism:** The IV administration of glucagon leads to an immediate rise in blood glucose. This results from an increase in the degradation of hepatic glycogen stores (see Chapter 11, V.) and an increase in gluconeogenesis (see Chapter 10, IV.). During the daytime, postprandial blood glucose levels are maintained primarily by hepatic glycogenolysis, with additional blood glucose provided by gluconeogenesis. Since the interprandial period is longer as we sleep and glycogen stores become more limited, gluconeogenesis increases providing more of the blood glucose source as the night progresses. Glucagon also inhibits glycolysis by decreasing levels of the PFK-1 allosteric activator, fructose 2,6-bisphosphate (see Chapter 8, V.).

2. **Effects on lipid metabolism:** The effects of glucagon on lipid metabolism are primarily demonstrated in the liver, adipocytes, and muscle. FA synthesis is inhibited by phosphorylation and subsequent inactivation of ACC by adenosine monophosphate (AMP)-activated protein kinase (see Chapter 16, III.). The resulting decrease in malonyl CoA production removes the inhibition on carnitine palmitoyltransferase-1, required for transport of long-chain FAs into the mitochondrial matrix for β-oxidation (see Chapter 16, IV.). As a result, the decrease in ACC activity and cytosolic FA synthesis leads to increased FA availability for β-oxidation in the mitochondrial matrix and an increase in FA oxidation for energy production. This demonstrates a shift from energy production from carbohydrate metabolism in the well-fed state to FA metabolism in the fasting state in certain tissues. Glucagon also plays a role in lipolysis in adipocytes, but the major activators of hormone-sensitive lipase (via phosphorylation by protein kinase A) are the catecholamines epinephrine and norepinephrine. The free FAs mobilized in the blood from adipocytes are taken up by liver and muscle tissues and oxidized to acetyl CoA by β-oxidation. The liver uses the acetyl CoA for energy and as a substrate for ketone body synthesis. The brain, which cannot use

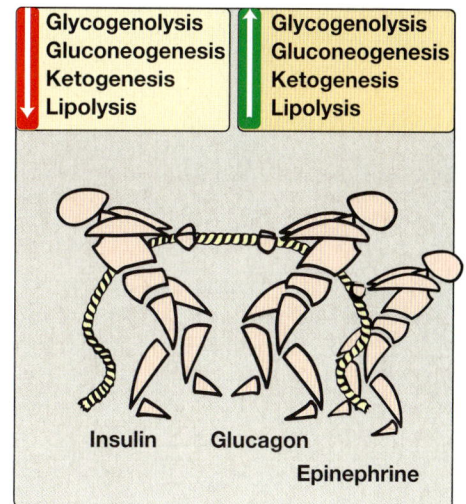

↓ Glycogenolysis Gluconeogenesis Ketogenesis Lipolysis	↑ Glycogenolysis Gluconeogenesis Ketogenesis Lipolysis

Insulin   Glucagon
Epinephrine

Figure 23.11
Opposing actions of insulin and glucagon plus epinephrine.

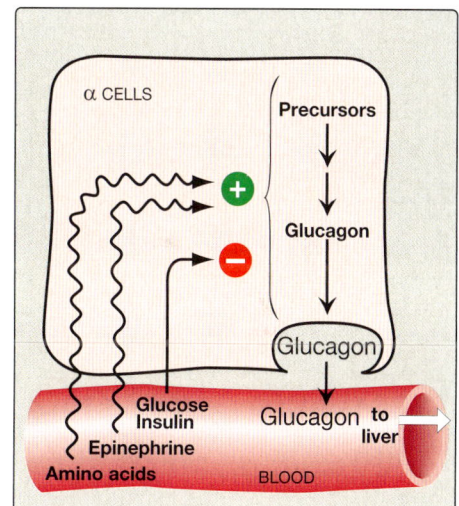

α CELLS

Precursors

Glucagon

Glucagon

Glucose
Insulin
Epinephrine
Amino acids   BLOOD

Glucagon to liver

Figure 23.12
Regulation of glucagon release from pancreatic α cells. Amino acids increase release of insulin and glucagon, whereas glucose increases release of insulin and decreases release of glucagon.

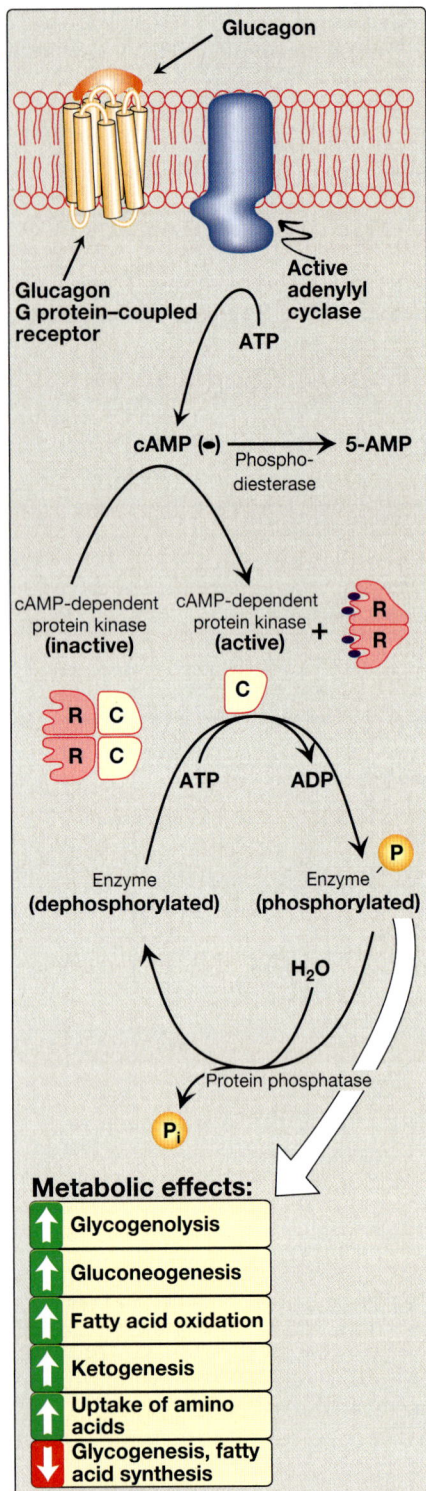

Figure 23.13
Mechanism of action of glucagon. (For clarity, G-protein activation of adenylyl cyclase has been omitted.) R, regulatory subunit; C, catalytic subunit; cAMP, cyclic adenosine monophosphate; ADP, adenosine diphosphate; **P**, phosphate.

FA for energy production, can use ketone bodies. Muscle cells will use the acetyl CoA and ketone bodies for energy. Glucagon and catecholamines also activate LL in cardiac and skeletal muscle tissues, to allow uptake of FA from VLDL complexes in the fasting state. Considering that glucagon stimulates GLUT-4 intracellular sequestration, and energy production from carbohydrates would be dependent on a decreasing supply of glycogen, it makes sense that muscle tissues would increase the utilization of FAs as an energy source in the fasting state. This is especially true for cardiac muscle, where up to 90% of energy production in the fasting state may come from β-oxidation.

3. **Effects on protein metabolism:** Glucagon increases uptake by the liver of amino acids supplied by muscle, resulting in increased availability of carbon skeletons for gluconeogenesis. As a consequence, plasma levels of amino acids are decreased.

D. Mechanism

Glucagon (and epinephrine) binds to high-affinity G protein–coupled receptors (GPCRs) on the cell membrane of hepatocytes. The GPCR for glucagon is distinct from the GPCRs that bind epinephrine. Epinephrine, not glucagon, receptors are found on skeletal muscle. Glucagon binding results in activation of adenylyl cyclase in the plasma membrane. This causes a rise in cyclic AMP (cAMP), which, in turn, activates cAMP-dependent protein kinase A and increases the phosphorylation of specific enzymes. This cascade results in the activation or inhibition of key regulatory enzymes involved in carbohydrate and lipid metabolism, as summarized in Figure 23.13. Glucagon, like insulin, also affects gene transcription (e.g., glucagon induces expression of phosphoenolpyruvate carboxykinase in gluconeogenesis). The overall effect of glucagon on metabolic processes includes increasing glycogenolysis, gluconeogenesis, FA mobilization and oxidation, ketogenesis, and amino acid uptake while decreasing glycogen and FA synthesis.

IV. HYPOGLYCEMIA

Hypoglycemia is characterized by (1) central nervous system (CNS) symptoms, including confusion, aberrant behavior, or coma; (2) a simultaneous blood glucose level no more than 50 mg/dL; and (3) symptoms being resolved within minutes following glucose administration (Fig. 23.14). Hypoglycemia is a medical emergency because the CNS has an absolute requirement for a continuous supply of blood-borne glucose to serve as a metabolic fuel. Transient hypoglycemia can cause cerebral dysfunction, whereas severe, prolonged hypoglycemia causes brain damage. Therefore, it is not surprising that the body has multiple overlapping mechanisms to prevent or correct hypoglycemia. The most important hormone changes in combating hypoglycemia are increased secretion of glucagon and catecholamines, combined with decreased insulin secretion.

A. Symptoms

The symptoms of hypoglycemia can be divided into two categories. Adrenergic (neurogenic, autonomic) symptoms, such as anxiety,

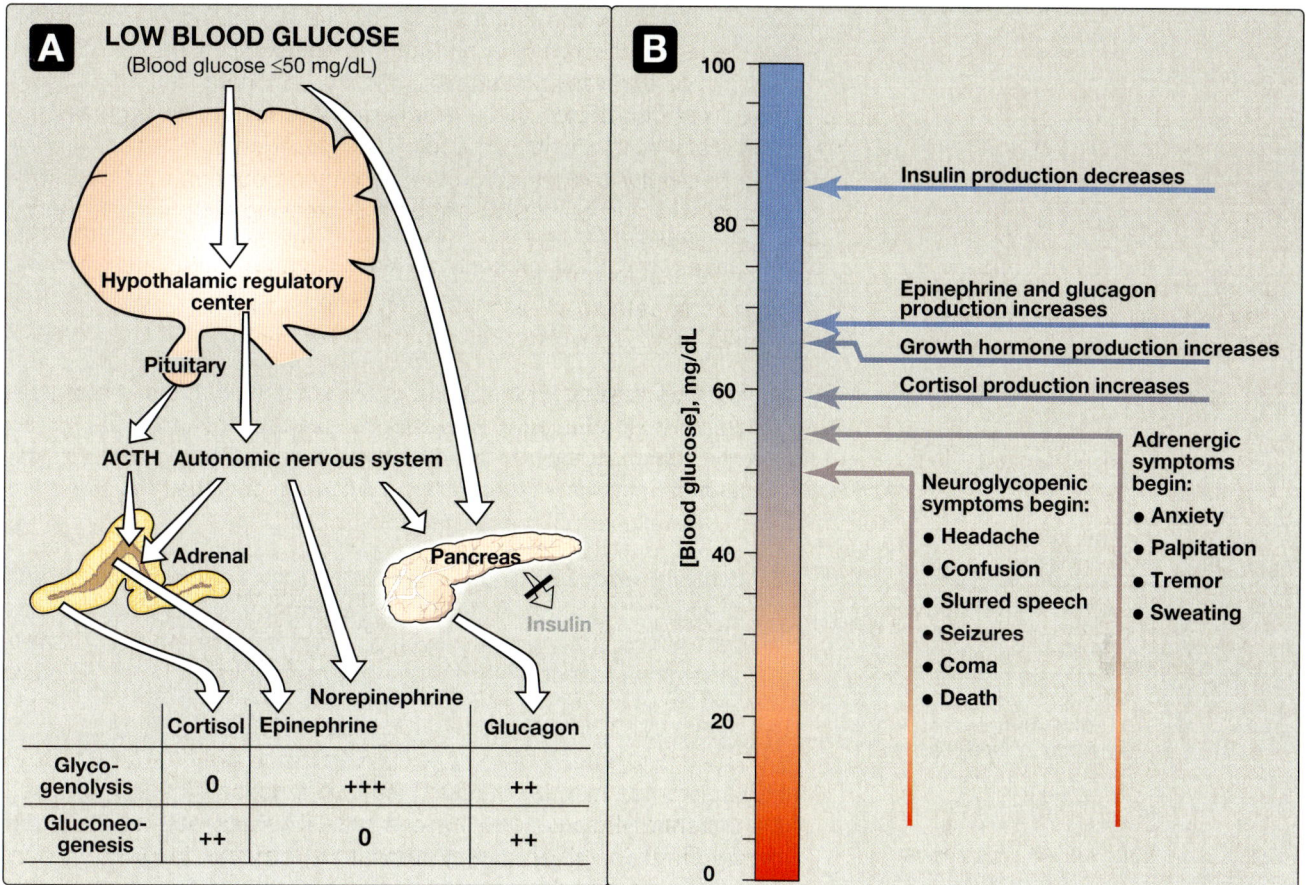

Figure 23.14
A. Actions of some of the glucoregulatory hormones in response to low blood glucose. **B**. Glycemic thresholds for the various responses to hypoglycemia. [Note: Normal fasting blood glucose is 70–99 mg/dL.] + = weak stimulation; ++ = moderate stimulation; +++ = strong stimulation; 0 = no effect; ACTH, adrenocorticotropic hormone.

palpitation, tremor, and sweating, are mediated by catecholamine release (primarily epinephrine) regulated by the hypothalamus in response to hypoglycemia. Adrenergic symptoms typically occur when blood glucose levels fall abruptly. The second category of hypoglycemic symptoms is neuroglycopenic. The impaired delivery of glucose to the brain (neuroglycopenia) results in impairment of brain function, causing headache, confusion, slurred speech, seizures, coma, and death. Neuroglycopenic symptoms often result from a gradual decline in blood glucose, often to levels less than 50 mg/dL. The slow decline in glucose deprives the CNS of fuel but fails to trigger an adequate adrenergic response.

B. Glucoregulatory systems

Humans have two overlapping glucose-regulating systems that are activated by hypoglycemia: (1) pancreatic α cells, which release glucagon, and (2) receptors in the hypothalamus, which respond to abnormally low concentrations of blood glucose. The hypothalamic glucoreceptors can trigger both the secretion of catecholamines (mediated by the sympathetic division of the autonomic nervous system) and release of adrenocorticotropic hormone (ACTH) and

GH by the anterior pituitary (Fig. 23.14). ACTH increases cortisol synthesis and secretion in the adrenal cortex (see Chapter 18, VII.). Glucagon, the catecholamines, cortisol, and GH are sometimes called the counterregulatory hormones because each opposes the action of insulin on glucose use.

1. **Glucagon and epinephrine:** Secretion of these counterregulatory hormones is most important in the acute, short-term regulation of blood glucose levels. Glucagon stimulates hepatic glycogenolysis and gluconeogenesis. Epinephrine promotes glycogenolysis and lipolysis. It inhibits insulin secretion, thereby preventing GLUT-4–mediated uptake of glucose by muscle and adipose tissues. Epinephrine assumes a critical role in hypoglycemia when glucagon secretion is deficient, for example, in the late stages of type 1 diabetes mellitus (see Chapter 25, II.). The prevention or correction of hypoglycemia fails when the secretion of both glucagon and epinephrine is deficient.

2. **Cortisol and GH:** These counterregulatory hormones are less important in the short-term maintenance of blood glucose concentrations. They do, however, play a role in the long-term (transcriptional) management of glucose metabolism.

C. Types

Hypoglycemia may be divided into four types: (1) insulin induced, (2) postprandial (sometimes called reactive hypoglycemia), (3) fasting hypoglycemia, and (4) alcohol related.

1. **Insulin-induced hypoglycemia:** Hypoglycemia occurs frequently in patients with diabetes who are receiving insulin treatment, particularly those striving to achieve tight control of blood glucose levels. Mild hypoglycemia in fully conscious patients is treated by oral administration of carbohydrates. Unconscious patients are typically given glucagon subcutaneously or intramuscularly (Fig. 23.15). Glucagon increases the blood glucose levels due to activation of hepatic glycogenolysis and gluconeogenesis.

2. **Postprandial hypoglycemia:** This is the second most common form of hypoglycemia. It is caused by an exaggerated insulin release following a meal, prompting transient hypoglycemia with mild adrenergic symptoms. The plasma glucose level returns to normal even if the patient is not fed. The only treatment usually required is that the patient eats frequent small meals rather than the usual three large meals.

3. **Fasting hypoglycemia:** Low blood glucose during fasting is rare but is more likely to present as a serious medical problem. Fasting hypoglycemia, which tends to produce neuroglycopenic symptoms, may result from a reduction in the rate of glucose production by hepatic glycogenolysis or gluconeogenesis. Thus, low blood glucose levels are often seen in patients with hepatocellular damage or adrenal insufficiency or in fasting individuals who have consumed large quantities of ethanol (see "Alcohol-related hypoglycemia"). Alternately, fasting hypoglycemia may

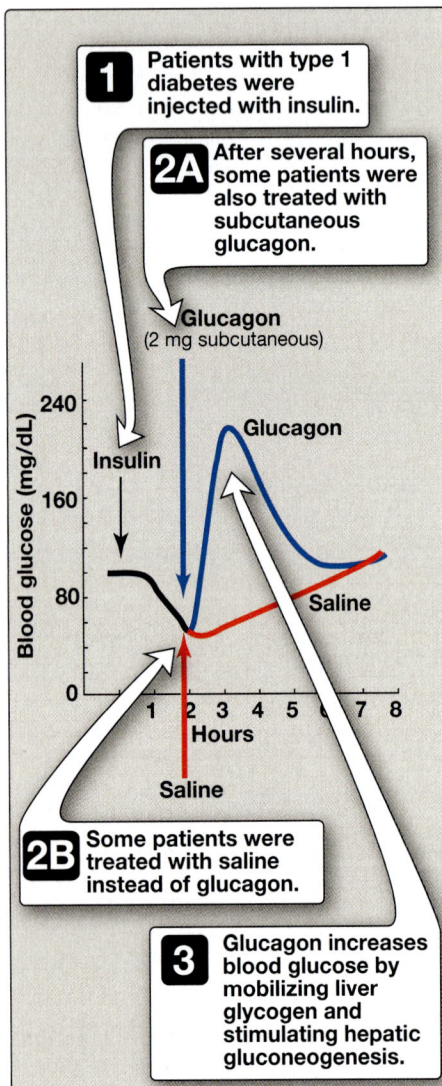

Figure 23.15
Reversal of insulin-induced hypoglycemia by administration of subcutaneous glucagon.

be the result of an increased rate of glucose use by the peripheral tissues because of the overproduction of insulin by rare pancreatic tumors. If left untreated, a patient with fasting hypoglycemia may lose consciousness and experience convulsions and coma. Certain inborn errors of metabolism (e.g., defects in FA oxidation) can result in fasting hypoglycemia.

4. **Alcohol-related hypoglycemia:** Alcohol (ethanol) is metabolized in the liver by two oxidation reactions (Fig. 23.16). Ethanol is first converted to acetaldehyde in the cytosol by zinc-containing alcohol dehydrogenase. Acetaldehyde is subsequently oxidized in the mitochondrial matrix to acetate by aldehyde dehydrogenase (ALDH). ALDH is inhibited by disulfiram, a drug that is used in aversion therapy for the treatment of chronic alcoholism. The resulting rise in acetaldehyde results in flushing, tachycardia, hyperventilation, and nausea. In each reaction, electrons are transferred to oxidized nicotinamide adenine dinucleotide (NAD^+), resulting in an increase in the ratio of the reduced form (NADH) to NAD^+. In alcohol-related hypoglycemia, the abundance of NADH favors the reduction of pyruvate to lactate and of oxaloacetate (OAA) to malate. Pyruvate and OAA are substrates in gluconeogenesis. Thus, the ethanol-mediated increase in NADH causes these gluconeogenic precursors to be diverted into alternate pathways, resulting in decreased synthesis of glucose. This can precipitate hypoglycemia, particularly in individuals who have depleted their stores of liver glycogen. Hypoglycemia can produce many of the behaviors associated with alcohol intoxication, such as agitation, impaired judgment, and combativeness. Therefore, alcohol consumption in vulnerable individuals (such as those who are fasting or have engaged in prolonged, strenuous exercise) can produce hypoglycemia that may contribute to the behavioral effects of alcohol.

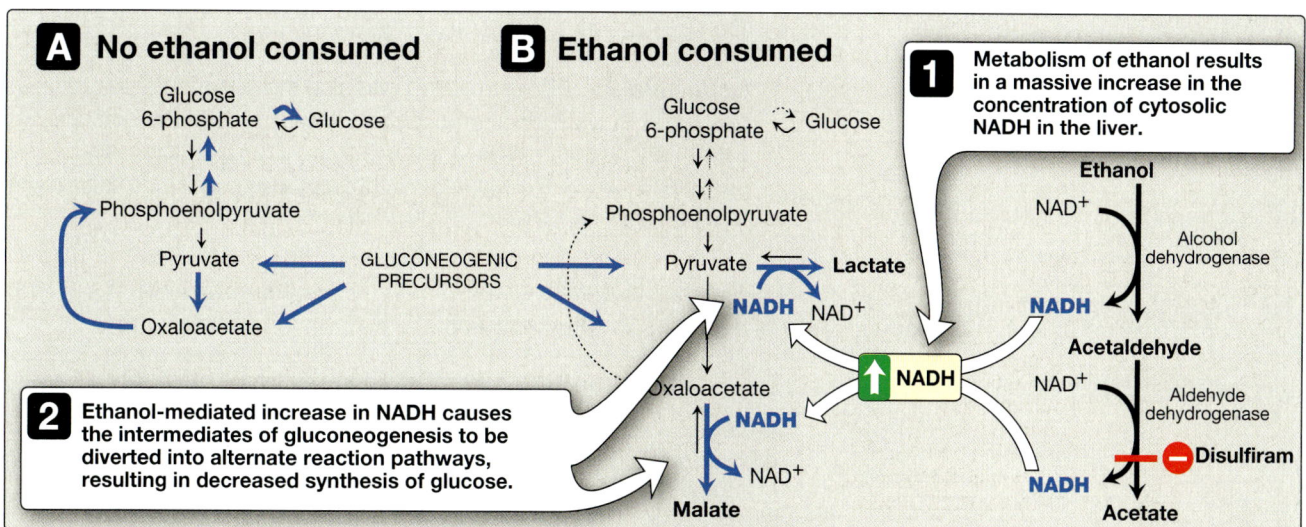

Figure 23.16
A. Normal gluconeogenesis in the absence of ethanol consumption. **B**. Inhibition of gluconeogenesis resulting from hepatic metabolism of ethanol. NAD(H), nicotinamide adenine dinucleotide.

Clinical Application 23.2: Other Alcohol-Related Metabolic Derangements

Other metabolic derangements result from excessive alcohol metabolism and the increased NADH/NAD$^+$ ratio. The effects are mediated through inhibition or reversal of oxidation reactions in which NAD$^+$ is the reaction electron acceptor cofactor. In addition to decreasing gluconeogenesis and causing hypoglycemia, several other reactions and pathways are affected, both in the cytosol and mitochondrial matrix. Lactic acid produced from pyruvate by lactate dehydrogenase can result in lactic acidosis. Lactic acidosis can cause a decrease in renal excretion of uric acid, resulting in hyperuricemia and gout-like symptoms. In glycolysis, the NADH/NAD$^+$ imbalance inhibits glyceraldehyde 3-phosphate dehydrogenase. Glyceraldehyde 3-phosphate is isomerized to dihydroxyacetone phosphate (DHAP). DHAP is then reduced to glycerol 3-phosphate by glycerol 3-phosphate dehydrogenase, driven by oxidation of NADH. β-Oxidation is also decreased since β-hydroxyacyl-CoA dehydrogenase requires NAD$^+$ as a cofactor. The decrease in FA oxidation and the presence of elevated glycerol 3-phosphate increases hepatic TAG synthesis coupled with impaired formation or release of VLDL complexes. This can result in hepatic steatosis (fatty liver) and hyperlipidemia. The NADH/NAD$^+$ imbalance also reverses the malate dehydrogenase reaction, decreasing the availability of OAA for the citric acid cycle (and gluconeogenesis). Acetyl-CoA, therefore, is diverted from the citric acid cycle toward hepatic ketone body formation (ketogenesis) and can result in alcoholic ketoacidosis. With continued alcohol consumption, alcoholic fatty liver can progress to alcoholic hepatitis and then to alcoholic cirrhosis. As liver damage increases, liver-specific enzymes are released into the blood, and hepatic functions will decrease. Laboratory results indicating hepatic dysfunction could include a plasma increase of liver-specific alanine and aspartate aminotransferases, alkaline phosphatase, and γ-glutamyl transferase; a decrease in blood urea nitrogen, total protein and albumin levels; and increased conjugated and unconjugated bilirubin (indicating hepatic jaundice), and bleeding time.

Because alcohol consumption can also increase the risk for hypoglycemia in patients using insulin, those in an intensive insulin treatment protocol (see Chapter 25, II.) are counseled about the increased risk of hypoglycemia that generally occurs many hours after alcohol ingestion.

V. CHAPTER SUMMARY

- The integration of **energy metabolism** is controlled primarily by **insulin** and the opposing actions of **glucagon** and catecholamines, particularly **epinephrine** (Fig. 23.17). Changes in the circulating levels of these hormones allow the body to store energy when food is abundant or to make stored energy available in times of **physiologic stress** (e.g., during survival crises, such as famine).

- **Insulin** is a peptide hormone produced by the β **cells** of the **islets of Langerhans** of the **pancreas**. It consists of disulfide-linked A and B chains. A rise in blood glucose is the most important signal for insulin **secretion**. The **catecholamines**, secreted in response to stress, trauma, or extreme exercise, inhibit insulin secretion.

- Insulin increases glucose uptake (by glucose transporters [**GLUT-4**] in muscle and adipose tissue) and the synthesis of **glycogen**, **protein**, and **triacylglycerol** (**TAG**); it is an **anabolic** hormone. These actions are mediated by binding to its membrane **tyrosine kinase receptor**. Binding initiates a cascade of cell-signaling responses, including phosphorylation of a family of proteins called **insulin receptor substrate** (**IRS**) **proteins**.

- **Glucagon** is a monomeric peptide hormone produced by the α **cells** of the pancreatic islets (both insulin and glucagon synthesis involve the formation of inactive precursors that are cleaved to form the active hormones). Glucagon, along with epinephrine, norepinephrine, cortisol, and growth hormone (the **counterregulatory hormones**), opposes many of the actions of insulin.

- Glucagon acts to maintain blood glucose during periods of potential hypoglycemia. Glucagon increases **glycogenolysis**, **gluconeogenesis**, **fatty acid oxidation**, **ketogenesis**, and **amino acid uptake**; it is a **catabolic** hormone. Glucagon secretion is stimulated by low blood glucose, amino acids, and catecholamines. Its secretion is inhibited by elevated blood glucose and by insulin.

- Glucagon binds to high-affinity **G protein–coupled receptors** (**GPCRs**) on the cell membrane of hepatocytes. Binding results in the activation of **adenylyl cyclase**, which produces the second messenger **cyclic adenosine monophosphate** (**cAMP**). Subsequent activation of **cAMP-dependent protein kinase A** results in the **phosphorylation**-mediated activation or inhibition of key regulatory enzymes involved in carbohydrate and lipid metabolism. Both insulin and glucagon affect **gene transcription**.

- **Hypoglycemia** is characterized by low blood glucose accompanied by **adrenergic** and **neuroglycopenic symptoms** that are rapidly resolved by the administration of glucose. Insulin-induced, postprandial, and fasting hypoglycemia result in the release of glucagon and epinephrine. The rise in the reduced form of **nicotinamide adenine dinucleotide** (**NADH**) that accompanies **ethanol** metabolism inhibits gluconeogenesis, leading to hypoglycemia in individuals with depleted stores. Alcohol consumption also increases the risk for hypoglycemia in patients using insulin. Chronic alcohol consumption can cause **fatty liver disease** and other metabolic pathway derangements including **inhibition of glycolysis** and **fatty acid oxidation, increased TAG synthesis, hepatic steatosis, hyperlipidemia, ketoacidosis, lactic acidosis**, and **hyperuricemia**.

Insulin

is a

Peptide hormone

secreted by

α Cells of the pancreas

secretion is stimulated by

↑ Blood glucose

secretion is inhibited by

↑ Epinephrine

causes

↑ Glucose uptake
Synthesis of
• Glycogen
• Protein
• Fat

↓ Glycogenolysis
Gluconeogenesis
Ketogenesis
Lipolysis

all mediated by

Activation of the insulin receptor

which leads to

Activation of receptor tyrosine kinase activity

which leads to

Phosphorylation of insulin receptor and IRSs

which leads to

Cascade of cell-signaling responses

which leads to

Phosphorylation and dephosphorylation of target proteins

also cause

Altered gene expression

Glucagon

is a

Peptide hormone

secreted by

α Cells of the pancreas

secretion is stimulated by

↓ Blood glucose

↑ Epinephrine

secretion is inhibited by

↑ Insulin

causes

↑ Glycogenolysis
Gluconeogenesis
Fatty acid oxidation
Ketogenesis
Uptake of amino acids

↓ Glycogenesis

all mediated by

Activation of the glucagon receptor

which leads to

Activation of adenylate cyclase

which leads to

Activation of protein kinases

which leads to

Phosphorylation of target proteins

Hypoglycemia

is characterized by

Central nervous system symptoms:
• Confusion
• Aberrant behavior
• Coma

and

Blood glucose ≤50 mg/dL

and

Rapid relief of symptoms following administration of glucose

occurs in

Patients receiving insulin treatment

Undernourished or fasted individuals consuming alcohol

Individuals with exaggerated postprandial insulin release

prompts

↑ Immediate secretion of
• Glucagon
• Epinephrine
• Norepinephrine

↓ Secretion of insulin

treated with

• Oral consumption of glucose in the conscious patient
• Subcutaneous or intramuscular injection of glucagon

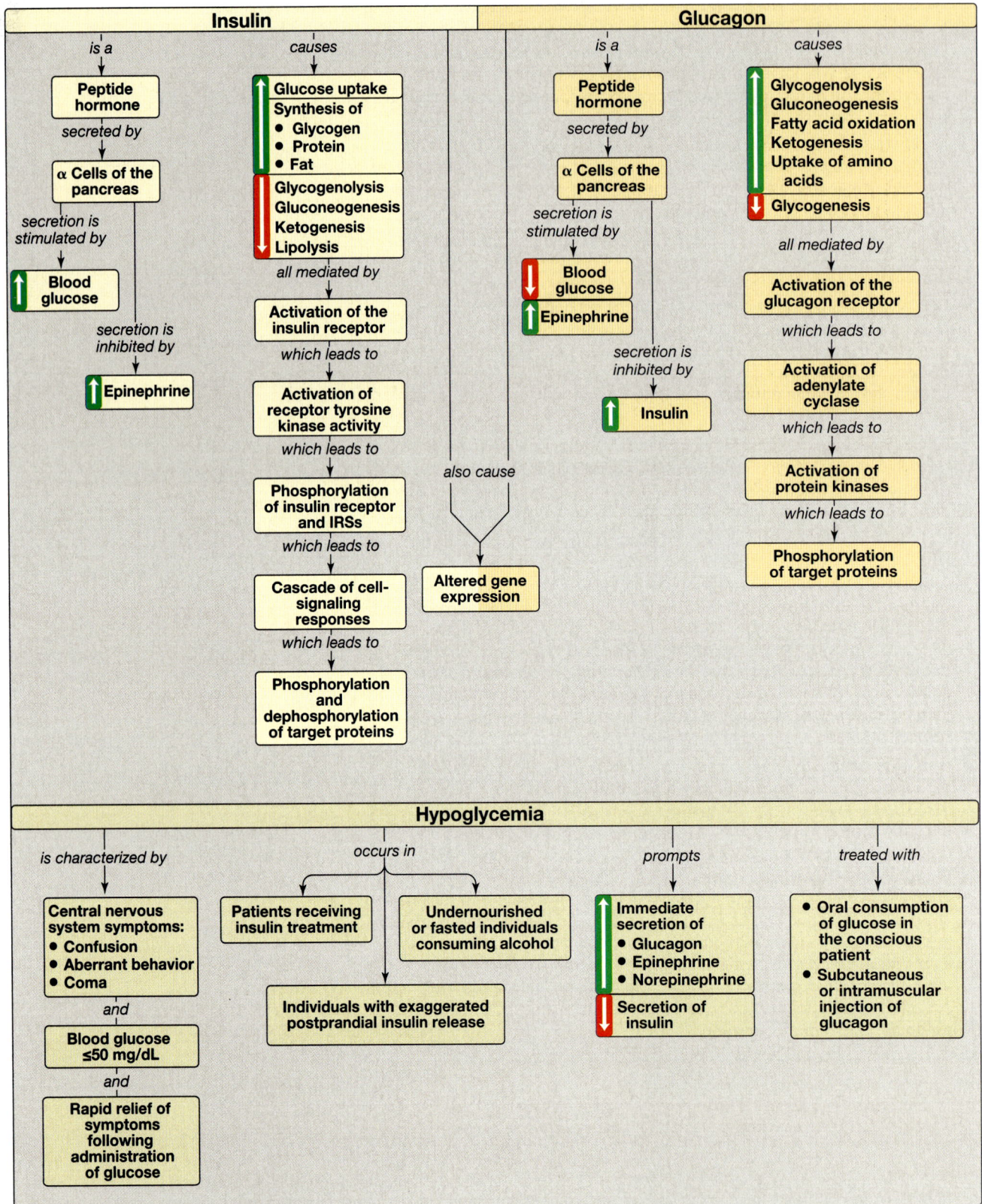

Figure 23.17
Key concept map for the metabolic effects of insulin and glucagon as well as hypoglycemia. IRSs, insulin receptor substrates.

Study Questions

Choose the ONE best answer.

23.1. Which of the following statements is true for insulin but not for glucagon?

  A. It is a peptide hormone secreted by pancreatic cells.
  B. Its actions are mediated by binding to a receptor found on the cell membrane of liver cells.
  C. Its effects include alterations in gene expression.
  D. Its secretion is decreased by the catecholamines.
  E. Its secretion is increased by amino acids.
  F. Its synthesis involves a nonfunctional precursor that gets cleaved to yield a functional molecule.

Correct answer = D. Secretion of insulin by pancreatic β cells is inhibited by the catecholamines, whereas glucagon secretion by the α cells is stimulated by them. All of the other statements are true for both insulin and glucagon.

23.2. In which one of the following tissues is glucose uptake into the cell insulin-dependent?

  A. Adipose
  B. Brain
  C. Liver
  D. Red blood cells
  E. Pancreas

Correct answer = A. The glucose transporter (GLUT-4) in adipose (and muscle) tissue is dependent on insulin. Insulin results in the movement of GLUT-4 from intracellular vesicles to the cell membrane. Glucagon can cause exocytosis of GLUT-4 from the plasma membrane to intracellular vesicles. The other tissues in the list contain other GLUT transporters that are independent of insulin because they are always located on the cell membrane.

23.3. A 39-year-old female is brought to the emergency room with weakness and dizziness. She recalls getting up early that morning to do her weekly errands and had skipped breakfast. She drank a cup of coffee for lunch and had nothing to eat during the day. She met with friends at 8 PM and had several drinks. As the evening progressed, she became weak and dizzy and was taken to the hospital. Laboratory tests revealed her blood glucose to be 45 mg/dL (normal = 70 to 99). She was given orange juice and immediately felt better. The biochemical basis of her alcohol-induced hypoglycemia is an increase in which of the following?

  A. Fatty acid oxidation
  B. Ratio of reduced NADH to oxidized NAD⁺ forms
  C. Oxaloacetate and pyruvate
  D. Use of acetyl coenzyme A in fatty acid synthesis
  E. Glycogen synthesis

Correct answer = B. The oxidation of ethanol to acetate by dehydrogenases is accompanied by the reduction of nicotinamide adenine dinucleotide (NAD⁺) to NADH. The rise in the NADH/NAD⁺ ratio shifts pyruvate to lactate and oxaloacetate (OAA) to malate, decreasing the availability of substrates for gluconeogenesis and resulting in hypoglycemia. The rise in NADH also reduces the NAD⁺ needed for fatty acid (FA) oxidation. The decrease in OAA shunts acetyl coenzyme A to ketogenesis. Inhibition of FA degradation results in re-esterification of FA into triacylglycerol that can result in fatty liver. Glycogen would not be synthesized under hypoglycemic conditions.

23.4. A patient is diagnosed with an insulinoma, a rare neuroendocrine tumor, the cells of which are derived primarily from pancreatic β cells. Which of the following would logically be characteristic of an insulinoma?

  A. Decreased body weight
  B. Decreased connecting peptide in the blood
  C. Decreased glucose in the blood
  D. Decreased insulin in the blood
  E. Decreased GLUT-4 activity

Correct answer = C. Insulinomas are characterized by constant production of insulin (and, therefore, of C-peptide) by the tumor cells. The increase in insulin drives glucose uptake by tissues such as muscle and adipose that have insulin-dependent GLUT-4 glucose transporters, resulting in hypoglycemia. However, the hypoglycemia is insufficient to suppress insulin production and secretion. Insulinomas, then, are characterized by increased blood insulin and decreased blood glucose. Insulin, as an anabolic hormone, results in weight gain.

23.5. In a patient with an even rarer glucagon-secreting tumor derived from the α cells of the pancreas, how would the presentation be expected to differ relative to the patient in Question 23.4?

A glucagon-secreting tumor of the pancreas (glucagonoma) would result in hyperglycemia, not hypoglycemia. The constant production of glucagon would result in constant gluconeogenesis, using amino acids from proteolysis as substrates. This results in loss of body weight.

23.6. In an individual who has fasted for more than 24 hours, which of the following normal tissue-specific metabolic outcomes would be expected?

A. Decrease in hepatic ketogenesis
B. Decrease in muscle FA oxidation
C. Increase in hepatic gluconeogenesis
D. Increase in hepatic glycogen storage
E. Increase in insulin to glucagon ratio

Correct answer = C. The fasting state will result in a low insulin-to-glucagon ratio. Higher glucagon levels will stimulate hepatic gluconeogenesis and glycogenolysis to help maintain blood glucose levels in the normal range and prevent hypoglycemia. In addition, glucagon signaling would result in the mobilization of free FA from adipocytes in the blood. This allows for more FA oxidation in muscle cells, especially in cardiac muscle. In the liver, FA oxidation provides acetyl-CoA as substrate for ketone body production. FA and ketone bodies can be used by certain tissues as an alternate carbon source, providing more glucose for the brain.

23.7. A runner is "carb loading" for a race the next day. Which of the following normal tissue-specific metabolic outcomes would be expected?

A. Decrease in insulin to glucagon ratio
B. Decrease in hepatic acetyl-CoA carboxylase activity
C. Decrease in hepatic carnitine palmitoyltransferase I activity
D. Increase in hepatic fructose 1,6-bisphosphatase activity
E. Increase in hepatic glycogen phosphorylase activity

Correct answer = C. Carbohydrate loading would indicate the well-fed state, resulting in a higher insulin-to-glucagon ratio. Higher insulin levels would result in activation of hepatic acetyl-CoA carboxylase, resulting in elevated malonyl-CoA levels for FA synthesis in the cytosol. Cytosolic malonyl-CoA increase would also inhibit hepatic carnitine palmitoyltransferase I activity, thereby inhibiting the carnitine shuttle and preventing FA import into the mitochondria for FA oxidation. In addition, insulin signaling will lead to inhibition of hepatic gluconeogenic glycogenolysis enzymes fructose 1,6-bisphosphatase and glycogen phosphorylase, respectively.

23.8. A patient with longstanding alcohol use disorder is seen in the emergency room with a significantly elevated blood alcohol level. Which of the following metabolic derangements could be expected?

A. Decrease in blood glucose levels
B. Decrease in ketogenesis
C. Decrease in lactic acid production
D. Increase in FA oxidation
E. Increase in renal urate excretion

Correct answer = A. Chronic elevated metabolism of ethanol leads to an increased NADH to NAD^+ ratio. Elevation of the reduced form NADH inhibits (or reverses) many reactions in many pathways where oxidized NAD^+ is a required cofactor. This cofactor imbalance leads to an increase in conversion of pyruvate to lactic acid by lactic acid dehydrogenase, causing a decrease in hepatic gluconeogenesis and hypoglycemia. The increase in lactic acid would also lead to decreased renal excretion of uric acid. Increased $NADH/NAD^+$ ratio also reverses malate dehydrogenase in the citric acid cycle and inhibits β hydroxyacyl-CoA dehydrogenase in β-oxidation. This leads to a decrease in FA oxidation and hepatic overproduction of ketone bodies from acetyl-CoA.

The Feed–Fast Cycle

24

I. OVERVIEW OF THE ABSORPTIVE STATE

The absorptive (well-fed) state is the 2- to 4-hour period after ingestion of a normal meal. During this interval, transient increases in plasma glucose, amino acids, and triacylglycerols (TAGs) occur, the latter primarily as components of chylomicrons synthesized and secreted by the intestinal mucosal cells (see Chapter 18, VI.). Pancreatic islet cells respond to gastrointestinal incretin secretion (see Chapter 23, II.) and the elevated level of glucose with increased secretion of insulin and decreased secretion of glucagon. The elevated insulin/glucagon ratio and the ready availability of circulating substrates make the absorptive state an anabolic period characterized by increased synthesis and storage of TAGs and glycogen to replenish fuel stores as well as increased synthesis of protein. During this absorptive period, virtually all tissues use glucose as a fuel, and the metabolic response of the body is dominated by alterations in the metabolism of liver, adipose tissue, skeletal muscle, and brain. In this chapter, an "organ map" is introduced that traces the movement of metabolites between tissues. The goal is to create an expanded and clinically useful vision of whole-body metabolism.

II. REGULATORY MECHANISMS

The flow of intermediates through metabolic pathways is controlled by four mechanisms: (1) the availability of substrates, (2) allosteric regulation of enzymes, (3) covalent modification of enzymes, and (4) induction–repression of enzyme synthesis, primarily through regulation of transcription. Although this scheme may at first seem redundant, each mechanism operates on a different timescale and allows the body to adapt to a wide variety of physiologic situations (Fig. 24.1). In the absorptive state, these regulatory mechanisms ensure that available nutrients are captured as glycogen, TAG, and protein.

A. Availability of substrates

In the absorptive phase, glucose is the predominant energy substrate for virtually all cell types. There are two main types of glucose transporters in cells, sodium–glucose linked transporters (SGLTs) and facilitated diffusion glucose transporters (GLUT), differing in terms of substrate specificity, tissue distribution and regulation. The GLUT transporter facilitates glucose uptake, and hexokinase traps glucose in the cell by phosphorylation, according to the relative kinetics (Michaelis constant [K_m]) of each enzyme. With high blood

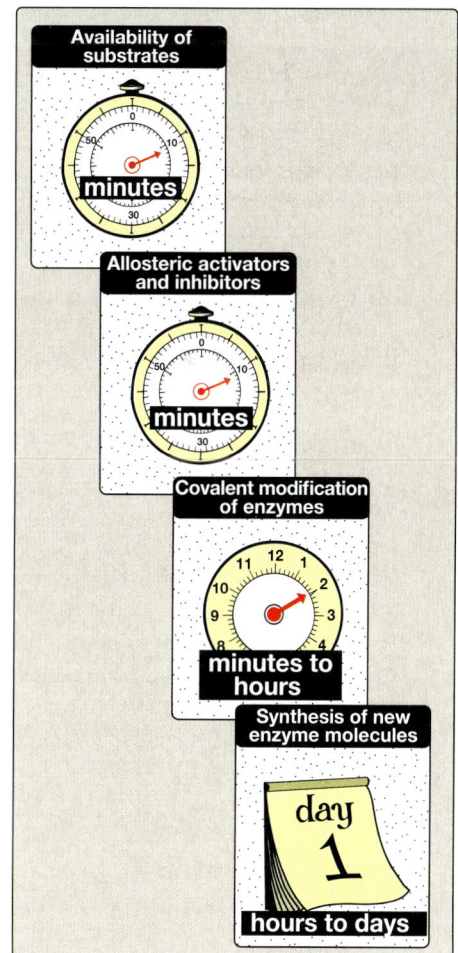

Availability of substrates — minutes

Allosteric activators and inhibitors — minutes

Covalent modification of enzymes — minutes to hours

Synthesis of new enzyme molecules — day 1 — hours to days

Figure 24.1
Control mechanisms of metabolism and some typical response times. [Note: Response times may vary according to the nature of the stimulus and from tissue to tissue.]

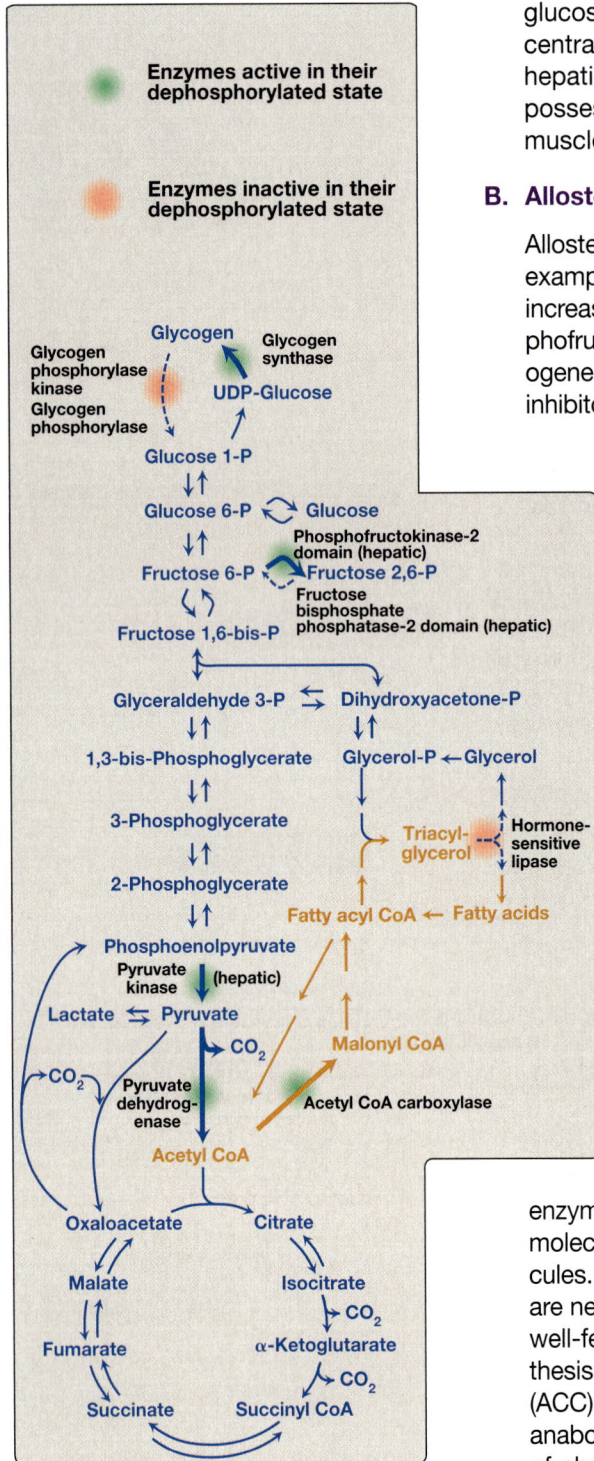

Figure 24.2
Important reactions of intermediary metabolism regulated by enzyme phosphorylation. **Blue text** = intermediates of carbohydrate metabolism; **brown text** = intermediates of lipid metabolism; P, phosphate; CoA, coenzyme A; CO_2, carbon dioxide.

glucose levels of the absorptive phase, the increased substrate concentration guarantees that there is even sufficient glucose for the hepatic isoforms, GLUT-2 and glucokinase (hexokinase IV), which possess higher K_m values (lower affinity) as compared to skeletal muscle isoforms, GLUT-4, and hexokinase I, respectively.

B. Allosteric effectors

Allosteric changes usually involve rate-determining reactions. For example, glycolysis in the liver is stimulated following a meal by an increase in fructose 2,6-bisphosphate, an allosteric activator of phosphofructokinase-1 ([PFK-1] see Chapter 8, V.). In contrast, gluconeogenesis is decreased by fructose 2,6-bisphosphate, an allosteric inhibitor of fructose 1,6-bisphosphatase (see Chapter 10, III.).

C. Covalent modification

The activity of many enzymes is regulated by the addition (via kinases in the fasting state, such as cyclic adenosine monophosphate [cAMP]–dependent protein kinase A [PKA] and adenosine monophosphate–activated protein kinase [AMPK]) or removal (via protein phosphatases in the well-fed absorptive state) of phosphate groups from specific serine, threonine, or tyrosine residues of the protein. In the absorptive state, most of the covalently regulated enzymes are in the dephosphorylated form and are active (Fig. 24.2). Three exceptions are glycogen phosphorylase kinase, glycogen phosphorylase (see Chapter 11, V.), and hormone-sensitive lipase ([HSL] see Chapter 16, IV.), which are inactive in their dephosphorylated form. In the liver, the dephosphorylated form of the bifunctional phosphofructokinase-2 (PFK-2) domain is active, increasing production of the allosteric activator fructose 2,6-bisphosphate (see Chapter 8, V.). The other domain, fructose 2,6-bisphosphatase, is inactive in the dephosphorylated form.

D. Induction and repression of enzyme synthesis

Increased (induction of) or decreased (repression of) enzyme synthesis leads to changes in the number of available enzyme molecules, rather than changing the activity of existing enzyme molecules. Enzymes subject to synthesis regulation are often those that are needed under specific physiologic conditions. For example, in the well-fed state, elevated insulin levels result in an increase in the synthesis of key enzymes, such as acetyl coenzyme A (CoA) carboxylase (ACC) and fatty acid (FA) synthase (see Chapter 16, III.), involved in anabolic metabolism. In the fasting state, glucagon induces expression of phosphoenolpyruvate carboxykinase (PEPCK) of gluconeogenesis (see Chapter 10, IV.). Therefore, both insulin and glucagon affect gene transcription in the absorptive well-fed and fasting states, respectively.

III. LIVER: NUTRIENT DISTRIBUTION CENTER

The liver is uniquely situated to process and distribute dietary nutrients because the venous drainage of the gut and pancreas passes through the hepatic portal vein before entry into the general circulation. Thus, after a meal, the liver is bathed in blood containing absorbed nutrients and elevated levels of insulin secreted by the pancreas. During the absorptive period, the liver takes up carbohydrates, lipids, and most amino acids. These nutrients are then metabolized, stored, or routed to other tissues. In this way, the liver smooths out potentially broad fluctuations in the availability of nutrients for the peripheral tissues.

A. Carbohydrate metabolism

The liver is normally a glucose-producing rather than a glucose-using organ. However, after a meal containing carbohydrates, the liver becomes a net consumer, retaining roughly 60 g of every 100 g of glucose presented by the portal system. This increased use reflects increased glucose uptake by the hepatocytes. Their insulin-independent glucose transporter (GLUT-2) has a high K_m for glucose and, therefore, takes up glucose only when blood glucose is high (see Chapter 8, IV.). Processes that are upregulated when hepatic glucose is increased include the following.

1. **Increased glucose phosphorylation:** With elevated blood glucose concentration and increased entry of glucose within hepatocytes by GLUT-2, glucokinase phosphorylates glucose to glucose 6-phosphate (Fig. 24.3). Glucokinase has a high K_m for glucose, is not subject to direct product inhibition, and has a sigmoidal reaction curve (see Chapter 8, V.).

2. **Increased glycogenesis:** The conversion of glucose 6-phosphate to glycogen is favored by the activation of glycogen synthase, both by dephosphorylation and by increased availability of glucose 6-phosphate, its positive allosteric effector (see Fig. 24.3).

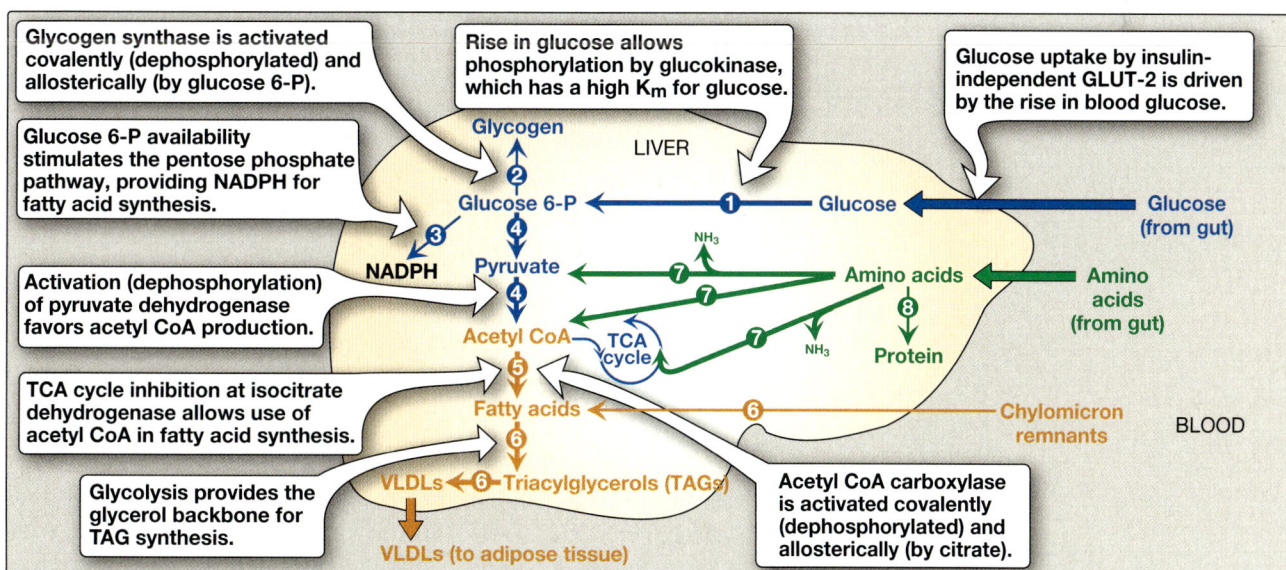

Figure 24.3
Major metabolic pathways in the liver in the absorptive state. [Note: The acetyl coenzyme A (CoA) is also used for cholesterol synthesis.] The *numbers in circles*, which appear both in the figure and in the text, indicate important pathways for carbohydrate, fat, or protein metabolism. **Blue text** = intermediates of carbohydrate metabolism; **brown text** = intermediates of lipid metabolism; **green text** = intermediates of protein metabolism; P, phosphate; TCA, tricarboxylic acid; VLDLs, very–low-density lipoproteins; GLUT, glucose transporter; NADPH, nicotinamide adenine dinucleotide phosphate; NH_3, ammonia.

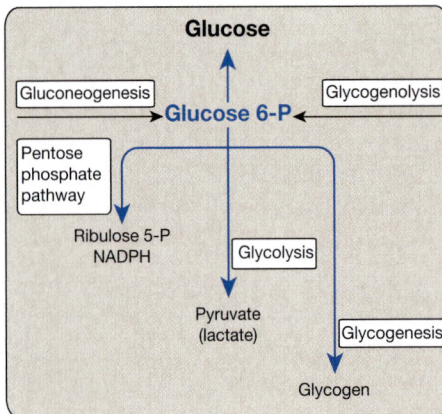

Glucose

Gluconeogenesis → **Glucose 6-P** ← Glycogenolysis

Pentose phosphate pathway

Ribulose 5-P
NADPH

Glycolysis

Pyruvate
(lactate)

Glycogenesis

Glycogen

Figure 24.4
Central role of glucose 6-phosphate in metabolism. [Note: The presence of glucose 6-phosphatase in the liver allows the production of free glucose from the glucose 6-phosphate produced in glycogenolysis and gluconeogenesis.] NADPH, nicotinamide adenine dinucleotide phosphate; P, phosphate.

3. **Increased pentose phosphate pathway activity:** The increased availability of glucose 6-phosphate, combined with the active use of reduced nicotinamide adenine dinucleotide phosphate (NADPH) in hepatic lipogenesis, stimulates the pentose phosphate pathway (see Chapter 13, II.). This pathway typically accounts for 5% to 10% of the glucose metabolized by the liver (see Fig. 24.3).

4. **Increased glycolysis:** In the liver, glycolysis is significant only during the absorptive period following a carbohydrate-rich meal. The conversion of glucose to pyruvate is stimulated by the elevated insulin/glucagon ratio that results in increased amounts of the regulated enzymes of glycolysis: glucokinase, PFK-1, and pyruvate kinase ([PK] see Chapter 8, VI.). Additionally, PFK-1 is allosterically activated by fructose 2,6-bisphosphate generated by the active (dephosphorylated) kinase domain of bifunctional PFK-2. PK is dephosphorylated and active. Pyruvate dehydrogenase (PDH), which converts pyruvate to acetyl CoA, is active (dephosphorylated) because pyruvate inhibits PDH kinase (see Fig. 24.3). The acetyl CoA is either used as a substrate for FA synthesis or is oxidized for energy in the tricarboxylic acid (TCA) cycle. Figure 24.4 summarizes the central role of glucose 6-phosphate in metabolism.

5. **Decreased glucose production:** While glycolysis and glycogenesis (pathways that promote glucose storage) are being stimulated in the liver in the absorptive state, gluconeogenesis and glycogenolysis (pathways that generate glucose) are being inhibited. Pyruvate carboxylase (PC), which catalyzes the first step in gluconeogenesis, is largely inactive because of low levels of acetyl CoA, its allosteric activator (see Chapter 10, III.). Acetyl CoA instead is being used as a substrate by ACC for FA synthesis (see Chapter 16, III.). The high insulin/glucagon ratio also favors inactivation of other gluconeogenic enzymes such as fructose 1,6-bisphosphatase by the allosteric inhibitor fructose 2,6-bisphosphate (see Fig. 8.17). Glycogenolysis is inhibited by dephosphorylation of glycogen phosphorylase and phosphorylase kinase (see Chapter 11, V.). The increased uptake and decreased hepatic production of blood glucose in the absorptive period also prevents hyperglycemia.

B. Fat metabolism

During the absorptive well-fed state, the liver will both synthesize FAs and take up excess dietary FAs and process these FAs as TAGs for export to other extrahepatic tissues in very–low-density lipoprotein (VLDL) complexes.

1. **Increased FA synthesis:** The liver is the primary site of *de novo* synthesis of FAs (see Fig. 24.3). FA synthesis, a cytosolic process, is favored in the absorptive period by availability of the substrates acetyl CoA (from glucose and amino acid metabolism) and NADPH (from glucose metabolism in the pentose phosphate pathway) and by the activation of ACC, both by dephosphorylation and by the presence of its allosteric activator, citrate. ACC catalyzes the formation of malonyl CoA from acetyl CoA,

the rate-limiting reaction for FA synthesis (see Chapter 16, III.). Malonyl CoA also inhibits carnitine palmitoyltransferase-I [CPT-I] of FA oxidation (see Chapter 16, IV.). Thus, citrate directly activates FA synthesis and indirectly inhibits FA degradation.

a. **Source of cytosolic acetyl CoA:** Pyruvate from aerobic glycolysis enters the mitochondria and is decarboxylated by PDH. The acetyl CoA product is combined with oxaloacetate (OAA) to form citrate via citrate synthase of the TCA cycle. When the TCA cycle is very active, ATP levels rise. ATP inhibits isocitrate dehydrogenase, leading to a buildup of citrate. Citrate leaves the mitochondria and enters the cytosol. Citrate is cleaved by ATP citrate lyase (induced by insulin), producing the acetyl CoA substrate of ACC plus OAA.

b. **Additional source of NADPH:** The OAA is reduced to malate, which is oxidatively decarboxylated to pyruvate by malic enzyme as NADPH is formed (see Chapter 16, III.).

2. **Increased triacylglycerol synthesis:** TAG synthesis is favored because fatty acyl CoAs are available both from *de novo* synthesis and from hydrolysis of TAGs from chylomicron remnants removed from the blood by hepatocytes (see Chapter 15, II.). Glycerol 3-phosphate, the backbone for TAG synthesis, is provided by glycolysis (see Chapter 16, III.). The liver packages these endogenous TAGs into VLDL complexes that are secreted into the blood for use by extrahepatic tissues, particularly adipose and muscle tissues (see Fig. 24.3).

C. Amino acid metabolism

The body cannot store amino acids. During the absorptive well-fed state, the liver degrades excess amino acids. In addition, it synthesizes and secretes needed proteins, some of which may have been degraded during the previous fasting period.

1. **Increased amino acid degradation:** In the absorptive period, more amino acids are present than the liver can use in the synthesis of proteins and other nitrogen-containing molecules. The surplus amino acids are not stored but are either released into the blood for other tissues to use in protein synthesis or deaminated, with the resulting carbon skeletons being degraded by the liver to pyruvate, acetyl CoA, or TCA cycle intermediates. These metabolites can be oxidized for energy or used in FA synthesis (see Fig. 24.3). The liver has limited capacity to initiate degradation of the branched-chain amino acids (BCAAs)—leucine, isoleucine, and valine. They pass through the liver essentially unchanged and are metabolized in muscle (see Chapter 20, III.).

2. **Increased protein synthesis:** The body does not store protein for energy in the same way that it maintains glycogen or TAG reserves. However, a transient increase in the synthesis of hepatic proteins does occur in the absorptive state, resulting in the replacement of any proteins that may have been degraded during the previous period of fasting (see Fig. 24.3).

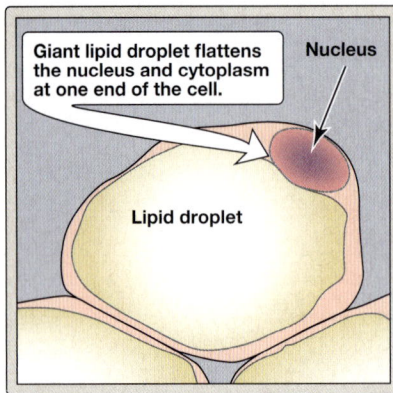

Figure 24.5
Colorized transmission electron micrograph of adipocytes.

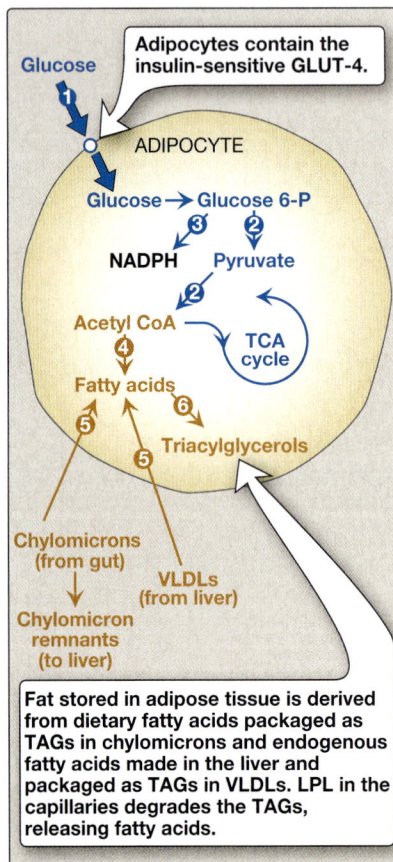

Figure 24.6
Major metabolic pathways in adipose tissue in the absorptive state. [Note: The numbers in the *circles*, which appear both in the figure and in the corresponding text, indicate important pathways for adipose tissue metabolism.] GLUT, glucose transporter; P, phosphate; NADPH, nicotinamide adenine dinucleotide; CoA, coenzyme A; TCA, tricarboxylic acid; TAGs, triacylglycerol; VLDLs, very–low-density lipoprotein; LPL, lipoprotein lipase.

IV. ADIPOSE TISSUE: ENERGY STORAGE DEPOT

Adipose tissue is second only to the liver in its ability to distribute fuel molecules. In a 70-kg man, white adipose tissue (WAT) weighs ~14 kg, or about half as much as the total muscle mass. Nearly the entire volume of each adipocyte in WAT can be occupied by a droplet of anhydrous, calorically dense TAGs (Fig. 24.5).

A. Carbohydrate metabolism

In the absorptive well-fed state, adipose tissue stores FAs as TAGs in fat droplets. The adipose tissue uses the glycolytic and pentose phosphate pathways to make glycerol 3-phosphate and NADPH, respectively, for the esterification of dietary FAs into TAGs for storage.

1. **Increased glucose transport:** Circulating insulin levels are elevated in the absorptive state, resulting in an influx of glucose into adipocytes via insulin-sensitive GLUT-4, recruited to the cell surface from intracellular vesicles (Fig. 24.6). The glucose is phosphorylated by hexokinase.

2. **Increased glycolysis:** The increased intracellular availability of glucose results in an enhanced rate of glycolysis (see Fig. 24.6). In adipose tissue, glycolysis serves a synthetic function by supplying glycerol 3-phosphate for TAG synthesis (see Chapter 16, III.). Unlike the liver, adipose tissue lacks glycerol kinase and cannot synthesize glycerol 3-phosphate from glycerol.

3. **Increased pentose phosphate pathway activity:** Adipose tissue can metabolize glucose by means of the pentose phosphate pathway, thereby producing NADPH, which is essential for FA synthesis (see Fig. 24.6). However, in humans, *de novo* synthesis is not a major source of FAs in adipose tissue, except when refeeding a previously fasted individual.

B. Fat metabolism

Most of the FAs added to the TAG stores of adipocytes after consumption of a lipid-containing meal are provided by the degradation of exogenous (dietary) TAGs in chylomicrons produced by the intestine, and from endogenous TAGs in VLDLs produced by the liver (see Fig. 24.6). The FAs are released from these lipoprotein complexes by lipoprotein lipase (LPL), an extracellular enzyme attached to the endothelial cells of capillary walls in many tissues, particularly adipose and muscle (see Chapter 18, VI.). In adipose tissue, LPL is upregulated by insulin. Thus, in the well-fed absorptive state, elevated levels of glucose and insulin favor storage of TAGs in adipose tissue (see Fig. 24.6). Elevated insulin also favors the dephosphorylated inactive form of HSL, thereby inhibiting lipolysis of stored TAGs in the well-fed state (see Chapter 16, IV.).

V. RESTING SKELETAL MUSCLE

Skeletal muscle accounts for ~40% of the body mass in individuals of healthy weight, and it can use glucose, amino acids, FAs, and ketone bodies as fuel. In the well-fed state, muscle takes up glucose via GLUT-4 (for energy and glycogen synthesis) and amino acids (for energy and protein synthesis). In contrast to liver, there is no covalent regulation of PFK-2 in skeletal muscle. However, in the cardiac isozyme, the kinase domain is activated by epinephrine-mediated phosphorylation (see Chapter 8, V.).

> Skeletal muscle is unique in being able to respond to substantial changes in the demand for ATP that accompanies contraction. At rest, muscle accounts for ~25% of the oxygen (O_2) consumption of the body, whereas during vigorous exercise, it is responsible for up to 90%. This underscores the fact that skeletal muscle, despite its potential for transient periods of anaerobic glycolysis, is an oxidative tissue.

A. Carbohydrate metabolism

In the absorptive well-fed state, muscle tissue uses glucose as its main energy source. It stores excess glucose in glycogen granules and synthesizes muscle proteins that may have been lost during the previous fasting state. Muscle is also the main tissue for metabolism of BCAAs.

1. **Increased glucose transport:** The transient increase in plasma glucose and insulin after a carbohydrate-rich meal leads to an increase in glucose transport into muscle cells (myocytes) by GLUT-4 (Fig. 24.7), thereby reducing blood glucose. Glucose is phosphorylated to glucose 6-phosphate by hexokinase and metabolized to meet the energy needs of myocytes.

2. **Increased glycogenesis:** The increased insulin/glucagon ratio and the availability of glucose 6-phosphate favor glycogen synthesis, particularly if glycogen stores have been depleted as a result of exercise or previous fasting (see Fig. 24.7). Glycogen synthase is active in the dephosphorylated state, whereas glycogen phosphorylase and phosphorylase kinase are both inactive in the dephosphorylated state (see Chapter 11, V.).

B. Fat metabolism

FAs are released from chylomicrons and VLDL by the action of LPL (see Chapter 18, VI.). However, FAs are of secondary importance as a fuel for resting muscle during the well-fed state, in which glucose is the primary source of energy. As a result, skeletal muscle secretes a lower or basal level of LPL in the absorptive phase. Cardiac muscle will always secrete LPL at a higher basal level compared to skeletal muscle. Cardiac muscle may obtain 50% to 60% of its energy from FAs in the absorptive phase. In the fasting state, this can increase to 90%.

C. Amino acid metabolism

1. **Increased protein synthesis:** An increase in amino acid uptake and protein synthesis occurs in the absorptive period after ingestion of a meal containing protein (see Fig. 24.7). This synthesis replaces protein degraded since the previous meal.

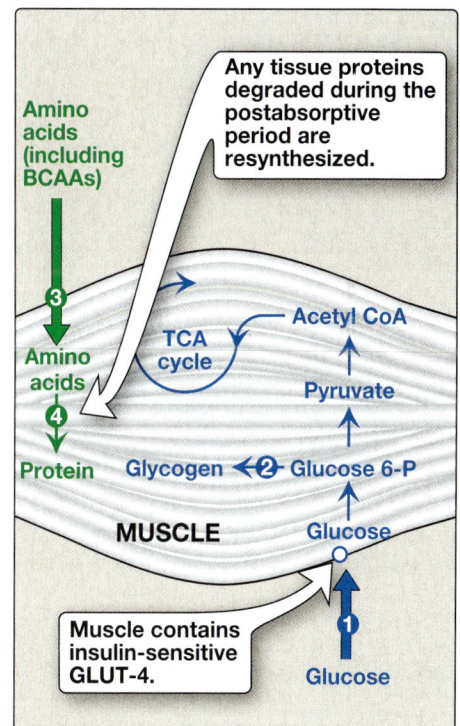

Figure 24.7
Major metabolic pathways in skeletal muscle in the absorptive state. [Note: The *numbers in circles*, which appear both in the figure and in the text, indicate important pathways for carbohydrate or protein metabolism.] CoA, coenzyme A; P, phosphate; GLUT, glucose transporter; BCAAs, branched-chain amino acids; TCA, tricarboxylic acid.

2. Increased BCAA uptake: Muscle is the principal site for degradation of the BCAAs because it contains the required transaminase (see Chapter 20, III.). The dietary BCAAs escape metabolism by the liver and are taken up by muscle, where they are used for protein synthesis (see Fig. 24.7) and as energy sources.

VI. BRAIN

Although contributing only 2% of the adult weight, the brain accounts for a consistent 20% of the basal O_2 consumption of the body at rest. Because the brain is vital to the proper functioning of all organs of the body, special priority is given to its fuel needs. To provide energy, substrates must be able to cross the endothelial cells that line the blood vessels in the brain (blood–brain barrier [BBB]). In the well-fed state, the brain exclusively uses glucose as a fuel (GLUT-1 of the BBB is insulin independent with a low K_m [1 to 2 mM]), completely oxidizing ~140 g/day to carbon dioxide and water. Because the brain contains no significant stores of glycogen, it is completely dependent on the availability of blood glucose (Fig. 24.8). If blood glucose levels fall below 50 mg/dL (normal fasting blood glucose is 70 to 99 mg/dL), cerebral function is impaired (see Fig. 23.14). The brain also lacks significant stores of TAGs, and the FAs circulating in the blood make little contribution to energy production for reasons that are unclear. The intertissue exchanges characteristic of the absorptive period are summarized in Figure 24.9.

VII. OVERVIEW OF THE FASTING STATE

Fasting begins if no food is ingested after the absorptive period. It may result from an inability to obtain food, the desire to lose weight rapidly, or clinical situations in which an individual cannot eat (e.g., because of trauma, surgery, cancer, or burns). In the absence of food, plasma levels of glucose, amino acids, and TAGs fall, triggering a decline in insulin secretion and an increase in glucagon, epinephrine, and cortisol secretion. The decreased insulin/counterregulatory hormone ratio and the decreased availability of circulating substrates make the postabsorptive period of nutrient deprivation a catabolic period characterized by degradation of TAGs, glycogen, and protein. This sets into motion an exchange of substrates among the liver, adipose tissue, skeletal muscle, and brain that is guided by two priorities: (1) the need to maintain adequate plasma levels of glucose to sustain energy metabolism in the brain, red blood cells, and other glucose-requiring tissues and (2) the need to mobilize FAs from TAGs in WAT for the synthesis and release of ketone bodies by the liver to supply energy to other tissues and spare body protein. As a result, blood glucose levels are maintained within a narrow range in the fasting state, while FA and ketone body levels increase. Maintaining glucose requires that the substrates for gluconeogenesis (such as pyruvate, alanine, and glycerol) be available.

A. Fuel stores

The metabolic fuels available in a normal 70-kg man at the beginning of a fast are shown in Figure 24.10. Observe the enormous caloric stores available in the form of TAGs compared with those contained

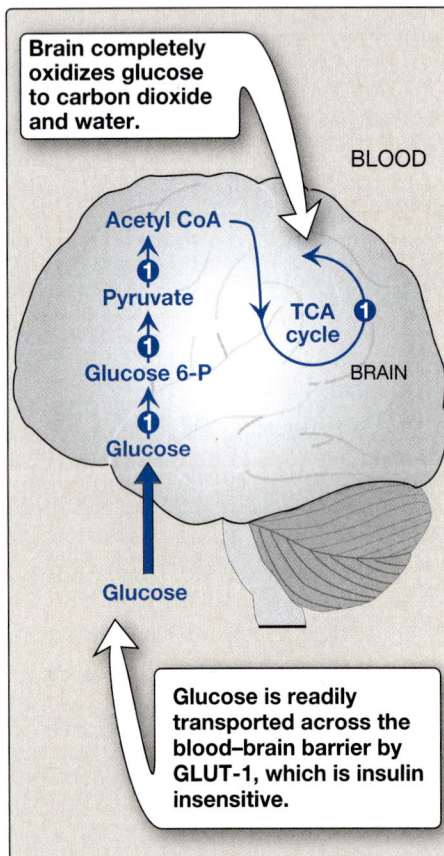

Figure 24.8
Major metabolic pathways in the brain in the absorptive state. [Note: The *numbers in circles*, which appear both in the figure and in the text, indicate important pathways for carbohydrate metabolism.] CoA, coenzyme A; TCA, tricarboxylic acid; P, phosphate; GLUT, glucose transporter.

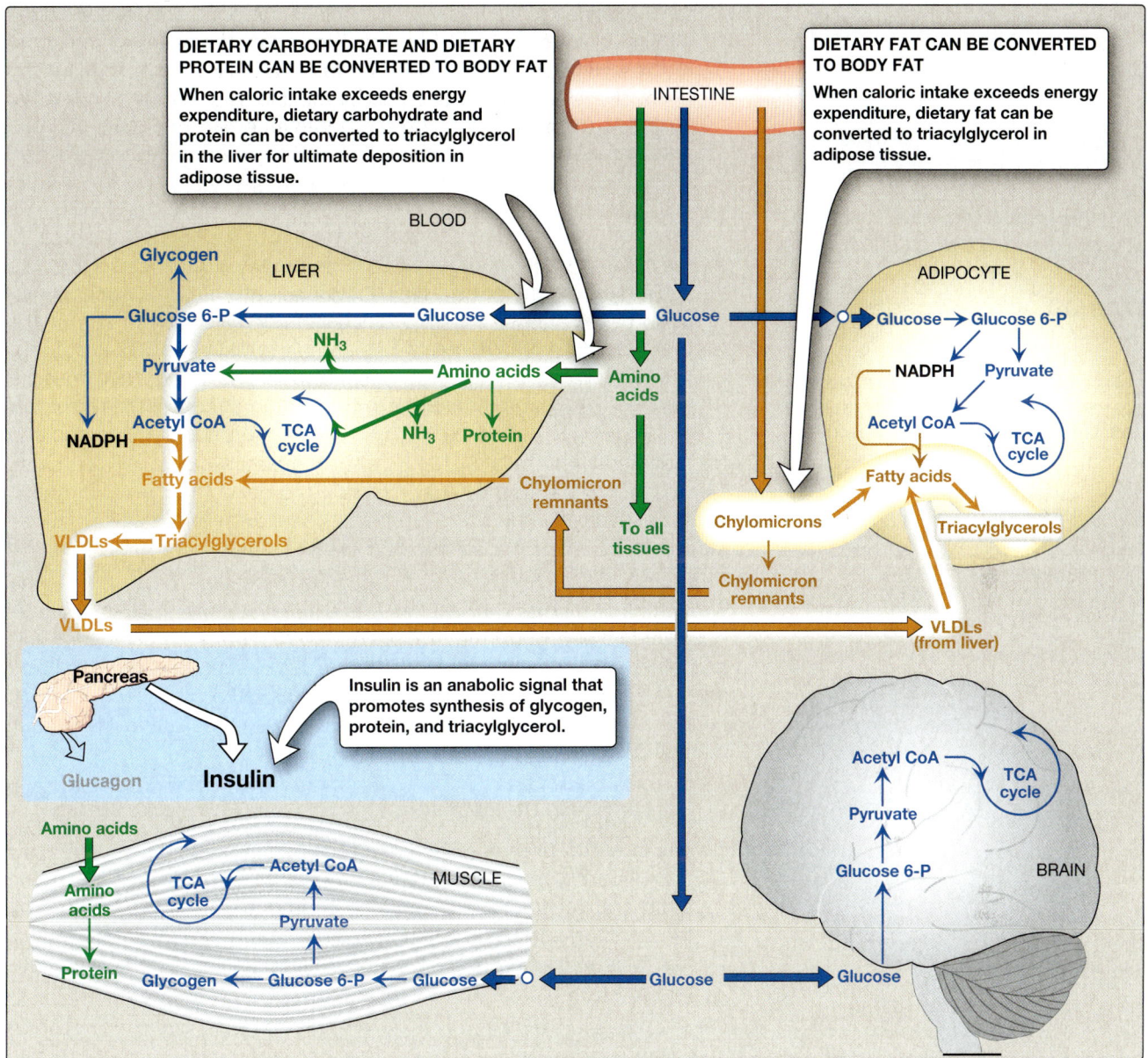

Figure 24.9
Intertissue relationships in the absorptive state and the hormonal signals that promote them. [Note: *Small circles* on the perimeter of muscle and the adipocyte indicate insulin-dependent glucose transporters.] P, phosphate; CoA, coenzyme A; NADPH, nicotinamide adenine dinucleotide phosphate; TCA, tricarboxylic acid; VLDLs, very–low-density lipoproteins.

in glycogen. Although protein is listed as an energy source, each protein also has a function unrelated to energy metabolism, as either a structural component of the body or as an enzyme. Therefore, only about one-third of the body's protein can be used for energy production without fatally compromising vital functions.

B. Enzymic changes

In the fasting state (as in the well-fed state), the flow of intermediates through the pathways of energy metabolism is controlled by four mechanisms: (1) the availability of substrates, (2) allosteric regulation

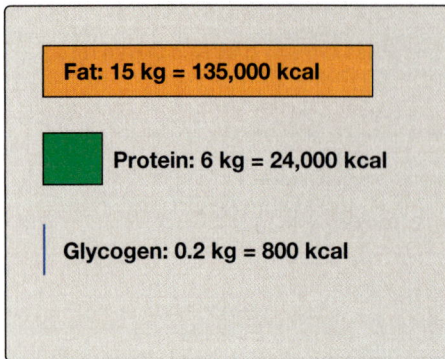

Fat: 15 kg = 135,000 kcal

Protein: 6 kg = 24,000 kcal

Glycogen: 0.2 kg = 800 kcal

Figure 24.10
Metabolic fuels present in a 70-kg man at the beginning of a fast. The fat stores are sufficient to meet energy needs for ~80 days.

of enzymes, (3) covalent modification of enzymes, and (4) induction–repression of enzyme synthesis. The metabolic changes observed in the fasting state are generally opposite those described for the absorptive state (see Fig. 24.9). For example, although most of the enzymes regulated by covalent modification are dephosphorylated and active in the well-fed state, they are phosphorylated and inactive in the fasting state. Three exceptions are glycogen phosphorylase, glycogen phosphorylase kinase (see Chapter 11, V.), and HSL (see Chapter 16, IV.), which are active in their phosphorylated states. In the fasting state, substrates are not provided by the diet but are available from the breakdown of stores and/or tissues, such as glycogenolysis with release of glucose from glycogen stores in the liver, lipolysis with release of FAs and glycerol from TAGs in adipose tissue, and proteolysis with release of amino acids from muscle. Recognition that the changes in the fasting state are the reciprocal of those in the well-fed state is helpful in understanding the ebb and flow of metabolism.

VIII. LIVER IN THE FASTING STATE

The primary role of the liver in the fasting state is maintenance of blood glucose through the production of glucose (from glycogenolysis and gluconeogenesis) for glucose-requiring tissues and the synthesis and distribution of ketone bodies for use by other tissues. Therefore, hepatic metabolism is distinguished from peripheral (or extrahepatic) metabolism.

A. Carbohydrate metabolism

The liver first uses glycogen degradation and then gluconeogenesis to maintain blood glucose levels to sustain energy metabolism of the brain and other glucose-requiring tissues in the fasting state. The presence of glucose 6-phosphatase in the liver allows the production of free glucose both from glycogenolysis and from gluconeogenesis (see Fig. 24.4).

1. **Increased glycogenolysis:** Figure 24.11 shows the sources of blood glucose during a typical three-meal day. After breakfast, lunch, and dinner, the primary source for blood glucose levels in the well-fed absorptive phase is from the diet, which peaks within the first 2 hours with a rise in insulin secretion. During this time, hepatic glycogen stores are replenished (see Fig. 24.3). As dietary blood glucose levels decline, there is increased secretion of glucagon and decreased secretion of insulin, representing the postabsorptive, fasting state. The increased glucagon/insulin ratio causes a rapid mobilization of hepatic glycogen stores (~80 g of glycogen from the well-fed state) because of PKA-mediated phosphorylation (and activation) of glycogen phosphorylase kinase that phosphorylates (and activates) glycogen phosphorylase (see Chapter 11, V.). Simultaneous phosphorylation of glycogen synthase inhibits glycogenesis. Gluconeogenesis contributes only a small percentage to blood glucose levels postabsorptive after breakfast and lunch, since the glycogen stores are sufficient during these shorter

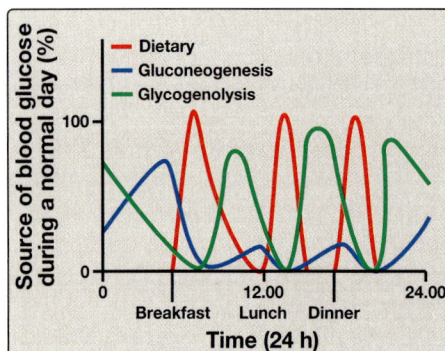

Figure 24.11
Sources of blood glucose during a typical three-meal day.

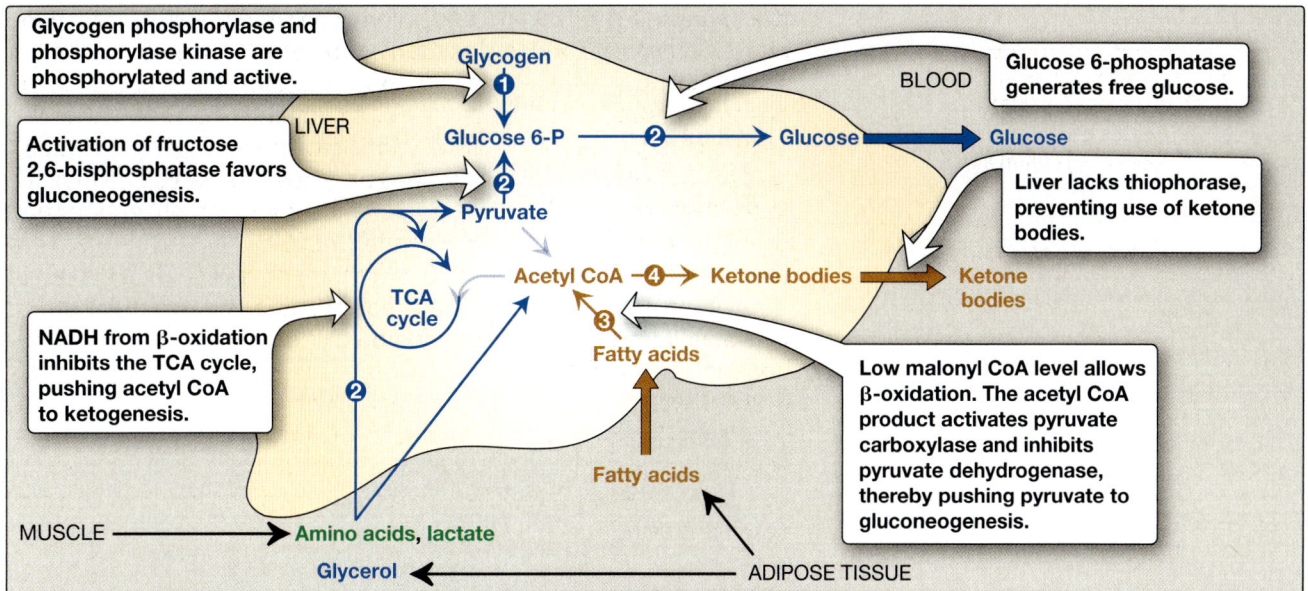

Figure 24.12
Major metabolic pathways in the liver during the fasting state. [Note: The *numbers in circles*, which appear both in the figure and in the corresponding citation in the text, indicate important metabolic pathways for carbohydrate or fat.] P, phosphate; CoA, coenzyme A; TCA, tricarboxylic acid; NADH, nicotinamide adenine dinucleotide.

interprandial periods. However hepatic glycogen stores are barely sufficient to maintain blood glucose levels during the longer time period (~12 hours) after dinner and fasting while we sleep. Late in the night or in early morning, as the major fraction of hepatic glycogen becomes depleted, gluconeogenesis becomes the primary source of blood glucose. If an individual continues fasting the following day, gluconeogenesis remains the main source for maintaining blood glucose levels. Figure 24.12 shows glycogen degradation as part of the overall metabolic response of the liver during the fasting state.

2. **Increased gluconeogenesis:** The synthesis of glucose and its release into the circulation are vital hepatic functions during short- and long-term fasting (see Fig. 24.12). The carbon skeletons for gluconeogenesis are derived primarily from glucogenic amino acids and lactate from muscle and glycerol from adipose tissue. Gluconeogenesis is favored by a decreased availability of the allosteric inhibitor fructose 2,6-bisphosphate and the subsequent activation of fructose 1,6-bisphosphatase, and by induction of PEPCK by glucagon (see Chapter 10, III.). Gluconeogenesis begins 4 to 6 hours after the last meal and becomes fully active as stores of liver glycogen are depleted (see Fig. 24.11). The decrease in fructose 2,6-bisphosphate simultaneously allosterically inhibits glycolysis at PFK-1 (see Chapter 8, V.).

B. Fat metabolism

The liver is active in maintaining blood glucose levels in the fasting state. It uses FA oxidation as its major source of energy production.

It also synthesizes ketone bodies, which can be used by the brain and other tissues for energy production.

1. **Increased FA oxidation:** The oxidation of FAs obtained from TAG hydrolysis in adipose tissue is the major source of energy in hepatic tissue in the fasting state (see Fig. 24.12). The fall in malonyl CoA because of phosphorylation (inactivation) of ACC by AMPK removes the inhibition on CPT-I, allowing β-oxidation to occur (see Chapter 16, IV.). FA oxidation generates nicotinamide adenine dinucleotide (NADH), flavin adenine dinucleotide ($FADH_2$), and acetyl CoA. The NADH inhibits the TCA cycle and shifts OAA to malate. This results in acetyl CoA being available for ketogenesis. The acetyl CoA is also an allosteric activator of PC and an allosteric inhibitor of PDH, thereby favoring use of pyruvate in gluconeogenesis (see Chapter 10, IV. and Fig. 10.9). Acetyl CoA is not a substrate for gluconeogenesis, in part because the PDH reaction is irreversible. Oxidation of NADH and $FADH_2$ coupled with oxidative phosphorylation supplies the energy required by the PC and PEPCK reactions of gluconeogenesis.

2. **Increased ketogenesis:** The liver is unique in being able to synthesize and release ketone bodies, primarily 3-hydroxybutyrate but also acetoacetate, for use as fuel by peripheral tissues. The liver itself (and peripheral tissues lacking mitochondria, e.g., RBCs) does not use ketone bodies as fuel because the liver lacks the mitochondrial enzyme thiophorase (see Chapter 16, V.). Ketogenesis, which starts during the first days of fasting, as shown in Figure 24.13, is favored when the concentration of acetyl CoA from FA oxidation exceeds the oxidative capacity of the TCA cycle. Ketogenesis releases CoA, ensuring its availability for continued FA oxidation. The availability of circulating water-soluble ketone bodies is important in the fasting state because they can be used for fuel by most tissues, including the brain, once their blood level is high enough. Ketone body concentration in blood increases from ~50 μM to ~6 mM in the fasting state. This reduces the need for gluconeogenesis from amino acid carbon skeletons, thus preserving essential protein (see Fig. 24.11). Ketogenesis as part of the overall hepatic response to the fasting state is shown in Figure 24.12. Ketone bodies are organic acids and, when present at high concentrations, can cause ketoacidosis.

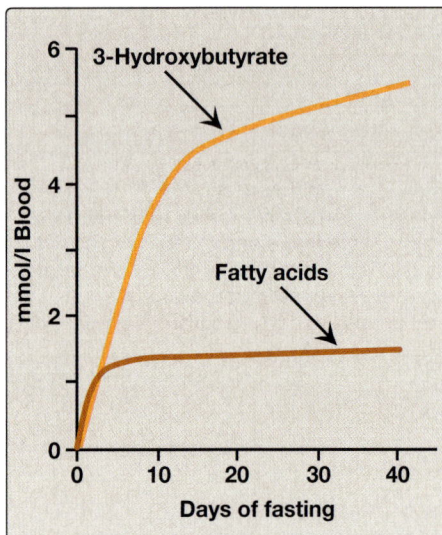

Figure 24.13
Concentrations of fatty acids and 3-hydroxybutyrate in the blood during the fasting state. [Note: 3-Hydroxybutyrate is made from the NADH-requiring reduction of acetoacetate.]

IX. ADIPOSE TISSUE IN THE FASTING STATE

A. Carbohydrate metabolism

Glucose transport by insulin-sensitive GLUT-4 into the adipocyte and its subsequent metabolism is decreased because of low levels of circulating insulin. This also results in decreased TAG synthesis.

B. Fat metabolism

Adipose tissue is second only to the liver in its ability to distribute energy compounds to the rest of the body in the fasting state. Adipose cells hydrolyze stored TAGs, providing free FAs in the blood to other tissues.

1. **Increased fat degradation:** The PKA-mediated phosphorylation and activation of HSL (see Chapter 16, IV.) and subsequent hydrolysis of stored fat (TAGs) are enhanced by the elevated catecholamines norepinephrine and epinephrine. These hormones, which are secreted from the sympathetic nerve endings in adipose tissue and/or from the adrenal medulla, are physiologically important activators of HSL (Fig. 24.14).

2. **Increased FA release:** FAs obtained from hydrolysis of TAGs stored in adipocytes are primarily released into the blood (see Fig. 24.14). Bound to albumin, they are transported to a variety of tissues for use as fuel. The glycerol produced from TAG degradation is used as a gluconeogenic precursor by the liver, which contains glycerol kinase. FAs can also be oxidized to acetyl CoA, which can enter the TCA cycle, thereby producing energy for the adipocyte.

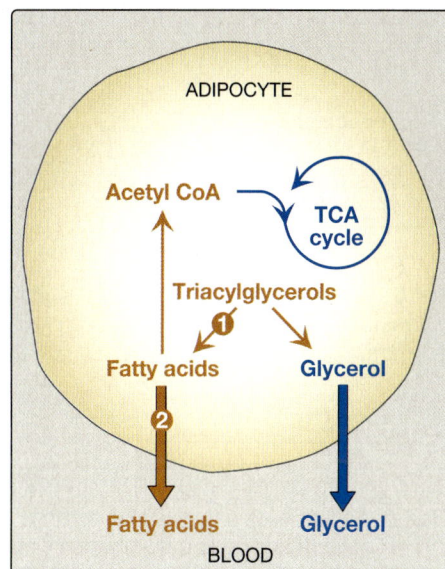

Figure 24.14
Major metabolic pathways in adipose tissue during the fasting state. [Note: The *numbers in the circles*, which appear both in the figure and in the corresponding citation in the text, indicate important pathways for fat metabolism.] CoA, coenzyme A; TCA, tricarboxylic acid.

> TAGs are produced by esterifying three FAs to glycerol 3-phosphate. The process of synthesizing glycerol 3-phosphate from substrates other than glucose is referred to as glyceroneogenesis (see Chapter 16, III.). Thiazolidinedione, an antihyperglycemic drug used in treating type 2 diabetes, functions to increase transcription of PEPCK in adipose tissues. This results in increasing glyceroneogenesis and TAG production in adipocytes. The subsequent reduction of plasma-free FAs improves insulin sensitivity, as muscle and other tissue types become more dependent on glucose oxidation for energy.

3. **Decreased FA uptake:** In the fasting state, the LPL activity of adipose tissue is low. Consequently, FAs from circulating TAGs within lipoprotein complexes are less available to adipose tissue in the fasting state than to muscle.

X. RESTING SKELETAL MUSCLE IN THE FASTING STATE

Resting muscle switches from glucose to FAs as its major fuel source in the fasting state. By contrast, exercising muscle initially uses creatine phosphate and its glycogen stores. During intense exercise, glucose 6-phosphate from glycogenolysis is converted to lactate by anaerobic glycolysis. The lactate is used by the liver for gluconeogenesis (Cori cycle; see Chapter 10, II. and Fig. 10.2). As these glycogen reserves are depleted, free FAs provided by the degradation of TAGs in adipose tissue become the dominant energy source. The contraction-based rise in AMP activates AMPK that phosphorylates and inactivates the muscle isozyme of ACC, decreasing malonyl CoA, enabling the carnitine shuttle CPT-1 and mitochondrial FA oxidation (see Chapter 16, IV.). Since muscle cells do not possess glucose 6-phosphatase, glucose 6-phosphate produced by muscle glycogenolysis in the fasting

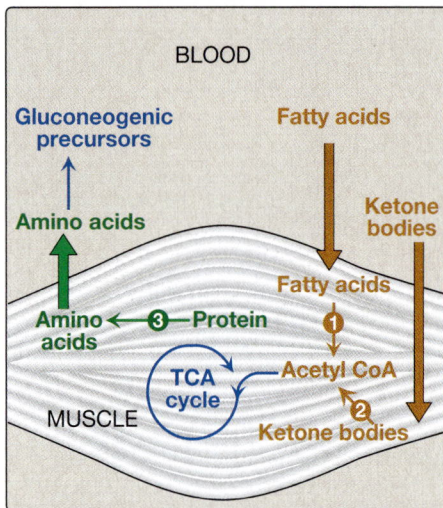

Figure 24.15
Major metabolic pathways in skeletal muscle during the fasting state. [Note: The *numbers in the circles*, which appear both in the figure and in the corresponding citation in the text, indicate important pathways for fat or protein metabolism.] CoA, coenzyme A; TCA, tricarboxylic acid.

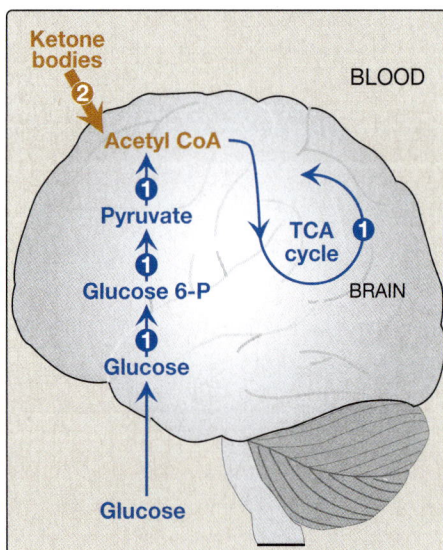

Figure 24.16
Major metabolic pathways in the brain during the fasting state. [Note: The *numbers in the circles*, which appear both in the figure and in the corresponding citation in the text, indicate important pathways for metabolism of fat or carbohydrates.] CoA, coenzyme A; TCA, tricarboxylic acid; P, phosphate.

state cannot be dephosphorylated or contribute to maintaining blood glucose levels.

A. Carbohydrate metabolism

Glucose transport into skeletal myocytes via insulin-sensitive GLUT-4 (see Chapter 8, IV.) and subsequent glucose metabolism are decreased because circulating insulin levels are low. Therefore, the glucose from hepatic gluconeogenesis is unavailable to muscle and adipose.

B. Lipid metabolism

Early in the fasting state, muscle tissue uses FAs from adipose tissue and ketone bodies from the liver as fuels (Fig. 24.15). In prolonged fasting, muscle decreases its use of ketone bodies (thus sparing them for the brain) and oxidizes FAs almost exclusively. Epinephrine signaling increases LPL expression in muscle cells, allowing cells to take up additional FAs from VLDL triglycerides in the fasting state. The acetyl CoA from FA oxidation indirectly inhibits PDH (by activation of PDH kinase). Pyruvate is transaminated to alanine and used by the liver for gluconeogenesis (glucose–alanine cycle; see Chapter 19, IV. and Fig. 19.13).

C. Protein metabolism

During the first few days of fasting, there is a rapid breakdown of muscle protein (e.g., glycolytic enzymes), providing amino acids that are used by the liver for gluconeogenesis (see Fig. 24.15). Because muscle does not have glucagon receptors, muscle proteolysis is initiated by a fall in insulin and sustained by a rise in glucocorticoids. Alanine and glutamine are quantitatively the most important glucogenic amino acids released from muscle. They are produced by the catabolism of BCAAs (see Chapter 19, IV. and Fig. 19.13). The glutamine is used as a fuel by enterocytes, for example, which send out alanine that is used in hepatic gluconeogenesis. In the second week of fasting, the rate of muscle proteolysis decreases, paralleling a decline in the need for glucose as a fuel for the brain, which has begun using ketone bodies as a source of energy.

XI. BRAIN IN THE FASTING STATE

During the early days of fasting, the brain continues to use only glucose as a fuel (Fig. 24.16). Blood glucose is maintained by hepatic gluconeogenesis from glucogenic precursors, such as amino acids from proteolysis and glycerol from lipolysis. In prolonged fasting (>2 to 3 weeks), plasma ketone bodies (see Fig. 24.12) reach significantly elevated levels and replace glucose as the primary fuel for the brain (Fig. 24.17; also see Fig. 24.16). This reduces the need for protein catabolism for gluconeogenesis: Ketone bodies spare glucose and, thus, muscle protein. As the duration of a fast extends from overnight to days to weeks, blood glucose levels initially drop and then are maintained at the lower level (65 to 70 mg/dL). The metabolic changes that occur during the fasting state ensure that all tissues have an adequate supply of fuel molecules. The response of the major tissues involved in energy metabolism during fasting is summarized in Figure 24.18.

XII. KIDNEY IN LONG-TERM FASTING

As fasting continues into early starvation and beyond, the kidney plays important roles. The renal cortex expresses the enzymes of gluconeogenesis, including glucose 6-phosphatase. In late fasting, ~50% of gluconeogenesis occurs here, although a portion of this glucose is used by the kidney itself. The kidney also provides compensation for the acidosis that accompanies the increased production of ketone bodies (organic acids). The glutamine released from the muscle's metabolism of BCAAs is taken up by the kidney and acted on by renal glutaminase and glutamate dehydrogenase (see Chapter 19, VI.), producing α-ketoglutarate, which can be used as a substrate for gluconeogenesis, plus ammonia (NH_3). The NH_3 picks up protons from ketone body dissociation and is excreted in the urine as ammonium (NH_4^+), thereby decreasing the acid load in the body (Fig. 24.19). Therefore, in long-term fasting, there is a switch from nitrogen disposal in the form of urea to disposal in the form of NH_4^+. As ketone body concentration rises, enterocytes, typically consumers of glutamine, become consumers of ketone bodies. This allows more glutamine to be available to the kidney.

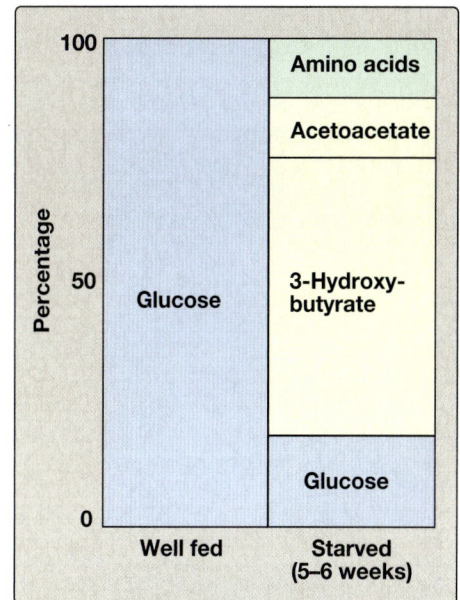

Figure 24.17
Fuel sources used by the brain to meet energy needs in the well-fed and starved states.

XIII. CHAPTER SUMMARY

- The flow of intermediates through metabolic pathways is controlled by **four regulatory mechanisms**: (1) availability of substrates, (2) allosteric activation and inhibition of enzymes, (3) covalent modification of enzymes, and (4) induction–repression of enzyme synthesis.

- In the **absorptive state**, the 2- to 4-hour period after ingestion of a meal, these mechanisms ensure that available nutrients are captured as **glycogen**, **triacylglycerols (TAGs)**, and **protein** (Fig. 24.20). During this interval, transient increases in plasma glucose, amino acids, and TAGs occur, the last primarily as components of **chylomicrons** synthesized by the intestinal mucosal cells.

- The **pancreas** responds to the elevated levels of glucose with an increased secretion of insulin and a decreased secretion of glucagon. The elevated **insulin/glucagon ratio** and the ready availability of circulating substrates make the absorptive state an **anabolic period** during which virtually all tissues use **glucose** as a fuel.

- In the absorptive phase, the **liver** replenishes its glycogen stores, replaces any needed hepatic proteins, and increases TAG synthesis. The latter are packaged in **very–low-density lipoprotein (VLDL) complexes**, which are exported to the peripheral tissues.

- In the absorptive phase in **adipose tissue**, **lipoprotein lipase increases** to provide dietary fatty acid (FA) from chylomicron complexes for TAG synthesis and storage. **Muscle** increases glycogen synthesis/storage, and protein synthesis to replace protein degraded since the previous meal. The **brain** uses glucose exclusively as a fuel.

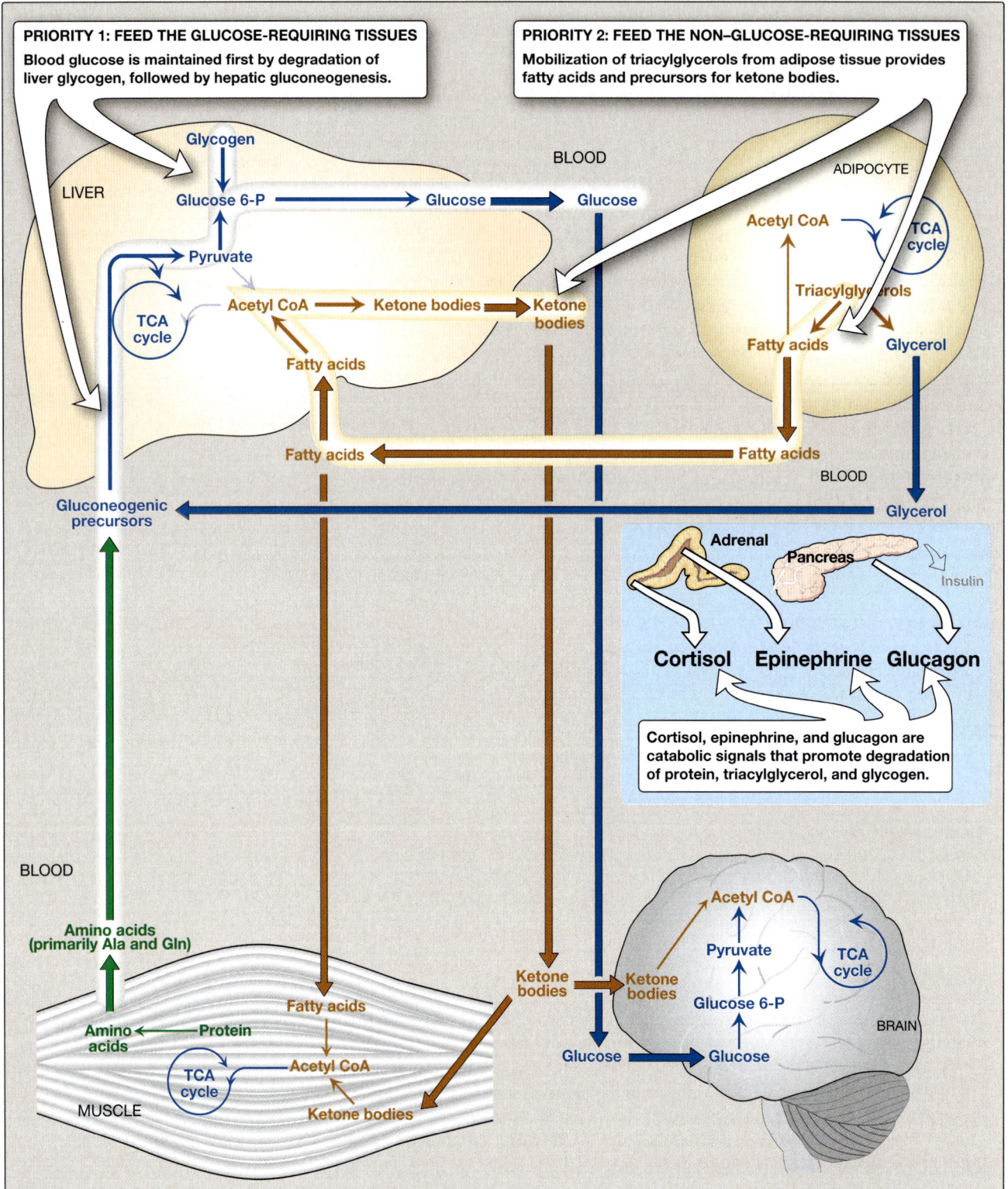

Figure 24.18
Intertissue relationships during the fasting state and the hormonal signals that promote them. P, phosphate; TCA, tricarboxylic acid; CoA, coenzyme A; Ala, alanine; Gln, glutamine.

- In the **fasting state**, plasma levels of glucose, amino acids, and TAGs fall, triggering a decline in insulin secretion and an increase in glucagon and **epinephrine** secretion. The decreased **insulin/counterregulatory hormone ratio** and the decreased availability of circulating substrates make the fasting state a **catabolic period**.

- The fasting state sets into motion an **exchange of substrates** among the liver, adipose tissue, skeletal muscle, and brain that is guided by two priorities: (1) the need to maintain adequate plasma levels of glucose to sustain energy metabolism of the brain and other glucose-requiring tissues and (2) the need to mobilize FAs from adipose tissue and release **ketone bodies** from liver to supply energy to other tissues.

- In the fasting state, the liver degrades glycogen and initiates **gluconeogenesis**, using increased **FA oxidation** to supply the energy and reducing equivalents needed for gluconeogenesis and the acetyl CoA building blocks for **ketogenesis**.

- In the fasting state, adipose tissue degrades stored TAGs by **hormone-sensitive lipase**, thus providing FAs to tissues and **glycerol** to the liver. The muscle can also use FAs as fuel as well as ketone bodies supplied by the liver. Lipoprotein lipase also increases to provide FAs to muscle from VLDL complexes. The liver uses the glycerol for gluconeogenesis.

- In the fasting state, **muscle protein** is degraded to supply amino acids for the liver to use in gluconeogenesis but decreases as ketone bodies increase, especially with more prolonged fasting. The brain can use both glucose and ketone bodies as fuels.

- From late fasting into starvation, the **kidneys** play important roles by synthesizing glucose and excreting the **protons** from ketone body dissociation as NH_4^+.

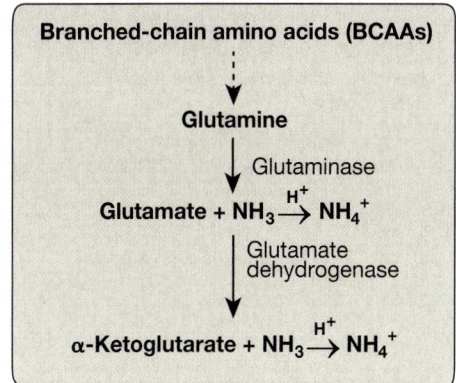

Branched-chain amino acids (BCAAs)

⋮

Glutamine

Glutaminase

$Glutamate + NH_3 \xrightarrow{H^+} NH_4^+$

Glutamate dehydrogenase

$\alpha\text{-Ketoglutarate} + NH_3 \xrightarrow{H^+} NH_4^+$

Figure 24.19
Use of glutamine from BCAA catabolism in muscle to generate ammonia (NH_3) used for the excretion of protons (H^+) as ammonium (NH_4^+) in the kidneys.

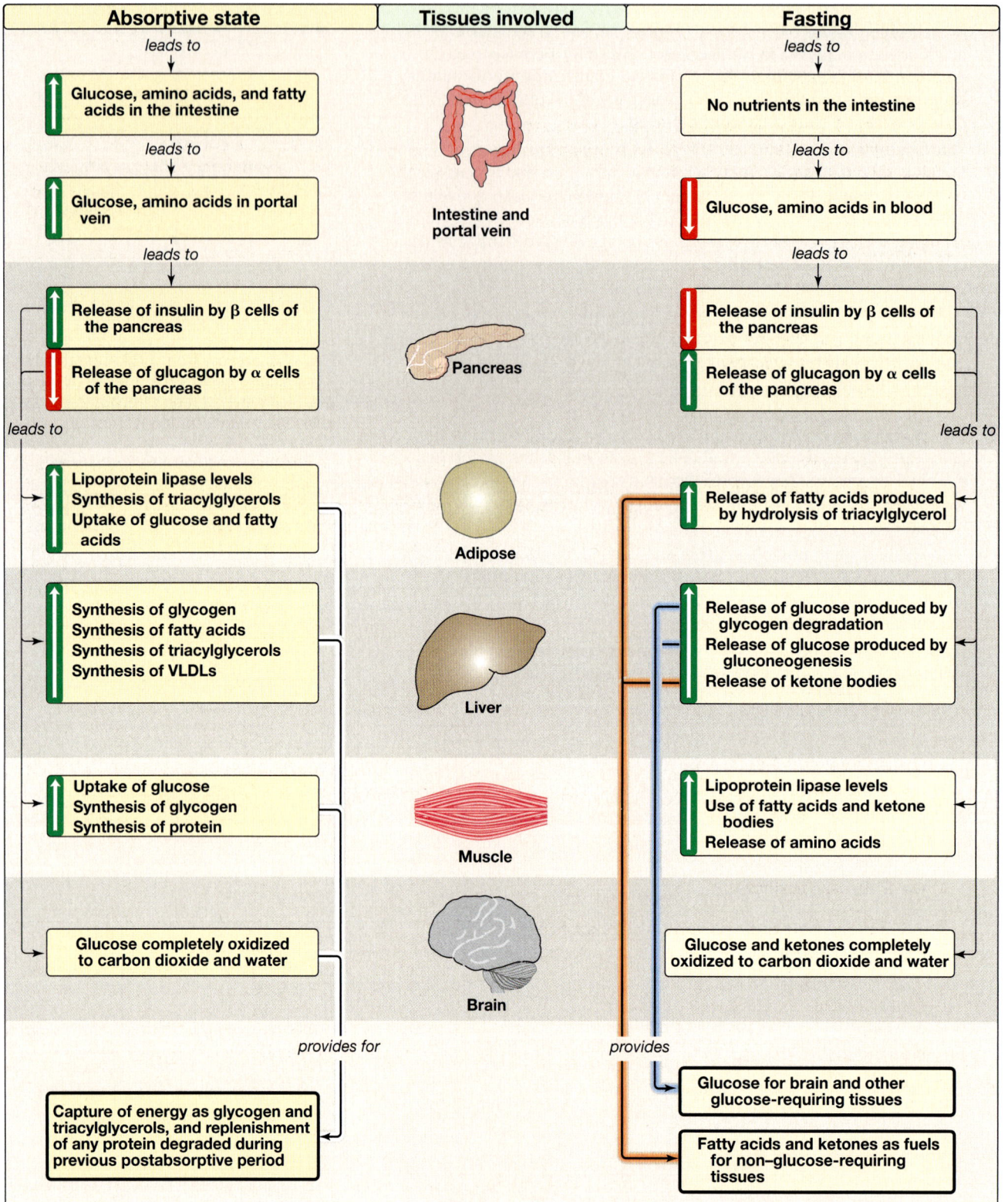

Absorptive state	Tissues involved	Fasting

Intestine and portal vein

Absorptive state:

↑ Glucose, amino acids, and fatty acids in the intestine

leads to

↑ Glucose, amino acids in portal vein

leads to

Fasting:

No nutrients in the intestine

leads to

↓ Glucose, amino acids in blood

leads to

Pancreas

↑ Release of insulin by β cells of the pancreas

↓ Release of glucagon by α cells of the pancreas

leads to

↓ Release of insulin by β cells of the pancreas

↑ Release of glucagon by α cells of the pancreas

leads to

Adipose

↑ Lipoprotein lipase levels
Synthesis of triacylglycerols
Uptake of glucose and fatty acids

↑ Release of fatty acids produced by hydrolysis of triacylglycerol

Liver

↑ Synthesis of glycogen
Synthesis of fatty acids
Synthesis of triacylglycerols
Synthesis of VLDLs

↑ Release of glucose produced by glycogen degradation
Release of glucose produced by gluconeogenesis
Release of ketone bodies

Muscle

↑ Uptake of glucose
Synthesis of glycogen
Synthesis of protein

↑ Lipoprotein lipase levels
Use of fatty acids and ketone bodies
Release of amino acids

Brain

Glucose completely oxidized to carbon dioxide and water

Glucose and ketones completely oxidized to carbon dioxide and water

provides for

provides

Capture of energy as glycogen and triacylglycerols, and replenishment of any protein degraded during previous postabsorptive period

Glucose for brain and other glucose-requiring tissues

Fatty acids and ketones as fuels for non–glucose-requiring tissues

Figure 24.20
Key concept map for the feed–fast cycle. VLDL, very–low-density lipoprotein.

Study Questions

Choose the ONE best answer.

24.1. Which of the following is elevated in plasma during the absorptive (well-fed) state as compared with the postabsorptive (fasting state) state?

A. Acetoacetate
B. Chylomicrons
C. Free fatty acids
D. Glucagon
E. Glycerol

Correct answer = B. Triacylglycerol-rich chylomicrons are synthesized in (and released from) the intestine following ingestion of a meal. Acetoacetate, free fatty acids, glucagon, and glycerol are elevated in the fasting state, not the absorptive state.

24.2. Which of the following statements concerning liver in the absorptive state is correct?

A. Fructose 2,6-bisphosphate is elevated
B. insulin stimulates the uptake of glucose
C. most covalently modified enzymes are in the phosphorylated state
D. oxidation of acetyl coenzyme A is increased
E. synthesis of glucokinase is repressed

Correct answer = A. The increased insulin and decreased glucagon levels characteristic of the absorptive state promote the synthesis of fructose 2,6-bisphosphate, which allosterically activates phosphofructokinase-1 of glycolysis while inhibiting fructose 1,6-bisphosphatase of gluconeogenesis. Most covalently modified enzymes are in the dephosphorylated state and are active, with the exception of glycogen phosphorylase kinase, glycogen phosphorylase, and hormone-sensitive lipase. Most of the acetyl coenzyme A is not oxidized in the well-fed state because it is being used in fatty acid synthesis. Uptake of glucose (by glucose transporter-2) into the liver is insulin independent. Synthesis of glucokinase is induced by insulin in the well-fed state.

24.3. Which of the following enzymes is phosphorylated and active in an individual who has been fasting for 12 hours?

A. Arginase
B. Carnitine palmitoyltransferase-I
C. Fatty acid synthase
D. Glycogen synthase
E. Hormone-sensitive lipase
F. Phosphofructokinase-1
G. Pyruvate dehydrogenase

Correct answer = E. Hormone-sensitive lipase of adipocytes is phosphorylated and activated by protein kinase A in response to epinephrine. Choices A, B, C, and F are not regulated covalently. Choices D and G are regulated covalently but are inactive if phosphorylated.

24.4. For a 70-kg male, in which period do ketone bodies supply the major portion of the brain's caloric needs?

A. Absorptive period
B. Overnight fast
C. 3-Day fast
D. 4-Week fast
E. 5-Month fast

Correct answer = D. Ketone bodies, made from the acetyl coenzyme A product of fatty acid oxidation, increase in the blood in the fasting state but must reach a critical level to cross the blood–brain barrier. Typically, this occurs in the second to third week of a fast. Fat stores in a 70-kg (~154-lb) man would not be able to supply his energy needs for 5 months.

24.5. In prolonged starvation, the kidney excretes ammonium (NH_4^+), in addition to urea, to remove excess nitrogen groups. Which of the following is also a main benefit of this NH_4^+ excretion?

   A. Decreases renal glutamine uptake
   B. Decreases ketoacidosis
   C. Increases enterocyte glutamine consumption
   D. Increases muscle alanine transamination
   E. Increases renal urea cycle capacity

Correct answer = B. In prolonged starvation, there is less proteolysis of muscle proteins for hepatic gluconeogenesis (via muscle alanine), and a corresponding increase of ketone body production. The kidney provides compensation for the acidosis that accompanies the increased ketone body production and dissociation. Glutamine consumption in intestinal enterocytes decreases and renal uptake of glutamine increases. Glutamine is converted by renal tissue to α-ketoglutarate, which can be used as a substrate for gluconeogenesis, plus ammonia (NH_3). The NH_3 picks up protons from ketone body dissociation and is excreted in the urine as ammonium (NH_4^+), thereby decreasing the acid load in the body (see Fig. 24.19). In long-term fasting, there is a decrease in nitrogen disposal by the urea cycle and an increase in the excretion of NH_4^+.

24.6. The diagram shows inputs to and outputs from pyruvate, a central molecule in energy metabolism. Which letter represents a reaction that requires biotin and is activated by acetyl coenzyme A?

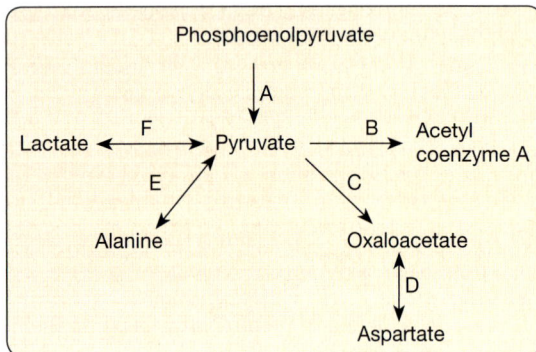

Correct answer = C. Pyruvate carboxylase, a mitochondrial enzyme of gluconeogenesis, requires biotin (and ATP) and is allosterically activated by acetyl coenzyme A from fatty acid oxidation. None of the other choices meets these criteria. A = pyruvate kinase; B = pyruvate dehydrogenase complex; D = aspartate aminotransferase; E = alanine aminotransferase; F = lactate dehydrogenase.

24.7. Which of the following best describes hormonal regulation of a metabolic reaction in the well-fed absorptive state?

   A. Decreased adipocyte lipoprotein lipase activity
   B. Decreased hepatic CPT-1 activity
   C. Decreased muscle GLUT-4 activity
   D. Increased adipocyte hormone-sensitive lipase activity
   E. Increased muscle glycogen phosphorylase activity

Correct answer = B. In the well-fed absorptive state, the insulin/counterregulatory hormone ratio is high, and blood glucose levels are elevated. As such, ACC is dephosphorylated (active), producing malonyl-CoA for FA synthesis. Malonyl-CoA inhibits CPT-1 on the mitochondrial surface to prevent FAs entering the mitochondrial matrix for β oxidation. Adipocytes would increase lipoprotein lipase activity, to allow hydrolysis to FAs from dietary TAGs in chylomicron complexes. Both muscle and adipocytes would increase GLUT-4 activity to import blood glucose. Adipocytes would decrease hormone-sensitive lipase. Muscle and liver would increase glycogen synthase, not glycogen phosphorylase, so as to synthesize and store glycogen.

24.8. Describe the tissue-specific regulation of lipoprotein lipase both in the well-fed absorptive and fasting states.

Lipoprotein lipase (LPL) is an extracellular enzyme attached to the endothelial capillary cells in many tissues, particularly adipose and muscle. When activated, LPL hydrolyzes FAs from TAG stores within circulating lipoprotein complexes, including dietary chylomicrons or hepatic VLDL complexes. In the well-fed absorptive state, the adipocytes increase their LPL levels, so as to obtain dietary FAs for storage within adipocytes as TAGs. Muscle would not induce LPL levels, as they utilize primarily glucose for energy (and to store it as glycogen). There would be a low basal level of LPL in muscle though, which is higher in cardiac muscle. In the fasting state, the adipocytes do not store FAs as TAGs, so the LPL level from adipocytes decreases. With the loss of GLUT-4 on the surface of muscle (and adipocytes), muscle cells switch from glucose to FAs as a potential energy source. As such muscle LPL levels increase, especially in cardiac muscle.

24.9. Which of the following best describes how blood glucose level is maintained in the fasting state?

A. decreased expression of phosphoenolpyruvate carboxykinase

B. glycerol precursor supplied from muscle

C. increased muscle glycogenolysis

D. inhibition of fructose 1,6-bisphosphatase

E. phosphorylation modification of key enzymes

Correct answer = E. In the fasting state, many key metabolic enzymes are regulated by covalent phosphorylation modification. In the liver, glycogen phosphorylase is activated by phosphorylation, allowing the hydrolysis of glycogen to glucose 6-phosphate. In the liver Glucose 6-phosphate is converted to glucose by glucose 6-phosphatase, allowing the maintenance of blood glucose from hepatic glycogen stores. Muscle glycogenolysis is also active in this manner, but muscle lacks glucose 6-phosphatase. This prevents the maintenance of blood glucose levels from muscle glycogen stores. Covalent phosphorylation modification of phosphofructokinase-2 (PFK-2) leads to a degradation of fructose 2,6-bisphosphate in the liver. A decrease in fructose 2,6-bisphosphate levels allosterically activates fructose 1,6-bisphosphatase-1 in gluconeogenesis. Glucagon leads to an increase in the transcriptional expression of the gluconeogenic enzyme phosphoenolpyruvate carboxykinase. Adipocytes hydrolyze TAGs to FAs and glycerol by hormone-sensitive lipase in the fasting state. Glycerol can be a gluconeogenic precursor for hepatocytes.

25 Diabetes Mellitus

I. OVERVIEW

Diabetes mellitus (diabetes) is not one disease but rather is a heterogeneous group of multifactorial, primarily polygenic syndromes characterized by elevated blood glucose caused by a relative or absolute deficiency in the hormone insulin. More than 38 million people in the United States (~11.6% of the population) have diabetes. Of this number, it is estimated that ~8 million are as yet undiagnosed. In addition, more than a third of U.S. adults are considered to have prediabetes, with the vast majority being unaware of their health status. Diabetes is the leading cause of adult blindness and amputation and a major cause of renal failure, nerve damage, heart attacks, and strokes. Diabetes is the seventh leading cause of death in the United States. Most cases of diabetes mellitus can be separated into two groups (Fig. 25.1): type 1 ([T1D],

Characteristics	Type 1 Diabetes	Type 2 Diabetes
AGE OF ONSET	Usually during childhood or puberty; symptoms develop rapidly	Frequently after age 35 years; symptoms develop gradually
NUTRITIONAL STATUS AT TIME OF DISEASE ONSET	Frequently undernourished	Obesity usually present
PREVALENCE	<10% of diagnosed diabetes	>90% of diagnosed diabetes
GENETIC PREDISPOSITION	Moderate	Very strong
DEFECT OR DEFICIENCY	β-Cell destruction, eliminating production of insulin	Insulin resistance combined with inability of β cells to produce appropriate quantities of insulin
FREQUENCY OF KETOSIS	Common	Rare
PLASMA INSULIN	Low to absent	High early in disease; low to absent in disease of long duration
ACUTE COMPLICATIONS	Ketoacidosis	Hyperosmolar hyperglycemic state
RESPONSE TO ORAL HYPOGLYCEMIC DRUGS	Unresponsive	Responsive
TREATMENT	Insulin always necessary	Diet, exercise, oral hypoglycemic drugs, insulin (may or may not be necessary); reduction of risk factors (weight reduction, smoking cessation, blood pressure control, treatment of dyslipidemia) essential to therapy

Figure 25.1
Comparison of type 1 and type 2 diabetes mellitus. [Note: The name of the disease reflects the clinical presentation of copious amounts of glucose-containing urine and is derived from the Greek word for "siphon" (*diabetes*) and the Latin word for "honey sweet" (*mellitus*).]

previously referred to as juvenile or juvenile-onset diabetes) and type 2 ([T2D], previously referred to as adult-onset diabetes). Both types of diabetes can affect adults and children. The incidence of disease onset at younger versus older ages was the reason for the previous terminology, but this has been updated in the new terminology to reflect our better understanding of the two types of diabetes. The incidence and prevalence of T2D is also increasing because of the aging of the U.S. population and the increasing prevalence of obesity and sedentary lifestyles (see Chapter 26, I.). The increase in children with T2D is particularly disturbing. At current rates, the number of individuals with T2D younger than age 20 years could increase 49% by 2050, and, if the rates increase, T2D cases in this age group could quadruple. A summary of both types of diabetes is shown in Figure 25.1.

II. TYPE 1 DIABETES

T1D constitutes ~5%, or 2 million of the ~38 million known cases of diabetes in the United States. The disease is characterized by an absolute deficiency of insulin caused by an autoimmune attack on the β islet cells of the pancreas. In T1D, the islets of Langerhans become infiltrated with activated T lymphocytes, leading to a condition called insulitis (immune cell infiltrates around and within the islets). Over a period of years, this autoimmune attack leads to gradual depletion of the β-cell population (Fig. 25.2). However, symptoms appear abruptly only after 80% to 90% of the β cells have been destroyed. At this point, the pancreas fails to respond adequately to ingestion of glucose, and insulin therapy is required to restore metabolic control and prevent life-threatening ketoacidosis. β-Cell destruction requires both a stimulus from the environment (such as a viral infection) and a genetic determinant that causes the β cells to be mistakenly identified as "nonself." Among monozygotic (identical) twins, if one sibling develops T1D, the other twin has only a 30% to 50% chance of developing the disease. This contrasts with T2D, in which the genetic influence is stronger with eventual development of the disease for virtually all monozygotic twinships. T1D can vary in incidence by ethnicity. In the United States, T1D is most common among non-Hispanic Caucasian/European American, followed by African American and Hispanic American populations. It is less common among Asian American populations, with Native Americans populations having the lowest incidence rates for T1D.

A. Diagnosis

The onset of T1D is sudden, occurring typically during childhood or puberty, hence the previous classification as "juvenile-onset diabetes." But with the recent increase in children diagnosed with T2D, the juvenile-onset diabetes classification has become less meaningful, leading to the updated T1D classification. Individuals with T1D can usually be recognized by the abrupt appearance of polyuria (frequent urination), polydipsia (excessive thirst), and polyphagia (excessive hunger), often triggered by physiologic stress such as an infection. These symptoms are usually accompanied by fatigue and weight loss. A clinical diagnosis of both types of diabetes is confirmed by a fasting blood glucose (FBG) level >125 mg/dL (normal = 70 to 99). Fasting is defined as no caloric intake for at least 8 hours, and is

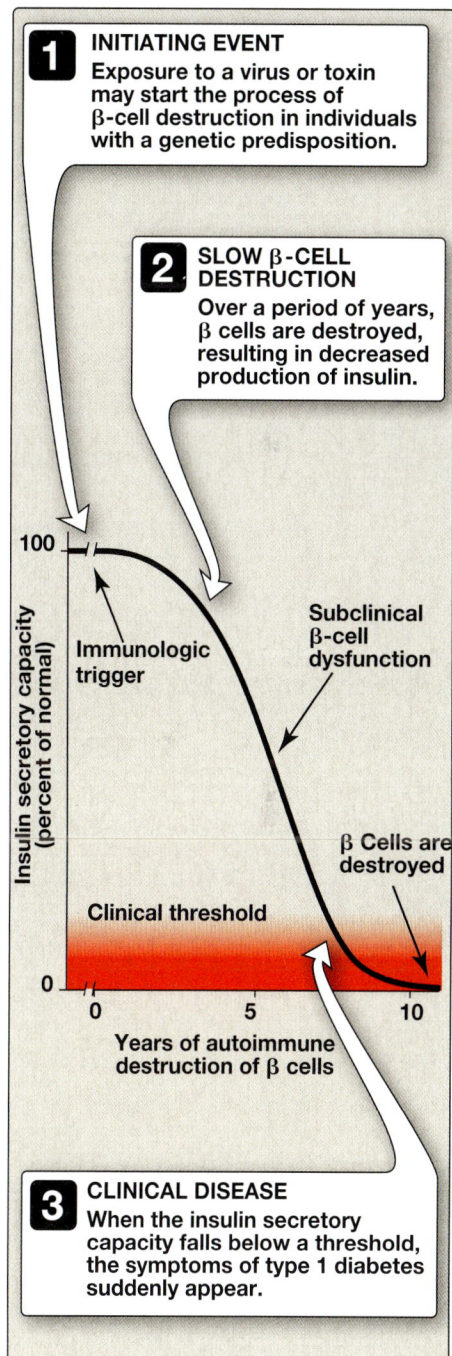

1 INITIATING EVENT
Exposure to a virus or toxin may start the process of β-cell destruction in individuals with a genetic predisposition.

2 SLOW β-CELL DESTRUCTION
Over a period of years, β cells are destroyed, resulting in decreased production of insulin.

Immunologic trigger

Subclinical β-cell dysfunction

β Cells are destroyed

Clinical threshold

Years of autoimmune destruction of β cells

3 CLINICAL DISEASE
When the insulin secretory capacity falls below a threshold, the symptoms of type 1 diabetes suddenly appear.

Figure 25.2
Insulin secretory capacity during onset of type 1 diabetes. [Note: Rate of autoimmune destruction of β cells may be faster or slower than shown.]

required for the FBG test. An FBG between 100 and 125 mg/dL can be categorized as an impaired FBG, or impaired glucose tolerance, and reflects either the functional absence of insulin (in T1D) or insulin resistance (in T2D). Individuals with impaired FBG (impaired glucose tolerance) have a blood glucose above normal, but not high enough to diagnose as diabetes. Such individuals are considered prediabetic and are at increased risk for later developing T2D. The measurement of the percent of glycated hemoglobin ([HbA$_{1c}$] see Chapter 3, II.) in the blood can both diagnose diabetes and assess overall glycemic control of patients with diabetes. The rate of formation of HbA$_{1c}$ is proportional to the average blood glucose concentration over the previous 2 months (the approximate half-life of a normal red blood cell). Normal HbA$_{1c}$ is less than 5.5%; impaired glucose tolerance or prediabetes range is 5.5% to 6.4%; diabetic diagnosis requires HbA$_{1c}$ levels at least 6.5%. HbA$_{1c}$ testing does not require fasting, since it measures the average blood glucose levels over the 2-month period. Diagnosis can also be made on the basis of a nonfasting (random) blood glucose level more than 200 mg/dL with consistent clinical symptoms, although a FBG or HbA$_{1c}$ test would likely be ordered to confirm the diagnosis of a nonfasting random blood glucose test. The oral glucose tolerance test, in which blood glucose is measured 2 hours after ingestion of a solution containing 75 g of glucose, is also used but is less convenient. It is most typically used to screen pregnant women for gestational diabetes early in the third trimester.

> When blood glucose is >180 mg/dL, the ability of renal sodium-dependent glucose transporters (SGLTs) to reclaim glucose is impaired, and glucose "spills" into urine. The loss of glucose into the urine is accompanied by water loss, resulting in the characteristic polyuria (with dehydration) and polydipsia diabetes symptoms.

B. Metabolic changes

The metabolic abnormalities of T1D, mainly hyperglycemia, ketonemia, and hypertriacylglycerolemia, result from an absolute functional deficiency of insulin that profoundly affects metabolism in three tissues: liver, skeletal muscle, and white adipose (Fig. 25.3).

1. **Hyperglycemia and ketonemia:** Elevated levels of blood glucose and ketone bodies are hallmarks of untreated T1D (see Fig. 25.3; also see Appendix Integrative Case 3). Hyperglycemia is caused by increased hepatic production of glucose via gluconeogenesis and glycogenolysis, combined with diminished peripheral utilization of glucose, due to downregulation of insulin-sensitive glucose transporter GLUT-4 on the surface of both muscle and adipose tissue (see Chapter 8, IV.). Ketonemia results from increased mobilization of fatty acids (FAs) from adipose tissue triacylglycerol (TAG), combined with accelerated hepatic FA β oxidation and synthesis of 3-hydroxybutyrate and acetoacetate (ketone bodies; see Chapter 16, V.). Acetyl coenzyme A from FA β oxidation is the substrate for ketogenesis and is an allosteric activator of pyruvate carboxylase, a gluconeogenic enzyme.

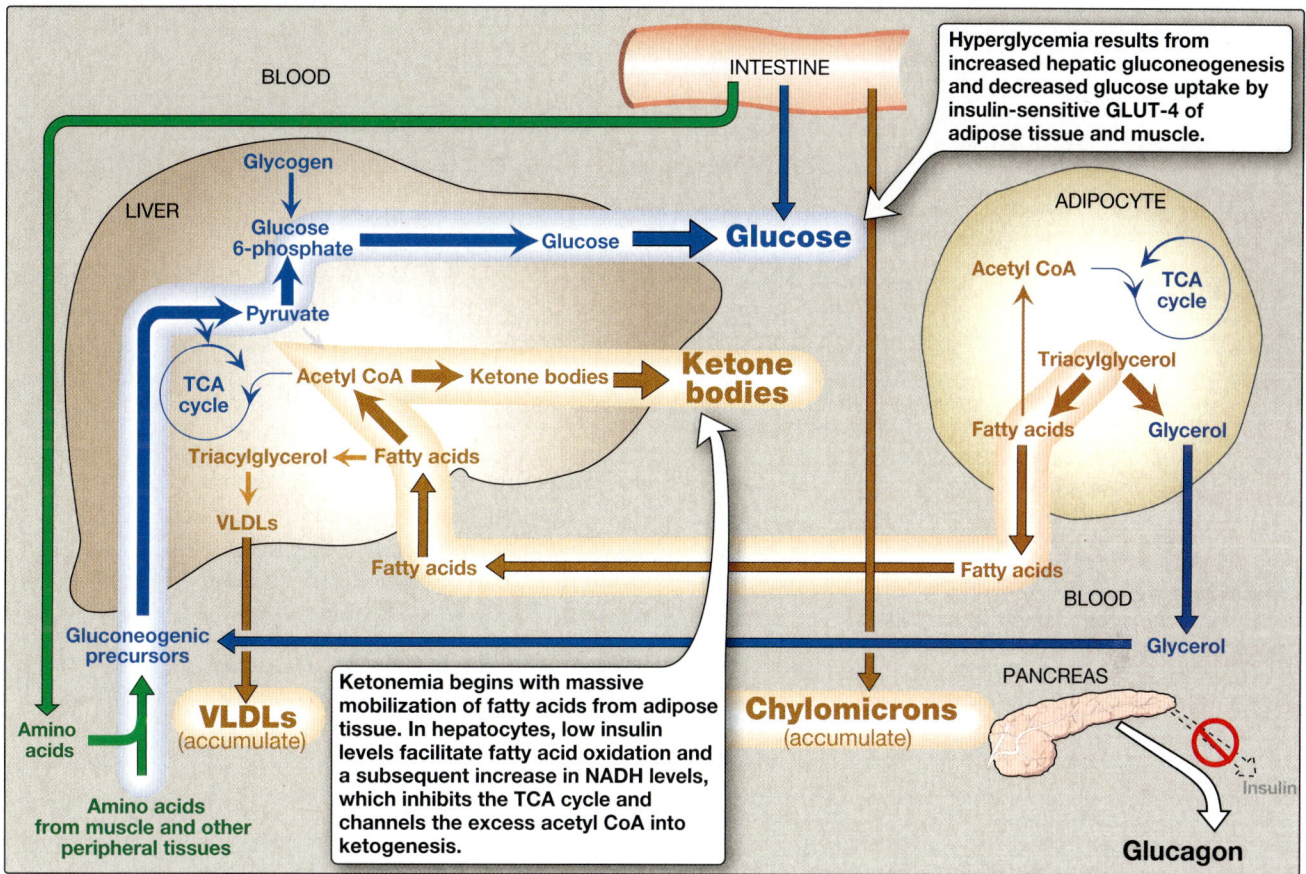

Figure 25.3
Intertissue relationships in type 1 diabetes. TCA, tricarboxylic acid; CoA, coenzyme A; VLDLs, very–low-density lipoproteins; GLUT, glucose transporter.

Diabetic ketoacidosis (DKA), a type of metabolic acidosis caused by an imbalance between ketone body production and use, occurs in 25% to 40% of those newly diagnosed with T1D and may recur if the patient becomes ill (most commonly with an infection) or does not comply with therapy. DKA is treated by replacing fluid and electrolytes and administering short-acting insulin to gradually correct hyperglycemia without precipitating hypoglycemia.

2. **Hypertriacylglycerolemia:** Not all of the FAs flooding the liver from the adipose tissue can be disposed of through oxidation and ketone body synthesis. These excess FAs are converted to TAG in the liver, which are packaged and secreted into the bloodstream in very–low-density lipoproteins ([VLDLs] see Chapter 18, VI.). Chylomicrons rich in dietary TAGs are packaged by intestinal mucosal cells and secreted into the bloodstream following a meal. Because lipoprotein lipase in the capillary beds of adipose tissue is low under diabetic conditions (synthesis of the enzyme is decreased when insulin levels are low), the blood plasma chylomicron and VLDL levels are elevated, resulting in hypertriacylglycerolemia (see Fig. 25.3).

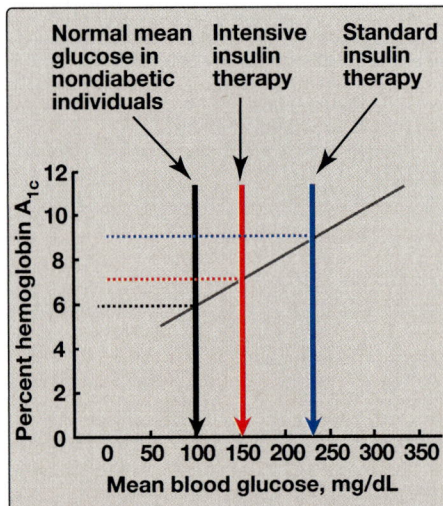

Figure 25.4
Correlation between mean blood glucose and percent hemoglobin A$_{1c}$ in patients with type 1 diabetes receiving intensive or standard insulin therapy. [Note: Nondiabetic individuals are included for comparison.]

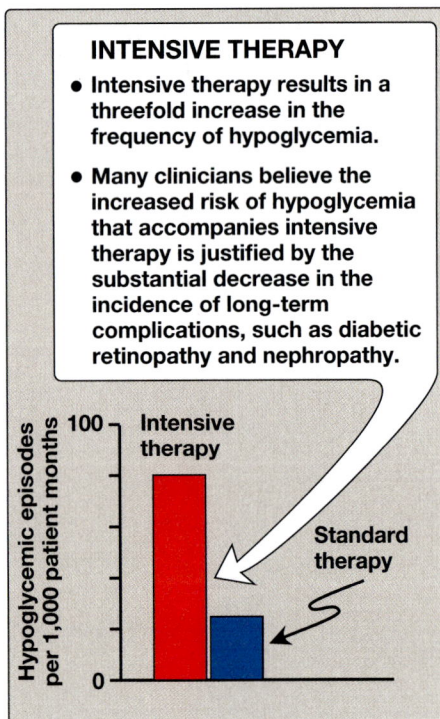

Figure 25.5
Effect of intensive therapy or standard therapy on hypoglycemic episodes in patient populations.

C. Treatment

Individuals with T1D have a functional absence of endogenous insulin, and therefore must rely on exogenous insulin delivered subcutaneously either by periodic injection or by continuous pump-assisted infusion. This will control the T1D metabolic derangements, hyperglycemia and ketonemia. Two types of therapeutic injection regimens are currently used, standard and intensive. [Note: Pump delivery is also considered intensive therapy.]

1. **Standard versus intensive treatment:** Standard treatment is typically two to three daily injections of recombinant human insulin. Mean blood glucose levels achieved by such treatment are typically 225 to 275 mg/dL, with a HbA$_{1c}$ level of 8% to 9% of the total hemoglobin (Fig. 25.4, blue arrow). In contrast to standard therapy, intensive treatment seeks to more closely normalize blood glucose through more frequent monitoring and subsequent injections of insulin, typically at least four times a day. Mean blood glucose levels of 150 mg/dL can be achieved, with HbA$_{1c}$ ~7% of the total hemoglobin (Fig. 25.4, red arrow). Normal mean blood glucose is ~100 mg/dL, and HbA$_{1c}$ is no more than 6% (Fig. 25.4, black arrow). Therefore, normalization of glucose values (euglycemia) is not achieved even in intensively treated patients. Nonetheless, patients on intensive therapy show at least a 50% reduction in the long-term microvascular complications of diabetes (i.e., retinopathy, nephropathy, and neuropathy) compared with patients receiving standard care. This confirms that the complications of diabetes are related to an elevation of plasma glucose.

2. **Complication of insulin therapy (hypoglycemia):** One of the therapeutic goals in cases of diabetes is to decrease blood glucose levels to minimize the development of long-term complications of the disease (see Section IV. for a discussion of the chronic complications of diabetes). However, the appropriate dosage of insulin is difficult to achieve. Hypoglycemia caused by excess insulin is the most common complication of insulin therapy, occurring in more than 90% of patients. The frequency of hypoglycemic episodes, seizures, and coma is particularly high with intensive treatment regimens designed to achieve tight control of blood glucose (Fig. 25.5). In healthy individuals, hypoglycemia triggers a compensatory secretion of counter-regulatory hormones, most notably glucagon and epinephrine, which promote hepatic production of glucose (see Chapter 23, IV. and Fig. 23.13). However, patients with T1D also develop a deficiency of glucagon secretion. This defect occurs early in the disease and is almost universally present 4 years after diagnosis. Therefore, these patients rely on epinephrine secretion to prevent severe hypoglycemia. However, as the disease progresses, T1D patients show diabetic autonomic neuropathy and impaired ability to secrete epinephrine in response to hypoglycemia. The combined deficiency of glucagon and epinephrine secretion creates a symptom-free condition sometimes called "hypoglycemia unawareness." Thus, patients with long-standing T1D are particularly vulnerable to hypoglycemia. Hypoglycemia can also be exacerbated by strenuous exercise. Because exercise

promotes glucose uptake into muscle and decreases the need for exogenous insulin, patients are advised to check blood glucose levels before or after intensive exercise to prevent or abort hypoglycemia.

3. **Contraindications for intensive therapy (tight control):** Children are not put on a program of tight control of blood glucose before age 8 years because of the risk that episodes of hypoglycemia may adversely affect brain development. Older adults typically do not go on tight control because hypoglycemia can cause strokes and heart attacks in this population. Also, the major goal of tight control is to prevent complications many years later. Tight control, then, is most worthwhile for otherwise healthy people who can expect to live at least 10 more years. For most nonpregnant adults with diabetes, the individual treatment strategies and goals are based on the duration of diabetes, age/life expectancy, and known comorbid conditions.

III. TYPE 2 DIABETES

T2D is the most common form of the disease, comprising ~95% of the U.S. population with diabetes. In the United States, T2D is most common among Hispanic Americans, Native Americans, and African Americans, followed by Asian Americans. Non-Hispanic Caucasian/European American populations have the lowest incidence rates for T2D. Typically, T2D develops gradually without obvious symptoms. The disease is often detected by routine screening tests. However, many individuals with T2D have symptoms of polyuria and polydipsia of several weeks' duration. Polyphagia may be present but is less common. T2D is characterized by a peripheral tissue insulin resistance resulting in hyperglycemia; insufficient or impaired compensatory insulin secretion; and, ultimately, β-cell failure (Fig. 25.6). Significantly, the tissue insulin resistance and insufficient compensatory insulin secretion are both required for hyperglycemia to increase to levels sufficient for diagnosis. Insulin therapy is not required to sustain life. However, in more than 90% of patients with T2D, insulin may eventually become required to control hyperglycemia and keep HbA$_{1c}$ lower than 7%. The eventual need for insulin therapy has eliminated the previous designation of T2D as non-insulin-dependent diabetes. The metabolic alterations observed in T2D are milder than those described for T1D, in part because insulin secretion in T2D, although inadequate, does restrain excessive ketogenesis and blunts the development of DKA. Insulin suppresses the release of glucagon (see Chapter 23, III.). Diagnosis is based on the presence of hyperglycemia as described above. The pathogenesis does not involve viruses or autoimmune antibodies and is not completely understood. An acute complication of T2D in the elderly is a hyperosmolar hyperglycemic nonketotic state characterized by severe hyperglycemia and dehydration and altered mental status.

A. Insulin resistance

Insulin resistance is the decreased ability of target tissues, such as the liver, white adipose, and skeletal muscle, to respond properly to normal (or elevated) circulating concentrations of insulin. For example,

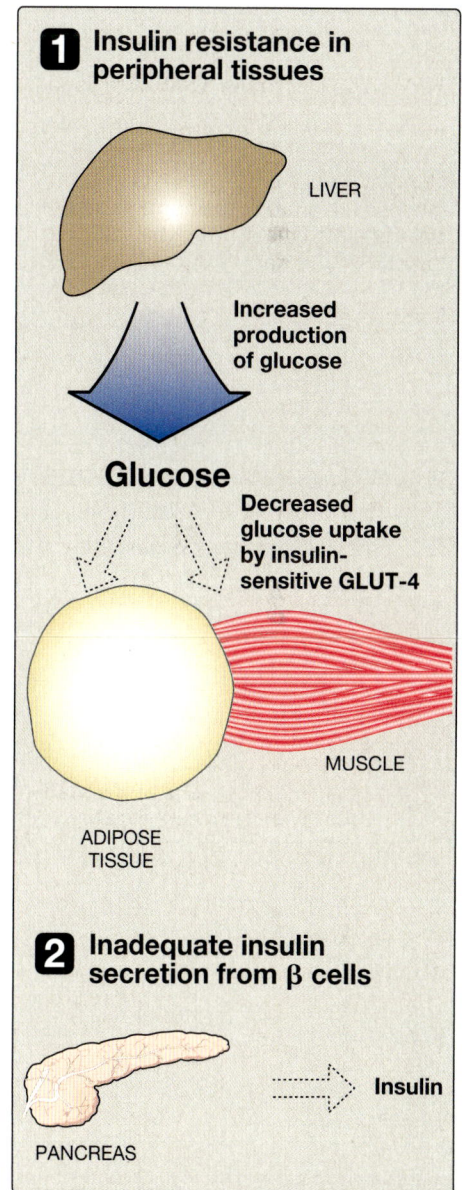

1 Insulin resistance in peripheral tissues

LIVER

Increased production of glucose

Glucose

Decreased glucose uptake by insulin-sensitive GLUT-4

MUSCLE

ADIPOSE TISSUE

2 Inadequate insulin secretion from β cells

Insulin

PANCREAS

Figure 25.6
Major factors contributing to hyperglycemia observed in type 2 diabetes. GLUT, glucose transporter.

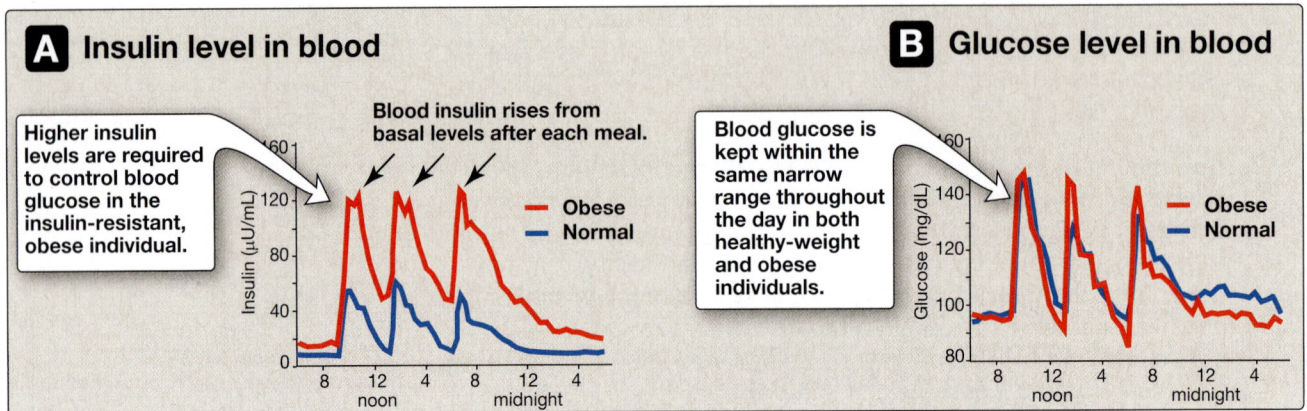

Figure 25.7
Daily blood insulin (**A**), and blood glucose (**B**), level changes in normal-weight and obese subjects.

insulin resistance is characterized by increased hepatic glucose production, decreased glucose uptake by muscle and adipose tissue, and increased adipose TAG lipolysis with an increase in serum free fatty acids (FFAs).

1. **Insulin resistance and obesity:** Although obesity is the most common cause of insulin resistance and increases the risk of developing T2D, most people with obesity and insulin resistance do not develop diabetes. In the absence of a defect in β-cell function, obese individuals can sufficiently compensate for the insulin resistance with elevated levels of insulin secretion. For example, Figure 25.7A shows that insulin secretion may be two to three times higher in obese patients than it is in individuals with leaner physiques. This higher insulin concentration compensates for the diminished effect of the hormone (as a result of insulin resistance) and produces blood glucose levels similar to those observed in lean individuals (Fig. 25.7B).

2. **Insulin resistance and T2D:** Insulin resistance alone will not lead to T2D. Rather, T2D develops in insulin-resistant individuals who also show impaired β-cell function. Insulin resistance and the subsequent risk for the development of T2D is commonly observed in individuals who have obesity, are physically inactive, are elderly, and/or are in the 3% to 5% of pregnant women who develop gestational diabetes. These patients are unable to sufficiently compensate for the insulin resistance with increased insulin release. Figure 25.8 shows the time course for the development of insulin resistance, hyperglycemia, and the loss of β-cell function. Prior to the age when a diagnosis of T2D is possible, an individual with insulin resistance can compensate by secreting higher than normal levels of insulin (>100% of normal). As a result, the individual can maintain glucose levels close to normal (although they may have glucose levels in the prediabetes range). At some point, the elevated secretion of insulin is no longer sufficient to fully compensate for the insulin resistance, and there is an increase in the blood glucose concentration above the diagnostic threshold (>125 mg/dL FBG or ≥6.5% HbA$_{1c}$). After the initial diagnosis, the β-cell defect may result in declining insulin secretion and worsening hyperglycemia. The decreasing insulin

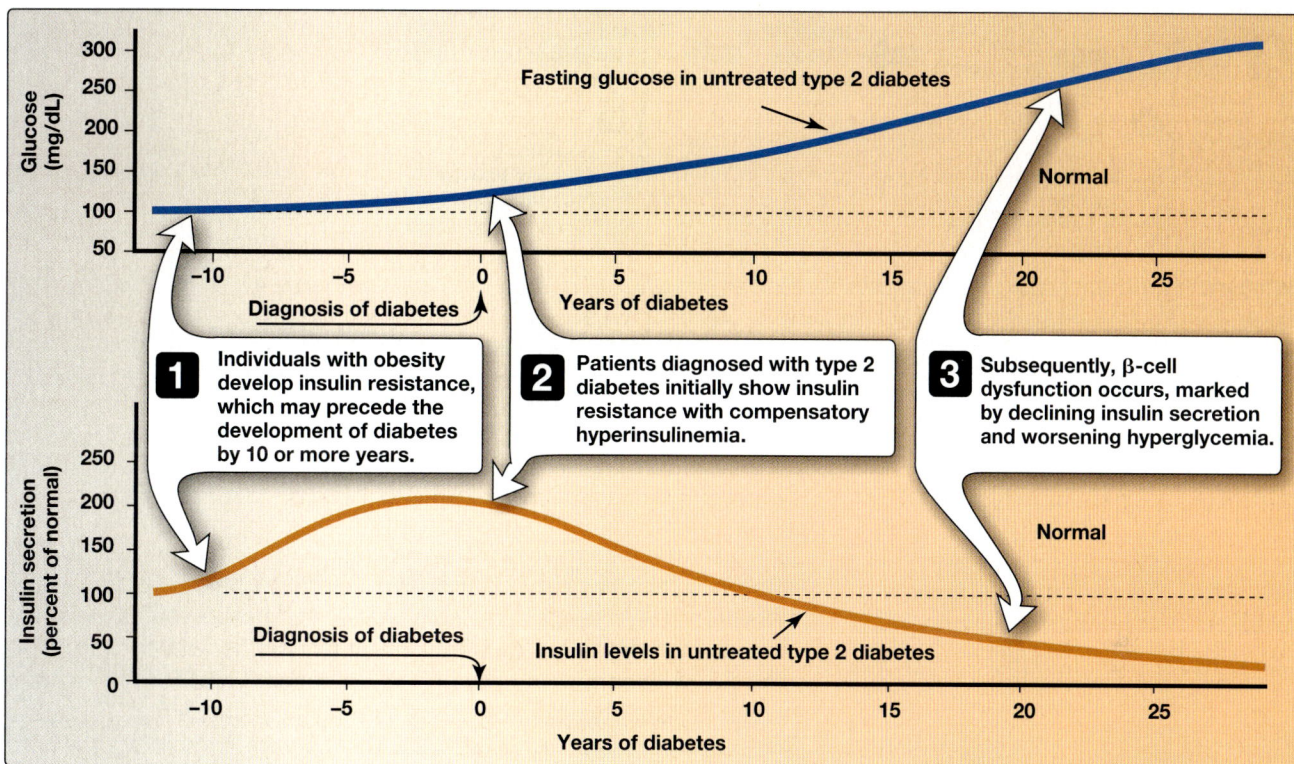

Figure 25.8
Progression of blood glucose and insulin levels in patients with type 2 diabetes.

secretion may eventually continue to be well below 100% normal. As this continues, decreasing insulin secretion may eventually dictate the need for exogenous insulin therapy.

3. **Causes of insulin resistance:** Insulin resistance increases with weight gain and decreases with weight loss. Excess adipose tissue (particularly in the abdomen) is key in the development of insulin resistance. Adipose is not simply an energy storage tissue, but also a secretory tissue. With obesity, changes in adipose secretions result in insulin resistance (Fig. 25.9). These include secretion of proinflammatory cytokines such as interleukin 6 and tumor necrosis factor-α by activated macrophages (inflammation is associated with insulin resistance); increased synthesis of leptin, a protein that acts on the brain to regulate food intake and energy expenditure (see Chapter 26, IV.); and decreased secretion of adiponectin (see Chapter 26, II.), a protein that reduces FFA levels in the blood, increases FA β oxidation and increases insulin sensitivity. The net result is chronic, low-grade inflammation and increased insulin resistance. One effect of insulin resistance is increased lipolysis and production of serum FFAs (see Fig. 25.9). FFA availability impairs insulin signaling, decreases muscle and adipose uptake of glucose contributing to hyperglycemia, and increases ectopic deposition of TAG in liver (hepatic steatosis). Steatosis results in nonalcoholic fatty liver disease (NAFLD). If accompanied by inflammation, a more serious condition, nonalcoholic steatohepatitis (NASH), can develop.

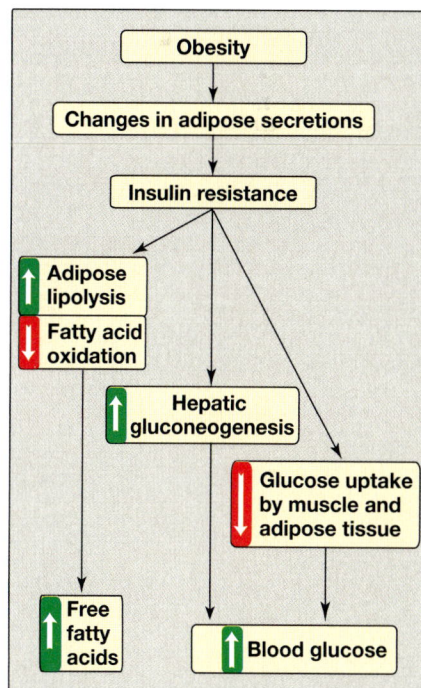

Figure 25.9
Obesity, insulin resistance, and hyperglycemia. [Note: Inflammation also is associated with insulin resistance.]

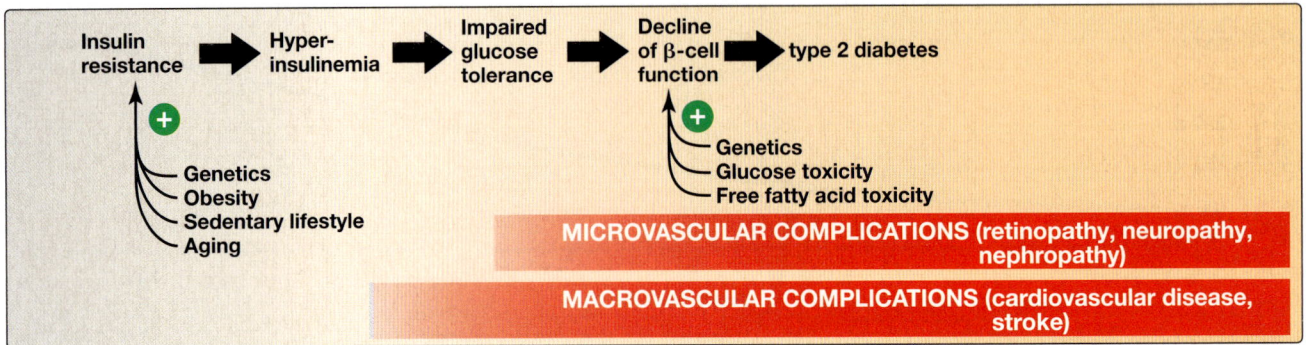

Figure 25.10
Typical progression of type 2 diabetes.

B. Dysfunctional β cells

In T2D, the pancreas initially retains β-cell capacity, resulting in insulin levels that vary from above normal to below normal. If the elevated insulin levels allow sufficient compensation of insulin resistance, hyperglycemia and T2D do not result. However, with time, the β cell becomes increasingly dysfunctional and fails to secrete enough insulin to correct the prevailing hyperglycemia. Thus, the natural progression of the disease results from a declining ability to fully compensate for the increasing insulin resistance, resulting in hyperglycemia levels that permit diagnosis (Fig. 25.10). Deterioration of β-cell function may be accelerated by the toxic effects of sustained hyperglycemia, elevated serum FFAs, and a proinflammatory environment. If insulin levels continue to decrease below a threshold, the patient may eventually require exogenous insulin therapy.

Clinical Application 25.1: Gestational Diabetes

Gestational diabetes mellitus (GDM) is defined as glucose intolerance of variable degree with an onset or first recognition during pregnancy. GDM accounts for 90% of cases of diabetes during pregnancy. GDM mimics T2D with maternal tissue insulin resistance with an insufficient insulin secretory compensatory response, resulting in maternal hyperglycemia. During pregnancy, placental hormones (estrogen, human placental lactogen, prolactin, cortisol, and progesterone) may affect maternal metabolic changes, including insulin resistance, decreased glucose uptake in muscle and adipose tissues, increase hepatic gluconeogenesis, and more. The resulting hyperglycemia provides excess glucose across the placenta to the developing fetus with an increase in fetal insulin secretion. Fetal hyperglycemia and hyperinsulinemia can have both short- and long-term effects on the developing fetus, including macrosomia. Factors that increase the risk for developing GDM include the age of the female during pregnancy (>35), previous pregnancies or family history with GDM, ethnicity (higher risk in Native American, African American, Hispanic American), obesity, unhealthy lifestyle, and more. GDM usually resolves itself upon delivery of the newborn, although GDM increases the risk that the mother may eventually develop T2D. Between 33% and 50% of patients with GDM progress to overt T2D within 5 years, depending on ethnicity (and other risk factors). Diagnosis of GDM involves an abnormal oral glucose tolerance test.

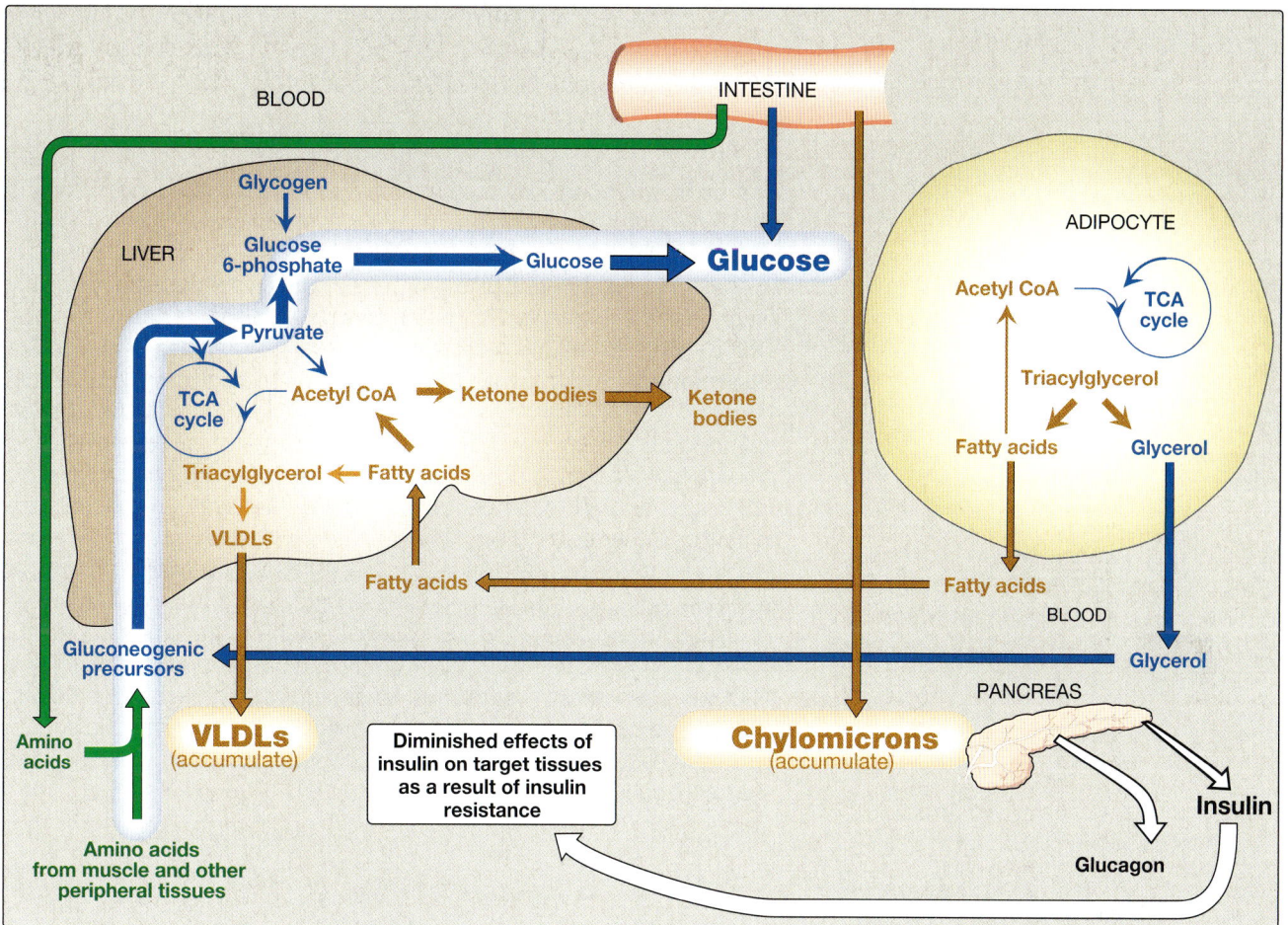

Figure 25.11
Intertissue relationships in type 2 diabetes. [Note: Ketogenesis is restrained as long as insulin action is adequate.] TCA, tricarboxylic acid; CoA, coenzyme A; VLDLs, very–low-density lipoproteins.

C. Metabolic changes

The abnormalities of glucose and TAG metabolism in T2D are the result of insulin resistance that occurs primarily in liver, skeletal muscle, and white adipose tissue (Fig. 25.11).

1. **Hyperglycemia:** Hyperglycemia is caused by increased hepatic production of glucose, combined with diminished use of glucose by muscle and adipose tissues (due to insulin resistance).

2. **Dyslipidemia:** In adipose tissue, TAG lipolysis is increased, resulting in an increase in serum FFAs. In the liver, FFA are converted to TAGs, which are packaged in VLDL and secreted into the bloodstream. Dietary TAG–rich chylomicrons are synthesized and secreted by the intestinal mucosal cells following a meal. Because lipoprotein TAG degradation catalyzed by lipoprotein lipase in adipose tissue is low in diabetes, the plasma chylomicron and VLDL levels are elevated, resulting in hypertriacylglycerolemia. Low levels of high-density lipoproteins are also associated with T2D, likely as a result of increased degradation. Significantly, ketonemia is usually minimal or absent in patients with T2D

(as compared to nonadherent T1D patients) because the presence of insulin, even with insulin resistance, restrains any significant increase in hepatic ketogenesis or the development of DKA.

D. Treatment

The goal in treating T2D is to maintain blood glucose concentrations within normal limits and to prevent the development of long-term complications. Weight reduction, exercise, and medical nutrition therapy (dietary modifications) can often correct the hyperglycemia of newly diagnosed T2D patients. Oral antihyperglycemic agents can be used by T2D patients to reduce blood glucose levels. Antihyperglycemics are also referred to as hypoglycemic agents, as hypoglycemia can result from their use. T2D antihyperglycemic agents include biguanides such as metformin (decreases hepatic gluconeogenesis), sulfonylureas and meglitinides (increase insulin secretion), thiazolidinediones (decrease FFA levels and increase peripheral insulin sensitivity), α-glucosidase inhibitors (decrease absorption of dietary carbohydrate), incretins (decrease glucagon secretion, increase insulin secretion, feeling of satiety), and SGLT inhibitors (decrease renal reabsorption of glucose), or insulin therapy may also be required to achieve satisfactory plasma glucose levels. Bariatric surgery in individuals with morbid obesity and T2D has been shown to result in disease remission in most patients, although remission may not be permanent. Antihyperglycemic agents and their tissue-specific effects on glucose and lipid metabolism are summarized in Figure 25.12.

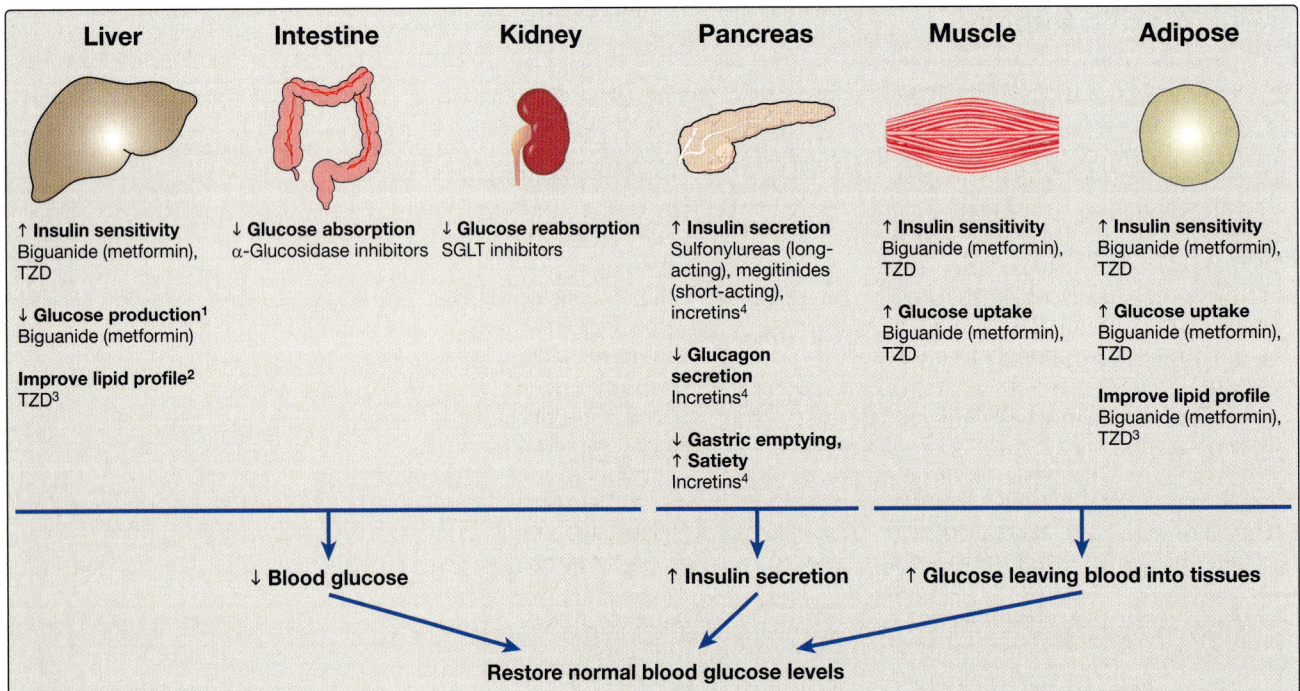

Figure 25.12
Antihyperglycemic agents and their tissue-specific effects on glucose and lipid metabolism in type 2 diabetes. **1.** Decrease hepatic gluconeogenesis and glycogenolysis. **2.** Combination of increasing HDL, decreasing triacylglycerides, and/or decreasing adipocyte lipolysis. **3.** Increase in adiponectin release from adipocytes, increased β-oxidation. **4.** Incretins and DPP4 inhibitors. TZD, thiazolidinediones; DPP4, dipeptidyl peptidase 4; SGLT, sodium-glucose cotransporter.

IV. CHRONIC EFFECTS AND PREVENTION

As noted, available therapies moderate the hyperglycemia of diabetes but fail to completely normalize metabolism. The long-standing elevation of blood glucose is associated with the chronic vascular complications of diabetes including cardiovascular disease (CVD) and stroke (macrovascular complications) as well as retinopathy, nephropathy, and neuropathy (microvascular complications). Intensive insulin treatment (see Section II. C. 3. and Fig. 25.5) delays the onset and slows the progression of some long-term complications. For example, the incidence of retinopathy decreases as control of blood glucose improves and HbA_{1c} levels decrease (Fig. 25.13). Data concerning the effect of tight control on CVD in T2D are less clear. The benefits of tight control of blood glucose outweigh the increased risk of severe hypoglycemia in most patients. How hyperglycemia causes the chronic complications of diabetes is unclear. In cells in which glucose uptake is not dependent on insulin, elevated blood glucose leads to increased intracellular glucose and its metabolites. For example, increased intracellular sorbitol contributes to cataract formation in diabetes (see Chapter 12, II.). Additionally, hyperglycemia promotes glycation of other cellular proteins in a reaction analogous to the formation of HbA_{1c}. These glycated proteins undergo additional reactions and become advanced glycation end products (AGEs) that mediate some of the early microvascular changes of diabetes and can reduce wound healing. Some AGEs bind to a membrane

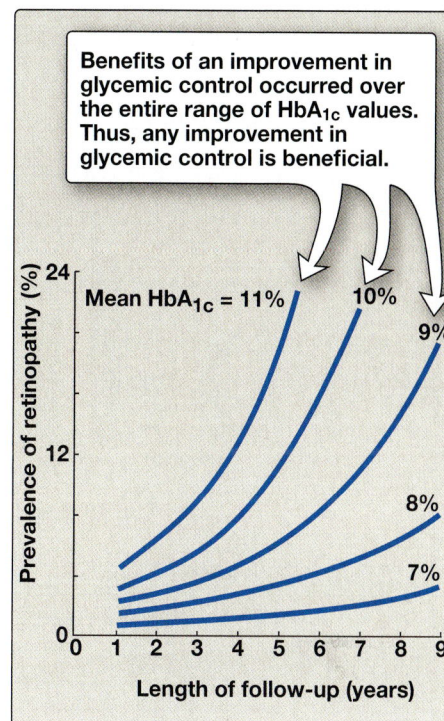

Benefits of an improvement in glycemic control occurred over the entire range of HbA_{1c} values. Thus, any improvement in glycemic control is beneficial.

Figure 25.13
Relationship of glycemic control and diabetic retinopathy. HbA_{1c}, glycated hemoglobin.

Clinical Application 25.2: Hyperosmolar Hyperglycemic Nonketotic Syndrome

Hyperosmolar hyperglycemic nonketotic syndrome ([HHNS] also referred to as hyperosmolar hyperglycemic nonketotic coma or hyperosmolar hyperglycemic state) is characterized by extreme hyperglycemia (often >600 mg/dL), hyperosmolarity (>320 mOsm/dL; normal = 275 to 299 mOsm/dL) and dehydration, without significant ketoacidosis. It usually presents in older patients with T2D and carries a higher mortality rate than DKA, estimated at approximately 15%. It is usually brought on by secondary events in addition to diabetes, including illness/infection, not taking enough fluids, or fluid loss (burns; stroke; recent surgery; or drugs, such as phenytoin, diazoxide, glucocorticoids, and diuretics). In patients with a pre-existing lack of, or resistance to insulin, a physiologic stress such as an acute illness can cause further net reduction in circulating insulin. The basic underlying mechanism of HHNS is a reduction in the effective circulating insulin with a concomitant elevation of counterregulatory hormones, such as glucagon, catecholamines, cortisol, and growth hormone. Decreased renal clearance and decreased peripheral utilization of glucose lead to extreme hyperglycemia. Hyperglycemia results in hyperosmolarity osmotic diuresis, an osmotic shift of fluid to the intravascular space, and intracellular dehydration. This diuresis leads to a life-threatening loss of electrolytes, such as sodium and potassium. Unlike T1D patients, patients with HHNS do not develop significant ketoacidosis, likely due to the availability of insulin in amounts sufficient to inhibit ketogenesis, but not sufficient to prevent hyperglycemia. Hyperosmolarity may also decrease lipolysis, limiting the amount of FFAs available for ketogenesis. Treatment includes restoring fluids and electrolyte balance, insulin to counteract hyperglycemia, glucose to combat hypoglycemia, and monitor for hyponatremia and hypokalemia.

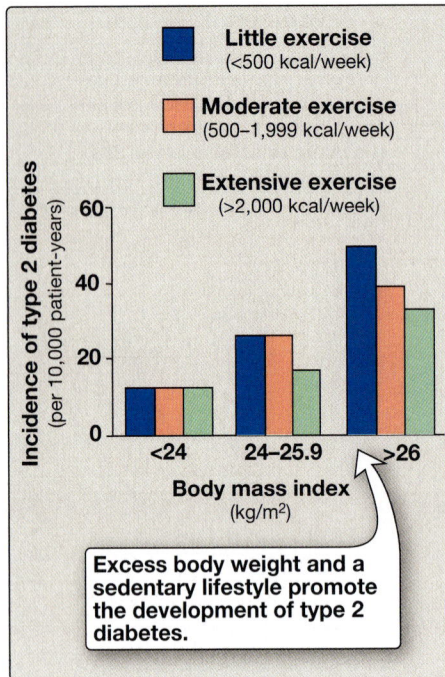

Figure 25.14
Effect of body mass index and exercise on the development of type 2 diabetes.

receptor (RAGE), causing the release of proinflammatory molecules. While there is currently no preventative treatment for T1D, the risk for T2D can be significantly decreased by a combined regimen of medical nutrition therapy, weight loss, exercise, and aggressive control of hypertension and dyslipidemias. For example, Figure 25.14 shows the incidence of disease in normal and overweight individuals with varying degrees of exercise.

V. CHAPTER SUMMARY

- **Diabetes mellitus** is a heterogeneous group of syndromes characterized by an **elevation** of **fasting blood glucose** (**FBG**) caused by a relative or absolute deficiency of insulin (Fig. 25.15).

- Diabetes is the leading cause of **adult blindness** and **amputation** and a major cause of **renal failure, nerve damage, heart attacks,** and **stroke**.

- Diabetes can be classified into two groups, **T1D** and **T2D**.

- T1D constitutes ~5% of >30 million cases of diabetes in the United States. The disease is characterized by an **absolute deficiency** of **insulin** caused by an **autoimmune attack** on the **pancreatic β cells**. This destruction requires an **environmental stimulus** (such as a viral infection) and a **genetic determinant** that causes the β cell to be mistakenly identified as "nonself." The **metabolic abnormalities** of T1D include **hyperglycemia, diabetic ketoacidosis** (**DKA**), and **hypertriacylglycerolemia** that result from a deficiency of insulin. Those with T1D must rely on **exogenous insulin** delivered subcutaneously to control hyperglycemia and ketoacidosis.

- **T2D** has a strong **genetic** component. It results from a combination of **insulin resistance** and **dysfunctional β cells**. Insulin resistance is the decreased ability of target tissues, such as liver, white adipose, and skeletal muscle, to respond properly to normal (or elevated) circulating concentrations of insulin. **Obesity** is the most common cause of insulin resistance. However, most people with obesity and insulin resistance do not develop diabetes. In the absence of a defect in β-cell function, obese individuals without diabetes can fully compensate for insulin resistance with elevated levels of insulin. Insulin resistance alone will not lead to T2D. Rather, T2D develops in insulin-resistant individuals who also show impaired β-cell function and are not able to compensate with sufficient insulin secretion to prevent hyperglycemia. The acute **metabolic alterations** observed in T2D are **milder** than those described for the completely insulin-dependent type 1 form of the disease, in part because insulin secretion in T2D, although inadequate, does restrain **ketogenesis** and blunts the development of DKA.

- Available treatments for diabetes moderate the hyperglycemia but fail to completely normalize metabolism. The long-standing elevation of blood glucose is associated with the **chronic complications** of diabetes including **cardiovascular disease** and **stroke** (**macrovascular**) as well as **retinopathy, nephropathy,** and **neuropathy** (**microvascular**).

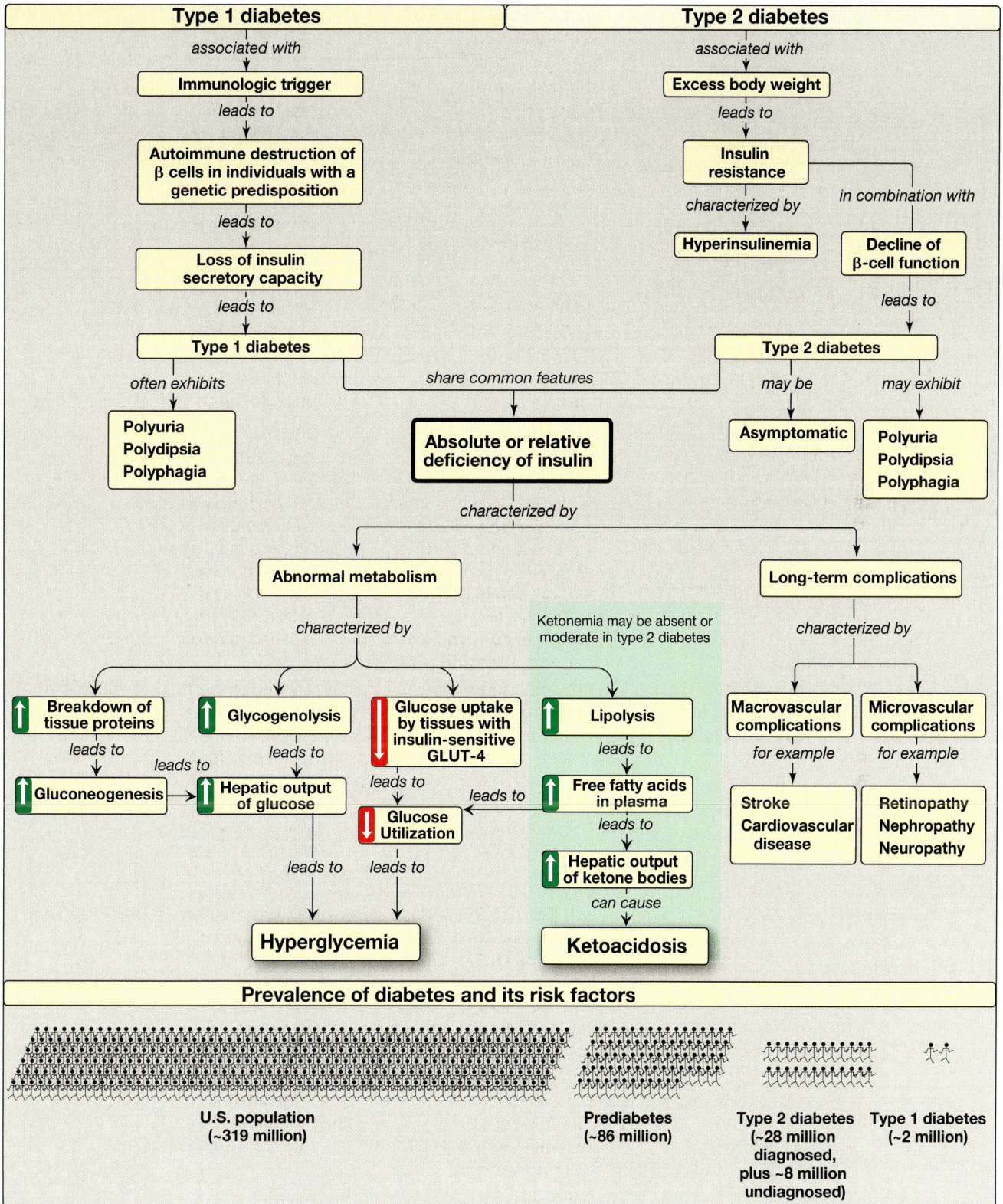

Figure 25.15
Key concept map for diabetes. [Note: Data are from 2014.] GLUT, glucose transporter.

Study Questions

Choose the ONE best answer.

25.1. Three patients being evaluated for gestational diabetes are given an oral glucose tolerance test. Based on the data shown, which patient is prediabetic?

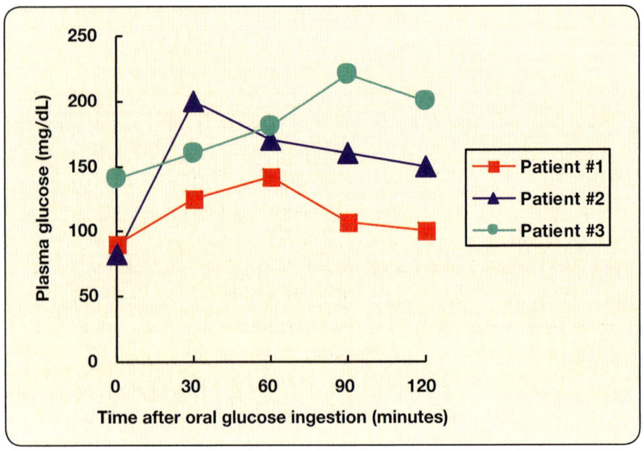

A. Patient #1
B. Patient #2
C. Patient #3
D. All
E. None

Correct answer = B. Patient #2 has a normal fasting blood glucose (FBG) but an impaired glucose tolerance (GT) as reflected in her blood glucose level at 2 hours and, so, is described as prediabetic. Patient #1 has a normal FBG and GT, whereas patient #3 has diabetes.

25.2. Relative or absolute lack of insulin in humans would result in which of the following reactions in the liver?

A. Decreased activity of hormone-sensitive lipase
B. Decreased gluconeogenesis from lactate
C. Decreased glycogenolysis
D. Increased formation of 3-hydroxybutyrate
E. Increased glycogenesis

Correct answer = D. Low insulin levels favor the liver producing ketone bodies, using acetyl coenzyme A generated by β oxidation of the fatty acids provided by hormone-sensitive lipase (HSL) in adipose tissue (not liver). Low insulin also causes activation of HSL, decreased glycogen synthesis, and increased gluconeogenesis and glycogenolysis.

25.3. Which of the following is characteristic of untreated diabetes regardless of the type?

A. Hyperglycemia
B. Ketoacidosis
C. Low levels of hemoglobin A_{1c}
D. Normal levels of C-peptide
E. Obesity

Correct answer = A. Elevated blood glucose occurs in type 1 diabetes (T1D) as a result of a lack of insulin. In type 2 diabetes (T2D), hyperglycemia is due to a defect in β-cell function and insulin resistance. The hyperglycemia results in elevated hemoglobin A_{1c} levels. Ketoacidosis is rare in T2D, whereas obesity is rare in T1D. C (connecting)-peptide is a measure of insulin synthesis. It would be virtually absent in T1D and initially increased then decreased in T2D.

25.4. An individual with obesity as well as T2D typically:

    A. benefits from receiving insulin about 6 hours after a meal.

    B. has a higher plasma level of glucagon than does a healthy individual.

    C. has a lower plasma level of insulin than does a normal individual early in the disease process.

    D. shows improvement in glucose tolerance if body weight is reduced.

    E. shows sudden onset of symptoms.

> Correct answer = D. Many individuals with type 2 diabetes are obese, and almost all show some improvement in blood glucose with weight reduction. Symptoms usually develop gradually. These patients have elevated insulin levels and usually do not require insulin (certainly not 6 hours after a meal) until late in the disease. Glucagon levels are typically normal or low.

For questions 25.5 to 25.9, match the antihyperglycemic drug with its major therapeutic mechanism of action.

    A. Decreases FFA levels and increases peripheral insulin sensitivity

    B. Increases insulin secretion

    C. Decrease dietary carbohydrate absorption

    D. Decreases hepatic gluconeogenesis

    E. Decreases renal reabsorption of glucose

25.5. α-Glucosidase inhibitors

25.6. Biguanides

25.7. SGLT inhibitors

25.8. Sulfonylureas

25.9. Thiazolidinediones

> Correct answers = C, D, E, B, A, respectively. α-Glucosidase inhibitors decrease intestinal dietary glucose absorption. Biguanides including metformin decrease hepatic glucose production. SGLT inhibitors decrease reabsorption of glucose in the kidneys, so excess glucose is excreted in the urine. Sulfonylureas (and meglitinides) are insulin secretagogues that stimulate secretion of insulin from the pancreas. Thiazolidinediones signal through peroxisome proliferator-activated receptors that mediate glyceroneogenesis and storage of FFA in adipocytes, thereby increasing insulin sensitivity in peripheral tissues.

26 Obesity

I. OVERVIEW

Obesity is a disorder of body weight regulatory systems characterized by an accumulation of excess body fat. In primitive societies, in which daily life required a high level of physical activity and food was only available intermittently, a genetic tendency favoring storage of excess calories as fat may have conferred survival value. Today, however, the sedentary lifestyle and abundance and wide variety of palatable, inexpensive foods in industrialized societies has undoubtedly contributed to an obesity epidemic. As adiposity has increased, so has the risk of developing associated diseases, such as type 2 diabetes (T2D), cardiovascular disease (CVD), hypertension, cancer, and arthritis. Particularly alarming is the explosion of obesity in children and adolescents, which has shown a threefold increase in prevalence over the last four decades. Approximately 1 in 5 children and adolescents ages 6 to 19 years are obese. In the United States, the lifetime risk of becoming overweight or obese is ~50% and 25%, respectively. Obesity has increased globally, and, by some estimates, there are more obese than undernourished individuals worldwide.

II. ASSESSMENT

Because the amount of body fat is difficult to measure directly, it is usually determined from an indirect measure, the body mass index (BMI), which has been shown to correlate with the amount of body fat in most individuals. A notable exception are athletes who have large amounts of lean muscle mass. Measuring the waist size with a tape measure is also used to screen for obesity, because this measurement reflects the amount of fat in the central abdominal area of the body. The presence of excess central fat is associated with an increased risk for morbidity and mortality, independent of the BMI. A waist size 40 inches or more for men and 35 inches or more for women is considered a risk factor.

A. Body mass index

The BMI (defined as weight in kg/[height in m]^2) provides a measure of relative weight, adjusted for height. This allows comparisons within and between populations. The healthy range for BMI is between 18.5 and 24.9. Individuals with a BMI between 25 and 29.9 are considered overweight, those with a BMI 30 and up are defined as obese, and a BMI more than 40 is considered severely (morbidly) obese (Fig. 26.1). These cutoffs are based on studies examining the

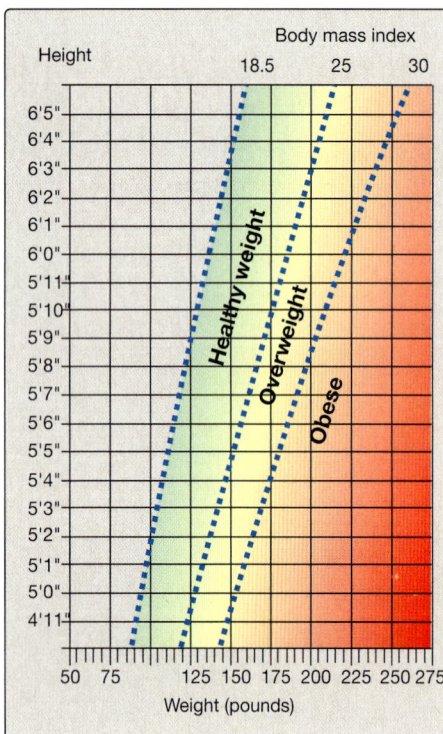

Figure 26.1
To use this body mass index (BMI) chart, find height in the left-hand column. Move across the row to weight. Height and weight intersect at the individual's BMI. [Note: To calculate BMI using pounds and inches, use BMI = weight in pounds/(height in inches)^2 × 703. Anyone >100-lb overweight is considered morbidly obese.]

relationship of BMI to premature death and are similar in men and women. Nearly two-thirds of U.S. adults are overweight, and more than one-third of those are obese. Children with a BMI for age above the 95th percentile are considered obese.

B. Anatomic differences in fat deposition

The anatomic distribution of body fat has a major influence on associated health risks. A waist/hip ratio (WHR) more than 0.8 for women and more than 1.0 for men is defined as android, apple-shaped, or upper-body obesity and is associated with more fat deposition in the trunk (Fig. 26.2A). In contrast, a lower WHR reflects a preponderance of fat distributed in the hips and thighs and is called gynoid, pear-shaped, or lower-body obesity. It is defined as a WHR of less than 0.8 for women and less than 1.0 for men. The pear shape, more commonly found in women, presents a much lower risk of metabolic disease, and some studies indicate that it may actually be protective. Thus, the clinician can use simple indices of body shape to identify those who may be at higher risk for metabolic diseases associated with obesity.

> About 80% to 90% of human body fat is stored in subcutaneous (subq) depots in the abdominal (upper body) and the gluteal–femoral (lower body) regions. The remaining 10% to 20% is in visceral depots located deep within the abdominal cavity (Fig. 26.2B). Excess fat in visceral and abdominal subq stores increases health risks associated with obesity.

C. Biochemical differences in regional fat depots

The regional types of fat described are biochemically different. Subq adipocytes from the lower body, particularly in women, are larger, very efficient at fat (triacylglycerol [TAG]) deposition, and tend to mobilize fatty acids (FAs) more slowly than subq adipocytes from the upper body. Visceral adipocytes are the most metabolically active. In obese individuals, both abdominal subq and visceral depots have high rates of lipolysis and contribute to increased availability of free fatty acids (FFAs). FFAs impair insulin signaling and are proinflammatory (see Chapter 25, III.). These metabolic differences may contribute to the higher health risk found in individuals with upper body (abdominal) obesity.

1. **Endocrine function:** White adipose tissue (WAT), once thought to be a passive reservoir of TAGs, is now known to play an active role in body weight regulatory systems. For example, the adipocyte is an endocrine cell that secretes a number of protein regulators (adipokines), such as the hormones leptin and adiponectin. Leptin regulates appetite as well as metabolism (see Section IV. A. 1.). Adiponectin reduces FFA levels in the blood (by increasing FA oxidation in muscles) and has been associated with improved lipid profiles; increased insulin sensitivity, resulting in better glycemic control; and reduced inflammation in patients with diabetes. Adiponectin levels decrease as body weight and obesity increases, whereas leptin levels increase.

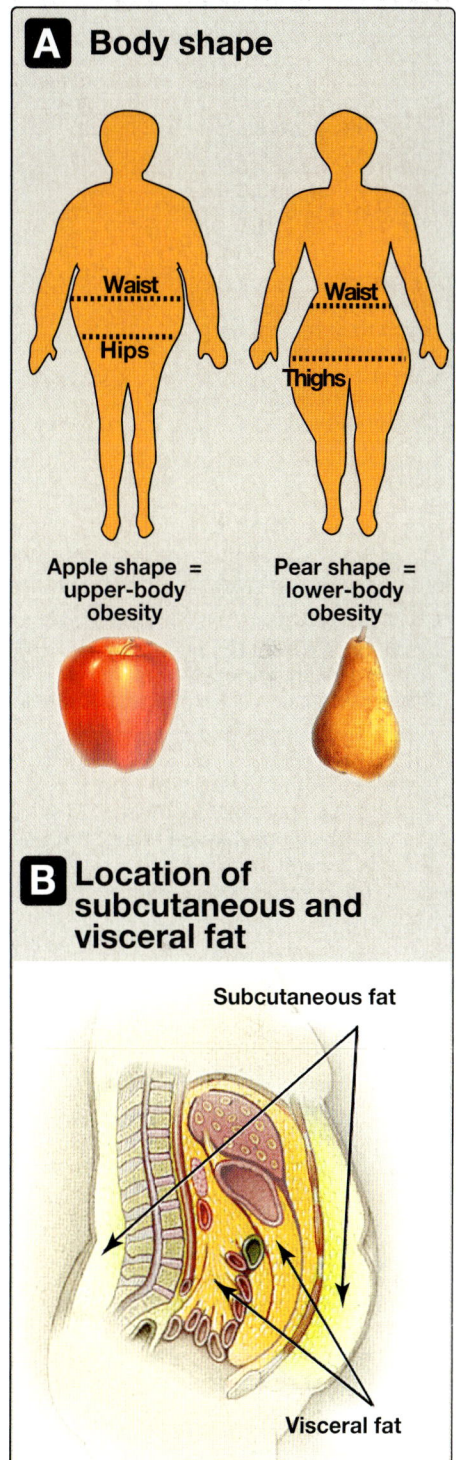

A Body shape

Apple shape = upper-body obesity

Pear shape = lower-body obesity

B Location of subcutaneous and visceral fat

Subcutaneous fat

Visceral fat

Figure 26.2
A. Individuals with upper-body obesity (**left**) have greater health risks than individuals with lower-body obesity (**right**). **B.** Visceral fat is located inside the abdominal cavity, packed in between the internal organs. Subcutaneous fat is found underneath the skin.

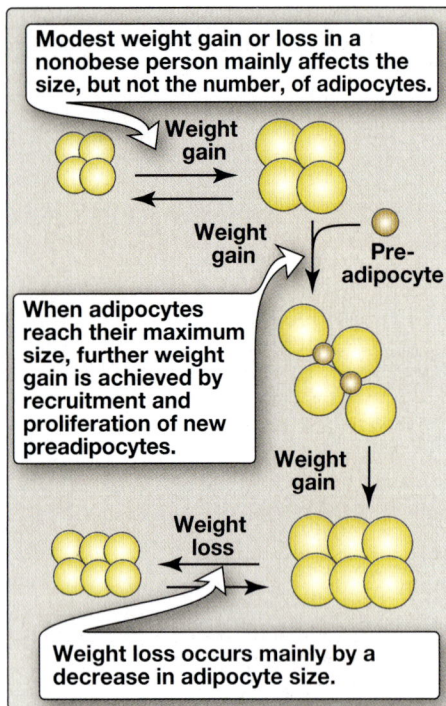

Figure 26.3
Hypertrophic (increased size) and hyperplastic (increased number) changes to adipocytes are thought to occur in severe obesity.

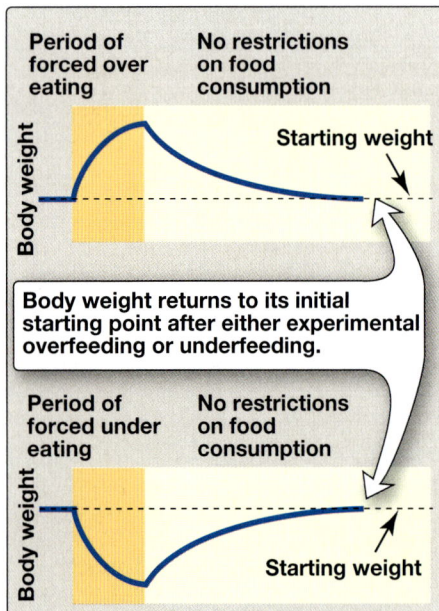

Figure 26.4
Weight changes following episodes of overfeeding or underfeeding followed by feeding with no restrictions.

2. Importance of portal circulation: With obesity, there is increased release of FFAs and secretion of proinflammatory cytokines, such as interleukin 6 (IL-6) and tumor necrosis factor-α (TNF-α), from adipose tissue. Cytokines are small proteins that regulate the immune system. One hypothesis for why abdominal adipose depots have such a large influence on metabolic dysfunction in obesity is that the FFAs and cytokines released from these depots enter the portal vein and, therefore, have direct access to the liver. In the liver, they may lead to insulin resistance (see Chapter 25, III.) and increased hepatic synthesis of TAGs, which are released as components of very–low-density lipoprotein particles and contribute to the hypertriacylglycerolemia associated with obesity. By contrast, FFAs from lower body subq adipose depots enter the general circulation, where they can be oxidized in muscle and, therefore, reach the liver in lower concentration.

D. Adipocyte size and number

As TAGs are stored, adipocytes can expand to an average of two to three times their normal volume (Fig. 26.3). However, the ability of fat cells to expand is limited. With prolonged overnutrition, preadipocytes within adipose tissue are stimulated to proliferate and differentiate into mature fat cells, increasing the number of adipocytes. Thus, most obesity is due to a combination of increased fat cell size (hypertrophy) and number (hyperplasia). Obese individuals can have up to five times the normal number of adipocytes. Like other tissues, the adipose tissue undergoes continuous remodeling. Contrary to early dogma, we now know that adipocytes can die. The estimated average lifespan of an adipocyte is 10 years. If excess calories cannot be accommodated within adipose tissue, the excess FAs "spill over" into other tissues, such as muscle and the liver. The amount of this ectopic fat is strongly associated with insulin resistance. With weight loss in an obese individual, the size of the fat cells is reduced, but the number is not usually affected. Thus, a normal amount of body fat is achieved by decreasing the size of the fat cell below normal. However, small fat cells are very efficient at reaccumulating fat, and this may drive appetite and weight regain.

III. BODY WEIGHT REGULATION

The body weight of most individuals tends to be relatively stable over time. This observation prompted the hypothesis that each individual has a biologically predetermined "set point" for body weight. The body attempts to add to adipose stores when the body weight falls below the set point and to lose adipose from stores when the body weight rises above the set point. Thus, the body defends the set point. For example, with weight loss, appetite increases and energy expenditure falls, whereas with overfeeding, appetite falls, and energy expenditure may slightly increase (Fig. 26.4). However, a strict set point model explains neither why some individuals fail to revert to their starting weight after a period of overeating nor the current epidemic of obesity.

A. Genetic contributions

It is now evident that genetic mechanisms play a major role in determining body weight.

1. **Biologic origin:** The importance of genetics as a determinant of obesity is indicated by the observation that children who are adopted usually show a body weight that correlates with their biologic rather than adoptive parents. Furthermore, identical twins have very similar BMIs, whether reared together or apart, and their BMIs are more similar than those of nonidentical, dizygotic twins.

2. **Mutations:** Rare, single-gene mutations can cause human obesity. For example, mutations in the gene for leptin (causing decreased production) or its receptor (decreased function) result in hyperphagia (increased appetite for and consumption of food) and severe obesity (Fig. 26.5). This underscores the importance of the leptin system in regulating human body weight (see Section IV.). Most humans with obesity have elevated leptin levels but are resistant to the appetite-regulating effects of this hormone.

B. Environmental and behavioral contributions

The epidemic of obesity occurring over the last several decades cannot be simply explained by changes in genetic factors, which are stable on this short time scale. Clearly, environmental factors, such as the ready availability of palatable, energy-dense foods, play a role. Furthermore, sedentary lifestyles decrease physical activity and enhance the tendency to gain weight. Eating behaviors, such as portion size, variety of foods consumed, an individual's food preferences, and the number of people present during eating, also influence food consumption. However, it is important to note that many individuals in this same environment do not become obese. The susceptibility to obesity appears to be explained, at least in part, by an interaction of an individual's genes and their environment and can be influenced by additional factors such as maternal under- or overnutrition that may "set" the body regulatory systems to defend a higher or lower level of body fat. Thus, epigenetic changes (see Chapter 33, IV.) likely influence the risk for obesity.

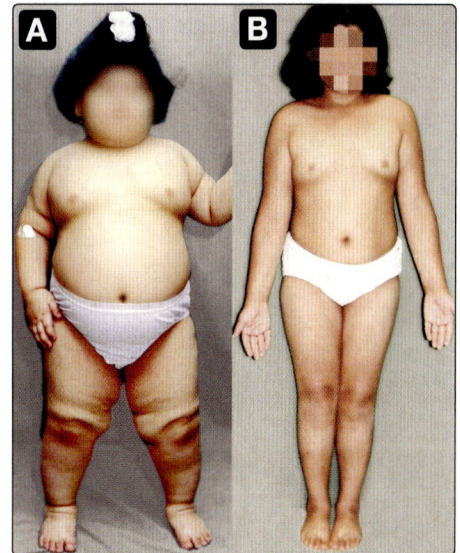

Figure 26.5
A. Patient with leptin deficiency before initiation of therapy at age 5 years.
B. Patient at age 9 years after 48 months of therapy with subcutaneous injections of recombinant leptin.

IV. MOLECULAR INFLUENCES

The cause of obesity can be summarized in a deceptively simple application of the first law of thermodynamics: obesity results when energy (caloric) intake exceeds energy expenditure. However, the mechanism underlying this imbalance involves a complex interaction of biochemical, neurologic, environmental, and psychological factors. The basic neural and humoral pathways that regulate appetite, energy expenditure, and body weight involve systems that regulate short-term food intake (meal to meal), and signals for the long-term (day to day, week to week, year to year) regulation of body weight (Fig. 26.6).

A. Long-term signals

Long-term signals reflect the status of fat (TAG) stores.

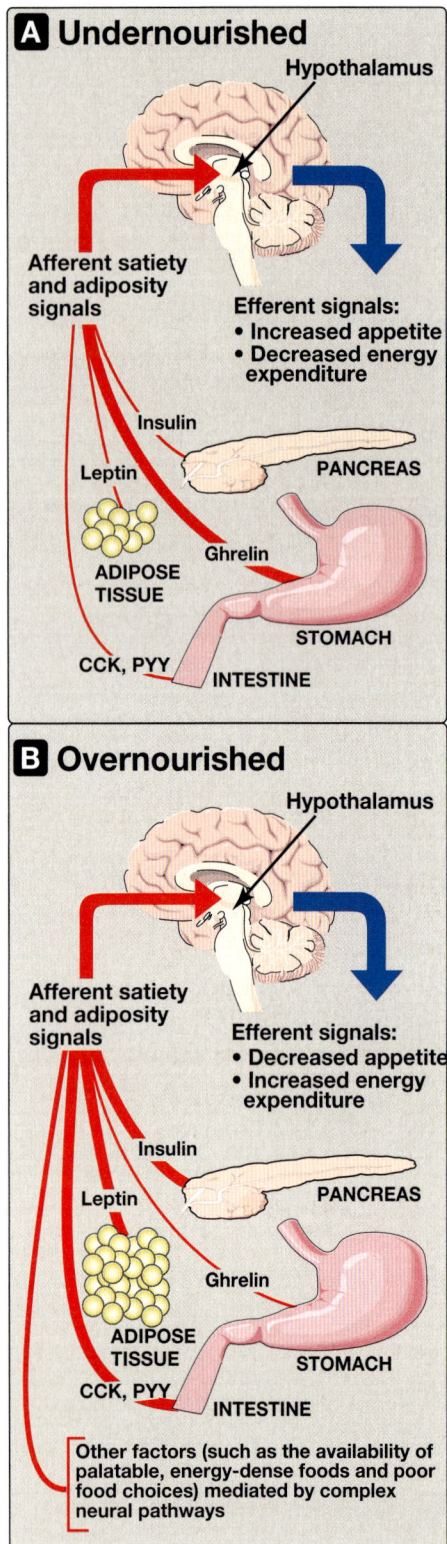

A Undernourished

Hypothalamus

Afferent satiety
and adiposity
signals

Efferent signals:
• Increased appetite
• Decreased energy
  expenditure

Insulin

Leptin

PANCREAS

ADIPOSE
TISSUE

Ghrelin

STOMACH

CCK, PYY

INTESTINE

B Overnourished

Hypothalamus

Afferent satiety
and adiposity
signals

Efferent signals:
• Decreased appetite
• Increased energy
  expenditure

Insulin

Leptin

PANCREAS

ADIPOSE
TISSUE

Ghrelin

STOMACH

CCK, PYY

INTESTINE

Other factors (such as the availability of
palatable, energy-dense foods and poor
food choices) mediated by complex
neural pathways

Figure 26.6
Some signals that influence appetite
and satiety in undernourished (**A**) and
overnourished (**B**) states. CCK, cholecystokinin; PYY, peptide YY.

1. **Leptin:** Leptin is an adipocyte peptide hormone that is made and secreted in proportion to the size of fat stores. It acts on the brain to regulate food intake and energy expenditure. When we consume more calories than we need, body fat increases, and leptin production by adipocytes increases. The body adapts by increasing energy use (increasing activity) and decreasing appetite (an anorexigenic effect). When body fat decreases, the opposite effects occur. Unfortunately, most obese individuals are leptin resistant, and the leptin system may be better at preventing weight loss than preventing weight gain. Leptin's effects are mediated through binding to receptors in the arcuate nucleus of the hypothalamus.

2. **Insulin:** Many individuals with obesity are also hyperinsuline-mic, as a compensatory mechanism to insulin resistance (see Chapter 25, III.). Like leptin, insulin acts on hypothalamic neurons to dampen appetite. See Chapter 23 for the effects of insulin on whole-body metabolism.

B. Short-term signals

Short-term signals from the gastrointestinal (GI) tract control hunger and satiety, which affect the size and number of meals over a time course of minutes to hours. In the absence of food intake (between meals), the stomach produces ghrelin, an orexigenic (appetite-stimulating) hormone that drives hunger. As food is consumed, GI hormones, including cholecystokinin and peptide YY, among others, induce satiety (an anorexigenic effect), thereby terminating eating, through actions on gastric emptying and neural signals to the hypothalamus. Within the hypothalamus, neuropeptides (such as orexigenic neuropeptide Y [NPY] and anorexigenic α-melanocyte–stimulating hormone [α-MSH]) and neurotransmitters (such as anorexigenic serotonin and dopamine) are important in regulating hunger and satiety. Long-term and short-term signals interact, insofar as leptin increases secretion of α-MSH and decreases secretion of NPY. Thus, many complex regulatory loops control the size and number of meals in relationship to the status of body fat stores. α-MSH, a cleavage product of proopiomelanocortin, binds to the melanocortin-4 receptor [MC4R]. Loss-of-function mutations to MC4R are associated with early-onset obesity.

V. METABOLIC EFFECTS

The primary metabolic effects of obesity include dyslipidemia, glucose intolerance, and insulin resistance expressed primarily in the liver, skeletal muscle, and adipose tissue. These metabolic abnormalities reflect molecular signals originating from the increased mass of adipocytes (see Figs. 25.9 and 26.6). Interestingly, about 30% of obese individuals do not show these metabolic abnormalities.

A. Metabolic syndrome

Abdominal obesity is associated with a cluster of metabolic abnormalities, including hyperglycemia, insulin resistance, hyperinsulinemia, atherogenic dyslipidemia (high levels of small low-density lipoprotein [LDL], low levels of high-density lipoprotein [HDL] and elevated TAGs), and hypertension, that is referred to as metabolic syndrome

(Fig. 26.7). It is a risk factor for developing CVD and T2D. The low-grade, chronic, systemic inflammation seen with obesity contributes to the pathogenesis of insulin resistance and T2D and likely plays a role in metabolic syndrome. In obesity, adipocytes release proinflammatory mediators such as IL-6 and TNF-α. Additionally, levels of adiponectin, which normally dampens inflammation and sensitizes tissues to insulin, are low.

B. Nonalcoholic liver disease

Obesity and insulin resistance is associated with increased lipolysis of WAT TAGs and FFAs in circulation. This leads to ectopic deposition of TAGs in the liver (hepatic steatosis) and results in an increased risk for nonalcoholic fatty liver disease ([NAFLD] see Chapter 25, III.).

VI. OBESITY AND HEALTH

Obesity is correlated with an increased risk of death (Fig. 26.8). Obesity is also a risk factor for a number of chronic conditions, including T2D, dyslipidemia, hypertension, CVD, some cancers, gallstones, arthritis, gout, pelvic floor disorders (e.g., urinary incontinence), NAFLD, and sleep apnea. The relationship between obesity and associated morbidities is stronger among individuals younger than age 55 years. After age 74 years, there is no longer an association between increased BMI and mortality. Obesity also has social consequences (e.g., stigmatization and discrimination). Weight loss in obese individuals leads to decreased blood pressure, plasma TAG, and blood glucose levels. HDL levels also increase.

VII. WEIGHT REDUCTION

Weight reduction can help reduce obesity complications. To achieve weight reduction, the patient must decrease energy intake or increase energy expenditure, although decreasing energy intake is thought to contribute more to inducing weight loss. Typically, a plan for weight reduction combines dietary change; increased physical activity; and behavioral modification, which can include nutrition education and meal planning, recording food intake through food diaries, modifying factors that lead to overeating, and relearning cues to satiety. Medications or surgery may be recommended. Once weight loss is achieved, weight maintenance is a separate process that requires vigilance because the majority of patients regain weight after they stop their weight-loss efforts.

A. Caloric restriction

Dieting is the most commonly practiced approach to weight control. Because 1 lb of adipose tissue corresponds to ~3,500 kcal, the effect that caloric restriction will have on the amount of adipose tissue can be estimated. Weight loss on calorie-restricted diets is determined primarily by caloric intake and not nutrient composition. However, compositional aspects can affect glycemic control and the blood lipid profile. Caloric restriction is ineffective over the long term for

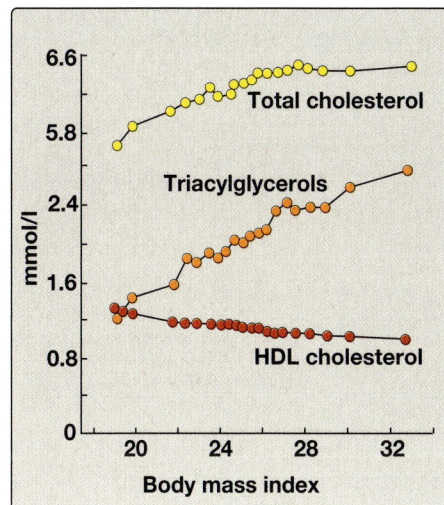

Figure 26.7
Body mass index and changes in blood lipids. HDL, high-density lipoprotein.

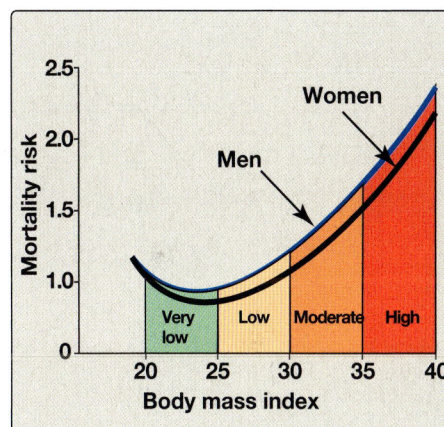

Figure 26.8
Body mass index and the relative risk of death.

many individuals. More than 90% of people who attempt to lose weight regain the lost weight when dietary intervention is suspended. Nonetheless, although few individuals will reach their ideal weight with treatment, weight losses of 10% of body weight over a 6-month period often reduce blood pressure and lipid levels and enhance control of T2D.

B. Physical activity

An increase in physical activity can create an energy deficit. Although adding exercise to a hypocaloric regimen may not produce a greater weight loss initially, exercise is a key component of programs directed at maintaining weight loss. In addition, physical activity increases cardiopulmonary fitness and reduces the risk of CVD, independently of weight loss. Persons who combine caloric restriction and exercise with behavioral treatment may expect to lose ~5% to 10% of initial body weight over a period of 4 to 6 months. Studies show that individuals who maintain their exercise program regain less weight after their initial weight loss.

C. Pharmacologic treatment

The U.S. Food and Drug Administration has approved several weight-loss medications for use in adults. They include orlistat (decreases absorption of dietary fat), lorcaserin, and phentermine in combination with topiramate (promote satiety through serotonin signaling), liraglutide or semaglutide (incretin mimetics; decrease appetite, slow gastric emptying, increase insulin secretion and decrease glucagon secretion by activating glucagon-like peptide 1 or gastric inhibitory polypeptide receptors; see Fig. 25.12), and bupropion in combination with naltrexone (increase metabolism by reducing appetite). Their effects on weight reduction tend to be modest. Pharmacologic activation of brown adipocytes is also being explored (see Chapter 6, VI.).

D. Surgical treatment

Gastric bypass and restriction surgeries are effective in causing weight loss in those who are severely obese. Through mechanisms that remain poorly understood, these operations greatly improve glycemic control in individuals with diabetes who are morbidly obese. Implantation of a device that electrically stimulates the vagus nerve to decrease food intake has also been approved.

VIII. CHAPTER SUMMARY

- **Obesity**, the accumulation of excess body fat, results when **energy (caloric) intake** exceeds **energy expenditure** (Fig. 26.9). Obesity is increasing in industrialized countries because of a reduction in daily energy expenditure and an increase in energy intake resulting from the increasing availability of palatable, inexpensive foods.
- **Body mass index (BMI)** is easy to determine and highly correlated to body fat. Nearly 69% of U.S. adults are **overweight** (BMI ≥25), and >33% of this group are **obese** (BMI ≥30).
- The anatomic distribution of body fat has a major influence on associated health risks. Excess fat located in the **abdomen** (upper body, apple shape), as reflected in **waist size**, is associated with greater risk for **hypertension**, **insulin resistance**, **diabetes**, **dyslipidemia**, and **coronary heart disease** as compared to fat located in the hips and thighs (lower body, pear shape).

- A person's weight is determined by genetic and environmental factors.
- **Appetite** is influenced by afferent, or incoming, signals (i.e., neural signals, circulating hormones such as **leptin**, and metabolites) that are integrated by the **hypothalamus**. These diverse signals prompt release of hypothalamic peptides (such as **neuropeptide Y [NPY]** and α-**melanocyte–stimulating hormone [α-MSH]**) and activate outgoing, efferent neural signals.
- Obesity is correlated with an increased risk of **death** and is also a risk factor for a number of **chronic conditions**.
- **Weight reduction** is achieved best with **negative energy balance**, that is, by decreasing caloric intake and increasing physical activity. Virtually all diets that limit particular groups of foods or macronutrients lead to short-term weight loss. Long-term maintenance of weight loss is difficult to achieve.
- Modest reduction in food intake occurs with **pharmacologic treatment**. **Surgical procedures**, such as gastric bypass, designed to limit food intake are an option for the severely obese patient who has not responded to other treatments.

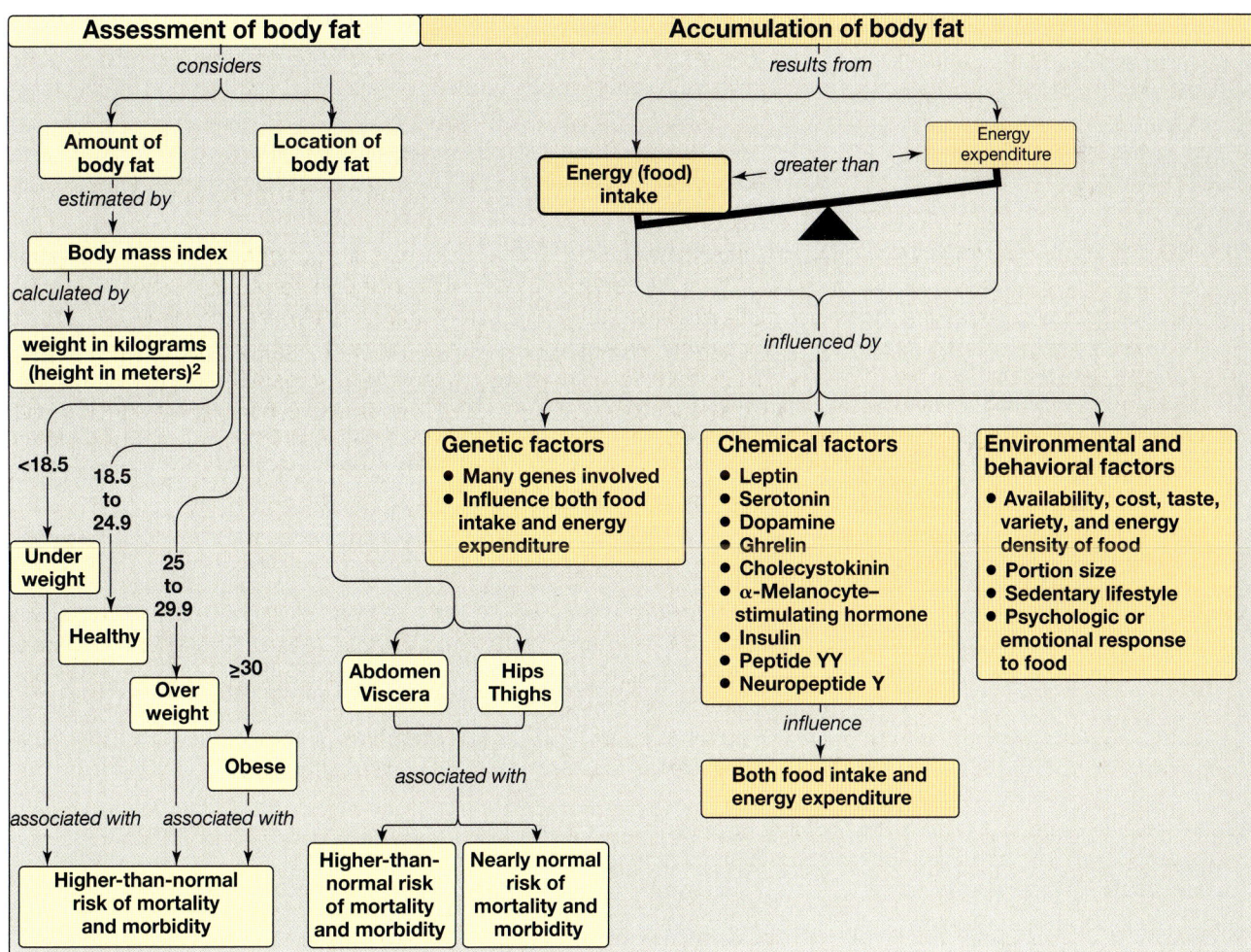

Figure 26.9
Key concept map for obesity. [Note: Body mass index may also be calculated by weight in pounds/(height in inches)$^2 \times 703$.]

Study Questions

Choose the ONE best answer.

For Questions 26.1 and 26.2, use the following scenario.

A 40-year-old female, 5-ft, 1-in (155-cm) tall and weighing 188 lb (85.5 kg), seeks your advice on how to lose weight. Her waist measured 41 in and her hips 39 in. The remainder of the physical examination and the blood laboratory data were all within the normal range. Her only child (who is age 14 years), her sister, and both of her parents are overweight. The patient recalls being overweight throughout her childhood and adolescence. Over the past 15 years, she had been on seven different diets for periods of 2 weeks to 3 months, losing from 5 to 25 lb each time. On discontinuation of the diets, she regained weight, returning to 185 to 190 lb.

26.1. Calculate and interpret the body mass index for the patient.

> Body mass index (BMI) = weight in kg/(height in m)^2 = $85.5/1.55^2$ = 35.6. Because her BMI is >30, the patient is classified as obese.

26.2. Which of the following statements best describes the patient?
- A. She has approximately the same number of adipocytes as an individual of normal weight, but each adipocyte is larger.
- B. She shows an apple pattern of fat distribution.
- C. She would be expected to show higher-than-normal levels of adiponectin.
- D. She would be expected to show lower-than-normal levels of circulating leptin.
- E. She would be expected to show lower-than-normal levels of circulating triacylglycerols (TAGs).

> Correct answer = B. Her waist/hip ratio (WHR) is 1.05 (41/39). Apple shape is defined as a WHR of >0.8 for women and >1.0 for men. Therefore, she has an apple pattern of fat distribution, more commonly seen in males. Compared with other women of the same body weight who have a gynoid (pear-shaped) fat pattern, her android fat pattern places her at greater risk for diabetes, hypertension, dyslipidemia, and coronary heart disease. Individuals with marked obesity and a history dating to early childhood have a fat depot made up of too many adipocytes, each fully loaded with TAGs. Plasma leptin levels are proportional to fat mass, suggesting that resistance to leptin, rather than its deficiency, occurs in human obesity. Adiponectin levels decrease with increasing fat mass. The elevated circulating free fatty acids characteristic of obesity are carried to the liver and converted to TAGs. The TAGs are released as components of very–low-density lipoproteins, resulting in elevated plasma TAG levels, or are stored in the liver, resulting in hepatic steatosis.

26.3. Which one of the following metabolic abnormalities is associated with abdominal obesity and metabolic syndrome?
- A. Higher-than-normal levels of glucose
- B. Higher-than-normal levels of HDL
- C. Lower-than-normal blood pressure
- D. Lower-than-normal levels of insulin
- E. Lower-than-normal levels of TAG

> Correct answer = A. Abdominal obesity is associated with metabolic syndrome, which is defined as a cluster of metabolic abnormalities including hyperglycemia, insulin resistance, hyperinsulinemia, atherogenic dyslipidemia (high levels of LDL, low levels of HDL, and elevated TAGs), and hypertension. Metabolic syndrome is a risk factor for developing CVD and T2D. Low-grade, chronic, systemic inflammation seen with obesity contributes to insulin resistance and T2D and likely plays a role in metabolic syndrome.

26.4. Which of the following statements about leptin is correct?

    A. Leptin expression and secretion is inversely proportional to the size of fat stores.

    B. Leptin signals to decrease appetite and increase energy expenditure.

    C. Leptin signals to decrease secretion of α-melanocyte–stimulating hormone (α-MSH).

    D. Leptin signals to increase secretion of adiponectin from adipocytes.

    E. Leptin signals to increase secretion of neuropeptide Y (NPY).

Correct answer = B. As we consume more calories than we need, body fat increases. Leptin levels increase, whereas adiponectin levels decrease as body weight increases. The body adapts to increasing energy by increasing energy use (increasing activity) and decreasing appetite (an anorexigenic effect). Leptin signaling increases secretion of anorexigenic α-MSH and decreases secretion of orexigenic NPY.

26.5. Which of the following statements best describes the relationship between obesity and nonalcoholic fatty liver disease?

    A. Chylomicrons increase delivery of FAs to the liver.

    B. Increased FFAs in circulation lead to deposition of TAGs in the liver.

    C. Increased NADH/NAD$^+$ ratio increases glycerol 3-phosphate dehydrogenase activity.

    D. Insulin resistance leads to decreased lipolysis of WAT TAG stores.

    E. Obesity leads to increased hepatic FA synthesis.

Correct answer = B. Obesity and insulin resistance are associated with increased lipolysis of WAT TAG and FFAs in circulation, not chylomicrons. This leads to ectopic deposition of TAG in the liver (hepatic steatosis) and results in an increased risk for nonalcoholic fatty liver disease (NAFLD). Obesity and insulin resistance would lead to a decrease in hepatic FA synthesis. An increase in NADH/NAD$^+$ ratio results from chronic alcohol use disorder.

26.6. Which of the following statements best describes the signaling influences on appetite, satiety, and weight in the under- or overnourished?

    A. Ghrelin, an anorexic signaling molecule, mediates increased satiety in the overnourished.

    B. Hyperinsulinemia in overweight individuals leads to increased hunger.

    C. Leptin expression is increased in the undernourished.

    D. Leptin signals to decrease energy expenditure in the undernourished.

    E. Obese individuals are often leptin resistant, making it harder to lose weight.

Correct answer = E. As we consume more calories than we need, body fat increases. Leptin levels increase proportionate to the size of the fat stores (low in undernourished, increased in over-nourished). In the overnourished, increased leptin results in increased satiety and an increase in energy expenditure as a mechanistic attempt to shift weight regulation back to a lower biologic set point. Unfortunately, most obese individuals are leptin resistant. This can have the effect of increasing appetite, decreasing energy expenditure, and making it harder to lose weight. Many overweight individuals are also hyperinsulinemic as a compensatory mechanism to insulin resistance. Hyperinsulinemia has an effect on the hypothalamus to dampen the appetite. Ghrelin is higher in the undernourished, and is an orexigenic (appetite-stimulating) hormone that drives hunger.

26.7. Which of the following statements best describes weight reduction?

    A. Caloric restriction alone is best to prevent weight gain after dietary restrictions are eventually suspended.

    B. Increasing energy expenditure is thought to contribute more to weight loss than decreasing energy intake.

    C. Incretin mimetics decreases absorption of dietary fats.

    D. Semaglutide increases gastric emptying, decreases insulin secretion, and increases glucagon secretion.

    E. Weight losses of 10% of body weight over a 6-month period often reduces blood pressure and lipid levels and enhances control of T2D.

Correct answer = E. Typically, a successful plan for weight reduction combines decreasing energy intake, increasing energy expenditure and behavioral modifications. Decreasing energy intake is thought to contribute more to inducing weight loss than just increasing energy expenditure. Individuals who combine caloric restriction and exercise with behavioral treatment may expect to lose ~5% to 10% of initial body weight over a period of 4 to 6 months, often with reduced blood pressure and lipid levels and enhanced control of T2D. Studies show that individuals who maintain their exercise program regain less weight after their initial weight loss. Orlistat decreases absorption of dietary fat, whereas incretin memetics such as semaglutide decrease appetite, slow gastric emptying, increase insulin secretion, and decrease glucagon secretion.

Nutrition: Overview and Macronutrients

27

I. OVERVIEW

Nutrients are the constituents of food necessary to sustain the normal functions of the body. All energy (calories) is provided by three classes of nutrients: fats, carbohydrates, and protein (Fig. 27.1). Because the intake of these energy-rich molecules is larger (g amounts) than that of the other dietary nutrients (mg to μg amounts), they are called macronutrients. Although alcohol is also an energy source, it is not a nutrient, and it interferes with growth, maintenance, and repair. This chapter focuses on the kinds and amounts of macronutrients needed to maintain optimal health and prevent chronic disease. Those nutrients needed in lesser amounts (mg or μg), vitamins and minerals, are called micronutrients and are considered in Chapters 28 and 29. The names macronutrient and micronutrients do not signify their relative importance, but rather denote their relative dietary intake requirements. A nutrient is a micronutrient when less than a gram is required daily.

II. DIETARY REFERENCE INTAKES

Committees of U.S. and Canadian experts organized by the Food and Nutrition Board of the Institute of Medicine of the National Academy of Sciences have compiled dietary reference intakes (DRIs), which are estimates of the amounts of nutrients required to prevent deficiencies and maintain optimal health and growth. The DRI expands on the recommended dietary allowances (RDAs), which have been published with periodic revisions since 1941. Unlike RDAs, DRIs establish upper limits on the consumption of some nutrients and incorporate the role of nutrients in lifelong health, going beyond mere prevention of deficiency diseases. Both the DRI and the RDA refer to long-term average daily nutrient intakes, because it is not necessary to consume the full RDA every day.

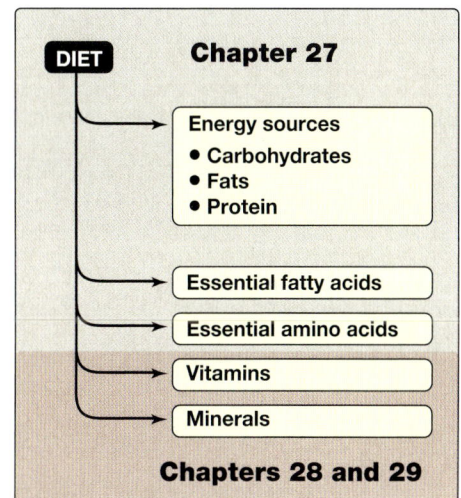

Figure 27.1
Essential nutrients obtained from the diet. [Note: Ethanol may provide a significant contribution to the daily caloric intake of some individuals.]

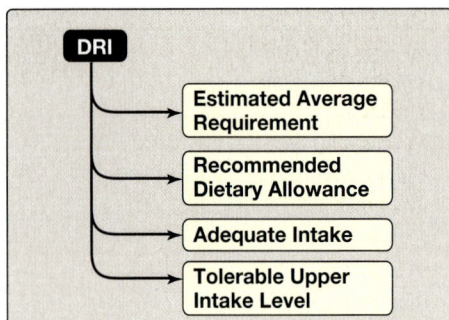

Figure 27.2
Components of dietary reference intakes (DRIs).

A. Definition

DRIs consist of four dietary reference standards for the intake of nutrients designated for specific life stage (age) groups, physiologic states, and gender (Fig. 27.2).

1. **Estimated average requirement:** The average daily nutrient intake level estimated to meet the requirement of 50% of the healthy individuals in a particular life stage and gender group is the estimated average requirement (EAR) (Fig. 27.3). It is useful in estimating the actual requirements in groups and individuals.

2. **RDA:** The RDA is the average daily nutrient intake level sufficient to meet the requirements of nearly all (97% to 98%) individuals in a particular life stage and gender group (Fig. 27.3). The RDA is not the minimal requirement for healthy individuals, but it is intentionally set to provide a margin of safety for most individuals. The EAR serves as the foundation for setting the RDA. If the standard deviation (SD) of the EAR is available, and the requirement for the nutrient is normally distributed, the RDA is set at 2 SDs above the EAR (i.e., $RDA = EAR + 2\ SD_{EAR}$).

3. **Adequate intake:** An adequate intake (AI) is set instead of an RDA if sufficient scientific evidence is not available to calculate an EAR or RDA. The AI is based on estimates of nutrient intake by a group (or groups) of apparently healthy people. For example, the AI for young infants, for whom human milk is the recommended sole source of food for the first 6 months, is based on the estimated daily mean nutrient intake supplied by human milk for healthy, full-term infants who are exclusively breastfed.

4. **Tolerable upper intake level:** The highest average daily nutrient intake level likely to pose no risk of adverse health effects to almost all individuals in the general population is the tolerable upper intake level (UL). As intake increases above the UL, potential risk of adverse effects may increase. The UL is useful because of the increased availability of fortified foods and the

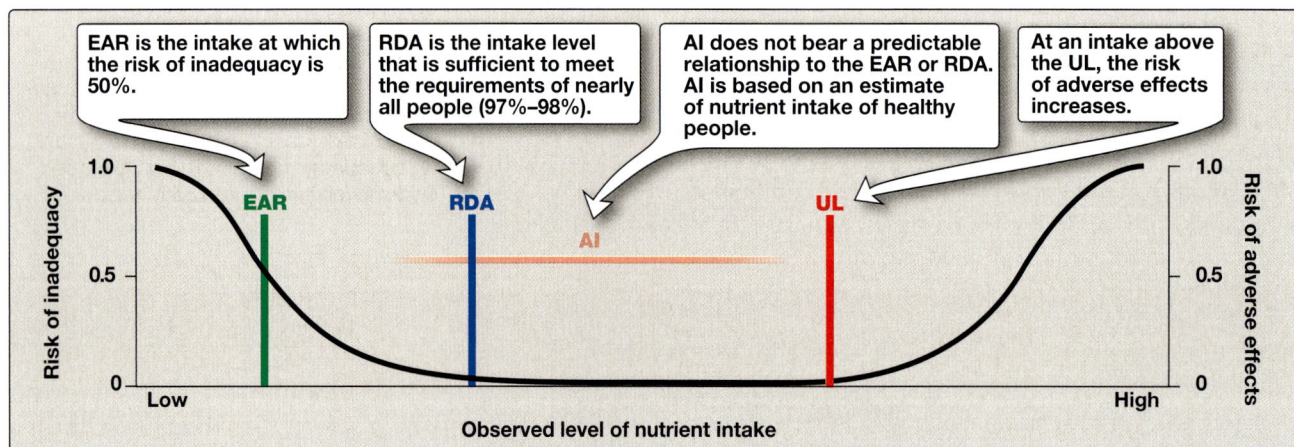

Figure 27.3
Comparison of dietary reference intake components. EAR, estimated average requirement; RDA, recommended dietary allowance; AI, adequate intake; UL, tolerable upper intake level.

increased use of dietary supplements. For some nutrients, there may be insufficient data on which to develop a UL.

B. Using the dietary reference intakes

Most nutrients have a set of DRIs (Fig. 27.4). Usually a nutrient has an EAR and a corresponding RDA. Most are set by age and gender and may be influenced by special factors, such as pregnancy and lactation in women (see Section IX.). When the data are not sufficient to estimate an EAR (or an RDA), an AI is designated. Intakes below the EAR need to be improved because the probability of adequacy is 50% or less (see Fig. 27.3). Intakes between the EAR and RDA likely need to be improved because the probability of adequacy is less than 98%, and intakes at or above the RDA can be considered adequate. Intakes above the AI can be considered adequate. Intakes between the UL and the RDA can be considered to have no risk for adverse effects. [Note: Because the DRI is designed to meet the nutritional needs of healthy individuals, it does not include any special needs of sick individuals.]

III. ENERGY REQUIREMENT IN HUMANS

The estimated energy requirement (EER) is the average dietary energy intake predicted to maintain an energy balance (i.e., the calories consumed are equal to the energy expended) in a healthy adult of a defined age, gender, and height whose weight and level of physical activity are consistent with good health. Differences in genetics, body composition, metabolism, and behavior make it difficult to accurately predict a person's caloric requirements. However, some simple approximations can provide useful estimates. For example, sedentary adults require ~30 kcal/kg/day to maintain body weight, moderately active adults require 35 kcal/kg/day, and very active adults require 40 kcal/kg/day.

A. Energy content of food

The energy content of food is calculated from the heat released by the total combustion of food in a calorimeter. The standard conversion factors for determining the metabolic caloric value of fat, protein, and carbohydrate are shown in Figure 27.5. A calorie is the amount of energy needed to raise the temperature of 1 gram of water by 1°C. A kilocalorie (kcal) is the amount of energy needed to raise 1,000 grams (1 kg) of water by 1°C. In the field of nutrition, 1,000-calorie units are known as kilocalories or Cal. In nutritional terms, this means that 1 g of carbohydrate is equivalent to 4 Cal (1 gram of carbohydrate is equivalent to 4 kcal of energy).

Note that the energy content of fat is more than twice that of carbohydrate or protein, whereas the energy content of ethanol is intermediate between those of fat and carbohydrate. [Note: The joule (J) is the International System of Units (SI) used for energy and is widely used in countries other than the United States. One cal = 4.2 J; 1 kcal (1 Cal, 1 food calorie) = 4.2 kJ. For uniformity, many scientists are promoting the use of joules rather than calories in the United States. However, kcal still predominates and is used throughout this text.]

MICRO-NUTRIENT	EAR, RDA, or AI	UL
Thiamine	EAR, RDA	—
Riboflavin	EAR, RDA	—
Niacin	EAR, RDA	UL
Vitamin B$_6$	EAR, RDA	UL
Folate	EAR, RDA	UL
Vitamin B$_{12}$	EAR, RDA	—
Pantothenic acid	AI	—
Biotin	AI	—
Choline	AI	UL
Vitamin C	EAR, RDA	UL
Vitamin A	EAR, RDA	UL
Vitamin D	EAR, RDA	UL
Vitamin E	EAR, RDA	UL
Vitamin K	AI	—
Boron	—	UL
Calcium	EAR, RDA	UL
Chromium	AI	—
Copper	EAR, RDA	UL
Fluoride	AI	UL
Iodine	EAR, RDA	UL
Iron	EAR, RDA	UL
Magnesium	EAR, RDA	UL
Manganese	AI	UL
Molybdenum	EAR, RDA	UL
Nickel	—	UL
Phosphorus	EAR, RDA	UL
Selenium	EAR, RDA	UL
Vanadium	—	UL
Zinc	EAR, RDA	UL

Figure 27.4
Dietary reference intakes for vitamins and minerals in individuals ages 1 year and older. [Note: A recommended daily allowance (RDA) has been set for carbohydrate and protein (macronutrients) but not for fat.] EAR, estimated average requirement; AI, adequate intake; UL, tolerable upper intake level; –, no value established.

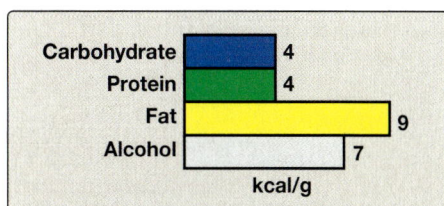

Macronutrient	kcal/g
Carbohydrate	4
Protein	4
Fat	9
Alcohol	7

Figure 27.5
Average energy available from macronutrients and alcohol.

SUBSTRATE	RQ
Carbohydrate	1.00
Protein	0.84
Fat	0.71

Figure 27.6
Respiratory quotient (RQ). [Note: For protein, the nitrogen is removed and excreted, and the α-keto acids are oxidized.]

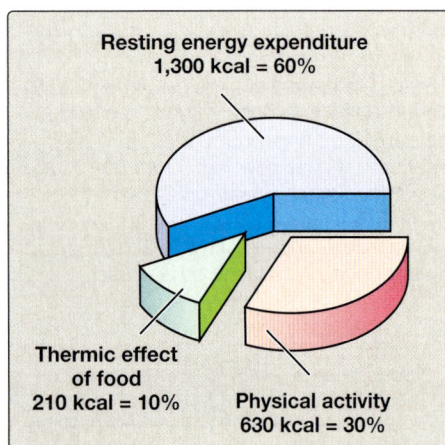

Figure 27.7
Estimated total energy expenditure in a healthy 20-year-old woman, 5 ft, 4 in (165 cm) tall, weighing 110 lb (50 kg), and engaged in light activity.

B. Use of food energy in the body

The energy generated by metabolism of the macronutrients is used for three energy-requiring processes that occur in the body: resting metabolic rate (RMR), physical activity, and the thermic effect of food. Another minor process that requires energy is thermogenesis. The number of kcal expended by these processes in a 24-hour period is the total energy expenditure (TEE).

1. **RMR:** RMR is the energy expended by an individual in a resting, postabsorptive state. It represents the energy required to carry out the normal body functions, such as respiration, blood flow, and ion transport. RMR can be determined by various methods such as calorimetry, doubly labeled water, or mathematical formulas. However, indirect calorimetry is the most commonly used method to quantify RMR by measuring oxygen (O_2) consumed or carbon dioxide (CO_2) produced. The ratio of CO_2 to O_2 is the respiratory quotient (RQ). It reflects the metabolic fuel or substrate being oxidized for energy in tissues (Fig. 27.6). RQ for carbohydrates, proteins and fats are 1.0, 0.84, and 0.71 respectively. For example, complete oxidation of glucose uses 6 O_2 and produces 6 CO_2; therefore, the ratio is 1. On the other hand, when oxidized, the most common fatty acid (FA), palmitate, uses 23 O_2 and produces 16 CO_2, hence the ratio of RQ = CO_2/O_2 = 0.7. An RQ close to 0.8 reflects the oxidation of the mixture of fat and carbohydrate in the diet.

   RMR also can be estimated using equations that include sex and age (RMR reflects lean muscle mass, which is highest in men and the young) as well as height and weight. A commonly used rough estimate is 1 kcal/kg/h for men and 0.9 kcal/kg/h for women. A basal metabolic rate (BMR) can be determined if more stringent environmental conditions are used, but it is not routinely done. RMR is ~10% higher than the BMR. In an adult, the 24-hour RMR, known as the resting energy expenditure (REE), is ~1,800 kcal for men (70 kg) and 1,300 kcal for women (50 kg). From 60% to 75% of the TEE in sedentary individuals is attributable to the REE (Fig. 27.7). [Note: Hospitalized individuals are commonly hypercatabolic due to various factors such as stress, illness, surgery, trauma, or infection. In such states, the body may break down muscle protein and other tissues at an accelerated rate to meet the increased energy demands of healing and recovery or to respond to the inflammation. As a result, RMR is multiplied by an injury factor that ranges from 1.0 (mild infection) to 2.0 (severe burns) in calculating their TEE.]

2. **Physical activity:** Muscular activity provides the greatest variation in the TEE. The amount of energy consumed depends on the duration and intensity of the exercise. This energy cost is expressed as a multiple of the RMR (range is 1.1 to >8.0) referred to as the physical activity ratio (PAR) or the metabolic equivalent of the task (MET). In general, a lightly active person requires ~30% to 50% more calories than the RMR (see Fig. 27.7), whereas a highly active individual may require at least 100% calories above the RMR.

3. **Thermic effect of food:** The production of heat by the body increases as much as 30% above the resting level during the

digestion and absorption of food. This is called the thermic effect of food, or diet-induced thermogenesis. The thermic response to food intake may amount to 5% to 10% of the TEE.

4. **Thermogenesis:** There are two types of thermogenesis: adaptive and nonexercise activity thermogenesis (NEAT). Adaptive thermogenesis is the regulated production of heat in response to environmental changes in temperature and diet, for example, shivering in response to cold. NEAT includes the common daily activities, such as fidgeting, walking to work, pacing while talking on the phone, and standing.

IV. ACCEPTABLE MACRONUTRIENT DISTRIBUTION RANGES

Acceptable macronutrient distribution ranges (AMDRs) are defined as a range of intakes for a particular macronutrient that is associated with reduced risk of chronic disease while providing adequate amounts of essential nutrients. The AMDR for adults is 45% to 65% of their total calories from carbohydrates, 20% to 35% from fat, and 10% to 35% from protein (Fig. 27.8). The biologic properties of dietary fat, carbohydrate, and protein are described next.

V. DIETARY FATS

The incidence of a number of chronic diseases is significantly influenced by the kinds and amounts of nutrients consumed (Fig. 27.9). High dietary intake of certain fats can lead to increased triglyceride levels, which is a risk factor for the incidence of coronary heart disease (CHD), but evidence linking dietary fat and the risk for cancer or obesity is much weaker.

> Earlier recommendations emphasized decreasing the total amount of dietary fat. Unfortunately, this resulted in increased consumption of refined grains and added sugars. Data now show that the type of fat is a more important risk factor than the total amount of fat.

A. Plasma lipids and coronary heart disease

Plasma cholesterol is derived from the diet or from endogenous biosynthesis. In either case, cholesterol is transported between the tissues in combination with protein and phospholipids as lipoproteins.

1. **Low-density and high-density lipoproteins:** The level of plasma cholesterol is not precisely regulated but, rather, varies in response to diet. Elevated levels of total cholesterol (hypercholesterolemia) result in an increased risk for CHD (Fig. 27.10). A much stronger correlation exists between CHD and the level of cholesterol in low-density lipoprotein ([LDL-C] see Chapter 18). As LDL-C increases, CHD increases. In contrast, elevated levels of high-density lipoprotein cholesterol (HDL-C) have been associated

MACRONUTRIENT	AMDR (percent of energy)
Fat	20–35
ω–6 Polyunsaturated fatty acids	5–10
ω–3 Polyunsaturated fatty acids	0.6–1.2*

Approximately 10% of the total fat can come from longer-chain, ω–3 or ω–6 fatty acids.

Carbohydrate	45–65

• RDA
  Men and women: 130 g/day

No more than 10% of total calories should come from added sugars.

Fiber
• AI
  Men: 38 g/day; women: 25 g/day

Protein	10–35

• RDA
  Men: 56 g/day; women: 46 g/day

Figure 27.8
Acceptable macronutrient distribution ranges (AMDRs) in adults. [Note: *A growing body of evidence suggests that higher levels of ω-3 polyunsaturated fatty acids provide protection against coronary heart disease.] RDA, recommended dietary allowance; AI, adequate intake.

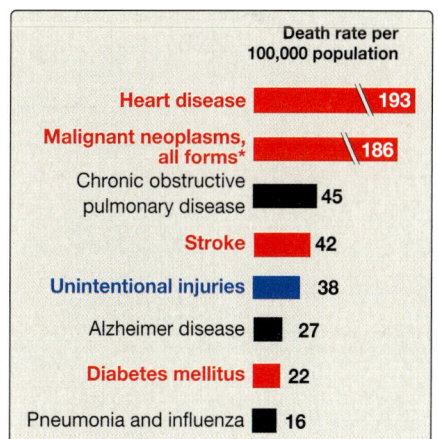

Death rate per 100,000 population

Heart disease	193
Malignant neoplasms, all forms*	186
Chronic obstructive pulmonary disease	45
Stroke	42
Unintentional injuries	38
Alzheimer disease	27
Diabetes mellitus	22
Pneumonia and influenza	16

Figure 27.9
Influence of nutrition on some common causes of death in the United States in the year 2010. *Red* indicates causes of death in which the diet plays a significant role. *Blue* indicates causes of death in which excessive alcohol consumption plays a part. [Note: *Diet plays a role in only some forms of cancer.]

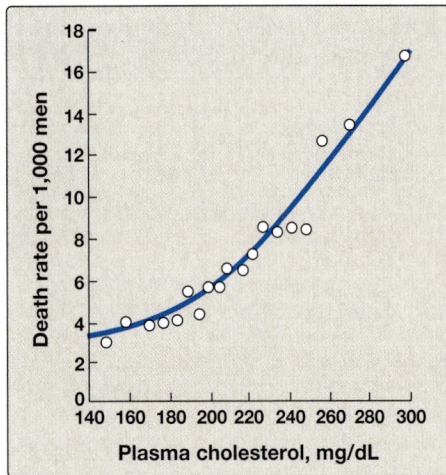

Figure 27.10
Correlation of the death rate from coronary heart disease with the concentration of plasma cholesterol. [Note: Data were obtained from a multiyear study of men with the death rate adjusted for age.]

with a decreased risk for heart disease (see Chapter 18). Elevated plasma triacylglycerol (TAG) is associated with CHD, but a causative relationship has yet to be demonstrated. Abnormal levels of plasma lipids (dyslipidemias) act in combination with smoking, obesity, sedentary lifestyle, insulin resistance, and other risk factors to increase the risk of CHD.

2. **Benefits of lowering plasma cholesterol:** Dietary or pharmacologic management of hypercholesterolemia has been shown to be effective in decreasing LDL-C, increasing HDL-C, and reducing the risk for cardiovascular events. The diet-induced changes in plasma cholesterol concentrations are modest, typically 10% to 20%, whereas treatment with statin drugs decreases plasma cholesterol by 30% to 60% (see Chapter 18, III. E. 5.). Dietary changes and drug treatment can also lower TAG.

B. Dietary fats and plasma lipids

TAGs are quantitatively the most important class of dietary fats. The influence of TAGs on blood lipids is determined by the chemical nature of their constituent FAs. The absence or presence and number of double bonds (saturated vs. mono- and polyunsaturated), the location of the double bonds (ω-6 vs. ω-3), and the *cis* versus *trans* configuration of the unsaturated FAs are the most important structural features that influence blood lipids. Figure 27.13 summarizes the effects of dietary fats.

1. **Saturated fats:** TAGs composed primarily of FAs whose hydrocarbon chains do not contain any double bonds are referred to as saturated fats. Consumption of saturated fats is positively associated with high levels of total plasma cholesterol and LDL-C and an increased risk of CHD. The main sources of saturated FAs are dairy and meat products, along with some vegetable oils, such as coconut and palm oils. These oils are staples in certain parts of the world, like Latin America and Asia, but are less common in U.S. diets. The current accepted dietary guidelines to reduce CVD risk include decreasing saturated fat intake to less than 10% of total caloric intake; substituting them with fish or plant sources, which are rich in polyunsaturated fatty acids (PUFAs), plant proteins, and complex carbohydrates; and avoiding *trans* FAs. It is important to note that replacing saturated fats with refined carbohydrates or meat protein may increase CVD risk.

> Saturated FAs with carbon chain lengths of 14 (myristic) and 16 (palmitic) are most potent in increasing the plasma cholesterol level. Stearic acid (18 carbons, found in many foods including chocolate) has little effect on blood cholesterol.

2. **Monounsaturated fats:** TAGs containing primarily FAs with one double bond are referred to as monounsaturated fats (MUFAs). FAs containing more than one double bond are PUFAs. MUFAs are generally obtained from plant-based oils. When substituted for saturated FAs in the diet, MUFAs lower both total plasma

cholesterol and LDL-C and maintain or increase HDL-C. This ability of MUFAs to favorably modify lipoprotein levels may explain, in part, the observation that Mediterranean cultures, with diets rich in olive oil that is high in monounsaturated oleic acid, show a low incidence of CHD. Although there is no AMDR for MUFAs, it is recommended that the fats in diet should be mostly unsaturated FAs (MUFAs and PUFAs).

a. **Mediterranean diet:** The Mediterranean diet is an example of a diet rich in MUFAs (from olive oil, olives, nuts, and fish) and PUFAs (from fish oils, plant oils, and some nuts) but low in saturated fat. For example, Figure 27.11 shows the composition of the Mediterranean diet in comparison with both a Western diet similar to that consumed in the United States and a typical low-fat diet. The Mediterranean diet contains seasonally fresh food, with an abundance of plant material, low amounts of red meat, and olive oil as the principal source of fat. The Mediterranean diet is associated with decreased plasma total cholesterol, LDL-C, and TAGs and increased HDL-C when compared with a typical Western diet higher in saturated fats.

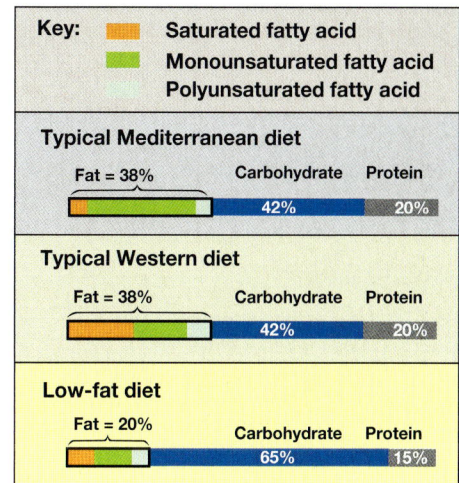

3. **Polyunsaturated fats:** TAGs containing primarily FAs with more than one double bond are referred to as polyunsaturated fats. The effects of PUFAs on cardiovascular disease are influenced by the location of the double bonds within the molecule.

a. **ω-6 Fatty acids:** These are long-chain PUFAs, with the first double bond beginning at the sixth bond position when starting from the methyl (ω) end of the FA molecule. They are also called n-6 fatty acids (see Chapter 16). Consumption of fats containing ω-6 PUFAs, principally linoleic acid (18:2 [9,12]), obtained from vegetable oils, lowers plasma cholesterol when substituted for saturated fats. Plasma LDL-C is lowered, but HDL-C, which protects against CHD, is also lowered, partially offsetting the benefits of lowering LDL-C. Nuts, avocados, olives, soybeans, and various oils (e.g., sunflower and corn oil) are common sources of these FAs. The AMDR for linoleic acid is 5% to 10%. The lower recommendation for intake of PUFAs relative to MUFAs is because of concern that free radical–mediated oxidation (peroxidation) of PUFAs may lead to deleterious products.

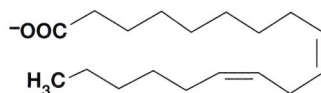

Linoleic acid
(18:2, ω-6)

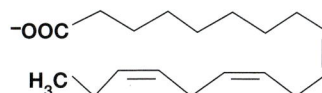

α-Linolenic acid
(18:3, ω-3)

b. **ω-3 Fatty acids:** These are long-chain PUFAs, with the first double bond beginning at the third bond position from the methyl (ω) end. Dietary ω-3 PUFAs suppress cardiac arrhythmias, reduce plasma TAGs, decrease the tendency for thrombosis, lower blood pressure, and substantially reduce the risk of cardiovascular mortality, but they have little effect on LDL-C or HDL-C levels. Evidence suggests that they

Figure 27.11
Composition of typical Mediterranean, Western, and low-fat diets.

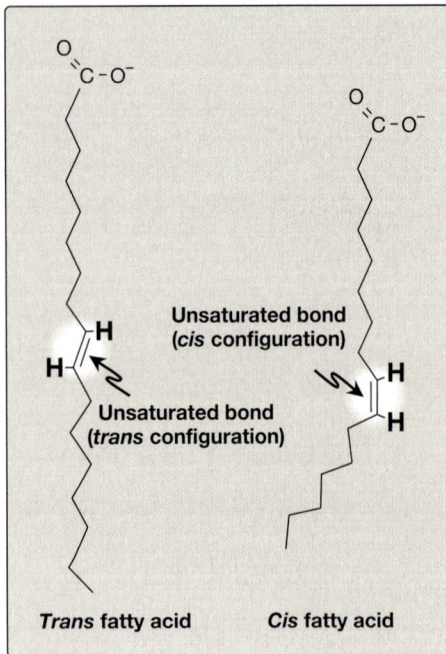

Figure 27.12
Structure of *cis* and *trans* fatty acids.

have anti-inflammatory effects. The ω-3 PUFAs, principally α-linolenic acid, 18:3(9,12,15), are found in plant oils, such as flaxseed and canola, and some nuts, such as walnuts. The AMDR for α-linolenic acid is 0.6% to 1.2%. Fish oil contains the long-chain ω-3 docosahexaenoic acid ([DHA] 22:6) and eicosapentaenoic acid ([EPA] 20:5). Two meals containing fatty fish (e.g., salmon, mackerel, anchovies, sardines, herring) per week are recommended. [Note: DHA is included in infant formulas to promote brain development.] Linoleic and α-linolenic acids are essential fatty acids (EFAs) required for membrane fluidity and synthesis of eicosanoids (see Chapter 17, VIII.). EFA deficiency, caused primarily by fat malabsorption, is characterized by scaly dermatitis as a result of the depletion of skin ceramides with long-chain FAs (see Chapter 17, III. F).

4. **Trans fatty acids:** *Trans* FAs are chemically classified as unsaturated FAs but behave more like saturated FAs in the body because they elevate LDL-C and lower HDL-C, thereby increasing the risk of CHD (Fig. 27.12). *Trans* FAs do not occur naturally in plants but occur in small amounts in animals. However, *trans* FAs are formed during the hydrogenation of vegetable oils (e.g., in the manufacture of margarine and partially hydrogenated vegetable oil). *Trans* FAs have historically been a major component of many commercial baked goods, such as cookies, and most deep-fried foods. Many manufacturers have reformulated their products to be free of *trans* fats. In 2006, the U.S. Food and Drug Administration required that Nutrition Facts labels (see Section VIII. B. 2.) display the *trans*-fat content of packaged food and has taken steps to eliminate artificial *trans* fats in processed, which took full effect in 2020.

5. **Dietary cholesterol:** Cholesterol is found only in animal products. The American Heart Association declared in 2015 that "there is

TYPE OF FAT	METABOLIC EFFECTS		EFFECTS ON DISEASE PREVENTION
Trans fatty acid	⬆ LDL ⬇ HDL		⬆ Incidence of coronary heart disease
Saturated fatty acid	⬆ LDL Little effect on HDL		⬆ Incidence of coronary heart disease; may increase risk of prostate, colon cancer
Monounsaturated fatty acid	⬇ LDL Maintain or increase HDL		⬇ Incidence of coronary heart disease
Polyunsaturated ω–6 fatty acids such as linoleic acid	⬇ LDL ⬇ HDL Provide arachidonic acid, which is an important precursor of prostaglandins and leukotrienes		⬇ Incidence of coronary heart disease
Polyunsaturated ω–3 fatty acids such as DHA	Little effect on LDL Little effect on HDL Suppress cardiac arrhythmias, reduce serum triacylglycerols, decrease the tendency for thrombosis, lower blood pressure, reduce inflammation		⬇ Incidence of coronary heart disease ⬇ Risk of sudden cardiac death

Figure 27.13
Effects of dietary fats. LDL, low-density lipoprotein; HDL, high-density lipoprotein; DHA, docosahexaenoic acid.

insufficient evidence to determine whether lowering dietary cholesterol reduces LDL-C" and the Dietary Guidelines Advisory Committee concluded that the "available evidence shows no appreciable relationship between consumption of dietary cholesterol and serum cholesterol."

C. Other dietary factors affecting coronary heart disease

In the past, the recommendation about alcohol was moderate consumption of alcohol (up to 1 drink/day for women and up to 2 drinks/day for men) decreases the risk of CHD, because there is a positive correlation between moderate alcohol (ethanol) consumption and the plasma concentration of HDL-C. However, in 2023, the World Health Organization published a statement that no level of alcohol consumption is safe for human health.

VI. DIETARY CARBOHYDRATES

The primary role of dietary carbohydrates is to provide energy. The obesity epidemic emerged in 1980s in western countries. During this same period, carbohydrate consumption has significantly increased (as fat consumption decreased), leading some observers to link obesity with carbohydrate consumption. However, obesity has also been related to increasingly inactive lifestyles and to calorie-dense foods served in expanded portion sizes. Carbohydrates are not inherently fattening.

A. Classification

Dietary carbohydrates are classified as simple sugars (monosaccharides and disaccharides), complex sugars (polysaccharides), and fiber.

1. **Monosaccharides:** Glucose and fructose are the principal monosaccharides found in food. Glucose is abundant in fruits, sweet corn, corn syrup, and honey. Free fructose is found together with free glucose in honey and fruits (e.g., apples).

   a. **High-fructose corn syrup:** High-fructose corn syrups (HFCSs) are prepared through enzymatic processing to convert glucose into fructose. Pure corn syrup (100% glucose) is then added to fructose to produce a desired sweetness. In the United States, HFCS 55 (containing 55% fructose and 42% glucose) is commonly used as a substitute for sucrose in beverages, including soft drinks, with HFCS 42 used in processed foods. The composition and metabolism of HFCS and sucrose are similar, the major difference being that HFCS is ingested as a mixture of monosaccharides (Fig. 27.14). Most studies have shown no significant difference between sucrose and HFCS meals in either postprandial glucose or insulin responses. [Note: The rise in the use of HFCS parallels the rise in obesity, but a substantial causal relationship has not been demonstrated due to the multifactorial nature of obesity, such as overall diet quality, physical inactivity, and genetic predisposition.]

2. **Disaccharides:** The most abundant disaccharides are sucrose (glucose + fructose), lactose (glucose + galactose), and maltose (glucose + glucose). Sucrose is ordinary table sugar and is

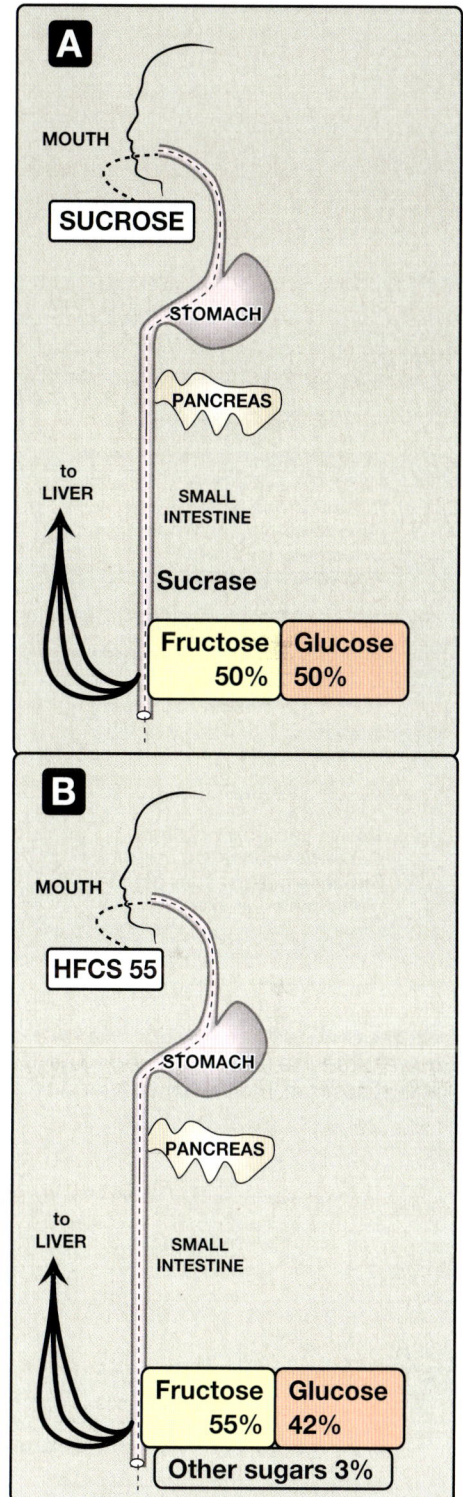

Figure 27.14
Digestion of sucrose, (**A**), or high-fructose corn syrup (HFCS) 55, (**B**), leads to absorption of glucose plus fructose.

abundant in molasses and maple syrup. Lactose is the principal sugar found in milk. Maltose is a product of enzymic digestion of glycogen and starches in the intestine. It is also found in significant quantities in beer and malt liquors as maltose is found in germinating grains. The term "sugar" refers to monosaccharides and disaccharides. "Added sugars" are those sugars and syrups (such as HFCS) added to foods during processing or preparation.

3. **Polysaccharides:** Complex carbohydrates are composed of oligosaccharides and polysaccharides, which are predominantly polymers of glucose. Examples of polysaccharides are starch, glycogen, and dietary fiber. Starch is an example of a complex carbohydrate found in abundance in plants. Common sources include wheat and other grains, potatoes, dried peas and beans (legumes), and vegetables.

4. **Fiber:** Dietary fiber, soluble and insoluble is the edible part of plants that is nondigestible, nonstarch carbohydrates and lignin (a noncarbohydrate polymer of aromatic alcohols). Dietary fiber provides little energy but has several beneficial effects. The AI for dietary fiber is 25 g/day for women and 38 g/day for men. However, most American diets are far lower in fiber at approximately 15 g/day. Insoluble fibers pass through the digestive track largely unchanged and adds bulk to the diet (Fig. 27.15). Fiber can absorb 10 to 15 times its own weight in water, drawing fluid into the lumen of the intestine and increasing bowel motility and promoting bowel movements (laxation). Fiber-rich diets decrease the risk for constipation, hemorrhoids, and diverticulosis. Examples of insoluble fiber are cellulose, hemicellulose, and lignin.

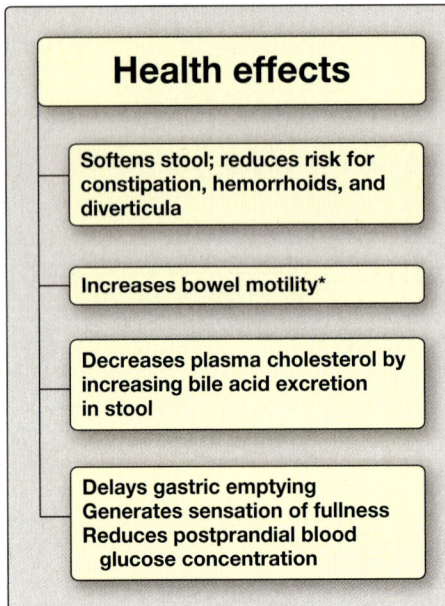

Figure 27.15
Actions of dietary fiber. [Note: *Increasing bowel motility decreases exposure time of the intestines to carcinogens.]

Although soluble fiber is resistant to digestion and absorption in the human small intestine, it is completely or partially fermented by the gut bacteria to short-chain fatty acids (SCFAs) in the large intestine. SCFAs play an essential role in regulating host metabolism, immune system, and cell proliferation. Examples of soluble fiber are pectins, β-glucans, gums, and mucilage. Soluble fiber delays gastric emptying and can result in a sensation of fullness (satiety). This delayed emptying also results in reduced spikes in blood glucose following a meal. Second, consumption of soluble fiber has been shown to lower LDL-C levels by increasing fecal bile acid excretion and interfering with dietary fats, cholesterol, and cholesterol reabsorption (see Chapter 18, V.). For example, diets rich (25 to 50 g/day) in the soluble fiber oat bran are associated with a modest, but significant, reduction in risk for CHD by lowering total cholesterol and LDL-C levels.

"Functional fiber" is the term used for isolated fiber that has proven health benefits such as commercially available fiber supplements. Introduction of fiber into the diet should be gradual as it can lead to abdominal discomfort, gas, diarrhea, and even constipation.

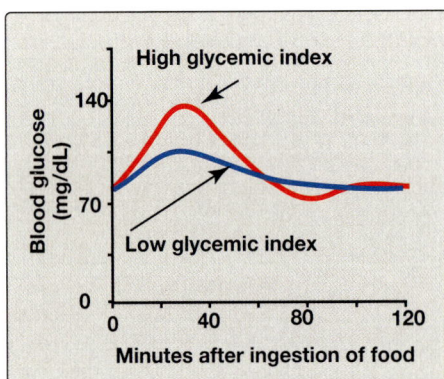

Figure 27.16
Blood glucose concentrations following ingestion of food with a low or high glycemic index (GI). [Note: GI is defined as the area under the blood glucose curve.]

B. Dietary carbohydrate and blood glucose

Some carbohydrate-containing foods produce a rapid rise followed by a steep fall in blood glucose concentration, whereas others result in a gradual rise followed by a slow decline (Fig. 27.16). Thus, they differ in their glycemic response (GR). [Note: Fiber blunts the GR.] The glycemic index (GI) ranks carbohydrate-rich foods on a scale of 0 to 100 based on the GR they cause relative to the GR caused by the same amount (50 g) of carbohydrate eaten in the form of white bread or glucose. A low GI is less than 55, whereas a high GI is at least 70. Evidence suggests that a low-GI diet improves glycemic control in individuals with diabetes. Food with a low GI tends to create a sense of satiety over a longer period of time and may be helpful in limiting caloric intake. Glycemic load (GL) takes into consideration not only GI but also the quantity of carbohydrates in a typical serving. Therefore, GL provides a more accurate picture of a food's impact on blood sugar levels and can be a more useful guide when managing diets for conditions such as diabetes. For example, carrots can have a high GI and a low GL, which means that the carbohydrates in carrots may cause a rapid increase in blood glucose but the actual amount of carbohydrates in a typical serving of carrots is low, so the overall effect on blood sugar (the GL) is small.

C. Carbohydrate requirements

Carbohydrates are not essential nutrients, because the carbon skeletons of most amino acids can be converted into glucose (see Chapter 20, II. A.). However, they supply essential nutrients such as vitamins and minerals. In addition, the absence of dietary carbohydrate leads to ketogenesis (see Chapter 16, V.) and degradation of body protein whose constituent amino acids provide carbon skeletons for gluconeogenesis (see Chapter 10, II. C.). The RDA for carbohydrate is set at 130 g/day for adults and children, based on the amount of glucose used by carbohydrate-dependent tissues, such as the brain and erythrocytes. However, this level of intake is usually exceeded. Adults should consume 45% to 65% of their total calories from carbohydrates. It is now recommended that added sugars represent no more than 10% of total energy intake because of concerns that they may displace nutrient-rich foods from the diet. [Note: Added sugars are associated with increased body weight and type 2 diabetes.]

D. Simple sugars and disease

Carbohydrates yield 4 kcal/g (the same as protein and less than half that of fat; see Fig. 27.5) and result in fat synthesis only when consumed in excess of the body's energy needs. There is no direct evidence that the consumption of simple sugars naturally present in food is harmful. Contrary to popular belief, carbohydrates are not inherently fattening; however, the type and quality of carbohydrates consumed can significantly influence weight management. In particular, processed and refined carbs with high GIs are more likely to contribute to weight gain. However, there is an association between sucrose consumption and dental caries, particularly in the absence of fluoride treatment (see Chapter 29, III. E.).

VII. DIETARY PROTEIN

The AMDR for protein is 10% to 35%. Dietary protein provides essential amino acids ([EAAs] see Fig. 20.2). Nine of the 20 amino acids needed for the synthesis of body proteins are essential (i.e., they cannot be synthesized in humans). In addition, unlike carbohydrates and fats, there is no specialized storage molecule in the body for excess proteins, such as glycogen for stored carbohydrates and TAG for stored fats in adipose tissue.

A. Protein quality

The quality of a dietary protein is a measure of its ability to provide the EAAs required for tissue maintenance. The protein digestibility–corrected amino acid score (PDCAAS) and the digestible indispensable amino acid score (DIAAS) are both methods used to evaluate the quality of protein in food. PDCAAS measures protein digestibility as a whole, using fecal rather than ileal digestibility as is the case for DIAAS. As a result, DIAAS provides a more accurate assessment of the protein's availability for use in the body. However, PDCAAS is still in use, primarily due to its historical precedence and the regulatory frameworks built around it.

1. **Proteins from animal sources:** Proteins from meat, poultry, milk, and fish have high quality because they contain all the EAAs in proportions similar to those required for synthesis of human tissue proteins (Fig. 27.17), and they are more readily digested. They are known as complete proteins. Gelatin prepared from animal collagen is an exception. It has a low biologic value as a result of deficiencies in several EAAs.

2. **Proteins from plant sources:** Plant proteins have lower quality than animal proteins since they have low amounts of more than one of the EAAs; therefore, they are called incomplete proteins. Proteins from different plant sources may be combined in such a way that the result is equivalent in nutritional value to animal protein. For example, wheat (lysine deficient but methionine rich) may be combined with kidney beans (methionine poor but lysine rich) to produce higher biologic value than either of the component proteins (Fig. 27.18). [Note: Animal proteins can also complement the biologic value of plant proteins.]

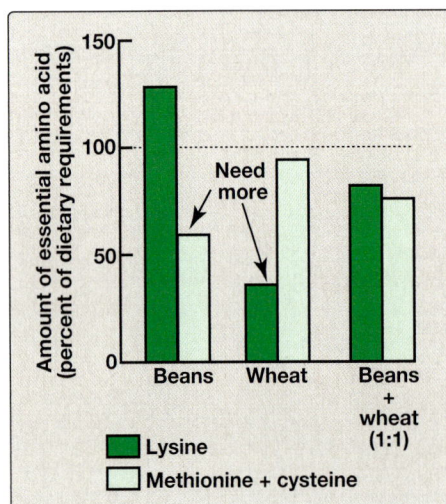

Source	PDCAAS value
Animal proteins	
Egg	1.00
Milk protein	1.00
Beef/poultry/fish	0.82–0.92
Gelatin	0.08
Plant proteins	
Soybean protein	1.00
Kidney beans	0.68
Whole wheat bread	0.40

Figure 27.17
Relative quality of some common dietary proteins. PDCAAS, protein digestibility–corrected amino acid score.

Figure 27.18
Combining two incomplete proteins that have complementary amino acid deficiencies results in a mixture with higher biologic value.

During periods of physiologic stress, such as illness, injury, or intense physical activity, the body's demand for certain amino acids may exceed its ability to produce them. As a result, these nonessential amino acids become "conditionally essential" and must be acquired from dietary sources. Examples include arginine, cysteine, glutamine, tyrosine, and proline. Ensuring AI of these amino acids during times of stress is important for effective recovery and maintaining health.

B. Nitrogen balance

Nitrogen balance occurs when the amount of nitrogen consumed equals that of the nitrogen excreted in the urine (primarily as urinary urea nitrogen [UUN]), sweat, and feces. Most healthy adults are normally in nitrogen balance. [Note: On average, there is 1 g nitrogen in 6.25 g protein.]

1. **Positive nitrogen balance:** This occurs when nitrogen intake exceeds nitrogen excretion. It is observed in situations in which tissue growth occurs, for example, in childhood, pregnancy, muscle

building, or during recovery from an emaciating illness as the body repairs and regenerates tissue.

2. **Negative nitrogen balance:** This occurs when nitrogen loss is greater than nitrogen intake. It is associated with inadequate dietary protein or lack of an EAA. Common causes include malnutrition and cachexia, and it can also arise from physiologic stresses, such as trauma, burns, illness, infection, or surgery.

> Nitrogen (N) balance (g N_{in} – g N_{out}) in a 24-hour period can be determined by the formula, N balance = protein intake in g/6.25 – (UUN + 4 g), where 4 g accounts for urinary loss in forms other than UUN plus loss in skin and feces.

C. Protein requirements

The amount of dietary protein required in the diet varies with its biologic value. The greater the proportion of animal protein in the diet, the less protein is required. The RDA for protein is computed for proteins of mixed biologic value at 0.8 g/kg of body weight for adults, or ~56 g of protein for a 70-kg individual. People who exercise strenuously on a regular basis may benefit from extra protein to maintain muscle mass, and a daily intake of ~1 g/kg has been recommended for athletes. Women who are pregnant or lactating require up to 30 g/day in addition to their basal requirements. To support growth, infants should consume 2 g/kg/day. Disease states influence protein needs. Protein restriction may be needed in kidney disease, whereas burns require increased protein intake.

1. **Consumption of excess protein:** There is no physiologic advantage to the consumption of more protein than the RDA. Protein consumed in excess of the body's needs is deaminated, and the resulting carbon skeletons are metabolized to provide energy or acetyl coenzyme A for FA synthesis. When excess protein is eliminated from the body as urinary nitrogen, it is often accompanied by increased urinary calcium, thereby increasing the risk of nephrolithiasis (kidney stones) and osteoporosis.

2. **Protein-sparing effect of carbohydrates:** The dietary protein requirement is influenced by the carbohydrate content of the diet. When the intake of carbohydrates is low, amino acids are deaminated to provide carbon skeletons for the synthesis of glucose that is needed as a fuel by the central nervous system. If carbohydrate intake is more than 130 g/day, substantial amounts of protein are metabolized to provide precursors for gluconeogenesis. Therefore, carbohydrate is considered to be "protein-sparing," because it allows amino acids to be used for repair and maintenance of tissue protein rather than for gluconeogenesis.

D. Protein-energy (calorie) malnutrition

In developed countries, protein-energy malnutrition (PEM), also known as protein-energy undernutrition (PEU), is most commonly seen in patients with medical conditions that decrease appetite or alter how nutrients are digested or absorbed or in hospitalized patients with major trauma or infections. Such highly catabolic patients frequently

Table 27.1: Physical Features of Extreme Protein-Energy Malnutrition (PEM) in Children

Type of PEM	Weight for Age (% Expected)	Weight for Height	Edema	Muscle and Fat Content
Kwashiorkor	60–80	Normal or ↓	Present	↓
Marasmus	<60	Markedly ↓	Absent	Markedly ↓

Note: The fatty liver and skin and hair changes of kwashiorkor are not seen in marasmus.

require intravenous (or parenteral) or tube-based (enteral) administration of nutrients. PEM may also be seen in children and older adults who are malnourished. In low-income countries, inadequate intake of protein and/or calories is the primary cause of PEM. Affected individuals show a variety of symptoms, including a depressed immune system with a reduced ability to resist infection. Death from secondary infection is common. PEM is a spectrum of degrees of malnutrition, and two extreme forms are kwashiorkor and marasmus (Table 27.1). Marasmic kwashiorkor has features of both forms.

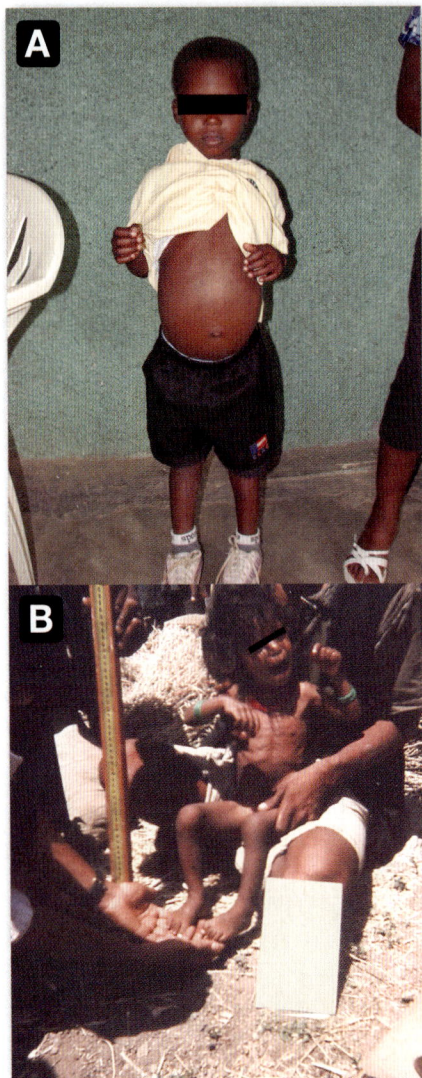

1. **Kwashiorkor:** Kwashiorkor occurs when protein deprivation is relatively greater than the reduction in total calories. Protein deprivation is associated with severely decreased synthesis of visceral protein. Kwashiorkor is commonly seen in low-income countries in children after weaning at about age 1 year, when their diet consists predominantly of carbohydrates. Typical symptoms include stunted growth, skin lesions, depigmented hair, anorexia, fatty liver, bilateral pitting edema, and decreased serum albumin concentration. Edema results from the lack of adequate blood proteins, primarily albumin, to maintain the distribution of water between blood and tissues. It may mask muscle and fat loss. Therefore, chronic malnutrition is reflected in the level of serum albumin. Because caloric intake from carbohydrates may be adequate, insulin levels suppress lipolysis and proteolysis. Kwashiorkor is, therefore, nonadapted malnutrition.

2. **Marasmus:** Marasmus occurs when calorie deprivation is relatively greater than the reduction in protein. It usually occurs in low-income countries in children younger than age 1 year when breast milk is supplemented or replaced with watery gruels of native cereals that are usually deficient in both protein and calories. Typical symptoms include arrested growth, extreme muscle wasting and depletion of subcutaneous fat (emaciation), weakness, and anemia (Fig. 27.19). Individuals with marasmus do not show the edema observed in kwashiorkor. Refeeding severely malnourished individuals can result in hypophosphatemia (see Chapter 29, II. A. 2.), because any available phosphate is used to phosphorylate carbohydrate intermediates. Milk is frequently given because it is rich in phosphate.

Cachexia, a wasting disorder characterized by loss of appetite and muscle atrophy (with or without increased lipolysis) that cannot be reversed by conventional nutritional support, is seen with a number of chronic diseases, such as cancer and chronic pulmonary and renal disease. It is associated with decreased treatment tolerance and response and decreased survival time. Cachexia is caused by various factors, including cytokines such as tumor necrosis factor (TNF)-α, interleukin (IL)-1β, IL-6, and interferon (IFN)-γ.

Figure 27.19
A. Child with kwashiorkor. Note the swollen belly and lower legs. **B.** Child with marasmus.

VIII. NUTRITION TOOLS

A set of tools has been developed that gives consumers information about what (and how much) they should eat as well as the nutritional content of the foods they do eat. Additional tools allow medical professionals to assess whether an individual's nutritional needs are being met.

A. MyPlate

MyPlate was designed by the U.S. Department of Agriculture (USDA) to graphically illustrate its recommendations as to what food groups and how much of each should be consumed daily. In MyPlate, the relative amounts of each of five food groups (vegetables, grains, protein, fruit, and dairy) are represented by the relative size of their section on the plate (Fig. 27.20). The number of servings depends on variables that include age and sex. The Healthy Eating Plate created by experts at Harvard School of Public Health and Harvard Medical School is also used. It differs from MyPlate, which recommends limiting milk and substituting with water. In addition, it recommends healthy oils and physical activity.

Figure 27.20
MyPlate.

B. Nutrition facts label

Most types of packaged goods are required to have a Nutrition Facts label, or "food label" that includes the size of a single serving, the Cal it provides, and the number of servings per container (Fig. 27.21). In addition, a percent daily value (%DV) is shown for most nutrients listed. The %DV is based on a 2,000-Cal diet for healthy adults.

1. **Percent daily value:** The %DV compares the amount of a given nutrient in a single serving of a product to the recommended daily intake for that nutrient. For example, the %DV for the micronutrients listed, as well as for total carbohydrates and fiber, are based on their recommended minimum daily intake. Thus, if the label lists 20% for calcium, one serving provides 20% of the minimum recommended amount of calcium needed each day. In contrast, the %DV for saturated fat, cholesterol, and sodium are based on their recommended maximum daily intake, and the %DV reflects what percentage of this maximum a serving provides. There is no %DV for protein because the recommended intake depends on body weight. "Sugars" represent mono- and disaccharides. The remainder of the carbohydrate (total carbohydrate − [fiber + sugars]) are the oligo- and polysaccharides.

2. **Nutrition Facts labels:** In 2014, the USDA proposed the following changes to the Nutrition Facts label: Added sugars, vitamin D, and potassium are to be included; vitamins A and C, total fat, and calories from fat are to be removed since the type of fat is more important than the amount; and serving size is to be adjusted to reflect the amounts people are now consuming. Additionally, design was changed to highlight key parts of the label (Fig. 27.22).

C. Nutrition assessment

Nutrition assessment evaluates nutritional status based on clinical information. It includes (but is not limited to) dietary history,

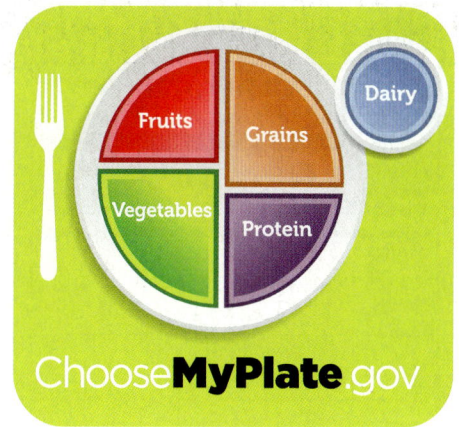

Figure 27.21
Nutrition Facts label (food label).

Nutrition Facts

8 servings per container

Serving size   2/3 cup (55 g)

Amount per 2/3 cup
Calories   **230**

% DV*		
12%	**Total Fat** 8 g	
5%	Saturated Fat 1 g	
	Trans Fat 0 g	
0%	**Cholesterol** 0 mg	
7%	**Sodium** 160 mg	
12%	**Total Carbs** 37 g	
14%	Dietary Fiber 4 g	
	Sugars 1 g	
	Added Sugars 0 g	
	Protein 3 g	
10%	**Vitamin D** 2 mcg	
20%	**Calcium** 260 mg	
45%	**Iron** 8 mg	
5%	**Potassium** 235 mg	

* Footnote on Daily Values (DV) and calories reference to be inserted here.

Figure 27.22
Nutrition Facts label showing changes proposed in 2014 for implementation by 2018.

anthropometric measures, and laboratory data. Assessment findings may result in medical nutrition therapy (MNT), which is an evidence-based medical approach for the treatment of certain medical conditions using a personalized nutrition plan. For example, MNT for hyperlipidemia involves reducing the quantity and types of fats and often calories in the diet.

1. **Dietary history:** This is a record of food intake over a period of time. For a food diary, the specific types and exact amounts of food eaten are recorded in "real time" (as soon as possible after eating) for a period of 3 to 7 days. Retrospective approaches include a food frequency questionnaire (e.g., what fruits were eaten and how often they were eaten in a typical day, week, or month) and a 24-hour recall of the specific foods and the amounts eaten in the last 24 hours.

2. **Anthropometric measures:** These physical measures of the body include (but are not limited to) weight, height, body mass index (indicator of obesity; see Chapter 26, II. A.), skin-fold thickness (indicator of subcutaneous fat), and waist circumference (indicator of abdominal fat; see Chapter 26, II.). [Note: Ideal body weight can be calculated using the Hamwi method: 106 lb (for males) or 100 lb (for females) for the first 5 ft of height + 5 lb for every inch over 5 ft, with an adjustment of −10% for a small frame and +10% for a large one.]

3. **Laboratory data:** These are obtained by tests performed on body fluids, tissues, and waste. They can include plasma LDL-C (for cardiovascular risk), fecal fat (for malabsorption), red blood cell indices (for vitamin deficiencies), and N balance and serum proteins (such as albumin and transthyretin [prealbumin]) for protein–energy status. [Note: These proteins are made in the liver and transport molecules such as FAs and thyroxine (see Chapter 29, IV. A.) through blood. Low albumin levels correlate with increased morbidity and mortality in hospitalized patients. The short half-life (2 to 3 days) of transthyretin as compared to that of albumin (20 days) has led to its use in monitoring the progress of hospitalized patients.]

> Nutritional insufficiency can be the result of inadequate nutrient intake (e.g., from an inability to eat, loss of appetite, or decreased availability), inadequate absorption, decreased utilization, increased excretion, or increased requirements.

IX. NUTRITION AND THE LIFE STAGES

Macronutrient energy sources, micronutrients, EFAs, and EAAs are required at every life stage. Additionally, each stage has specific nutrition needs.

A. Infancy, childhood, and adolescence

The rapid growth and development in infancy (birth to age 1 year) and childhood (age 1 year to adolescence) necessitate higher energy and protein needs relative to body size than are required in subsequent life stages. In adolescence, the marked increases in height and weight that occur increase nutritional needs. Growth charts are used to compare an individual's stature (height) and/or weight to the expected values for others of the same age (≤20 years) and sex (Fig. 27.23). They are based on data from large numbers of normal individuals over time. [Note: Deviations from the expected growth curve, as reflected in the crossing of two or more percentile lines, raise concern.]

1. **Infants:** Ideal infant nutrition is based on human breast milk because it provides calories and most micronutrients in amounts appropriate for the human infant. Carbohydrates, protein, and fat are present in a 7:3:1 ratio. In addition to the disaccharide lactose, human milk contains nearly 200 unique oligosaccharides. About 90% of the microbiota in the breastfed infant's intestine is represented by one type, *Bifidobacterium infantis*, which expresses all the enzymes needed to degrade these complex sugars. The sugars, in turn, act as prebiotics that support the growth of *B. infantis*, a probiotic. However, breast milk is low in vitamin D, and exclusively breastfed infants require vitamin D supplementation. Human milk provides antibodies and other proteins that reduce the risk of infection.

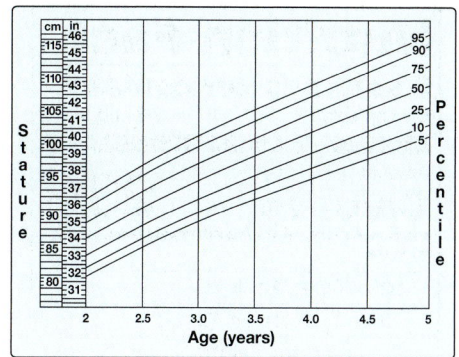

Figure 27.23
Clinical growth chart of stature-for-age for boys ages 2–5 years from the Centers for Disease Control and Prevention (CDC) (see https://www.cdc.gov/growthcharts/). Charts for girls are pink.

> The microbiota in and on the human body plus their genomes are referred to as the microbiome, which is acquired at birth from the environment and changes with the life stages. The gut microbiome influences host nutrition by facilitating processing of food consumed and is itself influenced by that food. Its relationship with undernutrition, obesity, and diabetes is under investigation.

2. **Children:** As with infants, children have increased need for calories and nutrients. The primary concerns in this stage, however, are deficiencies of iron and calcium.

3. **Adolescents:** In the teen years, increases in height and weight increase the need for calories, protein, calcium, iron, and phosphorus. Eating patterns in this stage can result in overconsumption of fat, sodium, and sugar and underconsumption of vitamin A, thiamine, and folic acid. Eating disorders and obesity are concerns in this age group.

B. Adulthood

Overnutrition is a concern in young adults, whereas malnutrition is a concern in older adults.

1. **Young adults:** Nutrition in young adults focuses on the maintenance of good health and the prevention of disease. The goal is a diet rich in plant-based foods (with a focus on fiber and whole grains), limited intake of saturated fat and *trans* FAs, and balanced intake of ω-3 and ω-6 PUFAs.

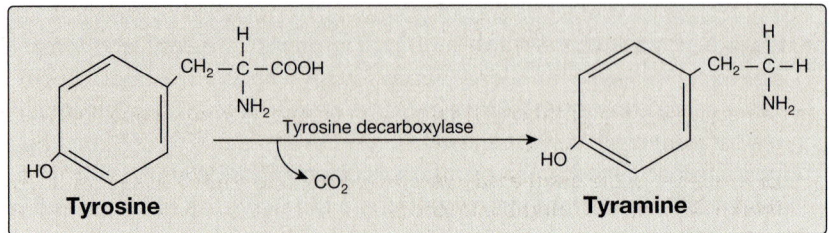

Figure 27.24
Decarboxylation of tyrosine to tyramine. CO_2, carbon dioxide.

2. **Pregnant or lactating women:** The requirements for calories, protein, and virtually all micronutrients increase in pregnancy and lactation. Supplementation with folic acid (to prevent neural tube defects; see Chapter 28, II. 2.), vitamin D, calcium, iron, iodine, and DHA is typically recommended.

3. **Older adults:** Aging increases the risk of malnutrition. Decreased appetite resulting from a reduced sense of taste (dysgeusia) and smell (hyposmia) decreases nutrient intake. Physical limitations, including problems with dentition, and psychosocial factors, such as isolation, may also play a role in reduced intake. Inadequate intake of protein, calcium, and vitamins D and B_{12} is common. Vitamin B_{12} deficiency can result from decreased absorption caused by achlorhydria (reduced stomach acid; see Chapter 28, III.). In aging, lean muscle mass decreases and fat increases, resulting in decreased RMR. Drug–nutrient interactions can occur at any life stage but are more common as the number of medications increases as in aging.

Clinical Application 27.1: Monoamine Oxidase Inhibitors

Monoamine oxidase inhibitors (MAOIs), used to treat depression (see Chapter 21, III. A. 4.) and early Parkinson disease, can interact with tyramine-containing foods. Tyramine is a monoamine derived from the decarboxylation of tyrosine during the curing, aging, or fermentation of food (Fig. 27.24). It causes the release of norepinephrine, increasing blood pressure, and heart rate. Patients who take MAOIs and consume such foods are at risk for a hypertensive crisis.

X. CHAPTER SUMMARY

- **Dietary reference intakes** (**DRIs**) provide estimates of the amounts of nutrients required to prevent deficiencies and maintain optimal health and growth.
- DRIs consist of the **estimated average requirement** (**EAR**), the **recommended dietary allowance** (**RDA**), the **adequate intake** (**AI**), and the **tolerable upper intake level** (**UL**).
- **EAR** is the average daily nutrient intake level estimated to meet the requirement of 50% of the healthy individuals in a particular life stage (age) and gender group.
- **RDA** is the average daily dietary intake level that is sufficient to meet the nutrient requirements of nearly all (97% to 98%) individuals in a life stage and gender group.
- **AI** is set instead of an RDA if sufficient scientific evidence is not available to calculate the RDA.
- **UL** is the highest average daily nutrient intake level likely to pose no risk of adverse health effects to almost all individuals in the general population.
- The energy generated by the metabolism of the **macronutrients** (9 kcal/g of fat and 4 kcal/g of protein or carbohydrate) is used for three energy-requiring processes that occur in the body: **resting metabolic rate**, **physical activity**, and the **thermic effect of food**.
- **Acceptable macronutrient distribution ranges** (**AMDRs**) are defined as the ranges of intake for a particular macronutrient that are associated with reduced risk of chronic disease while providing adequate amounts of essential nutrients.
- Adults should consume **45% to 65%** of their **total calories** from **carbohydrates**, **20% to 35%** from **fat**, and **10% to 35%** from **protein** (Fig. 27.25).
- Elevated levels of cholesterol in low-density lipoproteins (LDL-C) result in increased risk for **coronary heart disease** (**CHD**).
- Elevated levels of cholesterol in high-density lipoproteins (HDL-C) have been associated with a decreased risk for CHD.
- Dietary or drug treatment of **hypercholesterolemia** is effective in decreasing LDL-C, increasing HDL-C, and reducing the risk for CHD.
- Consumption of **saturated fats** is strongly associated with high levels of total plasma and LDL-C. When substituted for saturated fatty acids (FAs) in the diet, **monounsaturated fats** lower both total plasma cholesterol and LDL-C but maintain or increase HDL-C.
- Consumption of food containing ω-6 **polyunsaturated FAs** lowers plasma LDL-C, but HDL-C, which protects against CHD, is also lowered.
- Dietary ω-3 **polyunsaturated fats** suppress cardiac arrhythmias and reduce plasma triacylglycerols, decrease the tendency for thrombosis, and substantially reduce the risk of cardiovascular mortality.
- **Carbohydrates** provide **energy** and **fiber** to the diet. When they are consumed as part of a diet in which caloric intake is equal to energy expenditure, they do not promote obesity.
- Dietary **protein** provides **essential amino acids** (**EAAs**).
- **Protein quality** is a measure of its ability to provide the EAAs required for tissue maintenance. Proteins from animal sources, in general, have higher-quality protein than that derived from plants. However, proteins from different plant sources may be combined in such a way that the result is equivalent in nutritional value to animal protein.
- **Positive nitrogen** (**N**) **balance** occurs when N intake exceeds N excretion. It is observed in situations in which tissue growth occurs (e.g., childhood, pregnancy, or recovery from an emaciating illness).
- **Negative N balance** occurs when N losses are greater than N intake. It is associated with inadequate dietary protein; lack of an EAA; or during physiologic stresses such as trauma, burns, illness, or surgery.
- **Kwashiorkor** occurs when protein deprivation is relatively greater than the reduction in total calories. It is characterized by edema.
- **Marasmus** occurs when calorie deprivation is relatively greater than the reduction in protein. No edema is seen. Both are extreme forms of **protein-energy malnutrition** (**PEM**). **Nutrition Facts labels** give consumers information about the nutritional content of packaged foods.
- Medical assessment of nutritional status includes **dietary history**, **anthropometric measures**, and **laboratory data**. Each life stage has specific nutrition needs.
- **Growth charts** are used to monitor the growth pattern of an individual from birth through adolescence.
- **Drug–nutrient interactions** are of concern, especially in older adults.

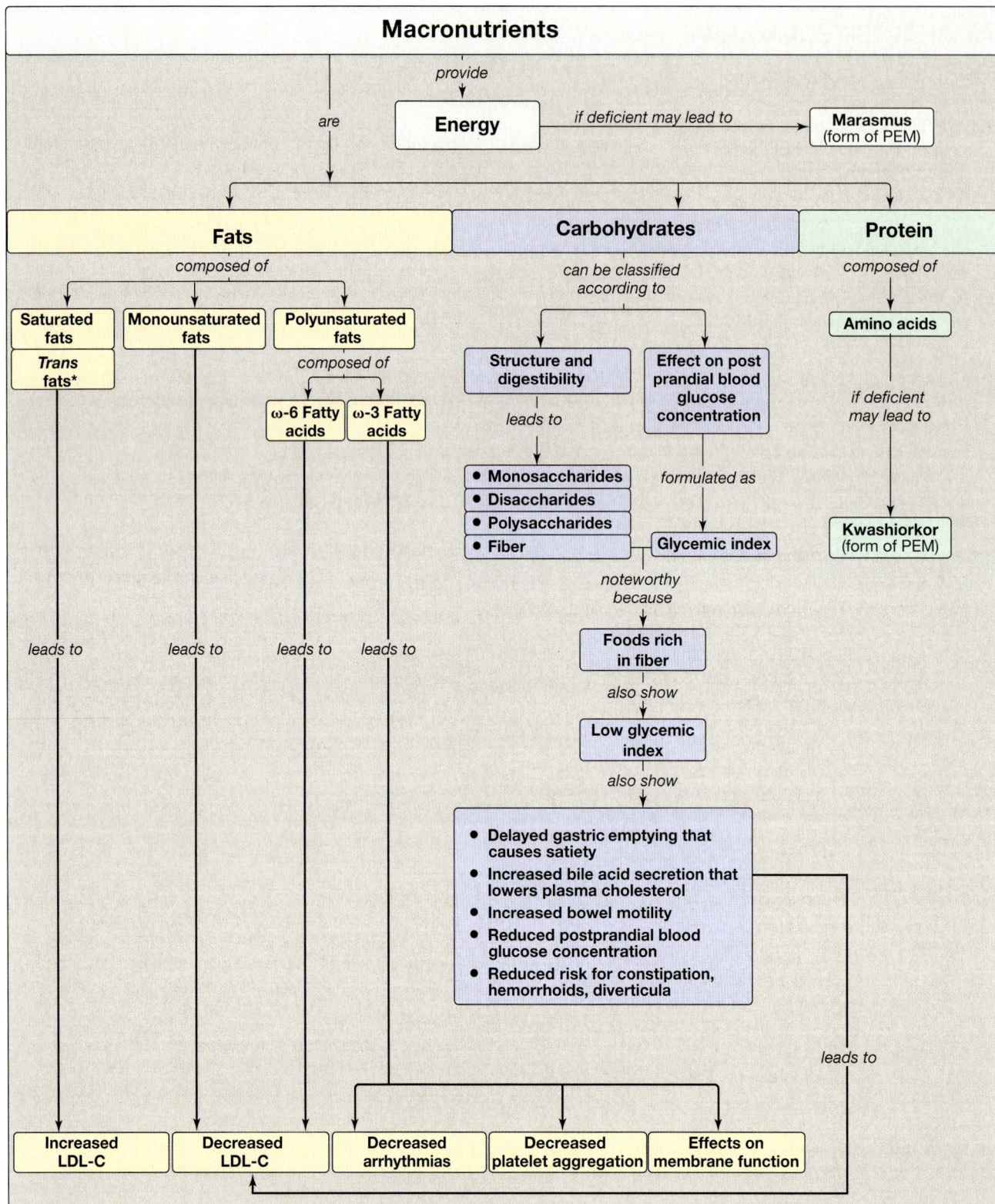

Figure 27.25
Key concept map for the macronutrients. [Note: *Trans* fatty acids are chemically classified as unsaturated.] PEM, protein-energy malnutrition; LDL, low-density lipoprotein; C, cholesterol.

Study Questions

Choose the ONE best answer.

27.1. Which of the statements would support a diagnosis of kwashiorkor? The children:

A. appear well nourished due to increased deposition of fat in adipose tissue.

B. display abdominal and peripheral edema.

C. have serum albumin levels above normal.

D. have markedly decreased weight for height.

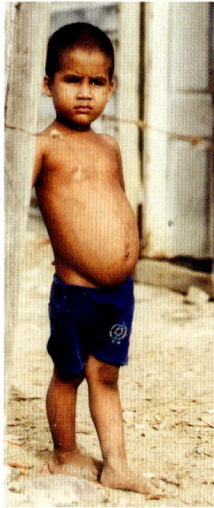

The correct answer = B. Kwashiorkor is caused by inadequate protein intake in the presence of fair-to-good energy (calorie) intake. Typical findings in a patient with kwashiorkor include abdominal and peripheral edema (note the swollen belly and legs) caused largely by a decreased serum albumin concentration. Body fat stores are depleted, but weight for height can be normal because of edema. Treatment includes a diet adequate in calories and protein.

27.2. Which of the following statements concerning dietary fat is correct?

A. Coconut oil is rich in monounsaturated fats, and olive oil is rich in saturated fats.

B. Fatty acids containing *trans* double bonds, unlike the naturally occurring *cis* isomers, raise high-density lipoprotein cholesterol levels.

C. The polyunsaturated fatty acids linoleic and linolenic acids are required components.

D. Triacylglycerols obtained from plants generally contain less unsaturated fatty acids than do those from animals.

Correct answer = C. Humans are unable to make linoleic and linolenic fatty acids. Consequently, these fatty acids are essential in the diet. Coconut oil is rich in saturated fats, and olive oil is rich in monounsaturated fats. *Trans* fatty acids raise plasma levels of low-density lipoprotein cholesterol, not high-density lipoprotein cholesterol. Triacylglycerols obtained from plants generally contain more unsaturated fatty acids than do those from animals.

27.3. If a 70-kg man is consuming a daily average of 275 g of carbohydrate, 75 g of protein, and 65 g of fat, which of the following conclusions can reasonably be drawn?

A. About 20% of his calories are derived from fats.

B. His diet contains a sufficient amount of fiber.

C. He is in nitrogen balance.

D. His macronutrients distribution conforms to current recommendations.

E. His total energy intake per day is ~3,000 kcal.

Correct answer = D. The total energy intake is (275 g carbohydrate × 4 kcal/g) + (75 g protein × 4 kcal/g) + (65 g fat × 9 kcal/g) = 1,100 + 300 + 585 = 1,985 total kcal/day. The percentage of calories derived from carbohydrate is 1,100/1,985 = 55, from protein is 300/1,985 = 15, and from fat is 585/1,985 = 30. These are very close to current recommendations. The amount of fiber or nitrogen balance cannot be deduced from the data presented. If the protein is of low biologic value, a negative nitrogen balance is possible.

27.4. In chronic bronchitis, excessive mucus production causes airway obstruction that results in hypoxemia (low blood oxygen level), impaired expiration, and hypercapnia (carbon dioxide retention). Why might a high-fat, low-carbohydrate diet be recommended for a patient with chronic obstructive pulmonary disease (COPD) caused by chronic bronchitis?

A. Fat contains more oxygen atoms relative to carbon or hydrogen atoms than do carbohydrates.

B. Fat is calorically less dense than carbohydrates.

C. Fat metabolism generates less carbon dioxide than carbohydrates.

D. The respiratory quotient (RQ) for fat is higher than the RQ for carbohydrates.

Correct answer = C. A treatment goal for COPD caused by acute bronchitis is to ensure appropriate nutrition without increasing the RQ, which is the ratio of carbon dioxide (CO_2) produced to oxygen consumed, thereby minimizing the production of CO_2. Less CO_2 is produced from the metabolism of fat (RQ = 0.7) than from the catabolism of carbohydrate (RQ = 1.0). Fat contains fewer oxygen atoms. Fat is calorically denser than is carbohydrate. [Note: RQ is determined by indirect calorimetry.]

27.5. A 32-year-old male who was rescued from a house fire was admitted to the hospital with burns over 45% of his body (severe burns). The patient weighs 154 lb (70 kg) and is 72 in (183 cm) tall. Which of the following is the best rapid estimate of the immediate daily caloric needs of this patient?

A. 1,345 kcal

B. 1,680 kcal

C. 2,690 kcal

D. 3,360 kcal

Correct answer = D. A commonly used rough estimate of the total energy expenditure (TEE) for men is 1 kcal/1 kg body weight/24 h. [Note: It is 0.8 kcal for females.] For this patient, that value is 1,680 kcal (1 kcal/kg/h × 24 h × 70 kg). In addition, an injury factor of 2 for severe burns must be included in the calculation: 1,680 kcal × 2 = 3,360 kcal.

27.6. Which of the following is the best advice to give a patient who asks about the notation "%DV" (percent daily value) on the Nutrition Facts label?

A. Select foods to achieve 100% daily value for each nutrient each day.

B. Select foods that have the highest percent daily value for all nutrients.

C. Select foods with a low percent daily value for the micronutrients.

D. Select foods with a low percent daily value for saturated fat.

Correct answer = D. The percent daily value (%DV) compares the amount of a given nutrient in a single serving of a product to the recommended daily intake for that nutrient. The %DV for the micronutrients listed on the label, as well as for total carbohydrates and fiber, is based on their recommended minimum daily intake, whereas the %DV for saturated fat, cholesterol, and sodium is based on their recommended maximum daily intake.

For Questions 27.7 and 27.8, use the following case.

A sedentary 50-year-old male weighing 176 lb (80 kg) requests a physical examination. He denies any health problems. Routine blood analysis is unremarkable except for plasma total cholesterol of 295 mg/dL (reference = <200 mg). The patient declines drug therapy for hypercholesterolemia. Analysis of a 1-day dietary recall shows:

Kilocalories	3,475 kcal	Cholesterol	822 mg
Protein	102 g	Saturated fat	69 g
Carbohydrate	383 g	Total fat	165 g
Fiber	6 g		

27.7. Decreasing which of the following dietary components would have the greatest effect in lowering the patient's plasma cholesterol?

A. Carbohydrates
B. Cholesterol
C. Fiber
D. Monounsaturated fat
E. Polyunsaturated fat
F. Saturated fat

> Correct answer = F. The intake of saturated fat most strongly influences plasma cholesterol in this diet. The patient is consuming a high-calorie, high-fat diet with 42% of the fat as saturated fat. The most important dietary recommendations are to lower total caloric intake, substitute monounsaturated and polyunsaturated fats for saturated fats, and increase dietary fiber. A decrease in dietary cholesterol would be helpful but is not a primary objective.

27.8. A 48-year-old male is recovering from an emaciating illness that resulted in significant muscle wasting. He has been cleared by his health care provider to begin a diet to help build lost muscle mass. Which of the following dietary modifications is the most appropriate recommendation?

A. Decrease protein intake to minimize renal load.
B. Increase intake of high-quality proteins to facilitate muscle repair and growth.
C. Increase fat intake to add caloric density to meals.
D. Decrease caloric intake to maintain current weight.

> After an emaciating illness, a patient needs to rebuild lost muscle mass. A diet rich in high-quality proteins helps ensure a supply of essential amino acids necessary for muscle protein synthesis, thereby promoting a positive nitrogen balance, which is required for anabolism and recovery.

28 Micronutrients: Vitamins

I. OVERVIEW

Vitamins are organic molecules that cannot be synthesized in adequate quantities by humans and, therefore, must be supplied by the diet. Nine vitamins (folic acid, cobalamin, ascorbic acid, pyridoxine, thiamine, niacin, riboflavin, biotin, and pantothenic acid) are classified as water soluble. Because they are readily excreted in the urine, toxicity is rare. However, deficiencies can develop quickly. Four vitamins (A, D, K, and E) are termed fat soluble (Fig. 28.1). They are released, absorbed, and transported (in chylomicrons; see Chapter 18, VI. B.) with dietary fat. They are stored in the liver and adipose tissue and are eliminated more slowly than water-soluble vitamins. In fact, consumption of vitamins A and D in excess of the dietary reference intakes (DRI) (see Chapter 27) can lead to the accumulation of toxic quantities of these compounds. Vitamins are required to perform specific cellular functions. For example, many water-soluble vitamins

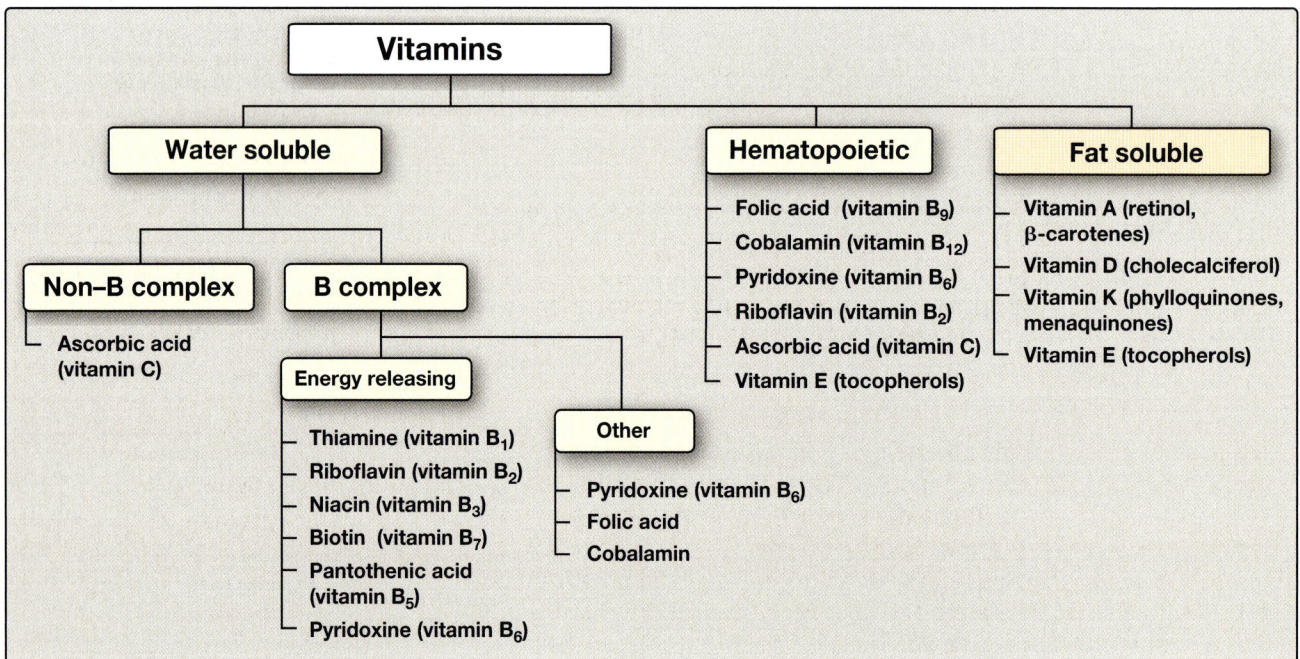

Figure 28.1
Classification of the vitamins. Because they are required in lesser amounts than the macronutrients (carbohydrate, protein, and lipid), vitamins are termed micronutrients.

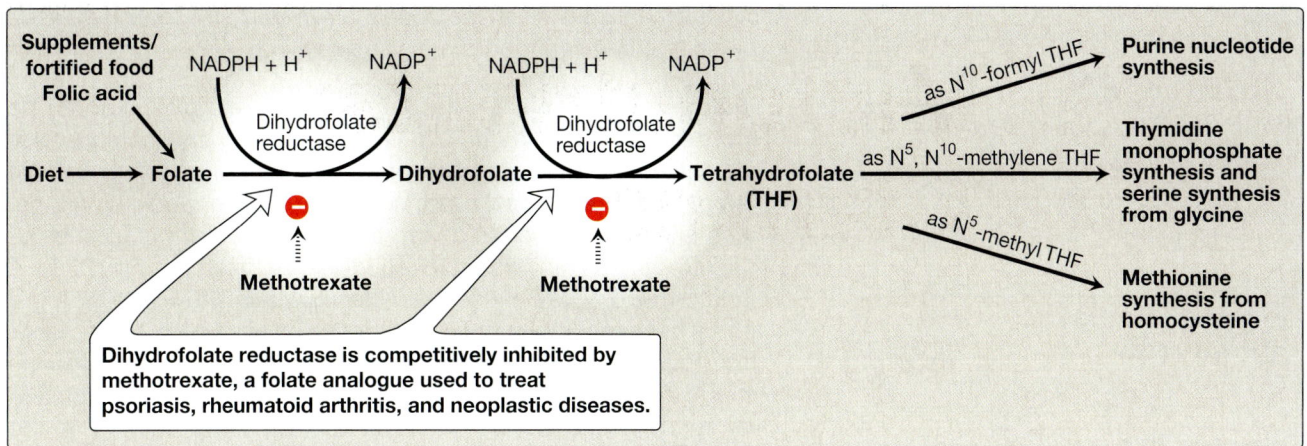

Figure 28.2
Production and use of tetrahydrofolate. NADP(H), nicotinamide adenine dinucleotide phosphate.

are precursors of coenzymes for the enzymes of intermediary metabolism. In contrast to the water-soluble vitamins, only one fat-soluble vitamin (vitamin K) has a coenzyme function.

II. FOLIC ACID (VITAMIN B₉)

Vitamin B_9 describes many forms of naturally occurring folate. Folic acid is the synthetic form of folate used in supplements and in fortification of foods. However, these two terms, folic acid and folate, are often used interchangeably. Folic acid plays a key role in one-carbon metabolism and is essential for the biosynthesis of several compounds. Folic acid deficiency is probably the most common vitamin deficiency in the United States, particularly among pregnant women and individuals with alcoholism. [Note: Leafy, dark-green vegetables are a good source of folic acid.]

A. Function

Tetrahydrofolate (THF), the reduced, coenzyme form of folate, receives one-carbon fragments from donors such as serine, glycine, and histidine and transfers them to intermediates in the synthesis of amino acids, purine nucleotides, and thymidine monophosphate (TMP), a pyrimidine nucleotide incorporated into DNA (Fig. 28.2).

B. Nutritional anemias

In anemia, the blood has a lower-than-normal concentration of hemoglobin, which results in a reduced ability to transport oxygen (O_2). Nutritional anemias (i.e., those caused by inadequate intake of one or more essential nutrients) can be classified according to red blood cell (RBC) size, or mean corpuscular volume ([MCV] Fig. 28.3). Microcytic anemia (MCV < normal), caused by lack of iron, is the most common form of nutritional anemia. The second major category of nutritional anemia, macrocytic (MCV > normal), results from a deficiency in folic acid or vitamin B_{12}. [Note: These macrocytic anemias are commonly called megaloblastic because a deficiency of either vitamin (or both) causes accumulation of large, immature RBC precursors,

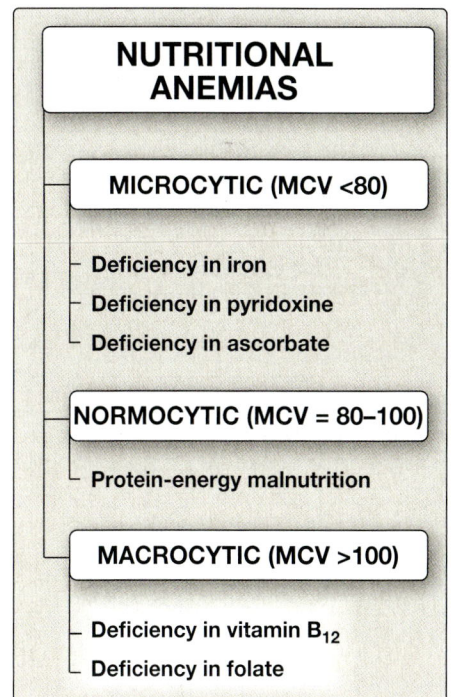

Figure 28.3
Classification of nutritional anemias by red cell size. The normal mean corpuscular volume (MCV) for people age >18 years is 80–100 µm³. [Note: Microcytic anemia is also seen with heavy metal (e.g., lead) poisoning.]

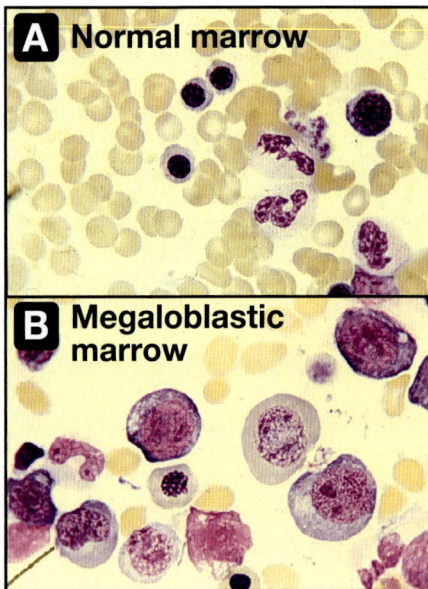

Figure 28.4
Bone marrow histology in healthy (**A**) and folate-deficient (**B**) individuals.

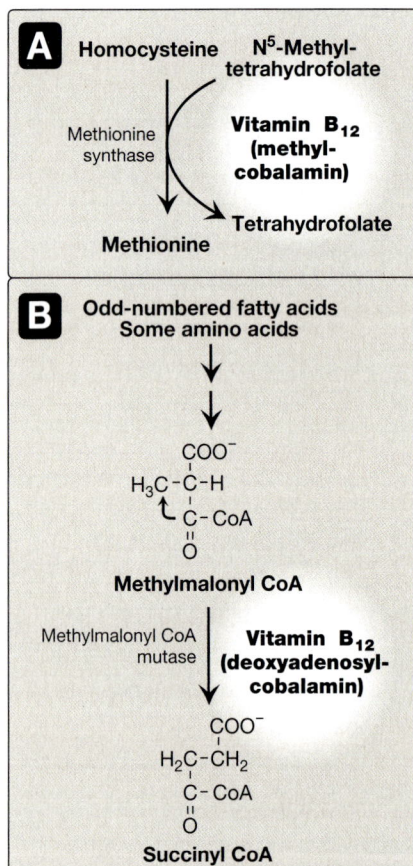

Figure 28.5
A, B. Reactions requiring coenzyme forms of vitamin B_{12}. CoA, coenzyme A.

known as megaloblasts, in the bone marrow and the blood (Fig. 28.4). Hypersegmented neutrophils are also seen.]

1. **Folate and anemia:** Inadequate serum levels of folate can be caused by increased demand (e.g., pregnancy and lactation; see Chapter 27, IX.), poor absorption caused by pathology of the small intestine, alcohol use disorder, or certain medications that can interfere with folate absorption or metabolism (e.g., methotrexate inhibits dihydrofolate reductase inhibitor; see Fig. 28.2). A folate-free diet can cause a deficiency within a few weeks. A primary consequence of folic acid deficiency is megaloblastic anemia (see Fig. 28.4), characterized by reduced synthesis of purine nucleotides and TMP. This leads to impaired DNA replication and arrest cell cycle leading to a reduced rate of cell division. At the same time, the production of hemoglobin in the cytoplasm proceeds without interruption, resulting in enlarged RBCs, which are termed macrocytic cells.

2. **Folate and neural tube defects (NTDs):** Spina bifida and anencephaly, the most common NTDs, affect ~3,000 pregnancies in the United States annually. Folic acid supplementation before conception and during the first trimester has been shown to significantly reduce NTDs. Therefore, all women of childbearing age are advised to consume 0.4 mg/day (400 µg/day) of folic acid to reduce the risk of having a pregnancy affected by an NTD and 10 times that amount if a previous pregnancy was affected. Adequate folate nutrition must occur at the time of conception because critical folate-dependent development occurs in the first weeks of fetal life, at a time when many women are not yet aware of their pregnancy. In 1998, the U.S. Food and Drug Administration authorized the fortification of cereal grain products with folic acid and also recommended folate supplementation in the form of pills resulting in a dietary supplementation of ~0.1 mg/day. This supplementation allows ~50% of all reproductive-aged women to receive 0.4 mg of folate from all sources.

III. COBALAMIN (VITAMIN B_{12})

Vitamin B_{12} is required in humans for two essential enzymatic reactions: remethylation of homocysteine (Hcy) to methionine and isomerization of methylmalonyl coenzyme A (CoA), which is produced during the degradation of some amino acids (isoleucine, valine, threonine, and methionine) and fatty acids (FAs) with odd numbers of carbon atoms (Fig. 28.5). When cobalamin is deficient, unusual (branched) FAs accumulate and become incorporated into cell membranes, including those of the central nervous system (CNS). This may account for some of the neurologic manifestations of vitamin B_{12} deficiency. [Note: Folic acid (as N^5-methyl THF) is also required in the remethylation of Hcy. Therefore, deficiency of B_{12} or folate results in elevated Hcy levels.]

A. Structure and coenzyme forms

Cobalamin contains a corrin ring system that resembles the porphyrin ring of heme (see Chapter 21), but differs in that two of the pyrrole

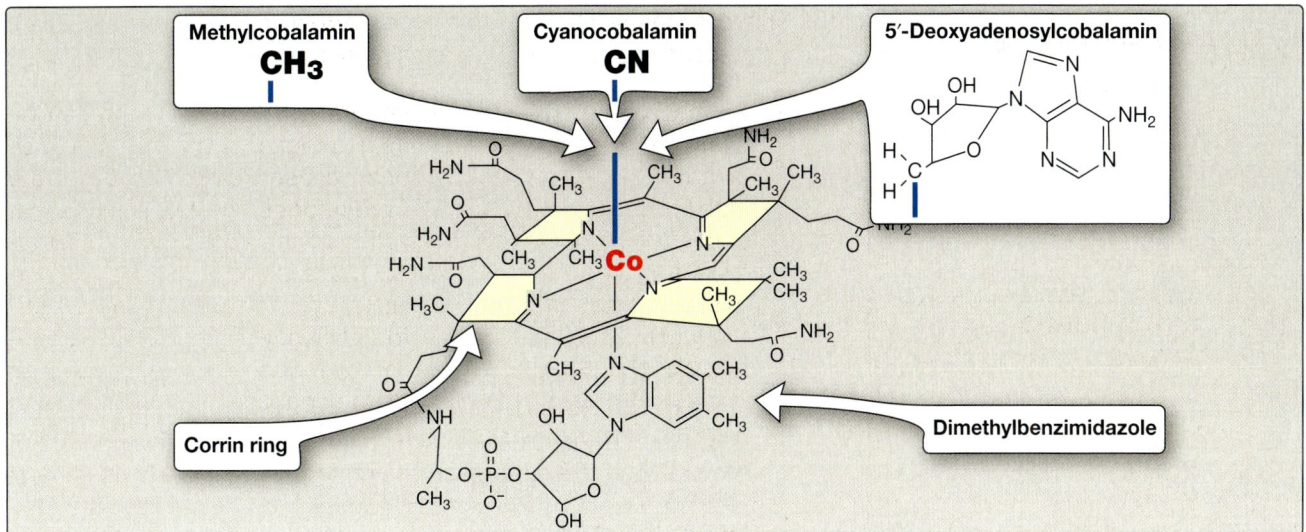

Figure 28.6
Structure of vitamin B$_{12}$ (cyanocobalamin) and its coenzyme forms (methylcobalamin and 5′-deoxyadenosylcobalamin).

rings are linked directly rather than through a methene bridge. Cobalt (see Chapter 29, IV.) is held in the center of the corrin ring by four coordination bonds with the nitrogens of the pyrrole groups. The remaining coordination bonds of the cobalt are with the nitrogen of 5,6-dimethylbenzimidazole and with cyanide in commercial preparations of the vitamin in the form of cyanocobalamin (Fig. 28.6). The physiologic coenzyme forms of cobalamin are 5′-deoxyadenosylcobalamin and methylcobalamin, in which cyanide is replaced with 5′-deoxyadenosine or a methyl group, respectively (see Fig. 28.6).

B. Distribution

Vitamin B$_{12}$ is synthesized only by microorganisms, and it is not present in plants. Animals obtain the vitamin preformed from their intestinal microbiota (see Chapter 27, IX. A.) or by eating foods derived from other animals. Cobalamin is present in appreciable amounts in liver, red meat, fish, eggs, dairy products, and fortified cereals.

C. Folate trap hypothesis

The effects of cobalamin deficiency are most pronounced in rapidly dividing cells, such as the erythropoietic tissue of bone marrow and the mucosal cells of the intestine. Such tissues need both the N^5,N^{10}-methylene and N^{10}-formyl forms of THF for the synthesis of nucleotides required for DNA replication (see Chapter 22.). However, in vitamin B$_{12}$ deficiency, the utilization of the N^5-methyl form of THF in the B$_{12}$-dependent methylation of Hcy to methionine is impaired. Because the methylated form cannot be converted directly to other forms of THF, folate is trapped in the N^5-methyl form, which accumulates. The levels of the other forms decrease. Thus, cobalamin deficiency leads to a deficiency of the THF forms needed in purine and TMP synthesis, resulting in the symptoms of megaloblastic anemia.

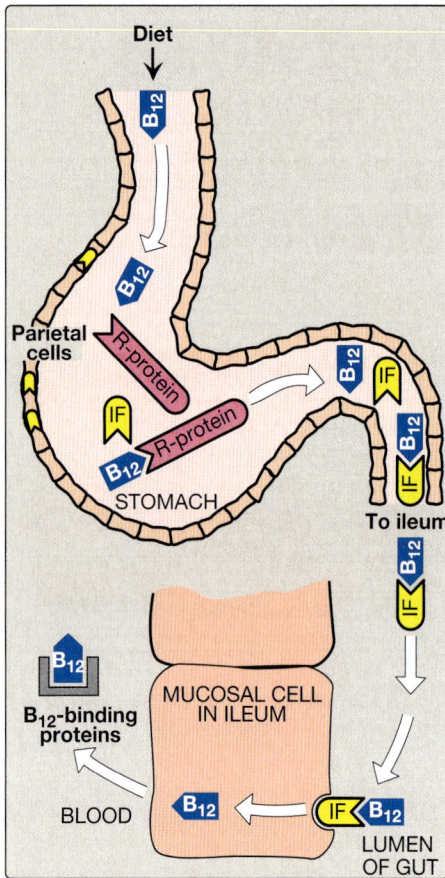

Figure 28.7
Absorption of vitamin B_{12}. [Note: Acid-dependent release of B_{12} from food is not shown.] IF, intrinsic factor.

D. Clinical indications for cobalamin

In contrast to other water-soluble vitamins, significant amounts (2 to 5 mg) of vitamin B_{12} are stored in the body. As a result, it may take several years for the clinical symptoms of B_{12} deficiency to develop as a result of decreased intake of the vitamin. [Note: Deficiency happens much more quickly (in months) if absorption is impaired. The Schilling test evaluates vitamin B_{12} absorption.] Vitamin B_{12} deficiency can be determined by the level of methylmalonic acid in blood, which is elevated in individuals with low intake or decreased absorption of the vitamin.

1. **Pernicious anemia:** Vitamin B_{12} deficiency is most commonly seen in patients who fail to absorb the vitamin from the intestine (Fig. 28.7). B_{12} is released from food in the acidic environment of the stomach. Malabsorption of cobalamin in older adults is most often due to reduced secretion of gastric acid (achlorhydria). Free vitamin B_{12} then binds a glycoprotein (R-protein or haptocorrin), and the complex moves into the intestine. Vitamin B_{12} is released from the R-protein by pancreatic enzymes and binds another glycoprotein, intrinsic factor (IF). The cobalamin–IF complex travels through the intestine and binds to a receptor (cubilin) on the surface of mucosal cells in the ileum. The cobalamin is transported into the mucosal cell and, subsequently, into the general circulation, where it is carried by its binding protein (transcobalamin). Vitamin B_{12} is taken up and stored in the liver, primarily. It is released into bile and efficiently reabsorbed in the ileum. Severe malabsorption of vitamin B_{12} leads to pernicious anemia. This disease is most commonly a result of an autoimmune destruction of the gastric parietal cells responsible for the synthesis of IF (lack of IF prevents vitamin B_{12} absorption). Patients who have had a partial or total gastrectomy become IF deficient and, therefore, vitamin B_{12} deficient. Individuals with cobalamin deficiency are usually anemic (folate recycling is impaired), and they show neuropsychiatric symptoms as the disease develops. The CNS effects are irreversible. Pernicious anemia requires lifelong treatment with either high-dose oral vitamin B_{12} or intramuscular injection of cyanocobalamin. Supplementation works even in the absence of IF because ~1% of vitamin B_{12} uptake is by IF-independent diffusion.

> Folic acid supplementation can partially reverse the hematologic abnormalities of vitamin B_{12} deficiency and, therefore, can mask a cobalamin deficiency. Thus, to prevent the later CNS effects of vitamin B_{12} deficiency, therapy for megaloblastic anemia is initiated with both vitamin B_{12} and folic acid until the cause of the anemia can be determined.

IV. ASCORBIC ACID (VITAMIN C)

The active form of vitamin C is ascorbic acid (Fig. 28.8). Its main function is as a reducing agent. Vitamin C is a coenzyme in hydroxylation reactions (e.g., hydroxylation of prolyl and lysyl residues in collagen, and hydroxylation of dopamine to norepinephrine in epinephrine synthesis), where its role is to keep the iron (Fe) of **hydroxylases** in the reduced, ferrous (Fe^{+2}) form. Thus, vitamin C is required for the maintenance of normal connective tissue as well as for wound healing. Vitamin C also facilitates the absorption of dietary nonheme iron from the intestine by reduction of the ferric form (Fe^{+3}) to the ferrous form (Fe^{+2}) (see Chapter 29, III. B.).

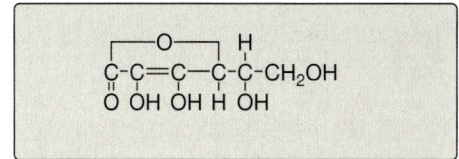

Figure 28.8
Structure of ascorbic acid.

A. Deficiency

Ascorbic acid deficiency results in scurvy, a disease characterized by sore and spongy gums, loose teeth, fragile blood vessels, hemorrhage, swollen joints, bone changes, and fatigue (Fig. 28.9). Many of the deficiency symptoms can be explained by the decreased hydroxylation of collagen, resulting in defective connective tissue. A microcytic anemia caused by decreased absorption of iron may also be seen.

B. Chronic disease prevention

Vitamin C is one of a group of nutrients that includes vitamin E (see Section XIV. and Chapter 13, IV. B. 2.) and β-carotene (see Section XI. A.), which are known as antioxidants. [Note: Vitamin C regenerates the functional, reduced form of vitamin E.] While it is commonly thought that vitamin C or E supplementation may reduce the incidence of some chronic diseases, there is no evidence to support these claims.

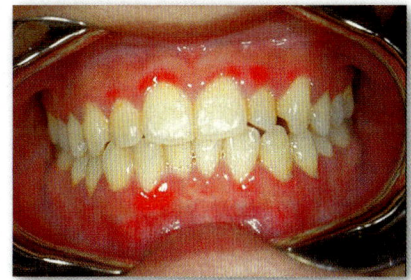

Figure 28.9
Oral manifestations in a patient with scurvy.

V. PYRIDOXINE (VITAMIN B$_6$)

Vitamin B$_6$ is a collective term for pyridoxine, pyridoxal, and pyridoxamine, all derivatives of pyridine. They differ only in the nature of the functional group attached to the ring (Fig. 28.10). Pyridoxine occurs primarily in plants, whereas pyridoxal and pyridoxamine are found in foods obtained from animals. All three compounds can serve as precursors of the biologically active coenzyme, pyridoxal phosphate (PLP). PLP functions as a coenzyme for a large number of enzymes, particularly those that catalyze reactions involving amino acids, for example, in the transsulfuration of Hcy to cysteine, and in the synthesis of dopamine and serotonin. [Note: PLP is also required by **glycogen phosphorylase** (see Chapter 11).]

Reaction type	Example
Transamination	Oxaloacetate + glutamate ⇌ aspartate + α-ketoglutarate
Deamination	Serine → pyruvate + NH$_3$
Decarboxylation	Histidine → histamine + CO$_2$
Condensation	Glycine + succinyl CoA → δ-aminolevulinic acid

A. Clinical indications for pyridoxine

Isoniazid, a drug commonly used to treat tuberculosis, can induce vitamin B$_6$ deficiency by forming an inactive derivative with PLP. Thus, vitamin B$_6$ supplementation is essential for some patients to prevent the development of neuropathy presenting with symptoms such as irritability, depression, confusion, and in severe cases, seizures. Otherwise, dietary deficiencies in pyridoxine are rare but have been observed in newborn infants fed formulas low in vitamin B$_6$, in women taking oral contraceptives, and in those with alcohol use disorder.

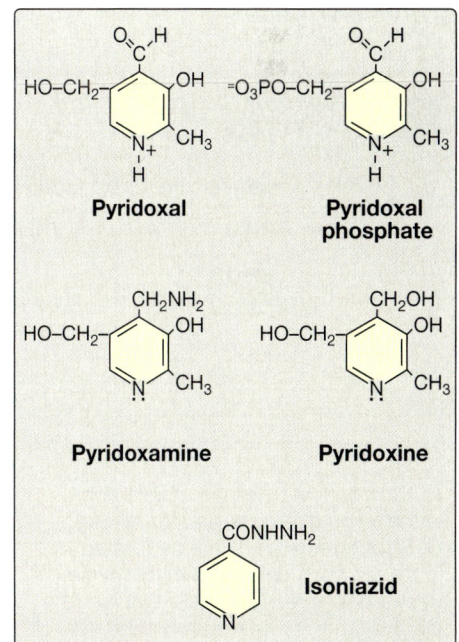

Figure 28.10
Structures of vitamin B$_6$ and the antituberculosis drug isoniazid.

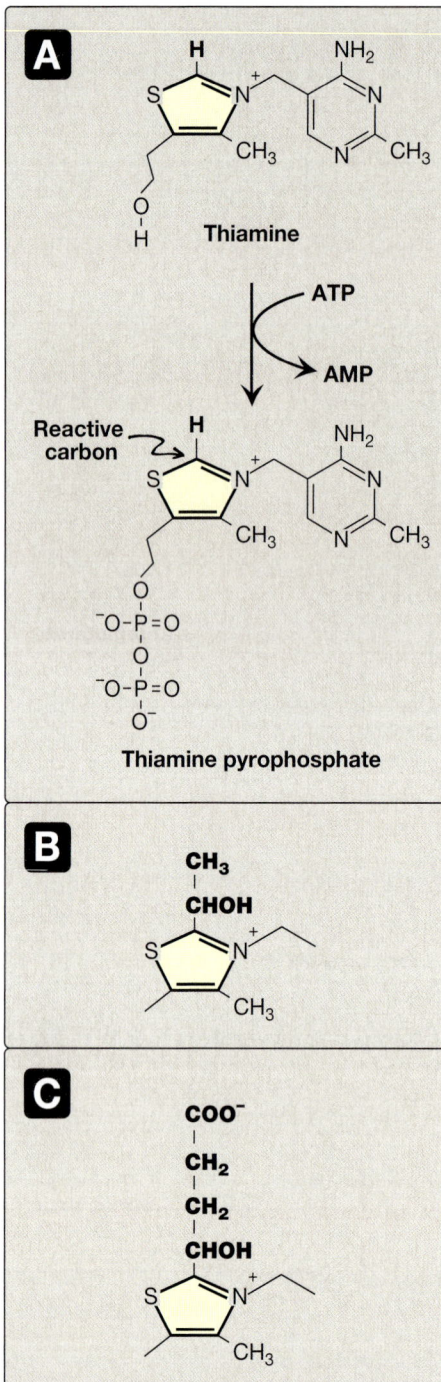

Figure 28.11
A. Structure of thiamine and its coenzyme form, thiamine pyrophosphate. **B.** Structure of intermediate formed in the reaction catalyzed by pyruvate dehydrogenase. **C.** Structure of intermediate formed in the reaction catalyzed by α-ketoglutarate dehydrogenase. AMP, adenosine monophosphate.

B. Toxicity

Vitamin B_6 is the only water-soluble vitamin with significant toxicity. Neurologic symptoms (sensory neuropathy) characterized by a loss of sensation in the extremities and sometimes difficulty coordinating movements (ataxia) occur at intakes above 500 mg/day, an amount nearly 400 times the recommended dietary allowance (RDA) and more than 5 times the tolerable upper intake limit (UL). (See Chapter 27 for a discussion of the RDA and UL.) Substantial improvement, but not complete recovery, occurs when the vitamin is discontinued.

VI. THIAMINE (VITAMIN B_1)

Thiamine pyrophosphate (TPP) is the biologically active form of the vitamin, formed by the transfer of a pyrophosphate group from ATP to thiamine (Fig. 28.11). TPP serves as a coenzyme in the formation or degradation of α-ketols by **transketolase** (Fig. 28.12A) and in the oxidative decarboxylation of α-keto acids (Fig. 28.12B).

A. Clinical indications for thiamine

The oxidative decarboxylation of pyruvate and α-ketoglutarate, which plays a key role in energy metabolism of most cells, is particularly important in CNS tissues. In thiamine deficiency, the activity of these two dehydrogenase-catalyzed reactions is decreased, resulting in decreased production of ATP and, therefore, impaired cellular function. TPP is also required by branched-chain α-keto acid dehydrogenase of muscle. [Note: It is the decarboxylase of each of these α-keto acid dehydrogenase multienzyme complexes that requires TPP.] Thiamine deficiency is diagnosed by an increase in erythrocyte transketolase activity observed with the addition of TPP.

1. **Beriberi:** This severe thiamine-deficiency syndrome is found in areas where there is severe malnutrition or in areas where starchy, low-thiamine food, such as polished rice is the major component of the diet. Adult beriberi is classified as dry (characterized by peripheral neuropathy, especially in the legs) or wet (characterized by edema because of dilated cardiomyopathy).

2. **Wernicke–Korsakoff syndrome:** In the United States, thiamine deficiency, which is seen primarily in association with chronic alcohol use disorder, is due to dietary insufficiency or impaired intestinal absorption of the vitamin. Some individuals with alcohol use disorder develop Wernicke–Korsakoff syndrome, a thiamine-deficiency state characterized by mental confusion, gait ataxia, nystagmus (a to-and-fro motion of the eyeballs), and ophthalmoplegia (weakness of eye muscles) with Wernicke encephalopathy as well as memory problems and hallucinations with Korsakoff dementia. The syndrome is treatable with thiamine supplementation, but recovery of memory is typically incomplete.

VII. NIACIN (VITAMIN B_3)

Niacin, or nicotinic acid, is a substituted pyridine derivative. The biologically active coenzyme forms are nicotinamide adenine dinucleotide (NAD^+) and its phosphorylated derivative, nicotinamide adenine dinucleotide phosphate (NADP⁺), as shown in Figure 28.13. Nicotinamide, a derivative of nicotinic acid that contains an amide instead of a carboxyl group, also occurs in the diet. Nicotinamide is readily deaminated in the body and, therefore, is nutritionally equivalent to nicotinic acid. NAD⁺ and NADP⁺ serve as coenzymes in oxidation–reduction reactions in which the coenzyme undergoes reduction of the pyridine ring by accepting two electrons from a hydride ion, as shown in Figure 28.14. The reduced forms of NAD⁺ and NADP⁺ are NADH and NADPH, respectively. A metabolite of tryptophan, quinolinate, can be converted to NAD(P). (In comparison, 60 mg of tryptophan = 1 mg of niacin.)

A. Distribution

Niacin is found in unrefined and enriched grains and cereal, milk, and lean meats (especially liver).

B. Clinical indications for niacin

1. **Deficiency:** A deficiency of niacin causes pellagra, a disease involving the skin, gastrointestinal tract, and CNS. The symptoms of pellagra progress through the three Ds: dermatitis (photosensitive), diarrhea, and dementia. If untreated, death (a fourth D) occurs. Hartnup disorder, characterized by defective absorption of tryptophan, can result in pellagra-like symptoms. Corn is low in both niacin and tryptophan, and corn-based diets can cause pellagra.

2. **Hyperlipidemia treatment:** Niacin at doses of 1.5 g/day, or 100 times the RDA, strongly inhibits lipolysis in adipose tissue, the primary producer of circulating free fatty acids (FFAs). The liver normally uses these circulating FFA as a major precursor for triacylglycerol (TAG) synthesis. Thus, niacin causes a decrease in liver TAG synthesis, which is required for very–low-density lipoprotein ([VLDL] see Chapter 18, VI. C.) production. In addition, niacin raises high-density lipoprotein and lowers Lp(a) levels (see Chapter 18, VI. F.). Low-density lipoprotein ([LDL] the cholesterol-rich lipoprotein) is derived from VLDL in the plasma. Thus, both plasma TAG (in VLDL) and

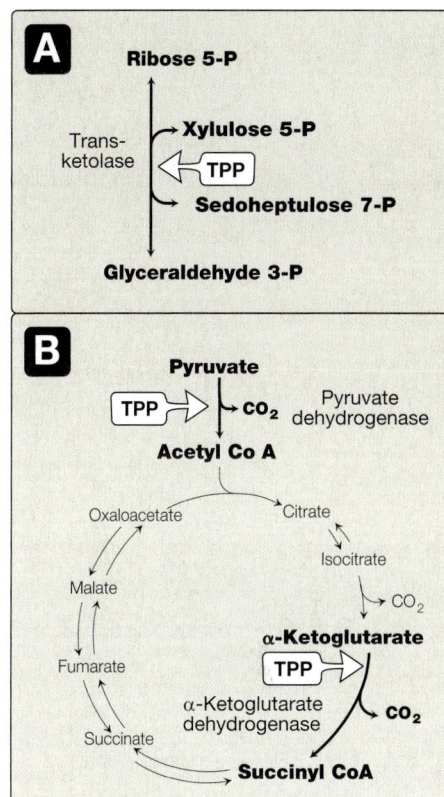

Figure 28.12
Reactions that use thiamine pyrophosphate (TPP) as coenzyme. **A.** Transketolase. **B.** Pyruvate dehydrogenase and α-ketoglutarate dehydrogenase. [Note: TPP is also used by branched-chain α-keto acid dehydrogenase.] P, phosphate; CoA, coenzyme A; CO_2, carbon dioxide.

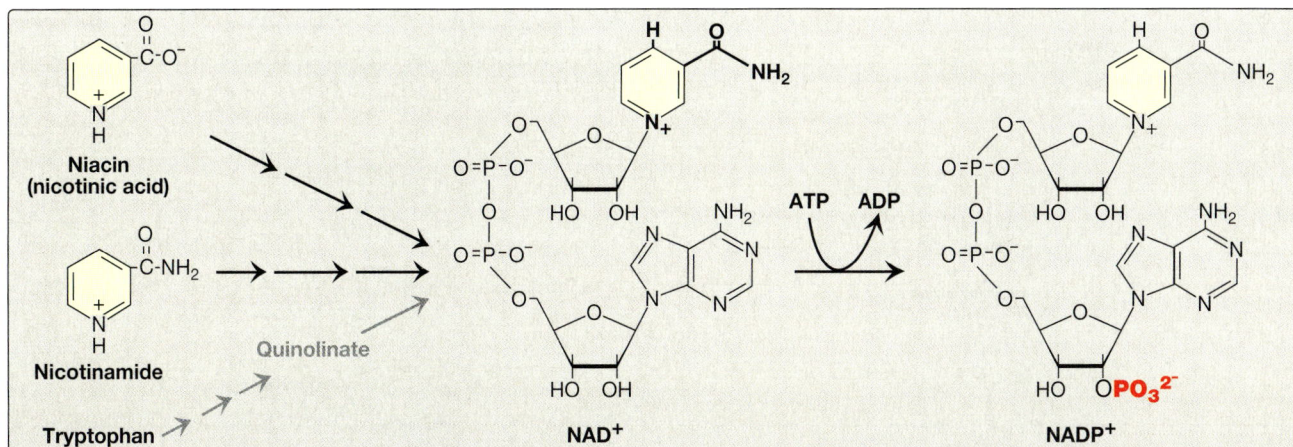

Figure 28.13
Structure and biosynthesis of oxidized nicotinamide adenine dinucleotide (NAD⁺) and nicotinamide adenine dinucleotide phosphate (NADP⁺). ADP, adenosine diphosphate.

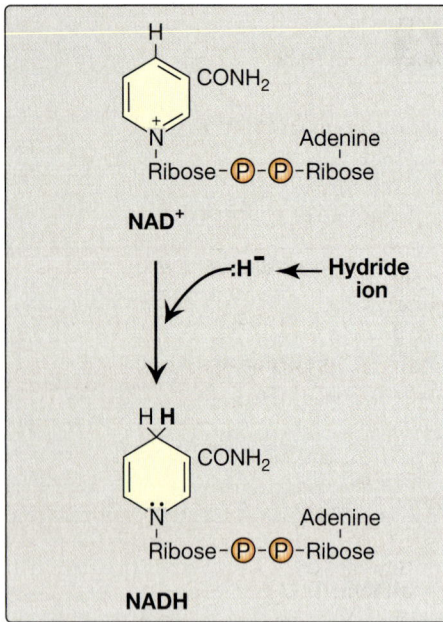

Figure 28.14
Reduction of oxidized nicotinamide adenine dinucleotide (NAD^+) to NADH. [Note: The hydride ion consists of a hydrogen (H) atom plus an electron.] Ⓟ, phosphate.

cholesterol (in LDL) are lowered. Therefore, niacin is particularly useful in the treatment of type IIb hyperlipoproteinemia, in which both VLDL and LDL are elevated. The high doses of niacin required can cause acute, prostaglandin-mediated flushing and hepatotoxicity, which can lead to jaundice, fatigue, and abdominal pain. Aspirin can reduce this side effect by inhibiting prostaglandin synthesis (see Chapter 17, VIII. A. 2.). Itching may also occur. Liver function tests (LFTs) should be monitored in patients taking high doses of niacin due to the risk of hepatotoxicity.

VIII. RIBOFLAVIN (VITAMIN B$_2$)

The two biologically active forms of vitamin B$_2$ are flavin mononucleotide (FMN) and flavin adenine dinucleotide (FAD), formed by the transfer of an adenosine monophosphate moiety from ATP to FMN (Fig. 28.15). FMN and FAD are each capable of reversibly accepting two hydrogen atoms, forming FMNH$_2$ or FADH$_2$ (fully reduced form of FMN and FAD), respectively. FMN and FAD are bound tightly, sometimes covalently, to flavoenzymes (e.g., NADH dehydrogenase [FMN] and succinate dehydrogenase [FAD]) that catalyze the oxidation or reduction of a substrate. Riboflavin deficiency is not associated with a major human disease, although it frequently accompanies other vitamin deficiencies. Symptoms of riboflavin deficiency include dermatitis, angular cheilitis (also termed cheilosis, or angular stomatitis), characterized by fissuring at the corners of the mouth, and glossitis, which presents as a smooth and dark appearance of the tongue. An easier way to remember riboflavin deficiency is by the "3 itises": dermatitis, cheilitis, and glossitis. Because riboflavin is light sensitive, phototherapy for hyperbilirubinemia (see Chapter 21, II. E. 2.) may require supplementation with the vitamin.)

IX. BIOTIN (VITAMIN B$_7$)

Biotin is a coenzyme in carboxylation reactions, in which it serves as a carrier of activated carbon dioxide (CO_2) for the mechanism of biotin-dependent carboxylations. Biotin is covalently bound to the ε-amino group

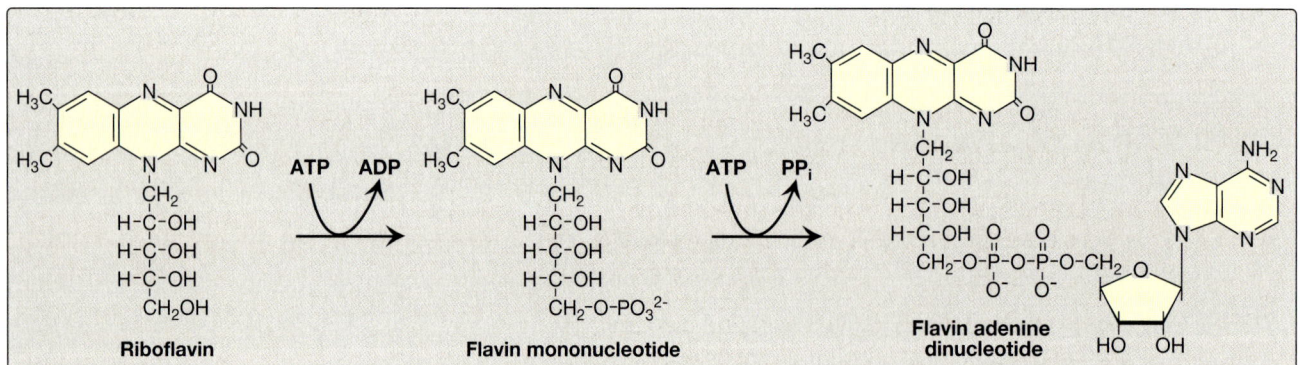

Figure 28.15
Structure and biosynthesis of the oxidized forms of flavin mononucleotide and flavin adenine dinucleotide. ADP, adenosine diphosphate; PP$_i$, pyrophosphate.

of lysine residues in biotin-dependent enzymes (Fig. 28.16). Biotin deficiency does not occur naturally because the vitamin is widely distributed in food. Also, a large percentage of the biotin requirement in humans is supplied by intestinal bacteria. However, the addition of raw egg white to the diet as a source of protein can induce symptoms of biotin deficiency, namely, dermatitis, hair loss, loss of appetite, and nausea. Raw egg white contains the glycoprotein avidin, which tightly binds biotin and prevents its absorption from the intestine. With a normal diet, however, it has been estimated that 20 eggs/day would be required to induce a deficiency syndrome. Inclusion of raw eggs in the diet is not recommended because of the possibility of salmonellosis caused by infection with *Salmonella enterica*.

> Multiple carboxylase deficiency results from decreased ability to add biotin to carboxylases during their synthesis or to remove it during their degradation. Treatment is biotin supplementation.

X. PANTOTHENIC ACID (VITAMIN B₅)

Pantothenic acid is a component of CoA, which functions in the transfer of acyl groups (Fig. 28.17). CoA contains a thiol group that carries acyl compounds as activated thiol esters. Examples of such structures are succinyl CoA, fatty acyl CoA, and acetyl CoA. Pantothenic acid is also a component of the acyl carrier protein domain of **FA synthase**. Eggs, liver, and yeast are the most important sources of pantothenic acid, although the vitamin is widely distributed. Pantothenic acid deficiency is not well characterized in humans, and no RDA has been established.

XI. VITAMIN A

Vitamin A is a fat-soluble vitamin that comes primarily from animal sources as retinol (preformed vitamin A), a retinoid. The retinoids, a family of structurally related molecules, are essential for vision, reproduction, growth, and maintenance of epithelial tissues. They also play a role in immune function. Retinoic acid, derived from the oxidation of retinol, mediates most of the actions of the retinoids, except for vision, which depends on retinal, the aldehyde derivative of retinol.

A. Structure

The retinoids include the natural forms of vitamin A, retinol and its metabolites, and synthetic forms (drugs; Fig. 28.18).

1. **Retinol:** A primary alcohol containing a β-ionone ring with an unsaturated side chain, retinol is found in animal tissues as a retinyl ester with long-chain FA. It is the storage form of vitamin A.

2. **Retinal:** This is the aldehyde derived from the oxidation of retinol. Retinal and retinol can readily be interconverted.

3. **Retinoic acid:** This is the acid derived from the oxidation of retinal. Retinoic acid cannot be reduced in the body and, therefore, cannot give rise to either retinal or retinol.

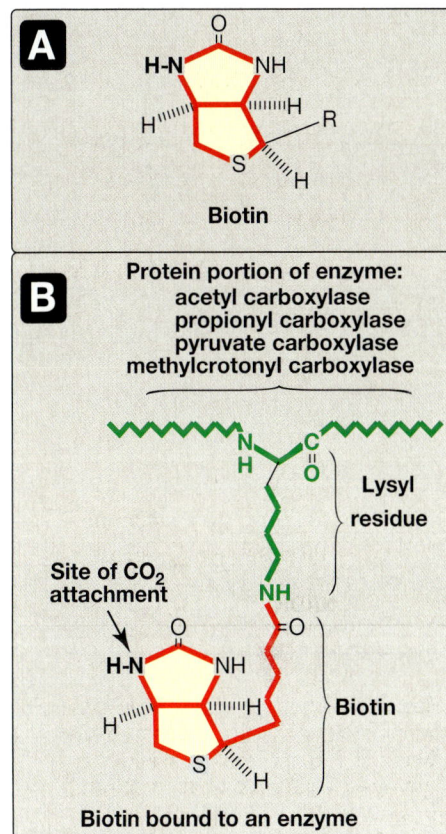

Biotin

Protein portion of enzyme:
acetyl carboxylase
propionyl carboxylase
pyruvate carboxylase
methylcrotonyl carboxylase

Lysyl residue

Site of CO_2 attachment

Biotin

Biotin bound to an enzyme

Figure 28.16
A. Structure of biotin. **B.** Biotin covalently bound to a lysyl residue of a biotin-dependent enzyme. CO_2, carbon dioxide.

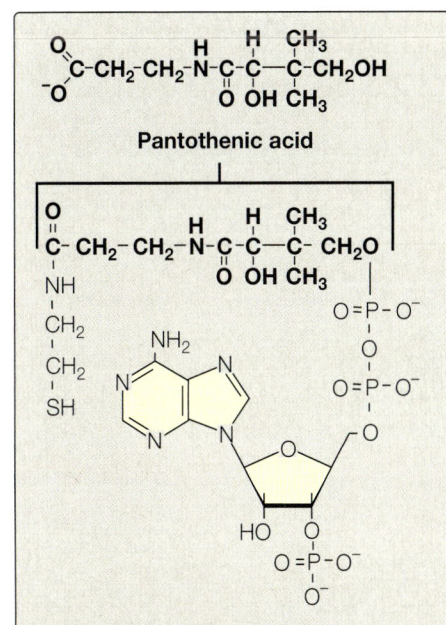

Pantothenic acid

Figure 28.17
Structure of coenzyme A.

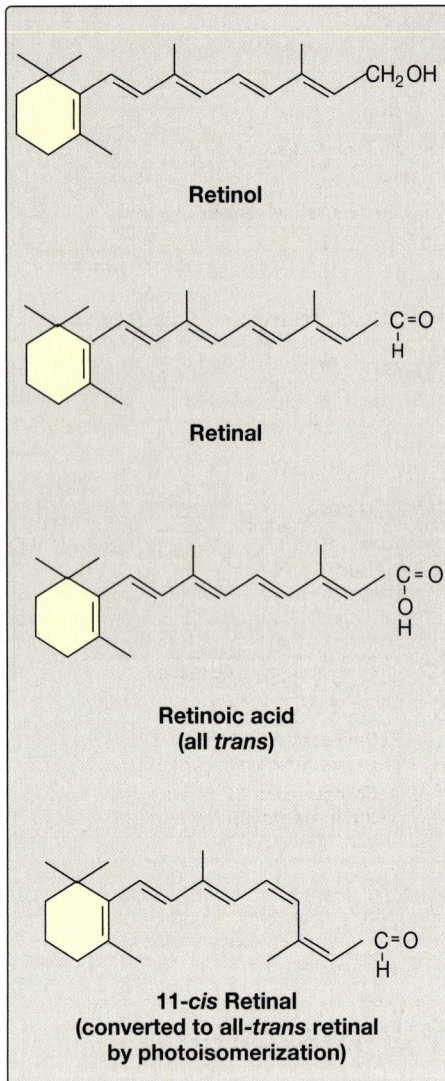

Figure 28.18
Structure of the retinoids.

4. **β-Carotene:** Plant foods contain β-carotene (provitamin A), which can be oxidatively and symmetrically cleaved in the intestine to yield two molecules of retinal. In humans, the conversion is inefficient, and the vitamin A activity of β-carotene is only about 1/12 that of retinol.

B. Absorption and transport to the liver

Retinyl esters from the diet are hydrolyzed in the intestinal mucosa, releasing retinol and FFAs (Fig. 28.19). Retinol derived from esters and from the reduction of retinal from β-carotene cleavage is re-esterified to long-chain FAs within the enterocytes and secreted as a component of chylomicrons into the lymphatic system. Retinyl esters contained in chylomicron remnants are taken up by, and stored in, the liver. Note that all fat-soluble vitamins are carried in chylomicrons.

C. Release from the liver

When needed, retinol is released from the liver and transported through the blood to extrahepatic tissues by retinol-binding protein complexed with transthyretin (see Fig. 28.19). The ternary complex binds to a transport protein on the surface of the cells of peripheral tissues, permitting retinol to enter. An intracellular retinol-binding protein carries retinol to sites in the nucleus where the vitamin regulates transcription in a manner analogous to that of steroid hormones.

D. Retinoic acid mechanism of action

Retinol is oxidized to retinoic acid. Retinoic acid binds with high affinity to specific receptor proteins (retinoic acid receptors [RARs]) present in the nucleus of target tissues such as epithelial cells (Fig. 28.20). The RAR proteins are part of the superfamily of transcriptional regulators that includes the nuclear receptors for steroid and thyroid hormones and vitamin D, all of which function in a similar way (see Chapter 18, VII. D.).

The activated retinoic acid–RAR complex binds to response elements on DNA and recruits activators or repressors to regulate retinoid-specific RNA synthesis, resulting in control of the production of specific proteins that mediate several physiologic functions. For example, retinoids control the expression of the gene for keratin in most epithelial tissues of the body.

E. Functions

Vitamin A is essential for vision; epithelial maintenance; immune response regulation; mucosal integrity; and growth, development, and reproduction.

1. **Visual cycle:** Vitamin A is a component of the visual pigments of rod and cone cells. Rhodopsin, the visual pigment of the rod cells in the retina, consists of 11-*cis* retinal bound to the protein opsin (see Fig. 28.19). When rhodopsin, a G protein–coupled receptor, is exposed to light, a series of photochemical isomerizations occurs, which results in the bleaching of rhodopsin and release of all-*trans* retinal and opsin. This process activates the G protein transducin, triggering a nerve impulse that is transmitted by

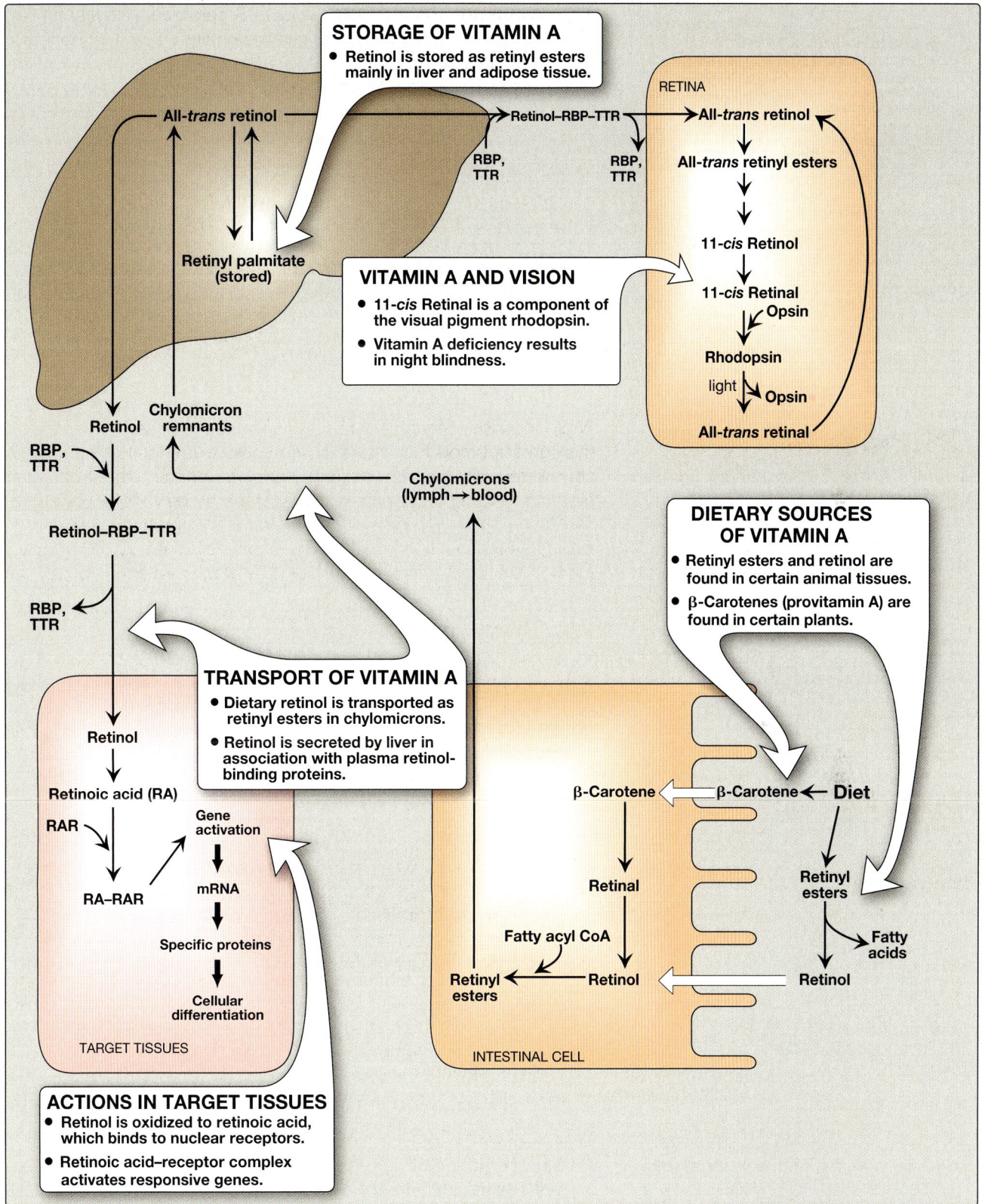

STORAGE OF VITAMIN A
- Retinol is stored as retinyl esters mainly in liver and adipose tissue.

VITAMIN A AND VISION
- 11-*cis* Retinal is a component of the visual pigment rhodopsin.
- Vitamin A deficiency results in night blindness.

DIETARY SOURCES OF VITAMIN A
- Retinyl esters and retinol are found in certain animal tissues.
- β-Carotenes (provitamin A) are found in certain plants.

TRANSPORT OF VITAMIN A
- Dietary retinol is transported as retinyl esters in chylomicrons.
- Retinol is secreted by liver in association with plasma retinol-binding proteins.

ACTIONS IN TARGET TISSUES
- Retinol is oxidized to retinoic acid, which binds to nuclear receptors.
- Retinoic acid–receptor complex activates responsive genes.

Figure 28.19
Absorption, transport, and storage of vitamin A and its derivatives. [Note: β-Carotene is a carotenoid, a plant pigment with antioxidant activity.] RBP, retinol-binding protein; TTR, transthyretin; RAR, retinoic acid receptor; CoA, coenzyme A; mRNA, messenger RNA.

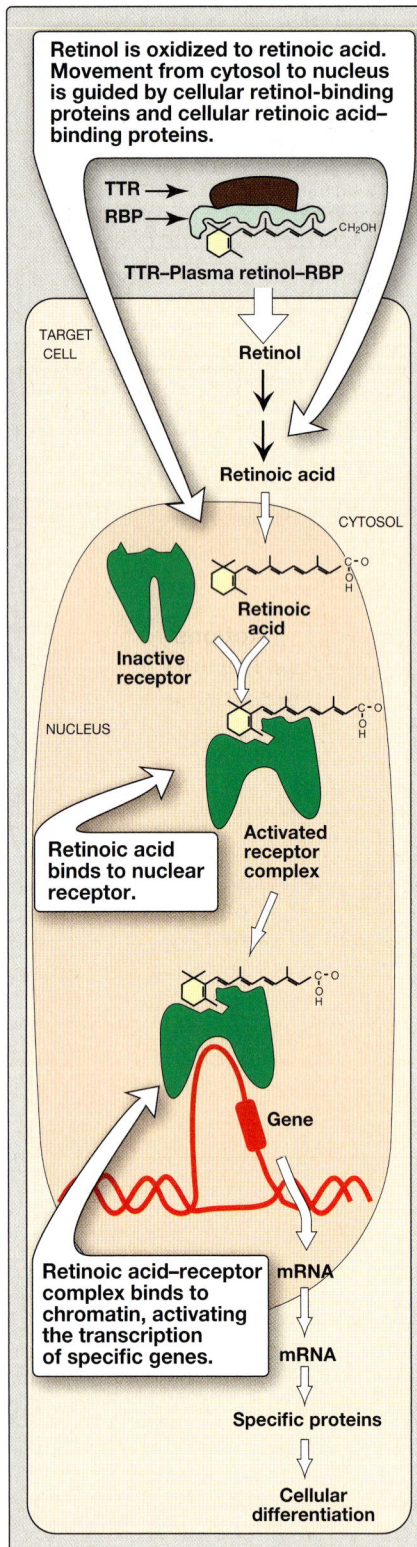

Figure 28.20
Action of the retinoids. [Note: Retinoic acid–receptor complex forms a dimer, but is shown as monomer for simplicity.] TTR, transthyretin; RBP, retinol-binding protein; mRNA, messenger RNA.

the optic nerve to the brain. Regeneration of rhodopsin requires isomerization of all-*trans* retinal back to 11-*cis* retinal. All-*trans* retinal is reduced to all-*trans* retinol, esterified, and isomerized to 11-*cis* retinol that is oxidized to 11-*cis* retinal. The latter combines with opsin to form rhodopsin, thus completing the cycle. Similar reactions are responsible for color vision in the cone cells.

2. **Epithelial cell maintenance:** Vitamin A is essential for normal differentiation of epithelial tissues and mucus secretion and, thus, supports the body's barrier-based defense against pathogens.

3. **Reproduction:** Retinol and retinal are essential for normal reproduction, supporting spermatogenesis in the male and preventing fetal resorption in the female. Retinoic acid is inactive in maintaining reproduction and in the visual cycle but promotes growth and differentiation of epithelial cells.

F. Distribution

Liver, kidney, cream, butter, and egg yolks are good sources of preformed vitamin A. Yellow, orange, and dark-green vegetables and fruits are good sources of the carotenes (provitamin A).

G. Requirement

The RDA for adults is 900 retinol activity equivalents (RAEs) for males and 700 RAEs for females. In comparison, 1 RAE = 1 μg of retinol, 12 μg of β-carotene, or 24 μg of other carotenoids.

H. Clinical indications for vitamin A

Although chemically related, retinoic acid and retinol have distinctly different therapeutic applications. Retinol and its carotenoid precursor are used as dietary supplements, whereas various forms of retinoic acid are useful in dermatology (Fig. 28.21).

1. **Deficiency:** Vitamin A, administered as retinol or retinyl esters, is used to treat patients who are deficient in the vitamin. Night blindness (nyctalopia) is one of the earliest signs of vitamin A deficiency. The visual threshold is increased, making it difficult to see in dim light. Prolonged deficiency leads to an irreversible loss in the number of visual cells. Severe deficiency leads to xerophthalmia, a pathologic dryness of the conjunctiva and cornea, caused, in part, by increased keratin synthesis. If untreated, xerophthalmia results in corneal ulceration and, ultimately, in blindness because of the formation of opaque scar tissue. The condition is most commonly seen in children in low-income tropical countries. More than 500,000 children worldwide are blinded each year by xerophthalmia caused by insufficient vitamin A in the diet.

2. **Skin conditions:** Dermatologic problems such as acne are effectively treated with retinoic acid or its derivatives (see Fig. 28.21). Mild cases of acne and skin aging are treated with tretinoin (all-*trans* retinoic acid). Tretinoin is too toxic for systemic (oral) administration in treating skin conditions, and its use is mainly confined to topical application. However, in patients with severe cystic acne unresponsive to conventional therapies, isotretinoin (13-*cis* retinoic

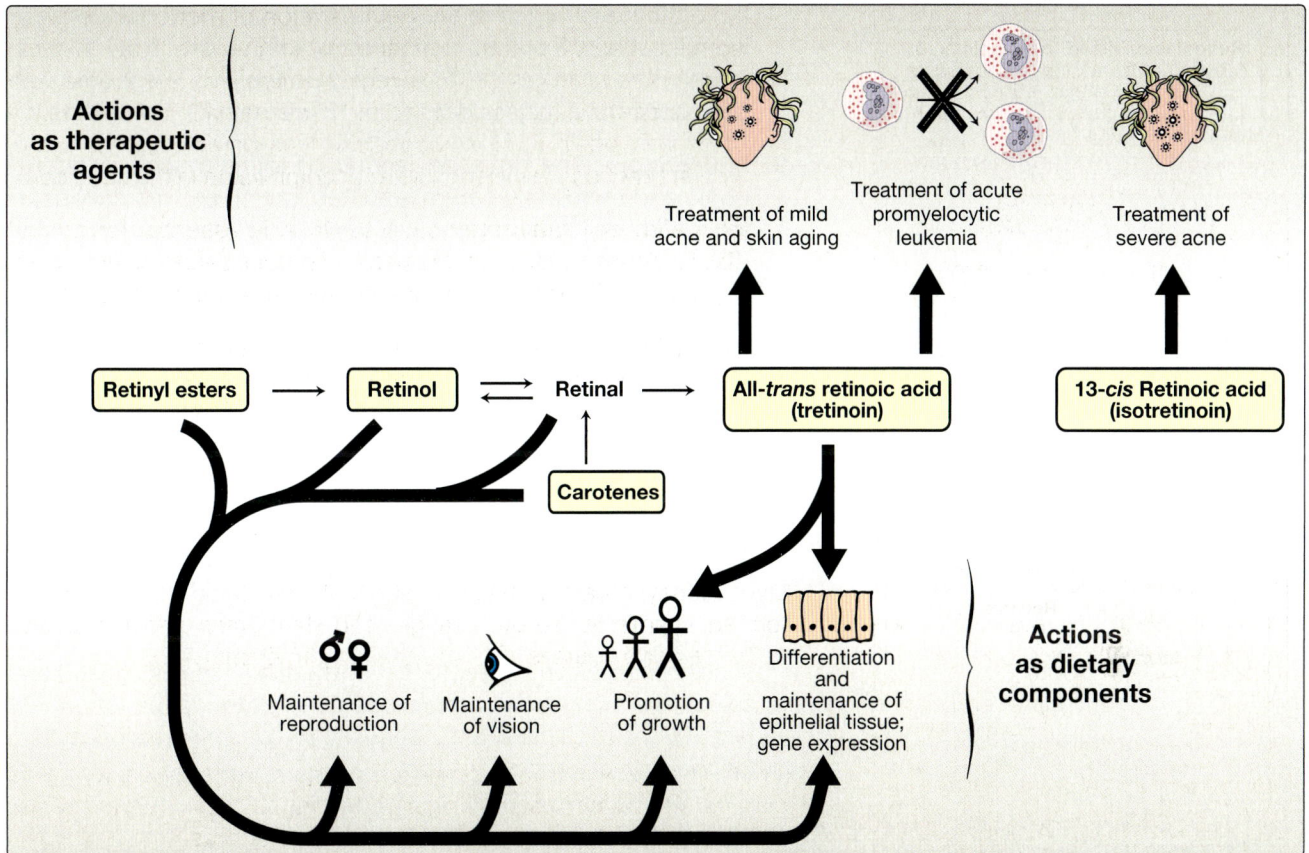

Figure 28.21
Summary of actions of retinoids. Compounds in boxes are available as dietary components or as pharmacologic agents.

acid) is administered orally. An oral synthetic retinoid is used to treat psoriasis. Oral tretinoin is also used in treating acute promyelocytic leukemia.

I. Retinoid toxicity

Retinoid toxicity, also known as hypervitaminosis A, typically results from the overconsumption of vitamin A through supplements or certain medications, leading to symptoms like dry, itchy skin; liver enlargement; and increased intracranial pressure. Isotretinoin, a synthetic analogue of retinoic acid used for severe acne treatment, is teratogenic and therefore contraindicated in women who are pregnant or may become pregnant and may increase the risk of cardiovascular disease with prolonged use.

1. **Vitamin A:** Excessive intake of vitamin A (but not carotene) produces a toxic syndrome called hypervitaminosis A. Amounts exceeding 7.5 mg/day of retinol should be avoided. Early signs of chronic hypervitaminosis A are reflected in the skin, which becomes dry and pruritic because of decreased keratin synthesis; in the liver, which becomes enlarged and can become cirrhotic; and in the CNS, where a rise in intracranial pressure may mimic the symptoms of a brain tumor. Pregnant women, in particular, should not ingest excessive quantities of vitamin A because of its potential

for teratogenesis causing congenital malformations in the developing fetus. UL is 3,000 μg preformed vitamin A/day. [Note: Vitamin A promotes bone growth. In excess, however, it is associated with decreased bone mineral density and increased risk of fractures.]

2. **Isotretinoin:** The drug, an isomer of retinoic acid, is teratogenic and absolutely contraindicated in women with childbearing potential unless they have severe, disfiguring cystic acne that is unresponsive to standard therapies. Pregnancy must be excluded before treatment begins, and birth control must be used. Prolonged treatment with isotretinoin can result in an increase in TAGs and cholesterol, providing some concern for an increased risk of cardiovascular disease (CVD).

XII. VITAMIN D

The D vitamins are a group of sterols that have a hormone-like function. The active molecule, 1,25-dihydroxycholecalciferol ([1,25-diOH-D$_3$] calcitriol), binds to intracellular receptor proteins. The 1,25-diOH-D$_3$–receptor complex interacts with response elements in the nuclear DNA of target cells in a manner similar to that of vitamin A (see Fig. 28.20) and either selectively stimulates or represses gene transcription. The most prominent actions of calcitriol are to regulate the serum levels of calcium and phosphorus.

A. Distribution

Vitamin D can be synthesized endogenously, or it can be obtained through the diet from plants or animal sources. Dietary vitamin D is essential for those with limited sun exposure.

1. **Endogenous vitamin precursor:** 7-Dehydrocholesterol, an intermediate in cholesterol synthesis, is converted to cholecalciferol in the dermis and epidermis of humans exposed to sunlight and transported to liver bound to vitamin D–binding protein.

2. **Diet:** Ergocalciferol (vitamin D$_2$), found in plants, and cholecalciferol (vitamin D$_3$), found in animal tissues, are sources of preformed vitamin D activity (Fig. 28.22). Vitamin D$_2$ and vitamin D$_3$ differ chemically only in the presence of an additional double-bond and methyl group in the plant sterol. Dietary vitamin D is packaged into chylomicrons. Preformed vitamin D is a dietary requirement only in individuals with limited exposure to sunlight.

B. Metabolism

The active form of vitamin D is calcitriol. It is synthesized from vitamins D$_2$ and D$_3$ via hydroxylation in the liver to produce 25-hydroxycholecalciferol ([25-OH-D$_3$]), then further converted in the kidney by 25-hydroxycholecalciferol 1-hydroxylase, with its activity tightly regulated by serum calcium and phosphate levels through the synthesis of parathyroid hormone (PTH).

1. **1,25-diOH-D3 formation:** Vitamins D$_2$ and D$_3$ are not biologically active but are converted *in vivo* to calcitriol, the active form of the D

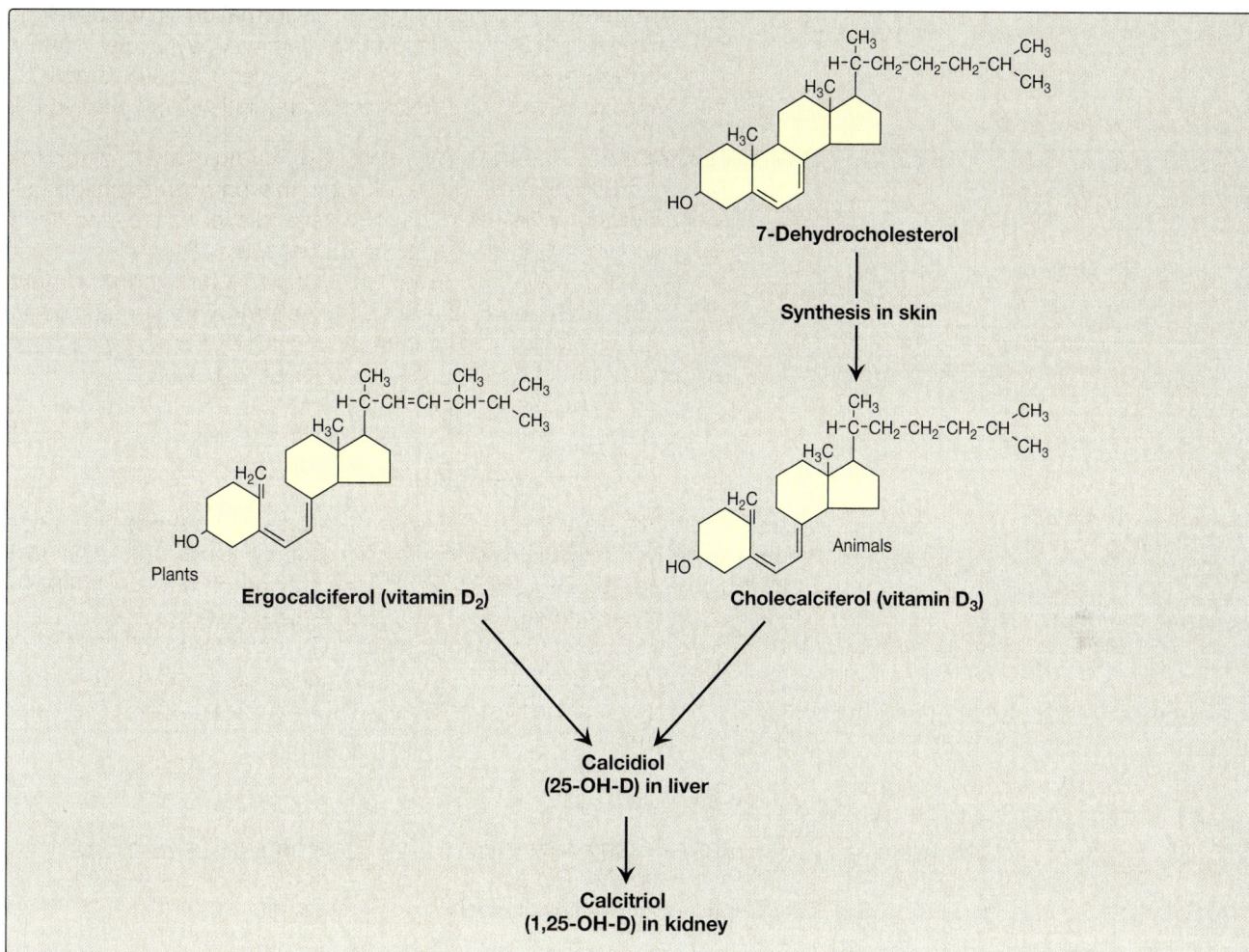

Figure 28.22
Sources of vitamin D. Vitamins D_2 and D_3 are first converted to calcidiol and then to calcitriol (active vitamin D). [Note: 7-Dehydrocholesterol (provitamin D_3) is decreased in the skin of older adults.]

vitamin, by two sequential hydroxylation reactions (Fig. 28.23). The first hydroxylation occurs at the 25 position and is catalyzed by a specific **25-hydroxylase** in the liver. The product of the reaction, 25-OH-D_3 (calcidiol), is the predominant form of vitamin D in the serum and the major storage form. 25-OH-D_3 is further hydroxylated at the 1 position by **25-hydroxycholecalciferol 1-hydroxylase** found primarily in the kidney, resulting in the formation of 1,25-diOH-D_3 (calcitriol). [Note: Both hydroxylases are cytochrome P450 proteins (see Chapter 13).]

2. **Hydroxylation regulation:** Calcitriol is the most potent vitamin D metabolite. Its formation is tightly regulated by the level of serum phosphate (PO_4^{3-}) and calcium ions (Ca^{2+}) as shown in Figure 28.24. **25-Hydroxycholecalciferol 1-hydroxylase** activity is increased directly by low serum PO_4^{3-} or indirectly by low serum Ca^{2+}, which triggers the secretion of PTH from the chief cells of the parathyroid gland. PTH upregulates the **1-hydroxylase**.

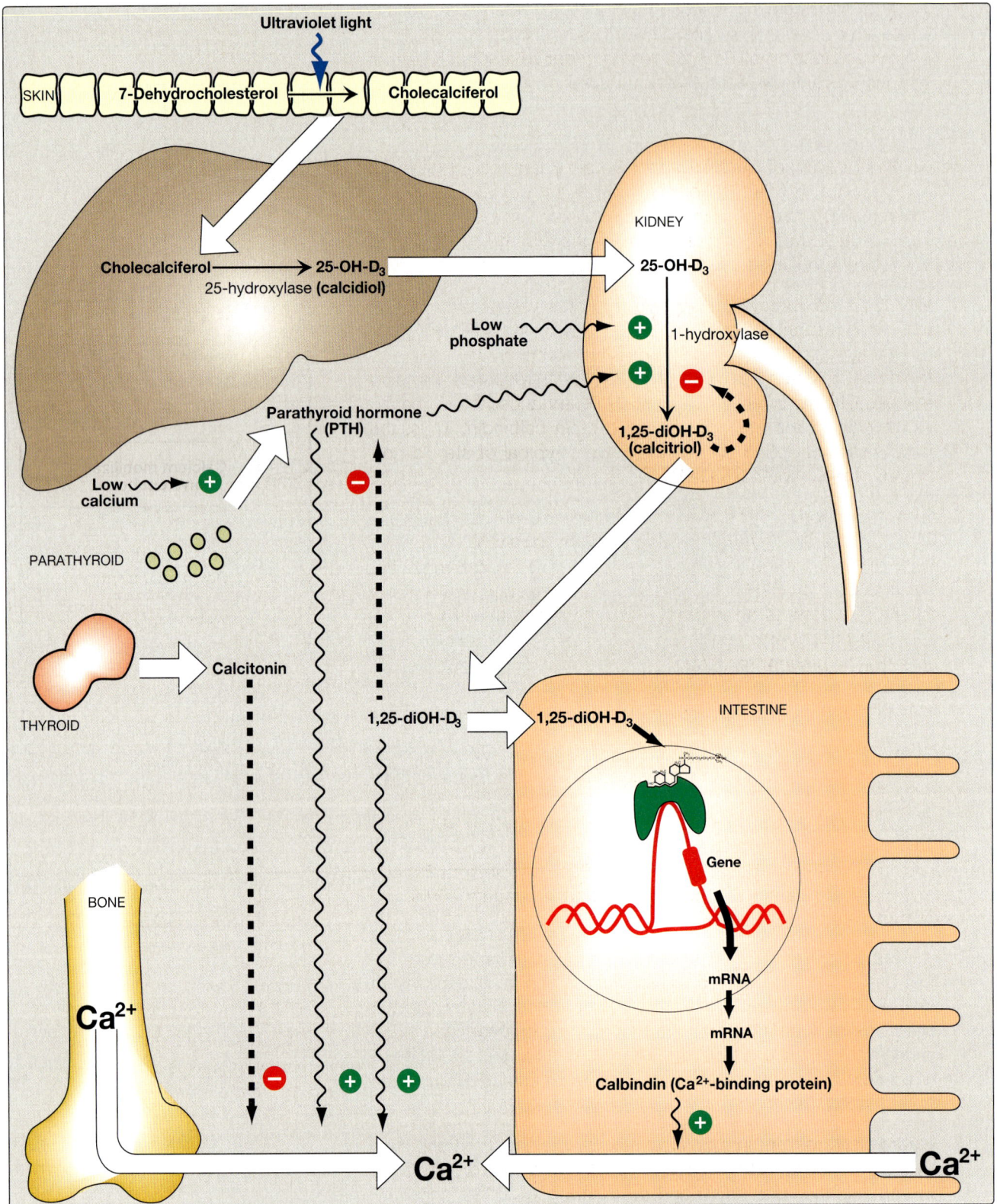

Figure 28.23
Metabolism and actions of vitamin D. [Note: Calcitonin, a thyroid hormone, decreases blood calcium (Ca^{2+}) by inhibiting mobilization from bone, absorption from the intestine, and reabsorption by the kidney. It opposes the actions of PTH.] mRNA, messenger RNA; 25-OH-D_3, 25-hydroxycholecalciferol; 1,25-diOH-D_3, 1,25-dihydroxycholecalciferol.

Thus, hypocalcemia caused by insufficient dietary Ca^{2+} results in elevated levels of serum 1,25-diOH-D_3. [Note: 1,25-diOH-D_3 inhibits expression of PTH, forming a negative-feedback loop. It also inhibits activity of the 1-hydroxylase.]

C. Function

The overall function of calcitriol is to maintain adequate serum levels of Ca^{2+}. It performs this function by (1) increasing uptake of Ca^{2+} by the intestine, (2) minimizing loss of Ca^{2+} by the kidney by increasing reabsorption, and (3) stimulating resorption (demineralization) of bone when blood Ca^{2+} is low (see Fig. 28.23).

1. **Effect on the intestine:** Calcitriol stimulates intestinal absorption of Ca^{2+} by first entering the intestinal cell and binding to a cytosolic receptor. The 1,25-diOH-D_3–receptor complex then moves to the nucleus, where it selectively interacts with response elements on the DNA. As a result, Ca^{2+} uptake is enhanced by increased expression of the calcium-binding protein calbindin. Thus, the mechanism of action of 1,25-diOH-D_3 is typical of steroid hormones (see Chapter 18, VII. D.).

2. **Effect on bone:** Bone is composed of collagen and crystals of $Ca_5(PO_4)_3OH$ (hydroxylapatite). When blood Ca^{2+} is low, 1,25-diOH-D_3 stimulates bone resorption by a process that is enhanced by PTH. The result is an increase in serum Ca^{2+}. Therefore, bone is an important reservoir of Ca^{2+} that can be mobilized to maintain serum levels. PTH and calcitriol also work together to prevent renal loss of Ca^{2+}.

D. Distribution and requirement

Vitamin D occurs naturally in fatty fish, liver, and egg yolk. Milk, unless it is artificially fortified, is not a good source. The RDA for individuals ages 1 to 70 years is 15 µg/day and 20 µg/day if older than age 70 years. Experts disagree, however, on the optimal level of vitamin D needed to maintain health. [Note: One µg vitamin D = 40 international units (IUs).] Because breast milk is a poor source of vitamin D, supplementation is recommended for breastfed infants.

E. Clinical indications for vitamin D

Vitamin D deficiency leads to rickets in children and osteomalacia in adults. Chronic kidney disease results in renal osteodystrophy, impairing the formation of active vitamin D and leading to bone demineralization. Hypoparathyroidism leads to hypocalcemia and is typically managed with vitamin D and calcium supplements.

1. **Nutritional rickets:** Vitamin D deficiency causes a net demineralization of bone, resulting in rickets in children and osteomalacia in adults (Fig. 28.25). Rickets is characterized by the continued formation of the collagen matrix of bone, but incomplete mineralization results in soft, pliable bones. In osteomalacia, demineralization of existing bones increases their susceptibility to fracture. Insufficient exposure to daylight and/or deficiencies

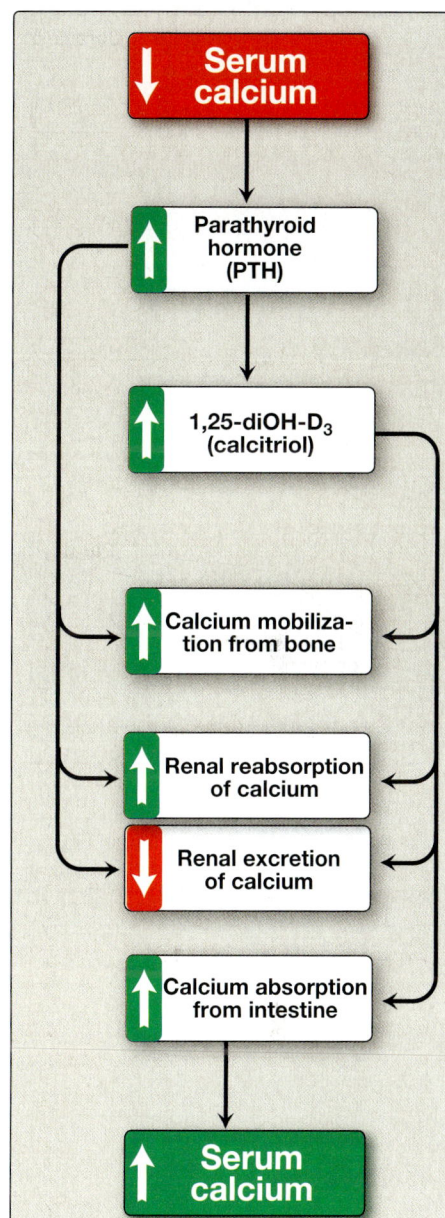

Figure 28.24
Response to low serum calcium. [Note: Calcitriol also increases intestinal absorption and renal reabsorption of phosphate. In contrast, PTH decreases renal reabsorption of phosphate.] 1,25-diOH-D_3, 1,25-dihydroxycholecalciferol.

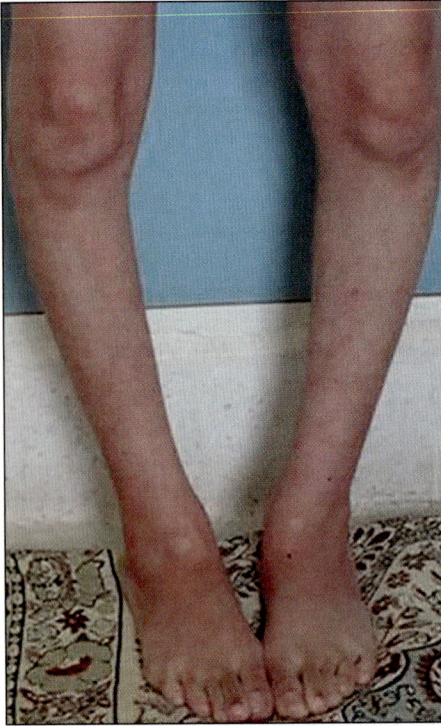

Figure 28.25
Bowed legs of a middle-age man with osteomalacia, a nutritional vitamin D deficiency that results in demineralization of the skeleton.

in vitamin D consumption occurs predominantly in infants and older adults. Vitamin D deficiency is more common in northern latitudes because less vitamin D synthesis occurs in the skin as a result of reduced exposure to ultraviolet light. Loss-of-function mutations in the vitamin D receptor result in hereditary vitamin D–deficient rickets.

2. **Renal osteodystrophy:** Chronic kidney disease causes decreased ability to form active vitamin D as well as increased retention of PO_4^{3-}, resulting in hyperphosphatemia and hypocalcemia. The low blood Ca^{2+} causes a rise in PTH and associated bone demineralization with release of Ca^{2+} and PO_4^{3-}. Supplementation with vitamin D is an effective therapy. However, supplementation must be accompanied by PO_4^{3-} reduction therapy to prevent further bone loss and precipitation of calcium phosphate crystals.

3. **Hypoparathyroidism:** Lack of PTH causes hypocalcemia and hyperphosphatemia. PTH increases phosphate excretion. Patients may be treated with vitamin D and calcium supplementation.

F. Toxicity

Like all fat-soluble vitamins, vitamin D can be stored in the body and is only slowly metabolized. High doses (100,000 IU for weeks or months) can cause loss of appetite, nausea, thirst, and weakness. Enhanced Ca^{2+} absorption and bone resorption result in hypercalcemia, which can lead to deposition of calcium salts in soft tissue (metastatic calcification). The UL is 100 µg/day (4,000 IU/day) for individuals ages 9 years and older, with a lower level for those younger than age 9 years. Toxicity is only seen with use of supplements. Excess vitamin D produced in the skin is converted to inactive forms.

XIII. VITAMIN K

The principal role of vitamin K is in the posttranslational modification of a number of proteins (most of which are involved with blood clotting), in which it serves as a coenzyme in the carboxylation of certain glutamic acid residues in these proteins. Vitamin K exists in several active forms, for example, in plants as phylloquinone (or vitamin K_1), and in intestinal bacteria as menaquinone (or vitamin K_2). A synthetic form of vitamin K, menadione, is able to be converted to K_2.

A. Function

Vitamin K is crucial for synthesizing γ-carboxyglutamate, Gla, necessary for the activation of specific coagulation factors in the liver, a process inhibited by warfarin. Additionally, Gla residues, which bind calcium, play roles in bone mineralization, reducing arterial calcification, and regulating blood clot formation.

1. **γ-Carboxyglutamate formation:** Vitamin K is essential for γ-carboxyglutamate formation during the posttranslational modification of certain coagulation factors—prothrombin (FII), FVII,

FIX, and FX—which are synthesized in the liver. A simple mnemonic to remember the coagulation factors that require vitamin K is "2+7 equals 9 not 10." Formation of the functional versions of these enzyme factors requires the vitamin K–dependent carboxylation of several glutamic acid residues to Gla residues (Fig. 28.26). The carboxylation reaction requires **γ-glutamyl carboxylase**, O_2, CO_2, and the hydroquinone form of vitamin K (which gets oxidized to the epoxide form). The formation of Gla residues is sensitive to inhibition by warfarin, a synthetic analog of vitamin K that inhibits **vitamin K epoxide reductase (VKOR)**, the enzyme required to regenerate the functional hydroquinone form of vitamin K.

2. **Prothrombin interaction with membranes:** Gla residues are good chelators of positively charged calcium ions, because of their two adjacent, negatively charged carboxylate groups. With prothrombin, for example, the prothrombin–calcium complex is able to bind to negatively charged membrane phospholipids on the surface of damaged endothelium and platelets. Attachment to membrane increases the rate at which the proteolytic conversion of prothrombin to thrombin can occur (Fig. 28.27).

3. **γ-Carboxyglutamate residues in other proteins:** Gla residues are also present in proteins other than those involved in forming a blood clot. For example, osteocalcin important for bone mineralization; matrix Gla protein (MPG), which reduces arterial calcification; and proteins C and S involved in limiting the formation of blood clots also undergo γ-carboxylation.

B. Distribution and requirement

Vitamin K is found in cabbage, kale, spinach, egg yolks, and liver. The adequate intake for vitamin K is 120 µg/day for adult males and 90 µg for adult females. There is also synthesis of the vitamin by the gut microbiota.

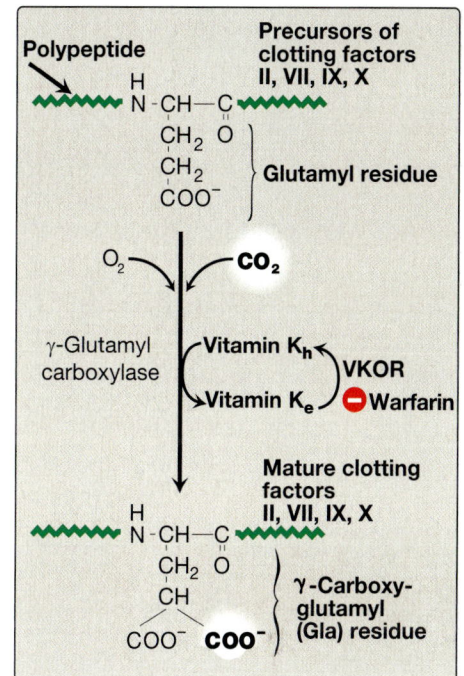

Figure 28.26
Carboxylation of glutamate to form γ-carboxyglutamate. H, hydroquinone; e, epoxide; VKOR, vitamin K epoxide reductase.

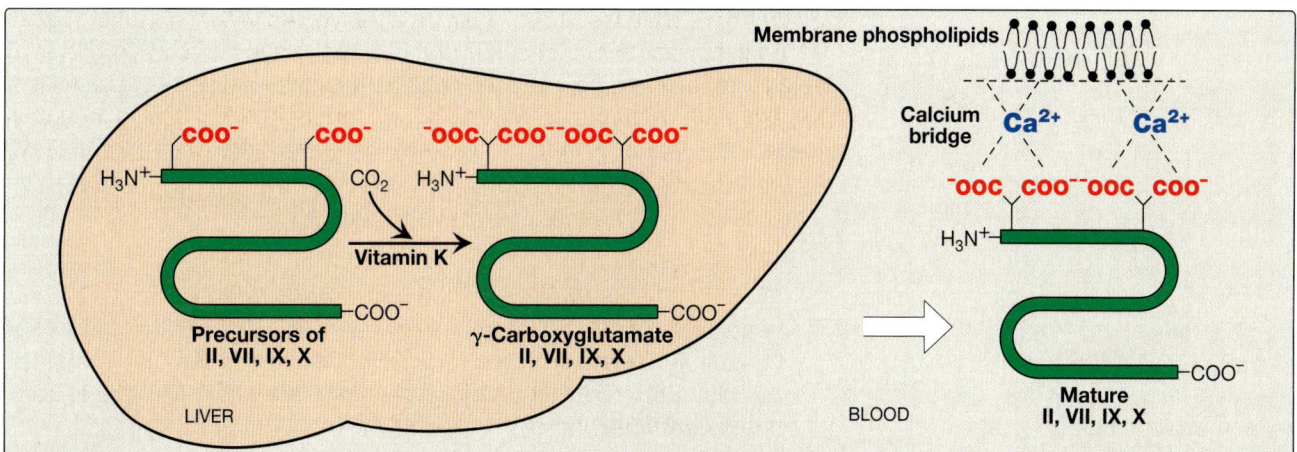

Figure 28.27
Role of vitamin K in blood coagulation. CO_2, carbon dioxide.

C. Clinical indications for vitamin K

Vitamin K deficiency is rare due to dietary sources and synthesis by gut bacteria but can occur with antibiotic use or in malnourished individuals, requiring supplementation to prevent bleeding and support bone health. Newborns, lacking intestinal bacteria and receiving insufficient vitamin K from human milk, are given a prophylactic dose to prevent hemorrhagic disease.

1. **Deficiency:** A true vitamin K deficiency is unusual because adequate amounts are generally obtained from the diet and produced by intestinal bacteria. If the bacterial population in the gut is decreased (e.g., by antibiotics), the amount of endogenously formed vitamin is decreased, and this can lead to hypoprothrombinemia in the marginally malnourished individual (e.g., a debilitated geriatric patient). This condition may require supplementation with vitamin K to correct the bleeding tendency. In addition, certain cephalosporin antibiotics (e.g., cefamandole) cause hypoprothrombinemia, apparently by a warfarin-like mechanism that inhibits VKOR. Consequently, their use in treatment is usually supplemented with vitamin K. Deficiency can also affect bone health.

2. **Deficiency in newborns:** Because newborns have sterile intestines, they initially lack the bacteria that synthesize vitamin K. Because human milk provides only about one fifth of the daily requirement for vitamin K, it is recommended that all newborns receive a single intramuscular dose of vitamin K as prophylaxis against hemorrhagic disease of the newborn.

D. Toxicity

Prolonged administration of large doses of menadione can produce hemolytic anemia and jaundice in the infant, because of toxic effects on the RBC membrane. Therefore, it is no longer used to treat vitamin K deficiency. No UL for the natural form has been set.

XIV. VITAMIN E

The E vitamins consist of eight naturally occurring tocopherols, of which α-tocopherol is the most active (Fig. 28.28). Vitamin E functions as an antioxidant in prevention of nonenzymic oxidations (e.g., oxidation of LDL and peroxidation of polyunsaturated FA by O_2 and free radicals). [Note: Vitamin C regenerates active vitamin E.]

A. Distribution and requirements

Vegetable oils are rich sources of vitamin E, whereas liver and eggs contain moderate amounts. The RDA for α-tocopherol is 15 mg/day for adults. The vitamin E requirement increases as the intake of polyunsaturated FAs increases to limit FA peroxidation.

B. Deficiency

Newborns, particularly preterm infants, have low reserves of vitamin E, but breast milk and infant formulas contain the vitamin. Deficiency

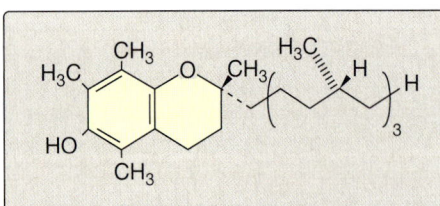

Figure 28.28
Structure of vitamin E (α-tocopherol).

in vitamin E can lead to hemolytic anemia due to the destruction of RBCs. Premature infants are especially susceptible to this condition and may require vitamin E supplementation to protect against oxidative stress, which can cause both hemolysis and retinopathy of prematurity. When observed in adults, deficiency is usually associated with defective lipid absorption or transport. [Note: Abetalipoproteinemia, caused by a defect in the formation of chylomicrons (and VLDL), results in vitamin E deficiency.]

C. Clinical indications for vitamin E

Vitamin E is not recommended for the prevention of chronic disease, such as CVD or cancer. Clinical trials using vitamin E supplementation have been uniformly disappointing. For example, subjects in the Alpha-Tocopherol, Beta-Carotene Cancer Prevention Study trial who received high doses of vitamin E not only lacked cardiovascular benefit but also had an increased incidence of stroke. [Note: Vitamins E and C are used to slow the progression of age-related macular degeneration.]

D. Toxicity

Vitamin E is the least toxic of the fat-soluble vitamins, and no toxicity has been observed at doses of 300 mg/day (UL = 1,000 mg/day).

> Populations consuming diets high in fruits and vegetables show decreased incidence of some chronic diseases. However, clinical trials have failed to show a definitive benefit from supplements of folic acid; vitamins A, C, or E; or antioxidant combinations for the prevention of cancer or CVD.

XV. CHAPTER SUMMARY

The vitamins are summarized in Figure 28.29.

VITAMIN	OTHER NAMES	ACTIVE FORM	FUNCTION
Vitamin B$_9$	Folic acid	Tetrahydro-folic acid	Transfer one-carbon units; synthesis of methionine, serine, purine nucleotides, and thymidine monophosphate
Vitamin B$_{12}$	Cobalamin	Methylcobalamin Deoxyadenosyl cobalamin	Coenzyme for reactions: Homocysteine $\rightarrow$ methionine Methylmalonyl CoA $\rightarrow$ succinyl CoA
Vitamin C	Ascorbic acid	Ascorbic acid	Antioxidant Dopamine $\rightarrow$ norepinephrine Coenzyme for hydroxylation reactions, for example: In procollagen: Proline $\rightarrow$ hydroxyproline Lysine $\rightarrow$ hydroxylysine
Vitamin B$_6$	Pyridoxine Pyridoxamine Pyridoxal	Pyridoxal phosphate	Coenzyme for enzymes, particularly in amino acid metabolism Heme synthesis
Vitamin B$_1$	Thiamine	Thiamine pyrophosphate	Coenzyme of enzymes catalyzing: Pyruvate $\rightarrow$ acetyl CoA α-Ketoglutarate $\rightarrow$ Succinyl CoA Ribose 5-P + xylulose 5-P $\rightarrow$ Sedoheptulose 7-P + Glyceraldehyde 3-P Branched-chain α-keto acid oxidation
Vitamin B$_3$	Niacin Nicotinic acid	NAD$^+$, NADP$^+$	Electron transfer
Vitamin B$_2$	Riboflavin	FMN, FAD	Electron transfer
Vitamin B$_7$	Biotin	Enzyme-bound biotin	Carboxylation reactions Pyruvate $\rightarrow$ oxaloacetate Acetyl CoA $\rightarrow$ malonyl CoA Propionyl Co $\rightarrow$ methylmalonyl CoA 3-Methylcrotonyl CoA $\rightarrow$ 3-methylglutaconyl CoA
Vitamin B$_5$	Pantothenic acid	Coenzyme A	Acyl carrier

WATER SOLUBLE

FAT SOLUBLE

VITAMIN	OTHER NAMES	ACTIVE FORM	FUNCTION
Vitamin A	Retinol Retinal Retinoic acid β-Carotene	Retinol Retinal Retinoic acid	Maintenance of reproduction Vision Promotion of growth Differentiation and maintenance of epithelial tissues Gene expression
Vitamin D	Cholecalciferol Ergocalciferol	1,25-Dihydroxy-cholecalciferol	Calcium uptake Gene expression
Vitamin K	Menadione Menaquinone Phylloquinone	Menadione Menaquinone Phylloquinone	γ-Carboxylation of glutamate residues in clotting and other proteins
Vitamin E	α-Tocopherol	Any of several tocopherol derivatives	Antioxidant

Figure 28.29
Summary of vitamins. [Note: Choline, like vitamin D, is considered an essential micronutrient in humans even though we are able to synthesize it.] P, phosphate; NAD(P), nicotinamide adenine dinucleotide (phosphate); FMN, flavin mononucleotide; FAD, flavin adenine dinucleotide; CoA, coenzyme A. (continued on the next page)

DEFICIENCY	SIGNS AND SYMPTOMS	TOXICITY	NOTES
Megaloblastic anemia Neural tube defects	Anemia Birth defects	None	Administration of high levels of folate can mask vitamin B_{12} deficiency
Pernicious anemia Dementia Spinal degeneration	Megaloblastic anemia Neuropsychiatric symptoms	None	Pernicious anemia is treated with intramuscular or high-dose oral vitamin B_{12}
Scurvy	Sore, spongy gums Loose teeth Poor wound healing Bleeding	None	Benefits of supplementation not established in controlled trials
Rare	Glossitis Neuropathy	Yes	Deficiency can be induced by isoniazid Sensory neuropathy occurs at high doses
Beriberi Wernicke–Korsakoff syndrome (most common in alcohol use disorder)	Peripheral neuropathy (dry form), edema and cardiomyopathy (wet form) Confusion, ataxia, memory loss, hallucinations, dysregulated eye movements	None	–
Pellagra	Dermatitis Diarrhea Dementia	None	High doses of niacin used to treat hyperlipidemia
Rare	Dermatitis Angular stomatitis	None	–
Rare	Dermatitis	None	Consumption of large amounts of raw egg whites (which contains a protein, avidin, that binds biotin) can induce a biotin deficiency
Rare	–	None	–
			WATER SOLUBLE
			FAT SOLUBLE
Night blindness Xerophthalmia Infertility Growth retardation	Increased visual threshold Dryness of cornea	Yes	β-Carotene not acutely toxic, but supplementation is not recommended Excess vitamin A can increase incidence of fractures
Rickets (in children) Osteomalacia (in adults)	Soft, pliable bones	Yes	Vitamin D is not a true vitamin because it can be synthesized in skin; application of sunscreen lotions or presence of dark skin color decreases this synthesis.
Newborn Rare in adults	Bleeding	Rare	Vitamin K produced by intestinal bacteria. Vitamin K deficiency common in newborns Intramuscular treatment with vitamin K is recommended at birth
Rare	Red blood cell fragility leads to hemolytic anemia	None	Benefits of supplementation for disease prevention not established in controlled trials

Figure 28.29
Summary of vitamins. (continued from the previous page)

Study Questions

Choose the ONE best answer.

For Questions 28.1–28.5, match the vitamin deficiency to the clinical consequence.

A. Folic acid
B. Niacin
C. Vitamin A
D. Vitamin B_{12}

E. Vitamin C
F. Vitamin D
G. Vitamin E
H. Vitamin K

28.1. Bleeding

28.2. Diarrhea and dermatitis

28.3. Neural tube defects

28.4. Night blindness (nyctalopia)

28.5. Sore, spongy gums and loose teeth

Correct answers = H, B, A, C, E. Vitamin K is required for formation of the γ-carboxyglutamate residues in several proteins required for blood clotting. Consequently, a deficiency of vitamin K results in a tendency to bleed. Niacin deficiency is characterized by the three Ds: diarrhea, dermatitis, and dementia (and death, a fourth D, if untreated). Folic acid deficiency can result in neural tube defects in the developing fetus. Night blindness is one of the first signs of vitamin A deficiency. Rod cells in the retina detect white and black images and work best in low light, for example, at night. Rhodopsin, the visual pigment of the rod cells, consists of 11-*cis* retinal bound to the protein opsin. Vitamin C is required for the hydroxylation of proline and lysine during collagen synthesis. Severe vitamin C deficiency (scurvy) results in defective connective tissue, characterized by sore and spongy gums, loose teeth, capillary fragility, anemia, and fatigue.

28.6. A 52-year-old woman presents with fatigue of several months' duration. Blood studies reveal a macrocytic anemia, reduced levels of hemoglobin, elevated levels of homocysteine, and normal levels of methylmalonic acid. Which of the following is most likely deficient in this patient?

A. Folic acid
B. Folic acid and vitamin B_{12}
C. Iron
D. Vitamin C

Correct answer = A. Macrocytic anemia is seen with deficiencies of folic acid, vitamin B_{12}, or both. Vitamin B_{12} is utilized in only two reactions in the body: the remethylation of homocysteine (Hcy) to methionine, which also requires folic acid (as tetrahydrofolate [THF]), and the isomerization of methylmalonyl coenzyme A to succinyl coenzyme A, which does not require THF. The elevated Hcy and normal methylmalonic acid levels in the patient's blood reflect a deficiency of folic acid as the cause of the macrocytic anemia. Iron deficiency causes microcytic anemia, as can vitamin C deficiency.

28.7. A 10-month-old infant, whose family recently relocated from the Northeast to the South of the United States, is being evaluated for the bowed appearance of her legs. The parents report that the child is still being breastfed and takes no supplements. Which one of the following statements concerning this girl's nutritional status is correct?

A. The girl's nutrition is likely adequate due to exclusive breastfeeding.
B. The girl may require vitamin D supplementation due to limited sun exposure.
C. The girl's leg condition is unrelated to her nutritional status.
D. Relocation to the south has likely resolved any previous vitamin deficiencies.

Correct answer = B. Vitamin D is required in the diet of individuals with limited exposure to sunlight, such as those living at northern latitudes and those with dark skin. Note that breast milk is low in vitamin D, and the lack of supplementation increases the risk of a deficiency.

28.8. A premature infant presents with hemolytic anemia and retinopathy of prematurity. Supplementation with which of the following could prevent of such complications?

    A. Vitamin C as an antioxidant and involving in collagen synthesis

    B. Vitamin K facilitating blood clotting

    C. Vitamin E as an antioxidant and stabilizing cell membranes

    D. Vitamin A essential for essential for vision

Correct answer: C. Hemolytic anemia, and retinopathy of prematurity is associated with oxidative damage. Premature infants are at a higher risk for vitamin E deficiency due to low stores at birth and the high levels of polyunsaturated fatty acids in their tissues, which are susceptible to oxidation. Vitamin C, while also an antioxidant, does not play the same specific role in preventing these conditions in premature infants. Vitamin K is essential for blood clotting but is not associated with preventing hemolysis or retinopathy. Vitamin A is crucial for vision, but its supplementation would not directly prevent hemolytic anemia or retinopathy of prematurity.

28.9. A 45-year-old woman presents to the Emergency Department with a 1-month history of progressive abdominal pain and watery diarrhea. She reports feelings of depression and frequent memory lapses. She has a history of alcohol use disorder. Physical examination shows a pigmented, scaly rash on sun-exposed areas of the face, neck, and hands. She states that the rash exacerbates with sunlight exposure. Considering her history and symptoms, which metabolic pathway is most likely impaired in this patient?

    A. Pyridoxal phosphate–dependent reactions

    B. Thiamin-dependent reactions

    C. FAD^+ coenzyme–dependent reactions

    D. NAD^+ and $NADP^+$ coenzyme–dependent reactions

The patient's symptoms of dermatitis (particularly in sun-exposed areas), diarrhea, and neurologic symptoms (depression and memory lapses), combined with a history of alcohol use disorder, suggest a deficiency of niacin (vitamin B_3). This deficiency leads to pellagra, characterized by the triad of dermatitis, diarrhea, and dementia. Niacin is essential for the formation of NAD^+ and $NADP^+$, which are coenzymes involved in numerous metabolic pathways, particularly those involving redox reactions.

28.10. Why might a deficiency of vitamin B_6 result in fasting hypoglycemia? Deficiency of what other vitamin could also result in hypoglycemia?

Vitamin B_6 is required for glycogen degradation by glycogen phosphorylase. A deficiency would result in fasting hypoglycemia. Additionally, a deficiency of biotin (required by pyruvate carboxylase of gluconeogenesis) would also result in fasting hypoglycemia.

29 Micronutrients: Minerals

I. OVERVIEW

Minerals are inorganic substances (elements) required in small amounts by the body. They function in a number of processes including formation of bones and teeth, fluid balance, nerve conduction, muscle contraction, signaling, and catalysis. [Note: Several minerals are essential enzyme cofactors.] Like the organic vitamins (see Chapter 28), minerals are micronutrients required in mg or μg amounts. Those required by adults in the largest amounts (>100 mg/day) are referred to as the macrominerals. Minerals required in amounts between 1 and 100 mg/day are the microminerals (trace minerals). Ultratrace minerals are required in amounts less than 1 mg/day (Fig. 29.1). [Note: The classification of specific minerals into these categories can vary among sources.] Mineral concentrations in the body are influenced by their rates of absorption and excretion.

MINERAL CLASSIFICATIONS	RDA (OR AI*) FOR ADULTS
MACROMINERALS	
Calcium (Ca^{2+})	1,000–2,000 mg
Chloride (Cl^-)	1,800–2,300 mg*
Magnesium (Mg^{2+})	310–420 mg
Phosphorus (P^+)	700 mg
Potassium (K)	4,700 mg*
Sodium (Na^+)	1,500 mg*
MICROMINERALS (TRACE)	
Chromium (Cr)	30–35 mg
Copper (Cu)	900 μg
Fluorine (as fluoride [F^-])	3–4 mg
Iron (Fe)	8–18 mg
Manganese (Mn)	1.8–2.3 mg*
Zinc (Zn)	8–11 mg
MICROMINERALS (ULTRATRACE)	
Iodine (I)	150 μg
Molybdenum (Mo)	45 μg
Selenium (Se)	55 μg

Figure 29.1
Classification of minerals and recommended amounts to be consumed/day by adults. [Note: *An adequate intake (AI) is set if insufficient scientific evidence is available to calculate a recommended dietary allowance (RDA).]

II. MACROMINERALS

The macrominerals include calcium (Ca^{2+}), phosphorus ([P] as inorganic phosphate [P_i, or PO_4^{3-}]), magnesium (Mg^{2+}), sodium (Na^+), chloride (Cl^-), and potassium (K^+). The free ionic forms are electrolytes.

A. Calcium and phosphorus

These macrominerals are considered together because they are components of hydroxylapatite ($Ca_5[PO_4]_3OH$), which makes up bones and teeth.

1. **Calcium:** Ca^{2+} is the most abundant mineral in the body, with ~98% being found in bones. The remainder is involved in a number of processes such as signaling, muscle contraction, and blood clotting. Ca^{2+} binds to a variety of proteins including calmodulin (see Chapter 11), **phospholipase A$_2$**, and **protein kinase C** and alters their activity. [Note: Calbindin is a vitamin D–induced intracellular Ca^{2+}-binding protein involved in Ca^{2+} absorption in the intestine.] Dairy products, many green vegetables (e.g., broccoli, but not spinach), and fortified orange juice are good dietary sources. Although dietary deficiency syndromes are unknown, average Ca^{2+} intake in the United States is insufficient for optimal bone health. Toxicity is seen only with supplements (Tolerable Upper Limit [UL] = 2,500 mg/day for adults). Hypercalcemia (elevated serum Ca^{2+})

can result from overproduction of parathyroid hormone (PTH). This may cause constipation and kidney stones. Hypocalcemia (low serum Ca^{2+}) can result from a deficiency of PTH or vitamin D. It can lead to bone demineralization (resorption). [Note: The hormonal regulation of serum Ca^{2+} level was presented in the vitamin D section of Chapter 28 and is reviewed in Section II. A. 3.]

> Bone mass increases from infancy through the early reproductive years and then shows an age-related loss in both men and women that increases the risk for fracture. This loss is greatest in individuals with certain genetic predispositions, such as those observed in some postmenopausal women. Studies have shown that concomitant supplementation with Ca^{2+} and vitamin D leads to an increase in bone mass density.

2. **Phosphorus:** Free phosphate (P_i) is the most abundant intracellular anion. However, 85% of the body's phosphorus is in the form of inorganic hydroxylapatite, with most of the remainder in intracellular organic compounds such as phospholipids, nucleic acids, ATP, and creatine phosphate. Phosphate is supplied as ATP for **kinases** and as P_i for **phosphorylases** (e.g., **glycogen phosphorylase**; see Chapter 11). [Note: Its addition by **kinases** or removal by **phosphatases** is an important means of covalent regulation of enzymes (see Chapter 24).] Phosphorus is widely distributed in food (milk is a good source), and dietary deficiency is rare. Hypophosphatemia can be caused by refeeding carbohydrates to malnourished patients (refeeding syndrome; see Chapter 27, VII. D.), overuse of aluminum-containing antacids (aluminum chelates P_i), and increased urinary loss in response to increased production of PTH. Muscle weakness is a common symptom. Hyperphosphatemia is caused primarily by decreased PTH levels. The excess P_i can combine with Ca^{2+} and form crystals that deposit in soft tissue (metastatic calcification). The Ca^{2+}/P_i ratio is important for bone formation (the ratio is ~2:1 in bone), and some experts are concerned that replacement of Ca^{2+}-rich milk by Ca^{2+}-poor, P_i-rich soft drinks can affect bone health.

3. **Hormonal regulation:** Serum levels of Ca^{2+} and P_i are primarily controlled by calcitriol (1,25-dihydroxycholecalciferol, the active form of vitamin D) and PTH, both of which respond to a decrease in serum Ca^{2+}. Calcitriol, produced by the kidneys, increases serum Ca^{2+} and P_i by increasing bone resorption and intestinal absorption and renal reabsorption of Ca^{2+} and P_i (Fig. 29.2). PTH (from the parathyroid glands) increases serum Ca^{2+} by increasing bone resorption, increasing renal reabsorption of Ca^{2+}, and activating the renal **1-hydroxylase** that produces calcitriol from calcidiol ([25-OH-D_3] Fig. 29.3). In contrast to calcitriol, PTH decreases P_i reabsorption in the kidneys, lowering serum P_i. [Note: High serum P_i increases PTH and decreases calcitriol.] A third hormone, calcitonin (from the C cells of the thyroid gland), responds to elevated serum Ca^{2+} levels by promoting bone mineralization and increasing renal excretion of Ca^{2+} (and P_i).

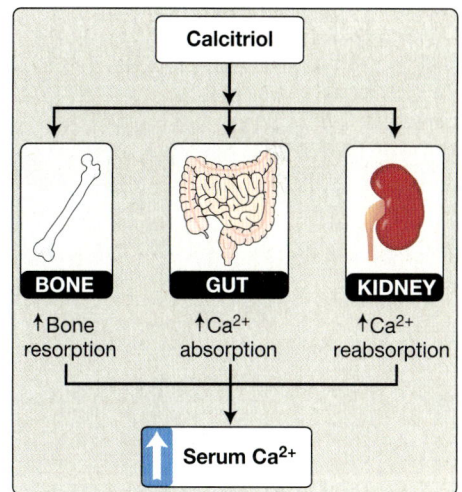

Figure 29.2
Effect of calcitriol on serum calcium (Ca^{2+}).

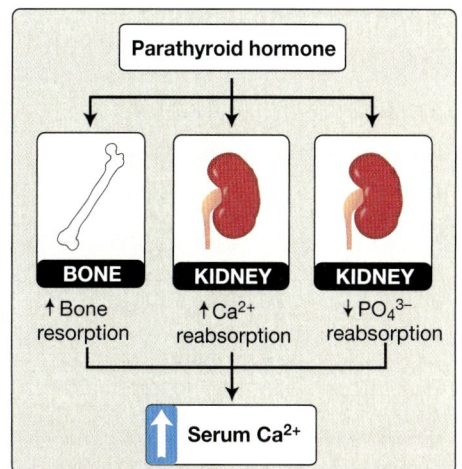

Figure 29.3
Effect of parathyroid hormone on serum calcium (Ca^{2+}). PO_4^{3-}, phosphate.

B. Magnesium

About 60% of the body's Mg^{2+} is in bone, but it accounts for just 1% of bone mass. The mineral is required by a variety of enzymatic reactions, including phosphorylation by **kinases** (Mg^{2+} binds the ATP cosubstrate) and phosphodiester bond formation by **DNA** and **RNA polymerases**. Mg^{2+} is widely distributed in foods, but the average intake in the United States is below the recommended level. Hypomagnesemia can result from decreased absorption or increased excretion of Mg^{2+}. Symptoms include hyperexcitability of skeletal muscles and nerves and cardiac arrhythmias. With hypermagnesemia, hypotension is seen. [Note: Magnesium sulfate is used in the treatment of preeclampsia, a hypertensive disorder of pregnancy.]

C. Sodium, chloride, and potassium

These macrominerals are considered together because they play important roles in several physiologic processes. For example, they maintain water balance, osmotic equilibrium, acid–base balance (pH), and the electrical gradients across cell membranes (membrane potential) that are essential for the functioning of neurons and myocytes.[1]

1. **Sodium and chloride:** Na^+ and Cl^- are primarily extracellular electrolytes. They are readily absorbed from foods containing salt (NaCl), much of which comes from processed foods. [Note: Na^+ is required for the intestinal absorption (and renal reabsorption) of glucose and galactose (see Chapter 7) and free amino acids by Na^+-linked transporters. Cl^- is used to form hydrochloric acid required for digestion (see Chapter 19, III. A. 1.).] In the United States, the average daily consumption of NaCl is one and a half to three times the Adequate Intake (AI) of 3.8 mg/day (UL = 5.8 g/day). Dietary deficiency is rare.

   a. **Hypertension:** Na^+ intake is related to blood pressure (BP). Ingestion of Na^+ stimulates thirst centers in the brain and secretion of antidiuretic hormone from the pituitary, leading to water retention. This results in an increase in plasma volume and, consequently, an increase in BP. Chronic hypertension can damage the heart, kidneys, and blood vessels. Modest reductions in Na^+ intake have been shown to result in modest reductions in BP.

   b. **Hyper- and hyponatremia:** Hypernatremia, typically caused by excess water loss, and hyponatremia, typically caused by decreased ability to excrete water, can result in severe brain damage. Chronic hyponatremia increases Ca^{2+} excretion and can result in osteoporosis (low bone mass).

2. **Potassium:** In contrast to Na^+, K^+ is primarily an intracellular electrolyte. The concentration differential of Na^+ and K^+ across the cell membrane is maintained by the **Na^+/K^+-ATPase** (Fig. 29.4). In contrast to Na^+ and Cl^-, K^+ (like Mg^{2+}) is insufficiently consumed in Western diets because its primary sources, fruits and vegetables, are not consumed in adequate amounts. Increasing dietary

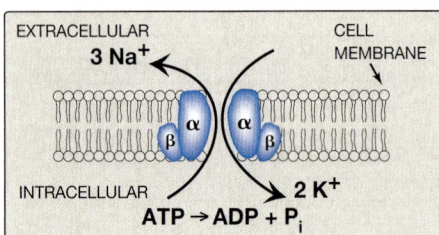

Figure 29.4
Na^+/K^+-ATPase. Na^+, sodium; K^+, potassium; ADP, adenosine diphosphate; P_i, phosphate.

---

[1]Note: These processes are discussed in *Lippincott® Illustrated Reviews: Physiology*, 3e.

K^+ decreases BP by increasing Na^+ excretion. There is a narrow range for normal serum K^+ levels, and even modest changes (up or down, resulting in hyper- or hypokalemia) can result in cardiac arrhythmias and skeletal muscle weakness. Hypokalemia can result from laxative overuse. No UL for K^+ has been established.

III. TRACE MINERALS (MICROMINERALS)

The trace minerals include copper (Cu), iron (Fe), manganese (Mn), and zinc (Zn). They are required by adults in amounts between 1 and 100 mg/day.

A. Copper

Cu is a key component of several enzymes that play critical functions in the body (Fig. 29.5). These include **ferroxidases** such as the **ceruloplasmin** and **hephaestin** involved in the oxidation of ferrous iron (Fe^{2+}) to the ferric form (Fe^{3+}) that is required for its intracellular storage or transport through blood (see Section III. B. 1.). Meat, shellfish, nuts, and whole grains are good dietary sources of Cu. Dietary deficiency is uncommon. If a deficiency does develop, anemia may be seen because of the effect on Fe metabolism. Toxicity from dietary sources is rare (UL = 10 mg/day). Menkes syndrome and Wilson disease are genetic causes of Cu deficiency and Cu overload, respectively.

1. **Menkes syndrome:** In Menkes syndrome, a rare X-linked disorder seen in ~1:140,000 X,Y males, efflux of dietary Cu out of intestinal enterocytes into the circulation by a Cu-transporting ATPase (ATP7A) is impaired. This results in systemic Cu deficiency. Consequently, urinary and serum-free (unbound) Cu are low, as is the concentration of **ceruloplasmin**, which carries more than 90% of the Cu in the circulation (Fig. 29.6). Progressive neurologic degeneration and connective tissue disorders are seen, as are changes to hair (to more tightly coiled hair). Parenteral administration of Cu has been used as a treatment with varying success. The mildest form of Menkes syndrome is called occipital horn syndrome.

2. **Wilson disease:** In Wilson disease, an autosomal-recessive disorder affecting 1:35,000 live births, efflux of Cu from the liver by **ATP7B** is impaired. Cu accumulates in the liver; leaks into the blood; and is deposited in the brain, eyes, kidneys, and skin. In contrast to Menkes syndrome, urinary and serum-free Cu are high (see Fig. 29.6). Hepatic dysfunction and neurologic and psychiatric symptoms are seen. Kayser–Fleischer rings (corneal deposits of Cu) may be present (Fig. 29.7). Life-long use of Cu-chelating agents, such as penicillamine, is the treatment. However, zinc therapy is an equally effective and safer alternative, offering low toxicity and fewer side effects at a lower cost.

The bioavailability (percent of the amount ingested that is able to be absorbed) of a mineral can be influenced by other minerals. For example, excess Zn decreases the absorption of Cu, and Cu is needed for the absorption of Fe.

Cu-REQUIRING ENZYME	FUNCTION
Cytochrome c oxidase	Transfers electrons from cytochrome c to oxygen in the ETC
Dopamine β-hydroxylase	Hydroxylates dopamine to norepinephrine
Ferroxidases	Oxidize iron
Lysyl oxidase	Forms cross-links in collagen and elastin
Tyrosinase	Synthesizes melanin
Superoxide dismutase (nonmitochondrial form; also requires zinc)	Converts superoxide to hydrogen peroxide

Figure 29.5
Examples of enzymes that require copper (Cu). ETC, electron transport chain.

VARIABLE	MENKES	WILSON
Whole-body Cu	Low	High
Free serum Cu	Low	High
Urinary Cu	Low	High
Inheritance	X-linked	AR
Cu-transporting ATPase affected	ATP7A	ATP7B

Figure 29.6
Comparison of Menkes syndrome and Wilson disease. Cu, copper; AR, autosomal recessive.

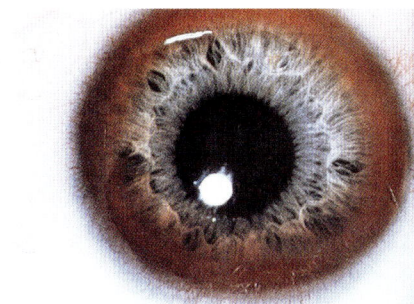

Figure 29.7
Kaiser–Fleischer rings.

B. Iron

The adult body typically contains 3 to 4 g of Fe. It is a component of many proteins, both catalytic (e.g., **hydroxylases** such as **prolyl hydroxylase** and noncatalytic). Iron can be linked to sulfur (S) as seen in the Fe-S proteins of the electron transport chain, or it can be part of the heme prosthetic group in proteins such as hemoglobin (~70% of all Fe), myoglobin, and the cytochromes. Free ionic Fe is toxic because it can cause production of the hydroxyl radical, a reactive oxygen species (ROS). Dietary Fe is available as Fe^{2+} in heme (animal sources) and Fe^{3+} in nonheme sources (plants). Heme iron is less abundant, but it is better absorbed. Meat, poultry, some shellfish, iron-fortified foods such as breakfast cereals and grains, lentils, and green leafy vegetables are good dietary sources of Fe. About 10% of ingested Fe is absorbed. This amount, ~1 to 2 mg/day, is sufficient to replace Fe lost from the body primarily by the sloughing of cells.

1. **Absorption, storage, and transport:** Intestinal uptake of heme is by a heme carrier protein (Fig. 29.8). Within the enterocytes, **heme oxygenase** releases Fe^{2+} from heme (see Chapter 21, II. D. 1.). Nonheme Fe is taken up via the apical membrane protein divalent metal ion transporter-1 (DMT-1). Vitamin C enhances absorption of nonheme Fe because it is the coenzyme for **duodenal cytochrome b (Dcytb)**, a **ferrireductase** that reduces Fe^{3+} to Fe^{2+}. Absorbed Fe^{2+} from heme and nonheme sources has two possible fates: it can be (1) oxidized to Fe^{3+} and stored by the intracellular protein ferritin (up to 4,500 Fe^{3+}/ferritin) or (2) transported out of the enterocyte by the basolateral membrane protein ferroportin, oxidized by the Cu-containing membrane protein **hephaestin**, as shown in Figure 29.8. In the plasma, ceruloplasmin, a Cu-binding protein with ferroxidase promote Fe^{2+} oxidation and taken up by the plasma transport protein transferrin (2 Fe^{3+}/transferrin).

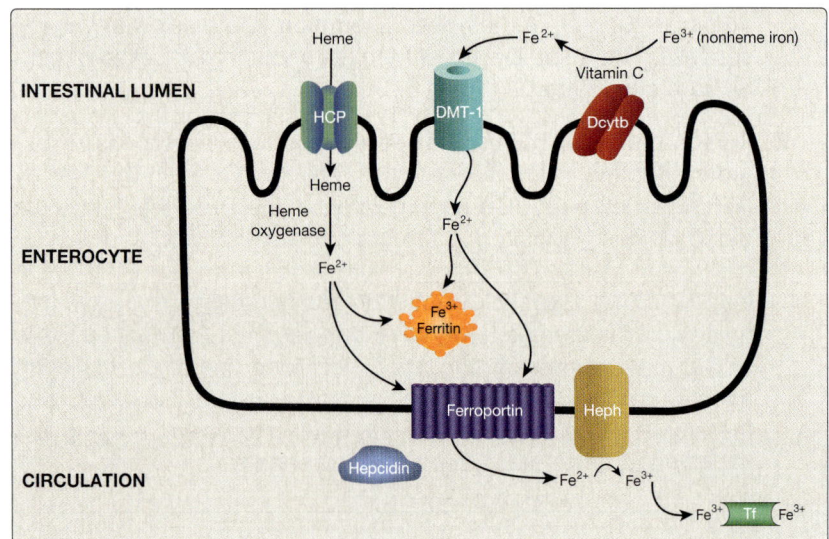

Figure 29.8
Absorption, storage, and transport of dietary iron (Fe). HCP, heme carrier protein; DMT, divalent metal ion transporter; Dcytb, duodenal cytochrome b (a ferrireductase); Heph, hephaestin; Tf, transferrin.

In normal individuals, transferrin (Tf) is about one-third saturated with Fe^{3+}. Ferroportin, the only known exporter of Fe from cells to the blood in humans, is regulated by the hepatic peptide hepcidin that induces internalization and lysosomal degradation of ferroportin. Therefore, hepcidin is the central molecule in Fe homeostasis. Transcription of hepcidin is suppressed when Fe is deficient.

2. **Recycling:** Macrophages phagocytose old and/or damaged red blood cells (RBCs), freeing heme Fe that is sent out of the cells via ferroportin, oxidized by **ceruloplasmin**, and transported by Tf as described above. This recycled Fe meets ~90% of our daily need, which is predominantly for erythropoiesis.

3. **Uptake:** Tf-bound Fe^{3+} from enterocytes and macrophages binds to receptors (TfRs) on erythroblasts and other Fe-requiring cells and is taken up by receptor-mediated endocytosis. The Fe^{3+} is released from Tf for use (or stored on ferritin), and the TfR (and Tf) is recycled in a process similar to the receptor-mediated endocytosis seen with low-density lipoprotein particles (see Chapter 18, VI. D. 1.). Regulation of the translation of the messenger RNA for ferritin and the TfR by iron regulatory proteins and iron-responsive elements is discussed in Chapter 33.

4. **Deficiency:** Fe deficiency can result in microcytic, hypochromic anemia, the most common anemia in the United States, as a result of decreased hemoglobin synthesis and, consequently, decreased RBC size (Fig. 29.9). Treatment is the administration of Fe in various ways depending on the severity of anemia.

5. **Excess:** Iron overload can arise often from hereditary hemochromatosis, but also from frequent blood transfusions, overuse of iron supplements, chronic liver diseases, and accidental ingestion. Acute Fe poisoning is the most common cause of poisoning deaths of children younger than age 6 years (UL = 40 mg/day for children, 45 mg/day for adults). Treatment is use of an Fe chelator. Hereditary hemochromatosis is an autosomal-recessive disorder of Fe overload found primarily in those of Northern European ancestry. It is most commonly caused by mutations to the *HFE* (high Fe) gene. Hyperpigmentation, presenting as a bronze coloring of the skin, can result from iron deposits. Additionally, these iron deposits may cause damage to the pancreas, potentially leading to diabetes, and can also affect the liver, a major storage site for iron. Damage to the heart may also occur due to the excessive iron accumulation. In hereditary hemochromatosis, serum Fe and Tf saturation are elevated. Treatment is phlebotomy or use of Fe chelators.

Fe overload is seen with mutations to proteins of Fe metabolism that result in inappropriately low levels of hepcidin. It can result in hemosiderosis (the deposition of hemosiderin, an intracellular, insoluble storage form of Fe).

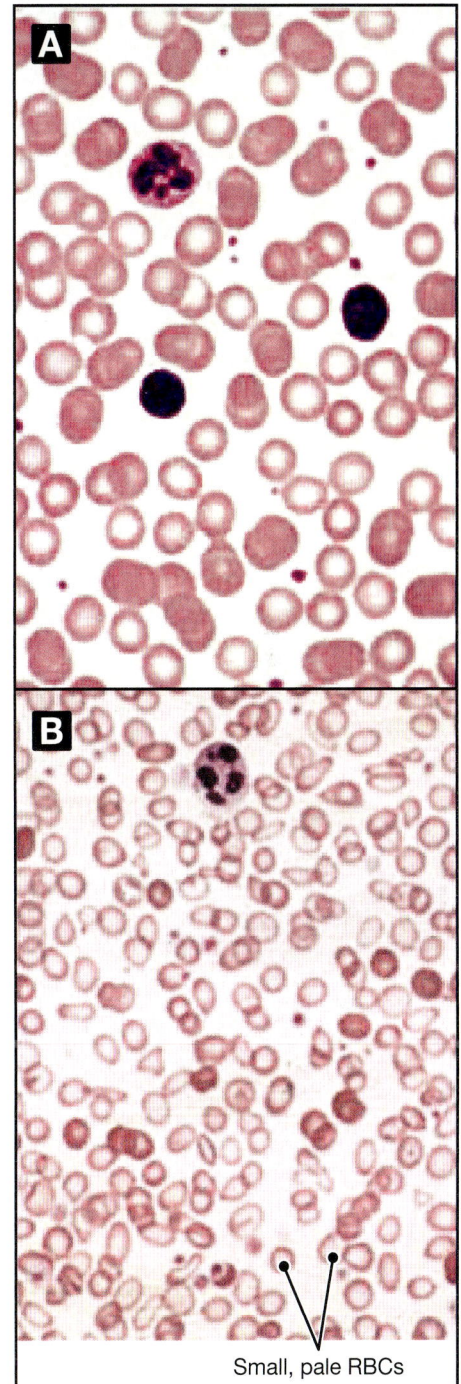

Figure 29.9
A. Normal red blood cells (RBCs).
B. Small (microcytic), pale (hypochromic) RBCs in microcytic anemia.

C. Manganese

Mn is important for the function of several enzymes (Fig. 29.10). Whole grains, legumes (e.g., beans and peas), nuts, and tea (especially green tea) are good sources of the mineral. Consequently, Mn

Mn-REQUIRING ENZYME	FUNCTION
Arginase-I	Hydrolyzes arginine to urea plus ornithine in the urea cycle
Glycosyltransferases	Transfer sugars in proteoglycan synthesis
Pyruvate carboxylase	Carboxylates pyruvate to OAA in gluconeogenesis
Superoxide dismutase (mitochondrial form)	Converts superoxide to hydrogen peroxide

Figure 29.10
Examples of enzymes that require manganese (Mn). OAA, oxaloacetate.

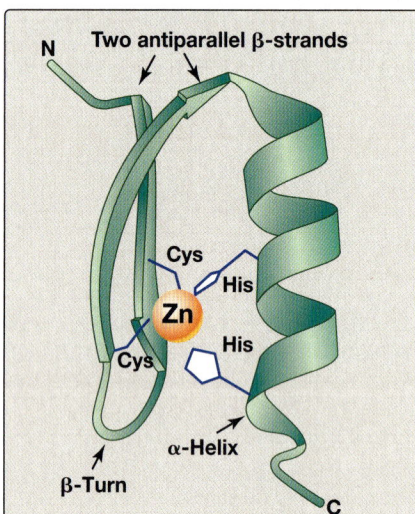

Figure 29.11
Zinc (Zn) finger is a common motif in proteins that bind DNA. Cys, cysteine; His, histidine.

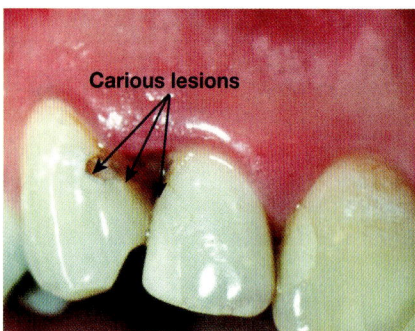

Figure 29.12
Dental caries (cavities).

deficiency in humans is rare. Toxicity from foods and/or supplements is also rare (UL = 11 mg/day for adults).

D. Zinc

Zn plays important structural and catalytic functions in the body. Zinc fingers are supersecondary structures in proteins (e.g., transcription factors) that bind to DNA and regulate gene expression (Fig. 29.11). Hundreds of enzymes require Zn for activity. Examples include **alcohol dehydrogenase**, which oxidizes ethanol to acetaldehyde (see Chapter 23, IV. C. 4.); **carbonic anhydrase**, which is important in the bicarbonate buffer system (see Chapter 3); (δ-Aminolevulinic dehydratase, aka porphobilinogen synthase) (porphobilinogen synthase) of heme synthesis, which is inhibited by lead (lead replaces the zinc; see Chapter 21, II. B. 2.); and the nonmitochondrial isoform of **superoxide dismutase** (**SOD**), which also requires Cu (see Fig. 29.5). Dietary sources of Zn include meat, fish, eggs, and dairy products. Phytates (phosphate storage molecules in plants such as grains, seeds, legumes, some nuts) irreversibly bind Zn in the intestine, decreasing its absorption, and can result in deficiency. [Note: Phytates may also bind Ca^{2+} and nonheme Fe.] Several drugs (e.g., penicillamine) chelate metals, and their use may cause Zn deficiency. Severe Zn deficiency is seen in acrodermatitis enteropathica, an autosomal-recessive disorder that arises due to a defect in the intestinal transporter for Zn. Symptoms include rashes around the orifices and in the limbs, slowed growth and development, diarrhea, and immune deficiencies. Vision problems may also occur because Zn is needed in the metabolism of vitamin A.

> Eukaryotic cells infected with bacteria can restrict availability of the essential micronutrients Fe, Mn, and Zn to the pathogens. This decreases the intracellular survival of the pathogen and is known as "nutritional immunity."

E. Other microminerals

Chromium (Cr) and fluorine (F) also play roles in the body. Cr potentiates the action of insulin by an unknown mechanism. It is found in fruits, vegetables, dairy products, and meat. F (as fluoride [F^-]) is added to water in many parts of the world to reduce the incidence of dental caries (Fig. 29.12). F^- replaces the hydroxyl group of hydroxylapatite, forming fluorapatite that is more resistant to the enamel-dissolving acid produced by mouth bacteria.

IV. ULTRATRACE MINERALS

The ultratrace minerals include iodine (I), selenium (Se), and molybdenum (Mo). They are required by adults in amounts less than 1 mg/day.

A. Iodine

Iodine is utilized in the synthesis of the thyroid hormones triiodothyronine (T_3) and thyroxine (T_4) that are required for development, growth, and metabolism. Circulating iodide (I^-) is taken up ("trapped")

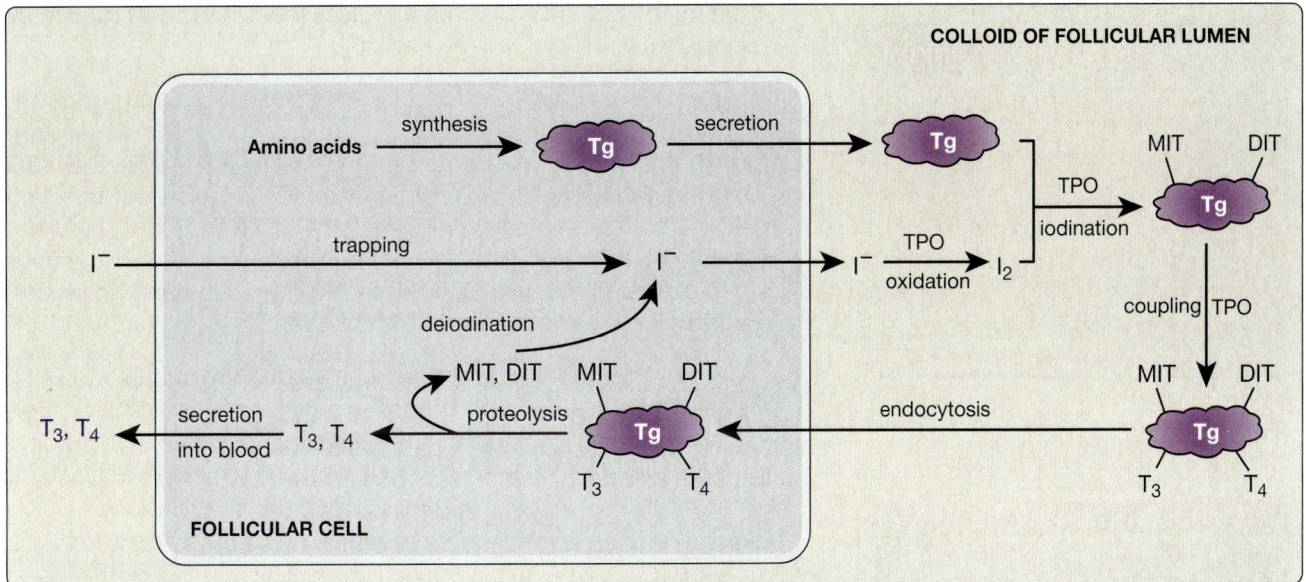

Figure 29.13
Thyroid hormone synthesis. Tg, thyroglobulin; I⁻, iodide; I_2, iodine; TPO, thyroperoxidase; MIT, monoiodinated tyrosine; DIT, diiodinated tyrosine; T_3, triiodothyronine; T_4, thyroxine.

and concentrated in the epithelial follicular cells of the thyroid gland. It then is sent into the colloid of the follicular lumen where it is oxidized to iodine (I_2) by **thyroperoxidase** (**TPO**), as shown in Figure 29.13. TPO then uses I_2 to iodinate selected tyrosine residues in thyroglobulin (Tg), forming monoiodinated tyrosine (MIT) and diiodinated tyrosine (DIT), as shown in Figure 29.14. [Note: Tg is synthesized and secreted into colloid by follicular cells.] The coupling of two DIT on Tg gives T_4, whereas coupling one MIT and one DIT gives T_3. The iodinated Tg is endocytosed and stored in follicular cells until needed, at which time it is proteolytically digested to release T_3 and T_4, which are secreted into the circulation (see Fig. 29.13). Under normal conditions, ~90% of secreted thyroid hormone is T_4 that is carried by transthyretin. In target tissues (e.g., the liver and developing brain), T_4 is converted to T_3 (the more active form) by Selenium (Se)-containing **deiodinases**. T_3 binds to a nuclear receptor that binds DNA at thyroid response elements and functions as a transcription factor. Thyroid hormone production is controlled by thyrotropin (thyroid-stimulating hormone [TSH]) from the anterior pituitary. TSH secretion is itself controlled by thyrotropin-releasing hormone (TRH) from the hypothalamus.

1. **Hypothyroidism:** Underingestion of iodine (I) can result in goiter, an enlargement of the thyroid in response to excessive stimulation by TSH, as shown in Figure 29.15. More severe deficiency results in hypothyroidism that is characterized by fatigue, weight gain, decreased thermogenesis, and decreased metabolic rate. If hormone deficiency occurs during fetal and infant development (congenital hypothyroidism), irreversible intellectual disability (formerly called "cretinism"), hearing loss, spasticity, and short stature can result. In the United States, dairy products, seafood, and meat are the primary sources of I. The use of iodized salt has greatly reduced dietary I deficiency. [Note: Autoimmune

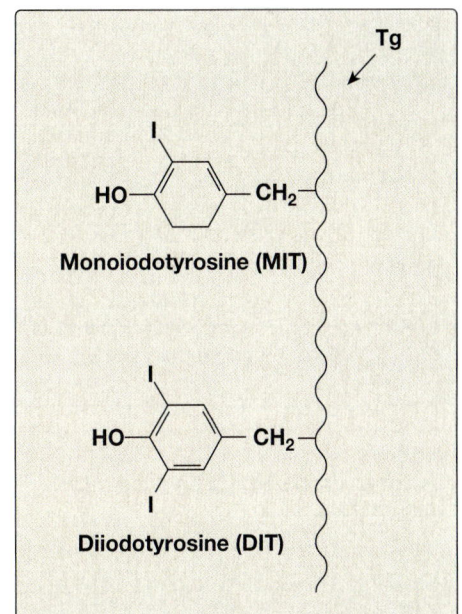

Figure 29.14
Iodination of thyroglobulin (Tg) with production of MIT and DIT.

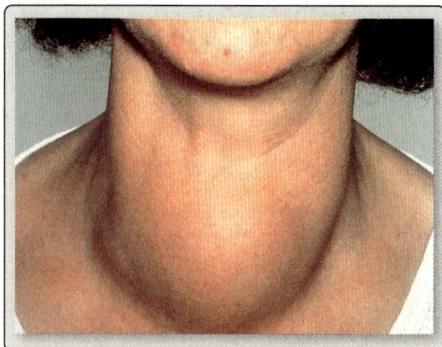

Figure 29.15
Goiter.

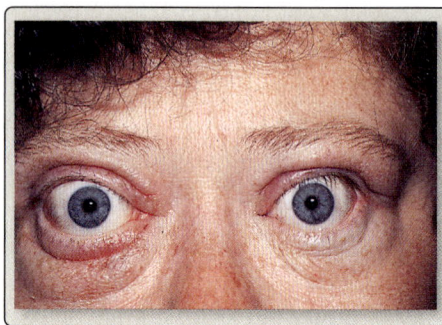

Figure 29.16
Exophthalmos.

Mo-REQUIRING ENZYME	FUNCTION
Aldehyde oxidase	Metabolizes drugs
Sulfite oxidase	Converts sulfite to sulfate in metabolism of the sulfur-containing amino acids methionine and cysteine
Xanthine oxidase	Oxidizes hypoxanthine to xanthine and xanthine to uric acid in purine degradation

Figure 29.17
Enzymes (oxidases) that require molybdenum (Mo).

destruction of TPO is a cause of Hashimoto thyroiditis (a primary hypothyroidism).]

2. **Hyperthyroidism:** This condition is the result of overproduction of thyroid hormone. Although it can be caused by overingestion of I-containing supplements (UL = 1.1 g/day for adults), the most common cause of hyperthyroidism is Graves disease, in which an antibody that mimics the effect of TSH is produced, resulting in dysregulated production of thyroid hormone. This can cause nervousness, weight loss, increased perspiration and heart rate, protruding eyes (exophthalmos; Fig. 29.16), and goiter.

3. **Goiter:** This condition denotes an enlarged thyroid gland, which does not inherently imply thyroid dysfunction. It signals an underlying condition causing abnormal growth of the thyroid. Goiter may develop in cases of hyperthyroidism, hypothyroidism, or even when hormone levels are normal (euthyroidism).

B. Selenium

Se is present in ~25 human proteins (selenoproteins) as a constituent of the amino acid selenocysteine, which is derived from serine (see Chapter 20, V. D. 1.). Selenoproteins include **glutathione peroxidase** that oxidizes glutathione in the reduction of hydrogen peroxide, a ROS, to water (see Chapter 13); **thioredoxin reductase** that reduces thioredoxin, a coenzyme of **ribonucleotide reductase** (see Chapter 22, IV. A.); and **deiodinases** that remove iodine from thyroid hormones. Meat, dairy products, and grains are important dietary sources. Keshan disease, first identified in China, is a cardiomyopathy caused by eating foods produced from Se-deficient soil. Selenosis describes toxicity caused by overingestion of supplements causes brittle nails and hair. Cutaneous and neurologic effects may also be seen (UL = 400 µg in adults).

C. Molybdenum

Molybdenum (Mo) functions as a cofactor for a small number of mammalian **oxidases** (Fig. 29.17). Legumes are important dietary sources. No dietary deficiency syndromes are known. Mo has low toxicity in humans (UL = 2 mg/day in adults).

Cobalt (Co), an ultratrace mineral, is a component of vitamin B_{12} (cobalamin; see Chapter 28, III.), which is required as methylcobalamin in the remethylation of homocysteine to methionine (see Chapter 20, III. E.) or adenosylcobalamin in the isomerization of methylmalonyl coenzyme A (CoA) to succinyl CoA. No Recommended Dietary Allowance or Daily Reference Intake (see Chapter 27, II. B.) has been established for Co.

V. CHAPTER SUMMARY

The minerals are summarized in Figure 29.18.

CLASSIFICATION	FUNCTION(S)	NOTES
Macrominerals: >100 mg/day for adults		
Calcium (Ca^{2+})	Component of hydroxylapatite ($Ca_5[PO_4]_3OH$) of bone and teeth, muscle contraction, signaling, blood clotting	Hypocalcemia with PTH or Vitamin D deficiency impairs calcium absorption, leading to low serum calcium levels. Hypercalcemia with increased PTH can cause excessive calcium release from bones, leading to a higher risk of kidney stones.
Chloride (Cl^-)	Fluid balance (along with Na, K), digestion	Dietary deficiency rare; overingested as NaCl
Magnesium (Mg^{2+})	Component (minor) of bone; regulates enzyme activity (binds substrate or enzyme)	Average U.S. intake is below recommended level; hyperexcitability and arrhythmias seen with hypomagnesemia; hypotension with hypermagnesemia
Phosphorus (P^+)	Component of hydroxylapatite of bone and teeth, energy storage, membrane structure, regulation	Dietary deficiency rare; hypophosphatemia with muscle weakness in refeeding syndrome, increased PTH, and use of aluminum-containing antacids; hyperphosphatemia with metastatic calcification in PTH deficiency
Potassium (K)	Membrane potential, blood pressure	Average U.S. intake is below recommended level; modest changes up or down in serum level result in arrhythmias and muscle weakness
Sodium (Na)	Membrane potential; blood volume and pressure; uptake of glucose, galactose, and amino acids	Dietary deficiency rare; overingested as NaCl; hyponatremia seen with excess water loss; hypernatremia with water retention
Microminerals (Trace): 1–100 mg/day		
Chromium (Cr)	Potentiates insulin action	Mechanism unknown
Copper (Cu)	Enzyme cofactor	Dietary deficiency rare; Menkes (genetic systemic Cu deficiency) and Wilson (genetic systemic Cu overload)
Fluorine (as fluoride [F^-])	Increases resistance to enamel-dissolving acid of mouth bacteria	Deficiency results in dental caries
Iron (Fe)	Enzyme cofactor, oxygen binding, Fe-S proteins	Dietary deficiency results in microcytic anemia; hereditary hemochromatosis, a genetic disease of Fe overload, with "bronze diabetes" (hyperglycemia, hyperpigmentation)
Manganese (Mn)	Enzyme cofactor	Dietary deficiency rare
Zinc (Zn)	Enzyme cofactor, protein structure (Zn finger)	Phytates and some drugs decrease absorption; severe deficiency (acrodermatitis enteropathica) with transporter defect
Microminerals (Ultratrace): <1 mg/day		
Iodine (I)	Thyroid hormone (T_3, T_4) synthesis	Underingestion causes goiter, hypothyroidism with fatigue, weight gain, and decreased metabolic rate; neurologic damage in congenital deficiency; hyperthyroidism (overproduction of T_3, T_4) in Graves disease
Molybdenum (Mo)	Enzyme cofactor	Dietary deficiency unknown
Selenium (Se)	Found (as selenocysteine) in selenoproteins	Dietary deficiency rare (Keshan disease with Se-deficient soil), toxicity from supplements

Figure 29.18
Summary of minerals. PTH, parathyroid hormone; Cl^-, chloride; S, sulfur; T_3, triiodothyronine; T_4, thyroxine.

Study Questions

For Questions 29.1–29.7, match the mineral to the most appropriate description.

A. Calcium
B. Chloride
C. Copper
D. Iodine
E. Iron
F. Magnesium
G. Manganese
H. Molybdenum
I. Phosphorus
J. Potassium
K. Selenium
L. Sodium
M. Zinc

29.1. Hereditary hemochromatosis develops due to excess deposition of which of the following minerals?

29.2. Which mineral is the major extracellular anion?

29.3. A decrease of which mineral is seen in refeeding syndrome and with overuse of aluminum-containing antacids?

29.4. Which mineral is a constituent of some amino acids found in proteins involved in antioxidant defense, thyroid hormone metabolism, and redox reactions?

29.5. Which mineral is required for the formation of a super-secondary protein structure that allows binding to DNA? (Its deficiency can result in dermatitis.)

29.6. Deficiency of which mineral can cause bone pain, tetany (intermittent muscle spasms), paresthesia (a "pins and needles" sensation), and an increased tendency to bleed?

29.7. Deficiency of which mineral can result in goiter and a decreased metabolic rate?

Correct answers = E, B, I, K, M, A, D. Hereditary hemochromatosis develops due to the excess deposition of iron in the tissues, which occur in hereditary hemochromatosis. The iron deposits can lead to a bronze coloring of the skin and can also cause damage to the pancreas, potentially leading to diabetes. Chloride is the major extracellular anion. [Note: Sodium is the major extracellular cation, potassium is the major intracellular cation, and phosphate is the major intracellular anion. The concentration differential across the membrane is maintained by active transport.] Carbohydrate metabolism involves the generation of phosphorylated intermediates. Refeeding severely malnourished individuals traps phosphate and results in hypophosphatemia. Muscle weakness is a common symptom. Selenocysteine, an amino acid formed from serine and selenium, is found in proteins (selenoproteins) such as glutathione peroxidase, deiodinases, and thioredoxin reductase. Zinc fingers are a type of structural motif found in proteins (e.g., transcription factors) that bind to DNA. Severe deficiency of zinc as a result of mutations to its intestinal transporter can result in acrodermatitis enteropathica, which is characterized by dermatitis, diarrhea, and alopecia. Calcium is required for bone mineralization, muscle contraction, nerve conduction, and blood clotting. Its deficiency will affect all of these processes. Thyroid hormones are iodinated tyrosines released by proteolytic digestion of thyroglobulin. Underconsumption of iodine causes enlargement of the thyroid in an attempt to increase hormone synthesis. Goiter can also result if too much hormone is made, as in Graves disease, or if too little is made, as in Hashimoto disease. Both are autoimmune diseases. Thyroid hormone increases the resting metabolic rate.

29.8. A 6-month-old infant presents with irritability and muscular spasms. Blood tests reveal hypocalcemia and decreased levels of parathyroid hormone (PTH). Which of the following laboratory findings most likely increased in this infant?

A. Bone resorption
B. Calcium reabsorption in the kidney
C. Serum calcitriol
D. Serum phosphate

Correct answer = D. PTH increases bone resorption (demineralization) resulting in the release of calcium and phosphate. It also increases the renal reabsorption of calcium, because PTH activates the renal hydroxylase that converts calcidiol to calcitriol. PTH also increases the renal excretion of phosphate. Calcium plays a key role in muscle contraction and nerve function. When calcium levels are low, nerve cells become more easily excited and can fire impulses excessively. This hyperexcitability leads to muscle spasms, which can manifest as twitching, cramps, or tetany.

29.9. A 28-year-old man is seen for evaluation of recent, severe, upper-right-quadrant pain. He also reports some difficulty with fine motor tasks. No jaundice is observed on physical examination. Laboratory testing is remarkable for elevated liver function tests (serum aspartate and alanine aminotransferases) and elevated urinary calcium and phosphate. Ophthalmology consult reveals Kayser–Fleischer rings in the cornea. The patient is started on penicillamine and zinc. Which of the following is the most likely diagnosis?

A. Graves disease
B. Hereditary hemochromatosis
C. Hypercalcemia
D. Hyperphosphatemia
E. Keshan disease
F. Menkes syndrome
G. Selenosis
H. Wilson disease

Correct answer = H. The patient has Wilson disease, an autosomal-recessive disorder that decreases copper efflux from the liver because of mutations to the hepatic copper transport protein ATP7B. Some copper leaks into the blood and is deposited in the brain, eyes, kidney, and skin. This results in liver and kidney damage, neurologic effects, and corneal changes caused by the excess copper. Administration of the metal chelator penicillamine is the treatment. [Note: Because zinc is also chelated, supplementation with zinc is common.] Graves disease results in hyperthyroidism. Hereditary hemochromatosis is a disorder of iron overload. Keshan disease is the result of selenium deficiency, whereas selenosis is caused by selenium excess. Menkes syndrome is the result of a systemic deficiency in copper as a result of mutations to ATP7A, an intestinal copper transport protein.

29.10. A 52-year-old woman is seen because of unplanned changes in the pigmentation of her skin that give her a tanned appearance. Physical examination shows hyperpigmentation, hepatomegaly, and mild scleral icterus. Laboratory tests are remarkable for elevated serum transaminases (liver function tests) and elevated fasting blood glucose. Results of other tests are pending. Which of the following is the most likely diagnosis?

A. Graves disease
B. Hereditary hemochromatosis
C. Hypercalcemia
D. Hyperphosphatemia
E. Keshan disease
F. Menkes syndrome
G. Selenosis
H. Wilson disease

Correct answer = B. The patient has hereditary hemochromatosis, a disease of iron overload that results from inappropriately low levels of hepcidin caused primarily by mutations to the *HFE* (high iron) gene. Hepcidin regulates ferroportin, the only known iron export protein in humans, by increasing its degradation. The increase in iron with hepcidin deficiency causes hyperpigmentation and hyperglycemia. Phlebotomy or use of iron chelators is the treatment. [Note: Pending lab tests would show an increase in serum iron and transferrin saturation.]

DNA Structure, Replication, and Repair

30

I. OVERVIEW

Nucleic acids are required for the storage and expression of genetic information. There are two chemically distinct types of nucleic acids: deoxyribonucleic acid (DNA) and ribonucleic acid ([RNA] see Chapter 31). DNA, the repository of genetic information (the genome), is present not only in chromosomes in the nucleus of eukaryotic organisms, but also in mitochondria and the chloroplasts of plants. Prokaryotic cells, which lack nuclei, have a single chromosome but may also contain nonchromosomal DNA in the form of plasmids. The genetic information found in DNA is copied and transmitted to daughter cells through DNA replication. Each cell type is specialized, expressing only those genes that are required for it to perform its role in maintaining the organism. The DNA contained in a fertilized egg encodes the information that directs the development of an organism. This development may involve the production of billions of cells. Therefore, DNA must be able not only to replicate precisely each time a cell divides, but also to have the information that it contains be selectively expressed and processed for production of the set of functional RNA and protein products needed for cellular function. Transcription (RNA synthesis) is the first stage in the expression of genetic information (see Chapter 31). Next, the code contained in the nucleotide sequence of messenger RNA molecules is translated (protein synthesis; see Chapter 32), thus completing gene expression. The regulation of gene expression is discussed in Chapter 33.

> The flow of information from DNA to RNA to protein is termed the "central dogma" of molecular biology (Fig. 30.1). It is descriptive of all organisms, with the exception of some viruses that have RNA as the repository of their genetic information.

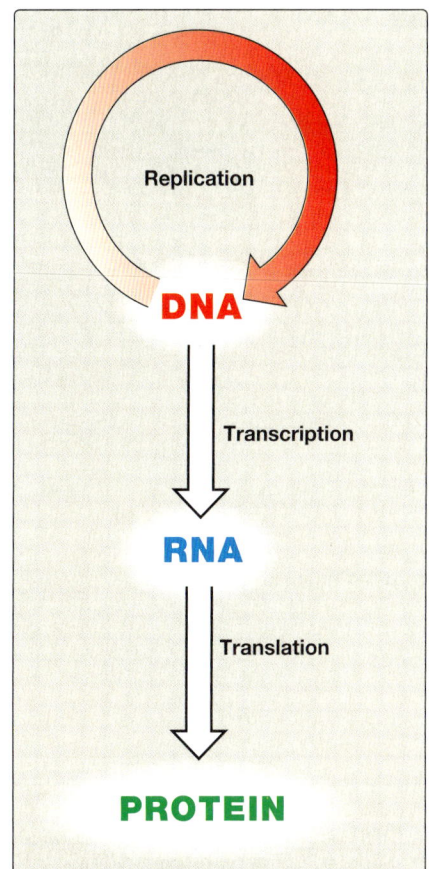

Figure 30.1
"Central dogma" of molecular biology.

II. DNA STRUCTURE

DNA is a polymer of deoxyribonucleoside monophosphates (dNMP), also called nucleotides, covalently linked by 3′-to-5′ phosphodiester bonds. With the exception of a few viruses that contain single-stranded DNA (ssDNA), DNA exists as a double-stranded molecule (dsDNA), in which the two strands wind around each other, forming a double helix. [Note: The sequence of the linked dNMP is primary structure, whereas the double helix is secondary structure.] In eukaryotic cells, DNA is found associated with various types of proteins (known collectively as nucleoprotein) present in the nucleus, whereas in prokaryotes, the protein–DNA complex is present in a non–membrane-bound region known as the nucleoid.

A. 3′-to-5′ Phosphodiester bonds

Phosphodiester bonds join the 3′-hydroxyl group of the deoxyribose of one nucleotide to the 5′-hydroxyl group of the deoxyribose of an adjacent nucleotide through a phosphoryl group (Fig. 30.2). The resulting long, unbranched chain has polarity, with both a 5′ end (the end with the free phosphate) and a 3′ end (the end with the free hydroxyl) that are not attached to other nucleotides. By convention, the bases located along the resulting deoxyribose-phosphate

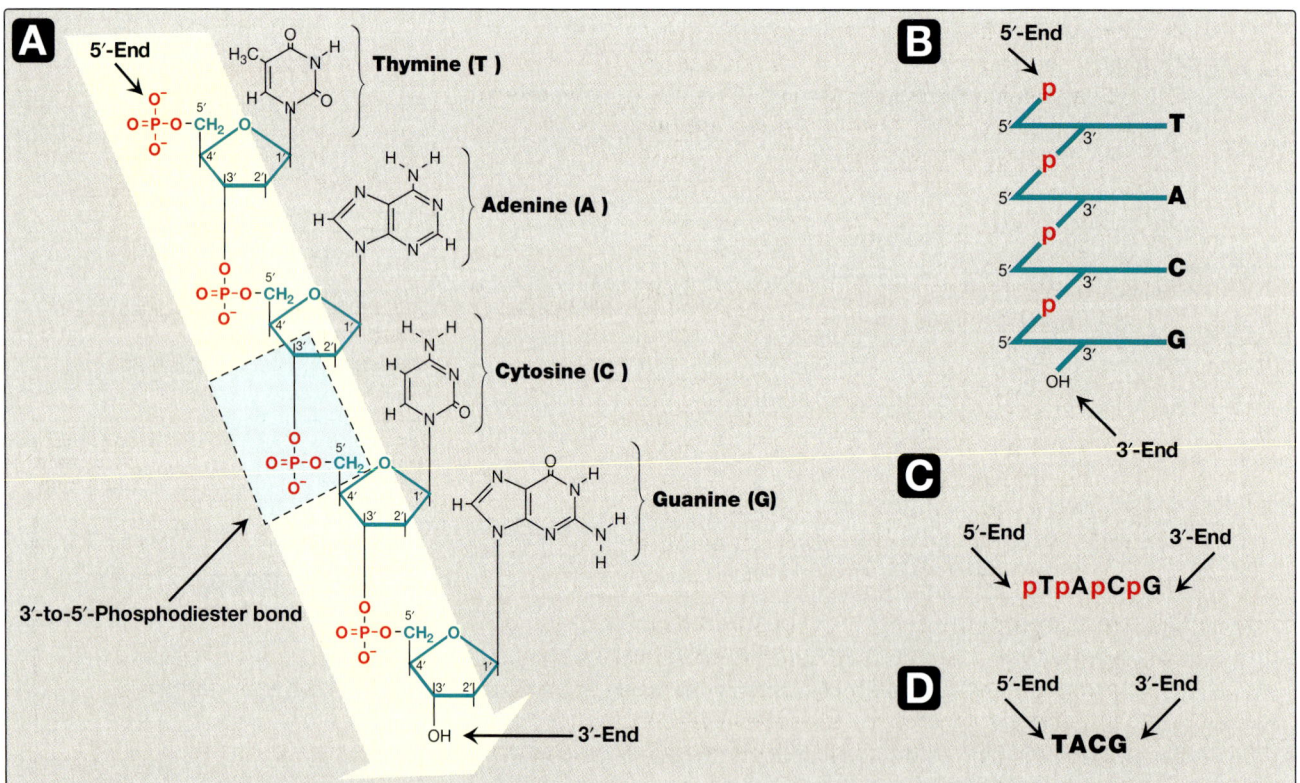

Figure 30.2
A. DNA with the nucleotide sequence shown written in the 5′→3′ direction. A 3′-to-5′ phosphodiester bond is shown highlighted in the *blue box*, and the deoxyribose-phosphate backbone is shaded in *yellow* with the *arrow* indicating the direction of DNA strand synthesis. **B.** DNA written in a more stylized form, emphasizing the deoxyribose-phosphate (p) backbone. **C.** A simpler representation of the nucleotide sequence. **D.** The simplest (and most common) representation. [Note: The nucleotide base sequence is assumed to be written in the 5′→3′ direction unless otherwise indicated.]

backbone are always written in sequence from the 5′ end of the chain to the 3′ end. For example, the sequence of the DNA shown in Figure 30.2A is written TACG and is read "thymine, adenine, cytosine, guanine." Phosphodiester linkages between nucleotides can be hydrolyzed enzymatically by a family of nucleases, deoxyribonucleases for DNA and ribonucleases for RNA, or cleaved hydrolytically by chemicals. [Note: Only RNA is cleaved by alkali.]

B. Double helix

In the double helix, the two chains are coiled around a common axis called the helical axis. The chains are paired in an antiparallel manner (i.e., the 5′ end of one strand is paired with the 3′ end of the other strand), as shown in Figure 30.3. In the DNA helix, the hydrophilic deoxyribose-phosphate backbone of each chain is on the outside of the molecule, whereas the hydrophobic bases are stacked inside. The overall structure resembles a twisted ladder. The spatial relationship between the two strands in the helix creates a major (wide) groove and a minor (narrow) groove. These grooves provide access for the binding of regulatory proteins to their specific recognition sequences along the DNA chain. [Note: Certain anticancer drugs, such as dactinomycin (actinomycin D), exert their cytotoxic effect by intercalating into the minor groove of the DNA double helix, thereby interfering with both DNA and RNA synthesis.]

1. **Base pairing:** The bases of one strand of DNA are paired with the bases of the second strand, so that an adenine (A) is always paired with a thymine (T), and a cytosine (C) is always paired with a guanine (G). [Note: The base pairs are perpendicular to the helical axis (see Fig. 30.3).] Therefore, one polynucleotide chain of the DNA double helix is always the complement of the other. Given the sequence of bases on one chain, the sequence of bases on the complementary chain can be determined (Fig. 30.4). [Note: The specific base-pairing in DNA leads to the Chargaff rule, which states that in any sample of dsDNA, the amount of A equals the amount of T, the amount of G equals the amount of C, and the total amount of purines (A + G) equals the total amount of pyrimidines (T + C).] The base pairs are held together by hydrogen bonds: two between A and T and three between G and C (Fig. 30.5). The base pairs are also stacked along the axis so that the planes of their rings are parallel. The hydrogen bonds of the base pairs, plus the hydrophobic interactions between the stacked bases, stabilize the structure of the double helix.

2. **DNA strand separation:** The two strands of the double helix separate when hydrogen bonds between the paired bases are disrupted. Disruption can occur in the laboratory if the pH of the DNA solution is altered so that the nucleotide bases ionize, or if the solution is heated. [Note: Covalent phosphodiester bonds are not broken by such treatment.] When DNA is heated, the temperature at which one-half of the helical structure is lost and single-stranded regions are formed is defined as the melting temperature (T_m). The loss of helical structure in DNA, called denaturation, can be monitored by measuring its absorbance at 260 nm. [Note: ssDNA has a higher relative absorbance at this wavelength than does dsDNA.] Because there are three hydrogen bonds between G and C but

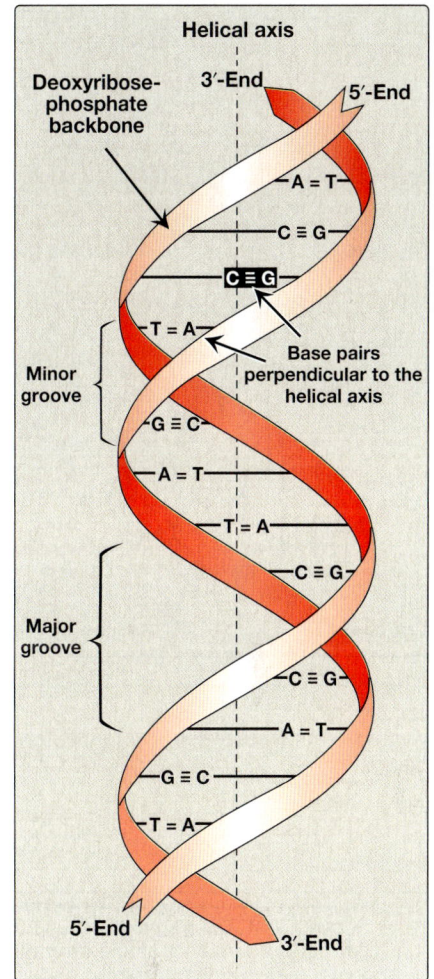

Figure 30.3
DNA double helix, illustrating some of its major structural features. T, thymine; A, adenine; C, cytosine; G, guanine.

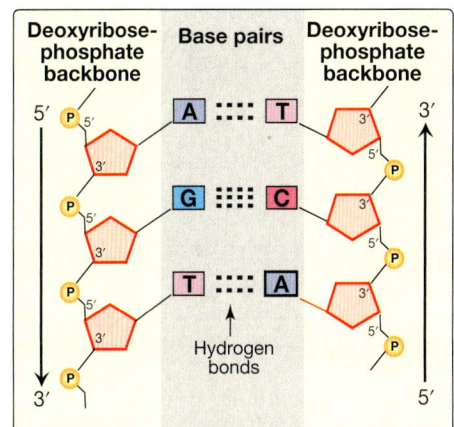

Figure 30.4
Two complementary DNA sequences. T, thymine; A, adenine; C, cytosine; G, guanine.

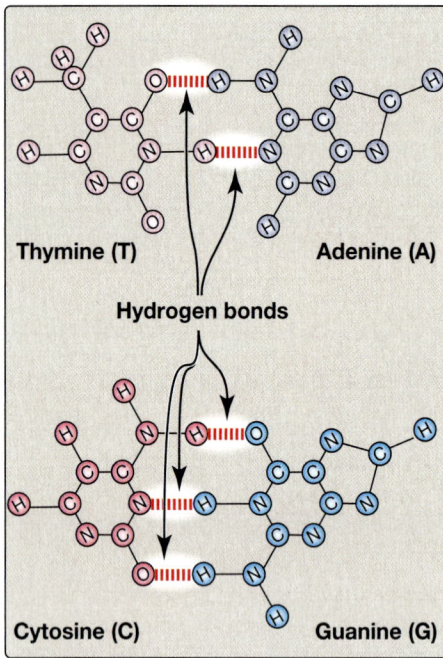

Figure 30.5
Hydrogen bonds between complementary bases.

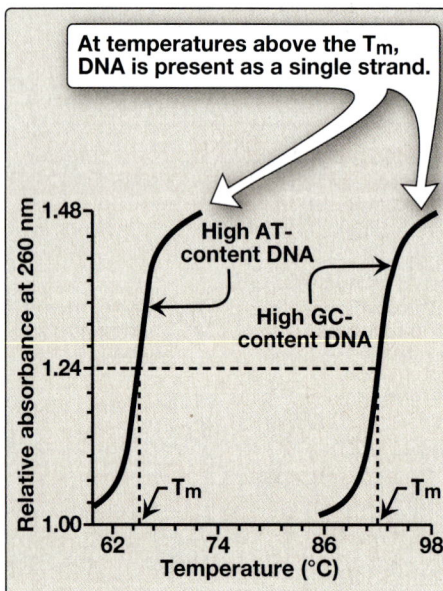

Figure 30.6
Melting temperatures (T_m) of DNA molecules with different nucleotide compositions.

only two between A and T, DNA that contains high concentrations of A and T denatures at a lower temperature than does G- and C-rich DNA (Fig. 30.6). If the DNA solution is cooled or titrated to neutral pH, complementary DNA strands can reform the double helix by the process called renaturation (or, reannealing). [Note: Separation of the two strands over short regions occurs during both DNA and RNA synthesis.]

3. **Structural forms:** There are three major structural forms of DNA: the B form (described by Watson and Crick in 1953), the A form, and the Z form. The B form is a right-handed helix with 10 bps per 360° turn (or twist) of the helix, and with the planes of the bases perpendicular to the helical axis. Chromosomal DNA is thought to consist primarily of B-DNA (Fig. 30.7 shows a space-filling model of B-DNA). The A form is produced by moderately dehydrating the B form. It is also a right-handed helix, but there are 11 bps per turn, and the planes of the base pairs are tilted 20° away from the perpendicular to the helical axis. The conformation found in DNA–RNA hybrids or RNA–RNA double-stranded regions is probably very close to the A form. Z-DNA is a left-handed helix that contains 12 bps per turn (see Fig. 30.7). [Note: The deoxyribose-phosphate backbone zigzags, hence, the name Z-DNA.] Stretches of Z-DNA can occur naturally in regions of DNA that have a sequence of alternating purines and pyrimidines (e.g., poly GC). Transitions between the B and Z helical forms of DNA may play a role in regulating gene expression.

C. Linear and circular DNA molecules

Each chromosome in the nucleus of a eukaryote consists of one long, linear molecule of dsDNA, which is bound by a complex mixture of proteins (histone and nonhistone) to form chromatin. Eukaryotes have closed, circular, dsDNA molecules in their mitochondria, as do plant chloroplasts. A prokaryotic organism typically contains a single, circular, dsDNA molecule. Each prokaryotic chromosome is associated with nonhistone proteins that help compact the DNA to form a nucleoid. In addition, most species of bacteria also contain small, circular, extrachromosomal DNA molecules called plasmids. Plasmid DNA carries genetic information and undergoes replication that may or may not be synchronized to chromosomal division. [Note: The use of plasmids as vectors in recombinant DNA technology is described in Chapter 34.]

> Plasmids may carry genes that convey antibiotic resistance to the host bacterium and may facilitate the transfer of genetic information from one bacterium to another.

@ III. STEPS IN PROKARYOTIC DNA REPLICATION

When the two strands of dsDNA are separated, each can serve as a template for the replication (synthesis) of a new complementary strand. This produces two daughter molecules, each of which contains two DNA strands (one old, one new) in an antiparallel orientation (see Fig. 30.3). This process is called semiconservative replication because, although the parental duplex is separated into two halves (and, therefore, is not conserved as an entity), each of the parental strands remains intact in one of the two new duplexes (Fig. 30.8). The enzymes involved in DNA replication are template-directed magnesium (Mg^{2+})-requiring polymerases that can synthesize the complementary sequence of each strand with extraordinary fidelity. The reactions described in this section were first known from studies of the bacterium *Escherichia coli*, and the description given below refers to the process in prokaryotes. DNA synthesis in higher organisms is more complex but involves the same types of mechanisms. In either case, initiation of DNA replication commits the cell to continue the process until the entire genome has been replicated.

A. Complementary strand separation

In order for the two complementary strands of the parental dsDNA to be replicated, they must first separate (or "melt") over a small region, because the polymerases use only ssDNA as a template. In prokaryotic organisms, DNA replication begins at a single, unique nucleotide sequence, a site called the origin of replication, or ori (oriC in *E. coli*), as shown in Figure 30.9A. [Note: This sequence is referred to as a consensus sequence, because the order of nucleotides at this site is essentially the same in different bacteria.] The ori includes short, AT-rich segments that facilitate melting. In eukaryotes, replication begins at multiple sites in each chromosome (Fig. 30.9B). Having multiple origins of replication provides a mechanism for rapidly replicating the great length of eukaryotic DNA molecules.

B. Replication fork formation

As the two strands unwind and separate, synthesis occurs at two replication forks that move away from the origin in opposite directions (bidirectionally), generating a replication bubble (see Fig. 30.9). [Note: The term "replication fork" derives from the Y-shaped structure in which the tines of the fork represent the separated strands (Fig. 30.10).]

1. **Required proteins:** Initiation of DNA replication requires the recognition of the origin (start site) by a group of proteins that form the prepriming complex. These proteins are responsible for melting at the ori, maintaining the separation of the parental strands, and unwinding the double helix ahead of the advancing replication fork. In *E. coli*, these proteins include the following.

   a. **DnaA protein:** DnaA protein initiates replication by binding to specific nucleotide sequences (DnaA boxes) within oriC. Binding causes an AT-rich region (the DNA unwinding element) in the origin to melt. Melting (strand separation) results in a short, localized region of ssDNA.

   b. **DNA helicases:** These enzymes bind to ssDNA near the replication fork and then move into the neighboring double-stranded region, forcing the strands apart (in effect, unwinding the double helix). Helicases require energy provided by ATP hydrolysis (see Fig. 30.10). Unwinding at the replication fork causes

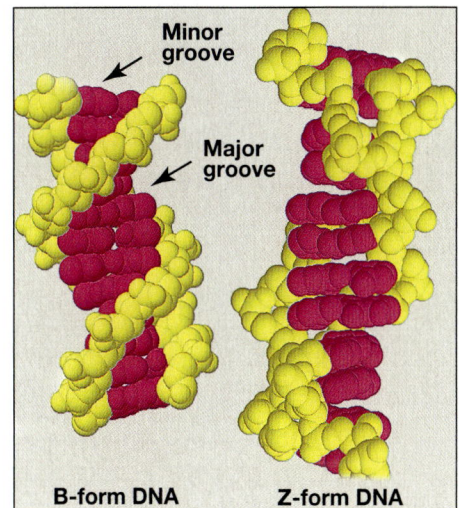

Figure 30.7
Structures of B-DNA and Z-DNA.

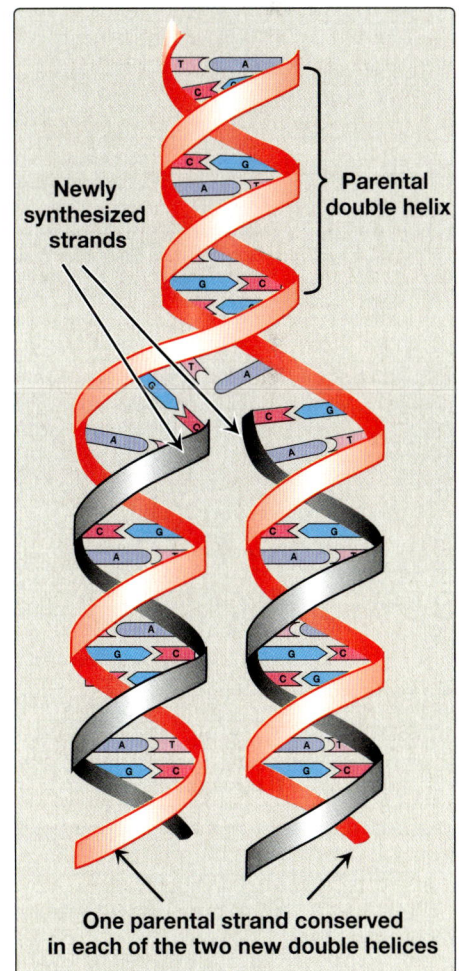

Figure 30.8
Semiconservative replication of DNA.

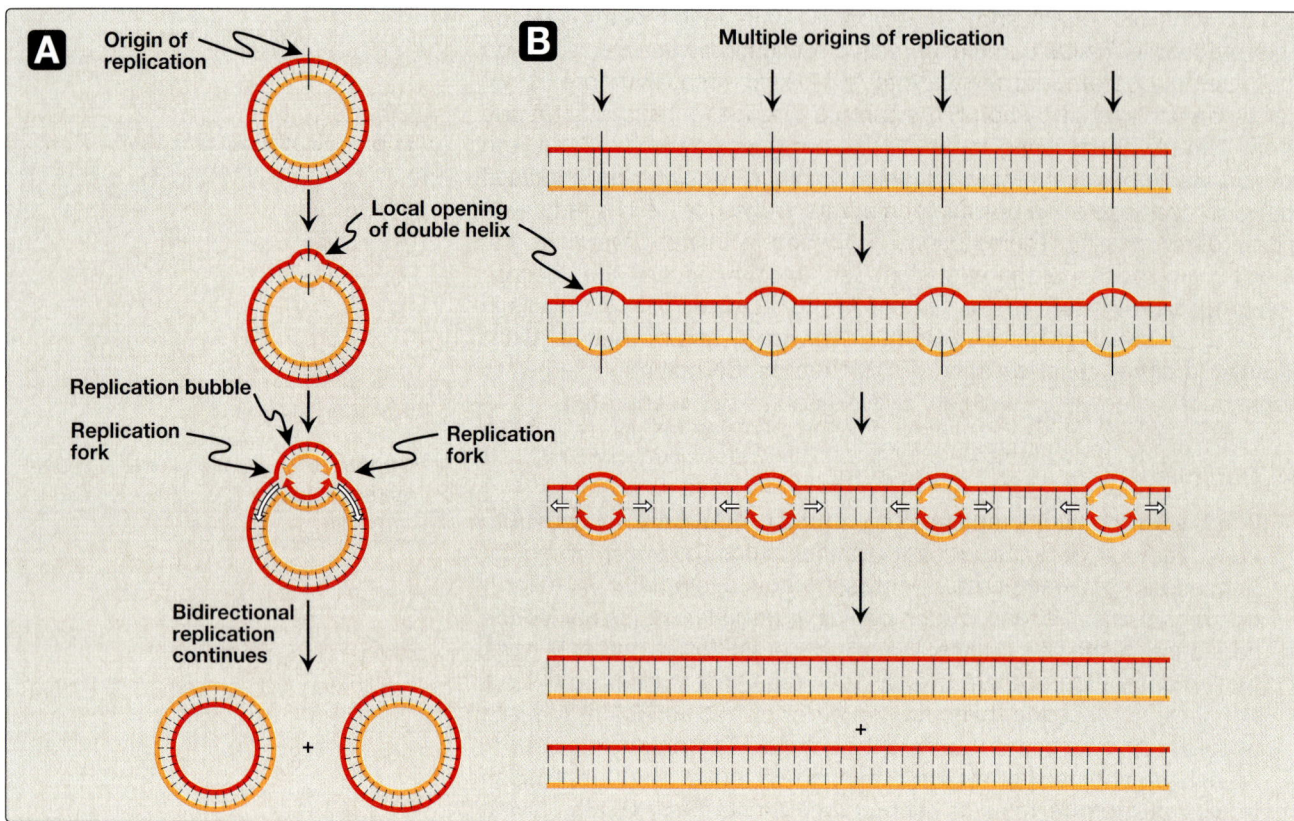

Figure 30.9
Replication of DNA: origins and replication forks. **A.** Small, circular prokaryotic DNA. **B.** Long, linear eukaryotic DNA.

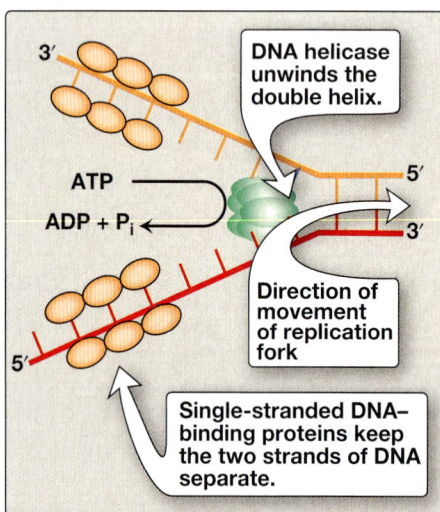

Figure 30.10
Proteins responsible for maintaining the separation of the parental strands and unwinding the double helix ahead of the advancing replication fork. ADP, adenosine diphosphate; P$_i$, inorganic phosphate.

supercoiling in other regions of the DNA molecule. [Note: DnaB is the principal helicase of replication in *E. coli*. Binding of this hexameric protein to DNA requires DnaC.] Supercoiling is a type of tertiary structure in which the double helix of a chromosome crosses over on itself one or more times to relieve torsional strain in the DNA molecule.

c. **Single-stranded DNA–binding protein:** This protein binds to the ssDNA generated by helicases (see Fig. 30.10). Binding is cooperative (i.e., the binding of one molecule of single-stranded binding [SSB] protein makes it easier for additional molecules of SSB protein to bind tightly to the DNA strand). The SSB proteins are not enzymes, but rather serve to shift the equilibrium between dsDNA and ssDNA in the direction of the single-stranded forms. These proteins not only keep the two strands of DNA separated in the area of the replication origin, thus providing the single-stranded template required by polymerases, but also protect the DNA from nucleases that degrade ssDNA.

2. **Solving the problem of supercoils:** Supercoiling can result from overwinding (positive supercoiling) or underwinding (negative supercoiling) of DNA. As the two strands of the double helix are separated, a problem is encountered, namely, the appearance of positive supercoils in the region of DNA ahead of the replication

fork (Fig. 30.11) and negative supercoils in the region behind the fork. The accumulating positive supercoils interfere with further separation of the DNA strands. [Note: Supercoiling can be demonstrated by tightly grasping one end of a helical telephone cord while twisting the other end. If the cord is twisted in the direction of tightening the coils, the cord will wrap around itself in space to form positive supercoils. If the cord is twisted in the direction of loosening the coils, the cord will wrap around itself in the opposite direction to form negative supercoils.] To solve this problem, there is a group of enzymes called DNA topoisomerases, which are responsible for removing supercoils in the helix by transiently cleaving one or both of the DNA strands.

a. **Type I DNA topoisomerases:** These enzymes reversibly cleave one strand of the double helix and form a covalent bond to the end of the nicked strand. They have both strand-cutting and strand-resealing activities. They do not require ATP, but rather appear to store the energy from the phosphodiester bond they cleave, reusing the energy to reseal the strand (Fig. 30.12). Each time an enzyme creates a transient nick in one DNA strand, it rotates around the intact DNA strand before resealing the nick, thus relieving (relaxing) accumulated supercoils. Type I topoisomerases relax negative supercoils (i.e., those that contain fewer turns of the helix than does relaxed DNA) in *E. coli* and both negative and positive supercoils (i.e., those that contain fewer or more turns of the helix than does relaxed DNA) in many prokaryotic cells (but not *E. coli*) and in eukaryotic cells.

b. **Type II DNA topoisomerases:** These enzymes bind tightly to the DNA double helix and make transient breaks in both strands. The enzyme then passes a second part of the DNA double helix through the break and, finally, reseals the break (Fig. 30.13). As a result, both negative and positive supercoils can be relieved by this ATP-requiring process. DNA gyrase, a type II topoisomerase found in bacteria and plants, has the unusual property of being able to introduce negative supercoils into circular DNA using energy from the hydrolysis of ATP. This facilitates the replication of DNA because the negative supercoils neutralize the positive supercoils introduced during opening of the double helix. It also aids in the transient strand separation required during transcription.

> Anticancer agents, such as the camptothecins, target human type I topoisomerases, whereas etoposide targets human type II topoisomerases. Bacterial DNA gyrase is a unique target of a group of antimicrobial agents called fluoroquinolones (e.g., ciprofloxacin).

C. Direction of DNA replication

The DNA polymerases (DNA pols) responsible for copying the DNA templates are only able to read the parental nucleotide sequences

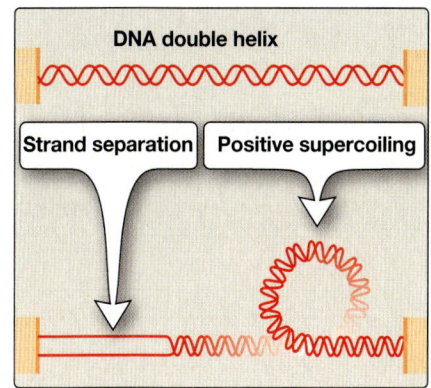

Figure 30.11
Positive supercoiling resulting from DNA strand separation.

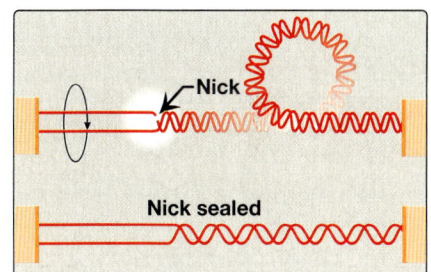

Figure 30.12
Action of type I DNA topoisomerases.

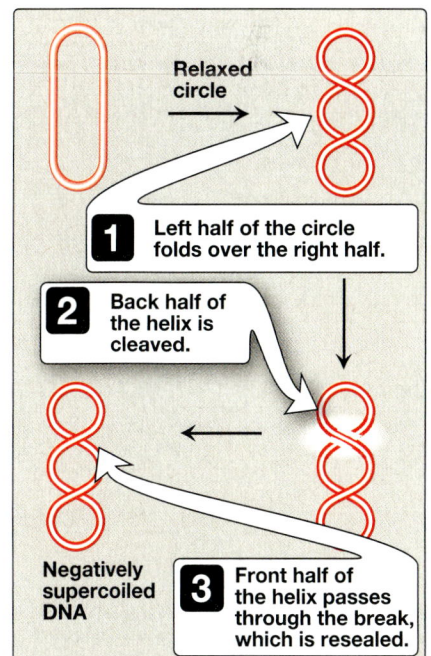

Figure 30.13
Action of type II DNA topoisomerase.

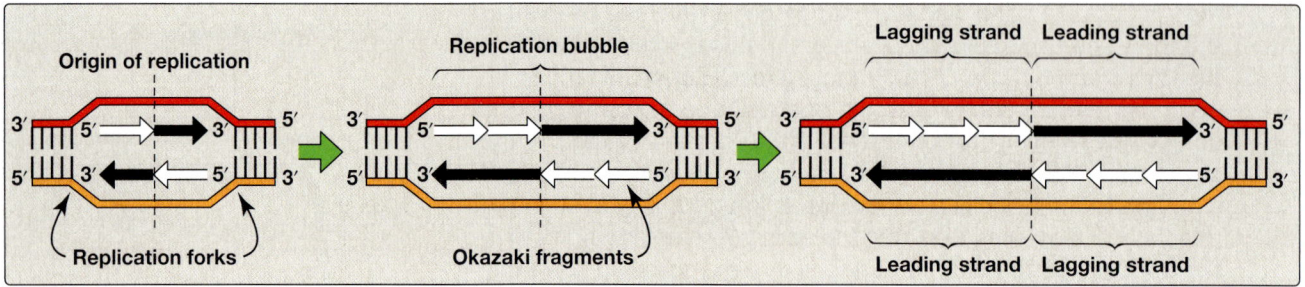

Figure 30.14
Semidiscontinuous synthesis of DNA. *Black arrows*, continuous synthesis; *white arrows*, discontinuous.

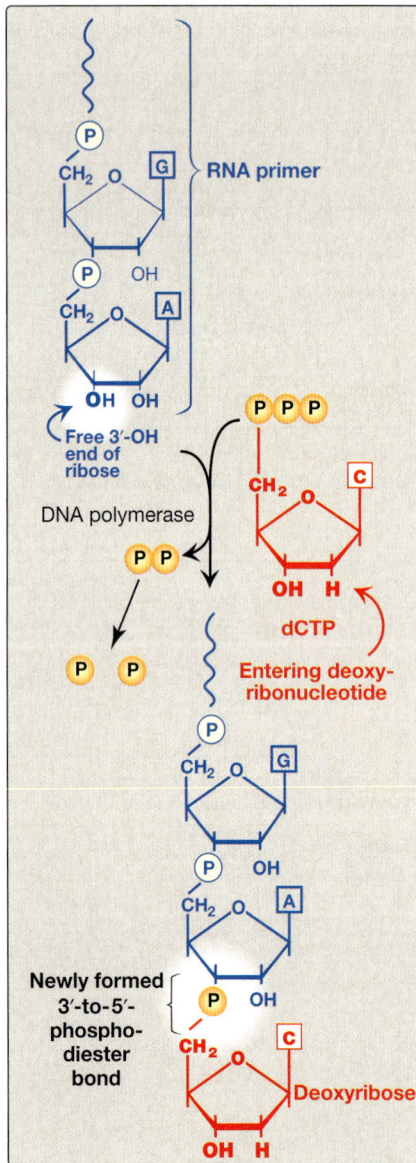

Figure 30.15
Use of an RNA primer to initiate DNA synthesis. Ⓟ and Ⓟ, phosphate; dCTP, deoxycytidine triphosphate.

in the 3'→5' direction, and they synthesize the new DNA strands only in the 5'→3' (antiparallel) direction. Therefore, beginning with one parental double helix, the two newly synthesized stretches of nucleotide chains must grow in opposite directions, one in the 5'→3' direction toward the replication fork and one in the 5'→3' direction away from the replication fork (Fig. 30.14). This feat is accomplished by a slightly different mechanism on each strand.

1. **Leading strand:** The strand that is being copied in the direction of the advancing replication fork is synthesized continuously and is called the leading strand.

2. **Lagging strand:** The strand that is being copied in the direction away from the replication fork is synthesized discontinuously, with small fragments of DNA being copied near the replication fork. These short stretches of discontinuous DNA, termed Okazaki fragments, are eventually joined (ligated) by ligase to become a single, continuous strand. The new strand of DNA produced by this mechanism is termed the lagging strand.

D. RNA primer

DNA pols cannot initiate synthesis of a complementary strand of DNA on a totally single-stranded template. Rather, they require an RNA primer, which is a short piece of RNA base paired to the DNA template, thereby forming a double-stranded DNA–RNA hybrid. The free hydroxyl group on the 3' end of the RNA primer serves as the first acceptor of a deoxynucleotide by action of a DNA pol (Fig. 30.15). [Note: Recall that glycogen synthase also requires a primer in the form of a short glycogen molecule.]

1. **Primase:** A specific RNA polymerase, called primase (DnaG), synthesizes the short stretches of RNA (~10 nucleotides long) that are complementary and antiparallel to the DNA template. In the resulting hybrid duplex, the uracil (U) in RNA pairs with A in DNA. As shown in Figure 30.16, these short RNA sequences are constantly being synthesized at the replication fork on the lagging strand, but only one RNA sequence at the origin of replication is required on the leading strand. The substrates for this process are 5'-ribonucleoside triphosphates, and pyrophosphate (PP_i) is released as each ribonucleoside monophosphate is added through formation of a 3'-to-5' phosphodiester

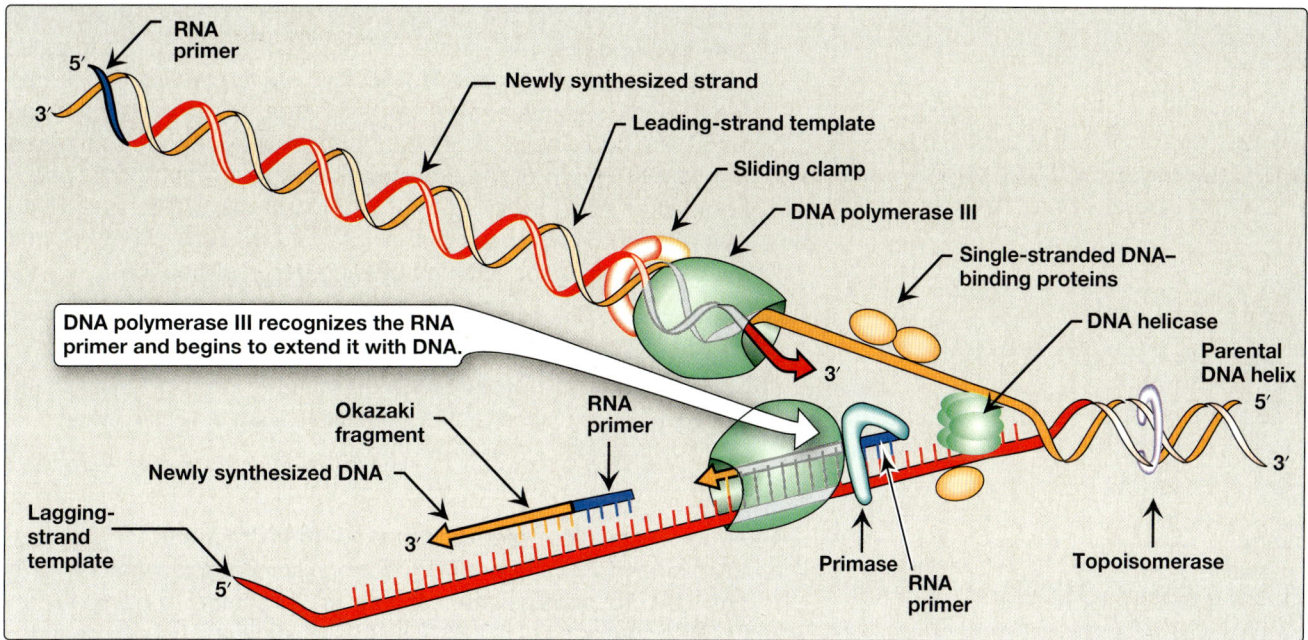

Figure 30.16
Elongation of the leading and lagging strands. [Note: The DNA sliding clamp is not shown for the lagging strand.]

bond. [Note: The RNA primer is later removed, as described in Section III. F.]

2. **Primosome:** The addition of primase converts the prepriming complex of proteins required for DNA strand separation to a primosome. The primosome makes the RNA primer required for leading-strand synthesis and initiates Okazaki fragment formation in discontinuous lagging-strand synthesis. As with DNA synthesis, the direction of synthesis of the primer is 5′→3′.

E. Chain elongation

Prokaryotic (and eukaryotic) DNA pols elongate a new DNA strand by adding deoxyribonucleotides, one at a time, to the 3′ end of the growing chain (see Fig. 30.16). The sequence of nucleotides that are added is dictated by the base sequence of the parental strand, which serves as a template. Incoming nucleotides, used in the synthesis of the new strand, pair with the bases of the template.

1. **DNA polymerase III:** DNA chain elongation is catalyzed by the multisubunit enzyme, DNA pol III. Using the 3′-hydroxyl group of the RNA primer as the acceptor of the first deoxyribonucleotide, DNA pol III begins to add nucleotides along the single-stranded template that specifies the sequence of bases in the newly synthesized chain. DNA pol III is a highly processive enzyme (i.e., it remains bound to the template strand as it moves along and does not diffuse away and then rebind before adding each new nucleotide). The processivity of DNA pol III is the result of the β-subunits of the holoenzyme forming a ring that encircles and moves along the template strand of the DNA, thus serving as a sliding DNA clamp. [Note: Clamp formation is facilitated by

a protein complex, the clamp loader, and ATP hydrolysis.] The new (daughter) strand grows in the 5'→3' direction, antiparallel to the parental strand (see Fig. 30.16). The nucleotide substrates are 5'-deoxyribonucleoside triphosphates. PP$_i$ is released when each new deoxynucleoside monophosphate is added to the free 3'-hydroxyl group of the growing chain through a 3'-to-5' phosphodiester bond (see Fig. 30.15). Hydrolysis of PP$_i$ to 2 P$_i$ by pyrophosphatase means that a total of two high-energy bonds are used to drive the addition of each deoxynucleotide.

> The production of PP$_i$ with subsequent hydrolysis to 2 P$_i$ is a common theme in biochemistry. Removal of the PP$_i$ drives the reaction that generates PP$_i$ in the forward direction, making it essentially irreversible.

All four substrates (deoxyadenosine triphosphate [dATP], deoxythymidine triphosphate [dTTP], deoxycytidine triphosphate [dCTP], and deoxyguanosine triphosphate [dGTP]) must be present for DNA elongation to occur. DNA synthesis stalls when the concentration of the nucleotide falls below the K$_m$ for the polymerase binding to the nucleotide.

2. **Proofreading newly synthesized DNA:** It is highly important for the survival of an organism that the nucleotide sequence of DNA be replicated with as few errors as possible. Misreading of the template sequence could result in deleterious, perhaps lethal, mutations. To ensure replication fidelity, DNA pol III has a proofreading activity (3'→5' exonuclease; Fig. 30.17) in addition

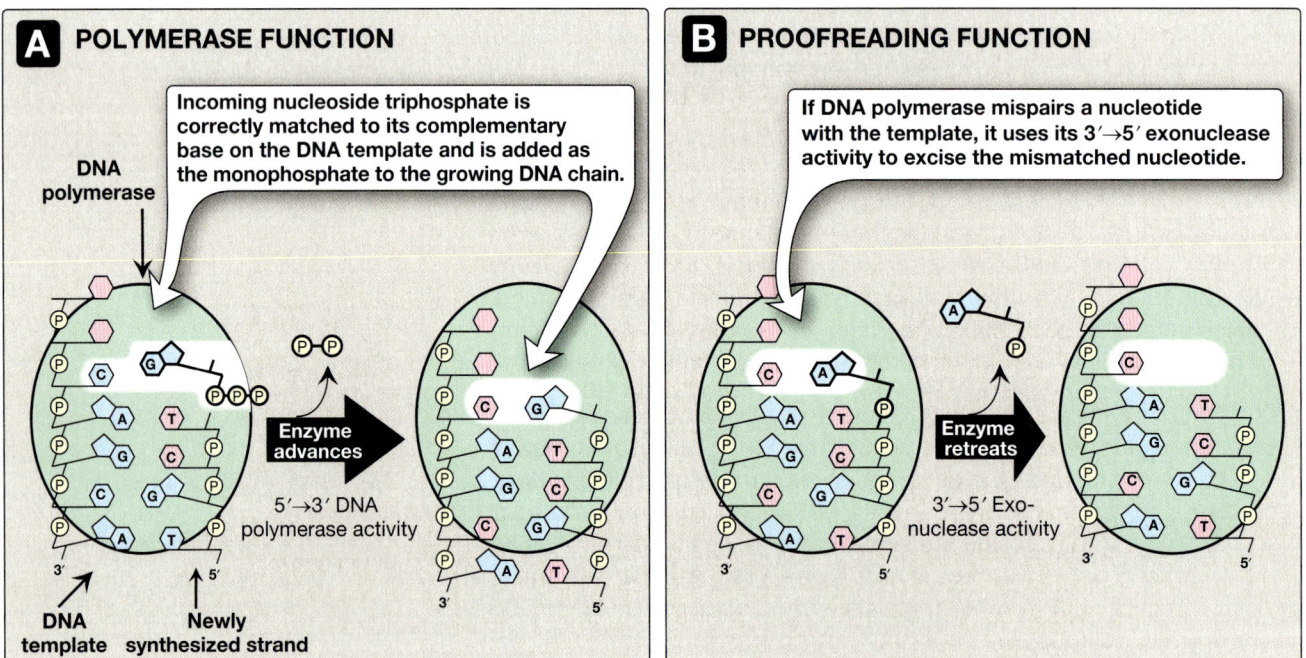

Figure 30.17
DNA polymerase III (**A**) 5'→3' polymerase activity for DNA strand synthesis and (**B**) 3'→5' exonuclease activity for DNA strand proofreading.

to its 5′→3′ polymerase activity. As each nucleotide is added to the chain, DNA pol III checks to make certain the base of the newly added nucleotide is, in fact, the complement of the base on the template strand. If it is not, the 3′→5′ exonuclease activity removes the erroneously added nucleotide in the direction opposite to polymerization. [Note: Because the exonuclease function of DNA pol III requires an improperly base-paired 3′-hydroxy terminus, it does not degrade correctly paired nucleotide sequences.] For example, if the template base is C and the enzyme inserts an A instead of a G into the new chain, the 3′→5′ exonuclease activity hydrolytically removes the misplaced nucleotide. The 5′→3′ polymerase activity then repeats the nucleotide addition step and inserts the correct nucleotide containing G (see Fig. 30.17). [Note: The 5′→3′ polymerase and 3′→5′ exonuclease domains are located on different subunits of DNA pol III.]

> Sickle cell anemia is caused by a single nucleotide change, an error of inserting a T in the place of an A, in the β-globin gene. This mutation results in an incorrect amino acid (a valine in the place of a glutamate) in the β-globin protein that alters the function of the protein in the red blood cell.

F. RNA primer excision and replacement by DNA

DNA pol III continues to synthesize DNA on the lagging strand until it nears the 5′ end of an RNA primer. When this occurs, the RNA is excised, and the gap between Okazaki fragments is filled by DNA pol I.

1. **5′→3′ Exonuclease activity:** In addition to having the 5′→3′ polymerase activity that synthesizes DNA and the 3′→5′ exonuclease activity that proofreads the newly synthesized DNA like DNA pol III, monomeric DNA pol I also has a 5′→3′ exonuclease activity that is able to hydrolytically remove the RNA primer. [Note: Exonucleases remove nucleotides from the end of the DNA chain, rather than cleaving the chain internally as do endonucleases (Fig. 30.18).] First, DNA pol I locates the space (nick) between the 3′ end of the DNA newly synthesized by DNA pol III and the 5′ end of the adjacent RNA primer. Next, DNA pol I hydrolytically removes the RNA nucleotides ahead of itself, moving in the 5′→3′ direction (5′→3′ exonuclease activity). As it removes ribonucleotides, DNA pol I replaces them with deoxyribonucleotides, synthesizing DNA in the 5′→3′ direction (5′→3′ polymerase activity). As it synthesizes the DNA, it also proofreads using its 3′→5′ exonuclease activity to remove errors. This removal/synthesis/proofreading continues until the RNA primer is totally degraded, and the gap is filled with DNA (Fig. 30.19). [Note: DNA pol I uses its 5′→3′ polymerase activity to fill in gaps generated during most types of DNA repair.]

2. **Comparison of 5′→3′ and 3′→5′ exonuclease activities:** The 5′→3′ exonuclease activity of DNA pol I allows the polymerase, moving 5′→3′, to hydrolytically remove one or more nucleotides

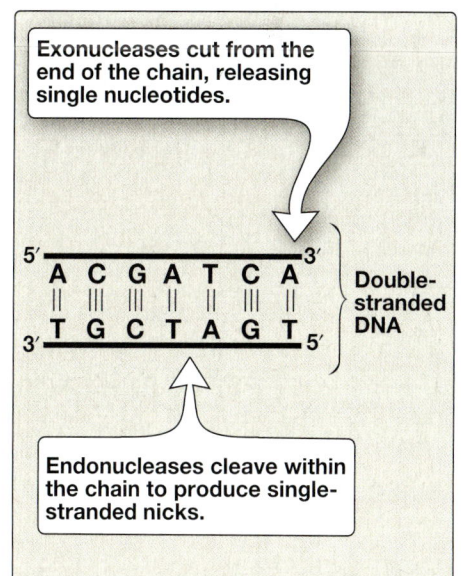

Exonucleases cut from the end of the chain, releasing single nucleotides.

5′ A C G A T C A 3′ Double-stranded DNA
3′ T G C T A G T 5′

Endonucleases cleave within the chain to produce single-stranded nicks.

Figure 30.18
Endonuclease versus exonuclease activity. [Note: Restriction endonucleases cleave both strands.]

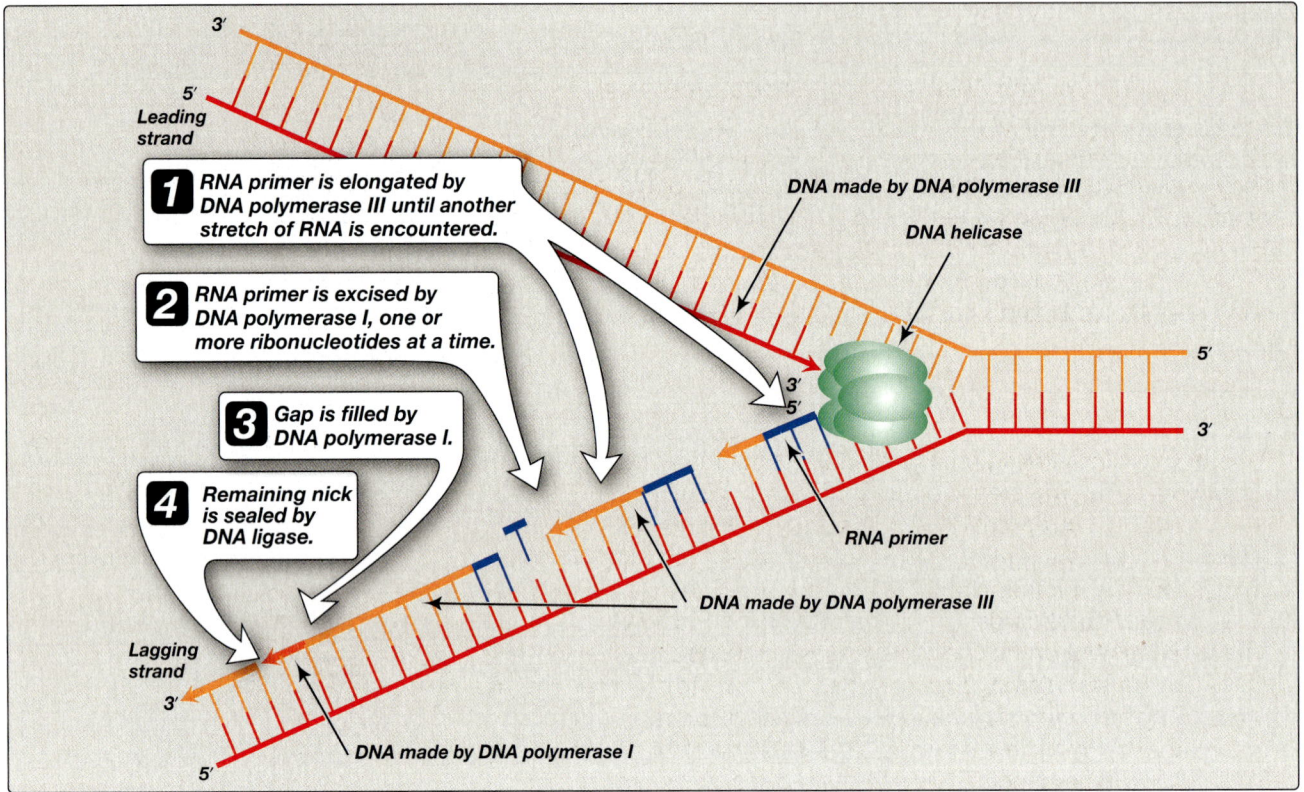

Figure 30.19
Removal of RNA primer and filling of the resulting gaps by DNA polymerase I.

at a time from the 5′ end of the ~10-nucleotide–long RNA primer. In contrast, the 3′→5′ exonuclease activity of DNA pol I and pol III allows these polymerases, moving 3′→5′, to hydrolytically remove one misplaced nucleotide at a time from the 3′ end of a growing DNA strand, increasing the fidelity of replication such that newly replicated DNA has no more than one error per 10^7 nucleotides.

G. DNA ligase

DNA pol can only catalyze phosphodiester bond formation between a DNA strand and a mononucleotide and cannot join two sections of a DNA strand. The final phosphodiester linkage between the 5′-phosphate group on the DNA synthesized by DNA pol III and the 3′-hydroxyl group on the DNA made by DNA pol I is catalyzed by DNA ligase (Fig. 30.20). The joining (ligation) of these two stretches of DNA requires energy, which, in most organisms, is provided by the cleavage of ATP to adenosine monophosphate + PP_i.

H. Termination

Replication termination in *E. coli* is mediated by sequence-specific binding of the protein, terminus utilization substance (Tus) to replication termination (Ter) sites on the DNA, stopping the movement of the replication fork.

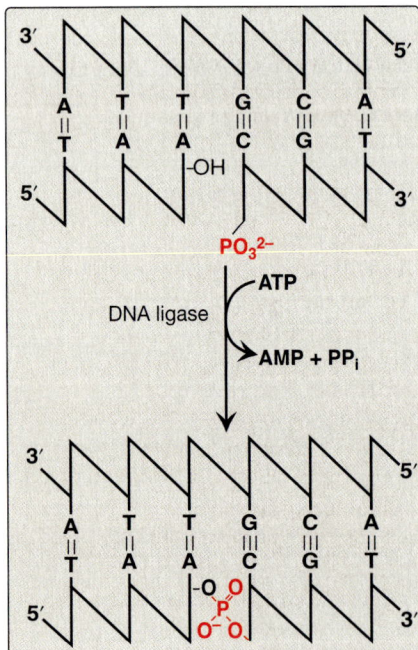

Figure 30.20
Formation of a phosphodiester bond by DNA ligase.

IV. EUKARYOTIC DNA REPLICATION

The process of eukaryotic DNA replication closely follows that of prokaryotic DNA synthesis. Some differences, such as the multiple origins of replication in eukaryotic cells versus single origins of replication in prokaryotes, have already been noted. Eukaryotic origin recognition proteins, ssDNA-binding proteins, and ATP-dependent DNA helicases have been identified, and their functions are analogous to those of the prokaryotic proteins previously discussed. In contrast, RNA primers are removed by RNase H and flap endonuclease 1 (FEN1) rather than by a DNA pol (Fig. 30.21).

A. Eukaryotic cell cycle

The events surrounding eukaryotic DNA replication and cell division (mitosis) are coordinated to produce the cell cycle (Fig. 30.22). The period preceding replication is called the gap 1 phase (G_1). DNA replication occurs during the synthesis (S) phase. Following DNA synthesis, there is another phase (G_2, or gap 2) before mitosis (M). Cells that have stopped dividing, such as mature T lymphocytes, are said to have gone out of the cell cycle into the G_0 phase. Such quiescent cells can be stimulated to reenter the G_1 phase to resume division. [Note: The cell cycle is controlled at a series of checkpoints that prevent entry into the next phase of the cycle until the preceding phase has been completed. Two key classes of proteins that control the progress of a cell through the cell cycle are the cyclins and cyclin-dependent kinases (Cdks).]

B. Eukaryotic DNA polymerases

At least five high-fidelity eukaryotic DNA pols have been identified and categorized on the basis of molecular weight, cellular location, sensitivity to inhibitors, and the templates or substrates on which they act. They are designated by Greek letters rather than by Roman numerals (Fig. 30.23).

1. **Pol α:** Pol α is a multisubunit enzyme. One subunit has primase activity, which initiates strand synthesis on the leading strand and at the beginning of each Okazaki fragment on the lagging strand. The primase subunit synthesizes a short RNA primer that is extended by the 5′→3′ polymerase activity of pol α, generating a short piece of DNA that is later extended by a more processive DNA pol such as pol ε or pol δ. [Note: Pol α is also referred to as pol α/primase.]

2. **Pol ε and pol δ:** Pol ε is recruited to complete DNA synthesis on the leading strand, whereas pol δ elongates the Okazaki fragments of the lagging strand, each using 3′→5′ exonuclease activity to proofread the newly synthesized DNA. [Note: DNA pol ε associates with proliferating cell nuclear antigen (PCNA), a protein that serves as a sliding DNA clamp in much the same way the β subunits of DNA pol III do in *E. coli*, thus ensuring high processivity.]

3. **Pol β and pol γ:** Pol β is involved in gap filling in DNA repair. Pol γ replicates mitochondrial DNA.

FUNCTION	PROTEIN(S)
Origin recognition	ORC
Helicase activity	MCM
ssDNA protection	RPA
Primer synthesis	Pol a/primase
Sliding clamp	PCNA
Primer removal	RNase H, FEN1

Figure 30.21
Proteins and their function in eukaryotic replication. ORC, origin recognition complex; MCM, minichromosome maintenance (complex); RPA, replication protein A; PCNA, proliferating cell nuclear antigen; FEN, flap endonuclease.

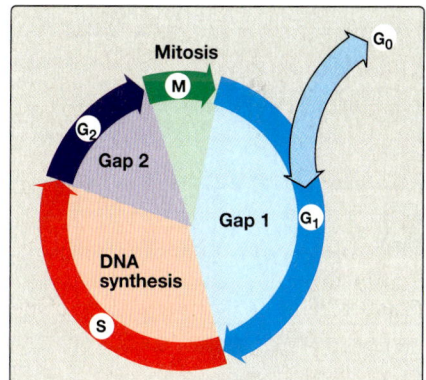

Figure 30.22
Eukaryotic cell cycle. [Note: Cells can leave the cell cycle and enter a reversible quiescent state called G_0.]

POLY-MERASE	FUNCTION	PROOF-READING*
Pol α (alpha)	• **Contains primase** • **Initiates DNA synthesis**	–
Pol β (beta)	• **Repair**	–
Pol δ (delta)	• **Elongates Okazaki fragments of the lagging strand**	+
Pol ε (epsilon)	• **Elongates the leading strand**	+
Pol γ (gamma)	• **Replicates mitochondrial DNA**	+

Figure 30.23
Activities of eukaryotic DNA polymerases (pol). [Note: The *asterisk* denotes 3′→5′ exonuclease activity.]

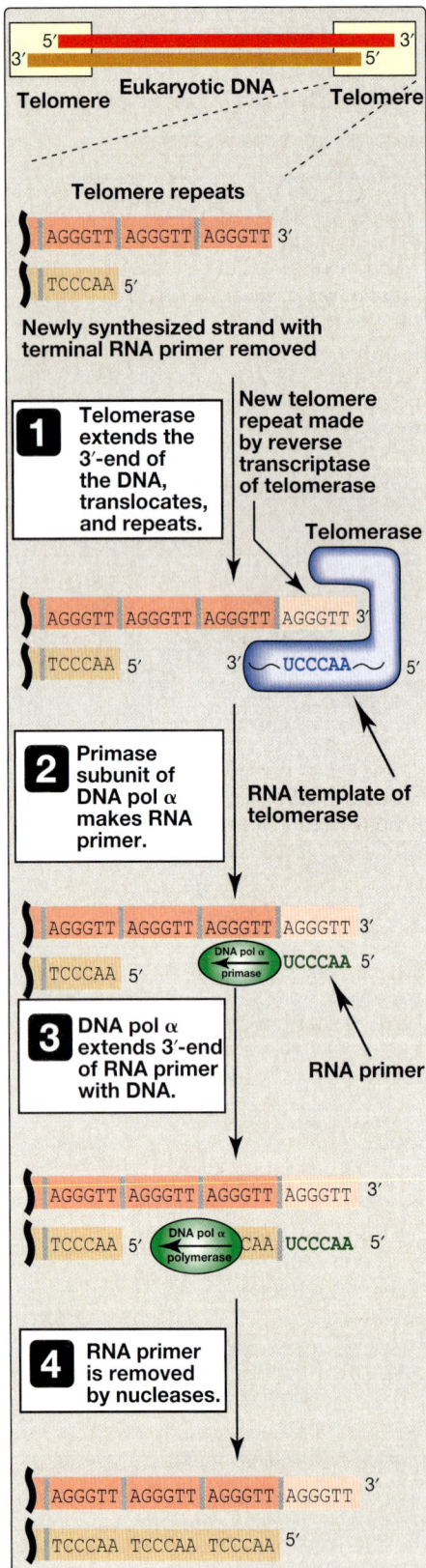

Figure 30.24
Mechanism of action of telomerase, a ribonucleoprotein. Pol, polymerase.

C. Telomeres

Telomeres are complexes of DNA, associated with proteins (collectively known as shelterin) located at the ends of linear chromosomes. They maintain the structural integrity of the chromosome, preventing attack by nucleases, and allow repair systems to distinguish a true end from a break in dsDNA. In humans, telomeric DNA consists of several thousand tandem repeats of a noncoding hexameric sequence, AGGGTT, base paired to the repeating AACCCT sequence. The strand with AGGGTT repeats (the "G-rich strand") is longer than its complementary strand with AACCCT repeats (the "C-rich strand"), leaving ssDNA a few hundred nucleotides in length at the 3′ end. The single-stranded region is thought to fold back on itself, forming a loop structure that is stabilized by protein.

1. **Telomere shortening:** Eukaryotic cells face a special problem in replicating the ends of their linear DNA molecules. Following removal of the RNA primer from the extreme 5′ end of the lagging strand, there is no way to fill in the remaining gap with DNA. Consequently, in most normal human somatic cells, telomeres shorten with each successive cell division. Once telomeres are shortened beyond some critical length, the cell is no longer able to divide and is said to be senescent. In germ cells and stem cells, as well as in cancer cells, telomeres do not shorten, and the cells do not senesce. This is a result of the activity of the ribonucleoprotein telomerase, which maintains telomeric length in these cells.

2. **Telomerase:** This complex contains a protein, TERT that acts as a reverse transcriptase and a short piece of RNA, TERC that acts as a template. The C-rich RNA template base pairs with the G-rich, single-stranded 3′ end of telomeric DNA (Fig. 30.24). The reverse transcriptase uses the RNA template to synthesize DNA in the usual 5′→3′ direction, extending the already longer 3′ end. Telomerase then translocates to the newly synthesized end, and the process is repeated. Once the G-rich strand has been lengthened, primase activity of DNA pol α can use it as a template to synthesize an RNA primer. The primer is extended by DNA pol α and then removed by nucleases.

> Telomeres may be viewed as mitotic clocks in that their length in most cells is inversely related to the number of times the cells have divided. The study of telomeres provides insight into the biology of normal aging, diseases of premature aging (the progerias), and cancer.

D. Reverse transcriptases

As seen with telomerase, reverse transcriptases are RNA-directed DNA pols. A reverse transcriptase is involved in the replication of retroviruses, such as human immunodeficiency virus (HIV). These viruses carry their genome in the form of ssRNA molecules. Following infection of a host cell, the viral enzyme reverse transcriptase uses the viral RNA as a template for the 5′→3′ synthesis of viral DNA, which then becomes integrated into host chromosomes. Reverse transcriptase activity is also seen with transposons, DNA elements that can move about the genome. In eukaryotes, most transposons are transcribed to RNA, the RNA is used as a template for DNA synthesis by a reverse transcriptase encoded by the transposon, and the DNA is randomly inserted into the genome. [Note: Transposons that involve an RNA intermediate are called retrotransposons or retroposons.]

E. DNA replication inhibition by nucleoside analogs

DNA chain growth can be blocked by the incorporation of certain nucleoside analogs that have been modified on the sugar portion (Fig. 30.25). For example, removal of the hydroxyl group from the 3' carbon of the deoxyribose ring as in 2',3' dideoxyinosine ([ddI] also known as didanosine), or conversion of the deoxyribose to another sugar, such as arabinose, prevents further chain elongation. By blocking DNA synthesis, these compounds slow the division of rapidly proliferating cancer cells and the replication of viruses. Cytosine arabinoside (cytarabine, or araC) has been used in anticancer chemotherapy, whereas adenine arabinoside (vidarabine, or araA) is an antiviral agent. Substitution on the sugar moiety, as seen in azidothymidine (AZT), also called zidovudine (ZDV), also terminates DNA chain elongation. [Note: These drugs are generally supplied as nucleosides, which are then converted to nucleotides by cellular kinases.]

V. EUKARYOTIC DNA ORGANIZATION

A typical (diploid) human somatic cell contains 46 chromosomes, whose total DNA is ~2 m long! It is difficult to imagine how such a large amount of genetic material can be effectively packaged into a volume the size of a cell nucleus so that it can be efficiently replicated, and its genetic information expressed. To do so requires the interaction of DNA with a large number of proteins, each of which performs a specific function in the ordered packaging of these long molecules of DNA. Eukaryotic DNA is associated with tightly bound basic proteins, called histones. These serve to order the DNA into fundamental structural units, called nucleosomes, which resemble beads on a string. Nucleosomes are further arranged into increasingly more complex structures that organize and condense the long DNA molecules into chromosomes that can be segregated during cell division. [Note: The complex of DNA and protein found inside the nuclei of eukaryotic cells is called chromatin.]

A. Histones and nucleosome formation

There are five classes of histones, designated H1, H2A, H2B, H3, and H4. These small, evolutionarily conserved proteins are positively charged at physiologic pH as a result of their high content of lysine and arginine. Because of their positive charge, they form ionic bonds with negatively charged DNA. Histones, along with ions such as Mg^{2+}, help neutralize the negatively charged DNA phosphate groups.

1. **Nucleosomes:** Two molecules each of H2A, H2B, H3, and H4 form the octameric core of the individual nucleosome "beads." Around this structural core, a segment of dsDNA is wound nearly

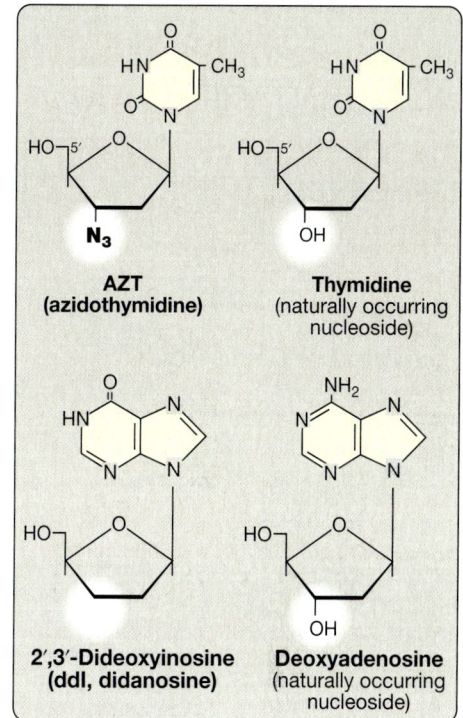

Figure 30.25
Examples of nucleoside analogs that lack a 3'-hydroxyl group. [Note: The ddI is converted to its active form (dideoxy ATP).]

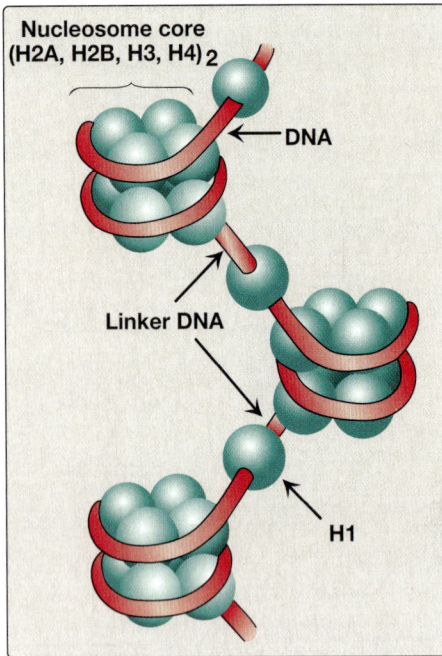

Figure 30.26
Organization of human DNA, illustrating the structure of nucleosomes. H, histone.

twice (Fig. 30.26). Winding eliminates a helical turn, causing negative supercoiling. [Note: The N-terminal ends of these histones can be acetylated, methylated, or phosphorylated. These reversible covalent modifications influence how tightly the histones bind to the DNA, thereby affecting the expression of specific genes. Histone modification is an example of epigenetics, or heritable changes in gene expression caused without alteration of the nucleotide sequence.] Neighboring nucleosomes are joined by linker DNA ~50 bps long. H1 is not found in the nucleosome core, but instead binds to the linker DNA chain between the nucleosome beads. H1 is the most tissue specific and species specific of the histones. It facilitates the packing of nucleosomes into more compact structures.

2. **Higher levels of organization:** Nucleosomes can be packed more tightly (stacked) to form a nucleofilament. This structure assumes the shape of a coil, often referred to as a 30-nm fiber. The fiber is organized into loops that are anchored by a nuclear scaffold containing several proteins. Additional levels of organization lead to the final chromosomal structure (Fig. 30.27).

B. Nucleosome fate during DNA replication

Parental nucleosomes are disassembled to allow access to DNA during replication. Once DNA is synthesized, nucleosomes form rapidly. Their histone proteins come both from *de novo* synthesis and from the transfer of parental histones.

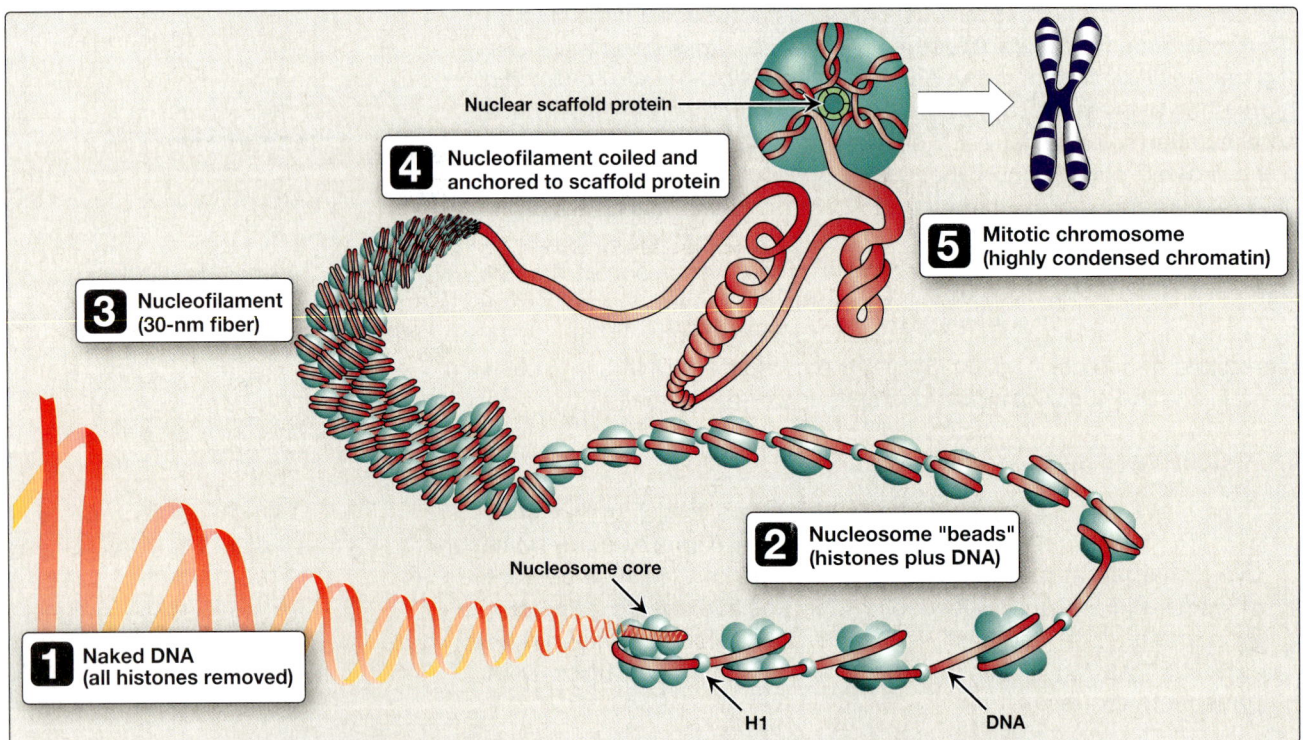

Figure 30.27
Structural organization of eukaryotic DNA. [Note: A 10^4 linear compaction is seen from 1–5.] H, histone.

@ VI. DNA REPAIR

Despite the elaborate proofreading system employed during DNA synthesis, errors (including incorrect base pairing or insertion of one to a few extra nucleotides) can occur. In addition, DNA is constantly being subjected to environmental insults that cause the alteration or removal of nucleotide bases. The damaging agents can be either chemicals (e.g., nitrous acid, which can deaminate bases) or radiation (e.g., nonionizing ultraviolet [UV] radiation from sunlight, which can fuse two pyrimidines adjacent to each other in the DNA, and high-energy ionizing radiation, which can cause double-strand breaks). Bases are also altered or lost spontaneously from mammalian DNA at a rate of many thousands per cell per day. If the damage is not repaired, a permanent change (mutation) is introduced that can result in any number of deleterious effects, including loss of control over the proliferation of the mutated cell, leading to cancer. Luckily, cells are remarkably efficient at repairing damage done to their DNA, especially when the damage affects only one or two bases at a location on the same strand of the DNA duplex. Most of the repair systems (which are called excision repair systems) involve recognition of the damage (lesion) on the DNA, removal, or excision of the damage, filling the gap left by excision using the undamaged, complementary strand as a template for DNA synthesis, and ligation to restore the continuity of the repaired strand. These excision repair systems remove one to tens of nucleotides. [Note: Repair synthesis of DNA can occur outside of the S phase.] Damage may also affect both strands of the DNA at location (e.g., double-strand breaks). These forms of damage are repaired by different repair systems than those removing damage to one strand.

A. Mismatch repair

Sometimes replication errors escape the proofreading activity during DNA synthesis, causing a mismatch of one to several bases. In *E. coli*, mismatch repair (MMR) is mediated by a group of proteins known as the Mut proteins (Fig. 30.28). Homologous proteins are present in humans. [Note: MMR occurs within minutes of replication and reduces the error rate of replication from 1 in 10^7 to 1 in 10^9 nucleotides.]

1. **Mismatched strand identification:** When a mismatch occurs, the Mut proteins that identify the mispaired nucleotide(s) must be able to discriminate between the correct strand and the strand with the mismatch. In prokaryotes, discrimination is based on the degree of methylation. GATC sequences, which are found once every thousand nucleotides, are methylated on the A residue by DNA adenine methylase (DAM). This methylation is not done immediately after synthesis, so the DNA is hemimethylated (i.e., the parental strand is methylated, but the daughter strand is not). The methylated parental strand is assumed to be correct, and it is the daughter strand that gets repaired. [Note: The exact mechanism by which the daughter strand is identified in eukaryotes is not yet known, but likely involves recognition of nicks in the newly synthesized strand.]

2. **Repair procedure:** When the strand containing the mismatch is identified, an endonuclease nicks the strand, and the mismatched nucleotide(s) is/are removed by an exonuclease. Additional

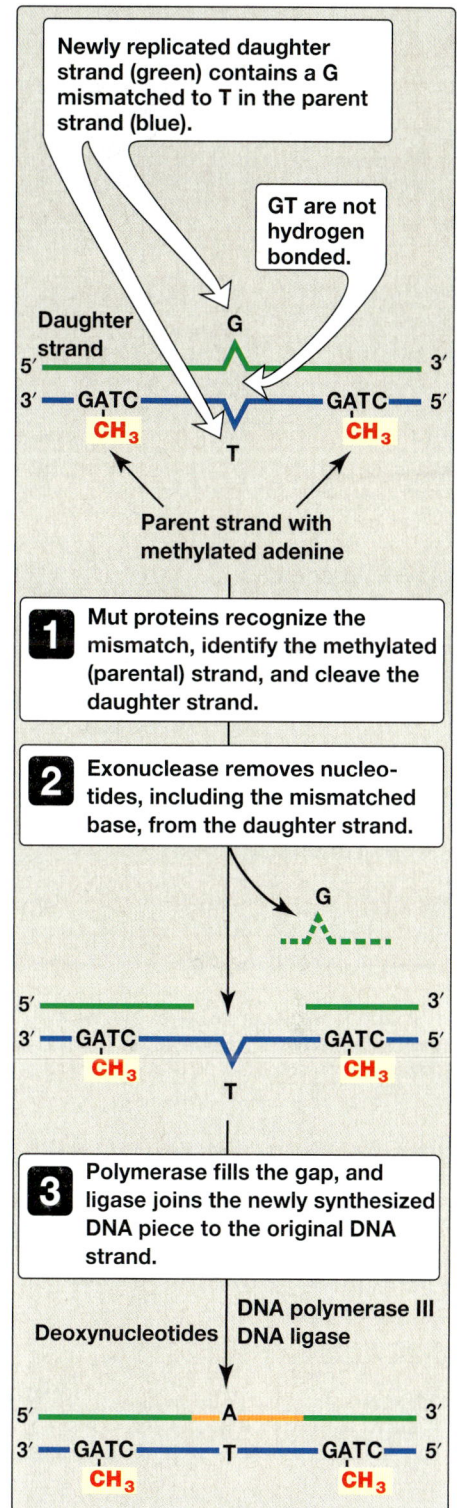

Figure 30.28
Methyl-directed mismatch repair in *Escherichia coli*. [Note: Mut S protein recognizes the mismatch and recruits Mut L. The complex activates Mut H, which cleaves the unmethylated (daughter) strand.]

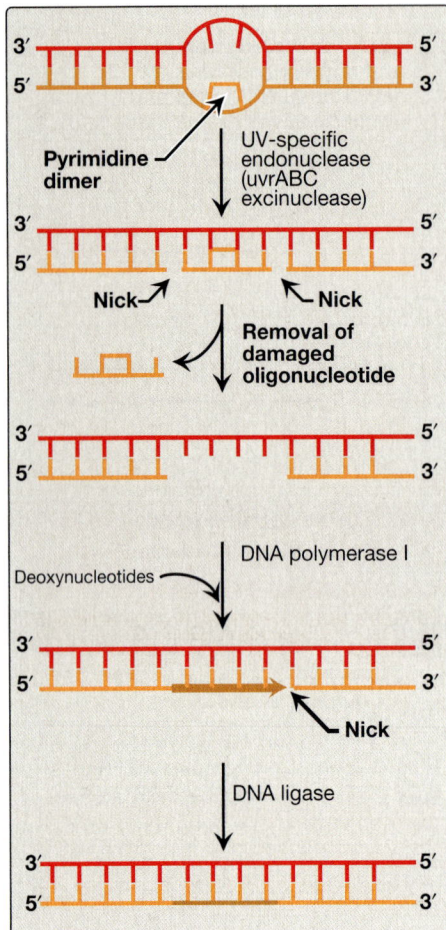

Figure 30.29
Nucleotide excision repair of pyrimidine dimers in *Escherichia coli* DNA. UV, ultraviolet.

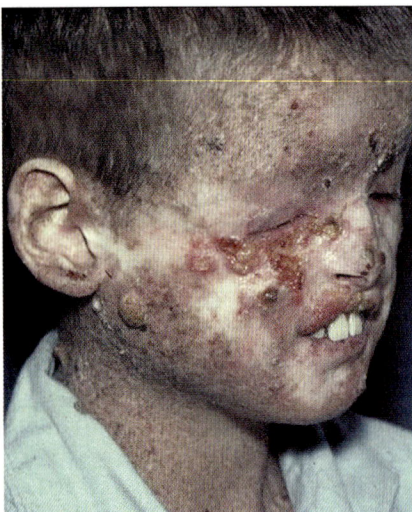

Figure 30.30
Patient with xeroderma pigmentosum.

nucleotides at the 5' and 3' ends of the mismatch are also removed. The gap left by the removal of the nucleotides is filled, using the sister strand as a template, by a DNA pol, typically DNA pol III. The 3' hydroxyl of the newly synthesized DNA is joined to the 5' phosphate of the remaining stretch of the original DNA strand by DNA ligase.

> Defects in the proteins involved in MMR in humans are associated with Lynch syndrome, also known as hereditary nonpolyposis colorectal cancer (HNPCC). Mutations in MSH2 and MLH1 (two human homologs of bacterial Mut proteins) account for 90% of cases. Although HNPCC confers an increased risk for developing colon cancer (as well as other cancers), only about 5% of all colon cancer is the result of mutations in MMR.

B. Nucleotide excision repair

Exposure of a cell to UV radiation can result in the covalent joining of two adjacent pyrimidines (usually Ts), producing a dimer. These intrastrand cross-links prevent DNA pol from replicating the DNA strand beyond the site of dimer formation. T dimers are excised in bacteria by UvrABC proteins in a process known as nucleotide excision repair (NER), as illustrated in Figure 30.29. The NER pathway is also present in humans (see Section VI. B. 2.). NER has two DNA damage recognition mechanisms, global genomic repair that finds damage throughout chromosomes and transcription-coupled repair that identifies DNA lesions encountered by RNA polymerases.

1. **Recognition and excision of UV-induced dimers:** A UV-specific endonuclease (called uvrABC excinuclease) recognizes the bulky dimer and cleaves the damaged strand on both the 5' side and 3' side of the lesion. A short oligonucleotide containing the dimer is excised, leaving a gap in the DNA strand. This gap is filled in using DNA pol I and DNA ligase. The human NER pathway uses additional proteins to remove pyrimidine dimers that form in skin cells and to repair DNA damage that is created by chemical exposure, such as G adducts caused by benzo[a]pyrene from cigarette smoke. NER occurs throughout the cell cycle.

2. **UV radiation and cancer:** In the rare, human genetic disease xeroderma pigmentosum (XP), an individual's skin cells cannot repair pyrimidine dimers caused by sunlight, resulting in extensive accumulation of mutations and, consequently, early and numerous skin cancers (Fig. 30.30). XP can be caused by defects in seven genes that code for the XP proteins required for NER of UV damage.

C. Base excision repair

DNA bases can be altered, either spontaneously, as is the case with C, which slowly undergoes deamination (the loss of its amino group) to form U, or by the action of deaminating or alkylating compounds. For example, nitrous acid, which is formed by the cell from precursors such as the nitrates, deaminates C, A (to hypoxanthine), and G (to xanthine). Dimethyl sulfate can alkylate (methylate) A. Bases can also be lost spontaneously by hydrolysis from the deoxyribose sugar backbone. For example, ~10,000 purine bases are lost this way per cell per day. Lesions involving base alterations or loss can be corrected by base excision repair ([BER] Fig. 30.31).

1. **Abnormal base removal:** In BER, abnormal bases, such as U, which can occur in DNA by either deamination of C or improper use of dUTP instead of dTTP during DNA synthesis, are recognized by specific DNA glycosylases that hydrolytically cleave them from the deoxyribose-phosphate backbone of the strand. This leaves an apyrimidinic site, or apurinic if a purine was removed, both referred to as AP sites.

2. **AP site recognition and repair:** Specific AP endonucleases recognize that a base is missing and initiate the process of excision and gap filling by making an endonucleolytic cut just to the 5′ side of the AP site. A deoxyribose phosphate lyase removes the single, base-free, sugar phosphate residue. DNA pol I and DNA ligase complete the repair process.

D. Double-strand break repair

Ionizing radiation, chemotherapeutic agents such as doxorubicin, and oxidative free radicals can cause double-strand breaks in DNA that can be lethal to the cell. [Note: Such breaks also occur naturally during genetic recombination.] dsDNA breaks cannot be corrected by the previously described strategy of excising the damage on one strand and using the undamaged strand as a template for replacing the missing nucleotide(s). Instead, they are repaired by one of two systems. The first is nonhomologous end joining (NHEJ), in which a group of proteins mediates the recognition, processing, and ligation of the ends of two DNA fragments. However, some DNA is lost in the process. Consequently, NHEJ is error prone and mutagenic. Defects in NHEJ are associated with a predisposition to cancer and immunodeficiency syndromes. The second repair system, homologous recombination (HR), uses the enzymes that normally perform genetic recombination between homologous chromosomes during meiosis. This system is much less error prone ("error free") than NHEJ because any DNA that was lost is replaced using homologous DNA as a template. HR occurs in late S and G_2 of the cell cycle, whereas NHEJ can occur anytime. [Note: Mutations to the proteins BRCA1 or BRCA2 (breast cancer 1 or 2), which are involved in HR, increase the risk for developing breast and ovarian cancer.]

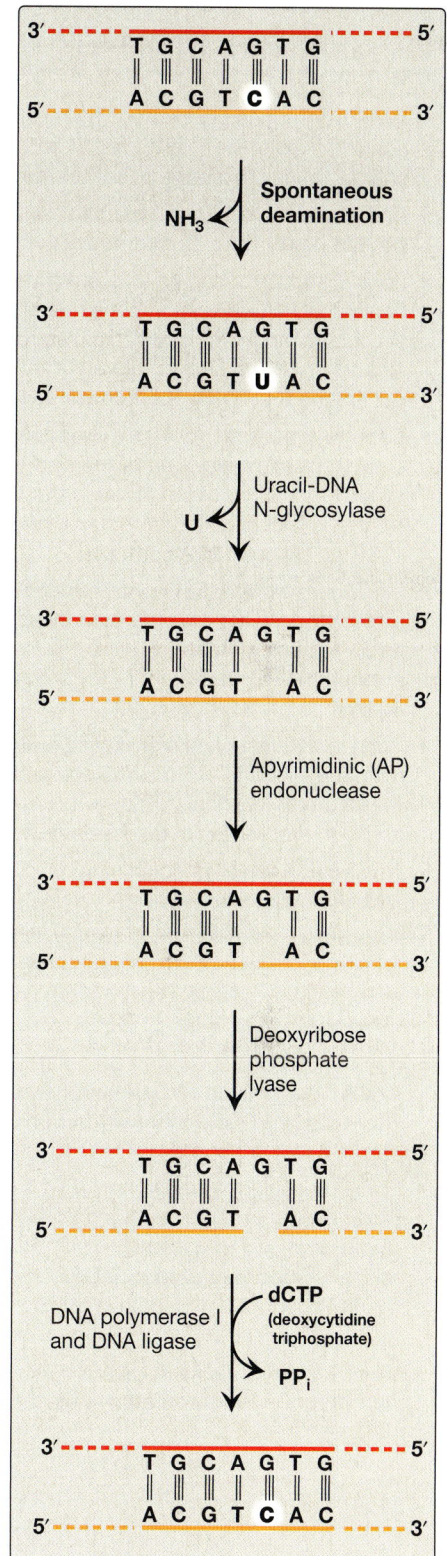

Figure 30.31
Correction of base alterations by base excision repair. C, cytosine; U, uracil; NH_3, ammonia; PP_i, pyrophosphate.

VII. CHAPTER SUMMARY

- DNA is composed of two polymers (chains) of deoxyribonucleoside monophosphates ([dNMP] nucleotides). Each chain has **polarity**, with both a 5′ end (free phosphate) and a 3′ end (free hydroxyl). The nucleotide sequences of the chains are read from the 5′ end to the 3′ end (Fig. 30.32).

- DNA exists as a double-stranded (ds)DNA, in which the two chains are paired in an **antiparallel** manner and form a **double helix**. Through hydrogen bonding, **A** pairs with **T**, and **C** pairs with **G**.

- Each DNA strand serves as a **template** for constructing a **complementary** daughter strand (**semiconservative replication**). DNA replication begins at an **origin of replication** where the two strands unwind and separate and synthesis occurs bidirectionally at two **replication forks** that move away from the origin. As **helicase separates the two strands**, positive **supercoils** are produced in the region of DNA ahead of the replication fork and negative supercoils behind the fork. **DNA topoisomerases types I** and **II** remove supercoils.

- **DNA polymerases (pols)** synthesize new DNA strands only in the **5′→3′** direction and require a short RNA **primer** created by **primase**. One of the new strands must grow in the 5′→3′ direction toward the replication fork (**leading strand**), while the other grows in the 5′→3′ direction away from the replication fork (**lagging strand**). Leading-strand synthesis is continuous from one RNA primer, whereas the lagging strand needs many primers (**discontinuous** synthesis involving **Okazaki fragments**).

- In *Escherichia coli*, DNA chain elongation is catalyzed by **DNA pol III**, using **5′-deoxyribonucleoside triphosphates** as substrates. The enzyme **proofreads** the newly synthesized DNA, removing terminal mismatched nucleotides with its **3′→5′ exonuclease** activity. RNA primers are removed by **DNA pol I**, using its **5′→3′ exonuclease** activity. This enzyme fills the gaps with DNA, proofreading as it synthesizes. The final phosphodiester linkage is catalyzed by **DNA ligase**.

- There are at least five high-fidelity **eukaryotic DNA pols**. Pol α is a multisubunit enzyme, one subunit of which is a **primase**. Pol α 5′→3′ polymerase activity adds a short piece of DNA to the RNA primer. **Pol ε** completes DNA synthesis on the leading strand, whereas **pol δ** elongates each lagging strand fragment. **Pol β** is involved with DNA repair, and **pol γ** replicates mitochondrial DNA. Pols ε, δ, and γ use 3′→5′ exonuclease activity to proofread.

- **Nucleoside analogs** containing modified sugars can be used to block DNA chain growth. They are useful in anticancer and antiviral chemotherapy.

- **Telomeres** are stretches of **highly repetitive DNA** that are bound by protein and protect the **ends** of linear chromosomes. As most cells divide and age, these sequences are shortened, contributing to senescence. In cells that do not senesce (e.g., germline and cancer cells), the ribonucleoprotein **telomerase** employs its protein component **reverse transcriptase** to extend the telomeres, using its **RNA** component as a **template**.

- Pairs of the positively charged histone proteins (H2A, H2B, H3, and H4) form an octameric structural core around which DNA is wrapped, creating a **nucleosome**. The DNA connecting the nucleosomes, called **linker DNA**, is bound to H1. Nucleosomes are packed more tightly to form a nucleofilament. Additional levels of organization create a chromosome.

- Three types of DNA repair correct most DNA damage in chromosomes: nucleotide excision repair (NER), base excision repair (BER), and mismatch repair (MMR). Each removes specific types of DNA damage (Fig. 30.33). NER removes **pyrimidine dimers** caused by ultraviolet radiation, BER replaces abnormal bases and AP sites, and MMR corrects mispaired bases caused by DNA pol errors. Defects in the **XP proteins** needed for NER in humans result in **xeroderma pigmentosum (XP)**. Defective MMR in humans, mainly caused by mutations in the *MSH2* and *MLH1* genes, is associated with **hereditary nonpolyposis colorectal cancer**.

- Double-strand breaks in DNA are repaired by **nonhomologous end joining** (error prone) and template-requiring **homologous recombination** ("error free").

DNA Structure

DNA Replication

consists of

Deoxyribose

Phosphate in a diester linkage

Bases
- Adenine
- Thymine
- Cytosine
- Guanine

in the form of

Deoxy-ribonucleotides

forming a

Double-stranded helical molecule

in which each strand has

Antiparallel orientation

creating

Polarity

in which the nucleotide sequence is read from the

5′-End — *to the* → **3′-End**

stabilized by

Hydrophobic interactions between stacked bases

Hydrogen bonds between base pairs

specifically

Adenine
Thymine ← *which pairs with*

and

Cytosine
Guanine ← *which pairs with*

Replication

Denaturation ← *as a result of* — **Single stranded** — *can become*

begins at an

Origin of replication

creating

Replication forks

where

Local separation of strands — *facilitated by* →
- **Helicase**
- **Single-stranded DNA–binding protein**

allows binding of

leads to → **Positive supercoils**

DNA polymerases

which synthesize

which are removed by

New DNA strands

only in the

DNA topoisomerases

5′→ 3′ direction — *extending* → **RNA primer**

resulting in

synthesized by

Bidirectional elongation

involving

Primase

Both strands of the double helix

which serve as

Templates

for constructing

Two complementary daughter strands — *a process called* → **Semiconservative replication**

consisting of

One leading strand

synthesized continuously in

5′→3′ direction toward the replication fork

requires

Only one RNA primer

One lagging strand

synthesized discontinuously in

5′→3′ direction away from the replication fork

requires

Many RNA primers

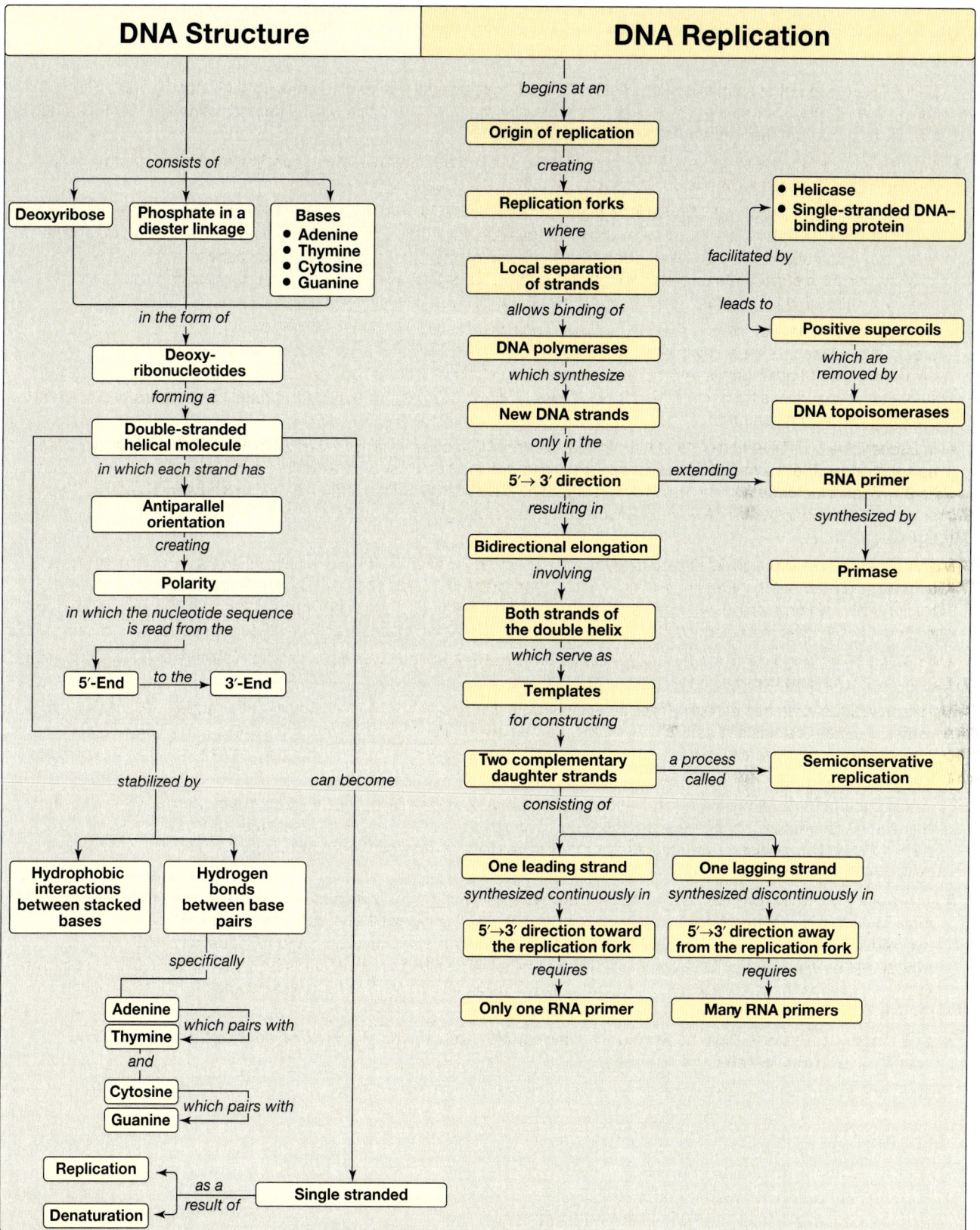

Figure 30.32
Key concept map for DNA structure and replication.

DNA Repair

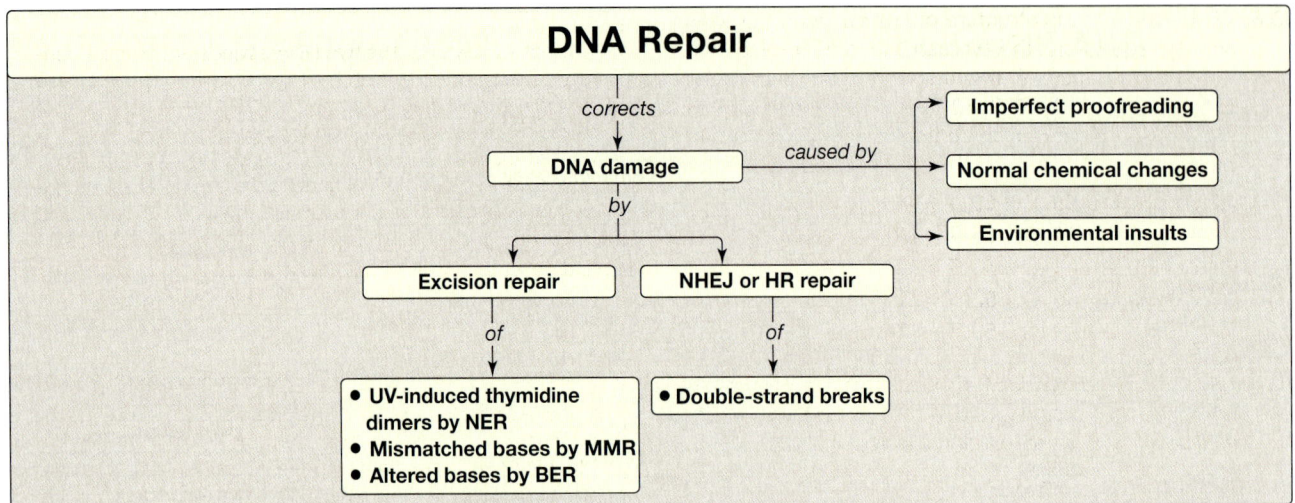

Figure 30.33
Key concept map for DNA repair. NHEJ, nonhomologous end joining; HR, homologous recombination; UV, ultraviolet; NER, nucleotide excision repair; MMR, mismatch repair; BER, base excision repair.

Study Questions

Choose the ONE best answer.

30.1. A 10-year-old female is brought by her parents to the dermatologist. She has many freckles on her face, neck, arms, and hands, and the parents report that she is unusually sensitive to sunlight. Two basal cell carcinomas are identified on her face. Which of the following processes is most likely to be defective in this patient?

A. Repair of double-strand breaks by error-prone homologous recombination

B. Removal of mismatched bases from the 3′ end of Okazaki fragments by a methyl-directed process

C. Removal of pyrimidine dimers from DNA by nucleotide excision repair

D. Removal of uracil from DNA by base excision repair

E. Removal of incorrectly paired nucleotides by Pol ε 3′→5′ exonuclease activity

> Correct answer = C. The sensitivity to sunlight, extensive freckling on parts of the body exposed to the sun, and presence of skin cancer at a young age indicate that the patient most likely has xeroderma pigmentosum (XP). These patients are deficient in any one of several XP proteins required for nucleotide excision repair of pyrimidine dimers in ultraviolet radiation–damaged DNA. Double-strand breaks are repaired by nonhomologous end joining (error prone) or homologous recombination ("error free"). Methylation is not used for strand discrimination in eukaryotic mismatch repair. Uracil is removed from DNA molecules by a specific glycosylase in base excision repair, but a defect in this process does not cause XP.

30.2. Telomeres are complexes of DNA and protein that protect the ends of linear chromosomes. In most normal human somatic cells, telomeres shorten with each division. In stem cells and cancer cells, however, telomeric length is maintained. In the synthesis of telomeres:

A. telomerase, a ribonucleoprotein, provides both the RNA and the protein needed for synthesis.

B. the RNA of telomerase serves as a primer.

C. the RNA of telomerase is a ribozyme.

D. the protein of telomerase is a DNA-directed DNA polymerase.

E. the shorter C-rich strand gets extended.

F. the direction of synthesis is 3′→5′.

> Correct answer = A. Telomerase is a ribonucleoprotein particle required for telomere maintenance. Telomerase contains an RNA that serves as the template, not the primer, for the synthesis of telomeric DNA by the reverse transcriptase of telomerase. Telomeric RNA has no catalytic activity. As a reverse transcriptase, telomerase synthesizes DNA using its RNA template and so is an RNA-directed DNA polymerase. The direction of synthesis, as with all DNA synthesis, is 5′→3′, and it is the 3′ end of the already longer G-rich strand that gets extended.

30.3. While studying the structure of a small gene that was sequenced during the Human Genome Project, an investigator notices that one strand of the DNA molecule contains 20 A, 25 G, 30 C, and 22 T. How many of each base is found in the complete double-stranded molecule?

    A. A = 40, G = 50, C = 60, T = 44
    B. A = 42, G = 55, C = 55, T = 42
    C. A = 44, G = 60, C = 50, T = 40
    D. A = 45, G = 45, C = 52, T = 52
    E. A = 50, G = 47, C = 50, T = 47

> Correct answer = B. The two DNA strands are complementary to each other, with A base paired with T and G base paired with C. So, for example, the 20 A on the first strand would be paired with 20 T on the second strand, the 25 G on the first strand would be paired with 25 C on the second strand, and so forth. When these are all added together, the correct numbers of each base are indicated in choice B. Note that, in the correct answer, A = T and G = C.

30.4. Based on the following 5′→3′ strand sequences, which DNA fragment will denature at the lowest temperature if all fragments are in solution at the same pH?

    A. AGGCCTCCAATA
    B. AGGGCTCCAATA
    C. AGGACTCCAATA
    D. ACGACTCCCATA
    E. AAGTTTGCAATA

> Correct answer = E. The melting temperature of DNA depends on its nucleotide composition. Base pairing between G and C creates three hydrogen bonds, while A and T base pairing makes only two hydrogen bonds. So, AT base pairs require less heat energy to "melt." This means that DNA with more AT base pairs will denature at a lower temperature. The structure of DNA bases can change if the pH of the solution changes. This will change the melting temperature of the DNA.

30.5. In human cells, DNA wraps around and binds to histone proteins to form nucleosomes. What property of the histone proteins allows them to bind tightly to DNA?

    A. Spherical tertiary structure
    B. Endonuclease activity and covalent bonding with end of a DNA strand
    C. Intercalation of aromatic amino acids into the DNA helix
    D. Methyl groups that create hydrophobic interactions
    E. Positive charges on amino acids

> Correct answer = E. Histones contain lysine and arginine amino acids that are positively charged at physiologic pH. These amino acids form ionic bonds with the negative phosphates in the phosphodiester backbone of DNA. Histones do not have endonuclease activity.

30.6. List the order in which the following enzymes participate in leading strand synthesis during prokaryotic replication.

    A. Ligase
    B. Polymerase I (3′→5′ exonuclease activity)
    C. Polymerase I (5′→3′ exonuclease activity)
    D. Polymerase I (5′→3′ polymerase activity)
    E. Polymerase III
    F. Primase

> Correct answer: F, E, C, D, B, A. Primase makes the RNA primer; polymerase (pol) III extends the primer with DNA (and proofreads); pol I removes the primer with its 5′→3′ exonuclease activity, fills in the gap with its 5′→3′ polymerase activity, and removes errors with its 3′→5′ exonuclease activity; and ligase makes the 5′-to-3′ phosphodiester bond that links the DNA made by pols I and III.

30.7. Dideoxynucleotides lack a 3′-hydroxyl group. Why would the incorporation of a dideoxynucleotide into DNA stop replication?

> The lack of the 3′-OH group prevents formation of the 3′ hydroxyl-to-5′-phosphate bond that links one nucleotide to the next in DNA.

31 RNA Structure, Synthesis, and Processing

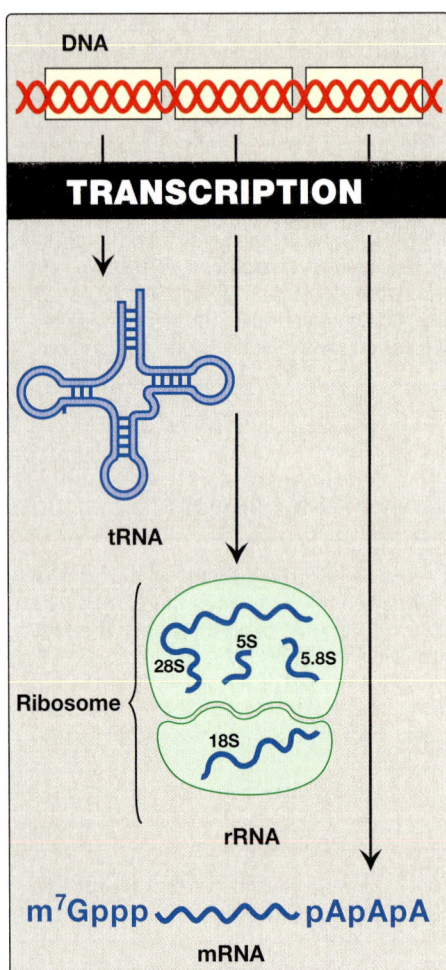

I. OVERVIEW

The genetic master plan of an organism is contained in the sequence of deoxyribonucleotides in its DNA. However, it is through ribonucleic acid (RNA), the "working copies" of DNA that the master plan is expressed (Fig. 31.1). The copying process, during which a DNA strand serves as a template for the synthesis of RNA, is called transcription. Transcription produces messenger RNA (mRNA), which is translated into sequences of amino acids (proteins), ribosomal RNA (rRNA), transfer RNA (tRNA), and additional RNA molecules that perform specialized structural, catalytic, and regulatory functions and are not translated. The untranslated RNA is called noncoding RNA (ncRNA). Therefore, the final product of gene expression can be RNA or protein, depending upon the gene. [Note: Only ~2% of the genome encodes proteins.] A central feature of transcription is that it is highly selective. For example, many transcripts are made from some regions of the DNA. In other regions, few or no transcripts are made. This selectivity is due, at least in part, to signals embedded in the nucleotide sequence of the DNA. These signals instruct the RNA polymerase (RNA pol) where to start, how often to start, and where to stop transcription. Several regulatory proteins are also involved in this selection process. The biochemical differentiation of an organism's tissues is ultimately a result of the selectivity of the transcription process. [Note: This selectivity of transcription is in contrast to the "all-or-none" nature of genomic replication.] Another important feature of transcription is that many primary RNA transcripts that are true copies of one of the two DNA strands may undergo various modifications, such as terminal additions, base modifications, trimming, and internal segment removal, which convert the inactive primary transcript into a functional molecule. The transcriptome is the complete set of RNA transcripts expressed by a genome.

II. RNA STRUCTURE

There are three major types of RNA that participate in the process of protein synthesis: rRNA, tRNA, and mRNA. Like DNA, these RNA are unbranched polymeric molecules composed of nucleoside monophosphates joined together by 3′-to-5′ phosphodiester bonds. However, they

Figure 31.1
Expression of genetic information by transcription. [Note: RNA shown are eukaryotic.] tRNA, transfer RNA; rRNA, ribosomal RNA; mRNA, messenger RNA; m⁷Gppp, 7-methylguanosine-triphosphate cap; pApApA, poly-A tail; p, phosphate.

510

differ from DNA in several ways. For example, they are considerably smaller than DNA, contain ribose instead of deoxyribose and uracil (U) instead of thymine (T), and exist as single strands that are capable of folding into complex structures. The three major types of RNA also differ from each other in size, function, and special structural modifications. In eukaryotes, additional small ncRNA molecules found in the nucleolus (small nucleolar RNA [snoRNA]), nucleus (small nuclear RNA [snRNA]), and cytoplasm (microRNA [miRNA]) perform specialized functions as described. [Note: In mammalian cells, a population of long ncRNA (lncRNA) molecules is produced at very low amounts. These lncRNA species appear to regulate gene expression in both the nucleus and cytoplasm.]

A. Ribosomal RNA

rRNA are found in association with several proteins as components of the ribosomes, the complex structures that serve as the sites for protein synthesis. Prokaryotic cells contain three distinct size species of rRNA (23S, 16S, and 5S, where S is the Svedberg unit for sedimentation rate that is determined by the size and shape of the particle), as shown in Figure 31.2. Eukaryotic cells contain four rRNA species (28S, 18S, 5.8S, and 5S) encoded by nuclear DNA and two rRNA species (12S and 16S) encoded by mitochondrial DNA. Together, rRNA make up ~80% of the total RNA in the cell. [Note: Some RNA function as catalysts, e.g., an rRNA in protein synthesis. RNA with catalytic activity is termed a ribozyme.]

B. Transfer RNA

tRNA are the smallest (4S) of the three major types of RNA molecules. There is at least one specific type of tRNA molecule for each of the 20 amino acids commonly found in proteins. Together, tRNA make up ~15% of the total RNA in the cell. The tRNA molecules contain a high percentage of unusual (modified) bases, for example, dihydrouracil (see Fig. 22.2), and have extensive intrachain base pairing that leads to characteristic secondary cloverleaf structure and tertiary structure (Fig. 31.3). Each tRNA serves as an adaptor molecule that carries its specific amino acid, covalently attached to its 3′ end, to the site of protein synthesis. There, it recognizes the genetic code sequence on an mRNA, which specifies the addition of that amino acid to the growing peptide chain. In eukaryotic cells, tRNA are encoded within both the nuclear and mitochondrial chromosomes.

The human mitochondrial chromosome carries 22 tRNA genes. Mutations in these genes can cause human disease. Mutations in the mitochondrial gene for tRNA^{Lys} are associated with myoclonic epilepsy (jerking muscle spasms) with ragged red fibers (MERRF), a disorder that affects skeletal muscle structure and function (myopathy), and with mitochondrial encephalomyopathy, lactic acidosis, and stroke-like episodes (MELAS), which affects the brain, nervous system, and muscles. MELAS is also caused by mutations in the mitochondrial tRNA^{Leu} gene.

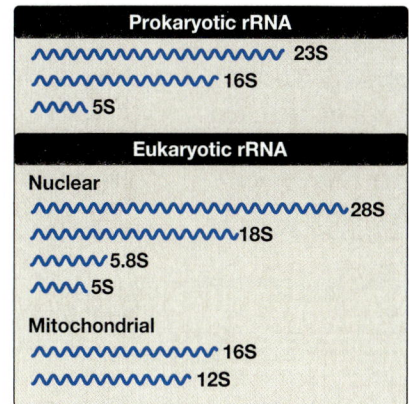

Figure 31.2
Prokaryotic and eukaryotic ribosomal RNA (rRNA). S, Svedberg unit.

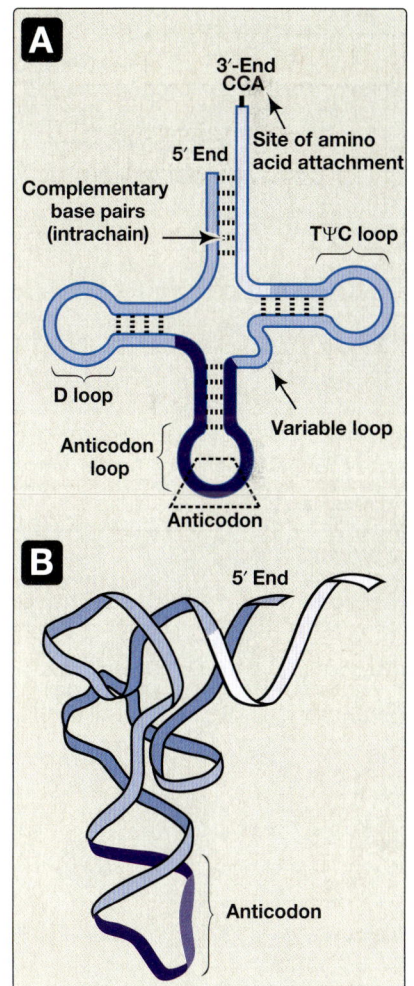

Figure 31.3
A. Characteristic transfer RNA (tRNA) secondary structure (cloverleaf).
B. Folded (tertiary) tRNA structure found in cells. D, dihydrouracil; Ψ, pseudouracil; T, thymine; C, cytosine; A, adenine.

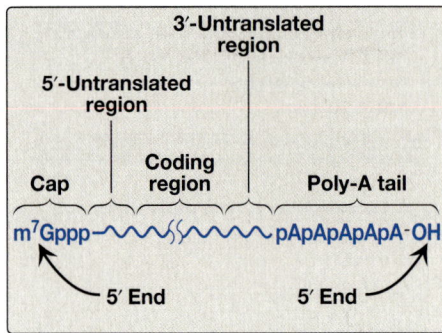

Figure 31.4
Structure of eukaryotic messenger RNA.
G, guanine; A, adenine.

C. Messenger RNA

mRNA comprises only ~5% of the RNA in a cell, yet is by far the most heterogeneous type of RNA in size and base sequence. mRNA is coding RNA in that it carries genetic information from DNA for use in protein synthesis. In eukaryotes, this involves transport of mRNA out of the nucleus and into the cytosol. An mRNA carrying information from more than one gene is polycistronic (cistron = gene). Polycistronic mRNA is characteristic of prokaryotes, mitochondria, some viruses, and plant chloroplasts. An mRNA carrying information from only one gene is monocistronic and is characteristic of eukaryotes. In addition to the protein-coding regions that can be translated, mRNA contains untranslated regions at its 5′ and 3′ ends (Fig. 31.4). Special structural characteristics of eukaryotic (but not prokaryotic) mRNA include a long sequence of adenine (A) nucleotides (a poly-A tail) on the 3′ end of the RNA, plus a cap on the 5′ end consisting of a molecule of 7-methylguanosine attached through an unusual (5′-to-5′) triphosphate linkage (see Fig. 31.17). The mechanisms for modifying mRNA to create these special structural characteristics are discussed.

III. PROKARYOTIC GENE TRANSCRIPTION

The structure of magnesium-requiring RNA pol, the signals that control transcription, and the varieties of modification that RNA transcripts can undergo differ among organisms, particularly from prokaryotes to eukaryotes. Therefore, the discussions of prokaryotic and eukaryotic transcription are presented separately.

A. Prokaryotic RNA polymerase

In bacteria, one species of RNA pol synthesizes all of the RNA except for the short RNA primers needed for DNA replication [Note: RNA primers are synthesized by the specialized, monomeric enzyme primase.] RNA pol is a multisubunit enzyme that recognizes a nucleotide sequence (the promoter region) at the beginning of a length of DNA that is to be transcribed. From the end of the promoter region, it makes a complementary RNA copy of the DNA template strand and then recognizes a DNA sequence (the termination region) that signals for the end of transcription. RNA is synthesized from its 5′ end to its 3′ end, antiparallel to its DNA template strand. The template is copied as it is in DNA synthesis, in which a guanine (G) on the DNA specifies a cytosine (C) in the RNA, a C specifies a G, a T specifies an A, but an A specifies a U instead of a T (Fig. 31.5). The RNA, then, is complementary to the DNA template (antisense, minus) strand and identical to the coding (sense, plus) strand, with U replacing T. Within the DNA molecule, regions of both strands can serve as templates for transcription. For a given gene, however, only one of the two DNA strands can be the template. Which strand is used is determined by the location of the promoter for that gene. Transcription by RNA pol involves a core enzyme and several auxiliary proteins.

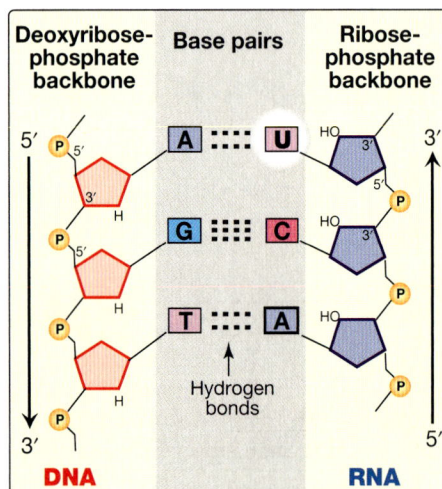

Figure 31.5
Antiparallel, complementary base pairs between DNA and RNA. T, thymine; A, adenine; C, cytosine; G, guanine; U, uracil.

1. **Core enzyme:** Five peptide subunits, 2 α, 1 β, 1 β′, and 1 Ω, are required for RNA pol function and form the core enzyme. These are necessary for RNA pol assembly (α, Ω), template binding (β′),

and 5′→3′ polymerase activity (β) (Fig. 31.6). However, this core enzyme lacks specificity (i.e., it cannot recognize the promoter region on the DNA template).

2. **Holoenzyme:** The σ subunit (sigma factor) enables RNA pol to recognize promoter regions on the DNA. The σ subunit plus the core enzyme make up the holoenzyme. Different σ factors recognize different groups of genes. The major σ factor in *Escherichia coli* is σ^{70}.

B. Steps in RNA synthesis

The process of transcription of a typical gene of *E. coli* can be divided into three phases: initiation, elongation, and termination. A transcription unit extends from the promoter to the termination region, and the initial product of transcription by RNA pol is termed the primary transcript.

1. **Initiation:** Transcription begins with the binding of the RNA pol holoenzyme to a region of the DNA known as the promoter, which is not transcribed. The prokaryotic promoter contains characteristic consensus sequences (Fig. 31.7). [Note: In consensus sequences, the base shown at each position is the base most frequently encountered at that position.] Those that are recognized by prokaryotic RNA pol σ factors include the following.

    a. **-35 Sequence:** A consensus sequence (5′-TTGACA-3′), centered about 35 bases to the left of the transcription start site (see Fig. 31.7), is the initial point of contact for the holoenzyme, and a closed complex is formed. [Note: By convention, the regulatory sequences that control transcription are designated by the 5′→3′ nucleotide sequence on the coding strand. A base in the promoter region is assigned a negative number if it occurs prior to (to the left of, toward the 5′ end of, or "upstream" of) the transcription start site. Therefore, the TTGACA sequence is centered at approximately base –35. The first base at the transcription start site is assigned a position of +1. There is no base designated "0."]

    b. **Pribnow box:** The holoenzyme moves and covers a second consensus sequence (5′-TATAAT-3′), centered at about –10 (see Fig. 31.7), which is the site of melting (unwinding) of a short stretch (~14 base pairs) of DNA. This initial melting

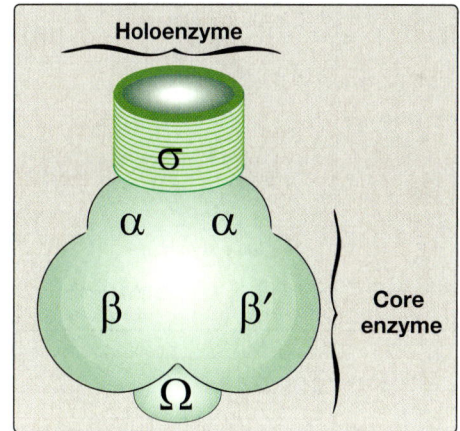

Figure 31.6
Components of prokaryotic RNA polymerase.

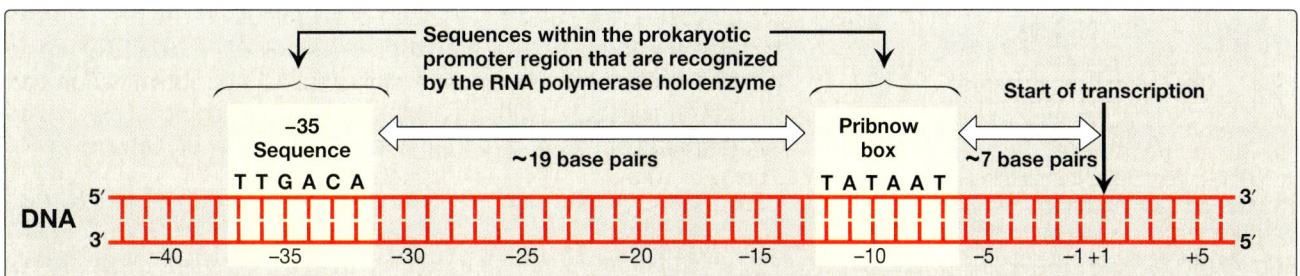

Figure 31.7
Structure of the prokaryotic promoter region. T, thymine; G, guanine; A, adenine; C, cytosine.

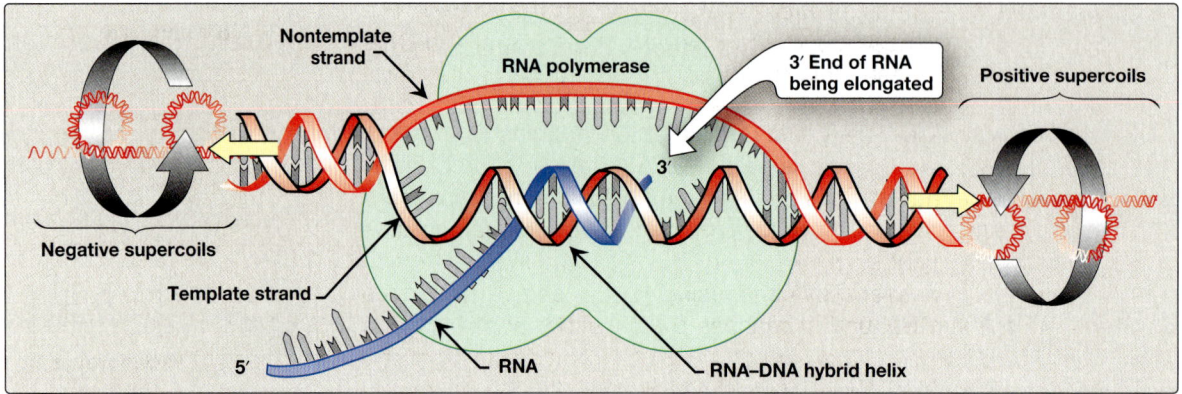

Figure 31.8
Local unwinding of DNA by RNA polymerase and formation of an open initiation complex (transcription bubble).

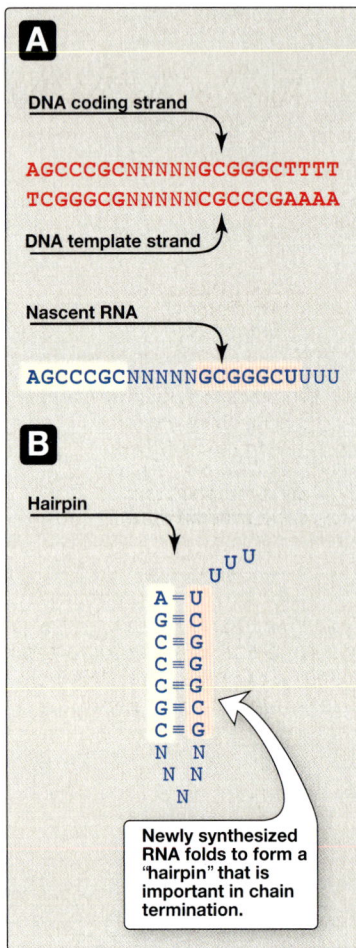

Figure 31.9
Rho-independent termination of prokaryotic transcription. **A.** DNA template sequence generates a self-complementary sequence in the nascent RNA. **B.** Hairpin structure formed by the RNA. N represents a noncomplementary base; A, adenine, T, thymine; G, guanine; C, cytosine; U, uracil.

converts the closed initiation complex to an open complex known as a transcription bubble. The Pribnow box consensus sequence is recognized by σ^{70}. [Note: A mutation in either the −10 or the −35 sequence can affect the transcription of the gene controlled by the mutant promoter.]

2. **Elongation:** Once the promoter has been recognized and bound by the holoenzyme, local unwinding of the DNA helix, mediated by the polymerase, continues (Fig. 31.8). [Note: Unwinding generates supercoils in the DNA that can be relieved by DNA topoisomerases.] RNA pol begins to synthesize a transcript of the DNA sequence, and several short pieces of RNA are made and discarded. The elongation phase begins when the transcript (typically starting with a purine) exceeds 10 nucleotides in length. Sigma factor is then released, and the core enzyme is able to leave (clear) the promoter and move along the template strand in a processive manner, serving as its own sliding clamp. During transcription, a short DNA–RNA hybrid helix is formed (see Fig. 31.8). Like DNA pol, RNA pol uses nucleoside triphosphates as substrates and releases pyrophosphate each time a nucleoside monophosphate is added to the growing chain. As with replication, transcription is always in the 5′→3′ direction. In contrast to DNA pol, RNA pol does not require a primer and does not have a 3′→5′ exonuclease domain for proofreading. Misincorporation of a ribonucleotide causes RNA pol to pause, backtrack, cleave the transcript, and restart. Nonetheless, transcription has a higher error rate than does replication.

3. **Termination:** The elongation of the single-stranded RNA chain continues until a termination signal is reached. Termination can be intrinsic (occur without additional proteins) or dependent upon the participation of a protein known as the ρ (rho) factor.

   a. **Rho independent:** For most prokaryotic genes, this type of termination requires a sequence in the DNA template for generating a sequence in the nascent (newly made) RNA that is self-complementary (Fig. 31.9). This allows the RNA to fold back on itself, forming a GC-rich stem (stabilized by hydrogen bonds) plus a loop. This structure is known as a "hairpin." Additionally, just beyond the hairpin, the RNA transcript contains a string of uracil residues (Us) at the 3′ end. The bonding of these Us to the complementary As of the DNA template is weak. This facilitates the separation of the newly synthesized RNA from its DNA template, as the double helix "zips up" behind the RNA pol.

b. **Rho dependent:** This requires the participation of the additional protein rho, which is a hexameric ATPase with helicase activity. Rho binds a C-rich rho utilization (rut) site near the 5′ end of the nascent RNA and, using its ATPase activity, moves along the RNA until it reaches the RNA pol paused at the termination site. The ATP-dependent helicase activity of rho separates the RNA–DNA hybrid helix, causing the release of the RNA.

4. **Antibiotics:** Some antibiotics prevent bacterial cell growth by inhibiting RNA synthesis. For example, rifampin (rifampicin) inhibits transcription initiation by binding to the β subunit of prokaryotic RNA pol and preventing chain growth beyond three nucleotides (Fig. 31.10). Rifampin is important in the treatment of tuberculosis. Dactinomycin (actinomycin D) was the first antibiotic to find therapeutic application in tumor chemotherapy. It inserts (intercalates) between the DNA bases and inhibits transcription initiation and elongation in tumor cells.

IV. EUKARYOTIC GENE TRANSCRIPTION

The transcription of eukaryotic genes is a far more complicated process than transcription in prokaryotes. Eukaryotic transcription involves separate polymerases for the synthesis of rRNA, tRNA, and mRNA. In addition, a large number of proteins called transcription factors (TFs) are involved. TFs bind to distinct sites (elements) on the DNA within the core promoter region, close (proximal) to it, or some distance away (distal). They are required for both the assembly of a transcription initiation complex at the promoter and the determination of which genes are to be transcribed. [Note: Each eukaryotic RNA pol has its own promoters and TFs that bind core promoter sequences.] For TFs to recognize and bind to their specific DNA sequences, the chromatin structure in that region must be decondensed (relaxed) to allow access to the DNA. The role of transcription in the regulation of gene expression is discussed in Chapter 33.

A. Chromatin structure and gene expression

The association of DNA with histones to form nucleosomes affects the ability of the transcription machinery to access the DNA to be transcribed. Most actively transcribed genes are found in a relatively decondensed form of chromatin called euchromatin, whereas most inactive segments of DNA are found in highly condensed heterochromatin. The interconversion of these forms is called chromatin remodeling. A major component of chromatin remodeling is the covalent modification of histones (e.g., the acetylation of lysine residues at the amino terminus of histone proteins), as shown in Figure 31.11. Acetylation, mediated by histone acetyltransferases (HATs), eliminates the positive charge on the lysine, thereby decreasing the interaction

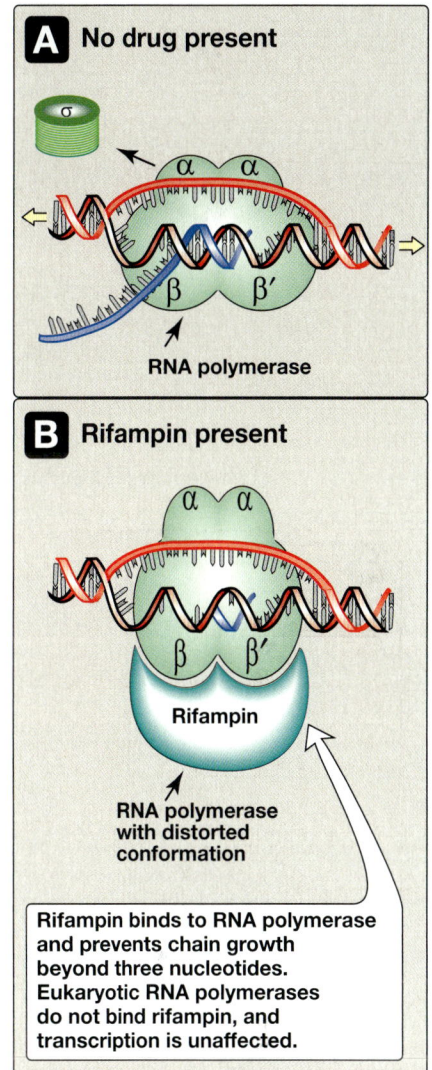

A No drug present

RNA polymerase

B Rifampin present

Rifampin

RNA polymerase with distorted conformation

Rifampin binds to RNA polymerase and prevents chain growth beyond three nucleotides. Eukaryotic RNA polymerases do not bind rifampin, and transcription is unaffected.

Figure 31.10
A. Prokaryotic transcript elongation by RNA polymerase with no drug present. **B.** Inhibition of prokaryotic RNA polymerase by rifampin (rifampicin).

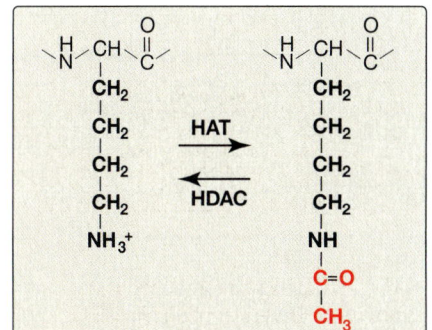

Figure 31.11
Acetylation/deacetylation of a lysine residue in a histone. Acetyl coenzyme A provides the acetyl group. HAT, histone acetyltransferase; HDAC, histone deacetylase.

of the histone with the negatively charged DNA. Removal of the acetyl group by histone deacetylases (HDACs) restores the positive charge and fosters stronger interactions between histones and DNA. [Note: The ATP-dependent repositioning of nucleosomes is also required to access DNA.]

B. Nuclear RNA polymerases

There are three distinct types of RNA pol in the nucleus of eukaryotic cells. All are large enzymes with multiple subunits. Each type of RNA pol recognizes particular genes. [Note: Mitochondria contain a single RNA pol that resembles the bacterial enzyme in its function. This enzyme is responsible for transcribing all genes on the mitochondrial chromosome.]

1. **RNA polymerase I:** This enzyme synthesizes the precursor of the 28S, 18S, and 5.8S rRNA within the nucleolus.

2. **RNA polymerase II:** This enzyme synthesizes the primary RNA transcripts in the nucleus that are processed to form mRNA. Mature mRNA is translated to make proteins. RNA pol II also synthesizes certain small ncRNA, such as snoRNA, snRNA, and miRNA, and lncRNA.

   a. **Promoters for RNA polymerase II:** In some genes transcribed by RNA pol II, a sequence of nucleotides (TATAAA) that is nearly identical to that of the Pribnow box is found centered ~25 nucleotides upstream of the transcription start site. This core promoter consensus sequence is called the TATA, or Hogness, box. In the majority of genes, however, no TATA box is present. Instead, different core promoter elements such as initiator (Inr) or downstream promoter element (DPE) are present (Fig. 31.12). [Note: No one consensus sequence is found in all core promoters.] Because these sequences are on the same molecule of DNA as the gene being transcribed, they are *cis*-acting. The sequences serve as binding sites for proteins known as general transcription factors (GTFs), which, in turn, interact with each other and with RNA pol II.

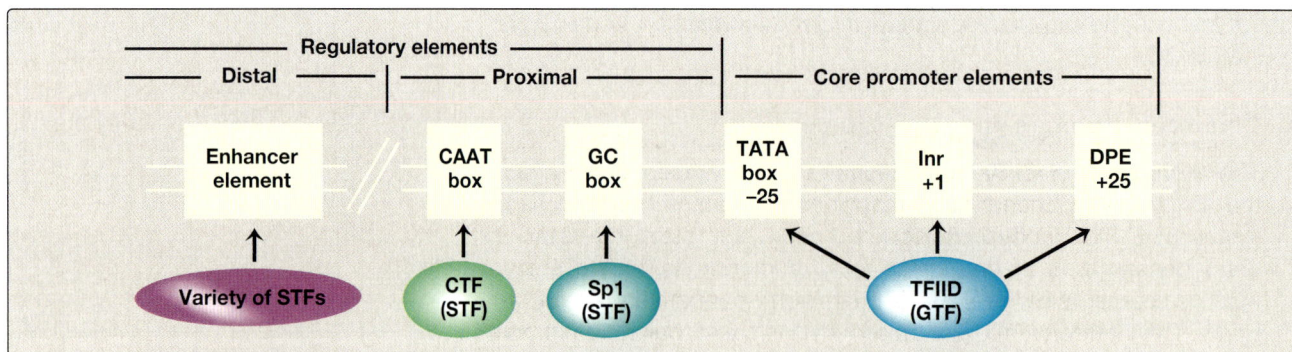

Figure 31.12
Eukaryotic gene *cis*-acting promoter and regulatory elements and their *trans*-acting general and specific transcription factors (GTFs and STFs, respectively). Inr, initiator; DPE, downstream promoter element.

b. **General transcription factors:** GTFs are the essential proteins required for recognition of the promoter, recruitment of RNA pol II to the promoter, formation of the preinitiation complex, and initiation of transcription at a basal level (Fig. 31.13A). GTFs are encoded by different genes, are synthesized in the cytosol, and diffuse (transit) to their sites of action, and so are *trans*-acting. In contrast to the prokaryotic holoenzyme, eukaryotic RNA pol II does not itself recognize and bind the promoter. Instead, transcription factor IID (TFIID), a GTF containing TATA-binding protein and TATA-associated factors, recognizes and binds the TATA box (and other core promoter elements). TFIIF, another GTF, brings the polymerase to the promoter. The helicase activity of TFIIH melts the DNA, and its kinase activity phosphorylates polymerase, allowing it to clear the promoter.

c. **Regulatory elements and transcriptional activators:** Additional consensus sequences lie upstream of the core promoter (see Fig. 31.12). Those close to the core promoter (within ~200 nucleotides) are the proximal regulatory elements, such as the CAAT and GC boxes. Those farther away are the distal regulatory elements such as enhancers (see Section IV. B. 2. d.). Proteins known as transcriptional activators or specific transcription factors (STFs) bind these regulatory elements. STFs bind to promoter proximal elements to regulate the frequency of transcription initiation and to distal elements to mediate the response to signals such as hormones and regulate which genes are expressed at a given point in time. A typical protein-coding eukaryotic gene has binding sites for many such factors. STFs have two binding domains. One is a DNA-binding domain, the other is a transcription activation domain that recruits the GTFs to the core promoter as well as coactivator proteins such as the HAT enzymes involved in chromatin modification. [Note: The mediator, a multisubunit coactivator of RNA pol II–catalyzed transcription, binds the polymerase, the GTF, and the STF and regulates transcription initiation.]

> **II** Transcriptional activators bind DNA through a variety of motifs, such as the helix-loop-helix, zinc finger, and leucine zipper.

d. **Role of enhancers:** Enhancers are distal regulatory elements that increase the rate of initiation of transcription by RNA pol II. Enhancers are typically on the same chromosome as the gene whose transcription they stimulate (Fig. 31.13B). However, they can (1) be located upstream (to the 5′ side) or downstream (to the 3′ side) of the transcription start site, (2) be close to or thousands of base pairs away from the promoter (Fig. 31.14), and (3) occur on either strand of the DNA. Enhancers contain DNA sequences called response elements that bind STFs. By bending or looping the DNA, STFs can interact with other TFs bound to a promoter and

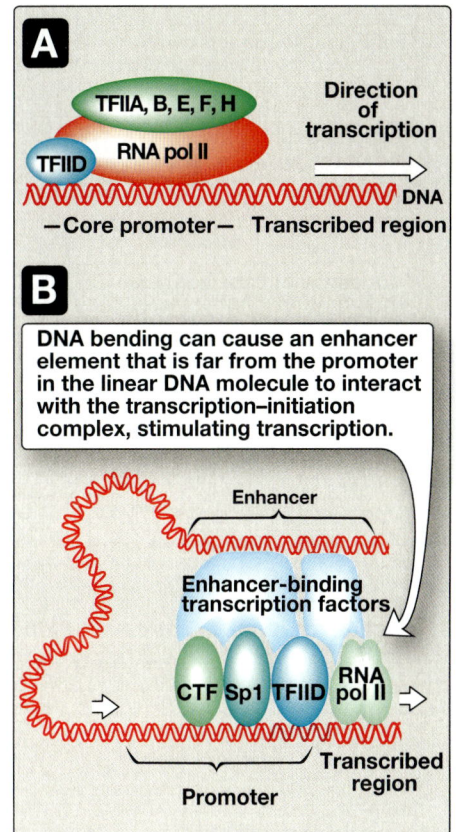

Figure 31.13
A. Association of the general transcription factors (TFII) and RNA polymerase II (RNA pol II) at the core promoter. [Note: The Roman numeral II denotes a TF for RNA pol II.] **B.** Enhancer stimulation of transcription. CTF, CAAT box transcription factor; Sp1, specificity factor-1.

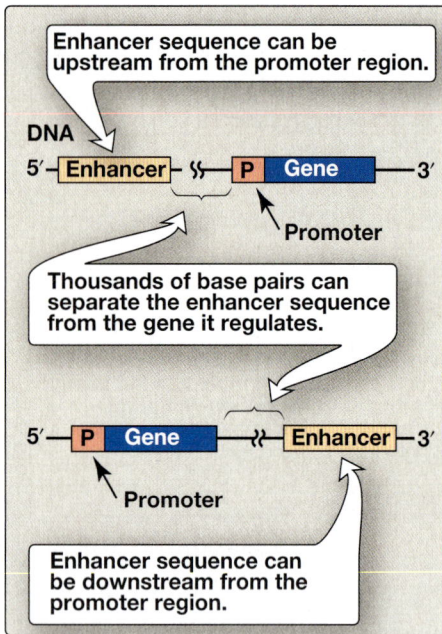

Figure 31.14
Some possible locations of enhancer sequences.

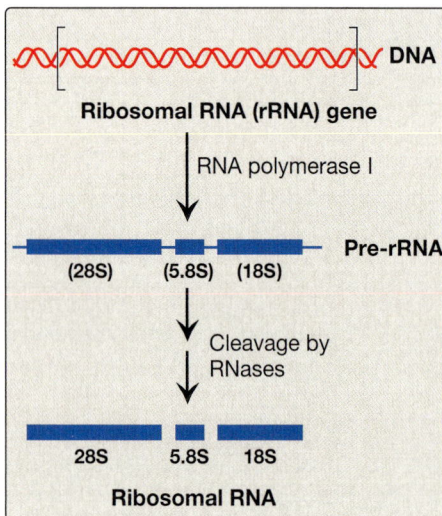

Figure 31.15
Posttranscriptional processing of eukaryotic ribosomal RNA by ribonucleases (RNases). S, Svedberg unit.

associated with RNA pol II, thereby stimulating transcription (see Fig. 31.13B). The mediator also binds enhancers. [Note: Although silencers are similar to enhancers in that they also can act over long distances, they reduce gene expression.]

e. **RNA polymerase II inhibitor:** α-Amanitin, a potent toxin produced by the poisonous mushroom *Amanita phalloides* (sometimes called the "death cap"), binds RNA pol II tightly and slows its movement, thereby inhibiting mRNA synthesis.

3. **RNA polymerase III:** This enzyme synthesizes tRNA, 5S rRNA, and some snRNA and snoRNA.

V. POSTTRANSCRIPTIONAL MODIFICATION OF RNA

A primary transcript is the initial, linear, RNA copy of a transcription unit (the segment of DNA between specific initiation and termination sequences). The primary transcripts of both prokaryotic and eukaryotic tRNA and rRNA are posttranscriptionally modified by cleavage of the original transcripts by ribonucleases. tRNA are further modified to help give each species its unique identity. In contrast, prokaryotic mRNA is generally identical to its primary transcript, whereas eukaryotic mRNA is extensively modified both co- and posttranscriptionally.

A. Ribosomal RNA

rRNA of both prokaryotic and eukaryotic cells is generated from long precursor molecules called pre-rRNA. The 23S, 16S, and 5S rRNA of prokaryotes are produced from a single pre-rRNA molecule, as are the 28S, 18S, and 5.8S rRNA of eukaryotes (Fig. 31.15). [Note: Eukaryotic 5S rRNA is synthesized by RNA pol III and modified separately.] The pre-rRNA are cleaved by ribonucleases to yield intermediate-sized pieces of rRNA, which are further processed (trimmed by exonucleases and modified at some bases and riboses) to produce the required RNA species. [Note: In eukaryotes, nuclear rRNA genes are found in long, tandem arrays and are transcribed and processed in the nucleolus, with base and sugar modifications facilitated by snoRNA. Mitochondrial 12S and 16S rRNA undergo base and sugar modifications by specific enzymes within the organelle.]

B. Transfer RNA

Both eukaryotic and prokaryotic tRNA are also made from longer precursor molecules that must be modified (Fig. 31.16). Sequences at both ends of the molecule are removed, and, if present, an intervening sequence intron is removed from the anticodon loop by nucleases. Other posttranscriptional modifications include addition of a –CCA sequence by nucleotidyltransferase to the 3′ terminal end of tRNA and modification of bases at specific positions to produce the unusual bases characteristic of tRNA.

C. Eukaryotic messenger RNA

The collection of all the primary transcripts synthesized in the nucleus by RNA pol II is known as heterogeneous nuclear RNA (hnRNA). The pre-mRNA components of hnRNA undergo extensive co- and

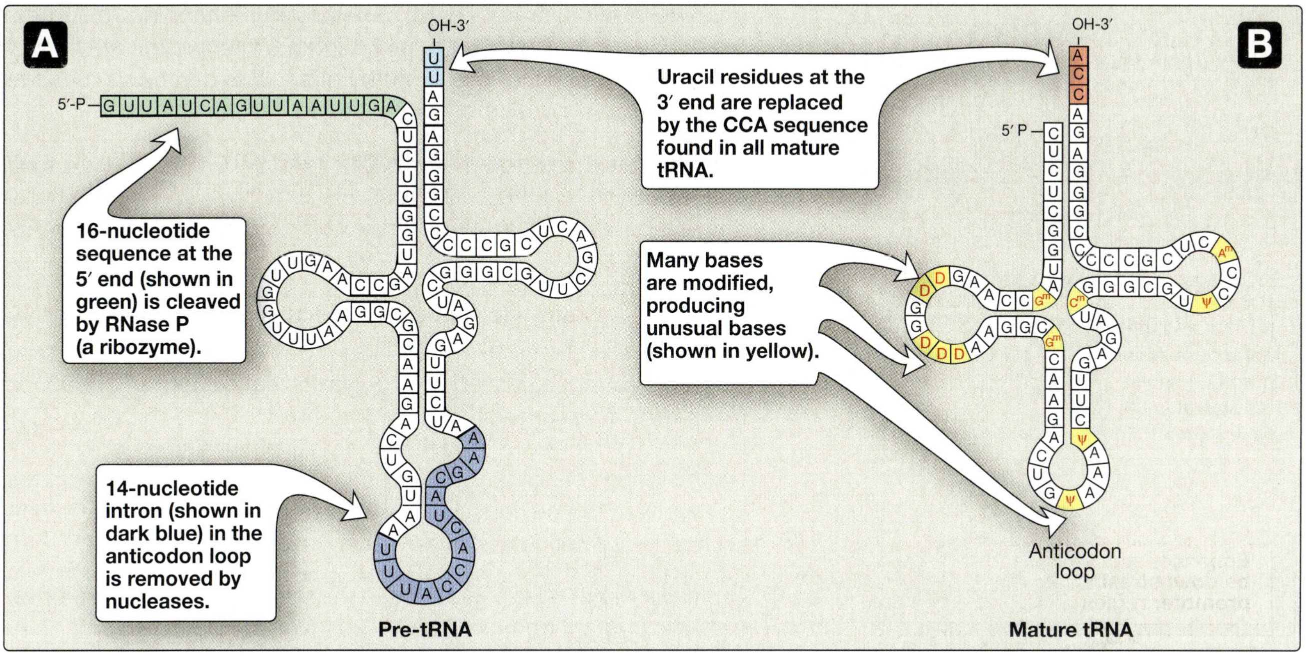

Figure 31.16
A. Precursor transfer RNA (pre-tRNA) transcript. **B.** Mature (functional) tRNA after posttranscriptional modification. Modified bases include D (dihydrouracil), ψ (pseudouracil), and ᵐ, which means that the base has been methylated.

posttranscriptional modification in the nucleus and become mature mRNA. These modifications usually include the following. [Note: Pol II itself recruits the proteins required for the modifications.]

1. **Addition of a 5′ cap:** This is the first of the processing reactions for pre-mRNA (Fig. 31.17). The cap is a 7-methylguanosine attached to the 5′ terminal end of the mRNA through an unusual 5′-to-5′ triphosphate linkage that is resistant to most nucleases. Creation of the cap requires removal of the γ-phosphoryl group from the 5′ triphosphate of the pre-mRNA, followed by addition of guanosine monophosphate (from guanosine triphosphate) by the nuclear enzyme guanylyltransferase. Methylation of this terminal G occurs in the cytosol and is catalyzed by

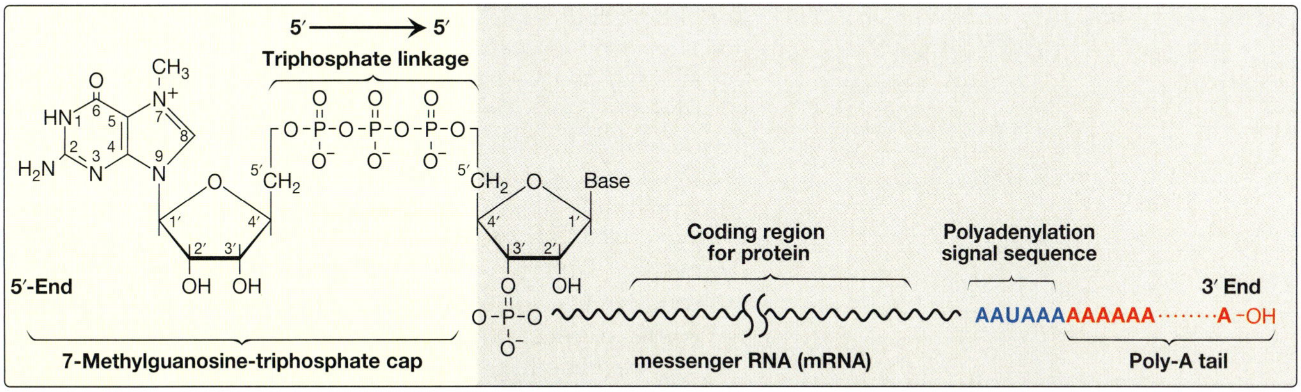

Figure 31.17
Posttranscriptional modification of mRNA showing the 7-methylguanosine cap and polyadenylate (poly-A) tail.

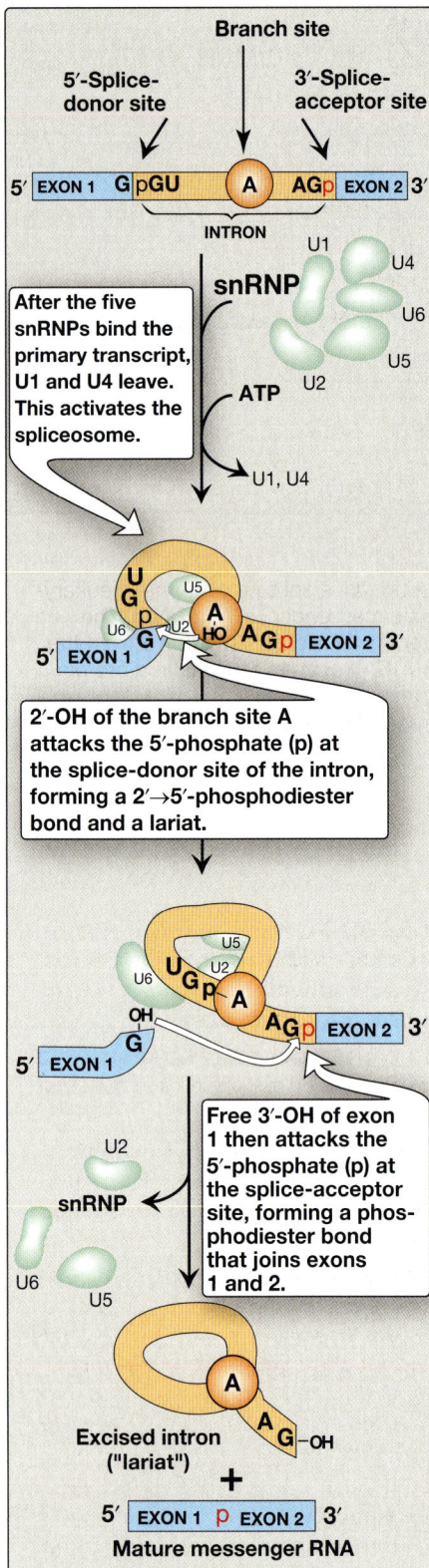

Figure 31.18
Splicing. [Note: U1 binds the 5′ donor site, and U2 binds the branch A and the 3′ acceptor site. Addition of U4–U6 completes the complex.] snRNP, small nuclear ribonucleoprotein particle.

guanine-7-methyltransferase. S-Adenosylmethionine (SAM) is the source of the methyl group. Additional methylation steps may occur. The addition of this 7-methylguanosine cap helps stabilize the mRNA and permits efficient initiation of translation.

2. **Addition of a 3′-poly-A tail:** Most eukaryotic mRNA (with several exceptions, including those for the histones and those in mitochondria) have a chain of 40 to 250 adenylates (adenosine monophosphates) attached to the 3′ end (see Fig. 31.17). This poly-A tail is not transcribed from the DNA but rather is added by the nuclear enzyme, polyadenylate polymerase, using ATP as the substrate. The pre-mRNA is cleaved downstream of a consensus sequence, called the polyadenylation signal sequence (AAUAAA), found near the 3′ end of the RNA, and the poly-A tail is added to the new 3′ end. The poly-A tail terminates eukaryotic transcription. In addition, it helps stabilize the mRNA, facilitates its exit from the nucleus, and aids in translation. After the mRNA enters the cytosol, the poly-A tail is gradually shortened.

3. **Splicing:** Maturation of eukaryotic mRNA usually involves removal from the primary transcript of RNA sequences (introns or intervening sequences) that do not code for protein. The remaining coding (expressed) sequences, the exons, are joined together to form the mature mRNA. The process of removing introns and joining exons is called splicing. The molecular complex that accomplishes these tasks is known as the spliceosome. A few eukaryotic primary transcripts contain no introns (e.g., those from histone genes and from mitochondrial DNA). Others contain a few introns, whereas some, such as the primary transcripts for the α-chains of collagen, contain more than 50 introns that must be removed.

   a. **Role of small nuclear RNA:** In association with multiple proteins, U-rich snRNA form five small nuclear ribonucleoprotein particles (snRNPs, or "snurps") designated as U1, U2, U4, U5, and U6 that mediate splicing. This group of snRNP facilitates the removal of introns by forming base pairs with the consensus sequences at each end of the intron (Fig. 31.18). [Note: In systemic lupus erythematosus, an autoimmune disease, patients produce antibodies against their own nuclear proteins such as snRNP.]

   b. **Mechanism:** The binding of snRNPs brings the sequences of neighboring exons into the correct alignment for splicing, allowing two transesterification reactions (catalyzed by the RNA of U2, U5, and U6) to occur. The 2′-OH group of an A nucleotide (known as the branch site A) in the intron attacks the phosphate at the 5′ end of the intron (splice donor site), forming an unusual 2′→5′ phosphodiester bond and creating a "lariat" structure (see Fig. 31.18). The newly freed 3′-OH of exon 1 attacks the 5′ phosphate at the splice acceptor site, forming a phosphodiester bond that joins exons 1 and 2. The excised intron is released as a lariat, which is typically degraded but may be a precursor for ncRNA such as snoRNA. [Note: The GU and AG sequences at the beginning and end, respectively, of introns are invariant. However, additional sequences are critical for splice-site recognition.] After introns have been removed and exons joined, the mature mRNA molecules pass into the cytosol through pores in the nuclear membrane. [Note: The introns in tRNA (see Fig. 31.16) are removed by a different mechanism.]

   c. **Effect of splice-site mutations:** Mutations at splice sites can lead to improper splicing and the production of aberrant proteins. It is estimated that at least 20% of all genetic diseases are a result of mutations that affect RNA splicing. For example, mutations that cause the incorrect splicing of β-globin mRNA are responsible for some cases of β-thalassemia, a disease in which the production of the β-globin protein is defective. Splice-site mutations can result in exons being skipped (removed) or introns being retained. These mutations can also activate cryptic splice sites, which are sites that contain the 5′ or 3′ consensus sequence but are not normally used.

4. **Alternative splicing:** The pre-mRNA molecules from more than 90% of human genes can be spliced in alternative ways in different tissues. Because this produces multiple variations of the mRNA and, therefore, of its protein product, it is a mechanism for producing a large, diverse set of proteins from a limited set of genes (Fig. 31.19). For example, the mRNA for tropomyosin (TM), an actin filament–binding protein of the cytoskeleton (and of the contractile apparatus in muscle cells), undergoes extensive tissue-specific, alternative splicing with production of multiple isoforms of the TM protein.

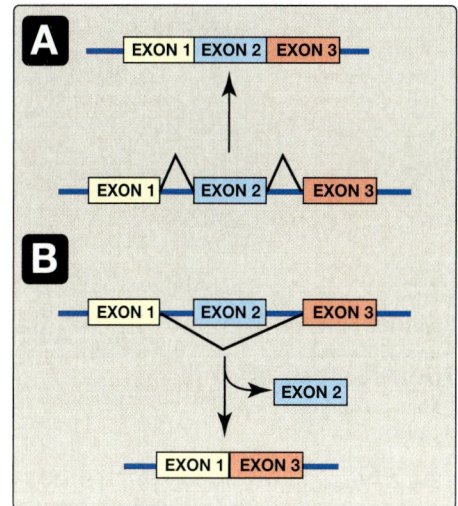

Figure 31.19
Alternative splicing patterns in eukaryotic messenger RNA (mRNA). The removal (skipping) of exon 2 from the mRNA in panel B results in a protein product that is different than the one made from the mRNA in panel A.

VI. CHAPTER SUMMARY

- Three major types of RNA participate in the process of protein synthesis: **ribosomal (rRNA)**, **transfer (tRNA)**, and **messenger (mRNA)**. RNA differs from DNA by containing **ribose** instead of deoxyribose and **U** instead of T. **rRNA** is a component of the **ribosomes**. **tRNA** serves as an **adaptor** molecule that carries a specific amino acid to the site of protein synthesis. **mRNA** (coding RNA) carries genetic information from DNA for use in protein synthesis.

- The process of RNA synthesis is called **transcription**. The enzyme that synthesizes RNA, **RNA polymerase (pol)**, uses **ribonucleoside triphosphates** as substrates for **5′→3′ polymerase activity**. In both prokaryotes and eukaryotes, RNA pol does not require a primer.

- In **prokaryotic** cells, the **core RNA pol enzyme** has five subunits (2 α, 1 β, 1 β′, and 1 Ω). The core enzyme requires an additional subunit, **sigma (σ) factor**, to recognize the nucleotide sequence (**promoter** region) in DNA. This region contains **consensus sequences** that are highly conserved and include the **–10 Pribnow box** and the **–35 sequence**. Another protein, **rho (ρ)**, is required for **termination** of transcription of some genes.

- In the **eukaryotic** cell nucleus, there are three distinct types of RNA pol. **RNA pol I** synthesizes the precursor of rRNA in the nucleolus. **RNA pol II** synthesizes the precursors for mRNA and some ncRNA, and **RNA pol III** synthesizes the precursors of tRNA and 5S rRNA. Core **promoters** for genes transcribed by **RNA pol II** contain *cis*-acting consensus sequences, such as the **TATA (Hogness) box**, which serve as binding sites for *trans*-acting general transcription factors (**GTFs**). Upstream of these are **proximal** regulatory elements, such as the CAAT and GC boxes, and **distal** regulatory elements, such as **enhancers**. **Specific transcription factors** ([STFs] transcriptional activators) and **mediator complex** bind these elements and regulate gene expression. Eukaryotic transcription requires that the **chromatin** be relaxed (decondensed) in a process known as **chromatin remodeling**.

- A **primary transcript** is a linear copy of a **transcription unit**, the segment of DNA between specific initiation and termination sequences. Prokaryotic mRNA is generally identical to its primary transcript, whereas eukaryotic **pre-mRNA** is extensively modified co- and posttranscriptionally. For example, a **7-methylguanosine cap** is attached to the 5′ end of the mRNA through a 5′-to-5′ linkage. A **long poly-A tail** is attached by polyadenylate polymerase to the 3′ end of most mRNA. Most eukaryotic mRNA also contains **intervening sequences** (**introns**) that must be removed for the mRNA to be functional. Their removal, as well as the joining of **expressed sequences** (**exons**), requires a **spliceosome** composed of **"snurps"** that mediate the process of **splicing**. Eukaryotic mRNA is **monocistronic**, containing information from just one gene, whereas prokaryotic mRNA is **polycistronic** (Fig. 31.20).

RNA Structure

consists of

| Ribose | Phosphate in a diester linkage | Bases |

consisting of

- Adenine (A)
- Uracil (U)
- Cytosine (C)
- Guanine (G)

producing

Ribonucleotides

polymers of which form

rRNA

Structural features include

- Associates with proteins
- Three size species in prokaryotes
- Four size species in eukaryotes
- Modified ribose and bases

functions as

Structural component of ribosomes

tRNA

Structural features include

- Unusual bases
- Extensive intra chain basepairing
- At least one specific type of molecule for each of the 20 amino acids found in proteins
- 3′-CCA sequence

functions as

Adaptor molecule that carries a specific amino acid to the ribosome–mRNA complex

Eukaryotic mRNA

Structural features include

- 3′-Poly-A tail
- 5′-Cap of 7-methyl-guanosine
- Monocistronic

functions as

Template for protein synthesis

Eukaryotic Transcription: DNA-Directed RNA Synthesis

consists of

Initiation

requires

Chromatin remodeling, binding of general transcription factors and RNA polymerase to core promoter sites up- or downstream of the transcription start site

which is facilitated by

Specific transcription factors bound to enhancer sequences and by the Mediator complex

Elongation

requires

Local unwinding of the DNA helix

followed by

Synthesis of a 5′→3′ RNA transcript coded for by the DNA template read in the 3′→5′ direction

Termination

requires

Different processes for the three types of RNA polymerase, with poly-A tailing used for RNA pol II

results in

Release of RNA polymerase and newly synthesized transcript from DNA

Co- and post transcriptional modification

for example

Trimming, addition of 3′-CCA, and base modification in pre-tRNA

Cleavage, trimming, and base/sugar modification in pre-rRNA

Addition of a 3′-poly-A tail and a 5′-7-methyl guanosine cap to pre-mRNA

Splicing of pre-mRNA to remove noncoding introns and join exons

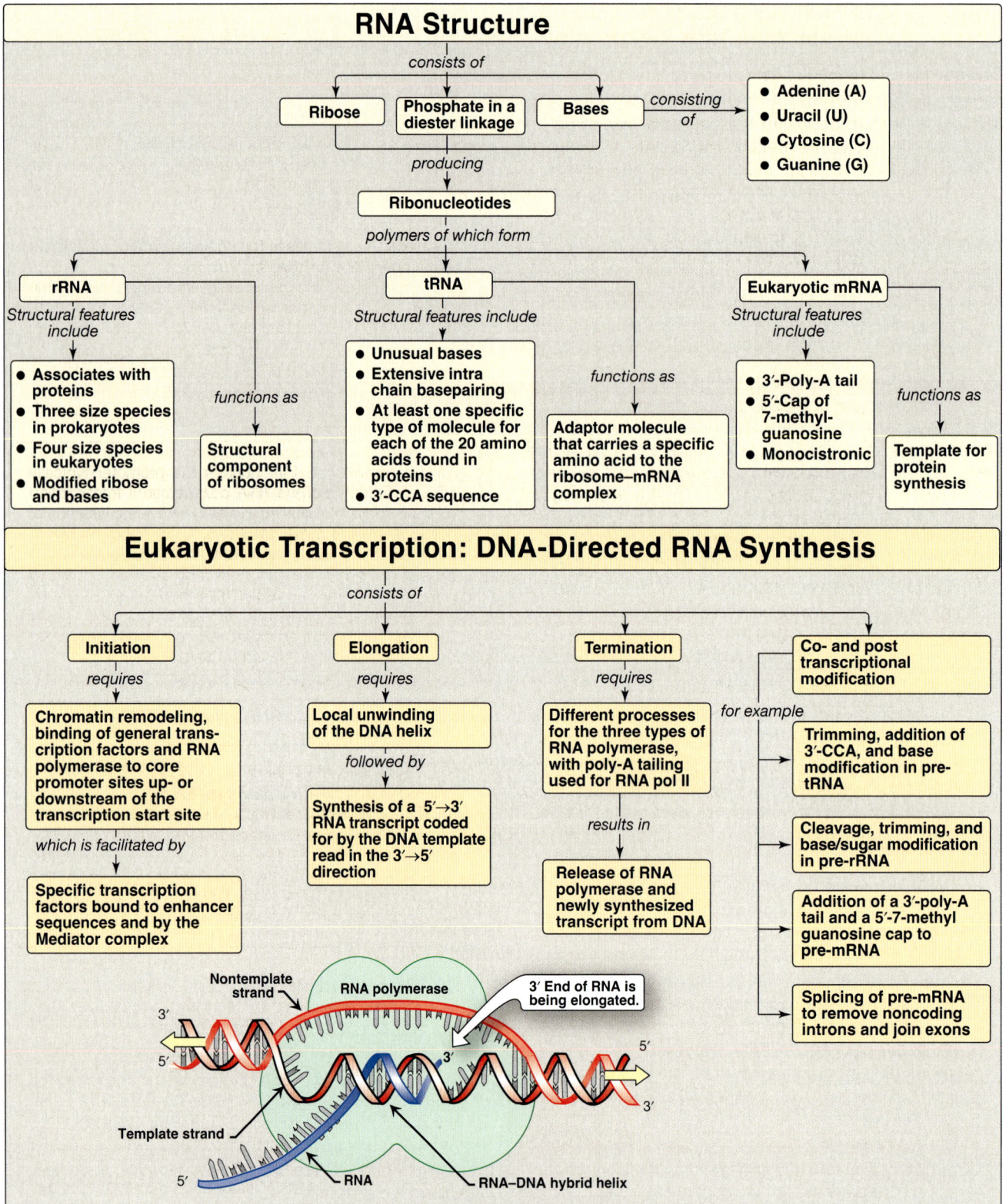

Nontemplate strand
RNA polymerase
3′ End of RNA is being elongated.
Template strand
RNA
RNA–DNA hybrid helix

Figure 31.20
Key concept map for RNA structure and synthesis. rRNA, ribosomal RNA; tRNA, transfer RNA; mRNA, messenger RNA.

Study Questions

Choose the ONE best answer.

31.1. An 8-month-old male with severe anemia is found to have β-thalassemia. Genetic analysis shows that one of his β-globin genes has a mutation that creates a new splice acceptor site 19 nucleotides upstream of the normal splice acceptor site of the first intron. Which of the following best describes the new messenger RNA molecule that can be produced from this mutant gene?

A. Exon 1 will be too short.

B. Exon 1 will be too long.

C. Exon 2 will be too short.

D. Exon 2 will be too long.

E. Exon 2 will be missing.

> Correct answer = D. Because the mutation creates an additional splice acceptor site (the 3′ end) upstream of the normal acceptor site of intron 1, the 19 nucleotides that are usually found at the 3′ end of the excised intron 1 lariat can remain behind as part of exon 2. The presence of these extra nucleotides in the coding region of the mutant messenger RNA (mRNA) molecule will prevent the ribosome from translating the message into a normal β-globin protein molecule. Those mRNA for which the normal splice site is used to remove the first intron will be normal, and their translation will produce normal β-globin protein.

31.2. A 4-year-old child who easily tires and has trouble walking is diagnosed with Duchenne muscular dystrophy, an X-linked recessive disorder. Genetic analysis shows that the patient's gene for the muscle protein dystrophin contains a mutation in its promoter region. Of the choices listed, which of the following would be the most likely to be defective due to this mutation?

A. Initiation of dystrophin transcription

B. Termination of dystrophin transcription

C. Capping of dystrophin messenger RNA

D. Splicing of dystrophin messenger RNA

E. Tailing of dystrophin messenger RNA

> Correct answer = A. Mutations in the promoter typically prevent formation of the RNA polymerase II transcription initiation complex, resulting in a decrease in the initiation of messenger RNA (mRNA) synthesis. A deficiency of dystrophin mRNA will result in a deficiency in the production of the dystrophin protein. Capping, splicing, and tailing defects are not a consequence of promoter mutations. They can, however, result in mRNA with decreased stability (capping and tailing defects) or an mRNA in which exons have been skipped (lost) or introns retained (splicing defects).

31.3. A mutation to which of the following sequences in eukaryotic messenger RNA (mRNA) would most likely affect the process by which the 3′-end polyadenylate (poly-A) tail is added to the mRNA?

A. AAUAAA

B. CAAT

C. CCA

D. GU… A … AG

E. TATAAA

> Correct answer = A. An endonuclease cleaves mRNA just downstream of this polyadenylation signal, creating a new 3′ end to which polyadenylate polymerase adds the poly-A tail using ATP as the substrate in a template-independent process. CAAT and TATAAA are sequences found in promoters for RNA polymerase II. CCA is added to the 3′ end of pre-transfer RNA by nucleotidyltransferase. GU…A… AG denotes an intron in eukaryotic pre-mRNA.

31.4. Which of the following protein factors identifies the promoter of protein-coding genes in eukaryotes?

A. Pribnow box

B. Rho

C. Sigma

D. TFIID

E. U1

> Correct answer = D. The general transcription factor TFIID recognizes and binds core promoter elements such as the TATA-like box in eukaryotic protein-coding genes. These genes are transcribed by RNA polymerase II. The Pribnow box is a *cis*-acting element in prokaryotic promoters. Rho is involved in the termination of prokaryotic transcription. Sigma is the subunit of prokaryotic RNA polymerase that recognizes and binds the prokaryotic promoter. U1 is a ribonucleoprotein involved in splicing of eukaryotic pre-mRNA.

31.5. A small RNA species is isolated from a eukaryotic cell. The RNA has an amino acid covalently linked to its 3'-end. Which of the following would be a posttranscriptional modification also found in this RNA species?

    A. Apurinic sites

    B. Dihydrouracil

    C. 7-Methylguanosine

    D. Pyrimidine dimers

    E. Xanthine

Correct answer = B. The RNA species is tRNA, which carries an amino acid to the site of protein synthesis. Some uracil bases of tRNA are modified after its transcription to produce dihydrouracil and pseudouracil. 7-Methylguanosine is a posttranscriptional modification of mRNA. Apurinic sites and pyrimidine dimers are forms of DNA damage. Xanthine is a degradation product of purine bases.

31.6. A scientist growing yeast cell cultures suspects that the cultures are contaminated with bacteria. Which of the following rRNA species could they detect in the culture to confirm that bacteria are present?

    A. 28S rRNA

    B. 23S rRNA

    C. 16S rRNA

    D. 12S rRNA

    E. 5S rRNA

Correct answer = B. 23S rRNA is unique to bacteria, which are prokaryotic cells. Yeast, which are eukaryotic cells, uniquely contain 28S and 12S rRNA species. 16S rRNA is found in bacteria and the mitochondria of eukaryotic cells. 5S rRNA is also found in both bacteria and yeast. So, only 23S rRNA would confirm that bacteria are in the culture.

31.7. What is the sequence (conventionally written) of the RNA product of the DNA template sequence, GATCTAC, also conventionally written?

Correct answer = 5'-GUAGAUC-3'. Nucleic acid sequences are conventionally written 5' to 3'. The template strand (5'-GATCTAC-3') is used as 3'-CATCTAG-5'. The RNA product is complementary to the template strand (and identical to the coding strand), with U replacing T.

Protein Synthesis

32

I. OVERVIEW

Genetic information, stored in the chromosomes and transmitted to daughter cells through DNA replication, is expressed through transcription to RNA and, in the case of messenger RNA (mRNA), subsequent translation into proteins (polypeptides) as shown in Figure 32.1. [Note: The proteome is the complete set of proteins expressed in a cell.] The process of protein synthesis is called translation, because the "language" of the nucleotide sequence on the mRNA is translated into the language of an amino acid sequence. Translation requires a genetic code, through which the information contained in the nucleotide sequence is expressed to produce a specific amino acid sequence. Any alteration in the nucleotide sequence may result in an incorrect amino acid being inserted into the protein, potentially causing disease or even death of the organism. Newly made immature (nascent) proteins undergo a number of processes to achieve their functional form. They must fold properly, otherwise misfolding can result in aggregation or degradation of the protein. Many proteins are covalently modified to alter their activities. Last, proteins are targeted to their final intra- or extracellular destinations by signals present in the proteins themselves.

II. GENETIC CODE

The genetic code is a "dictionary" that identifies the correspondence between a sequence of nucleotide bases and a sequence of amino acids. Each individual "word" in the code is composed of three nucleotide bases. These genetic "words" are called codons.

A. Codons

Codons are presented in the mRNA language of adenine (A), guanine (G), cytosine (C), and uracil (U). They are sequences of three nucleotides that are always written from the 5′ end to the 3′ end. The four mRNA nucleotide bases produce 64 different combinations of three-bases so that each codon is a "triplet code" for an amino acid, as shown in the table in Figure 32.2.

1. **How to translate a codon:** This table can be used to translate any codon and, thus, to determine which amino acids are coded for by an mRNA sequence. For example, the codon AUG codes for methionine ([Met] see Fig. 32.2). [Note: AUG is the start codon

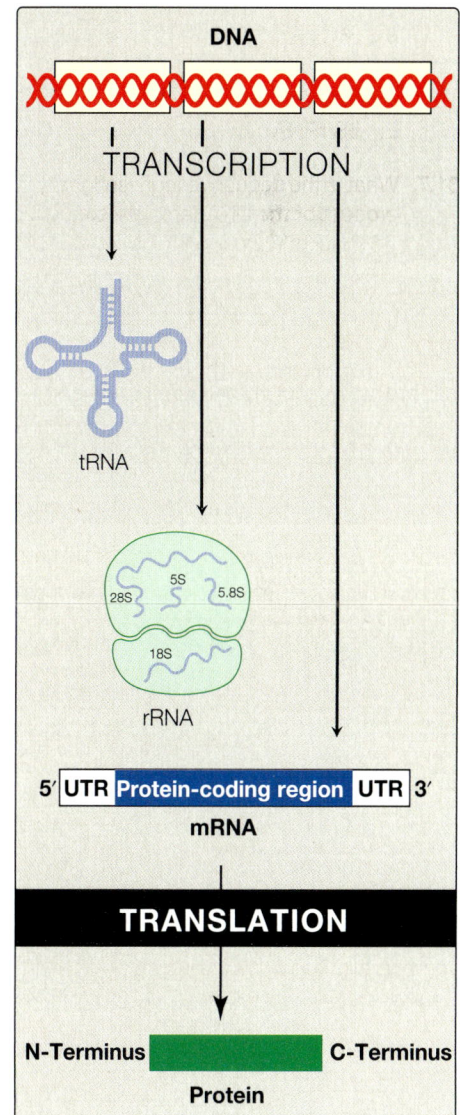

Figure 32.1
Protein synthesis or translation. tRNA, transfer RNA; rRNA, ribosomal RNA; mRNA, messenger RNA; UTR, untranslated region.

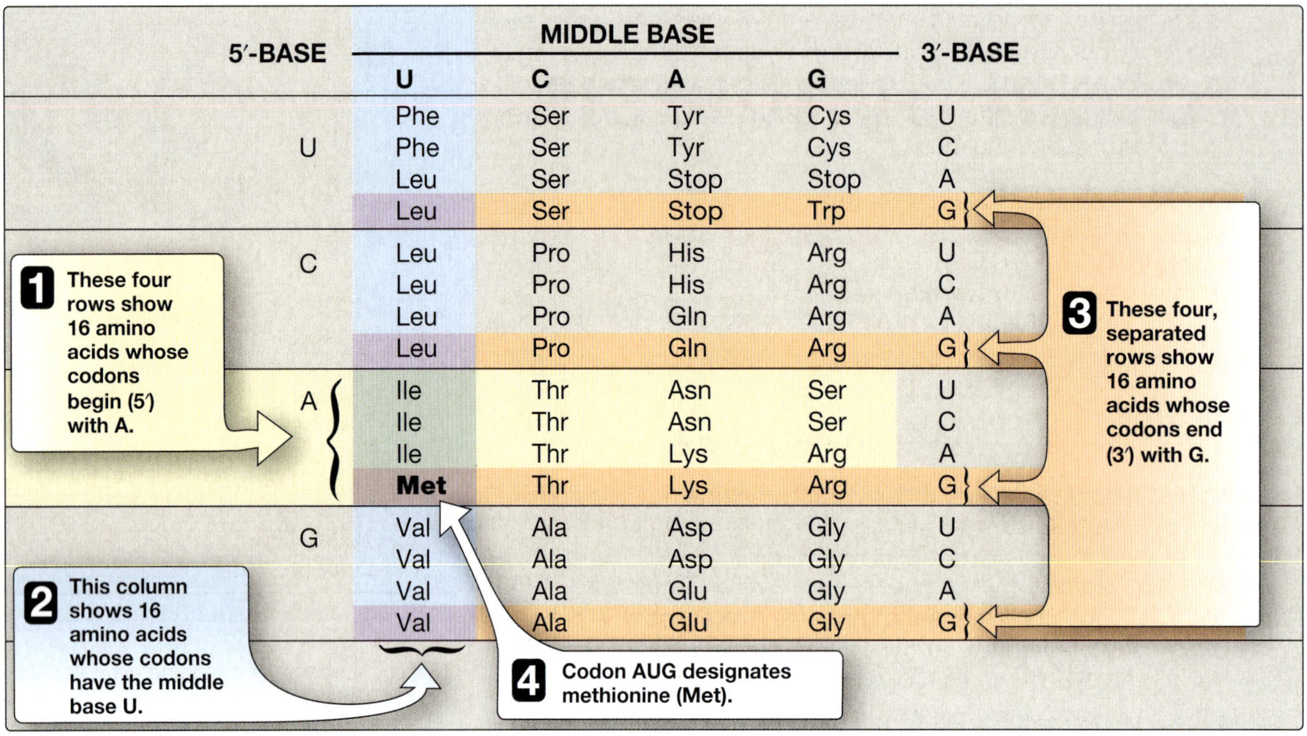

Figure 32.2
Use of the genetic code table to translate the codon AUG. A, adenine; G, guanine; C, cytosine; U, uracil. The three-letter abbreviations for many common amino acids are shown as examples.

that initiates protein translation.] Sixty-one of the 64 codons code for the 20 standard amino acids.

2. **Termination codons:** Three of the codons, UAA, UAG, and UGA, do not code for amino acids. Instead, these are termination codons, also called stop or nonsense codons. When one of these codons appears in an mRNA sequence, synthesis of the polypeptide coded for by that mRNA stops and translation terminates.

B. Characteristics

Usage of the genetic code is remarkably consistent throughout all living organisms. Characteristics of the genetic code include the following.

1. **Specificity:** The genetic code is specific (unambiguous), because a particular codon always codes for the same amino acid.

2. **Universality:** The genetic code is virtually universal insofar as its specificity has been conserved from very early stages of evolution, with only slight differences in the manner in which the code is translated. An exception occurs in mitochondria, where a few codons have meanings different than those shown in Figure 32.2. For example, in mitochondria UGA codes for tryptophan (Trp) instead of terminating translation.

3. **Degeneracy:** The genetic code is degenerate (sometimes called redundant). Although each codon corresponds to a single amino

acid, a given amino acid may have more than one triplet coding for it. For example, arginine (Arg) is specified by six different codons (see Fig. 32.2). Only Met and Trp have just one coding triplet. Most codons that code for the same amino acid differ only in the last base of the triplet.

4. **Nonoverlapping and continuous:** The genetic code is non-overlapping and continuous, meaning that the code is written as sequence of letters and is read from a fixed starting point as sets of three bases. For example, the code AGCUGGAUACAU is read as AGC UGG AUA CAU. The order of the codons in an mRNA that produces the correct sequence of amino acids in a protein is called the reading frame.

C. Consequences of altering the nucleotide sequence

Changing a single nucleotide base (a point mutation) in the coding region of an mRNA can lead to any one of three results (Fig. 32.3).

1. **Silent mutation:** The codon containing the changed base may code for the same amino acid. For example, if the serine (Ser) codon UCA is changed at the third base and becomes UCU, it still codes for Ser. This is termed a silent mutation.

2. **Missense mutation:** The codon containing the changed base may code for a different amino acid. For example, if the Ser codon UCA is changed at the first base and becomes CCA, it will code for a different amino acid (in this case, proline [Pro]). This is termed a missense mutation.

3. **Nonsense mutation:** Changing a base in a codon may create a termination codon. For example, if the Ser codon UCA is changed at the second base and becomes UAA, the new codon stops translation at that point and produces a shortened (truncated) protein. This is termed a nonsense mutation. [Note: The nonsense-mediated decay pathway degrades mRNA containing these mutations.]

4. **Other mutations:** Other mutations in mRNA can alter the amount or structure of the protein produced by translation.

   a. **Trinucleotide repeat expansion:** Occasionally, a sequence of three bases that is repeated in tandem will become amplified in number so that too many copies of the triplet occur. If this happens within the coding region of a gene, the protein will contain many extra copies of one amino acid. For example, expansion of the CAG codon in exon 1 of the gene for huntingtin protein leads to the insertion of many extra glutamine residues in the protein and causes the neurodegenerative disorder Huntington disease (Fig. 32.4). The additional glutamines result in an abnormally long protein that is cleaved, producing toxic fragments that aggregate in neurons. If the trinucleotide repeat expansion occurs in an untranslated region (UTR) of a gene, the result can be a decrease in the amount of protein produced, as seen in fragile X syndrome and myotonic dystrophy. In fragile X syndrome, the most common cause of intellectual disability in males, the expansion results in gene

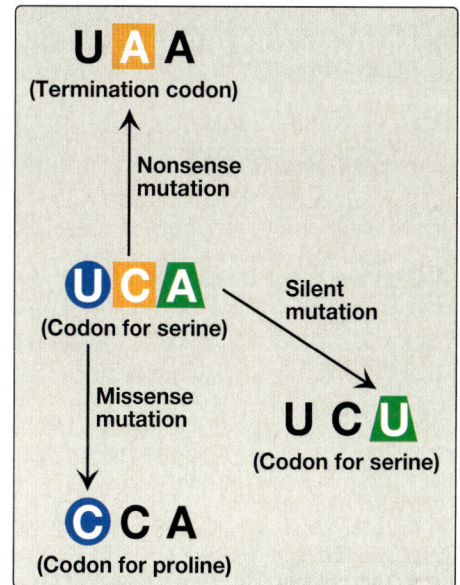

Figure 32.3
Possible effects of changing a single nucleotide base in the coding region of a messenger RNA. A, adenine; C, cytosine; U, uracil.

Huntington disease

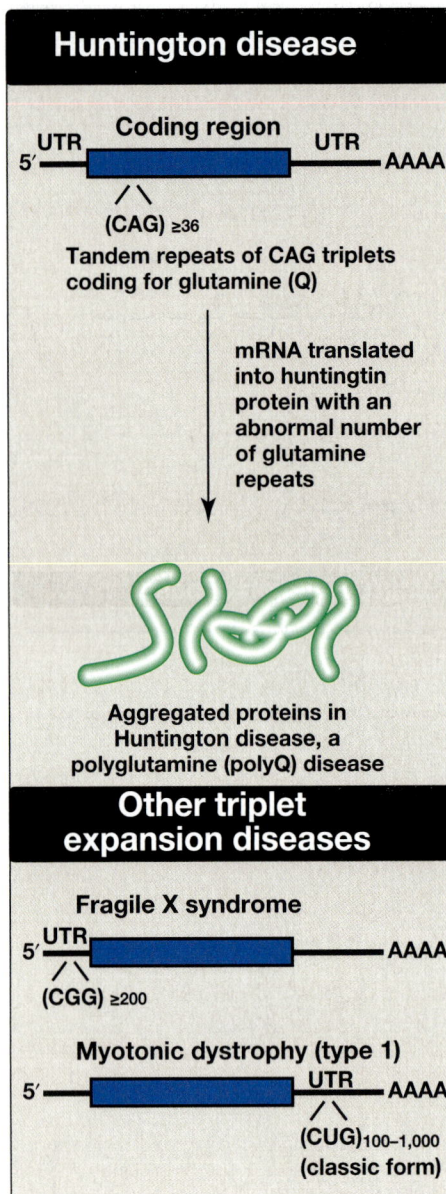

Coding region

UTR | Coding region | UTR
5′ ━━━ ▭ ━━━ AAAA

(CAG) ≥36

Tandem repeats of CAG triplets coding for glutamine (Q)

mRNA translated into huntingtin protein with an abnormal number of glutamine repeats

Aggregated proteins in Huntington disease, a polyglutamine (polyQ) disease

Other triplet expansion diseases

Fragile X syndrome

UTR
5′ ━━━ ▭ ━━━ AAAA

(CGG) ≥200

Myotonic dystrophy (type 1)

UTR
5′ ━━━ ▭ ━━━ AAAA

(CUG)100–1,000 (classic form)

Figure 32.4
Tandem triplet repeats in messenger RNA (mRNA) causing Huntington disease and other triplet expansion diseases. [Note: In unaffected individuals, the number of repeats in the huntingtin protein is <27; in fragile X mental retardation protein, it is 5–44; and in myotonic dystrophy protein kinase, it is 5–34.] UTR, untranslated region; A, adenine; C, cytosine; G, guanine; U, uracil; Q, single-letter abbreviation for glutamine.

silencing through DNA hypermethylation. More than 20 triplet expansion diseases are known.

b. **Splice-site mutations:** Mutations at splice sites can alter the way in which introns are removed from pre-mRNA molecules, producing aberrant proteins. For example, in myotonic dystrophy, a muscle disorder, gene silencing is the result of splicing alterations due to triplet expansion.

c. **Deletion and insertion mutations:** A loss (deletion) or addition (insertion) of one or more nucleotides from a gene can have different effects on the encoded protein. If one or two nucleotides are either deleted from or inserted into the coding region, the result is a frameshift mutation in the mRNA. The mutation alters the reading frame and can produce a protein product with a radically different amino acid sequence or even a truncated product due to the creation of a termination codon as the base sequence of the triplets change (Fig. 32.5). When three nucleotides are added or deleted, the effect on the protein depends on where the changes occur. If the three nucleotides are added within an existing codon sequence or are deleted from two adjacent codons, a frameshift happens. If three nucleotides are added between two codons, either a new amino acid is added into the protein, or a stop is generated that shortens the product. The deletion of a codon causes the loss of an amino acid. Loss or addition of three nucleotides may maintain the reading frame but can result in serious pathology. For example, cystic fibrosis (CF), an inherited disease that primarily affects the pulmonary and digestive systems, is most commonly caused by deletion of three nucleotides from the coding region of the *CFTR* gene. The deletion removes the triplet from the mRNA that codes for phenylalanine (Phe, or F) at position 508 in the CFTR protein. The loss of Phe at position 508 (ΔF508) prevents normal folding of the CFTR protein, leading to its destruction by the proteasome. CFTR normally functions as a chloride channel in epithelial cells, and its loss results in thick, sticky mucus in the lungs and pancreas, leading to lung infections and a digestive deficiency known as pancreatic insufficiency. In more than 70% of individuals with CF, the ΔF508 mutation is the cause of the disease.

III. COMPONENTS REQUIRED FOR TRANSLATION

A large number of components are required for the synthesis of a protein from the mRNA template carrying its specific sequence of codons. These include all the amino acids that are found in the finished product, transfer RNA (tRNA) for each of the amino acids, functional ribosomes, energy sources, and enzymes as well as noncatalytic protein factors needed for the initiation, elongation, and termination steps of polypeptide chain synthesis. [Note: The majority of translation in eukaryotic cells occurs in the cytosol, but mitochondrial DNA (mtDNA) encodes proteins that are synthesized in the organelle. So, all required components must be present in mitochondria.]

A. Amino acids

All the amino acids that eventually appear in the finished protein must be present at the time of protein synthesis. If one amino acid is missing, translation stops at the codon specifying that amino acid. [Note: This demonstrates the importance of having all the essential amino acids in sufficient quantities in the diet to ensure continued protein synthesis.]

B. Transfer RNA

At least one specific type of tRNA is required for each amino acid. In humans, there are at least 50 species of tRNA, whereas bacteria contain at least 30 species. [Note: The mitochondria of human cells encode 22 tRNA that participate in protein synthesis within the organelle.] Because there are only 20 different amino acids commonly carried by tRNA, some amino acids have more than one specific tRNA molecule. This is particularly true of those amino acids that are coded for by several codons. Two features of a tRNA are critical for codon translation into an amino acid, the amino acid attachment site and the anticodon.

1. **Amino acid attachment site:** Each tRNA molecule has an attachment site for a specific (cognate) amino acid at its 3′ end (Fig. 32.6). The carboxyl group of the amino acid is in an ester linkage with the 3′ hydroxyl of the ribose portion of the A nucleotide in the –CCA sequence at the 3′ end of the tRNA. [Note: A tRNA with a covalently attached (activated) amino acid is charged. Without an attached amino acid, it is uncharged.]

2. **Anticodon:** Each tRNA molecule also contains a three-base nucleotide sequence, the anticodon, which pairs with a specific codon on the mRNA (see Fig. 32.6). This codon specifies the insertion into the growing polypeptide chain of the amino acid carried by that tRNA.

C. Aminoacyl-tRNA synthetases

This family of 20 different enzymes is required for attachment of amino acids to their corresponding tRNA. Each member of this family recognizes a specific amino acid and all the tRNA that correspond to that amino acid (isoaccepting tRNA). There may be up to five isoaccepting tRNA per amino acid. Aminoacyl-tRNA synthetases catalyze a two-step reaction that results in the covalent attachment of the α-carboxyl group of an amino acid to the A in the –CCA sequence at the 3′ end of its corresponding tRNA. The overall reaction requires ATP, which is cleaved to adenosine monophosphate and inorganic pyrophosphate (PP$_i$), as shown in Figure 32.7. The extreme specificity of the synthetases in recognizing both the amino acid and its cognate tRNA contributes to the high fidelity of translation of the genetic message. In addition to their synthetic activity, the aminoacyl-tRNA synthetases have a proofreading, or editing activity that can remove an incorrect amino acid from the enzyme or the tRNA molecule.

D. Functionally competent ribosomes

As shown in Figure 32.8, ribosomes are large complexes of protein and ribosomal RNA (rRNA), in which rRNA predominates. They consist of

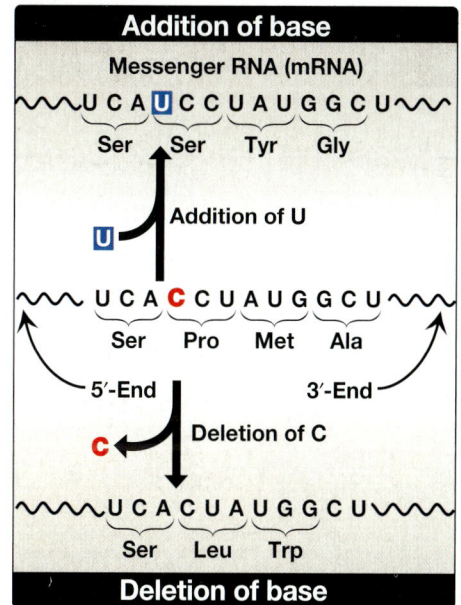

Figure 32.5
Frameshift mutations as a result of addition or deletion of a base can cause an alteration in the reading frame of mRNA. A, adenine; C, cytosine; G, guanine; U, uracil.

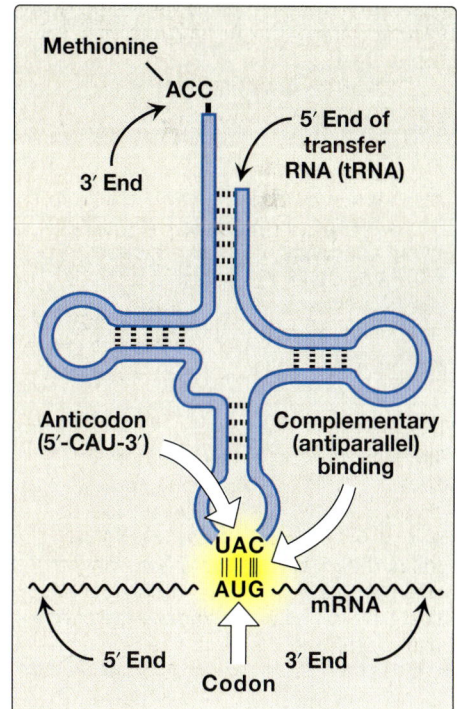

Figure 32.6
Complementary, antiparallel binding of the anticodon for methionyl-tRNA (CAU) to the messenger RNA (mRNA) codon for methionine (AUG), the initiation codon for translation.

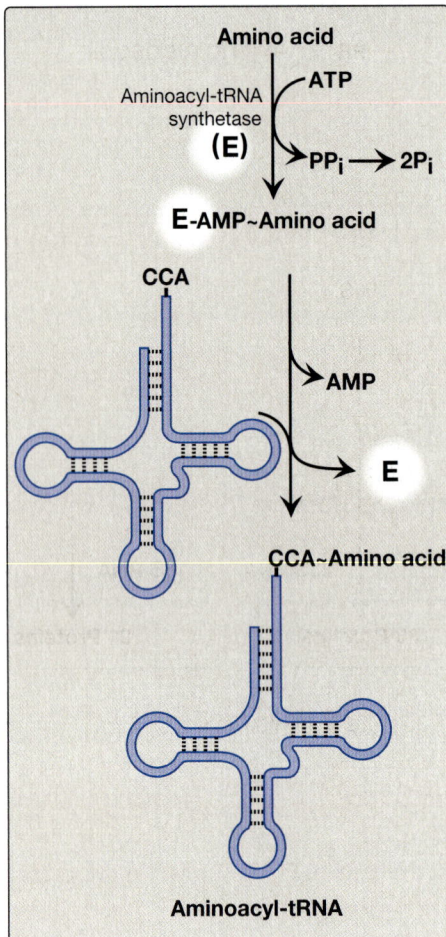

Figure 32.7
Attachment of a specific amino acid to its corresponding transfer RNA (tRNA) by an aminoacyl-tRNA synthetase. PP$_i$, pyrophosphate; P$_i$, inorganic phosphate; A, adenine; C, cytosine; AMP, adenosine monophosphate; ~, high-energy bond.

two subunits (one large and one small) whose relative sizes are given in terms of their sedimentation coefficients, or Svedberg (S) values. [Note: Because the S values are determined by both shape and size, their numeric values are not strictly additive (e.g., the prokaryotic 50S and 30S ribosomal subunits together form a 70S ribosome). The eukaryotic cytosolic 60S and 40S subunits form an 80S ribosome.] Prokaryotic and eukaryotic cytosolic ribosomes are similar in structure and serve the same function, namely, as the macromolecular complexes in which the synthesis of proteins occurs. [Note: The mtDNA encodes two rRNA, 16S rRNA and 12S rRNA (see Fig. 31.2). These rRNA form the large 39S and small 28S subunits, respectively, of the 55S ribosome that participates in protein synthesis in the organelle.]

1. **Ribosomal RNA:** As discussed, prokaryotic ribosomes contain three size species of rRNA, whereas cytosolic ribosomes in eukaryotes contain four (see Fig. 32.8). The rRNA are generated from a single pre-rRNA by the action of ribonucleases and the modification of some bases and ribose sugars.

The small ribosomal subunit binds mRNA and determines the accuracy of translation by ensuring correct base pairing between the mRNA codon and the tRNA anticodon. The large ribosomal subunit catalyzes formation of the peptide bonds that link amino acid residues in a protein.

2. **Ribosomal proteins:** Ribosomal proteins play a variety of roles in the structure and function of the ribosome and its interactions with other components of the translation system.

3. **A, P, and E sites:** The ribosome has three binding-sites for tRNA molecules: the A, P, and E sites, each of which extends over both subunits. Together, they cover three neighboring codons. During translation, the A site binds an incoming aminoacyl-tRNA as directed by the codon currently occupying this site. This codon specifies the next amino acid to be added to the growing peptide chain. The P site is occupied by peptidyl-tRNA. This tRNA carries the chain of amino acids that has already been synthesized. The E site is occupied by the empty tRNA as it is about to exit the ribosome. (See Fig. 32.13 for an illustration of the role of the A, P, and E sites in translation.)

4. **Cellular location:** In eukaryotic cells, the cytosolic ribosomes either are free or are in close association with the endoplasmic reticulum (which is then known as the rough endoplasmic reticulum [rER]). rER-associated ribosomes are responsible for synthesizing proteins (including glycoproteins) that are to be exported from the cell, incorporated into membranes, or imported into lysosomes (see Fig. 14.17 for an overview of the latter process). Cytosolic free ribosomes synthesize proteins required in the cytosol itself or destined for the nucleus, mitochondria, or peroxisomes. [Note: Mitochondria contain their own 55S ribosome that functions in the translation of the 13 proteins encoded in the mtDNA. These proteins are subunits of the complexes in

the electron transport chain (ETC). Most mitochondrial proteins, however, are encoded by nuclear DNA, synthesized completely in the cytosol, and then targeted to mitochondria.]

E. Protein factors

Initiation, elongation, and termination (or, release) factors are required for polypeptide synthesis. Some of these protein factors perform a catalytic function, whereas others appear to stabilize the synthetic machinery. [Note: A number of the factors are small, cytosolic G proteins and thus are active when bound to guanosine triphosphate (GTP) and inactive when bound to guanosine diphosphate (GDP). See Chapter 8 II. D. 1 and Fig. 8.7 for a review of membrane-associated G proteins.]

F. Energy sources

Cleavage of four high-energy bonds is required for the addition of one amino acid to the growing polypeptide chain: two from ATP in the aminoacyl-tRNA synthetase reaction, one in the removal of PP_i and one in the subsequent hydrolysis of the PP_i, to two P_i by pyrophosphatase, and two from GTP, one for binding the aminoacyl-tRNA to the A site and one for the translocation step (see Fig. 32.13). [Note: Additional ATP and GTP molecules are required for initiation in eukaryotes, whereas an additional GTP molecule is required for termination in both eukaryotes and prokaryotes.] Translation, then, is a major consumer of energy.

IV. CODON RECOGNITION BY TRANSFER RNA

Correct pairing of the codon in the mRNA with the anticodon of the tRNA is essential for accurate translation (see Fig. 32.6). Most tRNA (iso-accepting tRNA) recognize more than one codon for a given amino acid.

A. Antiparallel binding between codon and anticodon

Binding of the tRNA anticodon to the mRNA codon follows the rules of complementary and antiparallel binding, that is, the mRNA codon is read 5'→3' by an anticodon pairing in the opposite (3'→5') orientation (Fig. 32.9). [Note: Nucleotide sequences are always written in the 5' to 3' direction unless otherwise noted. Two nucleotide sequences orient in an antiparallel manner.]

B. Wobble hypothesis

The mechanism by which a tRNA can recognize more than one codon for a specific amino acid is described by the wobble hypothesis, which states that codon–anticodon pairing follows the traditional Watson–Crick rules (G pairs with C and A pairs with U) for the first two bases of the codon but can be less stringent for the last base. The base at the 5' end of the anticodon (the first base of the anticodon) is not as spatially defined as the other two bases. Movement of that first base allows nontraditional base pairing with the 3' base of the codon (the last base of the codon). This movement is called wobble and allows a single tRNA to recognize more than one codon.

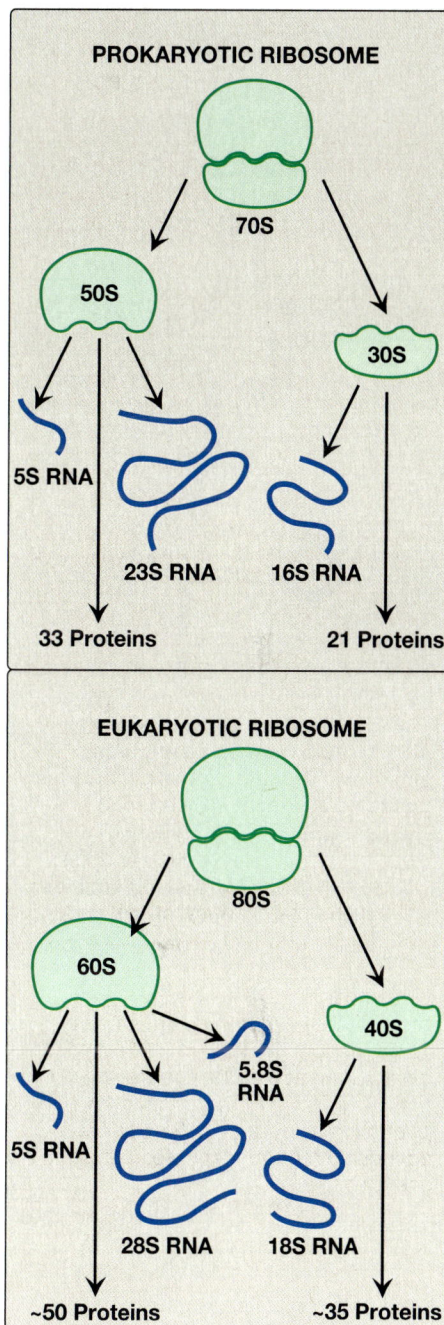

Figure 32.8
Ribosomal composition. [Note: The number of proteins in the eukaryotic ribosomal subunits varies somewhat from species to species.] S, Svedberg unit.

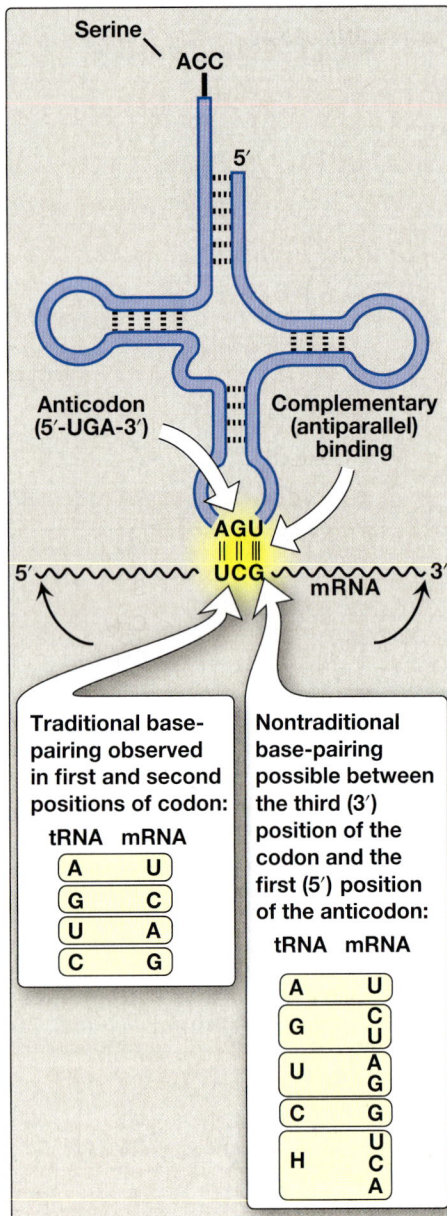

Serine

ACC

5′

Anticodon
(5′-UGA-3′)

Complementary
(antiparallel)
binding

AGU
‖ ‖ ‖
5′ ∿∿∿ UCG ∿∿∿ 3′
                    mRNA

Traditional base-pairing observed in first and second positions of codon:

tRNA	mRNA
A	U
G	C
U	A
C	G

Nontraditional base-pairing possible between the third (3′) position of the codon and the first (5′) position of the anticodon:

tRNA	mRNA
A	U
G	C, U
U	A, G
C	G
H	U, C, A

Figure 32.9
Wobble: Nontraditional base pairing between the 5′ nucleotide (first nucleotide) of the anticodon and the 3′ nucleotide (last nucleotide) of the codon. Hypoxanthine (H) is the product of adenine deamination and the base in the nucleotide inosine monophosphate (IMP). A, adenine; G, guanine; C, cytosine; U, uracil; tRNA, transfer RNA; mRNA, messenger RNA.

Examples of these flexible pairings are shown in Figure 32.9. The result of wobble is that a unique tRNA is not required to read each of the codons that code for amino acids.

V. STEPS IN TRANSLATION

The process of protein synthesis translates the 3-letter alphabet of nucleotide sequences on mRNA into the 20-letter alphabet of amino acids that constitute proteins. The mRNA is translated from its 5′ end to its 3′ end, producing a protein synthesized from its amino (N)-terminal end to its carboxyl (C)-terminal end. Prokaryotic mRNA often has several coding regions (i.e., they are polycistronic). Each coding region has its own initiation and termination codon and produces a separate species of polypeptide. This is also true for mitochondrial mRNA in eukaryotic cells. In contrast, mRNA made in the nucleus of eukaryotic cells has only one coding region (i.e., it is monocistronic). The process of translation is divided into three separate steps: initiation, elongation, and termination. Translation in the cytosol of eukaryotic cells resembles that of prokaryotic cells in most aspects. Individual differences are noted in the text. The steps of mitochondrial translation are generally conserved with those of prokaryotic cells and require protein factors distinct from those needed for translation in the eukaryotic cell cytosol. Mitochondrial translation will not be discussed in detail.

> Translation and transcription are temporally linked in prokaryotes and in eukaryotic mitochondria, with translation starting before transcription is completed. The nuclear membrane in eukaryotes physically separates the growing RNA transcripts in the nucleus from the ribosomes in the cytosol. So, mRNA must be exported from the nucleus for protein synthesis to occur.

A. Initiation

Initiation of protein synthesis involves the assembly of the components of the translation system before peptide-bond formation occurs. These components include the two ribosomal subunits, the mRNA to be translated, the aminoacyl-tRNA specified by the first codon in the message, GTP, and initiation factors (IFs) that facilitate the assembly of this initiation complex (see Fig. 32.13). [Note: In prokaryotes, three IFs are known (IF-1, IF-2, and IF-3), whereas in eukaryotes, there are many (designated eIF to indicate eukaryotic origin). Eukaryotes also require ATP for initiation.] The following are two mechanisms by which the ribosome recognizes the nucleotide sequence (AUG) that initiates translation.

1. **Shine–Dalgarno sequence:** In *Escherichia coli*, a purine-rich sequence of nucleotide bases, known as the Shine–Dalgarno (SD) sequence, is located 6 to 10 bases upstream of the initiating AUG codon on the mRNA molecule (i.e., near its 5′ end). The 16S rRNA component of the small (30S) ribosomal subunit has a nucleotide sequence near its 3′ end that is complementary to all or part of

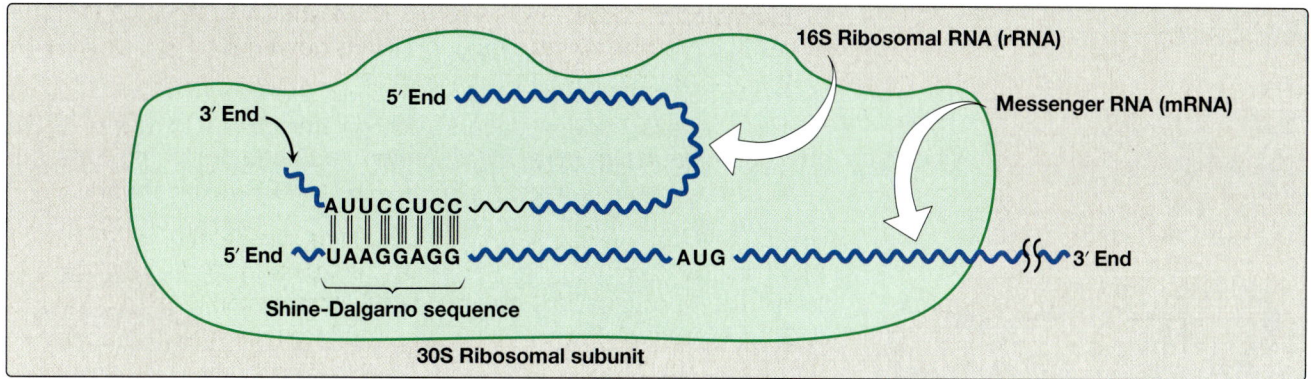

Figure 32.10
Complementary binding between prokaryotic mRNA Shine–Dalgarno sequence and 16S rRNA. S, Svedberg unit.

the SD sequence. Therefore, the 5′ end of the mRNA and the 3′ end of the 16S rRNA can form complementary base pairs, facilitating the positioning of the 30S subunit on the mRNA in close proximity to the initiating AUG codon (Fig. 32.10).

2. **5′ Cap:** Eukaryotic mRNA does not have SD sequences. In the cytosol of eukaryotes, the small (40S) ribosomal subunit (aided by members of the eIF-4 family of proteins) binds close to the cap structure at the 5′ end of the mRNA and moves 5′→3′ along the mRNA until it encounters the initiator AUG. This scanning process requires ATP. Cap-independent initiation can occur if the 40S subunit binds to an internal ribosome entry site close to the start codon. [Note: Interactions between the cap-binding eIF-4 proteins and the poly-A tail–binding proteins on eukaryotic mRNA mediate circularization of the mRNA and likely prevent the use of incompletely processed mRNA in cytosolic translation.]

3. **Initiation codon:** The initiating AUG is recognized by a special initiator tRNA (tRNA$_i$). Recognition is facilitated by IF-2-GTP in prokaryotes and eIF-2-GTP (plus additional eIF) in eukaryotes. The charged tRNA$_i$ is the only tRNA recognized by (e)IF-2 and the only tRNA to go directly to the P site on the small subunit. [Note: Base modifications distinguish tRNA$_i$ from the tRNA used for internal AUG codons.] In bacteria and mitochondria, tRNA$_i$ carries an N-formylated methionine (fMet), as shown in Figure 32.11. After Met is attached to tRNA$_i$, the formyl group is added by the enzyme transformylase, which uses N^{10}-formyl tetrahydrofolate as the carbon donor. In eukaryotes, the cytosolic tRNA$_i$ carries a Met that is not formylated. In both prokaryotic and eukaryotic cells, this N-terminal Met is usually removed before translation is completed. The large ribosomal subunit then joins the complex, and a functional ribosome is formed with the charged tRNA$_i$ in the P site. The A site is empty. [Note: Specific (e)IF function as antiassociation factors and prevent premature addition of the large subunit.] The GTP on (e)IF-2 gets hydrolyzed to GDP. In eukaryotes, the G nucleotide exchange factor eIF-2B facilitates the reactivation of eIF-2-GDP through replacement of GDP by GTP.

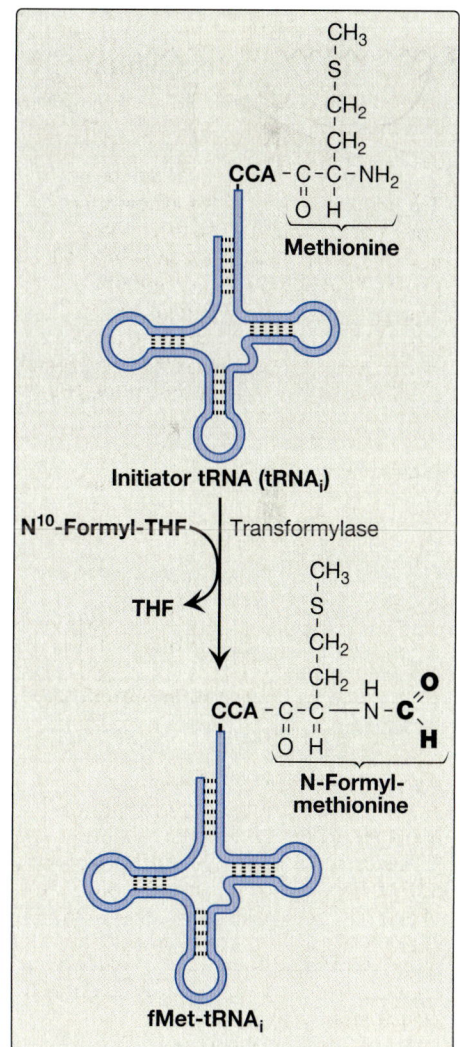

Figure 32.11
Generation of the initiator N-formylmethionyl-transfer RNA (fMet-tRNA$_i$). THF, tetrahydrofolate; C, cytosine; A, adenine.

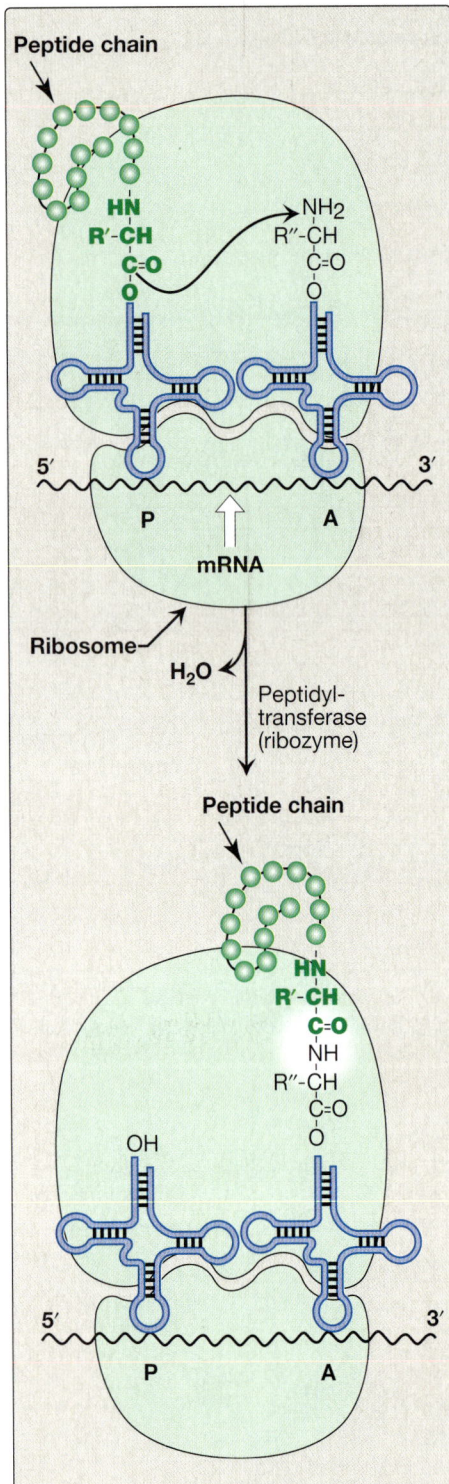

Figure 32.12
Formation of a peptide bond. Peptide bond formation results in transfer of the peptide on the transfer RNA (tRNA) in the P site to the amino acid on the tRNA in the A site (transpeptidation). mRNA, messenger RNA; R′, R″, different amino acid side chains.

B. Elongation

Elongation of the polypeptide involves the addition of amino acids to the carboxyl end of the growing chain. Delivery of the aminoacyl-tRNA whose codon appears next on the mRNA template in the ribosomal A site (a process known as decoding) is facilitated in *E. coli* by elongation factors EF-Tu-GTP and EF-Ts and requires GTP hydrolysis. [Note: In eukaryotes, comparable elongation factors are EF-1α-GTP and EF-1βγ. Both EF-Ts and EF-1βγ function in guanine nucleotide exchange.] Peptide bond formation between the α-carboxyl group of the amino acid in the P site and the α-amino group of the amino acid in the A site is catalyzed by peptidyl transferase, an activity intrinsic to an rRNA of the large subunit (Fig. 32.12). [Note: Because this rRNA catalyzes the reaction, it is a ribozyme.] After the peptide bond has been formed, the peptide on the tRNA at the P site is transferred to the amino acid on the tRNA at the A site, a process known as transpeptidation (Fig. 32.13). The ribosome then advances three nucleotides toward the 3′ end of the mRNA. This process is known as translocation and, in prokaryotes, requires the participation of EF-G-GTP (eukaryotes use EF-2-GTP) and GTP hydrolysis. Translocation causes movement of the uncharged tRNA from the P to the E site for release and movement of the peptidyl-tRNA from the A to the P site. The process is repeated until a termination codon is encountered. Several antibiotics target the initiation and elongation steps of prokaryotic protein synthesis. Their actions are summarized in Figure 32.13. [Note: Because of the length of most mRNA, more than one ribosome at a time can translate a message. Such a complex of one mRNA and a number of ribosomes is called a polysome, or polyribosome.]

C. Termination

Termination occurs when one of the three termination codons moves into the A site. These codons are recognized in *E. coli* by release factors: RF-1, which recognizes UAA and UAG, and RF-2, which recognizes UGA and UAA. The binding of these release factors results in hydrolysis of the bond linking the peptide to the tRNA at the P site, causing the nascent protein to be released from the ribosome. A third release factor, RF-3-GTP, then causes the release of RF-1 or RF-2 as GTP is hydrolyzed. [Note: Eukaryotes have a single release factor, eRF, for translation in the cytosol, which recognizes all three termination codons. A second factor, eRF-3, functions like the prokaryotic RF-3. See Figure 32.14 for a summary of the factors used in translation.] The newly synthesized polypeptide may undergo further modification as described below, and the ribosomal subunits, mRNA, tRNA, and protein factors can be recycled and used to synthesize another polypeptide. [Note: In prokaryotes, ribosome recycling factors mediate separation of the subunits. In eukaryotes, eRF and ATP hydrolysis are required.]

D. Translation regulation

Gene expression is most commonly regulated at the transcriptional level, but translation may also be regulated. An important mechanism by which this is achieved in eukaryotes is by covalent modification of eIF-2: Phosphorylated eIF-2 is inactive. In both eukaryotes and

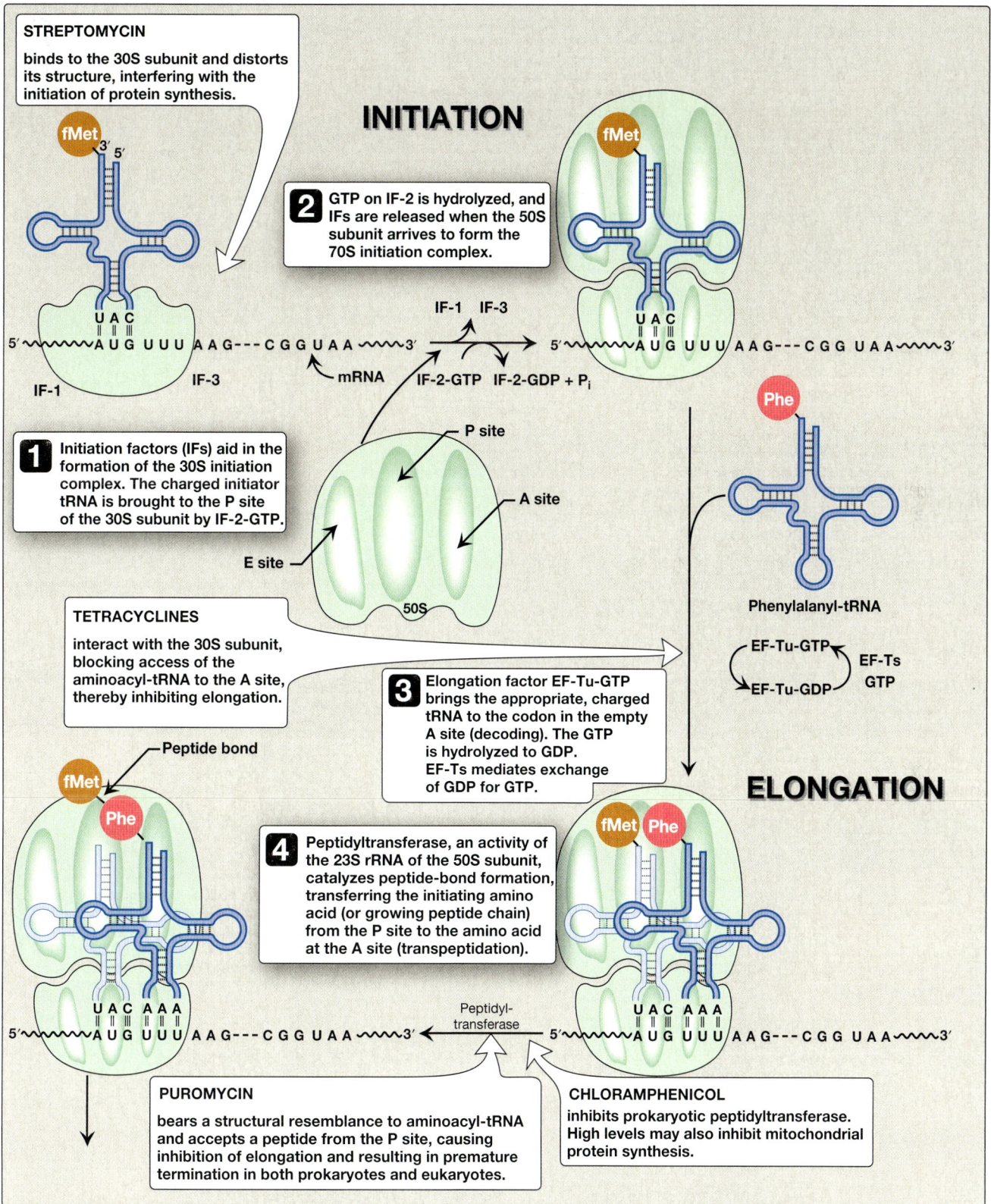

Figure 32.13
Steps in prokaryotic protein synthesis (translation), and their inhibition by antibiotics. [Note: EF-Ts is a guanine nucleotide exchange factor. It facilitates the removal of guanosine diphosphate (GDP) from EF-Tu, allowing its replacement by guanosine triphosphate (GTP). The eukaryotic equivalent is EF-1βγ.] fMet, formylated methionine; S, Svedberg unit; Phe, phenylalanine; Lys, lysine; Arg, arginine; tRNA, transfer RNA; mRNA, messenger RNA. (continued on the next page)

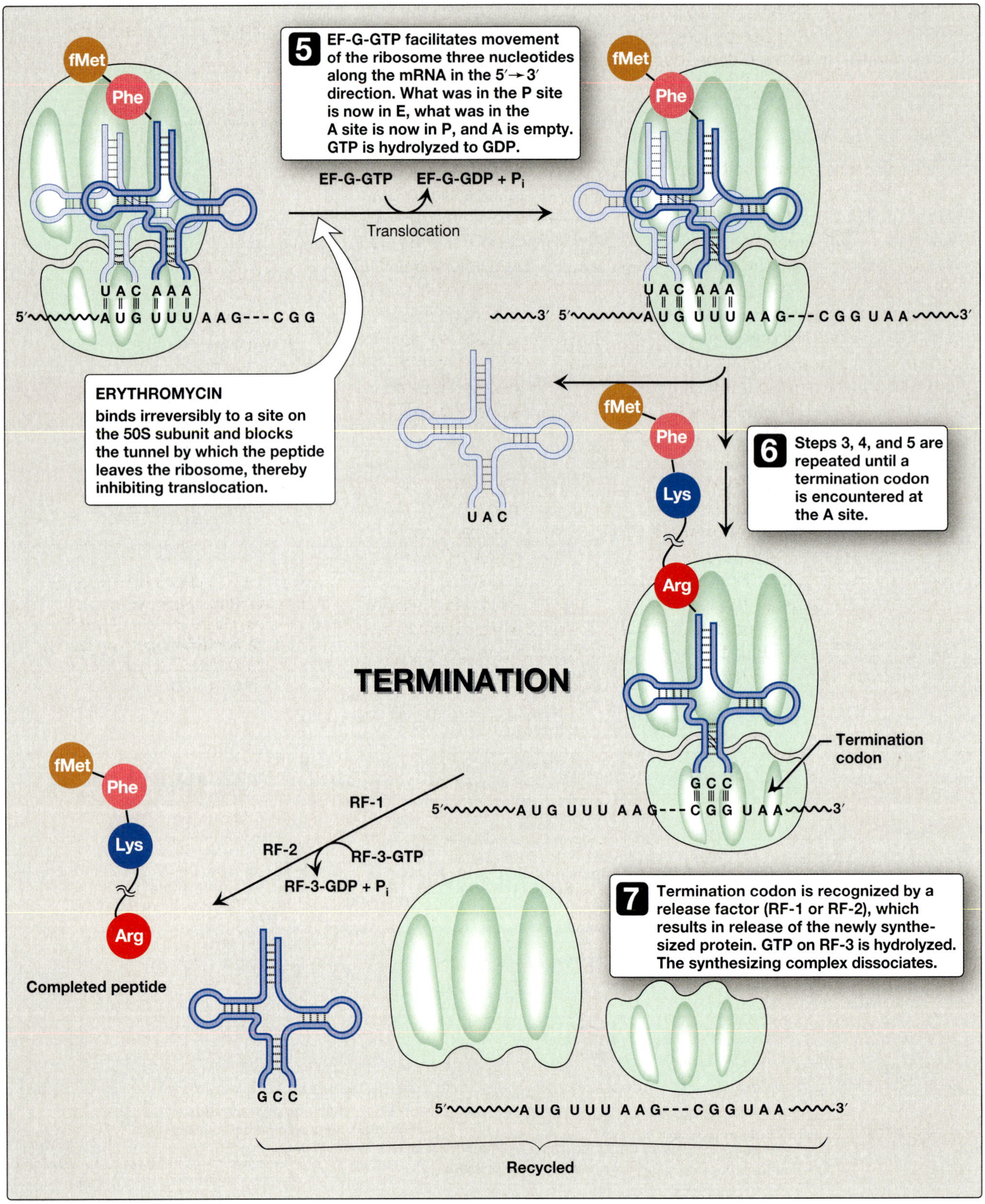

5 EF-G-GTP facilitates movement of the ribosome three nucleotides along the mRNA in the 5'→3' direction. What was in the P site is now in E, what was in the A site is now in P, and A is empty. GTP is hydrolyzed to GDP.

EF-G-GTP   EF-G-GDP + P$_i$

Translocation

ERYTHROMYCIN binds irreversibly to a site on the 50S subunit and blocks the tunnel by which the peptide leaves the ribosome, thereby inhibiting translocation.

6 Steps 3, 4, and 5 are repeated until a termination codon is encountered at the A site.

U A C A A A
5'~~~~AUG UUU AAG---CGG

U A C A A A
~~~3' 5'~~~AUG UUU AAG---CGG UAA~~~3'

U A C

fMet
Phe
Lys

Arg

TERMINATION

Termination codon

G C C
5'~~~~~AUG UUU AAG---CGG UAA~~~3'

RF-1

RF-2   RF-3-GTP

RF-3-GDP + P$_i$

7 Termination codon is recognized by a release factor (RF-1 or RF-2), which results in release of the newly synthesized protein. GTP on RF-3 is hydrolyzed. The synthesizing complex dissociates.

fMet
Phe
Lys
Arg

Completed peptide

G C C

5'~~~~~AUG UUU AAG---CGG UAA~~~3'

Recycled

Figure 32.13
[Note: In eukaryotes, diphtheria toxin inactivates EF-2 (the equivalent of prokaryotic EF-G), thereby inhibiting the translocation phase of elongation. Ricin, a toxin from castor beans, removes a specific A from the 28S ribosomal RNA (rRNA) in the large subunit of eukaryotic ribosomes, thereby inhibiting ribosomal function.] (continued from the previous page)

prokaryotes, regulation can also be achieved through proteins that bind mRNA and inhibit its use by blocking translation. [Note: Regulation of mitochondrial translation depends on nuclear encoded proteins and is less understood than cytosolic translation regulation.]

E. Protein targeting

Although most protein synthesis in eukaryotes is initiated in the cytoplasm, many proteins perform their functions within subcellular organelles or outside of the cell. Such proteins normally contain amino acid sequences that direct the proteins to their final locations. For example, secreted proteins are targeted during synthesis (cotranslational targeting) to the rER by the presence of an N-terminal hydrophobic signal sequence. The sequence is recognized by the signal-recognition particle (SRP), a ribonucleoprotein that binds the ribosome, halts elongation, and delivers the ribosome–peptide complex to an rER membrane channel (the translocon) via interaction with the SRP receptor. Translation resumes, the protein enters the rER lumen, and its signal sequence is cleaved (Fig. 32.15). The protein moves through the rER and the Golgi, is processed, packaged into vesicles, and secreted. Proteins targeted after synthesis (posttranslational) include nuclear proteins that contain an internal, short, basic nuclear localization signal; mitochondrial matrix proteins that contain an N-terminal, amphipathic, α-helical mitochondrial entry sequence; and peroxisomal proteins that contain a C-terminal tripeptide signal. [Note: Nuclear-encoded protein subunits of the ETC complexes are targeted to mitochondria. The mtDNA carries the code for additional protein subunits of the ETC complexes. Coordinated expression of these subunits is essential for oxidative phosphorylation and cellular energy production.

Cell	Factor	Function
Initiation		
Prok Euk	IF-2-GTP eIF-2-GTP	Bring charged initiating tRNA to P site
Prok Euk	IF-3 eIF-3	Prevent association of subunits
Elongation		
Prok Euk	EF-Tu-GTP EF1α-GTP	Bring all other charged tRNA to A site
Prok Euk	EF-Ts EF-1$\beta\gamma$	Guanine nucleotide exchange factors
Prok Euk	EF-G-GTP EF-2-GTP	Translocation
Termination		
Prok Euk	RF-1, 2 eRF	Recognize stop codons
Prok Euk	RF-3-GTP eRF-3-GTP	Release of other RF

Figure 32.14
Protein factors in the three stages of translation. Prok, prokaryotes; Euk, eukaryotes; tRNA, transfer RNA; IF, initiation factor; EF, elongation factor; RF, release factor; GTP, guanosine triphosphate.

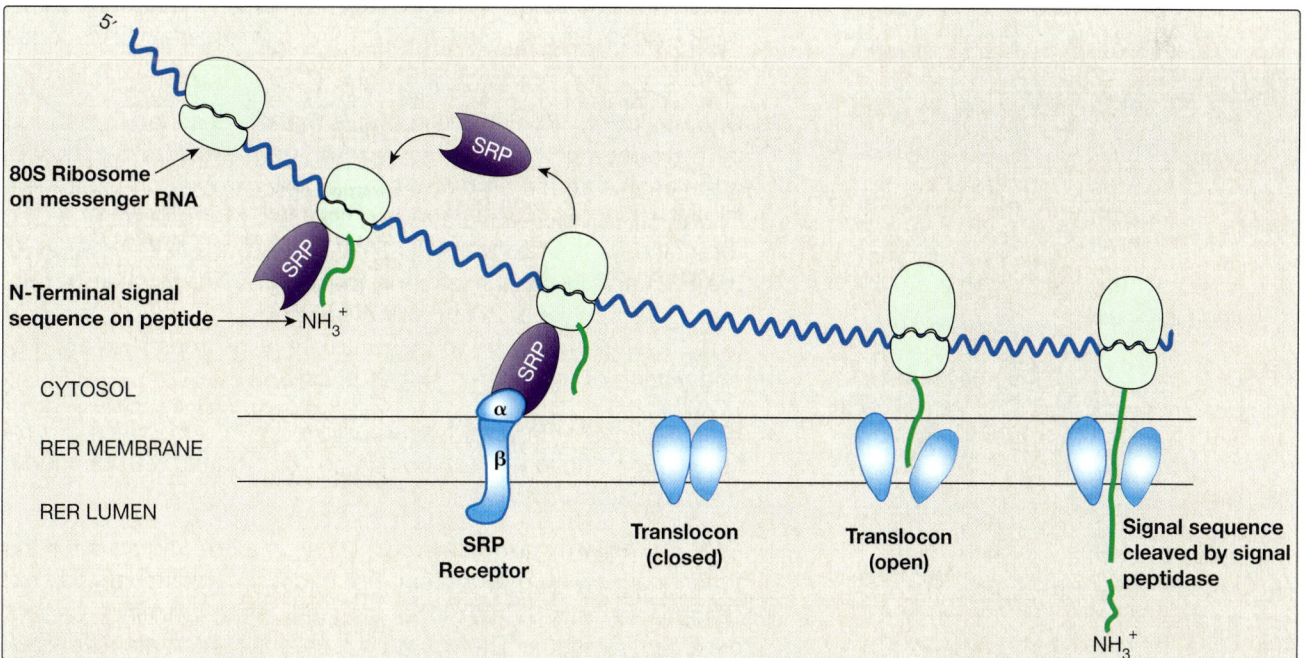

Figure 32.15
Cotranslational targeting of proteins to the rough endoplasmic reticulum (rER). SRP, signal recognition particle.

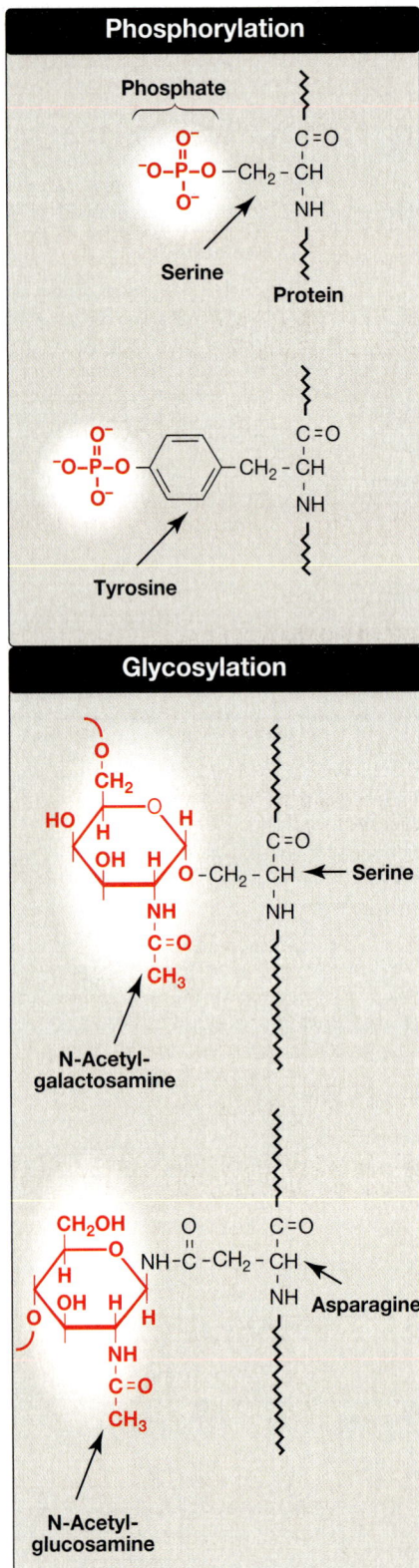

Phosphorylation

Phosphate

Serine

Protein

Tyrosine

Glycosylation

N-Acetyl-galactosamine

Serine

Asparagine

N-Acetyl-glucosamine

Figure 32.16
Covalent modification of some amino acid residues. (continued on the next page)

VI. CO- AND POSTTRANSLATIONAL MODIFICATIONS

Many polypeptides are covalently modified, either while they are still attached to the ribosome (cotranslational) or after their synthesis has been completed (posttranslational). These modifications may include removal of part of the translated sequence or the covalent addition of one or more chemical groups required for protein activity. Trimming of the protein and covalent modifications of amino acids may also enable proper folding and targeting of the protein.

A. Trimming

Many proteins destined for secretion are initially made as large, precursor molecules that are not functionally active. Portions of the protein must be removed by specialized endoproteases, resulting in the release of an active molecule. The cellular site of the cleavage reaction depends on the protein to be modified. Some precursor proteins are cleaved in the rER or the Golgi; others are cleaved in developing secretory vesicles (e.g., insulin; see Fig. 23.4); and still others, such as collagen, are cleaved after secretion.

B. Covalent attachments

Protein function can be affected by the covalent attachment of a variety of chemical groups (Fig. 32.16). Examples include the following.

1. **Phosphorylation:** Phosphorylation occurs on the hydroxyl groups of Ser, threonine, or, less frequently, tyrosine residues in a protein. It is catalyzed by one of a family of protein kinases and may be reversed by the action of protein phosphatases. The phosphorylation may increase or decrease the functional activity of the protein. Several examples of phosphorylation reactions have been previously discussed (e.g., see Chapter 11 for the regulation of glycogen synthesis and degradation).

2. **Glycosylation:** Many of the proteins that are destined to become part of a membrane or to be secreted from a cell have carbohydrate chains added en bloc to the amide nitrogen of asparagine (N linked) or built sequentially on the hydroxyl groups of a Ser, threonine, or hydroxylysine (O linked). N-glycosylation occurs in the RER and O-glycosylation in the Golgi. (The process of producing such glycoproteins was discussed.) N-Glycosylated acid hydrolases are targeted to the matrix of lysosomes by the phosphorylation of mannose residues at carbon 6.

3. **Hydroxylation:** Pro and lysine residues of the α-chains of collagen are extensively hydroxylated by vitamin C–dependent hydroxylases in the rER.

4. **Other covalent modifications:** These may be required for the functional activity of a protein. For example, additional carboxyl groups can be added to glutamate residues by vitamin K–dependent carboxylation. The resulting γ-carboxyglutamate (Gla) residues are essential for the activity of several of the blood-clotting proteins (see Chapter 35). Biotin is covalently bound to the ε-amino groups of lysine residues of biotin-dependent enzymes that

catalyze carboxylation reactions such as pyruvate carboxylase (see Fig. 10.3). Attachment of lipids, such as farnesyl groups, can help anchor proteins to membranes. Many eukaryotic proteins are cotranslationally acetylated at the N end. [Note: Reversible acetylation of histone proteins influences gene expression.]

C. Protein degradation

Proteins must fold to assume their functional, native state. Folding can be spontaneous (as a result of the primary structure) or facilitated by proteins known as chaperones. Proteins that are defective (e.g., misfolded) or destined for rapid turnover are often marked for destruction by ubiquitination, the covalent attachment of chains of a small, highly conserved protein called ubiquitin (see Fig. 19.3). Proteins marked in this way are rapidly degraded by the proteasome, which is a macromolecular, ATP-dependent, proteolytic system located in the cytosol. For example, misfolding of the CFTR protein results in its proteasomal degradation. [Note: If folding is impeded, unfolded proteins accumulate in the rER causing stress that triggers the unfolded protein response, in which the expression of chaperones is increased; global translation is decreased by eIF-2 phosphorylation; and the unfolded proteins are sent to the cytosol, ubiquitinated, and degraded in the proteasome by a process called ER-associated degradation.]

VII. CHAPTER SUMMARY

- **Codons** in are composed of three nucleotides in mRNA, which contains the bases **A**, **G**, **C**, and **U**. Codons are always written **5′→3′**.

- Of the 64 possible three-base combinations, 61 code for the 20 standard amino acids and 3 signal for the termination of protein synthesis (**translation**). In an organism, the genetic code is specific (each codon produces one amino acid) and degenerate (more than one codon can code for each amino acid).

- Altering the nucleotide sequence in a codon can cause **silent mutations** (the altered codon codes for the original amino acid), **missense mutations** (the altered codon codes for a different amino acid), or **nonsense mutations** (the altered codon is a termination codon). Frameshift mutations that result from the addition or deletion of a base can cause an alteration in the reading frame of mRNA.

- **Translation** of a protein from its mRNA requires all of the **amino acids** in the protein; the **tRNA** and **aminoacyl-tRNA synthetase** for each amino acid; fully competent **ribosomes** (70S in prokaryotes, 80S in the eukaryotic cell cytosol, and 55S in mitochondria); **protein factors** needed for initiation, elongation, and termination of protein synthesis; and **ATP** and **GTP** as energy sources.

- Ribosomes are large complexes of **protein** and **rRNA**. They consist of **two subunits**, 30S and 50S in prokaryotes, 40S and 60S in the eukaryotic cell cytosol, and 39S and 28S in mitochondria. Each ribosome has three tRNA-binding sites: the A, P, and E sites that cover three neighboring codons. The **A site** binds an **incoming aminoacyl-tRNA**, the **P site** is occupied by **peptidyl-tRNA**, and the **E site** is occupied by the **empty tRNA**.

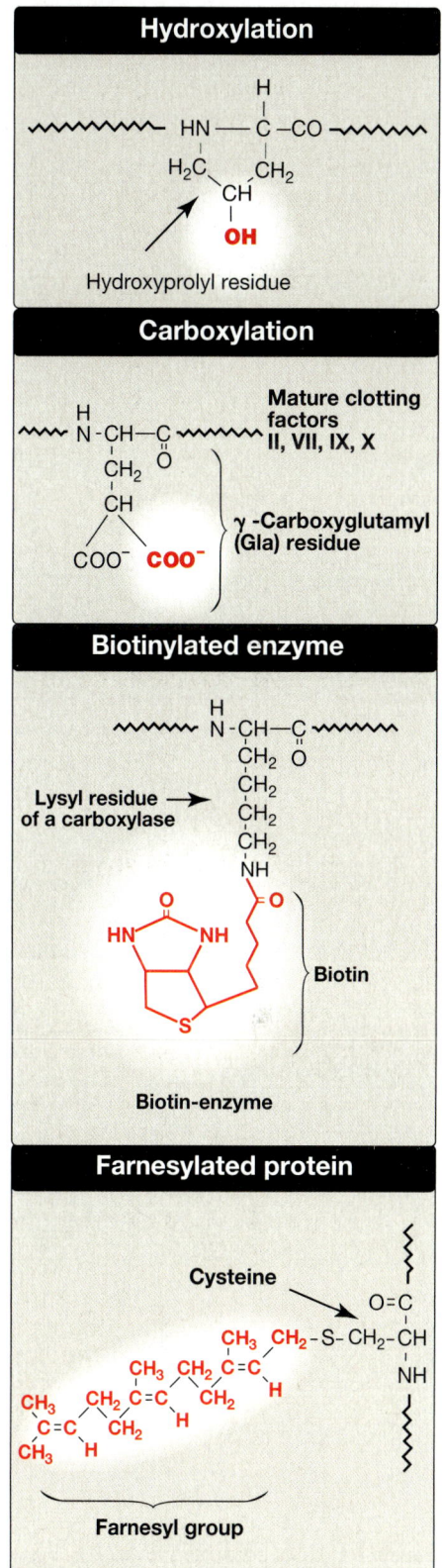

Figure 32.16
Covalent modification of some amino acid residues. (continued from the previous page)

- An mRNA codon is recognized by a tRNA **anticodon** following the rules of **complementarity** and **antiparallel** binding. The **wobble hypothesis** states that the first (5′) base of the anticodon is not as spatially constrained as the other two bases. Nontraditional base pairing may occur between the first (5′) anticodon base and the last (3′) base of the codon, thus allowing a single tRNA to recognize more than one codon for a specific amino acid.

- For the initiation of **translation**, an mRNA must associate with the small ribosomal subunit. The process requires **initiation factors (IFs)**. In **prokaryotes**, a purine-rich region of the mRNA (the **SD sequence**) base pairs with a complementary sequence on 16S rRNA, resulting in the positioning of the small subunit on the mRNA. In eukaryotes, this positioning is guided by the **5′ cap** of the mRNA, which is bound by proteins of the eIF-4 family. The **initiation codon** is **AUG. N-Formylmethionine** is the initiating amino acid in prokaryotes, whereas **methionine** is used in eukaryotes. The charged initiator tRNA (tRNA$_i$) is brought to the P site by **(e)IF-2**.

- **Elongation** (lengthening) of the polypeptide chain occurs by the addition of amino acids to its carboxyl end. **Elongation factors** facilitate the binding of the aminoacyl-tRNA to the A site as well as the movement of the ribosome along the mRNA. The formation of the peptide bond is catalyzed by **peptidyl transferase**, which is an activity intrinsic to the rRNA of the large subunit and, therefore, is a **ribozyme**. Following peptide bond formation, the ribosome advances along the mRNA in the **5′→3′ direction** to the next codon (**translocation**). Because of the length of most mRNA, more than one ribosome at a time can translate a message, forming a **polysome**.

- **Termination** begins when a termination codon moves into the A site and is recognized by **release factors**. The newly synthesized protein is released from the ribosomal complex, and the ribosome is dissociated from the mRNA.

- Numerous **antibiotics** interfere with the process of protein synthesis in prokaryotes.

- Polypeptide chains may be covalently modified during or after translation. Such modifications include amino acid **removal; phosphorylation,** which may activate or inactivate the protein; **glycosylation**, which plays a role in **protein targeting**; and **hydroxylation**, such as that seen in collagen.

- Protein targeting can be either **cotranslational** (as with secreted proteins) or **posttranslational** (as with mitochondrial matrix proteins).

- Proteins must **fold** to achieve their functional form. Folding can be spontaneous or facilitated by **chaperones**. Proteins that are defective (e.g., misfolded) or destined for rapid turnover are marked for destruction by the attachment of chains of a small, highly conserved protein called **ubiquitin**. Ubiquitinated proteins are rapidly degraded by a cytosolic complex known as the **proteasome** (Fig. 19.3).

Flow of genetic information

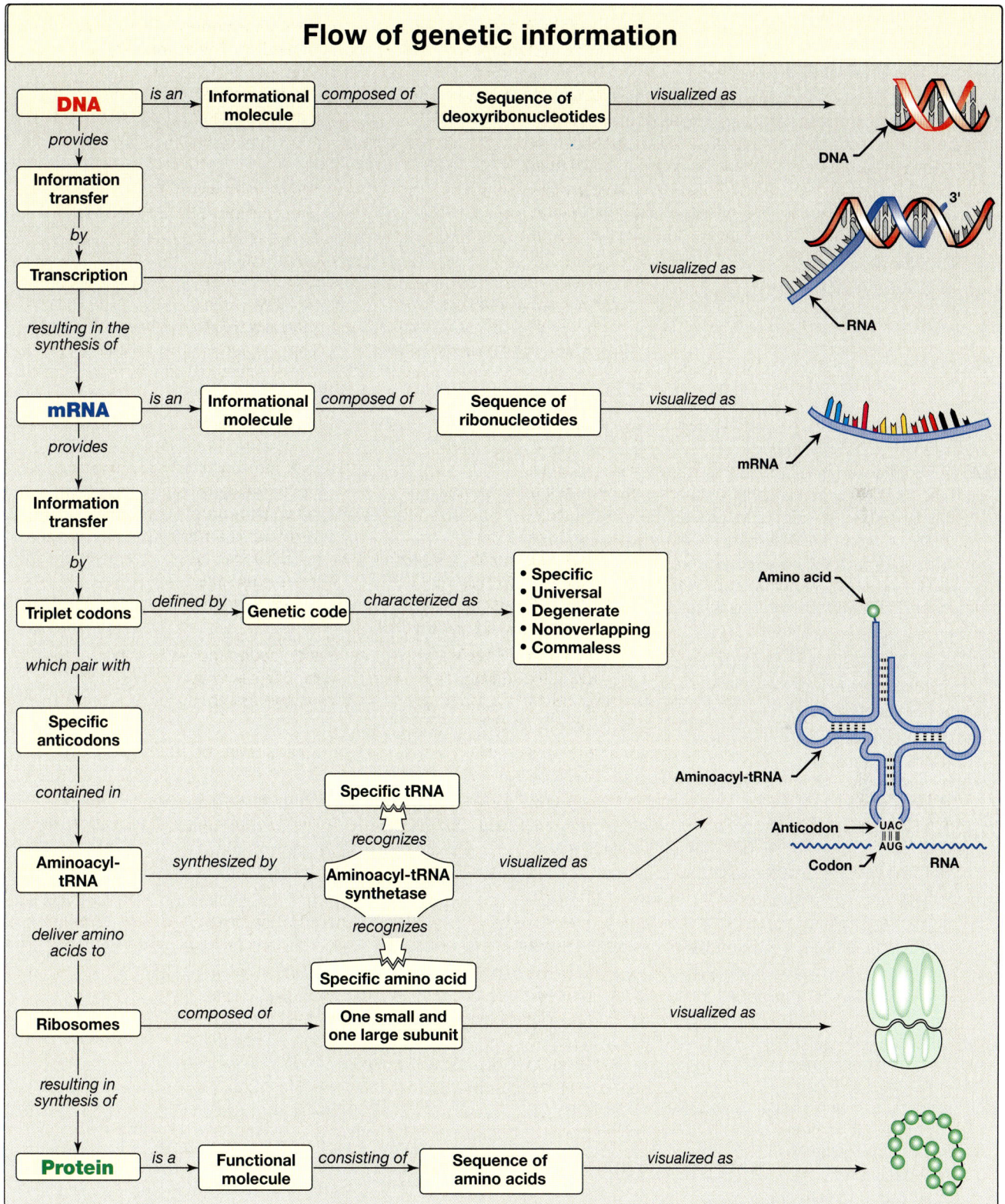

Figure 32.17
Key concept map for protein synthesis. mRNA, messenger RNA; tRNA, transfer RNA; A, adenine; G, guanine; C, cytosine; U, uracil.

Study Questions

Choose the ONE best answer.

32.1. A 20-year-old male with microcytic anemia is found to have an abnormal form of β-globin (Hemoglobin Constant Spring) that is 172 amino acids long, rather than the 141 found in the normal protein. Which of the following point mutations is consistent with this abnormality? Use Figure 32.2 to answer the question.

   A. CGA → UGA
   B. GAU → GAC
   C. GCA → GAA
   D. UAA → CAA
   E. UAA → UAG

> Correct answer = D. Mutating the normal termination (stop) codon from UAA to CAA in β-globin messenger RNA causes the ribosome to insert a glutamine at that point. It will continue extending the protein chain until it comes upon the next stop codon farther down the message, resulting in an abnormally long protein. The replacement of CGA (arginine) with UGA (stop) would cause the protein to be too short. GAU and GAC both code for aspartate and would cause no change in the protein. Changing GCA (alanine) to GAA (glutamate) would not change the size of the protein product. A change from UAA to UAG would simply change one termination codon for another and would have no effect on the protein.

32.2. A pharmaceutical company is studying a new antibiotic that inhibits bacterial protein synthesis. When this antibiotic is added to an *in vitro* protein synthesis system that is translating mRNA sequence AUGUUUUUUUAG, the only product formed is the dipeptide fMet-Phe. What step in protein synthesis is most likely inhibited by the antibiotic?

   A. Initiation
   B. Binding of tRNA to the ribosomal A site
   C. Peptidyl transferase activity
   D. Ribosomal translocation
   E. Termination

> Correct answer = D. Because fMet-Phe (formylated methionyl-phenylalanine) is made, the ribosomes must be able to complete initiation, bind Phe-tRNA to the A site, and use peptidyl transferase activity to form the first peptide bond. Because the ribosome is not able to proceed any further, ribosomal movement (translocation) is most likely the inhibited step. Therefore, the ribosome is stopped before it reaches the termination codon of this message.

32.3. A tRNA molecule that is supposed to carry cysteine (tRNA^Cys^) is mischarged, so that it actually carries alanine (Ala-tRNA^Cys^). Assuming no correction occurs, what would be the most likely fate of this alanine residue during protein synthesis?

   A. Alanine is incorporated into a protein.
   B. Cysteine is incorporated into a protein.
   C. Alanine is transferred to a tRNA^Ala^ in the E site of the ribosome.
   D. No protein synthesis occurs as alanine remains attached to the tRNA.
   E. Alanine is chemically converted to cysteine by cellular enzymes.

> Correct answer = A. Once an amino acid is attached to a tRNA molecule, only the anticodon of that tRNA determines the specificity of incorporation. Therefore, the incorrectly activated alanine will be incorporated into the protein at a position determined by a cysteine codon. A mischarged tRNA will cause a change in the protein that is not due to a mutation in the DNA.

32.4. In a patient with cystic fibrosis (CF) caused by the ΔF508 mutation, the mutant CF transmembrane conductance regulator (CFTR) protein folds incorrectly. The patient's cells modify this abnormal protein by attaching ubiquitin molecules to it. What is the fate of this modified CFTR protein?

   A. It is degraded by the proteasome.
   B. It is placed into storage vesicles.
   C. It is repaired by cellular enzymes.
   D. It is targeted to the lysosome.
   E. It is secreted from the cell.

> Correct answer = A. Ubiquitination usually marks old, damaged, or misfolded proteins for destruction by the cytosolic proteasome. There is no known cellular mechanism for repair of damaged proteins. Proteins are targeted to the matrix of the lysosome by a mannose 6-phosphate residue.

32.5. Many antimicrobials inhibit translation. Which of the following antimicrobials is correctly paired with its mechanism of action?

    A. Chloramphenicol inhibits transformylase.

    B. Erythromycin binds to the 60S ribosomal subunit.

    C. Puromycin inactivates elongation factor-2.

    D. Streptomycin binds to the 30S ribosomal subunit.

    E. Tetracyclines inhibit peptidyl transferase.

Correct answer = D. Streptomycin binds the 30S subunit and inhibits translation initiation. Chloramphenicol inhibits the peptidyl transferase activity of the 23S rRNA (ribozyme) of the 50S subunit. Erythromycin binds the 50S ribosomal subunit (60S denotes a eukaryote) and blocks the tunnel through which the peptide leaves the ribosome. Puromycin has structural similarity to aminoacyl-transfer RNA. It is incorporated into the growing chain, inhibits elongation, and results in premature termination in both prokaryotes and eukaryotes. Tetracyclines bind the 30S ribosomal subunit and block access to the A site, inhibiting elongation.

32.6. Translation of a synthetic polyribonucleotide containing the repeating sequence CAA in a cell-free protein-synthesizing system produces three homopolypeptides: polyglutamine, polyasparagine, and polythreonine. If the codons for glutamine and asparagine are CAA and AAC, respectively, which of the following triplets is the codon for threonine?

    A. AAC

    B. ACA

    C. CAA

    D. CAC

    E. CCA

Correct answer = B. The synthetic polynucleotide sequence of CAACAACAACAA… could be read by the *in vitro* protein-synthesizing system starting at the first C, the first A, or the second A (i.e., in any one of three reading frames). In the first case, the first triplet codon would be CAA, which codes glutamine; in the second case, the first triplet codon would be AAC, which codes for asparagine; in the last case, the first triplet codon would be ACA, which codes for threonine.

32.7. Which of the following is required for both prokaryotic and eukaryotic protein synthesis?

    A. Binding of the small ribosomal subunit to the Shine–Dalgarno sequence

    B. Formylated methionyl-transfer (t)RNA

    C. Movement of the messenger RNA out of the nucleus and into the cytoplasm

    D. Recognition of the 5′ cap by initiation factors

    E. Translocation of the peptidyl-tRNA from the A site to the P site

Correct answer = E. In both prokaryotes and eukaryotes, continued translation (elongation) requires movement of the peptidyl-tRNA from the A to the P site to allow the next aminoacyl-tRNA to enter the A site. Only prokaryotes have a Shine–Dalgarno sequence and use formylated methionine and only eukaryotes have a nucleus and co- and posttranscriptionally process their mRNA.

32.8. α1-Antitrypsin (AAT) deficiency can result in emphysema, a lung pathology, because the action of elastase, a serine protease, is unopposed. Deficiency of AAT in the lungs is the consequence of impaired secretion from the liver, the site of its synthesis. Proteins such as AAT that are destined to be secreted are best characterized by which of the following statements?

    A. Their synthesis is initiated on the smooth endoplasmic reticulum.

    B. They contain a mannose 6-phosphate targeting signal.

    C. They always contain methionine as the N-terminal amino acid.

    D. They are produced from translation products that have an N-terminal hydrophobic signal sequence.

    E. They contain no sugars with O-glycosidic linkages because their synthesis does not involve the Golgi.

Correct answer = D. Synthesis of secreted proteins is begun on free (cytosolic) ribosomes. As the N-terminal signal sequence of the peptide emerges from the ribosome, it is bound by the signal-recognition particle, taken to the rough endoplasmic reticulum (rER), threaded into the lumen, and cleaved as translation continues. The proteins move through the rER and the Golgi and undergo processing such as N-glycosylation (rER) and O-glycosylation (Golgi). In the Golgi, they are packaged in secretory vesicles and released from the cell. The smooth endoplasmic reticulum is associated with synthesis of lipids, not proteins, and has no ribosomes attached. Phosphorylation at carbon 6 of terminal mannose residues in glycoproteins targets these proteins (acid hydrolases) to lysosomes. The N-terminal methionine is removed from most proteins during processing.

32.9. Why is the genetic code described as both degenerate and unambiguous?

> A given amino acid can be coded for by more than one
> codon (degenerate code), but a given codon codes for just
> one particular amino acid (unambiguous code).

32.10. Explain how the genetic code in mitochondrial DNA
(mtDNA) differs from the universal genetic code
and provide an example of how this difference is
expressed in mitochondrial protein synthesis.

> The genetic code in mitochondria differs from the univer-
> sal genetic code because a few codons are read differ-
> ently. An example is the codon UGA, which signals for a
> stop in the universal code, but in mitochondria, it codes
> for the amino acid tryptophan. This exception reflects a
> variation in the genetic code of human cells that makes
> protein synthesis in mitochondria different from translation
> on cytoplasmic ribosomes.

Regulation of Gene Expression

33

I. OVERVIEW

Gene expression refers to the multistep process that ultimately results in the production of a functional gene product, either ribonucleic acid (RNA) or protein. The first step in gene expression, the use of deoxyribonucleic acid (DNA) for the synthesis of RNA (transcription), is the primary site of regulation in both prokaryotes and eukaryotes. In eukaryotes, however, gene expression also involves extensive posttranscriptional and posttranslational processes as well as actions that influence access to particular regions of the DNA. Each of these steps can be regulated to provide additional control over the kinds and amounts of functional products that are produced.

Not all genes are tightly regulated. In fact, genes described as "constitutive" encode products required for basic cellular functions and so are expressed at essentially a constant level. They are also known as "housekeeping" genes. Regulated genes, however, are expressed only under certain conditions. They may be expressed in all cells of the body or in only a subset of cells. The gene for fibrinogen alpha chain, for example, is expressed only in hepatocytes. The ability to regulate gene expression (i.e., to determine if, how much, and when particular gene products will be made) gives the cell control over structure and function. This is the basis for cellular differentiation, morphogenesis, and adaptability of any organism. Control of gene expression is best understood in prokaryotes, but many themes are repeated in eukaryotes. Figure 33.1 shows some of the sites where gene expression can be controlled.

II. REGULATORY SEQUENCES AND MOLECULES

Regulation of transcription, the initial step in all gene expression, is controlled by regulatory sequences of DNA that are usually embedded in the noncoding regions of the genome. The interaction between these DNA sequences and regulatory molecules, such as transcription factors, can induce or repress the transcriptional machinery, influencing the kinds and amounts of products that are produced. The regulatory DNA sequence elements are called *cis*-acting because they influence expression of genes that are located either downstream or upstream of the elements on the same chromosome. The regulatory molecules are called *trans*-acting because they can diffuse (transit) through the cell from their site of synthesis to their DNA-binding sites (Fig. 33.2). For example, a protein transcription factor (a *trans*-acting molecule) that

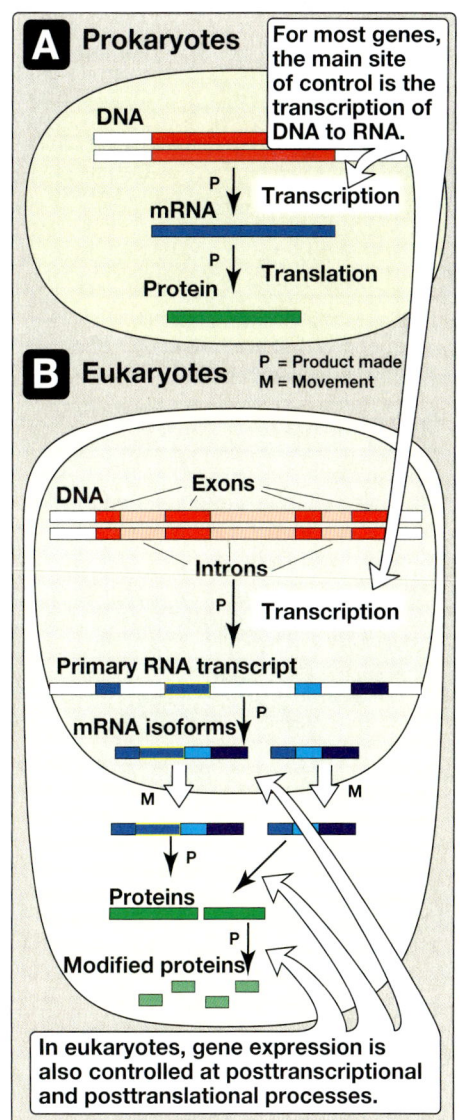

Figure 33.1
Control of gene expression. mRNA, messenger RNA.

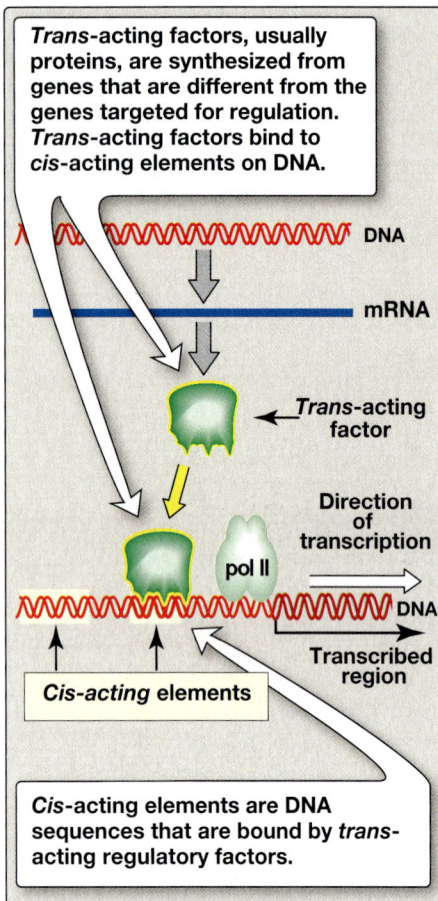

Trans-acting factors, usually proteins, are synthesized from genes that are different from the genes targeted for regulation. Trans-acting factors bind to cis-acting elements on DNA.

Cis-acting elements are DNA sequences that are bound by trans-acting regulatory factors.

Figure 33.2
Cis-acting elements and trans-acting factors. mRNA, messenger RNA; pol II, RNA polymerase II.

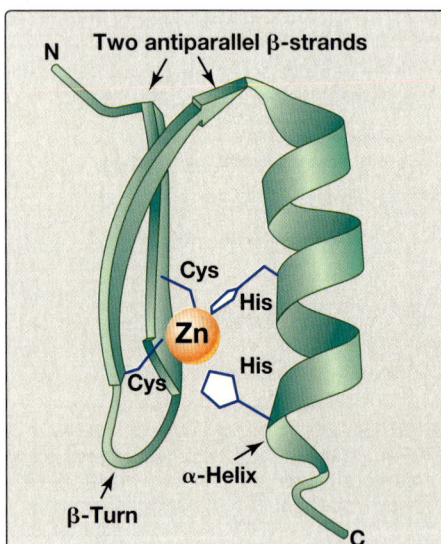

Figure 33.3
Zinc (Zn) finger is a common motif in proteins that bind DNA. Cys, cysteine; His, histidine.

regulates a gene on chromosome 6 might itself have been encoded by a gene on chromosome 11. The binding of proteins to DNA is through structural motifs in the protein such as the leucine zipper, helix-turn-helix, or zinc finger (Fig. 33.3).

III. REGULATION OF PROKARYOTIC GENE EXPRESSION

In prokaryotes such as the bacterium *Escherichia coli*, regulation of gene expression occurs primarily at the level of transcription and, in general, is mediated by the binding of *trans*-acting proteins to *cis*-acting regulatory elements on their single DNA molecule (chromosome). Transcriptional control in prokaryotes can involve the initiation or premature termination of transcription.

A. Messenger RNA transcription from bacterial operons

In bacteria, the structural genes that encode proteins involved in a particular metabolic pathway are often found sequentially grouped on the chromosome along with the *cis*-acting elements that regulate the transcription of these genes. The transcription product is a single polycistronic messenger RNA (mRNA). The genes are, thus, coordinately regulated (i.e., turned on or off as a unit). This entire package is referred to as an operon.

B. Operators in bacterial operons

Bacterial operons contain an operator, a segment of DNA that regulates the activity of the structural genes of the operon by reversibly binding a protein known as the repressor. If the operator is not bound by the repressor, RNA polymerase (RNA pol) binds the promoter, passes over the operator, and reaches the protein-coding genes that it transcribes to mRNA. If the repressor is bound to the operator, RNA pol is blocked and does not produce mRNA. As long as the repressor is bound to the operator, no mRNA (and, therefore, no proteins) are made. However, when an inducer molecule is present, it binds to the repressor, causing the repressor to change its conformation so that it no longer binds the operator. When this happens, RNA pol can initiate transcription. One of the best-understood examples is the inducible lactose (*lac*) operon of *E. coli* that illustrates both positive and negative regulation (Fig. 33.4).

C. Lactose operon

The *lac* operon contains the genes that code for three proteins involved in the catabolism of the disaccharide lactose: the *lacZ* gene codes for β-galactosidase, which hydrolyzes lactose to galactose and glucose; the *lacY* gene codes for a permease, which facilitates the movement of lactose into the cell; and the *lacA* gene codes for thiogalactoside transacetylase, which acetylates galactosides, lactosides, and glucosides. All of these proteins are maximally produced only when lactose is available to the cell and glucose, the preferred fuel for bacteria, is not. The regulatory portion of the operon is upstream of the three structural genes and consists of the promoter region where RNA pol binds and two additional sites, the operator (O) and the catabolite activator protein (CAP) sites, where regulatory proteins bind.

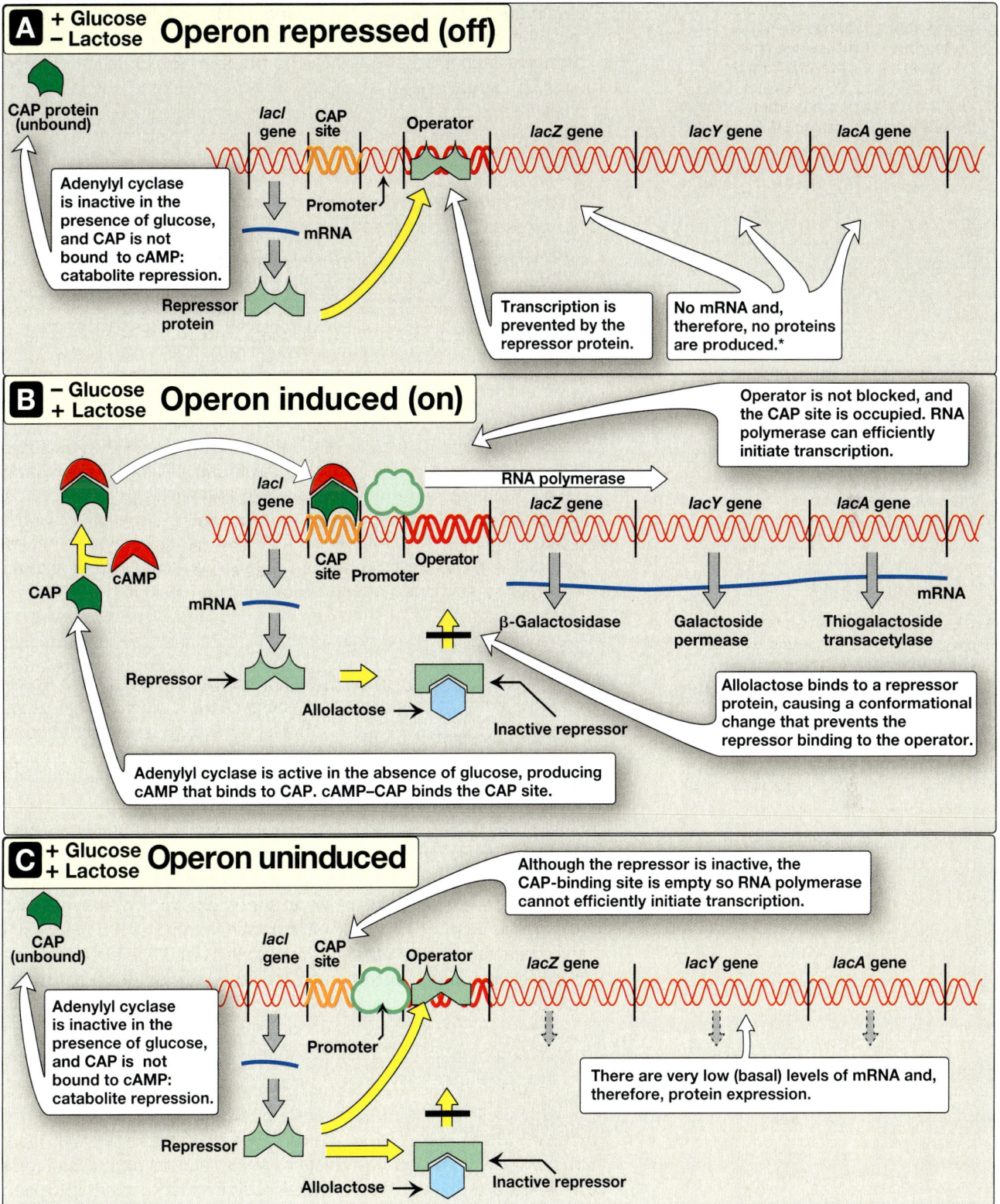

A + Glucose – Lactose **Operon repressed (off)**

CAP protein (unbound)

lacI gene
CAP site
Operator
lacZ gene
lacY gene
lacA gene

Adenylyl cyclase is inactive in the presence of glucose, and CAP is not bound to cAMP: catabolite repression.

Promoter
mRNA

Repressor protein

Transcription is prevented by the repressor protein.

No mRNA and, therefore, no proteins are produced.*

B – Glucose + Lactose **Operon induced (on)**

Operator is not blocked, and the CAP site is occupied. RNA polymerase can efficiently initiate transcription.

RNA polymerase

lacI gene
lacZ gene
lacY gene
lacA gene

cAMP
CAP site Promoter
Operator

CAP

mRNA
mRNA

β-Galactosidase
Galactoside permease
Thiogalactoside transacetylase

Repressor

Allolactose
Inactive repressor

Allolactose binds to a repressor protein, causing a conformational change that prevents the repressor binding to the operator.

Adenylyl cyclase is active in the absence of glucose, producing cAMP that binds to CAP. cAMP–CAP binds the CAP site.

C + Glucose + Lactose **Operon uninduced**

Although the repressor is inactive, the CAP-binding site is empty so RNA polymerase cannot efficiently initiate transcription.

CAP (unbound)

lacI gene
CAP site
Operator
lacZ gene
lacY gene
lacA gene

Adenylyl cyclase is inactive in the presence of glucose, and CAP is not bound to cAMP: catabolite repression.

Promoter

There are very low (basal) levels of mRNA and, therefore, protein expression.

Repressor

Allolactose
Inactive repressor

Figure 33.4
The lactose operon of *Escherichia coli* in the presence of (**A**) only glucose, (**B**) only lactose, and (**C**) both sugars. *[Note: Even when the operon has been turned off, the repressor transiently dissociates from the operator at a slow rate, allowing a very low level of expression. The synthesis of a few molecules of permease (and β-galactosidase) allows the organism to respond rapidly should glucose become unavailable.] CAP, catabolite activator protein; cAMP, cyclic adenosine monophosphate; mRNA, messenger RNA.

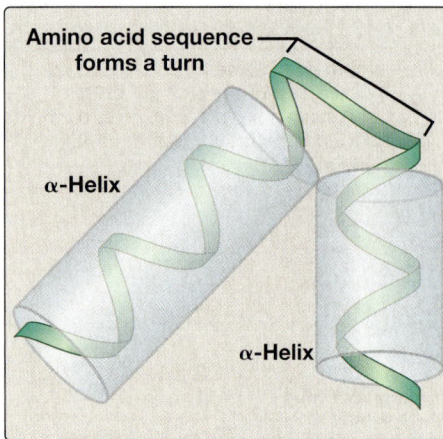

Figure 33.5
Helix-turn-helix motif of the *lac* repressor protein.

The *lacZ*, *lacY*, and *lacA* genes are maximally expressed only when the O site is empty and the CAP site is bound by a complex of cyclic adenosine monophosphate (cAMP) and the CAP, sometimes called the cAMP regulatory protein (CRP). A regulatory gene, the *lacI* gene, codes for the repressor protein (a *trans*-acting factor) that binds to the O site with high affinity. The *lacI* gene has its own promoter and is not part of the *lac* operon.

1. **When only glucose is available:** In this case, the *lac* operon is repressed (turned off). Repression is mediated by the repressor protein binding via a helix-turn-helix motif, as shown in Figure 33.5, to the O site, which is downstream of the promoter (see Fig. 33.4A). Binding of the repressor interferes with the binding of RNA pol to the promoter, thereby inhibiting transcription of the structural genes. This is an example of negative regulation.

2. **When only lactose is available:** In this case, the *lac* operon is induced (turned on and maximally expressed). A small amount of lactose is converted to an isomer, allolactose. This compound is an inducer that binds to the repressor protein, changing its conformation so that it can no longer bind to the O site. In the absence of glucose, adenylyl cyclase is active and makes cAMP that binds to the CAP. The cAMP–CAP *trans*-acting complex binds to the CAP site, causing RNA pol to initiate transcription with high efficiency at the promoter site (see Fig. 33.4B). This is an example of positive regulation. The transcript is a single polycistronic mRNA molecule that contains three sets of start and stop codons. Translation of the mRNA produces the three proteins that allow lactose to be used for energy production by the cell. In contrast to the inducible *lacZ*, *lacY*, and *lacA* genes, whose expression is regulated, the *lacI* gene is constitutive. Its gene product, the repressor protein, is always made and binds to the O site unless the inducer, allolactose, is present.

3. **When both glucose and lactose are available:** In this case, the *lac* operon is uninduced, and transcription is negligible, even if lactose is present at a high concentration. Adenylyl cyclase is inhibited in the presence of glucose (a process known as catabolite repression) so no cAMP–CAP complex forms, and the CAP site remains empty. Therefore, the RNA pol is unable to effectively initiate transcription, even though the repressor is not bound to the O site. Consequently, the three structural genes of the operon are expressed only at a very low (basal) level (see Fig. 33.4C). Induction (when only lactose is present) causes a 50-fold enhancement over basal-level expression.

D. Tryptophan operon

The tryptophan *(trp)* operon contains five structural genes that code for enzymes required for the synthesis of the amino acid tryptophan (Trp). As with the *lac* operon, the *trp* operon is subject to repression. However, for the trp operon, repression includes Trp itself binding to a repressor protein and facilitating the binding of the repressor to the operator: Trp is a corepressor. Because repression by Trp is not always complete, the *trp* operon, unlike the *lac* operon, is also regulated by a process known as attenuation. With attenuation,

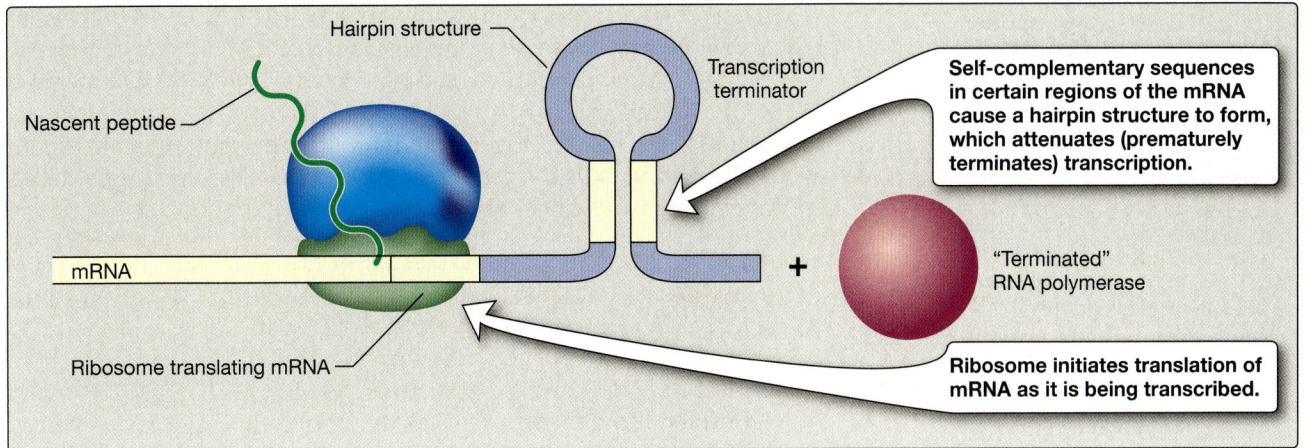

Figure 33.6
Attenuation of transcription of the *trp* operon when tryptophan is plentiful. mRNA, messenger RNA.

transcription is initiated but is terminated well before completion (Fig. 33.6). If Trp is plentiful, transcription initiation that escaped repression by Trp is attenuated (stopped) by the formation of an attenuator, a hairpin (stem-loop) structure in the mRNA similar to that seen in rho-independent termination (see Fig. 31.9). Recall that transcription and translation are temporally linked in prokaryotes, so attenuation also results in the formation of a truncated, non-functional peptide product that is rapidly degraded. If Trp becomes scarce, the operon is expressed. The 5′ end of the mRNA contains two adjacent codons for Trp. The lack of Trp causes ribosomes to stall at these codons, covering regions of the mRNA required for the formation of the attenuation hairpin. This prevents attenuation and allows transcription to continue.

> Transcriptional attenuation can occur in prokaryotes because translation of an mRNA begins before its synthesis is complete. This does not occur in eukaryotes because the presence of a membrane-bound nucleus spatially and temporally separates transcription and translation.

E. Coordination of transcription and translation

Although transcriptional regulation of mRNA production is primary in bacteria, regulation of ribosomal RNA (rRNA) and protein synthesis plays an important role in adaptation to environmental stress.

1. **Stringent response:** *E. coli* has seven operons that synthesize the rRNA needed for ribosome assembly, and each is regulated in response to changes in environmental conditions. Regulation in response to amino acid starvation is known as the stringent response. The binding of an uncharged transfer RNA (tRNA) to the A site of a ribosome triggers a series of events that leads to the production of an alarm molecule (an alarmone) that alerts the cell to stress that threatens cell survival. The alarmone in this case is guanosine 5′-diphosphate, 3′-diphosphate (ppGpp), an unusual

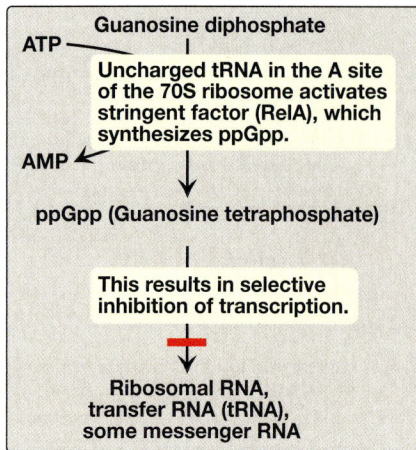

Figure 33.7
Regulation of transcription by the stringent response to amino acid starvation. S, Svedberg unit.

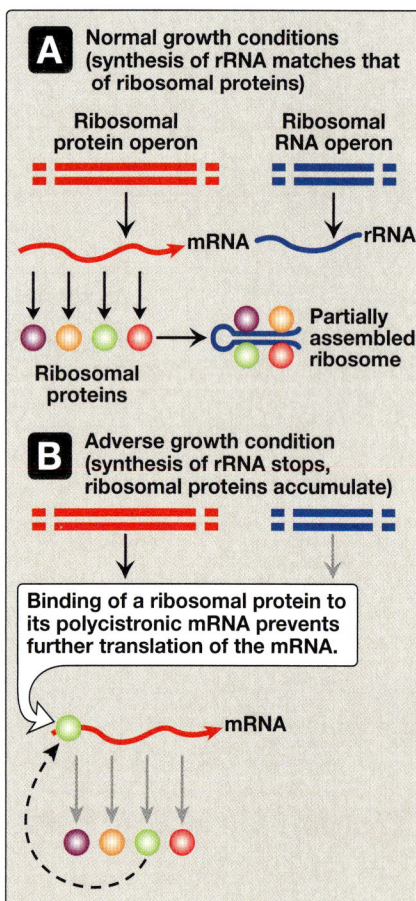

Figure 33.8
Regulation of translation by an excess of ribosomal proteins. mRNA, messenger RNA; rRNA, ribosomal RNA.

derivative of guanosine diphosphate (GDP). The synthesis of ppGpp is catalyzed by stringent factor (RelA), an enzyme physically associated with ribosomes. Elevated levels of ppGpp result in inhibition of rRNA synthesis (Fig. 33.7). ppGpp binds RNA pol and alters promoter selection through the use of different sigma factors for the polymerase. In addition to rRNA synthesis, tRNA synthesis and some mRNA synthesis (e.g., for ribosomal proteins [r-proteins]) are also inhibited. However, synthesis of mRNA for enzymes required for amino acid biosynthesis is not inhibited. The stringent response prevents the wasteful production of more ribosomes and promotes the production of needed amino acids when amino acids are scarce.

2. **Regulatory ribosomal proteins:** Operons for r-proteins can be inhibited by an excess of their own protein products. For each operon, one specific r-protein functions in the repression of translation of the polycistronic mRNA from that operon (Fig. 33.8). The r-protein does so by binding to the Shine–Dalgarno (SD) sequence located on the mRNA just upstream of the first initiating AUG codon (see Fig. 32.10) and acting as a physical impediment to the binding of the small ribosomal subunit to the SD sequence. Thus, one r-protein inhibits synthesis of all the r-proteins of the operon. This same r-protein also binds to rRNA and with a higher affinity than for mRNA. If the concentration of rRNA falls, the r-protein then is available to bind its own mRNA and inhibit its translation. This coordinated regulation keeps the synthesis of r-proteins in balance with the transcription of rRNA, so that each is present in appropriate amounts for the formation of ribosomes.

IV. REGULATION OF EUKARYOTIC GENE EXPRESSION

The higher degree of complexity of eukaryotic genomes, as well as the presence of a nuclear membrane, necessitates a wider range of regulatory processes. As with the prokaryotes, transcription is the primary site of regulation. Again, the theme of *trans*-acting factors binding to *cis*-acting elements is seen. Operons, however, are not found in eukaryotes, so these organisms must use alternate strategies to solve the problem of how to coordinately regulate all the genes required for a specific response. In eukaryotes, gene expression is also regulated at multiple levels other than transcription. For example, the major modes of posttranscriptional regulation at the mRNA level are alternative mRNA splicing and polyadenylation, control of mRNA stability, and control of translational efficiency. Additional regulation at the protein level occurs by mechanisms that modulate stability, processing, or targeting of the protein.

A. Coordinate regulation

Organisms with more than one chromosome need to coordinately regulate a group of genes to cause a particular response. An underlying theme occurs repeatedly: A *trans*-acting protein functions as a specific transcription factor (STF) that binds to a *cis*-acting regulatory consensus sequence (see Fig. 31.12) on each of the genes in the group even if they are on different chromosomes. Recall that the STF has a DNA-binding domain (DBD) and a transcription-activation domain (TAD). The TAD recruits coactivators, such as histone acetyltransferases, and the general transcription factors that, along with RNA pol, are required for formation of the transcription initiation complex at the promoter. Although the TAD recruits a variety of proteins, the specific effect of any one of them is dependent on the protein composition of the complex. This is known as combinatorial control. Two examples of coordinate regulation in eukaryotes are the galactose circuit and the hormone response system.

1. **Galactose circuit:** This regulatory scheme allows for the use of galactose when glucose is not available. In yeast, a unicellular organism, the genes required to metabolize galactose are on different chromosomes. Coordinated expression is mediated by the protein Gal4 (Gal = galactose), a STF that binds to a short regulatory DNA sequence upstream of each of the genes. The sequence is called the upstream activating sequence Gal (UAS$_{Gal}$). Binding of Gal4 to UAS$_{Gal}$ through zinc fingers in its DBD occurs in both the absence and presence of galactose. When the sugar is absent, the regulatory protein Gal80 binds Gal4 at its TAD, thereby inhibiting gene transcription (Fig. 33.9A). When present, galactose activates the Gal3 protein. Gal3 binds Gal80, thereby allowing Gal4 to activate transcription (Fig. 33.9B). When glucose is present, it prevents the use of galactose by inhibiting expression of Gal4 protein.

2. **Hormone-response system:** Hormone-response elements (HREs) are DNA regulatory sequences that bind *trans*-acting proteins and regulate gene expression in response to hormonal

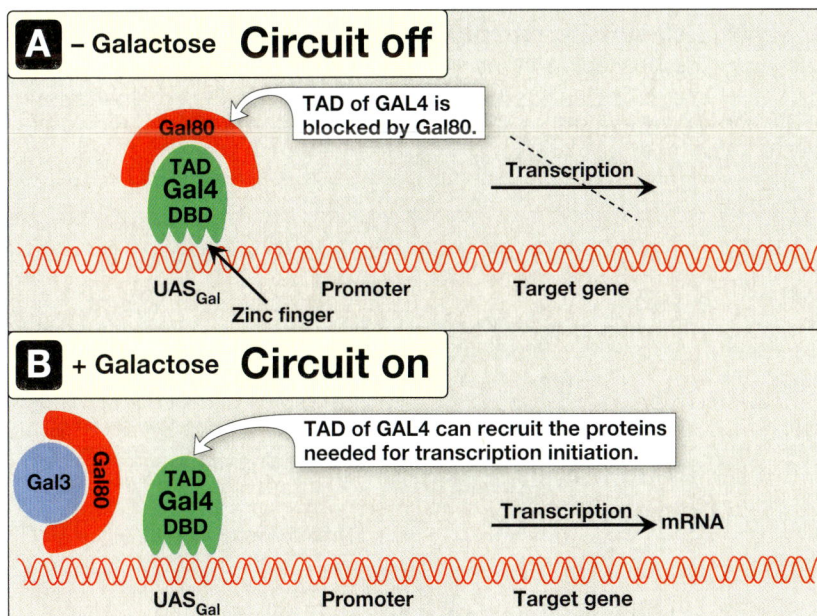

Figure 33.9
Regulation of galactose circuit in yeast in the (**A**) absence and (**B**) presence of galactose. [Note: Target genes, whether on the same or a different chromosome, each have an upstream activating sequence galactose (UAS$_{Gal}$).] TAD, transcription-activation domain; DBD, DNA-binding domain; mRNA, messenger RNA.

signals in multicellular organisms. Hormones bind to either intracellular (nuclear) receptors (e.g., steroid hormones; see Fig. 18.29) or cell-surface receptors (e.g., the peptide hormone glucagon; see Fig. 23.13).

a. **Intracellular receptors:** Members of the nuclear receptor superfamily, which includes the steroid hormone (glucocorticoids, mineralocorticoids, androgens, and estrogens), vitamin D, retinoic acid, and thyroid hormone receptors, function as STFs. In addition to domains for DNA-binding and transcriptional activation, these receptors also contain a ligand-binding domain. For example, the steroid hormone cortisol (a glucocorticoid) binds intracellular receptors at the ligand-binding domain (Fig. 33.10). Binding causes a conformational change in the receptor that activates it. The receptor–hormone complex enters the nucleus, dimerizes, and binds via a zinc finger motif to DNA at a regulatory element, the glucocorticoid-response element (GRE) that is an example of an HRE. Binding allows recruitment of coactivators to the TAD of the receptor and results in expression of cortisol-responsive genes, each of which is under the control of its own GRE. Binding of the receptor–hormone complex to the GRE allows coordinate expression of a group of target genes, even though these genes are on different chromosomes. The GRE can be located upstream or downstream of the genes it regulates and at great distances from them. The GRE, then, can function as an enhancer sequence (see Fig. 31.14). If the hormone–receptor complexes associate with repressors, they inhibit transcription.

b. **Cell-surface receptors:** These receptors include those for insulin, epinephrine, and glucagon. Glucagon, for example, is a peptide hormone that binds its G protein–coupled plasma membrane receptor on glucagon-responsive cells.

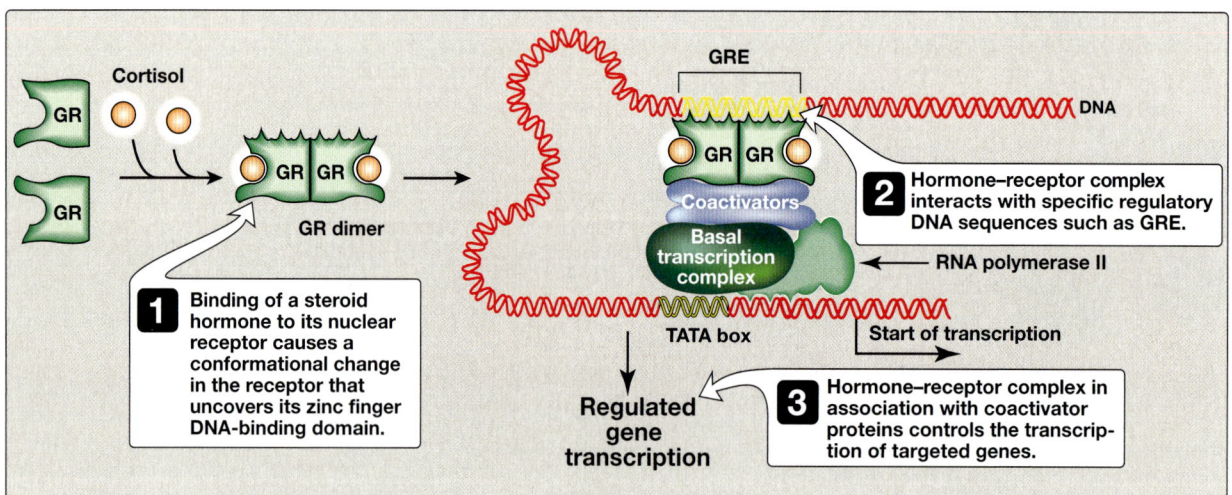

Figure 33.10
Transcriptional regulation by intracellular steroid hormone receptors. GRE, glucocorticoid response element; GR, glucocorticoid receptor.

This extracellular signal is then transduced to intracellular cAMP, a second messenger, which can affect protein expression (and activity) through protein kinase A–mediated phosphorylation (Fig. 33.11; also see Fig. 8.8). In response to a rise in cAMP, a *trans*-acting factor (cAMP response element–binding [CREB] protein) is phosphorylated and activated. Active CREB protein binds via a leucine zipper motif to a *cis*-acting regulatory element, the cAMP-response element (CRE), resulting in transcription of target genes with CREs in their promoters. An example of a gene upregulated by the cAMP/CRE/CREB system is that of phosphoenolpyruvate carboxykinase (PEPCK), a key enzyme of gluconeogenesis that is induced by glucagon and cortisol.

B. Messenger RNA processing and use

Eukaryotic mRNA undergoes several processing events before it is exported from the nucleus to the cytoplasm for use in protein synthesis. Capping at the 5′ end, polyadenylation at the 3′ end (see Fig. 31.17), and splicing (see Fig. 31.18) are essential for the production of a functional eukaryotic messenger from most pre-mRNA. Variations in splicing and polyadenylation can affect gene expression. In addition, messenger stability also affects gene expression.

1. **Alternative splicing:** Tissue-specific protein isoforms can be made from the same pre-mRNA through alternative splicing, which can involve exon skipping (loss), intron retention, and use of alternative splice donor or acceptor sites (Fig. 33.12). For example, the pre-mRNA for tropomyosin (TM) undergoes tissue-specific alternative splicing to yield a number of TM isoforms. More than 90% of all human genes undergo alternative splicing.

2. **Alternative polyadenylation:** Some pre-mRNA transcripts have more than one site for cleavage and polyadenylation. Alternative polyadenylation generates mRNA with different 3′ ends, altering the untranslated region (UTR) or the coding (translated) sequence. For example, this mechanism is used in the production of the membrane-bound and secreted forms of immunoglobulin M by naïve B cells during the adaptive immune response against a pathogen.

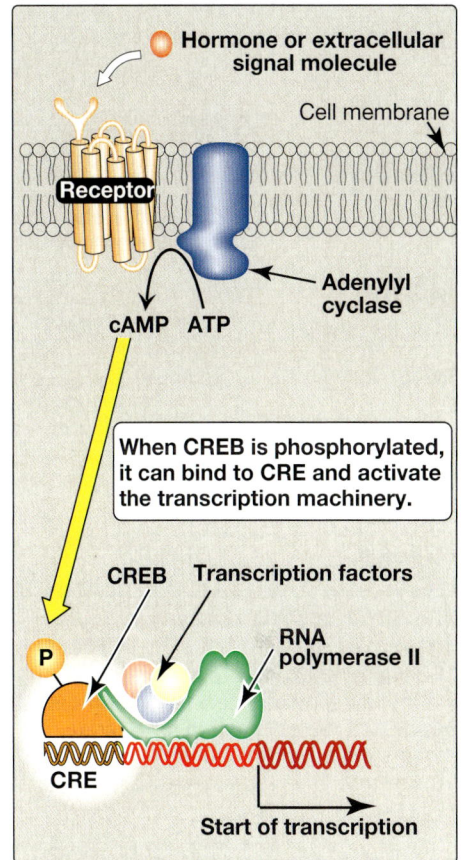

Figure 33.11
Transcriptional regulation by receptors located in the cell membrane. [Note: Cyclic adenosine monophosphate (cAMP) activates protein kinase A that phosphorylates cAMP response element–binding (CREB) protein.] CRE, cAMP response element.

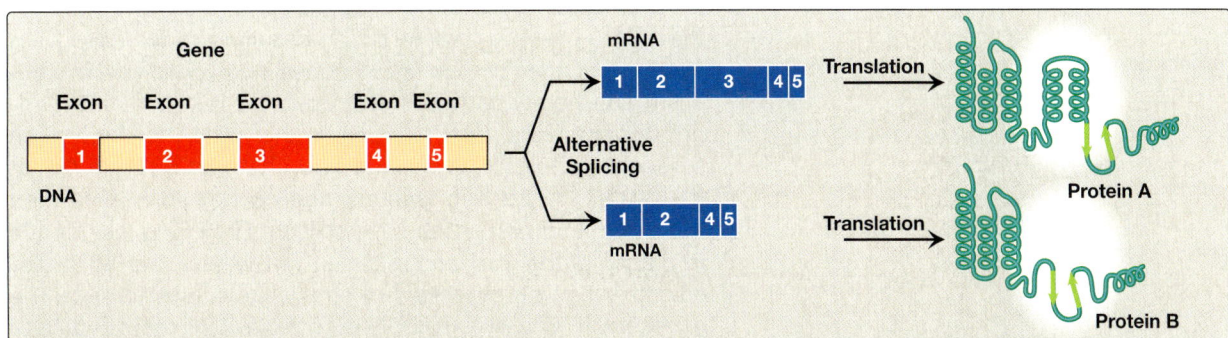

Figure 33.12
Tissue-specific alternative splicing produces different proteins, or isoforms, from a single gene. mRNA, messenger RNA.

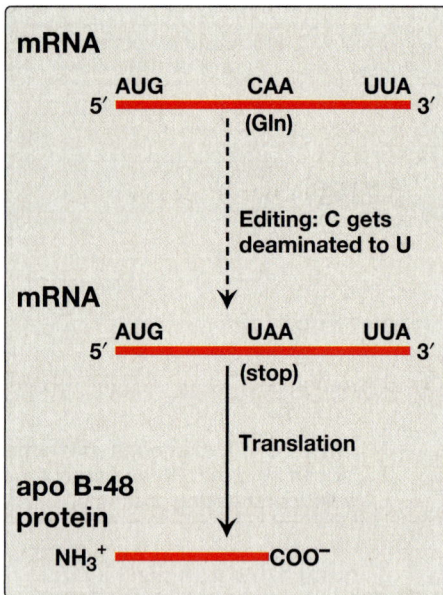

Figure 33.13
Editing of apolipoprotein (apo) B mRNA in the intestine and generation of the apo B-48 protein needed for chylomicron synthesis. Gln, glutamine; mRNA, messenger RNA; A, adenine; C, cytosine; G, guanine; U, uracil.

The use of alternative splicing and polyadenylation sites, as well as alternative transcription start sites explains, at least in part, how the ~20,000 to 25,000 genes in the human genome can give rise to well over 100,000 proteins.

3. **Messenger RNA editing:** Even after mRNA has been fully processed, it may undergo an additional posttranscriptional modification in which a base in the mRNA is altered. This is known as RNA editing. An important example in humans occurs with the transcript for apolipoprotein (apo) B, an essential component of chylomicrons (see Fig. 18.17) and very–low-density lipoproteins ([VLDLs] see Fig. 18.18). Apo B mRNA is made in the liver and the small intestine. Liver apo B mRNA produces the full-length protein, apo B-100, which is incorporated into VLDLs. However, in the intestine, apo B mRNA is edited such that the cytosine (C) base in the CAA codon for glutamine is enzymatically deaminated to uracil (U). This changes the sense codon CAA to the nonsense or stop codon UAA, as shown in Figure 33.13, and results in the translation of a shorter protein, apo B-48. Apo B-48 is made only in the intestine and is incorporated into chylomicrons.

4. **Messenger RNA stability:** How long an mRNA remains in the cytosol before it is degraded influences how much protein can be produced from it. Regulation of iron metabolism and the gene-silencing process of RNA interference (RNAi) illustrate the importance of mRNA stability in the regulation of gene expression.

    a. **Iron metabolism:** Transferrin (Tf) is a plasma protein that transports iron. Tf binds to a cell-surface transferrin receptor (TfR) that gets internalized and provides cells, such as erythroblasts in the bone marrow, with iron. The mRNA for the TfR has several *cis*-acting iron-responsive elements (IREs) in its 3′-UTR. IREs have a short stem-loop structure that can be bound by *trans*-acting iron regulatory proteins (IRPs), as shown in Figure 33.14. When the iron concentration in the cell is low, the IRPs bind to the 3′-IRE and stabilize the mRNA for TfR, allowing TfR synthesis. When intracellular iron levels are high, the IRPs dissociate. The lack of IRP bound to the mRNA hastens its destruction, resulting in decreased TfR synthesis. Other iron-sensitive genes involved in iron metabolism and heme synthesis are regulated in a different manner through IREs. One example is the mRNA for ferritin, an intracellular protein of iron storage, that has a single IRE in its 5′-UTR. When iron levels in the cell are low, IRPs bind the 5′-IRE and prevent the use of the mRNA, and less ferritin is made. When iron accumulates in the cell, the IRPs dissociate, allowing synthesis of ferritin molecules to store the excess iron. Another example is aminolevulinic acid synthase 2 (ALAS2), the regulated enzyme of heme synthesis in erythroblasts (see Fig. 21.3). ALAS2 mRNA also contains a 5′-IRE that controls its translation in a similar manner.

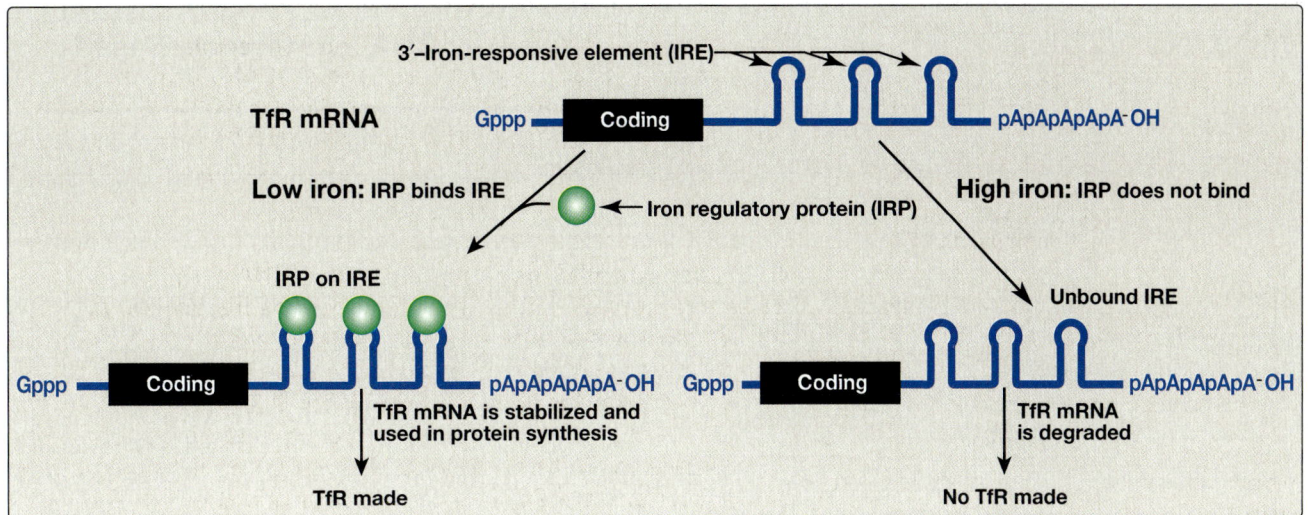

Figure 33.14
Regulation of transferrin receptor (TfR) synthesis. [Note: The IREs are located in the 3'-UTR (untranslated region) of TfR messenger RNA (mRNA).] Gppp, 7-methylguanosine cap; p(Ap)$_n$A-OH, polyadenylate tail.

b. **RNA interference:** RNAi is a mechanism of gene silencing through decreased expression of mRNA, either by repression of translation or by increased degradation. It plays a key role in such fundamental processes as cell proliferation, differentiation, and apoptosis. RNAi is mediated by short (~22 nucleotides), noncoding RNAs called microRNA (miRNA). miRNAs arise from far longer, genomically encoded nuclear transcripts, primary miRNAs (pri-miRNAs) that are partially processed in the nucleus to pre-miRNA by an endonuclease (Drosha), then transported to the cytoplasm. There, another endonuclease (Dicer) completes the processing and generates short, double-stranded miRNA. A single strand (the guide or antisense strand) of the miRNA associates with a cytosolic protein complex known as the RNA-induced silencing complex (RISC). The guide strand hybridizes with a complementary sequence in the 3'-UTR of a full-length target mRNA, bringing RISC to the mRNA. This can result in repression of translation of the mRNA or its degradation by an endonuclease (Argonaute/Ago/Slicer) of the RISC. The extent of complementarity appears to be the determining factor (Fig. 33.15). RNAi can also be triggered by the introduction of exogenous double-stranded short, interfering RNA (siRNA) into a cell, a process that has enormous therapeutic potential.

5. **Messenger RNA translation:** Regulation of gene expression can also occur at the level of mRNA translation. One mechanism by which translation is regulated is through phosphorylation of the eukaryotic translation initiation factor eIF-2 (Fig. 33.16). Phosphorylation of eIF-2 prevents its reactivation by inhibiting GDP–GTP exchange and thereby inhibits its function in translation initiation (see Fig. 32.13). The kinases responsible for eIF-2 phosphorylation are activated in response to environmental conditions, such as amino acid starvation, heme deficiency in

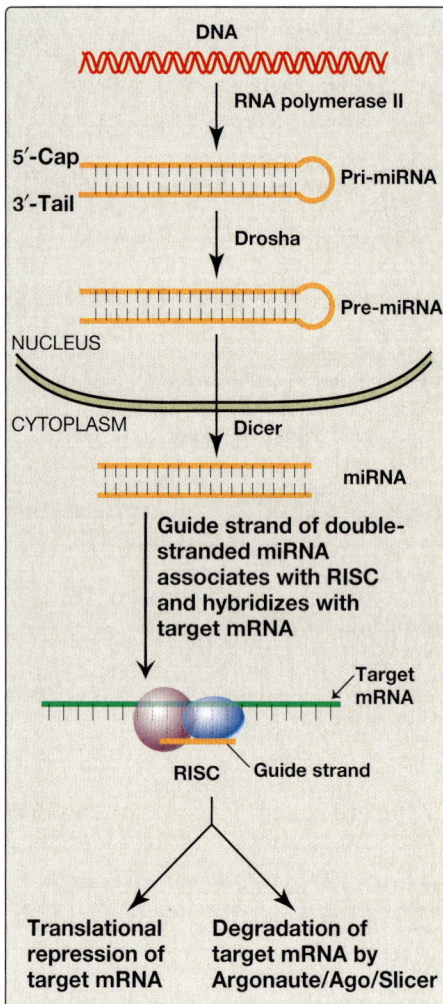

Figure 33.15
Biogenesis and actions of microRNA
(miRNA). [Note: The extent of complementarity between the target messenger RNA (mRNA) and the miRNA
determines the final outcome, with
perfect complementarity resulting in
mRNA degradation.] Pri, primary; RISC,
RNA-induced silencing complex.

Clinical Application 33.1: RNA Interference–Based Therapies

Several RNAi-based therapies have reached clinical application. Notable examples are four, siRNA-based agents, inclisiran, patisiran, lumasiran, and givosiran. The first three agents work by guiding RISC to the 3′-UTR of target mRNAs. Inclisiran prevents the synthesis of proprotein convertase subilisin-kinexin type 9 (PCSK9). PCSK9 plays a role in degrading the LDL receptor making its mRNA a prime target in successfully lowering LDL-C in patients with atherosclerotic disease, particularly those with heterozygous familial hypercholesterolemia (see Chapter 18, VI. D.). Patisiran is used to treat peripheral nerve disease (polyneuropathy) in patients with hereditary transthyretin-mediated amyloidosis (hATTR) caused by a mutation in the gene-encoding transthyretin (TTR). This drug prevents the production of abnormal TTR protein and reduces the buildup of amyloid deposits containing TTR that form in peripheral nerves and in the heart. Lumarsin targets hydroxyacid oxidase 1 mRNA to reduce the amount of the glycolate oxidase enzyme and its production of oxalate. This agent, which is most helpful for patients with primary hyperoxaluria, prevents the deposition of oxalate in the kidney, which causes urinary calcium oxalate stones (nephrolithiasis) and ultimately leads to kidney disease or even failure. The fourth agent, givosiran, guides RISC to a sequence within the coding region of the mRNA for aminolevulinate synthase 1 (ALAS1) to reduce the production of this enzyme in patients with acute hepatic porphyria. These individuals have debilitating symptoms caused by the accumulation of neurotoxic levels of aminolevulinic acid and porphobilinogen (see Chapter 21, II. C. 1.).

erythroblasts, the presence of double-stranded RNA (signaling viral infection), and the accumulation of misfolded proteins in the rough endoplasmic reticulum.

C. Regulation through variations in DNA

Gene expression in eukaryotes is also influenced by the accessibility of DNA to the transcriptional apparatus, the number of copies of genes, and the arrangement of DNA. Localized transitions between the B and Z forms of DNA (see Fig. 30.7) can also affect gene expression.

1. **Access to DNA:** In eukaryotes, DNA is found complexed with histone and nonhistone proteins to form chromatin (see Fig. 30.27). Transcriptionally active, decondensed chromatin (euchromatin) differs from the more condensed, inactive form (heterochromatin) in a number of ways. Active chromatin contains histone proteins that have been covalently modified at their amino terminal ends by reversible methylation, acetylation, or phosphorylation (see Chapter 31, IV. A. for a discussion of histone acetylation/deacetylation by histone acetyltransferase and histone deacetylase). Such modifications decrease the positive charge of these basic proteins, thereby decreasing the strength of their association with negatively charged DNA. This relaxes the nucleosome, allowing transcription factors access to specific regions on the DNA. Nucleosomes can also be repositioned, an ATP-requiring process that is part of chromatin remodeling. Another difference

between transcriptionally active and inactive chromatin is the extent of methylation of C bases in CG-rich regions (CpG islands) in the promoter region of many genes. Methylation is carried out by methyltransferases that use S-adenosylmethionine as the methyl donor (Fig. 33.17). Transcriptionally active genes are less methylated (hypomethylated) than their inactive counterparts, suggesting that DNA hypermethylation silences gene expression. Modification of histones and methylation of DNA are epigenetic in that they are heritable changes affecting DNA structure that alter gene expression without altering the base sequence. Epigenetic changes also occur in cancers making the enzymes responsible for the specific changes possible targets for cancer therapy.

2. **Gene copy number:** A change up or down in the number of copies of a gene can affect the amount of gene product produced. An increase in copy number (gene amplification) has contributed to increased genomic complexity and is a mechanism used to enhance protein production during stages of the developmental process in nonmammalian species, such as *Xenopus laevis* (African clawed frogs) and *Drosophilia melanogaster* (fruit flies). In mammals, however, gene amplification is associated with some diseases and is involved in the mechanism by which cells develop resistance to particular chemotherapeutic drugs. For example, the gene encoding the N-MYC transcription factor is amplified in the human brain cancer neuroblastoma. N-MYC interacts with chromatin remodeling factors and with other transcription factors to regulate target genes. Gene amplification causes an excess of N-MYC protein production that results in its action as an oncogene to dysregulate gene expression and promote tumor growth. Additionally, gene amplification is involved in resistance to the drug methotrexate, an inhibitor of the enzyme dihydrofolate reductase (DHFR). DHFR recycles dihydrofolate to tetrahydrofolate, which is required for the thymidylate synthase production of thymidine monophosphate (TMP) in the pyrimidine biosynthetic pathway (see Fig. 28.2). TMP is the essential precursor for thymidine triphosphate (TTP) that is necessary for DNA synthesis. The amplification of the *DHFR* gene results in the expression of more DHFR enzyme, which enables the cells exposed to methotrexate to survive because thymidine nucleotide production can continue in the presence of the drug.

3. **Arrangement of DNA:** The process by which immunoglobulins (antibodies) are produced by B lymphocytes involves permanent rearrangements of the DNA in these cells. The immunoglobulins (e.g., IgG) consist of two light and two heavy chains, with each chain containing regions of variable and constant amino acid sequence. The variable region is the result of somatic recombination of segments within both the light- and the heavy-chain genes. During B-lymphocyte development, single variable (V), diversity (D), and joining (J) gene segments are randomly selected and brought together through gene rearrangement to form a unique variable region (Fig. 33.18). This process allows the generation of 10^9 to 10^{11} different immunoglobulins from a single gene, providing the diversity needed for the recognition of an enormous number of antigens. DNA rearrangement may

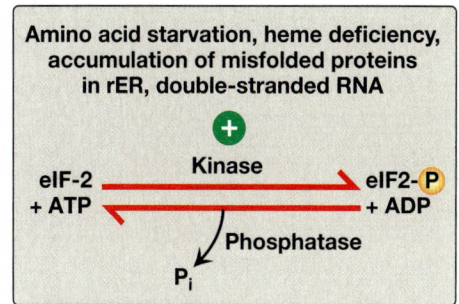

Figure 33.16
Regulation of translation initiation in eukaryotes by phosphorylation of eukaryotic translation initiation factor, eIF-2. rER, rough endoplasmic reticulum; ADP, adenosine diphosphate; P_i, inorganic phosphate; P, phosphate.

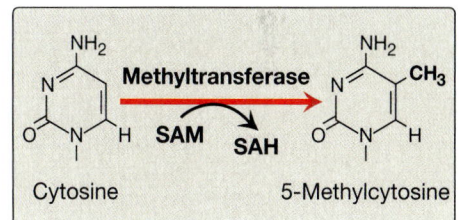

Figure 33.17
Methylation of cytosine in eukaryotic DNA. SAM, S-adenosylmethionine; SAH, S-adenosylhomocysteine.

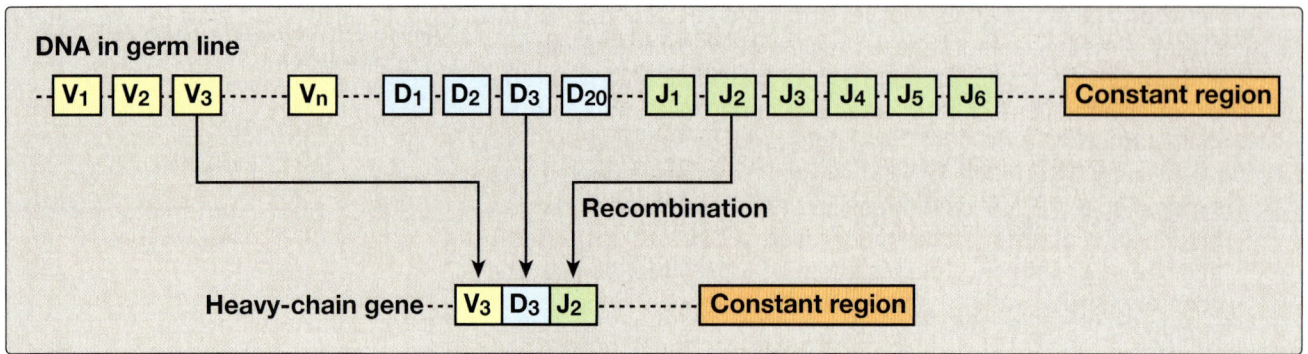

Figure 33.18
DNA rearrangements in the generation of immunoglobulins. V, variable; D, diversity; J, joining.

also lead to pathology. Translocation, an aberrant process by which two different chromosomes exchange DNA segments, is a mechanism associated with congenital syndromes and cancer development (see Fig. 34.13).

4. **Mobile DNA elements:** Transposons (Tns) are mobile segments of DNA that move in an essentially random manner from one site to another on the same or a different chromosome. Movement is mediated by transposase, an enzyme encoded by the Tn itself. Movement can be direct, in which transposase cuts out and then inserts the Tn at a new site, or replicative, in which the Tn is copied and the copy inserted elsewhere while the original remains in place. In eukaryotes, including humans, replicative transposition frequently involves an RNA intermediate made by a reverse transcriptase (see Chapter 30, IV. D.), in which case the Tn is called a retrotransposon. Transposition has contributed to structural variation in the genome but also has the potential to alter gene expression and even to cause disease. Tns comprise ~50% of the human genome, with retrotransposons accounting for 90% of Tns. Although the vast majority of these retrotransposons have lost the ability to move, some are still active. Their transposition is thought to be the basis for some rare cases of hemophilia A and Duchenne muscular dystrophy. In bacteria, plasmids (extrachromosomal circular DNA) may contain Tn-carrying antibiotic resistance genes. These genes can move from the plasmid to the bacterial chromosome so that the bacterium is resistant to one or more antimicrobial drugs even if the plasmid is lost from the cell. Bacteria may replicate then exchange plasmids during conjugation, further spreading drug resistance.

V. CHAPTER SUMMARY

- **Gene expression** produces a functional gene product (either RNA or protein).
- **Genes** can be either **constitutive** (always expressed) or **regulated** (expressed only under certain conditions).

- Regulation of gene expression occurs primarily at **transcription** in both prokaryotes and eukaryotes and is mediated through ***trans*-acting proteins** binding to ***cis*-acting regulatory DNA elements** (Fig. 33.19).

- In **eukaryotes**, regulation also occurs through DNA **modifications** and through **posttranscriptional** and **posttranslational processing**.

- In **prokaryotes**, the coordinate regulation of genes whose protein products are required for a particular process is achieved through **operons** (groups of functionally related genes sequentially arranged on the chromosome along with the regulatory elements that determine their transcription). Examples from *Escherichia coli* are the *lac* **operon** containing the *Z, Y,* and *A* structural genes involved in the catabolism of lactose, and the *trp* **operon,** which contains genes needed for the synthesis of Trp. The trp operon is also regulated by attenuation, in which mRNA synthesis that escaped repression by Trp is terminated before completion.

- In prokaryotes, transcription of **ribosomal RNA (rRNA)** and **transfer RNA (tRNA)** is selectively inhibited by the **stringent response** to amino acid starvation. **Translation** is also a site of prokaryotic gene regulation: Excess ribosomal (r)-proteins bind the **Shine–Dalgarno (SD) sequence** on their own polycistronic messenger RNA (mRNA), preventing ribosomes from binding.

- In eukaryotes, hormones coordinate the expression of groups of genes by binding to an intracellular receptor that acts as a *trans*-acting protein (as with steroid hormones) or to a cell surface receptor that initiates **second-messenger** signaling to activate a *trans*-acting protein (as with peptide hormones). In each case, the protein recognizes a specific response element and binds to the DNA sequence using structural motifs such as a **zinc finger** or a **leucine zipper**.

- **Co-** and **posttranscriptional regulation** is also seen in eukaryotes and includes **alternative mRNA splicing** and **polyadenylation**, mRNA **editing**, and variations in mRNA **stability**. **Transferrin receptor** synthesis is enhanced by mRNA stability when iron concentrations are low. **RNA interference** is used to control mRNA stability and translation and is the basis for a new class of therapeutic agents.

- Regulation at the **translational level** can be caused by the **phosphorylation** and inhibition of **eukaryotic initiation factor-2**. Gene expression in eukaryotes is also influenced by **accessibility** of DNA to the transcriptional apparatus (as seen with **epigenetic** changes to histone proteins), the gene copy number, and the **arrangement** of the DNA.

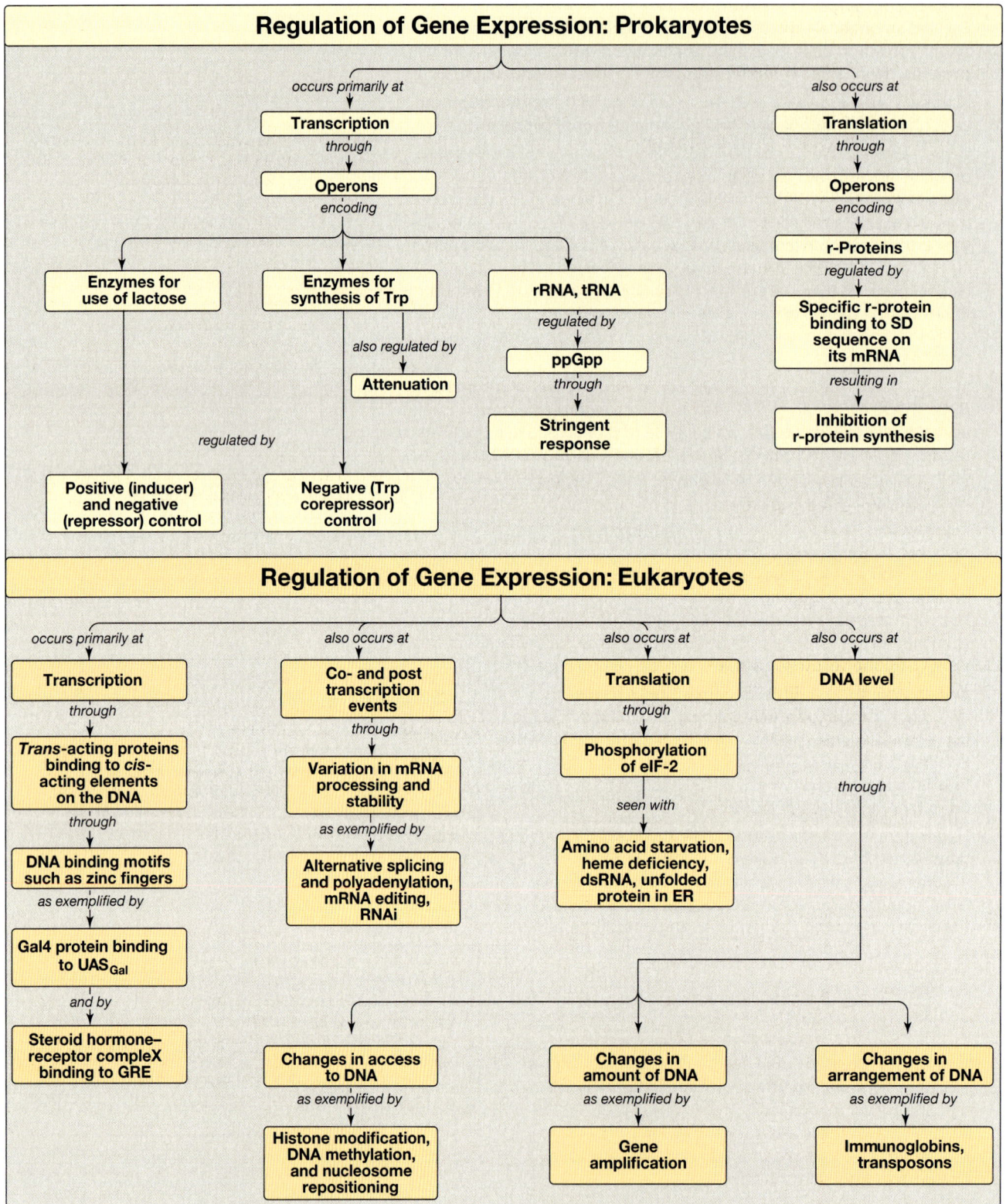

Figure 33.19
Summary of key concepts for the regulation of gene expression. Trp, tryptophan; rRNA, tRNA, mRNA, ribosomal, transfer, and messenger RNA, respectively; ppGpp, guanosine tetraphosphate; r-protein, ribosomal protein; SD, Shine–Dalgarno; Gal, galactose; UAS, upstream activating sequence; GRE, glucocorticoid response element; RNAi, RNA interference; eIF, eukaryotic initiation factor; ds, double stranded; ER, endoplasmic reticulum.

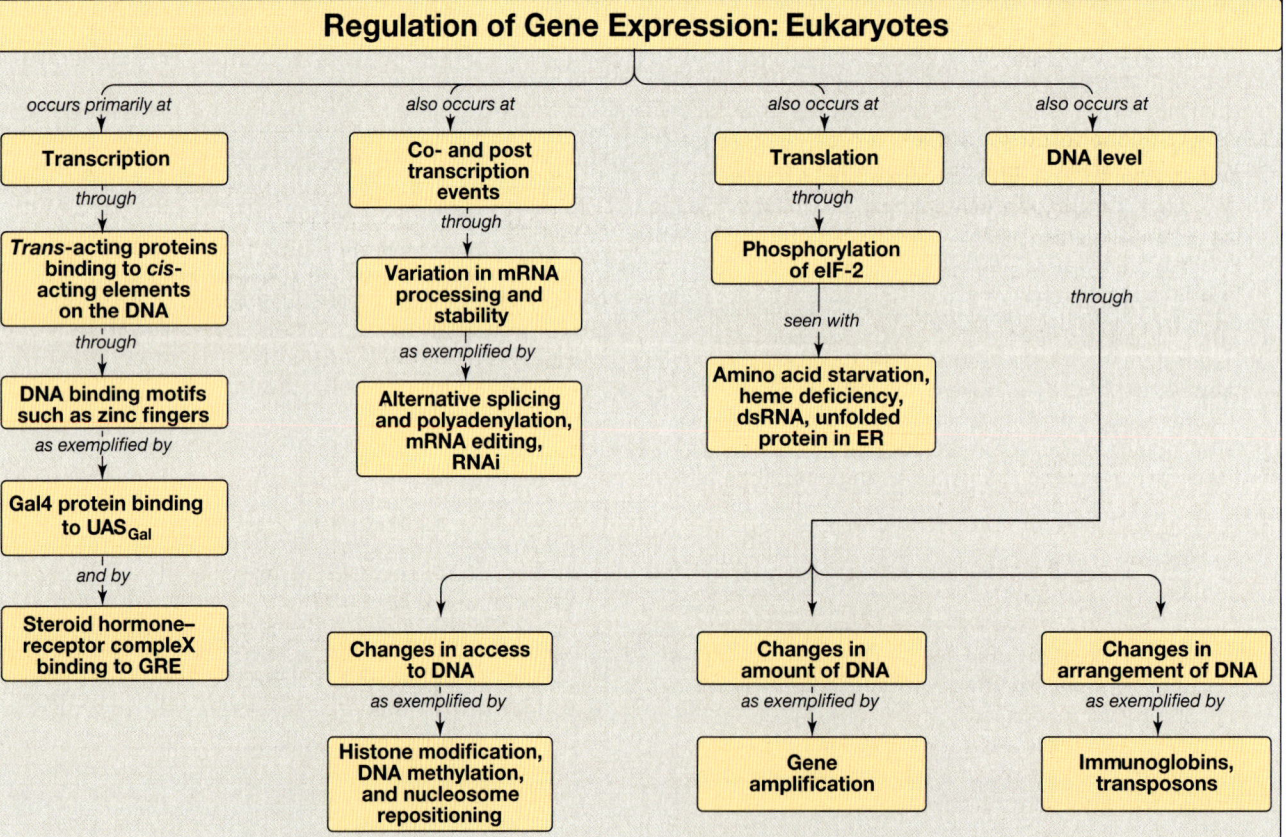

Study Questions

Choose the ONE best answer.

33.1. Which of the following mutations is most likely to result in reduced expression of the *lac* operon?

    A. cya⁻ (no adenylyl cyclase made)

    B. i⁻ (no repressor protein made)

    C. Oᶜ (operator cannot bind repressor protein)

    D. One resulting in impaired glucose uptake

    E. relA⁻ (no stringent response occurs)

> Correct answer = A. In the absence of glucose, adenylyl cyclase makes cyclic adenosine monophosphate (cAMP), which forms a complex with the catabolite activator protein (CAP). The cAMP–CAP complex binds the CAP site on the DNA, causing RNA polymerase to bind more efficiently to the *lac* operon promoter, thereby increasing expression of the operon. With cya⁻ mutations, adenylyl cyclase is not made, and so the operon is unable to be maximally expressed even when glucose is absent and lactose is present. The absence of a repressor protein or decreased ability of the repressor to bind the operator results in constitutive (essentially constant) expression of the *lac* operon.

33.2. Which of the following is best described as *cis*-acting?

    A. Cyclic adenosine monophosphate response element–binding protein

    B. Operator

    C. Repressor protein

    D. Thyroid hormone nuclear receptor

    E. Histone modification

> Correct answer = B. The operator is part of the DNA itself and so is *cis*-acting. The cyclic adenosine monophosphate response element–binding protein, repressor protein, and thyroid hormone nuclear receptor protein are molecules that diffuse (transit) to the DNA, bind, and affect the expression of that DNA and so are *trans*-acting.

33.3. Which of the following is the basis for the intestine-specific expression of apolipoprotein B-48?

    A. DNA rearrangement and loss

    B. DNA transposition

    C. RNA alternative splicing

    D. RNA editing

    E. RNA interference

> Correct answer = D. The production of apolipoprotein (apo) B-48 in the intestine and apo B-100 in liver is the result of RNA editing in the intestine, where a sense codon is changed to a nonsense codon by posttranscriptional deamination of cytosine to uracil. DNA rearrangement and transposition, as well as RNA interference and alternative splicing, do alter gene expression but are not the basis of apo B-48 tissue-specific production.

33.4. Which of the following is a likely consequence of the increased iron accumulation seen in patients with hemochromatosis?

    A. The messenger RNA for the transferrin receptor is stabilized by the binding of iron regulatory proteins to its 3′–iron-responsive elements.

    B. The messenger RNA for the transferrin receptor is not bound by iron regulatory proteins and is degraded.

    C. The messenger RNA for ferritin is not bound by iron regulatory proteins at its 5′–iron-responsive element and is translated.

    D. The messenger RNA for ferritin is bound by iron regulatory proteins and is not translated.

    E. Both B and C are correct.

> Correct answer = E. When iron levels in the body are high, as is seen with hemochromatosis, there is increased synthesis of the iron-storage molecule, ferritin, and decreased synthesis of the transferrin receptor (TfR) that mediates iron uptake by cells. These effects are the result of *cis*-acting iron-responsive elements not being bound by *trans*-acting iron regulatory proteins, resulting in degradation of the messenger RNA (mRNA) for TfR and increased translation of the mRNA for ferritin.

33.5. Patients with estrogen receptor–positive (hormone responsive) breast cancer may be treated with the drug tamoxifen, which binds the estrogen nuclear receptor without activating it. Which of the following is the most logical outcome of tamoxifen use?

    A. Increased acetylation of estrogen-responsive genes

    B. Increased growth of estrogen receptor–positive breast cancer cells

    C. Increased production of cyclic adenosine monophosphate

    D. Inhibition of the estrogen operon

    E. Inhibition of transcription of estrogen-responsive genes

Correct answer = E. Tamoxifen competes with estrogen for binding to the estrogen nuclear receptor. Tamoxifen fails to activate the receptor, preventing its binding to DNA sequences that upregulate expression of estrogen-responsive genes. Tamoxifen, then, blocks the growth-promoting effects of these genes and results in growth inhibition of estrogen-dependent breast cancer cells. Acetylation increases transcription by relaxing the nucleosome. Cyclic adenosine monophosphate is a regulatory signal mediated by cell-surface rather than nuclear receptors. Mammalian cells do not have operons.

33.6. The *ZYA* region of the *lac* operon will be maximally expressed if:

    A. cyclic adenosine monophosphate levels are low.

    B. glucose and lactose are both available.

    C. the attenuation stemloop is able to form.

    D. the CAP site is occupied.

    E. the Shine–Dalgarno sequence is not accessible.

Correct answer = D. It is only when glucose is gone, cyclic adenosine monophosphate (cAMP) levels are increased, the cAMP–catabolite activator protein (CAP) complex is bound to the CAP site, and lactose is available that the operon is maximally expressed (induced). If glucose is present, the operon is off as a result of catabolite repression. The *lac* operon is not regulated by attenuation, a mechanism for stopping transcription in some operons such as the *trp* operon.

33.7. X-chromosome inactivation is a process by which one of two X chromosomes in mammalian females is condensed and inactivated to prevent overexpression of X-linked genes. What would most likely be true about the degree of DNA methylation and histone acetylation on the inactivated X chromosome?

Cytosines in CpG islands would be hypermethylated, and histone proteins would be deacetylated. Both conditions are associated with decreased gene expression, and both are important in maintaining X inactivation.

Biotechnology and Human Disease

34

I. OVERVIEW

The human genome contains ~3 billion (10^9) base pairs (bps) of DNA that encode 20,000 to 25,000 protein-coding genes located on 23 chromosomes in the haploid genome. In 2003, the Human Genome Project revealed the first complete sequence of the human genome and revolutionized our ability to research human gene function, study regions of DNA traditionally thought to be inactive in human biology, and diagnose human disease. Before this project, many steps were required to study the genome because the exact DNA sequence was unknown. The three basic tools (restriction endonucleases, biologic cloning of DNA, and probes) for genome analysis are still in use today, but the complete genome sequence has enabled faster isolation and cloning (or amplification to make identical copies) of gene regions (Fig. 34.1). Today, the amplification of a DNA sequence is accomplished mainly through a molecular technique called the polymerase chain reaction (PCR). PCR produces millions of copies of DNA sequence within a few hours, a process that took weeks or longer using traditional biologic cloning vectors. PCR coupled with DNA sequencing has permitted the identification of variations in human DNA that give rise to disease. This knowledge has led to the more rapid diagnosis of disease, the emergence of treatments based on a patient's genetics or the genetics of cancerous tumors (a field called pharmacogenomics), and the success of gene therapy and gene editing approaches for treating several genetic disorders. Advances in technology have also made possible the rapid whole-genome sequencing of individual humans, animals, plants, and microbes. The following sections cover some applications of biotechnology in diagnosis and therapy in medicine.

II. RESTRICTION ENDONUCLEASES

Bacterial enzymes, called restriction endonucleases (restriction enzymes), which cleave double-stranded DNA (dsDNA) into smaller, more manageable fragments, are used to facilitate genetic analysis. Each restriction enzyme cleaves dsDNA at a specific nucleotide sequence (restriction site) and can be used to obtain precisely defined DNA segments called restriction fragments. These fragments can then be transferred to genetic tools called vectors for biologic DNA cloning or for the study of gene regulation or of gene products.

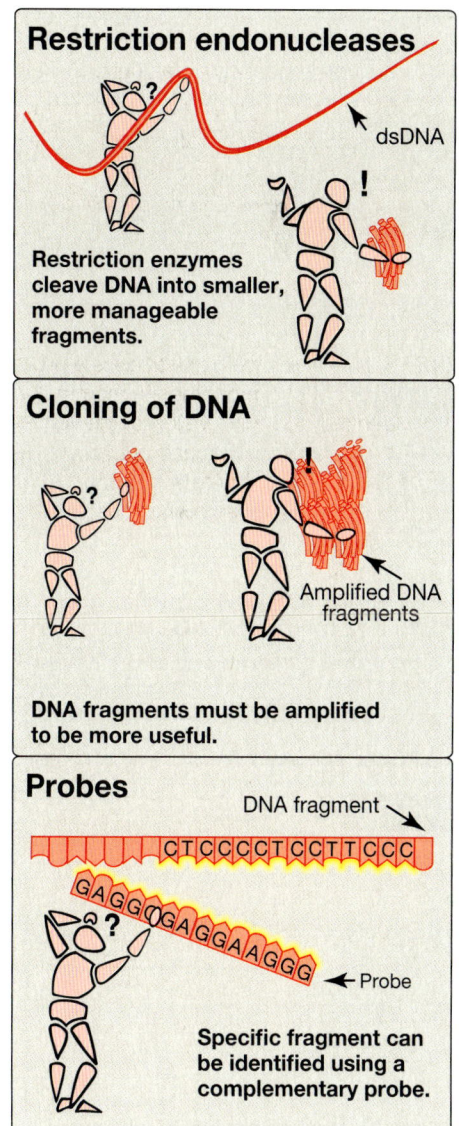

Restriction endonucleases

dsDNA

Restriction enzymes cleave DNA into smaller, more manageable fragments.

Cloning of DNA

Amplified DNA fragments

DNA fragments must be amplified to be more useful.

Probes

DNA fragment

CTCCCCTCCTTCCC

GAGGGGAGGAAGGG

Probe

Specific fragment can be identified using a complementary probe.

Figure 34.1
Three tools that facilitate analysis of human DNA. dsDNA, double-stranded DNA.

563

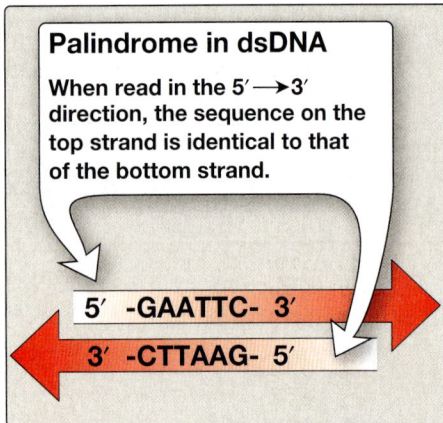

Figure 34.2
Recognition sequence of restriction endonuclease EcoRI. dsDNA, double-stranded DNA; A, adenine; C, cytosine; G, guanine; T, thymine.

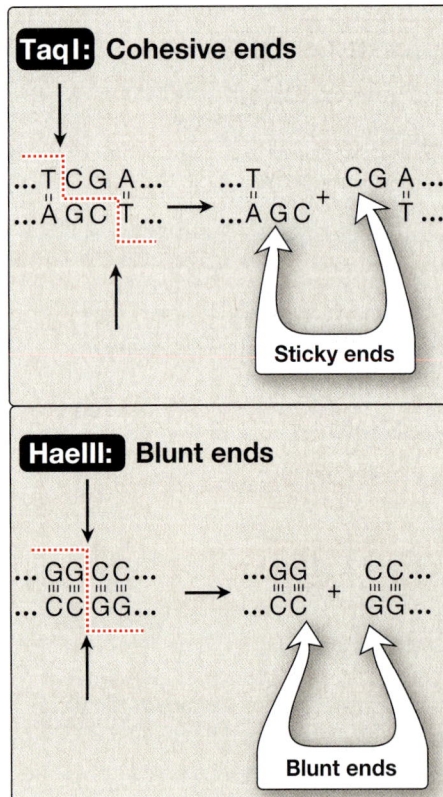

Figure 34.3
Specificity of TaqI and HaeIII restriction endonucleases. A, adenine; C, cytosine; G, guanine; T, thymine.

A. Specificity

Restriction endonucleases recognize restriction sites, which are short stretches of dsDNA (4 to 8 bps) that contain specific nucleotide sequences. These sequences, which differ for each restriction enzyme, are palindromes, that is, the nucleotide sequence on the two strands of DNA is identical if each is read in the 5′→3′ direction (Fig. 34.2). Restriction endonucleases cleave dsDNA into fragments of different sizes (restriction fragments) depending on the size of the restriction site sequence. An enzyme with a 4-bp restriction site has a greater probability of locating its restriction site in than does an enzyme with an 8-bp restriction site. Therefore, more restriction fragments are created from the same DNA target by enzymes with shorter restriction sites than those with longer restriction sequences. Hundreds of these enzymes, each having different cleavage specificities (varying in both nucleotide sequences and length of recognition sites) are commercially available for laboratory use.

In bacteria, restriction endonucleases limit (restrict) the expression of nonbacterial (foreign) DNA. Bacterial DNA is methylated at adenine bases, which protects the bacterial genome from being recognized. The endonucleases find their restriction sequences in foreign DNA and cleave the DNA into fragments. This process makes bacteria that have restriction endonucleases resistant to bacterial viruses called bacteriophages.

B. Nomenclature

A restriction enzyme is named according to the organism from which it was isolated. The first letter of the name is from the genus of the bacterium. The next two letters are from the name of the species. An additional letter indicates the type or strain (as needed), and a Roman numeral is appended to indicate the order in which the enzyme was discovered in that particular organism. For example, HaeIII is the third restriction endonuclease isolated from the bacterium *Haemophilus aegyptius*.

C. Sticky and blunt ends

Restriction enzymes cleave dsDNA so as to produce a 3′-hydroxyl group on one end and a 5′-phosphate group on the other. Some restriction endonucleases, such as TaqI, form staggered cuts that produce sticky or cohesive ends (i.e., the resulting DNA fragments have single-stranded regions that are complementary to each other), as shown in Figure 34.3. Other restriction endonucleases, such as HaeIII, produce fragments that have blunt ends that are entirely double stranded and, therefore, do not form hydrogen bonds with each other. Using the enzyme DNA ligase (see Fig. 30.20), sticky ends of a DNA fragment of interest can be covalently joined with other DNA fragments that have sticky ends produced by cleavage with the same restriction endonuclease (Fig. 34.4). A ligase encoded by bacteriophage T4 can covalently join blunt-ended fragments.

III. DNA CLONING AND SEQUENCING

DNA cloning is a process that produces many identical copies (amplification) of DNA. This may be accomplished by the introduction of the DNA of interest into a cell (biologic cloning), which will replicate the DNA. Alternatively, PCR, an *in vitro* technique, may be used to amplify specific segments of DNA from a tiny sample of genetic material. Biologic DNA cloning by cellular replication is mainly used if the genome sequence of the organism is not available or if the DNA sequence to be amplified is more than 20 kilobases (kb). PCR is a more sensitive approach but requires knowledge of sequence flanking the DNA segment to be amplified (see Section III. B.). For biologic cloning of a nucleotide sequence of interest, the total cellular DNA is first cleaved with a specific restriction enzyme, creating hundreds of thousands of fragments. Each of the resulting DNA fragments is joined to a DNA vector molecule (referred to as a cloning vector) to form a hybrid, or recombinant, DNA molecule. Each recombinant molecule is introduced into a single host cell (e.g., a bacterium), where it is replicated. [Note: The process of introducing foreign DNA into a cell is called transformation for bacteria and yeast and transfection for higher eukaryotes.] As the host cell multiplies, it creates a colony of cells in which every bacterium contains copies of the same inserted DNA fragment and is a "clone" of the original cell. The recombinant molecules can be released from the host cells by disruption of the cell membranes. The cloned DNA fragments can be cleaved from their vectors using the appropriate restriction endonuclease and then isolated by purification techniques. By this mechanism, many identical copies of the DNA of interest can be produced. PCR is the preferred amplification technique in medicine, where DNA may be obtained from samples of blood, saliva, amniotic fluid, or solid tissue, and in forensic science, where DNA from hair follicles or scrapings of fluids from fabric may require analysis. In medicine, PCR amplification allows for genetic analysis to detect prenatal disorders, diagnose a patient with an inherited disease, and develop patient-specific therapeutic strategies.

A. Biologic cloning vectors

A vector is a molecule of DNA to which the fragment of DNA to be cloned is joined. Essential properties of a vector include the (1) capacity for autonomous replication within a host cell, (2) presence of at least one specific nucleotide sequence recognized by a restriction endonuclease, and (3) presence of at least one gene (such as an antibiotic resistance gene) that confers survival to the host cells and allows for host cell selection. Commonly used vectors include plasmids and viruses. If a viral vector is used to deliver DNA into human cells, the process of introducing the DNA into the cells is called transduction. Other vectors facilitate replication of larger DNA segments or expression of gene products within the host cell from complementary DNA (cDNA, expression vectors).

1. **Prokaryotic plasmids:** Prokaryotic organisms typically contain single, large, circular chromosomes. In addition, most species of bacteria also normally contain small, circular, extrachromosomal DNA molecules called plasmids (Fig. 34.5). Plasmid DNA undergoes replication that may or may not be synchronized to chromosomal division. Plasmids may carry genes that convey antibiotic resistance to the host bacterium and may facilitate the transfer

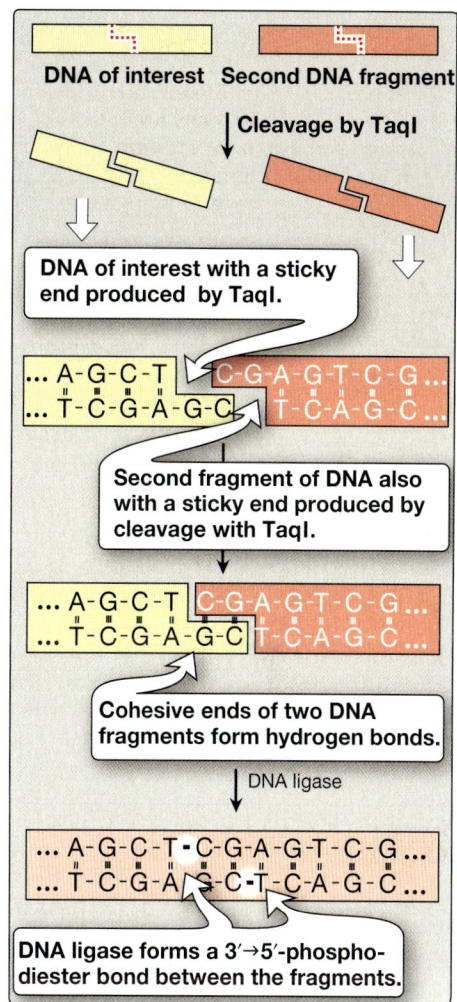

Figure 34.4
Formation of recombinant DNA from restriction fragments with sticky ends. A, adenine; C, cytosine; G, guanine; T, thymine.

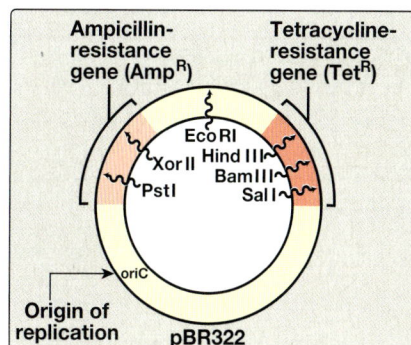

Figure 34.5
Partial map of pBR322 indicating the positions of its antibiotic resistance genes and 6 of the >40 unique sites recognized by specific restriction endonucleases. P, plasmid.

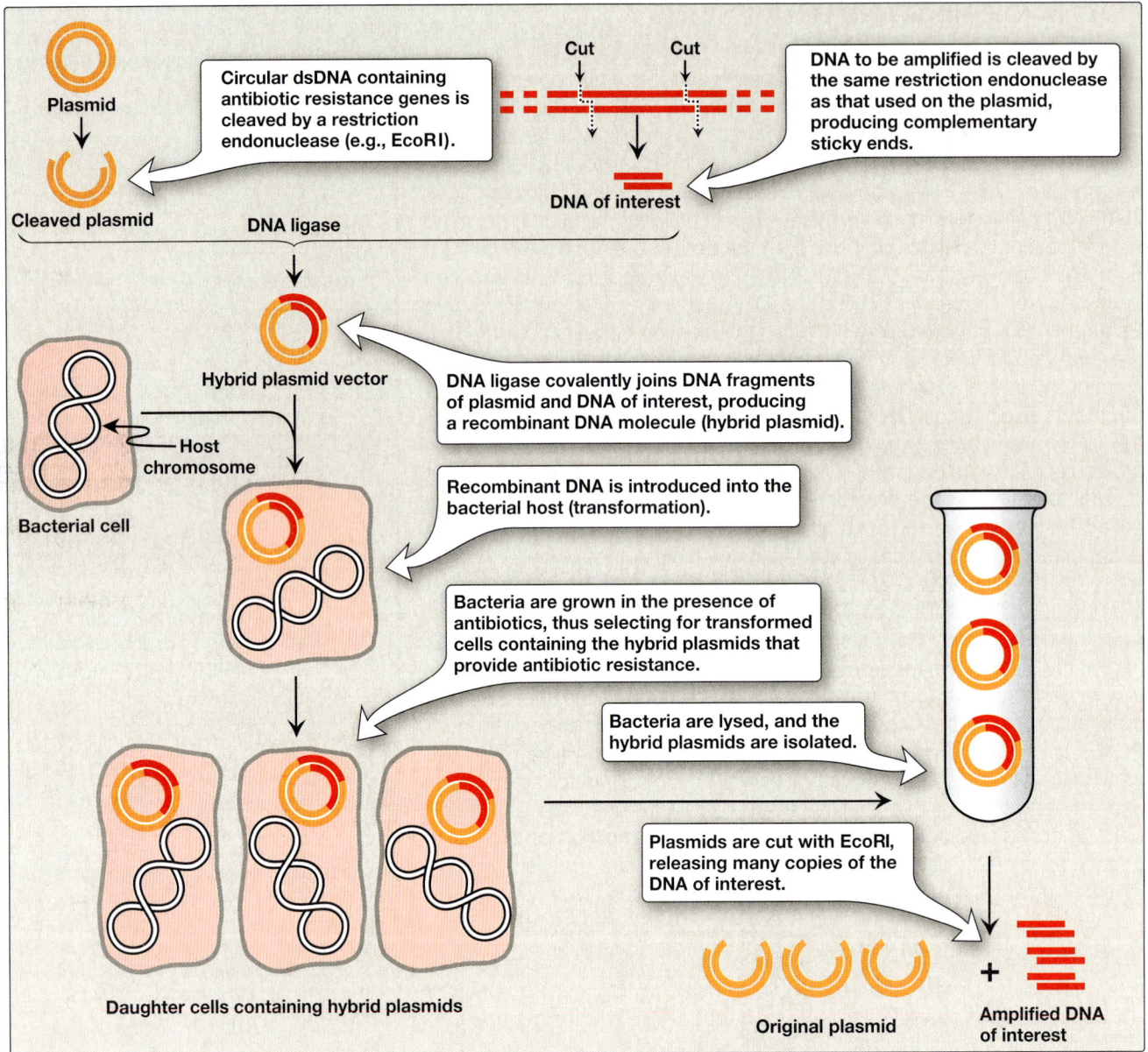

Figure 34.6
Summary of biologic gene cloning. [Note: Transformation is inefficient in that only a small percentage of the cells will contain the recombinant plasmid.] dsDNA, double-stranded DNA.

of genetic information from one bacterium to another. They can be readily isolated from bacterial cells and can be cleaved at specific sites by restriction endonucleases to allow for the insertion of up to 15 kb of foreign DNA (cut with the same restriction enzyme). The recombinant DNA molecule (hybrid plasmid) can be introduced into a bacterium, which can survive growth challenges, multiply, and produce numerous copies of the plasmid. If the plasmid vector provides antibiotic resistance, the bacteria will grow in the presence of antibiotics, thus selecting for cells containing the hybrid plasmids (Fig. 34.6). Artificial plasmids are routinely constructed. An example is the classic pBR322 (see Fig. 34.5), which contains an origin of replication, two antibiotic resistance genes, and more than 40 unique restriction sites. Use of plasmids is limited by the size of the DNA that can be inserted.

2. **Other cloning vectors:** In addition to the prokaryotic plasmids already described, naturally occurring viruses that infect bacteria (e.g., bacteriophage λ) or mammalian cells (e.g., retroviruses) as well as artificial constructs such as cosmids and bacterial or yeast artificial chromosomes (BAC and YAC, respectively), are used as cloning vectors because of their capacity to deliver foreign DNA into cells or to accommodate much larger DNA segments for replication. BAC and YAC can accept DNA inserts of 100 to 300 kb and 250 to 1,000 kb, respectively. Vectors such as these have aided molecular genetics research and therapeutics.

3. **Expression vectors:** Vectors that can be replicated in host cells and direct the expression of inserted DNA sequences in the host cells are called expression vectors. The inserted DNA is typically composed of sequence that is an exact copy of the mRNA for a gene. If a protein-coding gene of interest is expressed at a high level in a particular tissue, the mRNA transcribed from that gene is likely also present at high concentrations in the cells of that tissue. For example, reticulocyte mRNA is composed largely of molecules that code for the α-globin and β-globin chains of hemoglobin A (HbA). The mRNA in a cell can be isolated from transfer RNA and ribosomal RNA by the presence of its poly-A tail and then used as a template to make a cDNA strand to the mRNA using the enzyme reverse transcriptase (Fig. 34.7). Following RNA digestion with RNase H replication of the second DNA strand and ligation, the resulting cDNA are dsDNA molecules with one strand equivalent to the mRNA base sequence, lacking introns or other intervening sequence removed during RNA splicing. Specific cDNA molecules can be further amplified in additional cycles of PCR. Because cDNA lacks introns, it can be cloned into an expression vector for the synthesis of eukaryotic proteins by bacteria (Fig. 34.8). These special plasmids contain a bacterial promoter for transcription of the cDNA and a Shine–Dalgarno (SD) sequence that allows the bacterial ribosome to initiate translation of the resulting mRNA molecule. For example, the cDNA may be inserted downstream of the lac promoter and within the *lacZ* gene, such that the mRNA produced contains an SD sequence, a few codons for the bacterial protein, and all of the codons for the eukaryotic protein. This allows for more efficient expression and results in the production of a fusion protein. Therapeutic human insulin is made in bacteria through this technology. However, the extensive co- and posttranslational modifications required for most other human proteins (e.g., blood clotting factors) necessitates the use of eukaryotic, even mammalian, expression vectors and host cells.

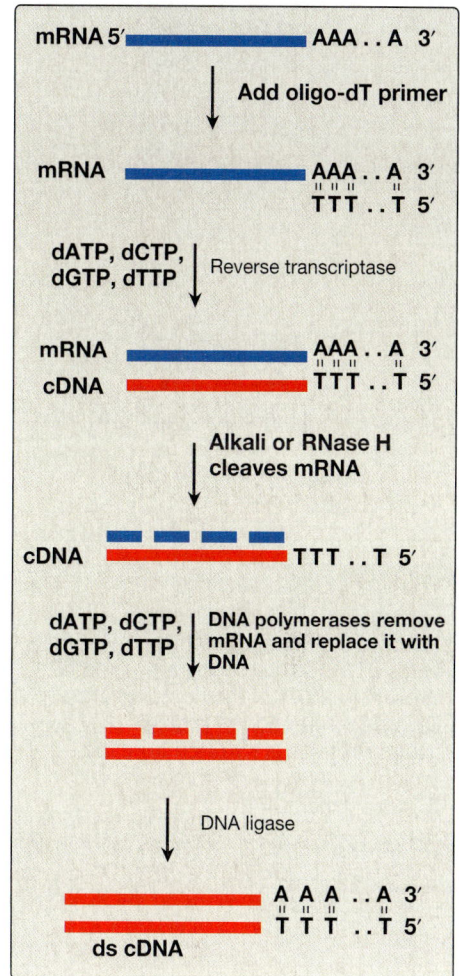

Figure 34.7
Synthesis of complementary DNA (cDNA) from messenger RNA (mRNA) using reverse transcriptase. Ligation of double-stranded (ds) DNA sequences containing a restriction site to each end allows biologic cloning of cDNA. [Note: DNA is resistant to alkaline hydrolysis.] dATP, dCTP, dGTP, dTTP, deoxyadenosine, deoxycytidine, deoxyguanosine, and deoxythymidine triphosphates.

@ B. Polymerase chain reaction

PCR is a rapid *in vitro* method for amplifying a selected DNA sequence that does not rely on the biologic (*in vivo*) cloning vectors. It is more sensitive than biologic cloning in that it can amplify a target sequence that makes up less than one part in a million of the total initial DNA sample. This sensitivity makes PCR a very common tool in research,

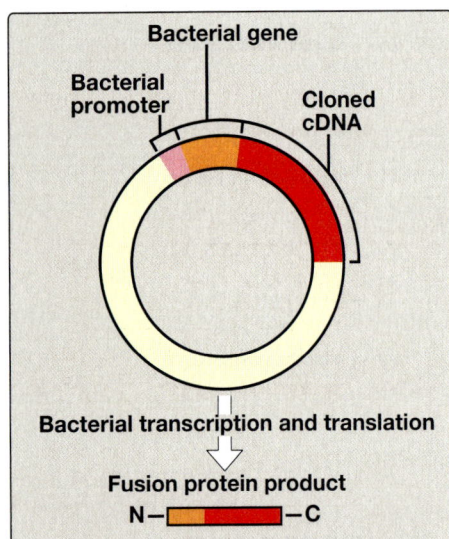

Bacterial gene

Bacterial promoter

Cloned cDNA

Bacterial transcription and translation

Fusion protein product

N —▭ —C

Figure 34.8
Expression vector. The product is a fusion protein that contains some amino acids of the bacterial protein ■ and all the amino acids of the complementary DNA (cDNA)-encoded protein ■. [Note: Proteins are written from the amino (N)-terminus to the carboxy (C)-terminus.]

forensic science, and clinical diagnostics. It can be used to amplify DNA sequences from any source, including viral, bacterial, plant, or animal, as long as some sequence information adjacent to the target sequence is known to construct a short primer for DNA polymerase (pol). In fact, PCR may be used to detect viruses within human blood cells. Restriction sites also may be incorporated into the PCR primer to enable insertion of the product into expression vectors for further analysis of gene products. PCR uses DNA pol to repetitively amplify targeted portions of genomic or cDNA through a four-step reaction cycle shown in Figure 34.9. Each cycle of amplification doubles the amount of DNA in the sample, leading to an exponential increase (2^n, where n = cycle number) in DNA with repeated cycles of amplification (Fig. 34.10). The amplified DNA products can then be sequenced or used in other analyses. The procedure for PCR requires the following five steps.

1. **Constructing primer:** It is not necessary to know the nucleotide sequence of the target DNA in the PCR method. However, it is necessary to know the nucleotide sequence of short segments on each side of the target DNA. These stretches, called flanking sequences, bracket the DNA sequence of interest. The nucleotide sequences of the flanking regions are used to construct two, single-stranded oligonucleotides, usually 20 to 35 nucleotides long, which are complementary to the respective flanking sequences. The 3′-hydroxyl end of each oligonucleotide points toward the target sequence (see Fig. 34.9). These synthetic oligonucleotides function as primers in PCR.

2. **Sample preparation:** A sample for analysis is prepared by adding the DNA (either genomic DNA or cDNA), the primers, an excess of deoxyribonucleoside triphosphates (dNTPs), and a heat-stable DNA pol to an appropriate buffer solution.

3. **Denaturing DNA:** The sample is heated to ~95 °C to separate the DNA into single strands (ssDNAs).

4. **Annealing primers:** The sample is cooled to ~50 °C, and the two primers (one for each strand) anneal to a complementary sequence on the ssDNA.

5. **Extending primers:** The sample temperature is raised to ~72 °C to initiate the synthesis of two new strands complementary to the original DNA strands. DNA pol adds nucleotides to the 3′-hydroxyl end of the primer, and strand growth extends in the 5′→3′ direction across the target DNA, making complementary copies of the target. The PCR products from this extension can be several thousand bps long. At the completion of one cycle of replication, the reaction mixture is heated again to separate the strands (of which there are now four). Each strand anneals to the complementary primer, and the step of primer extension is repeated. By using a heat-stable DNA pol (e.g., Taq from the bacterium *Thermus aquaticus* that normally lives at high temperatures), the polymerase is not denatured and, therefore, does not have to be added at each successive cycle. However, Taq lacks proofreading activity. Typically, 20 to 30 cycles are run during this process, amplifying the DNA by a million fold (2^{20}) to a billion fold (2^{30}). [Note: Each extension product includes a sequence at its 5′ end that is complementary to the primer (see Fig. 34.9). Thus, each newly synthesized strand

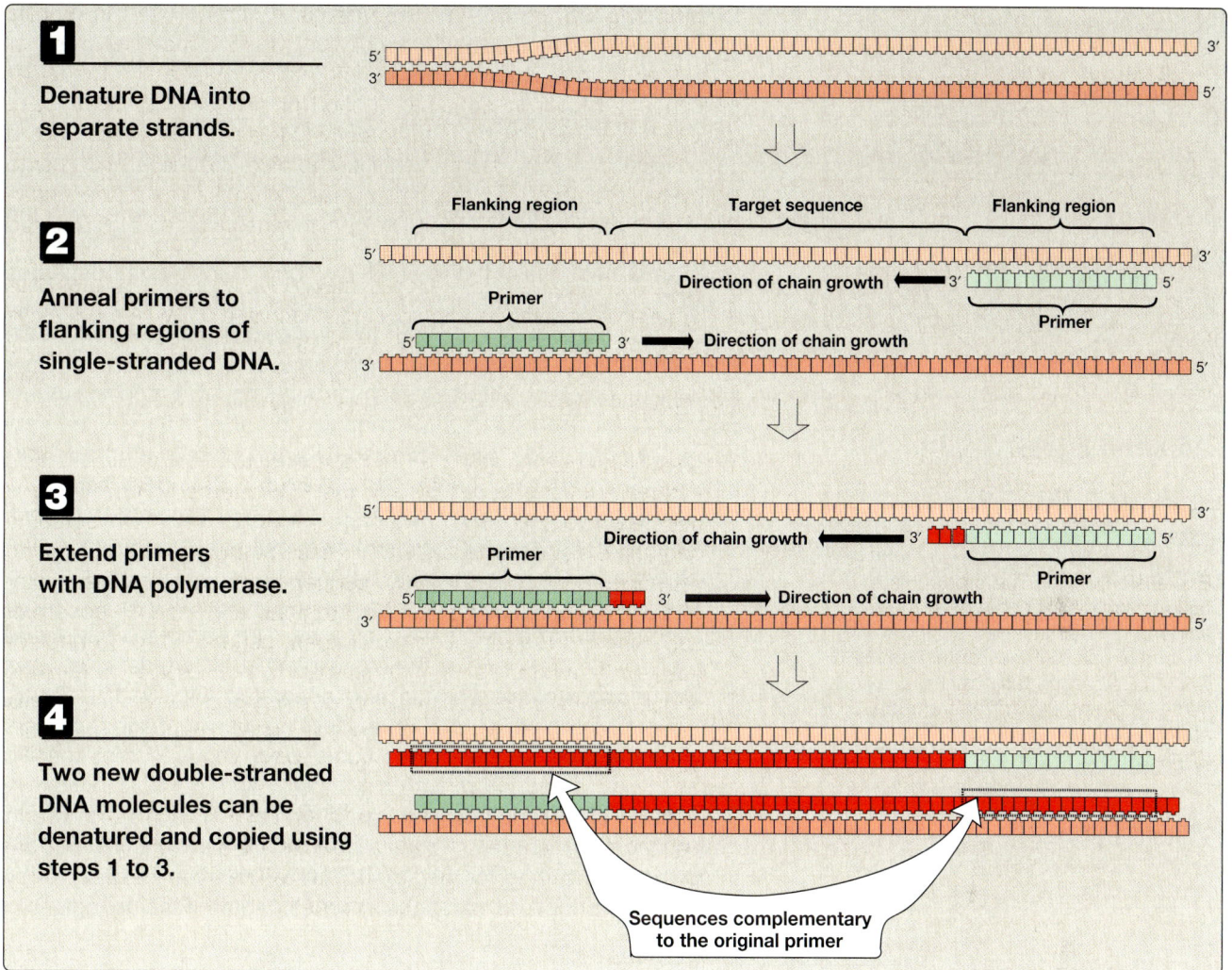

1 Denature DNA into separate strands.

2 Anneal primers to flanking regions of single-stranded DNA.

3 Extend primers with DNA polymerase.

4 Two new double-stranded DNA molecules can be denatured and copied using steps 1 to 3.

Figure 34.9
Steps (denature, anneal, extend) in one cycle of the polymerase chain reaction.

can act as a template for the successive cycles (see Fig. 34.10). This leads to an exponential increase in the amount of target DNA with each cycle, hence, the name "polymerase chain reaction."] Probes can be made during PCR by adding labeled nucleotides to the samples before the last few cycles.

Quantitative PCR (qPCR), also known as real-time PCR, allows quantification of the amount (copy number) of the target nucleic acid after each cycle of amplification (i.e., in real time) rather than at the end and is useful in determining viral load (amount of virus). A nucleic acid amplification test for viral load may be done when a person has experienced a high-risk exposure for HIV or has early symptoms HIV in order to diagnose infection and to guide antiretroviral treatment. Multiplex PCR uses several primers to amply multiple gene regions. This is particularly useful in detecting a loss of DNA sequence (deletions) in large genes, such as *CTFR*, which contains 27 exons encoding the cystic fibrosis transmembrane conductance regulator.

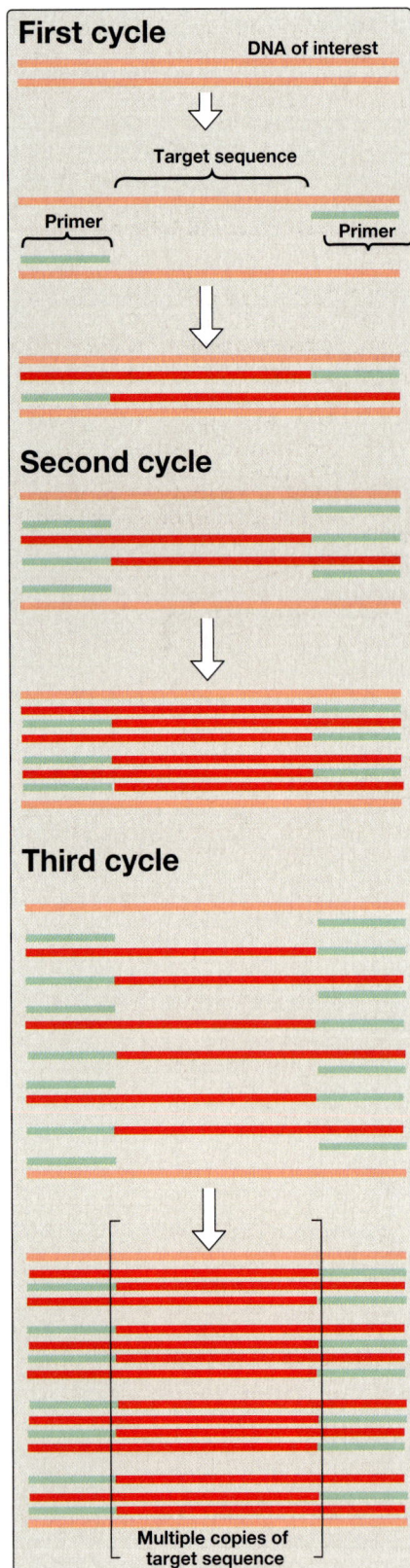

First cycle

DNA of interest

Target sequence

Primer

Primer

Second cycle

Third cycle

Multiple copies of
target sequence

Figure 34.10
Multiple cycles of the polymerase chain
reaction.

C. Sequencing cloned DNA fragments

The base sequence of DNA fragments that have been cloned can be determined. The original procedure for this purpose was the Sanger dideoxynucleotide chain termination method (Fig. 34.11). In this method, the ssDNA to be sequenced is used as the template for DNA synthesis by DNA pol. A radiolabeled primer complementary to the 3′ end of the target DNA is added to the sample, along with the four dNTPs. The sample is divided into four reaction tubes, and a small amount of one of the four dideoxyribonucleoside triphosphates (ddNTPs) is added to each tube. Because it contains no 3′-hydroxyl group, incorporation of a ddNMP terminates elongation at that point. The products of this reaction, then, consist of a mixture of DNA strands of different lengths, each terminating at a specific base. Separation of the various DNA products by size in an electric field using polyacrylamide gel electrophoresis, followed by autoradiography, yields a pattern of bands from which the DNA base sequence can be read. The shorter the fragment, the farther it travels on the gel, with the shortest fragment representing that which was made first (i.e., the 5′ end). In a more modern approach to the Sanger method, the four ddNTPs, each linked to a different fluorescent dye, are mixed with the ssDNA in a single reaction tube. The sample is separated by capillary electrophoresis, the fluorescent labels are detected, and a color readout of the sequence is generated (Fig. 34.12). Today, high throughput next-generation technologies using distinct primers allow for the rapid sequencing of hundreds to thousands of specific genes in only a few hours with high fidelity and low cost. Selective sequencing allows the interrogation of genes with variants known to cause inherited disorders. These advances are particularly important in newborn screening for single-nucleotide, insertion-deletion, copy number, and structural variations in genes or chromosome regions.

IV. PROBES

How can a specific DNA sequence of interest be detected in a mixture of genetic material? The answer lies in the use of a probe, a piece of ssDNA or RNA, labeled with a radioisotope, such as ^{32}P, or with a nonradioactive molecule, such as biotin linked to a fluorescent dye (a fluorochrome). The sequence of a probe is complementary to a sequence in the DNA of interest, called the target DNA. The detectable signal from the probe is used to identify target DNA among the bands produced from gel electrophoresis and even within the chromosomes of cells collected for molecular genetic analysis.

A. Hybridization to DNA

The utility of probes hinges on the process of hybridization (or annealing) in which a probe containing a complementary sequence binds a single-stranded sequence of a target DNA. Following gel electrophoresis, ssDNA, produced by alkaline denaturation of dsDNA within a gel, is first bound to a solid support, such as a nitrocellulose membrane. The immobilized DNA strands are prevented from self-annealing but are available for hybridization to the exogenous, labeled, single-stranded probe. When cellular chromosomes contain the target DNA, the chromosomes are denatured after being attached to a microscope slide so that probes added to the slide may find their target. Excess probe molecules that do not hybridize are removed by washing the membrane or slide with a buffered solution.

B. Target DNA detection

If the sequence of all or part of the target DNA is known, short, single-stranded oligonucleotide probes can be synthesized that are complementary to a small region of the gene of interest. If the sequence of the gene is unknown, the amino acid sequence of the protein, the final gene product, may be used to construct a nucleic acid probe using the genetic code as a guide. Because of the degeneracy of the genetic code, it is necessary to synthesize several oligonucleotides.

1 Single-stranded DNA of unknown sequence is used as a template.

2 Add primer and DNA polymerase + dATP, dGTP, dCTP, dTTP.

3 Split the sample into four tubes, and add one of the four dideoxynucleotides to each.

4 Synthesis proceeds until the dideoxynucleotide is incorporated into a DNA strand. DNA terminating in a dideoxynucleotide cannot be elongated because it lacks a 3'-OH, which is required for formation of a 3'→5' phosphodiester bond.

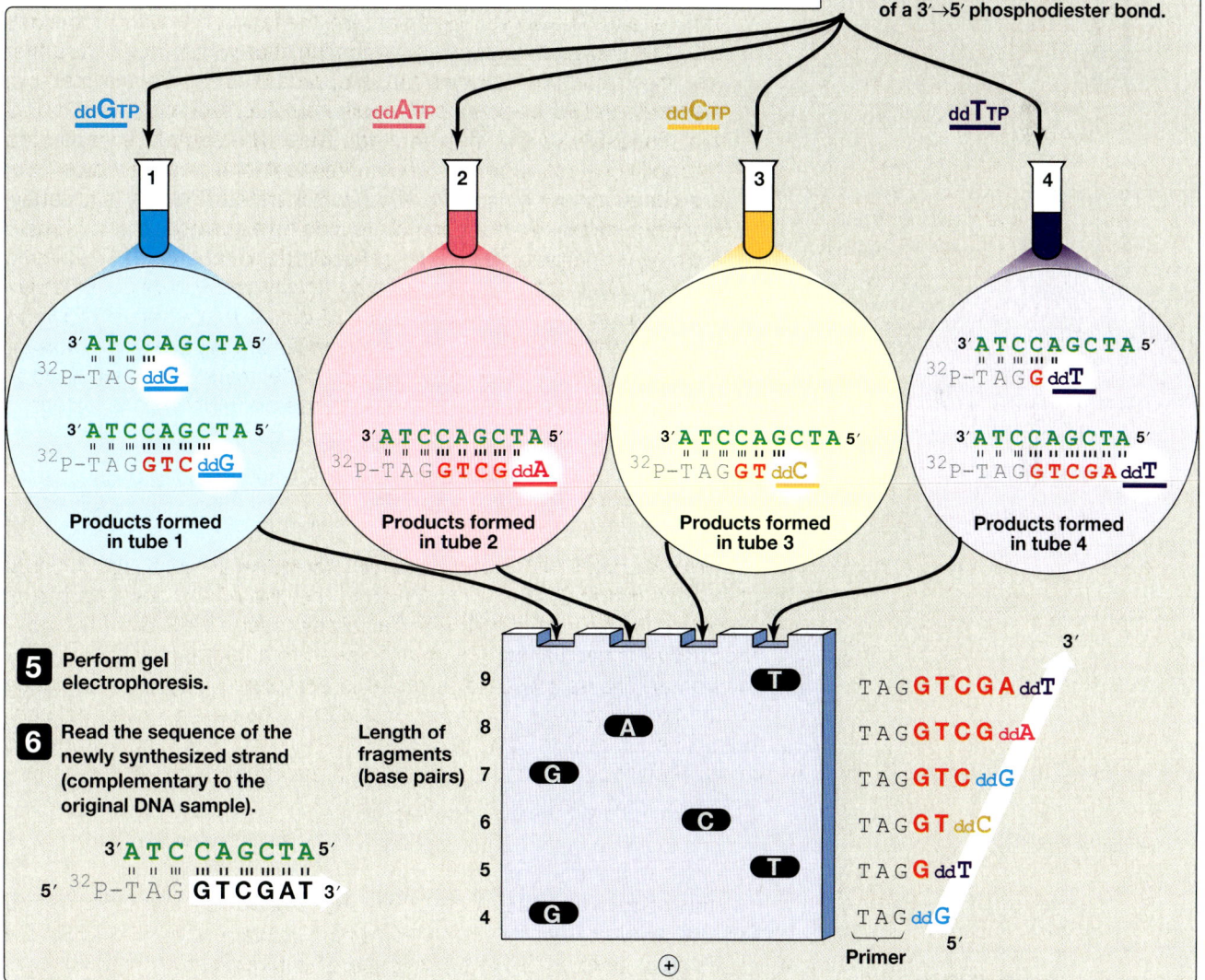

Figure 34.11
DNA sequencing by the Sanger dideoxynucleotide method. [Note: The original method utilized a radiolabeled primer. Fluorescent dye-labeled dideoxyribonucleoside triphosphates are now commonly used.] A, adenine; C, cytosine; G, guanine; T, thymine; d, deoxy; dd, dideoxy.

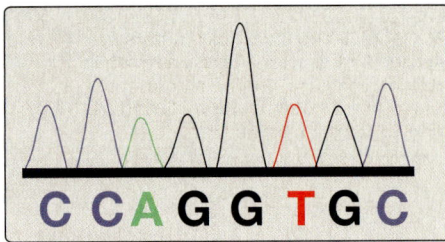

Figure 34.12
Color readout of a DNA sequence.

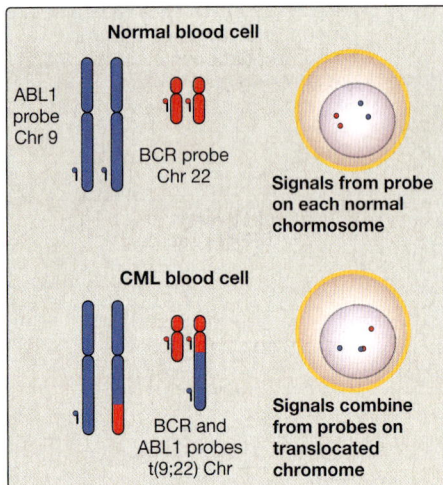

Figure 34.13
Fluorescence *in-situ* hybridization (FISH).

In contrast, cDNA probes, produced by reverse transcription of an RNA sample, contain many thousands of bases, and their binding to a target DNA with a single-base change is unaffected. Historically, radioactive isotopes, such as ^{32}P, were incorporated into nucleic acid probes so that the probes were detectable by exposure to x-ray film or phosphorimager screens. Now, nonradiolabeled probes are produced by chemically linking a detectable molecule to the nucleic acid. One of the most common linkages is based on the vitamin biotin. Biotin binds tightly to avidin, an abundant protein in chicken egg whites. A form of avidin produced by bacteria, streptavidin, also binds tightly to biotin. Streptavidin can be purified, attached to a fluorescent dye, then exposed to the biotin-linked probe. The fluorescent probe hybridized to its target DNA is detectable optically with great sensitivity. When fluorescently labeled probes are used to detect DNA or RNA sequences in cell or tissue preparations, the process is called fluorescence *in situ* hybridization (FISH). One use of FISH is the detection of chronic myeloid leukemia (CML) cells in blood samples. In CML, a portion of chromosome 9 containing the *ABL1* gene changes places with a section of chromosome 22 near the *BCR* gene. The new chromosome formed by this translocation of genetic material can be detected in the nucleus of cells using fluorescently labeled probes that bind to target sequences within the *BCR* and *ABL1* genes. The normal chromosome 9 is visible as the color of the ABL1 probe. The normal chromosome 22 is seen as the color of the BCR probe. The translocated chromosome containing sequence from both genes labels with both probes and has a mixed color signal (Fig. 34.13).

V. SOUTHERN BLOTTING

Southern blotting combines the use of restriction enzymes, electrophoresis, and DNA probes to generate, separate, and detect pieces of DNA. This technique is used when ample genomic DNA is available for analysis. In medicine and forensic science, Southern blotting has largely been replaced by PCR-based techniques that allow analysis of limited genetic material with specific, nonradioactive detection methods.

A. Procedure

This method, named after its inventor, Edward Southern, involves the following steps (Fig. 34.14). First, DNA is extracted from cells (e.g., a patient's white blood cells). Second, the DNA is cleaved into many fragments using a restriction enzyme. Third, the resulting fragments (all of which are negatively charged) are separated on the basis of size by electrophoresis. [Note: Because the large fragments move more slowly than the smaller ones, the lengths of the fragments, usually expressed as the number of bps, can be calculated from comparison of the positions of the fragments relative to standard fragments of known size in a DNA ladder.] The DNA fragments in the gel are denatured and transferred (blotted) to a nitrocellulose membrane for analysis. If the original DNA consists of the individual's entire genome, the enzymic digest contains at least 10^6 fragments. The gene of interest is on only one (or a few if the gene itself was fragmented) of these pieces of DNA. If all the DNA fragments

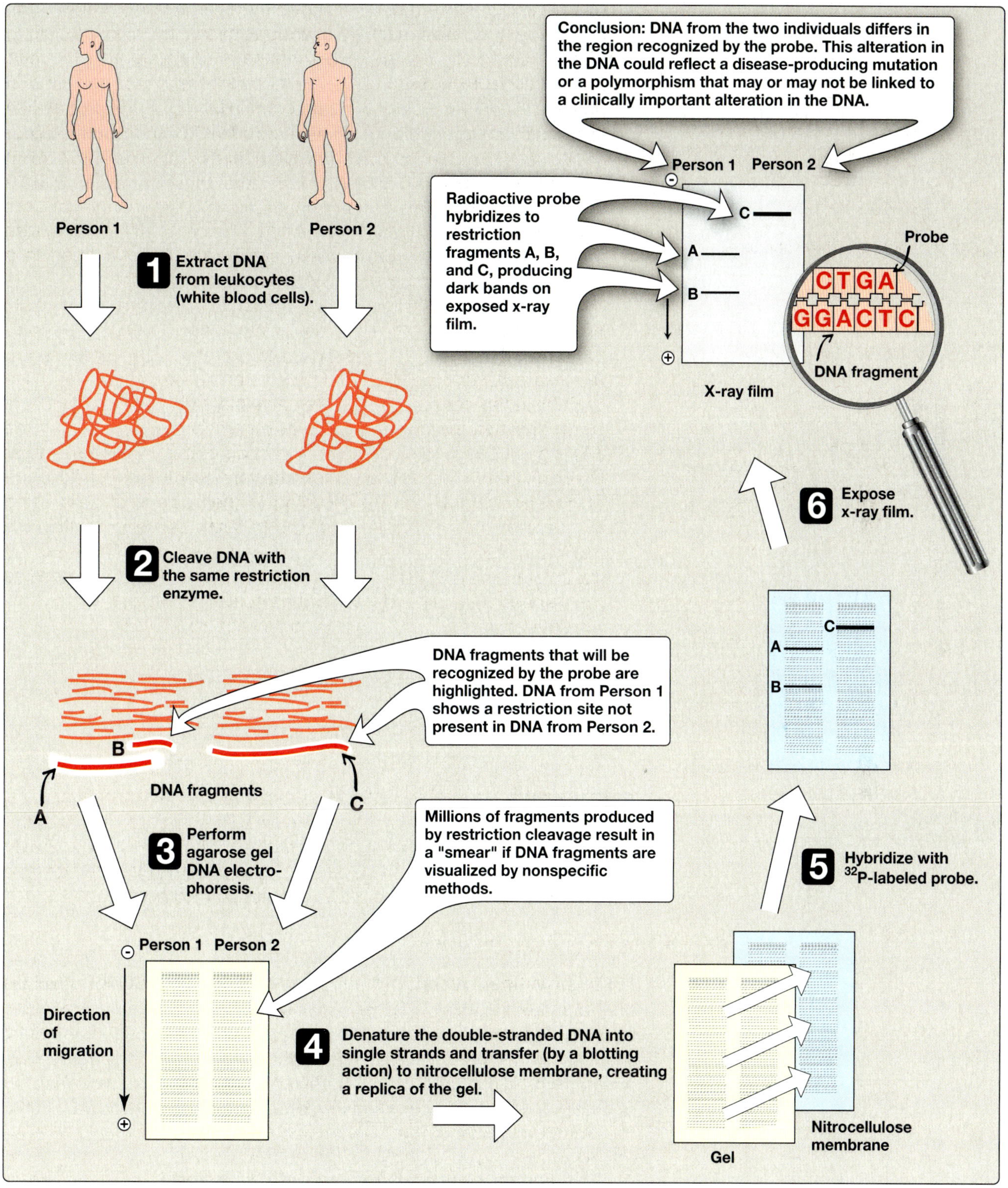

Figure 34.14
Southern blotting procedure. [Note: Nonradiolabeled probes are now commonly used.]

were visualized by a nonspecific technique, they would appear as an unresolved blur of overlapping bands. To avoid this, the last step in Southern blotting uses a probe to identify the DNA fragments of interest. The patterns observed on Southern blot analysis depend both on the specific restriction endonuclease and on the probe used to visualize the restriction fragments. Variants of the Southern blot have been facetiously named "northern" if RNA is being studied and "western" if protein is being studied, neither of which relates to anyone's name or to points of the compass.

B. Mutation detection

Southern blotting can detect DNA mutations such as large insertions or deletions, trinucleotide repeat expansions, and rearrangements of nucleotides. It can also detect point mutations (replacement of one nucleotide by another) that cause the loss or gain of restriction sites. Such mutations cause the pattern of bands to differ from those seen with a normal gene. Longer fragments are generated if a restriction site is lost. For example, in Figure 34.14, person 2 lacks a restriction site present in person 1. Alternatively, the point mutation may create a new cleavage site with the production of shorter fragments. If a specific gene region is amplified by PCR before restriction digestion, a fluorescent dye that binds to the DNA in the gel can be used for direct detection of the bands by ultraviolet light. The detection of DNA bands with fluorescent dyes requires that each band contain at least 100 nanograms (ng) of DNA.

VI. RESTRICTION FRAGMENT LENGTH POLYMORPHISM

It has been estimated that the genomes of any two unrelated people are 99.5% identical. With 6 billion bps in the diploid human genome, that represents variation in ~30 million bps. These genome variations are the result of mutations that lead to polymorphisms. A polymorphism is traditionally defined as a sequence variation at a given locus (allele) in more than 1% of a population. The change in genotype results in either no change in phenotype or a harmless change in phenotype, causes increased susceptibility to a disease, or, rarely, causes a disease. Polymorphisms primarily occur in the 98% of the genome that does not encode proteins (i.e., in introns and intergenic regions). A restriction fragment length polymorphism (RFLP) is a genetic variant that can be observed by cleaving the DNA into fragments (restriction fragments) with a restriction endonuclease. The length of the restriction fragments is altered if the variant alters the DNA sequence so as to create or abolish a restriction site. RFLP can be used to detect human genetic variations, for example, in prospective parents or in fetal tissue.

A. DNA variations resulting in restriction fragment length polymorphism

Two types of DNA variations commonly result in RFLP: single-base changes in the DNA sequence and tandem repeats of DNA sequences.

1. **Single-base changes:** About 90% of human genome variation comes in the form of single nucleotide polymorphisms (SNPs,

pronounced "snips"), that is, variations that involve just one base (Fig. 34.15). The substitution of one nucleotide at a restriction site can render the site unrecognizable by a particular restriction endonuclease. A new restriction site can also be created by the same mechanism. In either case, cleavage with an endonuclease results in fragments of lengths that differ from the normal and can be detected by DNA gel electrophoresis and Southern blotting, as shown in Figure 34.16, or by visualization of restriction fragments directly in the gel with fluorescent dye that binds to DNA (see Fig. 34.19). The altered restriction site can be either at the site of a disease-causing mutation (rare) or at a site some distance from the mutation. If the DNA of an individual has gained a restriction site by base substitution, then enzymatic cleavage yields at least one additional fragment. Conversely, if a mutation results in loss of a restriction site, fewer fragments are produced by enzymatic cleavage. This is most easily observed when a specific gene mutation is investigated, such as the specific phenylalanine hydroxylase (PAH) gene mutation causing phenylketonuria (PKU) in Figure 34.19.

2. **Tandem repeats:** Polymorphisms in chromosomal DNA can also arise from the presence of a variable number of tandem repeats (VNTRs), as shown in Figure 34.15. These are short sequences of DNA at scattered locations in the genome, repeated in tandem (one after another). The number of these repeat units varies from person to person but is unique for any given individual and, therefore, serves as a molecular "fingerprint." Cleavage by restriction enzymes yields fragments that vary in length depending on how many repeated segments are contained in the fragment (see Fig. 34.16). Many different VNTR loci have been identified and are extremely useful for DNA fingerprint analysis, such as in forensic and paternity cases. It is important to emphasize that these polymorphisms, whether SNPs or VNTRs, are simply markers, which, in most cases, have no known effect on the structure, function, or rate of production of any particular protein.

B. Tracing chromosomes from parent to offspring

An individual who is heterozygous for a polymorphism has a sequence variation in the DNA of one chromosome and not in the homologous chromosome (i.e., in the chromosome from the mother and not from the father, or *vice versa*). In such individuals, each chromosome can be traced from parent to offspring by determining the presence or absence of the polymorphism.

C. Genetic risk and diagnosis

A couple with a family history of severe genetic disease may want to know their risk of having an affected child. But, couples may not be aware of their family history before becoming pregnant. Prenatal genetic testing may be necessary to determine the presence of a disorder in a developing fetus if problems with growth are detected. Ultrasound allows visualization of the fetus to detect gross anatomic defects (e.g., neural tube defects [NTDs]) that are characteristic of particular genetic disorders. Amniocentesis, the sampling of amniotic

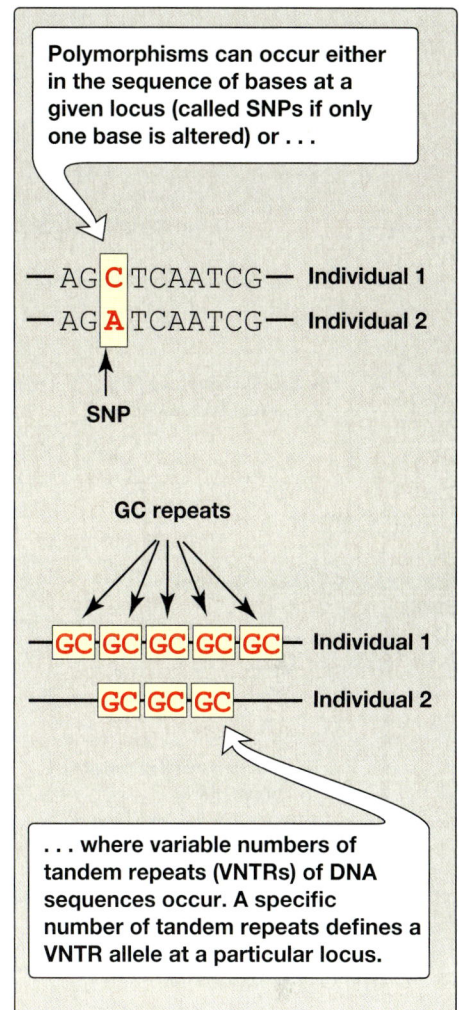

Figure 34.15
Common forms of genetic polymorphism. SNPs, single nucleotide polymorphisms; A, adenine; C, cytosine; G, guanine; T, thymine.

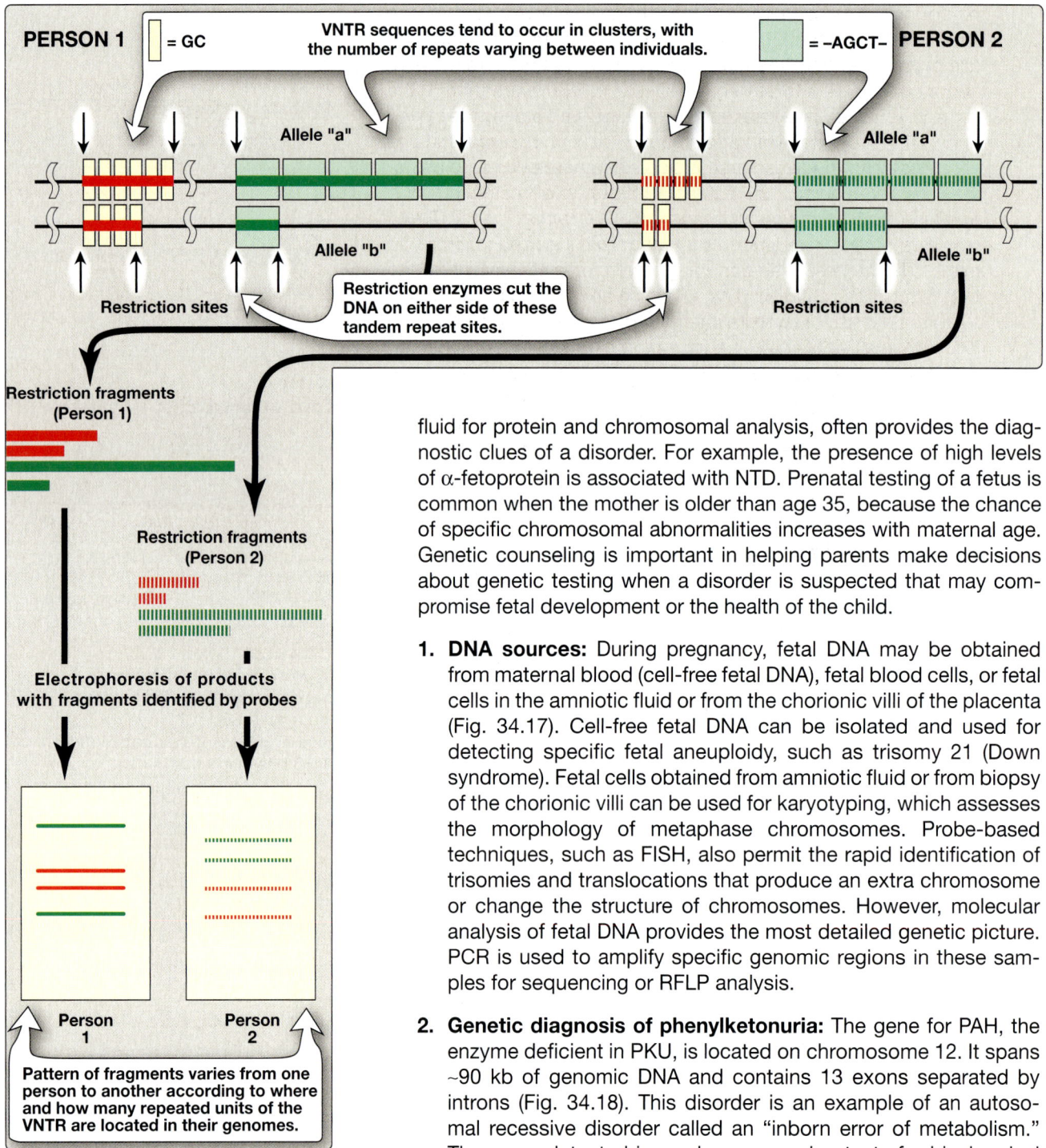

Figure 34.16
Restriction fragment length polymorphism of variable number tandem repeats (VNTRs). For each person, a pair of homologous chromosomes is shown.

fluid for protein and chromosomal analysis, often provides the diagnostic clues of a disorder. For example, the presence of high levels of α-fetoprotein is associated with NTD. Prenatal testing of a fetus is common when the mother is older than age 35, because the chance of specific chromosomal abnormalities increases with maternal age. Genetic counseling is important in helping parents make decisions about genetic testing when a disorder is suspected that may compromise fetal development or the health of the child.

1. **DNA sources:** During pregnancy, fetal DNA may be obtained from maternal blood (cell-free fetal DNA), fetal blood cells, or fetal cells in the amniotic fluid or from the chorionic villi of the placenta (Fig. 34.17). Cell-free fetal DNA can be isolated and used for detecting specific fetal aneuploidy, such as trisomy 21 (Down syndrome). Fetal cells obtained from amniotic fluid or from biopsy of the chorionic villi can be used for karyotyping, which assesses the morphology of metaphase chromosomes. Probe-based techniques, such as FISH, also permit the rapid identification of trisomies and translocations that produce an extra chromosome or change the structure of chromosomes. However, molecular analysis of fetal DNA provides the most detailed genetic picture. PCR is used to amplify specific genomic regions in these samples for sequencing or RFLP analysis.

2. **Genetic diagnosis of phenylketonuria:** The gene for PAH, the enzyme deficient in PKU, is located on chromosome 12. It spans ~90 kb of genomic DNA and contains 13 exons separated by introns (Fig. 34.18). This disorder is an example of an autosomal recessive disorder called an "inborn error of metabolism." These are detected in newborn screening tests for biochemical metabolites in a drop of blood collected from the infant soon after birth. Infants with PKU have very high levels of phenylalanine (Phe) and detectable levels of abnormal phenylketones (mainly phenylpyruvate) in their blood (see Chapter 20, VI. A.). The excessive Phe alters blood–brain barrier transport of amino acids, and phenylpyruvate inhibits pyruvate metabolism in the brain to cause myelination deficiencies and hinder neurologic development. So, rapid removal of Phe from the infant's diet and nutritional supplementation are necessary to prevent permanent

neurologic deficits. An infant with PKU has a mutation in the *PAH* gene on both chromosomes, and the specific mutation (or allele) inherited from the father and mother can be determined if the family decides to undergo genetic testing. More than 1,000 mutations in the *PAH* gene have been identified worldwide in persons with PKU. These mutations usually do not directly affect any restriction endonuclease recognition site and are easier to detect by DNA sequencing. However, a common allele of *PAH* in persons of European ancestry has a G-to-A mutation in exon 7 that removes a HinfI restriction site found in the normal *PAH* gene. [Note: The base substitution changes arginine at position 261 to glutamine in the PAH enzyme, reducing its function in converting Phe to tyrosine.] This allele can be identified in a family affected by PKU using PCR and RFLP analysis on epithelial cell DNA from a cheek swab (Fig. 34.19). Exon 7 of *PAH* is amplified from each person's cell sample by PCR. The DNA then is cleaved with the HinfI restriction enzyme and subjected to gel electrophoresis. Because PCR produces more than 100 ng of the *PAH* exon 7 DNA, a fluorescent dye that binds to DNA can be used to detect the DNA bands directly in the gel. The vertical arrow over the normal *PAH* gene indicates the HinfI cleavage site (GAGTC) that creates restriction fragments "a" and "b." The G-to-A transition at the first position eliminates the HinfI restriction site in the mutant *PAH* allele so the DNA remains full-length and is designated "m." The DNA migrates in the gel according to size during electrophoresis, so the full-length band appears at the top of the gel, and the restriction fragments migrate faster according to their decreasing size in base pairs. The heterozygous carriers in the pedigree, the father (I-1), the mother (I-2), and the infant's brother (II-2), have bands representing both the normal and mutant *PAH* exon 7. The infant with PKU (II-3) is homozygous for the mutant *PAH* allele and has only the full-length band. The infant's sister (II-1) has two normal *PAH* genes, so all of the DNA in her sample appears as restriction fragments "a" and "b." During genetic counseling, the couple should be informed that they have a 25% chance of having another child with PKU.

VII. GENE EXPRESSION ANALYSIS

The tools of biotechnology not only allow the study of gene structure, but also provide ways of analyzing the mRNA and protein products of gene expression.

A. Determining messenger RNA levels

mRNA levels are usually determined by the hybridization of labeled probes to either mRNA itself or to cDNA produced from mRNA by reverse transcriptase PCR, referred to as RT-PCR.

1. **Northern blots:** Northern blots are similar to Southern blots (see Fig. 34.14), except that the sample contains a mixture of mRNA molecules that are separated by electrophoresis, then transferred to a membrane and hybridized with a probe. The intensity of the detected bands gives a measure of the amount and size of the

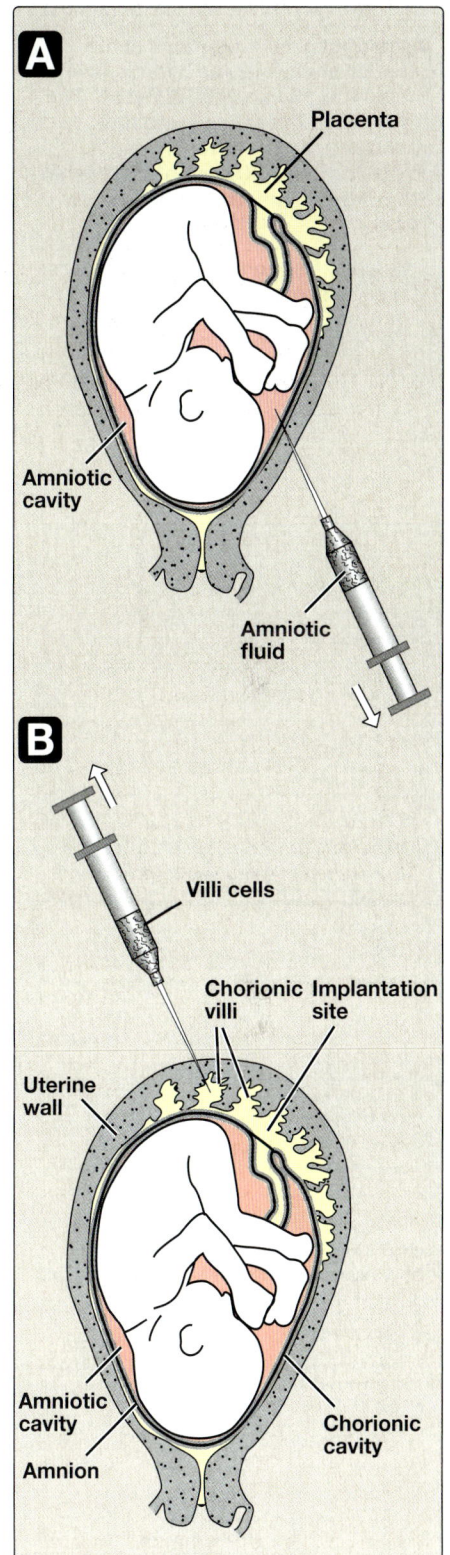

Figure 34.17
Sampling of fetal cells. **A.** Amniotic fluid. **B.** Chorionic villus.

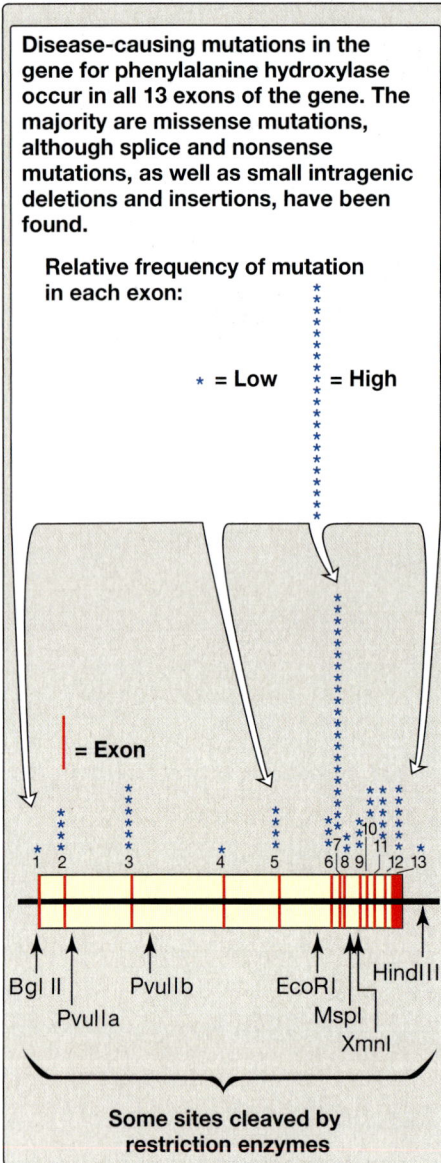

Figure 34.18
Gene for phenylalanine hydroxylase showing 13 exons, restriction sites, and some of the >500 mutations causing phenylketonuria.

target mRNA molecules in the sample. This technique has largely been replaced by faster and more sensitive methods.

2. **Microarrays:** DNA microarrays contain thousands of immobilized ssDNA sequences organized in an area no larger than a microscope slide. These microarrays are used to analyze a sample for the presence of gene variations or mutations (genotyping) or to determine the patterns of mRNA production (gene expression analysis), analyzing thousands of genes at the same time. For genotyping analysis, the sample is from genomic DNA. For expression analysis, the population of mRNA molecules from a particular cell type is converted to cDNA and labeled with a fluorescent tag (Fig. 34.20). This mixture is then exposed to a gene (or, DNA) chip, which is a glass slide or membrane containing thousands of tiny spots of DNA, each corresponding to a different gene. The amount of fluorescence bound to each spot is a measure of the amount of that particular mRNA in the sample. DNA microarrays are used to determine the differing patterns of gene expression in two different types of cells (e.g., normal and cancer cells; see Fig. 34.20). They can also be used to subclassify cancers, such as breast cancer, to optimize treatment. [Note: Microarrays involving proteins and the antibody (Ab) or other proteins that recognize them are being used to identify biomarkers to aid in the diagnosis, prognosis, and treatment of disease based on a patient's protein expression profile. Protein (and DNA) microarrays are important tools in the development of personalized (precision) medicine in which the treatment and/or prevention strategies consider the genetic, environmental, and lifestyle variations among individuals.]

B. Protein analysis

The kinds and amounts of proteins in cells do not always directly correspond to the amounts of mRNA present. Some mRNAs are translated more efficiently than others, and some proteins undergo posttranslational modification. When analyzing the abundance and interactions of a large number of proteins, automated methods involving a variety of techniques, such as mass spectrometry and two-dimensional electrophoresis, are used. When investigating one, or a limited number of proteins, labeled Abs are used to detect and quantify specific proteins and to determine posttranslational modifications. Protein detection assays using specific Abs are also used every day, at home, by people around the world.

1. **Enzyme-linked immunosorbent assays:** These assays (known as ELISAs) are performed in the wells of a microtiter dish. The antigen (protein) is bound to the plastic of the dish. The probe used consists of an Ab specific for the protein (such as the troponin T and I proteins in heart muscle cells that rise in blood following myocardial infarction) to be measured. The Ab is covalently bound to an enzyme, which will produce a colored product when exposed to its substrate. The amount of color produced is proportional to the amount of Ab present and, indirectly, to the amount of protein in a test sample.

2. **Western blots:** Western blots (also called immunoblots) are similar to Southern blots, except that proteins, instead of nucleic acid

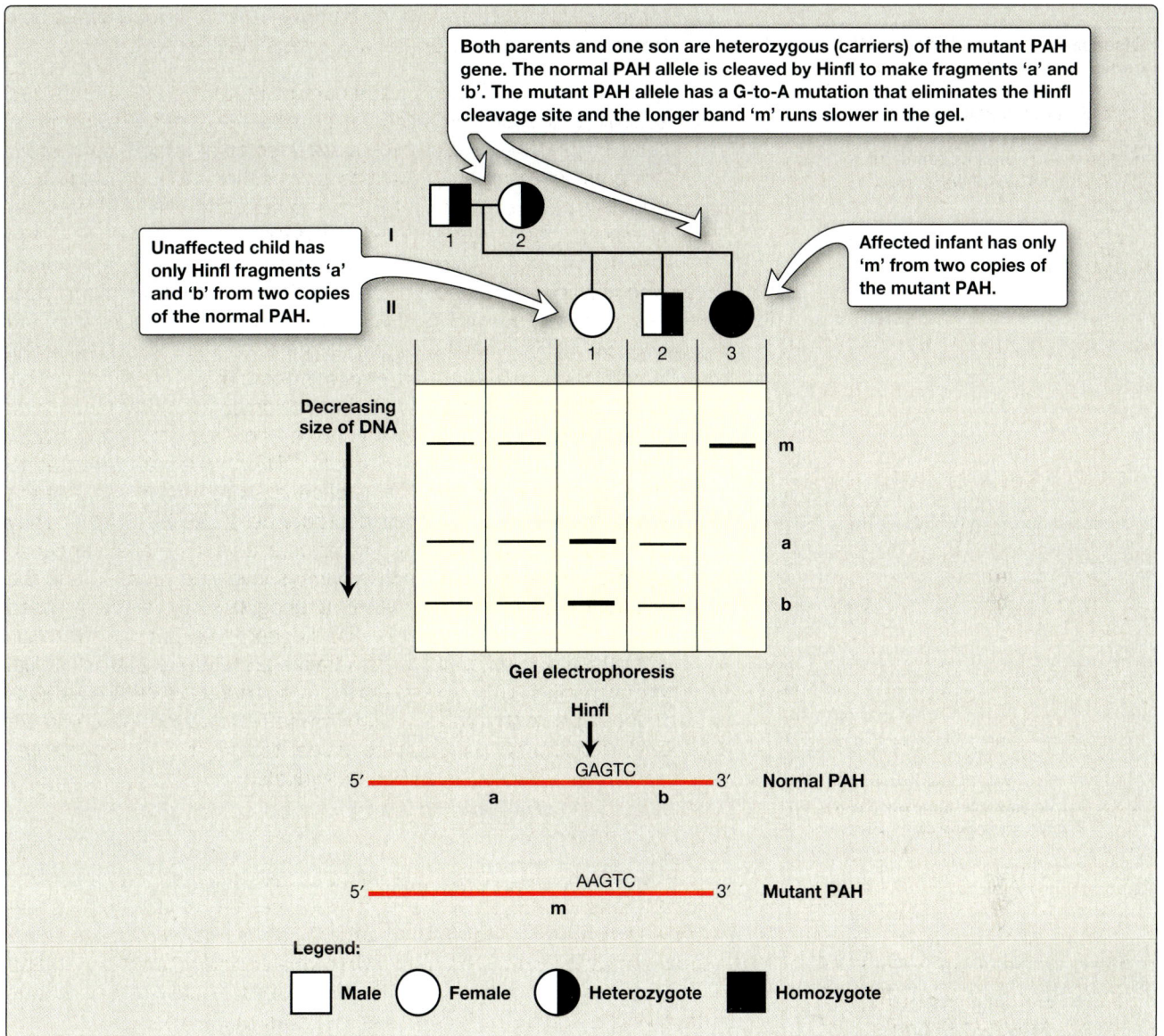

Figure 34.19
Analysis of restriction fragment length polymorphism in a family with a child affected by phenylketonuria (PKU), an autosomal recessive disease. The molecular defect in the gene for phenylalanine hydroxylase (PAH) in the family eliminates a restriction site.

molecules, are separated by electrophoresis and blotted (transferred) to a membrane. The probe is labeled Ab, which produces a band on the membrane at the location of its antigen.

3. **Lateral flow immunoassay:** Lateral flow immunoassays (LFIA) allow the rapid antibody detection of an antigen, such as the nucleocapsid (N) protein antigen from SARS-CoV-2, in a mixed sample, such as nasal mucus (see Fig. 34.21). Nasal mucus from a swab is applied to a "sample pad" saturated with buffer solution. Proteins in the mucus travel by capillary action across a "conjugate release pad" that contains an antibody to the SARS-CoV-2 N protein. The antibody to the N protein is conjugated to

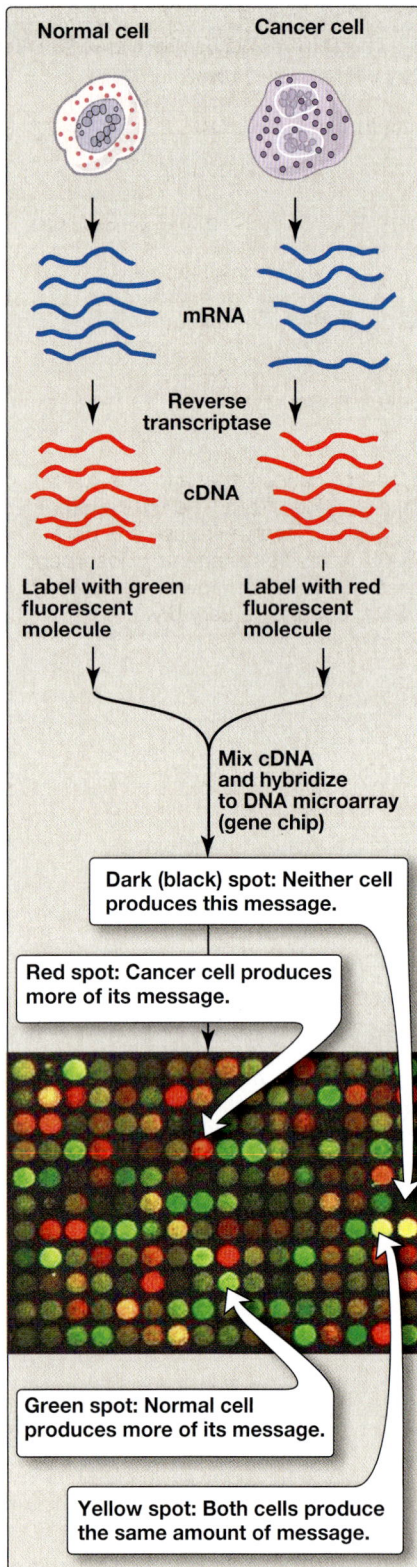

Figure 34.20
Microarray analysis of gene expression using DNA (gene) chips. [Note: Protein chips are also used.] mRNA, messenger RNA; cDNA, complementary DNA.

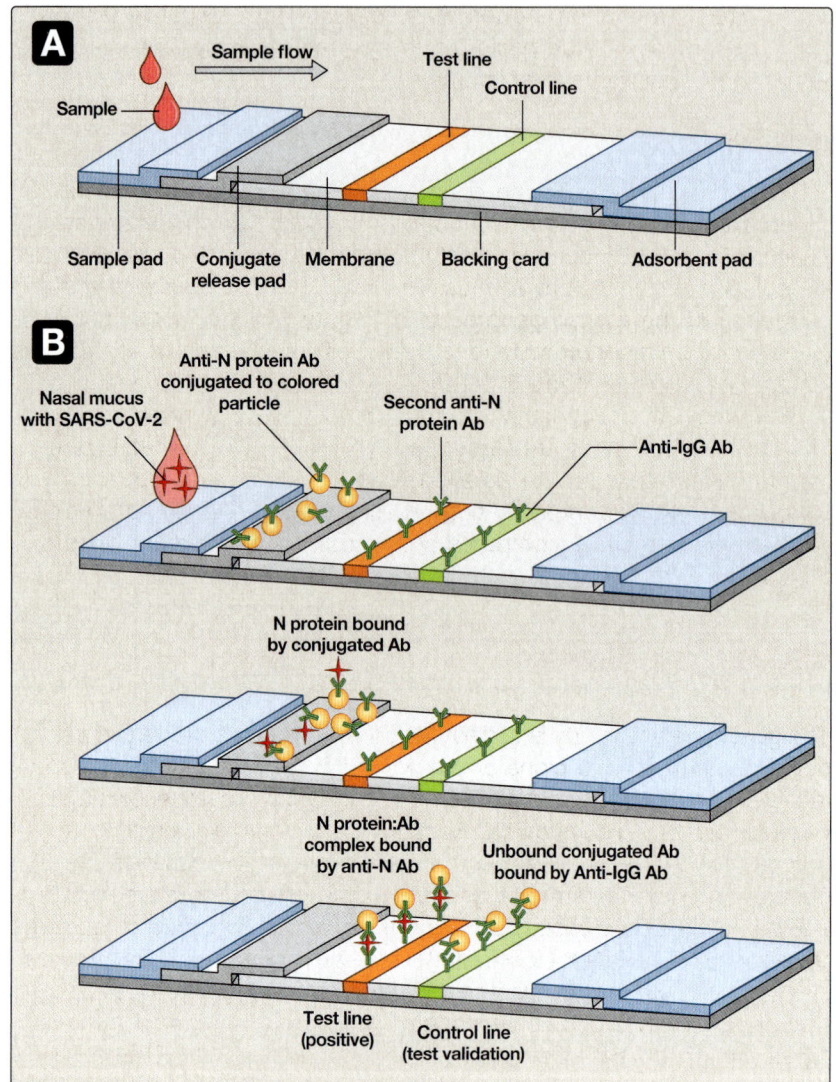

Figure 34.21
Lateral flow immunoassay for SARS-CoV-2.

a colored molecule that allows its visualization when it is concentrated in one area of the membrane. The N protein–antibody complex, free antibody, and other proteins in the sample move by capillary action along a membrane. The membrane contains two detection lines, a test line and a control line, over which the sample migrates. At the test line, a second antibody to another part of the N protein binds to it and stops the N protein–antibody complex. Other proteins and unbound conjugated antibodies in the sample continue toward the control line farther along the membrane. The control line has an antibody that binds and stops the unbound conjugated antibody (e.g., anti-IgG antibody) in the sample. The color of the conjugated antibody will intensify at the test line if the nasal mucus is infected with SARS-CoV-2 and also at the control line. The color at the control line confirms that the conjugated antibody to the N protein is abundant and should have bound to N protein if it was present in the nasal mucus.

This is an important control for confirming a negative test. This technology is the basis for the at-home testing kits for SARS-CoV-2 available to consumers.

C. Proteomics

The study of the proteome, or all the proteins expressed by a genome, including their relative abundance, distribution, posttranslational modifications, functions, and interactions with other macromolecules, is known as proteomics. The 20,000 to 25,000 protein-coding genes of the human genome translate into well over 100,000 proteins when posttranscriptional and posttranslational modifications are considered. Although a genome remains essentially unchanged, the amounts and types of proteins in any particular cell change dramatically as genes are turned on and off. [Note: Proteomics (and genomics) required the parallel development of bioinformatics, the computer-based organization, storage, and analysis of biologic data.] Figure 34.22 compares some of the analytic techniques discussed in this chapter.

VIII. GENE THERAPY

The goal of gene therapy is to treat disease through the delivery of a functional gene (typically a clone of the normal DNA sequence for the gene) into the somatic cells of a patient who has a defect in that gene as a result of a disease-causing mutation. Because somatic gene therapy changes only the targeted somatic cells, the change is not passed on to the next generation. [Note: In germline gene therapy, the germ cells are modified, and so the change is passed on. A long-standing moratorium on germline gene therapy is in effect worldwide.] There are two types of gene transfer: (1) ex vivo, in which cells from the patient are removed, transduced, and returned, and (2) in vivo, in which the cells are directly transduced. Both types require the use of a viral vector to deliver the DNA. Challenges of gene therapy include development of vectors, achievement of long-lived expression, and prevention of side effects such as an immune response. The first successful gene therapy involved two patients with severe combined immunodeficiency disease (SCID) caused by mutations to the gene for adenosine deaminase (ADA). It utilized mature T lymphocytes transduced ex vivo with a retroviral vector (Fig. 34.23). [Note: Human ADA cDNA is now used.]

Gene editing, as opposed to gene transfer, allows a mutated gene to be repaired. Combinations of DNA-binding molecules (proteins or RNA) and endonucleases are used to identify and cleave the mutated sequence. Cleavage activates homologous recombination repair of dsDNA breaks that integrates DNA containing the correct sequence into the gene. An endonuclease guided to a specific DNA sequence by a custom-designed RNA has been used in gene editing in human cells. The technique is based on (and named for) the prokaryotic CRISPR-Cas9 (clustered regularly interspaced short palindromic repeats [CRISPR]-associated protein) system that identifies and cleaves foreign DNA in bacterial cells. CRISPR-Cas9 technology has been developed for modifying the human β-globin gene in hematopoietic stem cells (Fig. 34.24). It has recently been approved for treating patients with sickle cell anemia.

TECHNIQUE	SAMPLE ANALYZED
Southern blot	DNA
Northern blot	RNA
Western blot	Protein
Microarray	cDNA or genomic DNA
	Protein
ELISA	Protein
LIFA	Protein

Figure 34.22
Techniques used to analyze DNA, RNA, and proteins. [Note: The three blotting techniques involve the use of a gel.] ELISA, enzyme-linked immunosorbent assay; LIFA, lateral flow immunoassay; cDNA, complementary DNA.

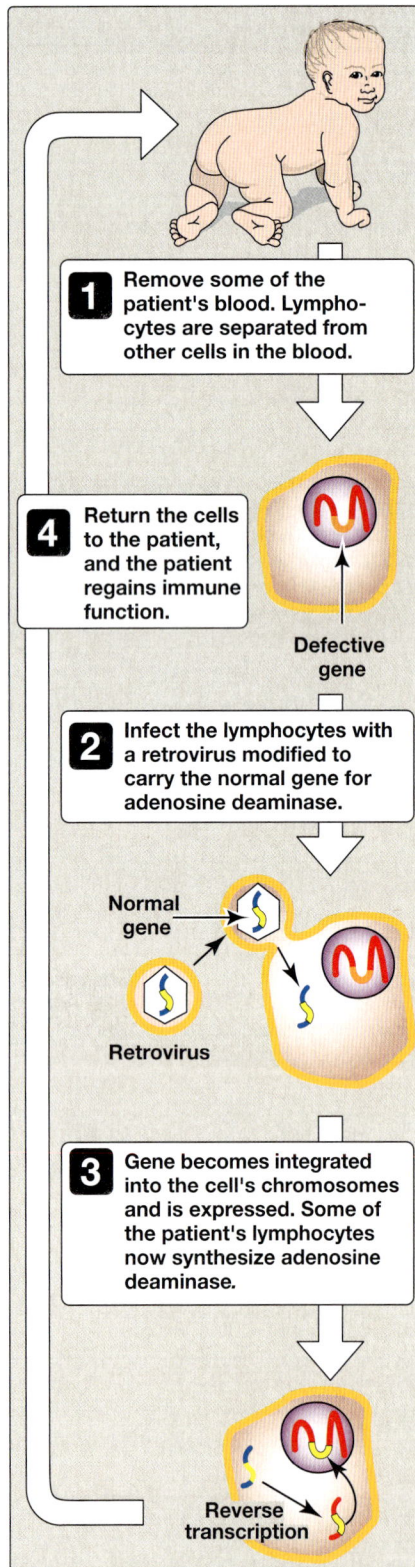

Figure 34.23
Gene therapy for severe combined immunodeficiency disease caused by adenosine deaminase deficiency. [Note: Bone marrow stem cells and a modified retroviral vector are now used.]

IX. TRANSGENIC ANIMALS

Transgenic animals can be produced by injecting a cloned foreign gene (a transgene) into a fertilized egg. If the gene randomly and stably integrates into a chromosome, it will be present in the germline of the resulting animal and can be passed from generation to generation. A giant mouse called "Supermouse" was produced in this way by injecting the gene for rat growth hormone into a fertilized mouse egg. Transgenic animals have been designed that produce therapeutic human proteins in their milk, a process called "pharming." Antithrombin, produced in the milk of transgenic goats, is used clinically to prevent blood clots in patients with a hereditary deficiency in the anticoagulation factor. If the functional transgene undergoes targeted (not random) insertion, a knockin (KI) animal that expresses the gene is created. Targeted insertion of a nonfunctional version of the transgene creates a knockout (KO) animal that does not express the gene. Such genetically engineered animals can serve as models for the study of a corresponding human disease.

X. CHAPTER SUMMARY

- **Restriction endonucleases** cleave double-stranded (ds)DNA at specific **palindromic** sequences (**restriction sites**). DNA fragments produced by cuts at the same **restriction site** can be ligated to form a **recombinant DNA molecule**.

- DNA cloning, which is the **amplification** (production of many copies) of DNA, requires a **recombinant DNA molecule** produced from a **vector** and a smaller DNA fragment of interest.

- Vectors are capable of **autonomous replication** within the host cell, contain at least one specific restriction site, and carry at least one gene, such as an **antibiotic resistance gene**, that confers the ability to select for host cells containing the vector. Prokaryotic **plasmids** can serve as vectors.

- **Complementary** (**c**)**DNA may** be produced from the mRNA molecules present in a cell. The cDNA for human genes can be cloned into an **expression vector** for the synthesis of proteins by bacteria or eukaryotes.

- **Polymerase chain reaction** (**PCR**) is a rapid method for **amplifying** a DNA sequence selected by the use of specific DNA primers that pair with flanking sequences. Some applications of the PCR technique include (1) comparison of a normal gene with a mutant form of the gene, (2) forensic analysis of DNA samples, (3) detection of low-abundance nucleic acid sequences, and (4) prenatal diagnosis and carrier detection of genetic disorders.

- **Southern blotting** can be used to detect specific sequences in DNA. The DNA is cleaved using a restriction endonuclease, after which the fragments are separated by **gel electrophoresis**, denatured within the gel, and transferred (blotted) onto a **nitrocellulose membrane** for analysis. The fragment of interest is detected using specific ssDNA or RNA probes.

- **Polymorphisms** (DNA sequence variations) in the human genome can arise from single-base changes or from tandem repeats. A polymorphism can serve as a genetic marker that can be followed through families.

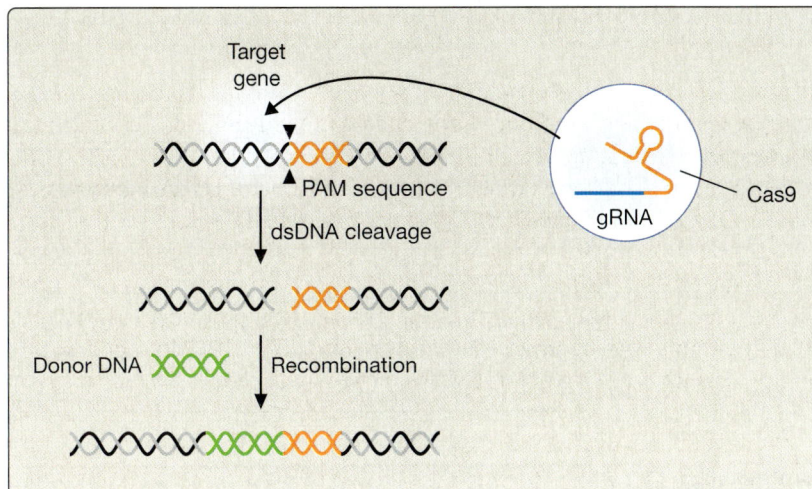

Figure 34.24
Simplified depiction of the CRISPR-Cas9 system. Hematopoietic stem cells are transfected with expression vectors encoding the Cas9 nuclease and a guide RNA (gRNA) along with a plasmid containing a donor DNA sequence. The Cas9 nuclease (*blue circle*) binds to the gRNA that directs Cas9 to the target gene. Cas9 creates a dsDNA break at a protospacer adjacent motif (PAM) sequence (*orange*). The donor DNA sequence (*green*) is inserted by recombination. This process edits the target gene removing a gene mutation or altering expression of a gene.

- A **restriction fragment length polymorphism** (**RFLP**) is a genetic variant that can be observed by cleaving the DNA of chromosomes with restriction endonucleases and separating the fragments by gel electrophoresis. A base substitution in one or more nucleotides at a restriction site can render the site unrecognizable by a particular restriction endonuclease or can create a new restriction site to produce DNA fragments of lengths differing from the expected fragments. **RFLP analysis** can be used to distinguish the chromosomes of individuals.

- The mRNA products of gene expression can be measured by **northern blots** and **microarrays**. In northern blot analysis, a sample containing a mixture of mRNA molecules are separated by electrophoresis, transferred to a membrane, and then hybridized to a radiolabeled probe for detection. In a **microarray**, the differing patterns of gene expression between two types of cells (e.g., normal and cancer cells) can be analyzed.

- **Enzyme-linked immunosorbent assays** (**ELISAs**), **western blots** (**immunoblots**), and **lateral flow immunoassays** are used to detect specific proteins using antibodies.

- **Proteomics** is the study of all proteins expressed by a genome.

- The goal of **gene therapy** is to replace a defective gene in a **somatic cell** with a normal cloned gene, whereas the goal of **gene editing** is to repair a mutated gene or to modify gene expression. CRISPR-Cas9 technology for modifying β-globin gene expression in hematopoietic stem cells is being used to treat patients with sickle cell anemia.

- Insertion of a foreign gene (transgene) into the germline of an animal creates a **transgenic animal** that can produce therapeutic proteins or serve as gene **knockin** or **knockout** models for human diseases.

Study Questions

Choose the ONE best answer.

34.1. HindIII is a restriction endonuclease. Which of the following is most likely to be the recognition sequence for this enzyme?

    A. AAGAAG
    B. AAGAGA
    C. AAGCTT
    D. AAGGAA
    E. AAGTTC

Correct answer = C. The vast majority of restriction endonucleases recognize palindromes in double-stranded DNA, and AAGCTT is the only palindrome among the choices. Because the sequence of only one DNA strand is given, the base sequence of the complementary strand must be determined. To be a palindrome, both strands must have the same sequence when read in the 5′→3′ direction. Thus, the complement of 5′-AAGCTT-3′ is also 5′-AAGCTT-3′.

34.2. A 38-year-old woman undergoes prenatal screening for chromosomal disorders in her developing fetus. Which of the following source of fetal DNA can be obtained with the least risk to the fetus?

    A. Amniotic fluid
    B. Chorionic villi
    C. Fetal blood
    D. Maternal blood

Correct answer = D. During pregnancy, fetal DNA enters the maternal circulation and can be obtained from a maternal blood sample. This is called "cell-free fetal DNA." Obtaining DNA from fetal cells in the amniotic fluid, the chorionic villi of the placenta, or the fetus require invasive procedures that pose a risk to the fetus.

34.3. A physician would like to determine the global patterns of gene expression in two different types of tumor cells in order to develop the most appropriate form of chemotherapy for each patient. Which of the following techniques would be most appropriate for this purpose?

    A. Enzyme-linked immunosorbent assay
    B. CRISPR-Cas9
    C. Microarray
    D. Northern blot
    E. Southern blot

Correct answer = C. Microarray analysis allows the determination of messenger RNA (mRNA) production (gene expression) from thousands of genes at once. A northern blot only measures mRNA production from one gene at a time. Enzyme-linked immunosorbent assay measure protein production (also gene expression) but only from one gene at a time. Southern blots are used to analyze DNA, not the products of DNA expression. CRISPR-Cas9 is a gene editing technique used to repair or replace defective genes in specific cell types.

34.4. A 2-week-old infant is diagnosed with a urea cycle defect. Enzymic analysis of the infant's lymphocytes showed no activity for ornithine transcarbamylase (OTC), an enzyme of the cycle. Molecular analysis revealed that the messenger RNA (mRNA) product of the gene for OTC was identical in length to that of a control. Which of the techniques listed below would be used to determine if OTC protein is produced from the mRNA?

    A. Dideoxy chain termination
    B. Northern blot
    C. Polymerase chain reaction
    D. Southern blot
    E. Western blot

Correct answer = E. Western blot allows analysis of the proteins present (expressed) in a particular cell or tissue. Southern blot is used for DNA analysis, whereas northern blot is used for mRNA analysis. Northern blot or reverse-transcriptase PCR (RT-PCR) may have been done to determine the length of the OTC mRNA. Dideoxy chain termination is used to sequence DNA.

Blood Clotting

35

I. OVERVIEW

Blood clotting (coagulation) is designed to rapidly stop bleeding from a damaged blood vessel in order to maintain a constant blood volume (hemostasis). Coagulation is accomplished through vasoconstriction and the formation of a clot (thrombus) that consists of a plug of platelets (primary hemostasis) and a meshwork of the protein fibrin (secondary hemostasis) that stabilizes the platelet plug. Clotting occurs in association with membranes on the surface of platelets and damaged blood vessels (Fig. 35.1). Processes to limit clot formation to the area of damage and remove the clot once vessel repair is underway also play essential roles in hemostasis. [Note: Separate discussions of the formation of the platelet plug and the fibrin meshwork facilitate presentation of these multistep, multicomponent processes. However, the two work together to maintain hemostasis.]

> If clotting occurs within an intact vessel such that the lumen is occluded and blood flow is impeded, the condition is called thrombosis. Serious tissue damage can occur due to a loss of the oxygen supplied by the blood. Thrombosis in an artery of the heart can cause a myocardial infarction (MI). The resulting damage to the heart muscle decreases the pumping ability of the heart and may cause death.

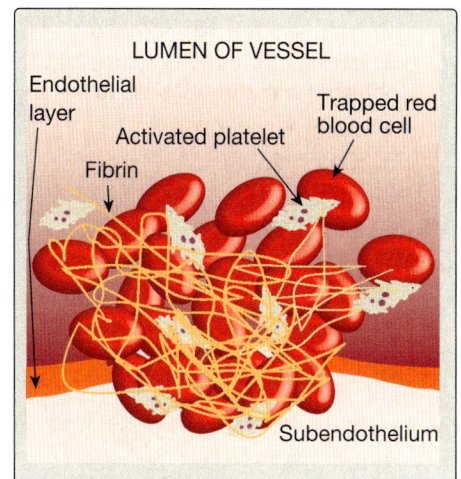

Figure 35.1
Blood clot formed by a plug of activated platelets and a meshwork of fibrin at the site of vessel injury.

II. SECONDARY HEMOSTASIS—FIBRIN MESHWORK FORMATION

The formation of the fibrin meshwork requires participation of platelets and involves two unique pathways, the extrinsic and intrinsic pathways that converge to form a common pathway (Fig. 35.2). In each pathway, the major components are proteins (called factors [F]) designated by Roman numerals. A few factors have additional names. For example, factor I (FI) is fibrinogen and factor II (FII) is prothrombin. The factors are glycoproteins that are synthesized and secreted primarily by the liver.

A. Proteolytic cascade

In response to a vascular injury, the factors, which are inactive zymogen proteases, are converted sequentially to an active form by proteolytic

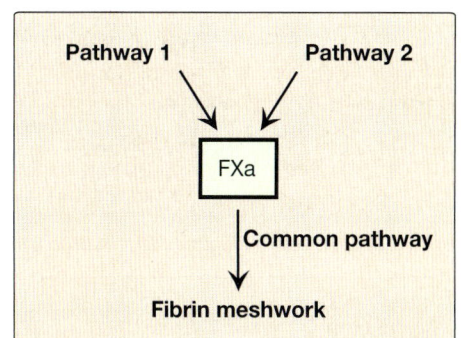

Figure 35.2
Three pathways involved in formation of the fibrin meshwork. F, factor; a, active.

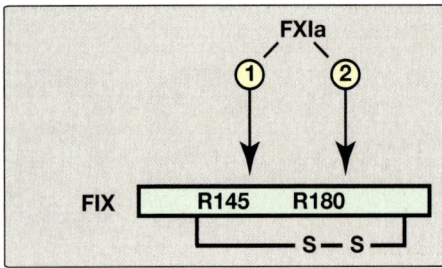

Figure 35.3
Activation of FIX via proteolysis by the serine protease FXIa. [Note: Activation can occur by conformational change for some of the factors.] F, factor; a, active; R, arginine.

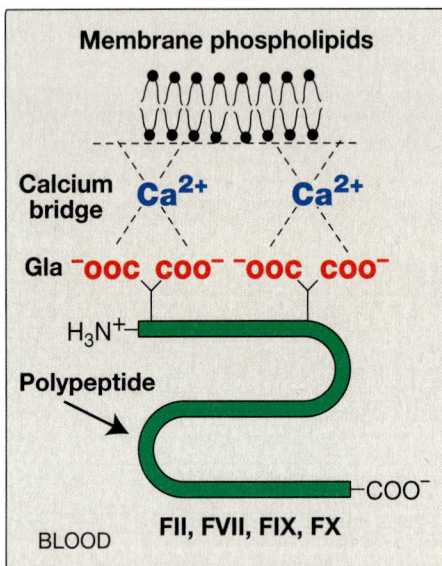

Figure 35.4
Ca^{2+} facilitates the binding of γ-carboxy-glutamate (Gla)-containing factors to membrane phospholipids. F, factor.

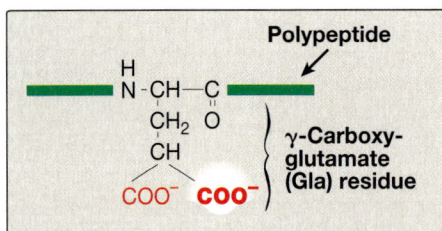

Figure 35.5
Gla residue.

cleavage. The protein product of one activation reaction initiates the next cleavage event in a cascade. The active form of an F is denoted by a lowercase "a" after the numeral. The active proteins FIIa (also called thrombin), FVIIa, FIXa, FXa, and FXIa, are enzymes of the serine protease family and cleave a peptide bond on the carboxyl side of an arginine or lysine residue in a polypeptide. For example, FIX is activated through cleavage at arginine 145 and arginine 180 by FXIa (Fig. 35.3). The proteolytic cascade results in enormous rate acceleration, because one active protease can produce many molecules of active product each of which, in turn, can activate many molecules of the next protein in the cascade. In some cases, activation can be caused by a conformational change in the protein in the absence of proteolysis. Nonproteolytic proteins also play a role as accessory proteins (cofactors) in the pathways. FIII (also called tissue factor [TF]), FV, and FVIII are the accessory proteins.

B. Role of phosphatidylserine and calcium

The presence of the negatively charged phospholipid phosphatidylserine (PS) and positively charged calcium ions (Ca^{2+}) accelerates the rate of some steps in the clotting cascade.

1. **Phosphatidylserine:** PS is located primarily on the intracellular (cytosolic) face of the plasma membrane. Its extracellular exposure signals injury to the endothelial cells that line blood vessels. PS is also exposed on the surface of activated platelets.

2. **Calcium ions:** Ca^{2+} binds the negatively charged γ-carboxyglutamate (Gla) residues present in four of the serine proteases of clotting (FII, FVII, FIX, and FX), facilitating the binding of these proteins to exposed phospholipids (Fig. 35.4). The Gla residues are good chelators of Ca^{2+} because of their two adjacent negatively charged carboxylate groups (Fig. 35.5). The use of Ca^{2+} chelating agents, such as sodium citrate, in blood-collecting tubes or bags prevents the blood from clotting.

C. Formation of γ-carboxyglutamate residues

γ-Carboxylation is a posttranslational modification in which 9 to 12 glutamate residues (at the amino [N]-terminus of the protein) become carboxylated at the γ carbon, thereby forming Gla residues. The process occurs within the rough endoplasmic reticulum (rER) of hepatocytes. γ-Carboxylation requires a protein substrate, oxygen (O_2), carbon dioxide (CO_2), γ-glutamyl carboxylase, and the hydroquinone form of vitamin K as a coenzyme (Fig. 35.6). In the reaction, the hydroquinone form of vitamin K is oxidized to its epoxide form as O_2 is reduced to water. Dietary vitamin K, a fat-soluble vitamin (see Chapter 28, XIII.), is reduced from the quinone form to the hydroquinone coenzyme form by the enzyme vitamin K–epoxide reductase ([VKOR] Fig. 35.7). The reductase, an integral protein of the rER membrane, is required to regenerate the functional hydroquinone form of vitamin K from the epoxide form generated in the γ-carboxylation reaction.

D. Pathways

Two pathways may initiate the formation of the fibrin meshwork: the extrinsic and the intrinsic pathways, which converge on the common

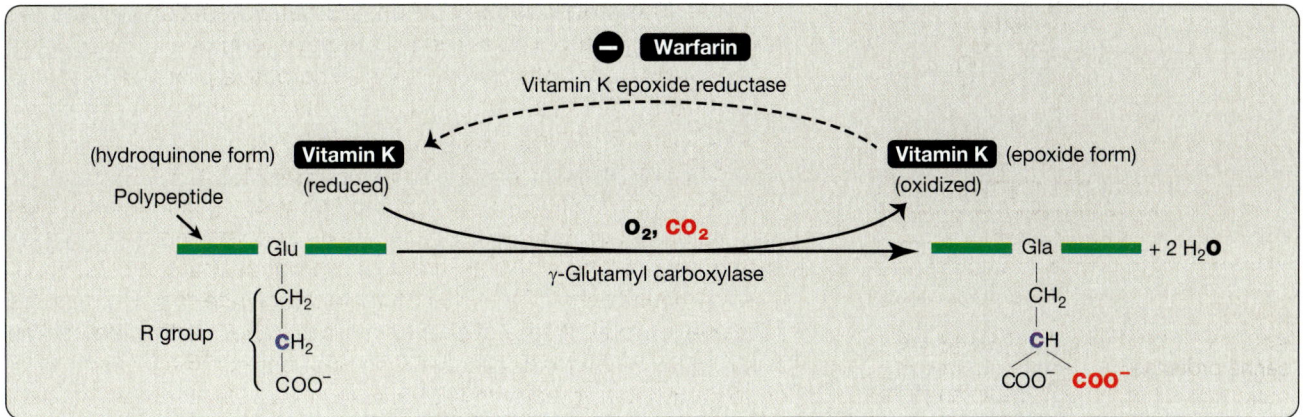

Figure 35.6
γ-Carboxylation of a glutamate (Glu) residue to γ-carboxyglutamate (Gla) by vitamin K–requiring γ-glutamyl carboxylase. The γ carbon is shown in *blue*. O_2, oxygen; CO_2, carbon dioxide.

Clinical Application 35.1: Warfarin Therapy

Warfarin, a synthetic analog of vitamin K, inhibits VKOR. Thus, warfarin and related drugs act as anticoagulants that inhibit clotting by functioning as vitamin K antagonists. Warfarin salts are used therapeutically to limit clot formation, but patients on warfarin therapy require careful monitoring of their blood clotting time to regulate their warfarin dose and prevent excessive bleeding. Vitamin K competes with warfarin so patients must also limit their intake of high vitamin K foods. Genetic differences (genotypes) in the gene for catalytic subunit 1 of VKOR (*VKORC1*) influence patient response to warfarin. For example, a polymorphism in the promoter region of the gene decreases gene expression, resulting in less VKOR being made, thereby necessitating a lower dose of warfarin to achieve a therapeutic level. The cytochrome P450 enzyme that metabolizes warfarin, CYP2C9, may contain several polymorphisms that also lower the therapeutic dose for individuals with these genetic variants. Therefore, the label for warfarin contains dosing recommendations based on the *VKORC1* and *CYP2C9* genotypes. The influence of genetics on an individual's response to drugs is known as pharmacogenetics.

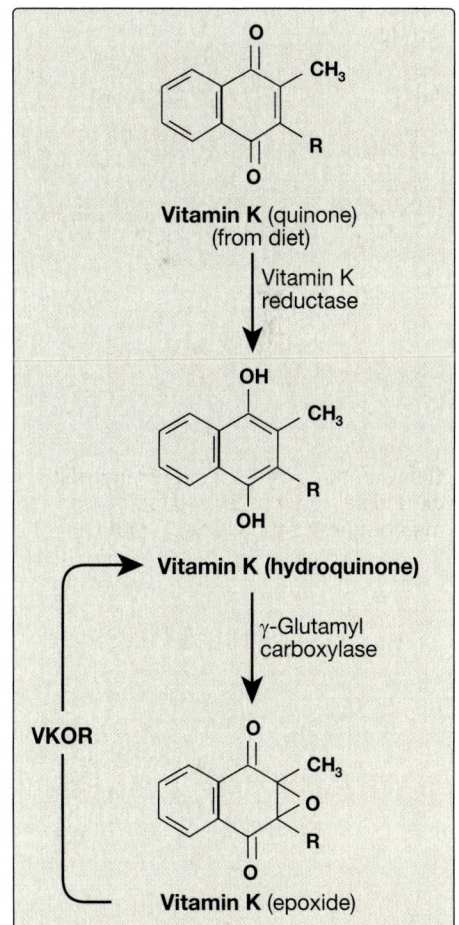

Figure 35.7
Vitamin K cycle. VKOR, vitamin K epoxide reductase.

pathway to create the fibrin clot. Production of FXa by both the extrinsic and intrinsic pathways triggers the common pathway (see Fig. 35.2).

1. **Extrinsic pathway:** This pathway involves TF, or FIII, a protein that is not ordinarily in the blood but becomes exposed when blood vessels are injured. TF is a transmembrane glycoprotein abundant in the perivascular cells of the subendothelium. It is an extravascular accessory protein and not a protease. Any injury that exposes TF to blood rapidly (within seconds) initiates the extrinsic or TF pathway. Once exposed, TF binds a circulating Gla-containing protein, FVII, activating it through conformational change. FVII also may be activated proteolytically by thrombin (see Section II. D. 3.) or by several other serine proteases. FVII–TF complex activation requires the presence

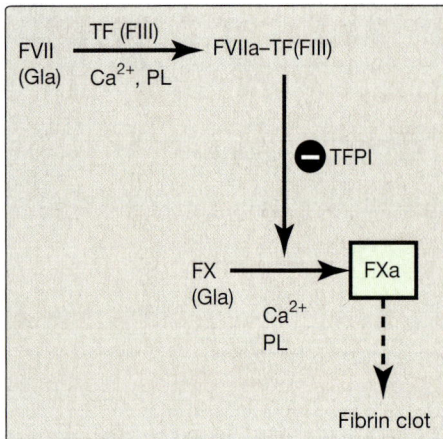

Figure 35.8
Extrinsic or tissue factor (TF) pathway. Binding of FVII to exposed TF (FIII) activates FVII. [Note: The pathway is quickly inhibited by tissue factor pathway inhibitor (TFPI).] F, factor; Gla, γ-carboxyglutamate; Ca^{2+}, calcium; PL, phospholipid; a, active.

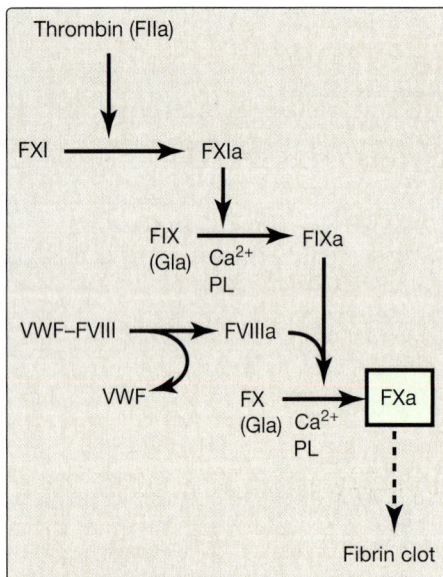

Figure 35.9
FX-activation phase of the intrinsic pathway. [Note: von Willebrand factor (VWF) stabilizes FVIII in the circulation.] Gla, γ-carboxyglutamate; PL, phospholipid; a, active; F, factor; Ca^{2+}, calcium.

of Ca^{2+} and phospholipids. The TF–FVIIa complex then binds and activates FX by proteolysis (Fig. 35.8). Therefore, activation of FX by the extrinsic pathway occurs in association with the perivascular cell membrane. FXa goes on to promote the common pathway activation of FII (prothrombin) to generate FIIa (thrombin). The extrinsic pathway is quickly inactivated by TF pathway inhibitor (TFPI) that, in an FXa-dependent process, binds to the TF–FVIIa complex and prevents further production of FXa.

2. **Intrinsic pathway:** All of the protein factors involved in the intrinsic pathway are present in the blood and are, therefore, intravascular. The sequence of events leading to the activation of FX to FXa by the intrinsic pathway is initiated by thrombin. Thrombin converts FXI to FXIa, which, in turn, activates FIX, a Gla-containing serine protease. FIXa combines with FVIIIa (a blood-borne accessory protein), and this complex activates FX, another Gla-containing serine protease (Fig. 35.9). The complex containing FIXa, FVIIIa, and FX forms on exposed negatively charged membrane regions, where FX is activated to FXa. This complex is sometimes referred to as Xase. Binding of the complex to membrane phospholipids requires Ca^{2+}.

> The inactivation of the extrinsic pathway by TFPI results in dependence on the intrinsic pathway for continued production of FXa. This explains why individuals with hemophilia bleed in an uncontrolled way even though they have an intact extrinsic pathway.

3. **Common pathway:** FXa produced by both the intrinsic and the extrinsic paths initiates the common pathway, a sequence of reactions that results in the generation of fibrin (FIa), as shown in

---

Clinical Application 35.2: Hemophilia

Hemophilia is a coagulopathy—a defect in the ability to clot. Hemophilia A, which accounts for 80% of all hemophilia, results from deficiency of FVIII, whereas deficiency of FIX causes hemophilia B. Both forms of hemophilia are X-linked since the genes for both factors are on the X chromosome. Each deficiency is characterized by decreased and delayed ability to clot and/or formation of abnormally friable (easily disrupted) clots. This can be manifested, for example, by bleeding into the joints (Fig. 35.10). The extent of the factor deficiency determines the severity of the disease. The primary treatment for hemophilia is factor replacement therapy using FVIII or FIX obtained from pooled human blood or from recombinant DNA technology. However, antibodies to the factors can develop reducing the effectiveness of the treatment. A newer therapeutic approach for hemophilia A involves administering a monoclonal antibody, emicizumab (Hemlibra), that mimics the function of FVIII and links FIX and FX promoting FX activation. An extremely rare bleeding disorder, referred to as hemophilia C, results from the autosomal inheritance of a deficiency of FXI.

Figure 35.11. FXa associates with FVa (a blood-borne accessory protein) and, in the presence of Ca^{2+} and phospholipids, forms a membrane-bound complex referred to as prothrombinase. The complex cleaves prothrombin (FII) to thrombin (FIIa), and FVa potentiates the proteolytic activity of FXa to control thrombin (FIIa) production. The binding of Ca^{2+} to the Gla residues in FII facilitates the binding of FII to the membrane and to the prothrombinase complex, with subsequent cleavage to FIIa. Cleavage excises the Gla-containing region, releasing thrombin (FIIa) from the membrane, thereby freeing it to activate fibrinogen (FI) in the blood. Notably, this is the only example of cleavage of a Gla protein that results in the release of a Gla-containing peptide. The peptide travels to the liver where it is thought to act as a signal for increased production of clotting proteins. Direct inhibitors of FXa are in clinical use as oral anticoagulants to prevent thrombosis and its consequences. In contrast to warfarin, they have a more rapid onset and shorter half-life and do not require routine monitoring.

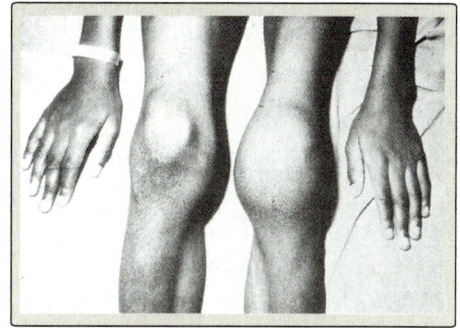

Figure 35.10
Acute bleeding into joint spaces (hemarthrosis) in an individual with hemophilia.

a. **Fibrinogen cleavage to fibrin:** Fibrinogen (sometimes referred to as FI) is a soluble glycoprotein made by the liver. It consists of dimers of three different polypeptide chains ($[\alpha\beta\gamma]_2$) held together at the N termini by disulfide bonds. The N termini of the α and β chains form "tufts" on the central of three globular domains (Fig. 35.12). The tufts are negatively charged and result in repulsion between fibrinogen molecules. Thrombin (FIIa) cleaves the charged tufts (releasing fibrinopeptides A and B), and FI becomes FIa (fibrin). As a result of the loss of charge, the fibrin monomers are able to noncovalently associate in a staggered array, and a soft (soluble) fibrin clot is formed.

A common point mutation (*G20210A*) in which an adenine (A) replaces a guanine (G) at nucleotide 20210 in the 3' untranslated region of the gene for prothrombin (FII) leads to increased levels of FII in the blood. This results in one type of thrombophilia, a condition characterized by an increased tendency for blood to clot and a two- to three-fold greater risk for venous thrombosis.

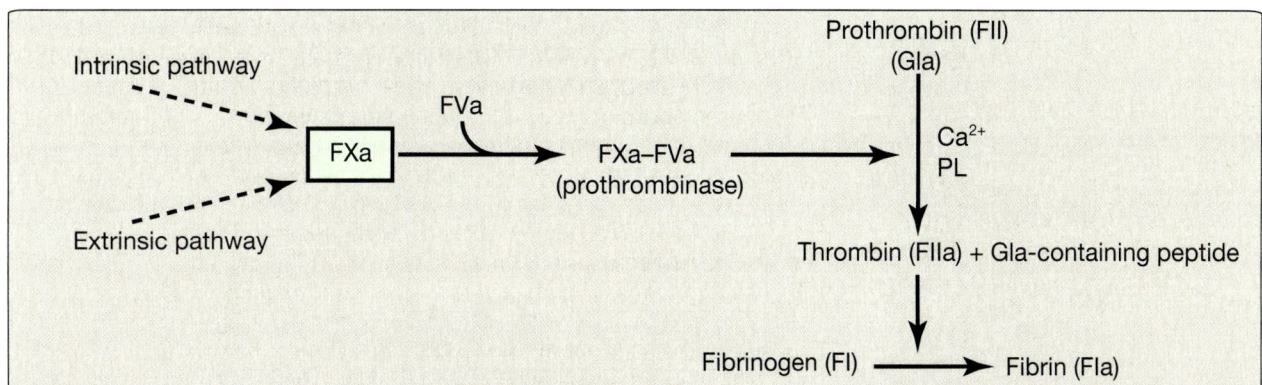

Figure 35.11
Generation of fibrin by FXa and the common pathway. F, factor; Gla, γ-carboxyglutamate; PL, phospholipid; a, active; Ca^{2+}, calcium.

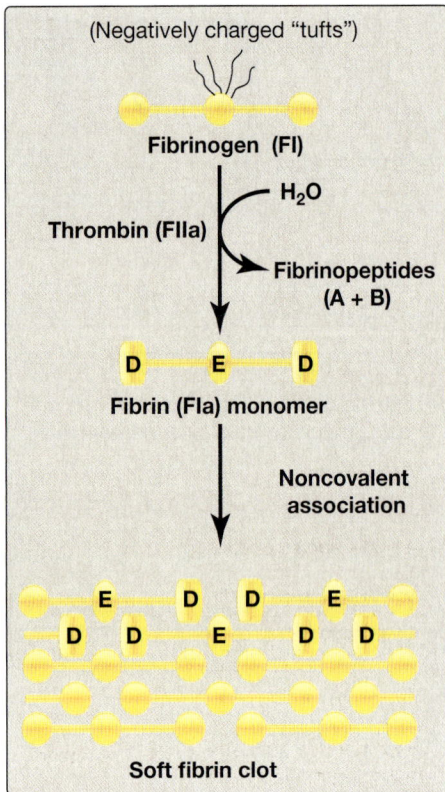

Figure 35.12
Conversion of fibrinogen to fibrin and formation of the soft fibrin clot. [Note: D and E refer to nodular domains on the protein.]

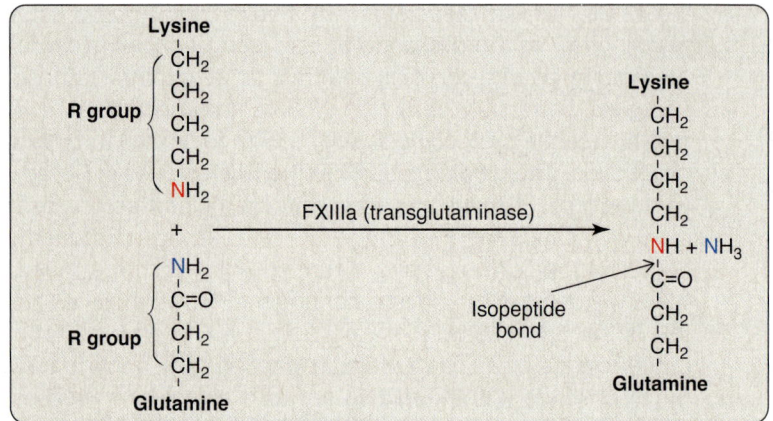

Figure 35.13
Cross-linking of fibrin. FXIIIa forms a covalent isopeptide bond between lysine and glutamine residues. F, factor; NH_3, ammonia.

b. Fibrin cross-linking: The associated fibrin molecules become covalently cross-linked, converting the soft clot to a hard (insoluble) clot. FXIIIa, a transglutaminase, covalently links the γ-carboxamide of a glutamine residue in one fibrin molecule to the ε-amino of a lysine residue in another through formation of an isopeptide bond and release of ammonia (Fig. 35.13). As one of its many functions, thrombin (FIIa) activates FXIII to promote fibrin cross-linking (Fig. 35.14).

c. Importance of thrombin: The extrinsic pathway produces the FXa that initiates the formation of thrombin (FIIa) from pro-thrombin and thereby promotes clotting. Thrombin (FIIa) then activates factors of the common (FV, FI, FXIII), intrinsic (FXI, FVIII), and extrinsic (FVII) pathways (Fig. 35.14). The intrinsic pathway amplifies and sustains clotting after the extrinsic pathway has been inhibited by TFPI. [Note: Dabigatran is a direct thrombin inhibitor (DTI) used clinically as an oral antico-agulant. Lepirudin, the recombinant form of a peptide secreted from the salivary gland of medicinal leeches, is a potent, injectable DTI used to treat patients with complications due to heparin therapy.] Additional cross talk between the pathways of clotting is achieved by the FVIIa–TF-mediated activation of the intrinsic pathway and the FXIIa-mediated activation of the extrinsic pathway. The complete picture of physiologic blood clotting via the formation of a hard fibrin clot is shown in Figure 35.15. The factors of the clotting cascade are shown organized by function in Figure 35.16.

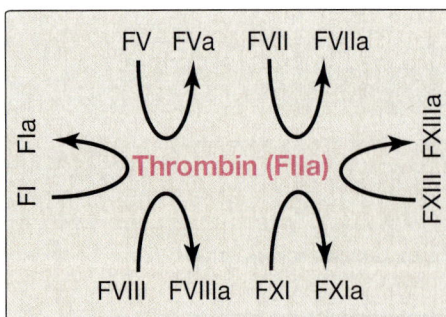

Figure 35.14
Importance of thrombin in formation of the fibrin clot. a, active; F, factor.

Clinical laboratory tests are available to evaluate the function of the extrinsic pathway through the common pathway (pro-thrombin time [PT] using thromboplastin and expressed as the international normalized ratio [INR]) and the intrinsic pathway through common pathway (activated partial thromboplastin time [aPTT]). Thromboplastin is a combination of phospholip-ids + TF. A derivative, partial thromboplastin, contains just the phospholipid portion because TF is not needed to activate the intrinsic pathway.

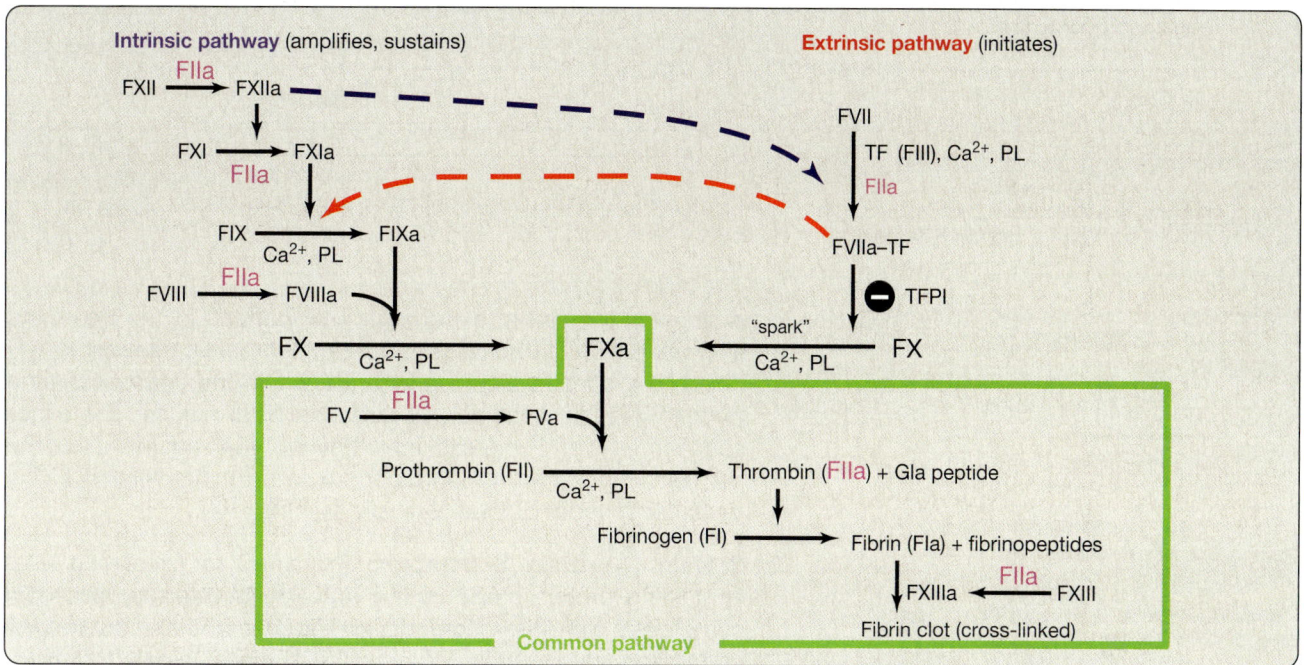

Figure 35.15
Complete picture of physiologic blood clotting via the formation of a cross-linked (hard) fibrin clot. a, active; F, factor; TF, tissue factor; TFPI, tissue factor pathway inhibitor; PL, phospholipid; Ca^{2+}, calcium; Gla, γ-carboxyglutamate.

III. LIMITING CLOTTING

The ability to limit clotting to areas of damage (anticoagulation) and remove clots once repair processes are underway (fibrinolysis) are exceedingly important aspects of hemostasis. These actions are performed by proteins that inactivate clotting factors either by binding to them and removing them from the blood or by degrading them and also by proteins that degrade the fibrin meshwork.

A. Inactivating proteins

Proteins synthesized by the liver and by the blood vessels themselves balance the need to form clots at sites of vessel injury with the need to limit their formation beyond the injured area.

1. **Antithrombin:** Antithrombin III ([ATIII] also referred to as antithrombin [AT]), is a serine protease inhibitor or "serpin" that is produced in the liver and circulates in the blood. It inactivates free thrombin (FIIa, a serine protease) by binding to it and carrying it to the liver, preventing it from participating in coagulation (Fig. 35.17). Specifically, thrombin binds to a reactive loop on ATIII and then cleaves a peptide bond in ATIII causing a conformational change that traps thrombin in a covalent complex. The affinity of ATIII for thrombin is greatly increased when ATIII is bound to heparan sulfate, an intracellular glycosaminoglycan released in response to injury by mast cells associated with blood vessels. ATIII also inactivates FXa and the other serine proteases of clotting, FIXa, FXIa, FXIIa, and the FVIIa–TF complex. Heparin, a therapeutic anticoagulant, is administered intravenously and

| Serine proteases |
| II, VII, IX, X, XI, XII |

| Gla-containing proteases |
| II, VII, IX, X |

| Accessory proteins |
| III, V, VIII |

Figure 35.16
Protein factors of the clotting cascade organized by function. The activated form would be denoted by an "a" after the numeral. [Note: Calcium is IV. There is no VI. I (fibrin) is neither a protease nor an accessory protein. XIII is a transglutaminase.] Gla, γ-carboxyglutamate.

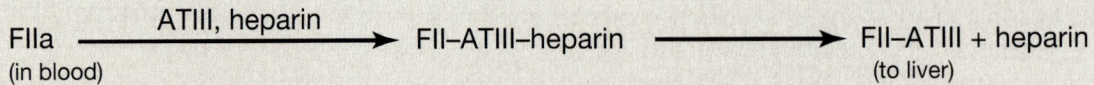

FIIa ———ATIII, heparin———▶ FII–ATIII–heparin ——————▶ FII–ATIII + heparin
(in blood)   (to liver)

Figure 35.17
Inactivation of FIIa (thrombin) by binding of antithrombin III (ATIII) and transport to the liver. [Note: Heparin increases the affinity of ATIII for FIIa.] a, active; F, factor.

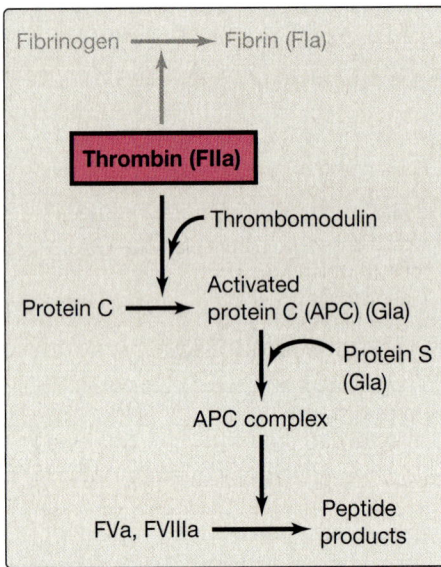

Figure 35.18
Formation and action of the APC complex. Gla, γ-carboxyglutamate; a, active; F, factor.

is commonly used in along with oral warfarin in the treatment and prevention of thrombosis. During therapeutic treatment, ATIII binds to a specific pentasaccharide within the oligosaccharide form of heparin. Inhibition of thrombin (FIIa) requires the oligosaccharide form, whereas inhibition of FXa requires only the pentasaccharide form. Fondaparinux, a synthetic version of the pentasaccharide, is used clinically to inhibit FXa.

2. **Protein C–protein S complex:** Protein C, a circulating Gla-containing protein made in the liver, is activated by thrombin complexed with thrombomodulin. Thrombomodulin, an integral membrane glycoprotein of endothelial cells, binds thrombin, thereby decreasing thrombin's affinity for fibrinogen and increasing its affinity for protein C. Protein C in complex with protein S, also a Gla-containing protein, forms the activated protein C (APC) complex that cleaves the accessory proteins FVa and FVIIIa, which are required for maximal activity of FXa (Fig. 35.18). Protein S helps anchor APC to the clot. Thrombomodulin, then, modulates the activity of thrombin, converting it from a protein of coagulation to a protein of anticoagulation, thereby limiting the extent of clotting.

B. Fibrinolysis

Clots are temporary patches that must be removed once wound repair has begun. The fibrin clot is cleaved by the protease plasmin to fibrin degradation products (Fig. 35.19). Measurement of D-dimer, a fibrin degradation product containing two cross-linked D domains,

---

Clinical Application 35.3: Thrombophilia

Thrombophilia (hypercoagulability) can result from the presence of FV Leiden, from excess production of thrombin (*G20210A* mutation), and from deficiencies of proteins C, S, and ATIII. Another form of thrombophilia is caused by antiphospholipid antibodies, which may be present in persons with autoimmune disorders, such as systemic lupus erythematosus. Factor V Leiden is a mutant form of FV with a glutamine substituted for arginine at position 506 and a resistance to APC. It is the most common inherited cause of thrombophilia in the United States. Heterozygotes are estimated to have a 7-fold increase in the risk for venous thrombosis, and homozygotes have up to a 50-fold increase. Women with FV Leiden are at even greater risk of thrombosis during pregnancy or when taking estrogen. A thrombus that forms in the deep veins of the leg (deep venous thrombosis [DVT]) can cause a pulmonary embolism (PE) if part or all of the clot breaks off, travels to the lungs, and blocks circulation.

can be used to assess the extent of clotting. Plasmin is a serine protease that is generated from plasminogen by plasminogen activators. Plasminogen, secreted by the liver into the circulation, binds to fibrin and is incorporated into clots as they form. Tissue plasminogen activator (TPA, t-PA), made by vascular endothelial cells and secreted in an inactive form in response to thrombin, becomes active (TPA$_a$) when bound to fibrin–plasminogen. Bound plasmin and TPA$_a$ are protected from their inhibitors, α_2-antiplasmin and plasminogen activator inhibitors, respectively. Once the fibrin clot is dissolved, plasmin and TPA$_a$ become available to their inhibitors. Therapeutic fibrinolysis can sometimes be achieved by treatment with commercially available TPA made by recombinant DNA techniques. Its main use is in treating ischemic stroke patients. Mechanical clot removal (thrombectomy) is more commonly used for treatment of MI. [Note: Urokinase is a plasminogen activator (u-PA) made in a variety of tissues and originally isolated from urine. Streptokinase (from bacteria) activates both free and fibrin-bound plasminogen.]

> Plasminogen contains structural motifs known as "kringle domains" that mediate protein–protein interactions. Because lipoprotein (a) (Lp[a]) also contains kringle domains, it competes with plasminogen for binding to Fla. The potential to inhibit fibrinolysis may be the basis for the association of elevated Lp(a) with increased risk for cardiovascular disease (see Chapter 18, VI. F.).

IV. PRIMARY HEMOSTASIS—PLATELET PLUG FORMATION

Platelets (thrombocytes) are small, anucleate fragments of megakaryocytes that adhere to exposed collagen of damaged endothelium, get activated, and aggregate to form a platelet plug (Fig. 35.20; also see Fig. 35.1). Formation of the platelet plug is referred to as primary hemostasis because it is the first response to bleeding. In a healthy adult, there are 150,000 to 450,000 platelets per µL of blood. They have a life span of up to 10 days, after which they are taken up by the liver and spleen and destroyed. Platelet number and activity are measured clinically to determine problems with platelet production or function that may affect clotting.

A. Adhesion

Adhesion of platelets to exposed collagen at the site of vessel injury is mediated by the protein von Willebrand factor (VWF), a glycoprotein secreted from endothelial cells and released from activated platelets. VWF binds to collagen, and platelets bind to VWF via glycoprotein Ib (GPIb), a component of a membrane receptor complex (GPIb–V–IX) on the platelet surface (Fig. 35.21). Binding to VWF stops the forward movement of platelets. [Note: Deficiency in the receptor for VWF results in Bernard–Soulier syndrome, a disorder of decreased platelet adhesion.] In addition to mediating the binding of platelets to collagen, VWF also binds to and stabilizes FVIII in the blood. Deficiency of VWF results in von Willebrand disease (VWD), the most common

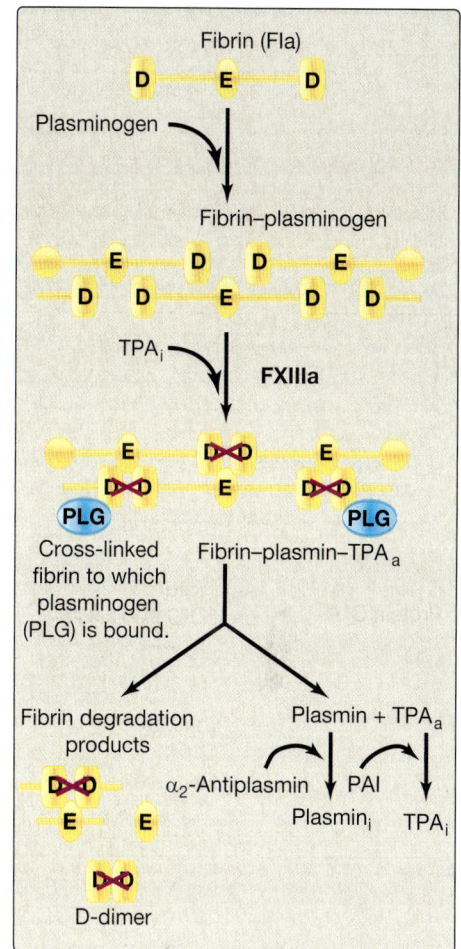

Figure 35.19
Fibrinolysis. Plasmin cleaves cross-linked fibrin into fibrin degradation products. Plasmin and TPA are released from the clot. TPA, tissue plasminogen activator; i, inactive; a, active; PAI, plasminogen activator inhibitor.

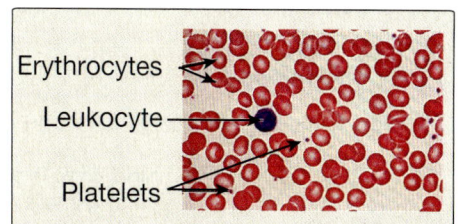

Figure 35.20
Size comparison of platelets, erythrocytes, and a leukocyte.

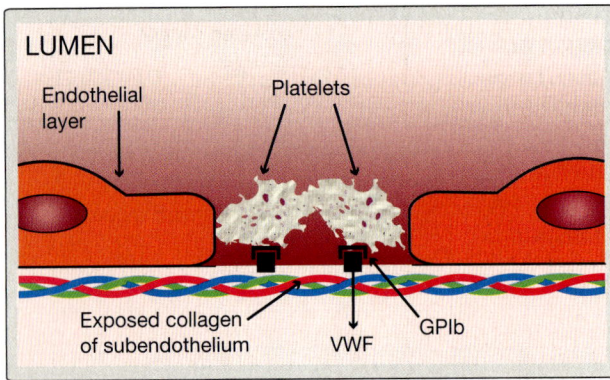

Figure 35.21
Binding of platelets via the glycoprotein Ib receptor (GPIb) to von Willebrand factor (VWF). VWF is bound to the exposed collagen at a site of injury.

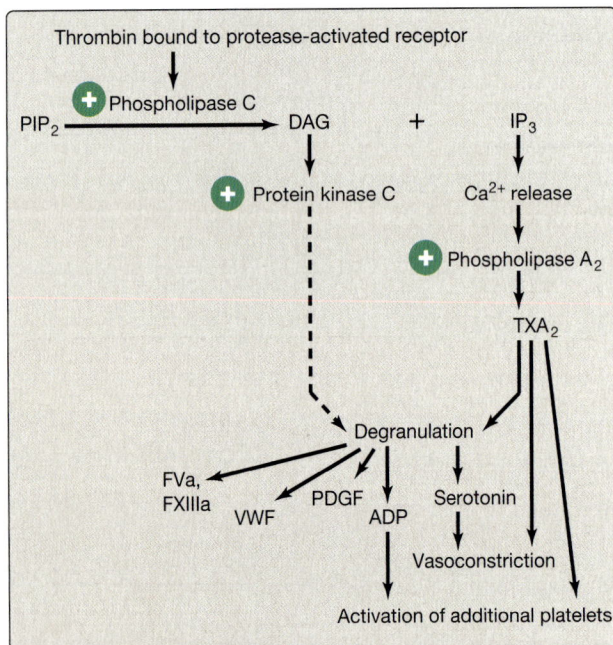

Figure 35.22
Platelet activation by thrombin. [Note: Protease-activated receptors are a type of G protein–coupled receptor.] PIP$_2$, phosphoinositol bisphosphate; DAG, diacylglycerol; IP$_3$, inositol trisphosphate; Ca^{2+}, calcium; TXA$_2$, thromboxane A$_2$; ADP, adenosine diphosphate; PDGF, platelet-derived growth factor; VWF, von Willebrand factor; F, factor.

inherited coagulopathy. VWD results from decreased binding of platelets to collagen and a deficiency in FVIII (due to increased degradation). Platelets can also bind directly to collagen via the membrane receptor glycoprotein VI (GPVI). Recall that damage to the vascular endothelium also exposes TF, initiating the extrinsic pathway of blood clotting and activation of FX (see Fig. 35.8).

B. Activation

Once adhered to areas of injury, platelets become activated. Platelet activation involves morphologic (shape) changes and degranulation, the process by which platelets secrete the contents of their α and δ (or, dense) storage granules. Activated platelets also expose PS on their surface. The externalization of PS is mediated by a Ca^{2+}-activated enzyme known as scramblase that disrupts the membrane phospholipid asymmetry created by flippases. The exposure of PS on the surface of activated platelets allows formation of the Xase complex (VIIIa, IXa, X, and Ca^{2+}) with subsequent formation of FXa and generation of thrombin (FIIa). Thrombin is the most potent platelet activator. Thrombin binds to and activates protease-activated receptors, a family of G protein–coupled receptor (GPCR), on the surface of platelets (Fig. 35.22). Thrombin is primarily associated with G$_q$ proteins, resulting in activation of phospholipase C and a rise in diacylglycerol (DAG) and inositol trisphosphate (IP$_3$). Recall that thrombomodulin on the surface of endothelial cells binds thrombin and thereby decreases thrombin availability for platelet activation (see Fig. 35.18).

1. **Degranulation:** DAG activates protein kinase C, a key event for degranulation. IP$_3$ causes the release of Ca^{2+} from dense granules. The Ca^{2+} activates phospholipase A$_2$, which cleaves membrane phospholipids to release arachidonic acid, the substrate for the synthesis of thromboxane A$_2$ (TXA$_2$) in activated platelets by cyclooxygenase-1 (COX-1) (see Chapter 17, VIII. B.). TXA$_2$ released from activated platelets causes vasoconstriction, augments degranulation, and binds to platelet GPCR, causing activation of additional platelets. Recall that aspirin irreversibly inhibits COX and, consequently, TXA$_2$ synthesis and is referred to as an antiplatelet drug. Degranulation also results in release of serotonin and adenosine diphosphate (ADP) from dense granules. Serotonin causes vasoconstriction. ADP binds to GPCR on the surface of platelets, activating additional platelets. Clopidogrel, an ADP-receptor antagonist, is used in combination with aspirin in dual antiplatelet therapy for patients who are at high risk for cardiovascular events.

Platelet-derived growth factor (involved in wound healing), VWF, FV, FXIII, and fibrinogen are among other proteins released from α-granules. A variety of cells, including endothelial cells and activated platelets, synthesize platelet-activating factor (PAF), an ether phospholipid, that binds PAF receptors (GPCR) on the surface of platelets and trigger platelet activation.

2. **Morphologic change:** The change in shape of activated platelets from discoidal to spherical with pseudopod-like processes that facilitate platelet–platelet and platelet–surface interactions (Fig. 35.23) is initiated by the release of Ca^{2+} from dense granules. Ca^{2+} bound to calmodulin mediates the activation of myosin light chain kinase that phosphorylates the myosin light chain, resulting in a major reorganization of the platelet cytoskeleton.

C. Aggregation

Activation causes dramatic changes in platelets that lead to their aggregation. Structural changes in a surface receptor (GPIIb/IIIa) expose binding sites for fibrinogen. Bound fibrinogen molecules link activated platelets to one another, with a single fibrinogen able to bind two platelets (Fig. 35.24). Fibrinogen is converted to fibrin by thrombin and then covalently cross linked by FXIIIa coming from both the blood and the platelets. Fibrin formation (secondary hemostasis) strengthens the platelet plug. Rare defects in the platelet receptor for fibrinogen result in Glanzmann thrombasthenia (decreased platelet function), whereas autoantibodies to this receptor are a cause of immune thrombocytopenia (decreased platelet number).

Unnecessary activation of platelets is prevented because (1) an intact vascular wall is separated from the blood by a monolayer of endothelial cells, preventing the contact of platelets with collagen; (2) endothelial cells synthesize prostaglandin I_2 ([PGI_2] or prostacyclin) and nitric oxide, each of which causes vasodilation; and (3) endothelial cells have a cell surface ADPase that converts ADP to adenosine monophosphate.

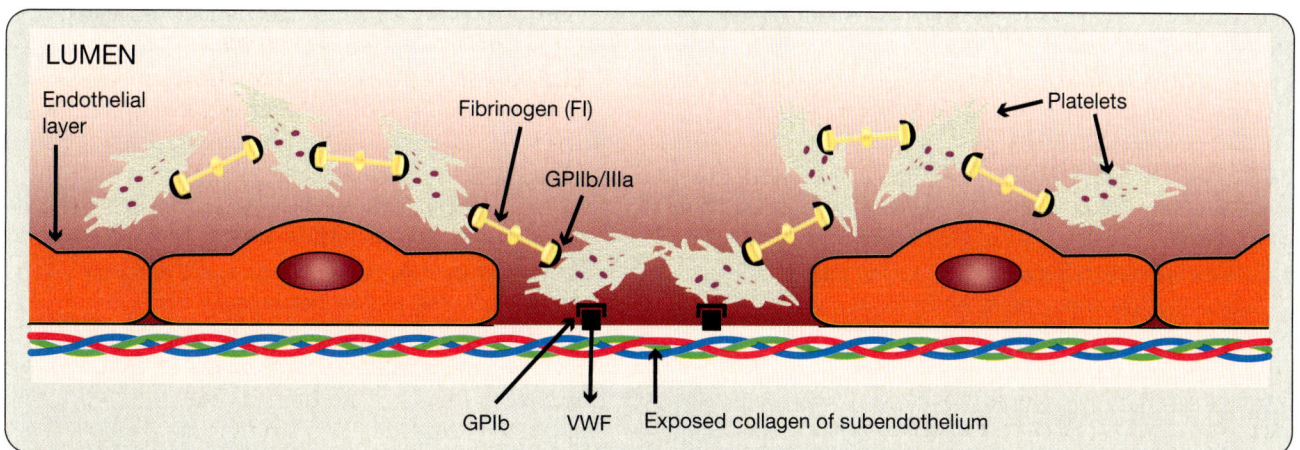

Figure 35.23
Activated platelets undergo calcium (Ca^{2+})-initiated shape change.

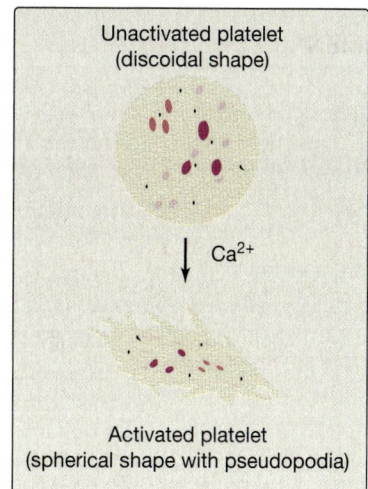

Figure 35.24
Linking of platelets by fibrinogen via the glycoprotein (GP) IIb/IIIa receptor. [Note: The shapes in the fibrinogen molecule represent the two D and one E domains.] GPIb, glycoprotein Ib receptor; VWF, von Willebrand factor.

V. CHAPTER SUMMARY

- **Blood clotting (coagulation)** rapidly stops bleeding from a damaged blood vessel in order to maintain a constant blood volume (**hemostasis**). Coagulation is accomplished through formation of a **clot (thrombus)** consisting of a plug of **platelets** and a meshwork of the protein **fibrin** (Fig. 35.25).

- The formation of the **fibrin meshwork** by the **clotting cascade** involves the **extrinsic** and **intrinsic pathways**, which include protein factors (F) that converge at **FXa** to form the **common pathway**. Many of the protein factors are **serine proteases.**

- **γ-Glutamyl carboxylase** and its coenzyme, the hydroquinone form of **vitamin K**, are required for the formation of γ-carboxyglutamate (Gla) residues in the clotting proteases **FII, FVII, FIX**, and **FX**. Calcium and Gla residues facilitate the binding of these proteins to negatively charged **phosphatidylserine (PS)** at the site of vessel damage and on the surface of platelets.

- In the carboxylase reaction, vitamin K gets oxidized to the nonfunctional epoxide form. **Warfarin**, a synthetic analog of vitamin K used clinically to reduce clotting, inhibits the enzyme **vitamin K–epoxide reductase (VKOR)** that regenerates the functional, reduced vitamin K.

- The extrinsic pathway is initiated by exposure of **FIII (TF)**, an **accessory protein** in vascular subendothelium. Circulating FVII binds to and is activated by TF forming **TF–FVIIa**, a complex that in turn activates FX by proteolysis. FXa allows **thrombin** (FIIa) production by the common pathway. Thrombin then activates components of the intrinsic pathway. The extrinsic pathway is rapidly inhibited by **TF pathway inhibitor (TFPI)**.

- The intrinsic pathway is initiated by **thrombin**, which activates FXI to FXIa. FXIa activates FIX to FIXa. FIXa then combines with FVIIIa, and the complex activates FX. FVIII deficiency results in **hemophilia A**, whereas FIX deficiency results in the less common **hemophilia B**.

- In the common pathway, FXa associates with **FVa** (an accessory protein), forming **prothrombinase** that cleaves **prothrombin (FII)** to **thrombin (FIIa)**. Thrombin then cleaves **fibrinogen (FI)** to **fibrin (FIa)**.

- Fibrin monomers associate, forming a **soluble (soft) fibrin clot,** and are then **cross-linked** by **FXIIIa**, forming an **insoluble (hard)** fibrin clot. The fibrin clot is cleaved (**fibrinolysis**) by the protein **plasmin,** a serine protease that is generated from **plasminogen** by **plasminogen activators** such as **TPA (or t-PA)**. Recombinant TPA is used therapeutically in ischemic stroke.

- The liver and blood vessels produce **anticoagulation** proteins that limit coagulation. **Antithrombin (AT)**, a serine protease inhibitor or **serpin**, is activated by heparan sulfate (or the anticoagulant drug heparin) and binds to and removes thrombin and FXa from the blood. **Protein C** is activated by the **thrombin–thrombomodulin** complex and then forms a complex with **protein S**, producing **activated protein C (APC)**. The APC complex cleaves the accessory proteins FVa and FVIIIa. **FV Leiden** is resistant to APC and causes the most common inherited **thrombophilic** condition in the United States.

- **Platelet plug** formation is initiated when a wound to a tissue damages blood vessels and exposes collagen in the vessel subendothelium to the vessel lumen. Platelets (thrombocytes) adhere to the exposed collagen through an interaction between GPIb on their surface and von Willebrand factor (VWF) that is bound to collagen in the subendothelium. Deficiency of VWF results in von Willebrand disease, the most common inherited coagulopathy.

- Once adhered, platelets are activated and then aggregate at the damaged site. Activation involves changes in shape and degranulation, the process by which platelets release the contents of their storage granules. Thrombin is the most potent activator of platelets.

- Activated platelets release substances that cause vasoconstriction, recruit and activate other platelets, and support the formation of a fibrin clot. Structural changes in the surface receptor GPIIb/IIIa expose binding sites for fibrinogen that links activated platelets together to make the initial loose plug of platelets (primary hemostasis).

- Fibrinogen is converted to fibrin by thrombin. Fibrin is cross-linked by FXIIIa coming both from the blood and from platelets. This strengthens the fibrin meshwork and stabilizes the platelet plug (secondary hemostasis).

- Disorders of platelets and coagulation proteins can impair the ability to clot. Prothrombin time and activated partial thromboplastin time are clinical laboratory tests used to evaluate the clotting cascade.

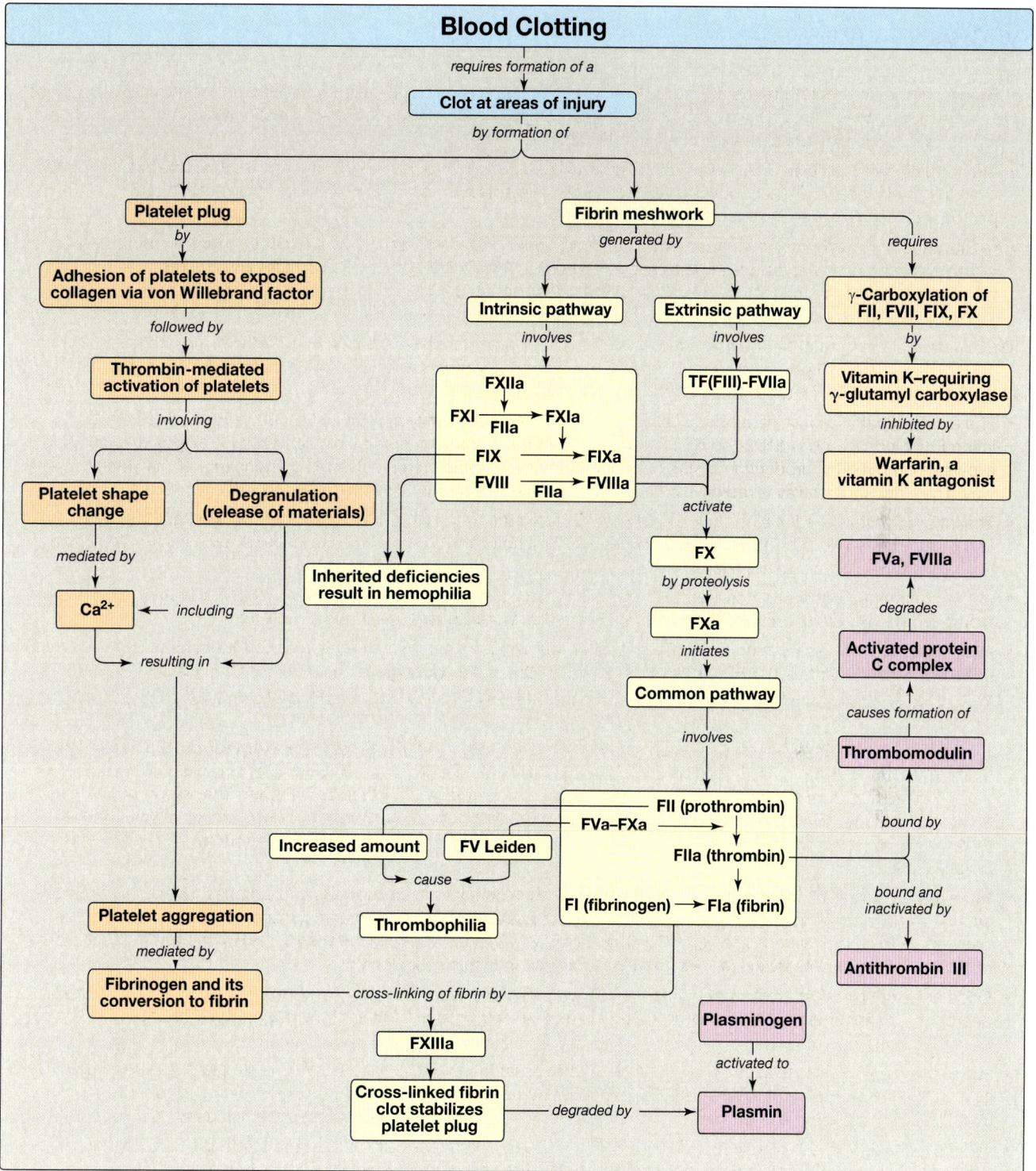

Blood Clotting

requires formation of a

Clot at areas of injury

by formation of

Platelet plug

by

Adhesion of platelets to exposed collagen via von Willebrand factor

followed by

Thrombin-mediated activation of platelets

involving

Platelet shape change

mediated by

Ca²⁺ ← *including* ← **Degranulation (release of materials)**

resulting in

Platelet aggregation

mediated by

Fibrinogen and its conversion to fibrin — *cross-linking of fibrin by* —

Fibrin meshwork

generated by

Intrinsic pathway

involves

FXIIa
FXI —FIIa→ FXIa
FIX → FIXa
FVIII —FIIa→ FVIIIa

Inherited deficiencies result in hemophilia

Extrinsic pathway

involves

TF(FIII)-FVIIa

activate

FX

by proteolysis

FXa

initiates

Common pathway

involves

FII (prothrombin)
FVa–FXa
FIIa (thrombin)
FI (fibrinogen) → FIa (fibrin)

Increased amount   **FV Leiden** → *cause* ←

Thrombophilia

FXIIIa

Cross-linked fibrin clot stabilizes platelet plug — *degraded by* → **Plasmin**

Plasminogen

activated to

Plasmin

requires

γ-Carboxylation of FII, FVII, FIX, FX

by

Vitamin K–requiring γ-glutamyl carboxylase

inhibited by

Warfarin, a vitamin K antagonist

FVa, FVIIIa

degrades

Activated protein C complex

causes formation of

Thrombomodulin

bound by

bound and inactivated by

Antithrombin III

Figure 35.25
Key concept map for blood clotting. a, active; F, factor; Ca²⁺, calcium.

Study Questions

Choose the ONE best answer.

For Questions 35.1–35.5, match the most appropriate protein factors (F) of clotting to the description.

A. FI   F. FVIII
B. FII   G. FIX
C. FIII   H. FX
D. FV   I. FXI
E. FVII   J. FXIII

35.1. This factor activates components of the intrinsic, extrinsic, and common pathways.

35.2. This factor converts the soluble clot to an insoluble clot.

35.3. This factor initiates the common pathway.

35.4. This factor is an accessory protein that potentiates the activity of factor Xa.

35.5. This factor is a γ-carboxyglutamate–containing serine protease of the extrinsic pathway.

Correct answers = B, J, H, D, E. Thrombin (FII) is formed in the common pathway and activates components in each of the three pathways of the clotting cascade. FXIII, a transglutaminase, covalently cross-links associated fibrin monomers, thereby converting a soluble clot to an insoluble one. The generation of FXa by the intrinsic and extrinsic pathways initiates the common pathway. FV increases the activity of FXa. It is one of three accessory (nonprotease) proteins. The others are FIII (TF) and FVIII (complexes with FIX to activate FX). FVII is a γ-carboxyglutamate–containing serine protease that complexes with FIII in the extrinsic pathway.

35.6. In which patient would prothrombin time (PT) be unaffected and activated partial thromboplastin time (aPTT) be prolonged?

A. Patient on aspirin therapy
B. Patient with end-stage liver disease
C. Patient with hemophilia
D. Patient with thrombocytopenia

Correct answer = C. PT measures the activity of the extrinsic through the common pathways, and aPTT measures the activity of the intrinsic through the common pathways. Patients with hemophilia are deficient in either FVIII (hemophilia A) or FIX (hemophilia B), components of the common pathway. They have an intact extrinsic pathway. Therefore, the PT is unaffected, and the aPTT is prolonged. Patients on aspirin therapy and those with thrombocytopenia have alterations in platelet function and number, respectively, and not in the proteins of the clotting cascade. Therefore, both the PT and the aPTT are unaffected. Patients with end-stage liver disease have decreased ability to synthesize clotting proteins. They show prolonged PT and aPTT.

35.7. Which of the following can be ruled out in a patient with thrombophilia?

A. Deficiency of antithrombin
B. Deficiency of FIX
C. Deficiency of protein C
D. Excess of prothrombin
E. Expression of FV Leiden

Correct answer = B. Symptomatic deficiencies in clotting factors will present with a decreased ability to clot (coagulopathy). Thrombophilia, however, is characterized by an increased tendency to clot. Choices A, C, D, and E result in thrombophilia.

35.8. Current guidelines for the treatment of patients with acute ischemic stroke (a stroke caused by a blood clot obstructing a vessel that supplies blood to the brain) include the recommendation that tissue plasminogen activator (TPA) be used shortly after the onset of symptoms. The basis of the recommendation for TPA is that it activates:

A. antithrombin.
B. the activated protein C complex.
C. the receptor for von Willebrand factor.
D. the serine protease that degrades fibrin.
E. thrombomodulin.

Correct answer = D. TPA converts plasminogen to plasmin. Plasmin (a serine protease) degrades the fibrin meshwork, removing the obstruction to blood flow. Antithrombin III in association with heparin binds thrombin and carries it to the liver, decreasing thrombin's availability in the blood. The activated protein C complex degrades the accessory proteins FV and FVIII. The platelet receptor for von Willebrand factor is not affected by TPA. Thrombomodulin binds thrombin and converts it from a protein of coagulation to one of anticoagulation by decreasing its activation of fibrinogen and increasing its activation of protein C.

35.9. The adhesion, activation, and aggregation of platelets provide the initial plug at the site of vessel injury. Which of the following statements concerning the formation of this platelet plug is correct?

A. Activated platelets undergo a shape change that decreases their surface area.
B. Formation of a platelet plug is prevented in intact vessels by the production of thromboxane A_2 by endothelial cells.
C. The activation phase requires production of cyclic adenosine monophosphate.
D. The adhesion phase is mediated by the binding of platelets to von Willebrand factor via glycoprotein Ib.
E. Thrombin activates platelets by binding to a protease-activated G protein–coupled receptor and causing activation of protein kinase A.

Correct answer = D. The adhesion phase of platelet plug formation is initiated by the binding of von Willebrand factor to a receptor (glycoprotein Ib) on the surface of platelets. A shape change from discoidal to spherical with pseudopodia increases the surface area of platelets. Thromboxane A_2 is made by platelets. It causes platelet activation and vasoconstriction. Adenosine diphosphate is released from activated platelets, and it itself activates platelets. Thrombin works primarily through receptors coupled to G_q proteins causing activation of phospholipase C.

35.10. Nephrotic syndrome is a kidney disease characterized by protein loss in the urine ($\geq$3 g/day) that is accompanied by edema. The loss of protein results in a hypercoagulable state. Excretion of which of the following proteins would explain the thrombophilia seen in the syndrome?

A. Antithrombin
B. FV
C. FVIII
D. Prothrombin

Correct answer = A. Antithrombin III (ATIII) inhibits the action of thrombin (FIIa), a Gla-containing protein of clotting that activates the extrinsic, intrinsic, and common pathways. Excretion of ATIII in nephrotic syndrome allows the actions of FIIa to continue, resulting in a hypercoagulable state. The other choices are proteins required for clotting. Their excretion in the urine would decrease clotting.

35.11. Blocking the action of which of the following proteins would be a rational therapy for hemophilia B?

A. FIX
B. FXIII
C. Protein C
D. Tissue factor pathway inhibitor

Correct answer = D. Hemophilia B is a coagulopathy caused by decreased thrombin production by the common pathway as a result of a deficiency in FIX of the intrinsic pathway. Because the extrinsic pathway also can result in thrombin production, blocking the inhibitor of this pathway (tissue factor pathway inhibitor) should, in principle, increase thrombin production.

35.12. The parents of a newborn are unsure whether to allow the injection of vitamin K that is recommended shortly after birth to prevent vitamin K deficiency bleeding, which is caused by the low levels of the vitamin in newborns. The activity of which of the following protein factors involved in clotting would be decreased in this patient if she does not receive the injection?

A. FV
B. FVII
C. FXI
D. FXIII

> Correct answer = B. FVII is a γ-carboxyglutamate (Gla)-containing protein of clotting. The creation of Gla residues by γ-glutamyl carboxylase requires vitamin K as a coenzyme. FII, FIX, and FX, as well as proteins C and S that limit clotting, also contain Gla residues. The other choices do not contain Gla residues.

35.13. Thrombin, produced in the common pathway of clotting, has both procoagulant and anticoagulant activities. Which of the following is an anticoagulant activity of thrombin?

A. Activating FXIII
B. Binding to thrombomodulin
C. Increasing nitric oxide production
D. Inhibiting FV and FVIII
E. Inhibiting platelet activation

> Correct answer = B. Thrombin bound to thrombomodulin activates protein C that degrades the accessory proteins FV and FVIII, thereby inhibiting clotting. Activation of FXIII by thrombin strengthens the fibrin clot. Nitric oxide, a vasodilator made by endothelial cells, decreases clot formation. It is not affected by thrombin. Thrombin is a powerful activator of platelets.

35.14. Which of the following pattern of results would be expected for a patient with a deficiency in FXIII?

A. Both prothrombin time and activated partial thromboplastin time are decreased.
B. Both prothrombin time and activated partial thromboplastin time are increased.
C. Both prothrombin time and activated partial thromboplastin time are unchanged.
D. Only prothrombin time is affected.
E. Only activated partial thromboplastin time is affected.

> Correct answer = C. FXIII is a transglutaminase that cross-links fibrin molecules in a soft clot to form a hard clot. Its deficiency does not affect the PT or aPTT tests. [Note: It is evaluated by a clot solubility test.]

35.15. Why do individuals with Scott syndrome, a rare disorder caused by mutations to scramblase in platelets, have a tendency to bleed?

> Scramblase moves phosphatidylserine (PS) from the cytosolic leaflet to the extracellular leaflet in the plasma membrane of platelets. This disrupts the asymmetrical localization of membrane phospholipids created by ATP-dependent flippases (move PS from extracellular to cytosolic leaflet) and flippases (move phosphatidylcholine [PC] in the opposite direction). Having PS on the outer face of platelet membranes provides a site for protein clotting factors to interact and activate thrombin. If scramblase is inactive, PS is not available to these factors, and bleeding results.

35.16. Several days after having had their home treated for an infestation of rats, the parents of a 3-year-old female become concerned that she might be ingesting the poison-containing pellets. After calling the Poison Hotline, they take her to the emergency department. Blood studies reveal a prolonged prothrombin and activated partial thromboplastin time and a decreased concentration of thrombin, FVII, FIX, and FX. Why might administration of vitamin K be a rational approach to the treatment of this patient?

> Many rodent poisons are super warfarins, drugs that have a long half-life in the body. Warfarin inhibits γ-carboxylation (production of γ-carboxyglutamate, or Gla, residues), and the clotting proteins reported as decreased are the Gla-containing proteases of the clotting cascade. [Note: Proteins C and S of anticlotting are also Gla-containing proteins.] Because warfarin functions as a vitamin K antagonist, administration of vitamin K is a rational approach to treatment.

APPENDIX
Clinical Cases

I. INTEGRATIVE CASES

Metabolic pathways, initially presented in isolation, are, in fact, linked to form an interconnected network. The following four integrative case studies illustrate how a perturbation in one process can result in perturbations in other processes of the network.

CASE 1: CHEST PAIN

Patient Presentation: A 35-year-old male with severe substernal chest pain of ~2 hours' duration is brought by ambulance to his local hospital at 5 AM. The pain is accompanied by dyspnea (shortness of breath), diaphoresis (sweating), and nausea.

Focused History: The patient reports episodes of exertional chest pain in the last few months, but they were less severe and of short duration. He smokes (2 to 3 packs per day), drinks alcohol only rarely, eats a "typical" diet, and walks with his wife most weekends. His blood pressure has been normal. Family history reveals that his father and paternal aunt died of heart disease at ages 45 and 39 years, respectively. His mother and younger (age 31 years) brother are said to be in good health.

Physical Examination (Pertinent Findings): He is pale and clammy and is in distress due to chest pain. Blood pressure and respiratory rate are elevated. Corneal arcus (lipid deposits on the periphery of his corneas; see left image) and xanthelasmas (under the skin on and around his eyelids; see right image) are noted. No xanthomas (deposits on his tendons) are detected.

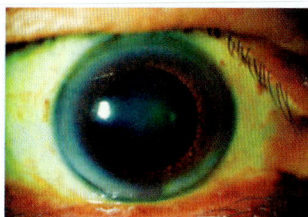

Corneal arcus   **Xanthelasmas**

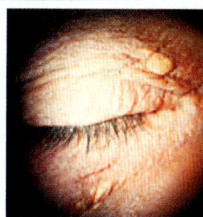

Pertinent Test Results: The patient's electrocardiogram is consistent with an acute myocardial infarction (MI). Angiography reveals areas of severe stenosis (narrowing) of several coronary arteries. Initial results from the clinical laboratory include the following:

	Patient	Reference Range
Troponin	0.5 ng/mL	<0.04 ng/mL
Total cholesterol	365 mg/dL (**H**)	<200
Low-density lipoprotein (LDL) cholesterol	304 mg/dL (**H**)	<130
High-density lipoprotein (HDL) cholesterol	38 mg/dL (**L**)	>45
Triglycerides (triacylglycerols)	115 mg/dL	<150

H, high; **L,** low. [Note: the patient had not eaten since the night before he was brought to the emergency department and prior to the blood draw.]

Diagnosis: Acute MI, the irreversible necrosis (death) of heart muscle secondary to ischemia (decreased blood supply), is caused by the occlusion (blockage) of a blood vessel most commonly by a blood clot (thrombus). The patient subsequently is determined to have heterozygous familial hypercholesterolemia (FH), also known as type IIa hyperlipidemia.

Immediate Treatment: He is given O_2, a vasodilator, and pain medication and undergoes a procedure to place stents to reestablish perfusion (restore blood flow to the heart).

Long-Term Treatment: Lipid-lowering drugs such as statins, daily aspirin, and counseling on nutrition, exercise, and smoking cessation would likely be part of the long-term treatment plan.

Prognosis: Patients with heterozygous FH have ~50% of the normal numbers of functional LDL receptors and hypercholesterolemia (two to three times normal) that puts them at high risk (>50% risk) for premature coronary heart disease (CHD). However, fewer than 5% of patients with hypercholesterolemia actually have FH.

Nutrition Nugget: Dietary recommendations for individuals with heterozygous FH include limiting saturated fats to <7% of total calories and cholesterol to <200 mg/day, substituting unsaturated fats for saturated fats, and adding soluble fiber (10 to 20 g/day) and plant sterols (2 g/day) for their hypocholesterolemic effects. Fiber increases cholesterol and bile acid (BA) excretion. This results in increased hepatic uptake of cholesterol-rich LDL to supply the substrate for BA synthesis. Phytosterols (plant sterols) decrease cholesterol absorption in the intestine.

Genetics Gem: FH is caused by hundreds of different mutations in the gene for the LDL receptor (on chromosome 19) that affect receptor expression and/or function. FH is an autosomal-dominant disease in which homozygotes are more seriously affected than heterozygotes. Heterozygous FH has an incidence of ~1:500 in the general population. It is associated with increased risk of cardiovascular disease. Genetic screening of the first-degree relatives of this patient would identify affected individuals for treatment.

REVIEW QUESTIONS: Choose the ONE best answer

RQ1. Triacylglycerols are glycerol-based lipids. Which of the following is also a glycerol-based lipid?
   A. Ganglioside GM_2
   B. Phosphatidylcholine
   C. Prostaglandin PGI_2
   D. Sphingomyelin
   E. Vitamin D

RQ2. Which of the following explains why statins are of benefit to patients with hypercholesterolemia?
   A. Decrease a rate-limiting and regulated step of *de novo* cholesterol biosynthesis
   B. Decrease expression of the gene for the LDL receptor
   C. Increase oxidation of cholesterol to $CO_2 + H_2O$
   D. Increase absorption of bile salts in the enterohepatic circulation
   E. Decrease cholesterol by increasing steroid hormone and vitamin D synthesis

RQ3. Statins are competitive inhibitors of hydroxymethylglutaryl coenzyme A reductase. Which of the following statements about the mechanism of action of statins is correct?
   A. Statins function as irreversible inhibitors.
   B. Statins cause an increase in both the apparent K_m and the V_{max}.
   C. Statins increase the apparent K_m and have no effect on the V_{max}.
   D. Statins decrease both the apparent K_m and the apparent V_{max}.
   E. Statins have no effect on the K_m and decrease the apparent V_{max}.

RQ4. Decreased tissue perfusion results in hypoxia (decreased O_2 availability). Which of the following is correct about hypoxia relative to normoxia?
   A. Electron transport is upregulated to provide protons for ATP synthesis.
   B. Ratio of NAD^+ to NADH increases.
   C. Pyruvate dehydrogenase complex is active.
   D. Process of substrate-level phosphorylation is increased in the cytosol.
   E. Tricarboxylic acid cycle is upregulated.

THOUGHT QUESTIONS

TQ1. Relative to an individual with familial defective LDL receptors, what would be the expected phenotype in an individual with familial defective apolipoprotein B-100? What would be the expected phenotype in an individual with apolipoprotein E4, the isoform that only poorly binds its receptor?

TQ2. Why was low-dose aspirin prescribed? **Hint:** What product of arachidonic acid metabolism is inhibited by aspirin?

TQ3. Heart muscle normally uses aerobic metabolism to meet its energy needs. However, in hypoxia, anaerobic glycolysis is increased. What allosteric activator of glycolysis is responsible for this effect? With hypoxia, what will be the end product of glycolysis?

TQ4. One of the reasons for encouraging smoking cessation and exercise for this patient is that these changes raise the level of HDL, and elevated HDL reduces the risk for CHD. How does a rise in HDL reduce the risk for CHD?

CASE 2: SEVERE FASTING HYPOGLYCEMIA

Patient Presentation: The patient is a 4-month-old male whose mother is concerned about the "twitching" movements he makes just before feedings. She tells the pediatrician that the movements started ~1 week ago, are most apparent in the morning, and disappear shortly after eating.

Focused History: The child was born at full term following a normal pregnancy and delivery. He appeared normal at birth. He has been at the 30th percentile for both weight and length since birth. His immunizations are up to date. He last ate a few hours ago.

Physical Examination (Pertinent Findings): The child appears sleepy and feels clammy to the touch. His respiratory rate is elevated. His temperature is normal. He has a protuberant, firm abdomen that appears to be nontender. His liver is palpable 4 cm below the right costal margin and is smooth.

Pertinent Test Results:

	Patient	Pediatric Reference Range
Glucose	50 mg/dL (**L**)	60–105
Lactate	3.4 mmol/L (**H**)	0.6–3.2
Uric acid	5.6 mg/dL (**H**)	2.4–5.4
Total cholesterol	220 mg/dL (**H**)	<170
Triglycerides (triacylglycerols)	280 mg/dL (**H**)	<90
pH	7.30 (**L**)	7.35–7.45
HCO_3^-	12 mEq/L (**L**)	19–25

H, high; **L,** low.

The child is sent to the regional children's hospital for further evaluation. Ultrasound studies confirm hepatomegaly and enlarged, symmetrical kidneys but no evidence of tumors. A liver biopsy is performed. The hepatocytes are enlarged. Staining reveals large amounts of lipid (primarily triacylglycerol) and carbohydrate. Liver glycogen is elevated in amount and normal in structure. Enzyme assay using liver homogenate treated with detergent reveals <10% of the normal activity of glucose 6-phosphatase, an enzyme of the endoplasmic reticular (ER) membrane in the liver and the kidneys.

Diagnosis: This child has glucose 6-phosphatase deficiency (glycogen storage disease [GSD] type Ia, von Gierke disease).

Treatment (Immediate): He was given glucose intravenously (IV), and his blood glucose level rose into the normal range. However, as the day progressed, it fell to well below normal. Administration of glucagon had no effect on blood glucose levels but increased blood lactate. His blood glucose levels were able to be maintained only by constant infusion of glucose.

Prognosis: Individuals with glucose 6-phosphatase deficiency develop hepatic adenomas starting in the second decade of life and are at increased risk for hepatic carcinoma. Kidney involvement can cause impaired tubular function resulting in acidosis, and glomerular function may also be impaired and can result in chronic kidney disease. Patients are at increased risk for developing gout, but this rarely occurs before puberty.

Nutrition Nugget: Long-term medical nutrition therapy for this child is designed to maintain his blood glucose levels in the normal range. Frequent (every 2 to 3 hours) daytime feedings that are rich in carbohydrate (provided by uncooked cornstarch that is slowly hydrolyzed) and nighttime nasogastric infusion of glucose are advised. Avoidance of fructose and galactose is recommended because they are metabolized to glycolytic intermediates and lactate, which can exacerbate the metabolic problems. Calcium and vitamin D supplements are prescribed.

Genetics Gem: GSD Ia is an autosomal-recessive disorder caused by >100 known mutations to the gene for glucose 6-phosphatase located on chromosome 17. It has an incidence of 1:100,000 and accounts for ~25% of all cases of GSD in the United States. It is one of the few genetic causes of hypoglycemia in newborns. GSD Ia is not routinely screened for in newborns. [Note: Deficiency of the translocase that moves glucose 6-phosphate into the ER is the cause of GSD Ib. Hypoglycemia and neutropenia are seen.]

REVIEW QUESTIONS: Choose the ONE best answer

RQ1. Which of the following explains hypoglycemia in the patient?
A. Unphosphorylated glucose cannot be produced by glycogenolysis or gluconeogenesis.
B. Glycogen phosphorylase is dephosphorylated and inactive; glycogen cannot be degraded.
C. Hormone-sensitive lipase is inactive; substrates for gluconeogenesis cannot be generated.
D. Decreased insulin/glucagon ratio upregulates glucose transporters in liver and kidneys.

RQ2. The patient was prescribed calcium supplements because chronic acidosis can cause bone demineralization, resulting in osteopenia. Which of the following explains why vitamin D was also prescribed?
A. Binds G_q protein–coupled membrane receptors and causes a rise in inositol trisphosphate
B. Cannot be synthesized by humans and, therefore, must be supplied in the diet
C. Is a fat-soluble vitamin that increases intestinal absorption of calcium
D. Acts as the coenzyme-prosthetic group for calbindin, a calcium transporter in the intestine

RQ3. The hepatomegaly and renomegaly seen in this child are primarily the result of an increase in the amount of glycogen stored in these organs. What is the basis for glycogen accumulation in these organs?
A. Glycolysis is downregulated, which pushes glucose to glycogenesis.
B. Increased oxidation of fatty acids spares glucose for glycogenesis.
C. Glucose 6-phosphate is an allosteric activator of glycogen synthase b.
D. The rise in the insulin/glucagon ratio favors glycogenesis.

RQ4. Glucose 6-phosphatase is an integral protein of the ER membrane. Which of the following statements about such proteins is correct?
A. If glycosylated, the carbohydrate portion of the protein extends into the cytosol.
B. They are synthesized on ribosomes that are free in the cytosol.
C. The membrane-spanning domain consists of hydrophilic amino acids.
D. The initial targeting signal is an amino-terminal hydrophobic signal sequence.

THOUGHT QUESTIONS

TQ1. What is the likely reason for the patient's twitching movements?
TQ2. Why was the liver homogenate treated with detergent? **Hint:** Think about where the enzyme is located.
TQ3. Why is this patient's blood glucose level unaffected by glucagon? **Hint:** What is the role of glucagon in normal individuals who experience a drop in blood glucose?
TQ4. Why are urate and lactate elevated in a disorder of glycogen metabolism? **Hint:** It is the result of a decrease in inorganic phosphate (P_i), but why is P_i decreased?
TQ5. Why are triacylglycerols and cholesterol elevated? **Hint:** Glucose is the primary carbon source for their synthesis. Why are ketone bodies not elevated?

CASE 3: HYPERGLYCEMIA AND HYPERKETONEMIA

Patient Presentation: A 40-year-old female was brought to the Emergency Department in a disoriented, confused state by her husband.

Focused History: Her husband reveals that the patient has had type 1 diabetes (T1D) for the last 24 years and this is her first medical emergency in 2 years.

Physical Examination (Pertinent Findings): The patient displayed signs of dehydration including dry mucous membranes and skin, poor skin turgor, and low blood pressure. She also had signs of acidosis such as deep, rapid breathing (Kussmaul respiration). Her breath had a faintly fruity odor. Her temperature was normal.

Pertinent Test Results: Results of blood tests performed by the clinical laboratory are shown below:

	Patient	Reference Range
Glucose	414 mg/dL (23 mmol/L) (**H**)	70–99 (3.9–5.5)
Blood urea nitrogen	8 mmol/L (**H**)	2.5–6.4
3-Hydroxybutyrate	350 mg/dL (**H**)	0–3
HCO_3^-	12 mmol/L (**L**)	22–28
Na^+	136 mmol/L	138–150
K^+	5.3 mmol/L	3.5–5.0
Cl^-	102 mmol/L	95–105
pH	7.1 (**L**)	7.35–7.45

H, high; **L,** low.

Microscopic examination of her urine revealed white blood cells, suspicious for a urinary tract infection (UTI), which was later confirmed by urine culture.

Diagnosis: This patient is in diabetic ketoacidosis (DKA) precipitated by a UTI. [Note: Diabetes increases the risk for infections such as UTI.]

Immediate Treatment: She was administered insulin. Rehydration was initiated with normal saline given IV. Blood glucose, ketone bodies, and electrolytes were measured periodically. Antibiotic treatment for her UTI was started.

Long-Term Treatment: Diabetes increases the risk for macrovascular complications including coronary artery disease and stroke and microvascular complications such as retinopathy, nephropathy, and neuropathy. Ongoing monitoring for these complications will be continued.

Prognosis: Diabetes is the seventh leading cause of death by disease in the United States. Individuals with diabetes have a reduced life expectancy relative to those without diabetes.

Nutrition Nugget: Monitoring total intake of carbohydrates is primary in blood glucose control. Carbohydrates should come from whole grains, vegetables, legumes, and fruits. Low-fat dairy products and nuts and fish rich in ω-3 fatty acids are encouraged. Intake of saturated and trans fats should be minimized.	**Genetics Gem:** Autoimmune destruction of pancreatic β-cells is characteristic of T1D. Of the genetic loci that confer risk for T1D, the human-leukocyte antigen (HLA) region on chromosome 6 has the strongest association. The majority of genes in the HLA region are involved in the immune response.

REVIEW QUESTIONS: Choose the ONE best answer

RQ1. Which of the following statements concerning T1D is correct?
A. Diagnosis can be made by measuring fasting blood glucose or glycated hemoglobin (HbA_{1c}).
B. During periods of stress, a patient's urine will likely be negative for reducing sugars.
C. T1D is associated with obesity and a sedentary lifestyle.
D. Characteristic metabolic abnormalities of T1D result from insensitivity to insulin.
E. Treatment with exogenous insulin allows normalization of blood glucose.

RQ2. Which of the following regarding ketone bodies is true?
A. They are made from acetyl coenzyme A (CoA) primarily produced by the oxidation of glucose.
B. They are utilized by many tissues, particularly the liver, after conversion to acetyl CoA.
C. They include acetoacetate, which imparts a fruity odor to the breath.
D. They require albumin for transport through the blood.
E. They are utilized in energy metabolism as organic acids that can add to the proton load of the body.

RQ3. Adipose lipolysis followed by β-oxidation of fatty acid (FA) products is required for the generation of ketone bodies. Which of the following statements concerning the generation and use of FA is correct?
  A. Mitochondrial β-oxidation of FA is inhibited by malonyl CoA.
  B. Production of FA from adipose lipolysis is upregulated by insulin.
  C. The acetyl CoA product of FA β-oxidation inhibits the use of pyruvate for gluconeogenesis.
  D. The β-oxidation of FA utilizes reducing equivalents generated by gluconeogenesis.
  E. The FAs produced by lipolysis are taken up by the brain and oxidized for energy.

THOUGHT QUESTIONS

TQ1. At admission, the patient was hypoinsulinemic, and she was given insulin. Why did her hypoinsulinemia result in hyperglycemia? **Hint:** What is the role of insulin in glucose metabolism?

TQ2. Why is there glucose in her urine (glucosuria)? How is the glucosuria related to her dehydrated state?

TQ3. Why is the majority of the acetyl CoA from FA β-oxidation being used for ketogenesis rather than being oxidized in the tricarboxylic acid cycle?

TQ4. Was she in positive or negative nitrogen balance when she was brought to the hospital?

TQ5. What response to the DKA is apparent in this patient? What response is likely occurring in the kidney? **Hint:** In addition to conversion to urea, how is toxic ammonia removed from the body?

TQ6. What would be true about the levels of ketone bodies and glucose during periods of physiologic stress in individuals with impaired FA oxidation?

CASE 4: HYPOGLYCEMIA, HYPERKETONEMIA, AND LIVER DYSFUNCTION

Patient Presentation: A 59-year-old male with slurred speech, ataxia (loss of skeletal muscle coordination), and abdominal pain was brought to the Emergency Department.

Focused History: This patient is known to the Emergency Department staff from previous visits. He has a 6-year history of chronic, excessive alcohol consumption. He is not known to take illicit drugs. At this visit, the patient reports that he has been drinking heavily in the past day or so. He cannot recall having eaten anything in that time but admits to vomiting, without evidence of recent bleeding.

Physical Examination (Pertinent Findings): The physical examination was remarkable for the patient's emaciated appearance. (His body mass index was later determined to be 17.5, which put him in the underweight category.) His facial cheeks were erythematous (red in color) due to dilated blood vessels in the skin (telangiectasia). Eye movement was normal. Neither icterus (jaundice) nor edema (swelling due to fluid retention) were seen. The liver was slightly enlarged. Bedside tests revealed hypoglycemia and hyperketonemia (as acetoacetate). Blood was drawn and sent to the clinical laboratory.

Pertinent Test Results:

	Patient	Reference Range
Ethanol	180 mg/dL (**H**)	(>80 considered positive for DUI)
Glucose	58 mg/dL (**L**)	70–99
Lactate	23 mg/dL (**H**)	5–15
Uric acid	7.0 mg/dL	2.5–8.0
3-Hydroxybutyrate	50 mg/dL (**H**)	0–3.0
Total bilirubin	1.5 mg/dL (**H**)	0.3–1.0
Direct (conjugated) bilirubin	0.5 mg/dL (**H**)	0.1–0.3
Albumin	3.0 g/dL (**L**)	3.5–5.8
Aspartate transaminase (AST)	130 U/L (**H**)	0–35
Alanine transaminase (ALT)	75 U/L (**H**)	0–35
Prothrombin time	15.5 s (**H**)	11.0–13.2

DUI, driving under the influence; **H,** high; **L,** low.

Additional Tests: Complete blood count (CBC) and blood smear revealed a macrocytic anemia (see right image). Folate and B_{12} levels were ordered (AK).

Diagnosis: The patient has alcohol use disorder and alcoholic ketoacidosis.

Treatment (Immediate): Thiamine and glucose were given IV.

Prognosis: Alcohol dependence is the third most common cause of preventable death in the United States. People with alcohol use disorder are at increased risk for vitamin deficiencies, liver cirrhosis, pancreatitis, gastrointestinal bleeding, and some cancers.

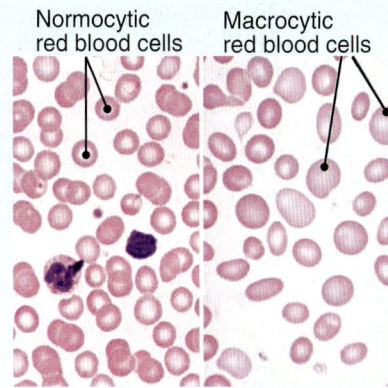

Normocytic red blood cells   Macrocytic red blood cells

Nutrition Nugget: Those with alcohol use disorder are at risk for vitamin deficiencies as a result of decreased intake and absorption. Thiamine (vitamin B_1) deficiency is common and can have serious consequences such as Wernicke–Korsakoff syndrome with its neurologic effects. Thiamine pyrophosphate (TPP), the coenzyme form, is required for the dehydrogenase-mediated oxidation of α-keto acids (such as pyruvate) as well as the transfer of two-carbon ketol groups by transketolase in the reversible sugar interconversions in the pentose phosphate pathway.

Genetics Gem: Acetaldehyde, the product of ethanol oxidation by the hepatic, cytosolic, nicotinamide adenine dinucleotide (NAD^+)-requiring enzyme alcohol dehydrogenase (ADH), is oxidized to acetate by the mitochondrial, NAD^+-requiring aldehyde dehydrogenase (ALDH2). Individuals of East Asian heritage often have a single nucleotide polymorphism (SNP) that renders ALDH2 essentially inactive. This results in aldehyde-induced facial flushing and mild to moderate intoxication after consumption of small amounts of ethanol.

REVIEW QUESTIONS: Choose the ONE best answer

RQ1. Many of the metabolic consequences of chronic excessive alcohol consumption seen in this patient are the result of an increase in the ratio of reduced nicotinamide adenine dinucleotide (NADH) to its oxidized form (NAD^+) in both the cytoplasm and mitochondria. Which of the following statements concerning the effects of the rise in mitochondrial NADH is correct?

A. Fatty acid oxidation is increased.

B. Gluconeogenesis is increased.

C. Lipolysis is inhibited.

D. The tricarboxylic acid cycle is inhibited.

E. The reduction of malate to oxaloacetate in the malate–aspartate shuttle is increased.

RQ2. Ethanol can also be oxidized by cytochrome P450 (CYP) enzymes, and CYP2E1 is an important example. CYP2E1, which is ethanol inducible, generates reactive oxygen species (ROS) in its metabolism of ethanol. Which of the following statements concerning the CYP proteins is correct?

A. CYP proteins are heme-containing dioxygenases.

B. CYP proteins of the inner mitochondrial membrane are involved in detoxification reactions.

C. CYP proteins of the smooth endoplasmic reticular membrane are involved in the synthesis of steroid hormones, bile acids, and calcitriol.

D. ROS such as hydrogen peroxide generated by CYP2E1 can be oxidized by glutathione peroxidase.

E. The pentose phosphate pathway is an important source of NADPH that provides the reducing equivalents needed for the regeneration of functional glutathione.

RQ3. Alcohol is known to modulate the levels of serotonin in the central nervous system, where the monoamine functions as a neurotransmitter. Which of the following statements about serotonin is correct?

A. It is associated with anxiety and depression.

B. It is degraded via methylation by monoamine oxidase.

C. It is released by activated platelets.

D. It is synthesized from tyrosine.

RQ4. Chronic, excessive consumption of alcohol is a leading cause of acute pancreatitis, a painful inflammatory condition that results from autodigestion of the gland by premature activation of pancreatic enzymes. Which of the following statements concerning the pancreas is correct?

    A. Autodigestion of the pancreas would be expected to result in a decrease in pancreatic proteins in the blood.

    B. In individuals who progress from acute to chronic pancreatitis, diabetes, and steatorrhea are expected findings.

    C. In response to secretin, the exocrine pancreas secretes protons to lower the pH in the intestinal lumen.

    D. Pancreatitis may also be seen in individuals with hypercholesterolemia.

THOUGHT QUESTIONS

TQ1. A. What effect does the rise in cytosolic NADH seen with ethanol metabolism have on glycolysis? **Hint:** What coenzyme is required in glycolysis?

    B. How does this relate to the fatty liver (hepatic steatosis) commonly seen in alcohol-dependent individuals?

TQ2. Why might individuals with a history of gouty attacks be advised to reduce their consumption of ethanol?

TQ3. Why might prothrombin time be affected in alcohol-dependent individuals?

TQ4. Folate and vitamin B_{12} deficiencies cause macrocytic anemia that may be seen in those with alcoholism. Why is it advisable to measure vitamin B_{12} levels before supplementing with folate in an individual with macrocytic anemia?

II. INTEGRATIVE CASE ANSWERS

CASE 1: Answers to Review Questions

RQ1. **Answer = B.** Phosphatidylcholine is a glycerol-based phospholipid derived from diacylglycerol phosphate (phosphatidic acid) and cytidine diphosphate-choline. Gangliosides are derived from ceramides, lipids with a sphingosine backbone. Prostaglandins (PGs) of the 2 series (such as PGI_2) are derived from the 20-carbon polyunsaturated FA arachidonic acid. Sphingomyelin is a sphingophospholipid derived from ceramide. Vitamin D is derived from an intermediate in the biosynthetic pathway for sterol cholesterol.

RQ2. **Answer = A.** Statins inhibit hydroxymethylglutaryl coenzyme A (HMG CoA) reductase, thereby preventing the nicotinamide adenine dinucleotide phosphate (NADPH)-dependent reduction of HMG CoA to mevalonate and decreasing cholesterol biosynthesis (see figure on next page). The decrease in cholesterol content caused by statins results in movement of the sterol regulatory element–binding protein-2 (SREBP-2) in complex with SREBP cleavage–activating protein (SCAP) from the endoplasmic reticular membrane to the Golgi membrane. SREBP-2 is cleaved, generating a transcription factor that moves to the nucleus and binds to the sterol regulatory element upstream of the genes for HMG CoA reductase and the low-density lipoprotein (LDL) receptor, increasing their expression. Humans are unable to degrade the steroid nucleus to $CO_2 + H_2O$. Bile acid (BA) sequestrants, such as cholestyramine, prevent the absorption of bile salts by the liver, thereby increasing their excretion. The liver then takes up cholesterol via the LDL receptor and uses it to make BA, thereby reducing blood cholesterol levels. Steroid hormones are synthesized from cholesterol, and vitamin D is synthesized in skin from an intermediate (7-dehydrocholesterol) in the cholesterol biosynthetic pathway. Therefore, inhibition of cholesterol synthesis would be expected to decrease their production as well.

2 Acetyl CoA
→ CoA
Acetoacetyl CoA (4 C)
Acetyl CoA → → CoA
HMG CoA (6 C)
2 NADPH + 2 H⁺ ⌐ *HMG CoA reductase*
2 NADP⁺ ⌐   ⊖ Statins
Mevalonate (6 C)
3 ATP ⌐
3 ADP + Pᵢ ⌐ → CO₂
(5 C) **Isopentenyl pyrophosphate** ↔ **Dimethylallyl pyrophosphate** (5 C)
→ PPᵢ
Geranyl pyrophosphate (10 C)
Isopentenyl pyrophosphate → → PPᵢ
Farnesyl pyrophosphate (15 C) ⇢ Ubiquinone, dolichol
                                    prenylation of proteins
Farnesyl pyrophosphate →
NADPH + H⁺ ⌐
NADP⁺ ⌐ → 2 PPᵢ
Squalene (30 C)
NADPH + H⁺ ⌐
NADP⁺ ⌐
Lanosterol (30 C) (cyclic)
                              sunlight
7-Dehydrocholesterol ————→ **Vitamin D**
Cholic acid *7-α-hydroxylase*
(bile acid) ⊖ Bile acids — **Cholesterol** (27 C) → **Steroid hormones**

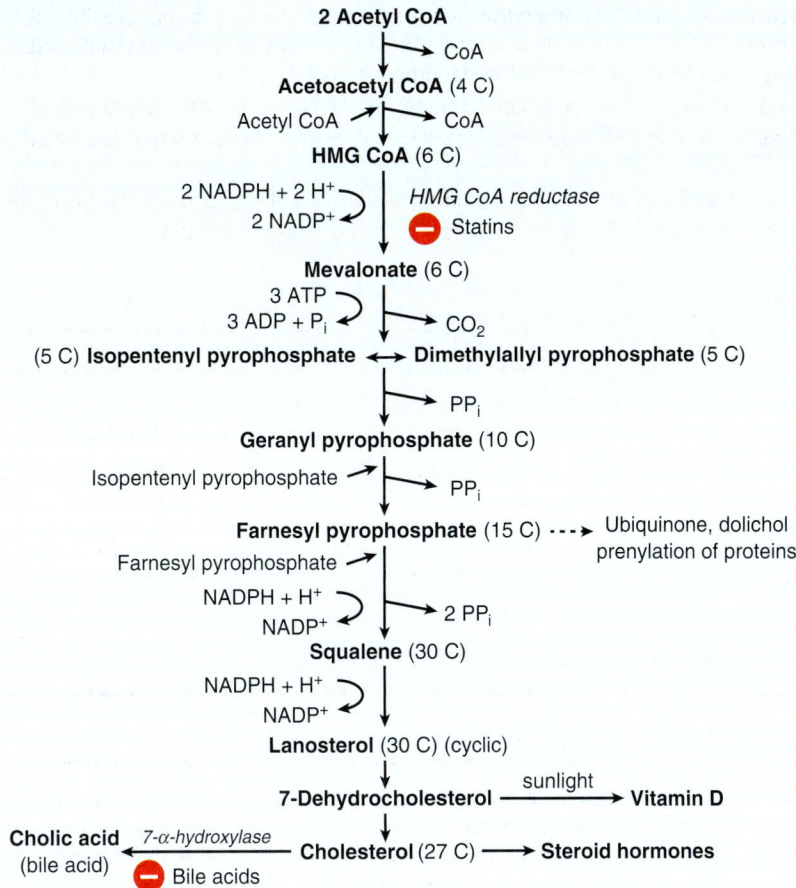

RQ3. **Answer = C.** Competitive inhibitors bind to the same site as the substrate (S) and prevent the S from binding. This results in an increase in the apparent K_m (Michaelis constant, or that S concentration that gives one-half of the maximal velocity [V_{max}]). However, because the inhibition can be reversed by adding additional substrate, the V_{max} is unchanged (see figure at right). It is noncompetitive inhibitors that decrease the apparent V_{max} and have no effect on K_m.

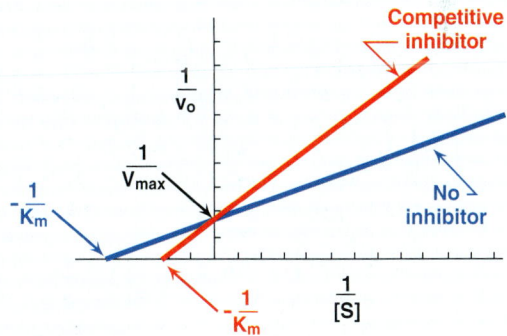

RQ4. **Answer = D.** In hypoxia, substrate-level phosphorylation in glycolysis provides ATP. Oxidative phosphorylation is inhibited by the lack of O_2. Because the rate of ATP synthesis by oxidative phosphorylation controls the rate of cellular respiration, electron transport is inhibited. The resulting rise in the ratio of the reduced form of nicotinamide adenine dinucleotide (NADH) to the oxidized form (NAD⁺) inhibits the tricarboxylic acid cycle and the PDH complex.

CASE 1: Answers to Thought Questions

TQ1. The phenotype would be the same. In familial defective apolipoprotein (apo) B-100, LDL receptors are normal in number and function, but the ligand for the receptor is altered such that binding to the receptor is decreased. Decreased ligand–receptor binding results in increased levels of LDL in the blood with hypercholesterolemia. [Note: The phenotype would be the same in individuals with a gain-of-function mutation to PCSK9, the protease that decreases recycling of the LDL receptor, thereby increasing its degradation.] With the apo E4 isoform, cholesterol-rich chylomicron remnants and intermediate-density lipoproteins (a.k.a., very–low-density lipoprotein [VLDL] remnants) would accumulate in blood.

TQ2. Aspirin irreversibly inhibits cyclooxygenase (COX) and, therefore, the synthesis of PGs, such as PGI_2 in vascular endothelial cells, and thromboxanes (TX), such as TXA_2 in activated platelets. TXA_2 promotes vasoconstriction and formation of a platelet plug, whereas PGI_2 inhibits these events. Because platelets are anucleate, they cannot overcome this inhibition by synthesizing more COX. However, endothelial cells have a nucleus. Aspirin, then, inhibits formation of blood clots by preventing production of TXA_2 for the life of the platelet.

TQ3. The decrease in ATP (as the result of a decrease in O_2 and, thus, a decrease in oxidative phosphorylation) causes an increase in adenosine monophosphate (AMP). AMP allosterically activates phosphofructokinase-1, the key regulated enzyme of glycolysis. The rise in glycolysis increases the production of ATP by substrate-level phosphorylation. It also increases the ratio of the reduced to oxidized forms of NAD. Under anaerobic conditions, pyruvate produced in glycolysis is reduced to lactate by lactate dehydrogenase as NADH is oxidized to NAD^+. NAD^+ is required for continued glycolysis. Because fewer ATP molecules are produced per molecule of substrate in substrate-level phosphorylation relative to oxidative phosphorylation, there is a compensatory increase in the rate of glycolysis under anaerobic conditions.

TQ4. High-density lipoprotein (HDL) functions in reverse cholesterol transport. It takes cholesterol from peripheral tissues (extrahepatic tissues such as lipid-laden arterial vascular smooth muscle cell or macrophages, a.k.a., foam cells) and brings it to the liver (see figure below). The ABCA1 transporter mediates the efflux of cholesterol to HDL. The cholesterol is esterified by extracellular lecithin–cholesterol acyltransferase (LCAT) that requires apo A-1 as a coenzyme. Some cholesteryl ester is transferred to VLDL by cholesteryl ester transfer protein (CETP) in exchange for triacylglycerol. The remainder is taken up by a scavenger receptor (SR-B1) on the surface of hepatocytes. The liver can use the cholesterol from HDL in the synthesis of BAs. Removal of cholesterol from lipid-laden foam cells prevents its accumulation (as cholesterol or cholesteryl ester), decreasing the risk of heart disease. [Note: In contrast, LDL carries cholesterol to peripheral tissues or back to the liver.]

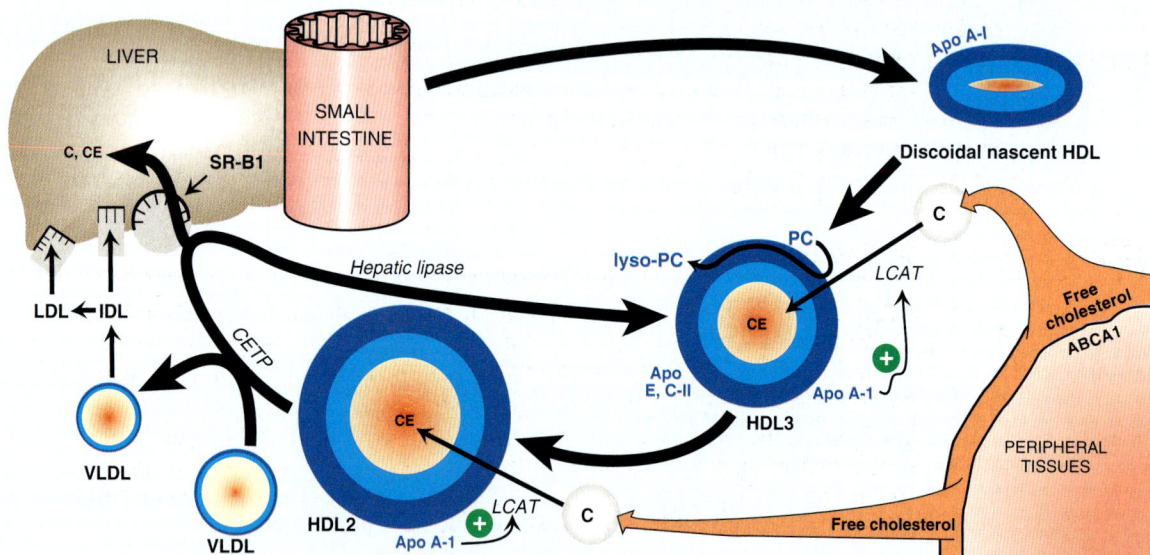

CASE 2: Answers to Review Questions

RQ1. **Answer = A.** Deficiency of glucose 6-phosphatase prevents the glucose 6-phosphate generated by glycogenolysis and gluconeogenesis from being dephosphorylated and released into the blood (see figure below). Blood glucose levels fall, and a severe, fasting hypoglycemia results. [Note: This patient's symptoms appeared only recently because, at age 4 months, his feedings are less frequent.] Hypoglycemia stimulates release of glucagon, which leads to phosphorylation and activation of glycogen phosphorylase kinase that phosphorylates and activates glycogen phosphorylase. Epinephrine is also released and leads to phosphorylation and activation of hormone-sensitive lipase. However, typical FAs cannot serve as substrates for gluconeogenesis. The glucose transporters in the liver and kidneys are insulin insensitive.

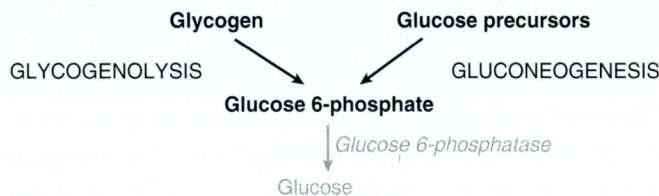

RQ2. **Answer = C.** Vitamin D is a fat-soluble vitamin that functions as a steroid hormone. In complex with its intracellular nuclear receptor, it increases transcription of the gene for calbindin, a calcium (Ca^{2+}) transporter protein in the intestine (see figure at right). Vitamin D does not bind to a membrane receptor and does not produce second messengers. It can be synthesized in the skin by the action of ultraviolet light on an intermediate of cholesterol synthesis, 7-dehydrocholesterol. Of the fat-soluble vitamins (A, D, E, and K), only K functions as a coenzyme.

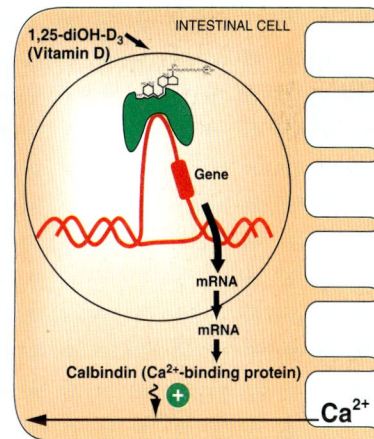

RQ3. **Answer = C.** Glucose 6-phosphate is a positive allosteric effector of the covalently inhibited (phosphorylated) glycogen synthase b. With the rise in glucose 6-phosphate, glycogen synthesis is activated, and glycogen stores are increased in both the liver and the kidneys. The increased availability of glucose 6-phosphate also drives glycolysis. The increase in glycolysis provides substrates for lipogenesis, thereby increasing synthesis of FA and triacylglycerols (TAG). In hypoglycemia, the insulin/glucagon ratio is low, not high.

RQ4. **Answer = D.** Membrane proteins are initially targeted to the endoplasmic reticulum (ER) by an amino-terminal hydrophobic signal sequence. Glycosylation is the most common posttranslational modification found in proteins. The glycosylated portion of membrane proteins is found on the extracellular face of the membrane. The membrane-spanning domain consists of ~22 hydrophobic amino acids. Proteins destined for secretion or for membranes, the ER lumen, Golgi, or lysosomes are synthesized on ribosomes associated with the ER.

CASE 2: Answers to Thought Questions

TQ1. The twitching is the result of the adrenergic response to hypoglycemia and is mediated by the rise in epinephrine. The adrenergic response includes tremor and sweating. Neuroglycopenia (impaired delivery of glucose to the brain) results in impairment of brain function that can lead to seizures, coma, and death. Neuroglycopenic symptoms develop if the hypoglycemia persists.

TQ2. Detergents are amphipathic molecules (i.e., they have both hydrophilic [polar] and hydrophobic [nonpolar] regions). Detergents solubilize membranes, thereby disrupting membrane structure. If the problem were the translocase needed to move the glucose 6-phosphate substrate into the ER, rather than the phosphatase, disruption of the ER membrane would allow the substrate access to the phosphatase.

TQ3. Glucagon, a peptide hormone released from pancreatic α-cells in hypoglycemia, binds its plasma membrane G protein–coupled receptor on hepatocytes. The α_s-subunit of the associated trimeric G protein is activated (guanosine diphosphate is replaced by guanosine triphosphate), separates from the β- and γ-subunits, and activates adenylyl cyclase that generates cyclic adenosine monophosphate (cAMP) from ATP. cAMP activates protein kinase A (PKA) that phosphorylates and activates glycogen phosphorylase kinase, which phosphorylates and activates glycogen phosphorylase. The phosphorylase degrades glycogen, generating glucose 1-phosphate that is converted to glucose 6-phosphate. With glucose 6-phosphatase deficiency, the degradative process stops here (see figure below). Consequently, administration of glucagon is unable to cause a rise in blood glucose. [Note: Epinephrine would be similarly ineffective.]

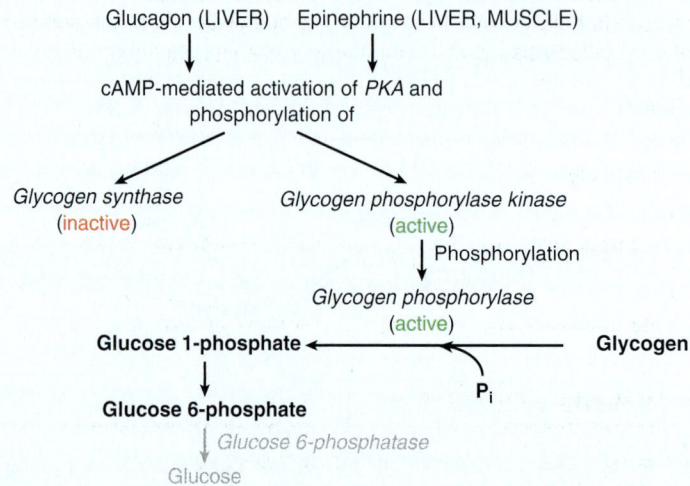

Glucagon (LIVER)   Epinephrine (LIVER, MUSCLE)

cAMP-mediated activation of *PKA* and
phosphorylation of

Glycogen synthase
(inactive)

Glycogen phosphorylase kinase
(active)
| Phosphorylation

Glycogen phosphorylase
(active)

Glucose 1-phosphate ◄────────── **Glycogen**
                                    P_i

Glucose 6-phosphate
   | *Glucose 6-phosphatase*
Glucose

TQ4. The availability of inorganic phosphate (P_i) is decreased because it is trapped as phosphorylated glycolytic intermediates as a result of the upregulation of glycolysis by the rise in glucose 6-phosphate. Urate is elevated because the trapping of P_i decreases the ability to phosphorylate adenosine diphosphate (ADP) to ATP, and the fall in ATP causes a rise in adenosine monophosphate (AMP). The AMP is degraded to urate. Additionally, the availability of glucose 6-phosphate drives the pentose phosphate pathway, resulting in a rise in ribose 5-phosphate (from ribulose 5-phosphate) and, consequently, a rise in purine synthesis. Nicotinamide adenine dinucleotide phosphate (NADPH) also rises. Purines made beyond need are degraded to urate (see figure below). [Note: The decrease in P_i reduces the activity of glycogen phosphorylase, resulting in increased storage of glycogen with a normal structure.] Lactate is elevated because the decrease in phosphorylation of ADP to ATP results in a decrease in cellular respiration (respiratory control) as a result of these processes being coupled. As a consequence, reduced nicotinamide adenine dinucleotide (NADH) from glycolysis cannot be oxidized by Complex I of the electron transport chain. Instead, it is oxidized by cytosolic lactate dehydrogenase with its coenzyme NADH as pyruvate is reduced to lactate. [Note: Pyruvate is increased as a result of the increase in glycolysis.] The lactate ionizes, releasing protons (H^+) and leading to a metabolic acidosis (low pH caused here by increased production of acid). Respiratory compensation causes an increased respiratory rate.

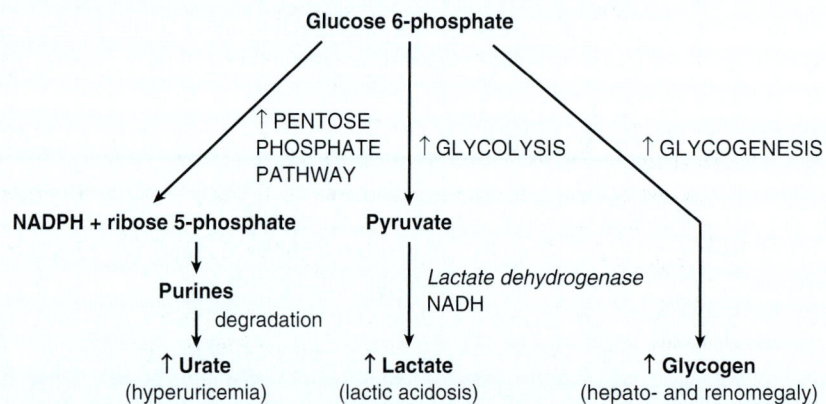

Glucose 6-phosphate

↑ PENTOSE
PHOSPHATE
PATHWAY

↑ GLYCOLYSIS

↑ GLYCOGENESIS

NADPH + ribose 5-phosphate   **Pyruvate**

Purines
   | degradation

Lactate dehydrogenase
NADH

↑ **Urate**
(hyperuricemia)

↑ **Lactate**
(lactic acidosis)

↑ **Glycogen**
(hepato- and renomegaly)

TQ5. Increased glycolysis results in increased availability of glycerol 3-phosphate for hepatic TAG synthesis. Additionally, some of the pyruvate generated in glycolysis will be oxidatively decarboxylated to acetyl coenzyme A (CoA). However, the tricarboxylic acid cycle is inhibited by the rise in NADH, and the acetyl CoA is transported to the cytosol as citrate. The rise of acetyl CoA in the cytosol results in increased fatty acid (FA) synthesis. Recall that citrate is an allosteric activator of acetyl CoA carboxylase (ACC). The malonyl product of ACC inhibits FA oxidation at the carnitine palmitoyltransferase I step. Because mitochondrial

FA oxidation generates the acetyl CoA substrate for hepatic ketogenesis, ketone body levels do not rise. The FA gets esterified to the glycerol backbone, resulting in an increase in TAG that gets sent out of the liver as components of very–low-density lipoproteins (VLDLs). [Note: The hypoglycemia results in release of epinephrine and the activation of TAG lipolysis with release of free FA into the blood. The FAs are oxidized, with the excess used in hepatic TAG synthesis.] The acetyl CoA is also a substrate for cholesterol synthesis. Thus, the increase in glycolysis results in the hyperlipidemia (see figure below).

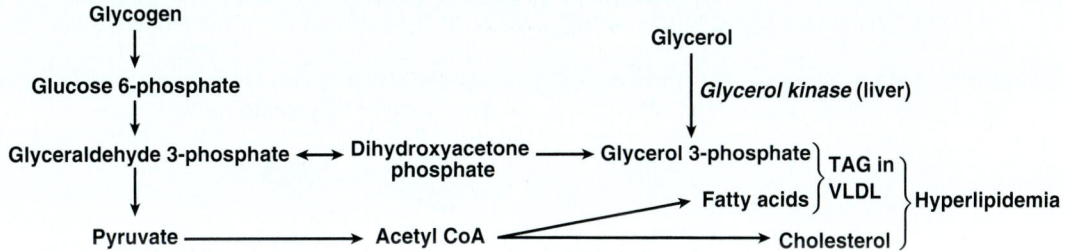

CASE 3: Answers to Review Questions

RQ1. **Correct answer = A.** Diabetes is characterized by hyperglycemia. Chronic hyperglycemia can result in the nonenzymatic glycosylation (glycation) of hemoglobin (Hb), producing HbA$_{1c}$. Therefore, measurement of glucose or HbA$_{1c}$ in the blood is used to diagnose diabetes. In response to physiologic stress (e.g., a urinary tract infection), secretion of counterregulatory hormones (such as the catecholamines) results in a rise in blood glucose. Glucose is a reducing sugar. It is type 2 diabetes (T2D) that is associated with obesity and a sedentary lifestyle and is caused by insensitivity to insulin (insulin resistance). T1D is caused by lack of insulin as a result of the autoimmune destruction of pancreatic β-cells. Even individuals on a program of tight glycemic control do not achieve euglycemia.

RQ2. **Correct answer = E.** The ketone bodies 3-hydroxybutyrate and acetoacetate are organic acids, and their ionization contributes to the proton load of the body. Ketone bodies are made in the mitochondria of liver cells using acetyl coenzyme A (CoA) generated primarily from the β-oxidation of fatty acids ([FAs]; see figure below). Because they are water soluble, they do not require a transporter. The liver cannot use them because it lacks the enzyme thiophorase, which moves CoA from succinyl CoA to acetoacetate for conversion to two molecules of acetyl CoA. It is the acetone released in the breath that can impart a fruity odor.

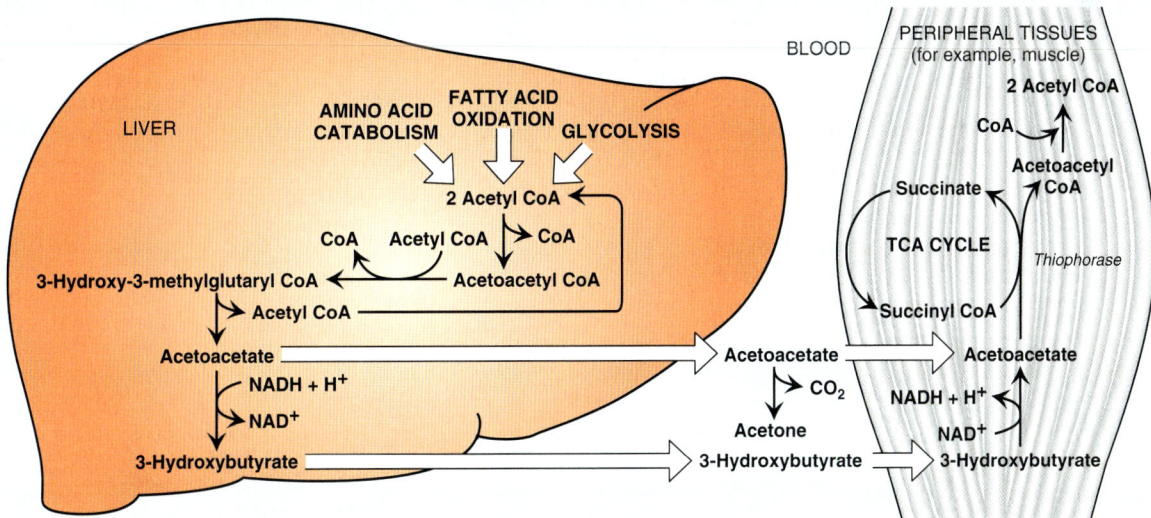

RQ3. **Correct answer = A.** Malonyl CoA, an intermediate of FA synthesis, inhibits FA β-oxidation through inhibition of carnitine palmitoyltransferase I. Lipolysis occurs when the insulin/counterregulatory hormone ratio decreases. Acetyl CoA, the product of FA β-oxidation, inhibits the pyruvate dehydrogenase (PDH) complex through activation of PDH kinase and activates pyruvate carboxylase. Acetyl CoA, then, pushes pyruvate to gluconeogenesis. β-Oxidation generates reduced nicotinamide adenine dinucleotide (NADH), the reducing equivalent required for gluconeogenesis. FAs are not readily catabolized for energy by the brain.

CASE 3: Answers to Thought Questions

TQ1. Hypoinsulinemia results in hyperglycemia because insulin is required for the uptake of blood glucose by muscle and adipose tissue. Their glucose transporter (GLUT-4) is insulin dependent in that insulin is required for movement of the transporter to the cell surface from intracellular storage sites. Insulin is also required to suppress hepatic gluconeogenesis. Insulin suppresses the release of glucagon from pancreatic α-cells. The resulting rise in the insulin/glucagon ratio results in the dephosphorylation and activation of the kinase domain of bifunctional phosphofructokinase-2 (PFK-2). The fructose 2,6-bisphosphate produced by PFK-2 activates phosphofructokinase-1 of glycolysis (see figure below). It also inhibits fructose 1,6-bisphosphatase (FBP-2), thereby inhibiting gluconeogenesis. With hypoinsulinemia, the failure to take up glucose from the blood while simultaneously sending it out into the blood results in hyperglycemia.

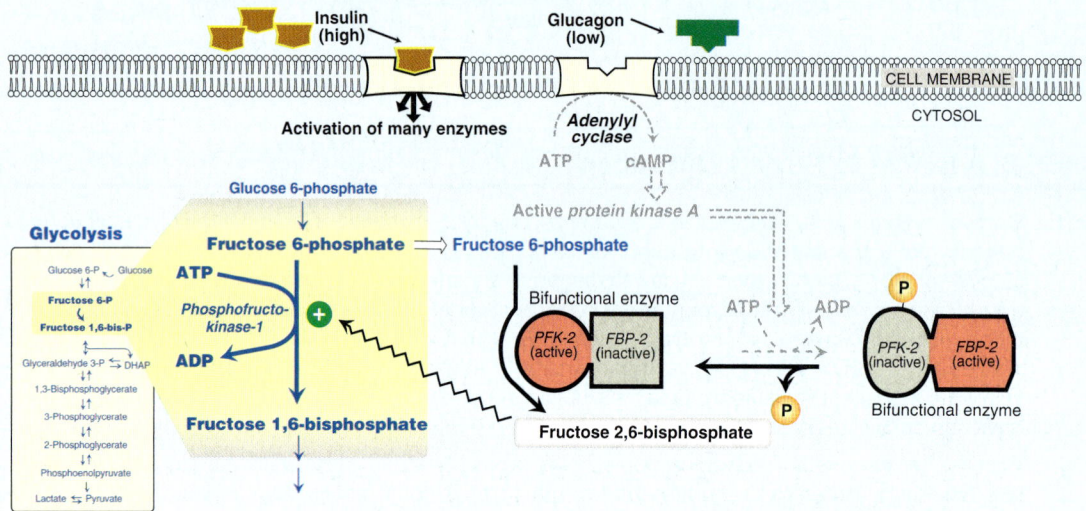

TQ2. The blood glucose level has exceeded the capacity of the kidney to reabsorb glucose (via a sodium-dependent glucose transporter [SGLT]). The high concentration of glucose in the urine osmotically draws water from the body. This causes increased urination (polyuria) with loss of water that results in dehydration.

TQ3. The NADH generated in FA β-oxidation inhibits the tricarboxylic acid (TCA) cycle at the three NADH-producing dehydrogenase steps. This shifts acetyl CoA away from oxidation in the TCA cycle and toward use as a substrate in hepatic ketogenesis.

TQ4. She was in negative nitrogen balance: More nitrogen was going out than coming in. This is reflected in the elevated blood urea nitrogen (BUN) level seen in the patient (see figure at right). [Note: The BUN value also reflects dehydration.] Muscle proteolysis and amino acid catabolism are occurring as a result of the fall in insulin. (Recall that skeletal muscle does not express the glucagon receptor.) Amino acid catabolism produces ammonia (NH_3), which is converted to urea by the hepatic urea cycle and sent into the blood. [Note: Urea in the urine is reported as urinary urea nitrogen.]

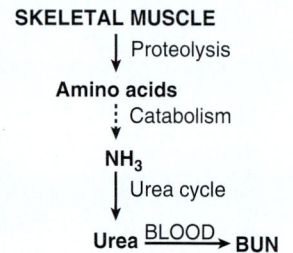

TQ5. The Kussmaul respiration seen in this patient is a respiratory response to the metabolic acidosis. Hyperventilation blows off CO_2 and water, reducing the concentration of protons (H^+) and bicarbonate (HCO_3^-) as reflected in the following equation:

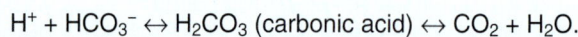

$$H^+ + HCO_3^- \leftrightarrow H_2CO_3 \text{ (carbonic acid)} \leftrightarrow CO_2 + H_2O.$$

The renal response includes, in part, the excretion of H^+ as ammonium (NH_4^+). Degradation of branched-chain amino acids in skeletal muscle results in the release of large amounts of glutamine (Gln) into the blood. The kidneys take up and catabolize the Gln, generating NH_3 in the process. The NH_3 is converted to NH_4^+ by secreted H^+ and is excreted (see figure at right). [Note: When ketone bodies are plentiful, enterocytes shift to using them as a fuel instead of Gln. This increases the amount of Gln going to the kidney.]

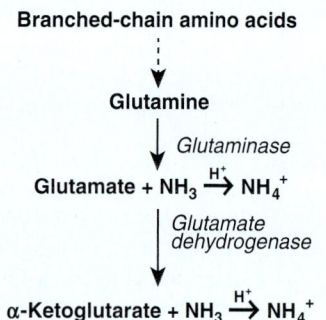

TQ6. Because FA β-oxidation supplies the acetyl CoA substrate for ketogenesis, impaired β-oxidation decreases the ability to make ketone bodies. Ketone bodies are an alternate to the use of glucose, and, thus, dependence on glucose increases. Because FA β-oxidation supplies the NADH and the nucleoside triphosphates needed for gluconeogenesis, glucose production decreases. The result is a hypoketotic hypoglycemia. Recall that this was seen with medium-chain acyl CoA dehydrogenase (MCAD) deficiency.

CASE 4: Answers to Review Questions

RQ1. **Answer = D.** The rise in reduced nicotinamide adenine dinucleotide (NADH) in the mitochondria decreases the tricarboxylic acid (TCA) cycle, FA oxidation, and gluconeogenesis. NADH inhibits the isocitrate dehydrogenase reaction, the key regulated step of the TCA cycle, and the α-ketoglutarate dehydrogenase reaction (see figure at right). It also favors the reduction of oxaloacetate (OAA) to malate (not malate to OAA), decreasing the availability of OAA for condensation with acetyl coenzyme A (CoA) in the TCA cycle and for gluconeogenesis. FA oxidation requires the oxidized form of nicotinamide adenine dinucleotide (NAD^+) for the 3-hydroxyacyl CoA dehydrogenase step and, thus, is inhibited by the rise in NADH. The decrease in FA oxidation decreases the production of ATP and acetyl CoA (the allosteric activator of pyruvate carboxylase) needed for gluconeogenesis. Lipolysis is activated in fasting as a consequence of the fall in insulin and the rise in catecholamines that result in activation of hormone-sensitive lipase.

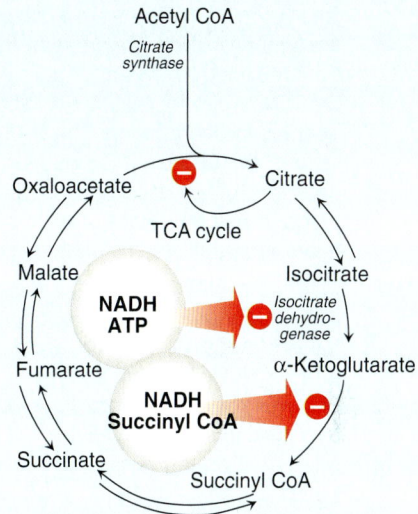

RQ2. **Answer = E.** The irreversible, oxidative portion of the pentose phosphate pathway provides the nicotinamide adenine dinucleotide phosphate (NADPH) that supplies the reducing equivalents needed for activity of cytochrome P450 (CYP) proteins and for the regeneration of functional (reduced) glutathione. It is also an important source of NADPH for reductive biosynthetic processes in the cytosol, such as FA and cholesterol synthesis. [Note: Malic enzyme is another source.] CYP proteins are monooxygenases (mixed-function oxidases). They incorporate one O atom from O_2 into the substrate as the other is reduced to water. It is the CYP proteins of the smooth endoplasmic reticular membrane that are involved in detoxification reactions. Those of the inner mitochondrial membrane are involved in the synthesis of steroid hormones, BAs, and vitamin D. Reactive oxygen species are reduced by glutathione peroxidase as glutathione is oxidized.

RQ3. **Answer = C.** Serotonin is released by activated platelets and causes vasoconstriction and platelet aggregation. [Note: Platelets do not synthesize serotonin, but they take up that which was made in the intestine and secreted into the blood.] Serotonin is associated with a feeling of well-being. It is degraded to 5-hydroxyindoleacetic acid by monoamine oxidase that catalyzes oxidative deamination. It is catechol-O-methyltransferase that catalyzes the methylation step in the degradation of the catecholamines. Serotonin is synthesized from tryptophan in a two-step process that utilizes tetrahydrobiopterin (BH_4)-requiring tryptophan hydroxylase and a pyridoxal phosphate (PLP)-requiring decarboxylase (see figure at right).

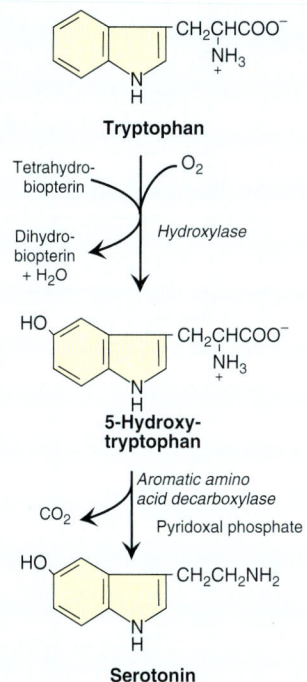

RQ4. **Answer = B.** The exocrine pancreas secretes enzymes required for the digestion of dietary carbohydrate, protein, and fat. The endocrine pancreas secretes the peptide hormones insulin and glucagon. Damage that affects the functions of the pancreas would lead to diabetes (decreased insulin) and steatorrhea (fatty stool), with the latter the consequence of maldigestion of dietary fat. As was seen with the rise of troponins in a myocardial infarction and transaminases in liver damage, loss of cellular integrity (as would be seen in autodigestion of the pancreas) results in proteins that normally are intracellular being found in higher-than-normal concentrations in the blood. Secretin causes the pancreas to release bicarbonate to raise the pH of the chyme coming to the intestine from the stomach. Pancreatic enzymes work best at neutral or slightly alkaline pH. Pancreatitis is seen in individuals with hypertriglyceridemia as a result of a deficiency in lipoprotein lipase or its coenzyme, apolipoprotein C-II.

CASE 4: Answers to Thought Questions

TQ1. A. The rise in cytosolic NADH seen with ethanol metabolism inhibits glycolysis. The glyceraldehyde 3-phosphate dehydrogenase step requires NAD^+, which gets reduced as glyceraldehyde 3-phosphate gets oxidized. With the rise in NADH, glyceraldehyde 3-phosphate accumulates.

B. Glyceraldehyde 3-phosphate from glycolysis is converted to glycerol 3-phosphate, the initial acceptor of FA in triacylglycerol (TAG) synthesis (see figure at right). FAs are available because of increased synthesis (from acetyl CoA, which is increased as a result of both increased production from the acetate product of acetaldehyde oxidation and decreased use in the TCA cycle), increased availability from lipolysis in adipose tissue, and decreased degradation. The TAG produced in the liver accumulate (due, in part, to decreased production of very–low-density lipoproteins) and cause fatty liver (steatosis). Hepatic steatosis is an early (and reversible) stage in alcohol-related liver disease. Subsequent stages are alcohol-related hepatitis (sometimes reversible) and cirrhosis (irreversible).

TQ2. The rise in NADH favors the reduction of pyruvate to lactate-by-lactate dehydrogenase. Lactate decreases the renal excretion of uric acid, thereby causing hyperuricemia, a necessary step in an acute gouty attack. [Note: The shift from pyruvate to lactate decreases the availability of pyruvate, a substrate for gluconeogenesis. This contributes to the hypoglycemia seen in AK.]

Glycerol 3-phosphate
Fatty acyl CoA ⟍ Acyltransferase
CoA ⟋
Lysophosphatidic acid
Fatty acyl CoA ⟍ Acyltransferase
CoA ⟋
Phosphatidic acid (DAG phosphate)
H_2O ⟍ Phosphatase
P_i ⟋
Diacylglycerol (DAG)
Fatty acyl CoA ⟍ Acyltransferase
CoA ⟋
Triacylglycerol (TAG)

TQ3. Prothrombin time (PT) measures the time it takes for plasma to clot after the addition of tissue factor, thereby allowing evaluation of the extrinsic (and common) pathways of coagulation. In the extrinsic pathway, Tissue Factor forms a complex with Factor VII and the complex is activated in a calcium (Ca^{2+})- and phospholipid (PL)-dependent process (see figure at right). Factor VII, like most of the proteins of clotting, is made by the liver. Alcohol-induced liver damage can decrease its synthesis. Additionally, Factor VII has a short half-life, and, as a γ-carboxyglutamate (Gla)-containing protein, its synthesis requires vitamin K. Poor nutrition can result in decreased availability of vitamin K and, therefore, decreased ability to clot. [Note: Severe liver disease results in prolonged PT and activated partial thromboplastin time, or aPPT.]

FVII (Gla) →[TF (FIII), Ca^{2+}, PL] [*FVIIa–TF/FIII*]
↓
FX (Gla) → *FXa*
⇣
Fibrin clot

TQ4. Administration of folate can mask a deficiency in vitamin B_{12} by reversing the hematologic manifestation (macrocytic anemia) of the deficiency. However, folate has no effect on the neurologic damage caused by B_{12} deficiency. Over time, then, the neurologic effects can become severe and irreversible. Thus, folate can mask a deficiency of B_{12} and prevent treatment until the neuropathy is apparent.

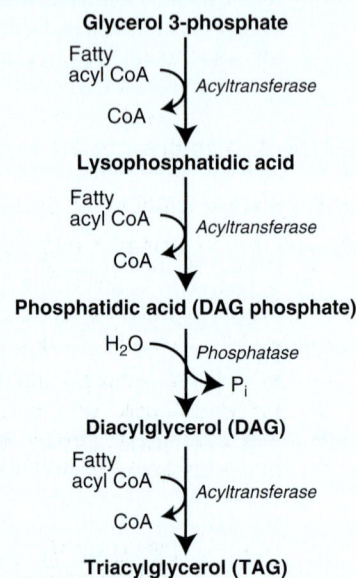

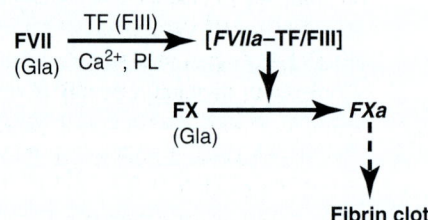

III. FOCUSED CASES

CASE 1: MICROCYTIC ANEMIA

Patient Presentation: A 24-year-old male is being evaluated as a follow-up to a pre-employment medical evaluation.

Focused History: He has no significant medical issues. His family history is unremarkable.

Pertinent Findings: The physical examination was normal. Routine analysis of his blood included the following results:

	Patient	Reference Range
Red blood cells	$4.8 \times 10^6/mm^3$	4.3–5.9
Hemoglobin	9.6 g/dL (**L**)	13.5–7.5 (men)
Mean corpuscular volume	70 μm^3 (**L**)	80–100
Serum iron	150 $\mu g/dL$	50–170

Based on the data, hemoglobin (Hb) electrophoresis was performed. The results are as follows:

	Patient	Reference Range
HbA	90% (L)	96–98
HbA_2	6% (H)	<3
HbF	4% (H)	<2

H, high; L, low. [Note: HbA includes HbA_{1c}.]

Diagnosis: This patient has β-thalassemia trait (β-thalassemia minor) that is causing a microcytic anemia (see image).

Treatment: None is required at this time. Patients are advised that iron supplements will not prevent their anemia.

Prognosis: β-Thalassemia trait does not cause mortality or significant morbidity. Patients should be informed of the genetic nature of their autosomal-recessive condition for family planning considerations because homozygous β-thalassemia (Cooley anemia) is a serious disorder.

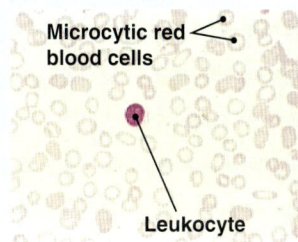

Microcytic red blood cells

Leukocyte

CASE-RELATED QUESTIONS: Choose the ONE best answer

Q1. Mutations to the gene for β-globin that result in decreased production of the protein are the cause of β-thalassemia. The mutations primarily affect gene transcription or posttranscriptional processing of the messenger RNA (mRNA) product. Which of the following statements concerning mRNA is correct?
   A. Eukaryotic mRNA is polycistronic.
   B. mRNA synthesis involves *trans*-acting factors binding to *cis*-acting elements.
   C. mRNA synthesis is terminated at the DNA base sequence thymine adenine guanine (TAG).
   D. Polyadenylation of the 5′-end of eukaryotic mRNA requires a methyl donor.
   E. Splicing of eukaryotic mRNA involves removal of exons and joining of introns.

Q2. HbA, a tetramer of 2α- and 2β-globin chains, delivers O_2 from the lungs to the tissues and protons and CO_2 from the tissues to the lungs. Increased concentration of which of the following will result in decreased O_2 delivery by HbA?
   A. 2,3-Bisphosphoglycerate
   B. Carbon dioxide
   C. Carbon monoxide
   D. Protons

Q3. What is the basis for the increase in HbA_2 and HbF (fetal Hb) in the β-thalassemias?

Q4. Why is the allele-specific oligonucleotide (ASO) hybridization technique useful in the diagnosis of all cases of sickle cell anemia but not all cases of β-thalassemia?

CASE 2: SKIN RASH

Patient Presentation: A 34-year-old female presents with a red, nonitchy rash on her left thigh along with flu-like symptoms.

Focused History: She reports that the rash first appeared a little over 2 weeks ago. It started out small but has gotten larger. She also thinks she is getting the flu because her muscles and joints ache (myalgia and arthralgia, respectively), and she has had a headache for the last few days. She reports that she and her husband took a camping trip last month.

Pertinent Findings: The physical examination is remarkable for the presence of a red, circular, flat lesion ~11 cm in size that resembles a bullseye (erythema migrans) (see image). She also has a low-grade fever.

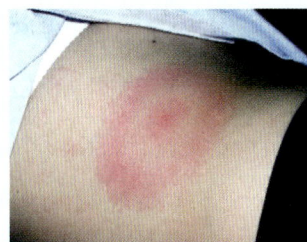

Diagnosis: The patient has Lyme disease caused by the bacterium *Borrelia burgdorferi*, which is transmitted by the bite of a tick in the genus *Ixodes*. Infected ticks are endemic in several regions of the United States.

Treatment: She is prescribed doxycycline, an antibiotic in the tetracycline family. Monitoring of the patient will continue until all symptoms have completely resolved. Blood is drawn for clinical laboratory tests.

Prognosis: Patients treated with the appropriate antibiotic in the early stages of Lyme disease typically recover quickly and completely.

CASE-RELATED QUESTIONS: Choose the ONE best answer

Q1. Antibiotics in the tetracycline class inhibit protein synthesis (translation) of prokaryotic mRNA at the initiation step. Which of the following statements about translation is correct?
   A. In eukaryotic translation, the initiating amino acid is formylated methionine.
   B. Only the charged initiating transfer RNA goes directly to the ribosomal A site.
   C. Peptidyltransferase is a ribozyme that forms the peptide bond between two amino acids.
   D. Prokaryotic translation can be inhibited by the phosphorylation of initiation factor 2.
   E. Termination of translation is independent of guanosine triphosphate hydrolysis.
   F. The Shine–Dalgarno sequence facilitates the binding of the large ribosomal subunit to mRNA.

Q2. The Centers for Disease Control and Prevention recommends a two-tier testing procedure for Lyme disease that involves a screening enzyme-linked immunosorbent assay (ELISA) followed by a confirmatory western blot analysis on any sample with a positive or equivocal ELISA result. Which of the following statements about these testing procedures is correct?
   A. Both techniques are used to detect specific mRNA.
   B. Both techniques involve the use of antibodies to detect proteins.
   C. ELISA requires the use of electrophoresis.
   D. Western blots require use of the polymerase chain reaction (PCR).

Q3. Why are eukaryotic cells unaffected by antibiotics in the tetracycline class?

CASE 3: BLOOD ON THE TOOTHBRUSH

Patient Presentation: A 34-year-old male who is experiencing homelessness presents for evaluation of bruising and bleeding gums.

Focused History: He has been living in a shelter. His diet mostly consists of cereal, coffee, and packaged snacks. Chewing is difficult.

Pertinent Findings: The physical examination was remarkable for the presence of swollen dark-colored gums (see image at right). Several of his teeth were loose, including one that anchors his dental bridge. Several black and blue marks (ecchymoses) were noted on the legs, and an unhealed sore was present on the right wrist. Inspection of his scalp revealed tiny red spots (petechiae) around some of the hair follicles. Blood was drawn for testing.

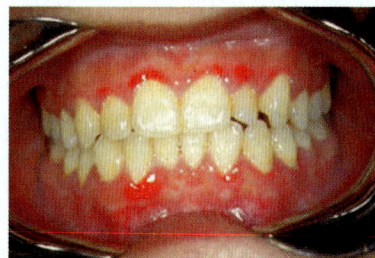

The results of blood tests are as follows:

	Patient	Reference Range
Red blood cells	$4.0 \times 10^6/mm^3$ (**L**)	4.3–5.9
Hemoglobin	10 g/dL (**L**)	13.5–17.5 (men)
Mean corpuscular volume	78 μm^3 (**L**)	80–100
Serum iron	40 μg/dL (**L**)	50–170
Serum ferritin	23 μg/L (**L**)	40–160 μg/L
Total iron-binding capacity	375 μg/dL (**H**)	300–360 μg/dL
Platelets	250×10^9/L	$150–350 \times 10^9$

The occult blood test was negative.

Results of follow-up tests (obtained several days after the appointment) included the following:

	Patient	Reference Range
Vitamin C (plasma)	0.16 mg/dL (L)	0.2–2

H, high; L, low.

Diagnosis: He has vitamin C deficiency with a microcytic, hypochromic anemia secondary to iron deficiency.

Treatment: He was prescribed vitamin C (as oral ascorbic acid) and iron (as oral ferrous sulfate) supplements. He will also be referred to social services.

Prognosis: The prognosis for recovery is good.

CASE-RELATED QUESTIONS: Choose the ONE best answer

Q1. Which of the following statements about vitamin C is correct?
   A. It is a competitive inhibitor of iron absorption in the intestine.
   B. It is a fat-soluble vitamin with a 3-month supply typically stored in adipose tissue.
   C. It is a coenzyme required for the hydroxylation of prolyl and lysyl residues in collagen.
   D. It is required for the cross-linking of collagen.

Q2. In contrast to the microcytic anemia characteristic of iron deficiency (common in older adults), a macrocytic anemia is seen with deficiencies of vitamin B_{12} and/or folic acid. These vitamin deficiencies are also common in persons experiencing homelessness. Which of the following statements concerning these vitamins is correct?
   A. An inability to absorb B_{12} results in pernicious anemia.
   B. Both vitamins cause changes in gene expression.
   C. Folic acid plays a key role in energy metabolism in most cells.
   D. Treatment with methotrexate can result in toxic levels of the coenzyme form of folic acid.
   E. Vitamin B_{12} is the coenzyme for amino acid deaminations, decarboxylations, and transaminations.

Q3. How do hemolytic anemias differ from nutritional anemias?

CASE 4: RAPID HEART RATE, HEADACHE, AND SWEATING

Patient Presentation: A 45-year-old female presents with concerns about sudden (paroxysmal), intense, brief episodes of headache, sweating (diaphoresis), and a racing heart (palpitations).

Focused History: She reports that the attacks started ~3 weeks ago. They last from 2 to 10 minutes, during which time she feels quite anxious. During the attacks, it feels as though her heart is skipping beats (arrhythmia). At first, she thought the attacks were related to recent stress at work and maybe even menopause. The last time it happened, she was in a pharmacy and had her blood pressure taken. She was told it was 165/110 mm Hg. The patient notes that she has lost weight (~8 lb) in this period even though her appetite has been good.

Pertinent Findings: The physical examination was remarkable for her thin, pale appearance. Blood pressure was elevated (150/100 mm Hg), as was the heart rate (110 to 120 beats/min). Based on her history, blood levels of normetanephrine and metanephrine were ordered. They were found to be elevated.

Diagnosis: She has a pheochromocytoma, a rare catecholamine-secreting tumor of the adrenal medulla.

Treatment: Imaging studies of the abdomen locate the tumor in her right adrenal gland and laparoscopic surgical removal of the tumor is performed. The tumor is found to be nonmalignant. Following surgery, her blood pressure returns to normal. Follow-up measurement of plasma metanephrines is performed 2 weeks later and is in the normal range.

Prognosis: The 5-year survival rate for nonmalignant pheochromocytomas is >95%.

CASE-RELATED QUESTIONS: Choose the ONE best answer

Q1. Pheochromocytomas secrete norepinephrine (NE) and epinephrine. Which of the following statements concerning the synthesis and degradation of these two biogenic amines is correct?
  A. The substrate for their synthesis is tryptophan, which is hydroxylated to 3,4-dihydroxyphenylalanine (DOPA) by tetrahydrobiopterin-requiring tryptophan hydroxylase.
  B. The conversion of DOPA to dopamine utilizes a pyridoxal phosphate–requiring carboxylase.
  C. The conversion of norepinephrine to epinephrine requires vitamin C.
  D. Degradation involves methylation by catechol-O-methyltransferase (COMT) and produces normetanephrine from NE and metanephrine from epinephrine.
  E. Normetanephrine and metanephrine are oxidatively deaminated to homovanillic acid by monoamine oxidase (MAO).

Q2. Which of the following statements concerning the actions of epinephrine and/or NE are correct?
  A. NE functions as a neurotransmitter and a hormone.
  B. They are initiated by autophosphorylation of select tyrosine residues in their receptors.
  C. They are mediated by binding to adrenergic receptors, a class of nuclear receptors.
  D. They result in the activation of glycogen and triacylglycerol synthesis.

Q3. NE bound to certain receptors causes vasoconstriction and an increase in blood pressure. Why might NE be used clinically in the treatment of septic shock?

CASE 5: SUN SENSITIVITY

Patient Presentation: A 6-year-old male is evaluated for freckle-like areas of hyperpigmentation on his face, neck, forearms, and lower legs.

Focused History: His father reports that the child has always been quite sensitive to the sun. His skin turns red (erythema) and his eyes hurt (photophobia) if he is exposed to the sun for any period of time.

Pertinent Findings: The physical examination was remarkable for the presence of thickened, scaly areas (actinic keratosis) and hyperpigmented areas on skin exposed to ultraviolet (UV) radiation from the sun. Small, dilated blood vessels (telangiectasia) were also seen. Tissue from several sites on his arms and legs was biopsied, and two were later determined to be squamous cell carcinomas.

Diagnosis: He has xeroderma pigmentosum, a rare defect in nucleotide excision repair of DNA.

Treatment: Avoidance of sun exposure is recommended. Protection from sunlight through use of sunscreens such as protective clothing that reflect UV radiation and chemicals that absorb it is essential. Frequent skin and eye examinations are recommended.

Prognosis: Most patients with xeroderma pigmentosum die at an early age from skin cancers. However, survival beyond middle age is possible.

CASE-RELATED QUESTIONS: Choose the ONE best answer

Q1. Which of the following regarding DNA repair is correct?
  A. DNA repair is performed only by eukaryotes.
  B. DNA repair of double-strand breaks is error free.
  C. DNA repair of mismatched bases involves repair of the parental strand.
  D. DAN repair of UV radiation–induced pyrimidine dimers involves removal of a short oligonucleotide containing the dimer.
  E. DNA repair of uracil produced by the deamination of cytosine requires the actions of endo- and exonucleases to remove the uracil base.

Q2. Which of the following regarding DNA synthesis or replication is correct?
  A. In both eukaryotes and prokaryotes, DNA replication requires an RNA primer.
  B. In eukaryotes, DNA replication requires condensation of chromatin.
  C. In prokaryotes, DNA replication is accomplished by a single DNA polymerase.
  D. DNA replication is initiated at random sites in the genome.
  E. DNA replication produces a polymer of deoxyribonucleoside monophosphates linked by $5' \rightarrow 3'$-phosphodiester bonds.

Q3. What is the difference between DNA proofreading and repair?

CASE 6: DARK URINE AND YELLOW SCLERAE

Patient Presentation: A 63-year-old male patient presents with fatigue and scleral icterus.

Focused History: He began treatment ~4 days ago with a sulfonamide antibiotic and a urinary analgesic for a UTI. He had been told that his urine would change color (become reddish) with the analgesic, but he reports that it has gotten darker (more brownish) over the last 2 days. Last night, his wife noticed that his eyes had a yellow tint. He says he feels as though he has no energy.

Pertinent Findings: The physical examination was remarkable for the patient's pale appearance, mild scleral icterus (jaundice), mild sple-nomegaly, and increased heart rate (tachycardia). His urine tested positive for hemoglobin (hemoglobinuria). A peripheral blood smear reveals a lower-than-normal number of red blood cells (RBC), with some containing precipitated hemoglobin (Heinz bodies; see image at right), and a higher-than-normal number of reticulocytes (immature RBC). Results of the CBC and blood chemistry tests are pending.

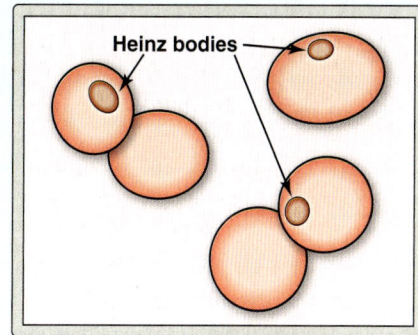

Heinz bodies

Diagnosis: This patient has glucose 6-phosphate dehydrogenase (G6PD) deficiency, an X-linked disorder that causes hemolysis (RBC lysis).

Treatment: G6PD deficiency can result in a hemolytic anemia in affected individuals exposed to oxidative agents, including infection, certain drugs, and fava beans. He will be switched to a different antibiotic and advised to avoid certain agents and to always report his condition to medical providers. He likely was not previously exposed to a strong oxidant stressor and was unaware he had this genetic defect.

Prognosis: In the absence of exposure to oxidative agents, G6PD deficiency does not cause mortality or significant morbidity.

CASE-RELATED QUESTIONS: Choose the ONE best answer

Q1. Which of the following statements concerning G6PD and the pentose phosphate pathway is correct?
   A. Deficiency of G6PD predominantly affects the liver.
   B. Deficiency of G6PD results in an inability to keep glutathione in its reduced form.
   C. The pentose phosphate pathway begins phosphorylated sugar interconversions followed by one final reversible reductive reaction.
   D. The NADPH produced in the pentose phosphate pathway is utilized in processes such nucleotide synthesis.

Q2. Blood chemistry tests revealed an elevation in the bilirubin level. Which of the following statements concerning bilirubin is correct?
   A. Hyperbilirubinemia results in deposition of bilirubin in the skin and sclerae resulting in jaundice.
   B. The solubility of bilirubin is increased by adding to two molecules of ascorbic acid in the bile duct.
   C. The conjugated form of bilirubin increases in the blood with a hemolytic anemia.
   D. Phototherapy can decrease the water solubility of excess bilirubin.

Q3. Why is urinary urobilinogen increased relative to normal in hemolytic jaundice and absent in obstructive jaundice?

CASE 7: JOINT PAIN

Patient Presentation: A 22-year-old male presents for follow-up 10 days after having been treated in the Emergency Department (ED) for severe inflammation at the base of his thumb.

Focused History: This was his first occurrence of severe joint pain. In the ED, he was given an anti-inflammatory medication. Fluid aspi-rated from the carpometacarpal joint of the thumb was negative for organisms but positive for needle-shaped monosodium urate (MSU) crystals (see image at right). The inflammatory symptoms have since resolved. He reports he is in good health otherwise, with no sig-nificant past medical history. His body mass index (BMI) is 31. No tophi (deposits of MSU crystals under the skin) were detected in the physical examination.

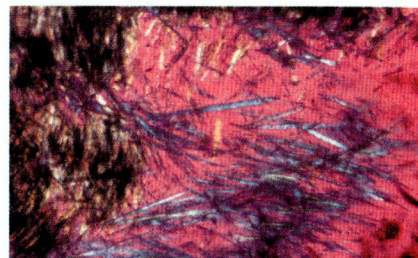

Pertinent Findings: Results on a 24-hour urine specimen and blood tests requested in advance of this visit revealed normal kidney function and uric acid section. His blood urate was 8.5 mg/dL (reference = 2.5 to 8.0). The unusually young age of presentation is suggestive of an enzymopathy of purine metabolism, and additional blood tests are ordered.

Diagnosis: The patient has gout (MSU crystal deposition disease), a type of inflammatory arthritis.

Treatment: He was given prescriptions for pain medication and for allopurinol and colchicine. The treatment goals are to reduce his blood urate levels to <6.0 mg/dL and to prevent additional attacks. He was advised to lose weight because being overweight or obese is a risk factor for gout. His BMI of 31 puts him in the obese category. He was also given written information on the association between diet and gout.

Prognosis: Gout increases the risk of developing renal stones. It is also associated with hypertension, diabetes, and heart disease.

CASE-RELATED QUESTIONS: Choose the ONE best answer

Q1. Allopurinol is converted in the body to oxypurinol, which functions as a noncompetitive inhibitor of an enzyme in purine metabolism. Which of the following statements concerning purine metabolism and its regulation is correct?

    A. As a noncompetitive inhibitor, oxypurinol increases the apparent K_m of the target enzyme.

    B. Colchicine inhibits xanthine oxidase, an enzyme of purine degradation.

    C. Glutamate provides two of the nitrogen atoms of the purine ring.

    D. In purine nucleotide synthesis, the nitrogenous base ring system is first constructed and then attached to ribose 5-phosphate.

    E. Oxypurinol inhibits the amidotransferase that initiates degradation of the purine ring system.

    F. Partial or complete enzymic deficiencies in the salvage of purine bases are characterized by hyperuricemia.

Q2. Which of the following statements is true of the pyrimidines?

    A. Carbamoyl phosphate synthetase I is the regulated enzymic activity in pyrimidine ring synthesis.

    B. Methotrexate decreases synthesis of the pyrimidine nucleotide thymidine monophosphate.

    C. Orotic aciduria is a pathology of pyrimidine degradation.

    D. Pyrimidine nucleotide synthesis is independent of 5-phosphoribosyl-1-pyrophosphate (PRPP).

Q3. The patient is subsequently shown to have a form of PRPP synthetase that shows increased enzymic activity. Why does this result in hyperuricemia?

CASE 8: NO BOWEL MOVEMENT

Patient Presentation: A 2-day-old female has not yet had a bowel movement.

Focused History: The infant was born at full term following a normal pregnancy and delivery. She appeared normal at birth. She is the first child of parents who are both in good health, with unremarkable family histories.

Pertinent Findings: The child has a distended abdomen. She recently vomited small amounts of bilious (green-colored) material.

Diagnosis: Meconium ileus (obstruction of the ileum by meconium, the first stool produced by newborns) was confirmed by abdominal x-rays. Because about 98% of full-term newborns with meconium ileus have cystic fibrosis (CF), this was the suspected diagnosis. Neonatal testing followed by genetic testing confirmed the diagnosis.

Treatment: The ileus was successfully treated without surgery. The family was referred to the CF center at the regional children's hospital.

Prognosis: CF is the most common life-limiting autosomal-recessive disease in persons of European descent and is seen in approximately 1/3,300 live births in the United States.

CASE-RELATED QUESTIONS: Choose the ONE best answer

Q1. Which of the following statements concerning CF is correct?
A. Clinical manifestations of CF are the consequence of chloride retention with increased water reabsorption that causes mucus on the epithelial surface to be excessively thick and sticky.
B. Excessive pancreatic secretion of insulin in CF commonly results in hypoglycemia.
D. Some *CFTR* mutations result in premature degradation of the CFTR protein through tagging with ubiquinone followed by proteasome-mediated proteolysis.
E. The most common mutation, ΔF508, results in the addition of an extra codon for phenylalanine (F) and is classified as a frameshift mutation.

Q2. The CFTR protein is an intrinsic plasma membrane glycoprotein. Which of the following is correct for the targeting of membrane proteins?
A. Processing includes trafficking to and through the Golgi.
B. Processing involves an amino-terminal signal sequence that is retained in the functional protein.
C. Processing occurs only after the protein has been completely synthesized and modified in the cytosol.
D. Requires the presence of mannose 6-phosphate residues within the protein.

Q3. Why might steatorrhea be seen with CF?

CASE 9: ELEVATED AMMONIA

Patient Presentation: A 40-hour-old male with signs of cerebral edema.

Focused History: The child was born at full term after a normal pregnancy and delivery. He appeared normal at birth. At age 36 hours, he became irritable, lethargic, and hypothermic. He fed only poorly and vomited. He also displayed tachypneic (rapid) breathing and neurologic posturing. At age 38 hours, he had a seizure.

Pertinent Findings: Respiratory alkalosis (increased pH, decreased CO_2 [hypocapnia]), increased ammonia, and decreased blood urea nitrogen were found. An amino acid screen revealed that argininosuccinate was increased >60-fold over baseline, and citrulline was increased 4-fold. Glutamine was elevated, and arginine (Arg) was decreased relative to normal.

Diagnosis: The patient has a urea cycle enzyme defect with neonatal onset.

Treatment: Hemodialysis was performed to remove ammonia. Sodium phenylacetate and sodium benzoate were administered to aid in excretion of waste nitrogen, as was Arg. Long-term treatment will include lifelong limitation of dietary protein; supplementation with essential amino acids; and administration of Arg, sodium phenylacetate, and sodium phenylbutyrate.

Prognosis: Survival into adulthood is possible. The degree of neurologic impairment is related to the degree and extent of the hyperammonemia.

CASE-RELATED QUESTIONS: Choose the ONE best answer

Q1. Based on the findings, which enzyme of the urea cycle is most likely to be deficient in this patient?
A. Arginase
B. Argininosuccinate lyase
C. Argininosuccinate synthetase
D. Carbamoyl phosphate synthetase I
E. Ornithine transcarbamoylase (OTC)

Q2. Why is Arg supplementation helpful in this case?

Q3. In individuals with partial (milder) deficiency of urea cycle enzymes, the level of which one of the following would be expected to be decreased during periods of physiologic stress?
A. Alanine
B. Ammonia
C. Glutamine
D. Insulin
E. pH

CASE 10: CALF PAIN

Patient Presentation: A 19-year-old female is being evaluated for pain and swelling in her right calf.

Focused History: 10 days ago, the patient had her spleen removed following a bicycle accident in which she fractured her tibial eminence, necessitating immobilization of the right knee. She has had a good recovery from the surgery. She is no longer taking pain medication but has continued her oral contraceptives (OCP).

Pertinent Findings: Her right calf is reddish in color (erythematous) and warm to the touch. It is visibly swollen. The left calf is normal in appearance and is without pain. An ultrasound is ordered.

Diagnosis: She has a deep venous thrombosis (DVT). OCP are a risk factor for DVT, as are surgery and immobilization.

Treatment (Immediate): Heparin is administered for anticoagulation.

Prognosis: In the 10 years following a DVT, about one-third of individuals have a recurrence.

CASE-RELATED QUESTIONS: Choose the ONE best answer

Q1. Which one of the following would increase the risk of thrombosis?
   A. Excess production of antithrombin
   B. Excess production of protein S
   C. Expression of FV Leiden
   D. Hypoprothrombinemia
   E. von Willebrand disease

Q2. Compare and contrast the actions of heparin and warfarin.

IV. FOCUSED CASES: ANSWERS TO CASE-BASED QUESTIONS

CASE 1: Anemia with β-Thalassemia Minor

Q1. **Answer = B.** Transcription (synthesis of single-stranded RNA from the template strand of double-stranded DNA) requires the binding of proteins (*trans*-acting factors) to sequences on the DNA (*cis*-acting elements). Eukaryotic mRNA is monocistronic because it contains information from just one gene (cistron). The base sequence TAG (thymine adenine guanine) in the coding strand of DNA is U(uracil) AG in the mRNA. UAG is a signal that terminates translation (protein synthesis), not transcription. It is formation of the 5′-cap of eukaryotic mRNA that requires methylation (using S-adenosylmethionine), not 3′-end polyadenylation. Splicing is the spliceosome-mediated process by which introns are removed from eukaryotic mRNA and exons joined.

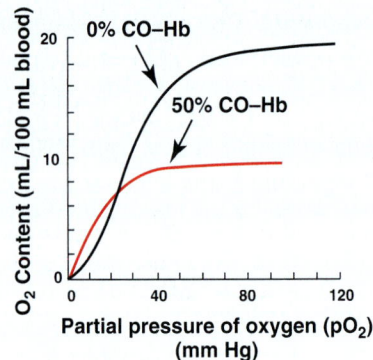

Q2. **Answer = C.** Carbon monoxide (CO) increases the affinity of hemoglobin (Hb)A for O_2, thereby decreasing the ability of HbA to offload O_2 in the tissues. CO stabilizes the R (relaxed) or oxygenated form and shifts the O_2 dissociation curve to the left, decreasing O_2 delivery (see figure at top right). The other choices decrease the affinity for O_2, stabilize the T (tense) or deoxygenated form, and cause a right shift in the curve.

Q3. HbA_2 and fetal Hb (HbF) do not contain β-globin. As β-globin production decreases, synthesis of HbA_2 ($\alpha_2\delta_2$) and HbF ($\alpha_2\gamma_2$) increases.

Q4. Sickle cell anemia is caused by a single point mutation (A→T) in the gene for β-globin that results in the replacement of glutamate by valine at the sixth amino acid position in the protein. Mutational analysis using ASO probes for that mutation (β^S) and for the normal sequence (β^A) is used in diagnosis (see figure at lower right). β-Thalassemia, in contrast, is caused by hundreds of different mutations. Mutational analysis using ASO probes can assess common mutations, including point

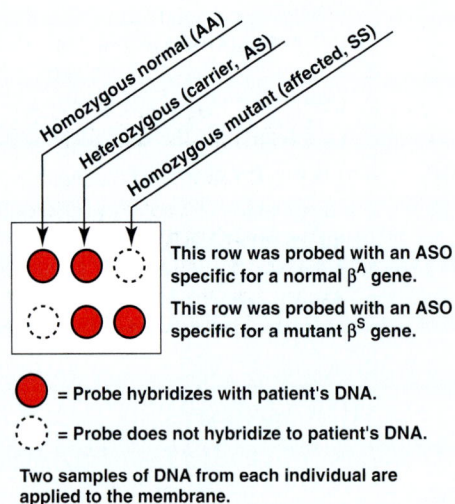

This row was probed with an ASO specific for a normal β^A gene.

This row was probed with an ASO specific for a mutant β^S gene.

◯ = Probe hybridizes with patient's DNA.

◌ = Probe does not hybridize to patient's DNA.

Two samples of DNA from each individual are applied to the membrane.

mutations, in at-risk populations (e.g., those of Mediterranean ancestry). β-Thalassemia is also known as Mediterranean anemia. However, less common mutations are often not included in the panel and can be detected only by DNA sequencing.

CASE 2: Skin Rash with Lyme Disease

Q1. **Answer = C.** Peptide-bond formation between the amino acid in the A site of the ribosome and the amino acid last added to the growing peptide in the P site is catalyzed by an RNA of the large ribosomal subunit. Any RNA with catalytic activity is referred to as a ribozyme (see figure below). Formylated methionine is used to initiate prokaryotic translation. The charged initiating transfer RNA (tRNA$_i$) is the only tRNA that goes directly to the P site, leaving the A site available for the tRNA carrying the next amino acid of the protein being made. Eukaryotic translation is inhibited by the phosphorylation of initiation factor 2 (eIF-2). The Shine–Dalgarno sequence is found in prokaryotic messenger RNA (mRNA) and facilitates the interaction of the mRNA with the small ribosomal subunit. In eukaryotes, the cap-binding proteins perform that task.

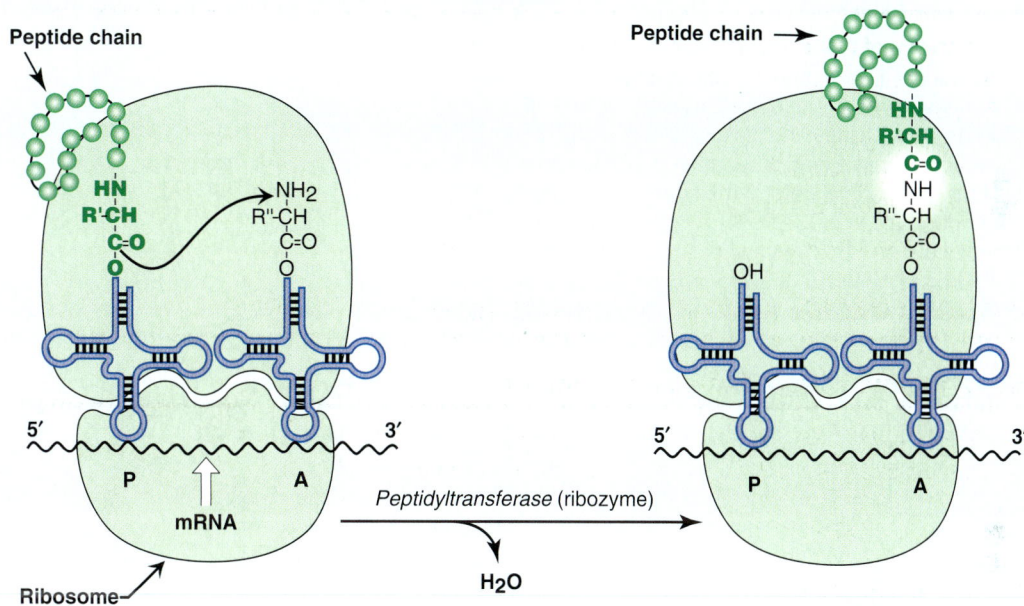

Q2. **Answer = B.** The ELISA and western blot are used to analyze proteins. Each makes use of antibodies to detect and quantify the protein of interest. It is western blots that utilize electrophoresis. The PCR is used to amplify DNA.

Q3. Antibiotics in the tetracycline family inhibit protein synthesis by binding to and blocking the A site of the small (30S) ribosomal subunit in prokaryotes. Tetracycline specifically interacts with the 16S ribosomal RNA (rRNA) component of the 30S subunit, inhibiting translation initiation. Eukaryotes do not contain 16S rRNA. Their small (40S) subunit contains 18S rRNA, which does not bind tetracycline.

CASE 3: Blood on the Toothbrush with Vitamin C Deficiency

Q1. **Answer = C.** Vitamin C (ascorbic acid) functions as a coenzyme in the hydroxylation of proline and lysine in the synthesis of collagen, a fibrous protein of the extracellular matrix. Vitamin C is also the coenzyme for duodenal cytochrome b (Dcytb) that reduces dietary iron from the ferric (Fe^{3+}) to the ferrous (Fe^{2+}) form that is required for absorption via the divalent metal transporter (DMT) of enterocytes (see figure on next page). With a deficiency of vitamin C, uptake of dietary iron is impaired and results in a microcytic, hypochromic anemia. As a water-soluble vitamin, vitamin C is not stored. Cross-linking of collagen by lysyl oxidase requires copper, not vitamin C.

INTESTINAL LUMEN

Heme Fe^{2+} Fe^{3+} (nonheme iron)

Vitamin C

HCP   DMT-1   Dcytb

ENTEROCYTE

Heme

Heme oxygenase Fe^{2+}

Fe^{2+}

Fe^{3+} Ferritin

Ferroportin   *Heph*

CIRCULATION

Hepcidin

Fe^{2+} Fe^{3+}

Fe^{3+} Tf Fe^{3+}

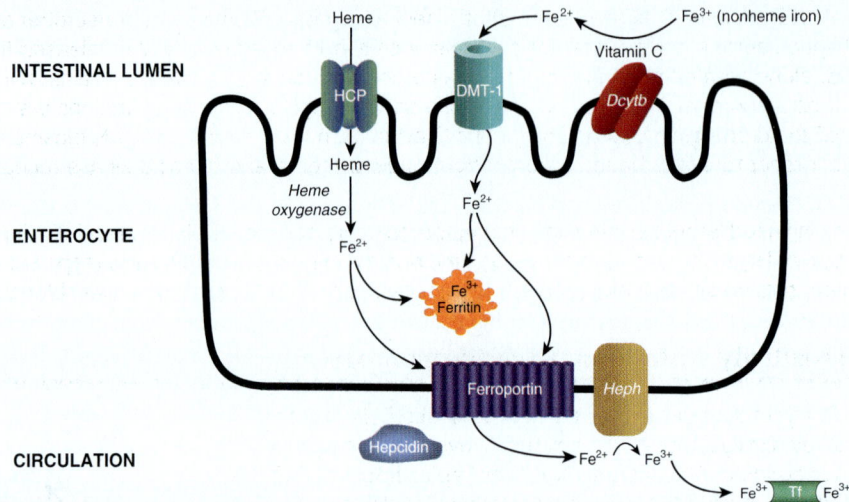

Q2. **Answer = A.** An inability to absorb vitamin B_{12} leads to pernicious anemia and is most commonly caused by decreased production of intrinsic factor (IF) by the parietal cells of the stomach (see figure at right). Vitamins D and A, in complex with their receptors, bind to DNA and alter gene expression. Thiamine (vitamin B_1) is a coenzyme in the oxidative decarboxylation of pyruvate and α-ketoglutarate and, therefore, is important in energy metabolism in most cells. Methotrexate inhibits dihydrofolate reductase, the enzyme that reduces dihydrofolate to tetrahydrofolate (THF), the functional coenzyme form of folate. This results in decreased availability of THF. It is pyridoxine (vitamin B_6) as pyridoxal phosphate that is the coenzyme for most reactions involving amino acids. [Note: Tetrahydrobiopterin is required by aromatic amino acid hydroxylases and nitric oxide synthases.]

Q3. Nutritional anemias are characterized by either increased red blood cell (RBC) size (folate and B_{12} deficiencies) or decreased RBC size (iron and vitamin C deficiencies). In hemolytic anemias, such as is seen in glucose 6-phosphate dehydrogenase and pyruvate kinase deficiencies and in sickle cell anemia, RBC size typically is normal, but RBC number is decreased and often involve RBC deformities or abnormalities.

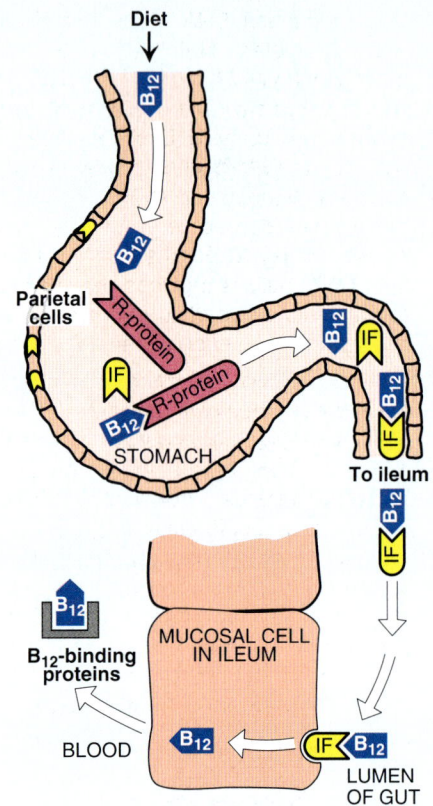

Diet

B_{12}

B_{12}

Parietal cells

R-protein

IF

B_{12}

IF

B_{12}

R-protein

B_{12}

IF B_{12}

STOMACH

To ileum

B_{12}

IF B_{12}

B_{12}

IF B_{12}

B_{12}

MUCOSAL CELL IN ILEUM

B_{12}-binding proteins

BLOOD

B_{12}

IF B_{12}

LUMEN OF GUT

CASE 4: Rapid Heart Rate, Headache, and Sweating with a Pheochromocytoma

Q1. **Answer = D.** Degradation of both epinephrine and NE involves methylation by catechol-O-methyltransferase (COMT) that produces normetanephrine from NE and metanephrine from epinephrine (see figure at right). Both of these products are deaminated to vanillylmandelic acid by MAO. The substrate for the synthesis of the catecholamines is tyrosine, which gets hydroxylated to DOPA by tetrahydrobiopterin-requiring tyrosine hydroxylase. DOPA is converted to dopamine by a pyridoxal phosphate–requiring decarboxylase. [Note: Most carboxylases require biotin.] NE is converted to epinephrine by methylation, and S-adenosylmethionine provides the methyl group.

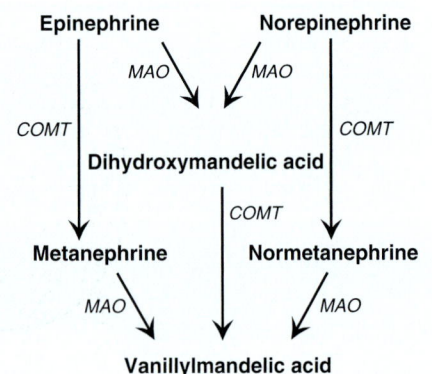

Epinephrine   **Norepinephrine**

MAO   *MAO*

COMT   *COMT*

Dihydroxymandelic acid

COMT

Metanephrine   **Normetanephrine**

MAO   *MAO*

Vanillylmandelic acid

Q2. **Answer = A.** NE released from the sympathetic nervous system functions as a neurotransmitter that acts on postsynaptic neurons and causes, for example, increased heart rate. It is also released from the adrenal medulla and, along with epinephrine, functions as a counterregulatory hormone against hypoglycemia that results in mobilization of stored fuels (e.g., glucose and triacylglycerols). These actions are mediated by the binding of NE to adrenergic receptors, which are G protein–coupled receptors of the plasma membrane, and not to nuclear receptors like those of steroid hormones or membrane tyrosine kinase receptors like that of insulin.

Q3. Septic shock is vasodilatory hypotension (low blood pressure caused by blood vessel dilation) resulting from the production of large amounts of nitric oxide by inducible nitric oxide synthase in response to infection. NE bound to receptors on smooth muscle cells causes vasoconstriction and, thus, raises blood pressure.

CASE 5: Sun Sensitivity with Xeroderma Pigmentosum

Q1. **Answer = D.** Pyrimidine dimers are the characteristic DNA lesions caused by ultraviolet (UV) radiation. Their repair involves the excision of an oligonucleotide containing the dimer and replacement of that oligonucleotide, a process known as nucleotide excision repair (NER). (See figure at right for a representation of the process in prokaryotes.) DNA repair systems are found in prokaryotes and eukaryotes. Nothing is error free, but the homologous recombination (HR) method of double-strand break repair is much less prone to error than is the nonhomologous end joining (NHEJ) method because any DNA that was lost is replaced. Mismatch-base repair (MMR) involves identification and repair of the newly synthesized (daughter) strand. In prokaryotes, the extent of strand methylation is used to discriminate between the strands. Base excision repair (BER), the mechanism by which uracil is removed from DNA, utilizes a glycosylase to remove the base, creating an apyrimidinic or apurinic (AP) site. The sugar phosphate is then removed by the actions of an endo- and exonuclease.

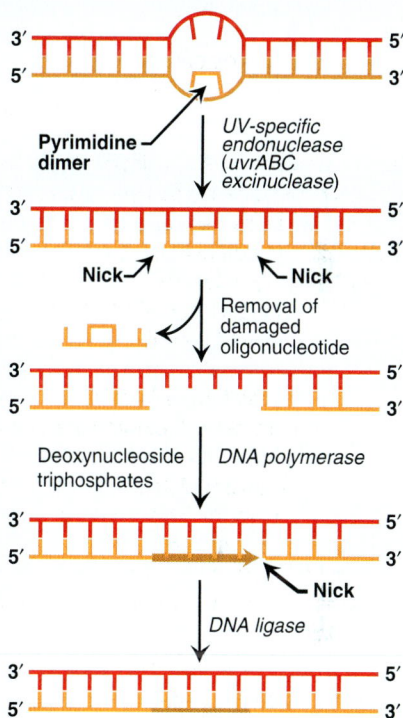

Q2. **Answer = A.** All replication requires an RNA primer because DNA polymerases (pol) cannot initiate DNA synthesis. The chromatin of eukaryotes gets decondensed (relaxed) for replication. Relaxation can be accomplished, for example, by acetylation via histone acetyltransferases. Prokaryotes have more than one DNA pol. For example, pol III extends the RNA primer with DNA, and pol I removes the primer and replaces it with DNA. Replication is initiated at specific locations (one in prokaryotes, many in eukaryotes) that are recognized by proteins (e.g., DnaA in prokaryotes). Deoxynucleoside monophosphates (dNMP) are joined by a phosphodiester bond that links the 3′-hydroxyl group of the last dNMP (deoxyribonucleoside monophosphate) added with the 5′-phosphate group of the incoming nucleotide, thereby forming a 3′→5′-phosphodiester bond as pyrophosphate is released.

Q3. Proofreading occurs during replication in the S (synthesis of DNA) phase of the cell cycle and involves the 3′→5′ exonuclease activity possessed by some DNA pol (see figure below). Because repair can occur independently of replication, it can be performed outside of the S phase.

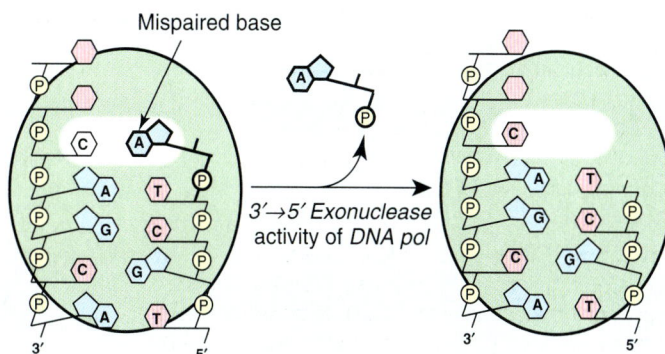

CASE 6: Dark Urine and Yellow Sclerae with Glucose 6-Phosphate Dehydrogenase Deficiency

Q1. **Answer = B.** Glutathione in its reduced form (G-SH) is an important antioxidant. NADPH produced via the pentose phosphate pathway is required to keep glutathione reduced. The selenium-containing enzyme glutathione peroxidase reduces hydrogen peroxide (H_2O_2, a reactive oxygen species) to water as glutathionine is oxidized (G-S-S-G). Reduced nicotinamide adenine dinucleotide phosphate (NADPH)-requiring glutathionine reductase regenerates G-SH from G-S-S-G (see Figure A). The NADPH is supplied by the oxidative reactions of the pentose phosphate pathway (see Figure B), which is regulated by the availability of NADPH at the glucose 6-phosphate dehydrogenase (G6PD)-catalyzed step (the first step). Deficiency of G6PD occurs in all cells, but the effects are seen mainly in red blood cells where the pentose phosphate pathway is the only source of NADPH. The pathway involves two irreversible oxidative reactions, each of which generates NADPH. The NADPH is used in reductive processes such as FA synthesis (not oxidation) as well as steroid hormone and cholesterol synthesis. Ribose, the 5 carbon sugar also produced in the pathway is used for nucleotide synthesis.

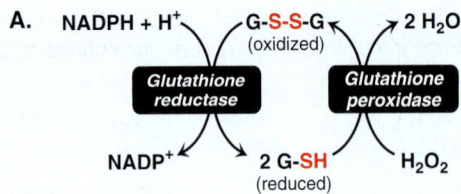

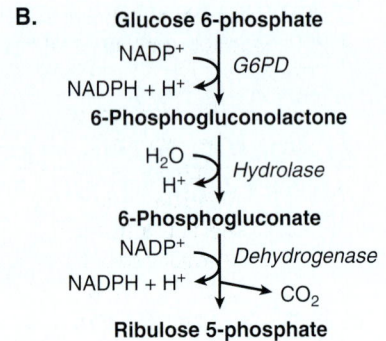

Q2. **Answer = A.** Jaundice (icterus) refers to the yellow color of the skin, nail beds, and sclerae that results from bilirubin deposition when the bilirubin level in the blood is elevated (hyperbilirubinemia; see Image C). Bilirubin has low solubility in aqueous solutions, and its solubility is increased by conjugation with uridine diphosphate–glucuronic acid in the liver, forming bilirubin diglucuronide or conjugated bilirubin (CB). In hemolytic conditions, such as G6PD deficiency, both CB and unconjugated bilirubin (UCB) are increased, but it is UCB that is found in the blood. CB is sent into the intestine. Phototherapy can be used to convert bilirubin to isomeric forms that are more water soluble and can be excreted. Bilirubin is the product of heme degradation in cells of the mononuclear phagocyte system, particularly in the liver and the spleen.

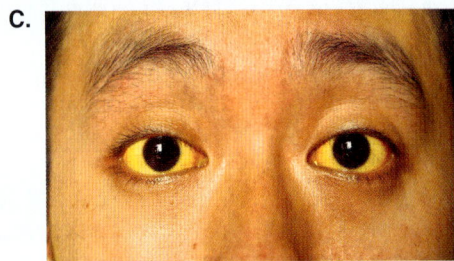

Q3. With hemolysis, more bilirubin is produced and conjugated. Higher than normal levels of conjugated bilirubin are sent to the intestine and converted to urobilinogen, some of which is reabsorbed, enters the portal blood, and travels to the kidney. Because the source of urinary urobilinogen is intestinal urobilinogen, urinary urobilinogen will be low in obstructive jaundice because intestinal urobilinogen will be low as a result of the obstruction of the common bile duct (see Figure D).

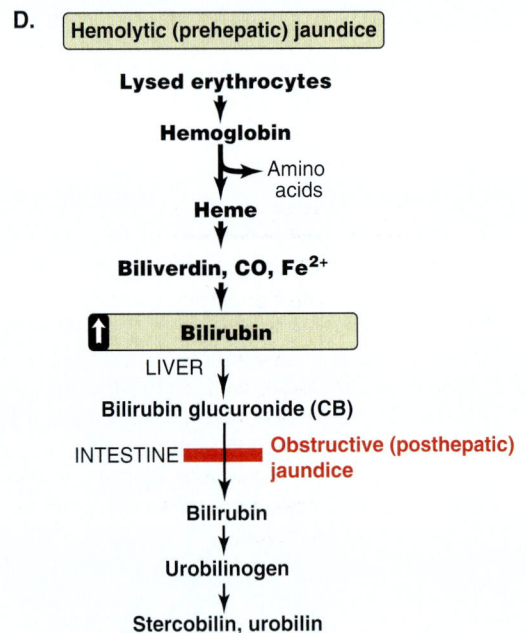

CASE 7: Joint Pain with Gout

Q1. **Answer = F.** Salvage of the purine bases hypoxanthine and guanine to the purine nucleotides inosine monophosphate (IMP) and guanosine monophosphate (GMP) by hypoxanthine-guanine phosphoribosyltransferase (HGPRT) requires 5-phosphoribosyl-1-pyrophosphate (PRPP) as the source of the ribose 1-phosphate. Salvage decreases the amount of substrate available for degradation to uric acid. Therefore, a deficiency in salvage results in hyperuricemia (see figure at right). Noncompetitive inhibitors such as oxypurinol have no effect on the Michaelis constant (K_m) but decrease the apparent maximal velocity (V_{max}). Colchicine is an anti-inflammatory drug. It has no effect on the enzymes of purine synthesis or degradation. Glutamine (not glutamate) is a nitrogen source for purine ring synthesis. In purine nucleotide synthesis, the purine ring system is constructed on the ribose 5-phosphate provided by PRPP.

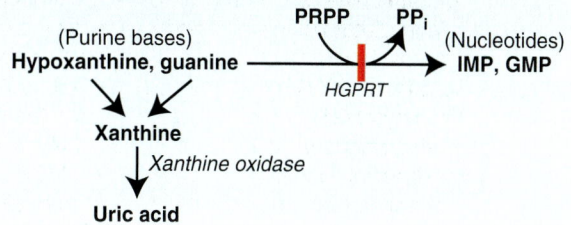

Allopurinol and its metabolite, oxypurinol, inhibit xanthine oxidase of purine degradation. The amidotransferase is the regulated enzyme of purine synthesis. Its activity is decreased by purine nucleotides and increased by PRPP.

Q2. **Answer = B.** Methotrexate inhibits dihydrofolate reductase, decreasing the availability of N^5,N^{10}-methylene tetrahydrofolate needed for synthesis of deoxythymidine monophosphate (dTMP) from deoxyuridine monophosphate (dUMP) by thymidylate synthase (see figure at right). Carbamoyl phosphate synthetase (CPS) II is the regulated enzymic activity of pyrimidine biosynthesis in humans. CPS I is an enzyme of the urea cycle. Orotic aciduria is a rare pathology of pyrimidine synthesis caused by a deficiency in one or both enzymic activities of bifunctional uridine monophosphate synthase. Pyrimidine nucleotide synthesis, like purine synthesis and salvage, requires PRPP.

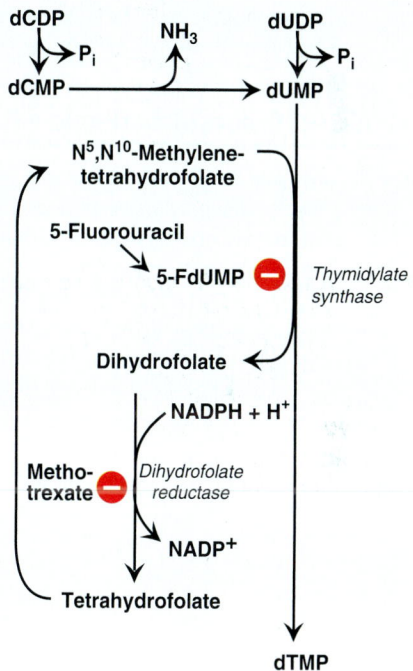

Q3. Increased activity of PRPP synthetase results in increased synthesis of PRPP. This results in an increase in purine nucleotide synthesis beyond need. The excess purine nucleotides get degraded to uric acid, thereby causing hyperuricemia.

CASE 8: No Bowel Movement with Cystic Fibrosis

Q1. **Answer = A.** The clinical manifestations of cystic fibrosis (CF) are the consequence of chloride retention with increased water absorption that causes mucus on an epithelial surface to be excessively thick and sticky. The result is pulmonary and gastrointestinal problems such as respiratory infection and impaired exocrine and endocrine pancreatic functions (pancreatic insufficiency). Impaired endocrine pancreatic function can result in diabetes with associated hyperglycemia. Some mutations do result in increased degradation of the CF transmembrane conductance regulator (CFTR) protein, but degradation is initiated by tagging the protein with ubiquitin, not ubiquinone. Frameshift mutations alter the reading frame through the addition or deletion of nucleotides by a number not divisible by three. Because the ΔF509 mutation is caused by the loss of three nucleotides that code for phenylalanine (F) at position 509 in the CFTR protein, it is not a frameshift mutation.

Q2. **Answer = A.** Targeting of proteins destined to function as components of the plasma membrane is an example of cotranslational targeting. It involves the initiation of translation on cytosolic ribosomes; recognition of the amino (N)-terminal signal sequence in the protein by the signal recognition particle; movement of the protein-synthesizing complex to the outer face of the membrane of the endoplasmic reticulum (ER); and continuation of protein synthesis, such that the protein is threaded into the lumen of the ER and packaged into vesicles that travel to and through the Golgi and eventually fuse with the plasma membrane. The N-terminal signal sequence is removed by a peptidase in the lumen of the ER. Mannose 6-phosphate is the signal that cotranslationally targets proteins to the matrix of the lysosome where they function as acid hydrolases.

Q3. The pancreatic insufficiency seen in some patients with CF results in a decreased ability to digest food, and digestion is required for absorption. Dietary fats move through the intestine and are excreted in the stool (see figure at right), which is foul smelling and bulky and may float. Patients are at risk for malnutrition and deficiencies in fat-soluble vitamins. Oral supplementation of pancreatic enzymes is the treatment.

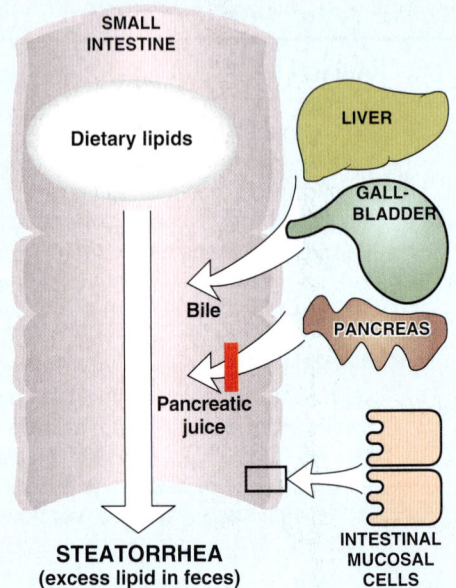

CASE 9: Hyperammonemia with a Urea Cycle Defect

Q1. **Answer = B.** Argininosuccinate lyase (ASL) cleaves argininosuccinate to arginine (Arg) and fumarate. The increase in argininosuccinate and citrulline and the decrease in Arg indicate a deficiency in ASL (see figure below). With arginase deficiency, Arg would be increased, not decreased. Additionally, with arginase deficiency, the hyperammonemia would be less severe because two nitrogens are excreted. Deficiency of argininosuccinate synthetase (ASS) would also cause an increase in citrulline, but argininosuccinate would be low to absent. Deficiency of carbamoyl phosphate synthetase (CPS) I is characterized by low levels of Arg and citrulline. Deficiency of OTC, the only X-linked enzyme of the urea cycle, would result in low levels of Arg and citrulline and elevated levels of urinary orotic acid. [Note: The orotic acid is elevated because the carbamoyl phosphate (CP) substrate of OTC is being used in the cytosol as a substrate for pyrimidine synthesis.]

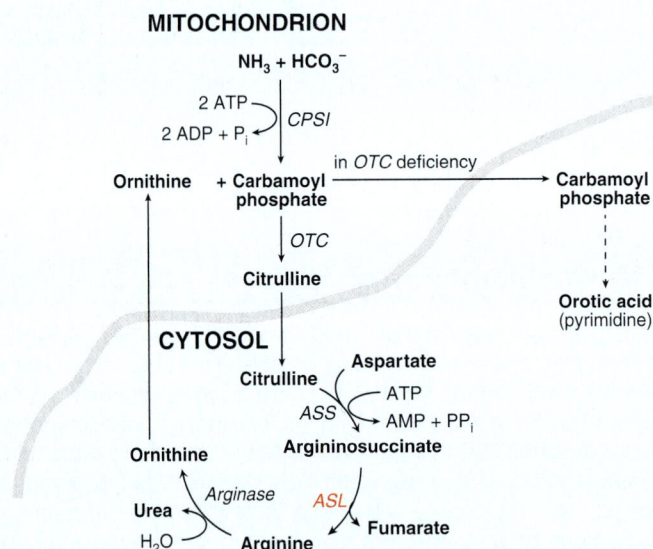

Q2. Arg supplementation is helpful because the Arg will be hydrolyzed to urea + ornithine by arginase. The ornithine will be combined with CP to form citrulline (see figure above). With ASL (and ASS) deficiency, citrulline accumulates and is excreted, thereby carrying waste nitrogen out of the body.

Q3. **Answer = D.** In individuals with milder (partial) deficiencies in the enzymes of the urea cycle, hyperammonemia may be triggered by physiologic stress (e.g., an illness or prolonged fasting) that decreases the insulin/counterregulatory hormone ratio. [Note: The degree of the hyperammonemia is usually less severe than that seen in the neonatal-onset forms.] The shift in the ratio results, in part, in skeletal muscle proteolysis, and the amino acids that are released get degraded. Degradation involves transamination by pyridoxal phosphate–requiring aminotransferases that generate the α-keto acid derivative of the amino acid + glutamate. The glutamate undergoes oxidative deamination to α-ketoglutarate and ammonia (NH_3) by glutamate dehydrogenase (GDH; see figure at right). [Note: GDH is unusual in that it uses both nicotinamide adenine dinucleotide (NAD) and nicotinamide adenine dinucleotide phosphate (NADP) as coenzymes.]

The NH_3, which is toxic, can be transported to the liver as glutamine (Gln) and alanine (Ala). The Gln is generated by the amination of glutamate by ATP-requiring glutamine synthetase. In the liver, the enzyme glutaminase removes the NH_3, which can be converted to urea by the urea cycle or excreted as ammonium (NH_4^+) (see figure at right). Gln, then, is a nontoxic vehicle of NH_3 transport in the blood. Ala is generated in skeletal muscle from the catabolism of the branched-chain amino acids (BCAA). In the liver, Ala is transaminated by alanine transaminase (ALT) to pyruvate (used in gluconeogenesis) and glutamate. Thus, Ala carries nitrogen to the liver for conversion to urea (see figure on next page). Therefore, defects in the urea cycle would result in an elevation in NH_3, Gln, and Ala. The elevated NH_3 drives respiration, and the hyperventilation causes a rise in pH (respiratory alkalosis). [Note: Hyperammonemia is toxic to the nervous system. Although the exact mechanisms are not completely understood, it is known that the metabolism of large amounts of NH_3 to Gln (in the astrocytes of the brain) results in osmotic effects that cause the brain to swell. Additionally, the rise in Gln decreases the availability of glutamate, an excitatory neurotransmitter.]

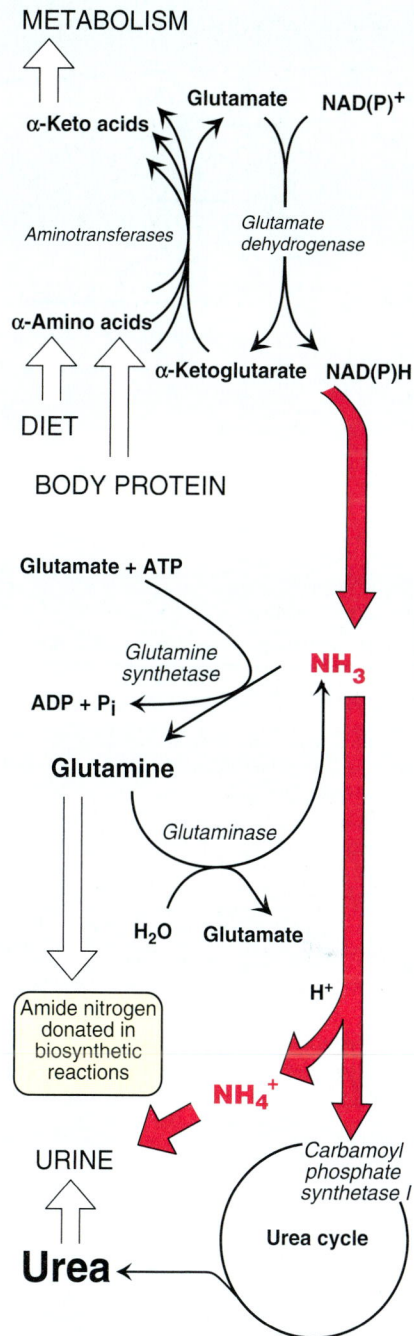

Intrinsic pathway (amplifies, sustains) — **Extrinsic pathway** (initiates)

Common pathway

CASE 10: Swollen, Painful Calf with Deep Venous Thrombosis

Q1. **Answer = C.** Factor V Leiden is a mutant form of Factor V that is resistant to proteolysis by the activated protein C complex. Decreased ability to degrade Factor V allows continued production of activated thrombin and leads to an increased risk of clot formation or thrombophilia. Antithrombin (ATIII) and protein S are proteins of anticoagulation. Increased, not decreased, production of prothrombin would result in thrombophilia. Deficiency of von Willebrand factor causes a coagulopathy or a deficiency in clotting through effects as a carrier of Factor VIII and also on platelet aggregation.

Q2. Heparin, a glycosaminoglycan, functions as an anticoagulant. It activates antithrombin (sometimes called ATIII), enabling antithrombin to inhibit thrombin and Factor Xa. Antithrombin functions by cleaving thrombin and Factor Xa, rendering them inactive. Warfarin, a synthetic analog of vitamin K, inhibits vitamin K, the coenzyme that is required for the γ-carboxylation of glutamate residues to γ-carboxyglutamate (Gla) residues in Factors II, VII, IX, and X (see figures below).

Index

Note: Page numbers followed by *f* indicate figures; those followed by *t* indicate tables. Also positional and configurational designations in chemical names (for example, "3-", "α", "*N*-", "ᴅ-") are ignored in alphabetization.

A

AAT. *See* α1-antitrypsin (AAT)
ABCG5/8 transporter, 257
Abdominal obesity, 418–419. *See also* Obesity
Abetalipoproteinemia, 205, 270, 469
ABO blood group, antigens, 243
Absorptive (well-fed) state
  adipose tissue, energy storage depot, 382
    carbohydrate metabolism, 382, 382*f*
    fat metabolism, 382, 382*f*
  brain in, metabolic pathways in, 384, 384*f*
  control mechanisms of metabolism, 377–378, 377*f*
  glucose in, 377
  key concept map for, 394*f*
  liver (nutrient distribution center) in, 378–382, 379*f*
    amino acid metabolism, 379*f*, 381
    carbohydrate metabolism, 378–380, 379*f*
    fat metabolism, 379*f*, 380–381
  overview of, 377, 378*f*
  regulatory mechanisms in, 377–378, 377*f*
    allosteric effectors, 378
    availability of substrates, 377–378
    covalent modification, 378, 378*f*
    induction and repression of enzyme synthesis, 378
  skeletal muscle in, 382–384
    amino acid metabolism, 383–384, 383*f*
    carbohydrate metabolism, 383
    fat metabolism, 383
ACC. *See* Acetyl CoA carboxylation (ACC)
Acceptable macronutrient distribution ranges (AMDRs), 429, 429*f*
  for adults, 429, 429*f*
  definition, 429
  for protein, 435
Acetic acid, 102*f*
  Henderson–Hasselbalch equation, 7
  titration curve of, 7, 7*f*
Acetoacetate, 227–228, 285*f*, 304–305, 304*f*, 309, 313*f*, 315*f*
  formation of, 305–306, 305*f*–306*f*, 309, 309*f*–310*f*, 317–318
  ketogenic amino acids and, 305, 305*f*
  in liver, 388, 388*f*
  use in peripheral tissues, 227, 227*f*
Acetoacetyl CoA, 309
  in amino acid metabolism, 309
  formation of, 309
Acetone, 227–229
  production of, 227, 227*f*
Acetylcholine, synthesis of, 237
Acetyl CoA carboxylation (ACC), 214, 214*f*
  ACC2, 222
  covalent regulation of, 214, 214*f*
  in FA synthesis, 213–214, 364
  inactive form of, 214
  to malonyl CoA, 213–214, 214*f*
    long-term regulation, 214, 214*f*
    short-term regulation, 214, 214*f*
Acetyl coenzyme A (CoA), 109, 109*f*, 137
  amino acids from, 309
  carboxylation to malonyl CoA, 213–214, 214*f*
  cytosolic, 213, 213*f*
  deficiency, 129, 129*f*
  in FA synthesis, 213

and gluconeogenesis, 131, 131*f*
  mitochondrial, production of, 213
  production, 127–130, 213, 213*f*
  regulation, 129, 129*f*
  succinyl, 306–309, 310*f*
  TCA cycle and, 127–130
    arsenic poisoning, 129–130
    coenzymes, 128, 128*f*
    deficiency, 129, 129*f*
    PDHC component enzymes, 128
    regulation, 129, 129*f*
N-acetylgalactosamine (GalNAc), 181, 184, 185*f*, 244
N-acetylglucosamine-6-sulfatase, deficiency of, 188*f*
N-acetylglucosamine (GlcNAc), 184, 185*f*
N-acetylglucosamine phosphotransferase, 194
N-acetylglutamate (NAG), 294, 296*f*
N-acetylglutamate synthase (NAGS), 296
N-acetylneuraminic acid (NANA), 185, 185*f*
Achlorhydria, 442, 452
Acid–base balance, disturbances in, 11*t*
Acid dissociation constant (K_a), 7
Acid hydrolase(s), 187
  deficiency of, 187
  in glycoprotein degradation, 187
  N-glycosylated, 538
  lysosomal, 194
    in glycoprotein degradation, 187
Acidic glycosphingolipids, 244, 244*f*
Acidic sugar synthesis, 185–186, 186*f*
  glucuronic acid, 185–186, 186*f*
  ʟ-iduronic acid, 186
Acid lipase(s), 200
Acne, 460–462
  cystic, 460–461, 461*f*
  retinoids for, 460
  tretinoin for, 460
Aconitase, 130, 130*f*
ACP. *See* Acyl carrier protein (ACP)
Acrodermatitis enteropathica, 480
ACTH. *See* Adrenocorticotropic hormone (ACTH)
Activated partial thromboplastin time (aPTT), 590
Activated protein C (APC) complex, 592, 592*f*
Activation energy (E_a), 63–64, 64*f*
Active site of chemistry, 64–65
  catalysis, 64
  transition-state stabilization, 64, 64*f*
  transition-state visualization, 64–65, 65*f*
Acute intermittent porphyria (AIP), 325, 325*f*, 327, 328*f*, 329
Acyclovir, 354
Acyl carrier protein (ACP), 214, 215*f*
Acyl CoA:cholesterol acyltransferase (ACAT), 204, 272*f*, 273, 273*f*
Acyl CoA dehydrogenase, 224, 226–227
Acyl CoA:diacylglycerol acyltransferase, 204, 204*f*
Acyl CoA:monoacylglycerol acyltransferase, 204
Acyl CoA oxidase, 226
Acyltransferase(s), 204
AD. *See* Alzheimer disease (AD)
Adaptive thermogenesis, 429

Addison disease, 277
Adenine (A), 341, 341*f*, 489, 512
Adenine arabinoside (vidarabine, or araA), 501
Adenine nucleotides, 85
Adenine phosphoribosyltransferase (APRT), 346, 346*f*
Adenosine deaminase (ADA), 349, 350*f*
  deficiency, 348, 351–352
  gene therapy for, 581, 582*f*
  in uric acid formation, 349, 350*f*
Adenosine diphosphate (ADP), 85, 85*f*
  to ATP, 89–91, 109, 109*f*
  and PDH complex activity, 127, 129*f*
  phosphorylation of, 89–93
  transport, 91
Adenosine kinase, 346
Adenosine monophosphate–activated protein kinase (AMPK), 214, 214*f*, 231*f*, 261, 261*f*, 378, 388–389
Adenosine monophosphate (AMP), 84, 160, 204*f*, 213*f*, 345–346, 345*f*
  to ATP ratio, 139, 139*f*
  cyclic, 112*f*, 113*f*, 119*f*, 152, 152*f*–154*f*, 174
  with sulfate group, 186
  synthesis of, 343
Adenosine triphosphate (ATP), 62, 84–85
  to ADP, 89–91, 109, 109*f*
  and aminoacyl-tRNA synthetase, 529, 530*f*
  apoptosis and, 93
  in catabolism, 109, 109*f*
  chemiosmotic hypothesis of, 89–91
    ATP synthase, 89–91, 90*f*
    proton pump, 89
  common intermediates, 84
  consumption, aerobic glycolysis, 112
  energy carried by, 84, 85*f*
  equivalent transport, reducing, 91–92, 92*f*
  exergonic hydrolysis of, 84, 84*f*
  and gluconeogenesis, 137–138, 138*f*
  in glycogenesis, 146–148
  in glycolytic pathway, 112–113, 113*f*
  inherited defects in oxidative phosphorylation, 92, 92*t*
  membrane transport systems, 91–92
  mitochondria and, 93
  phosphorylation of, 89–93
  for polyisoprenoid squalene, 258
  production, 119
  for protein synthesis, 515
  synthases, 89–91, 90*f*, 145
    coupling in OXPHOS, 89–90
    oligomycin, 90
    synthetic uncouplers, 91, 91*f*
    uncoupling proteins (UCPs), 90–91, 91*f*
  transport, 91
S-Adenosylhomocysteine (SAH), 307, 307*f*
  in epinephrine synthesis, 334
  formation of, 307, 307*f*
  hydrolysis of, 307*f*
S-Adenosylmethionine (SAM), 307, 307*f*, 310, 334, 334*f*, 520, 557
  activated methyl group of, transfer to methyl acceptors, 307
  in phosphatidylcholine synthesis, 238, 238*f*
  synthesis of, 308
Adenylate kinase, 85, 346

Figure Sources

Figure 2.12. Modified from Garrett RH, Grisham CM. *Biochemistry.* Saunders College Publishing; 1995:193. Figure 6.36.

Figure 2.13. From Dobson CM. Protein misfolding, evolution and disease. *Trends Biochem Sci.* 1999;24(9):329–332, Figure 3.

Figure 3.1. Illustration, Irving Geis. Image from the Irving Geis Collection, Howard Hughes Medical Institute. Rights owned by HHMI. Not to be reproduced without permission.

Figure 3.20. Photo From Fizkes/Shutterstock.com

Figure 3.21B. Courtesy of Whit Fisher, MD.

Figure 4.3. From Pawlina W. *Histology: A Text and Atlas.* 9th ed. Wolters Kluwer; 2024.

Figure 4.4. Modified from Yurchenco PD, Birk DE, Mecham RP, eds. *Extracellular Matrix Assembly and Structure.* Academic Press; 1994.

Figure 4.8. From Council ML, Sheinbein D, Cornelius LA; *The Washington Manual of Dermatology Diagnostics.* Wolters Kluwer Health; 2016.

Figure 4.10. From Gru AA. *Pediatric Dermatopathology and Dermatology.* Wolters Kluwer; 2019, Figure 9-1A.

Figure 4.11. From Radiograph from Jorde LB, Carey JC, Bamshad MJ, et al. *Medical Genetics.* 2nd ed. Mosby; 1999. http://medgen.genetics.utah.edu/index.htm

Figure 17.13. From *Urbana Atlas of Pathology.* University of Illinois College of Medicine at Urbana-Champaign. Image number 26.

Figure 17.20. Kumar V, Hagler H, Schneider N. *Interactive Case Study Companion to Robbins Pathologic Basis of Disease.* 6th ed. W B Saunders Co Ltd; 2003.

Figure 18.9. Based on https://commons.wikimedia.org/wiki/File:Cholic_acid.jpg

Figure 18.13. From Husain AN, Stocker JT, Dehner LP. *Stocker and Dehner's Pediatric Pathology.* 4th ed. Wolters Kluwer; 2016, Figure 15–62D.

Figure 20.21. Prostock-studio/Shutterstock.

Figure 20.23. A: From Bullough PG. *Orthopaedic Pathology.* 5th ed. Mosby, Inc.; 2010. Figure 11–31. **B:** Modified from Vigorita VJ. *Orthopaedic Pathology.* 3rd ed. Wolters Kluwer; 2016, Figure 16–53B.

Figure 21.6. Image provided by Stedman's.

Figure 21.7. From Rich MW. Porphyria cutanea tarda. *Postgrad Med.* 1999;105:208–214.

Figure 21.14. Phototake.

Figure 22.16. From Ballantyne JC, Fishman SM, Rathmell JP. *Bonica's Management of Pain.* 5th ed. Wolters Kluwer; 2019, Figure 34-10.

Figure 22.18. From Rubin E, Reisner HM. *Principles of Rubin's Pathology.* 7th ed. Wolters Kluwer; 2019. Figure 22–43D.

Figure 23.2. Courtesy of Gwen V. Childs, PhD. http://www.cytochemistry.net/

Figure 23.14. Cryer PE, Fisher JN, Shamoon H. Hypoglycemia. *Diabetes Care.* 1994;17:734–753. Copyright and all rights reserved. Material from this publication has been used with the permission of American Diabetes Association.

Figure 24.11. Data from Baynes JW, Dominiczak MH. *Medical Biochemistry.* 4th ed. Saunders; 2014.

Figure 26.5. From Gibson W, Farooqi IS, Moreau M, et al. Hypoglycemia. *J Clin Endocrinol Metab.* 2004;89(10):4821 by permission of Oxford University Press.

Figure 27.19. A: Courtesy of Corey Heitz, MD. **B:** From Centers for Disease Control and Prevention. Public Health Image Library. Atlanta, GA.

Figure for Question 27.1. Shutterstock ID: 1176895957. Photo by Nelson Bastidas.

Figure 28.4. Courtesy of Matthews JH. Queen's University Department of Medicine, Division of Hematology/Oncology, Kingston, Canada.

Figure 30.7. Courtesy of Nolan J. Associate Professor of Biology, School of Science and Technology, Room A1394, Georgia Gwinnett College.

Figure 35.10. From Foerster J, Lee G, Lukens J, et al. *Wintrobe's Clinical Hematology.* 10th ed. Lippincott Williams & Wilkins; 1998.

Figure 35.20. From Cohen BJ, Taylor JJ. *Memmler's the Human Body in Health and Disease.* 10th ed. Lippincott Williams & Wilkins; 2005.

Appendix, Integrative Cases, Case 1 Figures. From Gold DH, Weingeist TA. *Color Atlas of the Eye in Systemic Disease.* Lippincott Williams & Wilkins; 2001.

Appendix, Focused Cases, Case 2 Figure. From Goodheart HP. *Goodheart's Photographs of Common Skin Disorders.* 2nd ed. Lippincott Williams & Wilkins; 2003.

Appendix, Focused Cases, Case 6 Figure C. From Zay Nyi Nyi/Shutterstock.com

Appendix, Focused Cases, Case 7 Figure. From Rubin E, Reisner HM. *Principles of Rubin's Pathology.* 7th ed. Wolters Kluwer; 2019. Figure 22–43D.